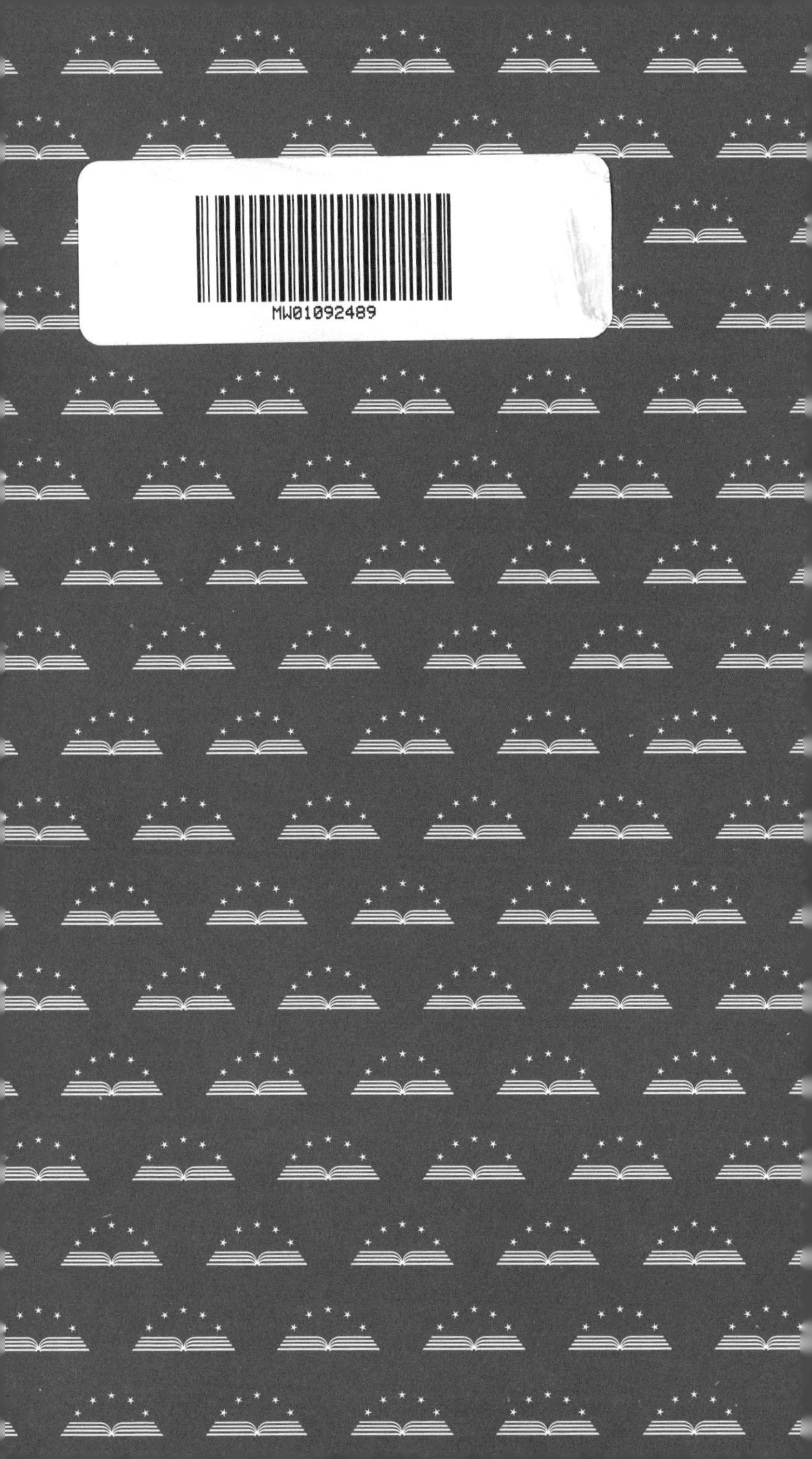
MW01092489

Library of America, a nonprofit organization, champions our nation's cultural heritage by publishing America's greatest writing in authoritative new editions and providing resources for readers to explore this rich, living legacy.

WENDELL BERRY

Wendell Berry

ESSAYS 1993–2017

INCLUDING

Life Is a Miracle

AND SELECTIONS FROM

Sex, Economy, Freedom & Community

Another Turn of the Crank

Citizenship Papers

The Way of Ignorance

What Matters?

Imagination in Place

It All Turns on Affection

Our Only World

The Art of Loading Brush

Jack Shoemaker, *editor*

THE LIBRARY OF AMERICA

WENDELL BERRY: ESSAYS 1993–2017

Published in the United States by Library of America.
Visit our website at www.loa.org.

All texts reprinted by arrangement with Counterpoint Press.

This paper exceeds the requirements of
ANSI/NISO Z39.48–1992 (Permanence of Paper).

Distributed to the trade in the United States
by Penguin Random House Inc.
and in Canada by Penguin Random House Canada Ltd.

Library of Congress Control Number: 2018952819
ISBN 978-1-59853-608-9

First Printing
The Library of America—317

Manufactured in the United States of America

Wendell Berry: Essays 1993–2017
is published with support from

THE GOULD FAMILY FOUNDATION

WALTER E. ROBB

To Mary and Den
my children
with thanks

Contents

FROM

SEX, ECONOMY, FREEDOM & COMMUNITY (1993)

Conservation and Local Economy

In our relation to the land, we are ruled by a number of terms and limits set not by anyone's preference but by nature and by human nature:

I. Land that is used will be ruined unless it is properly cared for.

II. Land cannot be properly cared for by people who do not know it intimately, who do not know how to care for it, who are not strongly motivated to care for it, and who cannot afford to care for it.

III. People cannot be adequately motivated to care for land by general principles or by incentives that are merely economic—that is, they won't care for it merely because they think they should or merely because somebody pays them.

IV. People are motivated to care for land to the extent that their interest in it is direct, dependable, and permanent.

V. They will be motivated to care for the land if they can reasonably expect to live on it as long as they live. They will be more strongly motivated if they can reasonably expect that their children and grandchildren will live on it as long as they live. In other words, there must be a mutuality of belonging: they must feel that the land belongs to them, that they belong to it, and that this belonging is a settled and unthreatened fact.

VI. But such belonging must be appropriately limited. This is the indispensable qualification of the idea of land ownership. It is well understood that ownership is an incentive to care. But there is a limit to how much land can be owned before an owner is unable to take proper care of it. The need for attention increases with the intensity of use. But the *quality* of attention decreases as acreage increases.

VII. A nation will destroy its land and therefore itself if it does not foster in every possible way the sort of thrifty, prosperous, permanent rural households and communities that have the desire, the skills, and the means to care properly for the land they are using.

In an age notoriously impatient of restraints, such a list of rules will hardly be welcome, but that these *are* the rules of land use I have no doubt. I am convinced of their authenticity both by common wisdom and by my own experience and observation. The rules exist; the penalties for breaking them are obvious and severe; the failure of land stewardship in this country is the result of a general disregard for all of them.

As proof of this failure, there is no need to recite again the statistics of land ruination. The gullies and other damages are there to be seen. Very little of our land that is being used—for logging, mining, or farming—is being well used. Much of our land has never been well used. Those of us who know what we are looking at know that this is true. And after observing the worsening condition of our land, we have only to raise our eyes a little to see the worsening condition of those who are using the land and who are entrusted with its care. We must accept as a fact that by now, our country (as opposed to our nation) is characteristically in decline. War, depression, inflation, usury, the attitudes of the industrial economy, social and educational fashions—all have taken their toll. For a long time, the news from everywhere in rural America has been almost unrelievedly bad: bankruptcy, foreclosure, depression, suicide, the departure of the young, the loneliness of the old, soil loss, soil degradation, chemical pollution, the loss of genetic and specific diversity, the extinction or threatened extinction of species, the depletion of aquifers, stream degradation, the loss of wilderness, strip mining, clear-cutting, population loss, the loss of supporting economies, the deaths of towns. Rural American communities, economies, and ways of life that in 1945 were thriving and, though imperfect, full of promise for an authentic human settlement of our land are now as effectively destroyed as the Jewish communities of Poland; the means of destruction were not so blatantly evil, but they have proved just as thorough. The news of rural decline and devastation has been accompanied, to be sure, by a chorus of professional, institutional, and governmental optimists, who continue to insist that all is well, that we are making things worse only as a way of making things better, that farmers who failed are merely "inefficient producers" for whose failure the country is better off, that money and technology will fill the gaps, that government will fill the gaps, that science will soon free us from our regrettable dependence on the soil. We have heard that it is good business and good labor economics to destroy the last remnants of American wilderness. We have heard that the rural population is actually growing because city people are moving to the country and commuters are replacing farmers. We have heard that the rural economy can be repaired by moving the urban economy out into the country and by replacing rural work with work in factories and offices. And all the while the real conditions of the rural land and the rural people have been getting worse.

Of the general condition of the American countryside, my own community will serve well enough as an example. The town of Port Royal, Kentucky, has a population of about one hundred people. The town came into existence as a trading center, serving the farms in a few square miles of hilly country on the west side of the Kentucky River. It has never been much bigger than it is now. But whereas now it is held together by habit or convenience, once it was held together by a complex local economy. In my mother's childhood, in the years before World War I, there were sixteen business and professional enterprises in the town, all serving the town and the surrounding farms. By the time of my own childhood, in the years before World War II, the number had been reduced to twelve, but the town and its tributary landscape were still alive as a community and as an economy. Now, counting the post office, the town has five enterprises, one of which does not serve the local community. There is now no market for farm produce in the town or within forty miles. We no longer have a garage or repair shop of any kind. We have had no doctor for forty years and no school for thirty. Now, as a local economy and therefore as a community, Port Royal is dying.

What does the death of a community, a local economy, cost its members? And what does it cost the country? So far as I know, we have no economists who are interested in such costs. Nevertheless, when you must drive ten or twenty or more miles to reach a doctor or a school or a mechanic or to find parts for farm machinery, the costs exist, and they are increasing. As they increase, they make the economy of every farm and household less tenable.

As people leave the community or, remaining in the place, drop out of the local economy, as the urban-industrial economy more and more usurps the local economy, as the scale and speed of work increase, care declines. As care declines, the natural supports of the human economy and community also decline, for whatever is used, is used destructively.

We in Port Royal are part of an agricultural region surrounded by cities that import much of their food from distant places. Though we urgently need crops that can be substituted for tobacco, we produce practically no vegetables or other foods for consumption in our region. Having no local food economy, we produce a less and less diverse food supply for the general market. This condition implies and virtually requires the abuse of our land and our people, and they are abused. We are also part of a region that is abundantly and diversely forested, and we have no forest economy. We have no local wood products industry. This makes it almost certain that our woodlands and their owners will be abused, and they are abused.

We provide, moreover, a great deal of recreation for our urban neighbors—hunting, fishing, boating, and the like—and we have the capacity to provide more. But for this we receive little or nothing, and sometimes we suffer damage.

In our region, furthermore, there has been no public effort to preserve the least scrap of land in its pristine condition. And the last decade or so of agricultural depression has caused much logging of the few stands of mature forest in private hands. Now, if we want our descendants to know what the original forest was like—that is, to know the original nature of our land—we must start from scratch and grow the necessary examples over the next two or three hundred years.

My part of rural America is, in short, a colony, like every other part of rural America. Almost the whole landscape of this country—from the exhausted cotton fields of the plantation South to the eroding wheatlands of the Palouse, from the strip mines of Appalachia to the clear-cuts of the Pacific slope—is in the power of an absentee economy, once national and now increasingly international, that is without limit in its greed and without mercy in its exploitation of land and people. Between the prosperity of this vast centralizing economy and the prosperity of any local economy or locality, there is now a radical disconnection. The accounting that measures the wealth of corporations, great banks, and national treasuries takes no measure of the civic or economic or natural health of places like Port Royal, Kentucky; Harpster, Ohio; Indianola, Iowa; Matfield Green, Kansas; Wolf Hole, Arizona; or Nevada City, California—and it does not intend to do so.

In 1912, according to William Allen White, "the county in the United States with the largest assessed valuation was Marion County, Kansas. . . . Marion County happened to have a larger per capita of bank deposits than any other American county. . . . Yet no man in Marion County was rated as a millionaire, but the jails and poorhouses were practically empty. The great per capita of wealth was actually distributed among the people who earned it." This, of course, is the realization of that dream that is sometimes called Jeffersonian but is really the dream of the economically oppressed throughout human history. And because this was a rural county, White was not talking just about bank accounts; he was talking about real capital—usable property. That era and that dream are now long past. Now the national economy, which is increasingly a global economy, no longer prospers by the prosperity of the land and people but by their exploitation.

The Civil War made America safe for the moguls of the railroads and of the mineral and timber industries who wanted to be free to exploit the countryside. The work of these industries and their successors is now almost complete. They have dispossessed, disinherited, and moved into the urban economy almost the entire citizenry; they have defaced and plundered the countryside. And now this great corporate enterprise, thoroughly uprooted and internationalized, is moving toward the exploitation of the whole world under the shibboleths of "globalization," "free trade," and "new world order." The proposed revisions in the General Agreement on Tariffs and Trade are intended solely to further this exploitation. The aim is simply and unabashedly to bring every scrap of productive land and every worker on the planet under corporate control.

The voices of the countryside, the voices appealing for respect for the land and for rural community, have simply not been heard in the centers of wealth, power, and knowledge. The centers have decreed that the voice of the countryside shall be that of Snuffy Smith or Li'l Abner, and only that voice have they been willing to hear.

"The business of America is business," a prophet of our era too correctly said. Two corollaries are clearly implied: that the business of the American government is to serve, protect, and defend business; and that the business of the American people is to serve the government, which means to serve business. The costs of this state of things are incalculable. To start with, people in great numbers—because of their perception that the government serves not the country or the people but the corporate economy—do not vote. Our leaders, therefore, are now in the curious—and hardly legitimate—position of asking a very substantial number of people to cheer for, pay for, and perhaps die for a government that they have not voted for.

But when the interests of local communities and economies are relentlessly subordinated to the interests of "business," then two further catastrophes inevitably result. First, the people are increasingly estranged from the native wealth, health, knowledge, and pleasure of their country. And, second, the country itself is destroyed.

It is not possible to look at the present condition of our land and people and find support for optimism. We must not fool ourselves. It is altogether conceivable that we may go right along with this business of "business," with our curious religious faith in technological progress, with our glorification of our own greed and violence always rationalized by our indignation at the greed and violence of others, until our land, our world, and ourselves are utterly destroyed. We know from history that massive human failure is possible. It is foolish to assume that we will save ourselves from any fate that we have made possible simply because we have the conceit to call ourselves *Homo sapiens*.

On the other hand, we want to be hopeful, and hope is one of our duties. A part of our obligation to our own being and to our descendants is to study our life and our condition, searching always for the authentic underpinnings of hope. And if we look, these underpinnings can still be found.

For one thing, though we have caused the earth to be seriously diseased, it is not yet without health. The earth we have before us now is still abounding and beautiful. We must learn again to see that present world for what it is. The health of nature is the primary ground of hope—if we can find the humility and wisdom to accept nature as our teacher. The pattern of land stewardship is set by nature. This is why we must have stable rural economies and communities; we must keep alive in every place the human knowledge of the nature of that place. Nature is the best farmer and forester, for she does not destroy the land in order to make it productive. And so in our wish to preserve our land, we are not without the necessary lessons, nor are we without instruction, in our cultural and religious tradition, necessary to learn those lessons.

But we have not only the example of nature; we have still, though few and widely scattered, sufficient examples of competent and loving human stewardship of the earth. We have, too, our own desire to be healthy in a healthy world. Surely, most of us still have, somewhere within us, the fundamental human wish to die in a world in which we have been glad to live. And we *are*, in spite of much evidence to the contrary, somewhat sapient. We *can* think—if we will. If we know carefully enough who, what, and where we are, and if we keep the scale of our work small enough, we can think responsibly.

These assets are not the gigantic, technical, and costly equipment that we tend to think we need, but they are enough. They are, in fact, God's plenty. Because we have these assets, which are the supports of our legitimate hope, we can start from where we are, with what we have, and imagine and work for the healings that are necessary.

But we must begin by giving up any idea that we can bring about these healings without fundamental changes in the way we think and live. We face a choice that is starkly simple: we must change or be changed. If we fail to change for the better, then we will be changed for the worse. We cannot blunder our way into health by the same sad and foolish hopes by which we have blundered into disease. We must see that the standardless aims of industrial communism and industrial capitalism equally have failed. The aims of productivity, profitability, efficiency, limitless growth, limitless wealth, limitless power, limitless mechanization and automation can enrich and empower the few (for a while), but they will sooner or later ruin us all. The gross national product and the corporate bottom line are utterly meaningless as measures of the prosperity or health of the country.

If we want to succeed in our dearest aims and hopes as a people, we must understand that we cannot proceed any further without standards, and we must see that ultimately the standards are not set by us but by nature. We must see that it is foolish, sinful, and suicidal to destroy the health of nature for the sake of an economy that is really not an economy at all but merely a financial system, one that is unnatural, undemocratic, sacrilegious, and ephemeral. We must see the error of our effort to live by fire, by burning the world in order to live in it. There is no plainer symptom of our insanity than our avowed intention to maintain by fire an unlimited economic growth. Fire destroys what nourishes it and so in fact imposes severe limits on any growth associated with it. The true source and analogue of our economic life is the economy of plants, which never exceeds natural limits, never grows beyond the power of its place to support it, produces no waste, and enriches and preserves itself by death and decay. We must learn to grow like a tree, not like a fire. We must repudiate what Edward Abbey called "the ideology of the cancer cell": the idiotic ideology of "unlimited economic growth" that pushes blindly toward the limitation of massive catastrophe.

We must give up also our superstitious conviction that we can contrive technological solutions to all our problems. Soil loss, for example, is a problem that embarrasses all of our technological pretensions. If soil were all being lost in a huge slab somewhere, that would appeal to the would-be heroes of "science and technology," who might conceivably engineer a glamorous, large, and speedy solution—however many new problems they might cause in doing so. But soil is not usually lost in slabs or heaps of magnificent tonnage. It is lost a little at a time over millions of acres by the careless acts of millions of people. It cannot be saved by heroic feats of gigantic technology but only by millions of small acts and restraints, conditioned by small fidelities, skills, and desires. Soil loss is ultimately a cultural problem; it will be corrected only by cultural solutions.

The aims of production, profit, efficiency, economic growth, and technological progress imply, as I have said, no social or ecological standards, and in practice they submit to none. But there is another set of aims that does imply a standard, and these aims are freedom (which is pretty much a synonym for personal and local self-sufficiency), pleasure (that is, our gladness to be alive), and longevity or sustainability (by which we signify our wish that human freedom and pleasure may last). The standard implied by all of these aims is health. They depend ultimately and inescapably on the health of nature; the idea that freedom and pleasure can last long in a diseased world is preposterous. But these good things depend also on the health of human culture, and human culture is to a considerable extent the knowledge of economic and other domestic procedures—that is, ways of work, pleasure, and education—that preserve the health of nature.

In talking about health, we have thus begun to talk about community. But we must take care to see how this standard of health enlarges and clarifies the idea of community. If we speak of a *healthy* community, we cannot be speaking of a community that is merely human. We are talking about a neighborhood of humans in a place, plus the place itself: its soil, its water, its air, and all the families and tribes of the nonhuman creatures that belong to it. If the place is well preserved, if its entire membership, natural and human, is present in it, and if the human economy is in practical harmony with the nature of the place, then the community is healthy. A diseased community will be suffering natural losses that become, in turn, human losses. A healthy community is sustainable; it is, within reasonable limits, self-sufficient and, within reasonable limits, self-determined—that is, free of tyranny.

Community, then, is an indispensable term in any discussion of the connection between people and land. A healthy community is a form that includes all the local things that are connected by the larger, ultimately mysterious form of the Creation. In speaking of community, then, we are speaking of a complex connection not only among human beings or between humans and their homeland but also between the human economy and nature, between forest or prairie and field or orchard, and between troublesome creatures and pleasant ones. *All* neighbors are included.

From the standpoint of such a community, any form of land abuse—a clear-cut, a strip mine, an overplowed or overgrazed field—is as alien and as threatening as it would be from the standpoint of an ecosystem. From such a standpoint, it would be plain that land abuse reduces the possibilities of local life, just as do chain stores, absentee owners, and consolidated schools.

One obvious advantage of such an idea of community is that it provides a common ground and a common goal between conservationists and small-scale land users. The long-standing division between conservationists and farmers, ranchers, and other private small-business people is distressing because it is to a considerable extent false. It is readily apparent that the economic forces that threaten the health of ecosystems and the survival of species are equally threatening to economic democracy and the survival of human neighborhoods.

I believe that the most necessary question now—for conservationists, for small-scale farmers, ranchers, and businesspeople, for politicians interested in the survival of democracy, and for consumers—is this: What must be the economy of a healthy community based in agriculture or forestry? It *cannot* be the present colonial economy in which only "raw materials" are exported and *all* necessities and pleasures are imported. To be healthy, land-based communities will need to add value to local products, they will need to supply local demand, and they will need to be reasonably self-sufficient in food, energy, pleasure, and other basic requirements.

Once a person understands the necessity of healthy local communities and community economies, it becomes easy to imagine a range of reforms that might bring them into being.

It is at least conceivable that useful changes might be started or helped along by consumer demand in the cities. There is, for example, already evidence of a growing concern among urban consumers about the quality and the purity of food. Once this demand grows extensive and competent enough, it will have the power to change agriculture—if there is enough left of agriculture, by then, to be changed.

It is even conceivable that our people in Washington might make decisions tending toward sustainability and self-sufficiency in local economies. The federal government could do much to help, if it would. Its mere acknowledgment that problems exist would be a promising start.

But let us admit that urban consumers are not going to be well informed about their economic sources very soon and that a federal administration enlightened about the needs and problems of the countryside is not an immediate prospect.

The real improvements then must come, to a considerable extent, from the local communities themselves. We need local revision of our methods of land use and production. We need to study and work together to reduce scale, reduce overhead, reduce industrial dependencies; we need to market and process local products locally; we need to bring local economies into harmony with local ecosystems so that we can live and work with pleasure in the same places indefinitely; we need to substitute ourselves, our neighborhoods, our local resources, for expensive imported goods and services; we need to increase cooperation among all local economic entities: households, farms, factories, banks, consumers, and suppliers. If we are serious about reducing government and the burdens of government, then we need to do so by returning economic self-determination to the people. And we must not do this by inviting destructive industries to provide "jobs" in the community; we must do it by fostering economic democracy. For example, as much as possible of the food that is consumed locally ought to be locally produced on small farms, and then processed in small, non-polluting plants that are locally owned. We must do everything possible to provide to ordinary citizens the opportunity to own a small, usable share of the country. In that way, we will put local capital to work locally, not to exploit and destroy the land but to use it well. This is not work just for the privileged, the well-positioned, the wealthy, and the powerful. It is work for everybody.

I acknowledge that to advocate such reforms is to advocate a kind of secession—not a secession of armed violence but a quiet secession by which people find the practical means and the strength of spirit to remove themselves from an economy that is exploiting them and destroying their homeland. The great, greedy, indifferent national and international economy is killing rural America, just as it is killing America's cities—it is killing our country. Experience has shown that there is no use in appealing to this economy for mercy toward the earth or toward any human community. All true patriots must find ways of opposing it.

Conservation Is Good Work

THERE ARE, as nearly as I can make out, three kinds of conservation currently operating. The first is the preservation of places that are grandly wild or "scenic" or in some other way spectacular. The second is what is called "conservation of natural resources"—that is, of the things of nature that we intend to use: soil, water, timber, and minerals. The third is what you might call industrial troubleshooting: the attempt to limit or stop or remedy the most flagrant abuses of the industrial system. All three kinds of conservation are inadequate, both separately and together.

Right at the heart of American conservation, from the beginning, has been the preservation of spectacular places. The typical American park is in a place that is "breathtakingly" beautiful or wonderful and of little apparent economic value. Mountains, canyons, deserts, spectacular landforms, geysers, waterfalls—these are the stuff of parks. There is, significantly, no prairie national park. Wilderness preserves, as Dave Foreman points out, tend to include much "rock and ice" and little marketable timber. Farmable land, in general, has tempted nobody to make a park. Wes Jackson has commented with some anxiety on the people who charge blindly across Kansas and eastern Colorado, headed for the mountains west of Denver. These are nature lovers and sightseers, but they are utterly oblivious of or bored by the rich natural and human history of the Plains. The point of Wes Jackson's anxiety is that the love of nature that limits itself to the love of places that are "scenic" is implicitly dangerous. It is dangerous because it tends to exclude unscenic places from nature and from the respect that we sometimes accord to nature. This is why so much of the landscape that is productively used is also abused; it is used solely according to standards dictated by the financial system and not at all according to standards dictated by the nature of the place. Moreover, as we are beginning to see, it is going to be extremely difficult to make enough parks to preserve vulnerable species and the health of ecosystems or large watersheds.

"Natural resources," the part of nature that we are going

to use, is the part outside the parks and preserves (which, of course, we also use). But "conservation of natural resources" is now in confusion because it is a concept that has received much lip service but not much thought or practice. Part of the confusion is caused by thinking of "natural resources" as belonging to one category when, in fact, they belong to two: surface resources, like soils and forests, which can be preserved in use; and underground resources, like coal or oil, which cannot be. The one way to conserve the minable fuels and materials that use inevitably exhausts is to limit use. At present, we have no intention of limiting such use, and so we cannot say that we are at all interested in the conservation of exhaustible resources. Surface or renewable resources, on the other hand, can be preserved in use so that their yield is indefinitely sustainable.

Sustainability is a hopeful concept not only because it is a present necessity but because it has a history. We know, for example, that some agricultural soils have been preserved in continuous use for several thousand years. We know, moreover, that it is possible to improve soil in use. And it is clear that a forest can be used in such a way that it remains a forest, with its biological communities intact and its soil undamaged, while producing a yield of timber. But the methods by which exhaustible resources are extracted and used have set the pattern also for the use of sustainable resources, with the result that now soils and forests are not merely being used but are being used up, exactly as coal seams are used up.

Since the sustainable use of renewable resources depends on the existence of settled, small local economies and communities capable of preserving the local knowledge necessary for good farming and forestry, it is obvious that there is no simple, easy, or quick answer to the problem of the exhaustion of sustainable resources. We probably are not going to be able to conserve natural resources so long as our extraction and use of the goods of nature are wasteful and improperly scaled, or so long as these resources are owned or controlled by absentees, or so long as the standard of extraction and use is profitability rather than the health of natural and human communities.

Because we are living in an era of ecological crisis, it is understandable that much of our attention, anxiety, and energy is focused on exceptional cases, the outrages and extreme abuses of the industrial economy: global warming, the global assault on the last remnants of wilderness, the extinction of species, oil spills, chemical spills, Love Canal, Bhopal, Chernobyl, the burning oil fields of Kuwait. But a conservation effort that concentrates only on the extremes of industrial abuse tends to suggest that the only abuses are the extreme ones when, in fact, the earth is probably suffering more from many small abuses than from a few large ones. By treating the spectacular abuses as exceptional, the powers that be would like to keep us from seeing that the industrial system (capitalist or communist or socialist) is in itself and by necessity of all of its assumptions extremely dangerous and damaging and that it exists to support an extremely dangerous and damaging way of life. The large abuses exist within and because of a pattern of smaller abuses.

Much of the Sacramento River is now dead because a carload of agricultural poison was spilled into it. The powers that be would like us to believe that this colossal "accident" was an exception in a general pattern of safe use. Diluted and used according to the instructions on the label, they will tell us, the product that was spilled is harmless. They neglect to acknowledge any of the implications that surround the accident: that if this product is to be used in dilution almost everywhere, it will have to be manufactured, stored, and transported in concentration somewhere; that even in "harmless" dilution, such chemicals are getting into the water, the air, the rain, and into the bodies of animals and people; that when such a product is distributed to the general public, it will inevitably be spilled in its concentrated form in large or small quantities and that such "accidents" are anticipated, discounted as "acceptable risk," and charged to nature and society by the powers that be; that such chemicals are needed, in the first place, because the scale, the methods, and the economy of American agriculture are all monstrously out of kilter; that such chemicals are used to replace the work and intelligence of people; and that such a deformed agriculture is made necessary, in the first place, by the public's demand for a diet that is at once cheap and luxurious—too cheap to support adequate agricultural communities or good agricultural methods or good maintenance of agricultural land and yet so goofily self-indulgent as to demand, in every season, out-of-season foods produced by earth-destroying machines and chemicals.

We tend to forget, too, in our just and necessary outrage at the government-led attack on public lands and the last large tracts of wilderness, that for the very same reasons and to the profit of the very same people, thousands of woodlots are being abusively and wastefully logged.

Here, then, are three kinds of conservation, all of them urgently necessary and all of them failing. Conservationists have won enough victories to give them heart and hope and a kind of accreditation, but they know better than anybody how immense and how baffling their task has become. For all their efforts, our soils and waters, forests and grasslands are being used up. Kinds of creatures, kinds of human life, good natural and human possibilities are being destroyed. Nothing now exists anywhere on earth that is not under threat of human destruction. Poisons are everywhere. Junk is everywhere.

These dangers are large and public, and they inevitably cause us to think of changing public policy. This is good, so far as it goes. There should be no relenting in our efforts to influence politics and politicians. But in the name of honesty and sanity we must recognize the limits of politics. It is, after all, much easier to improve a policy than it is to improve the community the policy attempts to affect. And it is also probable that some changes required by conservation cannot be politically made and that some necessary changes will have to be made by the governed without the help or approval of the government.

I must admit here that my experience over more than twenty years as part of an effort to influence agricultural policy has not been encouraging. Our arguments directed at the government and the universities by now remind me of the ant crawling up the buttocks of the elephant with love on his mind. We have not made much impression. My conclusion, I imagine, is the same as the ant's, for these great projects, once undertaken, are hard to abandon: we have got to get more radical.

However destructive may be the policies of the government and the methods and products of the corporations, the root of the problem is always to be found in private life. We must learn to see that every problem that concerns us as conservationists always leads straight to the question of how we live. The world is being destroyed, no doubt about it, by the greed of the rich and powerful. It is also being destroyed by popular demand. There are not enough rich and powerful people to consume the whole world; for that, the rich and powerful need the help of countless ordinary people. We acquiesce in the wastefulness and destructiveness of the national and global economics by acquiescing in the wastefulness and destructiveness of our own households and communities. If conservation is to have a hope of succeeding, then conservationists, while continuing their effort to change public life, are going to have to begin the effort to change private life as well.

The problems we are worried about are caused not just by other people but by ourselves. And this realization should lead directly to two more. The first is that solving these problems is not work merely for so-called environmental organizations and agencies but also for individuals, families, and local communities. We are used to hearing about turning off unused lights, putting a brick in the toilet tank, using water-saving shower heads, setting the thermostat low, sharing rides, and so forth—pretty dull stuff. But I'm talking about actual jobs of work that are interesting because they require intelligence and because they are accomplished in response to interesting questions: What are the principles of household economy, and how can they be applied under present circumstances? What are the principles of a neighborhood or a local economy, and how can they be applied under present circumstances? What do people already possess in their minds and bodies, in their families and neighborhoods, in their dwellings and in their local landscape, that can replace what is now being supplied by our consumptive and predatory so-called economy? What can we supply to ourselves cheaply or for nothing that we are now paying dearly for? To answer such questions requires more intelligence and involves more pleasure than all the technological breakthroughs of the last two hundred years.

Second, the realization that we ourselves, in our daily economic life, are causing the problems we are trying to solve ought to show us the inadequacy of the language we are using to talk about our connection to the world. The idea that we live in something called "the environment," for instance, is utterly preposterous. This word came into use because of the pretentiousness of learned experts who were embarrassed by the religious associations of "Creation" and who thought "world" too mundane. But "environment" means that which surrounds or encircles us; it means a world separate from ourselves, outside us. The real state of things, of course, is far more complex and intimate and interesting than that. The world that environs us, that is around us, is also within us. We are made of it; we eat, drink, and breathe it; it is bone of our bone and flesh of our flesh. It is also a Creation, a holy mystery, made for and to some extent by creatures, some but by no means all of whom are humans. This world, this Creation, belongs in a limited sense to us, for we may rightfully require certain things of it—the things necessary to keep us fully alive as the kind of creature we are—but we also belong to it, and it makes certain rightful claims on us: that we care properly for it, that we leave it undiminished not just to our children but to all the creatures who will live in it after us. None of this intimacy and responsibility is conveyed by the word *environment.*

That word is a typical product of the old dualism that is at the root of most of our ecological destructiveness. So, of course, is "biocentrism." If life is at the center, what is at the periphery? And for that matter, *where* is the periphery? "Deep ecology," another bifurcating term, implies that there is, a couple of layers up, a shallow ecology that is not so good—or that an ecosystem is a sort of layer cake with the icing on the bottom. Not only is this language incapable of giving a description of our relation to the world; it is also academic, artificial, and pretentious. It is the sort of language used by a visiting expert who does not want the local people to ask any questions. (I am myself an anthropobiotheointerpenetrist and a gastrointero-environmentalist, but I am careful to say so only in the company of other experts.)

No settled family or community has ever called its home place an "environment." None has ever called its feeling for its home place "biocentric" or "anthropocentric." None has ever thought of its connection to its home place as "ecological," deep or shallow. The concepts and insights of the ecologists are of great usefulness in our predicament, and we can hardly escape the need to speak of "ecology" and "ecosystems." But the terms themselves are culturally sterile. They come from the juiceless, abstract intellectuality of the universities which was invented to disconnect, displace, and disembody the mind. The real names of the environment are the names of rivers and river valleys; creeks, ridges, and mountains; towns and cities; lakes, woodlands, lanes, roads, creatures, and people.

And the real name of our connection to this everywhere different and differently named earth is "work." We are connected by work even to the places where we don't work, for all places are connected; it is clear by now that we cannot exempt one place from our ruin of another. The name of our *proper* connection to the earth is "good work," for good work involves much giving of honor. It honors the source of its materials; it honors the place where it is done; it honors the art by which it is done; it honors the thing that it makes and the user of the made thing. Good work is always modestly scaled, for it cannot ignore either the nature of individual places or the differences between places, and it always involves a sort of religious humility, for not everything is known. Good work can be defined only in particularity, for it must be defined a little differently for every one of the places and every one of the workers on the earth.

The name of our present society's connection to the earth is "bad work"—work that is only generally and crudely defined, that enacts a dependence that is ill understood, that enacts no affection and gives no honor. Every one of us is to some extent guilty of this bad work. This guilt does not mean that we must indulge in a lot of breast-beating and confession; it means only that there is much good work to be done by every one of us and that we must begin to do it. All of us are responsible for bad work, not so much because we do it ourselves (though we all do it) as because we have it done for us by other people.

Here we are bound to see our difficulty as almost overwhelming. What proxies have we issued, and to whom, to use the earth on our behalf? How, in this global economy, are we to render anything like an accurate geographical account of our personal economies? How do we take our lives from this earth that we are so anxious to protect and restore to health?

Most of us get almost all the things we need by buying them; most of us know only vaguely, if at all, where those things come from; and most of us know not at all what damage is involved in their production. We are almost entirely dependent on an economy of which we are almost entirely ignorant. The provenance, for example, not only of the food we buy at the store but of the chemicals, fuels, metals, and other materials necessary to grow, harvest, transport, process, and package that food is almost necessarily a mystery to us. To know the full economic history of a head of supermarket cauliflower would require an immense job of research. To be so completely and so ignorantly dependent on the present abusive food economy certainly defines us as earth abusers. It also defines us as potential victims.

Living as we now do in almost complete dependence on a global economy, we are put inevitably into a position of ignorance and irresponsibility. No one can know the whole globe. We can connect ourselves to the globe as a whole only by means of a global economy that, without knowing the earth, plunders it for us. The global economy (like the national economy before it) operates on the superstition that the deficiencies or needs or wishes of one place may safely be met by the ruination of another place. To build houses here, we clear-cut the forests there. To have air-conditioning here, we strip-mine the mountains there. To drive our cars here, we sink our oil wells there. It is an absentee economy. Most people aren't using or destroying what they can see. If we cannot see our garbage or the grave we have dug with our energy proxies, then we assume that all is well. The issues of carrying capacity and population remain abstract and not very threatening to most people for the same reason. If this nation or region cannot feed its population, then food can be imported from other nations or regions. All the critical questions affecting our use of the earth are left to be answered by "the market" or the law of supply and demand, which proposes no limit on either supply or demand. An economy without limits is an economy without discipline.

Conservationists of all kinds would agree, I think, that no discipline, public or private, is implied by the industrial economy and that none is practiced by it. The implicit wish of the industrial economy is that producers might be wasteful, shoddy, and irresponsible and that consumers might be gullible, extravagant, and irresponsible. To fulfill this wish, the industrial economy employs an immense corps of hireling politicians, publicists, lobbyists, admen, and adwomen. The consequent ruin is notorious: we have been talking about it for generations; it brought conservation into being. And conservationists have learned very well how to address this ruin as a public problem. There is now no end to the meetings and publications in which the horrifying statistics are recited, usually with the conclusion that pressure should be put on the government to do something. Often, this pressure has been applied, and the government has done something. But the government has not done enough and may never do enough. It is likely that the government *cannot* do enough.

The government's disinclination to do more than it does is explained, of course, by the government's bought-and-paid-for servitude to interests that do not want it to do more. But there may also be a limit of another kind: a government that could do enough, assuming it had the will, would almost certainly be a government radically and unpleasantly different from the one prescribed by our Constitution. A government undertaking to protect all of nature that is now abused or threatened would have to take total control of the country. Police and bureaucrats—and opportunities for malfeasance—would be everywhere. To wish only for a public or a political solution to the problem of conservation may be to wish for a solution as bad as the problem and still be unable to solve it.

The way out of this dilemma is to understand the ruin of nature as a problem that is both public and private. The failure of public discipline in matters of economy is only the other face of the failure of private discipline. If we have worked at the issues of public policy so long and so exclusively as to bring political limits into sight, then let us turn—not instead but also—to issues of private economy and see how far we can go in that direction. It is a direction that may take us further and produce more satisfactory and lasting results than the direction of policy.

The dilemma of private economic responsibility, as I have said, is that we have allowed our suppliers to enlarge our economic boundaries so far that we cannot be responsible for our effects on the world. The only remedy for this that I can see is to draw in our economic boundaries, shorten our supply lines, so as to permit us literally to know where we are economically. The closer we live to the ground that we live from, the more we will know about our economic life; the more we know about our economic life, the more able we will be to take responsibility for it. The way to bring discipline into one's personal or household or community economy is to limit one's economic geography.

This obviously opens up an agenda almost as daunting as the political agenda. The difference—a consoling one—is that when we try to influence policy, only large jobs must be done; whereas when we seek to reform private economies, the work is necessarily modest, and it can be started by anybody anywhere. What is required is the formation of local economic strategies and eventually of local economies—by which to resist abuses of natural and human communities by the larger economy. And, of course, in talking about the formation of local economies capable of using an earthly place without ruining it, we are talking about the reformation of people; we are talking about reviving good work as an economic force.

If we think of this task of rebuilding local economies as one large task that must be done in a hurry, then we will again be overwhelmed and will want the government to do it. If, on the other hand, we define the task as beginning the reformation of our private or household economies, then the way is plain. What we must do is use well the considerable power we have as consumers: the power of choice. We can choose to buy or not to buy, and we can choose what to buy. The standard by which we choose must be the health of the community—and by that we must mean the *whole* community: ourselves, the place where we live, and all the humans and other creatures who live there with us. In a healthy community, people will be richer in their neighbors, in neighborhood, in the health and pleasure of neighborhood, than in their bank accounts. It is better, therefore, even if the cost is greater, to buy near at hand than to buy at a distance. It is better to buy from a small, privately owned local store than from a chain store. It is better to buy a good product than a bad one. Do not buy anything you don't need. Do as much as you can for yourself. If you cannot do something for yourself, see if you have a neighbor who can do it for you. Do everything you can to see that your money stays as long as possible in the local community. If you have money to invest, try to invest it locally, both to help the local community and to keep from helping the larger economy that is destroying local communities. Begin to ask yourself how your money could be put at minimal interest into the hands of a young person who wants to start a farm, a store, a shop, or a small business that the community needs. This agenda can be followed by individuals and single families. If it is followed by people in groups—churches, conservation organizations, neighborhood associations, groups of small farmers, and the like—the possibilities multiply and the effects will be larger.

The economic system that most affects the health of the world and that may be most subject to consumer influence is that of food. And the issue of food provides an excellent example of what I am talking about. If you want to reform your own food economy, you can make a start without anybody's permission or help. If you have a place to do it, grow some food for yourself. Growing some of your own food gives you pleasure, exercise, knowledge, sales resistance, and standards. Your own food, if you grow it the right way, will taste good and so will cause you to wish to buy food that tastes good. So far as you can, buy food that is locally grown. Tell your grocer that you are interested in locally grown food. If you can't find locally grown food in stores, then see if you can deal directly with a local farmer. The value of this, for conservationists, is that when consumers are acquainted and friendly with their producers, they can influence production. They can know the land on which their food is produced. They can refuse to buy food that is produced with dangerous chemicals or by other destructive practices. As these connections develop, local agriculture will diversify, become more healthy and more stable, employ more people. As local demand increases and becomes more knowledgeable, small food-processing industries will enter the local economy. Everything that is done by the standard of community health will make new possibilities for good work —that is for the responsible use of the world.

The forest economy is not so obviously subject to consumer influence, but such influence is sorely needed. Both the forests themselves and their human communities suffer for the want of local forest economies—properly scaled wood-products industries that would be the basis of stable communities and would provide local incentives for the good use of the forest. People who see that they must depend on the forest for generations, in a complex local forest economy, will want the forest to last and be healthy; they will *not* want to see all the marketable timber ripped out of it as fast as possible. Both forest and farm communities would benefit from technologies that could be locally supplied and maintained. Draft horses, for example, are better than large machines, both for the woods and for the local economy.

The economy of recreation has hardly been touched as an issue of local economy and conservation, though conservationists and consumers alike have much to gain from making it such an issue. At present, there is an almost complete disconnection between the economic use of privately owned farm and forest land and its use for recreation. Such land is now much used by urban people for hunting and fishing, but mainly without benefit to the landowners, who therefore receive no incentive from this use to preserve wildlife habitat or to take the best care of their woodlands and stream margins. They need to receive such incentives. It is not beyond reason that public funds might be given to private landowners to preserve and enhance the recreational value—that is, the wildness—of their land. But since governments are unlikely to do this soon, the incentives need to be provided by consumer and conservation groups working in cooperation with farm groups. The rule of the food economy ought to apply to the recreation economy: find your pleasure and your rest as near home as possible. In Kentucky, for example, we have hundreds of miles of woodland stretching continuously along the sides of our creek and river valleys. Why should conservation and outdoor groups not pay an appropriate price to farmers to maintain hiking trails and campsites and preserve the forests in such places? The money that would carry a family to a vacation in a distant national park could thus be kept at home and used to help the local economy and protect the local countryside.

The point of all this is the use of local buying power, local gumption, and local affection to see that the best care is taken of the local land. This sort of effort would bridge the gap, now so destructive, between the conservationists and the small farmers and ranchers, and that would be one of its great political benefits. But the fundamental benefit would be to the world and ourselves. We would begin to protect the world not just by conserving it but also by living in it.

Christianity and the Survival of Creation

I CONFESS that I have not invariably been comfortable in front of a pulpit; I have never been comfortable behind one. To be behind a pulpit is always a forcible reminder to me that I am an essayist and, in many ways, a dissenter. An essayist is, literally, a writer who attempts to tell the truth. Preachers must resign themselves to being either right or wrong; an essayist, when proved wrong, may claim to have been "just practicing." An essayist is privileged to speak without institutional authorization. A dissenter, of course, must speak without privilege.

I want to begin with a problem: namely, that the culpability of Christianity in the destruction of the natural world and the uselessness of Christianity in any effort to correct that destruction are now established clichés of the conservation movement. This is a problem for two reasons.

First, the indictment of Christianity by the anti-Christian conservationists is, in many respects, just. For instance, the complicity of Christian priests, preachers, and missionaries in the cultural destruction and the economic exploitation of the primary peoples of the Western Hemisphere, as of traditional cultures around the world, is notorious. Throughout the five hundred years since Columbus's first landfall in the Bahamas, the evangelist has walked beside the conqueror and the merchant, too often blandly assuming that their causes were the same. Christian organizations, to this day, remain largely indifferent to the rape and plunder of the world and of its traditional cultures. It is hardly too much to say that most Christian organizations are as happily indifferent to the ecological, cultural, and religious implications of industrial economics as are most industrial organizations. The certified Christian seems just as likely as anyone else to join the military-industrial conspiracy to murder Creation.

The conservationist indictment of Christianity is a problem, second, because, however just it may be, it does not come from an adequate understanding of the Bible and the cultural traditions that descend from the Bible. The anti-Christian conservationists characteristically deal with the Bible by waving it off. And this dismissal conceals, as such dismissals are apt to do, an ignorance that invalidates it. The Bible is an inspired book written by human hands; as such, it is certainly subject to criticism. But the anti-Christian environmentalists have not mastered the first rule of the criticism of books: you have to read them before you criticize them. Our predicament now, I believe, requires us to learn to read and understand the Bible in the light of the present fact of Creation. This would seem to be a requirement both for Christians and for everyone concerned, but it entails a long work of true criticism—that is, of careful and judicious study, not dismissal. It entails, furthermore, the making of very precise distinctions between biblical instruction and the behavior of those peoples supposed to have been biblically instructed.

I cannot pretend, obviously, to have made so meticulous a study; even if I were capable of it, I would not live long enough to do it. But I have attempted to read the Bible with these issues in mind, and I see some virtually catastrophic discrepancies between biblical instruction and Christian behavior. I don't mean disreputable Christian behavior, either. The discrepancies I see are between biblical instruction and allegedly respectable Christian behavior.

If because of these discrepancies Christianity were dismissible, there would, of course, be no problem. We could simply dismiss it, along with the twenty centuries of unsatisfactory history attached to it, and start setting things to rights. The problem emerges only when we ask, Where then would we turn for instruction? We might, let us suppose, turn to another religion—a recourse that is sometimes suggested by the anti-Christian conservationists. Buddhism, for example, is certainly a religion that could guide us toward a right respect for the natural world, our fellow humans, and our fellow creatures. I owe a considerable debt myself to Buddhism and Buddhists. But there are an enormous number of people—and I am one of them—whose native religion, for better or worse, is Christianity. We were born to it; we began to learn about it before we became conscious; it is, whatever we think of it, an intimate belonging of our being; it informs our consciousness, our language, and our dreams. We can turn away from it or against it, but that will only bind us tightly to a reduced version of it. A better possibility is that this, our native religion, should survive and renew itself so that it may become as largely and truly instructive as we need it to be. On such a survival and renewal of the Christian religion may depend the survival of the Creation that is its subject.

II

If we read the Bible, keeping in mind the desirability of those two survivals—of Christianity and the Creation—we are apt to discover several things about which modern Christian organizations have kept remarkably quiet or to which they have paid little attention.

We will discover that we humans do not own the world or any part of it: "The earth is the Lord's, and the fulness thereof: the world and they that dwell therein." There is in our human law, undeniably, the concept and right of "land ownership." But this, I think, is merely an expedient to safeguard the mutual belonging of people and places without which there can be no lasting and conserving human communities. This right of human ownership is limited by mortality and by natural constraints on human attention and responsibility; it quickly becomes abusive when used to justify large accumulations of "real estate," and perhaps for that reason such large accumulations are forbidden in the twenty-fifth chapter of Leviticus. In biblical terms, the "landowner" is the guest and steward of God: "The land is mine; for ye are strangers and sojourners with me."

We will discover that God made not only the parts of Creation that we humans understand and approve but all of it: "All things were made by him; and without him was not anything made that was made." And so we must credit God with the making of biting and stinging insects, poisonous serpents, weeds, poisonous weeds, dangerous beasts, and disease-causing microorganisms. That we may disapprove of these things does not mean that God is in error or that He ceded some of the work of Creation to Satan; it means that we are deficient in wholeness, harmony, and understanding—that is, we are "fallen."

We will discover that God found the world, as He made it, to be good, that He made it for His pleasure, and that He continues to love it and to find it worthy, despite its reduction and corruption by us. People who quote John 3:16 as an easy formula for getting to Heaven neglect to see the great difficulty implied in the statement that the advent of Christ was made possible by God's love for the world—not God's love for Heaven or for the world as it might be but for the world as it was and is. Belief in Christ is thus dependent on prior belief in the inherent goodness—the lovability—of the world.

We will discover that the Creation is not in any sense independent of the Creator, the result of a primal creative act long over and done with, but is the continuous, constant participation of all creatures in the being of God. Elihu said to Job that if God "gather unto himself his spirit and his breath; all flesh shall perish together." And Psalm 104 says, "Thou sendest forth thy spirit, they are created." Creation is thus God's presence in creatures. The Greek Orthodox theologian Philip Sherrard has written that "Creation is nothing less than the manifestation of God's hidden Being." This means that we and all other creatures live by a sanctity that is inexpressibly intimate, for to every creature, the gift of life is a portion of the breath and spirit of God. As the poet George Herbert put it:

> Thou art in small things great, not small in any . . .
> For thou art infinite in one and all.

We will discover that for these reasons our destruction of nature is not just bad stewardship, or stupid economics, or a betrayal of family responsibility; it is the most horrid blasphemy. It is flinging God's gifts into His face, as if they were of no worth beyond that assigned to them by our destruction of them. To Dante, "despising Nature and her goodness" was a violence against God. We have no entitlement from the Bible to exterminate or permanently destroy or hold in contempt anything on the earth or in the heavens above it or in the waters beneath it. We have the right to use the gifts of nature but not to ruin or waste them. We have the right to use what we need but no more, which is why the Bible forbids usury and great accumulations of property. The usurer, Dante said, "condemns Nature . . . for he puts his hope elsewhere."

William Blake was biblically correct, then, when he said that "everything that lives is holy." And Blake's great commentator Kathleen Raine was correct both biblically and historically when she said that "the sense of the holiness of life is the human norm."

The Bible leaves no doubt at all about the sanctity of the act of world-making, or of the world that was made, or of creaturely or bodily life in this world. We are holy creatures living among other holy creatures in a world that is holy. Some people know this, and some do not. Nobody, of course, knows it all the time. But what keeps it from being far better known than it is? Why is it apparently unknown to millions of professed students of the Bible? How can modern Christianity have so solemnly folded its hands while so much of the work of God was and is being destroyed?

III

Obviously, "the sense of the holiness of life" is not compatible with an exploitive economy. You cannot know that life is holy if you are content to live from economic practices that daily destroy life and diminish its possibility. And many if not most Christian organizations now appear to be perfectly at peace with the military-industrial economy and its "scientific" destruction of life. Surely, if we are to remain free and if we are to remain true to our religious inheritance, we must maintain a separation between church and state. But if we are to maintain any sense or coherence or meaning in our lives, we cannot tolerate the present utter disconnection between religion and economy. By "economy" I do not mean "economics," which is the study of money-making, but rather the ways of human housekeeping, the ways by which the human household is situated and maintained within the household of nature. To be uninterested in economy is to be uninterested in the practice of religion; it is to be uninterested in culture and in character. Probably the most urgent question now faced by people who would adhere to the Bible is this: What sort of economy would be responsible to the holiness of life? What, for Christians, would be the economy, the practices and the restraints, of "right livelihood"? I do not believe that organized Christianity now has any idea. I think its idea of a Christian economy is no more or less than the industrial economy—which is an economy firmly founded on the seven deadly sins and the breaking of all ten of the Ten Commandments. Obviously, if Christianity is going to survive as more than a respecter and comforter of profitable iniquities, then Christians, regardless of their organizations, are going to have to interest themselves in economy—which is to say, in nature and in work. They are going to have to give workable answers to those who say we cannot live without this economy that is destroying us and our world, who see the murder of Creation as the only way of life.

The holiness of life is obscured to modern Christians also by the idea that the only holy place is the built church. This idea may be more taken for granted than taught; nevertheless, Christians are encouraged from childhood to think of the church building as "God's house," and most of them could think of their houses or farms or shops or factories as holy places only with great effort and embarrassment. It is understandably difficult for modern Americans to think of their dwellings and workplaces as holy, because most of these are, in fact, places of desecration, deeply involved in the ruin of Creation.

The idea of the exclusive holiness of church buildings is, of course, wildly incompatible with the idea, which the churches also teach, that God is present in all places to hear prayers. It is incompatible with Scripture. The idea that a human artifact could contain or confine God was explicitly repudiated by Solomon in his prayer at the dedication of the Temple: "Behold, the heaven and the heaven of heavens cannot contain thee: how much less this house that I have builded?" And these words of Solomon were remembered a thousand years later by Saint Paul, preaching at Athens:

> God that made the world and all things therein, seeing that he is lord of heaven and earth, dwelleth not in temples made with hands . . .
>
> For in him we live, and move, and have our being; as certain also of your own poets have said.

Idolatry always reduces to the worship of something "made with hands," something confined within the terms of human work and human comprehension. Thus, Solomon and Saint Paul both insisted on the largeness and the at-largeness of God, setting Him free, so to speak, from *ideas* about Him. He is not to be fenced in, under human control, like some domestic creature; He is the wildest being in existence. The presence of His spirit in us is our wildness, our oneness with the wilderness of Creation. That is why subduing the things of nature to human purposes is so dangerous and why it so often results in evil, in separation and desecration. It is why the poets of our tradition so often have given nature the role not only of mother or grandmother but of the highest earthly teacher and judge, a figure of mystery and great power. Jesus' own specifications for his church have nothing at all to do with masonry and carpentry but only with people; his church is "where two or three are gathered together in my name."

The Bible gives exhaustive (and sometimes exhausting) attention to the organization of religion: the building and rebuilding of the Temple; its furnishings; the orders, duties, and paraphernalia of the priesthood; the orders of rituals and ceremonies. But that does not disguise the fact that the most significant religious events recounted in the Bible do not occur in "temples made with hands." The most important religion in that book is unorganized and is sometimes profoundly disruptive of organization. From Abraham to Jesus, the most important people are not priests but shepherds, soldiers, property owners, workers, housewives, queens and kings, manservants and maidservants, fishermen, prisoners, whores, even bureaucrats. The great visionary encounters did not take place in temples but in sheep pastures, in the desert, in the wilderness, on mountains, on the shores of rivers and the sea, in the middle of the sea, in prisons. And however strenuously the divine voice prescribed rites and observances, it just as strenuously repudiated them when they were taken to *be* religion:

> Your new moons and your appointed feasts my soul hateth: they are a trouble unto me; I am weary to bear them.
>
> And when you spread forth your hands, I will hide mine eyes from you: yea, when you make many prayers, I will not hear: your hands are full of blood.
>
> Wash you, make you clean; put away the evil of your doings from before mine eyes; cease to do evil;
>
> Learn to do well; seek judgment, relieve the oppressed, judge the fatherless, plead for the widow.

Religion, according to this view, is less to be celebrated in rituals than practiced in the world.

I don't think it is enough appreciated how much an outdoor book the Bible is. It is a "hypaethral book," such as Thoreau talked about—a book open to the sky. It is best read and understood outdoors, and the farther outdoors the better. Or that has been my experience of it. Passages that within walls seem improbable or incredible, outdoors seem merely natural. This is because outdoors we are confronted everywhere with wonders; we see that the miraculous is not extraordinary but the common mode of existence. It is our daily bread. Whoever really has considered the lilies of the field or the birds of the air and pondered the improbability of their existence in this warm world within the cold and empty stellar distances will hardly balk at the turning of water into wine—which was, after all, a very small miracle. We forget the greater and still continuing miracle by which water (with soil and sunlight) is turned into grapes.

It is clearly impossible to assign holiness exclusively to the built church without denying holiness to the rest of Creation, which is then said to be "secular." The world, which God looked at and found entirely good, we find none too good to pollute entirely and destroy piecemeal. The church, then, becomes a kind of preserve of "holiness," from which certified lovers of God assault and plunder the "secular" earth.

Not only does this repudiate God's approval of His work; it refuses also to honor the Bible's explicit instruction to regard the works of the Creation as God's revelation of Himself. The assignation of holiness exclusively to the built church is therefore logically accompanied by the assignation of revelation exclusively to the Bible. But Psalm 19 begins, "The heavens declare the glory of God; and the firmament sheweth his handiwork." The word of God has been revealed in facts from the moment of the third verse of the first chapter of Genesis: "Let there be light: and there was light." And Saint Paul states the rule: "The invisible things of him from the creation of the world are clearly seen, being understood by the things that are made." Yet from this free, generous, and sensible view of things, we come to the idolatry of the book: the idea that nothing is true that cannot be (and has not been already) written.

The misuse of the Bible thus logically accompanies the abuse of nature: if you are going to destroy creatures without respect, you will want to reduce them to "materiality"; you will want to deny that there is spirit or truth in them, just as you will want to believe that the only holy creatures, the only creatures with souls, are humans—or even only Christian humans.

By denying spirit and truth to the nonhuman Creation, modern proponents of religion have legitimized a form of blasphemy without which the nature- and culture-destroying machinery of the industrial economy could not have been built—that is, they have legitimized bad work. Good human work honors God's work. Good work uses no thing without respect, both for what it is in itself and for its origin. It uses neither tool nor material that it does not respect and that it does not love. It honors nature as a great mystery and power, as an indispensable teacher, and as the inescapable judge of all work of human hands. It does not dissociate life and work, or pleasure and work, or love and work, or usefulness and beauty. To work without pleasure or affection, to make a product that is not both useful and beautiful, is to dishonor God, nature, the thing that is made, and whomever it is made for. This is blasphemy: to make shoddy work of the work of God. But such is not possible when the entire Creation is understood as holy and when the works of God are understood as embodying and thus revealing His spirit.

In the Bible we find none of the industrialist's contempt or hatred for nature. We find, instead, a poetry of awe and reverence and profound cherishing, as in these verses from Moses' valedictory blessing of the twelve tribes:

> And of Joseph he said, Blessed of the Lord be his land, for the precious things of heaven, for the dew, and for the deep that croucheth beneath,
>
> And for the precious fruits brought forth by the sun, and for the precious things put forth by the moon,
>
> And for the chief things of the ancient mountains, and for the precious things of the lasting hills,
>
> And for the precious things of the earth and fullness thereof, and for the good will of him that dwelt in the bush.

IV

I have been talking, of course, about a dualism that manifests itself in several ways: as a cleavage, a radical discontinuity, between Creator and creature, spirit and matter, religion and nature, religion and economy, worship and work, and so on. This dualism, I think, is the most destructive disease that afflicts us. In its best-known, its most dangerous, and perhaps its fundamental version, it is the dualism of body and soul. This is an issue as difficult as it is important, and so to deal with it we should start at the beginning.

The crucial test is probably Genesis 2:7, which gives the process by which Adam was created: "The Lord God formed man of the dust of the ground, and breathed into his nostrils the breath of life: and man became a living soul." My mind, like most people's, has been deeply influenced by dualism, and I can see how dualistic minds deal with this verse. They conclude that the formula for man-making is man = body + soul. But that conclusion cannot be derived, except by violence, from Genesis 2:7, which is not dualistic. The formula given in Genesis 2:7 is not man = body + soul; the formula there is soul = dust + breath. According to this verse, God did not make a body and put a soul into it, like a letter into an envelope. He formed man of dust; then, by breathing His breath into it, He made the dust live. The dust, formed as man and made to live, did not *embody* a soul; it *became* a soul. "Soul" here refers to the whole creature. Humanity is thus presented to us, in Adam, not as a creature of two discrete parts temporarily glued together but as a single mystery.

We can see how easy it is to fall into the dualism of body and soul when talking about the inescapable worldly dualities of good and evil or time and eternity. And we can see how easy it is, when Jesus asks, "For what is a man profited, if he shall gain the whole world, and lose his own soul?" to assume that he is condemning the world and appreciating the disembodied soul. But if we give to "soul" here the sense that it has in Genesis 2:7, we see that he is doing no such thing. He is warning that in pursuit of so-called material possessions, we can lose our understanding of ourselves as "living souls"—that is, as creatures of God, members of the holy community of Creation. We can lose the possibility of the atonement of that membership. For we are free, if we choose, to make a duality of our one living soul by disowning the breath of God that is our fundamental bond with one another and with other creatures.

But we can make the same duality by disowning the dust. The breath of God is only one of the divine gifts that make us living souls; the other is the dust. Most of our modern troubles come from our misunderstanding and misvaluation of this dust. Forgetting that the dust, too, is a creature of the Creator, made by the sending forth of His spirit, we have presumed to decide that the dust is "low." We have presumed to say that we are made of two parts: a body and a soul, the body being "low" because made of dust, and the soul "high." By thus valuing these two supposed-to-be parts, we inevitably throw them into competition with each other, like two corporations. The "spiritual" view, of course, has been that the body, in Yeats's phrase, must be "bruised to pleasure soul." And the "secular" version of the same dualism has been that the body, along with the rest of the "material" world, must give way before the advance of the human mind. The dominant religious view, for a long time, has been that the body is a kind of scrip issued by the Great Company Store in the Sky, which can be cashed in to redeem the soul but is otherwise worthless. And the predictable result has been a human creature able to appreciate or tolerate only the "spiritual" (or mental) part of Creation and full of semiconscious hatred of the "physical" or "natural" part, which it is ready and willing to destroy for "salvation," for profit, for "victory," or for fun. This madness constitutes the norm of modern humanity and of modern Christianity.

But to despise the body or mistreat it for the sake of the "soul" is not just to burn one's house for the insurance, nor is it just self-hatred of the most deep and dangerous sort. It is yet another blasphemy. It is to make nothing—and worse than nothing—of the great Something in which we live and move and have our being.

When we hate and abuse the body and its earthly life and joy for Heaven's sake, what do we expect? That out of this life that we have presumed to despise and this world that we have presumed to destroy, we would somehow salvage a soul capable of eternal bliss? And what do we expect when with equal and opposite ingratitude, we try to make of the finite body an infinite reservoir of dispirited and meaningless pleasures?

Times may come, of course, when the life of the body must be denied or sacrificed, times when the whole world must literally be lost for the sake of one's life as a "living soul." But such sacrifice, by people who truly respect and revere the life of the earth and its Creator, does not denounce or degrade the body but rather exalts it and acknowledges its holiness. Such sacrifice is a refusal to allow the body to serve what is unworthy of it.

V

If we credit the Bible's description of the relationship between Creator and Creation, then we cannot deny the spiritual importance of our economic life. Then we must see how religious issues lead to issues of economy and how issues of economy lead to issues of art. By "art" I mean all the ways by which humans make the things they need. If we understand that no artist—no maker—can work except by reworking the works of Creation, then we see that by our work we reveal what we think of the works of God. How we take our lives from this world, how we work, what work we do, how well we use the materials we use, and what we do with them after we have used them—all these are questions of the highest and gravest religious significance. In answering them, we practice, or do not practice, our religion.

The significance—and ultimately the quality—of the work we do is determined by our understanding of the story in which we are taking part.

If we think of ourselves as merely biological creatures, whose story is determined by genetics or environment or history or economics or technology, then, however pleasant or painful the part we play, it cannot matter much. Its significance is that of mere self-concern. "It is a tale / Told by an idiot, full of sound and fury, / Signifying nothing," as Macbeth says when he has "supp'd full with horrors" and is "aweary of the sun."

If we think of ourselves as lofty souls trapped temporarily in lowly bodies in a dispirited, desperate, unlovable world that we must despise for Heaven's sake, then what have we done for this question of significance? If we divide reality into two parts, spiritual and material, and hold (as the Bible does *not* hold) that only the spiritual is good or desirable, then our relation to the material Creation becomes arbitrary, having only the quantitative or mercenary value that we have, in fact and for this reason, assigned to it. Thus, we become the judges and inevitably the destroyers of a world we did not make and that we are bidden to understand as a divine gift. It is impossible to see how good work might be accomplished by people who think that our life in this world either signifies nothing or has only a negative significance.

If, on the other hand, we believe that we are living souls, God's dust and God's breath, acting our parts among other creatures all made of the same dust and breath as ourselves; and if we understand that we are free, within the obvious limits of mortal human life, to do evil or good to ourselves and to the other creatures—then all our acts have a supreme significance. If it is true that we are living souls and morally free, then all of us are artists. All of us are makers, within mortal terms and limits, of our lives, of one another's lives, of things we need and use.

This, Ananda Coomaraswamy wrote, is "the normal view," which "assumes . . . not that the artist is a special kind of man, but that every man who is not a mere idler or parasite is necessarily some special kind of artist." But since even mere idlers and parasites may be said to work inescapably, by proxy or influence, it might be better to say that everybody is an artist—either good or bad, responsible or irresponsible. Any life, by working or not working, by working well or poorly, inescapably changes other lives and so changes the world. This is why our division of the "fine arts" from "craftsmanship," and "craftsmanship" from "labor," is so arbitrary, meaningless, and destructive. As Walter Shewring rightly said, both "the plowman and the potter have a cosmic function." And bad art in any trade dishonors and damages Creation.

If we think of ourselves as living souls, immortal creatures, living in the midst of a Creation that is mostly mysterious, and if we see that everything we make or do cannot help but have an everlasting significance for ourselves, for others, and for the world, then we see why some religious teachers have understood work as a form of prayer. We see why the old poets invoked the muse. And we know why George Herbert prayed, in his poem "Mattens":

> Teach me thy love to know;
> That this new light, which now I see,
> May both the work and workman show.

Work connects us both to Creation and to eternity. This is the reason also for Mother Ann Lee's famous instruction: "Do all your work as though you had a thousand years to live on earth, and as you would if you knew you must die tomorrow."

Explaining "the perfection, order, and illumination" of the artistry of Shaker furniture makers, Coomaraswamy wrote, "All tradition has seen in the Master Craftsman of the Universe the exemplar of the human artist or 'maker by art,' and we are told to be 'perfect, *even as* your Father in heaven is perfect.'" Searching out the lesson, for us, of the Shakers' humble, impersonal, perfect artistry, which refused the modern divorce of utility and beauty, he wrote, "Unfortunately, we do not desire to be such as the Shaker was; we do not propose to 'work as though we had a thousand years to live, and as though we were to die tomorrow.' Just as we desire peace but not the things that make for peace, so we desire art but not the things that make for art. . . . we have the art that we deserve. If the sight of it puts us to shame, it is with ourselves that the reformation must begin."

Any genuine effort to "re-form" our arts, our ways of making, must take thought of "the things that make art." We must see that no art begins in itself; it begins in other arts, in attitudes and ideas antecedent to any art, in nature, and in inspiration. If we look at the great artistic traditions, as it is necessary to do, we will see that they have never been divorced either from religion or from economy. The possibility of an entirely secular art and of works of art that are spiritless or ugly or useless is not a possibility that has been among us for very long. Traditionally, the arts have been ways of making that have placed a just value on their materials or subjects, on the uses and the users of the things made by art, and on the artists themselves. They have, that is, been ways of giving honor to the works of God. The great artistic traditions have had nothing to do with what we call "self-expression." They have not been destructive of privacy or exploitive of private life. Though they have certainly originated things and employed genius, they have no affinity with the modern cults of originality and genius. Coomaraswamy, a good guide as always, makes an indispensable distinction between genius in the modern sense and craftsmanship: "Genius inhabits a world of its own. The master craftsman lives in a world inhabited by other men; he has neighbors." The arts, traditionally, belong to the neighborhood. They are the means by which the neighborhood lives, works, remembers, worships, and enjoys itself.

But most important of all, now, is to see that the artistic traditions understood every art primarily as a skill or craft and ultimately as a service to fellow creatures and to God. An artist's first duty, according to this view, is technical. It is assumed that one will have talents, materials, subjects—perhaps even genius or inspiration or vision. But these are traditionally understood not as personal properties with which one may do as one chooses but as gifts of God or nature that must be honored in use. One does not dare to use these things without the skill to use them well. As Dante said of his own art, "far worse than in vain does he leave the shore . . . who fishes for the truth and has not the art." To use gifts less than well is to dishonor them and their Giver. There is no material or subject in Creation that in using, we are excused from using well; there is no work in which we are excused from being able and responsible artists.

VI

In denying the holiness of the body and of the so-called physical reality of the world—and in denying support to the good economy, the good work, by which alone the Creation can receive due honor—modern Christianity generally has cut itself off from both nature and culture. It has no serious or competent interest in biology or ecology. And it is equally uninterested in the arts by which humankind connects itself to nature. It manifests no awareness of the specifically Christian cultural lineages that connect us to our past. There is, for example, a splendid heritage of Christian poetry in English that most church members live and die without reading or hearing or hearing about. Most sermons are preached without any awareness at all that the making of sermons is an art that has at times been magnificent. Most modern churches look like they were built by robots without reference to the heritage of church architecture or respect for the place; they embody no awareness that work can be worship. Most religious music now attests to the general assumption that religion is no more than a vaguely pious (and vaguely romantic) emotion.

Modern Christianity, then, has become as specialized in its organizations as other modern organizations, wholly concentrated on the industrial shibboleths of "growth," counting its success in numbers, and on the very strange enterprise of "saving" the individual, isolated, and disembodied soul. Having witnessed and abetted the dismemberment of the households, both human and natural, by which we have our being as creatures of God, as living souls, and having made light of the great feast and festival of Creation to which we were bidden as living souls, the modern church presumes to be able to save the soul as an eternal piece of private property. It presumes moreover to save the souls of people in other countries and religious traditions, who are often saner and more religious than we are. And always the emphasis is on the individual soul. Some Christian spokespeople give the impression that the highest Christian bliss would be to get to Heaven and find that you are the only one there—that you were right and all the others wrong. Whatever its twentieth-century dress, modern Christianity as I know it is still at bottom the religion of Miss Watson, intent on a dull and superstitious rigmarole by which supposedly we can avoid going to "the bad place" and instead go to "the good place." One can hardly help sympathizing with Huck Finn when he says, "I made up my mind I wouldn't try for it."

Despite its protests to the contrary, modern Christianity has become willy-nilly the religion of the state and the economic status quo. Because it has been so exclusively dedicated to incanting anemic souls into Heaven, it has been made the tool of much earthly villainy. It has, for the most part, stood silently by while a predatory economy has ravaged the world, destroyed its natural beauty and health, divided and plundered its human communities and households. It has flown the flag and chanted the slogans of empire. It has assumed with the economists that "economic forces" automatically work for good and has assumed with the industrialists and militarists that technology determines history. It has assumed with almost everybody that "progress" is good, that it is good to be modern and up with the times. It has admired Caesar and comforted him in his depredations and defaults. But in its de facto alliance with Caesar, Christianity connives directly in the murder of Creation. For in these days, Caesar is no longer a mere destroyer of armies, cities, and nations. He is a contradicter of the fundamental miracle of life. A part of the normal practice of his power is his willingness to destroy the world. He prays, he says, and churches everywhere compliantly pray with him. But he is praying to a God whose works he is prepared at any moment to destroy. What could be more wicked than that, or more mad?

The religion of the Bible, on the contrary, is a religion of the state and the status quo only in brief moments. In practice, it is a religion for the correction equally of people and of kings. And Christ's life, from the manger to the cross, was an affront to the established powers of his time, just as it is to the established powers of our time. Much is made in churches of the "good news" of the Gospels. Less is said of the Gospels' bad news, which is that Jesus would have been horrified by just about every "Christian" government the world has ever seen. He would be horrified by our government and its works, and it would be horrified by him. Surely no sane and thoughtful person can imagine any government of our time sitting comfortably at the feet of Jesus while he is saying, "Love your enemies, bless them that curse you, do good to them that hate you, and pray for them that despitefully use you and persecute you." In fact, we know that one of the businesses of governments, "Christian" or not, has been to reenact the crucifixion. It has happened again and again and again. In *A Time for Trumpets*, his history of the Battle of the Bulge, Charles B. MacDonald tells how the SS Colonel Joachim Peiper was forced to withdraw from a bombarded château near the town of La Gleize, leaving behind a number of severely wounded soldiers of both armies. "Also left behind," MacDonald wrote, "on a whitewashed wall of one of the rooms in the basement was a charcoal drawing of Christ, thorns on his head, tears on his cheeks—whether drawn by a German or an American nobody would ever know." This is not an image that belongs to history but rather one that judges it.

Sex, Economy, Freedom, and Community

"It all turns on affection now," said Margaret. "Affection. Don't you see?"

—E. M. Forster, *Howards End*

THE SEXUAL HARASSMENT phase of the Clarence Thomas hearing was handled by the news media as if it were anomalous and surprising. In fact, it was only an unusually spectacular revelation of the destructiveness of a process that has been well established and well respected for at least two hundred years—the process, that is, of community disintegration. This process has been well established and well respected for so long, of course, because it has been immensely profitable to those in a position to profit. The surprise and dismay occasioned by the Thomas hearing were not caused by the gossip involved (for that, the media had prepared us very well) but by the inescapable message that this process of disintegration, so little acknowledged by politicians and commentators, can be severely and perhaps illimitably destructive.

In the government-sponsored quarrel between Clarence Thomas and Anita Hill, public life collided with private life in a way that could not have been resolved and that could only have been damaging. The event was depressing and fearful both because of its violations of due process and justice and because it was an attempt to deal publicly with a problem for which there is no public solution. It embroiled the United States Senate in the impossible task of adjudicating alleged offenses that had occurred in private, of which there were no witnesses and no evidence. If the hearing was a "lynching," as Clarence Thomas said it was, that was because it dealt a public punishment to an unconvicted and unindicted victim. But it was a peculiar lynching, all the same, for it dealt the punishment equally to the accuser. It was not a hearing, much less a trial; it was a story-telling contest that was not winnable by either participant.

Its only result was damage to all participants and to the nation. Public life obviously cannot be conducted in that way, and neither can private life. It was a public procedure that degenerated into a private quarrel. It was a private quarrel that became a public catastrophe.

Sexual harassment, like most sexual conduct, is extremely dangerous as a public issue. A public issue, properly speaking, can only be an issue about which the public can confidently know. Because most sexual conduct is private, occurring only between two people, there are typically no witnesses. Apart from the possibility of a confession, the public can know about it only as a probably unjudgeable contest of stories. (In those rare instances when a sexual offense occurs before reliable witnesses, then, of course, it is a legitimate public issue.)

Does this mean that sexual conduct is *only* private in its interest and meaning? It certainly does not. For if there is no satisfactory way to deal publicly with sexual issues, there is also no satisfactory way to deal with them in mere privacy. To make sense of sexual issues or of sex itself, a third term, a third entity, has to intervene between public and private. For sex is not and cannot be any individual's "own business," nor is it merely the private concern of any couple. Sex, like any other necessary, precious, and volatile power that is commonly held, is everybody's business. A way must be found to entitle everybody's legitimate interest in it without either violating its essential privacy or allowing its unrestrained energies to reduce necessary public procedures to the level of a private quarrel. For sexual problems and potentialities that have a more-than-private interest, what is needed are common or shared forms and solutions that are not, in the usual sense, public.

The indispensable form that can intervene between public and private interests is that of community. The concerns of public and private, republic and citizen, necessary as they are, are not adequate for the shaping of human life. Community alone, as principle and as fact, can raise the standards of local health (ecological, economic, social, and spiritual) without which the other two interests will destroy one another.

By community, I mean the commonwealth and common interests, commonly understood, of people living together in a place and wishing to continue to do so. To put it another way, community is a locally understood interdependence of local people, local culture, local economy, and local nature. (Community, of course, is an idea that can extend itself beyond the local, but it only does so metaphorically. The idea of a national or global community is meaningless apart from the realization of local communities.) Lacking the interest of or in such a community, private life becomes merely a sort of reserve in which individuals defend their "right" to act as they please and attempt to limit or destroy the "rights" of other individuals to act as they please.

A community identifies itself by an understood mutuality of interests. But it lives and acts by the common virtues of trust, goodwill, forbearance, self-restraint, compassion, and forgiveness. If it hopes to continue long as a community, it will wish to—and will have to—encourage respect for all its members, human and natural. It will encourage respect for all stations and occupations. Such a community has the power—not invariably but as a rule—to enforce decency without litigation. It has the power, that is, to influence behavior. And it exercises this power not by coercion or violence but by teaching the young and by preserving stories and songs that tell (among other things) what works and what does not work in a given place.

Such a community is (among other things) a set of arrangements between men and women. These arrangements include marriage, family structure, divisions of work and authority, and responsibility for the instruction of children and young people. These arrangements exist, in part, to reduce the volatility and the danger of sex—to preserve its energy, its beauty, and its pleasure; to preserve and clarify its power to join not just husband and wife to one another but parents to children, families to the community, the community to nature; to ensure, so far as possible, that the inheritors of sexuality, as they come of age, will be worthy of it.

But the life of a community is more vulnerable than public life. A community cannot be made or preserved apart from the loyalty and affection of its members and the respect and goodwill of the people outside it. And for a long time, these conditions have not been met. As the technological, economic, and political means of exploitation have expanded, communities have been more and more victimized by opportunists outside themselves. And as the salesmen, saleswomen, advertisers, and propagandists of the industrial economy have become more ubiquitous and more adept at seduction, communities have lost the loyalty and affection of their members. The community, wherever you look, is being destroyed by the desires and ambitions of both private and public life, which for want of the intervention of community interests are also destroying one another. Community life is by definition a life of cooperation and responsibility. Private life and public life, without the disciplines of community interest, necessarily gravitate toward competition and exploitation. As private life casts off all community restraints in the interest of economic exploitation or ambition or self-realization or whatever, the communal supports of public life also and by the same stroke are undercut, and public life becomes simply the arena of unrestrained private ambition and greed.

As our communities have disintegrated from external predation and internal disaffection, we have changed from a society whose ideal of justice was trust and fairness among people who knew each other into a society whose ideal of justice is public litigation, breeding distrust even among people who know each other.

Once it has shrugged off the interests and claims of the community, the public language of sexuality comes directly under the influence of private lust, ambition, and greed and becomes inadequate to deal with the real issues and problems of sexuality. The public dialogue degenerates into a stupefying and useless contest between so-called liberation and so-called morality. The real issues and problems, as they are experienced and suffered in people's lives, cannot be talked about. The public language can deal, however awkwardly and perhaps uselessly, with pornography, sexual hygiene, contraception, sexual harassment, rape, and so on. But it cannot talk about respect, responsibility, sexual discipline, fidelity, or the practice of love. "Sexual education," carried on in this public language, is and can only be a dispirited description of the working of a sort of anatomical machinery—and this is a sexuality that is neither erotic nor social nor sacramental but rather a cold-blooded, abstract procedure that is finally not even imaginable.

The conventional public opposition of "liberal" and "conservative" is, here as elsewhere, perfectly useless. The "conservatives" promote the family as a sort of public icon, but they will not promote the economic integrity of the household or the community, which are the mainstays of family life. Under the sponsorship of "conservative" presidencies, the economy of the modern household, which once required the father to work away from home—a development that was bad enough —now requires the mother to work away from home, as well. And this development has the wholehearted endorsement of "liberals," who see the mother thus forced to spend her days away from her home and children as "liberated"—though nobody has yet seen the fathers thus forced away as "liberated." Some feminists are thus in the curious position of opposing the mistreatment of women and yet advocating their participation in an economy in which everything is mistreated.

The "conservatives" more or less attack homosexuality, abortion, and pornography, and the "liberals" more or less defend them. Neither party will oppose sexual promiscuity. The "liberals" will not oppose promiscuity because they do not wish to appear intolerant of "individual liberty." The "conservatives" will not oppose promiscuity because sexual discipline would reduce the profits of corporations, which in their advertisements and entertainments encourage sexual self-indulgence as a way of selling merchandise.

The public discussion of sexual issues has thus degenerated into a poor attempt to equivocate between private lusts and public emergencies. Nowhere in public life (that is, in the public life that counts: the discussions of political and corporate leaders) is there an attempt to respond to community needs in the language of community interest.

And although we seem more and more inclined to look on education, even as it teaches less and is more overcome by violence, as the solution to all our problems (thus delaying the solution for a generation), there is really not much use in looking to education for the help we need. For education has become increasingly useless as it has become increasingly public. Real education is determined by community needs, not by public tests. Nor is community interest or community need going to receive much help from television and the other public media. Television is the greatest disrespecter and exploiter of sexuality that the world has ever seen; even if the network executives decide to promote "safe sex" and the use of condoms, they will not cease to pimp for the exceedingly profitable "sexual revolution." It is, in fact, the nature of the electronic media to blur and finally destroy all distinctions between public and community. Television has greatly accelerated the process, begun long ago, by which many communities have been atomized and congealed into one public. Nor is government a likely source of help. As political leaders have squirmed free of the claims and responsibilities of community life, public life has become their private preserve. The public political voice has become increasingly the voice of a conscious and self-serving duplicity: it is now, for instance, merely typical that a political leader can speak of "the preciousness of all life" while armed for the annihilation of all life. And the right of privacy, without the intervening claims and responsibilities of community life, has moved from the individual to the government and assumed the name of "official secrecy." Whose liberation is that?

In fact, there is no one to speak for the community interest except those people who wish to adhere to community principles. The community, in other words, must speak in its own interest. It must learn to defend itself. And in its self-defense, it may use the many powerful arguments provided for it by the failures of the private and public aims that have so nearly destroyed it.

The defenders of community should point out, for example, that for the joining of men and women there need to be many forms that only a community can provide. If you destroy the ideal of the "gentle man" and remove from men all expectations of courtesy and consideration toward women and children, you have prepared the way for an epidemic of rape and abuse. If you depreciate the sanctity and solemnity of marriage, not just as a bond between two people but as a bond between those two people and their forebears, their children, and their neighbors, then you have prepared the way for an epidemic of divorce, child neglect, community ruin, and loneliness. If you destroy the economies of household and community, then you destroy the bonds of mutual usefulness and practical dependence without which the other bonds will not hold.

If these and all other community-made arrangements between men and women are removed, if the only arrangements left between them are those of sex and sexual politics, instinct and polity without culture, then sex and politics are headed not only toward many kinds of private and public suffering but toward the destruction of justice, as in the confrontation of Clarence Thomas and Anita Hill.

II

But to deal with community disintegration as merely a matter of sexual disorder, however destructive that disorder may be, is misleading. The problem is far more complex than I have been able to suggest so far. There is much more to be said on the issue of sex, and I will return to it, but for the sake of both truth and clarity I must first examine other issues that are related and, in some respects, analogous to the issue of sex.

It is certain, as I have already said, that communities are destroyed both from within and from without: by internal disaffection and external exploitation. It is certain, too, that there have always been people who have become estranged from their communities for reasons of honest difference or disagreement. But it can be argued that community disintegration typically is begun by an aggression of some sort from the outside and that in modern times the typical aggression has been economic. The destruction of the community begins when its economy is made—not *dependent* (for no community has ever been entirely independent)—but *subject* to a larger external economy. As an example, we could probably do no better than the following account of the destruction of the local wool economy of the parish of Hawkshead in the Lake District of England:

> The . . . reason for the decline of the customary tenant must be sought in the introduction of machinery towards the end of the eighteenth century, which extinguished not only the local spinning and weaving, but was also the deathblow of the local market. Before this time, idleness at a fellside farm was unknown, for clothes and even linen were home-made, and all spare time was occupied by the youths in carding wool, while the girls spun the "garn" with distaff and wheel. . . . The sale of the yarn to the local weavers, and at the local market, brought important profits to the dalesman, so that it not only kept all hands busy, but put money

> into his pocket. But the introduction of machinery for looms and for spinning, and consequent outside demand for fleeces instead of yarn and woven material, threw idle not only half of the family, but the local hand-weavers, who were no doubt younger sons of the same stock. Thus idleness took the place of thrift and industry among a naturally industrious class, for the sons and daughters of the 'statesmen, often too proud to go out to service, became useless encumbrances on the estates. Then came the improvement in agricultural methods [that is, technological innovations], which the 'statesman could not afford to keep abreast of . . . What else could take place but that which did? The estates became mortgaged and were sold, and the rich manufacturers, whose villas are on the margin of Windermere, have often enough among their servants the actual descendants of the old 'statesmen, whose manufactures they first usurped and whose estates they afterwards absorbed.

This paragraph sets forth the pattern of industrial exploitation of a locality and a local economy, a pattern that has prevailed for two hundred years. The industrialization of the eastern Kentucky coal fields early in the present century, though more violent, followed this pattern exactly. A decentralized, fairly independent local economy was absorbed and destroyed by an aggressive, monetarily powerful outside economy. And like the displaced farmers, spinners, and weavers of Hawkshead, the once-independent mountaineers of eastern Kentucky became the wage-earning servants of those who had dispossessed their parents, sometimes digging the very coal that their families had once owned and had sold for as little per acre as the pittance the companies paid per day. By now, there is hardly a rural neighborhood or town in the United States that has not suffered some version of this process.

The same process is destroying local economies and cultures all over the world. Of Ladakh, for example, Helena Norberg-Hodge writes:

> In the traditional culture, villagers provided for their basic needs without money. They had developed skills that enabled them to grow barley at 12,000 feet and to manage yaks and other animals at even higher elevations.

> People knew how to build houses with their own hands from the materials of the immediate surroundings. The only thing they actually needed from outside the region was salt, for which they traded. They used money in only a limited way, mainly for luxuries.
>
> Now, suddenly, as part of the international money economy, Ladakhis find themselves ever more dependent—even for vital needs—on a system that is controlled by faraway forces. They are vulnerable to decisions made by people who do not even know that Ladakh exists. . . . For two thousand years in Ladakh, a kilo of barley has been a kilo of barley, but now you cannot be sure of its value.

This, I think, speaks for itself: if you are dependent on people who do not know you, who control the value of your necessities, you are not free, and you are not safe.

The industrial revolution has thus made universal the colonialist principle that has proved to be ruinous beyond measure: the assumption that it is permissible to ruin one place or culture for the sake of another. Thus justified or excused, the industrial economy grows in power and thrives on its damages to local economies, communities, and places. Meanwhile, politicians and bureaucrats measure the economic prosperity of their nations according to the burgeoning wealth of the industrial interests, not according to the success or failure of small local economies or the reduction and often hopeless servitude of local people. The self-congratulation of the industrialists and their political minions has continued unabated to this day. And yet it is a fact that the industrialists of Hawkshead, like all those elsewhere and since, have lived off the public, just as surely as do the despised clients of "welfare." They have lived off a public of industrially destroyed communities, and they have not compensated for this destruction by their ostentatious contributions to the art, culture, and education of the professional class, or by their "charities" to the poor. Nor is this state of things ameliorated by efforts to enable a local population to "participate" in the global economy, by "education" or any other means. It is true that local individuals, depending on their capital, intelligence, cunning, or influence, may be able to "participate" to their apparent advantage, but local communities and places can "participate" only as victims. The global economy does not exist to help the communities and localities of the globe. It exists to siphon the wealth of those communities and places into a few bank accounts. To this economy, democracy and the values of the religious traditions mean absolutely nothing. And those who wish to help communities to survive had better understand that a merely political freedom means little within a totalitarian economy. Ms. Norberg-Hodge says, of the relatively new influence of the global economy on Ladakh, that

> Increasingly, people are locked into an economic system that pumps resources out of the periphery into the center—from the nonindustrialized to the industrialized parts of the world, from the countryside to the city, from the poor to the rich. Often, these resources end up back where they came from as commercial products . . . at prices that the poor can no longer afford.

This is an apt description not just of what is happening in Ladakh but of what is happening in my own rural county and in every other rural county in the United States.

The situation in the wool economy of Hawkshead at the end of the eighteenth century was the same as that which, a little later, caused the brief uprising of those workers in England who were called Luddites. These were people who dared to assert that there were needs and values that justly took precedence over industrialization; they were people who rejected the determinism of technological innovation and economic exploitation. In them, the community attempted to speak for itself and defend itself. It happened that Lord Byron's maiden speech in the House of Lords, on February 27, 1812, dealt with the uprising of the Luddites, and this, in part, is what he said:

> By the adoption of one species of [weaving] frame in particular, one man performed the work of many, and the superfluous laborers were thrown out of employment. Yet it is to be observed, that the work thus executed was inferior in quality; not marketable at home, and merely hurried over with a view to exportation. . . .

> The rejected workmen . . . conceived themselves to be sacrificed to improvements in mechanism. In the foolishness of their hearts they imagined that the maintenance and well-doing of the industrious poor were objects of greater importance than the enrichment of a few individuals by any improvement, in the implements of trade, which threw the workmen out of employment, and rendered the laborer unworthy of his hire.

The Luddites did, in fact, revolt not only against their own economic oppression but also against the poor quality of the machine work that had replaced them. And though they destroyed machinery, they "abstained from bloodshed or violence against living beings, until in 1812 a band of them was shot down by soldiers." Their movement was suppressed by "severe repressive legislation" and by "many hangings and transportations."

The Luddites thus asserted the precedence of community needs over technological innovation and monetary profit, and they were dealt with in a way that seems merely inevitable in the light of subsequent history. In the years since, the only group that I know of that has successfully, so far, made the community the standard of technological innovation has been the Amish. The Amish have differed from the Luddites in that they have not destroyed but merely declined to use the technologies that they perceive as threatening to their community. And this has been possible because the Amish are an agrarian people. The Luddites could not have refused the machinery that they destroyed; the machinery had refused them.

The victory of industrialism over Luddism was thus overwhelming and unconditional; it was undoubtedly the most complete, significant, and lasting victory of modern times. And so one must wonder at the intensity with which any suggestion of Luddism still is feared and hated. To this day, if you say you would be willing to forbid, restrict, or reduce the use of technological devices in order to protect the community—or to protect the good health of nature on which the community depends—you will be called a Luddite, and it will not be a compliment. To say that the community is more important than machines is certainly Christian and certainly democratic, but it is also Luddism and therefore not to be tolerated.

Technological determinism, then, has triumphed. By now the rhetoric of technological determinism has become thoroughly mixed with the rhetoric of national mysticism so that a political leader may confidently say that it is "our destiny" to go to the moon or Mars or wherever we may go for the profit of those who will provide the transportation. Because this "destiny" has become the all-shaping superstition of our time and is therefore never debated in public, we have difficulty seeing that the triumph of technological determinism is the defeat of community. But it is a fact that in both private conversation and public dialogue, the community has neither status nor standing from which to plead in its own defense. There is no denying, of course, that "community" ranks with "family," "our land," and "our beloved country" as an icon of the public vocabulary; everybody is for it, and this means nothing. If individuals or groups have the temerity to oppose an actual item on the agenda of technological process because it will damage a community, the powers that be will think them guilty of Luddism, sedition, and perhaps insanity. Local and community organizations, of course, do at times prevail over the would-be "developers," but the status of these victories remains tentative. The powers that be, and the media as well, treat such victories as anomalies, never as the work of a legitimate "side" in the public dialogue.

Some time ago, for example, my friends in the Community Farm Alliance asked me to write a newspaper piece opposing the Bush administration's efforts to revise the General Agreement on Tariffs and Trade; the proposed revisions would destroy all barriers and restrictions to international trade in food, thus threatening the already precarious survival of farm communities in the industrial countries and also the local availability of food in Third World countries. I wrote the article and sent it to the editor of the op-ed page of the *New York Times*, who had invited me to make such submissions—and who rejected my article on the ground that the subject was "not sexy." I conclude that if the opponents of the GATT revisions should win, their victory will have an effect limited to the issue—that is, such a victory would not cause their point of view to be acknowledged, much less represented, among the powers that be, and the public standing of community interests would be no higher than before.

The triumph of the industrial economy is the fall of community. But the fall of community reveals how precious and how necessary community is. For when community falls, so must fall all the things that only community life can engender and protect: the care of the old, the care and education of children, family life, neighborly work, the handing down of memory, the care of the earth, respect for nature and the lives of wild creatures. All of these things have been damaged by the rule of industrialism, but of all the damaged things probably the most precious and the most damaged is sexual love. For sexual love is the heart of community life. Sexual love is the force that in our bodily life connects us most intimately to the Creation, to the fertility of the world, to farming and the care of animals. It brings us into the dance that holds the community together and joins it to its place.

In dealing with community, as in dealing with everything else, the industrial economy goes for the nucleus. It does this because it wants the power to cause fundamental change. To make sex the preferred bait of commerce may seem merely the obvious thing to do, once greed is granted its now conventional priority as a motive. But this could happen only after a probably instinctive sense of the sanctity and dignity of the body—the sense of its having been "fearfully and wonderfully made" had been destroyed. Once this ancient reticence had been broken down, then the come-on of the pimp could be instituted as the universal spiel of the marketplace; everything could be sold on the promise of instant, innocent sexual gratification, "no strings attached." Sexual energy cannot be made publicly available for commercial use—that is, prostituted—without destroying all of its communal or cultural forms: forms of courtship, marriage, family life, household economy, and so on. The devaluation of sexuality, like the devaluation of a monetary currency, destroys its correspondence to other values.

In the wake of the Thomas-Hill catastrophe in Washington, the *New York Times Sunday Magazine* contained a skin lotion advertisement that displayed a photograph of the naked torso of a woman. From a feminist point of view, this headless and footless body represents the male chauvinist's sexual ideal: a woman who cannot think and cannot escape. From a point of view somewhat more comprehensive—the point of view of community—it represents also the commercial ideal of the industrial economy: the completely seducible consumer, unable either to judge or to resist.

The headlessness of this lotionable lady suggests also another telling indication of the devaluation of sexual love in modern times—that is, the gravitation of attention from the countenance, especially the eyes, to the specifically sexual anatomy. The difference, of course, is that the countenance is both physical and spiritual. There is much testimony to this in the poetic tradition and elsewhere. Looking into one another's eyes, lovers recognize their encounter as a meeting not merely of two bodies but of two living souls. In one another's eyes, moreover, they see themselves reflected not narcissistically but as singular beings, separate and small, far inferior to the creature that they together make.

In this meeting of eyes, there is an acknowledgment that love is more than sex:

> 'This Ecstasy doth unperplex,'
> We said, 'and tell us what we love;
> We see by this it was not sex;
> We see we saw not what did move.'

These lines are from John Donne's poem "The Ecstasy," in which the lovers have been joined by the "double string" of their mutual gaze. This is not a disembodied love. Far from it. For love is finally seen in the poem as "that subtle knot, which makes us man" by joining body and soul together, just as it joins the two lovers. Sexual love is thus understood as both fact and mystery, physical motion and spiritual motive. That this complex love should be reduced simply to sex has always seemed a fearful thing to the poets. As late as our own century, for example, we have these lines from Wallace Stevens:

> If sex were all, then every trembling hand
> Could make us squeak, like dolls, the wished-for words.

The fear of the reduction of love to sex is obviously subject to distortion. At worst, this fear has caused a kind of joyless prudery. But at best, it has been a just fear, implying both an appropriate awe and a reluctance that humans should become the puppets of a single instinct. Such a puppetry is possible, and its sign, in modern terms, is the body as product, made delectably consumable by the application of lotions and other commercial liquids. But there is a higher, juster love of which the sign is the meeting of the eyes.

In the poetry that I know, the most graceful and the richest testimony to the power of the eyes is in act three, scene two, of *The Merchant of Venice*, where Portia says to Bassanio:

> Beshrew your eyes,
> They have o'erlook'd me and divided me,
> One half of me is yours, the other half yours,—
> Mine own I would say: but if mine then yours,
> And so all yours.

Portia is no headless woman. She is the brightest, most articulate character in the play. Witty and charming as these lines are, they are not frivolous. They attest to the sexual and the spiritual power of a look, which has just begun an endless conversation between two living souls. This speech of hers is as powerful as it is because she knows exactly what she is doing. What she is doing is giving herself away. She has entered into a situation in which she must find her life by losing it. She is glad, and she is frightened. She is speaking joyfully and fearfully of the self's suddenly irresistible wish to be given away. And this is an unconditional giving, on which, she knows, time and mortality will impose their inescapable conditions; she will have remembered the marriage ceremony with its warnings of difficulty, poverty, sickness, and death. There is nothing "safe" about this. This love has no place to happen except in this world, where it cannot be made safe. And this scene finally becomes a sort of wedding, in which Portia says to Bassanio, "Myself, and what is mine, to you and yours / Is now converted. . . . I give them with this ring."

There is no "sexism" or "double standard" in this exchange of looks and what it leads to. The look of which Portia speaks could have had such power only because it answered a look of hers; she has, in the same moment, "o'erlook'd" Bassanio. She is not speaking as a woman submitting to the power of a man, much less to seduction. She is speaking as one of a pair who are submitting to the redemptive power of love. Responding to Portia's pledge of herself to him, Bassanio cannot say much;

he is not so eloquent as she, and his mind has become "a wild of nothing, save of joy." He says only:

> when this ring
> Parts from this finger, then parts life from hence,—
> O then be bold to say Bassanio's dead!

But that is enough. He is saying, as his actual wedding will require him to say, that he will be true to Portia until death. But he is also saying, as she has said, that he is dead to his old life and can now live only in the new life of their love. He might have said, with Dante, "Of a surety I have now set my feet on that point of life, beyond the which he must not pass who would return."

This is a play about the precedence of affection and fidelity over profit, and so it is a play about the order of community. Lovers must not, like usurers, live for themselves alone. They must finally turn from their gaze at one another back toward the community. If they had only themselves to consider, lovers would not need to marry, but they must think of others and of other things. They say their vows to the community as much as to one another, and the community gathers around them to hear and to wish them well, on their behalf and on its own. It gathers around them because it understands how necessary, how joyful, and how fearful this joining is. These lovers, pledging themselves to one another "until death," are giving themselves away, and they are joined by this as no law or contract could ever join them. Lovers, then, "die" into their union with one another as a soul "dies" into its union with God. And so here, at the very heart of community life, we find not something to sell as in the public market but this momentous giving. If the community cannot protect this giving, it can protect nothing—and our time is proving that this is so.

We thus can see that there are two kinds of human economy. There is the kind of economy that exists to protect the "right" of profit, as does our present public economy; this sort of economy will inevitably gravitate toward protection of the "rights" of those who profit most. Our present public economy is really a political system that safeguards the private exploitation of the public wealth and health. The other kind of economy exists for the protection of gifts, beginning with the "giving in marriage," and this is the economy of community, which now has been nearly destroyed by the public economy.

There are two kinds of sexuality that correspond to the two kinds of economy. The sexuality of community life, whatever its inevitable vagaries, is centered on marriage, which joins two living souls as closely as, in this world, they can be joined. This joining of two who know, love, and trust one another brings them in the same breath into the freedom of sexual consent and into the fullest earthly realization of the image of God. From their joining, other living souls come into being, and with them great responsibilities that are unending, fearful, and joyful. The marriage of two lovers joins them to one another, to forebears, to descendants, to the community, to Heaven and earth. It is the fundamental connection without which nothing holds, and trust is its necessity.

Our present sexual conduct, on the other hand, having "liberated" itself from the several trusts of community life, is public, like our present economy. It has forsaken trust, for it rests on the easy giving and breaking of promises. And having forsaken trust, it has predictably become political. In private life, as in public, we are attempting to correct bad character and low motives by law and by litigation. "Losing kindness," as Lao-tzu said, "they turn to justness." The superstition of the anger of our current sexual politics, as of other kinds of anger, is that somewhere along the trajectory of any quarrel a tribunal will be reached that will hear all complaints and find for the plaintiff; the verdict will be that the defendant is entirely wrong, the plaintiff entirely right and entirely righteous. This, of course, is not going to happen. And because such "justice" cannot happen, litigation only prolongs itself. The difficulty is that marriage, family life, friendship, neighborhood, and other personal connections do not depend exclusively or even primarily on justice—though, of course, they all must try for it. They depend also on trust, patience, respect, mutual help, forgiveness—in other words, the *practice* of love, as opposed to the mere *feeling* of love.

As soon as the parties to a marriage or a friendship begin to require strict justice of each other, then that marriage or friendship begins to be destroyed, for there is no way to adjudicate the competing claims of a personal quarrel. And so these relationships do not dissolve into litigation, really; they dissolve into a feud, an endless exchange of accusations and retributions. If the two parties have not the grace to forgive the inevitable offenses of close connection, the next best thing is separation and silence. But why should separation have come to be the virtually conventional outcome of close relationships in our society? The proper question, perhaps, is not why we have so much divorce, but why we are so unforgiving. The answer, perhaps, is that, though we still recognize the feeling of love, we have forgotten how to practice love when we don't feel it.

Because of our determination to separate sex from the practice of love in marriage and in family and community life, our public sexual morality is confused, sentimental, bitter, complexly destructive, and hypocritical. It begins with the idea of "sexual liberation": whatever people desire is "natural" and all right, men and women are not different but merely equal, and all desires are equal. If a man wants to sit down while a pregnant woman is standing or walk through a heavy door and let it slam in a woman's face, that is all right. Divorce on an epidemic scale is all right; child abandonment by one parent or another is all right; it is regrettable but still pretty much all right if a divorced parent neglects or refuses to pay child support; promiscuity is all right; adultery is all right. Promiscuity among teenagers is pretty much all right, for "that's the way it is"; abortion as birth control is all right; the prostitution of sex in advertisements and public entertainment is all right. But then, far down this road of freedom, we decide that a few lines ought to be drawn. Child molestation, we wish to say, is not all right, nor is sexual violence, nor is sexual harassment, nor is pregnancy among unmarried teenagers. We are also against venereal diseases, the diseases of promiscuity, though we tend to think that they are the government's responsibility, not ours.

In this cult of liberated sexuality, "free" of courtesy, ceremony, responsibility, and restraint, dependent on litigation and expert advice, there is much that is human, sad to say, but there is no sense or sanity. Trying to draw the line where we are trying to draw it, between carelessness and brutality, is like insisting that falling is flying—until you hit the ground—and then trying to outlaw hitting the ground. The pretentious, fantastical, and solemn idiocy of the public sexual code could not be better exemplified than by the now-ubiquitous phrase "sexual partner," which denies all that is implied by the names of "husband" or "wife" or even "lover." It denies anyone's responsibility for the consequences of sex. With one's "sexual partner," it is now understood, one must practice "safe sex" —that is, one must protect oneself, not one's partner or the children that may come of the "partnership."

But the worst hypocrisy of all is in the failure of the sexual libertarians to come to the defense of sexually liberated politicians. The public applies strenuously to public officials a sexual morality that it no longer applies to anyone privately and that it does not apply to other liberated public figures, such as movie stars, artists, athletes, and business tycoons. The prurient squeamishness with which the public and the public media poke into the lives of politicians is surely not an expectable result of liberation. But this paradox is not the only one. According to its claims, sexual liberation ought logically to have brought in a time of "naturalness," ease, and candor between men and women. It has, on the contrary, filled the country with sexual self-consciousness, uncertainty, and fear. Women, though they may dress as if the sexual millennium had arrived, hurry along our city streets and public corridors with their eyes averted, like hunted animals. "Eye contact," once the very signature of our humanity, has become a danger. The meeting ground between men and women, which ought to be safeguarded by trust, has become a place of suspicion, competition, and violence. One no longer goes there asking how instinct may be ramified in affection and loyalty; now one asks how instinct may be indulged with the least risk to personal safety.

Seeking to "free" sexual love from its old communal restraints, we have "freed" it also from its meaning, its responsibility, and its exaltation. And we have made it more dangerous. "Sexual liberation" is as much a fraud and as great a failure as the "peaceful atom." We are now living in a sexual atmosphere so polluted and embittered that women must look on virtually any man as a potential assailant, and a man must look on virtually any woman as a potential accuser. The idea that this situation can be corrected by the courts and the police only compounds the disorder and the danger. And in the midst of this acid rainfall of predation and recrimination, we presume to teach our young people that sex can be made "safe"—by the use, inevitably, of purchased drugs and devices. What a lie! Sex was never safe, and it is less safe now than it has ever been.

What we are actually teaching the young is an illusion of thoughtless freedom and purchasable safety, which encourages them to tamper prematurely, disrespectfully, and dangerously with a great power. Just as the public economy encourages people to spend money and waste the world, so the public sexual code encourages people to be spendthrifts and squanderers of sex. The basis of true community and household economy, on the other hand, is thrift. The basis of community sexuality is respect for everything that is involved—and respect, here as everywhere, implies discipline. By their common principles of extravagance and undisciplined freedom, our public economy and our public sexuality are exploiting and spending moral capital built up by centuries of community life—exactly as industrial agriculture has been exploiting and spending the natural capital built up over thousands of years in the soil.

In sex, as in other things, we have liberated fantasy but killed imagination, and so have sealed ourselves in selfishness and loneliness. Fantasy is of the solitary self, and it cannot lead us away from ourselves. It is by imagination that we cross over the differences between ourselves and other beings and thus learn compassion, forbearance, mercy, forgiveness, sympathy, and love—the virtues without which neither we nor the world can live.

Starting with economic brutality, we have arrived at sexual brutality. Those who affirm the one and deplore the other will have to explain how we might logically have arrived anywhere else. Sexual lovemaking between humans is not and cannot be the thoughtless, instinctual coupling of animals; it is not "recreation"; it is not "safe." It is the strongest prompting and the greatest joy that young people are likely to experience. Because it is so powerful, it is risky, not just because of the famous dangers of venereal disease and "unwanted pregnancy" but also because it involves and requires a giving away of the self that if not honored and reciprocated, inevitably reduces dignity and self-respect. The invitation to give oneself away is not, except for the extremely ignorant or the extremely foolish, an easy one to accept.

Perhaps the current revulsion against sexual harassment may be the beginning of a renewal of sexual responsibility and self-respect. It must, at any rate, be the beginning of a repudiation of the idea that sex among us is merely natural. If men and women are merely animals, it is hard to see how sexual harassment could have become an issue, for such harassment is no more than the instinctive procedure of male animals, who openly harass females, usually by unabashed physical display and contact; it is their way of asking who is and who is not in estrus. Women would not think such behavior offensive if we had not, for thousands of years, understood ourselves as specifically human beings—creatures who, if in some ways animal-like, are in other ways God-like. In asking men to feel shame and to restrain themselves—which one would not ask of an animal—women are implicitly asking to be treated as human beings in that full sense, as living souls made in the image of God. But any humans who wish to be treated and to treat others according to that definition must understand that this is not a kindness that can be conferred by a public economy or by a public government or by a public people. It can only be conferred on its members by a community.

III

Much of the modern assault on community life has been conducted within the justification and protection of the idea of freedom. Thus, it is necessary to try to see how the themes of freedom and community have intersected.

The idea of freedom, as Americans understand it, owes its existence to the inevitability that people will disagree. It is a way of guaranteeing to individuals and to political bodies the right to be different from one another. A specifically American freedom began with our wish to assert our differences from England, and its principles were then worked out in the effort to deal with differences among the states. The result is the Bill of Rights, of which the cornerstone is the freedom of speech. This freedom is not only the basic guarantee of political liberty but it also obligates public officials and private citizens alike to acknowledge the inherent dignity and worth of individual people. It exists only as an absolute; if it can be infringed at all, then probably it can be destroyed entirely.

But if it is an absolute, it is a peculiar and troubling one. It is not an absolute in the sense that a law of nature is. It is not absolute even as the moral law is. One person alone can uphold the moral law, but one person alone cannot uphold the freedom of speech. The freedom of speech is a public absolute, and it can remain absolute only so long as a sufficient segment of the public believes that it is and consents to uphold it. It is an absolute that can be destroyed by public opinion. This is where the danger lies. If this freedom is abused and if a sufficient segment of the public becomes sufficiently resentful of the abuses, then the freedom will be revoked. It is a freedom, therefore, that depends directly on responsibility. And so the First Amendment alone is not a sufficient guarantee of the freedom of speech.

As we now speak of it, freedom is almost always understood as a public idea having to do with the liberties of individuals. The public dialogue about freedom almost always has to do with the efforts of one group or another to wrest these individual liberties from the government or to protect them from another group. In this situation, it is inevitable that freedom will be understood as an issue of power. This is perhaps as necessary as it is unavoidable. But power is not the only issue related to freedom.

From another point of view, not necessarily incompatible, freedom has long been understood as the consequence of knowing the truth. When Jesus said to his followers, "Ye shall know the truth, and the truth shall make you free," he was not talking primarily about politics, but the political applicability of the statement has been obvious for a long time, especially to advocates of democracy. According to this line of thought, freedom of speech is necessary to political health and sanity because it permits speech—the public dialogue—to correct itself. Thomas Jefferson had this in mind when he said in his first inaugural address, "If there be any among us who would wish to dissolve this Union or to change its republican form, let them stand undisturbed as monuments of the safety with which error of opinion may be tolerated where reason is left free to combat it." The often-cited "freedom to be wrong" is thus a valid freedom, but it is a poor thing by itself; its validity comes from the recognition that error is real, identifiable as such, dangerous to freedom as to much else, and controvertible. The freedom to be wrong is valid, in other words, because it is the unexcisable other half of the freedom to be right. If freedom is understood as merely the privilege of the unconcerned and uncommitted to muddle about in error, then freedom will certainly destroy itself.

But to define freedom only as a public privilege of private citizens is finally inadequate to the job of protecting freedom. It leaves the issue too public and too private. It fails to provide a circumstance for those private satisfactions and responsibilities without which freedom is both pointless and fragile. Here as elsewhere, we need to interpose between the public and the private interests a third interest: that of the community. When there is no forcible assertion of the interest of community, public freedom becomes a sort of refuge for escapees from the moral law—those who hold that there is, in Mary McGrory's words, "no ethical transgression except an indictable one."

Public laws are meant for a public, and they vary, sometimes radically, according to forms of government. The moral law, which is remarkably consistent from one culture to another, has to do with community life. It tells us how we should treat relatives and neighbors and, by metaphorical extension, strangers. The aim of the moral law is the integrity and longevity of the community, just as the aim of public law is the integrity and longevity of a political body. Sometimes, the identities of community and political body are nearly the same, and in that case public laws are not necessary because there is, strictly speaking, no "public." As I understand the term, *public* means simply all the people, apart from any personal responsibility or belonging. A public building, for example, is a building which everyone may use but to which no one belongs, which belongs to everyone but not to anyone in particular, and for which no one is responsible except "public employees." A community, unlike a public, has to do first of all with belonging; it is a group of people who belong to one another and to their place. We would say, "We belong to our community," but never "We belong to our public."

I don't know when the concept of "the public," in our sense, emerged from the concept of "the people." But I am aware that there have been human situations in which the concept of "the public" was simply unnecessary. It is not quite possible, for example, to think of the Bushmen or the Eskimos as "publics" or of any parts of their homelands as "public places." And in the traditional rural villages of England there was no public place but rather a "common." A public, I suppose, becomes necessary when a political body grows so large as to include several divergent communities.

A public government, with public laws and a public system of justice, founded on democratic suffrage, is in principle a good thing. Ideally, it makes possible a just and peaceable settlement of contentions arising between communities. It also makes it possible for a mistreated member of a community to appeal for justice outside the community. But obviously such a government can fall short of its purpose. When a public government becomes identified with a public economy, a public culture, and public fashions of thought, it can become the tool of a public process of nationalism or "globalization" that is oblivious of local differences and therefore destructive of communities.

"Public" and "community," then, are different—perhaps radically different—concepts that under certain circumstances are compatible but that, in the present economic and technological monoculture, tend to be at odds. A community, when it is alive and well, is centered on the household—the family place and economy—and the household is centered on marriage. A public, when it is working in the best way—that is, as a political body intent on justice—is centered on the individual. Community and public alike, then, are founded on respect—the one on respect for the family, the other on respect for the individual. Both forms of respect are deeply traditional, and they are not fundamentally incompatible. But they are different, and that difference, once it is instituted in general assumptions, can be the source of much damage and much danger.

A household, according to its nature, will seek to protect and prolong its own life, and since it will readily perceive its inability to survive alone, it will seek to join its life to the life of a community. A young person, coming of age in a healthy household and community, will understand her or his life in terms of membership and service. But in a public increasingly disaffected and turned away from community, it is clear that individuals must be increasingly disinclined to identify themselves in such a way.

The individual, unlike the household and the community, always has two ways to turn: she or he may turn either toward the household and the community, to receive membership and to give service, or toward the relatively unconditional life of the public, in which one is free to pursue self-realization, self-aggrandizement, self-interest, self-fulfillment, self-enrichment, self-promotion, and so on. The problem is that—unlike a married couple, a household, or a community—one individual represents no fecundity, no continuity, and no harmony. The individual life implies no standard of behavior or responsibility.

I am indebted to Judith Weissman for the perception that there are two kinds of freedom: the freedom of the community and the freedom of the individual. The freedom of the community is the more fundamental and the more complex. A community confers on its members the freedoms implicit in familiarity, mutual respect, mutual affection, and mutual help; it gives freedom its proper aims; and it prescribes or shows the responsibilities without which no one can be legitimately free, or free for very long. But to confer freedom or any other benefits on its members, a community must also be free from outside pressure or coercion. It must, in other words, be so far as possible the cause of its own changes; it must change in response to its own changing needs and local circumstances, not in response to motives, powers, or fashions coming from elsewhere. The freedom of the individual, by contrast, has been construed customarily as a license to pursue any legal self-interest at large and at will in the domain of public liberties and opportunities.

These two kinds of freedom, so understood, are clearly at odds. In modern times, the dominant freedom has been that of the individual, and Judith Weissman believes—correctly, I think—that this self-centered freedom is still the aim of contemporary liberation movements:

> The liberation of the individual self for fulfillments, discoveries, pleasures, and joys, and the definition of oppression as mental and emotional constraints . . . this combination existing at the heart of Shelley's Romantic radicalism remains basically unchanged in later feminist writers. . . .

Freedom defined strictly as individual freedom tends to see itself as an escape from the constraints of community life

—constraints necessarily implied by consideration for the nature of a place; by consideration for the needs and feelings of neighbors; by kindness to strangers; by respect for the privacy, dignity, and propriety of individual lives; by affection for a place, its people, and its nonhuman creatures; and by the duty to teach the young.

But certain liberationist intellectuals are not the only ones who have demanded this sort of freedom. Almost everybody now demands it, as she or he has been taught to do by the schools, by the various forms of public entertainment, and by salespeople, advertisers, and other public representatives of the industrial economy. People are instructed to free themselves of all restrictions, restraints, and scruples in order to fulfill themselves as individuals to the utmost extent that the law allows. Moreover, we treat corporations as "persons"—an abuse of a metaphor if ever there was one!—and allow to them the same liberation from community obligations that we allow to individuals.

But there is a paradox in all this, and it is as cruel as it is obvious: as the emphasis on individual liberty has increased, the liberty and power of most individuals has declined. Most people are now finding that they are free to make very few significant choices. It is becoming steadily harder for ordinary people —the unrich, the unprivileged—to choose a kind of work for which they have a preference, a talent, or a vocation, to choose where they will live, to choose to work (or to live) at home, or even to choose to raise their own children. And most individuals ("liberated" or not) choose to conform not to local ways and conditions but to a rootless and placeless monoculture of commercial expectations and products. We try to be "emotionally self-sufficient" at the same time that we are entirely and helplessly dependent for our "happiness" on an economy that abuses us along with everything else. We want the liberty of divorce from spouses and independence from family and friends, yet we remain indissolubly married to a hundred corporations that regard us at best as captives and at worst as prey. The net result of our much-asserted individualism appears to be that we have become "free" for the sake of not much self-fulfillment at all.

However frustrated, disappointed, and unfulfilled it may be, the pursuit of self-liberation is still the strongest force now operating in our society. It is the dominant purpose not only of those feminists whose individualism troubles Judith Weissman but also of virtually the entire population; it determines the ethics of the professional class; it defines increasingly the ambitions of politicians and other public servants. This purpose is publicly sanctioned and publicly supported, and it operates invariably to the detriment of community life and community values.

All the institutions that "serve the community" are publicly oriented: the schools, governments and government agencies, the professions, the corporations. Even the churches, though they may have community memberships, do not concern themselves with issues of local economy and local ecology on which community health and integrity must depend. Nor do the people in charge of these institutions think of themselves as members of communities. They are itinerant, in fact or in spirit, as their careers require them to be. These various public servants all have tended to impose on the local place and the local people programs, purposes, procedures, technologies, and values that originated elsewhere. Typically, these "services" involve a condescension to and a contempt for local life that are implicit in all the assumptions—woven into the very fabric—of the industrial economy.

A community, especially if it is a rural community, is understood by its public servants as provincial, backward and benighted, unmodern, unprogressive, unlike "us," and therefore in need of whatever changes are proposed for it by outside interests (to the profit of the outside interests). Anyone who thinks of herself or himself as a member of such a community will sooner or later see that the community is under attack morally as well as economically. And this attack masquerades invariably as altruism: the community must be plundered, expropriated, or morally offended for its own good—but its good is invariably defined by the interest of the invader. The community is not asked whether or not it wishes to be changed, or how it wishes to be changed, or what it wishes to be changed into. The community is deemed to be backward and provincial, it is taught to believe and to regret that it is backward and provincial, and it is thereby taught to welcome the purposes of its invaders.

I have already discussed at some length the honored practice of community destruction by economic invasion. Now it will be useful to look at an example of the analogous practice of moral invasion. In 1989, Actors Theater of Louisville presented the premiere performance of Arthur Kopit's play *Bone-the-Fish*. The Louisville *Courier-Journal* welcomed Mr. Kopit and his play to town with the headline: "Arthur Kopit plans to offend almost everyone." In the accompanying article, Mr. Kopit is quoted as saying of his play:

> I am immodestly proud that it is written in consistently bad taste. It's about vile people who do vile things. They are totally loathsome, and I love them all. . . . I'm almost positive that it has something to offend everyone.

The writer of the article explains that "Kopit wrote 'Bone-the-Fish' out of a counterculture impulse, as a reaction against a complacency that he finds is corrupting American life."

My interest here is not in the quality or the point of Mr. Kopit's play, which I did not see (because I do not willingly subject myself to offense). I am interested in the article about him and his play merely as an example of the conventionality of the artistic intention to offend—and of the complacency of the public willingness not to be offended but passively to accept offense. Here we see the famous playwright coming from the center of culture to a provincial city, declaring his intention to "offend almost everyone," and here we see the local drama critic deferentially explaining the moral purpose of this intention. But the playwright makes three rather curious assumptions: (1) that the Louisville theater audience may be supposed (without proof) to be complacent and corrupt; (2) that they therefore deserve to be offended; and (3) that being offended will make them less complacent and less corrupt.

That Louisville theatergoers are more complacent and corrupt than theatergoers elsewhere (or than Mr. Kopit) is not an issue, for there is no evidence. That they deserve to be offended is not an issue for the same reason. That anyone's complacency and corruption can be corrected by being offended in a theater is merely a contradiction in terms, for people who are corrupted by complacency are by definition not likely to take offense. People who do take offense will be either fundamentally decent or aggressively corrupt. People who are fundamentally decent do not deserve to be offended and cannot be instructed by offense. People who are aggressively corrupt would perhaps see the offense but would not accept it. Mr. Kopit's preferred audience is therefore one that will applaud his audacity and pay no attention at all to his avowed didactic purpose—and this perhaps explains his love for "vile people."

If one thinks of Louisville merely as a public, there is not much of an issue here. Mr. Kopit's play is free speech, protected by the First Amendment, and that is that. If, however, one thinks of Louisville as a community or even as potentially a community, then the issue is sizable, and it is difficult. A public, as I have already suggested, is a rather odd thing; I can't think of anything else that is like it. A community is another matter, for it exists within a system of analogies or likenesses that clarify and amplify its meaning. A healthy community is like an ecosystem, and it includes—or it makes itself harmoniously a part of its local ecosystem. It is also like a household; it is the household of its place, and it includes the households of many families, human and nonhuman. And to extend Saint Paul's famous metaphor by only a little, a healthy community is like a body, for its members mutually support and serve one another.

If a community, then, is like a household, what are we to make of the artist whose intention is to offend? Would I welcome into my house any stranger who came, proud of his bad taste, professing his love for vile people and proposing to offend almost everyone? I would not, and I do not know anybody who would. To do so would contradict self-respect and respect for loved ones. By the same token, I cannot see that a community is under any obligation to welcome such a person. The public, so far as I can see, has no right to require a community to submit to or support statements that offend it.

I know that for a century or so many artists and writers have felt it was their duty—a mark of their honesty and courage—to offend their audience. But if the artist has a duty to offend, does not the audience therefore have a duty to be offended? If the public has a duty to protect speech that is offensive to the community, does not the community have the duty to respond, to be offended, and so defend itself against the offense? A community, as a part of a public, has no right to silence publicly protected speech, but it certainly has a right not to listen and to refuse its patronage to speech that it finds offensive. It is remarkable, however, that many writers and artists appear to be unable to accept this obvious and necessary limitation on their public freedom; they seem to think that freedom entitles them not only to be offensive but also to be approved and subsidized by the people whom they have offended.

These people believe, moreover, that any community attempt to remove a book from a reading list in a public school is censorship and a violation of the freedom of speech. The situation here involves what may be a hopeless conflict of freedoms. A teacher in a public school ought to be free to exercise his or her freedom of speech in choosing what books to teach and in deciding what to say about them. (This, to my mind, would certainly include the right to teach that the Bible is the word of God and the right to teach that it is not.) But the families of a community surely must be allowed an equal freedom to determine the education of their children. How free are parents who have no choice but to turn their children over to the influence of whatever the public will prescribe or tolerate? They obviously are not free at all. The only solution is trust between a community and its teachers, who will therefore teach as members of the community—a trust that in a time of community disintegration is perhaps not possible. And so the public presses its invasion deeper and deeper into community life under the justification of a freedom far too simply understood. It is now altogether possible for a teacher who is forbidden to teach the Bible to teach some other book that is not morally acceptable to the community, perhaps in order to improve the community by shocking or offending it. It is therefore possible that the future of community life in this country may depend on private schools and home schooling.

Does my objection to the intention to offend and the idea of improvement by offense mean that I believe it is invariably wrong to offend or that I think community and public life do not need improving? Obviously not. I do not mean at all to slight the issues of honesty and of artistic integrity that are involved. But I would distinguish between the intention to offend and the willingness to risk offending. Honesty and artistic integrity do not require anyone to intend to give offense, though they certainly may cause offense. The intention to offend, it seems to me, identifies the would-be offender as a public person. I cannot imagine anyone who is a member of a community who would purposely or gladly or proudly offend it, though I know very well that honesty might require one to do so.

Here we are verging on a distinction that had better be explicitly made. There is a significant difference between works of art made to be the vital possessions of a community (existing or not) and those made merely as offerings to the public. Some artists, and I am one of them, wish to live and work within a community, or within the hope of community, in a given place. Others wish to live and work outside the claims of any community, and these now appear to be an overwhelming majority. There is a difference between these two kinds of artists but not necessarily a division. The division comes when the public art begins to conventionalize an antipathy to community life and to the moral standards that enable and protect community life, as our public art has now done. Mr. Kopit's expressed eagerness to offend a local audience he does not know is representative of this antipathy. Our public art now communicates a conventional prejudice against old people, history, parental authority, religious faith, sexual discipline, manual work, rural people and rural life, anything local or small or inexpensive. At its worst, it glamorizes or glorifies drugs, promiscuity, pornography, violence, and blasphemy. Any threat to suppress or limit these public expressions will provoke much support for the freedom of speech. I concur in this. But as a community artist, I would like to go beyond my advocacy of the freedom of speech to deplore some of the uses that are made of it, and I wish that more of my fellow artists would do so as well.

I wish that artists and all advocates and beneficiaries of the First Amendment would begin to ask, for instance, how the individual can be liberated by disobeying the moral law, when the community obviously can be liberated only by obeying it. I wish that they would consider the probability that there is a direct relation between the public antipathy to community life and local ("provincial") places and the industrial destruction of communities and places. I wish, furthermore, that they could see that artists who make offensiveness an artistic or didactic procedure are drawing on a moral capital that they may be using up. A public is shockable or offendable only to the extent that it is already uncomplacent and uncorrupt—to the extent, in other words, that it is a community or remembers being one. What happens after the audience becomes used to being shocked and is therefore no longer shockable—as is apparently near to being the case with the television audience? What if offenses become stimulants—either to imitate the offenses or to avenge them? And what is the difference between the artist who wishes to offend the "provincials" and the industrialist or developer who wishes to dispossess them or convert them into a "labor force"?

The idea that people can be improved by being offended will finally have to meet the idea (espoused some of the time by some of the same people) that books, popular songs, movies, television shows, sex videos, and so on are "just fiction" or "just art" and therefore exist "for their own sake" and have no influence. To argue that works of art are "only" fictions or self-expressions and therefore cannot cause bad behavior is to argue also that they cannot cause good behavior. It is, moreover, to make an absolute division between art and life, experience and life, mind and body—a division that is intolerable to anyone who is at all serious about being a human or a member of a community or even a citizen.

Ananda Coomaraswamy, who had exhaustive knowledge of the traditional uses of art, wrote that "the purpose of any art . . . is to teach, to delight, *and above all to move*" (my emphasis). Of course art moves us! To assume otherwise not only contradicts the common assumption of teachers and writers from the earliest times almost until now; it contradicts everybody's experience. A cathedral, to mention only one of the most obvious examples, is a work of art made to cause a movement toward God, and this is in part a physical movement required by the building's structure and symbolism. But all works of any power move us, in both body and mind, from the most exalted music or poetry to the simplest dance tune. In fact, a dance tune is as good an example as a cathedral. An influence is cast over us, and we are moved. If we see that the influence is bad, we may be moved to reject it, but that is a second movement; it occurs only after we have felt the power of the influence.

People do not patronize the makers of pornographic films and sex videos because they are dispassionate appreciators of bad art; they do so because they wish to be moved. Perhaps the makers of pornographic films do not care what their products move their patrons to do. But if they do care, they are writing a check on moral capital to which they do not contribute. They trust that people who are moved by their work will not be moved to sexual harassment or child molestation or rape. They are banking heavily on the moral decency of their customers. And so are all of us who defend the freedom of speech. We are trusting—and not comfortably—that people who come under the influence of the sexual pandering, the greed, the commercial seductions, the moral oversimplification, the brutality, and the violence of our modern public arts will yet somehow remain under the influence of Moses and Jesus. I don't see how anyone can extend this trust without opposing in every way short of suppression the abuses and insults that are protected by it. The more a society comes to be divided in its assumptions and values, the more necessary public freedom becomes. But the more necessary public freedom becomes, the more necessary community responsibility becomes. This connection is unrelenting. And we should not forget that the finest works of art make a community of sorts of their audience. They do not divide people or justify or flatter their divisions; they define our commonwealth, and they enlarge it.

The health of a free public—especially that of a large nation under a representative government—depends on distrust. Thomas Jefferson thought so, and I believe he was right. In subscribing, generation after generation, to our Constitution, we extend to one another and to our government a trust that would be foolish if there were any better alternative. It is a breathtaking act of faith. And this trust is always so near to being misapplied that it cannot be maintained without distrust. People would fail it worse than they do if it were not for the constant vigilance and correction of distrust.

But a community makes itself up in more intimate circumstances than a public. And the health of a community depends absolutely on trust. A community knows itself and knows its place in a way that is impossible for a public (a nation, say, or a state). A community does not come together by a covenant, by a conscientious granting of trust. It exists by proximity, by neighborhood; it knows face to face, and it trusts as it knows. It learns, in the course of time and experience, what and who can be trusted. It knows that some of its members are untrustworthy, and it can be tolerant, because to know in this matter is to be safe. A community member can be trusted to be untrustworthy and so can be included. (A community can trust its liars to be liars, for example, and so enjoy them.) But if a community withholds trust, it withholds membership. If it cannot trust, it cannot exist.

One of the essential trusts of community life is that which holds marriages and families together. Another trust is that neighbors will help one another. Another is that privacy will be respected, especially the privacy of personal feeling and the privacy of relationships. All these trusts are absolutely essential, and all are somewhat fragile. But the most fragile, the most vulnerable to public invasion, is the trust that protects privacy. And in our time privacy has been the trust that has been most subjected to public invasion.

I am referring not just to the pryings and snoopings of our secret government, which contradict all that our public government claims to stand for, but also to those by now conventional publications of private grief, of violence to strangers, of the sexual coupling of strangers—all of which allow the indulgence of curiosity without sympathy. These all share in the evil of careless or malicious gossip; like careless or malicious gossip of any other kind, they destroy community by destroying respect for personal dignity and by destroying compassion. It is clear that no self-respecting human being or community would tolerate for a moment the representation of brutality or murder, on television or anywhere else, in such a way as to allow no compassion for the victim. But worst of all—and, I believe, involved in all—is the public prostitution of sex in guises of freedom ranging from the clinical to the commercial, from the artistic to the statistical.

One of the boasts of our century is that its artists—not to mention its psychologists, therapists, anthropologists, sociologists, statisticians, and pornographers—have pried open the bedroom door at last and shown us sexual love for what it "really" is. We have, we assume, cracked the shell of sexual privacy. The resulting implication that the shell is easily cracked disguises the probability that the shell is, in fact, not crackable at all and that what we have seen displayed is not private or intimate sex, not sexual love, but sex reduced, degraded, oversimplified, and misrepresented by the very intention to display it. Sex publicly displayed is public sex. Sex observed is not private or intimate and cannot be.

Could a voyeur conceivably crack the shell? No, for voyeurs are the most handicapped of all the sexual observers; they know only what they see. True intimacy, even assuming that it can be observed, cannot be known by an outsider and cannot be shown. An artist who undertakes to show the most intimate union of lovers—even assuming that the artist is one of the lovers—can only represent what she or he alone thinks it is. The intimacy, the union itself, remains unobserved. One cannot enter into this intimacy and watch it at the same time, any more than the mind can think about itself while it thinks about something else.

Is sexual love, then, not a legitimate subject of the imagination? It is. But the work of the imagination does not require that the shell be cracked. From Homer to Shakespeare, from the Bible to Jane Austen, we have many imaginings of the intimacy and power of sexual love that have respected absolutely its essential privacy and thus have preserved its intimacy and honored its dignity.

The essential and inscrutable privacy of sexual love is the sign both of its mystery and sanctity and of its humorousness. It is mysterious because the couple who are in it are lost in it. It is their profoundest experience of the being of the world and of their being in it and is at the same time an obliviousness to the world. This lostness of people in sexual love tends to be funny to people who are outside it. But having subscribed to the superstition that we have stripped away all privacy—and mystery and sanctity—from sex, we have become oddly humorless about it. Most people, for example, no longer seem to be aware of the absurdity of sexual vanity. Most people apparently see the sexual pretension and posturing of popular singers, athletes, and movie stars as some kind of high achievement, not the laughable inanity that it really is. Sexual arrogance, on the other hand, is not funny. It is dangerous, and there are some signs that our society has begun to recognize the danger. What it has not recognized is that the publication of sexual privacy is not only fraudulent but often also a kind of sexual arrogance, and a dangerous one.

Does this danger mean that any explicit representation of sexual lovemaking is inevitably wrong? It does not. But it means that such representations *can* be wrong and that when they are wrong, they are destructive.

The danger, I would suggest, is not in the representation but in the reductiveness that is the risk of representation and that is involved in most representations. What is so fearfully arrogant and destructive is the implication that what is represented, or representable, is all there is. In the best representations, I think, there would be a stylization or incompleteness that would convey the artist's honest acknowledgment that this is not all.

The best representations are surrounded and imbued with the light of imagination, so that they make one aware, with profound sympathy, of the two lives, not just the two bodies, that are involved; they make one aware also of the difficulty of full and open sexual consent between two people and of the history and the trust that are necessary to make possible that consent. Without such history and trust, sex is brutal, no matter what species is involved.

When sexual lovemaking is shown in art, one can respond intelligently to it by means of a handful of questions: Are the lovers represented as merely "physical" bodies or as two living souls? Does the representation make it possible to see why Eros has been understood not as an instinct or a "drive" but as a god? Are we asked to see this act as existing in and of and for itself or as joined to the great cycle of fertility and mortality? Does it belong to nature and to culture? Can we imagine this sweetness continuing on through the joys and difficulties of homemaking, the births and the upbringing of children, the deaths of parents and friends—through disagreements, hardships, quarrels, aging, and death? Does it encourage us to forget or to remember that "certainly it must some time come to pass that the very gentle Beatrice will die"?

And finally we must ask how the modern representations of lovemaking that we find in movies, books, paintings, sculptures, and on television measure up to the best love scenes that we know. The best love scene that I know is not explicitly sexual. It is the last scene of *The Winter's Tale*, in which Shakespeare brings onto the same stage, into the one light, young love in its astounding beauty, ardor, and hope, and old love with its mortal wrongs astoundingly graced and forgiven.

The relevance of such imagining is urgently practical; it is the propriety or justness that holds art and the world together. To represent sex without this fullness of imagination is to foreshadow the degradation and destruction of all that is not imagined. Just as the ruin of farmers, farming, and farmland may be predicted from a society's failure to imagine food in all its meanings and connections, so the failure to imagine sex in all its power and sanctity is to prepare the ruin of family and community life and of much else. In order to expose the privacy of sex, we have made of it another industrial specialization, leaving it naked not only of clothes and of customary discretions and courtesies but also of all its cultural and natural connections.

There are, we must realize, kinds of nakedness that are significantly and sometimes ominously different from each other. To know this, we have only to study the examples that are before us. There is—and who can ever forget it?—the nakedness of the photographs of prisoners in Hitler's death camps. This is the nakedness of absolute exposure to mechanical politics, politics gravitating toward the unimagining "efficiency" of machinery. I remember also a photograph of a naked small child running terrified down a dirt road in Vietnam, showing the body's absolute exposure to the indifference of air war, the appropriate technology of mechanical politics.

There is also the nakedness in advertising, in the worst kinds of fashionable or commercial art. This is the nakedness of free-market sexuality, the nakedness that is possible only in a society in which price is the only index of worth.

The nakedness of the death camps and of mechanical war denotes an absolute loss of dignity. In advertising, novels, and movies, the nakedness sometimes denotes a very significant and a very dangerous loss of dignity. Where the body has no dignity, where the sanctity of its own mystery and privacy is not recognized by a surrounding and protecting community, there can be no freedom. To destroy the dignity of the body—the dignity of any and every body—is to prepare the way for the enslaver, the rapist, the torturer, the user of cannon fodder. The nakedness or near-nakedness of some tribal peoples (I judge from the photographs that I have seen) is, in contrast, always dignified, and this dignity rests on a trust so complex and comprehensive as to be virtually unimaginable to us. The public nakedness of our own society involves no trust but only an exploitiveness that is inescapably economic and greedy. It is an abandonment of the self to self-exploitation and to exploitation by others.

There is also the nakedness of innocence, as, for example, in Degas's *Seated Bather Drying Herself*, in the Metropolitan Museum of Art, in which the body is shown in the unaware and unregarded coherence and mystery of its own being. This quality D. H. Lawrence saw and celebrated:

> People were bathing and posturing themselves on the beach
> and all was dreary, great robot limbs, robot breasts
> robot voices, robot even the gay umbrellas.
> But a woman, shy and alone, was washing herself under
> a tap
> and the glimmer of the presence of the gods was like lilies
> and like water-lilies.

Finally, there is the nakedness of sexual candor. However easy or casual nakedness may have been made by public freedom, the nakedness of sexual candor is not possible except within the culturally delineated conditions that establish and maintain trust. And it is utterly private. It can be suggested in art but not represented. Any effort to represent it, I suspect, will inevitably be bogus. What must we do to earn the freedom of being unguardedly and innocently naked to someone? Our own and other cultures suggest that we must do a lot. We must make promises and keep them. We must assume many fearful responsibilities and do much work. We must build the household of trust.

It is the community, not the public, that is the protector of the possibility of this candor, just as it is the protector of other tender, vulnerable, and precious things—the childhood of children, for example, and the fertility of fields. These protections are left to the community, for they can be protected only by affection and by intimate knowledge, which are beyond the capacities of the public and beyond the power of the private citizen.

IV

If the word *community* is to mean or amount to anything, it must refer to a place (in its natural integrity) and its people. It must refer to a placed people. Since there obviously can be no cultural relationship that is uniform between a nation and a continent, "community" must mean a people locally placed and a people, moreover, not too numerous to have a common knowledge of themselves and of their place. Because places differ from one another and because people will differ somewhat according to the characters of their places, if we think of a nation as an assemblage of many communities, we are necessarily thinking of some sort of pluralism.

There is, in fact, a good deal of talk about pluralism these days, but most of it that I have heard is fashionable, superficial, and virtually worthless. It does not foresee or advocate a plurality of settled communities but is only a sort of indifferent charity toward a plurality of aggrieved groups and individuals. It attempts to deal liberally—that is, by the superficial courtesies of tolerance and egalitarianism—with a confusion of claims.

The social and cultural pluralism that some now see as a goal is a public of destroyed communities. Wherever it exists, it is the result of centuries of imperialism. The modern industrial urban centers are "pluralistic" because they are full of refugees from destroyed communities, destroyed community economies, disintegrated local cultures, and ruined local ecosystems. The pluralists who see this state of affairs as some sort of improvement or as the beginning of "global culture" are being historically perverse, as well as politically naive. They wish to regard liberally and tolerantly the diverse, sometimes competing claims and complaints of a rootless society, and yet they continue to tolerate also the ideals and goals of the industrialism that caused the uprooting. They affirm the pluralism of a society formed by the uprooting of cultures at the same time that they regard the fierce self-defense of still-rooted cultures as "fundamentalism," for which they have no tolerance at all. They look with wistful indulgence and envy at the ruined or damaged American Indian cultures so long as those cultures remain passively a part of our plurality, forgetting that these cultures, too, were once "fundamentalist" in their self-defense. And when these cultures again attempt self-defense—when they again assert the inseparability of culture and place—they are opposed by this pluralistic society as self-righteously as ever. The tolerance of this sort of pluralism extends always to the uprooted and passive, never to the rooted and active.

The trouble with the various movements of rights and liberties that have passed among us in the last thirty years is that they have all been too exclusive and so have degenerated too readily into special pleading. They have, separately, asked us to stop exploiting racial minorities or women or nature, and they have been, separately, right to do so. But they have not, separately or together, come to the realization that we live in a society that exploits, first, everything that is not ourselves and then, inevitably, ourselves. To ask, within this general onslaught, that we should honor the dignity of this or that group is to ask that we should swim up a waterfall.

Any group that takes itself, its culture, and its values seriously enough to try to separate, or to remain separate, from the industrial line of march will be, to say the least, unwelcome in the plurality. The tolerance of these doctrinaire pluralists always runs aground on religion. You may be fascinated by religion, you may study it, anthropologize and psychoanalyze about it, collect and catalogue its artifacts, but you had better not believe in it. You may put into "the canon" the holy books of any group, but you had better not think them holy. The shallowness and hypocrisy of this tolerance is exposed by its utter failure to extend itself to the suffering people of Iraq, who are, by the standards of this tolerance, fundamentalist, backward, unprogressive, and in general not like "us."

The problem with this form of pluralism is that it has no authentic standard; its standard simply is what one group or another may want at the moment. Its professed freedom is not that of community life but that of a political group acting on the pattern of individualism. To get farther toward a practicable freedom, the group must measure itself and its wants by standards external to itself. I assume that these standards must be both cultural and ecological. If people wish to be free, then they must preserve the culture that makes for political freedom, and they must preserve the health of the world.

There is an insistently practical question that any person and any group seriously interested in freedom must ask: Can land and people be preserved anywhere by means of a culture that is in the usual sense pluralistic? E. M. Forster, writing *Howards End* in the first decade of this century, doubted that they could. Nothing that has happened in the intervening eighty-odd years diminishes that doubt, and much that has happened confirms it.

A culture capable of preserving land and people can be made only within a relatively stable and enduring relationship between a local people and its place. Community cultures made in this way would necessarily differ, and sometimes radically so, from one place to another, because places differ. This is the true and necessary pluralism. There can, I think, be no national policy of pluralism or multiculturalism but only these pluralities of local cultures. And if these cultures are of any value and worthy of any respect, they will not be elective—not determined by mere wishes—but will be formed in response to local nature and local needs.

At present, the rhetoric of racial and cultural pluralism works against the possibility of a pluralism of settled communities, exactly as do the assumptions and the practices of national and global economies. So long as we try to think of ourselves as African Americans or European Americans or Asian Americans, we will never settle anywhere. For an authentic community is made less in reference to who we are than to where we are. I cannot farm my farm as a European American—or as an American, or as a Kentuckian—but only as a person belonging to the place itself. If I am to use it well and live on it authentically, I cannot do so by knowing where my ancestors came from (which, except for one great-grandfather, I do not know and probably can never know); I can do so only by knowing where I am, what the nature of the place permits me to do here, and who and what are here with me. To know these things, I must ask the place. A knowledge of foreign cultures is useful, perhaps indispensable, to me in my effort to settle here, but it cannot tell me where I am.

That there should be peace, commerce, and biological and cultural outcrosses among local cultures is obviously desirable and probably necessary as well. But such a state of things would be radically unlike what is now called pluralism. To start with, a plurality of settled communities could not be preserved by the present-day pluralists' easy assumption that all cultures are equal or of equal value and capable of surviving together by tolerance. The idea of equality is a good one, so long as it means "equality before the law." Beyond that, the idea becomes squishy and sentimental because of manifest inequalities of all kinds. It makes no sense, for example, to equate equality with freedom. The two concepts must be joined precisely and within strict limits if their association is to make any sense at all. Equality, in certain circumstances, is anything but free. If we have equality and nothing else—no compassion, no magnanimity, no courtesy, no sense of mutual obligation and dependence, no imagination—then power and wealth will have their way; brutality will rule. A general and indiscriminate egalitarianism is free-market culture, which, like free-market economics, tends toward a general and destructive uniformity. And tolerance, in association with such egalitarianism, is a way of ignoring the reality of significant differences. If I merely tolerate my neighbors on the assumption that all of us are equal, that means I can take no interest in the question of which ones of us are right and which ones are wrong; it means that I am denying the community the use of my intelligence and judgment; it means that I am not prepared to defer to those whose abilities are superior to mine, or to help those whose condition is worse; it means that I can be as self-centered as I please.

In order to survive, a plurality of true communities would require not egalitarianism and tolerance but knowledge, an understanding of the necessity of local differences, and respect. Respect, I think, always implies imagination—the ability to see one another, across our inevitable differences, as living souls.

FROM
ANOTHER TURN OF THE CRANK
(1995)

Farming and the Global Economy

WE HAVE BEEN repeatedly warned that we cannot know where we wish to go if we do not know where we have been. And so let us start by remembering a little history.

As late as World War II, our farms were predominantly solar powered. That is, the work was accomplished principally by human beings and horses and mules. These creatures were empowered by solar energy, which was collected, for the most part, on the farms where they worked and so was pretty cheaply available to the farmer.

However, American farms had not become as self-sufficient in fertility as they should have been—or many of them had not. They were still drawing, without sufficient repayment, against an account of natural fertility accumulated over thousands of years beneath the native forest trees and prairie grasses.

The agriculture we had at the time of World War II was nevertheless often pretty good, and it was promising. In many parts of our country we had begun to have established agricultural communities, each with its own local knowledge, memory, and tradition. Some of our farming practices had become well adapted to local conditions. The best traditional practices of the Midwest, for example, are still used by the Amish with considerable success in terms of both economy and ecology.

Now that the issue of sustainability has arisen so urgently, and in fact so transformingly, we can see that the correct agricultural agenda following World War II would have been to continue and refine the already established connection between our farms and the sun and to correct, where necessary, the fertility deficit. There can be no question, now, that that is what we should have done.

It was, notoriously, not what we did. Instead, the adopted agenda called for a shift from the cheap, clean, and, for all practical purposes, limitless energy of the sun to the expensive, filthy, and limited energy of the fossil fuels. It called for the massive use of chemical fertilizers to offset the destruction of topsoil and the depletion of natural fertility. It called also for the displacement of nearly the entire farming population and the replacement of their labor and good farming practices by machines and toxic chemicals. This agenda has succeeded in its aims, but to the benefit of no one and nothing except the corporations that have supplied the necessary machines, fuels, and chemicals—and the corporations that have bought cheap and sold high the products that, as a result of this agenda, have been increasingly expensive for farmers to produce.

The farmers have not benefited—not, at least, as a class—for as a result of this agenda they have become one of the smallest and most threatened of all our minorities. Many farmers, sad to say, have subscribed to this agenda and its economic assumptions, believing that they would not be its victims. But millions, in fact, have been its victims—not farmers alone but also their supporters and dependents in our rural communities.

The people who benefit from this state of affairs have been at pains to convince us that the agricultural practices and policies that have almost annihilated the farming population have greatly benefited the population of food consumers. But more and more consumers are now becoming aware that our supposed abundance of cheap and healthful food is to a considerable extent illusory. They are beginning to see that the social, ecological, and even the economic costs of such "cheap food" are, in fact, great. They are beginning to see that a system of food production that is dependent on massive applications of drugs and chemicals cannot, by definition, produce "pure food." And they are beginning to see that a kind of agriculture that involves unprecedented erosion and depletion of soil, unprecedented waste of water, and unprecedented destruction of the farm population cannot by any accommodation of sense or fantasy be called "sustainable."

From the point of view, then, of the farmer, the ecologist, and the consumer, the need to reform our ways of farming is now both obvious and imperative. We need to adapt our farming much more sensitively to the nature of the places where the farming is done. We need to make our farming practices and our food economy subject to standards set not by the industrial system but by the health of ecosystems and of human communities.

The immediate difficulty in even thinking about agricultural reform is that we are rapidly running out of farmers. The tragedy of this decline is not just in its numbers; it is also in the fact that these farming people, assuming we will ever recognize our need to replace them, cannot be replaced anything like as quickly or easily as they have been dispensed with. Contrary to popular assumption, good farmers are not in any simple way part of the "labor force." Good farmers, like good musicians, must be raised to the trade.

The severe reduction of our farming population may signify nothing to our national government, but the members of country communities feel the significance of it—and the threat of it—every day. Eventually urban consumers will feel these things, too. Every day farmers feel the oppression of their long-standing problems: overproduction, low prices, and high costs. Farmers sell on a market that because of overproduction is characteristically depressed, and they buy their supplies on a market that is characteristically inflated—which is necessarily a recipe for failure, because farmers do not control either market. If they will not control production and if they will not reduce their dependence on purchased supplies, then they will keep on failing.

The survival of farmers, then, requires two complementary efforts. The first is entirely up to the farmers, who must learn —or learn again—to farm in ways that minimize their dependence on industrial supplies. They must diversify, using both plants and animals. They must produce, on their farms, as much of the required fertility and energy as they can. So far as they can, they must replace purchased goods and services with natural health and diversity and with their own intelligence. To increase production by increasing costs, as farmers have been doing for the last half century, is not only unintelligent; it is crazy. If farmers do not wish to cooperate any longer in their own destruction, then they will have to reduce their dependence on those global economic forces that intend and approve and profit from the destruction of farmers, and they will have to increase their dependence on local nature and local intelligence.

The second effort involves cooperation between local farmers and local consumers. If farmers hope to exercise any control over their markets, in a time when a global economy and global transportation make it possible for the products of any region to be undersold by the products of any other region, then they will have to look to local markets. The long-broken connections between towns and cities and their surrounding landscapes will have to be restored. There is much promise and much hope in such a restoration. But farmers must understand that this requires an economics of cooperation rather than competition. They must understand also that such an economy sooner or later will require some rational means of production control.

If communities of farmers and consumers wish to promote a sustainable, safe, reasonably inexpensive supply of good food, then they must see that the best, the safest, and most dependable source of food for a city is not the global economy, with its extreme vulnerabilities and extravagant transportation costs, but its own surrounding countryside. It is, in every way, in the best interest of urban consumers to be surrounded by productive land, well farmed and well maintained by thriving farm families in thriving farm communities.

If a safe, sustainable local food economy appeals to some of us as a goal that we would like to work for, then we must be careful to recognize not only the great power of the interests arrayed against us but also our own weakness. The hope for such a food economy as we desire is represented by no political party and is spoken for by no national public officials of any consequence. Our national political leaders do not know what we are talking about, and they are without the local affections and allegiances that would permit them to learn what we are talking about.

But we should also understand that our predicament is not without precedent; it is approximately the same as that of the proponents of American independence at the time of the Stamp Act—and with one difference in our favor: in order to do the work that we must do, we do not need a national organization. What we must do is simple: we must shorten the distance that our food is transported so that we are eating more and more from local supplies, more and more to the benefit of local farmers, and more and more to the satisfaction of local consumers. This can be done by cooperation among small organizations: conservation groups, churches, neighborhood associations, consumer co-ops, local merchants, local independent banks, and organizations of small farmers. It also can be done by cooperation between individual producers and consumers. We should not be discouraged to find that local food economies can grow only gradually; it is better that they should grow gradually. But as they grow they will bring about a significant return of power, wealth, and health to the people.

One thing at least should be obvious to us all: the whole human population of the world cannot live on imported food. Some people somewhere are going to have to grow the food. And wherever food is grown the growing of it will raise the same two questions: How do you preserve the land in use? And how do you preserve the people who use the land?

The farther the food is transported, the harder it will be to answer those questions correctly. The correct answers will not come as the inevitable by-products of the aims, policies, and procedures of international trade, free or unfree. They cannot be legislated or imposed by international or national or state agencies. They can only be supplied locally, by skilled and highly motivated local farmers meeting as directly as possible the needs of informed local consumers.

Conserving Forest Communities

I LIVE IN Henry County, near the lower end of the Kentucky River Valley, on a small farm that is half woodland. Starting from my back door, I could walk for days and never leave the woods except to cross the roads. Though Henry County is known as a farming county, 25 percent of it is wooded. From the hillside behind my house I can see thousands of acres of trees in the counties of Henry, Owen, and Carroll.

Most of the trees are standing on steep slopes of the river and creek valleys that were cleared and plowed at intervals from the early years of settlement until about the time of World War II. These are rich woodlands nevertheless. The soil, though not so deep as it once was, is healing from agricultural abuse and, because of the forest cover, is increasing in fertility. The plant communities consist of some cedar and a great diversity of hardwoods, shrubs, and wildflowers.

The history of these now-forested slopes over the last two centuries can be characterized as a cyclic alternation of abuse and neglect. Their best hope, so far, has been neglect—though even neglect has often involved their degradation by livestock grazing. So far, almost nobody has tried to figure out or has even wondered what might be the best use and the best care for such places. Often the trees have been regarded merely as obstructions to row cropping, which, because of the steepness of the terrain, has necessarily caused severe soil losses from water erosion. If an accounting is ever done, we will be shocked to learn how much ecological capital this kind of farming required for an almost negligible economic return: thousands of years of soil building were squandered on a few crops of corn or tobacco.

In my part of Kentucky, as in other parts, we never developed a local forest economy, and I think this was because of our preoccupation with tobacco. In the wintertime when farmers in New England, for example, employed themselves in the woods, our people went to their stripping rooms. Though in the earliest times we depended on the maple groves for syrup and sugar, we did not do so for very long. In this century, the fossil fuels weaned most of our households from firewood. For those reasons and others, we have never very consistently or very competently regarded trees as an economic resource.

And so as I look at my home landscape, I am happy to see that I am to a considerable extent a forest dweller. But I am unhappy to remember every time I look—for the landscape itself reminds me—that I am a dweller in a forest for which there is, properly speaking, no local forest culture and no local forest economy. That is to say that I live in a threatened forest.

Such woodlands as I have been describing are now mostly ignored so long as they are young. After the trees have reached marketable size, especially in a time of agricultural depression, the landowners come under pressure to sell them. And then the old cycle is repeated, as neglect is once more superseded by abuse. The salable trees are marked, and the tract of timber is sold to somebody who may have no connection, economic or otherwise, to the local community. The trees are likely to be felled and dragged from the woods in ways that do more damage than necessary to the land and the young trees. The skidder may take the logs straight upslope, leaving scars that (depending on how they catch the runoff) will be slow to heal or will turn into gullies that will never heal. There is no local *interest* connecting the woods workers to the woods. They do not regard the forest as a permanent resource but rather as a purchased "crop" that must be "harvested" as quickly and as cheaply as possible.

The economy of this kind of forestry is apt to be as deplorable as its ecology. More than likely only the prime log of each tree is taken—that is, the felled tree is cut in two below the first sizable branch, leaving many board feet in short logs (that would be readily usable, say, if there were small local woodworking shops) as well as many cords of firewood. The trees thus carelessly harvested will most likely leave the local community and the state as sawlogs or, at best, rough lumber. The only local economic benefit may well be the single check paid by the timber company to the landowner.

But the small landowners themselves may not receive the optimum benefit, for the prevailing assumptions and economic conditions encourage or require them to sell all their marketable trees at the same time. Unless the landowner is also a logger with the know-how and the means of cutting timber and removing it from the woods, the small, privately owned woodland is not likely to be considered a source of steady income, producing a few trees every year or every few years. For most such landowners in Kentucky, a timber sale may be thinkable only once or twice in a lifetime.

Furthermore, such landowners now must, as a matter of course, sell their timber on a market in which they have no influence, in which the power is held almost exclusively by the buyer. The sellers, of course, may choose not to sell—but only if they can *afford* not to sell. The private owners of Kentucky woodlands are in much the same fix that Kentucky tobacco producers were in before the time of the Burley Tobacco Growers Co-operative Association—and in much the same fix as most American farmers today. They cannot go to market except by putting themselves at the mercy of the market. This is a matter of no little significance and concern in a rural state in which 90.9 percent of the forestland "is owned by approximately 440,000 nonindustrial private owners, whose average holding is 26 acres."

I have been describing one version of present-day commercial forestry in Kentucky—what might be called the casual version.

But we also have in view another version. This is the big-money, large-scale corporate version. It involves the building of a large factory in a forested region, predictably accompanied by political advertisements about "job creation" and "improving the local economy." This factory, instead of sawing trees into boards, will reduce them to pulp for making paper, or it will grind or shred them and make boards or prefabricated architectural components by gluing together the resulting chips or strands.

Obviously, there are some advantages to these methods. Pulping or shredding can certainly use more of a tree than, say, a conventional sawmill. The laminated-strand process can make good building material out of low-quality trees. And there is no denying our society's need for paper and for building materials.

But from the point of view of either the forest or the local human community, there are also a number of problems associated with this kind of operation.

The fundamental problem is that it is costly and large in scale. It is therefore beyond the reach of small rural communities and so will be run inevitably for the benefit not of the local people but of absentee investors. And because of its cost and size, a large wood-products factory establishes in the local forest an enormous appetite for trees.

The very efficiency of a shredding mill—its ability to use small or low-quality trees—necessarily predisposes it to clear-cutting rather than to selective and sustained production. And a well-known inclination of such industries is toward forest monocultures, which do not have the ecological stability of natural forests.

As Kentuckians know from plenty of experience, nonexploitive relationships between large industries and small communities are extremely rare, if they exist at all. A large industrial operation might conceivably be established upon the most generous and forbearing principles of forestry and with the most benevolent intentions toward the local people. But we must remember that this large operation involves a large investment. And experience has taught us that large investments tend to take precedence over ecosystems and communities. In a time of economic adversity, the community and the forest will be sacrificed before the factory will be. The ideal of such operations is maximum profit to the owners or shareholders, who are not likely to be members of the local community. This means what it has always meant: labor and materials must be procured as cheaply as possible, and real human and ecological costs must be "externalized"—charged to taxpayers or to the future.

And so Kentucky forestry, at present, is mainly of two kinds: the casual and careless logging that is hardly more than an afterthought of farming, and the large-scale exploitation of the forest by absentee owners of corporations. Neither kind is satisfactory, by any responsible measure, in a state whose major natural resources will always be its productive soils and whose landscape today is one-half forested.

Kentucky has 12,700,000 acres of forest—almost 20,000 square miles. Very little of this is mature forest; nearly all of the old-growth timber had been cut down by 1940. Kentucky woodlands are nevertheless a valuable economic resource, supporting at present a wood-products industry with an annual payroll of $300 million and employing about 25,000 people. In addition, our forestlands contribute significantly to Kentucky's attractiveness to tourists, hunters, fishermen, and campers. They contribute indirectly to the economy by protecting our watersheds and our health.

But however valuable our forests may be now, they are nothing like so valuable as they can become. If we use the young forests we have now in the best way and if we properly care for them, they will continue to increase in board footage, in health, and in beauty for several more human generations. But already we are running into problems that can severely limit the value and usefulness of this resource to our people, because we have neglected to learn to practice good forest stewardship.

Moreover, we have never understood that the only appropriate human response to a diversified forest ecosystem is a diversified local forest economy. We have failed so far to imagine and put in place the sort of small-scale, locally owned logging and wood-products industries that would be the best guarantors of the long-term good use and good care of our forests. At present, it is estimated that up to 70 percent of the timber production of our forests leaves the state as logs or as raw lumber.

Lest you think that the situation and the problems I have outlined are of interest only to "tree huggers," let me remind you that during most of the history of our state, our rural landscapes and our rural communities have been in bondage to an economic colonialism that has exploited and misused both land and people. This exploitation has tended to become more severe with the growth of industrial technology. It has been most severe and most obvious in the coalfields of eastern Kentucky, but it has been felt and has produced its dire effects everywhere. With few exceptions our country people, generation after generation, have been providers of cheap fuels and raw materials to be used or manufactured in other places and to the profit of other people. They have added no value to what they have produced, and they have gone onto the markets without protection. They have sold their labor, their mineral rights, their crops, their livestock, and their trees with the understanding that the offered price was the price that they must take. Except for the tobacco program and the coal miners' union, rural Kentuckians have generally been a people without an asking price. We have developed the psychology of a subject people, willing to take whatever we have been offered and to believe whatever we have been told by our self-designated "superiors."

Now, with the two staple economies of coal and tobacco in doubt, we ask, "What can we turn to?" This is a question for every Kentuckian, but immediately it is a question for the rural communities. It is a question we may have to hold before ourselves for a long time, because the answer is going to be complex and difficult. If, however, as a part of the answer, we say, "Timber," I believe we will be right.

But we must be careful. In the past we have too often merely trusted that the corporate economy or the government would dispose of natural resources in a way that would be best for the land and the people. I hope we will not do that again. That trust has too often been catastrophically misplaced. From now on we should disbelieve that any corporation ever comes to any rural place to do it good, to "create jobs," or to bring to the local people the benefits of the so-called free market. It will be a tragedy if the members of Kentucky's rural communities ever again allow themselves passively to be sold off as providers of cheap goods and cheap labor. To put the bounty and the health of our land, our only commonwealth, into the hands of people who do not live on it and share its fate will always be an error. For whatever determines the fortune of the land determines also the fortune of the people. If the history of Kentucky teaches anything, it teaches that.

But the peculiarity of our history, so far, is that we have not had to learn the lesson. When the Old World races settled here, they saw a natural abundance so vast they could not imagine that it could be exhausted or ruined. Because it was vast and because virtually a whole continent was opening to the west, many of our forebears felt free to use the land carelessly and to justify their carelessness on the assumption that they could escape what they ruined. That early regardlessness of consequence infected our character, and so far it has dominated the political and economic life of our state. So far, for every Kentuckian, like Harry Caudill, willing to speak of the natural limits within which we have been living all along, there have been many who have wished only to fill their pockets and move on, leaving their ecological debts to be paid by somebody else's children.

But by this time, the era of cut-and-run economics *ought* to be finished. Such an economy cannot be rationally defended or even apologized for. The proofs of its immense folly, heartlessness, and destructiveness are everywhere. Its failure as a way of dealing with the natural world and human society can no longer be sanely denied. That this economic system persists and grows larger and stronger in spite of its evident failure has nothing to do with rationality or, for that matter, with evidence. It persists because, embodied now in multinational corporations, it has discovered a terrifying truth: If you can control a people's economy, you don't need to worry about its politics; its politics have become irrelevant. If you control people's choices as to whether or not they will work, and where they will work, and what they will do, and how well they will do it, and what they will eat and wear, and the genetic makeup of their crops and animals, and what they will do for amusement, then why should you worry about freedom of speech? In a totalitarian economy, any "political liberties" that the people might retain would simply cease to matter. If, as is often the case already, nobody can be elected who is not wealthy, and if nobody can be wealthy without dependence on the corporate economy, then what is your vote worth? The citizen thus becomes an economic subject.

A totalitarian economy might "correct" itself, of course, by a total catastrophe—total explosion or total contamination or total ecological exhaustion. A far better correction, however, would be a cumulative process by which states, regions, communities, households, or even individuals would begin to work toward economic self-determination and an appropriate measure of local independence. Such a course of action would involve us in a renewal of thought about our history and our predicament. We must ask again whether or not we really want to be a free people. We must consider again the linkages between land and landownership and land use and liberty. And we must ask, as we have not very seriously asked before, what are the best ways to use and to care for our land, our neighbors, and our natural resources.

If economists ever pay attention to such matters, they may find that as the scale of an enterprise increases, its standards become more and more simple, and it answers fewer and fewer needs in the local community. For example, in the summer of 1982, according to an article in *California Forestry Notes*, three men, using five horses, removed 400,780 board feet from a 35.5-acre tract in Latour State Forest. This was a "thinning operation." Two of the men worked full time as teamsters, using two horses each; one man felled the trees and did some skidding with a single horse. The job required sixty-four days. It was profitable both for the state forest and for the operator. During the sixty-four days the skidders barked a total of eight trees, only one of which was damaged badly enough to require removal. Soil disturbance in the course of the operation was rated as "slight."

At the end of this article the author estimates that a tractor could have removed the logs two and a half times as fast as the horses. And thus he implies a question that he does not attempt to answer: Is it better for two men and four horses to work sixty-four days, or for one man and one machine to do the same work in twenty-five and a half days? Assuming that the workers would all be from the local community, it is clear that the community, a timber company, and a manufacturer of mechanical skidders would answer that question in different ways. The timber company and the manufacturer would answer on the basis of a purely economic efficiency: the need to produce the greatest volume, hence the greatest profit, in the shortest time. The community, on the contrary—and just as much as a matter of self-interest—might reasonably prefer the way of working that employed the most people for the longest time and did the least damage to the forest and the soil. The community might conclude that the machine, in addition to the ecological costs of its manufacture and use, not only replaced the work of one man but more than halved the working time of another. From the point of view of the community, it is *not* an improvement when the number of employed workers is reduced by the introduction of labor-saving machinery.

This question of which technology is better is one that our society has almost never thought to ask on behalf of the local community. It is clear nevertheless that the corporate standard of judgment, in this instance as in others, is radically oversimplified, and that the community standard is sufficiently complex. By using more people to do better work, the economic need is met, but so are other needs that are social and ecological, cultural and religious.

We can safely predict that for a long time there are going to be people in places of power who will want to solve our local problems by inviting in some great multinational corporation. They will want to put millions of dollars of public money into an "incentive package" to make it worthwhile for the corporation to pay low wages for our labor and to pay low prices for, let us say, our timber. It is well understood that nothing so excites the glands of a free-market capitalist as the offer of a government subsidy.

But before we agree again to so radical a measure, producing maximum profits to people who live elsewhere and minimal, expensive benefits to ourselves and our neighbors, we ought to ask if we cannot contrive local solutions for our local problems, and if the local solutions might not be the best ones. It is not enough merely to argue against a renewal of the old colonial economy. We must have something else competently in mind.

If we don't want to subject our forests to the rule of absentee exploiters, then we must ask what kind of forest economy we would like to have. By "we" I mean all the people of our state, of course, but I mean also, and especially, the people of our state's rural counties and towns and neighborhoods.

Obviously, I cannot speak for anybody but myself. But as a citizen of this state and a member of one of its rural communities, I would like to offer a description of what I believe would be a good forest economy. The following are not my own ideas, as you will see, but come from the work of many people who have put first in their thoughts the survival and the good health of their communities.

A good forest economy, like any other good land-based economy, would aim to join the local human community and the local natural community or ecosystem together as conservingly and as healthfully as possible.

A good forest economy would therefore be a local economy, and the forest economy of a state or region would therefore be a decentralized economy. The only reason to centralize such an economy is to concentrate its profits into the fewest hands. A good forest economy would be owned locally. It would afford a decent livelihood to local people. And it would propose to serve local needs and fill local demands first, before seeking markets elsewhere.

A good forest economy would preserve the local forest in its native diversity, quality, health, abundance, and beauty. It would recognize no distinction between its own prosperity and the prosperity of the forest ecosystem. A good forest economy would function in part as a sort of lobby for the good use of the forest.

A good forest economy would be properly scaled. Individual enterprises would be no bigger than necessary to ensure the best work and the best livelihood for workers. The ruling purpose would be to do the work with the least possible disturbance to the local ecosystem and the local human community. Keeping the scale reasonably small is good for the forest. Only a local, small-scale forest economy would permit, for example, the timely and selective logging of small woodlots.

Another benefit of smallness of scale is that it preserves economic democracy and the right of private property. Property boundaries, as we should always remember, are human conventions, useful for defining not only privileges but also responsibilities, so that use may always be accompanied by knowledge, affection, care, and skill. Such boundaries exist only because the society as a whole agrees to their existence. If the right of landownership is used only to protect an owner's wish to abuse or destroy the land, upon which the community's welfare ultimately depends, then society's interest in maintaining the convention understandably declines. And so in the interest of democracy and property rights, there is much to be gained by keeping especially the land-based industries small.

A good forest economy would be locally complex. People

in the local community would be employed in forest management, logging, and sawmilling, in a variety of value-adding small factories and shops, and in satellite or supporting industries. The local community, that is, would be enabled by its economy to realize the maximum income from its local resource. This is the opposite of a colonial economy. It would answer unequivocally the question, To *whom* is the value added?

Furthermore, a local forest economy, living by the measure of local economic health, might be led to some surprising alterations of logging technology. For example, it would almost certainly have to look again at the use of draft animals in logging. This would not only be kinder to the forest but would also be another way of elaborating the economy locally, requiring lower investment and less spending outside the community.

A good forest economy would make good forestry attractive to landowners, providing income from recreational uses of their woodlands, markets for forest products other than timber, and so on.

A good forest economy would obviously need to be much interested in local education. It would, of course, need to pass on to its children the large culture's inheritance of book learning. But also, both at home and in school, it would want its children to acquire a competent knowledge of local geography, ecology, history, natural history, and of local songs and stories. And it would want a system of apprenticeships, constantly preparing young people to carry on the local work in the best way.

All along, I have been implying that a good forest economy would be a limited economy. It would be limited in scale and limited by the several things it would not do. But it would be limited also by the necessity to leave some wilderness tracts of significant acreage unused. Because of its inclination to be proud and greedy, human character needs this practical deference toward things greater than itself; this is, I think, a religious deference. Also, for reasons of self-interest and our own survival, we need wilderness as a standard. Wilderness gives us the indispensable pattern and measure of sustainability.

To assure myself that what I have described as a good forest economy is a real possibility, I went to visit the tribal forest of the Menominee Indians in northern Wisconsin. In closing, I want to say what I learned about that forest—from reading; from talking with Marshall Pecore, the forest manager, and others; and from seeing for myself.

The Menominee originally inhabited a territory of perhaps ten million acres in Wisconsin and northern Michigan. By the middle of the nineteenth century, as the country was taken up by white settlers, the tribal holding had been reduced to 235,000 acres, 220,000 acres of which were forested.

The leaders understood that if the Menominee were to live, they would have to give up their old life of hunting and gathering and make timber from their forest a major staple of their livelihood; they understood also that if the Menominee were to survive as a people, they would have to preserve the forest while they lived from it. And so in 1854 they started logging, having first instituted measures to ensure that neither the original nature nor the productive capacity of the forest would be destroyed by their work. Now, 140 years later, Menominee forest management has become technically sophisticated, but it is still rooted in cultural tradition, and its goal has remained exactly the same: to preserve the identification of the human community with the forest, and to give an absolute priority to the forest's ecological integrity. The result, in comparison to the all-too-common results of land use in the United States, is astonishing. In 1854, when logging was begun, the forest contained an estimated billion and a half board feet of standing timber. No records exist for the first thirteen years, but from 1865 to 1988 the forest yielded two billion board feet. And today, after 140 years of continuous logging, the forest still is believed to contain a billion and a half board feet of standing timber. Over those 140 years, the average diameter of the trees has been reduced by only one half of one inch—and that by design, for the foresters want fewer large hemlocks.

About 20 percent of the forest is managed in even-aged stands of aspen and jack pine, which are harvested by clear-cutting and which regenerate naturally. The rest of the forest is divided into 109 compartments, to each of which the foresters return every fifteen years to select trees for cutting. Their rule is to cut the worst and leave the best. That is, the loggers remove only those trees that are unlikely to survive for another fifteen years, those that are stunted or otherwise defective, and those that need to be removed in order to improve the stand. Old trees that are healthy and still growing are left uncut. As a result, this is an old forest, containing, for example, 350-year-old hemlocks, as well as cedars that are probably older. The average age of harvested maples is 140 to 180 years.

In support of this highly selective cutting, the forest is kept under constant study and evaluation. And loggers in the forest are strictly regulated and supervised. Even though the topography of the forest is comparatively level, skidders must be small and rubber-tired. Loggers must use permanent skid trails. And all logging contractors must attend training sessions.

The Menominee forest economy currently employs—in forest management, logging, milling, and other work—215 tribe members, or nearly 16 percent of the adult population of the reservation. As the Menominee themselves know, this is not enough; the economy of the forest needs to be more diverse. Its products at present are sawed lumber, logs, veneer logs, pulpwood, and "specialty woods" such as paneling and moldings. More value-adding industries are needed, and the Menominee are working on the problem. One knowledgeable observer has estimated that "they could probably turn twice the profit with half the land under management if they used more secondary processing."

Kentuckians looking for the pattern of a good local forest economy would have to conclude, I think, that the Menominee example is not complex enough, but that in all other ways it is excellent. We have much to learn from it. The paramount lesson undoubtedly is that the Menominee forest economy is as successful as it is because it is not understood primarily as an economy. Everybody I talked to on my visit urged me to understand that the forest is the basis of a culture and that the unrelenting cultural imperative has been to keep the forest intact—to preserve its productivity and the diversity of its trees, both in species and in age. The goal has always been a diverse, old, healthy, beautiful, productive, community-supporting forest that is home not only to its wild inhabitants but also to its human community. To secure this goal, the Menominee, following the dictates of their culture, have always done their work bearing in mind the needs of the seventh generation of their descendants.

And so, to complete my description of a good forest economy, I must add that it would be a long-term economy. Our modern economy is still essentially a crop-year economy—as though industrialism had founded itself upon the principles of the worst sort of agriculture. The ideal of the industrial economy is to shorten as much as possible the interval separating investment and payoff; it wants to make things fast, especially money. But even the slightest acquaintance with the vital statistics of trees places us in another kind of world. A forest makes things slowly; a good forest economy would therefore be a patient economy. It would also be an unselfish one, for good foresters must always look toward harvests that they will not live to reap.

Health Is Membership

FROM OUR CONSTANT and increasing concerns about health, you can tell how seriously diseased we are. Health, as we may remember from at least some of the days of our youth, is at once wholeness and a kind of unconsciousness. Disease (dis-ease), on the contrary, makes us conscious not only of the state of our health but of the division of our bodies and our world into parts.

The word "health," in fact, comes from the same Indo-European root as "heal," "whole," and "holy." To be healthy is literally to be whole; to heal is to make whole. I don't think mortal healers should be credited with the power to make holy. But I have no doubt that such healers are properly obliged to acknowledge and respect the holiness embodied in all creatures, or that our healing involves the preservation in us of the spirit and the breath of God.

If we were lucky enough as children to be surrounded by grown-ups who loved us, then our sense of wholeness is not just the sense of completeness in ourselves but also is the sense of belonging to others and to our place; it is an unconscious awareness of community, of having in common. It may be that this double sense of singular integrity and of communal belonging is our personal standard of health for as long as we live. Anyhow, we seem to know instinctively that health is not divided.

Of course, growing up and growing older as fallen creatures in a fallen world can only instruct us painfully in division and disintegration. This is the stuff of consciousness and experience. But if our culture works in us as it should, then we do not age merely into disintegration and division, but that very experience begins our education, leading us into knowledge of wholeness and of holiness. I am describing here the story of Job, of Lazarus, of the lame man at the pool of Bethesda, of Milton's Samson, of King Lear. If our culture works in us as it should, our experience is balanced by education; we are led out of our lonely suffering and are made whole.

In the present age of the world, disintegration and division, isolation and suffering seem to have overwhelmed us. The balance between experience and education has been overthrown; we are lost in experience, and so-called education is leading us nowhere. We have diseases aplenty. As if that were not enough, we are suffering an almost universal hypochondria. Half the energy of the medical industry, one suspects, may now be devoted to "examinations" or "tests"—to see if, though apparently well, we may not be latently or insidiously diseased.

If you are going to deal with the issue of health in the modern world, you are going to have to deal with much absurdity. It is not clear, for example, why death should increasingly be looked upon as a curable disease, an abnormality, by a society that increasingly looks upon life as insupportably painful and/or meaningless. Even more startling is the realization that the modern medical industry faithfully imitates disease in the way that it isolates us and parcels us out. If, for example, intense and persistent pain causes you to pay attention only to your stomach, then you must leave home, community, and family and go to a sometimes distant clinic or hospital, where you will be cared for by a specialist who will pay attention only to your stomach.

Or consider the announcement by the Associated Press on February 9, 1994, that "the incidence of cancer is up among all ages, and researchers speculated that environmental exposure to cancer-causing substances other than cigarettes may be partly to blame." This bit of news is offered as a surprise, never mind that the environment (so called) has been known to be polluted and toxic for many years. The blame obviously falls on that idiotic term "the environment," which refers to a world that surrounds us but is presumably different from us and distant from us. Our laboratories have proved long ago that cigarette smoke gets inside us, but if "the environment" surrounds us, how does *it* wind up inside us? So much for division as a working principle of health.

This, plainly, is a view of health that is severely reductive. It is, to begin with, almost fanatically individualistic. The body is seen as a defective or potentially defective machine, singular, solitary, and displaced, without love, solace, or pleasure. Its health excludes unhealthy cigarettes but does not exclude unhealthy food, water, and air. One may presumably be healthy in a disintegrated family or community or in a destroyed or poisoned ecosystem.

So far, I have been implying my beliefs at every turn. Now I had better state them openly.

I take literally the statement in the Gospel of John that God loves the world. I believe that the world was created and approved by love, that it subsists, coheres, and endures by love, and that, insofar as it is redeemable, it can be redeemed only by love. I believe that divine love, incarnate and indwelling in the world, summons the world always toward wholeness, which ultimately is reconciliation and atonement with God.

I believe that health is wholeness. For many years I have returned again and again to the work of the English agriculturist Sir Albert Howard, who said, in *The Soil and Health*, that "the whole problem of health in soil, plant, animal, and man [is] one great subject."

I am moreover a Luddite, in what I take to be the true and appropriate sense. I am not "against technology" so much as I am for community. When the choice is between the health of a community and technological innovation, I choose the health of the community. I would unhesitatingly destroy a machine before I would allow the machine to destroy my community.

I believe that the community—in the fullest sense: a place and all its creatures—is the smallest unit of health and that to speak of the health of an isolated individual is a contradiction in terms.

We speak now of "spirituality and healing" as if the only way to render a proper religious respect to the body is somehow to treat it "spiritually." It could be argued just as appropriately (and perhaps less dangerously) that the way to respect the body fully is to honor fully its materiality. In saying this, I intend no reduction. I do not doubt the reality of the experience and knowledge we call "spiritual" any more than I doubt the reality of so-called physical experience and knowledge; I recognize the rough utility of these terms. But I strongly doubt the advantage, and even the possibility, of separating these two realities.

What I'm arguing against here is not complexity or mystery but dualism. I would like to purge my own mind and language of such terms as "spiritual," "physical," "metaphysical," and "transcendental"—all of which imply that the Creation is divided into "levels" that can readily be peeled apart and judged by human beings. I believe that the Creation is one continuous fabric comprehending simultaneously what we mean by "spirit" and what we mean by "matter."

Our bodies are involved in the world. Their needs and desires and pleasures are physical. Our bodies hunger and thirst, yearn toward other bodies, grow tired and seek rest, rise up rested, eager to exert themselves. All these desires may be satisfied with honor to the body and its maker, but only if much else besides the individual body is brought into consideration. We have long known that individual desires must not be made the standard of their own satisfaction. We must consider the body's manifold connections to other bodies and to the world. The body, "fearfully and wonderfully made," is ultimately mysterious both in itself and in its dependences. Our bodies live, the Bible says, by the spirit and the breath of God, but it does not say how this is so. We are not going to *know* about this.

The distinction between the physical and the spiritual is, I believe, false. A much more valid distinction, and one that we need urgently to learn to make, is that between the organic and the mechanical. To argue this—as I am going to do—puts me in the minority, I know, but it does not make me unique. In *The Idea of a Christian Society*, T. S. Eliot wrote, "We may say that religion, as distinguished from modern paganism, implies a life in conformity with nature. It may be observed that the natural life and the supernatural life have a conformity to each other which neither has with the mechanistic life."

Still, I wonder if our persistent wish to deal spiritually with physical things does not come either from the feeling that physical things are "low" and unworthy or from the fear, especially when speaking of affection, that "physical" will be taken to mean "sexual."

The *New York Review of Books* of February 3, 1994, for example, carried a review of the correspondence of William and

Henry James along with a photograph of the two brothers standing together with William's arm around Henry's shoulders. Apropos of this picture, the reviewer, John Bayley, wrote that "their closeness of affection was undoubted and even took on occasion a quasi-physical form." It is Mr. Bayley's qualifier, "quasi-physical," that sticks in one's mind. What can he have meant by it? Is this prurience masquerading as squeamishness, or vice versa? Does Mr. Bayley feel a need to assure his psychologically sophisticated readers that even though these brothers touched one another familiarly, they were not homosexual lovers?

The phrase involves at least some version of the old dualism of spirit and body or mind and body that has caused us so much suffering and trouble and that raises such troubling questions for anybody who is interested in health. If you love your brother and if you and your brother are living creatures, how could your love for him not be physical? Not spiritual or mental only, not "quasi-physical," but physical. How could you not take a simple pleasure in putting your arm around him?

Out of the same dualism comes our confusion about the body's proper involvement in the world. People seriously interested in health will finally have to question our society's longstanding goals of convenience and effortlessness. What is the point of "labor saving" if by making work effortless we make it poor, and if by doing poor work we weaken our bodies and lose conviviality and health?

We are now pretty clearly involved in a crisis of health, one of the wonders of which is its immense profitability both to those who cause it and to those who propose to cure it. That the illness may prove incurable, except by catastrophe, is suggested by our economic dependence on it. Think, for example, of how readily our solutions become problems and our cures pollutants. To cure one disease, we need another. The causes, of course, are numerous and complicated, but all of them, I think, can be traced back to the old idea that our bodies are not very important except when they give us pleasure (usually, now, to somebody's profit) or when they hurt (now, almost invariably, to somebody's profit).

This dualism inevitably reduces physical reality, and it does so by removing its mystery from it, by dividing it absolutely from what dualistic thinkers have understood as spiritual or mental reality.

A reduction that is merely theoretical might be harmless enough, I suppose, but theories find ways of getting into action. The theory of the relative unimportance of physical reality has put itself into action by means of a metaphor by which the body (along with the world itself) is understood as a machine. According to this metaphor—which is now in constant general use—the human heart, for example, is no longer understood as the center of our emotional life or even as an organ that pumps; it is understood as "a pump," having somewhat the same function as a fuel pump in an automobile.

If the body is a machine for living and working, then it must follow that the mind is a machine for thinking. The "progress" here is the reduction of mind to brain and then of brain to computer. This reduction implies and requires the reduction of knowledge to "information." It requires, in fact, the reduction of everything to numbers and mathematical operations.

This metaphor of the machine bears heavily upon the question of what we mean by health and by healing. The problem is that like any metaphor, it is accurate only in some respects. A girl is only in some respects like a red rose; a heart is only in some respects like a pump. This means that a metaphor must be controlled by a sort of humorous intelligence, always mindful of the exact limits within which the comparison is meaningful. When a metaphor begins to control intelligence, as this one of the machine has done for a long time, then we must look for costly distortions and absurdities.

Of course, the body in most ways is not at all like a machine. Like all living creatures and unlike a machine, the body is not formally self-contained; its boundaries and outlines are not so exactly fixed. The body alone is not, properly speaking, a body. Divided from its sources of air, food, drink, clothing, shelter, and companionship, a body is, properly speaking, a cadaver, whereas a machine by itself, shut down or out of fuel, is still a machine. Merely as an organism (leaving aside issues of mind and spirit) the body lives and moves and has its being, minute by minute, by an interinvolvement with other bodies and other creatures, living and unliving, that is too complex to diagram or describe. It is, moreover, under the influence of thought and feeling. It does not live by “fuel” alone.

A mind, probably, is even less like a computer than a body is like a machine. As far as I am able to understand it, a mind is not even much like a brain. Insofar as it is usable for thought, for the association of thought with feeling, for the association of thoughts and feelings with words, for the connections between words and things, words and acts, thought and memory, a mind seems to be in constant need of reminding. A mind unreminded would be no mind at all. This phenomenon of reminding shows the extensiveness of mind—how intricately it is involved with sensation, emotion, memory, tradition, communal life, known landscapes, and so on. How you could locate a mind within its full extent, among all its subjects and necessities, I don’t know, but obviously it cannot be located within a brain or a computer.

To see better what a mind is (or is not), we might consider the difference between what we mean by knowledge and what the computer now requires us to mean by “information.” Knowledge refers to the ability to do or say the right thing at the right time; we would not speak of somebody who does the wrong thing at the wrong time as “knowledgeable.” People who perform well as musicians, athletes, teachers, or farmers are people of knowledge. And such examples tell us much about the nature of knowledge. Knowledge is formal, and it informs speech and action. It is instantaneous; it is present and available when and where it is needed.

“Information,” which once meant that which forms or fashions from within, now means merely “data.” However organized this data may be, it is not shapely or formal or in the true sense in-forming. It is not present where it is needed; if you have to “access” it, you don’t have it. Whereas knowledge moves and forms acts, information is inert. You cannot imagine a debater or a quarterback or a musician performing by “accessing information.” A computer chock full of such information is no more admirable than a head or a book chock full of it.

The difference, then, between information and knowledge is something like the difference between a dictionary and somebody’s language.

Where the art and science of healing are concerned, the

machine metaphor works to enforce a division that falsifies the process of healing because it falsifies the nature of the creature needing to be healed. If the body is a machine, then its diseases can be healed by a sort of mechanical tinkering, without reference to anything outside the body itself. This applies, with obvious differences, to the mind; people are assumed to be individually sane or insane. And so we return to the utter anomaly of a creature that is healthy within itself.

The modern hospital, where most of us receive our strictest lessons in the nature of industrial medicine, undoubtedly does well at surgery and other procedures that permit the body and its parts to be treated as separate things. But when you try to think of it as a place of healing—of reconnecting and making whole—then the hospital reveals the disarray of the medical industry's thinking about health.

In healing, the body is restored to itself. It begins to live again by its own powers and instincts, to the extent that it can do so. To the extent that it can do so, it goes free of drugs and mechanical helps. Its appetites return. It relishes food and rest. The patient is restored to family and friends, home and community and work.

This process has a certain naturalness and inevitability, like that by which a child grows up, but industrial medicine seems to grasp it only tentatively and awkwardly. For example, any ordinary person would assume that a place of healing would put a premium upon rest, but hospitals are notoriously difficult to sleep in. They are noisy all night, and the routine interventions go on relentlessly. The body is treated as a machine that does not need to rest.

You would think also that a place dedicated to healing and health would make much of food. But here is where the disconnections of the industrial system and the displacement of industrial humanity are most radical. Sir Albert Howard saw accurately that the issue of human health is inseparable from the health of the soil, and he saw too that we humans must responsibly occupy our place in the cycle of birth, growth, maturity, death, and decay, which is the health of the world. Aside from our own mortal involvement, food is our fundamental connection to that cycle. But probably most of the complaints you hear about hospitals have to do with the food, which, according to the testimony I have heard, tends to range from unappetizing to sickening. Food is treated as another unpleasant substance to inject. And this is a shame. For in addition to the obvious nutritional link between food and health, food can be a pleasure. People who are sick are often troubled or depressed, and mealtimes offer three opportunities a day when patients could easily be offered something to look forward to. Nothing is more pleasing or heartening than a plate of nourishing, tasty, beautiful food artfully and lovingly prepared. Anything less is unhealthy, as well as a desecration.

Why should rest and food and ecological health not be the basic principles of our art and science of healing? Is it because the basic principles already are technology and drugs? Are we confronting some fundamental incompatibility between mechanical efficiency and organic health? I don't know. I only know that sleeping in a hospital is like sleeping in a factory and that the medical industry makes only the most tenuous connection between health and food and no connection between health and the soil. Industrial medicine is as little interested in ecological health as is industrial agriculture.

A further problem, and an equally serious one, is that illness, in addition to being a bodily disaster, is now also an economic disaster. This is so whether or not the patient is insured. It is a disaster for us all, all the time, because we all know that personally or collectively, we cannot continue to pay for cures that continue to get more expensive. The economic disturbance that now inundates the problem of illness may turn out to be the profoundest illness of all. How can we get well if we are worried sick about money?

I wish it were not the fate of this essay to be filled with questions, but questions now seem the inescapable end of any line of thought about health and healing. Here are several more:

1. Can our present medical industry produce an adequate definition of health? My own guess is that it cannot do so. Like industrial agriculture, industrial medicine has depended increasingly on specialist methodology, mechanical technology, and chemicals; thus, its point of reference has become more and more its own technical prowess and less and less the health of creatures and habitats. I don't expect this problem to be

solved in the universities, which have never addressed, much less solved, the problem of health in agriculture. And I don't expect it to be solved by the government.

2. How can cheapness be included in the criteria of medical experimentation and performance? And why has it not been included before now? I believe that the problem here is again that of the medical industry's fixation on specialization, technology, and chemistry. As a result, the modern "health care system" has become a way of marketing industrial products, exactly like modern agriculture, impoverishing those who pay and enriching those who are paid. It is, in other words, an industry such as industries have always been.

3. Why is it that medical strictures and recommendations so often work in favor of food processors and against food producers? Why, for example, do we so strongly favor the pasteurization of milk to health and cleanliness in milk production? (Gene Logsdon correctly says that the motive here "is monopoly, not consumer health.")

4. Why do we so strongly prefer a fat-free or a germ-free diet to a chemical-free diet? Why does the medical industry strenuously oppose the use of tobacco, yet complacently accept the massive use of antibiotics and other drugs in meat animals and of poisons on food crops? How much longer can it cling to the superstition of bodily health in a polluted world?

5. How can adequate medical and health care, including disease prevention, be included in the structure and economy of a community? How, for example, can a community and its doctors be included in the same culture, the same knowledge, and the same fate, so that they will live as fellow citizens, sharers in a common wealth, members of one another?

II

It is clear by now that this essay cannot hope to be complete; the problems are too large and my knowledge too small. What I have to offer is an association of thoughts and questions wandering somewhat at random and somewhat lost within the experience of modern diseases and the often bewildering industry that undertakes to cure them. In my ignorance and bewilderment, I am fairly representative of those who go, or go with loved ones, to doctors' offices and hospitals. What I have written so far comes from my various efforts to make as much sense as I can of that experience. But now I had better turn to the experience itself.

On January 3, 1994, my brother John had a severe heart attack while he was out by himself on his farm, moving a feed trough. He managed to get to the house and telephone a friend, who sent the emergency rescue squad.

The rescue squad and the emergency room staff at a local hospital certainly saved my brother's life. He was later moved to a hospital in Louisville, where a surgeon performed a double-bypass operation on his heart. After three weeks John returned home. He still has a life to live and work to do. He has been restored to himself and to the world.

He and those who love him have a considerable debt to the medical industry, as represented by two hospitals, several doctors and nurses, many drugs and many machines. This is a debt that I cheerfully acknowledge. But I am obliged to say also that my experience of the hospital during John's stay was troubled by much conflict of feeling and a good many unresolved questions, and I know that I am not alone in this.

In the hospital what I will call the world of love meets the world of efficiency—the world, that is, of specialization, machinery, and abstract procedure. Or, rather, I should say that these two worlds come together in the hospital but do not meet. During those weeks when John was in the hospital, it seemed to me that he had come from the world of love and that the family members, neighbors, and friends who at various times were there with him came there to represent that world and to preserve his connection with it. It seemed to me that the hospital was another kind of world altogether.

When I said early in this essay that we live in a world that was created and exists and is redeemable by love, I did not mean to sentimentalize it. For this is also a fallen world. It involves error and disease, ignorance and partiality, sin and death. If this world is a place where we may learn of our involvement in immortal love, as I believe it is, still such learning is only possible here because that love involves us so inescapably in the limits, sufferings, and sorrows of mortality.

*

Like divine love, earthly love seeks plenitude; it longs for the full membership to be present and to be joined. Unlike divine love, earthly love does not have the power, the knowledge, or the will to achieve what it longs for. The story of human love on this earth is a story by which this love reveals and even validates itself by its failures to be complete and comprehensive and effective enough. When this love enters a hospital, it brings with it a terrifying history of defeat, but it comes nevertheless confident of itself, for its existence and the power of its longing have been proved over and over again even by its defeat. In the face of illness, the threat of death, and death itself, it insists unabashedly on its own presence, understanding by its persistence through defeat that it is superior to whatever happens.

The world of efficiency ignores both loves, earthly and divine, because by definition it must reduce experience to computation, particularity to abstraction, and mystery to a small comprehensibility. Efficiency, in our present sense of the word, allies itself inevitably with machinery, as Neil Postman demonstrates in his useful book *Technopoly*. "Machines," he says, "eliminate complexity, doubt, and ambiguity. They work swiftly, they are standardized, and they provide us with numbers that you can see and calculate with." To reason, the advantages are obvious, and probably no reasonable person would wish to reject them out of hand.

And yet love obstinately answers that no loved one is standardized. A body, love insists, is neither a spirit nor a machine; it is not a picture, a diagram, a chart, a graph, an anatomy; it is not an explanation; it is not a law. It is precisely and uniquely what it is. It belongs to the world of love, which is a world of living creatures, natural orders and cycles, many small, fragile lights in the dark.

In dealing with problems of agriculture, I had thought much about the difference between creatures and machines. But I had never so clearly understood and felt that difference as when John was in recovery after his heart surgery, when he was attached to many machines and was dependent for breath on a respirator. It was impossible then not to see that the breathing of a machine, like all machine work, is unvarying, an oblivious regularity, whereas the breathing of a creature is ever changing, exquisitely responsive to events both inside and outside the body, to thoughts and emotions. A machine makes breaths as a machine makes buttons, all the same, but every breath of a creature is itself a creature, like no other, inestimably precious.

Logically, in plenitude some things ought to be expendable. Industrial economics has always believed this: abundance justifies waste. This is one of the dominant superstitions of American history—and of the history of colonialism everywhere. Expendability is also an assumption of the world of efficiency, which is why that world deals so compulsively in percentages of efficacy and safety.

But this sort of logic is absolutely alien to the world of love. To the claim that a certain drug or procedure would save 99 percent of all cancer patients or that a certain pollutant would be safe for 99 percent of a population, love, unembarrassed, would respond, "What about the one percent?"

There is nothing rational or perhaps even defensible about this, but it is nonetheless one of the strongest strands of our religious tradition—it is probably the most essential strand—according to which a shepherd, owning a hundred sheep and having lost one, does not say, "I have saved 99 percent of my sheep," but rather, "I have lost one," and he goes and searches for the one. And if the sheep in that parable may seem to be only a metaphor, then go on to the Gospel of Luke, where the principle is flatly set forth again and where the sparrows stand not for human beings but for all creatures: "Are not five sparrows sold for two farthings, and not one of them is forgotten before God?" And John Donne had in mind a sort of equation and not a mere metaphor when he wrote, "If a clod be washed away by the sea, Europe is the less, as well as if a promontory were, as well as if a manor of thy friend's or of thine own were. Any man's death diminishes me."

It is reassuring to see ecology moving toward a similar idea of the order of things. If an ecosystem loses one of its native species, we now know that we cannot speak of it as itself minus one species. An ecosystem minus one species is a different ecosystem. Just so, each of us is made by—or, one might better say, made as—a set of unique associations with unique persons, places, and things. The world of love does not admit the principle of the interchangeability of parts.

When John was in intensive care after his surgery, his wife, Carol, was standing by his bed, grieving and afraid. Wanting to reassure her, the nurse said, "Nothing is happening to him that doesn't happen to everybody."

And Carol replied, "I'm not everybody's wife."

In the world of love, things separated by efficiency and specialization strive to come back together. And yet love must confront death, and accept it, and learn from it. Only in confronting death can earthly love learn its true extent, its immortality. Any definition of health that is not silly must include death. The world of love includes death, suffers it, and triumphs over it. The world of efficiency is defeated by death; at death, all its instruments and procedures stop. The world of love continues, and of this grief is the proof.

In the hospital, love cannot forget death. But like love, death is in the hospital but not of it. Like love, fear and grief feel out of place in the hospital. How could they be included in its efficient procedures and mechanisms? Where a clear, small order is fervently maintained, fear and grief bring the threat of large disorder.

And so these two incompatible worlds might also be designated by the terms "amateur" and "professional"—amateur, in the literal sense of lover, one who participates for love; and professional in the modern sense of one who performs highly specialized or technical procedures for pay. The amateur is excluded from the professional "field."

For the amateur, in the hospital or in almost any other encounter with the medical industry, the overriding experience is that of being excluded from knowledge—of being unable, in other words, to make or participate in anything resembling an "informed decision." Of course, whether doctors make informed decisions in the hospital is a matter of debate. For in the hospital even the professionals are involved in experience; experimentation has been left far behind. Experience, as all amateurs know, is not predictable, and in experience there are no replications or "controls"; there is nothing with which to compare the result. Once one decision has been made, we have destroyed the opportunity to know what would have happened if another decision had been made. That is to say that medicine is an exact science until applied; application involves intuition, a sense of probability, "gut feeling," guesswork, and error.

In medicine, as in many modern disciplines, the amateur is divided from the professional by perhaps unbridgeable differences of knowledge and of language. An "informed decision" is really not even imaginable for most medical patients and their families, who have no competent understanding of either the patient's illness or the recommended medical or surgical procedure. Moreover, patients and their families are not likely to know the doctor, the surgeon, or any of the other people on whom the patient's life will depend. In the hospital, amateurs are more than likely to be proceeding entirely upon faith—and this is a peculiar and scary faith, for it must be placed not in a god but in mere people, mere procedures, mere chemicals, and mere machines.

It was only after my brother had been taken into surgery, I think, that the family understood the extremity of this deed of faith. We had decided—or John had decided and we had concurred—on the basis of the best advice available. But once he was separated from us, we felt the burden of our ignorance. We had not known what we were doing, and one of our difficulties now was the feeling that we had utterly given him up to what we did not know. John himself spoke out of this sense of abandonment and helplessness in the intensive care unit, when he said, "I don't know what they're going to do to me or for me or with me."

As we waited and reports came at long intervals from the operating room, other realizations followed. We realized that under the circumstances, we could not be told the truth. We would not know, ever, the worries and surprises that came to the surgeon during his work. We would not know the critical moments or the fears. If the surgeon did any part of his work ineptly or made a mistake, we would not know it. We realized, moreover, that if we were told the truth, we would have no way of knowing that the truth was what it was.

We realized that when the emissaries from the operating room assured us that everything was "normal" or "routine," they were referring to the procedure and not the patient. Even as amateurs—perhaps *because* we were amateurs—we knew that what was happening was not normal or routine for John or for us.

*

That these two worlds are so radically divided does not mean that people cannot cross between them. I do not know how an amateur can cross over into the professional world; that does not seem very probable. But that professional people can cross back into the amateur world, I know from much evidence. During John's stay in the hospital there were many moments in which doctors and nurses—especially nurses!—allowed or caused the professional relationship to become a meeting between two human beings, and these moments were invariably moving.

The most moving, to me, happened in the waiting room during John's surgery. From time to time a nurse from the operating room would come in to tell Carol what was happening. Carol, from politeness or bravery or both, always stood to receive the news, which always left us somewhat encouraged and somewhat doubtful. Carol's difficulty was that she had to suffer the ordeal not only as a wife but as one who had been a trained nurse. She knew, from her own education and experience, in how limited a sense open-heart surgery could be said to be normal or routine.

Finally, toward the end of our wait, two nurses came in. The operation, they said, had been a success. They explained again what had been done. And then they said that after the completion of the bypasses, the surgeon had found it necessary to insert a "balloon pump" into the aorta to assist the heart. This possibility had never been mentioned, nobody was prepared for it, and Carol was sorely disappointed and upset. The two young women attempted to reassure her, mainly by repeating things they had already said. And then there was a long moment when they just looked at her. It was such a look as parents sometimes give to a sick or suffering child, when they themselves have begun to need the comfort they are trying to give.

And then one of the nurses said, "Do you need a hug?"

"Yes," Carol said.

And the nurse gave her a hug.

Which brings us to a starting place.

LIFE IS A MIRACLE

An Essay Against Modern Superstition
(2000)

In memory:
Lionel Basney (1946–1999)

"We are not getting something for nothing.
We are getting nothing for everything."

CONTENTS

Thy life's a miracle. Speak yet again.
King Lear, IV, vi, 55

I. Ignorance

THE EXPRESSED dissatisfaction of some scientists with the dangerous oversimplifications of commercialized science has encouraged me to hope that this dissatisfaction will run its full course. These scientists, I hope, will not stop with some attempt at a merely theoretical or technical "correction," but will press on toward a new, or a renewed, propriety in the study and the use of the living world.

No such change is foreseeable in the terms of the presently dominant mechanical explanations of things. Such a change is imaginable only if we are willing to risk an unfashionable recourse to our cultural tradition. Human hope may always have resided in our ability, in time of need, to return to our cultural landmarks and reorient ourselves.

One of the principal landmarks of the course of my own life is Shakespeare's tragedy of *King Lear*. Over the last forty-five years I have returned to *King Lear* many times. Among the effects of that play—on me, and I think on anybody who reads it closely—is the recognition that in all our attempts to renew or correct ourselves, to shake off despair and have hope, our starting place is always and only our experience. We can begin (and we must always be beginning) only where our history has so far brought us, with what we have done.

Lately my thoughts about the inevitably commercial genetic manipulations already in effect or contemplated have sent me back to *King Lear* again. The whole play is about kindness, both in the usual sense, and in the sense of truth-to-kind, naturalness, or knowing the limits of our specifically *human* nature. But this issue is dealt with most explicitly in an episode of the subplot, in which the Earl of Gloucester is recalled from despair so that he may die in his full humanity.

The old earl has been blinded in retribution for his loyalty to the king, and in this fate he sees a kind of justice for, as he says, "I stumbled when I saw." He, like Lear, is guilty of hubris or presumption, of treating life as knowable, predictable, and within his control. He has falsely accused and driven away his loyal son, Edgar. Exiled and under sentence of death, Edgar has disguised himself as a madman and beggar. He becomes, in that role, the guide of his blinded father, who asks to be led to Dover where he intends to kill himself by leaping off a cliff. Edgar's task is to save his father from despair, and he succeeds, for Gloucester dies at last " 'Twixt two extremes of passion, joy and grief. . . ." He dies, that is, within the proper bounds of the human estate. Edgar does not want his father to give up on life. To give up on life is to pass beyond the possibility of change or redemption. And so he does not lead his father to the cliff's verge, but only *tells* him he has done so. Gloucester renounces the world, blesses Edgar, his supposedly absent son, and, according to the stage direction, "Falls forward and swoons."

When he returns to consciousness, Edgar now speaks to him in the guise of a passer-by at the bottom of the cliff, from which he pretends to have seen Gloucester fall. Here he assumes explicitly the role of spiritual guide to his father.

Gloucester, dismayed to find himself still alive, attempts to refuse help: "Away, and let me die."

And then Edgar, after an interval of several lines in which he represents himself as a stranger, speaks the filial (and fatherly) line about which my thoughts have gathered:

Thy life's a miracle. Speak yet again.

This is the line that calls Gloucester back—out of hubris, and the damage and despair that invariably follow—into the properly subordinated human life of grief and joy, where change and redemption are possible.

The power of that line read in the welter of innovation and speculation of the bioengineers will no doubt be obvious. One immediately recognizes that suicide is not the only way to give up on life. We know that creatures and kinds of creatures can be killed, deliberately or inadvertently. And most farmers know that any creature that is sold has in a sense been given up on; there is a big difference between selling this year's lamb crop, which is, as such, all that it can be, and selling the breeding flock or the farm, which hold the immanence of a limitless promise.

A little harder to compass is the danger that we can give up on life also by presuming to "understand" it—that is by reducing it to the *terms* of our understanding and by treating it as predictable or mechanical. The most radical influence of reductive science has been the virtually universal adoption of the idea that the world, its creatures, and all the parts of its creatures are machines—that is, that there is no difference between creature and artifice, birth and manufacture, thought and computation. Our language, wherever it is used, is now almost invariably conditioned by the assumption that fleshly bodies are machines full of mechanisms, fully compatible with the mechanisms of medicine, industry, and commerce; and that minds are computers fully compatible with electronic technology.

This may have begun as a metaphor, but in the language as it is used (and as it affects industrial practice) it has evolved from metaphor through equation to identification. And this usage institutionalizes the human wish, or the sin of wishing, that life might be, or might be made to be, predictable.

I have read of Werner Heisenberg's principle that "Wherever one treats living organisms as physicochemical systems they must necessarily behave as such." I am not competent to have an opinion about the truth of that. I do feel able to say that whenever one treats living organisms as machines they must necessarily be *perceived* to behave as such. And I can see that the proposition is reversible: Whenever one perceives living organisms as machines they must necessarily be treated as such. William Blake made the same point earlier in this age of reduction and affliction:

> What seems to Be, Is, To those to whom
> It seems to Be, & is productive of the most dreadful
> Consequences to those to whom it seems to Be . . .

For quite a while it has been possible for a free and thoughtful person to see that to treat life as mechanical or predictable or understandable is to reduce it. Now, almost suddenly, it is becoming clear that to reduce life to the scope of our understanding (whatever "model" we use) is inevitably to enslave it, make property of it, and put it up for sale.

This is to give up on life, to carry it beyond change and redemption, and to increase the proximity of despair.

Cloning—to use the most obvious example—is not a way to improve sheep. On the contrary, it is a way to stall the sheep's lineage and make it unimprovable. No true breeder could consent to it, for true breeders have their farm and their market in mind, and always are trying to breed a better sheep. Cloning, besides being a new method of sheep-stealing, is only a pathetic attempt to make sheep predictable. But this is an affront to reality. As any shepherd would know, the scientist who thinks he has made sheep predictable has only made himself eligible to be outsmarted.

The same sort of limitation and depreciation is involved in the proposed cloning of fetuses for body parts, and in other extreme measures for prolonging individual lives. No individual life is an end in itself. One can live fully only by participating fully in the succession of the generations, in death as well as in life. Some would say (and I am one of them) that we can live fully only by making ourselves as answerable to the claims of eternity as to those of time.

The problem, as it appears to me, is that we are using the wrong language. The language we use to speak of the world and its creatures, including ourselves, has gained a certain analytical power (along with a lot of expertish pomp) but has lost much of its power to designate *what* is being analyzed or to convey any respect or care or affection or devotion toward it. As a result we have a lot of genuinely concerned people calling upon us to "save" a world which their language simultaneously reduces to an assemblage of perfectly featureless and dispirited "ecosystems," "organisms," "environments," "mechanisms," and the like. It is impossible to prefigure the salvation of the world in the same language by which the world has been dismembered and defaced.

By almost any standard, it seems to me, the reclassification of the world from creature to machine must involve at least a perilous reduction of moral complexity. So must the shift in our attitude toward the creation from reverence to understanding. So must the shift in our perceived relationship to nature from that of steward to that of absolute owner, manager, and engineer. So even must our permutation of "holy" to "holistic."

At this point I can only declare myself. I think that the poet and scholar Kathleen Raine was correct in reminding us that life, like holiness, can be known only by being experienced. To experience it is not to "figure it out" or even to understand it, but to suffer it and rejoice in it as it is. In suffering it and rejoicing in it as it is, we know that we do not and cannot understand it completely. We know, moreover, that we do not wish to have it appropriated by somebody's claim to have understood it. Though we have life, it is beyond us. We do not know how we have it, or why. We do not know what is going to happen to it, or to us. It is not predictable; though we can destroy it, we cannot make it. It cannot, except by reduction and the grave risk of damage, be controlled. It is, as Blake said, holy. To think otherwise is to enslave life, and to make, not humanity, but a few humans its predictably inept masters.

We need a new Emancipation Proclamation, not for a specific race or species, but for life itself—and that, I believe, is precisely what Edgar urges upon his once presumptuous and now desperate father:

> Thy life's a miracle. Speak yet again.

Gloucester's attempted suicide is really an attempt to recover control over his life—a control he believes (mistakenly) that he once had and has lost:

> O you mighty gods!
> This world I do renounce, and in your sights
> Shake patiently my great affliction off.

The nature of his despair is delineated in his belief that he can control his life by killing himself, which is a paradox we will meet again three and a half centuries later at the extremity of industrial warfare when we believed that we could "save" by means of destruction.

Later, under the guidance of his son, Gloucester prays a prayer that is exactly opposite to his previous one—

> You ever-gentle gods, take my breath from me;
> Let not my worser spirit tempt me again
> To die before you please

—in which he renounces control over his life. He has given up his life as an understood possession, and has taken it back as miracle and mystery. And his reclamation as a human being is acknowledged in Edgar's response: "Well pray you, father."

It seems clear that humans cannot significantly reduce or mitigate the dangers inherent in their use of life by accumulating more information or better theories or by achieving greater predictability or more caution in their scientific and industrial work. To treat life as less than a miracle is to give up on it.

*

I am aware how brash this commentary will seem, coming from me, who have no competence or learning in science. The issue I am attempting to deal with, however, is not knowledge but ignorance. In ignorance I believe I may pronounce myself a fair expert.

One of our problems is that we humans cannot live without acting; we *have* to act. Moreover, we *have* to act on the basis of what we know, and what we know is incomplete. What we have come to know so far is demonstrably incomplete, since we keep on learning more, and there seems little reason to think that our knowledge will become significantly more complete. The mystery surrounding our life probably is not significantly reducible. And so the question of how to act in ignorance is paramount.

Our history enables us to suppose that it may be all right to act on the basis of incomplete knowledge *if* our culture has an effective way of telling us that our knowledge is incomplete, and also of telling us how to act in our state of ignorance. We may go so far as to say that it is all right to act on the basis of sure knowledge, since our studies and our experience have given us knowledge that seems to be pretty sure. But apparently it is dangerous to act on the assumption that sure knowledge is complete knowledge—or on the assumption that our knowledge will increase fast enough to outrace the bad consequences of the arrogant use of incomplete knowledge. To trust "progress" or our putative "genius" to solve all the problems that we cause is worse than bad science; it is bad religion.

A second human problem is that evil exists and is an ever-present and lively possibility. We know that malevolence is always ready to appropriate the means that we have intended for good. For example, the technical means that have industrialized agriculture, making it (by very limited standards) more efficient and productive and easy, have also made it more toxic, more violent, and more vulnerable—have made it, in fact, far less dependable if not less predictable than it used to be.

One kind of evil certainly is the willingness to destroy what we cannot make—life, for instance—and we have greatly enlarged our means of doing that. And what are we to do? Must we let evil and our implication in it drive us to despair?

The present course of reductive science—as when we allow agriculture to be invaded by the technology of war and the economics of industrialism—*is* driving us to despair, as witness the incidence of suicide among farmers.

If we lack the cultural means to keep incomplete knowledge from becoming the basis of arrogant and dangerous behavior, then the intellectual disciplines themselves become dangerous. What is the point of the further study of nature if that leads to the further destruction of nature? To study the "purpose" of the organ within the organism or of the organism within the ecosystem is *still* reductive if we do so with the assumption that we will or can finally figure it out. This simply captures the world as the subject of present or future "understanding," which will become the basis of further industrial and commercial optimism, which will become the basis of further exploitation and destruction of communities, ecosystems, and local cultures.

I am not of course proposing an end to science and other intellectual disciplines, but rather a change of standards and goals. The standards of our behavior must be derived, not from the capability of technology, but from the nature of places and communities. We must shift the priority from production to local adaptation, from innovation to familiarity, from power to elegance, from costliness to thrift. We must learn to think about propriety in scale and design, as determined by human and ecological health. By such changes we might again make our work an answer to despair.

II. *Propriety*

MY GENERAL concern is with what I take to be the increasing inability of the scientific, artistic, and religious disciplines to help us address the issue of propriety in our thoughts and acts. "Propriety" is an old term, even an old-fashioned one, and is not much in favor. Its value is in its reference to the fact that we are not alone. The idea of propriety makes an issue of the fittingness of our conduct to our place or circumstances, even to our hopes. It acknowledges the always-pressing realities of context and of influence; we cannot speak or act or live out of context. Our life inescapably affects other lives, which inescapably affect our life. We are being measured, in other words, by a standard that we did not make and cannot destroy. It is by that standard, and only by that standard, that we know we are in a crisis in our relationship to nature. The term "environmental crisis," crude and inexact as it is, acknowledges that we have invoked this standard and have measured ourselves by it. A civilization that is destroying all of its sources in nature has raised starkly the issue of propriety, whether or not it wishes to have done so.

Propriety is the antithesis of individualism. To raise the issue of propriety is to deny that any individual's wish is the ultimate measure of the world. The issue presents itself as a set of questions: Where are we? (This question applies, with as much particularity as human competence will allow, to all of the world's millions of small localities.) Who are we? (The proper answer to this question depends on where we are and where we have been, and it includes history.) What is our condition? (This is a *practical* question.) What are our abilities? (This also is a practical question. It refers to abilities that are *proven*, not to abilities that are theoretical or potential, such as "aptitude" or I.Q.) What appropriately may we do in our own interest *here*? (And this question submits to the standard of the health of the place.) These questions address themselves to all the disciplines, but they do not call for specialized answers. They cannot, I think, be answered by specialists—or not, at least, by specialists in isolation from one another.

To ask such questions seriously now is not quite absurd, for the questions are valid and urgent, but it is nonetheless to risk a sort of comedy, for the questions are as foreign to our sciences and arts as presently practiced, and to our institutions of government, learning, and religion, as they are to the global corporations whose existence depends upon their (and our) willingness to ignore any such questions.

All of the disciplines are increasingly identifiable as professionalisms, which are increasingly conformable to the aims and standards of industrialism. All of the disciplines are failing the test of propriety because they are failing the test of locality. The professionals of the disciplines don't *care* where they are. Though they are inescapably in context, they assume or pretend that they think and work without context. They subscribe to the preeminence of the mind and (logically from that) of the career. The questions of propriety, calling as they must for local answers, call necessarily for *small* answers. But small local answers are now as far beneath the notice of professionalism as of commercialism. Professionalism aspires to *big* answers that will make headlines, money, and promotions. It longs, moreover, for answers that are uniform and universal—the same styles, explanations, routines, tools, methods, models, beliefs, amusements, etc., for everybody everywhere. And like the corporations, whose appetite for "growth" seems now ungovernable, the institutions of government, education, and religion are now all too likely to measure their success in terms of size and number. All the institutions seem to have learned to imitate the organizational structures and to adopt the values and aims of industrial corporations. It is astonishing to realize how quickly and shamelessly doctors and lawyers and even college professors have taken to drumming up trade, and how readily hospitals, once run according to the laws of healing, mercy, and charity, have submitted to the laws of professionalism, industrial methodology, careerism, and profit.

This is happening to all the disciplines, but because science is the most influential category of the disciplines, and increasingly has set the pattern for the rest, we must be concerned first of all with science. Stephen Edelglass, Georg Maier, Hans Gebert, and John Davy in their book, *The Marriage of Sense and Thought*, wrote that "Science now functions in society rather as the Church did in the Middle Ages." What kind of religion science is, and how it works as such, are questions we will have to deal with.

One used to hear a great deal about "pure science." The universities, one was given to understand, were full of scientists who were disinterestedly pursuing truth. "Pure science" did not permit the scientist to ask so crude and pragmatic a question as *why* this or that truth was being pursued; it was just assumed, not only that to know the truth was good, but that, once the truth was discovered, it would somehow be *used* for good. This is a singularly naive view of science (as it would be of any human enterprise), but it survived at least into the early days of space exploration, when a lot of aficionados of so-called high technology assumed that NASA existed to sponsor voyages of pure discovery: to learn whatever might be learned, to take pictures of the earth and other planets, and to provide extremely expensive mystical experiences to astronauts. Some people believed that this enterprise was really a sort of spiritual quest, and would always remain above the gross concerns of, for example, the military-industrial complex. It would promote instead a renewed tenderness toward our "planet" by such devices as pictures of half of said planet, taken at a distance that reduced it to a blue bauble something like a Christmas tree ornament. In our foolish insistence on substituting technology for vision, we forget that we are not the first to have seen "the whole earth" from such a distance. Dante saw it from a higher level of human accomplishment, and at far less economic and ecological cost, several hundred years before NASA.

The possibility of pure science was significantly diminished, surely, by the time early scientists had invented metallurgy and then gunpowder, and it diminished steadily from then on. By now, when the possibilities of application have so enormously multiplied and the greed of corporations has grown so elaborate that they wish to patent discoveries before they have been discovered, it appears safest to assume that all sciences are "applied." Science may at times have been altruistically applied. But even such nominally altruistic sciences as medicine and plant-breeding have now become so deeply interpenetrated with economics and politics that their motives are at best mixed with, and at worst replaced by, the motives of corporations and governments. If nothing else, the increasing costliness of the practice of conventional science, and its consequent dependence on large grants or investments, would mitigate against its purity. One can only assume that pure science now needs to move fast (and beg hard) to keep its skirts from being lifted by the ever randy and handy corporate giants.

As I have already confessed, I am not at all a scientist. And yet, like every human inhabitant of the modern world, I have experienced many of the effects (costs and benefits) of science; I have received a great deal of hearsay of it; and I know that I am always under its influence and at its mercy. Though I am unable to comment on its methods or the truth of its discoveries, I am nonetheless appropriately interested in its motives—in what it thinks it is doing and in how it justifies itself. I agree with the proposition that science (or "science-and-technology") has now become a sort of religion; I am aware also that in many ways it rules over us. I want to know by what power it has crowned and mitered itself.

I believe it is generally agreed that "science" means knowledge of a special kind: factual knowledge that can be proven by measure, that can stand up to empirical testing. Scientific knowledge is the hard cash of the modern economy of thought; its worth is constant, no matter who has it; its value is not derived from belief or opinion or speculation or desire. Once established, it cannot be argued about.

Science has to do, famously, with theory. "Theory," at root, is related to the word "theater"; it has to do with watching, with observation. A scientific theory is an aid to observation. It involves assumptions that appear to be consistent with known facts. It is not proven; it is useful because it may lead to evidence or to proofs.

Science also involves prediction. Prediction is a highly disciplined concept when it is used in relation to the methodology of proof: A thing is true only if it is *predictably* true; a thing is true, not because it is true now, but because it is true always. But in the hands of such "scientists" as meteorologists and economists, whose putative usefulness depends directly upon their ability to predict and whose predictions are frequently wrong, the meaning of prediction begins to slide from science toward journalism. The same slide occurs when scientists, on the basis of early results, predict the success of a course of experimentation. Alert readers of newspapers will certainly have noticed the frequency of reports that scientists "may have" discovered something or other, or that new data "may prove" something or other. Journalists, and apparently some scientists also, are partial to news stories beginning "Scientists foresee" or "Scientists predict."

This seems to come from abuse of faith, which is another essential attribute of science. There is a sort of scientific faith that is legitimate. It is hard to see how the work of science could be done if scientists did not have faith in the workability and soundness of their methods. This is not faith of the highest sort, obviously, but is akin to the unproven confidence with which we non-scientists face the unknowns of our own workdays. But under various suasions of profession and personality, this legitimate faith in scientific methodology seems to veer off into a kind of religious faith in the power of science to know all things and solve all problems, whereupon the scientist may become an evangelist and go forth to save the world.

This religification and evangelizing of science, in defiance of scientific principles, is now commonplace and is widely accepted or tolerated by people who are not scientists. We really seem to have conceded to scientists, to the extent of their own regrettable willingness to occupy it, the place once occupied by the prophets and priests of religion. This can have happened only because of a general abdication of our responsibility to be critical and, above all, self-critical.

Why is there not a robust, profoundly questioning criticism of science within the scientific disciplines? One reason, I assume, is that such self-criticism, especially in public, would be considered "unprofessional." Another reason is that the modern sciences, working always in such proximity to "application," are simply too lucrative or too potentially lucrative to be self-critical. The professions increasingly have adopted the standards and thought patterns of business: If you're making money, what can be wrong? The criticism of science most familiar to ordinary citizens is more than likely to take the form of a public protest against some ruinous local manifestation of applied science. The most ubiquitous and unignorable result of modern chemistry, for example, is pollution, but typically this result is dealt with by ordinary citizens, not by chemists.

In 1959, C. P. Snow spoke of science as having an "automatic corrective." At that time, maybe, one could reasonably suppose that "pure" science, safely withdrawn from application, might by its own processes of experimentation and proof more or less automatically correct itself. By now we know that the applied sciences are subject to no such corrective. The scale of experimentation has become too greatly enlarged, for now science may be said to be conducting many of its experiments on the scale of the world. Among the results are Chernobyl, the ozone hole, the acceleration of species extinction, and universal pollution.

If there are critics of science in the governments and the bureaucracies, they are largely inaudible. In the universities, the scientists generally proceed from promotion to promotion and from grant to grant, leaving few recorded moments of conscience or professional self-doubt; and the professors of the humanities seem for the most part merely to be abashed by the sciences, deferring to their certainties, adopting their values, admiring their wealth, and longing even to imitate their methodology and their jargon. The journalists think it intellectually chic to stand open-mouthed before any wonder of science whatsoever. The media, cultivating their mediocrity, seem quite comfortably unaware that many of the calamities from which science is expected to save the world were caused in the first place by science—which meanwhile is busy propagating further calamities, hailed now as wonders, from which later it will undertake to save the world. Nobody, so far as I have heard, is attempting to figure out how much of the progress resulting from this enterprise is *net*. It is as if a whole population has been genetically deprived of the ability to subtract.

I know that there are some scientists who are speaking and writing sound criticism of science or of scientific abuses of science, but these people seem to have the status of dissidents or heretics; they are not accepted as partners in a necessary dialogue. Typically, their criticisms and objections are not even answered. (If you are making money and have power, why debate?) In short, the scientific critics of science are not effective. That there has been no effective criticism of science is demonstrated, for instance, by science's failure to attend to the possibility of small-scale or cheap or low-energy or ecologically benign technologies. Most applications of science to our problems result in large payments to large corporations and in damages to ecosystems and communities. These eventually will have to be subtracted (but not, if they can help it, by the inventors or manufacturers) from whatever has been gained.

III. On Edward O. Wilson's Consilience

APPARENTLY everywhere in the "developed world" human communities and their natural and cultural supports are being destroyed, not by natural calamities or "acts of God" or invasion by foreign enemies, but by a sort of legalized vandalism known as "the economy." The economy now famously depends upon the authority and the applicable knowledge of science. It would therefore be useful to say what is the character of this science that has benefited us in so many ways, and yet has cost us so dearly and exacted from us such deference and such questionable permissions.

Since I am not able to conduct any kind of survey, I will focus my study upon a single book: *Consilience*, by Edward O. Wilson. I am aware of the several objections to treating any one book as representative, but I am encouraged to do so, not only by the advantage of economy, but also by my belief that Mr. Wilson's assumptions are widely shared both by his colleagues and by non-scientists, and that there is no idea in his book that would be surprising to any fairly regular reader of articles on science in a daily newspaper. I think, in short, that despite his pretensions to iconoclasm, Mr. Wilson speaks for a popular scientific orthodoxy. His book reads as though it was written to confirm the popular belief that science is entirely good, that it leads to unlimited progress, and that it has (or will have) all the answers.

I am interested in Mr. Wilson's book also because, like me, he is a conservationist. We have perhaps some other things in common, but we have differences too, and these concern me just as much and (as will presently be evident) more seriously and at greater length. Our fundamental difference may be that he is a university man through and through, and I have always been most comfortable out of school. Whereas Mr. Wilson apparently is satisfied with the modern university's commitment to departmented specialization, professional standards, industry-sponsored research, and a scheme of promotion and tenure based upon publication, I am distrustful of that commitment and think it has done harm, both to learning and to the world.

Obviously, I have no authority from which to question Mr. Wilson's scientific knowledge, which I believe to be great and admirable, as human knowledge goes. My interest, rather, is in his attitudes toward what he knows and (more important) what he doesn't know. It is in these attitudes, I believe, that he is most conventional, and in them he most conforms to the values and the psychology of industrialism.

To show how Mr. Wilson stands on the issue of knowledge and ignorance, I will catalogue here his fundamental biases and assumptions as they appear in his book, observing when he is being properly scientific, and when he is not. *Consilience* is in effect a scientific credo; its opinions are plainly stated.

I. MATERIALISM

Mr. Wilson is, to begin with, a materialist. He believes that this is "a lawful material world," all the laws of which can be explained and understood empirically, and are subject to scientific proofs. He holds that "all tangible phenomena, from the birth of stars to the workings of social institutions, are based on material processes that are ultimately reducible . . . to the laws of physics."

Science is an enterprise of materiality, dealing in empirical proofs, in the tangible, the measurable, and the countable. And so, in terms of procedure, there can be no objection to Mr. Wilson's materialism; he is a scientist, and from a scientist we require truths that are materially verifiable. But as a doctrine of belief, materialism takes him into several kinds of trouble. These troubles show up pretty plainly against his obviously genuine concern for conservation of the natural world.

A minor problem, perhaps, is the tendency of materialism to objectify the world, dividing it from the "objective observer" who studies it. The world thus becomes "the environment," a word which Mr. Wilson uses repeatedly when speaking of conservation, and which means "surroundings," a place that one is *in* but not *of.* The question raised by this objectifying procedure and its vocabulary is whether the problems of conservation can be accurately defined by an objective observer who observes at an intellectual remove, forgetting that he eats, drinks, and breathes the so-called environment.

A more serious problem is this: A theoretical materialism so strictly principled as Mr. Wilson's is inescapably deterministic. We and our works and acts, he holds, are determined by our genes, which are determined by the laws of biology, which are determined ultimately by the laws of physics. He sees that this directly contradicts the idea of free will, which even as a scientist he seems unwilling to give up, and which as a conservationist he cannot afford to give up. He deals with this dilemma oddly and inconsistently.

First, he says that we have, and need, "the illusion of free will" which, he says further, is "biologically adaptive." I have read his sentences several times, hoping to find that I have misunderstood them, but I am afraid that I understand them. He is saying that there is an evolutionary advantage in illusion. The proposition that our ancestors survived because they were foolish enough to believe an illusion is certainly optimistic, but it does not seem very probable. And what are we to think of a materialism that can be used to validate an illusion? Mr. Wilson nevertheless insists upon his point; in another place he speaks of "self-deception" as granting to our species the "adaptive edge."

Later, in discussing the need for conservation, Mr. Wilson affirms the Enlightenment belief that we can "choose wisely." How a wise choice can be made on the basis of an illusory freedom of will is impossible to conceive, and Mr. Wilson wisely chooses not to try to conceive it.

2. MATERIALISM AND MYSTERY

Furthermore, as internally awry and inconsistent as it is, Mr. Wilson makes of his materialism a little platform from which to look down upon (as he thinks) and patronize the opposition. He is a *militant* materialist, with a doctrinaire intolerance for any sort of mystery—or, for that matter, any sort of ambiguity or uncertainty. He understands mystery as attributable entirely to human ignorance, and thereby appropriates it for the future of human science; in his formula, the unknown = the to-be-known. I will have more to say about this presently. For now, it is enough merely to notice that he has no ability to confront mystery (or even the unknown) as such, and therefore has learned none of the lessons that humans have always learned when they have confronted mystery as such. His book is an exercise in a sort of academic hubris.

If modern science is a religion, then one of its presiding deities must be Sherlock Holmes. To the modern scientist as to the great detective, every mystery is a problem, and every problem can be solved. A mystery can exist only because of human ignorance, and human ignorance is always remediable. The appropriate response is not deference or respect, let alone reverence, but pursuit of "the answer."

This pursuit, however, is properly scientific only so long as the mystery is empirically or rationally solvable. When a scientist denies or belittles a mystery that cannot be solved, then he or she is no longer within the bounds of science.

Thus, when Mr. Wilson asserts that *Paradise Lost* owes nothing to "God's guidance of Milton's thoughts, as the poet himself believed," he is talking far beyond the reach of proof. He does not consider that *Paradise Lost* is the poem it is because Milton was a man faithful and humble enough to invoke the assistance of the "Heav'nly Muse." The only empirical truth available here is that Milton, believing, wrote *Paradise Lost*, and that Mr. Wilson, disbelieving, wrote *Consilience*, a book of a different order.

Believing that whatever is intangible does not exist, Mr. Wilson like many materialists, atheists, rationalists, realists, etc., thinks he has struck a killing blow against religious faith when he has asked to see its evidence. But of course religious faith *begins* with the discovery that there is no "evidence." There is no argument or trail of evidence or course of experimentation that can connect unbelief and belief.

By insisting upon so narrow a definition of reality, Mr. Wilson does not defeat religion, but only misunderstands it. He does not appreciate, because he cannot suspect, the possibility that religious faith may be a way of knowing things that cannot otherwise be known. Misunderstanding religion, he often appropriates and misuses such words as *transcendent*, *create*, *archetype*, *reverence*, and *sacred*.

In his chapter on "Ethics and Religion," he has "constructed a debate" between "the transcendentalist" (a straw man) and "the empiricist" (a stuffed shirt). The empiricist, having found the transcendentalist's simple-minded argument (rigged for defeat by Mr. Wilson) to be lacking in "objective evidence" and "statistical proofs," then proceeds to make the same sort of irrational swerve that Mr. Wilson himself made previously in affirming "the illusion of free will." Empiricism, the empiricist concedes, "is bloodless. People need . . . the poetry of affirmation. . . . It would be a sorry day if we abandoned our venerated sacral traditions. It would be a tragic misreading of history to expunge *under God* from the American Pledge of Allegiance. Whether atheists or true believers, let oaths be taken with hand on the Bible. . . . Call upon priests and ministers and rabbis to bless civil ceremony with prayer . . ." All this, Mr. Wilson calls "the presence of poetry."

"But to share reverence," the empiricist continues hopefully, "is not to surrender the precious self. . . ." A language can hardly endure this sort of abuse. It is impossible to tell what Mr. Wilson may mean by "share reverence," but to *feel* reverence, to *be* reverent, is exactly to surrender the "precious self," and is nothing else.

And then the empiricist administers his coup de grâce: "We can be proud as a species because, having discovered that we are alone, we owe the gods very little." This would be a noble blasphemy, like that of Job's wife, if the empiricist or Mr. Wilson believed in the gods, but neither one of them does. It is only a weary little cliché of a too familiar "scientific" iconoclasm—hubris without a bang.

But this materialism raises a question that *Consilience* does not acknowledge. If at last "all tangible phenomena" are empirically reduced to the laws of physics, then we will merely have completed a circle. We will have arrived again at the question that preceded Genesis: Where did the physical world come from? And physics of course can have no answer.

The principal point to be made, so far, is that Mr. Wilson's initially reputable materialism has led him far beyond objective evidence and statistical proofs and into what looks very much like poppycock. More examples will follow.

3. IMPERIALISM

Mr. Wilson's scientific "faith" (as he sometimes calls it) is in the ultimate empirical explainability of everything—that is, the "consilience" of all the disciplines "by the linking of facts and fact-based theory . . . to create a common groundwork of explanation." He concedes from time to time that he may be wrong, but this doubt is a mere gesture, entirely absent from the passages in which he is elatedly confident that he is right. His humility is only etiquette, not a conviction or even an attitude. It is not involved in his thought, and his thought is unaffected by it. His book contains, moreover, several passages in which he writes with candor of the great difficulties and perhaps the impossibilities of the consilience he wishes for, but these also do not dampen or qualify his passionate conviction. Waving aside ignorance and mystery and human limitation as merely illusory or irrelevant, he claims not only all knowledge but all future knowledge and everything unknown as the property of science. So confident is he in his power to know that his book abounds in one-sentence definitions of, for examples, the mind, consciousness, meaning, mood, creativity, insanity, art, science, the self (all of these are in one chapter, "The Mind"). The definitions themselves are mostly jargon, singularly unhelpful—"What we call *meaning* is the linkage among the neural networks created by the spreading excitation that enlarges imagery and engages emotion"—but their purpose is clear enough: It is to take possession of the subject in the name of the idea of consilience.

This idea is explicitly imperialistic, and it is implicitly tyrannical. Mr. Wilson is perfectly frank about his territorial ambitions. He wishes to see all the disciplines linked or unified—but strictly on the basis of science. Non-scientists are not invited to the negotiations, or at least they are not to participate on their own terms: "The key to the exchange between [science and the arts] is . . . reinvigoration of interpretation with the knowledge of science and its proprietary sense of the future." And if you have any doubt of the political and economic implications of modern science and of Mr. Wilson's advocacy, consider the following: "Governmental and private patrons of the brain scientists, like royal geographic commissions of past centuries, are generous. They know that history can be made by a single sighting of coastland, where inland lies virgin land and the future lineaments of empire."

Mr. Wilson's book scarcely acknowledges the existence of politics, and perhaps for good reason, for the putative ability to explain everything along with the denial of religion (or the appropriation of its appearances) is a property of political tyranny. So is the belief that one's explanations will save the world from some great threat. And Mr. Wilson himself, in his conviction that everything that is not a science *ought* to be and *will* be, shows himself to be a man with a fiercely proprietary mind and dire intentions toward the unenlightened.

The logic of his position is clear, and it is most disquieting. Would not, for example, any social theory be almost inescapably totalitarian if it were to be (in Mr. Wilson's terms) general, consilient, and predictive? Or if it *thought* it was, for the supposition would be just as dangerous to freedom as the fact. To suppose that the theory was predictive would prepare the way for a bloody "exogenous shock" that it did not predict. Mr. Wilson, in fact, alludes briefly to such possibilities in his chapter "The Enlightenment," but he does not confront the issue; he races past it and within three pages he is praising the Enlightenment's "new freedom": "It waved aside everything, every form of religious and civil authority, every imaginable fear, to give precedence to the ethic of free inquiry." The direction of Mr. Wilson's consilience, prescribed and announced, is toward an empirical dogma of dead certainty, "a common groundwork of explanation," the terms of which would exclude whatever cannot be empirically explained. Mere "humanitarianism" would do the rest: If you know with certainty what is true, should you not *enforce* the truth? If you see your poor subjects struggling and suffering in their error, how can you rightly forbear to impose the necessary corrections? Having spied out the coast, why should you not extend inland the "lineaments of empire"?

And this, though with some of us it would not be all right, would be incontestable if we had good reason to believe that everything could be explained on the terms proposed, which fortunately we do not. The only science we have or can have is *human* science; it has human limits and is involved always with human ignorance and human error. It is a fact that the solutions invented or discovered by science have tended to lead to new problems or to become problems themselves. Scientists discovered how to use nuclear energy to solve some problems, but any use of it is enormously dangerous to us all, and scientists have not discovered what to do with the waste. (They have not discovered what to do with old tires.) The availability of antibiotics leads to the overuse of antibiotics. And so on. Our daily lives are a daily mockery of our scientific pretensions. We are learning to know precisely the location of our genes, but significant numbers of us don't know the whereabouts of our children. Science does not seem to be lighting the way; we seem rather to be leapfrogging into the dark along series of scientific solutions, which become problems, which call for further solutions, which science is always eager to supply, and which it sometimes cannot supply. Sometimes it fails us infamously and fearfully. The so-called Y2K problem—the failure to manufacture computers capable of recognizing the year 2000—could have been prevented by perfectly ordinary human foresight, as limited as that is. The coming of the year 2000 could have been foretold by every child old enough to count. But among all the scientists who helped to develop "computer science," all who taught it in great universities, all who used it in their work, all who evangelized it as the answer to every intellectual problem, apparently no one with authority foresaw the Y2K fiasco until almost too late.

For a long time we humans have fairly successfully (but not invariably) avoided error within our systems of thought, but the systems themselves have often proved to be wrong. That is, our systems have made it possible (within the limits of the systems) to be consistent, but they have not preserved us from error. Our experience suggests that they cannot preserve us from error. Should we regret this? Probably not, since it is always the errors of our systems that have released us (so far) from the tyranny of our systems.

The presently dominant system of thought—which we should call, not "science," but "science-technology-and-industry"—has produced an unsurprising number of errors and an unsurprising number of failures. It is hard to see how our systems of thought could be other than fallible, once we grant that they cannot be contrived except by fallible creatures; fallibility is an infection in us that we inevitably communicate to our works. Who does not know this? Most of us begin every day with some kind of plan, and every day we see that plan altered or foiled because it necessarily lacks the scope of our nature and character, let alone that of reality. Our view of the world and even of our own experience is always to some degree distorted, oversimplified, or reduced, and so is varyingly liable to be in error. That we are in error means that our plans or systems tend to suffer the interference of bad surprises—and, let us not forget, also of good surprises. It is impossible to argue that we can know empirically anything that is beyond our mental capacity. What we understand has necessarily been limited by the limits of our understanding. Mr. Wilson predicts approvingly that "The world henceforth will be run by synthesizers, people able to put together the right information at the right time, think critically about it, and make important choices wisely." Synthesis, he says, is "holism." He does not acknowledge that the synthesis he is talking about will be neither whole nor holy, but rather an artifact made of parts that we have isolated and in our fashion understood ("We murder to dissect") and put together again in a way we understand.

The fallibility of a human system of thought is always the result of incompleteness. In order to include some things, we invariably exclude others. We can't include everything because we don't know everything; we can't comprehend what comprehends us. The incompleteness of a system is rarely if ever perceptible to those who made it or to those who benefit from it. To those who are excluded from it, the incompleteness of a system is, or eventually becomes, plain enough. One weakness of the present system, which Mr. Wilson does not mind, is that it excludes all inscrutable and ineffable things, including the life history of the human soul.

Another weakness, which Mr. Wilson does mind, is that it excludes the principle or the standard of ecological health. Science-technology-and-industry has enabled us to be precise (apparently) in describing objects that are extremely small and near or extremely large and far away. It has failed utterly to provide us with even adequate descriptions of the places and communities we live in—probably because it *cannot* do so.

There are scientists, one must suppose, who know all about atoms or molecules or genes, or galaxies or planets or stars, but who do not know where they are geographically, historically, or ecologically. Our schools are turning out millions of graduates who do not know, in this sense, where they are. Certain lamentable results predictably follow. Mr. Wilson thinks the present system can correct itself merely by enlarging its present claims. I think the present system can correct itself only by conscientiously trying to include what it has so far excluded —which, of course, would make it an entirely different system with entirely different claims.

If one has a science that is manifestly incomplete—that is surrounded by mystery—and yet one believes with passionate intensity that it should, can, and will complete itself by means of a consilience of all the disciplines, then as an immediate and necessary effect one subjects one's language to a heavy strain. Mr. Wilson's project calls for a language of great assurance, but he is writing inescapably about what he does not know, and so his language often is necessarily tentative. The future that his thesis forces him to try to see into is finally as obscure to him as it is to the rest of us. His writing about consilience is always under the sway of conditional verbs, of protestations of faith, of "if" and "until" and "likely" and "perhaps." And so it is not scientific; it is not theorizing; it is only a fairly ordinary kind of human supposing, guided only by certain popular and professional prejudices.

But the most unscientific and the most disturbing thing about this book is Mr. Wilson's appropriation of whatever is unknown. He does this by variations on the themes of "until" or "not yet." He cannot bring himself to say that scientists do not know something; he must say that they do not know it *yet*; he must say that one thing cannot be known *until* another thing is known. He says repeatedly things like this: "The belief in the possibility of consilience beyond science and across the great branches of learning is not yet science. . . . It cannot be proved with logic from first principles or grounded in any definitive set of empirical tests, at least not by any yet conceived." This "not yet" forthrightly appropriates mystery as future knowledge. It takes possession of life and the future of life in the names of its would-be explainers—and, it follows, of its would-be exploiters. As soon as a mystery is scheduled for solution, it is no longer a mystery; it is a problem. The most tyrannic of all reductions has thus been accomplished; a self-aggrandizing science has thus asserted its "proprietary sense of the future."

The practical result of such language is a sort of moral blindness. We cannot derive sound thinking about propriety of scale and conduct from the proposition that what we need to know we do not know "yet." Such an idea simply overrides the issue of limits. Without a lively recognition of our own limits —chiefly of our knowledge and of our ability to know—we cannot even approach the issue of the limits of nature.

What is the possibility of "consilience beyond science and across the great branches of learning"?

We don't know yet.

Why do the innocent suffer?

We don't know yet.

I am not proposing, of course, that mental work of any sort can do without hypothesis or theory or any other way of articulating one's sense of possibility. Because our knowledge is discontinuous and we are ignorant of the future, we must grant the necessity of some manner of supposing from one point of evidence to another. At the same time it seems only fair to insist that this process should not be extended indefinitely without evidence, and that the points of evidence ought to be reasonably close together. In shallow water one may not risk much by postulating the existence of an as yet invisible stepping stone just beneath the surface. But if the water is deep and swift, one should not start across if some of the stepping stones are hypothetical. It is absurd to accumulate enormous quantities of nuclear waste, telling ourselves that we don't know *yet* how to dispose of it. We might face the future a good deal more confidently now if nuclear scientists had had the humility and the candor to say simply, We don't know.

Clearly, there ought to be a limit beyond which we cease to hedge our ignorance with promises to "continue to study the problem." Scrupulous minds, in this age as in any other, not only must be constrained occasionally to confess ignorance, but also must continue to live with the old proposition that some things are not knowable. *Consilience*, at any rate, shows us a man's mind leaping with exuberant confidence from one merely conjectural stepping stone to another, oblivious of the rushing waters.

4. REDUCTIONISM

Reductionism, like materialism, has uses that are appropriate, and it also can be used inappropriately. It is appropriately used as a way (one way) of understanding what is empirically known or empirically knowable. When it becomes merely an intellectual "position" confronting what is not empirically known or knowable, then it becomes very quickly absurd, and also grossly desensitizing and false. Like materialism, reductionism belongs legitimately to science; as an article of belief, it causes trouble.

According to Mr. Wilson, "Science . . . is the *organized, systematic enterprise that gathers knowledge about the world and condenses the knowledge into testable laws and principles*" [his italics]. He says further that "The cutting edge of science is reductionism, the breaking apart of nature into its natural constituents." And reductionism has "a deeper agenda," which is "to fold the laws and principles of each level of organization into those at more general, hence more fundamental levels. Its strong form is total consilience, which holds that nature is organized by simple universal laws of physics to which all other laws and principles can eventually be reduced." Toward the end of his book, Mr. Wilson adds the following: "There is abundant evidence to support and none absolutely to refute the proposition that consilient explanations are congenial to the entirety of the great branches of learning."

Mr. Wilson's definitions of science and reductionism, granting him his prejudices, seem to me perfectly appropriate. His definition of consilience, however, like his exposition of it, becomes more contestable the farther it goes.

There obviously is a necessary usefulness in the processes of reduction. They are indispensable to scientists—and to the rest of us as well. It is valuable (sometimes) to know the parts of a thing and how they are joined together, to know what things do and do not have in common, and to know the laws or principles by which things cohere, live, and act. Such inquiries are native to human thought and work.

But reductionism also has one inherent limitation that is paramount, and that is abstraction: its tendency to allow the particular to be absorbed or obscured by the general. It is a curious paradox of science that its empirical knowledge of the material world gives rise to abstractions such as statistical averages which have no materiality and exist only as ideas. There is, empirically speaking, no average and no type. Between the species and the specimen the creature itself, the individual creature, is lost. Having been classified, dissected, and explained, the creature has disappeared into its class, anatomy, and explanation. The tendency is to equate the creature (or its habitat) with one's formalized knowledge of it. Mr. Wilson is somewhat aware of this problem for he insists upon the importance of "synthesis and integration." But he does not acknowledge that synthesis and integration are merely parts of an explanation, which is invariably and inevitably less than the thing explained. The synthesizing and integrating scientist is only ordering and making sense of as much as he knows. He is not making whole that which he has taken apart, and he should not claim credit for putting together what was already together.

The uniqueness of an individual creature is inherent, not in its physical or behavioral anomalies, but in its *life*. Its life is not its "life history," the typical cycle of members of its species from conception to reproduction to death. Its life is all that happens to it in its place. Its wholeness is inherent in its life, not in its physiology or biology. This wholeness of creatures and places together is never going to be apparent to an intelligence coldly determined to be empirical or objective. It shows itself to affection and familiarity.

The frequent insultingness of modern (scientific-technological-industrial) medicine is precisely its inclination to regard individual patients apart from their lives, as representatives or specimens of their age, sex, pathology, economic status, or some other category. The specialist to whom you have been "referred" may never have seen you before, may know nothing about you, and may never see you again, and yet he or she presumes to know exactly what is wrong with you. The same insultingness is now also a commonplace of politics, which treats individuals as representatives of racial, sexual, geographic, economic, ideological, and other categories, each with typical faults, complaints, rights, or virtues.

Science speaks properly a language of abstraction and abstract categories when it is properly trying to sort out and put in order the things it knows. But it often assumes improperly that it has said—or known—enough when it has spoken of "the cell" or "the organism," "the genome" or "the ecosystem" and given the correct scientific classification and name. Carried too far, this is a language of false specification and pretentious exactitude, never escaping either abstraction or the cold-heartedness of abstraction.

The giveaway is that even scientists do not speak of their loved ones in categorical terms as "a woman," "a man," "a child," or "a case." Affection requires us to break out of the abstractions, the categories, and confront the creature itself in its life in its place. The importance of this for Mr. Wilson's (and my) cause of conservation can hardly be overstated. For things cannot survive as categories but only as individual creatures living uniquely where they live.

We know enough of our own history by now to be aware that people *exploit* what they have merely concluded to be of value, but they *defend* what they love. To defend what we love we need a particularizing language, for we love what we particularly know. The abstract, "objective," impersonal, dispassionate language of science can, in fact, help us to know certain things, and to know some things with certainty. It can help us, for instance, to know the value of species and of species diversity. But it cannot replace, and it cannot become, the language of familiarity, reverence, and affection by which things of value ultimately are protected.

The abstractions of science are too readily assimilable to the abstractions of industry and commerce, which see everything as interchangeable with or replaceable by something else. There is a kind of egalitarianism which holds that any two things equal in price are equal in value, and that nothing is better than anything that may profitably or fashionably replace it. Forest = field = parking lot; if the price of alteration is right, then there is no point in quibbling over differences. One place is as good as another, one use is as good as another, one life is as good as another—if the price is right. Thus political sentimentality metamorphoses into commercial indifference or aggression. This is the industrial doctrine of the interchangeability of parts, and we apply it to places, to creatures, and to our fellow humans as if it were the law of the world, using all the while a sort of middling language, imitated from the sciences, that cannot speak of heaven or earth, but only of concepts. This is a rhetoric of nowhere, which forbids a passionate interest in, let alone a love of, anything in particular.

Directly opposed to this reduction or abstraction of things is the idea of the preciousness of individual lives and places. This does not come from science, but from our cultural and religious traditions. It is not derived, and it is not derivable, from any notion of egalitarianism. If all are equal, none can be precious. (And perhaps it is necessary to stop here to say that this ancient delight in the individuality of creatures is not the same thing as what we now mean by "individualism." It is the opposite. Individualism, in present practice, refers to the supposed "right" of an individual to act alone, in disregard of other individuals.)

We now have the phenomenon of "mitigation banking" by which a developer may purchase the "right" to spoil one place by preserving another. Science can measure and balance acreages in this way just as cold-heartedly as commerce; developers involved in such trading undoubtedly have the assistance of ecologists. Nothing insists that one place is not interchangeable with another except affection. If the people who live in such places and love them cannot protect them, nobody can.

It is not quite imaginable that people will exert themselves greatly to defend creatures and places that they have dispassionately studied. It is altogether imaginable that they will greatly exert themselves to defend creatures and places that they have involved in their lives and invested their lives in—and of course I know that many scientists make this sort of commitment.

I have been working this morning in front of a window where I have been at work on many mornings for thirty-seven years. Though I have been busy, today as always I have been aware of what has been happening beyond the window. The ground is whitened by patches of melting snow. The river, swollen with the runoff, is swift and muddy. I saw four wood ducks riding the current, apparently for fun. A great blue heron was fishing, standing in water up to his belly feathers. Through binoculars I saw him stoop forward, catch, and swallow a fish. At the feeder on the window sill, goldfinches, titmice, chickadees, nuthatches, and cardinals have been busy at a heap of free (to them) sunflower seeds. A flock of crows has found something newsworthy in the cornfield across the river. The woodpeckers are at work, and so are the squirrels. Sometimes from this outlook I have seen wonders: deer swimming across, wild turkeys feeding, a pair of newly fledged owls, otters at play, a coyote taking a stroll, a hummingbird feeding her young, a peregrine falcon eating a snake. When the trees are not in leaf, I can see the wooded slopes on both sides of the valley. I have known this place all my life. I long to protect it and the creatures who belong to it. During the thirty-seven years I have been at work here, I have been thinking a good part of the time about how to protect it. This is a small, fragile place, a slender strip of woodland between the river and the road. I know that in two hours a bulldozer could make it unrecognizable to me, and perfectly recognizable to every "developer."

The one thing that I know above all is that even to hope to protect it, I have got to break out of all the categories and confront it as it is; I must be present in its presence. I know at least some of the categories and value them and have found them useful. But here I am in my life, and I know I am not here as a representative white male American human, nor are the birds and animals and plants here as representatives of their sex or species. We all have our ways, forms, and habits. We all are what we are partly because we are here and not in another place. Some of us are mobile; some of us (such as the trees) have to be content merely to be flexible. All of us who are mobile are required by happenstance and circumstance and accident to make choices that are not instinctive, and that force us out of categories into our lives here and now. Even the trees are under this particularizing influence of place and time. Each one, responding to happenstance and circumstance and accident, has assumed a shape not quite like that of any other tree of its kind. The trees stand rooted in their mysteriously determined places, no place quite like any other, in strange finality.

The birds and animals have their nests in holes and burrows and crotches, each one's place a little unlike any other in the world—and so is the nest my mate and I have made.

In all of the thirty-seven years I have worked here, I have been trying to learn a language particular enough to speak of this place as it is and of my being here as I am. My success, as I well know, has been poor enough, and yet I am glad of the effort, for it has helped me to make, and to remember always, the distinction between reduction and the thing reduced. I know the usefulness of reductive language. To know that I am "a white male American human," that a red bird with black wings is "a scarlet tanager," that a tree with white bark is "a sycamore," that this is "a riparian plant community"—all that is helpful to a necessary kind of thought. But when I try to make my language more particular, I see that the life of this place is always emerging beyond expectation or prediction or typicality, that it is unique, given to the world minute by minute, only once, never to be repeated. And then is when I see that this life is a miracle, absolutely worth having, absolutely worth saving.

We are alive within mystery, by miracle. "Life," wrote Erwin Chargaff, "is the continual intervention of the inexplicable." We have more than we can know. We know more than we can say. The constructions of language (which is to say the constructions of thought) are formed *within* experience, not the other way around. Finally we live beyond words, as also we live beyond computation and beyond theory. There is no reason whatever to assume that the languages of science are less limited than other languages. Perhaps we should wish that after the processes of reduction, scientists would return, not to the processes of synthesis and integration, but to the world of our creatureliness and affection, our joy and grief, that precedes and (so far) survives all of our processes.

5. CREATURES AS MACHINES

There is a reduction, by now more formulaic than procedural, that seems endemic to modern science, and from science it has spread everywhere. I mean the definition or identification of the world and all its creatures as "machines." This is one of the fundamental assertions in *Consilience*. The Enlightenment thinkers, Mr. Wilson says, encouraged us to "think of the world as God's machine"—"if you still insist on a divine intervention." A little later he says, "People, after all, are just extremely complicated machines." Further on, he says that we are "organic machines," and that "an organism is a machine."

This machine business may once have had meaning. It may have been a way of asserting belief in the integrity of Creation and the physical coherence of creatures; it may have been a way of insisting on the indispensability of part to whole. The machine, in other words, had a certain usefulness as a *metaphor*. But the legitimacy of a metaphor depends upon our understanding of its limits. A friend of mine remembered an aunt who noted, correctly, that when Jesus said "I am the door," He did not mean that He had hinges and a knob. We must be careful to remember that a profession is not altogether like a field, or a camera altogether like a room, or a pedigree altogether like a crane's foot.

When a metaphor is construed as an equation, it is out of control; when it is construed as an identity, it is preposterous. If we are to assume that our language means anything at all, then the world is not a machine, and neither is an organism. A machine, to state only the greatest and most obvious difference, is a human artifact, and a world or an organism is not.

But Mr. Wilson, like many others, is fond of this error, and he carries it further. He says, first, that "the brain is a machine," and then he says that "the mind . . . is the brain at work." To lock the subject into this definition, and so forestall any difference of opinion or the introduction of any contradictory evidence, he says, "The surest way to grasp complexity in the brain, as in any other biological system, is to think of it as an engineering problem."

The proposed theory of human mentality, then, is a simple formula: mind = brain = machine. That is to say that the mind is singular, material, and altogether what it is in itself—just as a machine (once made) is a machine per se, singular and material. This is pretty close to what I would call the Tarzan theory of the mind, invented by Edgar Rice Burroughs. The Tarzan theory holds that a human, raised entirely by apes, would have a mind nonetheless fully human: a human brain, in (so to speak) a social and cultural vacuum, would still function on its own as a human mind.

But this raises an interesting question: Is there such a thing as a mind which is merely a brain which is a machine? Would one have a mind if one had no body, or no body except for a brain (whether or not it is a machine)—if one had no sense organs, no hands, no ability to move or speak, no sensory pains or pleasures, no appetites, no bodily needs? If we grant (for the sake of argument) that such may be theoretically possible, we must concede at the same time that it is not imaginable, and for the most literal of reasons: Such a mind could contain no image.

And now let us grant the mind a body and all that the body brings with it and implies: sense, imagery, motion, desire. Another question arises: Would such a mind alone be a mind? Would one mind alone, alone and therefore lacking a language and any need for signs or signification, be recognizable as a mind? Again, we may grant that this is theoretically possible (though I see no reason why we should). But if we encountered such a mind, how would we know that it was a mind?

Suppose, then, that we put two embodied minds together, making them male and female, and yet deny them a habitat and any familiarity of place and time. Now we would be approaching something recognizably a mind, because we would have two. There would need to be some sort of language. These two minds would be different, they would have desires toward one another, they would need to negotiate. But without a dwelling place this language would be too poverty-stricken and crude to be called human. We can grant, as before, a theoretical possibility to this sort of mind, at the same time being forced to grant also that it would not be recognizably a human mind. (All of us have seen, and most of us have been, adolescent lovers who had nothing to talk about but themselves, and we know how intelligent *that* is.)

But we have been begging the question all along with our grantings of theoretical possibility, for one can't be a brain without a body, or a body (for very long) without a familiar homeland. To have one mind you have got to have at least two (and undoubtedly many more) and a world. We could call this the Adam and Eve theory of the mind. The correct formula, in fact, is more like this: mind = brain + body + world + local dwelling place + community + history. "History" here would mean not just documented events but the whole heritage of culture, language, memory, tools, and skills. Mind in this definition has become hard to locate in an organ, organism, or place. It has become an immaterial presence or possibility that is capable of being embodied and placed.

And here the difference between organisms and machines becomes clearer. The idea of a mind and the idea of an organism are not separate ideas. Or we could say that they are momentarily separable—but only momentarily—for the purposes of thought. Every living creature embodies enough mind to know how to be itself and survive in its place, else it cannot live. A machine embodies none of the mind that made it, and it has nothing of an organism's dependence on its world and community and place and history. A machine, if shot into outer space never to return, would simply go on and on being a machine; after it ran out of fuel or traveled beyond guidance, it would still be a machine. A human mind, necessarily embodied, if shot into outer space never to return, would die as soon as it went beyond its sustaining connections and references.

How far from home can a mind go and still be a mind? Probably no scientist has yet made this measurement, but we can answer confidently: Not too far. How far can a machine go from home (supposing, for the sake of argument, that a machine has a home) and still be a machine? Theoretically, if it is not destroyed, it can go on forever.

If the mind is incomplete without a home, without familiar associations and points of reference outside itself, then it becomes possible to argue that the longer the mind of an individual or a community is at home the better it may become. But this "better" implies the willingness and the ability to practice the virtues of domestic economy: frugality, continuous household maintenance and repair, neighborliness, good husbandry of soil and water, ecosystem and watershed.

If the mind fails in this earthly housekeeping, it can only get worse. How much can a mind diminish its culture, its community and its geography—how much topsoil, how many species can it lose—and still be a mind?

*

If the mind is as complexly formed as I have suggested, then it seems unlikely that the mind of an individual can be the origin of intelligence or truth, any more than an individual brain can be. And I fail to see how an individual brain alone can have any originating power whatsoever.

Of the material origin of intelligence or truth, or even of mind, any answer given will lead only to another question. To settle the matter, one would have to *see* experimentally the point at which some physical activity or excitement of the brain, alone, transformed itself into an original idea.

But how can an idea, which is not material, have a material origin? "Average," for example, is an idea which partakes of none of the physical properties of the things that are averaged. Materialism itself is an idea, just as immaterial as any other. And Edward O. Wilson, despite his materialism, shows himself to be a man as interested in ideas as in the material world.

If ideas are not material, how can they have a material origin? If they are not material in origin, how can their origin be explained by materialist science? This is the major fault line of Mr. Wilson's book: His interest in explaining the origin of things whose authentic existence is denied by the terms of the proposed investigation.

Anybody who thinks that this "scientific" reduction of creatures to machines is merely an issue to be pondered by academic intellectuals is in need of a second thought. I suppose that there are no religious implications in this reductionism, for if you think creatures are machines, you have no religion. For artists who do not think of themselves as machines, there is one artistic implication: Don't be mechanical. But the implications for politics and conservation are profound.

It is evident to us all by now that modern totalitarian governments become more mechanical as they become more total. Under *any* political system there is always a tendency to expect the government to work with mechanical "efficiency"—that is, with speed and no redundancy. (Mechanical efficiency always "externalizes" inefficiencies, such as exhaust fumes, but still one can understand the temptation.) Our system, however, which claims freedom as its purpose, involves several powerful concepts that tend to retard the speed and efficiency of government and to make it unmechanical: the ideas of government by consent of the governed, of minority rights, of checks and balances, of trial by jury, of appellate courts, and so on. If we were to implement politically the idea that creatures are machines, we would lose all of those precious impediments to mechanical efficiency in government. The basis of our rights and liberties would be undermined. If people are machines, what is wrong, for example, with slavery? Why should a machine wish to be free? Why should a large machine honor a small machine's quaint protestations that it has thoughts or feelings or affections or aspirations?

It is not beside the point to remember that our government at times has seen fit to look upon the prosperity of many small producers and manufacturers as a political and economic good, and so has placed appropriate restraints upon the mechanical efficiencies of monopolists and foreign competitors. It is not mechanically efficient to recognize that unrestrained competition between an individual farmer or storekeeper and a great corporation is neither democratic nor fair. I suppose that our so-called conservatives have at least no inconsistency to apologize for; they have espoused the "freedom" of the corporations and their "global economy," and they have no conflicting inhibitions in favor of democracy and fairness. The "liberals," on the other hand, have made political correctness the measure of their social policy at the same time that they advance the economic determinism of the conservatives. Reconciling these "positions" is not rationally possible; you cannot preserve the traditional rights and liberties of a democracy by the mechanical principles of economic totalitarianism.

But for the time being (may it be short) the corporations thrive, and they are doing so at the expense of everything else. Their dogma of the survival of the wealthiest (i.e. mechanical efficiency) is the dominant intellectual fashion. A letter to the *New York Times* of July 8, 1999 stated it perfectly: "While change is difficult for those affected, the larger, more efficient business organization will eventually emerge and industry consolidation will occur to the benefit of the many." When you read or hear those words "larger" and "more efficient" you may expect soon to encounter the word "inevitable," and this letter writer conformed exactly to the rule: "We should not try to prevent the inevitable consolidation of the farming industry." This way of talking is now commonplace among supposedly intelligent people, and it has only one motive: the avoidance of difficult thought. Or one might as well say that the motive is the avoidance of thought, for that use of the word "inevitable" obviates the need to consider any alternative, and a person confronting only a single possibility is well beyond any need to think. The message is: "The machine is coming. If you are small and in the way, you must lie down and be run over." So high a level of mental activity is readily achieved by terrapins.

The reduction of creatures to machines is in principle directly opposed to the effort of conservation. It is, in the first place, part and parcel of the determinism that derives from materialism. Conservation depends upon our ability to make qualitative choices affecting our influence on the ecosystems we live in or from. Machines can make no such choices, and neither, presumably, can creatures who are machines. If we are machines, we can only do as we are bidden to do by the mechanical laws of our mechanical nature. By what determinism we *regret* our involvement in our mechanical devastations of the natural world has not been explained by Mr. Wilson or (so far as I know) by anybody.

But suppose we *don't* subscribe to this determinism. Suppose we don't believe that creatures are machines. Then we must see the extent to which conservation has been hampered by this idea, whether consciously advocated by scientists or thoughtlessly mouthed about in the media and in classrooms. The widespread belief that creatures *are* machines obviously makes it difficult to form an advocacy for creatures *against* machines. To confuse or conflate creatures with machines not only makes it impossible to see the differences between them; it also masks the conflict between creatures and machines that under industrialism has resulted so far in an almost continuous sequence of victories of machines over creatures.

To say as much puts me on difficult ground, I know. To confess, these days, that you think some things are more important than machines is almost sure to bring you face to face with somebody who will accuse you of being "against technology" —against, that is, "the larger, more efficient business organization" that will emerge inevitably "to the benefit of the many."

And so I would like to be as plain as possible. What I am against—and without a minute's hesitation or apology—is our slovenly willingness to allow machines and the idea of the machine to prescribe the terms and conditions of the lives of creatures, which we have allowed increasingly for the last two centuries, and are still allowing, at an incalculable cost to other creatures *and to ourselves.* If we state the problem that way, then we can see that the way to correct our error, and so deliver ourselves from our own destructiveness, is to quit using our technological capability as the reference point and standard of our economic life. We will instead have to measure our economy by the health of the ecosystems and human communities where we do our work.

It is easy for me to imagine that the next great division of the world will be between people who wish to live as creatures and people who wish to live as machines.

6. ORIGINALITY AND THE "TWO CULTURES"

If one of the deities or mythological prototypes of modern science is Sherlock Holmes, another, surely, is the pioneering navigator or land discoverer: Christopher Columbus or Daniel Boone. Mr. Wilson's book returns to this image again and again. He says that "Original discovery is everything." And he speaks of "new terrain," "the frontier," "the mother lode," "virgin soil," "the growing edge," "the cutting edge," and "virgin land." He speaks of scientists as "prospectors," as navigators who "steer for blue water, abandoning sight of land for a while" and (in several places) as explorers of unmapped territory.

This figure of the heroic discoverer, so prominent in the mind of so eminent a scientist, dominates as well the languages of scientific journalism and propaganda. It defines, one guesses, the ambition or secret hope of most scientists, industrial technologists, and product developers: to go where nobody has previously gone, to do what nobody has ever done.

There is nothing intrinsically wrong with heroic discovery. However, it is as much subject to criticism as anything else. That is to say that it may be either good or bad, depending on what is discovered and what use is made of it. Intelligence minimally requires us to consider the possibility that we might well have done without some discoveries, and that there might be two opinions from different perspectives about any given discovery—for example, the opinion of Cortés, and that of Montezuma. Perhaps intelligence requires us to consider even that some unexplored territory had better be treated as forbidden territory.

As a personal ambition, heroic discovery has obvious risks, even for the heroically gifted. The greatest risk is that one will trade one's life—all the ordinary satisfactions of homeland and family life—for the sake of a hope not ordinarily realizable. William Butler Yeats saw these possibilities as mutually exclusive and the choice between them as inescapable; he was wrong, I think, and yet he was undoubtedly right about the cost of choosing work over life:

> The intellect of man is forced to choose
> Perfection of the life, or of the work,
> And if it take the second must refuse
> A heavenly mansion, raging in the dark.
>
> When all that story's finished, what's the news?
> In luck or out the toil has left its mark:
> That old perplexity an empty purse,
> Or the day's vanity, the night's remorse.

Mr. Wilson believes that the desire to seek out "virgin land and the future lineaments of empire" is "basic to human nature." Maybe so, but it seems more likely to be basic to the nature of only some humans, among whom have been some of the worst. Its cultural justification is to be found in works of romantic individualism and self-glorification such as Tennyson's "Ulysses," in which the hero (rehabilitated from Dante's *Inferno*) yearns toward "that untraveled world," and desires

> To follow knowledge like a sinking star
> Beyond the utmost bound of human thought.

Dante, as Tennyson did not say, found this Ulysses in Hell among the Evil Counsellors. And we, in making a cultural ideal of the same heroic ambition, see only the good that we believe is inevitably in it, forgetting how much it may partake of adolescent fantasy, adult megalomania, and intellectual snobbery, or how closely allied it is to our continuing history of imperialism and colonialism.

As a *norm* of expectation or ambition, then, heroic discovery is potentially ruinous, and maybe insane. It is one of the versions of our obsession with "getting to the top." Unlike the culture of the European Middle Ages, which honored the vocations of the learned teacher, the country parson, and the plowman as well as that of the knight, or the culture of Japan in the Edo period which ranked the farmer and the craftsman above the merchant, our own culture places an absolute premium upon various kinds of stardom. This degrades and impoverishes ordinary life, ordinary work, and ordinary experience. It depreciates and underpays the work of the primary producers of goods, and of the performers of all kinds of essential but unglamorous jobs and duties. The inevitable practical results are that most work is now poorly done; great cultural and natural resources are neglected, wasted, or abused; the land and its creatures are destroyed; and the citizenry is poorly taught, poorly governed, and poorly served.

Moreover, in education, to place so exclusive an emphasis upon "high achievement" is to lie to one's students. Versions of Mr. Wilson's "original discovery is everything" are now commonly handed out in public schools. The goal of education-as-job-training, which is now the dominant pedagogical idea, is a high professional salary. Young people are being told, "You can be anything you want to be." Every student is given to understand that he or she is being prepared for "leadership." All of this is a lie. Original discovery is *not* everything. You don't, for instance, have to be an original discoverer in order to be a good science teacher. A high professional salary is *not* everything. You *can't* be everything you want to be; nobody can. Everybody *can't* be a leader; not everybody even wants to be. And these lies are not innocent. They lead to disappointment. They lead good young people to think that if they have an ordinary job, if they work with their hands, if they are farmers or housewives or mechanics or carpenters, they are no good.

C. P. Snow, in his 1959 lecture "The Two Cultures," alluded to this problem. His comment occurs in a footnote, but it is nevertheless maybe the most troubling insight of his lecture. One consequence of industrialism, he said, "is that there are no people left, clever, competent and resigned to a humble job. . . . Postal services, railway services, are likely slowly to deteriorate just because the people who once ran them are now being educated for different things."

Snow's general argument was that in Britain and other countries of the West, the literary and scientific cultures had become separated by lack of common knowledge and a common language. His lecture, which I think was never first-rate, is still provocative and useful; it was controversial from the start, and the events of the last forty years certainly have imposed questions and qualifications upon it. But his point, if not his bias, is still valid. The "two cultures" remain divided; they are divided both between and within themselves; and this state of things is still regrettable.

Mr. Wilson appears to accept Snow's argument without qualification, since it justifies his own project for reuniting all the intellectual disciplines by means of consilience. But on this issue of the primacy of originality and innovation, of the cutting edge and the unmapped territory, the scientific and the literary cultures appear now to be pretty much in agreement. The two cultures don't exactly meet under this heading, but here at least they overlap.

This agreement is to a considerable extent the result of the absorption of all the disciplines into the organization (and the value system) of the modern, corporatized university, and of the literary culture's envy of the power, wealth, and prestige of the scientific culture within that organization. Given the present structure of incentives and rewards, it is perhaps only natural that non-sciences would aspire to become sciences, and that non-scientists would aspire to be, like scientists, heroes of original discovery (or at least of "the liberation of the human spirit"), scouting the frontiers of human knowledge or experience, wielding the cutting edge of some social science or some critical theory or some "revolutionary" art.

If there is an economy of the life of the mind—as I assume there has to be, for the life of the mind involves the distribution of limited amounts of time, energy, and attention—then that economy, like any other, subsists upon the making of critical choices. You can't think, read, research, study, learn, or teach everything. To choose one thing is to choose against many things. To know some things well is to know other things not so well, or not at all. Knowledge is always surrounded by ignorance. We are, moreover, differently talented and are called by different vocations. All this explains, and to some extent justifies, any system of specialization in work or study. One cannot sensibly choose against specialization because, if for no other reason, all of us by nature are to some degree specialized. There can be no objection in principle to organizing a university as a convocation of specialties and specialists; that is what a university is bound to be.

But some serious questions remain, the most serious of which I would put this way: Can this convocation of specialists, who have been "called together" to learn and teach, actually come together? In other words, can the convocation become a conversation? For that, the convocation would have to have a common purpose, a common standard, and a common language. It would have to understand itself as a part, for better or worse, of the surrounding community. For reasons both selfish and altruistic, it would have to make the good health of its community the primary purpose of all its work. If that were the avowed purpose, then all the members and branches of the university would have to converse with one another, and their various professional standards would have to submit to the one standard of the community's health.

This has not happened in our universities. The opposite, in fact, is happening. Unlike the English agriculturist Sir Albert Howard, who moved from his specialty, mycology, to the "one great subject" of health, the modern university specialist moves ever away from health toward the utter departmentalization and disintegration of the life of the mind and of communities. The various specialties are moving ever outward from any center of interest or common ground, becoming ever farther apart, and ever more unintelligible to one another. Among the causes, I think, none is more prominent than the by now ubiquitous and nearly exclusive emphasis upon originality and innovation. This emphasis, operating within the "channels" of administration, affects in the most direct and practical ways all the lives within the university. It imposes the choice of work over life, exacting not only the personal costs spoken of in Yeats's poem, but very substantial costs to the community as well. And these are costs that can be accounted.

"Over the years," Mr. Wilson writes, "I have been presumptuous enough to counsel new Ph.D.'s in biology as follows: If you choose an academic career you will need forty hours a week to perform teaching and administrative duties, another twenty hours on top of that to conduct respectable research, and still another twenty hours to accomplish really important research." Mr. Wilson is thus prescribing to the young a normative work week of eighty hours. Since he mentions no days off, let us assume that he is speaking of seven workdays of about 11½ hours each, lasting, say, from eight o'clock in the morning until 7:30 at night, or until eight at night if we allow half an hour for lunch. There are 168 hours in a seven-day week. Eighty from 168 leaves 88 hours. If the young Ph.D. sleeps eight hours a night, that takes another 56 hours, leaving 32 hours, or about 4½ hours per day. In that 4½ hours he or she must eat, keep clean, shop, do domestic chores, commute, read, care for his or her (unfortunate) children, etc. The time left over may presumably be used for amusement and for taking part in family and community life.

I suppose we ought to yield a certain admiration to such a dedicated life of work and sacrifice. It is certain that all of us have benefited from such effort on the part of some people. But it is just as certain that we have been damaged and are threatened by similar effort on the part of other people. In fact, most people can be driven to such an extent only by a kind of professional or careerist panic. Young Ph.D.'s or assistant professors find themselves in a man-made evolutionary crisis known in the universities as "publish or perish." To survive they must "produce," for the fundamental academic fact of life is that (as Mr. Wilson puts it) "a discovery does not exist until it is safely reviewed and in print." All "tenure track" professors in all universities, at least until tenure, are under life-or-death pressure to "find their way to a publishable conclusion." If a tree falls in the absence of a refereed journal or a foundation, does it make a sound? The answer, in the opinion of the imitation corporate executives who now run our universities, is no.

This academic Darwinism inflicts severe penalties both upon those who survive and upon those who perish. Both must submit to an absolute economic system which values their lives strictly according to their "productivity"—which is to say that they submit to a form of slavery. Both must submit, at least until tenure, to a university-prescribed regimen of life in which time = work = original discovery = career, thus assuring the ascendancy of professional standards in the minds of the young, and the eclipse of any standard of any other kind. The modern university thus enforces obedience, not to the academic ideal of learning and teaching what is true, as a community of teachers and scholars passing on to the young the knowledge of the old, but obedience rather to the industrial economic ideals of high productivity and constant innovation. The problem here is not that we should object to hard work and exacting study, which any school might appropriately expect, but that we certainly can find reason to object to turning schools into factories, and to making originality or innovation the exclusive goal and measure of so much effort. Mr. Wilson in his counsel to the young is, in fact, helping to perpetuate a system of education that conforms exactly to the demands of the economic system the effects of which, as a conservationist, he so much regrets. The "cutting edge" is not critical or radical or intellectually adventurous. The cutting edge of science is now fundamentally the same as the cutting edge of product development. The university emphasis upon productivity and innovation is inherently conventional and self-protective. It is part and parcel of the status quo. The goal is innovation but not difference. The system exists to prevent "academic freedom" from causing unhappy surprises to corporations, governments, or university administrators.

The present conformity between science and the industrial economy is virtually required by the costliness of the favored kinds of scientific research and the consequent dependence of scientists on patronage. Mr. Wilson writes that "Science, like art, and as always through history, follows patronage." Even I know of some scientists whose work did not "follow patronage," and there certainly have been artists whose work did not, but this statement seems generally true of modern science. Mr. Wilson elaborates; there is, he says, "a cardinal principle in the conduct of scientific research: Find a paradigm for which you can raise money and attack with every method of analysis at your disposal." This principle, in effect, makes the patron the prescriber of the work to be done. It would seem to eliminate the scientist as a person or community member who would judge whether or not the work *ought* to be done. It removes the scientist from the human and ecological circumstances in which the work will have its effect, and which should provide one of the standards by which the work is to be judged; the scientist is thus isolated, by this principle of following patronage, in a career with a budget. What this has to do with the vaunted aim of pursuing truth cannot be determined until one knows where the money comes from and what the donor expects. The donor will determine what truth (and how much) will be pursued, and how far, and to what effect. The scientist, having succeeded or failed with one paradigm, will then presumably be free to find another, and another patron.

The young Ph.D.'s who work eighty hours a week in a system devoted to "really important research" are not going to have time to know their community, let alone to wonder about the possible effects of their work upon its health. They are not going to have time to confront the problems invariably raised by innovation, or to perform the necessary criticism. Nor, in reality, will they have time to know their students very well, or to teach very well. If your educational system gives the preponderance of its rewards (promotions, salary increases, tenure, publication, prizes, grants) to "original discovery" and "really important research," then to the same extent it discourages teaching. It is simply a matter of fact that if teachers know that their careers and their livelihoods depend almost entirely on research, then most will steal time from teaching to give to research—exactly as any rational person would expect.

Teaching, anyhow, cannot do well under the cult of innovation. Devotion to the new enforces a devaluation and dismissal of the old, which is necessarily the subject of teaching. Even if its goal is innovation, science does not *consist* of innovation; it consists of what has been done, what is so far known, what has been thought—just like the so-called humanities. And here we meet a strange and difficult question that may be uniquely modern: Can the past be taught, can it even be known, by people who have no respect for it? If you believe in the absolute superiority of the new, can you learn and teach anything identifiable as old? Here, as before, Mr. Wilson speaks from an entirely conventional point of view. He takes seriously no history before the Enlightenment, which he believes began the era of modern science. Of "prescientific cultures" he makes short work: "they are wrong, always wrong." They know nothing about "the real world," but can only "invent ingenious speculations and myths." And: "Without the instruments and accumulated knowledge of the natural sciences—physics, chemistry, and biology—humans are trapped in a cognitive prison. They are like intelligent fish born in a deep, shadowed pool." I think (or, anyhow, hope) he does not realize how merciless this is—for he has thus flipped away most human history, most human lives, and most of the human cultural inheritance—or how small and dull a world it leaves him in. To escape the "cognitive prison" of religion and mythology, he has consigned himself to the prison of materialist and reductive cognition. *Consilience*, exactly like Genesis, explains only what it is capable of explaining. But, unlike Genesis, it concedes nothing to mystery; it simply rules out or blots out whatever it can't explain or doesn't like. One thing it blots out is the damage that this intellectual complacency and condescension has done and is doing still to prescientific cultures and their homelands around the world.

"I mean no disrespect," Mr. Wilson says, "when I say that prescientific people, regardless of their innate genius, could never guess the nature of physical reality beyond the tiny sphere attainable by unaided common sense. . . . No shaman's spell or fast upon a sacred mountain can summon the electromagnetic spectrum. Prophets of the great religions were kept unaware of its existence, not because of a secretive god but because they lacked the hard-won knowledge of physics." It seems only courteous to inquire at this point if anybody, ever before, has had the originality to propose that the prophets *needed* to know about the electromagnetic spectrum? One may imagine a little play, as follows:

> Isaiah *(finger in the air and somewhat oblivious of the historical superiority of the modern audience)*: The voice said, Cry. And he said, What shall I cry? All flesh is grass, and all the goodliness thereof is as the flower of the field . . .
>
> Edward O. Wilson *(somewhat impressed, but nonetheless determined to do his bit for "evolutionary progress")*: But . . . But, sir! Are you aware of the existence of the electromagnetic spectrum?

CURTAIN

Even as a believer in "the potential of indefinite human progress," Mr. Wilson can be properly humble, when he has the notion. For instance, he says that "evolutionary progress is an obvious reality" if we mean by it "the production through time of increasingly complex and controlling organisms and societies, in at least some lines of descent, with regression always a possibility. . . ." One notes with gratitude this consent to the possibility of regression, but in fact Mr. Wilson is not much impressed or detained by any such possibility. Later on, returning to the subject of "preliterate humans," he concedes: "We are all still primitives compared to what we might become." There follows an avowal of the largeness of human ignorance, which, characteristically, he hastens past and quickly forgets: "*Yet* [my emphasis] the great gaps in knowledge are beginning to be filled . . . knowledge continues to expand globally. . . . Any trained person can retrieve and augment any part of it. . . . The explanations can be joined in space from molecule to ecosystem, and in time from microsecond to millennium." And then, speaking as if his program has already been completed, he says, "Now, with science and the arts combined [in consilience], we have it all." Here there is no functioning doubt or question, no live sense of the possibility of regression, no acknowledgment of the possibility that knowledge, if it can be accumulated, can also be lost. There is no hint that knowledge can be misused.

Though his head sometimes tells him that such concessions

should be made, his heart never does. In his heart, he is in agreement with the apparent majority of the public who now believe that the new inevitably replaces or invalidates the old, because the new, coming from an ever-growing fund of data, is inevitably better than the old. The rails of the future have been laid by genetic (or technological or economic) determination, and as we move forward we destroy justly and properly the rails of the past. This is strong, easeful, and reassuring doctrine, so long as one does not count its costs or number its losses.

If under the demands of a university system obsessively concentrated upon originality and innovation, science serves progress, industry, and the corporate economy, then the literary culture (to use the phrase of C. P. Snow) gives its tacit approval to the program of science-technology-and-industry and, itself, serves nothing—except, perhaps, for certain politically correct ideologies that could be as well served anywhere else. It serves less and less even the cause of literacy, which the university system has made the specialty of the English department, and which the English department has made the specialty of the freshman English program. The university as a whole gives no support to the cause of literacy. If technical or workmanly competence in writing is required only by teachers of freshman English, and virtually all other teachers either require no written work or grade what they do require "on content," then the message is unmistakable: Competence in writing does not matter and is not necessary. And that is what most students believe.

The English department, working under the same dire pressure to "produce" publishable books and articles as every other department, must assign the teaching of English composition to graduate students and a few specialists in the methodology and technology of composition-teaching, while most of the regular faculty concentrate on matters exalted far above grammar and punctuation and sentence structure. One result, as I know from my own experience and observation, is the certification of public school English teachers who do not necessarily know how to construct a coherent English sentence, or punctuate it, or make its nouns and verbs agree, or spell its common English words. Nor is it by any means certain that these certified English teachers will have a sense of literary history and tradition. Another result is the virtual languagelessness of many professional journalists and "communicators." Another is the increasing editorial slovenliness of newspapers and publishing companies.

The cult of progress and the new, along with the pressure to originate, innovate, publish, and attract students, has made the English department as nervously susceptible to fashion as a flock of teenagers. The academic "profession" of literature seems now to be merely tumbling from one critical or ideological fad to another, constantly "revolutionizing" itself in pathetic imitation of the "revolutionary" sciences, issuing all the while a series of passionless, jargonizing, "publishable" but hardly readable articles and books, in which a pretentious obscurity and dullness masquerade as profundity. And this, I think, is not easily definable as the fault of anybody in particular. It is the fault of a bad system—howbeit one that most people in it don't to all appearances object to, and one that nobody in it has effectively objected to. The university's convocation of the disciplines is not a conversation; it is incapable of criticizing itself. One of the most dangerous effects of the specialist system is to externalize its critics, and thus deprive them of standing.

Originality and innovation in science may be a danger to the community, because newness is not inherently good, and because the scientific disciplines use only professional standards in judging their work. There is no real criticism. (Ezra Pound has reminded us in *ABC of Reading* that, at root, to criticize is to choose.) Nobody seems able to subtract the negative results of scientific "advances" from the positive. Not many modern scientists would say, with Erwin Chargaff, that "all great scientific discoveries . . . carry . . . an irreversible loss of something that mankind cannot afford to lose." But, then, Chargaff had confronted fully the implication of modern science in the bombing of Hiroshima and Nagasaki and "the German extermination factories." He wrote: "The Nazi experiment in eugenics . . . was the outgrowth of the same kind of mechanistic thinking that, in an outwardly very different form, contributed to what most people would consider the glories of modern science."

In the literary culture, the preponderant aim of originality and innovation strikes directly at the community by granting precedence to intellectual fashion, and so depreciating literacy and literature and the cultural inheritance. The new supposedly dazzles the old out of existence, and people of our era are encouraged to pity their ancestors who had not the good fortune to be as we are. The cult of originality, however, seems to have produced about the expectable amount of bad work, but not much that is truly original. Instead, it has produced fashions and uniforms of originality. Political correctness becomes the intellectual and literary cutting edge.

One worries that the cultists of the new and original think they are doing what Ezra Pound told them to do. In fact, Pound did say that writers should "make it new," but that was probably as traditional an instruction as he ever gave. It is a statement perhaps too easy to understand as a flippant rejection of the old, but that is not what he meant. Pound used, to begin with, the verb "make," and he, like virtually every poet until recently, knew that our word "poet" came from a Greek word meaning "maker." To make, one must know how to make. And how does one learn? By reading. To Pound, how to write was the same question as how to read. To learn to write one must learn both a considerable portion of what has been written and *how* it was written. And so the first reference of the pronoun in "make it new" is the literary inheritance; one must renew the means of literature, which is to say the literary tradition, by making it newly applicable to contemporary needs and occasions. The new must come from the old, for where else would you get it? Not, anyhow, from contempt for the old, or from ambition. Pound's work, at its sanest, was always a testing of the usefulness of what he had read.

But I think he meant much more than that. There are passages in *ABC of Reading* that can be understood as glosses on "make it new." "A classic," Pound wrote there, "is classic not because it conforms to certain structural rules, or fits certain definitions. . . . It is classic because of a certain eternal and irrepressible freshness." And furthermore: "Great literature is simply language charged with meaning to the utmost possible degree." (Note that he does not use the adjective "new" in that sentence.) And furthermore: "Literature is news that STAYS news." The business of literature, then, is to renew not only itself but also our sense of the perennial newness of the world and of our experience; it is to renew our sense of the newness of what is eternally new.

Ananda Coomaraswamy, who was a more systematic student of artistic tradition than Pound, also wrote usefully on the subject of making it new: "There can be no property in ideas, because these are gifts of the Spirit, and not to be confused with talents. . . . No matter how many times [it] may already have been 'applied' by others, whoever conforms himself to an idea and makes it his own, will be working originally, but not so if he is expressing only his own ideals or opinions." And perhaps even more helpfully he wrote that "when there is realization, when the themes are felt and art *lives*, it is of no moment whether . . . the themes are new or old." It should be fairly clear that a culture has taken a downward step when it forsakes the always difficult artistry that renews what is neither new nor old and replaces it with an artistry that merely exploits what is fashionably or adventitiously "new," or merely displays the "originality" of the artist.

Scientists who believe that "original discovery is everything" justify their work by the "freedom of scientific inquiry," just as would-be originators and innovators in the literary culture justify their work by the "freedom of speech" or "academic freedom." Ambition in the arts and the sciences, for several generations now, has conventionally surrounded itself by talk of freedom. But surely it is no dispraise of freedom to point out that it does not exist spontaneously or alone. The hard and binding requirement that freedom must answer, if it is to last, or if in any meaningful sense it is to exist, is that of responsibility. For a long time the originators and innovators of the two cultures have made extravagant use of freedom, and in the process have built up a large debt to responsibility, little of which has been paid, and for most of which there is not even a promissory note.

The debt can be paid only by thought, work, deference, and affection given to the integrity of our ecological and cultural life. The condition which that integrity (or that one-time integrity) imposes on human work and human freedom is that everything we do has an effect or an influence. But it is generally true to say that among the originators of the modern era there has been no flinching before effects, for the purpose of the originators (as understood by themselves) has been the origination of causes only. This is the moral absurdity of specialization driven to the limit. The effects are understood simply as the causes of other original work by other specialists. And thus we have assumed that all problems merely lead to solutions, an article of pathological faith.

All along, the enterprise of science-industry-and-technology has been accompanied by a tradition of objection. Blake's revulsion at the "dark Satanic Mills" and Wordsworth's perception that "we murder to dissect" have been handed down through a succession of lives and works, and among the inheritors have been scientists as well as artists. The worry, I think, has always been that in our ever-accelerating effort to explain, control, use, and sell the world we would destroy the wholeness and the sanctity of all that which it is our highest obligation to "make new."

On the day after Hitler's troops marched into Prague, the Scottish poet Edwin Muir, then living in that city, wrote in his journal a note that recalls a similar lamentation of Montaigne about four hundred years earlier: "So many goodly citties ransacked and razed," Montaigne wrote; "so many nations destroyed and made desolate; so infinite millions of harmelesse people of all sexes, states and ages, massacred, ravaged and put to the sword; and the richest, the fairest and the best part of the world topsiturvied, ruined and defaced for the traffick of Pearles and Pepper: Oh mechanicall victories, oh base conquest." Muir wrote: "Think of all the native tribes and peoples, all the simple indigenous forms of life which Britain trampled upon, corrupted, destroyed . . . in the name of commercial progress. All these things, once valuable, once human, are now dead and rotten. The nineteenth century thought that machinery was a moral force and would make men better. How could the steam-engine make men better? Hitler marching into Prague is connected with all this. If I look back over the last hundred years it seems to me that we have lost more than we have gained, that what we have lost was valuable, and that what we have gained is trifling, for what we have lost was old and what we have gained is merely new."

Laboring in the shadow of the scientific apocalypse of World War II, C. S. Lewis wrote: "Dreams of the far future destiny of man were dragging up from its shallow and unquiet grave the old dream of Man as God. The very experiences of the dissecting room and the pathological laboratory were breeding a conviction that the stifling of all deep-set repugnances was the first essential for progress."

Albert Howard, at about the same time rethinking the role of science in agriculture, wrote: "It is a severe question, but one which imposes itself as a matter of public conscience, whether agricultural research in adopting the esoteric attitude, in putting itself above the public and above the farmer whom it professes to serve, in taking refuge in the abstruse heaven of the higher mathematics, has not subconsciously been trying to cover up what must be regarded as a period of ineptitude and of the most colossal failure. Authority has abandoned the task of illuminating the laws of Nature, has forfeited the position of the friendly judge, scarcely now ventures even to adopt the tone of the earnest advocate: it has sunk to the inferior and petty work of photographing the corpse. . . ."

And in the autobiographical meditation written when he was old, looking back over the course of science from the time of his own crisis of conscience at the revelations of World War II, Erwin Chargaff wrote: "The wonderful, inconceivably intricate tapestry is being taken apart strand by strand; each thread is being pulled out, torn up, and analyzed; and at the end even the memory of the design is lost and can no longer be recalled."

It is not easily dismissable that virtually from the beginning of the progress of science-technology-and-industry that we call the Industrial Revolution, while some have been confidently predicting that science, going ahead as it has gone, would solve all problems and answer all questions, others have been in mourning. Among these mourners have been people of the highest intelligence and education, who were speaking, not from nostalgia or reaction or superstitious dread, but from knowledge, hard thought, and the promptings of culture.

What were they afraid of? What were their "deep-set repugnances"? What did they mourn? Without exception, I think, what they feared, what they found repugnant, was the violation of life by an oversimplifying, feelingless utilitarianism; they feared the destruction of the living integrity of creatures, places, communities, cultures, and human souls; they feared the loss of the old prescriptive definition of humankind, according to which we are neither gods nor beasts, though partaking of the nature of both. What they mourned was the progressive death of the earth.

Wes Jackson of the Land Institute said once, thinking of the nuclear power and genetic engineering industries, "We ought to stay out of the nuclei." I remember that because I felt that he was voicing, not scientific intelligence, but a wise instinct: an intuition, common enough among human beings, that some things are and ought to be forbidden to us, off-limits, unthinkable, foreign, *properly* strange. I remember it furthermore because my own instinctive wish was to "stay out of the nuclei," and, as I well knew, this wish amounted exactly to nothing. One can hardly find a better example of modern science as a public predicament. For modern scientists work with everybody's proxy, whether or not that proxy has been given. A good many people, presumably, would have chosen to "stay out of the nuclei," but that was a choice they did not have. When a few scientists decided to go in, they decided for everybody. This "freedom of scientific inquiry" was immediately transformed into the freedom of corporate and/or governmental exploitation. And so the freedom of the originators and exploiters has become, in effect, the abduction and imprisonment of all the rest of us. Adam was the first, but not the last, to choose for the whole human race.

The specialist system, using only professional standards, thus isolates and overwhelmingly empowers the specialist as the only authorizer of his work—she alone is made the sole moral judge of the need or reason for her work. This solitary assumption of moral authority, of course, must *precede* the acceptance of patronage. Originality as a professional virtue gives far too much importance and power to originators, and at the same time isolates them socially and morally.

The specialist within the literary culture is isolated, it seems to me, in precisely the same way, and in fact *wishes* to be so isolated. The effects of the work of the literary specialists are not, of course, so directly practical as those of the scientists, but they are in the long run a part of the same disintegration, and are equally serious.

That the arts have been envious of the prestige, the drama, and the glamour of innovative science is suggested by the long-enduring vogue of "experimental art." "Experiment" is a word that seems displaced and uncomfortable outside of science; in science, I suppose, a failed experiment is still science, but in art a failed experiment, whatever else it may be, is not art. Misnomer or not, "experimentation" in the arts certainly bespeaks a hankering among artists for the heroism of life on the "cutting edge."

The science closest to art (in the opinion, anyhow, of many artists) is psychology and especially psychoanalysis. The study of the "psyche" is not a very exact science, but its subject matter is indigenous to the arts, and it is not hard to understand how attractive among artists have been the psychological theories of consciousness and "the unconscious." The idea of imitating in writing the "stream of consciousness" occurred early to novelists, and the psychoanalysts carried on and encouraged the artists' age-old fascination with dreams.

Maybe because modern artists took so many promptings from psychology, the scientific goal of "original discovery" became in art, and particularly in literature, the goal of original disclosure. It seems generally true that in the twentieth century writers' interest in personal life and in the inward life of persons became more intense, intimate, and in certain ways more articulate than before. The difference, roughly, is that between Tennyson's "Ulysses" and Eliot's "The Love Song of J. Alfred Prufrock." There is a new keenness, or a new kind of keenness, in understanding how people understand themselves.

Along with this interest in the intimate, inward histories of fictional persons has come (in what relation of cause or effect, I don't know) an interest in disclosing the private lives of real persons. Such disclosures are now conventional and commonplace in biography, in "confessional poetry," and in "fictionalized" accounts of actual lives and events. The most inward life is laid open, the most intimate details are shown, exactly as in dissection (which means cutting apart) or autopsy (which means seeing for yourself). One of the paramount originations of the modern literary culture is the discovery that privacy is penetrable and publishable, and that publication is not likely to be legally actionable. In fiction and poetry, in biography, in journalism and the entertainment industry, and finally in politics, the cutting edge for most of the twentieth century has been the dis-covering of the intimate, the secret, the sexual, the private, and the obscene. And this process of exposure has been carried on in the name of freedom by people priding themselves on their courage.

Has it required courage? So long as it involved legal or professional penalties, it most certainly did require courage. But now that the penalties have been removed, no courage is necessary. Public sexual revelations and public obscenity are now merely clichés, part of the uniform behavior of modish nonconformity and fashionable bad manners, but always performed by people who wish to be thought courageous.

Has it increased freedom? Well, of course people have become more free when they have earned or taken or been given the right to do what they previously were forbidden to do. But, as always, the worth of freedom depends upon how it is used. The value of freedom is probably not intrinsic and is certainly not limitless. It is generally understood by people who think about it that freedom can be abused, and that it rests, in the long run, on a common understanding of fairness: One should not increase one's freedom by reducing somebody else's.

I would question also the worth of freedom from what C. S. Lewis called our "deep-set repugnances," among which I would include our native and proper repugnance against nosiness, against having our privacy invaded. This has to do, I think, with our rightful fear of being misunderstood or too simply understood, or of having our profoundest experience misvalued. This, surely, is one of the reasons for Christ's insistence on the privacy of prayer. It is a part of our deepest and most precious integrity that we should speak (if we wish) for ourselves. We do not want self-appointed spokesmen for our souls. Sex and worship especially are inward to us, and they are especially fragile as possessions. Their nature is to be shared, and yet it is dangerous to speak of them carelessly. To speak of them carelessly is to violate yet another nucleus that ought to be sacrosanct.

Our present idea of freedom in science is too often reducible to thoughtlessness of consequence. Freedom in the arts frequently looks like mere carelessness in self-exposure or in exposing others. In both science and art there is a principled resistance to any suggestion that the specialist, within his or her work, might be subject or subordinate to anything.

On October 19, 1998, in New York City, the Authors Guild and the Authors Guild Foundation held a panel discussion, "Whose Life Is It, Anyway?"—a transcript of which was published in the *Authors Guild Bulletin*. The panelists, Cynthia Ozick, David Leavitt, Janna Malamud Smith, and Judy Collins, "addressed the moral, ethical and artistic implications of the writer's appropriation of others' lives and experiences." Parts of their conversation are illustrative of the problem of freedom, and I am going to quote from it at some length. Several of my ellipses indicate large omissions.

Cynthia Ozick said: "I could not fathom that fiction might not be an arena of total freedom. . . . I remember sitting on the edge of a bed with my mother-in-law, explaining how I'd been keeping a diary since 1953, and that everyone was in it. She was terrifically disturbed. 'No, no,' she cried, 'erase it, you can't have it, you mustn't do this.' And instantly I realized . . . that no writer could ever take that view. Life becomes real only through having been written. . . . Inevitably, writers are responsible for wounds and hurts—but the writer must say, I don't care, I don't give a damn. . . ."

David Leavitt quoted with approval a writing teacher who once told a class, "For every writer it is a rite of passage to write the story after which a member of your family will no longer speak to you." And later Mr. Leavitt states his credo: "I say anything goes in fiction—anything goes. If you start to take away bit by bit the rights of writers doing what they want, what you end up eroding is your own freedom."

Ms. Ozick agreed that "anything goes." But then she made an exception: "Yet I do have certain lines of limits. . . . I would not admire—I would strenuously object to—a novel which took a Holocaust-denial point of view. . . . But that's an extreme issue. For the writer . . . it's only make-believe, it's the world of enchantment. Make-believe and enchantment can't really harm anyone."

Janna Malamud Smith, politely on the contrary, said this: "When rationalizing their exposure of others, writers tend to claim two values as having overriding worth. One is the aesthetic goal of telling the story well. There's often a feeling that writing beautifully is an ultimate good, that telling a tale very well compensates any harm it might do to its subjects. The second virtue writers tend to honor is outing the truth. We take seriously the job of looking behind hypocrisy and social facade. . . . We like to believe there is a version of the truth that is superior and that we can state it. These are serious premises. . . . But I think they thrive best when they are occasionally pruned by opposing values. . . . The reason people feel betrayed when they find themselves in people's books is this: Intimacy . . . works because you are allowed to do things in a friendship, in a love relationship, that you can't do in public. So when the private things intimacy has allowed you to expose are suddenly made public, that is a legitimate reason for a feeling of profound betrayal. . . . The fact is that betrayals are a real thing."

Later, Ms. Smith speaks of the possibility that people might be "led astray by romantic novels," and she says, "Influence is real, and probably we need to think about that as well."

It certainly is true that writers are burdened with the responsibility to bear witness to such truth as they have seen or think they have seen. The responsibility goes with the trade: There is no value in telling anything if one does not try with all one's might to tell the truth. If human beings could be utterly confident of their ability to tell—or know—the truth, then the problem would be at least smaller, though a writer's insistence upon "total freedom" to tell the truth about other people would still be questionable. Since human beings can be wrong, since even with the best intentions they may know falsely and tell falsehoods, to say that "anything goes" and to leave it at that is far too simple.

My intention here is certainly not to promote any abridgment of the freedoms of speech and inquiry, though I believe that those freedoms are now being pretty severely abused and that the abuse of freedom threatens its survival. But I agree with Janna Malamud Smith that betrayal and influence are real and must be thought about.

I don't believe that the connection between art and life can ever be finally or even very satisfactorily resolved, any more than can be the connection between science and life. We join ourselves to the living world by the artifacts of art and science —by made things. And we are always going to be at least somewhat at fault, because we are ignorant and fallible and small; the living world is larger and more complex than our works. Because we must be always correcting our errors, art and science always need to be free to shift their ground and start again. The unendable, the necessarily ongoing problem of justice to the world and to one another thus enforces practically the requirement of freedom. And this freedom can survive, I believe, only by being well used.

What "well used" may mean was clearly shown by Richard C. Strohman in an article of April 1, 1999 in *The Daily Californian*. Mr. Strohman was writing in defense of "unimpeded science." He was worried about the costs to science of "new initiatives for university-corporate alliances" in the development of biotechnology. The costs, apparently, will be the familiar ones of too much specialization and of a falsifying oversimplification. Mr. Strohman wrote: "The corporate need for technology dedicated to specific products will . . . must . . . subvert the scientific need for unimpeded research. In academic biology the technological need to define complex behavior in terms of simple causality subverts the need for a wider, more complex research context.

"Here then is the real danger of the university-corporate 'merger' . . . a corporate need that must repress new ways of seeing nature."

And so science too must be concerned with "making it new" and with renewing itself, and now it must do so for a reason both new and urgent: to see that nature escapes the corporations which are newly empowered to oversimplify it, commodify it, and put it up for sale. Mr. Strohman's answer, I think, is the correct one: Enlarge the context of the work.

Freedom in both science and art probably depends upon enlarging the context of our work, increasing (rather than decreasing) the number of considerations we allow to bear upon it. This is because the ultimate context of our work is the world, which is always larger than the context of our thought. And so to complicate the consideration of freedom in literature by the considerations of betrayal and influence is not a diminishment of literature or freedom; it is a *just* enlargement of the context of work. If we could faithfully commit ourselves to the principle that nothing whatever can safely be said to lie outside the context of our work, then artists and scientists would have to be ready at any time to see that they have been wrong and to start again, making yet larger the context of the work. *That* is true freedom. It means simply that beyond all error we can begin again; redemption is possible. From this principle also we can make our way to critical judgments of an amplitude beyond specialization and professionalism: Work that diminishes the possibility of a new start, of "making it new," is bad work.

Janna Malamud Smith said, "You don't trade betrayal for writing . . ." That is not a simple statement, because to be a writer is not a simple predicament. There is a constant relationship, though never altogether settled and never altogether clear, between imagination and reality. If you are a fiction writer, you may, at one extreme, tell a story that is almost the story of something that actually happened; at the opposite extreme, you may tell a story that you have almost entirely imagined. But what you have imagined will always be somewhat informed by what you have actually known, and your actual knowing will always be somewhat informed by imagination. The extremes of reality and imagination, within the limits of human experience, are never pure. And so there is always some risk of betrayal. It is possible to allow imagination to abuse reality; it is possible by imagination to violate a real intimacy—and this leaves aside the possibility of deliberately tattling for meanness or revenge or some version of success. It is always possible too that imagination may be debased by a false or too narrow understanding of what is real.

Both imagination and a competent sense of reality are necessary to our life, and they necessarily discipline one another. Only imagination, for example, can give our home landscape and community a presence in our minds that is a sort of vision at once geographical and historical, practical and protective, affectionate and hopeful. But if that vision is not repeatedly corrected by a fairly accurate sense of reality, if the vision becomes fantastical or merely wishful, then both we and the landscape fall into danger; we may destroy the landscape, or the landscape (especially if damaged by us in our illusion) may destroy us.

To speak of betrayal as a possibility in literature is one way of acknowledging this necessary and inescapable tension between imagination and reality. Fiction can abuse reality by violating intimacy or confidence or privacy, or by being wrong. Not least among the offenses of literary artists is their frequent indifference to facts of history or natural history or ways of work.

To be careless of such betrayals is to reduce one's subject to the status of "raw material"—exactly as the ubiquitous enterprise of science-technology-and-industry reduces *its* subjects. The parallels of value and attitude among contemporary arts and sciences and the industrial economy are obvious, are ratified by convention, and are almost unnoticed by the would-be pioneers and heroes of the cutting edges. But if one lives, as I do, in a rural place, which is to say in the midst of other people's "raw material," then one does notice. And if one notices, then one knows that artistic and scientific betrayals are real and are serious. They are an affront to one's subject, and they endanger it.

Too much disclosure of the intimate, the secret, the sexual, the private, and the obscene is accomplished by mentioning or representing or picturing but not imagining. To represent the intimacy of desire or of grief without the art that compels one to imagine these things as the events of lives and of shared lives is actually to misrepresent them. This is the "objectivity" of the schools and the professions, which allows a university or a corporation to look at the community—its *own* community—as one looks at a distant landscape through fog. This sort of objectivity functions in art much the same as in science; it obstructs compassion; it obscures the particularity of creatures and places. In both, it is a failure of imagination.

Journalism and the electronic media, for example, routinely exhibit representations or disclosures of intimate emotion as objects of curiosity, as intrinsically interesting, or as proofs of artistic or journalistic courage. The perennial act of cutting-edge enterprise in reporting is to shove a camera or a microphone into the face of a grieving woman. But what is the qualitative difference between the man who cold-heartedly shoots another and the photographer who cold-heartedly photographs the corpse or the grieving widow? Are they not simply two parts of the same epidemic failure of imagination, which is to say a failure of compassion and of community life?

Such exposures do not make us free, and they do not increase our knowledge. They only compound human cruelty by a self-induced numbness to the suffering of others and to our common suffering.

To be indifferent to hurts given by one's writing to its human subjects, which exactly parallels the scientific-industrial indifference to the suffering of animal or human subjects of exploitation or experimentation—to say "I don't care, I don't give a damn"—is a betrayal not only of the subject of writing, which is invariably our common life, our neighborhood, but also of imagination itself. It is a refusal to be compassionate, a denial of the vital link between imagination and compassion. How can such a betrayal not impair one's ability to know the truth and to make art?

The world and its neighborhoods, natural and human, are not passively the subjects of art, any more than they are passively the subjects of science-industry-and-technology. They are affected by all that we do. And they respond. The world does not exist merely to be written about, any more than it exists merely to be studied. It is real, before and after human work. What we write is finally to be measured by the health of what we write about. What we think we know affects the health of the thing we think we know.

The problem of influence also is real, and it is inescapable. Ms. Ozick acknowledges as much when she wishes to exempt the Holocaust from her credo of not giving a damn, and so she undermines all her affirmation of artistic superiority and autonomy. To say that writing about the Holocaust may be influential is to say that writing may be influential, period. Who can deny that writing about Jews with contempt may cause them to be treated with contempt, or with violence? But the history of oppression forbids us to limit that liability to the Jews. We treat people, places, and things in accordance with the way we perceive them, and literature influences our perceptions. To leave aside more fashionable examples, who can deny that the history of coal mining in the southern Appalachians has been under the influence of writers who have written of the mountaineers as "briars" or "hillbillies"? And who wishes to say that our long exploitation and finally our virtual destruction of our farm population has not been influenced by generations of writers who have represented farmers as "yokels" who live in the "sticks" and do "mind-numbing work"? We can't deny that writing has an influence unless we can also deny that stereotyping and character assassination have an influence.

The question for art, then, is exactly the same as the question for science: Can it properly subordinate itself to concerns that are larger than its own? Can it judge itself by standards that are higher and more comprehensive than professional standards? The issue is the old one of propriety. Is every artist and every scientist to be "free" to work as if his or her discipline were the only one, or the dominant one? Or is it possible still to see one's work as occurring within a larger and ultimately a mysterious pattern of causes and influences? If we can see that we are mutually dependent upon one another and upon that mosaic of natural and human neighborhoods we call "the world," then it should not be too hard to see that there ought to be responsible connections between science and the knowledge of how to live, and between art and the art of living, and that there is always, inescapably, acknowledged or not, a complex connection between art and science.

7. PROGRESS WITHOUT SUBTRACTION

The task of thinking about Mr. Wilson's book is made difficult at every point by his adherence to the rather simple-minded popular doctrine of mechanical or automatic progress. His book perhaps was written as a defense of that doctrine. He believes, with the Enlightenment thinkers, in "the potential of indefinite human progress." He affirms the necessity to speak of "evolutionary progress." In spite of his perfunctory acknowledgment of the possibility of regression he speaks twice of the "Ratchet of Progress." He says that "humanity accepted the Ratchet of Progress" as if to suggest that we had a choice, but he doesn't say what we might have accepted instead. The practical effect of his belief in the inevitability of progress is to make him a poor critic of his own thought. In fact, for all his enthusiasm, he is a rather passive consumer of scientific platitudes. His idea of progress, for example, is both starkly deterministic (it is "evolutionary" and a "ratchet") and hazily romantic: Modern science, he says, is "driven by the faith that if we dream, press to discover, explain, and dream again, thereby plunging repeatedly into new terrain, the world will somehow come clearer and we will grasp the true strangeness of the universe. And the strangeness will all prove to be connected and make sense." Later, in his very sobering appraisal of our destruction of "the environment," he says, "We must plunge ahead and make the best of it, worried but confident of success. . . ."

This is utterly baffling. If our future is already determined by "evolutionary progress" and the "ratchet" is in place, there is no use in "plunging" anywhere—unless it is to exercise our "biologically adaptive" "illusion of free will." But if free will is an illusion, to what purpose do we make the world clearer? What Mr. Wilson evidently means by "plunging" is merely going ahead as we are going with our "really important research" and "following patronage." It is hard to see how any of this could be encouraging or useful to a conservationist.

If regression really is a possibility, then should we not watch for the signs of it? And should we not attempt to subtract regression from progression to get at least an approximate notion of net gain or net loss? Mr. Wilson concedes that people forget and die, but he says that "knowledge continues to expand globally while passing from one generation to the next." But in fact as knowledge expands globally it is being lost locally. This is the paramount truth of the modern history of rural places everywhere in the world. And it is the gravest problem of land use: Modern humans typically are using places whose nature they have never known and whose history they have forgotten; thus ignorant, they almost necessarily abuse what they use. If science has sponsored both an immensity of knowledge and an immensity of violence, what is the gain? If we "grasp the true strangeness of the universe" but forget how to farm, what is the gain?

Such questions, seriously asked and intelligently answered, lead directly to choices that people have the ability to make, but no such possibility is suggested in *Consilience.*

In Edward O. Wilson's view, the world is not a place where we all make in our daily lives intelligent or unintelligent choices affecting the future of the world. It is, rather, a place where the most genetically favored and the most richly subsidized scientists determine the future by "plunging ahead," each isolated in his or her vision of "new terrain," and each cut off from any restraining affection for old terrain.

Why should we trust them?

IV. Reduction and Religion

It is clearly bad for the sciences and the arts to be divided into "two cultures." It is bad for scientists to be working without a sense of obligation to cultural tradition. It is bad for artists and scholars in the humanities to be working without a sense of obligation to the world beyond the artifacts of culture. It is bad for both of these cultures to be operating strictly according to "professional standards," without local affection or community responsibility, much less any vision of an eternal order to which we all are subordinate and under obligation. It is even worse that we are actually confronting, not just "two cultures," but a whole ragbag of disciplines and professions, each with its own jargon more or less unintelligible to the others, and all saying of the rest of the world, "That is not my field."

The badness of all this is manifested first in the loss even of the pretense of intellectual or academic community. This is a loss increasingly ominous because intellectual engagement among the disciplines, across the lines of the specializations —that is to say *real* conversation—would enlarge the context of work; it would press thought toward a just complexity; it would work as a system of checks and balances, introducing criticism that would reach beyond the professional standards. Without such a vigorous conversation originating in the universities and emanating from them, we get what we've got: sciences that spread their effects upon the world as if the world were no more than an experimental laboratory; arts and "humanities" as unmindful of their influence as if the world did not exist; institutions of learning whose chief purpose is to acquire funds and be administered by administrators; governments whose chief purpose is to provide offices to members of political parties.

The ultimate manifestation of this incoherence is loss of trust—loss, moreover, of the entire cultural pattern by which we understand what it means to give and receive trust. The general assumption now is that everybody is working in his or her own interest and will continue to do so until checked by somebody whose self-interest is more powerful. That nobody now trusts the politicians or their governments is probably the noisiest of present facts. More quietly, people are withdrawing their trust from the professions, the corporations, the education system, the religious institutions, the medical industry. Perhaps no expert has yet assigned a quantitative value to trust; it is nonetheless certain that when we have finished subtracting trust from all we think we have gained, not much will be left.

And so it certainly is desirable—it probably is necessary—that the arts and the sciences should cease to be "two cultures" and become fully communicating, if not always fully cooperating, parts of one culture. (I believe, as I will show, that this culture when it comes will be in fact a mosaic of cultures, based upon every community's recognition that all its members have a common ground, and that this ground is the ground under their feet.) I have, therefore, not the slightest inclination to disagree with Mr. Wilson's wish for a "linkage of the arts and humanities." With his goal of "consilience," though I sympathize, I do not agree.

I do not agree because I do not think it is possible. I do not think it is possible because, as he defines it, it would impose the scientific methodology of reductionism upon cultural properties, such as religion and the arts, that are inherently alien to it, and that are often expressly resistant to reduction of any kind. Consilience, Mr. Wilson says, is "literally a 'jumping together' of knowledge by the linking of facts and fact-based theory across disciplines to create a common groundwork of explanation." And: "The only way either to establish or to refute consilience is by methods developed in the natural sciences—not . . . an effort led by scientists, or frozen in mathematical abstraction, but rather one allegiant to the habits of thought that have worked so well in exploring the material universe." The project of consilience, then, is not for scientists only, but it is only for science.

Whether or not science, religion, and the arts can be linked on "a common groundwork of explanation" depends upon a further question: Can religion and the arts be explained in the same way that science can be, or can they, in any comprehensive way, be explained at all? And this, it seems to me, depends upon another question that is even more important: Is knowledge by definition explainable, or is there such a thing as unexplainable knowledge?

I have in mind three statements that seem to me to test this issue of knowledge and explainability:

At the end of *King Lear*, the broken-hearted old king comes in with his faithful daughter Cordelia dead in his arms. He says: "Thou'lt come no more, / Never never never never never."

In II Samuel 18:33, David the king has just been told that his son, who has been his enemy, is dead. The King says: "O my son Absalom, my son, my son Absalom! would God I had died for thee, O Absalom, my son, my son!"

After the battle of Gettysburg, General Lee was overheard saying to himself, "Too bad! Too bad! Oh, too bad!"

These outcries "out of the depths" certainly express knowledge, and precisely too. They communicate knowledge. But the knowledge they convey cannot be proved, demonstrated, or explained; it cannot be taught or learned. These utterances are not "self-explanatory." They are as far as possible unlike what we now call "information." One either does or does not know what they mean. The idea of explaining them to someone who does not know is merely laughable.

Statements of religious faith seem to me to be of the same general kind. Job says: "I know that my redeemer liveth, and that he shall stand at the latter day upon the earth: And though . . . worms destroy this body, yet in my flesh shall I see God. . . ." This statement rests upon no evidence, no proof. It is not in any respectable sense a theory. Job calls it knowledge: He "knows" that what he says is true. A great many people who have read these verses have agreed; they too have known that this is so.

"The empiricist" in Mr. Wilson's chapter on "Ethics and Religion" would find Job's knowledge readily explainable as a "beneficent" falsehood, supported by no "objective evidence" or "statistical proofs." Mr. Wilson himself understands it as a genetically implanted "urge": "Perhaps . . . it can all eventually be explained as brain circuitry and deep, genetic history." People follow religion, he says, because it is "easier" than empiricism, the lab evidently being harder to bear than the cross. Mr. Wilson forgets, in calling attention to religion's want of statistical proofs, that empiricism can supply no statistical disproofs. His explanation of religion rather tends to prove that it is not explainable. God and the devices of human understanding are not the same subject.

Suppose, granting the hopelessness of empirical proof, that you took Job's statement of faith as seriously as Mr. Wilson wishes you to take empiricism; how, then, could you explain it to Mr. Wilson? It seems to me that you would have to concede —and here empirical evidence is available—that it could not be done.

His statement of his own "position" brings no clarification; though it is a statement of a faith somewhat less than scientific, for it has no proofs, it carefully does not touch the issue of religious faith: "I am an empiricist. On religion I lean toward deism but consider its proof largely *a problem* in astrophysics. The existence of a cosmological God who created the universe (as envisioned by deism) is *possible*, and *may eventually* be settled, *perhaps* by forms of material evidence *not yet* imagined. Or the matter may be forever beyond human reach. In contrast . . . the existence of a biological God, one who directs organic evolution and intervenes in human affairs (as envisioned by theism) is *increasingly* contravened by biology and the brain sciences." My emphases call attention to the extreme tentativeness of the thought. Mr. Wilson concedes on the same page, "I may be wrong," but that very concession exposes the hopelessness of the argument that he is proposing to settle by consilience. How could he be "proven" wrong? The faith of an empirical deist will probably have to wait a good while for proof or disproof by astrophysics. About as long, I imagine, as it will take the "increasing" evidence of biology and the brain sciences to culminate in empirical disproof of theism.

What is the difference between an "empirical" faith so hedged about and religious faith? One difference, to use Edwin Muir's terms, is that whereas religious faith is old, the empirical faith is merely new. A second difference is that religious faith has lived to grow old because to hundreds of generations it has appeared to rest upon a knowledge that is not empirical, whereas the empirical faith, as its language shows, rests only upon speculation.

There is no reason, as I hope and believe, that science and

religion might not live together in amity and peace, so long as they both acknowledge their real differences and each remains within its own competence. Religion, that is, should not attempt to dispute what science has actually proved; and science should not claim to know what it does not know, it should not confuse theory and knowledge, and it should disavow any claim on what is empirically unknowable.

The two cannot be reconciled by Mr. Wilson's consilience because consilience requires the acceptance of empiricism as a ruling dogma or orthodoxy, denying standing or consideration to any thought not subject to empirical proof. His proposed consilience, by attempting to impose on art and religion the methods and values of reductive science, would prolong the disunity and disintegration it is meant to heal. Like a naive politician, Mr. Wilson thinks he has found a way to reconcile two sides without realizing that his way is one of the sides. There is simply no reason for any person of faith to discuss consilience with Mr. Wilson. One cannot, in honesty, propose to reconcile Heaven and Earth by denying the existence of Heaven.

The danger of this sort of reconciliation, as twentieth-century politics has shown, is that whatever proposes to invalidate or abolish religion (and this is what consilience pretty openly proposes) is in fact attempting to put itself in religion's place. Science-as-religion is clearly a potent threat to freedom. Beyond that, it endangers real science. Science can function as religion only by making two unscientific claims: that it will *eventually* know everything, and that it will *eventually* solve all human problems. And here it is enough to note that at times Mr. Wilson allows the term "science" to become altogether too elastic.

Religion, as empiricists must finally grant, deals with a reality beyond the reach of empiricism. This larger reality does not manifest itself in the manner of laboratory results or in the manner of a newspaper front page. Christ does not come down from the cross and confound his tormentors, as good a movie as that would make. God does not speak loudly from Heaven in the most popular modern languages for all to hear. (If He did, we would have no need for science, or religion either.) It is nevertheless true that people believe in the existence of this larger reality, and accept religious truth as knowledge, because of their *experience.* John Milton, to whom Mr. Wilson so easily condescends, is only one of many poets in our tradition who wrote of an unevident reality, and who invoked the muse for aid in so great a task. The walls of the rational, empirical world are famously porous. What come through are dreams, imaginings, inspirations, visions, revelations. There is no use in stooping over these with a magnifying lens. Beyond any earthly reason we experience beauty in excess of use, justice in excess of anger, mercy in excess of justice, love in excess of deserving or fulfillment. We have known evil beyond imagining and seemingly beyond intention. We have known compassion and forgiveness beyond measure. And all of this is in excess of what Mr. Wilson means by "religion" and of what he means by "ethics."

Religion, it seems to me, has dealt with this reality clumsily enough, and that is why the history of a religion and its organizations is so frequently a blight on its teachings. But religion at least attempts to deal with religious experience on its own terms; it does not try to explain it by terms that are fundamentally alien to it. For thousands of years, for example, people (who were not dummies) have supposed that dreams come from outside the waking world, speaking to us at least some of the time, and however unclearly, of a reality beyond that world. Hamlet speaks for a lot of people, and very much to my point, when he says, "I could be bounded in a nutshell and count myself a king of infinite space were it not that I have bad dreams." The same, of course, is true of good dreams. Mr. Wilson says, typically, that "dreaming is a kind of insanity, a rush of visions, largely unconnected to reality . . . arbitrary in content . . . very likely a side effect of the reorganization and editing of information in the memory banks of the brain." Something of the sort, of course, may be said of inspiration, imagination, beauty, justice, mercy, and love—which consilience would require us to understand as mere strategies of survival encoded in our genes. But this kind of reduction is sufficiently answered by the fact that these things, thus explained, are no longer even conceptually what they were. Reduction does not necessarily limit itself to compacting and organizing knowledge; it also has the power to change what is known.

But biblical religion (which is the only religion that Mr. Wilson talks about) is also explicitly against reductionism. Mr. Wilson's spokesman "the empiricist" hauls out, as if he had thought of it himself, the most popular "environmental" cliché about Christianity: "With a second life waiting, suffering can be endured—especially in other people. The natural environment can be used up." This little platitude has passed from mouth to mouth for years, chewable but not swallowable. It is untrue. Nobody who has actually read the Gospels could believe it. It ignores the very point of the Incarnation. It ignores Christ's unfailing compassion for sufferers, whom He healed, one by one, as they came or were carried to Him. And there is nowhere in the Bible a single line that gives or implies a permission to "use up" the "natural environment."

On the contrary, the Bible says that between all creatures and God there is an absolute intimacy. All flesh lives by the spirit and breath of God. We "live, and move, and have our being" in God. In the Gospels it is a principle of faith that God's love for the world includes *every* creature individually, not just races or species. God knows of the fall of every sparrow; He has numbered "the very hairs of your head." Edgar was being perfectly scriptural when he said to his father, "Thy life's a miracle," and so was William Blake when he said that "everything that lives is holy." Julian of Norwich also was following scripture when she said that God "wants us to know that not only does he care for great and noble things, but equally for little and small, lowly and simple things as well." Stephanie Mills is witness to the survival of this tradition when she writes: "*A Sand County Almanac* is suffused with affection for distinct beings. . . ."

No attentive reader of the Bible can fail to see the writers' alertness to the individuality of things. The characters of humans are sharply observed and are appreciated for their unique qualities. And surely nobody, having read of him once, can forget the warhorse in Job 39:25, who "saith among the trumpets, Ha, ha." I don't know where you could find characterizations more deft and astute than those in the story of the resurrection in John 20:1–17. And again and again the biblical writers write of their pleasure and wonder in the "manifold" works of God, all keenly observed.

People who blame the Bible for the modern destruction of nature have failed to see its delight in the variety and individuality of creatures and its insistence upon their holiness. But that delight—in, say, the final chapters of Job or the 104th psalm—is far more useful to the cause of conservation than the undifferentiating abstractions of science. Empiricists fail to see how the language of religion (and I mean such language as I have quoted, not pulpit clichés) can speak of a non-empirical reality and convey knowledge, and how it can instruct those who use it in good faith. Reverence gives standing to creatures, and to our perception of them, just as the law gives standing to a citizen. Certain things appear only in certain lights. "The gods' presence in the world," Herakleitos said, "goes unnoticed by men who do not believe in the gods." To define knowledge as merely empirical is to limit one's ability to know; it enfeebles one's ability to feel and think.

We have come face to face with a paradox that we had better notice. Mr. Wilson's materialism is theoretical and reductionistic, tending, in his idea of consilience, toward "unity." People of faith, on the other hand, have always believed in the unity of truth in God, whose works are endlessly and countlessly various. There is a world of difference between this humanly unknowable unity of truth and Mr. Wilson's theoretical unity of knowledge, which supposes that mere humans can know, in some definitive or final way, the truth. And the results are wonderfully different: Acceptance of the mystery of unitary truth in God leads to glorification of the multiplicity of His works, whereas Mr. Wilson's goal of a cognitive unity produced by science leads to abstraction and reduction, the opposite of which is not synthesis. The principle that is opposite to reduction—and, when necessary, its sufficient answer—is God's love for all things, for each thing for its own sake and not for its category.

V. Reduction and Art

BY "THE ARTS" Mr. Wilson means "the creative arts, the personal productions of literature, visual arts, drama, music, and dance marked by those qualities which . . . we call the true and the beautiful." He says further that "The defining quality of the arts is the expression of the human condition by mood and feeling, calling into play all the senses, evoking both order and disorder." And he makes a strict distinction between science and the arts: "While biology has an important part to play in scholarly interpretation, the creative arts themselves can never be locked in by this or any other discipline of science. The reason is that the exclusive role of the arts is the transmission of the intricate details of human experience by artifice to intensify aesthetic and emotional response. Works of art communicate feeling directly from mind to mind, with no intent to explain why the impact occurs. In this defining quality, the arts are the antithesis of science.

"When addressing human nature, science is coarse-grained and encompassing, as opposed to the arts, which are fine-grained and interstitial. That is, science aims to create principles and use them in human biology to define the diagnostic qualities of the species; the arts use fine details to flesh out and make strikingly clear by implication those same qualities."

These proposed differences notwithstanding, Mr. Wilson argues that science and the arts can be brought into alignment or unity by "consilient explanation." The means of consilience is to be interpretation, which is "the logical channel of consilient explanation between science and the arts." Two questions about the arts are "the central concern of interpretation": "where they come from in both history and personal experience, and how their essential qualities of truth and beauty are to be described through ordinary language." Interpretation of the arts needs to be reinvigorated "with the knowledge of science and its proprietary sense of the future." Mr. Wilson expects that, thus reinvigorated, interpretation will finally show (whether theoretically or by proof is not clear to me) that the arts originate in "an inborn human nature"—that is, in "the material processes of the human mind." Again, the mind is equated with the brain; the consilient explanation of the arts depends upon the explanation of the brain:

"If the brain is ever to be charted and an enduring theory of the arts created as a part of the enterprise, it will be by stepwise and consilient contributions from the brain sciences, psychology, and evolutionary biology. And if during this process the creative mind is to be understood, it will need collaboration between scientists and humanities scholars.

"The collaboration, now in its early stages, is likely to conclude that innovation is a concrete biological process founded upon an intricacy of nerve circuitry and neuro-transmitter release."

Great artists are genetically gifted, not by "singular neurobiological traits," but rather "by a quantitative edge in powers shared in smaller degree with those less gifted," and this quantitative edge produces works that are "qualitatively new." Art is to be accounted for both by genetic evolution and by cultural evolution, but cultural evolution is under the sway of "epigenetic rules of human nature" that draw creative minds toward "certain thoughts and behavior," which, in turn, "bias cultural evolution toward the invention of archetypes, the widely recurring abstractions and core narratives that are dominant themes in the arts." The most enduring works of art are those that are truest to their origins in human nature: "It follows that even the greatest works of art might be understood fundamentally with knowledge of the biologically evolved epigenetic rules that guided them."

Having tried conscientiously to summarize Mr. Wilson's "working hypothesis" of "the biological origin of the arts," I find a residue of statements that I don't understand well enough to include in my summary. For example, he says, "*The arts are innately focused toward certain forms and themes but are otherwise freely constructed*" [his italics]. If the forms and themes, especially the forms, are determined by innate predisposition, then it is not clear how much latitude there can be for freedom of construction. What is called for, apparently, is a sample analysis of a work of art, showing what is "innate" and what is "freely constructed," and how the innate and the free can be conjoined in a work that is "qualitatively new" when innovation has already been described as "a concrete biological process."

Nor am I able to understand the statement that the quality of the arts "is measured by . . . the precision of their adherence to human nature." If the forms and themes of the arts are determined by "an inborn human nature" ("the material processes of the human mind"), then how could they not adhere to it? In a naturally determined system, how can anything happen, or how is anything conceivable, that is unnatural? We need now an example of a work of art that does not adhere to human nature—which, if produced, would testify to the authenticity of that freedom of will which Mr. Wilson has said is illusory. But in a system of biological determinism, how does the issue of quality arise in the first place? If everything is originated biologically and free will is an illusion, then what we get is what we've got, qualitative standards are irrelevant, and critical judgment also is an illusion.

But even the parts of Mr. Wilson's "working hypothesis" that I am able to comprehend are frequently in error.

He is much mistaken, to begin with, in his wish to limit the arts to "expression of the human condition by mood and feeling" and to "aesthetic and emotional response." The arts, of course, "express" by their native means: words, colors, shapes, sounds, etc. They also include knowledge. They can instruct. Literature, at least, can convey facts, adduce evidence, and make arguments. *Paradise Lost*, which is the only work of literature that Mr. Wilson discusses at length, is for his thesis particularly unfortunate. Milton's purpose in that poem was avowedly *not* to express the human condition by mood and feeling. His purpose was, as he said, to "assert Eternal Providence, / And justifie the wayes of God to men." His poem is, among much else, a great argument. If you read *Paradise Lost*, you will certainly be obliged to feel and to experience moods and to respond aesthetically and emotionally, but you will also have to employ all of your mind to think and comprehend and to make critical judgments. Milton would have been indignant at the suggestion that his art was in any exclusive way "the antithesis of science."

Mr. Wilson would like to exclude science from art, which is easy to do, maybe, in theory, but harder in practice, when one considers how much the arts have been influenced by science and how often science has provided the subject matter of art. It would be a daunting critical exercise to subtract astronomy from *The Divine Comedy*, or biology from *Walden*.

But he would also like to exclude art from science. Though he speaks of the need for "collaboration between scientists and humanities scholars," it is hard to see what use he would have for the humanities scholars, except maybe to provide a little bibliography. His "working hypothesis" of "the biological origin of the arts" is strictly a scientific hypothesis, and it proposes only scientific tasks. Mr. Wilson's councils obviously could not include any humanities scholars who might, for example, take seriously Milton's faith, or his poetic purpose, or his invocation to the Heavenly Muse. The humanities scholars of choice would be those who would affirm Mr. Wilson's materialism, in which case Milton (and a host of other artists) would not be represented or would be misrepresented.

Since Mr. Wilson sees the arts as products of "gene-culture coevolution," he naturally sees them as serving the cause of "survival and reproduction." I am happy to concede him this point. Though I am not much impressed by evolution as the ultimate explanation of life, I am altogether convinced that the arts have helped us to survive and reproduce; to believe otherwise, I would have to deny the existence and the efficacy of love songs. But species survival alone does not adequately account for the existence of the arts, and (if quality is an issue) it does not provide an adequate standard of art criticism. "Survival value," it seems to me, must deal in minimums, since any species dependent upon maximums would be too vulnerable to survive. The human race has survived because of its ability to survive famine, not because of its ability to survive feasts. Survival is possible at minimal levels—in poverty, exile, concentration camps—and this ability merely to persist and endure undoubtedly owes much to instinct, to "inborn human nature," unlearned. But surviving is not the same thing, it is not as high an accomplishment, as the desire to go on living one's own life after surviving, say, defeat or famine or poverty or illness or grief. To live at a high level, desiring and aspiring throughout a human lifetime with its inevitable griefs and troubles, requires culture that, beyond any genetic determination or epigenetic rules, must be deliberately taught and learned. Obviously, the desire to live at a high level can have "survival value" also. Nevertheless, the desire to survive and the desire to live are two different desires, and the second is more conscious, more deliberate, more a matter of education and cultural choice than the first.

Mr. Wilson speaks of human nature as if it were *only* inborn, a product only of evolution. And so he has little choice but to speak of art in the same way. His fundamental error, in proposing his consilience of science and art, is his assumption that works of art are properties of nature in the same way that organisms are. (He thus extends his reductive formula to read: work of art = organism = machine.) He understands works of art as the products of "talent," not as artifacts, not as things made by arts which exist by being taught and learned. Once, he says that the masters of the arts have "exceptional knowledge" and "technical skill," but nowhere does he speak of the cultural continuum by which such knowledge and skill are kept alive and handed down. He is interested almost exclusively in the artists' "talent" and their "intuitive grasp of inborn human nature." He does not understand the arts as ways of making or works of art as made things. He asks two questions about the arts: where they come from and how their qualities can be described; he does not ask how they are made. And so he can think of the arts and human nature merely as "natural." He thinks of human nature as "inborn," not as both inborn and to be learned from (among other things) works of art. If human nature (and therefore all its manifestations, such as the arts) is merely natural or inborn, then it is merely a subject of study; no standards of judgment are necessary. If human nature is also the product of learning and is to some extent made by art, then critical judgment is both possible and necessary, and we must deal with issues of will and choice. We are ready to ask, for example, what may be the effect of our cultural and artistic choices upon the natural world. And at this point we can see the error of segregating the "fine" or "creative" arts from the arts that are practical or economic. Why should our universities sponsor an active criticism of the fine arts (by specialized or professional standards, ignoring their effect on the world) but no criticism of farming or forestry or mining or manufacturing? This question, of course, can be answered by a crude evolutionism—those who survive do not bite the corporations that feed them—but it ought to give some anxiety to a conservationist.

Finally, if innovation (the "qualitatively new") is a primary requirement for art, then why are we still interested in works that are no longer new?

Can science and the arts be "linked" by "a common groundwork of explanation"? The answer depends upon the extent to which the arts are reducible to explanation. Mr. Wilson's project of consilience depends upon his assumption that works of art can be rendered into "interpretations" that can then be aligned with the laws of biology and ultimately with the laws of physics. He assumes, in other words, that a sufficient response to a work of art is to "interpret" it, and moreover that the resulting interpretation is as good as or is equal to the work of art. He says that "criticism can be as inspired and idiosyncratic as the work it addresses." (It is consistent with his view of things that he should both deny the possibility of inspiration and use "inspired" as a term of praise, but how he reconciles the supposed intellectual virtue of idiosyncrasy with his zeal for reduction to laws and principles is not clear.) To propose that the value of a work of art lies in its interpretation is to propose further that it is of interest only as an instance or specimen and that it can be not only explained but explained away.

And of course this would be all right if works of art were so constructed as to have extractable meanings or principles or laws. The problem is that they are not so constructed, and in this way they are in fact much like organisms. A chickadee is not constructed to exemplify the principles of its anatomy or the laws of aerodynamics or the life history of its species, and it has not been explained when these things have been extracted (or subtracted) from it.

For a while, in thinking of this question, I proposed to myself that the only things really explainable are explanations. That is not quite true, but it is near enough to the truth that I am unwilling to forget it.

What can be explained? Experiments, ideas, patterns, cause-effect relationships and connections *within defined limits*, anything that can be calculated, graphed, or diagrammed. And yet explanation changes whatever is explained into something explainable. Explanation is reductive, not comprehensive; most of the time, when you have explained something, you discover leftovers. An explanation is a bucket, not a well.

What can't be explained? I don't think creatures can be explained. I don't think lives can be explained. What we know about creatures and lives must be pictured or told or sung or danced. And I don't think pictures or stories or songs or dances can be explained. The arts are indispensable precisely because they are so nearly antithetical to explanation.

The arts are constitutionally resistant to the reduction that Mr. Wilson wishes to subject them to. This resistance manifests itself in two ways: Art insists upon the irreducibility of its subjects; and works of art, as objects, are by nature not reducible.

The power of art tends to be an individuating power, and that tendency is itself an affirmation of the value of individuals and of individuality. It is true that in our literature we have some allegories such as the play *Everyman* and *Pilgrim's Progress*, in which the characters represent abstractions, but this genre, though it contains important works, is a minor one. The dominant flow of our artistic tradition rises from the Bible and from Homer's epics, great works of individualization, which pause to delineate the characters not just of heroes and seers but also of children, housewives, bureaucrats, prostitutes, and tax collectors, of swineherds and old nurses and animals. This tradition, both sacred and democratic, has given us Odysseus and Penelope, Eumaios and Eurykleia, King David and Mary Magdalene, the Wife of Bath, Dante and Virgil, King Lear and Rosalind, Corin the shepherd and Falstaff, Tom Jones, Emma Woodhouse and Mr. Knightley, Captain Ahab, Huckleberry Finn, Tess of the D'Urbervilles, Leopold and Molly Bloom, Joe Christmas and Lena Grove. However much these characters may "stand for" us humans in our quests, flights, trials, and follies, they are each also intransigently themselves, and are valued as such. They all come out of the common fund of human experience, and so we recognize them, but not one of them is the same as anybody else. This tradition gives us the true-to-life portraits of shepherds in Flemish nativity scenes. It is realized pointedly in many a painting of the Virgin, in which she is represented both as the mother of Christ and as whatever ordinary girl posed for the artist.

The truest tendency of art is toward the exaltation, not the reduction, of its subjects. The highest art, as William Blake said, is able

> To see a World in a Grain of Sand
> And a Heaven in a Wild Flower

To paint a convincing portrait of the Virgin is to realize that for Christ to be born into this world He had to have a human mother. To write believably of a pilgrimage from Hell to Heaven, or of the transfiguring destitution of Gloucester and Lear, is to require time to remember eternity. Mr. Wilson's science, on the contrary, cannot see a world in a grain of sand. It can classify, name, and (within limits) explain a grain of sand, and divide it into ever smaller parts. There is no reason to say that this work is not admirable, valuable, or useful. But there is reason to say that it is not equivalent to, and it does not replace, the imagination of William Blake. Blake's lines remind us again of the miraculousness of life. This news has been delivered to us time after time in our long tradition. It cannot be proved. It only can be told or shown.

All art that rises above competence insists upon the irreducibility of its subjects, its materials, and its finished works. It makes things that are inherently valuable in themselves and are not interchangeable with other things. To a merely competent carpenter, one sound board may be pretty much the same as another. But to a fine carpenter or cabinetmaker, every board is unique. The better artist a woodworker becomes, the more aware he or she becomes of the individuality of boards and of the differences between them. The increase of art accounts for the increase of perception.

In the same way, to be competent a farmer must know the nature of species and breeds of animals. But the better the farmer, the more aware he or she is of the animals' individuality. "Every one is different," you hear the good stockmen say. "No two are alike." The ideal of livestock breeding over the centuries has not been to produce clones. Recognition of "type" is certainly important. But paramount is the ability to recognize the outstanding individual.

The plainest and most emphatic denunciation of critical reductionism, and one that is generally ignored by critics, is the "Notice" posted at the beginning of *Huckleberry Finn*: "Persons attempting to find a motive in this narrative will be prosecuted; persons attempting to find a moral in it will be banished; persons attempting to find a plot in it will be shot." Mark Twain's point, I think, is not that his book had no motive or moral or plot, but rather that its motive, its moral, and its plot were peculiar to itself as a whole, and could be conveyed only by itself as a whole. The motive, the moral, and the plot were not to be extracted and studied piecemeal like the organs of a laboratory frog. And the reason for this is plain: The value of *Huckleberry Finn* is not in its motive or moral or plot, but in its language. The book is valuable because it is a story *told*, not a story explained.

Or the problem of reduction in art may be illustrated by the problem of translating a poem from one language to another. The problem is that the poetry is in the language, or is the language. We can certainly translate the "sense" of one language to another, but the question of how to translate a poem from one language to another is the same question as how to translate a language from one language to another, which cannot be done. And so translators of poetry must accept failure as the primary condition of their work. They must settle, at best, for second best: Their translations must succeed or fail as new poems in their own language which at the same time serve as approximations or shadows of the original poems. Nobody, I think, has ever believed that there was an equation between a translation and the original.

You cannot translate a poem into an explanation, any more than you can translate a poem into a painting or a painting into a piece of music or a piece of music into a walking stick. A work of art says what it says in the only way it can be said. Beauty, for example, cannot be interpreted. It is not an empirically verifiable fact; it is not a quantity. Artists and critics and teachers and students certainly ought to notice that some things are beautiful and some are not; they ought to ask, and learn if they can, the difference between beauty and ugliness; they should learn how beautiful things are made and how things are made beautiful; but they might as well not ask what are the equivalents of beauty in ideas or pulse rates or dollars or "ordinary language." To believe that the arts can be interpreted so as to make them consilient with biology or physics is about equivalent to the belief that literary classics can survive as comic books or movies.

The truth too, as it appears in art, cannot be extracted as an idea or paraphrase. If we didn't, to start with, feel that a work of art was true, we wouldn't bother with it, or not for long. I don't think we stand before Gerard David's *Annunciation*, in the Metropolitan Museum, as speculators of the truth. We don't say, Can this be true? or, Might it have happened this way? Either we see that in the painting it is happening, or we don't. If we don't, we pass on by. Of course, if we assent to the painting, and if we are responsible people, we finally must ask if it is true. We must measure it against our knowledge of other paintings and other visions of holy things. We must ask if we are being fooled, or are fooling ourselves. Nevertheless, the painting must be accepted or rejected as itself, not on the basis of interpretation or our opinion of Luke 1:28–35. The painting says what it says in the only way it can be said.

I don't mean at all to say that criticism is impossible, or that it cannot be useful. Obviously, we need to talk about works of art. We must test our ways of knowing about them. We must learn them and teach them and describe them and study the ways they are made. We must compare them with one another, and evaluate them by whatever standards we can make applicable. But a work of criticism is not equivalent to a work of art and cannot replace it. The English departments and the biology departments and all other would-be consilient departments can spend the next millennium interpreting *King Lear*, and at the end of all that work the interpretation will still be one kind of thing and *King Lear* will still be a thing of a different kind. And so Mr. Wilson's idea that the arts' "essential qualities of truth and beauty [can be] described through ordinary language" is not merely off the subject, as might first appear; it is a violation of sense. It is saying, in effect, that extraordinary language can be described in ordinary language. This is too flimsy a scaffold to hold up much in the way of art criticism.

The question remains, Is it science? If "science" means proven knowledge or a methodology for proving knowledge, then Mr. Wilson's chapter on "The Arts and Their Interpretation" is no closer to science than it is to art criticism. In the chapter, he makes this startling confession: "Gene-culture coevolution is, I believe, the underlying process by which the brain evolved and the arts originated. It is the conceivable means most consistent with the joint findings of the brain sciences, psychology, and evolutionary biology. Still, *direct* evidence with reference to the arts is slender." On this "I believe," this "conceivable," and this "slender" evidence, Mr. Wilson's enormous speculation teeters, like the Balanced Rock. It is not a reassuring place for a picnic.

VI. A Conversation Out of School

THE DISCIPLINES are different from one another, each distinct in itself, and rightly so. Science and art are neither fundamental nor immutable. They are not life or the world. They are tools. The arts and the sciences are our kit of cultural tools. Science cannot replace art or religion for the same reason that you cannot loosen a nut with a saw or cut a board in two with a wrench. The first question about the disciplines is not how they originated but how and for what they are to be used.

But if the sciences and the arts are divided into "two cultures," or into many subcultures, they are nobody's kit of tools. They are not the subjects of one conversation. They cannot be used in collaboration. And if they cannot be gathered together in one culture by consilience—which, on the evidence of Mr. Wilson's book, is not probable—then what can gather them together?

The only reason, really, that we need this kit of tools is to build and maintain our dwelling here on earth. (Those who wish to live or do business in other worlds should be free to depart, but not to return.) Our dwelling here is the proper work of culture. If the tools can be used collaboratively, then maybe we can find what are the appropriate standards for our work and can then build a good and lasting dwelling—which actually would be a diversity of dwellings suited to the diversity of homelands. If the tools cannot be so used, then they will be used to destroy such dwellings as we have accomplished so far, and our homelands as well.

To begin to think of the possibility of collaboration among the disciplines, we must realize that the "two cultures" exist as such because both of them belong to the one culture of division and dislocation, opposition and competition, which is to say the culture of colonialism and industrialism. This culture has steadily increased the dependence of individuals, regions, and nations upon larger and larger collective economies at the same time that it has thrown individuals, regions, and nations into a competitiveness with one another that is limitlessly destructive and demeaning. This state of universal competition understands the world as an anti-pattern in which each thing is opposed to every other thing, and it destroys the self-sufficiency of all places—households, farms, communities, regions, nations—even as it destroys the self-sufficiency of the world.

The collective economy is run for the benefit of a decreasing number of increasingly wealthy corporations. These corporations understand their "global economy" as a producer of money, not of goods. The goods of the world such as topsoil or forests must decline so that the money may increase. To facilitate this process, the corporations patronize the disciplines, chiefly the sciences, but some of the money, as "philanthropy," trickles down upon the arts. The brokers of this patronage are the universities, which are the organizers of the disciplines in our time. Since the universities are always a-building and are always in need of money, they accept the economy's fundamental principle of the opposition of money to goods. Having thus accepted as real the world as an anti-pattern of competing opposites, it is merely inevitable that they should organize learning, not as a conversation of collaborating disciplines, but as an anti-system of opposed and competing divisions. They have departmented our one great responsibility to live ably and generously into a nest of irresponsibilities. The sciences are sectioned like a stockyard the better to serve the corporations. The so-called humanities, which might have supplied at least a corrective or chastening remembrance of the good that humans have sometimes accomplished, have been dismembered into utter fecklessness, turning out "communicators" who have nothing to say and "educators" who have nothing to teach.

Must we reconcile ourselves to this cultural disintegration, this cacophony of the disciplines? Is it possible, failing consilience, to bring the arts and the sciences into healthful coherence and community of purpose? Edward O. Wilson would like to consiliate art and science on the terms of science—wrongly, I believe. The correct response is not to substitute the terms of art, or to look about for some hardly imaginable compromise.

The correct response, I think, is to ask if science and art are inherently at odds with one another. It seems obvious that they are not. To see that they are not may require extracurricular thought, but once we have cracked the crust of academic convention we can see that "science" means knowing and that "art" means doing, and that one is meaningless without the other. Out of school, the two are commonly inter-involved and naturally cooperative in the same person—a farmer, say, or a woodworker—who knows and does, both at the same time. It may be more or less possible to know and do nothing, but it is not possible to do and know nothing. One does as one knows. It is not possible to imagine a farmer who does not use both science and art.

It is also obvious that there is no insuperable natural or inherent division between scientists and artists—at least there is none outside of the academic pigeonholes. It is possible for a scientist and an artist to take part in the same conversation. On this subject I can speak from experience. I have been for the greater part of my life an artist of sorts, a cottage industrialist of literature, and for the past nineteen years I have been involved in a conversation with Wes Jackson, who is a scientist, a plant geneticist, and co-founder of The Land Institute in Salina, Kansas. This conversation has been, from the beginning, an alliance and a friendship. To me, it has been an indispensable source of instruction and a continuous testing of my thoughts. I can't speak for Wes, of course, and I make no claims as to the quality of our talk; the point here is only that we have been able to talk to each other out of our supposedly estranged disciplines, making our disciplines in the process useful to one another. In many meetings, telephone calls, notes, and letters during nineteen years, we have almost always had questions, and sometimes have had answers, for each other.

One of the most interesting facts about our conversation, from the standpoint of this essay, is that we were not prepared for it by our schooling. Wes's Ph.D. in genetics and my M.A. in English were not designed to give us things to say to each other. Because of his knowledge of the Bible and various works of literature, Wes was better prepared to talk to me than I to talk to him. Before I met him, I had been for perhaps fifteen years under the influence of the writings of the English agricultural scientist, Sir Albert Howard, and this was all that enabled me to understand Wes's germinal idea that, to be enduring, agriculture must imitate the local processes of nature. Though Wes is a writer (we both are essayists), I am not a scientist. I am perfectly ignorant of some things that Wes knows perfectly. While I have been writing, in addition to essays about agriculture, a series of fragments of the history of an imagined rural community, Wes has been at work on a project to renew agriculture by the development of perennial grain crops, in imitation of the native prairie plant communities, thereby reducing the amount of plowing necessary for food production, thereby reducing our presently ruinous rates of soil erosion. Obviously, two men so divergently occupied, and so divided by education, cannot talk together by any notion of the unification or consilience of their disciplines. And so what has made our conversation possible? The list of reasons amounts in implication to a fairly complete criticism of the present organization of the disciplines and the assumptions of that organization:

1. Though Wes and I were specialized, and maybe too much so, by our formal schooling, that schooling was superimposed upon an earlier, older education that we have in common. We both were raised in agrarian families. We were taught as children to know, respect, and love farming. From childhood until now, our thoughts about agriculture have been informed and conditioned by the actual work of farming, and this is work that we *like*.

2. Though we both have taught in universities, I more than Wes, neither of us has made a life in a university or in a "university community." We have lived in the countryside, among farming people, and have been involved in farming. In our minds, the problems we have talked about have always had the aspect of particular places and people, intimately known and cared about.

3. In ways sometimes different and sometimes the same, we have been at work on the same problem: how to change from a culture and a system of agriculture that destroy land and people to a culture and a system able to conserve both.

4. Because good land use involves both science and art (knowing and doing) and cannot be understood or practiced as either alone, we have had no illusions about the self-sufficiency or the adequacy of either of our disciplines. Our conversation has been between two parts of an always uncompleted whole.

It has lasted so long partly, of course, because it has been enjoyable, but also because it has been necessary.

5. The questions that have concerned us have been the same, and all of them raise the most practical issues of propriety: How can land and people be well used? What is good use? What is good knowledge, good thought, good work? How can one become genuinely and honorably native to one's place? This last question (the terms of which are set forth in Wes's book *Becoming Native to This Place*) is paramount. Asking it removes one permanently from the "two cultures" of careerist artists and scientists.

6. Neither of us believes that either art or science can be "neutral." Influence and consequence are inescapable. History continues. You cannot serve both God and Mammon, and you cannot work without serving one or the other.

7. Though each of us possesses the specialized vocabulary of his discipline, our conversation uses such talk only when necessary. We both can speak common English. Each of us, moreover, can speak a local English that is a source both of pleasure and exactitude. Our conversation is always striving to be local and particular. It is full of proper nouns, names of places and people. This subject of language is of the greatest importance, and I will have to return to it.

VII. *Toward a Change of Standards*

I HAVE JUST implied that a scientist and an artist may have to live and work outside the university if they are to have a sustained, mutually instructive, and effective conversation. Some will argue with this, and so be it. I will only point out that the modern university is organized to divide the disciplines; that universities pay little or no attention to the local and earthly effects of the work that is done in them; and that in the universities one discipline is rarely called upon to answer questions that might be asked of it by another discipline. If the universities sponsored an authentic conversation among the disciplines, then, for example, the colleges of agriculture would long ago have been brought under questioning by the college of arts and sciences or of medicine. A vital, functioning intellectual community *could* not sponsor patterns of land use that are increasingly toxic, violent, and destructive of rural communities.

I don't at all mean to suggest that I know how to reorganize the disciplines. I don't know how to do that, and I doubt that anybody does. But I feel no hesitation in saying that the standards and goals of the disciplines need to be changed. It used to be that we thought of the disciplines as ways of being useful to ourselves, for we needed to earn a living, but also and more importantly we thought of them as ways of being useful to one another. As long as the idea of vocation was still viable among us, I don't believe it was ever understood that a person was "called" to be rich or powerful or even successful. People were taught the disciplines at home or in school for two reasons: to enable them to live and work both as self-sustaining individuals and as useful members of their communities, and to see that the disciplines themselves survived the passing of the generations.

Now we seem to have replaced the ideas of responsible community membership, of cultural survival, and even of usefulness, with the idea of professionalism. Professional education proceeds according to ideas of professional competence and according to professional standards, and this explains the decline in education from ideals of service and good work, citizenship and membership, to mere "job training" or "career preparation." The context of professionalism is not a place or a community but a career, and this explains the phenomenon of "social mobility" and all the evils that proceed from it. The religion of professionalism is progress, and this means that, in spite of its vocal bias in favor of practicality and realism, professionalism forsakes both past and present in favor of the future, which is never present or practical or real. Professionalism is always offering up the past and the present as sacrifices to the future, in which all our problems will be solved and our tears wiped away—and which, being the future, never arrives. The future is always free of past limitations and present demands, always stocked with newer merchandise than any presently available, always promising that what we are going to have is better than what we have. The future is the utopia of academic thought, for virtually anything is hypothetically possible there; and it is the always-expanding frontier of the industrial economy, the fictive real estate against which losses are debited and to which failures are exiled. The future is not anticipated or provided for, but is only bought or sold. The present is ever diminished by this buying and selling of shares in the future that rightfully are owned by the unborn.

Wallace Stegner knew, both from his personal experience and from his long study of his region, that the two cultures of the American West are not those of the sciences and the arts, but rather those of the two human kinds that he called "boomers" and "stickers," the boomers being "those who pillage and run," and the stickers "those who settle, and love the life they have made and the place they have made it in." This applies to our country as a whole, and maybe to all of Western civilization in modern times. The first boomers were the oceanic navigators of the European Renaissance. They were gold seekers. All boomers have been gold seekers. They are would-be Midases who want to turn all things into gold: plants and animals, trees, food and drink, soil and water and air, life itself, even the future.

The sticker theme has so far managed to survive, and to preserve in memory and even in practice the ancient human gifts of reverence, fidelity, neighborliness, and stewardship. But unquestionably the dominant theme of modern history has been that of the boomer. It is no surprise that the predominant arts and sciences of the modern era have been boomer arts and boomer sciences.

The collaboration of boomer science with the boomer mentality of the industrial corporations has imposed upon us a state of virtually total economy in which it is the destiny of every creature (humans not excepted) to have a price and to be sold. In a total economy, all materials, creatures, and ideas become commodities, interchangeable and disposable. People become commodities along with everything else. Only such an economy could seek to impose upon the world's abounding geographic and creaturely diversity the tyranny of technological and genetic monoculture. Only in such an economy could "life forms" be patented, or the renewability of nature and culture be destroyed. Monsanto's aptly named "terminator gene"—which, implanted in seed sold by Monsanto, would cause the next generation of seed to be sterile—is as grave an indicator of totalitarian purpose as a concentration camp.

The complicity of the arts and humanities in this conquest is readily apparent in the enthusiasm with which the disciplines, schools, and libraries have accepted their ever-growing dependence (at public expense) on electronic technologies that are, in fact, as all of history shows, not necessary to learning or teaching, and which have produced no perceptible improvement in either. This was accomplished virtually without a dissenting voice, without criticism, without regard even for the economic cost. It is the clearest demonstration so far that the cult of originality and innovation is in fact a crowd of conformists, tramping on one another's heels for fear of being the last to buy whatever is for sale.

With the same ardor, and in more or less the same stampede, this crowd has mastered a slang of personal "liberation" that has done little for real freedom (which requires perception of authentic differences and distinctions), but has set many free from their rightful obligations and responsibilities—to, for example, their spouses and their children. The arts, especially in their well-paying popular versions, have become adept as permission givers for this sort of freedom. But there is too close a kinship between the personal freedom from reverence, fidelity, neighborliness, and stewardship and the corporate freedom to pollute and exterminate. When, if ever, the accounting is properly done, many of our present "liberties" and "necessities" will be seen to owe too much to the exploitation of "cheap" labor, raw materials, energy, and food.

The dominant story of our age, undoubtedly, is that of adultery and divorce. This is true both literally and figuratively: The dominant *tendency* of our age is the breaking of faith and the making of divisions among things that once were joined. This story obviously must be told by somebody. Perhaps, in one form or another, it must be told (because it must be experienced) by everybody. But how has it been told, and how ought it be told? This is a critical question, but not a question merely for art criticism. The story can be told in a way that clarifies, that makes imaginable and compassionable, the suffering and the costs; or it can be told in a way that seems to grant an easy permission and absolution to adultery and divorce. Can literature, for example, be written according to standards that are not merely literary? Obviously it can. And it had better be.

Suppose, then, that we should change the standards, as in fact some scientists and some artists already are attempting to do. Suppose that the ultimate standard of our work were to be, not professionalism and profitability, but the health and durability of human and natural communities. Suppose we learned to ask of any proposed innovation the question that so far only the Amish have been wise enough to ask: What will this do to our community? Suppose we attempted the authentic multiculturalism of adapting our ways of life to the nature of the places where we live. Suppose, in short, that we should take seriously the proposition that our arts and sciences have the power to help us adapt and survive. What then?

Well, we certainly would have a healthier, prettier, more diverse and interesting world, a world less toxic and explosive, than we have now.

And how might this come about? Again, I have to say that I don't know. I don't like or trust large, official programs of improvement, and I don't want to appear to be inviting any such thing. But perhaps there is no harm in making suggestions, if I acknowledge that the suggestions are only mine, and if I make sure that my suggestions apply primarily to the thinking, work, and conduct of individuals. Here is my list:

1. Rather than the present economic hierarchy of the professions, which results in the denigration and undercompensation of essential jobs of work, particularly in the economies of land use, we should think and work toward an appropriate subordination of all the disciplines to the health of creatures, places, and communities. A science or an art, for example, that served settlement rather than the exploitation of "frontiers" would be subordinate to reverence, fidelity, neighborliness, and stewardship, to affection and delight. It would aim to keep our creatureliness intact.

2. We should banish from our speech and writing any use of the word "machine" as an explanation or definition of anything that is not a machine. Our understanding of creatures and our use of them are *not* improved by calling them machines.

3. We should abandon the idea that this world and our human life in it can be brought by science to some sort of mechanical perfection or predictability. We are creatures whose intelligence and knowledge are not invariably equal to our circumstances. The radii of knowledge have only pushed back —and enlarged—the circumference of mystery. We live in a world famous for its ability both to surprise us and to deceive us. We are prone to err, ignorantly or foolishly or intentionally or maliciously. One of the oddest things about us is the interdependency of our virtues and our faults. Our moral code depends on our shortcomings as much as our knowledge. It is only when we confess our ignorance that we can see our need for "the law and the prophets." It is only because we err and are ignorant that we make promises, which we keep, not because we are smart, but because we are faithful.

4. We should give up the frontier and its boomer "ethics" of greed, cunning, and violence, and, so near too late, accept settlement as our goal. Wes Jackson says that our schools now have only one major, upward mobility, and that we need to

offer a major in homecoming. I agree, and would only add that a part of the sense of "homecoming" must be home*making*, for we now must begin sometimes with remnants, sometimes with ruins.

5. We need to require from our teachers, researchers, and leaders—and attempt for ourselves—a responsible accounting of technological progress. What have we gained by computers, for example, *after* we have subtracted the ecological costs of making them, using them, and throwing them away, the value of lost time and work when "the computers are down," and the enormous economic cost of the "Y2K" correction?

6. We ought conscientiously to reduce our tolerance for ugliness. Why, if we are in fact "progressing," should so much expense and effort have resulted in so much ugliness? We ought to begin to ask ourselves what are the limits—of scale, speed, and probably expense as well—beyond which human work is bound to be ugly.

7. We should recognize the insufficiency, to our life here among living creatures, of the abstract categories of reductionist thought. Resist classification! Without some use of abstraction, thought is incoherent or unintelligible, perhaps unthinkable. But abstraction alone is merely dead. And here we return again to the crucial issue of language.

In our public dialogue (such as it is) we are now using many valuable words that are losing their power of reference, and have as a consequence become abstract, merely gestures. I have in mind words such as "patriotism," "freedom," "equality," and "rights," or "nature," "human," "wild," and "sustainable." We could make a longish list of words such as these, which we often use without thought or feeling, just to show which side we suppose we want to be on. This situation calls for language that is not sloganish and rhetorical, but rather is capable of reference, specification, precision, and refinement—a language never far from experience and example. In the work of the great poets, the heavenly and the earthly are not abstract, but are *present*; the language of those poets is whole and precise. We are mistaken to think that we can increase our earthly knowledge by ignoring Heaven, or become more intelligent by giving up Dante or condescending to Milton. The middling, politically correct language of the professions is incapable either of reverence or familiarity; it is headless and footless, loveless, a language of nowhere.

I believe that this need for a whole, vital, particularizing language applies just as strongly to the sciences as to the arts and humanities. For the human necessity is not just to know, but also to cherish and protect the things that are known, and to know the things that can be known only by cherishing. If we are to protect the world's multitude of places and creatures, then we must know them, not just conceptually but imaginatively as well. They must be pictured in the mind and in memory; they must be known with affection, "by heart," so that in seeing or remembering them the heart may be said to "sing," to make a music peculiar to its recognition of each particular place or creature that it knows well. I am remembering here the importance to Confucius of "the tones given off by the heart." To know imaginatively is to know intimately, particularly, precisely, gratefully, reverently, and with affection.

In *Consilience*, Edward O. Wilson says, "Today the entire planet has become home ground," but that is a conceptual statement only, and doubtful as such. No human has ever known, let alone imagined, the entire planet. And even in an age of "world travel," none of us lives on the entire planet; in fact, owing to so much mobility, a lot of people (as some of them will tell you) don't live anywhere. But if we are to know any part of the planet intimately, particularly, precisely, and with affection, then we must live somewhere in particular for a long time. We must be able to call up to the mind's eye by name a lot of local places, people, creatures, and things.

One of the most significant costs of the economic destruction of farm populations is the loss of local memory, local history, and local names. Field names, for instance, even such colorless names as "the front field" and "the back field," are vital signs of a culture. If the arts and the sciences ever waken from their rapture of academic specialization, they will make themselves at home in places they have helped to spoil, and set about reconstructing histories and remembering names.

8. We should value familiarity above innovation. Boomer scientists and artists want to discover (so to speak) a place where they have not been. Sticker scientists and artists want to know where they are. There is no reason that familiarity cannot be a goal just as worthy, demanding, and exciting as innovation—or, as I would argue, much more so. It would certainly give worthwhile employment to more people. And in fact its boundaries are much larger. Innovation is limited always by human ingenuity and human means; familiarity is limited only by the limits of life. The real infinitude of experience is in familiarity.

My own experience has shown me that it is possible to live in and attentively study the same small place decade after decade, and find that it ceaselessly evades and exceeds comprehension. There is nothing that it can be reduced to, because "it" is always, and not predictably, changing. It is never the same two days running, and the better one pays attention the more aware one becomes of these differences. Living and working in the place day by day, one is continuously revising one's knowledge of it, continuously being surprised by it and in error about it. And even if the place stayed the same, one would be getting older and growing in memory and experience, and would need for that reason alone to work from revision to revision. One knows one's place, that is to say, only within limits, and the limits are in one's mind, not in the place. This is a description of life in time in the world. A place, apart from our now always possible destruction of it, is inexhaustible. It *cannot* be altogether known, seen, understood, or appreciated.

That Cézanne returned many times to paint Mont Sainte-Victoire, or that William Carlos Williams spent a long life writing about Rutherford, New Jersey, does not mean that those places are now exhausted as subjects. It means that they are inexhaustible. There are many examples of this. One that I have kept in mind for nearly forty years, to define a hope and a consolation, is that of the French entomologist Jean Henri Fabre. Too poor to travel, Fabre spent the last thirty-odd years of his life studying the insects and other creatures of his small *harmas* near Sérignan, "a patch of pebbles enclosed within four walls." And surely his enthusiasm lasted so long because he studied the living creatures in their—and his—dwelling place.

There is nothing intrinsically wrong with an interest in discovery and innovation. It only becomes wrong when it is thought to be the norm of culture and of intellectual life. As such it is in the first place misleading. As I have already suggested, the effort of familiarity is always leading to discovery and the new, just as do the quests of explorers and "original researchers." The difference is that innovation for its own sake, and especially now when it so directly serves the market, is disruptive of human settlement, whereas the revelations of familiarity elaborate the local cultural pattern and tend toward settlement, which they also prevent from becoming static.

If local adaptation is important, as I believe it unquestionably is, then we must undertake, in both science and art, the effort of familiarity. In doing so, we will confront the endlessness of human knowledge, work, and experience. But we should not mislead ourselves: We will confront mystery too. There is more to the world, and to our own work in it, than we are going to know.

One of the best studies of local adaptation that I know is a book by George Sturt, *The Wheelwright's Shop*, which looks at traditional farm carts and wagons as products of "the age-long effort of Englishmen to fit themselves close and ever closer into England." Sturt understands these vehicles as distinctly local products, whose form and fabric evolved in a long, only partly conscious give and take between the people and the landscape. The accommodation inherent in the design was elegant, though not in any sure sense explainable:

> But where begin to describe so efficient an organism, in which all the parts interacted until it is hard to say which was modified first, to meet which other? Was it to suit the horses or the ruts, the loading or the turning, that the front wheels had to have a diameter of about four feet? Or was there something in the average height of a carter, or in the skill of wheel-makers, that fixed these dimensions? One only knew that, by a wonderful compromise, all these points had been provided for in the country tradition of fore-wheels for a waggon. And so all through. Was it to suit these same wheels that the front of a waggon was slightly curved up, or was that

> done in consideration of the loads, and the circumstance merely taken advantage of for the wheels? I could not tell. I cannot tell. I only know that in these and a hundred details every well-built farm-waggon (of whatever variety) was like an organism, reflecting in every curve and dimension some special need of its own countryside, or, perhaps, some special difficulty attending wheelwrights with the local timber.

This is the way a locally adapted culture works. Over a long time it learns to conform its artifacts to the local landscape, local circumstances, and local needs. This is exactly opposite to the way of industrialism, which forces the locality to conform to industrial artifacts, always with the most dreadful consequences to the locality. Having in the industrial age exchanged, as Sturt says, "local needs . . . for cosmopolitan wishes," we are a long way now from the saving elegance that his book recalls. And the most resolute and expensive projects of discovery and innovation on the part of science-technology-and-industry cannot take us there. Only a long, patient, loving effort of familiarity can do that.

Hanging his project from an "if," as usual, Edward O. Wilson looks forward in *Consilience* to "a Magellanic voyage that eventually encircles the whole of reality." This presumably will be the long end run that will carry us and "the environment" over the goal line of survival.

But in an earlier book, *Biophilia*, Mr. Wilson set forth a far different conclusion, one which he has now evidently repudiated, but which I wish to affirm: "That the naturalist's journey has only begun and for all intents and purposes will go on forever. That it is possible to spend a lifetime in a magellanic voyage around the trunk of a single tree."

A single tree? Well, life is a miracle and therefore infinitely of interest everywhere. We have perhaps sufficient testimony, from artists and scientists both, that if we watch, refine our intelligence and our attention, curb our greed and our pride, work with care, have faith, a single tree might be enough.

VIII. *Some Notes in Conclusion*

IN THE PROCESS that carries knowledge from the laboratory to the market there is not enough fear. And in the history of that process there has not been adequate accounting.

Richard Strohman has made clear what is objectionable about the infusion of money from biotechnology corporations into the universities: These grants will press university scientists in the direction of product development; an interest in product development increases the emphasis upon predictability (you cannot market, or not for long, an unpredictable product); but an undue emphasis upon predictability will tend to narrow the context of experimentation, making the product, in the context of the world, unpredictable in effect and influence.

This process has a historical analogue in the introduction of the internal combustion engine into agriculture. In the commercial workshops tractors had only to pass the test of mechanical correctness: They had to start and run more or less predictably. In the context of the world, however, these machines had effects and exerted influences that far surpassed their merely mechanical limits. They replaced agriculture's old dependence on the free energy of the sun with a dependence on purchased energy; in general, they increased farming's dependence on a supply economy that farmers cannot control or influence; over the years, these dependences have radically oversimplified the patterns of farming, replacing diversity with monoculture, crop rotation with continuous tillage, and human labor with machines and chemicals; they have replaced (in Wes Jackson's words) nature's wisdom with human cleverness; they have caused widespread, profound social and cultural disruption. All these changes are still in progress. Whatever the technological or quantitative gains, this industrialization of farming has been costly, and it will continue to be. Most of the costs have been "externalized"—that is, charged to nature or the public or the future.

The response to this in the land grant universities has been applause.

*

The faculties and administrations of universities are inexcusably bewildered between the superstition that knowledge is invariably good and the fact that it can be monetarily valuable and also dangerous.

There is in fact no reason to think that the professions are self-correcting, or that new knowledge necessarily compensates for old error.

*

The time is past, if ever there was such a time, when you can just discover knowledge and turn it loose in the world and assume that you have done good.

This, to me, is a sign of the incompleteness of science in itself—which is a sign of the need for a strenuous conversation among all the branches of learning. This is a conversation that the universities have failed to produce, and in fact have obstructed.

*

If we were as fearful of our knowledge and our power as in our ignorance we ought to be—and as our cultural and religious traditions instruct us to be—then we would be trying to reconnect the disciplines both within the universities and in the conduct of the professions.

*

In our present economic predicament, ethics, ecology, environmental law, etc. won't *as specialties* have much corrective force. They will be used to rationalize what is wrong.

*

The anti-smoking campaign, by its insistent reference to the expensiveness to government and society of death by smoking, has raised a question that it has not answered: What is the best and cheapest disease to die from, and how can the best and cheapest disease best be promoted?

*

An idea of health that does not generously and gracefully accommodate the fact of death is obviously incomplete. The crudest manifestation of modern medicine is its routine, stubborn, and finally cruel resistance to death. This comes of the refusal to accept death not only as part of health, which it demonstrably is, but also as a great mystery both in itself and as a part of the mystery that surrounds us all our lives. The medical industry's resistance is only sometimes an instance of scientific heroism; sometimes it is the fear of what we don't know anything about.

*

Science can teach us and help us to resist death, but it can't teach us to prepare for death or to die well.

The question of how you want to die is somewhat fantastical but nonetheless it is one that all the living need to consider, one that belongs to the issue of health, and one that health science can't answer.

Do you want to die at home with your people "in blessed peace around you," which is the death Tiresias foresaw for Odysseus and the one Homer seems to recommend?

Or do you want to die in the hands of the best medical professionals wherever they are?

Such questions may seem irrelevant until you realize that they define two very different lives.

*

The refusal of modern medicine to confront the deaths of its patients is only a function of its refusal to confront the unique and unempirically precious lives of its patients.

Analogous to that is modern agriculture's refusal to confront the life of a good or healthy farm as a cultural artifact, unique in place and character, complex in form, mysterious in its sources, and many times more fascinating and precious than the "unit of production" to which it has been reduced by the economics of "agribusiness" and the colleges of agriculture.

*

My worry is only partly about science-as-Pandora, an activity of questioning or curiosity which cannot undo the harm that it may do and sometimes does. (Science has armed us but it cannot disarm us.)

I am worried also about the *application* of science, which I think is generally crude. This cannot be solved merely by keeping the context of research as large as it actually is in application—which, of course, would be a sensible precaution. However large the context, however generous the acknowledgment of context, the results of the research still cannot be applied *both* generally and sensitively. Finally it is "brought home" to a specific community of persons and creatures in a specific place. If it is then applied in its abstract or generalized or marketable form, it will obscure the uniquenesses of the subject persons or creatures or places, or of their community, and this sort of application is almost invariably destructive.

The only remedy I can see is for scientists (and artists also) to understand and imagine themselves as members of, and sharers in, the fate of affected communities. Our schools now encourage people to regard as mere privileges the power and influence that they call leadership. But leadership without membership is a terrible thing.

*

The issue of the application of science is a political issue. If the science is applied only by, or can only be applied by, a government or a large corporation, then it is tyrannical. If it is to be applied, it should be applied locally by local people on a local scale, using the health of the locality as the standard of application and judgment.

The use of science by or upon people who do not understand it is always potentially tyrannical, and it is always dangerous.

*

Applying knowledge—scientific or otherwise—is an art. An artist is somebody who knows what to put where, and when to put it. A good artist is one who applies knowledge skillfully and sensitively to the particular creatures and places of the world. Good farmers and good architects and good doctors are the most obvious examples, but the same potential is in all the arts. This is why it is such a shame to see the so-called fine arts elaborating themselves as academic or professional specialties without reference to much of anything, perhaps in imitation of the supposedly pure sciences.

*

The modern scientific enterprise apparently is directed toward the goal of complete knowledge. But if you had complete knowledge, if you knew everything, could you then act? Could you apply what you knew, or would you be paralyzed by a surplus of considerations? If you were to map within a circle all possible relationships among all the points along its circumference, you would end up with a black circle—an engorgement of "information" that would not be knowledge, but rather the practical equivalent of the blank circle you began with.

Thus the proposition that it would be good to know everything is probably false. The real question that is always to be addressed is the one that arises from our state of ignorance: How does one act well—sensitively, compassionately, without irreparable damage—on the basis of *partial* knowledge?

Perhaps the most proper, and the most natural, response to our state of ignorance is not haste to increase the amount of available information, or even to increase knowledge, but rather a lively and convivial engagement with the issues of form, elegance, and kindness. These issues of "sustainability" are both scientific and artistic.

*

The problem that we confront in our sciences and arts, as I understand it, is not a problem of information but a problem of ignorance. Or, to put it another way, the problem is not primarily one of mass; it is a problem of form.

It is out of our confrontation with our ignorance that we come to the problem of form. The ignorant must hope, and with study they may come to know, that it is possible to achieve forms that partake of wholeness and even holiness and make sense, even though one does not know everything.

*

Good artists are people who can stick things together so that they stay stuck. They know how to gather things into formal arrangements that are intelligible, memorable, and lasting. Good forms confer health upon the things that they gather together. Farms, families, and communities are forms of art just as are poems, paintings, and symphonies. None of these things would exist if we did not make them. We can make them either well or poorly; this choice is another thing that we make.

*

Always informing our practice of artistic form is our sense of the formality of creation. This "sense" is not knowledge of the empirical or scientific sort. It does not tend toward any sort of description. It is the perfectly assimilated, perfectly forgotten knowledge by which all creatures live in their places.

*

Overhanging all our thought and work is the question of how certain of itself human knowledge can be.

I learned from the theologian Philip Sherrard to ask this question: If things are evolving, and if human consciousness is evolving along with everything else, where do we find a standpoint from which to understand the whole process?

To make the same point in a more practical way, let us take that ubiquitous and misleading word "environment"—which, as used, proposes that reality is composed of a creature and its surroundings. But if, as in fact we know, the creature is not only *in* its environment but *of* it, and if the relationship between creature and environment is mutually formative, and if this relationship is a process that cannot be stopped short of the creature's death, then how can we get outside the relationship in order to predict with certainty the effects of our participation?

Religion begins with such questions. But even reason can see that they define the issues of propriety and scale. If we can't know with final certainty what we are doing, then reason cautions us to be humble and patient, to keep the scale small, to be careful, to go slow.

*

In speaking of the reductionism of modern science, we should not forget that the primary reductionism is in the assumption that human experience or human meaning can be adequately represented in any human language. This assumption is false.

To show what I mean, I will give the example that is most immediate to my mind:

My grandson, who is four years old, is now following his father and me over some of the same countryside that I followed my father and grandfather over. When his time comes, my grandson will choose as he must, but so far all of us have been farmers. I know from my grandfather that when he was a child he too followed his father in this way, hearing and seeing, not knowing yet that the most essential part of his education had begun.

And so in this familiar spectacle of a small boy tagging along behind his father across the fields, we are part of a long procession, five generations of which I have seen, issuing out of generations lost to memory, going back, for all I know, across previous landscapes and the whole history of farming.

Modern humans tend to believe that whatever is known can be recorded in books or on tapes or on computer discs and then again learned by those artificial means.

But it is increasingly plain to me that the meaning, the cultural significance, even the practical value, of this sort of family procession across a landscape can be known but not told. These things, though they have a public value, do not have a public meaning; they are too specific to a particular small place and its history. This is exactly the tragedy in the modern displacement of people and cultures.

That such things can be known but not told can be shown by answering a simple question: *Who* knows the meaning, the cultural significance, and the practical value of this rural family's generational procession across its native landscape? The answer is not so simple as the question: No one person ever will know all the answer. My grandson certainly does not know it. And my son does not, though he has positioned himself to learn some of it, should he be so blessed.

I am the one who (to some extent) knows, though I know also that I cannot tell it to anyone living. I am in the middle now between my grandfather and my father, who are alive in my memory, and my son and my grandson, who are alive in my sight.

If my son, after thirty more years have passed, has the good pleasure of seeing his own child and grandchild in that procession, then he will know something like what I now know.

This living procession through time in a place *is* the record by which such knowledge survives and is conveyed. When the procession ends, so does the knowledge.

FROM

CITIZENSHIP PAPERS

(2003)

A Citizen's Response to "The National Security Strategy of the United States of America"

> The constituent parts of a state are obliged to hold their public faith with each other, and with all those who derive any serious interest under their engagements, as much as the whole state is bound to keep its faith with separate communities. Otherwise competence and power would soon be confounded, and no law be left but the will of a prevailing force.
>
> —Edmund Burke, *On the Revolution in France*

> America! America!
> God mend thine every flaw,
> Confirm thy soul in self control,
> Thy liberty in law.
>
> —Katharine Lee Bates, "America the Beautiful"

> To seek for peace by way of war is the same as to seek for chastity by way of fornication.
>
> —Anonymous theologian of the first century

THE NEW "National Security Strategy" published by the White House in September 2002, if carried out, would amount to a radical revision of the political character of our nation. This document was conceived in reaction to the terrorist attacks of September 11, 2001. Its central and most significant statement is this:

> While the United States will constantly strive to enlist the support of the international community, we will not hesitate to act alone, if necessary, to exercise our right of self-defense by acting preemptively against such terrorists . . .

A democratic citizen, properly uneasy, must deal here first of all with the question, Who is this "we"? It is not the "we" of the Declaration of Independence, which referred to a small group of signatories bound by the conviction that "governments [derive] their just powers from the consent of the governed." And it is not the "we" of the Constitution, which refers to "*the people* [my emphasis] of the United States."

Because of what is implied by the commitment to act alone and preemptively, this "we" of the new strategy can refer only to the president. It is a royal "we." A head of state, preparing to act alone in starting a preemptive war, will need to justify his intention by secret information, and will need to plan in secret and execute his plan without forewarning. A preemptive attack widely known and discussed, as in a democratic polity, would risk being preempted by a preemptive attack by the other side. The idea of a government acting alone in preemptive war is inherently undemocratic, for it does not require or even permit the president to obtain the consent of the governed. As a policy, this new strategy depends on the acquiescence of a public kept fearful and ignorant, subject to manipulation by the executive power, and on the compliance of an intimidated and office-dependent legislature. Even within the narrow logic of warfare, there is a substantial difference between a defensive action, for which the reason would be publicly known, and a preemptive or aggressive action, for which the reason would be known only by a few at the center of power. The responsibilities of the president obviously are not mine, and so I hesitate to doubt absolutely the necessity of governmental secrecy. But I feel no hesitation in saying that to the extent that a government is secret, it cannot be democratic or its people free. By this new doctrine, the president alone may start a war against any nation at any time, and with no more forewarning than preceded the Japanese attack on Pearl Harbor.

Would-be participating citizens of a democratic nation, unwilling to have their consent coerced or taken for granted, therefore have no choice but to remove themselves from the illegitimate constraints of this "we" in as immediate and public a way as possible.

But as this document and its supporters insist, we have now entered a new era when acts of war may be carried out not only by nations and "rogue nations," but also by individuals using weapons of mass destruction, and this requires us to give up some measure of freedom in return for some increase of security. The lives of every one of us may at any time be in jeopardy.

Even so, we need to ask: What does real security require of us? What does true patriotism require of us? What does freedom require of us?

The alleged justification for this new strategy is the recent emergence in the United States of international terrorism. But why the events of September 11, 2001, horrifying as they were, should have called for a radical new investiture of power in the executive branch is not clear.

"The National Security Strategy" defines terrorism as "premeditated, politically motivated violence perpetrated against innocents." This is truly a distinct kind of violence, but it is a kind old and familiar, even in the United States. All that was really new about the events of September 11, 2001, was that they raised the scale of such violence to that of "legitimate" modern warfare.

To imply by the word "terrorism" that this sort of terror is the work exclusively of "terrorists" is misleading. The "legitimate" warfare of technologically advanced nations likewise is premeditated, politically motivated violence perpetrated against innocents. The distinction between the *intention* to perpetrate violence against innocents, as in "terrorism," and the *willingness* to do so, as in "war," is not a source of comfort. We know also that modern war, like ancient war, often involves intentional violence against innocents.

A more correct definition of "terrorism" would be this: violence perpetrated unexpectedly without the authorization of a national government. Violence perpetrated unexpectedly *with* such authorization is not "terrorism" but "war." If a nation perpetrates violence officially—whether to bomb an enemy airfield or a hospital—it is not guilty of "terrorism." But there is no need to hesitate over the difference between "terrorism" and any violence or threat of violence that is terrifying. "The National Security Strategy" wishes to cause "terrorism" to be seen "in the same light as slavery, piracy, or genocide"—but not in the same light as war. It accepts and affirms the legitimacy of war.

This document concedes that "we are menaced less by fleets and armies than by catastrophic technologies in the hands of the embittered few." And yet our government, with our permission, continues to manufacture, stockpile, and trade in these catastrophic technologies, including nuclear, chemical, and biological weapons. In nuclear or biological warfare, in which we know we cannot limit effects, how do we distinguish our enemies from our friends—or our enemies from ourselves? Does this not bring us exactly to the madness of terrorists who kill themselves in order to kill others?

The official definition of "terrorism," then, is far too exclusive if we seriously wish to free the world of the terrors induced by human violence. But let us suppose that our opposition to terror could be justly or wisely limited to a "war against terrorism." How effective might such a war be?

The war against terrorism is not, strictly speaking, a war against nations, even though it has already involved international war in Afghanistan and presidential threats against other nations. This is a war against "the embittered few"—"thousands of trained terrorists"—who are "at large" among many thousands and even millions of others who are, in the language of this document, "innocents," and thus deserving of our protection.

Hunting these terrorists down will be like combing lice out of a head of hair. Unless we are willing to kill innocents in order to kill the guilty—unless we are willing to blow our neighbor's head off, or blow our own head off, to get rid of the lice—the need to be lethal will be impeded constantly by the need to be careful. Because of the inherent difficulties and because we must suppose a new supply of villains to be always in the making, we can expect the war on terrorism to be more or less endless, endlessly costly and endlessly supportive of a thriving bureaucracy.

Unless, that is, we should become willing to ask why, and to do something about the causes. Why do people become terrorists? Such a question is often dismissed as evidence of "liberal softness" toward malefactors. But that is not necessarily the case. Such a question may also arise from the recognition that problems have causes. There is, however, no acknowledgment in "The National Security Strategy" that terrorism might have a cause that could possibly be discovered and possibly remedied. "The embittered few," it seems, are merely "evil."

II

Much of the obscurity of our effort so far against terrorism originates in the now official idea that the enemy is evil and that we are (therefore) good, which is the precise mirror image of the official idea of the terrorists.

The epigraph of Part III of "The National Security Strategy" contains this sentence from President Bush's speech at the National Cathedral on September 14, 2001: "But our responsibility to history is already clear: to answer these attacks and rid the world of evil." A government, committing its nation to rid the world of evil, is assuming necessarily that it and its nation are good.

But the proposition that anything so multiple and large as a nation can be good is an insult to common sense. It is also dangerous, because it precludes any attempt at self-criticism or self-correction; it precludes public dialogue. It leads us far indeed from the traditions of religion and democracy that are intended to measure and so to sustain our efforts to be good. "There is none good but one, that is, God," Christ said. Also: "He that is without sin among you, let him first cast a stone at her." And Thomas Jefferson justified general education by the obligation of citizens to be critical of their government: "for nothing can keep it right but their own vigilant *and distrustful* [my emphasis] superintendence." An inescapable requirement of true patriotism, love for one's land, is a vigilant distrust of any determinative power, elected or unelected, that may preside over it.

And so it is not without reason or precedent that a would-be participating citizen should point out that in addition to evils originating abroad and supposedly correctable by catastrophic technologies in "legitimate" hands, we have an agenda of domestic evils, not only those that properly self-aware humans can find in their own hearts, but also several that are indigenous to our history as a nation: issues of economic and social justice, and issues related to the continuing and worsening maladjustment between our economy and our land.

There are kinds of violence that have nothing directly to do with unofficial or official warfare but are accepted as normal to our economic life. I mean such things as toxic pollution, land destruction, soil erosion, and the destruction of biological diversity, and of the degradation of ecological supports of agriculture. To anybody with a normal concern for health and sanity, these "externalized costs" are terrible and are terrifying.

I don't wish to make light of the threats and dangers that now confront us. There can be no doubt of the reality of terrorism as defined and understood by "The National Security Strategy," or of the seriousness of our situation, or of our need for security. But frightening as all this is, it does not relieve us of the responsibility to be as intelligent, principled, and practical as we can be. To rouse the public's anxiety about foreign terror while ignoring domestic terror, and to fail to ask if these terrors are in any way related, is wrong.

It is understandable that we should have reacted to the attacks of September 11, 2001, by curtailment of civil rights, by defiance of laws, and by resort to overwhelming force, for those actions are the ready products of fear and hasty thought. But they cannot protect us against the destruction of our own land by ourselves. They cannot protect us against the selfishness, wastefulness, and greed that we have legitimized here as economic virtues, and have taught to the world. They cannot protect us against our government's long-standing disdain for any form of self-sufficiency or thrift, or against the consequent dependence, which for the present at least is inescapable, on foreign supplies such as oil from the Middle East.

And they cannot protect us from what may prove to be the greatest danger of all: the estrangement of our people from one another and from our land. Increasingly, Americans—including, notoriously, their politicians—are not *from* anywhere. And so they have in this "homeland," which their government now seeks to make secure on their behalf, no home *place* that they are strongly moved to know or love or use well or protect.

*

It is no wonder that "The National Security Strategy," growing as it does out of unresolved contradictions in our domestic life, should attempt to compound a foreign policy out of contradictory principles.

There is, first of all, the contradiction of peace and war, or of war as the means of achieving and preserving peace. This document affirms peace; it also affirms peace as the justification of war and war as the means of peace—and thus it perpetuates a hallowed absurdity. But implicit in its assertion of this (and, by implication, any other) nation's right to act alone in its own interest is an acceptance of war as a permanent condition. Either way, it is cynical to invoke the ideas of cooperation, community, peace, freedom, justice, dignity, and the rule of law (as this document repeatedly does), and then proceed to assert one's intention to act alone in making war. One cannot reduce terror by holding over the world the threat of what it most fears.

All the things we supposedly want to secure are thus subverted by our proposed means of securing them. Edmund Burke recognized this contradiction and was wary of it: "Laws are commanded to hold their tongues amongst arms; and tribunals fall to the ground with the peace they are no longer able to uphold."

This is a contradiction not reconcilable except by a self-righteousness almost inconceivably naive. The authors write that "We will . . . use our foreign aid to promote freedom and support those who struggle *non-violently* [my emphasis] for it"; and they observe that

> In pursuing advanced military capabilities that can threaten its neighbors . . . China is following an outdated path that, in the end, will hamper its own pursuit of national greatness. In time, China will find that social and political freedom is the only source of that greatness.

Thus we come to the authors' implicit definition of "rogue state": any nation pursuing national greatness by advanced military capabilities that can threaten its neighbors—any nation, that is, except *our* nation.

If you think our displeasure with "rogue states" might have any underpinning in international law, then you will be disappointed to learn that

> We will take the actions necessary to ensure that our efforts to meet our global security commitments and protect Americans are not impaired by the potential for investigations, inquiry, or prosecution by the International Criminal Court (ICC), whose jurisdiction does not extend to Americans and which we do not accept.

The rule of law in the world, then, is to be upheld by a nation that has declared itself to be above the law. A childish hypocrisy here assumes the dignity of a nation's foreign policy. But if we perceive an illegitimacy in the catastrophic weapons and *ad lib* warfare of other nations, how can we not perceive the same illegitimacy in our own?

The contradiction between peace and war implies, of course, at every point a contradiction between security and war. We wish, this document says, to be cooperative with other nations, but the authors seem not to realize how rigidly our diplomacy and our offers to cooperate will be qualified by this new threat of overwhelming force to be used merely at the president's pleasure. We cannot hope to be secure when our government has declared, by its announced readiness "to act alone," its willingness to be everybody's enemy.

III

A further contradiction is that between war and commerce. This issue arises first of all in the war economy, which unsurprisingly regards war as a business and weapons as merchandise. However nationalistic may be the doctrine of "The National Security Strategy," the fact is that the business of warfare and the weapons trade have been thoroughly internationalized. Saddam Hussein possesses weapons of mass destruction, for example, partly because we sold him such weapons and the means of making them back when (madman or not) he was our "friend." But the internationalization of the weapons trade is a result inherent in international trade itself. It is a part of globalization. Mr. Bush's addition of this Security Strategy to the previous bipartisan commitment to globalization exposes an American dementia that has not been so plainly displayed before.

The America Whose Business Is Business has been internationalizing its economy in haste (for bad reasons, and with little foresight), looking everywhere for "trading partners," cheap labor, and tax shelters. Meanwhile, the America Whose Business Is National Defense is withdrawing from the world in haste (for bad reasons, with little foresight), threatening left and right, repudiating agreements, and angering friends. The problem of participating in the Global Economy for the benefit of Washington's corporate sponsors while maintaining a nationalist belligerence and an isolationist morality calls for superhuman intelligence in the Secretary of Commerce. The problem of "acting alone" in an international war while maintaining simultaneously our ability to import the foreign goods (for instance, oil) on which we have become dependent even militarily will call, likewise, for overtopping genius in the Secretary of Defense.

"The National Security Strategy" devotes a whole chapter to the president's resolve to "ignite a new era of global economic growth through free markets and free trade." But such a project cannot be wedded, even theoretically, to his commitment to a militarist nationalism ever prepared "to act alone." One must wonder when the government's corporate sponsors will see the contradiction and require the nation to assume a more humble posture in the presence of the global economy.

The conflict in future-abuse between this document and the sales talk of the corporations is stark, and it is pretty absurd. On the one hand, we have the future as a consumer's paradise in which everybody will be able to buy comfort, convenience, and happiness. On the other hand, we have this government's new future in which terrible things are bound to happen if we don't do terrible things in the present—which, of course, will make terrible things even more likely to happen in the future.

After World War II, we hoped the world might be united for the sake of peacemaking. Now the world is being "globalized" for the sake of trade and the so-called free market—for the sake, that is, of plundering the world for cheap labor, cheap energy, and cheap materials. How nations, let alone regions and communities, are to shape and protect themselves within this "global economy" is far from clear. Nor is it clear how the global economy can hope to survive the wars of nations.

If a nation cannot be "good" in any simple or incontestable way, then what can it reasonably be that is better than bad?

A nation can be charitable, as we can say with some confidence, for we need not look beyond our own for an example. Our nation, sometimes, has been charitable toward its own people; it has been kind to the elderly, the sick, the unemployed, and others unable to help themselves. And sometimes it has been charitable toward other nations, as when we helped even our onetime enemies to recover from World War II. But "charity" does not refer only to the institutional or governmental help we give to the "less fortunate." The word means "love." The commandment to "Love your enemies" suggests that charity must be without limit; it must include everything. A nation's charity must come from the heart and the imagination of its people. It requires us ultimately to see the world as a community of all the creatures, a community which, to be possessed by any, must be shared by all.

Perhaps that is only a better way of saying that a nation can be civilized. To be civil is to conduct oneself as a responsible citizen, honoring the lives and the rights of others. Our courts and jails are filled with the uncivil, who have presumed to act alone in their own interest. And we well know that incivility is now almost conventionally in business among us.

A nation can be independent, as our founders instructed us. If a nation cannot within reasonable measure be independent, it is hard to see how its existence might be justified. Though independence may at times require some sort of self-defense, it cannot be maintained by defiance of other nations or by making war against them. It can be maintained only by the most practical economic self-reliance. At the very least, a nation should be able sustainably to feed, clothe, and shelter its citizens, using its own sources and by its own work.

And of course that requires a nation to be, in the truest sense, patriotic: Its citizens must love their land with a knowing, intelligent, sustaining, and protective love. They must not, for any price, destroy its beauty, its health, or its productivity.

And they must not allow their patriotism to be degraded to a mere loyalty to symbols or any present set of officials.

A nation also can abide under the rule of law. Since the Alien and Sedition Laws of 1798, in times of national stress or emergency there have been arguments for the abridgement of citizenship under the Constitution. But the weakness of those arguments is in their invariable implication that a democracy such as ours can work only in the most favorable circumstances. If constitutional guarantees of rights and immunities cannot be maintained in unfavorable circumstances, what is their point or value? Their value in fact originates in the acknowledgment of their usefulness in the times of greatest difficulty and to those in greatest need, as does the value of international law.

It is impossible to think that constitutional government can be suspended in a time of danger, in deference to the greater "efficiency" of centralized power, and then easily or quickly restored. Efficiency may be a political virtue, but only if strictly limited. Our Constitution, by its separation of powers and its system of checks and balances, acts as a restraint upon efficiency by denying exclusive power to any branch of the government. The logic of governmental efficiency, unchecked, runs straight on, not only to dictatorship, but also to torture, assassination, and other abominations.

Such aims as charity, civility, independence, true patriotism, and lawfulness a nation of imperfect human beings may reasonably adopt as its standards. And we may conclude reasonably, rightly, and with no touch of self-contempt, that by those standards we are less charitable, less civil, less independent, less patriotic, and less law-abiding than we might be, and than we need to be. And do these shortcomings relate to the president's perception that we are less secure than we need to be? We would be extremely foolish to suppose otherwise.

One might reasonably assume that a policy of national security would advocate from the start various practical measures to conserve and to use frugally the nation's resources, the objects of this husbandry being a reduction in the nation's dependence on imports and a reduction in the competition between nations for necessary goods. One might reasonably expect the virtues of stewardship, thrift, self-sufficiency, and neighborliness to receive a certain precedence in the advocacy of political leaders. Since the country, to make itself secure, may be required to rely on itself, one might reasonably expect a due concern for the health and longevity of its soils, forests, and watersheds, its natural and its human communities, its domestic economy, and the natural systems on which that economy inescapably depends.

Such a concern could come only from perceiving the contradiction between national security and the present global economy, but there is no such perception in "The National Security Strategy." This document, ignoring all conflicts, proposes to go straight ahead with both projects: national security, which it defines forthrightly as isolationist, domineering, and violent; and the global economy, which it defines as international humanitarianism. It does allow that there is a connection between our national security and the economic condition of "the rest of the world," and that the extreme poverty of much of the rest of the world is "neither just nor stable." And of course one can only agree. But the authors assume that economic wrongs can be righted merely by "economic development" and the "free market," that it is the nature of these things to cure poverty, and that they will not be impeded by terrorism or a war against terrorism or a preemptive war for the security of one nation. These are articles of a faith available only to a politically sheltered economic elite.

As for conservation here and elsewhere, the authors provide a list of proposals that is short, vague or ambiguous, incomplete, and rather wildly miscellaneous:

> We will incorporate labor and environmental concerns into U.S. trade negotiations . . .
>
> We will . . . expand the sources and types of global energy . . .
>
> We will . . . develop cleaner and more energy efficient technologies.
>
> Economic growth should be accompanied by global efforts to *stabilize* [my emphasis] greenhouse gas concentrations . . .

> Our overall objective is to reduce America's greenhouse gas emissions relative to the size of our economy, cutting such emissions per unit of economic activity by 18 percent over the next 10 years . . .

But the only energy technologies specifically promoted here are those of "clean coal" and nuclear power—for both of which there is a strong corporate advocacy and a strong conservationist criticism or opposition. There is no mention of land loss, of soil erosion, of pollution of land, air, and water, or of the various threats to biological diversity—all problems of generally (and scientifically) recognized gravity.

Agriculture, which is involved with all the problems listed above, and with several others—and which is the economic activity most clearly and directly related to national security, if one grants that we all must eat—receives scant and superficial treatment amounting to dismissal. The document proposes only:

1. "a global effort to address new technology, science, and health regulations that needlessly impede farm exports and improved agriculture." This refers, without saying so, to the growing consumer resistance to genetically modified food. A global effort to overcome this resistance would help, not farmers and not consumers, but global agribusiness corporations such as Monsanto.
2. "transitional safeguards which we have used in the agricultural sector." This refers to government subsidies, which ultimately help the agribusiness corporations, not farmers.
3. promotion of "new technologies, including biotechnology, [which] have enormous potential to improve crop yields in developing countries while using fewer pesticides and less water." This is offered (as usual and questionably) as the solution to hunger, but its immediate benefit would be to the corporate suppliers of the technologies.

This is not an agriculture policy, let alone a national security strategy. It has the blindness, arrogance, and foolishness that are characteristic of top-down thinking by politicians and academic experts, assuming that "improved agriculture" would inevitably be the result of catering to the agribusiness corporations, and that national food security can be achieved merely by going on as before. It does not address any agricultural problem as such, and it ignores the vulnerability of our present food system—dependent as it is on genetically impoverished monocultures, cheap petroleum, cheap long-distance transportation, and cheap farm labor—to many kinds of disruption by "the embittered few," who, in the event of such disruption, would quickly become the embittered many. On eroding, ecologically degraded, increasingly toxic landscapes, worked by failing or subsidy-dependent farmers and by the cheap labor of migrants, we have erected the tottering tower of "agribusiness," which prospers and "feeds the world" (incompletely and temporarily) by undermining its own foundations.

But all our military strength, all our police, all our technologies and strategies of suspicion and surveillance cannot make us secure if we lose our ability to farm, or if we squander our forests, or if we exhaust or poison our water sources.

A policy of preemptive war that rests, as this one does, on such flimsy domestic underpinnings and on too great a concentration of power in the presidency, obviously risks or invites correction. And, as we know, correction can come by three means:

1. By strenuous public debate. But this requires a strong, independent political opposition, which at present we do not have. The country now contains many individuals and groups seriously troubled by issues of civil rights, food, health, agriculture, economy, peace, and conservation. These people have much in common, but they have no strong political voice, because few politicians have seen fit to speak for them.
2. By failure, as by some serious disruption of our food or transportation or energy systems. This might cause a principled and serious public debate, which might place us on a more stable economic and political footing. But we are a large nation, highly centralized in almost every way, and without much self-sufficiency, either national

or regional. Any failure, therefore, will be large and not nearly so easy to correct as to prevent.

3. By citizens' initiative. Responsibilities abandoned by the government properly are assumed by the people. An example of this is the already well-established movement for local economies, typically beginning with food. There is much hope in this effort, provided that it continues to grow. And we have, surviving from the Vietnam War, a surprisingly strong and numerous peace movement. As we learned from the Vietnam experience, the only effective answer to a secretive and unresponsive government is a citizens' revolt. The revolt against the war in Vietnam was nonviolent, effective, and finally successful. Such a correction is better than no correction, but it is far from ideal—far less to be preferred than correction by public debate. A citizens' revolt necessarily comes too late. And if it is not peaceable and responsibly led, it could easily destroy the things it is meant to save.

IV

The present administration has adopted a sort of official Christianity, and it obviously wishes to be regarded as Christian. But "Christian" war has always been a problem, best solved by avoiding any attempt to reconcile policies of national or imperial militarism with anything Christ said or did. The Christian gospel is a summons to peace, calling for justice beyond anger, mercy beyond justice, forgiveness beyond mercy, love beyond forgiveness. It would require a most agile interpreter to justify hatred and war by means of the Gospels, in which we are bidden to love our enemies, bless those who curse us, do good to those who hate us, and pray for those who despise and persecute us.

This peaceability has grown more practical—it has gained "survival value"—as industrial warfare has developed increasingly catastrophic weapons, which are abominable to our government, so far, only when other governments possess them. But since the end of World War II, when the terrors of industrial war had been fully revealed, many people and, by fits and starts, many governments have recognized that peace is not just a desirable condition, as was thought before, but is a practical necessity. It has become less and less thinkable that we might have a living and a livable world, or that we might have livable lives or any lives at all, if we do not make the world capable of peace.

And yet we have not learned to think of peace apart from war. We have received many teachings about peace and peaceability in biblical and other religious traditions, but we have marginalized those teachings, have made them abnormal, in deference to the great norm of violence and conflict. We wait, still, until we face terrifying dangers and the necessity to choose among bad alternatives, and then we think again of peace, and again we fight a war to secure it.

At the end of the war, if we have won it, we declare peace; we congratulate ourselves on our victory; we marvel at the newly proved efficiency of our latest, most "sophisticated" weapons; we ignore the cost in lives, materials, and property, in suffering and disease, in damage to the natural world; we ignore the inevitable residue of resentment and hatred; and we go on as before, having, as we think, successfully defended our way of life.

That is pretty much the story of our victory in the Gulf War of 1991. In the years between that victory and September 11, 2001, we did not alter our thinking about peace and war, which is to say that we thought much about war and little about peace; we continued to punish the defeated people of Iraq and their children; we made no effort to reduce our dependence on the oil we import from other, potentially belligerent countries; we made no improvement in our charity toward the rest of the world; we made no motion toward greater economic self-reliance; and we continued our extensive and often irreversible damages to our own land. We appear to have assumed merely that our victory confirmed our manifest destiny to be the richest, most powerful, most wasteful nation in the world. After the catastrophe of September 11, it again became clear to us how good it would be to be at peace, to have no enemies, to have no needless death to mourn. And then, our need for war following with the customary swift and deadly logic our need for peace, we took up the customary obsession with the evil of other people.

And now we are stirring up the question whether or not Islam is a warlike religion, ignoring the question, much more urgent for us, whether or not Christianity is a warlike religion. There is no hope in this. Islam, Judaism, Christianity—all have been warlike religions. All have tried to make peace and rid the world of evil by fighting wars. This has not worked. It is never going to work. The failure belongs inescapably to all of these religions insofar as they have been warlike, and to acknowledge this failure is the duty of all of them. It is the duty of all of them to see that it is wrong to destroy the world, or risk destroying it, to get rid of its evil.

It is useless to try to adjudicate a long-standing animosity by asking who started it or who is the most wrong. The only sufficient answer is to give up the animosity and try forgiveness, to try to love our enemies and to talk to them and (if we pray) to pray for them. If we can't do any of that, then we must begin again by trying to imagine our enemies' children, who, like *our* children, are in mortal danger because of enmity that they did not cause.

We can no longer afford to confuse peaceability with passivity. Authentic peace is no more passive than war. Like war, it calls for discipline and intelligence and strength of character, though it calls also for higher principles and aims. If we are serious about peace, then we must work for it as ardently, seriously, continuously, carefully, and bravely as we have ever prepared for war.

Thoughts in the Presence of Fear

I. The time will soon come when we will not be able to remember the horrors of September 11 without remembering also the unquestioning technological and economic optimism that ended on that day.

II. This optimism rested on the proposition that we were living in a "new world order" and a "new economy" that would "grow" on and on, bringing a prosperity of which every new increment would be "unprecedented."

III. The dominant politicians, corporate officers, and investors who believed this proposition did not acknowledge that the prosperity was limited to a tiny percentage of the world's people, and to an ever smaller number of people even in the United States; that it was founded upon the oppressive labor of poor people all over the world; and that its ecological costs increasingly threatened all life, including the lives of the supposedly prosperous.

IV. The "developed" nations had given to the "free market" the status of a religion, and were sacrificing to it their farmers, farmlands, and rural communities, their forests, wetlands, and prairies, their ecosystems and watersheds. They had accepted universal pollution and global warming as normal costs of doing business.

V. There was, as a consequence, a growing worldwide effort on behalf of economic decentralization, economic justice, and ecological responsibility. We must recognize that the events of September 11 make this effort more necessary than ever. We citizens of the industrial countries must continue the labor of self-criticism and self-correction. We must recognize our mistakes.

VI. The paramount doctrine of the economic and technological euphoria of recent decades has been that everything depends on innovation. It was understood as desirable, and even as necessary, that we should go on and on from one technological innovation to the next, which would cause the economy to "grow" and make everything better and better. This of course implied at every point a hatred of the past, of all things inherited and free. All things superseded in our progress of innovations, whatever their value might have been, were discounted as of no value at all.

VII. We did not anticipate anything like what has now happened. We did not foresee that all our sequence of innovations might be at once overridden by a greater one: the invention of a new kind of war that would turn our previous innovations against us, discovering and exploiting the debits and the dangers that we had ignored. We never considered the possibility that we might be trapped in the webwork of communication and transport that was supposed to make us free.

VIII. Nor did we foresee that the weaponry and the war science that we marketed and taught to the world would become available, not just to recognized national governments which possess so uncannily the power to legitimate large-scale violence, but also to "rogue nations," dissident or fanatical groups, and individuals—whose violence, though never worse than that of nations, is judged by the nations to be illegitimate.

IX. We had apparently accepted the belief that technology is only good; that it cannot serve evil as well as good; that it cannot serve our enemies as well as ourselves; that it cannot be used to destroy what is good, including our homelands and our lives.

X. We had accepted too the corollary belief that an economy (either as a money economy or as a life-support system) that is global in extent, technologically complex, and centralized is invulnerable to terrorism, sabotage, or war, and that it is protectable by "national defense."

XI. We now have a clear, inescapable choice that we must make. We can continue to promote a global economic system of unlimited "free trade" among corporations, held together by long and highly vulnerable lines of communication and supply, but *now* recognizing that such a system will have to be protected by a hugely expensive police force that will be worldwide, whether maintained by one nation or several or all, and that such a police force will be effective precisely to the extent that it oversways the freedom and privacy of the citizens of every nation.

XII. Or we can promote a decentralized world economy that would have the aim of assuring to every nation and region a *local* self-sufficiency in life-supporting goods. This would not eliminate international trade, but it would tend toward a trade in surpluses after local needs have been met.

XIII. One of the gravest dangers to us now, second only to further terrorist attacks against our people, is that we will attempt to go on as before with the corporate program of global "free trade," whatever the cost in freedom and civil rights, without self-questioning or self-criticism or public debate.

XIV. This is why the substitution of rhetoric for thought, always a temptation in a national crisis, must be resisted by officials and citizens alike. It is hard for ordinary citizens to know what is actually happening in Washington in a time of such great trouble; for all we know, serious and difficult thought may be taking place there. But the talk that we are hearing from politicians, bureaucrats, and commentators has so far tended to reduce the complex problems now facing us to issues of unity, security, normality, and retaliation.

XV. National self-righteousness, like personal self-righteousness, is a mistake. It is misleading. It is a sign of weakness. Any war that we may make now against terrorism will come as a new installment in a history of war in which we have fully participated. We are not innocent of making war against civilian populations. The modern doctrine of such warfare was set forth and enacted by General William Tecumseh Sherman, who held that a civilian population could be declared guilty and rightly subjected to military punishment. We have never repudiated that doctrine.

XVI. It is a mistake also—as events since September 11 have shown—to suppose that a government can promote and participate in a global economy and at the same time act exclusively in its own interest by abrogating its international treaties and standing aloof from international cooperation on moral issues.

XVII. And surely, in our country, under our Constitution, it is a fundamental error to suppose that any crisis or emergency can justify any form of political oppression. Since September 11, far too many public voices have presumed to "speak for us" in saying that Americans will gladly accept a reduction of freedom in exchange for greater "security." Some would, maybe. But some others would accept a reduction in security (and in global trade) far more willingly than they would accept any abridgement of our constitutional rights.

XVIII. In a time such as this, when we have been seriously and most cruelly hurt by those who hate us, and when we must consider ourselves to be gravely threatened by those same people, it is hard to speak of the ways of peace and to remember that Christ enjoined us to love our enemies, but this is no less necessary for being difficult.

XIX. Even now we dare not forget that since the attack on Pearl Harbor—to which the present attack has been often and not usefully compared—we humans have suffered an almost uninterrupted sequence of wars, none of which have brought peace or made us more peaceable.

XX. The aim and result of war necessarily are not peace but victory, and any victory won by violence necessarily justifies the violence that won it and leads to further violence. If we are serious about innovation, must we not conclude that we need something new to replace our perpetual "war to end war"?

XXI. What leads to peace is not violence but peaceableness, which is not passivity, but an alert, informed, practiced, and active state of being. We should recognize that while we have extravagantly subsidized the means of war, we have almost totally neglected the ways of peaceableness. We have, for example, several national military academies, but not one peace academy. We have ignored the teachings and the examples of Christ, Gandhi, Martin Luther King, and other peaceable leaders. And here we have an inescapable duty to notice also that war is profitable, whereas the means of peaceableness, being cheap or free, make no money.

XXII. The key to peaceableness is continuous practice. It is wrong to suppose that we can exploit and impoverish the poorer countries, while arming them and instructing them in the newest means of war, and then reasonably expect them to be peaceable.

XXIII. We must not again allow public emotion or the public media to caricature our enemies. If our enemies are now to be some nations of Islam, then we should undertake to *know* those enemies. Our schools should begin to teach the histories, cultures, arts, and languages of the Islamic nations. And our leaders should have the humility and the wisdom to ask the reasons some of those people have for hating us.

XXIV. Starting with the economies of food and farming, we should promote at home and encourage abroad the ideal of local self-sufficiency. We should recognize that this is the surest, the safest, and the cheapest way for the world to live. We should not countenance the loss or destruction of any local capacity to produce necessary goods.

XXV. We should reconsider and renew and extend our efforts to protect the natural foundations of the human economy: soil, water, and air. We should protect every intact ecosystem and watershed that we have left, and begin restoration of those that have been damaged.

XXVI. The complexity of our present trouble suggests as never before that we need to change our present concept of education. Education is not properly an industry, and its proper use is not to serve industries, either by job-training or by industry-subsidized research. Its proper use is to enable citizens to live lives that are economically, politically, socially, and culturally responsible. This cannot be done by gathering or "accessing" what we now call "information"—which is to say facts without context and therefore without priority. A proper education enables young people to put their lives in order, which means knowing what things are more important than other things; it means putting first things first.

XXVII. The first thing we must begin to teach our children (and learn ourselves) is that we cannot spend and consume endlessly. We have got to learn to save and conserve. We do need a "new economy," but one that is founded on thrift and care, on saving and conserving, not on excess and waste. An economy based on waste is inherently and hopelessly violent, and war is its inevitable by-product. We need a peaceable economy.

The Failure of War

IF YOU KNOW even as little history as I know, it is hard not to doubt the efficacy of modern war as a solution to any problem except that of retribution—the "justice" of exchanging one damage for another, which results only in doubling (and continuing) the damage and the suffering.

Apologists for war will immediately insist that war answers the problem of national self-defense. But the doubter, in reply, will ask to what extent the *cost* even of a successful war of national defense—in life, money, material goods, health, and (inevitably) freedom—may amount to a national defeat. And national defense in war always involves *some* degree of national defeat. This paradox has been with us from the very beginning of our republic. Militarization in defense of freedom reduces the freedom of the defenders. There is a fundamental inconsistency between war and freedom.

In asking such a question, the doubter will be mindful also that in a modern war, fought with modern weapons and on the modern scale, neither side can limit to "the enemy" or "the enemy country" the damage that it does. These wars damage the world. We know enough by now to know that you cannot damage a part of the world without damaging all of it. Modern war has not only made it impossible to kill "combatants" without killing "noncombatants," it has made it impossible to damage your enemy without damaging yourself. You cannot kill your enemy's women and children without offering your own women and children to the selfsame possibility. We (and, inevitably, others) have prepared ourselves to destroy our enemy by destroying the entire world—including, of course, ourselves.

That many have considered the increasing unacceptability of modern warfare is shown by the language of the propaganda surrounding modern war. Modern wars have characteristically been fought to end war. They have been fought in the name of peace. Our most terrible weapons have been made, ostensibly, to preserve and assure the peace of the world. "All we want is peace," we say, as we increase relentlessly our capacity to make war.

And yet at the end of a century in which we have fought two wars to end war and several more to prevent war and preserve peace, and in which scientific and technological progress has made war ever more terrible and less controllable in its effects, we still, by policy, give no consideration, and no countenance, to nonviolent means of national defense. We do indeed make much of diplomacy and diplomatic relations, but by diplomacy we mean invariably ultimatums for peace backed by the threat of war. It is always understood that we stand ready to kill those with whom we are "peacefully negotiating."

Our century of war, militarism, and political terror has unsurprisingly produced great—and successful!—advocates of true peace, among whom Mohandas K. Gandhi and Martin Luther King, Jr., are paramount examples. The considerable success that they achieved testifies to the presence, in the midst of violence, of an authentic and powerful desire for peace and, more important, of the proven will to make the necessary sacrifices. But so far as our government is concerned, these men and their great and authenticating accomplishments might as well never have existed. To achieve peace by peaceable means is not yet our goal. We cling to the hopeless paradox of making peace by making war.

Which is to say that we cling, in our public life, to a brutal hypocrisy. In our century of almost universal violence of humans against fellow humans and against our natural and cultural commonwealth, hypocrisy has been inescapable because our opposition to violence has been selective or merely fashionable. Some of us who approve of our monstrous military budget and our peacekeeping wars nonetheless deplore "domestic violence" and think that our society can be pacified by "gun control." Some of us are against capital punishment but for abortion. Some of us are against abortion but for capital punishment. Most of us, whatever our stand on preserving the lives of the thoughtlessly conceived unborn, thoughtlessly participate in an economy that steals from all the unborn.

One does not have to know very much or think very far in order to see the moral absurdity upon which we have erected our sanctioned enterprises of violence. Abortion-as-birth-control is justified as a "right," which can establish itself only by denying all the rights of another person, which is the most primitive intent of warfare. Capital punishment sinks us all to the same level of primal belligerence, at which an act of violence is avenged by another act of violence. What the justifiers of these wrongs ignore is the fact—as well established by the history of feuds or the history of anger as by the history of war—that violence breeds violence. Acts of violence committed in "justice" or in affirmation of "rights" or in defense of "peace" do not end violence. They prepare and justify its continuation.

The most dangerous superstition of the parties of violence is the idea that sanctioned violence can prevent or control unsanctioned violence. But if violence is "just" in one instance, as determined by the state, why, by a merely logical extension, might it not also be "just" in another instance, as determined by an individual? How can a society that justifies capital punishment and warfare prevent its justifications from being extended to assassination and terrorism? If a government perceives that some causes are so important as to justify the killing of children, how can it hope to prevent the contagion of its logic from spreading to its citizens—or to its citizens' children? If you so devalue human life that the accidentally conceived unborn may be permissibly killed, how do you keep that permission from being assumed by someone who has made the same judgment against the born?

I am aware of the difficulty of assigning psychological causes to acts of violence. Psychological causes abound. But here I am talking about the power of example, precedent, and reason.

If we give to these small absurdities the magnitude of international relations, we produce, unsurprisingly, some much larger absurdities. What could be more absurd than our attitude of high moral outrage against other nations for manufacturing the selfsame weapons that we manufacture? The difference, as our leaders say, is that we will use these weapons virtuously whereas our enemies will use them maliciously—a proposition that too readily conforms to a proposition of much less dignity: We will use them in *our* interest, whereas our enemies will use them in *theirs.* The issue of virtue in war is as obscure, ambiguous, and troubling as Abraham Lincoln found to be the issue of prayer in war: "Both [the North and the South] read the same Bible, and pray to the same God; and each invokes His aid against the other. . . . The prayers of both could not be answered—that of neither could be answered fully."

But recent American wars, having been both "foreign" and "limited," have been fought under a second illusion even more dangerous than the illusion of perfect virtue: We are assuming, and are encouraged by our leaders to assume, that, aside from the sacrifice of life, no personal sacrifice is required. In "foreign" wars, we do not directly experience the damage that we inflict upon the enemy. We hear and see this damage reported in the news, but we are not affected, and we don't mind. These limited, "foreign" wars require that some of our young people will be killed or crippled, and that some families will grieve, but these "casualties" are so widely distributed among our population as hardly to be noticed. Otherwise, we do not feel ourselves to be involved. We pay taxes to support the war, but that is nothing new, for we pay war taxes also in time of "peace." We experience no shortages, we suffer no rationing, we endure no limitations. We earn, borrow, spend, and consume in wartime as in peacetime.

And of course no sacrifice is required of those large economic interests that now principally constitute our "economy." No corporation will be required to submit to any limitation or to sacrifice a dollar. On the contrary, war is understood by some as the great cure-all and opportunity of our corporate economy. War ended the Great Depression of the 1930s, and we have maintained a war economy—an economy, one might justly say, of general violence—ever since, sacrificing to it an enormous economic and ecological wealth, including, as designated victims, the farmers and the industrial working class.

And so great costs are involved in our fixation on war, but the costs are "externalized" as "acceptable losses." And here we see how progress in war, progress in technology, and progress in the industrial economy are parallel to one another—or, very often, are merely identical.

Romantic nationalists, which is to say most apologists for war, always imply in their public speeches a mathematics or an accounting that cannot be performed. Thus by its suffering in the Civil War, the North is said to have "paid for" the emancipation of the slaves and the preservation of the Union. Thus we may speak of our liberty as having been "bought" by the bloodshed of patriots. I am fully aware of the truth in such statements. I know that I am one of many who have benefited from painful sacrifices made by other people, and I would not like to be ungrateful. Moreover, I am a patriot myself, and I know that the time may come for any of us when we must make extreme sacrifices for the sake of liberty.

But still I am suspicious of this kind of accounting. For one reason, it is necessarily done by the living on behalf of the dead. And I think we must be careful about too easily accepting, or being too easily grateful for, sacrifices made by others, especially if we have made none ourselves. For another reason, though our leaders in war always assume that there is an acceptable price, there is never a previously stated level of acceptability. The acceptable price, finally, is whatever is paid.

It is easy to see the similarity between this accounting of the price of war and our usual accounting of "the price of progress." We seem to have agreed that whatever has been (or will be) paid for so-called progress is an acceptable price. If that price includes the diminishment of privacy and the increase of government secrecy, so be it. If it means a radical reduction in the number of small businesses and the virtual destruction of the farm population, so be it. If it means cultural and ecological impoverishment, so be it. If it means the devastation of whole regions by extractive industries, so be it. If it means that a mere handful of people should own more billions of wealth than is owned by all of the world's poor, so be it.

But let us have the candor to acknowledge that what we call "the economy" or "the free market" is less and less distinguishable from warfare. For about half of this century we worried about world conquest by international communism. Now with less worry (so far) we are witnessing world conquest by international capitalism. Though its political means are milder (so far) than those of communism, this newly internationalized capitalism may prove even more destructive of human cultures and communities, of freedom, and of nature. Its tendency is just as much toward total dominance and control. Confronting this conquest, ratified and licensed by the new international trade agreements, no place and no community in the world may consider itself safe from some form of plunder. More and more people all over the world are recognizing that this is so, and they are saying that world conquest of any kind is wrong, period.

They are doing more than that. They are saying that *local* conquest also is wrong, and wherever it is taking place local people are joining together to oppose it. All over my own state of Kentucky this opposition is growing—from the west, where the exiled people of the Land Between the Lakes are struggling to save their confiscated homeland from bureaucratic degradation, to the east, where the native people of the mountains are still struggling to preserve their land from destruction by absentee corporations.

To have an economy that is warlike, that aims at conquest, and that destroys virtually everything that it is dependent on, placing no value on the health of nature or of human communities, is absurd enough. It is even more absurd that this economy, that in some respects is so much at one with our military industries and programs, is in other respects directly in conflict with our professed aim of national defense.

It seems only reasonable, only sane, to suppose that a gigantic program of preparedness for national defense would be founded, first of all, upon a principle of national and even regional economic independence. A nation determined to defend itself and its freedoms should be prepared, and always preparing, to live from its own resources and from the work and the skills of its own people. It should carefully husband and conserve those resources, justly compensate that work, and rigorously cultivate and nurture those skills. But that is not what we are doing in the United States today. What we are doing, as we prepare for and prosecute wars allegedly for national defense, is squandering in the most prodigal manner the natural and human resources of the nation.

At present, in the face of declining finite sources of fossil fuel energy, we have virtually no energy policy, either for conservation or for the development of safe and clean alternative sources. At present, our energy policy simply is to use all that we have. At present, moreover, in the face of a growing population needing to be fed, we have virtually no policy for land conservation, and *no* policy of just compensation to the primary producers of food. At present, our agricultural policy is to use up everything that we have, while depending increasingly on imported food, energy, technology, and labor.

Those are just two examples of our general indifference to our own needs. We thus are elaborating a direct and surely a dangerous contradiction between our militant nationalism and our espousal of the international "free market" ideology. How are we going to defend our freedoms (this is a question both for militarists and for pacifists) when we must import our necessities from international suppliers who have no concern or respect for our freedoms? What would happen if in the course of a war of national defense we were to be cut off from our foreign sources of supply? What would happen if, in a war of national defense, military necessity required us to attack or blockade our foreign suppliers? We have already fought one energy war allegedly in national defense. If our present policies of economic indifference continue, we may face wars for other commodities: food or water or shoes or steel or textiles.

What can we do to free ourselves of this absurdity?

With that question my difficulty declares itself, for I do not ask it as a teacher knowing the answer. I ask it knowing that by doing so I describe my own dilemma. The news media, the industrial economy, perhaps human nature as well, prompt us to want quick, neat answers to our questions, but I don't think my question has a quick, neat answer.

Obviously, we would be less absurd if we took better care of one another and of all our fellow creatures. We would be less absurd if we founded our public policies upon an honest description of our needs and our predicament, rather than upon fantastical descriptions of our wishes. We would be less absurd if our leaders would consider in good faith the proven alternatives to violence.

Such things are easy to say. But finally we must face this daunting question, not as a nation or a group, but as individual persons—as ourselves. We are disposed, somewhat by culture and somewhat by nature, to solve our problems by violence—by maximum force relentlessly applied—and even to enjoy doing so. And yet by now all of us must at least have suspected that our right to live, to be free, and to be at peace is not guaranteed by any act of violence. It can be guaranteed only by our willingness that all other persons should live, be free, and be at peace—and by our willingness to use or give our own lives to make that possible. To be incapable of such willingness is merely to resign ourselves to the absurdity we are in; and yet, if you are like me, you are unsure to what extent you are capable of it.

It appears then that the answer to my question may be only another question. But maybe we can take some encouragement from that. Maybe, if our questions lead to other questions, that is a sign that we are asking the right ones.

Here is the other question that the predicament of modern warfare forces upon us: How many deaths of other people's children by bombing or starvation are we willing to accept in order that we may be free, affluent, and (supposedly) at peace? To that question I answer pretty quickly: *None.* And I know that I am not the only one who would give that answer: Please. No children. Don't kill any children for *my* benefit.

If that is our answer, then we must know that we have not come to rest. Far from it. For now surely we must feel ourselves swarmed about with more questions that are urgent, personal, and intimidating. But perhaps also we feel ourselves beginning to be free, facing at last in our own lives the greatest challenge ever laid before us, the most comprehensive vision of human progress, the best advice and the least obeyed:

"Love your enemies, bless them that curse you, do good to them that hate you, and pray for them which despitefully use you, and persecute you;

"That ye may be the children of your Father which is in Heaven: for He maketh His sun to rise on the evil and the good, and sendeth rain on the just and on the unjust."

(1999)

Postscript

I have been advised that "most readers" will object to my treatment of abortion in this essay. I don't know that most readers will object, but I am sure that some will. And so I will deal with this subject more plainly.

The issue of abortion, so far as I understand it, involves two questions: Is it killing? And what is killed?

It is killing, of course. To kill is the express purpose of the procedure.

What is killed is usually described by apologists for abortion as "a fetus," as if that term names a distinct kind or species of being. But what this being might be, if it is not a human being, is not clear. Generally, pregnant women have thought and spoken of the beings in their wombs as "babies." The attempt to make a categorical distinction between a baby living in the womb and a baby living in the world is as tenuous as would be an attempt to make such a distinction between a living child and a living adult. No living creature is "viable" independently of an enveloping life-support system.

If the creature in the womb is a living human being, and so far also an innocent one, then it is wrong to treat it as an enemy. If we are worried about the effects of treating fellow humans as enemies or enemies of society eligible to be killed, how do we justify treating an innocent fellow human as an enemy-in-the-womb?

As for the "right to control one's own body," I am all for that. But implicit in that right is the responsibility to control one's body in such a way as to avoid dealing irresponsibly or violently or murderously with other bodies.

Women and men generally have understood that when they have conceived a child they have relinquished a significant measure of their independence, and that henceforth they must control their bodies in the interest of the child.

(2003)

In Distrust of Movements

I MUST BURDEN MY READERS as I have burdened myself with the knowledge that I speak from a local, some might say a provincial, point of view. When I try to identify myself to myself I realize that, in my most immediate reasons and affections, I am less than an American, less than a Kentuckian, less even than a Henry Countian, but am a man most involved with and concerned about my family, my neighbors, and the land that is daily under my feet. It is this involvement that defines my citizenship in the larger entities. And so I will remember, and I ask you to remember, that I am not trying to say what is thinkable everywhere, but rather what it is possible to think on the westward bank of the lower Kentucky River in the summer of 1998.

Over the last twenty-five or thirty years I have been making and remaking different versions of the same argument. It is not "my" argument, really, but rather one that I inherited from a long line of familial, neighborly, literary, and scientific ancestors. We could call it "the agrarian argument." This argument can be summed up in as many ways as it can be made. One way to sum it up is to say that we humans can escape neither our dependence on nature nor our responsibility to nature—and that, precisely because of this condition of dependence *and* responsibility, we are also dependent upon and responsible for human culture.

Food, as I have argued at length, is both a natural (which is to say a divine) gift and a cultural product. Because we must *use* land and water and plants and animals to produce food, we are at once dependent on and responsible to what we use. We must know both how to use and how to care for what we use. This knowledge is the basis of human culture. If we do not know how to adapt our desires, our methods, and our technology to the nature of the places in which we are working, so as to make them productive *and to keep them so*, that is a cultural failure of the grossest and most dangerous kind. Poverty and starvation also can be cultural products—if the culture is wrong.

Though this argument, in my keeping, has lengthened and acquired branches, in its main assumptions it has stayed the same. What has changed—and I say this with a good deal of wonder and with much thankfulness—is the audience. Perhaps the audience will always include people who are not listening, or people who think the agrarian argument is merely an anachronism, a form of entertainment, or a nuisance to be waved away. But increasingly the audience also includes people who take this argument seriously, because they are involved in one or more of the tasks of agrarianism. They are trying to maintain a practical foothold on the earth for themselves or their families or their communities. They are trying to preserve and develop local land-based economies. They are trying to preserve or restore the health of local communities and ecosystems and watersheds. They are opposing the attempt of the great corporations to own and control all of Creation.

In short, the agrarian argument now has a significant number of friends. As the political and ecological abuses of the so-called global economy become more noticeable and more threatening, the agrarian argument is going to have more friends than it has now. This being so, maybe the advocate's task needs to change. Maybe now, instead of merely propounding (and repeating) the agrarian argument, the advocate must also try to see that this argument does not win friends too easily. I think, myself, that this is the case. The tasks of agrarianism that we have undertaken are not going to be finished for a long time. To preserve the remnants of agrarian life, to oppose the abuses of industrial land use and finally correct them, and to develop the locally adapted economies and cultures that are necessary to our survival will require many lifetimes of dedicated work. This work does not need friends with illusions. And so I would like to speak—in a friendly way, of course—out of my distrust of "movements."

I have had with my friend Wes Jackson a number of useful conversations about the necessity of getting out of movements—even movements that have seemed necessary and dear to us—when they have lapsed into self-righteousness and self-betrayal, as movements seem almost invariably to do. People in movements too readily learn to deny to others the rights and privileges they demand for themselves. They too easily become unable to mean their own language, as when a "peace movement" becomes violent. They often become too specialized, as if they cannot help taking refuge in the pinhole vision of the industrial intellectuals. They almost always fail to be radical enough, dealing finally in effects rather than causes. Or they deal with single issues or single solutions, as if to assure themselves that they will not be radical enough.

And so I must declare my dissatisfaction with movements to promote soil conservation or clean water or clean air or wilderness preservation or sustainable agriculture or community health or the welfare of children. Worthy as these and other goals may be, they cannot be achieved alone. They cannot be responsibly advocated alone. I am dissatisfied with such efforts because they are too specialized, they are not comprehensive enough, they are not radical enough, they virtually predict their own failure by implying that we can remedy or control effects while leaving the causes in place. Ultimately, I think, they are insincere; they propose that the trouble is caused by *other* people; they would like to change policy but not behavior.

The worst danger may be that a movement will lose its language either to its own confusion about meaning and practice, or to preemption by its enemies. I remember, for example, my naïve confusion at learning that it was possible for advocates of organic agriculture to look upon the "organic method" as an end in itself. To me, organic farming was attractive both as a way of conserving nature and as a strategy of survival for small farmers. Imagine my surprise in discovering that there could be huge "organic" monocultures. And so I was somewhat prepared for the recent attempt of the United States Department of Agriculture to appropriate the "organic" label for food irradiation, genetic engineering, and other desecrations by the corporate food economy. Once we allow our language to mean anything that anybody wants it to mean, it becomes impossible to mean what we say. When "homemade" ceases to mean neither more nor less than "made at home," then it means anything, which is to say that it means nothing. The same decay is at work on words such as "conservation," "sustainable," "safe," "natural," "healthful," "sanitary," and "organic." The use of such words now requires the most exacting control of context and the use immediately of illustrative examples.

Real organic gardeners and farmers who market their produce locally are finding that, to a lot of people, "organic" means something like "trustworthy." And so, for a while, it will be useful for us to talk about the meaning and the economic usefulness of trust and trustworthiness. But we must be careful. Sooner or later, Trust Us Global Foods, Inc., will be upon us, advertising safe, sanitary, natural food irradiation. And then we must be prepared to raise another standard and move on.

As you see, I have good reasons for declining to name the movement I think I am a part of. I call it The Nameless Movement for Better Ways of Doing—which I hope is too long and uncute to be used as a bumper sticker. I know that movements tend to die with their names and slogans, and I believe that this Nameless Movement needs to live on and on. I am reconciled to the likelihood that from time to time it will name itself and have slogans, but I am not going to use its slogans or call it by any of its names. After this, I intend to stop calling it The Nameless Movement for Better Ways of Doing, for fear it will become the NMBWD and acquire a headquarters and a budget and an inventory of T-shirts covered with language that in a few years will be mere spelling.

Let us suppose, then, that we have a Nameless Movement for Better Land Use and that we know we must try to keep it active, responsive, and intelligent for a long time. What must we do?

What we must do above all, I think, is try to see the problem in its full size and difficulty. If we are concerned about land abuse, then we must see that this is an economic problem. Every economy is, by definition, a land-using economy. If we are using our land wrong, then something is wrong with our economy. This is difficult. It becomes more difficult when we recognize that, in modern times, every one of us is a member of the economy of everybody else. Every one of us has given many proxies to the economy to use the land (and the air, the water, and other natural gifts) on our behalf. Adequately supervising those proxies is at present impossible; withdrawing them is for virtually all of us, as things now stand, unthinkable.

But if we are concerned about land abuse, we have begun an extensive work of economic criticism. Study of the history of land use (and any local history will do) informs us that we have had for a long time an economy that thrives by undermining its own foundations. Industrialism, which is the name of our economy, and which is now virtually the only economy of the world, has been from its beginnings in a state of riot. It is based squarely upon the principle of violence toward everything on which it depends, and it has not mattered whether the form of industrialism was communist or capitalist; the violence toward nature, human communities, traditional agricultures, and local economies has been constant. The bad news is coming in from all over the world. Can such an economy somehow be fixed without being radically changed? I don't think it can.

The Captains of Industry have always counseled the rest of us to "be realistic." Let us, therefore, be realistic. Is it realistic to assume that the present economy would be just fine if only it would stop poisoning the earth, air, and water, or if only it would stop soil erosion, or if only it would stop degrading watersheds and forest ecosystems, or if only it would stop seducing children, or if only it would quit buying politicians, or if only it would give women and favored minorities an equitable share of the loot? Realism, I think, is a very limited program, but it informs us at least that we should not look for bird eggs in a cuckoo clock.

Or we can show the hopelessness of single-issue causes and single-issue movements by following a line of thought such as this: We need a continuous supply of uncontaminated water. Therefore, we need (among other things) soil-and-water-conserving ways of agriculture and forestry that are not dependent on monoculture, toxic chemicals, or the indifference and violence that always accompany big-scale industrial enterprises on the land. Therefore, we need diversified, small-scale land economies that are dependent on people. Therefore, we need people with the knowledge, skills, motives, and attitudes required by diversified, small-scale land economies. And all this is clear and comfortable enough, until we recognize the question we have come to: *Where are the people?*

Well, all of us who live in the suffering rural landscapes of the United States know that most people are available to those landscapes only recreationally. We see them bicycling or boating or hiking or camping or hunting or fishing or driving along and looking around. They do not, in Mary Austin's phrase, "summer and winter with the land." They are unacquainted with the land's human and natural economies. Though people have not progressed beyond the need to eat food and drink water and wear clothes and live in houses, most people have progressed beyond the domestic arts—the husbandry and wifery of the world—by which those needful things are produced and conserved. In fact, the comparative few who still practice that necessary husbandry and wifery often are inclined to apologize for doing so, having been carefully taught in our education system that those arts are degrading and unworthy of people's talents. Educated minds, in the modern era, are unlikely to know anything about food and drink or clothing and shelter. In merely taking these things for granted, the modern educated mind reveals itself also to be as superstitious a mind as ever has existed in the world. What could be more superstitious than the idea that money brings forth food?

I am not suggesting, of course, that everybody ought to be a farmer or a forester. Heaven forbid! I *am* suggesting that most people now are living on the far side of a broken connection, and that this is potentially catastrophic. Most people are now fed, clothed, and sheltered from sources, in nature and in the work of other people, toward which they feel no gratitude and exercise no responsibility.

We are involved now in a profound failure of imagination. Most of us cannot imagine the wheat beyond the bread, or the farmer beyond the wheat, or the farm beyond the farmer, or the history (human or natural) beyond the farm. Most people cannot imagine the forest and the forest economy that produced their houses and furniture and paper; or the landscapes, the streams, and the weather that fill their pitchers and bathtubs and swimming pools with water. Most people appear to assume that when they have paid their money for these things they have entirely met their obligations. And that is, in fact, the conventional economic assumption. The problem is that it is possible to starve under the rule of the conventional economic assumption; some people are starving now under the rule of that assumption.

Money does not bring forth food. Neither does the technology of the food system. Food comes from nature and from the work of people. If the supply of food is to be continuous for a long time, then people must work in harmony with nature. That means that people must find the right answers to a lot of questions. The same rules apply to forestry and the possibility of a continuous supply of forest products.

People grow the food that people eat. People produce the lumber that people use. People care properly or improperly for the forests and the farms that are the sources of those goods. People are necessarily at both ends of the process. The economy, always obsessed with its need to sell products, thinks obsessively and exclusively of the consumer. It mostly takes for granted or ignores those who do the damaging or the restorative and preserving work of agriculture and forestry. The economy pays poorly for this work, with the unsurprising result that the work is mostly done poorly. But here we must ask a very realistic economic question: Can we afford to have this work done poorly? Those of us who know something about land stewardship know that we cannot afford to pay poorly for it, because that means simply that we will not get it. And we know that we cannot afford land use without land stewardship.

One way we could describe the task ahead of us is by saying that we need to enlarge the consciousness and the conscience of the economy. Our economy needs to know—and care—what it is doing. This is revolutionary, of course, if you have a taste for revolution, but it is also merely a matter of common sense. How could anybody seriously object to the possibility that the economy might eventually come to know what it is doing?

Undoubtedly some people will want to start a movement to bring this about. They probably will call it the Movement to Teach the Economy What It Is Doing—the MTEWIID. Despite my very considerable uneasiness, I will agree to participate, but on three conditions.

My first condition is that this movement should begin by giving up all hope and belief in piecemeal, one-shot solutions. The present scientific quest for odorless hog manure should give us sufficient proof that the specialist is no longer with us. Even now, after centuries of reductionist propaganda, the world is still intricate and vast, as dark as it is light, a place of mystery, where we cannot do one thing without doing many things, or put two things together without putting many things together.

Water quality, for example, cannot be improved without improving farming and forestry, but farming and forestry cannot be improved without improving the education of consumers —and so on.

The proper business of a human economy is to make one whole thing of ourselves and this world. To make ourselves into a practical wholeness with the land under our feet is maybe not altogether possible—how would *we* know?—but, as a goal, it at least carries us beyond *hubris*, beyond the utterly groundless assumption that we can subdivide our present great failure into a thousand separate problems that can be fixed by a thousand task forces of academic and bureaucratic specialists. That program has been given more than a fair chance to prove itself, and we ought to know by now that it won't work.

My second condition is that the people in this movement (the MTEWIID) should take full responsibility for themselves as members of the economy. If we are going to teach the economy what it is doing, then we need to learn what *we* are doing. This is going to have to be a private movement as well as a public one. If it is unrealistic to expect wasteful industries to be conservers, then obviously we must lead in part the public life of complainers, petitioners, protesters, advocates and supporters of stricter regulations and saner policies. But that is not enough. If it is unrealistic to expect a bad economy to try to become a good one, then *we* must go to work to build a good economy. It is appropriate that this duty should fall to us, for good economic behavior is more possible for us than it is for the great corporations with their miseducated managers and their greedy and oblivious stockholders. Because it is possible for us, we must try in every way we can to make good economic sense in our own lives, in our households, and in our communities. We must do more for ourselves and our neighbors. We must learn to spend our money with our friends and not with our enemies. But to do this, it is necessary to renew local economies, and revive the domestic arts. In seeking to change our economic use of the world, we are seeking inescapably to change our lives. The outward harmony that we desire between our economy and the world depends finally upon an inward harmony between our own hearts and the creative spirit that is the life of all creatures, a spirit as near us as our flesh and yet forever beyond the measures of this obsessively measuring age. We can grow good wheat and make good bread only if we understand that we do not live by bread alone.

My third condition is that this movement should content itself to be poor. We need to find cheap solutions, solutions within the reach of everybody, and the availability of a lot of money prevents the discovery of cheap solutions. The solutions of modern medicine and modern agriculture are all staggeringly expensive, and this is caused in part, and maybe altogether, by the availability of huge sums of money for medical and agricultural research.

Too much money, moreover, attracts administrators and experts as sugar attracts ants—look at what is happening in our universities. We should not envy rich movements that are organized and led by an alternative bureaucracy living on the problems it is supposed to solve. We want a movement that is a movement because it is advanced by all its members in their daily lives.

Now, having completed this very formidable list of the problems and difficulties, fears and fearful hopes that lie ahead of us, I am relieved to see that I have been preparing myself all along to end by saying something cheerful. What I have been talking about is the possibility of renewing human respect for this earth and all the good, useful, and beautiful things that come from it. I have made it clear, I hope, that I don't think this respect can be adequately enacted or conveyed by tipping our hats to nature or by representing natural loveliness in art or by prayers of thanksgiving or by preserving tracts of wilderness—though I recommend all those things. The respect I mean can be given only by using well the world's goods that are given to us. This good use, which renews respect—which is the only currency, so to speak, of respect—also renews our pleasure. The callings and disciplines that I have spoken of as the domestic arts are stationed all along the way from the farm to the prepared dinner, from the forest to the dinner table, from stewardship of the land to hospitality to friends and strangers. These arts are as demanding and gratifying, as instructive and as pleasing as the so-called fine arts. To learn them, to practice them, to honor and reward them is, I believe, our profoundest calling. Our reward is that they will enrich our lives and make us glad.

The Total Economy

LET US BEGIN by assuming what appears to be true: that the so-called environmental crisis is now pretty well established as a fact of our age. The problems of pollution, species extinction, loss of wilderness, loss of farmland, and loss of topsoil may still be ignored or scoffed at, but they are not denied. Concern for these problems has acquired a certain standing, a measure of discussability, in the media and in some scientific, academic, and religious institutions.

This is good, of course; obviously, we can't hope to solve these problems without an increase of public awareness and concern. But in an age burdened with "publicity," we have to be aware also that as issues rise into popularity they rise also into the danger of oversimplification. To speak of this danger is especially necessary in confronting the destructiveness of our relationship to nature, which is the result, in the first place, of gross oversimplification.

The "environmental crisis" has happened because the human household or economy is in conflict at almost every point with the household of nature. We have built our household on the assumption that the natural household is simple and can be simply used. We have assumed increasingly over the last five hundred years that nature is merely a supply of "raw materials," and that we may safely possess those materials merely by taking them. This taking, as our technical means have increased, has involved always less reverence or respect, less gratitude, less local knowledge, and less skill. Our methodologies of land use have strayed from our old sympathetic attempts to imitate natural processes, and have come more and more to resemble the methodology of mining, even as mining itself has become more powerful technologically and more brutal.

And so we will be wrong if we attempt to correct what we perceive as "environmental" problems without correcting the economic oversimplification that caused them. This oversimplification is now either a matter of corporate behavior or of behavior under the influence of corporate behavior. This is sufficiently clear to many of us. What is not sufficiently clear, perhaps to any of us, is the extent of our complicity, as individuals and especially as individual consumers, in the behavior of the corporations.

What has happened is that most people in our country, and apparently most people in the "developed" world, have given proxies to the corporations to produce and provide *all* of their food, clothing, and shelter. Moreover, they are rapidly increasing their proxies to corporations or governments to provide entertainment, education, child care, care of the sick and the elderly, and many other kinds of "service" that once were carried on informally and inexpensively by individuals or households or communities. Our major economic practice, in short, is to delegate the practice to others.

The danger now is that those who are concerned will believe that the solution to the "environmental crisis" can be merely political—that the problems, being large, can be solved by large solutions generated by a few people to whom we will give our proxies to police the economic proxies that we have already given. The danger, in other words, is that people will think they have made a sufficient change if they have altered their "values," or had a "change of heart," or experienced a "spiritual awakening," and that such a change in passive consumers will necessarily cause appropriate changes in the public experts, politicians, and corporate executives to whom they have granted their political and economic proxies.

The trouble with this is that a proper concern for nature and our use of nature must be practiced, not by our proxy-holders, but by ourselves. A change of heart or of values without a practice is only another pointless luxury of a passively consumptive way of life. The "environmental crisis," in fact, can be solved only if people, individually and in their communities, recover responsibility for their thoughtlessly given proxies. If people begin the effort to take back into their own power a significant portion of their economic responsibility, then their inevitable first discovery is that the "environmental crisis" is no such thing; it is not a crisis of our environs or surroundings; it is a crisis of our lives as individuals, as family members, as community members, and as citizens. We have an "environmental crisis" because *we* have consented to an economy in which by eating, drinking, working, resting, traveling, and enjoying ourselves we are destroying the natural, the God-given, world.

We live, as we must sooner or later recognize, in an era of sentimental economics and, consequently, of sentimental politics. Sentimental communism holds in effect that everybody and everything should suffer for the good of "the many" who, though miserable in the present, will be happy in the future for exactly the same reasons that they are miserable in the present.

Sentimental capitalism is not so different from sentimental communism as the corporate and political powers claim to suppose. Sentimental capitalism holds in effect that everything small, local, private, personal, natural, good, and beautiful must be sacrificed in the interest of the "free market" and the great corporations, which will bring unprecedented security and happiness to "the many"—in, of course, the future.

These forms of political economy may be described as sentimental because they depend absolutely upon a political faith for which there is no justification. They seek to preserve the gullibility of the people by issuing a cold check on a fund of political virtue that does not exist. Communism and "free-market" capitalism both are modern versions of oligarchy. In their propaganda, both justify violent means by good ends, which always are put beyond reach by the violence of the means. The trick is to define the end vaguely—"the greatest good of the greatest number" or "the benefit of the many"—and keep it at a distance. For example, the United States government's agricultural policy, or non-policy, since 1952 has merely consented to the farmers' predicament of high costs and low prices; it has never envisioned or advocated in particular the prosperity of farmers or of farmland, but has only promised "cheap food" to consumers and "survival" to the "larger and more efficient" farmers who supposedly could adapt to and endure the attrition of high costs and low prices. And after each inevitable wave of farm failures and the inevitable enlargement of the destitution and degradation of the countryside, there have been the inevitable reassurances from government propagandists and university experts that American agriculture was now more efficient and that everybody would be better off in the future.

The fraudulence of these oligarchic forms of economy is in their principle of displacing whatever good they recognize (as well as their debts) from the present to the future. Their success depends upon persuading people, first, that whatever they have now is no good, and, second, that the promised good is certain to be achieved in the future. This obviously contradicts the principle—common, I believe, to all the religious traditions—that if ever we are going to do good to one another, then the time to do it is now; we are to receive no reward for promising to do it in the future. And both communism and capitalism have found such principles to be a great embarrassment. If you are presently occupied in destroying every good thing in sight in order to do good in the future, it is inconvenient to have people saying things like "Love thy neighbor as thyself" or "Sentient beings are numberless, I vow to save them." Communists and capitalists alike, "liberal" capitalists and "conservative" capitalists alike, have needed to replace religion with some form of determinism, so that they can say to their victims, "I'm doing this because I can't do otherwise. It is not my fault. It is inevitable." This is a lie, obviously, and religious organizations have too often consented to it.

The idea of an economy based upon several kinds of ruin may seem a contradiction in terms, but in fact such an economy is possible, as we see. It is possible, however, on one implacable condition: The only future good that it assuredly leads to is that it will destroy itself. And how does it disguise this outcome from its subjects, its short-term beneficiaries, and its victims? It does so by false accounting. It substitutes for the real economy, by which we build and maintain (or do not maintain) our household, a symbolic economy of money, which in the long run, because of the self-interested manipulations of the "controlling interests," cannot symbolize or account for anything but itself. And so we have before us the spectacle of unprecedented "prosperity" and "economic growth" in a land of degraded farms, forests, ecosystems, and watersheds, polluted air, failing families, and perishing communities.

This moral and economic absurdity exists for the sake of the allegedly "free" market, the single principle of which is this: Commodities will be produced wherever they can be produced at the lowest cost and consumed wherever they will bring the highest price. To make too cheap and sell too high has always been the program of industrial capitalism. The global "free market" is merely capitalism's so far successful attempt to enlarge the geographic scope of its greed, and moreover to give to its greed the status of a "right" within its presumptive territory. The global "free market" is free to the corporations precisely because it dissolves the boundaries of the old national colonialisms, and replaces them with a new colonialism without restraints or boundaries. It is pretty much as if all the rabbits have now been forbidden to have holes, thereby "freeing" the hounds.

The "right" of a corporation to exercise its economic power without restraint is construed, by the partisans of the "free market," as a form of freedom, a political liberty implied presumably by the right of individual citizens to own and use property.

But the "free market" idea introduces into government a sanction of an inequality that is not implicit in any idea of democratic liberty: namely that the "free market" is freest to those who have the most money, and is not free at all to those with little or no money. Wal-Mart, for example, as a large corporation "freely" competing against local, privately owned businesses, has virtually all the freedom, and its small competitors virtually none.

To make too cheap and sell too high, there are two requirements. One is that you must have a lot of consumers with surplus money and unlimited wants. For the time being, there are plenty of these consumers in the "developed" countries. The problem, for the time being easily solved, is simply to keep them relatively affluent and dependent on purchased supplies.

The other requirement is that the market for labor and raw materials should remain depressed relative to the market for retail commodities. This means that the supply of workers should exceed demand, and that the land-using economies should be allowed or encouraged to overproduce.

To keep the cost of labor low, it is necessary first to entice or force country people everywhere in the world to move into the cities—in the manner prescribed by the Committee for Economic Development after World War II—and, second, to continue to introduce labor-replacing technology. In this way it is possible to maintain a "pool" of people who are in the threatful position of being mere consumers, landless and poor, and who therefore are eager to go to work for low wages—precisely the condition of migrant farm workers in the United States.

To cause the land-using economies to overproduce is even simpler. The farmers and other workers in the world's land-using economies, by and large, are not organized. They are therefore unable to control production in order to secure just prices. Individual producers must go individually to the market and take for their produce simply whatever they are paid. They have no power to bargain or to make demands. Increasingly, they must sell, not to neighbors or to neighboring towns and cities, but to large and remote corporations. There is no competition among the buyers (supposing there is more than one), who *are* organized and are "free" to exploit the advantage of low prices. Low prices encourage overproduction, as producers attempt to make up their losses "on volume," and overproduction inevitably makes for low prices. The land-using economies thus spiral downward as the money economy of the exploiters spirals upward. If economic attrition in the land-using population becomes so severe as to threaten production, then governments can subsidize production without production controls, which necessarily will encourage overproduction, which will lower prices—and so the subsidy to rural producers becomes, in effect, a subsidy to the purchasing corporations. In the land-using economies, production is further cheapened by destroying, with low prices and low standards of quality, the cultural imperatives for good work and land stewardship.

This sort of exploitation, long familiar in the foreign and domestic colonialism of modern nations, has now become "the global economy," which is the property of a few supranational corporations. The economic theory used to justify the global economy in its "free market" version is, again, perfectly groundless and sentimental. The idea is that what is good for the corporations will sooner or later—though not of course immediately—be good for everybody.

That sentimentality is based, in turn, upon a fantasy: the proposition that the great corporations, in "freely" competing with one another for raw materials, labor, and market share, will drive each other indefinitely, not only toward greater "efficiencies" of manufacture, but also toward higher bids for raw materials and labor and lower prices to consumers. As a result, all the world's people will be economically secure—in the future. It would be hard to object to such a proposition if only it were true.

But one knows, in the first place, that "efficiency" in manufacture always means reducing labor costs by replacing workers with cheaper workers or with machines.

In the second place, the "law of competition" does *not* imply that many competitors will compete indefinitely. The law of competition is a simple paradox: Competition destroys competition. The law of competition implies that many competitors, competing on the "free market" without restraint, will ultimately and inevitably reduce the number of competitors to one. The law of competition, in short, is the law of war.

In the third place, the global economy is based upon cheap long-distance transportation, without which it is not possible to move goods from the point of cheapest origin to the point of highest sale. And cheap long-distance transportation is the basis of the idea that regions and nations should abandon any measure of economic self-sufficiency in order to specialize in production for export of the few commodities, or the single commodity, that can be most cheaply produced. Whatever may be said for the "efficiency" of such a system, its result (and, I assume, its purpose) is to destroy local production capacities, local diversity, and local economic independence. It destroys the economic security that it promises to make.

This idea of a global "free market" economy, despite its obvious moral flaws and its dangerous practical weaknesses, is now the ruling orthodoxy of the age. Its propaganda is subscribed to and distributed by most political leaders, editorial writers, and other "opinion makers." The powers that be, while continuing to budget huge sums for "national defense," have apparently abandoned any idea of national or local self-sufficiency, even in food. They also have given up the idea that a national or local government might justly place restraints upon economic activity in order to protect its land and its people.

The global economy is now institutionalized in the World Trade Organization, which was set up, without election anywhere, to rule international trade on behalf of the "free market"—which is to say on behalf of the supranational corporations—and to *over*rule, in secret sessions, any national or regional law that conflicts with the "free market." The corporate program of global "free trade" and the presence of the World Trade Organization have legitimized extreme forms of expert thought. We are told confidently that if Kentucky loses its milk-producing capacity to Wisconsin (and if Wisconsin's is lost to California), that will be a "success story." Experts such as Stephen C. Blank, of the University of California, Davis, have recommended that "developed" countries, such as the United States and the United Kingdom, where food can no longer be produced cheaply enough, should give up agriculture altogether.

The folly at the root of this foolish economy began with the idea that a corporation should be regarded, legally, as "a person." But the limitless destructiveness of this economy comes about precisely because a corporation is *not* a person. A corporation, essentially, is a pile of money to which a number of persons have sold their moral allegiance. Unlike a person, a corporation does not age. It does not arrive, as most persons finally do, at a realization of the shortness and smallness of human lives; it does not come to see the future as the lifetime of the children and grandchildren of anybody in particular. It can experience no personal hope or remorse, no change of heart. It cannot humble itself. It goes about its business as if it were immortal, with the single purpose of becoming a bigger pile of money. The stockholders essentially are usurers, people who "let their money work for them," expecting high pay in return for causing others to work for low pay. The World Trade Organization enlarges the old idea of the corporation-as-person by giving the global corporate economy the status of a super-government with the power to overrule nations.

I don't mean to say, of course, that all corporate executives and stockholders are bad people. I am only saying that all of them are very seriously implicated in a bad economy.

Unsurprisingly, among people who wish to preserve things other than money—for instance, every region's native capacity to produce essential goods—there is a growing perception that the global "free market" economy is inherently an enemy to the natural world, to human health and freedom, to industrial workers, and to farmers and others in the land-use economies; and, furthermore, that it is inherently an enemy to good work and good economic practice.

I believe that this perception is correct and that it can be shown to be correct merely by listing the assumptions implicit in the idea that corporations should be "free" to buy low and sell high in the world at large. These assumptions, so far as I can make them out, are as follows:

1. That there is no conflict between the "free market" and political freedom, and no connection between political democracy and economic democracy.
2. That there can be no conflict between economic advantage and economic justice.
3. That there is no conflict between greed and ecological or bodily health.
4. That there is no conflict between self-interest and public service.
5. That it is all right for a nation's or a region's subsistence to be foreign-based, dependent on long-distance transport, and entirely controlled by corporations.
6. That the loss or destruction of the capacity anywhere to produce necessary goods does not matter and involves no cost.
7. That, therefore, wars over commodities—our recent Gulf War, for example—are legitimate and permanent economic functions.
8. That this sort of sanctioned violence is justified also by the predominance of centralized systems of production, supply, communications, and transportation which are extremely vulnerable not only to acts of war between nations, but also to sabotage and terrorism.
9. That it is all right for poor people in poor countries to work at poor wages to produce goods for export to affluent people in rich countries.
10. That there is no danger and no cost in the proliferation of exotic pests, vermin, weeds, and diseases that accompany international trade, and that increase with the volume of trade.

11. That an economy is a machine, of which people are merely the interchangeable parts. One has no choice but to do the work (if any) that the economy prescribes, and to accept the prescribed wage.
12. That, therefore, vocation is a dead issue. One does not do the work that one chooses to do because one is called to it by Heaven or by one's natural abilities, but does instead the work that is determined and imposed by the economy. Any work is all right as long as one gets paid for it. (This assumption explains the prevailing "liberal" and "conservative" indifference toward displaced workers, farmers, and small-business people.)
13. That stable and preserving relationships among people, places, and things do not matter and are of no worth.
14. That cultures and religions have no legitimate practical or economic concerns.

These assumptions clearly prefigure a condition of total economy. A total economy is one in which everything—"life forms," for instance, or the "right to pollute"—is "private property" and has a price and is for sale. In a total economy, significant and sometimes critical choices that once belonged to individuals or communities become the property of corporations. A total economy, operating internationally, necessarily shrinks the powers of state and national governments, not only because those governments have signed over significant powers to an international bureaucracy or because political leaders become the paid hacks of the corporations, but also because political processes—and especially democratic processes—are too slow to react to unrestrained economic and technological development on a global scale. And when state and national governments begin to act in effect as agents of the global economy, selling their people for low wages and their people's products for low prices, then the rights and liberties of citizenship must necessarily shrink. A total economy is an unrestrained taking of profits from the disintegration of nations, communities, households, landscapes, and ecosystems. It licenses symbolic or artificial wealth to "grow" by means of the destruction of the real wealth of all the world.

Among the many costs of the total economy, the loss of the

principle of vocation is probably the most symptomatic and, from a cultural standpoint, the most critical. It is by the replacement of vocation with economic determinism that the exterior workings of a total economy destroy human character and culture also from the inside.

In an essay on the origin of civilization in traditional cultures, Ananda Coomaraswamy wrote that "the principle of justice is the same throughout. . . . [It is] that each member of the community should perform the task for which he is fitted by nature. . . ." The two ideas, justice and vocation, are inseparable. That is why Coomaraswamy spoke of industrialism as "the mammon of injustice," incompatible with civilization. It is by way of the practice of vocation that sanctity and reverence enter into the human economy. It was thus possible for traditional cultures to conceive that "to work is to pray."

Aware of industrialism's potential for destruction, as well as the considerable political danger of great concentrations of wealth and power in industrial corporations, American leaders developed, and for a while used, certain means of limiting and restraining such concentrations, and of somewhat equitably distributing wealth and property. The means were: laws against trusts and monopolies, the principle of collective bargaining, the concept of 100 percent parity between the land-using and the manufacturing economies, and the progressive income tax. And to protect domestic producers and production capacities, it is possible for governments to impose tariffs on cheap imported goods. These means are justified by the government's obligation to protect the lives, livelihoods, and freedoms of its citizens. There is, then, no necessity that requires our government to sacrifice the livelihoods of our small farmers, small-business people, and workers, along with our domestic economic independence, to the global "free market." But now all of these means are either weakened or in disuse. The global economy is intended as a means of subverting them.

In default of government protections against the total economy of the supranational corporations, people are where they have been many times before: in danger of losing their economic security and their freedom, both at once. But at the same time the means of defending themselves belongs to them in the form of a venerable principle: Powers not exercised by government return to the people. If the government does not propose to protect the lives, the livelihoods, and the freedoms of its people, then the people must think about protecting themselves.

How are they to protect themselves? There seems, really, to be only one way, and that is to develop and put into practice the idea of a local economy—something that growing numbers of people are now doing. For several good reasons, they are beginning with the idea of a local food economy. People are trying to find ways to shorten the distance between producers and consumers, to make the connections between the two more direct, and to make this local economic activity a benefit to the local community. They are trying to learn to use the consumer economies of local towns and cities to preserve the livelihoods of local farm families and farm communities. They want to use the local economy to give consumers an influence over the kind and quality of their food, and to preserve and enhance the local landscapes. They want to give everybody in the local community a direct, long-term interest in the prosperity, health, and beauty of their homeland. This is the only way presently available to make the total economy less total. It was once the only way to make a national or a colonial economy less total, but now the necessity is greater.

I am assuming that there is a valid line of thought leading from the idea of the total economy to the idea of a local economy. I assume that the first thought may be a recognition of one's ignorance and vulnerability as a consumer in the total economy. As such a consumer, one does not know the history of the products one uses. Where, exactly, did they come from? Who produced them? What toxins were used in their production? What were the human and ecological costs of producing and then of disposing of them? One sees that such questions cannot be answered easily, and perhaps not at all. Though one is shopping amid an astonishing variety of products, one is denied certain significant choices. In such a state of economic ignorance it is not possible to choose products that were produced locally or with reasonable kindness toward people and toward nature. Nor is it possible for such consumers to influence production for the better. Consumers who feel a prompting toward land stewardship find that in this economy they can have no stewardly practice. To be a consumer in the total economy, one must agree to be totally ignorant, totally passive, and totally dependent on distant supplies and self-interested suppliers.

And then, perhaps, one begins to *see* from a local point of view. One begins to ask, What is here, what is in my neighborhood, what is in me, that can lead to something better? From a local point of view, one can see that a global "free market" economy is possible only if nations and localities accept or ignore the inherent weakness of a production economy based on exports and a consumer economy based on imports. An export economy is beyond local influence, and so is an import economy. And cheap long-distance transport is possible only if granted cheap fuel, international peace, control of terrorism, prevention of sabotage, and the solvency of the international economy.

Perhaps also one begins to see the difference between a small local business that must share the fate of the local community and a large absentee corporation that is set up to escape the fate of the local community by ruining the local community.

So far as I can see, the idea of a local economy rests upon only two principles: neighborhood and subsistence.

In a viable neighborhood, neighbors ask themselves what they can do or provide for one another, and they find answers that they and their place can afford. This, and nothing else, is the *practice* of neighborhood. This practice must be, in part, charitable, but it must also be economic, and the economic part must be equitable; there is a significant charity in just prices.

Of course, everything needed locally cannot be produced locally. But a viable neighborhood is a community, and a viable community is made up of neighbors who cherish and protect what they have in common. This is the principle of subsistence. A viable community, like a viable farm, protects its own production capacities. It does not import products that it can produce for itself. And it does not export local products until local needs have been met. The economic products of a viable community are understood either as belonging to the community's subsistence or as surplus, and only the surplus is considered to be marketable abroad. A community, if it is to be viable, cannot think of producing solely for export, and it cannot permit importers to use cheaper labor and goods from other places to destroy the local capacity to produce goods that are needed locally. In charity, moreover, it must refuse to import goods that are produced at the cost of human or ecological degradation elsewhere. This principle of subsistence applies not just to localities, but to regions and nations as well.

The principles of neighborhood and subsistence will be disparaged by the globalists as "protectionism"—and that is exactly what it is. It is a protectionism that is just and sound, because it protects local producers and is the best assurance of adequate supplies to local consumers. And the idea that local needs should be met first and only surpluses exported does *not* imply any prejudice against charity toward people in other places or trade with them. The principle of neighborhood at home always implies the principle of charity abroad. And the principle of subsistence is in fact the best guarantee of giveable or marketable surpluses. This kind of protection is not "isolationism."

Albert Schweitzer, who knew well the economic situation in the colonies of Africa, wrote about seventy years ago: "Whenever the timber trade is good, permanent famine reigns in the Ogowe region, because the villagers abandon their farms to fell as many trees as possible." We should notice especially that the goal of production was "as many . . . as possible." And Schweitzer made my point exactly: "These people could achieve true wealth if they could develop their agriculture and trade to meet their own needs." Instead they produced timber for export to "the world market," which made them dependent upon imported goods that they bought with money earned from their exports. They gave up their local means of subsistence, and imposed the false standard of a foreign demand ("as many trees as possible") upon their forests. They thus became helplessly dependent on an economy over which they had no control.

Such was the fate of the native people under the African colonialism of Schweitzer's time. Such is, and can only be, the fate of everybody under the global colonialism of our time. Schweitzer's description of the colonial economy of the Ogowe region is in principle not different from the rural economy in Kentucky or Iowa or Wyoming now. A total economy, for all practical purposes, is a total government. The "free trade," which from the standpoint of the corporate economy brings "unprecedented economic growth," from the standpoint of the land and its local populations, and ultimately from the standpoint of the cities, is destruction and slavery. Without prosperous local economies, the people have no power and the land no voice.

Two Minds

HUMAN ORDERS—scientific, artistic, social, economic, and political—are fictions. They are untrue, not because they necessarily are false, but because they necessarily are incomplete. All of our human orders, however inclusive we may try to make them, turn out to be to some degree exclusive. And so we are always being surprised by something we find, too late, that we have excluded. Think of almost any political revolution or freedom movement or the ozone hole or mad cow disease or the events of September 11, 2001.

The present order, thus surprised, is then required to accommodate new knowledge and thus to be reordered. Thomas Kuhn described this process in *The Structure of Scientific Revolutions.*

But these surprises and changes obviously have their effect also on individual lives and on whole cultures. All of our fictions labor under an ever-failing need to be true. And this means that they labor under an obligation to be continuously revised.

Or, to put it another way, we humans necessarily make pictures in our minds of our places and our world. But we can do this only by selection, putting some things into the picture and leaving the rest out. And so we live in two landscapes, one superimposed upon the other.

First there is the cultural landscape made up of our own knowledge of where we are, of landmarks and memories, of patterns of use and travel, of remindings and meanings. The cultural landscape, among other things, is a pattern of exchanges of work, goods, and comforts among neighbors. It is the country we have in mind.

And then there is the actual landscape, which we can never fully know, which is always going to be to some degree a mystery, from time to time surprising us. These two landscapes are necessarily and irremediably different from each other. But there is danger in their difference; they can become *too* different. If the cultural landscape becomes too different from the actual landscape, then we will make practical errors that will be destructive of the actual landscape or of ourselves or of both.

You can learn this from the study of any landscape that is inhabited by humans, or from teachers such as Barry Lopez and Gary Nabhan, who have written on the traditional economies of the Arctic and the American Southwest. It is easier to understand, perhaps, when thinking about extreme landscapes: If the cultural landscapes come to be too much at odds with the actual landscapes of the Arctic or the desert, the penalties are apt to be swift and lethal. In more forgiving landscapes they will (perhaps) be slower, but finally just as dangerous.

In some temperate and well-watered areas, we humans have applied the most extreme industrial methods of landscape destruction. By disregarding the cultural landscape, and all values and protections that accrue therefrom to the actual landscape, the strip miners entirely destroy the actual landscape. Cropland erosion, caused by a serious incongruity between the cultural and the actual landscapes, is a slower form of destruction than strip mining, but given enough time, it too can be entirely destructive.

At present, in the United States and in much of the rest of the world, most of the cultural landscapes that still exist are hodgepodges of failing local memories, money-making schemes, ignorant plans, bucolic fantasies, misinformation, and the random facts that we now call "information." This is compounded by the outright destruction of innumerable burial sites and other sacred places, of natural and historical landmarks, and of entire actual landscapes. Moreover, we have enormous and increasing numbers of people who have no home landscape, though in every one of their economic acts they are affecting the actual landscapes of the world, mostly for the worse. This is a situation unprecedentedly disorderly and dangerous.

To be disconnected from any actual landscape is to be, in the practical or economic sense, without a home. To have no country carefully and practically in mind is to be without a culture. In such a situation, culture becomes purposeless and arbitrary, dividing into "popular culture," determined by commerce, advertising, and fashion, and "high culture," which is either social affectation, displaced cultural memory, or the merely aesthetic pursuits of artists and art lovers.

We are thus involved in a kind of lostness in which most people are participating more or less unconsciously in the destruction of the natural world, which is to say, the sources of their own lives. They are doing this unconsciously because they see or do very little of the actual destruction themselves, and they don't know, because they have no way to learn, how they are involved. At the same time, many of the same people fear and mourn the destruction, which they can't stop because they have no practical understanding of its causes.

Conservationists, scientists, philosophers, and others are telling us daily and hourly that our species is now behaving with colossal irrationality and that we had better become more rational. I agree as to the dimensions and danger of our irrationality. As to the possibility of curing it by rationality, or at least by the rationality of the rationalists, I have some doubts.

The trouble is not just in the way we are thinking; it is also in the way we, or anyhow we in the affluent parts of the world, are living. And it is going to be hard to define anybody's living as a series of simple choices between irrationality and rationality. Moreover, this is supposedly an age of reason; we are encouraged to believe that the governments and corporations of the affluent parts of the world are run by rational people using rational processes to make rational decisions. The dominant faith of the world in our time is in rationality. That in an age of reason, the human race, or the most wealthy and powerful parts of it, should be behaving with colossal irrationality ought to make us wonder if reason alone can lead us to do what is right.

It is often proposed, nowadays, that if we would only get rid of religion and other leftovers from our primitive past and become enlightened by scientific rationalism, we could invent the new values and ethics that are needed to preserve the natural world. This proposal is perfectly reasonable, and perfectly doubtful. It supposes that we can empirically know and rationally understand everything involved, which is exactly the supposition that has underwritten our transgressions against the natural world in the first place.

*

Obviously we need to use our intelligence. But how much intelligence have we got? And what sort of intelligence is it that we have? And how, at its best, does human intelligence work? In order to try to answer these questions I am going to suppose for a while that there are two different kinds of human mind: the Rational Mind and another, which, for want of a better term, I will call the Sympathetic Mind. I will say now, and try to keep myself reminded, that these terms are going to appear to be allegorical, too neat and too separate —though I need to say also that their separation was not invented by me.

The Rational Mind, without being anywhere perfectly embodied, is the mind all of us are supposed to be trying to have. It is the mind that the most powerful and influential people *think* they have. Our schools exist mainly to educate and propagate and authorize the Rational Mind. The Rational Mind is objective, analytical, and empirical; it makes itself up only by considering facts; it pursues truth by experimentation; it is uncorrupted by preconception, received authority, religious belief, or feeling. Its ideal products are the proven fact, the accurate prediction, and the "informed decision." It is, you might say, the official mind of science, industry, and government.

The Sympathetic Mind differs from the Rational Mind, not by being unreasonable, but by refusing to limit knowledge or reality to the scope of reason or factuality or experimentation, and by making reason the servant of things it considers precedent and higher.

The Rational Mind is motivated by the fear of being misled, of being wrong. Its purpose is to exclude everything that cannot empirically or experimentally be proven to be a fact.

The Sympathetic Mind is motivated by fear of error of a very different kind: the error of carelessness, of being unloving. Its purpose is to be considerate of whatever is present, to leave nothing out.

The Rational Mind is exclusive; the Sympathetic Mind, however failingly, wishes to be inclusive.

These two types certainly don't exhaust the taxonomy of minds. They are merely the two that the intellectual fashions of our age have most deliberately separated and thrown into opposition.

My purpose here is to argue in defense of the Sympathetic Mind. But my objection is not to the use of reason or to reasonability. I am objecting to the exclusiveness of the Rational Mind, which has limited itself to a selection of mental functions such as the empirical methodologies of analysis and experimentation and the attitudes of objectivity and realism. In order to go into business on its own, it has in effect withdrawn from all of human life that involves feeling, affection, familiarity, reverence, faith, and loyalty. The separability of the Rational Mind is not only the dominant fiction but also the master superstition of the modern age.

The Sympathetic Mind is under the influence of certain inborn or at least fundamental likes and dislikes. Its impulse is toward wholeness. It is moved by affection for its home place, the local topography, the local memories, and the local creatures. It hates estrangement, dismemberment, and disfigurement. The Rational Mind tolerates all these things "in pursuit of truth" or in pursuit of money—which, in modern practice, have become nearly the same pursuit.

I am objecting to the failure of the rationalist enterprise of "objective science" or "pure science" or "the disinterested pursuit of truth" to prevent massive damage both to nature and to human economy. The Rational Mind does not confess its complicity in the equation: knowledge = power = money = damage. Even so, the alliance of academic science, government, and the corporate economy, and their unifying pattern of sanctions and rewards, is obvious enough. We have resisted, so far, a state religion, but we are in danger of having both a corporate state and a state science, which some people, in both the sciences and the arts, would like to establish as a state religion.

The Rational Mind is the lowest common denominator of the government–corporation–university axis. It is the fiction that makes high intellectual ability the unquestioning servant of bad work and bad law.

Under the reign of the Rational Mind, there is no firewall between contemporary science and contemporary industry or economic development. It is entirely imaginable, for instance, that a young person might go into biology because of love for plants and animals. But such a young person had better be careful, for there is nothing to prevent knowledge gained for love of the creatures from being used to destroy them for the love of money.

Now some biologists, who have striven all their lives to embody perfectly the Rational Mind, have become concerned, even passionately concerned, about the loss of "biological diversity," and they are determined to do something about it. This is usually presented as a merely logical development from ignorance to realization to action. But so far it is only comedy. The Rational Mind, which has been destroying biological diversity by "figuring out" some things, now proposes to save what is left of biological diversity by "figuring out" some more things. It does what it has always done before: It defines the problem as a big problem calling for a big solution; it calls in the world-class experts; it invokes science, technology, and large grants of money; it propagandizes and organizes and "gears up for a major effort." The comedy here is in the failure of these rationalists to see that as soon as they have become passionately concerned they have stepped outside the dry, objective, geometrical territory claimed by the Rational Mind, and have entered the still mysterious homeland of the Sympathetic Mind, watered by unpredictable rains and by real sweat and real tears.

The Sympathetic Mind would not forget that so-called environmental problems have causes that are in part political and therefore have remedies that are in part political. But it would not try to solve these problems merely by large-scale political protections of "the environment." It knows that they must be solved ultimately by correcting the way people use their home places and local landscapes. Politically, but also by local economic improvements, it would stop colonialism in all its forms, domestic and foreign, corporate and governmental. Its first political principle is that landscapes should not be used by people who do not live in them and share their fate. If that principle were strictly applied, we would have far less need for the principle of "environmental protection."

The Sympathetic Mind understands the vital importance of the cultural landscape. The Rational Mind, by contrast, honors no cultural landscape, and therefore has no protective loyalty or affection for any actual landscape.

The definitive practical aim of the Sympathetic Mind is to adapt local economies to local landscapes. This is necessarily the work of local cultures. It cannot be done as a world-scale *feat* of science, industry, and government. This will seem a bitter bite to the optimists of scientific rationalism, which is scornful of limits and proud of its usurpations. But the science of the Sympathetic Mind is occupied precisely with the study of limits, both natural and human.

The Rational Mind does not work from any sense of geographical whereabouts or social connection or from any basis in cultural tradition or principle or character. It does not see itself as existing or working within a context. The Rational Mind doesn't think there is a context until it gets there. Its principle is to be "objective"—which is to say, unremembering and disloyal. It works within narrow mental boundaries that it draws for itself, as directed by the requirements of its profession or academic specialty or its ambition or its desire for power or profit, thus allowing for the "trade-off" and the "externalization" of costs and effects. Even when working outdoors, it is an indoor mind.

The Sympathetic Mind, even when working indoors, is an outdoor mind. It lives within an abounding and unbounded reality, always partly mysterious, in which everything matters, in which we humans are therefore returned to our ancient need for thanksgiving, prayer, and propitiation, in which we meet again and again the ancient question: How does one become worthy to use what must be used?

Whereas the Rational Mind is the mind of analysis, explanation, and manipulation, the Sympathetic Mind is the mind of our creatureliness.

Creatureliness denotes what Wallace Stevens called "the instinctive integrations which are the reasons for living." In our creatureliness we forget the little or much that we know about the optic nerve and the light-sensitive cell, and *we see*; we forget whatever we know about the physiology of the brain, and *we think*; we forget what we know of anatomy, the nervous system, the gastrointestinal tract, and *we work, eat, and sleep.*

We forget the theories and therapies of "human relationships," and we merely love the people we love, and even try to love the others. If we have any sense, we forget the fashionable determinisms, and we tell our children, "Be good. Be careful. Mind your manners. Be kind."

The Sympathetic Mind leaves the world whole, or it attempts always to do so. It looks upon people and other creatures as whole beings. It does not parcel them out into functions and uses.

The Rational Mind, by contrast, has rested its work for a long time on the proposition that all creatures are machines. This works as a sort of strainer to eliminate impurities such as affection, familiarity, and loyalty from the pursuit of knowledge, power, and profit. This machine-system assures the objectivity of the Rational Mind, which is itself understood as a machine, but it fails to account for a number of things, including the Rational Mind's own worries and enthusiasms. Why should a machine be bothered by the extinction of other machines? Would even an "intelligent" computer grieve over the disappearance of the Carolina parakeet?

The Rational Mind is preoccupied with the search for a sure way to avoid risk, loss, and suffering. For the Rational Mind, experience is likely to consist of a sequence of bad surprises and therefore must be booked as a "loss." That is why, to rationalists, the past and the present are so readily expendable or destructible in favor of the future, the era of no loss.

But the Sympathetic Mind accepts loss and suffering as the price, willingly paid, of its sympathy and affection—its wholeness.

The Rational Mind attempts endlessly to inform itself against its ruin by facts, experiments, projections, scoutings of "alternatives," hedgings against the unknown.

The Sympathetic Mind is informed by experience, by tradition-borne stories of the experiences of others, by familiarity, by compassion, by commitment, by faith.

The Sympathetic Mind is preeminently a faithful mind, taking knowingly and willingly the risks required by faith. The Rational Mind, ever in need of certainty, is always in doubt, always looking for a better way, asking, testing, disbelieving in everything but its own sufficiency to its own needs, which its experience and its own methods continually disprove. It is a skeptical, fearful, suspicious mind, and always a disappointed one, awaiting the supreme truth or discovery it expects of itself, which of itself it cannot provide.

To show how these two minds work, let us place them within the dilemma of a familiar story. Here is the parable of the lost sheep from the Gospel of St. Matthew: "If a man have an hundred sheep, and one of them be gone astray, doth he not leave the ninety and nine, and goeth into the mountains, and seeketh that which is gone astray? And if so be that he find it, verily I say unto you, he rejoiceth more of that sheep, than of the ninety and nine which went not astray."

This parable is the product of an eminently sympathetic mind, but for the moment that need not distract us. The dilemma is practical enough, and we can see readily how the two kinds of mind would deal with it.

The rationalist, we may be sure, has a hundred sheep because he has a plan for that many. The one who has gone astray has escaped not only from the flock but also from the plan. That this particular sheep should stray off in this particular place at this particular time, though it is perfectly in keeping with the nature of sheep and the nature of the world, is not at all in keeping with a rational plan. What is to be done? Well, it certainly would not be rational to leave the ninety and nine, exposed as they would then be to further whims of nature, in order to search for the one. Wouldn't it be best to consider the lost sheep a "trade-off" for the safety of the ninety-nine? Having thus agreed to his loss, the doctrinaire rationalist would then work his way through a series of reasonable questions. What would be an "acceptable risk"? What would be an "acceptable loss"? Would it not be good to do some experiments to determine how often sheep may be expected to get lost? If one sheep is likely to get lost every so often, then would it not be better to have perhaps 110 sheep? Or should one insure the flock against such expectable losses? The annual insurance premium would equal the market value of how many sheep? What is likely to be the cost of the labor of looking for one lost sheep after quitting time? How much time spent looking would equal the market value of the lost sheep? Should not one think of splicing a few firefly genes into one's sheep so that strayed sheep would glow in the dark? And so on.

But (leaving aside the theological import of the parable) the shepherd is a shepherd because he embodies the Sympathetic Mind. Because he is a man of sympathy, a man devoted to the care of sheep, a man who knows the nature of sheep and of the world, the shepherd of the parable is not surprised or baffled by his problem. He does not hang back to argue over risks, trade-offs, actuarial data, or market values. He does not quibble over fractions. He goes without hesitating to hunt for the lost sheep because he has committed himself to the care of the whole hundred, because he understands his work as the fulfillment of his whole trust, because he loves the sheep, and because he knows or imagines what it is to be lost. He does what he does on behalf of the whole flock because he wants to preserve himself as a whole shepherd.

He also does what he does because he has a particular affection for that particular sheep. To the Rational Mind, all sheep are the same; any one is the same as any other. They are interchangeable, like coins or clones or machine parts or members of "the work force." To the Sympathetic Mind, each one is different from every other. Each one is an individual whose value is never entirely reducible to market value.

The Rational Mind can and will rationalize any trade-off. The Sympathetic Mind can rationalize none. Thus we have not only the parable of the ninety and nine, but also the Buddhist vow to save all sentient beings. The parable and the vow are utterly alien to the rationalism of modern science, politics, and industry. To the Rational Mind, they "don't make sense" because they deal with hardship and risk merely by acknowledgment and acceptance. Their very point is to require a human being's suffering to involve itself in the suffering of other creatures, including that of other human beings.

The Rational Mind conceives of itself as eminently practical, and is given to boasting about its competence in dealing with "reality." But if you want to hire somebody to take care of your hundred sheep, I think you had better look past the "animal scientist" and hire the shepherd of the parable, if you can still find him anywhere. For it will continue to be more reasonable, from the point of view of the Rational Mind, to trade off the lost sheep for the sake of the sheep you have left—until only one is left.

If you think I have allowed my argument to carry me entirely into fantasy and irrelevance, then let me quote an up-to-date story that follows pretty closely the outline of Christ's parable. This is from an article by Bernard E. Rollin in *Christian Century*, December 19–26, 2001, p. 26:

> A young man was working for a company that operated a large, total-confinement swine farm. One day he detected symptoms of a disease among some of the feeder pigs. As a teen, he had raised pigs himself . . . so he knew how to treat the animals. But the company's policy was to kill any diseased animals with a blow to the head—the profit margin was considered too low to allow for treatment of individual animals. So the employee decided to come in on his own time, with his own medicine, and he cured the animals. The management's response was to fire him on the spot for violating company policy.

The young worker in the hog factory is a direct cultural descendant of the shepherd in the parable, just about opposite and perhaps incomprehensible to the "practical" rationalist. But the practical implications are still the same. Would you rather have your pigs cared for by a young man who had compassion for them or by one who would indifferently knock them in the head? Which of the two would be most likely to prevent the disease in the first place? Compassion, of course, is the crux of the issue. For "company policy" must exclude compassion; if compassion were to be admitted to consideration, such a "farm" could not exist. And yet one imagines that even the hardheaded realists of "management" must occasionally violate company policy by wondering at night what they would do if *all* the pigs got sick. (I suppose they would kill them all, collect the insurance, and move on. Perhaps all that has been foreseen and prepared for in the business plan, and there is no need, after all, to lie awake and worry.)

But what of the compassionate young man? The next sentence of Mr. Rollin's account says: "Soon the young man left agriculture for good. . . ." We need to pause here to try to understand the significance of his departure.

Like a strip mine, a hog factory exists in utter indifference to the landscape. Its purpose, as an animal *factory*, is to exclude from consideration both the nature of the place where it is and the nature of hogs. That it is a factory means that it could be in *any* place, and that the hog is a "unit of production." But the young man evidently was farm-raised. He evidently had in his mind at least the memory of an actual place and at least the remnants of its cultural landscape. In that landscape, things were respected according to their nature, which made compassion possible when their nature was violated. That this young man was fired from his job for showing compassion is strictly logical, for the explicit purpose of the hog factory is to violate nature. And then, logically enough, the young man "left agriculture for good." But when you exclude compassion from agriculture, what have you done? Have you not removed something ultimately of the greatest practical worth? I believe so. But this is one of the Rational Mind's world-scale experiments that has not yet been completed.

The Sympathetic Mind is a freedom-loving mind because it knows, given the inevitable discrepancies between the cultural and the actual landscapes, that everybody involved must be free to change. The idea that science and industry and government can discover for the rest of us the ultimate truths of nature and human nature, which then can be infallibly used to regulate our life, is wrong. The true work of the sciences and the arts is to keep all of us moving, in our own lives in our own places, between the cultural and the actual landscapes, making the always necessary and the forever unfinished corrections.

When the Rational Mind establishes a "farm," the result is bad farming. There is a remarkable difference between a hog factory, which exists only for the sake of its economic product, and a good farm, which exists for many reasons, including the pleasure of the farm family, their affection for their home, their satisfaction in their good work—in short, their patriotism. Such a farm yields its economic product as a sort of side effect of the health of a flourishing place in which things live according to their nature. The hog factory attempts to be a totally rational, which is to say a totally economic, enterprise. It strips away from animal life and human work every purpose, every benefit too, that is not economic. It comes about as the result of a long effort on the part of "scientific agriculture" to remove the Sympathetic Mind from all agricultural landscapes and replace it with the Rational Mind. And so good-bye to the shepherd of the parable, and to compassionate young men who leave agriculture for good. Good-bye to the cultural landscape. Good-bye to the actual landscape. These have all been dispensed with by the Rational Mind, to be replaced by a totalitarian economy with its neat, logical concepts of world-as-factory and life-as-commodity. This is an economy excluding all decisions but "informed decisions," purporting to reduce the possibility of loss.

Nothing so entices and burdens the Rational Mind as its need, and its self-imposed responsibility, to make "informed decisions." It is certainly possible for a mind to be informed—in several ways, too. And it is certainly possible for an informed mind to make decisions on the basis of all that has informed it. But that such decisions are "informed decisions"—in the sense that "informed decisions" are predictably right, or even that they are reliably better than uninformed decisions—is open to doubt.

The ideal of the "informed decision" forces "decision makers" into a thicket of facts, figures, studies, tests, and "projections." It requires long and uneasy pondering of "cost-benefit ratios"—the costs and benefits, often, of abominations. The problem is that decisions all have to do with the future, and all the actual knowledge we have is of the past. It is impossible to make a decision, however well-informed it may be, that is assuredly right, because it is impossible to know what will happen. It is only possible to know or guess that some things *may* happen, and many things that have happened have not been foreseen.

Moreover, having made an "informed decision," even one that turns out well, there is no way absolutely to determine whether or not it was a better decision than another decision that one might have made instead. It is not possible to compare a decision that one made with a decision that one did not make. There are no "controls," no "replication plots," in experience.

The Rational Mind is under relentless pressure to justify governmental and corporate acts on an ever-increasing scale of power, extent, and influence. Given that pressure, it may be not so very surprising that the Rational Mind should have a remarkable tendency toward superstition. Reaching their mental limits, which, as humans, they must do soon enough, the rationalists begin to base their thinking on principles that are sometimes astonishingly unsound: "Creatures are machines" or "Knowledge is good" or "Growth is good" or "Science will find the answer" or "A rising tide lifts all boats." Or they approach the future with a stupefying array of computers, models, statistics, projections, calculations, cost-benefit analyses, experts, and even better computers—which of course cannot foretell the end of a horse race any better than Bertie Wooster.

And so the great weakness of the Rational Mind, contrary to its protestations, is a sort of carelessness or abandonment that takes the form of high-stakes gambling—as when, with optimism and fanfare, without foreknowledge or self-doubt or caution, nuclear physicists or chemists or genetic engineers release their products into the whole world, making the whole world their laboratory.

Or the great innovators and decision makers build huge airplanes whose loads of fuel make them, in effect, flying bombs. And they build the World Trade Center, forgetting apparently the B-25 bomber that crashed into the seventy-ninth floor of the Empire State Building in 1945. And then on September 11, 2001, some enemies—of a kind we well knew we had and evidently had decided to ignore—captured two huge airplanes and flew them, as bombs, into the two towers of the World Trade Center. In retrospect, we may doubt that these shaping decisions were properly informed, just as we may doubt that the expensive "intelligence" that is supposed to foresee and prevent such disasters is sufficiently intelligent.

The decisions, if the great innovators and decision makers were given to reading poetry, might have been informed by

James Laughlin's poem "Above the City," which was written soon after the B-25 crashed into the Empire State Building:

> You know our office on the 18th
> floor of the Salmon Tower looks
> right out on the
>
> Empire State & it just happened
> we were finishing up some
> late invoices on
>
> a new book that Saturday morning
> when a bomber roared through the
> mist and crashed
>
> flames poured from the windows
> into the drifting clouds & sirens
> screamed down in
>
> the streets below it was unearthly
> but you know the strangest thing
> we realized that
>
> none of us were much surprised be-
> cause we'd always known that those
> two Paragons of
>
> Progress sooner or later would per-
> form before our eyes this demon-
> stration of their
> true relationship

It is tempting now to call this poem "prophetic." But it is so only in the sense that it is insightful; it perceives the implicit contradiction between tall buildings and airplanes. This contradiction was readily apparent also to the terrorists of September 11, but evidently invisible within the mist of technological euphoria that had surrounded the great innovators and decision makers.

In the several dimensions of its horror the destruction of the World Trade Center exceeds imagination, and that tells us something. But as a physical event it is as comprehensible as 1 + 1, and that tells us something else.

*

Now that terrorism has established itself among us as an inescapable consideration, even the great decision makers are beginning to see that we are surrounded by the results of great decisions not adequately informed. We have built many nuclear power plants, each one a potential catastrophe, that will have to be protected, not only against their inherent liabilities and dangers, but against terrorist attack. And we have made, in effect, one thing of our food supply system, and that will have to be protected (if possible) from bioterrorism. These are by no means the only examples of the way we have exposed ourselves to catastrophic harm and great expense by our informed, rational acceptance of the normalcy of bigness and centralization.

After September 11, it can no longer be believed that science, technology, and industry are only good or that they serve only one "side." That never has been more than a progressivist and commercial superstition. Any power that belongs to one side belongs, for worse as well as better, to all sides, as indifferent as the sun that rises "on the evil and on the good." Only in the narrowest view of history can the scientists who worked on the nuclear bomb be said to have worked for democracy and freedom. They worked inescapably also for the enemies of democracy and freedom. If terrorists get possession of a nuclear bomb and use it, then the scientists of the bomb will be seen to have worked also for terrorism. There is (so far as I can now see) nothing at all that the Rational Mind can do, after the fact, to make this truth less true or less frightening. This predicament cries out for a different kind of mind before and after the fact: a mind faithful and compassionate that will not rationalize about the "good use" of destructive power, but will repudiate *any* use of it.

Freedom also is neutral, of course, and serves evil as well as good. But freedom rests on the power of good—by free speech, for instance—to correct evil. A great destructive power simply prevents this small decency of freedom. There is no way to correct a nuclear explosion.

In the midst of the dangers of the Rational Mind's achievement of bigness and centralization, the Sympathetic Mind is as hard-pressed as a pacifist in the midst of a war. There is no greater violence that ends violence, and no greater bigness with which to solve the problems of bigness. All that the Sympathetic Mind can do is maintain its difference, preserve its own integrity, and attempt to see the possibility of something better.

The Sympathetic Mind, as the mind of our creatureliness, accepts life in this world for what it is: mortal, partial, fallible, complexly dependent, entailing many responsibilities toward ourselves, our places, and our fellow beings. Above all, it understands itself as limited. It knows without embarrassment its own irreducible ignorance, especially of the future. It deals with the issue of the future, not by knowing what is going to happen, but by knowing—within limits—what to expect, and what should be required, of itself, of its neighbors, and of its place. Its decisions are informed by its culture, its experience, its understanding of nature. Because it is aware of its limits and its ignorance, it is alert to issues of scale. The Sympathetic Mind knows from experience—not with the brain only, but with the body—that danger increases with height, temperature, speed, and power. It knows by common sense and instinct that the way to protect a building from being hit by an airplane is to make it shorter; that the way to keep a nuclear power plant from becoming a weapon is not to build it; that the way to increase the security of a national food supply is to increase the agricultural self-sufficiency of states, regions, and local communities.

Because it is the mind of our wholeness, our involvement with all things beyond ourselves, the Sympathetic Mind is alert as well to the issues of propriety, of the fittingness of our artifacts to their places and to our own circumstances, needs, and hopes. It is preoccupied, in other words, with the fidelity or the truthfulness of the cultural landscape to the actual landscape.

I know I am not the only one reminded by the World Trade Center of the Tower of Babel: "let us build . . . a tower, whose top may reach unto heaven; and let us make us a name, lest we be scattered abroad upon the face of the whole earth." All extremely tall buildings have made me think of the Tower of Babel, and this started a long time before September 11, 2001, for reasons that have become much clearer to me now that "those two Paragons of / progress" have demonstrated again "their true relationship."

Like all such gigantic buildings, from Babel onward, the World Trade Center was built without reference to its own landscape or to any other. And the reason in this instance is not far to find. The World Trade Center had no reference to a landscape because world trade, as now practiced, has none. World trade now exists to exploit indifferently the landscapes of the world, and to gather the profits to centers whence they may be distributed to the world's wealthiest people. World trade needs centers precisely to prevent the world's wealth from being "scattered abroad upon the face of the whole earth." Such centers, like the "global free market" and the "global village," are utopias, "no-places." They need to be no-places, because they respect no places and are loyal to no place.

As for the problem of building on Manhattan Island, the Rational Mind has reduced that also to a simple economic principle: Land is expensive but air is cheap; therefore, build in the air. In the early 1960s my family and I lived on the Lower West Side, not far uptown from what would become the site of the World Trade Center. The area was run down, already under the judgment of "development." But once, obviously, it had been a coherent, thriving local neighborhood of residential apartments and flats, small shops and stores, where merchants and customers knew one another and neighbors were known to neighbors. Walking from our building to the Battery was a pleasant thing to do because one had the sense of being in a real place that kept both the signs of its old human history and the memory of its geographical identity. The last time I went there, the place had been utterly dis-placed by the World Trade Center.

Exactly the same feat of displacement is characteristic of the air transportation industry, which exists to free travel from all considerations of place. Air travel reduces place to space in order to traverse it in the shortest possible time. And like gigantic buildings, gigantic airports must destroy their places and become no-places in order to exist.

People of the modern world, who have accepted the dominance and the value system of the Rational Mind, do not object, it seems, to this displacement, or to the consequent disconnection of themselves from neighborhoods and from the landscapes that support them, or to their own anonymity within crowds of strangers. These things, according to cliché, free one from the suffocating intimacies of rural or small town life. And yet we now are obliged to notice that placelessness, centralization, gigantic scale, crowdedness, and anonymity are conditions virtually made to order for terrorists.

It is wrong to say, as some always do, that catastrophes are "acts of God" or divine punishments. But it is not wrong to ask if they may not be the result of our misreading of reality or our own nature, and if some correction may not be needed. My own belief is that the Rational Mind has been performing impressively within the narrowly drawn boundaries of what it provably knows, but it has been doing badly in dealing with the things of which it is ignorant: the future, the mysterious wholeness and multiplicity of the natural world, the needs of human souls, and even the real bases of the human economy in nature, skill, kindness, and trust. Increasingly, it seeks to justify itself with intellectual superstitions, public falsehoods, secrecy, and mistaken hopes, responding to its failures and bad surprises with (as the terrorists intend) terror and with ever grosser applications of power.

But the Rational Mind is caught, nevertheless, in cross-purposes that are becoming harder to ignore. It is altogether probable that there is an executive of an air-polluting industry who has a beloved child who suffers from asthma caused by air pollution. In such a situation the Sympathetic Mind cries, "Stop! Change your life! Quit your job! At least try to discover the cause of the harm and *do* something about it!" And here the Rational Mind must either give way to the Sympathetic Mind, or it must recite the conventional excuse that is a confession of its failure: "There is nothing to be done. This is the way things are. It is inevitable."

The same sort of contradiction now exists between national security and the global economy. Our government, having long ago abandoned any thought of economic self-sufficiency, having ceded a significant measure of national sovereignty to the World Trade Organization, and now terrified by terrorism, is obliged to police the global economy against the transportation of contraband weapons, which can be detected if the meshes of the surveillance network are fine enough, and also against the transportation of diseases, which cannot be detected. This too will be excused, at least for a while, by the plea of inevitability, never mind that this is the result of a conflict of policies and of "informed decisions." Meanwhile, there is probably no landscape in the world that is not threatened with abuse or destruction as a result of somebody's notion of trade or somebody's notion of security.

When the Rational Mind undertakes to work on a large scale, it works clumsily. It inevitably does damage, and it cannot exempt even itself or its own from the damage it does. You cannot help to pollute the world's only atmosphere and exempt your asthmatic child. You cannot make allies and enemies of the same people at the same time. Finally the idea of the trade-off fails. When the proposed trade-off is on the scale of the whole world—the natural world for world trade, world peace for national security—it can fail only into world disaster.

The Rational Mind, while spectacularly succeeding in some things, fails completely when it tries to deal in materialist terms with the part of reality that is spiritual. Religion and the language of religion deal approximately and awkwardly enough with this reality, but the Rational Mind, though it apparently cannot resist the attempt, cannot deal with it at all.

But most of the most important laws for the conduct of human life probably are religious in origin—laws such as these: Be merciful, be forgiving, love your neighbors, be hospitable to strangers, be kind to other creatures, take care of the helpless, love your enemies. We must, in short, love and care for one another and the other creatures. We are allowed to make no exceptions. Every person's obligation toward the Creation is summed up in two words from Genesis 2:15: "Keep it."

It is impossible, I believe, to make a neat thing of this set of instructions. It is impossible to disentangle its various obligations into a list of discrete items. Selfishness, or even "enlightened self-interest," cannot find a place to poke in its awl. One's obligation to oneself cannot be isolated from one's obligation to everything else. The whole thing is balanced on the verb *to love*. Love for oneself finds its only efficacy in love for everything else. Even loving one's enemy has become a strategy of self-love as the technology of death has grown greater. And this the terrorists have discovered and have accepted: The death of your enemy is your own death. The whole network of interdependence and obligation is a neatly set trap. Love does not let us escape from it; it turns the trap itself into the means and fact of our only freedom.

This condition of lawfulness and this set of laws did not originate in the Rational Mind, and could not have done so. The Rational Mind reduces our complex obligation to care for one another to issues of justice, forgetting the readiness with which we and our governments reduce justice, in turn, to revenge; and forgetting that even justice is intolerable without mercy, forgiveness, and love.

Justice is a rational procedure. Mercy is not a procedure and it is not rational. It is a kind of freedom that comes from sympathy, which is to say imagination—the *felt* knowledge of what it is to be another person or another creature. It is free because it does not have to be just. Justice is desirable, of course, but it is virtually the opposite of mercy. Mercy, says the Epistle of James, "rejoiceth against judgment."

As for the law requiring us to "keep" the given or the natural world (to go in search of the lost sheep, to save all sentient beings), the Rational Mind, despite the reasonable arguments made by some ecologists and biologists, cannot cover that distance either. In response to the proposition that we are responsible for the health of all the world, the Rational Mind begins to insist upon exceptions and trade-offs. It begins to designate the profit-yielding parts of the world that may "safely" be destroyed, and those unprofitable parts that may be preserved as "natural" or "wild." It divides the domestic from the wild, the human from the natural. It conceives of a natural place as a place where no humans live. Places where humans live are not natural, and the nature of such places must be reduced to comprehensibility, which is to say destroyed as they naturally are. The Rational Mind, convinced of the need to preserve "biological diversity," wants to preserve it in "nature preserves." It cannot conceive or tolerate the possibility of preserving biological diversity in the whole world, or of an economic harmony between humans and a world that by nature exceeds human comprehension.

It is because of the world's ultimately indecipherable webwork of vital connections, dependences, and obligations, and because ultimately our response to it must be loving beyond knowing, that the works of the Rational Mind are ultimately disappointing even to some rationalists.

When the Rational Mind fails not only into bewilderment but into irrationality and catastrophe, as it repeatedly does, that is because it has so isolated itself within its exclusive terms that it goes beyond its limits without knowing it.

Finally the human mind must accept the limits of sympathy, which paradoxically will enlarge it beyond the limits of rationality, but nevertheless will limit it. It must find its freedom and its satisfaction by working within its limits, on a scale much smaller than the Rational Mind will easily accept, for the Rational Mind continually longs to extend its limits by technology. But the safe competence of human work extends no further, ever, than our ability to think and love at the same time.

Obviously, we *can* work on a gigantic scale, but just as obviously we cannot foresee the gigantic catastrophes to which gigantic works are vulnerable, any more than we can foresee the natural and human consequences of such work. We can develop a global economy, but only on the conditions that it will not be loving in its effects on its human and natural sources, and that it will risk global economic collapse. We can build gigantic works of architecture too, but only with the likelihood that the gathering of the economic means to do so will generate somewhere the will to destroy what we have built.

The efficacy of a law is in the ability of people to obey it. The larger the scale of work, the smaller will be the number of people who can obey the law that we should be loving toward the world, even those places and creatures that we must use. You will see the problem if you imagine that you are one of the many, or if you *are* one of the many, who can find no work except in a destructive industry. Whether or not it is economic slavery to have no choice of jobs, it certainly is moral slavery to have no choice but to do what is wrong.

And so conservationists have not done enough when they conserve wilderness or biological diversity. They also must conserve the possibilities of peace and good work, and to do that they must help to make a good economy. To succeed, they must help to give more and more people everywhere in the world the opportunity to do work that is both a living and a loving. This, I think, cannot be accomplished by the Rational Mind. It will require the full employment of the Sympathetic Mind—*all* the little intelligence we have.

The Whole Horse

> This modern mind sees only half of the horse—that half which may become a dynamo, or an automobile, or any other horsepowered machine. If this mind had much respect for the full-dimensioned, grass-eating horse, it would never have invented the engine which represents only half of him. The religious mind, on the other hand, has this respect; it wants the whole horse, and it will be satisfied with nothing less. I should say a religious mind that requires more than a half-religion.
>
> —Allen Tate, "Remarks on the Southern Religion," in *I'll Take My Stand*

One of the primary results—and one of the primary needs—of industrialism is the separation of people and places and products from their histories. To the extent that we participate in the industrial economy, we do not know the histories of our meals or of our habitats or of our families. This is an economy, and in fact a culture, of the one-night stand. "I had a good time," says the industrial lover, "but don't ask me my last name." Just so, the industrial eater says to the svelte industrial hog, "We'll be together at breakfast. I don't want to see you before then, and I won't care to remember you afterwards."

In this condition, we have many commodities, but little satisfaction, little sense of the sufficiency of anything. The scarcity of satisfaction makes of our many commodities, in fact, an infinite series of commodities, the new commodities invariably promising greater satisfaction than the older ones. And so we can say that the industrial economy's most-marketed commodity is satisfaction, and that this commodity, which is repeatedly promised, bought, and paid for, is never delivered. On the other hand, people who have much satisfaction do not need many commodities.

The persistent want of satisfaction is directly and complexly related to the dissociation of ourselves and all our goods from our and their histories. If things do not last, are not made to last, they can have no histories, and we who use these things can have no memories. We buy new stuff on the promise of satisfaction because we have forgot the promised satisfaction for which we bought our old stuff. One of the procedures of the industrial economy is to reduce the longevity of materials. For example, wood, which well made into buildings and furniture and well cared for can last hundreds of years, is now routinely manufactured into products that last twenty-five years. We do not cherish the memory of shoddy and transitory objects, and so we do not remember them. That is to say that we do not invest in them the lasting respect and admiration that make for satisfaction.

The problem of our dissatisfaction with all the things that we use is not correctable within the terms of the economy that produces those things. At present, it is virtually impossible for us to know the economic history or the ecological cost of the products we buy; the origins of the products are typically too distant and too scattered and the processes of trade, manufacture, transportation, and marketing too complicated. There are, moreover, too many reasons for the industrial suppliers of these products not to want their histories to be known.

When there is no reliable accounting and therefore no competent knowledge of the economic and ecological effects of our lives, we cannot live lives that are economically and ecologically responsible. This is the problem that has frustrated, and to a considerable extent undermined, the American conservation effort from the beginning. It is ultimately futile to plead and protest and lobby in favor of public ecological responsibility while, in virtually every act of our private lives, we endorse and support an economic system that is by intention, and perhaps by necessity, ecologically irresponsible.

If the industrial economy is not correctable within or by its own terms, then obviously what is required for correction is a countervailing economic idea. And the most significant weakness of the conservation movement is its failure to produce or espouse an economic idea capable of correcting the economic idea of the industrialists. Somewhere near the heart of the conservation effort as we have known it is the romantic assumption that, if we have become alienated from nature, we can become unalienated by making nature the subject of contemplation or art, ignoring the fact that we live necessarily in and from nature—ignoring, in other words, all the economic issues that are involved. Walt Whitman could say, "I think I could turn and live with animals" as if he did not know that, in fact, we do live with animals, and that the terms of our relation to them are inescapably established by our economic use of their and our world. So long as we live, we are going to be living with skylarks, nightingales, daffodils, waterfowl, streams, forests, mountains, and all the other creatures that romantic poets and artists have yearned toward. And by the way we live we will determine whether or not those creatures will live.

That this nature-romanticism of the nineteenth century ignores economic facts and relationships has not prevented it from setting the agenda for modern conservation groups. This agenda has rarely included the economics of land use, without which the conservation effort becomes almost inevitably long on sentiment and short on practicality. The giveaway is that when conservationists try to be practical they are likely to defend the "sustainable use of natural resources" with the argument that this will make the industrial economy sustainable. A further giveaway is that the longer the industrial economy lasts in its present form, the further it will demonstrate its ultimate impossibility: Every human in the world cannot, now or ever, own the whole catalogue of shoddy, high-energy industrial products, which cannot be sustainably made or used. Moreover, the longer the industrial economy lasts, the more it will eat away the possibility of a better economy.

The conservation effort has at least brought under suspicion the general relativism of our age. Anybody who has studied with care the issues of conservation knows that our acts are being measured by a real, absolute, and unyielding standard that was invented by no human. Our acts that are not in harmony with nature are inevitably and sometimes irremediably destructive. The standard exists. But having no opposing economic idea, conservationists have had great difficulty in applying the standard.

What, then, is the countervailing idea by which we might correct the industrial idea? We will not have to look hard to find it, for there is only one, and that is agrarianism. Our major difficulty (and danger) will be in attempting to deal with agrarianism as "an idea"—agrarianism is primarily a practice, a set of attitudes, a loyalty, and a passion; it is an idea only secondarily and at a remove. To use merely the handiest example: I was raised by agrarians, my bias and point of view from my earliest childhood were agrarian, and yet I never heard agrarianism defined, or even so much as named, until I was a sophomore in college. I am well aware of the danger in defining things, but if I am going to talk about agrarianism, I am going to have to define it. The definition that follows is derived both from agrarian writers, ancient and modern, and from the unliterary and sometimes illiterate agrarians who have been my teachers.

The fundamental difference between industrialism and agrarianism is this: Whereas industrialism is a way of thought based on monetary capital and technology, agrarianism is a way of thought based on land.

Agrarianism, furthermore, is a culture at the same time that it is an economy. Industrialism is an economy before it is a culture. Industrial culture is an accidental by-product of the ubiquitous effort to sell unnecessary products for more than they are worth.

An agrarian economy rises up from the fields, woods, and streams—from the complex of soils, slopes, weathers, connections, influences, and exchanges that we mean when we speak, for example, of the local community or the local watershed. The agrarian mind is therefore not regional or national, let alone global, but local. It must know on intimate terms the local plants and animals and local soils; it must know local possibilities and impossibilities, opportunities and hazards. It depends and insists on knowing very particular local histories and biographies.

Because a mind so placed meets again and again the necessity for work to be good, the agrarian mind is less interested in abstract quantities than in particular qualities. It feels threatened and sickened when it hears people and creatures and places spoken of as labor, management, capital, and raw material. It is not at all impressed by the industrial legendry of gross national products, or of the numbers sold and dollars earned by gigantic corporations. It is interested—and forever fascinated—by questions leading toward the accomplishment of good work: What is the best location for a particular building or fence?

What is the best way to plow *this* field? What is the best course for a skid road in *this* woodland? Should *this* tree be cut or spared? What are the best breeds and types of livestock for *this* farm?—questions which cannot be answered in the abstract, and which yearn not toward quantity but toward elegance. Agrarianism can never become abstract because it has to be practiced in order to exist.

And though this mind is local, almost absolutely placed, little attracted to mobility either upward or lateral, it is not provincial; it is too taken up and fascinated by its work to feel inferior to any other mind in any other place.

An agrarian economy is always a subsistence economy before it is a market economy. The center of an agrarian farm is the household. The function of the household economy is to assure that the farm family lives from the farm so far as possible. It is the subsistence part of the agrarian economy that assures its stability and its survival. A subsistence economy necessarily is highly diversified, and it characteristically has involved hunting and gathering as well as farming and gardening. These activities bind people to their local landscape by close, complex interests and economic ties. The industrial economy alienates people from the native landscape precisely by breaking these direct practical ties and introducing distant dependences.

Agrarian people of the present, knowing that the land must be well cared for if anything is to last, understand the need for a settled connection, not just between farmers and their farms, but between urban people and their surrounding and tributary landscapes. Because the knowledge and know-how of good caretaking must be handed down to children, agrarians recognize the necessity of preserving the coherence of families and communities.

The stability, coherence, and longevity of human occupation require that the land should be divided among many owners and users. The central figure of agrarian thought has invariably been the small owner or smallholder who maintains a significant measure of economic self-determination on a small acreage. The scale and independence of such holdings imply two things that agrarians see as desirable: intimate care in the use of the land, and political democracy resting upon the indispensable foundation of economic democracy.

A major characteristic of the agrarian mind is a longing for independence—that is, for an appropriate degree of personal and local self-sufficiency. Agrarians wish to earn and deserve what they have. They do not wish to live by piracy, beggary, charity, or luck.

In the written record of agrarianism, there is a continually recurring affirmation of nature as the final judge, lawgiver, and pattern-maker of and for the human use of the earth. We can trace the lineage of this thought in the West through the writings of Virgil, Spenser, Shakespeare, Pope, Jefferson, and on into the work of the twentieth-century agriculturists and scientists J. Russell Smith, Liberty Hyde Bailey, Albert Howard, Wes Jackson, John Todd, and others. The idea is variously stated: We should not work until we have looked and seen where we are; we should honor Nature not only as our mother or grandmother, but as our teacher and judge; we should "let the forest judge"; we should "consult the Genius of the Place"; we should make the farming fit the farm; we should carry over into the cultivated field the diversity and coherence of the native forest or prairie. And this way of thinking is surely allied to that of the medieval scholars and architects who saw the building of a cathedral as a symbol or analogue of the creation of the world. The agrarian mind is, at bottom, a religious mind. It subscribes to Allen Tate's doctrine of "the whole horse." It prefers the Creation itself to the powers and quantities to which it can be reduced. And this is a mind completely different from that which sees creatures as machines, minds as computers, soil fertility as chemistry, or agrarianism as an idea. John Haines has written that "the eternal task of the artist and the poet, the historian and the scholar . . . is to find the means to reconcile what are two separate and yet inseparable histories, Nature and Culture. To the extent that we can do this, the 'world' makes sense to us and can be lived in." I would add only that this applies also to the farmer, the forester, the scientist, and others.

The agrarian mind begins with the love of fields and ramifies in good farming, good cooking, good eating, and gratitude to God. Exactly analogous to the agrarian mind is the sylvan mind that begins with the love of forests and ramifies in good forestry, good woodworking, good carpentry, and gratitude to God. These two kinds of mind readily intersect; neither ever intersects with the industrial-economic mind. The industrial-economic mind begins with ingratitude, and ramifies in the destruction of farms and forests. The "lowly" and "menial" arts of farm and forest are mostly taken for granted or ignored by the culture of the "fine arts" and by "spiritual" religions; they are taken for granted or ignored or held in contempt by the powers of the industrial economy. But in fact they are inescapably the foundation of human life and culture, and their adepts are capable of as deep satisfactions and as high attainments as anybody else.

Having, so to speak, laid industrialism and agrarianism side by side, implying a preference for the latter, I will be confronted by two questions that I had better go ahead and answer.

The first is whether or not agrarianism is simply a "phase" that we humans had to go through and then leave behind in order to get onto the track of technological progress toward ever greater happiness. The answer is that although industrialism has certainly conquered agrarianism, and has very nearly destroyed it altogether, it is also true that in every one of its uses of the natural world industrialism is in the process of catastrophic failure. Industry is now desperately shifting—by means of genetic engineering, global colonialism, and other contrivances—to prolong its control of our farms and forests, but the failure nonetheless continues. It is not possible to argue sanely in favor of soil erosion, water pollution, genetic impoverishment, and the destruction of rural communities and local economies. Industrialism, unchecked by the affections and concerns of agrarianism, becomes monstrous. And this is because of a weakness identified by the Twelve Southerners of *I'll Take My Stand* in their "Statement of Principles": Under the rule of industrialism "the remedies proposed . . . are always homeopathic." That is to say that industrialism always proposes to correct its errors and excesses by more industrialization.

The second question is whether or not by espousing the revival of agrarianism we will commit the famous sin of "turning back the clock." The answer to that, for present-day North Americans, is fairly simple. The overriding impulse of agrarianism is toward the local adaptation of economies and cultures. Agrarian people wish to adapt the farming to the farm and the forestry to the forest. At times and in places we latter-day

Americans have come close to accomplishing this goal, and we have a few surviving examples, but it is generally true that we are much further from local adaptation now than we were fifty years ago. We never yet have developed stable, sustainable, locally adapted land-based economies. The good rural enterprises and communities that we will find in our past have been almost constantly under threat from the colonialism, first foreign and then domestic, and now "global," which has so far dominated our history, and which has been institutionalized for a long time in the industrial economy. The possibility of an authentically settled country still lies ahead of us.

If we wish to look ahead, we will see not only in the United States but in the world two economic programs that conform pretty exactly to the aims of industrialism and agrarianism as I have described them.

The first is the effort to globalize the industrial economy, not merely by the expansionist programs of supra-national corporations within themselves, but also by means of government-sponsored international trade agreements, the most prominent of which is the World Trade Organization Agreement, which institutionalizes the industrial ambition to use, sell, or destroy every acre and every creature of the world.

The World Trade Organization gives the lie to the industrialist conservatives' professed abhorrence of big government. The cause of big government, after all, is big business. The power to do large-scale damage, which is gladly assumed by every large-scale industrial enterprise, calls naturally and logically for government regulation, which of course the corporations object to. But we have a good deal of evidence also that the leaders of big business actively desire and promote big government. They and their political allies, while ostensibly working to "downsize" government, continue to promote government helps and "incentives" to large corporations; and, however absurdly, they adhere to their notion that a small government, taxing only the working people, can maintain a big highway system, a big military establishment, a big space program, and big government contracts.

But the most damaging evidence is the World Trade Organization itself, which is in effect a global government, with power to enforce the decisions of the collective against national laws that conflict with it. The coming of the World Trade Organization was foretold seventy years ago in the "Statement of Principles" of *I'll Take My Stand*, which said that "the true Sovietists or Communists . . . are the industrialists themselves. They would have the government set up an economic super-organization, which in turn would become the government." The agrarians of *I'll Take My Stand* did not foresee this because they were fortune-tellers, but because they had perceived accurately the character and motive of the industrial economy.

The second program, counter to the first, is composed of many small efforts to preserve or improve or establish local economies. These efforts on the part of nonindustrial or agrarian conservatives, local patriots, are taking place in countries both affluent and poor all over the world.

Whereas the corporate sponsors of the World Trade Organization, in order to promote their ambitions, have required only the hazy glamour of such phrases as "the global economy," "the global context," and "globalization," the local economists use a much more diverse and particularizing vocabulary that you can actually think with: "community," "ecosystem," "watershed," "place," "homeland," "family," "household."

And whereas the global economists advocate a world-government-by-economic-bureaucracy, which would destroy local adaptation everywhere by ignoring the uniqueness of every place, the local economists found their work upon respect for such uniqueness. Places differ from one another, the local economists say, therefore we must behave with unique consideration in each one; the ability to tender an appropriate practical regard and respect to each place in its difference is a kind of freedom; the inability to do so is a kind of tyranny. The global economists are the great centralizers of our time. The local economists, who have so far attracted the support of no prominent politician, are the true decentralizers and downsizers, for they seek an appropriate degree of self-determination and independence for localities. They seem to be moving toward a radical and necessary revision of our idea of a city. They are learning to see the city, not just as a built and paved municipality set apart by "city limits" to live by trade and transportation from the world at large, but rather as a part of a local community which includes also the city's rural neighbors, its surrounding landscape and its watershed, on which it might depend for at least some of its necessities, and for the health of which it might exercise a competent concern and responsibility.

At this point, I want to say point blank what I hope is already clear: Though agrarianism proposes that everybody has agrarian responsibilities, it does not propose that everybody should be a farmer or that we do not need cities. Nor does it propose that every product should be a necessity. Furthermore, any thinkable human economy would have to grant to manufacturing an appropriate and honorable place. Agrarians would insist only that any manufacturing enterprise should be formed and scaled to fit the local landscape, the local ecosystem, and the local community, and that it should be locally owned and employ local people. They would insist, in other words, that the shop or factory owner should not be an outsider, but rather a sharer in the fate of the place and its community. The deciders should have to live with the results of their decisions.

Between these two programs—the industrial and the agrarian, the global and the local—the most critical difference is that of knowledge. The global economy institutionalizes a global ignorance, in which producers and consumers cannot know or care about one another, and in which the histories of all products will be lost. In such a circumstance, the degradation of products and places, producers and consumers is inevitable.

But in a sound local economy, in which producers and consumers are neighbors, nature will become the standard of work and production. Consumers who understand their economy will not tolerate the destruction of the local soil or ecosystem or watershed as a cost of production. Only a healthy local economy can keep nature and work together in the consciousness of the community. Only such a community can restore history to economics.

I will not be altogether surprised to be told that I have set forth here a line of thought that is attractive but hopeless. A number of critics have advised me of this, out of their charity, as if I might have written of my hopes for forty years without giving a thought to hopelessness. Hope, of course, is always accompanied by the fear of hopelessness, which is a legitimate fear.

And so I would like to conclude by confronting directly the issue of hope. My hope is most seriously challenged by the fact of decline, of loss. The things that I have tried to defend are less numerous and worse off now than when I started, but in this I am only like all other conservationists. All of us have been fighting a battle that on average we are losing, and I doubt that there is any use in reviewing the statistical proofs. The point—the only interesting point—is that we have not quit. Ours is not a fight that you can stay in very long if you look on victory as a sign of triumph or on loss as a sign of defeat. We have not quit because we are not hopeless.

My own aim is not hopelessness. I am not looking for reasons to give up. I am looking for reasons to keep on. In outlining here the concerns of agrarianism, I have intended to show how the effort of conservation could be enlarged and strengthened.

What agrarian principles implicitly propose—and what I explicitly propose in advocating those principles at this time—is a revolt of local small producers and local consumers against the global industrialism of the corporations. Do I think that there is a hope that such a revolt can survive and succeed, and that it can have a significant influence upon our lives and our world?

Yes, I do. And to be as plain as possible, let me just say what I know. I know from friends and neighbors and from my own family that it is now possible for farmers to sell at a premium to local customers such products as organic vegetables, organic beef and lamb, and pasture-raised chickens. This market is being made by the exceptional goodness and freshness of the food, by the wish of urban consumers to support their farming neighbors, and by the excesses and abuses of the corporate food industry.

This is the pattern of an economic revolt that is not only possible but is happening. It is happening for two reasons: First, as the scale of industrial agriculture increases, so does the scale of its abuses, and it is hard to hide large-scale abuses from consumers. It is virtually impossible now for intelligent consumers to be ignorant of the heartlessness and nastiness of animal confinement operations and their excessive use of antibiotics, of the use of hormones in meat and milk production, of the stenches and pollutants of pig and poultry factories, of the use of toxic chemicals and the waste of soil and soil health in industrial row-cropping, of the mysterious or disturbing or threatening practices associated with industrial food storage, preservation, and processing. Second, as the food industries focus more and more on gigantic global opportunities, they cannot help but overlook small local opportunities, as is made plain by the proliferation of "community-supported agriculture," farmers markets, health food stores, and so on. In fact, there are some markets that the great corporations by definition cannot supply. The market for so-called organic food, for example, is really a market for good, fresh, trustworthy food, food from producers known and trusted by consumers, and such food cannot be produced by a global corporation.

But the food economy is only one example. It is also possible to think of good local forest economies. And in the face of much neglect, it is possible to think of local small business economies—some of them related to the local economies of farm and forest—supported by locally owned, community-oriented banks.

What do these efforts of local economy have to do with conservation as we know it? The answer, I believe, is *everything*. The conservation movement, as I said earlier, has a conservation program; it has a preservation program; it has a rather sporadic health-protection program; but it has no economic program, and because it has no economic program it has the status of something exterior to daily life, surviving by emergency, like an ambulance service. In saying this, I do not mean to belittle the importance of protest, litigation, lobbying, legislation, large-scale organization—all of which I believe in and support. I am saying simply that we must do more. We must confront, on the ground, and each of us at home, the economic assumptions in which the problems of conservation originate.

We have got to remember that the great destructiveness of the industrial age comes from a division, a sort of divorce, in our economy, and therefore in our consciousness, between production and consumption. Of this radical division of functions we can say, without much fear of oversimplifying, that the aim of industrial producers is to sell as much as possible and that the aim of industrial consumers is to buy as much as possible. We need only to add that the aim of both producer and consumer is to be so far as possible carefree. Because of various pressures, governments have learned to coerce from producers some grudging concern for the health and solvency of consumers. No way has been found to coerce from consumers any consideration for the methods and sources of production.

What alerts consumers to the outrages of producers is typically some kind of loss or threat of loss. We see that in dividing consumption from production we have lost the function of conserving. Conserving is no longer an integral part of the economy of the producer or the consumer. Neither the producer nor the consumer any longer says, "I must be careful of this so that it will last." The working assumption of both is that where there is some, there must be more. If they can't get what they need in one place, they will find it in another. That is why conservation is now a separate concern and a separate effort.

But experience seems increasingly to be driving us out of the categories of producer and consumer and into the categories of citizen, family member, and community member, in all of which we have an inescapable interest in making things last. And here is where I think the conservation movement (I mean that movement that has defined itself as the defender of wilderness and the natural world) can involve itself in the fundamental issues of economy and land use, and in the process gain strength for its original causes.

I would like my fellow conservationists to notice how many people and organizations are now working to save something of value—not just wilderness places, wild rivers, wildlife habitat, species diversity, water quality, and air quality, but also agricultural land, family farms and ranches, communities, children and childhood, local schools, local economies, local food markets, livestock breeds and domestic plant varieties, fine old buildings, scenic roads, and so on. I would like my fellow conservationists to understand also that there is hardly a small farm or ranch or locally owned restaurant or store or shop or business anywhere that is not struggling to conserve itself.

All of these people, who are fighting sometimes lonely battles to keep things of value that they cannot bear to lose, are the conservation movement's natural allies. Most of them have the same enemies as the conservation movement. There is no necessary conflict among them. Thinking of them, in their great variety, in the essential likeness of their motives and concerns, one thinks of the possibility of a defined community of interest among them all, a shared stewardship of all the diversity of good things that are needed for the health and abundance of the world.

I don't suppose that this will be easy, given especially the history of conflict between conservationists and land users. I only suppose that it is necessary. Conservationists can't conserve everything that needs conserving without joining the effort to use well the agricultural lands, the forests, and the waters that we must use. To enlarge the areas protected from use without at the same time enlarging the areas of *good* use is a mistake. To have no large areas of protected old-growth forest would be folly, as most of us would agree. But it is also folly to have come this far in our history without a single working model of a thoroughly diversified and integrated, ecologically sound, local forest economy. That such an economy is possible is indicated by many imperfect or incomplete examples, but we need desperately to put the pieces together in one place—and then in every place.

The most tragic conflict in the history of conservation is that between the conservationists and the farmers and ranchers. It is tragic because it is unnecessary. There is no irresolvable conflict here, but the conflict that exists can be resolved only on the basis of a common understanding of good practice. Here again we need to foster and study working models: farms and ranches that are knowledgeably striving to bring economic practice into line with ecological reality, and local food economies in which consumers conscientiously support the best land stewardship.

We know better than to expect very soon a working model of a conserving global corporation. But we must begin to expect —and we must, as conservationists, begin working for, and in—working models of nature-conserving local economies. These are possible now. Good and able people are working hard to develop them now. They need the full support of the conservation movement now. Conservationists need to go to these people, ask what they can do to help, and then help. A little later, having helped, they can in turn ask for help.

The Agrarian Standard

THE UNSETTLING OF AMERICA was published twenty-five years ago; it is still in print and is still being read. As its author, I am tempted to be glad of this, and yet, if I believe what I said in that book, and I still do, then I should be anything but glad. The book would have had a far happier fate if it could have been disproved or made obsolete years ago.

It remains true because the conditions it describes and opposes, the abuses of farmland and farming people, have persisted and become worse over the last twenty-five years. In 2002 we have less than half the number of farmers in the United States that we had in 1977. Our farm communities are far worse off now than they were then. Our soil erosion rates continue to be unsustainably high. We continue to pollute our soils and streams with agricultural poisons. We continue to lose farmland to urban development of the most wasteful sort. The large agribusiness corporations that were mainly national in 1977 are now global, and are replacing the world's agricultural diversity, useful primarily to farmers and local consumers, with bioengineered and patented monocultures that are merely profitable to corporations. The purpose of this new global economy, as Vandana Shiva has rightly said, is to replace "food democracy" with a worldwide "food dictatorship."

To be an agrarian writer in such a time is an odd experience. One keeps writing essays and speeches that one would prefer not to write, that one wishes would prove unnecessary, that one hopes nobody will have any need for in twenty-five years. My life as an agrarian writer has certainly involved me in such confusions, but I have never doubted for a minute the importance of the hope I have tried to serve: the hope that we might become a healthy people in a healthy land.

We agrarians are involved in a hard, long, momentous contest, in which we are so far, and by a considerable margin, the losers. What we have undertaken to defend is the complex accomplishment of knowledge, cultural memory, skill, self-mastery, good sense, and fundamental decency—the high and indispensable art—for which we probably can find no better name than "good farming." I mean farming as defined by agrarianism as opposed to farming as defined by industrialism: farming as the proper use and care of an immeasurable gift.

I believe that this contest between industrialism and agrarianism now defines the most fundamental human difference, for it divides not just two nearly opposite concepts of agriculture and land use, but also two nearly opposite ways of understanding ourselves, our fellow creatures, and our world.

The way of industrialism is the way of the machine. To the industrial mind, a machine is not merely an instrument for doing work or amusing ourselves or making war; it is an explanation of the world and of life. The machine's entirely comprehensible articulation of parts defines the acceptable meanings of our experience, and it prescribes the kinds of meanings the industrial scientists and scholars expect to discover. These meanings have to do with nomenclature, classification, and rather short lineages of causation. Because industrialism cannot understand living things except as machines, and can grant them no value that is not utilitarian, it conceives of farming and forestry as forms of mining; it cannot use the land without abusing it.

Industrialism prescribes an economy that is placeless and displacing. It does not distinguish one place from another. It applies its methods and technologies indiscriminately in the American East and the American West, in the United States and in India. It thus continues the economy of colonialism. The shift of colonial power from European monarchy to global corporation is perhaps the dominant theme of modern history. All along—from the European colonization of Africa, Asia, and the New World, to the domestic colonialism of American industries, to the colonization of the entire rural world by the global corporations—it has been the same story of the gathering of an exploitive economic power into the hands of a few people who are alien to the places and the people they exploit. Such an economy is bound to destroy locally adapted agrarian economies everywhere it goes, simply because it is too ignorant not to do so. And it has succeeded precisely to the extent that it has been able to inculcate the same ignorance in workers and consumers. A part of the function of industrial education is to preserve and protect this ignorance.

To the corporate and political and academic servants of global industrialism, the small family farm and the small farming community are not known, are not imaginable, and are therefore unthinkable, except as damaging stereotypes. The people of "the cutting edge" in science, business, education, and politics have no patience with the local love, local loyalty, and local knowledge that make people truly native to their places and therefore good caretakers of their places. This is why one of the primary principles in industrialism has always been to get the worker away from home. From the beginning it has been destructive of home employment and home economies. The office or the factory or the institution is the place for work. The economic function of the household has been increasingly the consumption of purchased goods. Under industrialism, the farm too has become increasingly consumptive, and farms fail as the costs of consumption overpower the income from production.

The idea of people working at home, as family members, as neighbors, as natives and citizens of their places, is as repugnant to the industrial mind as the idea of self-employment. The industrial mind is an organizational mind, and I think this mind is deeply disturbed and threatened by the existence of people who have no boss. This may be why people with such minds, as they approach the top of the political hierarchy, so readily sell themselves to "special interests." They cannot bear to be unbossed. They cannot stand the lonely work of making up their own minds.

The industrial contempt for anything small, rural, or natural translates into contempt for uncentralized economic systems, any sort of local self-sufficiency in food or other necessities. The industrial "solution" for such systems is to increase the scale of work and trade. It is to bring Big Ideas, Big Money, and Big Technology into small rural communities, economies, and ecosystems—the brought-in industry and the experts being invariably alien to and contemptuous of the places to which they are brought in. There is never any question of propriety, of adapting the thought or the purpose or the technology to the place.

The result is that problems correctable on a small scale are replaced by large-scale problems for which there are no large-scale corrections. Meanwhile, the large-scale enterprise has reduced or destroyed the possibility of small-scale corrections. This exactly describes our present agriculture. Forcing all agricultural localities to conform to economic conditions imposed from afar by a few large corporations has caused problems of the largest possible scale, such as soil loss, genetic impoverishment, and ground-water pollution, which are correctable only by an agriculture of locally adapted, solar-powered, diversified small farms—a correction that, after a half century of industrial agriculture, will be difficult to achieve.

The industrial economy thus is inherently violent. It impoverishes one place in order to be extravagant in another, true to its colonialist ambition. A part of the "externalized" cost of this is war after war.

Industrialism begins with technological invention. But agrarianism begins with givens: land, plants, animals, weather, hunger, and the birthright knowledge of agriculture. Industrialists are always ready to ignore, sell, or destroy the past in order to gain the entirely unprecedented wealth, comfort, and happiness supposedly to be found in the future. Agrarian farmers know that their very identity depends on their willingness to receive gratefully, use responsibly, and hand down intact an inheritance, both natural and cultural, from the past. Agrarians understand themselves as the users and caretakers of some things they did not make, and of some things that they cannot make.

I said a while ago that to agrarianism farming is the proper use and care of an immeasurable gift. The shortest way to understand this, I suppose, is the religious way. Among the commonplaces of the Bible, for example, are the admonitions that the world was made and approved by God, that it belongs to Him, and that its good things come to us from Him as gifts. Beyond those ideas is the idea that the whole Creation exists only by participating in the life of God, sharing in His being, breathing His breath. "The world," Gerard Manley Hopkins said, "is charged with the grandeur of God." Such thoughts seem strange to us now, and what has estranged us from them is our economy. The industrial economy could not have been derived from such thoughts any more than it could have been derived from the golden rule.

If we believed that the existence of the world is rooted in mystery and in sanctity, then we would have a different economy. It would still be an economy of use, necessarily, but it would be an economy also of return. The economy would have to accommodate the need to be worthy of the gifts we receive and use, and this would involve a return of propitiation, praise, gratitude, responsibility, good use, good care, and a proper regard for future generations. What is most conspicuously absent from the industrial economy and industrial culture is this idea of return. Industrial humans relate themselves to the world and its creatures by fairly direct acts of violence. Mostly we take without asking, use without respect or gratitude, and give nothing in return. Our economy's most voluminous product is waste—valuable materials irrecoverably misplaced, or randomly discharged as poisons.

To perceive the world and our life in it as gifts originating in sanctity is to see our human economy as a continuing moral crisis. Our life of need and work forces us inescapably to use in time things belonging to eternity, and to assign finite values to things already recognized as infinitely valuable. This is a fearful predicament. It calls for prudence, humility, good work, propriety of scale. It calls for the complex responsibilities of caretaking and giving-back that we mean by "stewardship." To all of this the idea of the immeasurable value of the resource is central.

We can get to the same idea by a way a little more economic and practical, and this is by following through our literature the ancient theme of the small farmer or husbandman who leads an abundant life on a scrap of land often described as cast-off or poor. This figure makes his first literary appearance, so far as I know, in Virgil's Fourth Georgic:

I saw a man,
An old Cilician, who occupied
An acre or two of land that no one wanted,
A patch not worth the ploughing, unrewarding

For flocks, unfit for vineyards; he however
By planting here and there among the scrub
Cabbages or white lilies and verbena
And flimsy poppies, fancied himself a king
In wealth, and coming home late in the evening
Loaded his board with unbought delicacies.

Virgil's old squatter, I am sure, is a literary outcropping of an agrarian theme that has been carried from earliest times until now mostly in family or folk tradition, not in writing, though other such people can be found in books. Wherever found, they don't vary by much from Virgil's prototype. They don't have or require a lot of land, and the land they have is often marginal. They practice subsistence agriculture, which has been much derided by agricultural economists and other learned people of the industrial age, and they always associate frugality with abundance.

In my various travels, I have seen a number of small homesteads like that of Virgil's old farmer, situated on "land that no one wanted" and yet abundantly productive of food, pleasure, and other goods. And especially in my younger days, I was used to hearing farmers of a certain kind say, "They may run me out, but they won't starve me out" or "I may get shot, but I'm not going to starve." Even now, if they cared, I think agricultural economists could find small farmers who have prospered, not by "getting big," but by practicing the ancient rules of thrift and subsistence, by accepting the limits of their small farms, and by knowing well the value of having a little land.

How do we come at the value of a little land? We do so, following this strand of agrarian thought, by reference to the value of *no* land. Agrarians value land because somewhere back in the history of their consciousness is the memory of being landless. This memory is implicit, in Virgil's poem, in the old farmer's happy acceptance of "an acre or two of land that no one wanted." If you have no land you have nothing: no food, no shelter, no warmth, no freedom, no life. If we remember this, we know that all economies begin to lie as soon as they assign a fixed value to land. People who have been landless know that the land is invaluable; it is worth everything. Preagricultural humans, of course, knew this too. And so, evidently, do the animals. It is a fearful thing to be without a "territory." Whatever the market may say, the worth of the land is what it always was: It is worth what food, clothing, shelter, and freedom are worth; it is worth what life is worth. This perception moved the settlers from the Old World into the New. Most of our American ancestors came here because they knew what it was to be landless; to be landless was to be threatened by want and also by enslavement. Coming here, they bore the ancestral memory of serfdom. Under feudalism, the few who owned the land owned also, by an inescapable political logic, the people who worked the land.

Thomas Jefferson, who knew all these things, obviously was thinking of them when he wrote in 1785 that "it is not too soon to provide by every possible means that as few as possible shall be without a little portion of land. The small landholders are the most precious part of a state. . . ." He was saying, two years before the adoption of our Constitution, that a democratic state and democratic liberties depend upon democratic ownership of the land. He was already anticipating and fearing the division of our people into settlers, the people who wanted "a little portion of land" as a home, and, virtually opposite to those, the consolidators and exploiters of the land and the land's wealth, who would not be restrained by what Jefferson called "the natural affection of the human mind." He wrote as he did in 1785 because he feared exactly the political theory that we now have: the idea that government exists to guarantee the right of the most wealthy to own or control the land without limit.

In any consideration of agrarianism, this issue of limitation is critical. Agrarian farmers see, accept, and live within their limits. They understand and agree to the proposition that there is "this much and no more." Everything that happens on an agrarian farm is determined or conditioned by the understanding that there is only so much land, so much water in the cistern, so much hay in the barn, so much corn in the crib, so much firewood in the shed, so much food in the cellar or freezer, so much strength in the back and arms—and no more. This is the understanding that induces thrift, family coherence, neighborliness, local economies. Within accepted limits, these virtues become necessities. The agrarian sense of abundance comes from the experienced possibility of frugality and renewal within limits.

This is exactly opposite to the industrial idea that abundance comes from the violation of limits by personal mobility, extractive machinery, long-distance transport, and scientific or technological breakthroughs. If we use up the good possibilities in this place, we will import goods from some other place, or we will go to some other place. If nature releases her wealth too slowly, we will take it by force. If we make the world too toxic for honeybees, some compound brain, Monsanto perhaps, will invent tiny robots that will fly about, pollinating flowers and making honey.

To be landless in an industrial society obviously is not at all times to be jobless and homeless. But the ability of the industrial economy to provide jobs and homes depends on prosperity, and on a very shaky kind of prosperity too. It depends on "growth" of the wrong things such as roads and dumps and poisons—on what Edward Abbey called "the ideology of the cancer cell"—and on greed with purchasing power. In the absence of growth, greed, and affluence, the dependents of an industrial economy too easily suffer the consequences of having no land: joblessness, homelessness, and want. This is not a theory. We have seen it happen.

I don't think that being landed necessarily means owning land. It does mean being connected to a home landscape from which one may live by the interactions of a local economy and without the routine intervention of governments, corporations, or charities.

In our time it is useless and probably wrong to suppose that a great many urban people ought to go out into the countryside and become homesteaders or farmers. But it is not useless or wrong to suppose that urban people have agricultural responsibilities that they should try to meet. And in fact this is happening. The agrarian population among us is growing, and by no means is it made up merely of some farmers and some country people. It includes urban gardeners, urban consumers who are buying food from local farmers, organizers of local food economies, consumers who have grown doubtful of the healthfulness, the trustworthiness, and the dependability of the corporate food system—people, in other words, who understand what it means to be landless.

Apologists for industrial agriculture rely on two arguments. In one of them, they say that the industrialization of agriculture, and its dominance by corporations, has been "inevitable." It has come about and it continues by the agency of economic and technological determinism. There has been simply nothing that anybody could do about it.

The other argument is that industrial agriculture has come about by choice, inspired by compassion and generosity. Seeing the shadow of mass starvation looming over the world, the food conglomerates, the machinery companies, the chemical companies, the seed companies, and the other suppliers of "purchased inputs" have done all that they have done in order to solve "the problem of hunger" and to "feed the world."

We need to notice, first, that these two arguments, often used and perhaps believed by the same people, exactly contradict each other. Second, though supposedly it has been imposed upon the world by economic and technological forces beyond human control, industrial agriculture has been pretty consistently devastating to nature, to farmers, and to rural communities, at the same time that it has been highly profitable to the agribusiness corporations, which have submitted not quite reluctantly to its "inevitability." And, third, tearful over human suffering as they always have been, the agribusiness corporations have maintained a religious faith in the profitability of their charity. They have instructed the world that it is better for people to buy food from the corporate global economy than to raise it for themselves. What is the proper solution to hunger? Not food from the local landscape, but industrial development. After decades of such innovative thought, hunger is still a worldwide calamity.

The primary question for the corporations, and so necessarily for us, is not how the world will be fed, but who will control the land, and therefore the wealth, of the world. If the world's people accept the industrial premises that favor bigness, centralization, and (for a few people) high profitability, then the corporations will control all of the world's land and all of its wealth. If, on the contrary, the world's people might again see the advantages of local economies, in which people live, so far as they are able to do so, from their home landscapes, and work patiently toward that end, eliminating waste and the cruelties of landlessness and homelessness, then I think they might reasonably hope to solve "the problem of hunger," and several other problems as well.

But do the people of the world, allured by TV, supermarkets, and big cars, or by dreams thereof, *want* to live from their home landscapes? *Could* they do so, if they wanted to? Those are hard questions, not readily answerable by anybody. Throughout the industrial decades, people have become increasingly and more numerously ignorant of the issues of land use, of food, clothing, and shelter. What would they do, and what *could* they do, if they were forced by war or some other calamity to live from their home landscapes?

It is a fact, well attested but little noticed, that our extensive, mobile, highly centralized system of industrial agriculture is extremely vulnerable to acts of terrorism. It will be hard to protect an agriculture of genetically impoverished monocultures that is entirely dependent on cheap petroleum and long-distance transportation. We know too that the great corporations, which grow and act so far beyond the restraint of "the natural affections of the human mind," are vulnerable to the natural depravities of the human mind, such as greed, arrogance, and fraud.

The agricultural industrialists like to say that their agrarian opponents are merely sentimental defenders of ways of farming that are hopelessly old-fashioned, justly dying out. Or they say that their opponents are the victims, as Richard Lewontin put it, of "a false nostalgia for a way of life that never existed." But these are not criticisms. They are insults.

For agrarians, the correct response is to stand confidently on our fundamental premise, which is both democratic and ecological: The land is a gift of immeasurable value. If it is a gift, then it is a gift to all the living in all time. To withhold it from some is finally to destroy it for all. For a few powerful people to own or control it all, or decide its fate, is wrong.

From that premise we go directly to the question that begins the agrarian agenda and is the discipline of all agrarian practice: What is the best way to use land? Agrarians know that this question necessarily has many answers, not just one. We are not asking what is the best way to farm everywhere in the world, or everywhere in the United States, or everywhere in Kentucky or Iowa. We are asking what is the best way to farm in each one of the world's numberless places, as defined by topography, soil type, climate, ecology, history, culture, and local need. And we know that the standard cannot be determined only by market demand or productivity or profitability or technological capability, or by any other single measure, however important it may be. The agrarian standard, inescapably, is local adaptation, which requires bringing local nature, local people, local economy, and local culture into a practical and enduring harmony.

Conservationist and Agrarian

I AM A CONSERVATIONIST and a farmer, a wilderness advocate and an agrarian. I am in favor of the world's wildness, not only because I like it, but also because I think it is necessary to the world's life and to our own. For the same reason, I want to preserve the natural health and integrity of the world's economic landscapes, which is to say that I want the world's farmers, ranchers, and foresters to live in stable, locally adapted, resource-preserving communities, and I want them to thrive.

One thing that this means is that I have spent my life on two losing sides. As long as I have been conscious, the great causes of agrarianism and conservation, despite local victories, have suffered an accumulation of losses, some of them probably irreparable—while the third side, that of the land-exploiting corporations, has appeared to grow ever richer. I say "appeared" because I think their wealth is illusory. Their capitalism is based, finally, not on the resources of nature, which it is recklessly destroying, but on fantasy. Not long ago I heard an economist say, "If the consumer ever stops living beyond his means, we'll have a recession." And so the two sides of nature and the rural communities are being defeated by a third side that will eventually be found to have defeated itself.

Perhaps in order to survive its inherent absurdity, the third side is asserting its power as never before: by its control of politics, of public education, and of the news media; by its dominance of science; and by biotechnology, which it is commercializing with unprecedented haste and aggression in order to control totally the world's land-using economies and its food supply. This massive ascendancy of corporate power over democratic process is probably the most ominous development since the end of World War II, and for the most part "the free world" seems to be regarding it as merely normal.

My sorrow in having been for so long on two losing sides has been compounded by knowing that those two sides have been in conflict, not only with their common enemy, the third side, but also, and by now almost conventionally, with each other. And I am further aggrieved in understanding that everybody on my two sides is deeply implicated in the sins and in the fate of the self-destructive third side.

As a part of my own effort to think better, I decided not long ago that I would not endorse any more wilderness preservation projects that do not seek also to improve the health of the surrounding economic landscapes and human communities. One of my reasons is that I don't think we can preserve either wildness or wilderness areas if we can't preserve the economic landscapes and the people who use them. This has put me into discomfort with some of my conservation friends, but that discomfort only balances the discomfort I feel when farmers or ranchers identify me as an "environmentalist," both because I dislike the term and because I sympathize with farmers and ranchers.

Whatever its difficulties, my decision to cooperate no longer in the separation of the wild and the domestic has helped me to see more clearly the compatibility and even the coherence of my two allegiances. The dualism of domestic and wild is, after all, mostly false, and it is misleading. It has obscured for us the domesticity of the wild creatures. More important, it has obscured the absolute dependence of human domesticity upon the wildness that supports it and in fact permeates it. In suffering the now-common accusation that humans are "anthropocentric" (ugly word), we forget that the wild sheep and the wild wolves are respectively ovicentric and lupocentric. The world, we may say, is wild, and all the creatures are homemakers within it, practicing domesticity: mating, raising young, seeking food and comfort. Likewise, though the wild sheep and the farm-bred sheep are in some ways unlike in their domesticities, we forget too easily that if the "domestic" sheep become too unwild, as some occasionally do, they become uneconomic and useless: They have reproductive problems, conformation problems, and so on. Domesticity and wildness are in fact intimately connected. What is utterly alien to both is corporate industrialism—a dis-placed economic life that is without affection for the places where it is lived and without respect for the materials it uses.

The question we must deal with is not whether the domestic and the wild are separate or can be separated; it is how, in the human economy, their indissoluble and necessary connection can be properly maintained.

But to say that wildness and domesticity are not separate, and that we humans are to a large extent responsible for the proper maintenance of their relationship, is to come under a heavy responsibility to be practical. I have two thoroughly practical questions on my mind.

The first is: Why should conservationists have a positive interest in, for example, farming? There are lots of reasons, but the plainest is: Conservationists eat. To be interested in food but not in food production is clearly absurd. Urban conservationists may feel entitled to be unconcerned about food production because they are not farmers. But they can't be let off so easily, for they all are farming by proxy. They can eat only if land is farmed on their behalf by somebody somewhere in some fashion. If conservationists will attempt to resume responsibility for their need to eat, they will be led back fairly directly to all their previous concerns for the welfare of nature.

Do conservationists, then, wish to eat well or poorly? Would they like their food supply to be secure from one year to the next? Would they like their food to be free of poisons, antibiotics, alien genes, and other contaminants? Would they like a significant portion of it to be fresh? Would they like it to come to them at the lowest possible ecological cost? The answers, if responsibly given, will influence production, will influence land use, will determine the configuration and the health of landscapes.

If conservationists merely eat whatever the supermarket provides and the government allows, they are giving economic support to all-out industrial food production: to animal factories; to the depletion of soil, rivers, and aquifers; to crop monocultures and the consequent losses of biological and genetic diversity; to the pollution, toxicity, and overmedication that are the inevitable accompaniments of all-out industrial food production; to a food system based on long-distance transportation and the consequent waste of petroleum and the spread of pests and diseases; and to the division of the countryside into ever-larger farms and ever-larger fields receiving always less human affection and human care.

If, on the other hand, conservationists are willing to insist on having the best food, produced in the best way, as close to their homes as possible, and if they are willing to learn to judge the quality of food and food production, then they are going to give economic support to an entirely different kind of land use in an entirely different landscape. This landscape will have a higher ratio of caretakers to acres, of care to use. It will be at once more domestic and more wild than the industrial landscape. Can increasing the number of farms and farmers in an agricultural landscape enhance the quality of that landscape as wildlife habitat? Can it increase what we might call the wilderness value of that landscape? It *can* do so, and the determining factor would be diversity. Don't forget that we are talking about a landscape that is changing in response to an increase in local consumer demand for local food. Imagine a modern agricultural landscape devoted mainly to corn and soybeans and to animal factories. And then imagine its neighboring city developing a demand for good, locally grown food. To meet that demand, local farming would have to diversify.

If that demand is serious, if it is taken seriously, if it comes from informed and permanently committed consumers, if it promises the necessary economic support, then that radically oversimplified landscape will change. The crop monocultures and animal factories will give way to the mixed farming of plants and animals. Pastured flocks and herds of meat animals, dairy herds, and poultry flocks will return, requiring, of course, pastures and hayfields. If the urban consumers would extend their competent concern for the farming economy to include the forest economy and its diversity of products, that would improve the quality and care, and increase the acreage, of farm woodlands. And we should not forget the possibility that good farmers might, for their own instruction and pleasure, preserve patches of woodland unused. As the meadows and woodlands flourished in the landscape, so would the wild birds and animals. The acreages devoted to corn and soybeans, grown principally as livestock feed or as raw materials for industry, would diminish in favor of the fruits and vegetables required by human dinner tables.

As the acreage under perennial cover increased, soil erosion would decrease and the water-holding capacity of the soil would increase. Creeks and rivers would grow cleaner and their flow more constant. As farms diversified, they would tend to become smaller because complexity and work increase with diversity, and so the landscape would acquire more owners. As the number of farmers and the diversity of their farms increased, the toxicity of agriculture would decrease—insofar as agricultural chemicals are used to replace labor and to defray the biological costs of monoculture. As food production became decentralized, animal wastes would be dispersed, and would be absorbed and retained in the soil as nutrients rather than flowing away as waste and as pollutants. The details of such a transformation could be elaborated almost endlessly. To make short work of it here, we could just say that a dangerously oversimplified landscape would become healthfully complex, both economically and ecologically.

Moreover, since we are talking about a city that would be living in large measure from its local fields and forests, we are talking also about a local economy of decentralized, small, nonpolluting value-adding factories and shops that would be scaled to fit into the landscape with the least ecological or social disruption. And thus we can also credit to this economy an increase in independent small businesses, in self-employment, and a decrease in the combustible fuel needed for transportation and (I believe) for production.

Such an economy is technically possible, there can be no doubt of that; we have the necessary methods and equipment. The capacity of nature to accommodate, and even to cooperate in, such an economy is also undoubtable; we have the necessary historical examples. This is not, from nature's point of view, a pipe dream.

What *is* doubtable, or at least unproven, is the capacity of modern humans to choose, make, and maintain such an economy. For at least half a century we have taken for granted that the methods of farming could safely be determined by the mechanisms of industry, and that the economies of farming could safely be determined by the economic interests of industrial corporations. We are now running rapidly to the end of the possibility of that assumption. The social, ecological, and even the economic costs have become too great, and the costs are still increasing, all over the world.

Now we must try to envision an agriculture founded, not on mechanical principles, but on the principles of biology and ecology. Sir Albert Howard and Wes Jackson have argued at length for such a change of standards. If you want to farm sustainably, they have told us, then you have got to make your farming conform to the natural laws that govern the local ecosystem. You have got to farm with both plants and animals in as great a diversity as possible, you have got to conserve fertility, recycle wastes, keep the ground covered, and so on. Or, as J. Russell Smith put it seventy years ago, you have got to "fit the farming to the land"—not to the available technology or the market, as important as those considerations are, but to the land. It is necessary, in short, to maintain a proper connection between the domestic and the wild. The paramount standard by which the work is to be judged is the health of the place where the work is done.

But this is not a transformation that we can just drift into, as we drift in and out of fashions, and it is not one that we should wait to be forced into by large-scale ecological breakdown. It won't happen if a lot of people—consumers and producers, city people and country people, conservationists and land users—don't get together deliberately to make it happen.

Those are some of the reasons why conservationists should take an interest in farming and make common cause with good farmers. Now I must get on to the second of my practical questions.

Why should farmers be conservationists? Or maybe I had better ask why *are* good farmers conservationists? The farmer lives and works in the meeting place of nature and the human economy, the place where the need for conservation is most obvious and most urgent. Farmers either fit their farming to their farms, conform to the laws of nature, and keep the natural powers and services intact—or they do not. If they do not, then they increase the ecological deficit that is being charged to the future. (I had better admit that some farmers do increase the ecological deficit, but they are not the farmers I am talking about. I am not asking conservationists to support destructive ways of farming.)

Good farmers, who take seriously their duties as stewards of Creation and of their land's inheritors, contribute to the welfare of society in more ways than society usually acknowledges, or even knows. These farmers produce valuable goods, of course; but they also conserve soil, they conserve water, they conserve wildlife, they conserve open space, they conserve scenery.

All that is merely what farmers *ought* to do. But since our present society's first standard in all things is profit and it loves to dwell on "economic reality," I can't resist a glance at these good farmers in their economic circumstances, for these farmers will be poorly paid for the goods they produce, and for the services they render to conservation they will not be paid at all. Good farmers today may market products of high quality and perform well all the services I have listed, and *still* be unable to afford health insurance, and *still* find themselves mercilessly caricatured in the public media as rural simpletons, hicks, or rednecks. And then they hear the voices of the "economic realists": "Get big or get out. Sell out and go to town. Adapt or die." We have had fifty years of such realism in agriculture, and the result has been more and more large-scale monocultures and factory farms, with their ever larger social and ecological —and ultimately economic—costs.

Why do good farmers farm well for poor pay and work as good stewards of nature for no pay, many of them, moreover, having no hope that their farms will be farmed by their children (for the reasons given) or that they will be farmed by anybody?

Well, I was raised by farmers, have farmed myself, and have in turn raised two farmers—which suggests to me that I may know something about farmers, and also that I don't know very much. But over the years I along with a lot of other people have wondered, "Why do they do it?" Why do farmers farm, given their economic adversities on top of the many frustrations and difficulties normal to farming? And always the answer is: "Love. They must do it for love." Farmers farm for the love of farming. They love to watch and nurture the growth of plants. They love to live in the presence of animals. They love to work outdoors. They love the weather, maybe even when it is making them miserable. They love to live where they work and to work where they live. If the scale of their farming is small enough, they like to work in the company of their children and with the help of their children. They love the measure of independence that farm life can still provide. I have an idea that a lot of farmers have gone to a lot of trouble merely to be self-employed, to live at least a part of their lives without a boss.

And so the first thing farmers as conservationists must try to conserve is their love of farming and their love of independence. Of course they can conserve these things only by handing them down, by passing them on to their children, or to *somebody's* children. Perhaps the most urgent task for all of us who want to eat well and to keep eating is to encourage farm-raised children to take up farming. And we must recognize that this only can be done economically. Farm children are not encouraged by watching their parents take their products to market only to have them stolen at prices less than the cost of production.

But farmers obviously are responsible for conserving much more than agrarian skills and attitudes. I have already told why farmers should be, as much as any conservationists, conservers of the wildness of the world—and that is their inescapable dependence on nature. Good farmers, I believe, recognize a difference that is fundamental between what is natural and what is man-made. They know that if you treat a farm as a factory and living creatures as machines, or if you tolerate the idea of "engineering" organisms, then you are on your way to something destructive and, sooner or later, too expensive. To treat creatures as machines is an error with large practical implications.

Good farmers know too that nature can be an economic ally. Natural fertility is cheaper, often in the short run, always in the long run, than purchased fertility. Natural health, inbred and nurtured, is cheaper than pharmaceuticals and chemicals. Solar energy—if you know how to capture and use it: in grass, say, and the bodies of animals—is cheaper than petroleum. The highly industrialized factory farm is entirely dependent on "purchased inputs." The agrarian farm, well integrated into the natural systems that support it, runs to an economically significant extent on resources and supplies that are free.

It is now commonly assumed that when humans took to agriculture they gave up hunting and gathering. But hunting and gathering remained until recently an integral and lively part of my own region's traditional farming life. People hunted for wild game; they fished the ponds and streams; they gathered wild greens in the spring, hickory nuts and walnuts in the fall; they picked wild berries and other fruits; they prospected for wild honey. Some of the most memorable, and least regrettable, nights of my own youth were spent in coon hunting with farmers. There is no denying that these activities contributed to the economy of farm households, but a further fact is that they were pleasures; they were wilderness pleasures, not greatly different from the pleasures pursued by conservationists and wilderness lovers. As I was always aware, my friends the coon hunters were not motivated just by the wish to tree coons and listen to hounds and listen to each other, all of which were sufficiently attractive; they were coon hunters also because they wanted to be afoot in the woods at night. Most of the farmers I have known, and certainly the most interesting ones, have had the capacity to ramble about outdoors for the mere happiness of it, alert to the doings of the creatures, amused by the sight of a fox catching grasshoppers, or by the puzzle of wild tracks in the snow.

As the countryside has depopulated and the remaining farmers have come under greater stress, these wilderness pleasures have fallen away. But they have not yet been altogether abandoned; they represent something probably essential to the character of the best farming, and they should be remembered and revived.

Those, then, are some reasons why good farmers are conservationists, and why all farmers ought to be.

What I have been trying to do is to define a congruity or community of interest between farmers and conservationists who are not farmers. To name the interests that these two groups have in common, and to observe, as I did at the beginning, that they also have common enemies, is to raise a question that is becoming increasingly urgent: Why don't the two groups publicly and forcefully agree on the things they agree on, and make an effort to cooperate? I don't mean to belittle their disagreements, which I acknowledge to be important. Nevertheless, cooperation is now necessary, and it is possible. If Kentucky tobacco farmers can meet with antismoking groups, draw up a set of "core principles" to which they all agree, and then support those principles, something of the sort surely could happen between conservationists and certain land-using enterprises: family farms and ranches, small-scale, locally owned forestry and forest products industries, and perhaps others. Something of the sort, in fact, is beginning to happen, but so far the efforts are too small and too scattered. The larger organizations on both sides need to take an interest and get involved.

If these two sides, which need to cooperate, have so far been at odds, what is the problem? The problem, I think, is economic. The small land users, on the one hand, are struggling so hard to survive in an economy controlled by the corporations that they are distracted from their own economy's actual basis in nature. They also have not paid enough attention to the difference between their always threatened local economies and the apparently thriving corporate economy that is exploiting them.

On the other hand, the mostly urban conservationists, who mostly are ignorant of the economic adversities of, say, family-scale farming or ranching, have paid far too little attention to the connection between *their* economic life and the despoliation of nature. They have trouble seeing that the bad farming and forestry practices that they oppose as conservationists are done on their behalf, and with their consent implied in the economic proxies they have given as consumers.

These clearly are serious problems. Both of them indicate that the industrial economy is not a true description of economic reality, and moreover that this economy has been wonderfully successful in getting its falsehoods believed. Too many land users and too many conservationists seem to have accepted the doctrine that the availability of goods is determined by the availability of cash, or credit, and by the market. In other words, they have accepted the idea always implicit in the arguments of the land-exploiting corporations: that there can be, and that there is, a safe disconnection between economy and ecology, between human domesticity and the wild world. Industrializing farmers have too readily assumed that the nature of their land could safely be subordinated to the capability of their technology, and that conservation could safely be left to conservationists. Conservationists have too readily assumed that the integrity of the natural world could be preserved mainly by preserving tracts of wilderness, and that the nature and nurture of the economic landscapes could safely be left to agribusiness, the timber industry, debt-ridden farmers and ranchers, and migrant laborers.

To me, it appears that these two sides are as divided as they are because each is clinging to its own version of a common economic error. How can this be corrected? I don't think it can be, so long as each of the two sides remains closed up in its own conversation. I think the two sides need to enter into *one* conversation. They have got to talk to one another. Conservationists have got to know and deal competently with the methods and economics of land use. Land users have got to recognize the urgency, even the economic urgency, of the requirements of conservation.

Failing this, these two sides will simply concede an easy victory to their common enemy, the third side, the corporate totalitarianism which is now rapidly consolidating as "the global economy" and which will utterly dominate both the natural world and its human communities.

FROM

THE WAY OF IGNORANCE

(2005)

Secrecy vs. Rights

On January 12, 2004, according to David Stout of the *New York Times*, the Supreme Court refused to reconsider a lower court's ruling "that the Justice Department was within its rights in refusing to identify more than 700 people, most of those Arabs or Muslims, arrested for immigration violations in connection with the attacks" of September 11, 2001. Mr. Stout usefully reduced the case to its gist by writing that it "pitted two fundamental values against each other—the right of the public to know details of how its government operates and the government's need to keep some information secret to protect national security."

The Supreme Court thus decided, in effect, to support the Justice Department's policy of holding prisoners in secret for the sake of national security, but it did not resolve the conflict between "the right of the public to know" and "the government's need" for secrecy.

The first thing we notice, of course, is that secret imprisonment necessarily denies to the prisoner any semblance of due process. But more is involved than that, and I want to try to say what more is involved.

The rights of the people are openly declared (not granted, but affirmed) in the founding documents of our nation. These rights were not understood as given, and therefore retractable, by the government at its discretion; they were understood, rather, as entitlements originating in "the laws of nature and of nature's God." The government's need for secrecy, by contrast, is a need that can be defined only by the government, and only in secret.

A government's wish to rule in secret on its own initiative and authority is perfectly understandable; this is a merely human weakness. *Of course* government officials would like to keep some information secret for reasons of national security, as well as for many other reasons that may readily be imagined. It is nevertheless true that a government's wish to govern in secret is the same as the wish to govern tyrannically, as we are shown by the secret imprisonment of the seven hundred.

Perhaps it is always the tendency of those in power to wish to rule autocratically, and this is what our founders feared and provided against. Aside from any issue of faith or theory, there is a practical reason for ascribing human rights to "the laws of nature and of nature's God." Rights that originate beyond and above governmental power cannot justly be abridged or revoked by a government.

In the United States the major political dialogue has been about rights: Can the fundamental human rights, set forth in the Declaration of Independence and the Constitution as belonging equally to all, be justifiably withheld from some? Over the course of our history until now, we have decided that they cannot be. We have not decided this without reason. The most persuasive argument of the Civil Rights Movement, for example, was that you cannot withhold civil rights from some people and still confidently guarantee the same rights to others; the denial turned against one group at one time can, at another time, be turned against another group. Injustices condoned against a native racial minority may with the same impunity be used against Arab and Muslim immigrants, or against dissident white protestants. Governments, the Declaration says, derive "their just powers from the consent of the governed." If the governed accept or allow a government that is arbitrary, self-authorized, and self-justified, then, whatever may be the fate of national security, mere citizens will no longer be secure in their rights.

There is no reason, now or ever, to make light of what we are now calling "national security." People want, naturally enough, "to be secure in their persons, houses, papers, and effects," not only against "unreasonable searches and seizures," but also against violence, whether domestic or foreign in origin. That is why we threw off the arbitrary rule of England and instituted a government of our own; we wished to be secure, which is to say that we wished not to be governed except by our own consent.

We may say then that national security has been our concern from the beginning, and that from the beginning the designated purpose of the nation was to secure the rights to which its citizens individually were entitled by the laws of nature and of nature's God. The two kinds of security were understood as one. If the security of the nation ceases to imply the security of all its inhabitants in their God-given rights, then, according to the Declaration of Independence, that nation's government will have delegitimized itself and should be replaced. That is not the wild idea of somebody in the current crop of left-wingers or right-wingers, but merely what the Declaration says, and we have been living with it unobjectingly for going on 228 years.

People's wish to be safe is undoubtedly one of the paramount concerns of politics and government. We want to be safe because we have perceived accurately that we live under threat of many dangers. We expect, rightly, that the government should give us reasonable protections against at least some of those dangers, especially the ones against which we can only be protected collectively. And we have to reckon with the likelihood of circumstances in which these protections may be supplied only selectively and incompletely.

The Bill of Rights was written in anticipation both of the tendency of government to usurp the rights of individuals and of circumstances in which those rights will be hard to preserve. It would always be reasonable to foresee times of stress between the government's obligation to safeguard the lives of people and its obligation to respect their rights. This is the reason for appending a Bill of Rights to the Constitution.

However, the Bill of Rights insists that the rights of individual persons must be unfailingly respected in all circumstances excepting only one, which is set forth in Article V: "No person shall be held to answer for a capital, or otherwise infamous crime, unless on a presentment or indictment of a grand jury, except in cases arising in the land or naval forces, or in the militia, when in actual service in time of war or public danger . . ." This exception is made, apparently, to validate military courts. But Article V expressly disdains to limit its application merely to citizens; it says "No person," not "No citizen." And having stated the exception, it returns promptly to the inclusive language it began with: "nor shall any person . . . be deprived of life, liberty, or property, without due process of law . . ."

It is clear, then, that secret imprisonment by the government of *any person*, citizen or immigrant or alien or enemy, is necessarily a denial of due process of law, insofar as no person can at the same time be secretly imprisoned and publicly tried. (We now have only the government's word that the number of prisoners is, or was, seven hundred.)

A further point is somewhat harder to state, but I think it is no less obvious. Terrorism, against which the government without formally declaring war says we are "at war," at the same time that it seeks to circumvent the legal conventions of war, has certainly precipitated a time of public danger. How badly frightened the general public may be by this state of affairs is a question hard to answer. But the federal government and the courts have given evidence that they, as the terrorists intended, are badly frightened. They are so badly frightened as to believe that they have no choice but to sacrifice the rights of persons in deference to the government's need for secrecy. But it is an error to believe that these two "fundamental values" can somehow be justly "balanced" by the government or the courts, or that the people can judge responsibly between their rights, which they can easily know, and a proclaimed "need," which the government so far forbids them to know. The Constitution, anyhow, does not provide for its own suspension by the fearful in a time of war and public danger.

Contempt for Small Places

NEWSPAPER EDITORIALS deplore such human-caused degradations of the oceans as the Gulf of Mexico's "dead zone," and reporters describe practices like "mountain removal" mining in eastern Kentucky. Some day we may finally understand the connections.

The health of the oceans depends on the health of rivers; the health of rivers depends on the health of small streams; the health of small streams depends on the health of their watersheds. The health of the water is exactly the same as the health of the land; the health of small places is exactly the same as the health of large places. As we know, disease is hard to confine. Because natural law is in force everywhere, infections move.

We cannot immunize the continents and the oceans against our contempt for small places and small streams. Small destructions add up, and finally they are understood collectively as large destructions. Excessive nutrient runoff from farms and animal factories in the Mississippi watershed has caused, in the Gulf of Mexico, a hypoxic or "dead zone" of five or six thousand square miles. In forty-odd years, strip mining in the Appalachian coal fields, culminating in mountain removal, has gone far toward the destruction of a whole region, with untold damage to the region's people, to watersheds, and to the waters downstream.

There is not a more exemplary history of our contempt for small places than that of Eastern Kentucky coal mining, which has enriched many absentee corporate shareholders and left the region impoverished and defaced. Coal industry representatives are now defending mountain removal—and its attendant damage to forests, streams, wells, dwellings, roads, and community life—by saying that in "10, 15, 20 years" the land will be restored, and that such mining has "created the [level] land" needed for further industrial development.

But when you remove a mountain you also remove the topsoil and the forest, and you do immeasurable violence to the ecosystem and the watershed. These things are not to be restored in ten or twenty years, or in ten or twenty hundred years. As for the manufacture of level places for industrial development, the supply has already far exceeded any foreseeable demand. And the devastation continues.

The contradictions in the state's effort "to balance the competing interests" were stated as follows by Ewell Balltrip, director of the Kentucky Appalachian Commission: "If you don't have mining, you don't have an economy, and if you don't have an economy you don't have a way for the people to live. But if you don't have environmental quality, you won't create the kind of place where people want to live."

Yes. And if the clearly foreseeable result is a region of flat industrial sites where nobody wants to live, we need a better economy.

Compromise, Hell!

WE ARE DESTROYING OUR COUNTRY—I mean our country itself, our land. This is a terrible thing to know, but it is not a reason for despair unless we decide to continue the destruction. If we decide to continue the destruction, that will not be because we have no other choice. This destruction is not necessary. It is not inevitable, except that by our submissiveness we make it so.

We Americans are not usually thought to be a submissive people, but of course we are. Why else would we allow our country to be destroyed? Why else would we be rewarding its destroyers? Why else would we all—by proxies we have given to greedy corporations and corrupt politicians—be participating in its destruction? Most of us are still too sane to piss in our own cistern, but we allow others to do so, and we reward them for it. We reward them so well, in fact, that those who piss in our cistern are wealthier than the rest of us.

How do we submit? By not being radical enough. Or by not being thorough enough, which is the same thing.

Since the beginning of the conservation effort in our country, conservationists have too often believed that we could protect the land without protecting the people. This has begun to change, but for a while yet we will have to reckon with the old assumption that we can preserve the natural world by protecting wilderness areas while we neglect or destroy the economic landscapes—the farms and ranches and working forests—and the people who use them. That assumption is understandable in view of the worsening threats to wilderness areas, but it is wrong. If conservationists hope to save even the wild lands and wild creatures, they are going to have to address issues of economy, which is to say issues of the health of the landscapes and the towns and cities where we do our work, and the quality of that work, and the well-being of the people who do the work.

Governments seem to be making the opposite error, believing that the people can be adequately protected without protecting the land. And here I am not talking about parties or party doctrines, but about the dominant political assumption.

Sooner or later, governments will have to recognize that if the land does not prosper, nothing else can prosper for very long. We can have no industry or trade or wealth or security if we don't uphold the health of the land and the people and the people's work.

It is merely a fact that the land, here and everywhere, is suffering. We have the "dead zone" in the Gulf of Mexico and undrinkable water to attest to the toxicity of our agriculture. We know that we are carelessly and wastefully logging our forests. We know that soil erosion, air and water pollution, urban sprawl, the proliferation of highways and garbage are making our lives always less pleasant, less healthful, less sustainable, and our dwelling places more ugly.

Nearly forty years ago my state of Kentucky, like other coal-producing states, began an effort to regulate strip mining. While that effort has continued, and has imposed certain requirements of "reclamation," strip mining has become steadily more destructive of the land and the land's future. We are now permitting the destruction of entire mountains and entire watersheds. No war, so far, has done such extensive or such permanent damage. If we know that coal is an exhaustible resource, whereas the forests over it are with proper use inexhaustible, and that strip mining destroys the forest virtually forever, how can we permit this destruction? If we honor at all that fragile creature the topsoil, so long in the making, so miraculously made, so indispensable to all life, how can we destroy it? If we believe, as so many of us profess to do, that the Earth is God's property and is full of His glory, how can we do harm to any part of it?

In Kentucky, as in other unfortunate states, and again at great public cost, we have allowed—in fact we have officially encouraged—the establishment of the confined animal-feeding industry, which exploits and abuses everything involved: the land, its people, the animals, and the consumers. If we love our country, as so many of us profess to do, how can we so desecrate it?

But the economic damage is not confined just to our farms and forests. For the sake of "job creation" in Kentucky, and in other backward states, we have lavished public money on corporations that come in and stay only so long as they can exploit people here more cheaply than elsewhere. The general purpose of the present economy is to exploit, not to foster or conserve.

Look carefully, if you doubt me, at the centers of the larger towns in virtually every part of our country. You will find that they are economically dead or dying. Good buildings that used to house needful, useful, locally owned small businesses of all kinds are now empty or have evolved into junk stores or antique shops. But look at the houses, the churches, the commercial buildings, the courthouse, and you will see that more often than not they are comely and well made. And then go look at the corporate outskirts: the chain stores, the fast-food joints, the food-and-fuel stores that no longer can be called service stations, the motels. Try to find something comely or well made there.

What is the difference? The difference is that the old town centers were built by people who were proud of their place and who realized a particular value in living there. The old buildings look good because they were built by people who respected themselves and wanted the respect of their neighbors. The corporate outskirts, on the contrary, were built by people who manifestly take no pride in the place, see no value in lives lived there, and recognize no neighbors. The only value they see in the place is the money that can be siphoned out of it to more fortunate places—that is, to the wealthier suburbs of the larger cities.

Can we actually suppose that we are wasting, polluting, and making ugly this beautiful land for the sake of patriotism and the love of God? Perhaps some of us would like to think so, but in fact this destruction is taking place because we have allowed ourselves to believe, and to live, a mated pair of economic lies: that nothing has a value that is not assigned to it by the market, and that the economic life of our communities can safely be handed over to the great corporations.

We citizens have a large responsibility for our delusion and our destructiveness, and I don't want to minimize that. But I don't want to minimize, either, the large responsibility that is borne by government.

It is commonly understood that governments are instituted to provide certain protections that citizens individually cannot provide for themselves. But governments have tended to assume that this responsibility can be fulfilled mainly by the police and the military services. They have used their regulatory powers reluctantly and often poorly. Our governments have only occasionally recognized the need of land and people to be protected against economic violence. It is true that economic violence is not always as swift, and is rarely as bloody, as the violence of war, but it can be devastating nonetheless. Acts of economic aggression can destroy a landscape or a community or the center of a town or city, and they routinely do so.

Such damage is justified by its corporate perpetrators and their political abettors in the name of the "free market" and "free enterprise," but this is a freedom that makes greed the dominant economic virtue, and it destroys the freedom of other people along with their communities and livelihoods. There are such things as economic weapons of massive destruction. We have allowed them to be used against us, not just by public submission and regulatory malfeasance, but also by public subsidies, incentives, and sufferances impossible to justify.

We have failed to acknowledge this threat and to act in our own defense. As a result, our once-beautiful and bountiful countryside has long been a colony of the coal, timber, and agribusiness corporations, yielding an immense wealth of energy and raw materials at an immense cost to our land and our land's people. Because of that failure also, our towns and cities have been gutted by the likes of Wal-Mart, which have had the permitted luxury of destroying locally owned small businesses by means of volume discounts.

Because as individuals or even as communities we cannot protect ourselves against these aggressions, we need our state and national governments to protect us. As the poor deserve as much justice from our courts as the rich, so the small farmer and the small merchant deserve the same economic justice, the same freedom in the market, as big farmers and chain stores. They should not suffer ruin merely because their rich competitors can afford (for a while) to undersell them.

Furthermore, to permit the smaller enterprises always to be ruined by false advantages, either at home or in the global economy, is ultimately to destroy local, regional, and even national capabilities of producing vital supplies such as food and textiles. It is impossible to understand, let alone justify, a government's willingness to allow the human sources of necessary goods to be destroyed by the "freedom" of this corporate anarchy. It is equally impossible to understand how a government can permit, and even subsidize, the destruction of the land or of the land's productivity. Somehow we have lost or discarded any controlling sense of the interdependence of the Earth and the human capacity to use it well. The governmental obligation to protect these economic resources, inseparably human and natural, is the same as the obligation to protect us from hunger or from foreign invaders. In result, there is no difference between a domestic threat to the sources of our life and a foreign one.

It appears that we have fallen into the habit of compromising on issues that should not, and in fact cannot, be compromised. I have an idea that a large number of us, including even a large number of politicians, believe that it is wrong to destroy the Earth. But we have powerful political opponents who insist that an Earth-destroying economy is justified by freedom and profit. And so we compromise by agreeing to permit the destruction only of parts of the Earth, or to permit the Earth to be destroyed a little at a time—like the famous three-legged pig that was too well loved to be eaten all at once.

The logic of this sort of compromising is clear, and it is clearly fatal. If we continue to be economically dependent on destroying parts of the Earth, then eventually we will destroy it all.

So long a complaint accumulates a debt to hope, and I would like to end with hope. To do so I need only repeat something I said at the beginning: Our destructiveness has not been, and it is not, inevitable. People who use that excuse are morally incompetent, they are cowardly, and they are lazy. Humans don't have to live by destroying the sources of their life. People can change; they can learn to do better. All of us, regardless of party, can be moved by love of our land to rise above the greed and contempt of our land's exploiters. This of course leads to practical problems, and I will offer a short list of practical suggestions.

We have got to learn better to respect ourselves and our dwelling places. We need to quit thinking of rural America as a colony. Too much of the economic history of our land has been that of the export of fuel, food, and raw materials that have been destructively and too cheaply produced. We must reaffirm the economic value of good stewardship and good work. For that we will need better accounting than we have had so far.

We need to reconsider the idea of solving our economic problems by "bringing in industry." Every state government appears to be scheming to lure in a large corporation from somewhere else by "tax incentives" and other squanderings of the people's money. We ought to suspend that practice until we are sure that in every state we have made the most and the best of what is already there. We need to build the local economies of our communities and regions by adding value to local products and marketing them locally before we seek markets elsewhere.

We need to confront honestly the issue of scale. Bigness has a charm and a drama that are seductive, especially to politicians and financiers; but bigness promotes greed, indifference, and damage, and often bigness is not necessary. You may need a large corporation to run an airline or to manufacture cars, but you don't need a large corporation to raise a chicken or a hog. You don't need a large corporation to process local food or local timber and market it locally.

And, finally, we need to give an absolute priority to caring well for our land—for every bit of it. There should be no compromise with the destruction of the land or of anything else that we cannot replace. We have been too tolerant of politicians who, entrusted with our country's defense, become the agents of our country's destroyers, compromising on its ruin.

And so I will end this by quoting my fellow Kentuckian, a great patriot and an indomitable foe of strip mining, the late Joe Begley of Blackey: "Compromise, hell!"

Charlie Fisher

I DON'T IMAGINE Charlie Fisher told me everything he has done, but in the day and a half I spent with him I did find out that he was raised on a truck farm, that for a while he rode bulls and exhibited a trick horse on the rodeo circuit, that as a young man he worked for a dairyman, and that later he had a dairy farm of his own. His interest in logging and in working horses began while he was a hired hand in the dairy. In the winter, between milkings, he and his elderly employer spent their time in the woods at opposite ends of a crosscut saw—which, Charlie says, made him tireder than it made the old dairyman. They cut some big timber and dragged out the logs with horses. The old dairyman saw that Charlie liked working horses and was good at it. And so it was that he became both a teamster and a logger.

Though he tried other employment, those two early interests stayed with him, and he has spent many years in logging with horses. There were times when he worked alone, cutting and skidding out the logs by himself. Later, his son, David, began to work with him, skidding out the logs while Charlie cut. David, who is now twenty-two, virtually grew up in the woods. He started skidding logs with a team when he was nine, and he is still working with his father, as both teamster and log cutter.

Nine years ago, near Andover in northeast Ohio, Charlie Fisher and Jeff Green formed a company, Valley Veneer, which involves both a logging operation and a sawmill. Charlie buys the standing timber, marks the trees that are to be cut, and supervises the logging crews, while Jeff keeps things going at the mill and markets the lumber.

The mill employs eight or nine hands, and it saws three million board feet a year. It provides a local market for local timber. This obviously is good for the economy of the Andover neighborhood, but it also is good for the forest. By establishing the mill, Charlie and Jeff have invested in the neighborhood and formed a permanent connection to it, and so they have an inescapable interest in preserving the productivity of the local forest. Thus a local forest economy, if it is complex enough, will tend almost naturally to act as a conserver of the local forest ecosystem. Valley Veneer, according to Charlie and Jeff, has been warmly received into the neighborhood. The company deals with the only locally-owned bank in the area. The bankers have been not only cooperative but also friendly, at times offering more help than Charlie and Jeff asked for.

The mill yard is the neatest I have ever seen. The logs are sorted and ricked according to species. Veneer logs are laid down separately with one end resting on a pole, so that they can be readily examined by buyers. The mill crew is skillful in salvaging good lumber from damaged or inferior trees. This is extremely important, as is Jeff's marketing of lumber from inferior species such as soft maple, for it means that the cutting in the woods is never limited to the best trees. Charlie marks the trees, knowing that whatever the woodland can properly yield—soft maple or fine furniture-quality cherry or trees damaged by disease or wind—can be sawed into boards and sold. The mill seemed to me an extraordinarily efficient place, where nothing of value is wasted. Twenty percent of the slabs are sold for firewood; the rest go to the chipper and are used for pulp. The sawdust is sold to farmers, who use it as bedding for animals.

The woods operation—Charlie's end of the business—consists of three logging crews, each made up of one log cutter and two teamsters. Each of the teamsters works two horses on a logging cart or "logging arch." And so Charlie routinely employs nine men and twelve horses. At times, the cutter also will do some skidding, and this increases the number of teams in use. The three crews will usually be at work at three different sites.

Mostly they log small, privately-owned woodlots within a radius of forty or fifty miles. Charlie recently counted up and found that he had logged 366 different tracts of timber in the last three years. And there are certain advantages to working on this scale. In a horse logging operation, it is best to limit the skidding distance to five or six hundred feet, though Charlie says they sometimes increase it to a thousand, and they can go somewhat farther in winter when snow or freezing weather reduces the friction. Big tracts, however, involve longer distances, and eventually it becomes necessary either to build a road for the truck or to use a bulldozer to move the logs from where the teamsters yard them in the woods to a second yarding place accessible from the highway. For this purpose, in addition to a log truck equipped with a hydraulic loading boom, Valley Veneer owns two bulldozers, one equipped with a fork, one with a blade, and both with winches. Even so, about 98 percent of the logs are moved with horses.

The logging crews work the year round and in all weather except pouring rain. The teamsters, who furnish their own teams and equipment, receive forty dollars per thousand board feet. Two of his teamsters, Charlie says, make more than thirty thousand dollars a year each.

The logging arch, in comparison to a mechanical skidder, is a very forthright piece of equipment. Like the forecart that is widely used for fieldwork, it is simply a way to provide a drawbar for a team of horses. There are a number of differences in design, but the major one is that the logging arch's drawbar is welded on edge-up and has slots instead of holes. The slots are made so as to catch and hold the links of a log chain. Each cart carries an eighteen-foot chain with a grab hook at each end. Four metal hooks (which Charlie calls "log grabs," but which are also called "J-hooks" or "logging dogs") are linked to rings and strung on the chain, thus permitting the cart to draw as many as four logs at a time. The chain can also be used at full-length if necessary to reach a hard-to-get-to-log. Larger logs require the use of tongs, which the teamsters also carry with them, or two grabs driven into the log on opposite sides. The carts are equipped also with a cant hook and a "skipper" with which to drive the grabs into the log and knock them out again.

The slotted drawbar permits the chain to be handily readjusted as the horses work a log into position for skidding. When the log is ready to go, it is chained as closely as possible to the drawbar, so that when the horses tighten, the fore end of the log is raised off the ground. This is the major efficiency of the logging arch: By thus raising the log, the arch both keeps it from digging and reduces its friction against the ground by more than half.

We watched a team drag out a twelve-foot log containing about 330 board feet. They were well-loaded but were not straining. Charlie says that a team can handle up to five or six hundred board feet. For bigger logs, they use an additional team or a bulldozer. A good teamster can skid 3000 to 3500 board feet a day in small logs. The trick, Charlie says, is to know what your horses can do, and then see that they do that much on every pull. Overload, and you're resting too much. Underload, and you're wasting energy and time. The important thing is to keep loaded and keep moving.

Charlie Fisher is a man of long experience in the woods and extensive knowledge of the timber business and of logging technology. He has no prejudice against mechanical equipment as such, but uses it readily according to need; for a time, during his thirties, he used mechanical skidders. That this man greatly prefers horses for use in the woods is therefore of considerable interest. I asked him to explain.

His first reason, and the most important, is one I'd heard before from draft horsemen: "I've always liked horses." Charlie and David are clearly the sort of men who can't quite live without horses. Between them, they own six excellent, very large Belgian geldings and two Belgian mares. Charlie, as he explained, owns three and a half horses, and David four and a half. The two halves, fortunately, belong to the same horse, which Charlie and David own in partnership. Charlie has long been an enthusiastic participant in pulling contests, and David has followed in his father's footsteps in the arena as in the woods. Last season, David participated in twenty-three contests and Charlie in five, which for him was many fewer than usual. Charlie and his wife, Becky, showed us several shelves crowded with trophies, many of which were David's. It looked to me like they are going to need more shelves. Charlie and Becky are very proud of David, who is an accomplished logger and horseman. David, Charlie says, is an exceptionally quiet hand with a team—unlike Charlie, who confessed, "I holler." Since they would have the horses anyhow, Charlie said, they might as well put them to work in the woods, which keeps them fit and allows them to earn their keep.

Charlie's second reason for using horses in the woods, almost as important as the first, is that he likes the woods, and horses leave the woods in better condition than a skidder. A team and a logging arch require a much narrower roadway than a skidder; unlike a skidder, they don't bark trees; and they leave their skidding trails far less deeply rutted. "The horse," Charlie says, "will always be the answer to good logging in a woods."

A third attractive feature of the horse economy in the woods is that the horse logger both earns and spends his money in the local community, whereas the mechanical skidder siphons money away from the community and into the hands of large corporate suppliers. Moreover, the horse logger's kinder treatment of the woods will, in the long run, yield an economic benefit.

And, finally, horses work far more cheaply and cost far less than a skidder, thus requiring fewer trees to be cut per acre, and so permitting the horse logger to be more selective and conservative.

(Another issue involved in the use of horses for work is that of energy efficiency. Legs are more efficient than wheels over rough ground—something that will quickly be apparent to you if you try riding a bicycle over a plowed field.)

Well ahead of the logging crews, Charlie goes into the woods to mark the trees that are to be cut. Except when he is working for a "developer" who is going to clear the land, Charlie never buys or marks trees with the idea of taking every one that is marketable. His purpose is to select a number of trees, often those that need cutting because they are diseased or damaged or otherwise inferior, which will provide a reasonable income to landowner and logger alike, without destroying the wood-making capacity of the forest. The point can best be understood by considering the difference between a year's growth added to a tree fourteen inches in diameter and that added to a tree four inches in diameter. Clear-cutting or any other kind of cutting that removes all the trees of any appreciable size radically reduces the wood-making capacity of the forest. After such a cutting, in Charlie's part of the country, it will be sixty to a hundred years before another cutting can be made. Of a clear-cut woodland that adjoined one of his own tracts, Charlie said, "In fifty years there still won't be a decent log in it."

Charlie does not believe that such practices are good for the forest or the people—or, ultimately, for the timber business. He stated his interest forthrightly in economic terms, but his is the right kind of economics: "I hope maybe there'll be trees here for my son to cut in ten or twenty years." If you don't overdo the cutting, he says, a woodland can yield a cash crop every ten to fifteen years. We looked at one tract of twenty acres on which Charlie had marked about 160 trees and written the owner a check for $23,000. Charlie described this as "a young piece of timber," and he said that it "definitely" could be logged again in ten years—at which time he could both take more and leave more good trees than he will take and leave at this cutting.

Owners of wooded land should consider carefully the economics of this twenty-acre tract. If it is selectively and carefully logged every ten years, as Charlie says it can be, then every acre will earn $1,150 every ten years, or $115 per year. And this comes to the landowner without expense or effort. (These particular figures, of course, apply only to this particular woodlot. Some tracts might be more productive, others less.)

We looked at marked woodlands, at woodlands presently being logged, and finally, at the end of the second day of our visit, at a woodland that one of Charlie's crews had logged three years ago. The last, a stand predominantly of hard and soft maples, provided convincing evidence of the good sense of Charlie's kind of forestry. Very few of the remaining trees had been damaged by trees felled during the logging. I saw not a single tree that had been barked by a skidded log. The skid trails had completely healed over; there was no sign of erosion. And, most striking, the woodland was still ecologically intact. It was still a diverse, uneven-aged stand of trees, many of which were over sixteen inches in diameter. We made a photograph of three trees, standing fairly close together, which varied in diameter from seventeen to twenty-one inches. After logging, the forest is still a forest, and it will go on making wood virtually without interruption or diminishment. It seems perfectly reasonable to think that, if several generations of owners were so inclined, this sort of forestry could eventually result in an "old growth" forest that would have produced a steady income for two hundred years.

I was impressed by a good many things during my visit with Charlie Fisher, but what impressed me most is the way that Charlie's kind of logging achieves a complex fairness or justice to the several interests that are involved: the woods, the landowner, the timber company, the woods crews and their horses.

Charlie buys standing trees, and he marks every tree he buys. Within a fairly narrow margin of error, Charlie knows what he is buying, and the landowner knows what he is getting paid for. When Charlie goes in to mark the trees, he is thinking not just about what he will take, but also about what he will leave. He sees the forest as it is, and he sees the forest as it will be when the logging job is finished. I think he sees it too as it will be in ten or fifteen or twenty years, when David or another logger will return to it. By this long-term care, he serves the forest and the landowner as well as himself. As he marks the trees he is thinking also of the logging crew that will soon be there. He marks each tree that is to be cut with a slash of red paint. Sometimes, where he has seen a leaning deadfall or a dead limb or a flaw in the trunk, he paints an arrow above the slash, and this means "Look up!" The horses, like the men, are carefully borne in mind. Everywhere, the aim is to do the work in the best and the safest way.

Moreover, these are not competing interests, but seem rather to merge into one another. Thus one of Charlie's economic standards—"I hope maybe there'll be trees here for my son to cut in ten or twenty years"—becomes, in application, an ecological standard. And the ecological standard becomes, again, an economic standard as it proves to be good for business.

Most landowners, Charlie says, care how their woodlands are logged. Though they may need the income from their trees, they don't want to sacrifice the health or beauty of their woods in order to get it. Charlie's way of logging recommends itself to such people; he does not need to advertise. As we were driving away from his house on the morning of our second day, one of the neighbors waved us to a stop. This man makes his living selling firewood, and he had learned of two people who wanted their woodlands logged by a horse logger. That is the way business comes to him, Charlie said. Like other horse loggers, he has all the work he can do, and more. It has been ten years since he has had to hunt for woodlots to log. He said, "Everybody else has buyers out running the roads, looking for timber." But he can't buy all that he is offered.

I don't know that I have ever met a man with more enthusiasms than Charlie Fisher. I have mentioned already his abounding interest in his family, in forestry, and in working and pulling horses, but I have neglected to say that he is also a coon hunter. This seems to me a most revealing detail. Here is a man who makes his living by walking the woods all day, and who then entertains himself by walking the woods at night.

He told me that he had a list of several things he had planned to do when he retired, but that now, at sixty-six, he is busier than ever.

"Well," I said, "you seem to be enjoying it."

"Oh," he said, "I *love* it!"

The Way of Ignorance

In order to arrive at what you do not know
You must go by a way which is the way of ignorance.
T. S. Eliot, "East Coker"

OUR PURPOSE HERE is to worry about the predominance of the supposition, in a time of great technological power, that humans either know enough already, or can learn enough soon enough, to foresee and forestall any bad consequences of their use of that power. This supposition is typified by Richard Dawkins's assertion, in an open letter to the Prince of Wales, that "our brains . . . are big enough to see into the future and plot long-term consequences."

When we consider how often and how recently our most advanced experts have been wrong about the future, and how often the future has shown up sooner than expected with bad news about our past, Mr. Dawkins's assessment of our ability to know is revealed as a superstition of the most primitive sort. We recognize it also as our old friend hubris, ungodly ignorance disguised as godly arrogance. Ignorance plus arrogance plus greed sponsors "better living with chemistry," and produces the ozone hole and the dead zone in the Gulf of Mexico. A modern science (chemistry or nuclear physics or molecular biology) "applied" by ignorant arrogance resembles much too closely an automobile being driven by a six-year-old or a loaded pistol in the hands of a monkey. Arrogant ignorance promotes a global economy while ignoring the global exchange of pests and diseases that must inevitably accompany it. Arrogant ignorance makes war without a thought of peace.

We identify arrogant ignorance by its willingness to work on too big a scale, and thus to put too much at risk. It fails to foresee bad consequences not only because some of the consequences of all acts are inherently unforeseeable, but also because the arrogantly ignorant often are blinded by money invested; they cannot afford to foresee bad consequences.

*

Except to the arrogantly ignorant, ignorance is not a simple subject. It is perhaps as difficult for ignorance to be aware of itself as it is for awareness to be aware of itself. One can hardly begin to think about ignorance without seeing that it is available in several varieties, and so I will offer a brief taxonomy.

There is, to begin with, the kind of ignorance we may consider to be inherent. This is ignorance of all that we cannot know because of the kind of mind we have—which, I will note in passing, is neither a computer nor exclusively a brain, and which certainly is not omniscient. We cannot, for example, know the whole of which we and our minds are parts. The English poet and critic Kathleen Raine wrote that "we cannot imagine how the world might appear if we did not possess the groundwork of knowledge which we do possess; nor can we in the nature of things imagine how reality would appear in the light of knowledge which we do not possess."

A part of our inherent ignorance, and surely a most formidable encumbrance to those who presume to know the future, is our ignorance of the past. We know almost nothing of our history as it was actually lived. We know little of the lives even of our parents. We have forgotten almost everything that has happened to ourselves. The easy assumption that we have remembered the most important people and events and have preserved the most valuable evidence is immediately trumped by our inability to know what we have forgotten.

There are several other kinds of ignorance that are not inherent in our nature but come instead from weaknesses of character. Paramount among these is the willful ignorance that refuses to honor as knowledge anything not subject to empirical proof. We could just as well call it materialist ignorance. This ignorance rejects useful knowledge such as traditions of imagination and religion, and so it comes across as narrow-mindedness. We have the materialist culture that afflicts us now because a world exclusively material is the kind of world most readily used and abused by the kind of mind the materialists think they have. To this kind of mind, there is no longer a legitimate wonder. Wonder has been replaced by a research agenda, which is still a world away from demonstrating the impropriety of wonder. The materialist conservationists need to tell us how a materialist culture can justify its contempt and destructiveness of material goods.

A related kind of ignorance, also self-induced, is moral ignorance, the invariable excuse of which is objectivity. One of the purposes of objectivity, in practice, is to avoid coming to a moral conclusion. Objectivity, considered a mark of great learning and the highest enlightenment, loves to identify itself by such pronouncements as the following: "You may be right, but on the other hand so may your opponent," or "Everything is relative," or "Whatever is happening is inevitable," or "Let me be the devil's advocate." (The part of devil's advocate is surely one of the most sought after in all the precincts of the modern intellect. Anywhere you go to speak in defense of something worthwhile, you are apt to encounter a smiling savant writhing in the estrus of objectivity: "Let me play the devil's advocate for a moment." As if the devil's point of view will not otherwise be adequately represented.)

There is also ignorance as false confidence, or polymathic ignorance. This is the ignorance of people who know "all about" history or its "long-term consequences" in the future. And this is closely akin to self-righteous ignorance, which is the failure to know oneself. Ignorance of one's self and confident knowledge of the past and future often are the same thing.

Fearful ignorance is the opposite of confident ignorance. People keep themselves ignorant for fear of the strange or the different or the unknown, for fear of disproof or of unpleasant or tragic knowledge, for fear of stirring up suspicion and opposition, or for fear of fear itself. A good example is the United States Department of Agriculture's panic-stricken monopoly of inadequate meat inspections. And there is the related ignorance that comes from laziness, which is the fear of effort and difficulty. Learning often is not fun, and this is well-known to all the ignorant except for a few "educators."

And finally there are for-profit ignorance, which is maintained by withholding knowledge, as in advertising, and for-power ignorance, which is maintained by government secrecy and public lies.

Kinds of ignorance (and there must be more than I have

named) may thus be sorted out. But having sorted them out, one must scramble them back together again by acknowledging that all of them can be at work in the same mind at the same time, and in my opinion they frequently are.

I may be talking too much at large here, but I am going to say that a list of kinds of ignorance comprises half a description of a human mind. The other half, then, would be supplied by a list of kinds of knowledge.

At the head of that list let us put the empirical or provable knowledge of the materialists. This is the knowledge of dead certainty or dead facts, some of which at least are undoubtedly valuable, undoubtedly useful, but at best this is static, smallish knowledge that always is what it always was, and it is rather dull. A fact may thrill us once, but not twice. Once available, it is easy game; we might call it sitting-duck knowledge. This knowledge becomes interesting again when it enters experience by way of use.

And so, as second, let us put knowledge as experience. This is useful knowledge, but it involves uncertainty and risk. How do you know if it is going to rain, or when an animal is going to bolt or attack? Because the event has not yet happened, there is no empirical answer; you may not have time to calculate the statistical probability even on the fastest computer. You will have to rely on experience, which will increase your chance of being right. But then you also may be wrong.

The experience of many people over a long time is traditional knowledge. This is the common knowledge of a culture, which it seems that few of us any longer have. To have a culture, mostly the same people have to live mostly in the same place for a long time. Traditional knowledge is knowledge that has been remembered or recorded, handed down, pondered, corrected, practiced, and refined over a long time.

A related kind of knowledge is made available by the religious traditions and is not otherwise available. If you premise the falsehood of such knowledge, as the materialists do, then of course you don't have it and your opinion of it is worthless.

There also are kinds of knowledge that seem to be more strictly inward. Instinct is inborn knowledge: how to suck, bite, and swallow; how to run away from danger instead of toward it. And perhaps the prepositions refer to knowledge that is more or less instinctive: up, down, in, out, etc.

Intuition is knowledge as recognition, a way of knowing without proof. We know the truth of the Book of Job by intuition.

What we call conscience is knowledge of the difference between right and wrong. Whether or not this is learned, most people have it, and they appear to get it early. Some of the worst malefactors and hypocrites have it in full; how else could they fake it so well? But we should remember that some worthy people have believed conscience to be innate, an "inner light."

Inspiration, I believe, is another kind of knowledge or way of knowing, though I don't know how this could be proved. One can say in support only that poets such as Homer, Dante, and Milton seriously believed in it, and that people do at times surpass themselves, performing better than all you know of them has led you to expect. Imagination, in the highest sense, is inspiration. Gifts arrive from sources that cannot be empirically located.

Sympathy gives us an intimate knowledge of other people and other creatures that can come in no other way. So does affection. The knowledge that comes by sympathy and affection is little noticed—the materialists, I assume, are unable to notice it—but in my opinion it cannot be overvalued.

Everybody who has done physical work or danced or played a game of skill is aware of the difference between knowing how and being able. This difference I would call bodily knowledge.

And finally, to be safe, we had better recognize that there is such a thing as counterfeit knowledge or plausible falsehood.

I would say that these taxonomies of mine are more or less reasonable; I certainly would not claim that they are scientific. My only assured claim is that any consideration of ignorance and knowledge ought to be at least as complex as this attempt of mine. We are a complex species—organisms surely, but also living souls—who are involved in a life-or-death negotiation, even more complex, with our earthly circumstances, which are complex beyond our ability to guess, let alone know. In dealing with those circumstances, in trying "to see into the future and plot long-term consequences," the human mind is neither capacious enough nor exact nor dependable. We are encumbered by an inherent ignorance perhaps not significantly reducible, as well as by proclivities to ignorance of other kinds, and our ways of knowing, though impressive within human limits, have the power to lead us beyond our limits, beyond foresight and precaution, and out of control.

What I have said so far characterizes the personal minds of individual humans. But because of a certain kind of arrogant ignorance, and because of the gigantic scale of work permitted and even required by powerful technologies, we are not safe in dealing merely with personal or human minds. We are obliged to deal also with a kind of mind that I will call corporate, although it is also political and institutional. This is a mind that is compound and abstract, materialist, reductionist, greedy, and radically utilitarian. Assuming as some of us sometimes do that two heads are better than one, it ought to be axiomatic that the corporate mind is better than any personal mind, but it can in fact be much worse—not least in its apparently limitless ability to cause problems that it cannot solve, and that may be unsolvable. The corporate mind is remarkably narrow. It claims to utilize only empirical knowledge—the preferred term is "sound science," reducible ultimately to the "bottom line" of profit or power—and because this rules out any explicit recourse to experience or tradition or any kind of inward knowledge such as conscience, this mind is readily susceptible to every kind of ignorance and is perhaps naturally predisposed to counterfeit knowledge. It comes to its work equipped with factual knowledge and perhaps also with knowledge skillfully counterfeited, but without recourse to any of those knowledges that enable us to deal appropriately with mystery or with human limits. It has no humbling knowledge. The corporate mind is arrogantly ignorant by definition.

Ignorance, arrogance, narrowness of mind, incomplete knowledge, and counterfeit knowledge are of concern to us because they are dangerous; they cause destruction. When united with great power, they cause great destruction. They have caused far too much destruction already, too often of irreplaceable things. Now, reasonably enough, we are asking if it is possible, if it is even thinkable, that the destruction can be stopped. To some people's surprise, we are again backed up against the fact that knowledge is not in any simple way good. We have often been a destructive species, we are more destructive now than we have ever been, and this, in perfect accordance with ancient warnings, is because of our ignorant and arrogant use of knowledge.

Before going further, we had better ask what it is that we humans need to know. We need to know many things, of course, and many kinds of things. But let us be merely practical for the time being and say that we need to know who we are, where we are, and what we must do to live. These questions do not refer to discrete categories of knowledge. We are not likely to be able to answer one of them without answering the other two. And all three must be well answered before we can answer well a further practical question that is now pressing urgently upon us: How can we work without doing irreparable damage to the world and its creatures, including ourselves? Or: How can we live without destroying the sources of our life?

These questions are perfectly honorable, we may even say that they are perfectly obvious, and yet we have much cause to believe that the corporate mind never asks any of them. It does not care who it is, for it is not anybody; it is a mind perfectly disembodied. It does not care where it is so long as its present location yields a greater advantage than any other. It will do anything at all that is necessary, not merely to live, but to aggrandize itself. And it charges its damages indifferently to the public, to nature, and to the future.

The corporate mind at work overthrows all the virtues of the personal mind, or it throws them out of account. The corporate mind knows no affection, no desire that is not greedy, no local or personal loyalty, no sympathy or reverence or gratitude, no temperance or thrift or self-restraint. It does not observe the first responsibility of intelligence, which is to know when you don't know or when you are being unintelligent. Try to imagine an official standing up in the high councils of a global corporation or a great public institution to say, "We have grown too big," or "We now have more power than we can responsibly use," or "We must treat our employees as our neighbors," or "We must count ourselves as members of this community," or "We must preserve the ecological integrity of our work places," or "Let us do unto others as we would have them do unto us"—and you will see what I mean.

The corporate mind, on the contrary, justifies and encourages the personal mind in its worst faults and weaknesses, such as greed and servility, and frees it of any need to worry about long-term consequences. For these reliefs, nowadays, the corporate mind is apt to express noisily its gratitude to God.

But now I must hasten to acknowledge that there are some corporations that do not simply incorporate what I am calling the corporate mind. Whether the number of these is increasing or not, I don't know. These organizations, I believe, tend to have hometowns and to count themselves as participants in the local economy and as members of the local community.

I would not apply to science any stricture that I would not apply to the arts, but science now calls for special attention because it has contributed so largely to modern abuses of the natural world, and because of its enormous prestige. Our concern here has to do immediately with the complacency of many scientists. It cannot be denied that science, in its inevitable applications, has given unprecedented extremes of scale to the technologies of land use, manufacturing, and war, and to their bad effects. One response to the manifest implication of science in certain kinds of destruction is to say that we need more science, or more and better science. I am inclined to honor this proposition, if I am allowed to add that we also need more than science.

But I am not at all inclined to honor the proposition that "science is self-correcting" when it implies that science is thus made somehow "safe." Science is no more safe than any other kind of knowledge. And especially it is not safe in the context of its gigantic applications by the corporate mind. Nor is it safe in the context of its own progressivist optimism. The idea, common enough among the universities and their ideological progeny, that one's work, whatever it is, will be beneficently disposed by the market or the hidden hand or evolution or some other obscure force is an example of counterfeit knowledge.

The obvious immediate question is, How *soon* can science correct itself? Can it correct itself soon enough to prevent or correct the real damage of its errors? The answer is that it cannot correct itself soon enough. Scientists who have made a plausible "breakthrough" hasten to tell the world, including of course the corporations. And while science may have corrected itself, it is not necessarily able to correct its results or its influence.

We must grant of course that science in its laboratories may be well under control. Scientists in laboratories did not cause the ozone hole or the hypoxic zones or acid rain or Chernobyl or Bhopal or Love Canal. It is when knowledge is corporatized, commercialized, and applied that it goes out of control. Can science, then, make itself responsible by issuing appropriate warnings with its knowledge? No, because the users are under no obligation to heed or respect the warning. If the knowledge is conformable to the needs of profit or power, the warning will be ignored, as we know. We are not excused by the doctrine of scientific self-correction from worrying about the influence of science on the corporate mind, and about the influence of the corporate mind on the minds of consumers and users. Humans in general have got to worry about the origins of the permission we have given ourselves to do large-scale damage. That permission is our problem, for by it we have made our ignorance arrogant and given it immeasurable power to do harm. We are killing our world on the theory that it was never alive but is only an accidental concatenation of materials and mechanical processes. We are killing one another and ourselves on the same theory. If life has no standing as mystery or miracle or gift, then what signifies the difference between it and death?

To state the problem more practically, we can say that the ignorant use of knowledge allows power to override the question of scale, because it overrides respect for the integrity of local ecosystems, which respect alone can determine the appropriate scale of human work. Without propriety of scale, and the acceptance of limits which that implies, there can be no form —and here we reunite science and art. We live and prosper by form, which is the power of creatures and artifacts to be made whole within their proper limits. Without formal restraints, power necessarily becomes inordinate and destructive. This is why the poet David Jones wrote in the midst of World War II that "man as artist hungers and thirsts after form." Inordinate size has of itself the power to exclude much knowledge.

*

What can we do? Anybody who goes on so long about a problem is rightly expected to have something to say about a solution. One is expected to "end on a positive note," and I mean to do that. But I also mean to be careful. The question, What can we do? especially when the problem is large, implies the expectation of a large solution.

I have no large solution to offer. There is, as maybe we all have noticed, a conspicuous shortage of large-scale corrections for problems that have large-scale causes. Our damages to watersheds and ecosystems will have to be corrected one farm, one forest, one acre at a time. The aftermath of a bombing has to be dealt with one corpse, one wound at a time. And so the first temptation to avoid is the call for some sort of revolution. To imagine that destructive power might be made harmless by gathering enough power to destroy it is of course perfectly futile. William Butler Yeats said as much in his poem "The Great Day":

> Hurrah for revolution and more cannon shot!
> A beggar upon horseback lashes a beggar on foot.
> Hurrah for revolution and cannon come again!
> The beggars have changed places, but the lash goes on.

Arrogance cannot be cured by greater arrogance, or ignorance by greater ignorance. To counter the ignorant use of knowledge and power we have, I am afraid, only a proper humility, and this is laughable. But it is only partly laughable. In his political pastoral "Build Soil," as if responding to Yeats, Robert Frost has one of his rustics say,

> I bid you to a one-man revolution—
> The only revolution that is coming.

If we find the consequences of our arrogant ignorance to be humbling, and we are humbled, then we have at hand the first fact of hope: We can change ourselves. We, each of us severally, can remove our minds from the corporate ignorance and arrogance that is leading the world to destruction; we can honestly confront our ignorance and our need; we can take guidance from the knowledge we most authentically possess, from experience, from tradition, and from the inward promptings of affection, conscience, decency, compassion, even inspiration.

This change can be called by several names—change of heart,

rebirth, metanoia, enlightenment—and it belongs, I think, to all the religions, but I like the practical way it is defined in the Confucian *Great Digest*. This is from Ezra Pound's translation:

> The men of old wanting to clarify and diffuse throughout the empire that light which comes from looking straight into the heart and then acting, first set up good government in their own states; wanting good government in their states, they first established order in their own families; wanting order in the home, they first disciplined themselves; desiring self-discipline, they rectified their own hearts; and wanting to rectify their hearts, they sought precise verbal definitions of their inarticulate thoughts [the tones given off by the heart]; wishing to attain precise verbal definitions, they set to extend their knowledge to the utmost.

This curriculum does not rule out science—it does not rule out knowledge of any kind—but it begins with the recognition of ignorance and of need, of being in a bad situation.

If the ability to change oneself is the first fact of hope, then the second surely must be an honest assessment of the badness of our situation. Our situation is extremely bad, as I have said, and optimism cannot either improve it or make it look better. But there is hope in seeing it as it is. And here I need to quote Kathleen Raine again. This is a passage written in the aftermath of World War II, and she is thinking of T. S. Eliot's poem *The Waste Land*, written in the aftermath of World War I. In *The Waste Land*, Eliot bears unflinching witness to the disease of our time: We are living the death of our culture and our world. The poem's ruling metaphor is that of a waterless land perishing for rain, an image that becomes more poignant as we pump down the aquifers and dry up or pollute the rivers.

> But Eliot [Kathleen Raine said] has shown us what the world is very apt to forget, that the statement of a terrible truth has a kind of healing power. In his stern vision of the hell that lies about us . . . , there is a quality of grave consolation. In his statement of the worst, Eliot has always implied the whole extent of the reality of which that worst is only one part.

Honesty is good, then, not just because it is a virtue, but for a practical reason: It can give us an accurate description of our problem, and it can set the problem precisely in its context.

Honesty, of course, is not a solution. As I have already said, I don't think there are solutions commensurate with our problems. I think the great problems call for many small solutions. But for that possibility to attain sufficient standing among us, we need not only to put the problems in context but also to learn to put our work in context. And here is where we turn back from our ambitions to consult both the local ecosystem and the cultural instructions conveyed to us by religion and the arts. All the arts and sciences need to be made answerable to standards higher than those of any art or science. Scientists and artists must understand that they can honor their gifts and fulfill their obligations only by living and working as human beings and community members rather than as specialists. What this may involve may not be predictable even by scientists. But the best advice may have been given by Hippocrates: "As to diseases make a habit of two things—to help, or at least, to do no harm."

The wish to help, especially if it is profitable to do so, may be in human nature, and everybody wants to be a hero. To help, or to try to help, requires only knowledge; one needs to know promising remedies and how to apply them. But to do no harm involves a whole culture, and a culture very different from industrialism. It involves, at the minimum, compassion and humility and caution. The person who will undertake to help without doing harm is going to be a person of some complexity, not easily pleased, probably not a hero, probably not a billionaire.

The corporate approach to agriculture or manufacturing or medicine or war increasingly undertakes to help at the risk of harm, sometimes of great harm. And once the risk of harm is appraised as "acceptable," the result often is absurdity: We destroy a village in order to save it; we destroy freedom in order to save it; we destroy the world in order to live in it.

The apostles of the corporate mind say, with a large implicit compliment to themselves, that you cannot succeed without risking failure. And they allude to such examples as that of the Wright brothers. They don't see that the issue of risk raises directly the issue of scale. Risk, like everything else, has an appropriate scale. By propriety of scale we limit the possible damages of the risks we take. If we cannot control scale so as to limit the effects, then we should not take the risk. From this, it is clear that some risks simply should not be taken. Some experiments should not be made. If a Wright brother wishes to risk failure, then he observes a fundamental decency in risking it alone. If the Wright airplane had crashed into a house and killed a child, the corporate mind, considering the future profitability of aviation, would count that an "acceptable" risk and loss. One can only reply that the corporate mind does not have the householder's or the parent's point of view.

I am aware that invoking personal decency, personal humility, as the solution to a vast risk taken on our behalf by corporate industrialism is not going to suit everybody. Some will find it an insult to their sense of proportion, others to their sense of drama. I am offended by it myself, and I wish I could do better. But having looked about, I have been unable to convince myself that there is a better solution or one that has a better chance of working.

I am trying to follow what T. S. Eliot called "the way of ignorance," for I think that is the way that is appropriate for the ignorant. I think Eliot meant us to understand that the way of ignorance is the way recommended by all the great teachers. It was certainly the way recommended by Confucius, for who but the ignorant would set out to extend their knowledge to the utmost? Who but the knowingly ignorant would know there is an "utmost" to knowledge?

But we take the way of ignorance also as a courtesy toward reality. Eliot wrote in "East Coker":

> The knowledge imposes a pattern, and falsifies,
> For the pattern is new in every moment
> And every moment is a new and shocking
> Valuation of all we have been.

This certainly describes the ignorance inherent in the human condition, an ignorance we justly feel as tragic. But it also is a way of acknowledging the uniqueness of every individual creature, deserving respect, and the uniqueness of every moment, deserving wonder. Life in time involves a great freshness that is falsified by what we already know.

And of course the way of ignorance is the way of faith. If enough of us will accept "the wisdom of humility," giving due honor to the ever-renewing pattern, accepting each moment's "new and shocking/Valuation of all we have been," then the corporate mind as we now have it will be shaken, and it will cease to exist as its members dissent and withdraw from it.

Quantity vs. Form

MY FAMILY AND I had a good friend I will call Lily. Lily was industrious and generous, a good neighbor. She was especially well-loved by her neighbors' children and grandchildren, though she had no children of her own. She lived a long time, surviving her husband by many years. At last, permanently ill and debilitated, she had to leave the small house that she and her husband had bought in their latter years and go to the nursing home. My brother, who was her lawyer, never until then much needed, arranged for the sale of her house and all her worldly goods.

I went to visit her a day or two after the sale. She was bedfast, sick and in some pain, but perfectly clear in her mind. We talked of the past and of several of our old neighbors, long gone. And then, speaking of the sale of her possessions, she said, "I'm all finished now. Everything is done."

She said this so cheerfully that I asked her, "Lily, is it a load off your mind?"

She said, "Well, Wendell, it hurt me. I laid here the night when I knew it was all gone, and I could *see* it all, all the things I'd cared for so long. But, yes, it is a load off my mind."

I was so moved and impressed by what she said that I wrote it down. She had lived her life and met her hardships bravely and cheerfully, and now she faced her death fully aware and responsible and with what seemed to me a completed grace. I didn't then and I don't now see how she could have been more admirable.

The last time I saw Lily she was in the hospital, where the inevitable course of her illness had taken her. By then, in addition to a seriously afflicted heart, she had not recovered from a bout of pneumonia, and because of osteoporosis she had several broken bones. She was as ill probably as a living creature can be and in great pain. She was dying. But in talking with the resident physician, I discovered that he had taken her off her pain medication to increase her appetite in the hope, he said, of "getting her back on her feet."

And so a life in every sense complete had to suffer at its end this addendum of useless and meaningless pain. I don't think this episode is unusual or anomalous at the present time. The doctor's stupidity and cowardice are in fact much mitigated by being perfectly conventional. The medical industry now instructs us all that longevity is a good in itself. Plain facts and simple mercy, moreover, are readily obscured by the supposed altruism of the intent to "heal."

I am obliged now to say that I am by no means an advocate of euthanasia or "assisted suicide." My purpose here is only to notice that the ideal of a whole or a complete life, as expressed in Psalm 128 or in Tiresias' foretelling of the death of Odysseus, now appears to have been replaced by the ideal merely of a *long* life. And I do not believe that these two ideals can be reconciled.

As a man growing old, I have not been able to free my mind of the story of Lily's last days or of other stories like it that I know, and I have not been able to think of them without fear. This fear is only somewhat personal. It is also a cultural fear, the fear that something valuable and necessary to our life is being lost.

To clarify my thoughts I have been in need of some further example, and recently the associations of reading led me to Robert Southey's account of the Battle of Trafalgar in his biography of Lord Nelson. I am by conviction a pacifist, but that does not prevent me from being moved and instructed by the story of a military hero. What impresses me in Southey's account is the substantial evidence that Nelson went into the battle both expecting and fully prepared to die.

He expected to die because he had refused any suggestion that he should enter the battle in disguise in order to save himself. Instead, he would wear, Southey wrote, "as usual, his Admiral's frockcoat, bearing on the left breast four stars of the different orders with which he was invested." He thus made himself the prime target of the engagement; he would live and fight as himself, though it meant that he would die unmistakably as himself. As for his decorations: "In honor I gained them, and in honor I will die with them." And before the battle he wrote out a prayer, asking for a British victory but also for humanity afterward toward the enemy. "For myself individually," he wrote, "I commit my life to Him who made me . . ."

At the end of his account, published in 1813, eight years after the battle, Southey wrote of the admiral's death a verdict undoubtedly not so remarkable then as the succeeding two centuries have made it: "There was reason to suppose, from the appearance upon opening the body, that in the course of nature he might have attained, like his father, to a good old age. Yet he cannot be said to have fallen prematurely whose work was done . . ."

Nelson was killed at the age of forty-seven, which would seem to us in our time to be a life cut "tragically short." But Southey credited to that life a formal completeness that had little to do with its extent and much to do with its accomplishments and with Nelson's own sense of its completeness: "Thank God, I have done my duty."

The issue of the form of a lived life is difficult, for the form as opposed to the measurable extent of a life has as much to do with inward consciousness as with verifiable marks left on the world. But we are already in the thick of the problem when we have noticed that there does seem to be such a thing as a good life; that a good life consists, in part at least, of doing well; and that this possibility is an ancient one, having apparently little to do with the progress of science or how much a person knows. And so we must ask how it is that one does not have to know everything in order to do well.

The answer, apparently, is that one does so by accepting formal constraints. We are excused from the necessity of creating the universe, and most of us will not have even to command a fleet in a great battle. We come to form, we in-form our lives, by accepting the obvious limits imposed by our talents and circumstances, by nature and mortality, and thus by getting the scale right. Form permits us to live and work gracefully within our limits.

In *The Soil and Health*, his light-giving book of 1947, Sir Albert Howard wrote:

> It needs a more refined perception to recognize throughout this stupendous wealth of varying shapes and forms

> the principle of stability. Yet this principle dominates. It dominates by means of an ever-recurring cycle, a cycle which, repeating itself silently and ceaselessly, ensures the continuation of living matter. This cycle is constituted of the successive and repeated processes of birth, growth, maturity, death, and decay.

Following, as he said, "an eastern religion," Howard speaks of this cycle as "The Wheel of Life." The life of nature depends upon the uninterrupted turning of this wheel. Howard's work rested upon his conviction, obviously correct, that a farm needed to incorporate within its own working the entire revolution of the wheel of life, so that it too might remain endlessly alive and productive by obeying "Nature's law of return." When, thirty years ago, I wrote in a poem, "The farm is an infinite form," this is what I meant.

The wheel of life is a form. It is a natural form, and it can become an artistic form insofar as the art of farming and the work of a farm can be made to conform to it. It can be made a form also of the art of living, but that, I think, requires an additional step. The wheel of human—that is, of *fully* human—life would consist over the generations of birth, growth, maturity, *ripeness*, death, and decay.

"Ripeness" is implicit in the examples of Lord Nelson and my friend Lily, but the term itself comes from act V, scene 2, of *King Lear*, in which Edgar says to his father:

> Men must endure
> Their going hence, even as their coming hither.
> Ripeness is all.

By "ripeness" Edgar means a perfect readiness for death, and his sentence echoes "The readiness is all" in act V, scene 2, of *Hamlet*. In the wheel of human life, "ripeness" adds to the idea of biological growth the growth in a living soul of the knowledge of time and eternity in preparation for death. And after the addition of "ripeness," "decay" acquires the further sense of the "plowing in" of experience and memory, building up the cultural humus. The art of living thus is practiced not only by individuals, but by whole communities or societies. It is the work of the long-term education of a people. Its purpose, we may say, is to make life conform gracefully both to its natural course and to its worldly limits. And this is in fulfillment of what Vermont Chief Justice Jeffrey L. Amestoy says is "our common responsibility . . . to imagine humanity the heart can recognize."

What is or what should be the goal of our life and work? This is a fearful question and it ought to be fearfully answered. Probably it should not be answered for anybody in particular by anybody else in particular. But the ancient norm or ideal seems to have been a life in which you perceived your calling, faithfully followed it, and did your work with satisfaction; married, made a home, and raised a family; associated generously with neighbors; ate and drank with pleasure the produce of your local landscape; grew old seeing yourself replaced by your children or younger neighbors, but continuing in old age to be useful; and finally died a good or a holy death surrounded by loved ones.

Now we seem to have lost any such thought of a completed life. We no longer imagine death as an appropriate end or as a welcome deliverance from pain or grief or weariness. Death now apparently is understood, and especially by those who have placed themselves in charge of it, as a punishment for growing old, to be delayed at any cost.

We seem to be living now with the single expectation that there should and will always be more of everything, including "life expectancy." This insatiable desire for more is the result of an overwhelming sense of incompleteness, which is the result of the insatiable desire for more. This is the wheel of death. It is the revolving of this wheel that now drives technological progress. The more superficial and unsatisfying our lives become, the faster we need to progress. When you are skating on thin ice, speed up.

The medical industry's invariable unction about life-saving, healing, and the extended life expectancy badly needs a meeting on open ground with tragedy, absurdity, and moral horror. To wish for a longer life is to wish implicitly for an extension of the possibility that one's life may become a burden or even a curse. And what are we to think when a criminal becomes a medical emergency by the beneficence of nature, is accorded the full panoply of technological mercy, and is soon back in practice? The moral horror comes when the suffering or dementia of an overly extended life is reduced to another statistical verification of the "miracle" of modern medicine; or when a mental disease, such as the inability to face death or an ungovernable greed for more of everything, is exploited for profit.

Perhaps there is nobody now who has not benefited in some urgently personal way from the technology of the modern medical industry. To disregard the benefit is a falsehood, and to be ungrateful is inexcusable. But even gratitude does not free us of the obligation to be critical when criticism is needed. And there can be no doubt that the rapid development of industrial technology in medicine—and, as I am about to show, in agriculture—is much in need of criticism. We need to study with great concern the effects of introducing the mechanical and chemical procedures and quantitative standards of industry into the organic world and into the care of creatures. If this has given us benefits, it has also charged us and our world with costs that, typically, have been ignored by the accountants of progress. There is never, at best, an exact fit between the organic world and industrial technology. At worst, there is contradiction, opposition, and serious damage.

Industrial technology tends to obscure or destroy the sense of appropriate scale and of propriety of application. The standard of performance tends to be set by the capacity of the technology rather than the individual nature of places and creatures. Industrial technology, instead of adapting itself to life, attempts to adapt life to itself by treating life as merely a mechanical or chemical process, and thus it inhibits the operation of love, imagination, familiarity, compassion, fear, and awe. It reduces responsibility to routine, and work to "processing." It destroys the worker's knowledge of what is being worked upon.

II

The opposition of quantity and form in agriculture is not so immediately painful as in medicine, but it is more obvious. The medical industry has lifted the "norm" of life expectancy out of reach by proposing to extend longevity ad infinitum. Likewise, agricultural science, agribusiness, and the food industry propose to increase production ad infinitum, and this is their only aim. They will increase production by any means and at any cost, even at the cost of future productivity, for they have no functioning idea of ecological or agricultural or human limits. And since the agricultural economy is controlled by agribusiness and the food industry, their fixation on quantity is too easily communicated to farmers.

The art of farming, as I said earlier, fashions the farm's cycle of productivity so that it conforms to the wheel of life. That is Sir Albert Howard's language. In Wes Jackson's language, the art of farming is to mimic on the farm the self-renewing processes of the local ecosystem. But that is not all. The art of farming is also the art of living on a farm. The form of a farm is partly in its embodied consciousness of ecological obligation, and thus in its annual cycle of work, but it is also in the arrangement of fields and buildings in relation to the life of the farm's human family whose focus is the household. There is thus a convergence or even a coincidence between the form of a farm and the form of a farming life. The art of sustaining fertility and the art of living on a farm are mutually enhancing and mutually reinforcing.

A long view of an old agricultural landscape, in America and even more in Europe, would show how fencerows and fields have conformed over time both to natural topography and to human use, and how the location of dwellings, barns, and outbuildings reveals the established daily and seasonal patterns of work. In talking now about such farms in such landscapes, we are talking mostly about the past. Such farms were highly diversified and formally complex, and sometimes they were impressively sensitive (though perhaps no farm can ever be sensitive enough) both to the requirements of the place and to human need. The pursuit of higher and higher productivity has replaced those complex forms with the form (if it can be called that) of a straight line. The minimal formality of the straight line is even further attenuated because the line really is an arrow pointing toward nothing at all that is present, but toward the goal of even more production in the future. The line, it is proposed, will go on and on from one record yield to another.

And the line of this determination is marked on the ground by longer and longer rows, which is to say larger and larger farms.

The exclusive standard of productivity destroys the formal integrity of a farm just as the exclusive standard of longevity destroys the formal integrity of a life. The quest for higher and higher production on farms leads almost inevitably to specialization, ignoring the natural impulsion toward diversity; specialization in turn obliterates local proprieties of scale and proportion and obscures any sense of human connection. Driven by fashion, debt, and bad science, the desire for more overrides completely the idea of a home or a home place or a home economy or a home community. The desire for quantity replaces the desire for wholeness or holiness or health. The sense of right proportion and scale cannot survive the loss of the sense of relationship, of the parts to one another and to the whole. The result, inevitably, is ugliness, violence, and waste.

Those of us who have watched, and have cared, have seen the old diverse and complex farm homesteads dissolving into an oversimplified, overcapitalized, market-determined agriculture that destroys farms and farmers. The fences, the fencerow plants and animals, the woodlots, the ponds and wetlands, the pastures and hayfields, the grassed waterways all disappear. The farm buildings go from disuse to neglect to decay and finally to fire and the bulldozer. The farmhouse is rented, dishonored, neglected until it too goes down and disappears. A neighborhood of home places, a diverse and comely farmed landscape, is thus replaced by a mechanical and chemical, entirely-patented agricultural desert. And this is a typical reductionist blunder, the success story of a sort of materialist fundamentalism.

By indulging a limitless desire for a supposedly limitless quantity, one gives up all the things that are most desirable. One abandons any hope of the formal completeness, grace, and beauty that come only by subordinating one's life to the whole of which it is a part, and thus one is condemned to the life of a fragment, forever unfinished and incomplete, forever greedy. One loses, that is, the sense of human life as an artifact, a part made imaginatively whole.

Renewing Husbandry

I REMEMBER WELL a summer morning in about 1950 when my father sent a hired man with a McCormick High Gear No. 9 mowing machine and a team of mules to the field I was mowing with our nearly new Farmall A. That memory is a landmark in my mind and my history. I had been born into the way of farming represented by the mule team, and I loved it. I knew irresistibly that the mules were good ones. They were stepping along beautifully at a rate of speed in fact only a little slower than mine. But now I saw them suddenly from the vantage point of the tractor, and I remember how fiercely I resented their slowness. I saw them as "in my way." For those who have had no similar experience, I was feeling exactly the outrage and the low-grade superiority of a hot-rodder caught behind an aged dawdler in urban traffic. It is undoubtedly significant that in the summer of 1950 I passed my sixteenth birthday and became eligible to solve all my problems by driving an automobile.

This is not an exceptional or a remarkably dramatic bit of history. I recite it here to confirm that the industrialization of agriculture is a part of my familiar experience. I don't have the privilege of looking at it as an outsider. It is not incomprehensible to me. The burden of this essay, on the contrary, is that the industrialization of agriculture is a grand oversimplification, too readily comprehensible, to me and to everybody else.

We were mowing that morning, the teamster with his mules and I with the tractor, in the field behind the barn on my father's home place, where he and before him his father had been born, and where his father had died in February of 1946. The old way of farming was intact in my grandfather's mind until the day he died at eighty-two. He had worked mules all his life, understood them thoroughly, and loved the good ones passionately. He knew tractors only from a distance, he had seen only a few of them, and he rejected them out of hand because he thought, correctly, that they compacted the soil.

Even so, four years after his death his grandson's sudden resentment of the "slow" mule team foretold what history would bear out: The tractor would stay and the mules would go. Year after year, agriculture would be adapted more and more to the technology and the processes of industry and to the rule of industrial economics. This transformation occurred with astonishing speed because, by the measures it set for itself, it was wonderfully successful. It "saved labor," it conferred the prestige of modernity, and it was highly productive.

Though I never entirely departed from farming or at least from thoughts of farming, and my affection for my homeland remained strong, during the fourteen years after 1950 I was much away from home and was not giving to farming the close and continuous attention I have given to it in the forty years since.

In 1964 my family and I returned to Kentucky, and in a year were settled on a hillside farm in my native community, where we have continued to live. Perhaps because I was a returned traveler intending to stay, I now saw the place more clearly than before. I saw it critically, too, for it was evident at once that the human life of the place, the life of the farms and the farming community, was in decline. The old self-sufficient way of farming was passing away. The economic prosperity that had visited the farmers briefly during World War II and for a few years afterward had ended. The little towns that once had been social and economic centers, thronged with country people on Saturdays and Saturday nights, were losing out to the bigger towns and the cities. The rural neighborhoods, once held together by common memories, common work, and the sharing of help, had begun to dissolve. There were no longer local markets for chickens or eggs or cream. The spring lamb industry, once a staple of the region, was gone. The tractors and other mechanical devices certainly were saving the labor of the farmers and farm hands who had moved away, but those who had stayed were working harder and longer than ever.

Because I remembered with affection and respect my grandparents and other country people of their generation, and because I had admirable friends and neighbors with whom I was again farming, I began to ask what was happening, and why. I began to ask what would be the effects on the land, on the community, on the natural world, and on the art of farming. And these questions have occupied me steadily ever since.

The effects of this process of industrialization have become so apparent, so numerous, so favorable to the agribusiness corporations, and so unfavorable to everything else, that by now the questions troubling me and a few others in the 1960s and 1970s are being asked everywhere.

There are no doubt many ways of accounting for this change, but for convenience and brevity I am going to attribute it to the emergence of context as an issue. It has become increasingly clear that the way we farm affects the local community, and that the economy of the local community affects the way we farm; that the way we farm affects the health and integrity of the local ecosystem, and that the farm is intricately dependent, even economically, upon the health of the local ecosystem. We can no longer pretend that agriculture is a sort of economic machine with interchangeable parts, the same everywhere, determined by "market forces" and independent of everything else. We are not farming in a specialist capsule or a professionalist department; we are farming in the world, in a webwork of dependences and influences more intricate than we will ever understand. It has become clear, in short, that we have been running our fundamental economic enterprise by the wrong rules. We were wrong to assume that agriculture could be adequately defined by reductionist science and determinist economics.

If you can keep the context narrow enough (and the accounting period short enough), then the industrial criteria of labor saving and high productivity seem to work well. But the old rules of ecological coherence and of community life have remained in effect. The costs of ignoring them have accumulated, until now the boundaries of our reductive and mechanical explanations have collapsed. Their collapse reveals, plainly for all to see, the ecological and social damages that they were meant to conceal. It will seem paradoxical to some that the national and global corporate economies have narrowed the context for thinking about agriculture, but it is merely the truth. Those large economies, in their understanding and in their accounting, have excluded any concern for the land and the people. Now, in the midst of much unnecessary human and ecological damage, we are facing the necessity of a new start in agriculture.

*

And so it is not possible to look back at the tableau of team and tractor on that morning in 1950 and see it as I saw it then. That is not because I have changed, though obviously I have; it is because, in the fifty-four years since then, history and the law of consequence have widened the context of the scene as circles widen on water around a thrown stone.

My impatience at the slowness of the mules, I think, was a fairly representative emotion. I thought I was witnessing a contest of machine against organism, which the machine was bound to win. I did not see that the team arrived at the field that morning from the history of farming and from the farm itself, whereas the tractor arrived from almost an opposite history, and by means of a process reaching a long way beyond that farm or any farm. It took me a long time to understand that the team belonged to the farm and was directly supportable by it, whereas the tractor belonged to an economy that would remain alien to agriculture, functioning entirely by means of distant supplies and long supply lines. The tractor's arrival had signaled, among other things, agriculture's shift from an almost exclusive dependence on free solar energy to a total dependence on costly fossil fuel. But in 1950, like most people at that time, I was years away from the first inkling of the limits of the supply of cheap fuel.

We had entered an era of limitlessness, or the illusion thereof, and this in itself is a sort of wonder. My grandfather lived a life of limits, both suffered and strictly observed, in a world of limits. I learned much of that world from him and others, and then I changed; I entered the world of labor-saving machines and of limitless cheap fossil fuel. It would take me years of reading, thought, and experience to learn again that in this world limits are not only inescapable but indispensable.

My purpose here is not to disturb the question of the use of draft animals in agriculture—though I doubt that it will sleep indefinitely. I want instead to talk about the tractor as an influence. The means we use to do our work almost certainly affect the way we look at the world. If the fragment of autobiography I began with means anything, it means that my transformation from a boy who had so far grown up driving a team to a boy driving a tractor was a sight-changing experience.

Brought up as a teamster but now driving a tractor, a boy almost suddenly, almost perforce, sees the farm in a different way: as ground to be got over by a means entirely different, at an entirely different cost. The team, like the boy, would grow weary, but that weariness has all at once been subtracted, and the boy is now divided from the ground by the absence of a living connection that enforced sympathy as a practical good. The tractor can work at maximum speed hour after hour without tiring. There is no longer a reason to remember the shady spots where it was good to stop and rest. Tirelessness and speed enforce a second, more perilous change in the way the boy sees the farm: Seeing it as ground to be got over as fast as possible and, ideally, without stopping, he has taken on the psychology of a traveler by interstate highway or by air. The focus of his attention has shifted from the place to the technology.

I now suspect that if we work with machines the world will seem to us to be a machine, but if we work with living creatures the world will appear to us as a living creature. Be that as it may, mechanical farming certainly makes it easy to think mechanically about the land and its creatures. It makes it easy to think mechanically even about oneself, and the tirelessness of tractors brought a new depth of weariness into human experience, at a cost to health and family life that has not been fully accounted.

Once one's farm and one's thoughts have been sufficiently mechanized, industrial agriculture's focus on production, as opposed to maintenance or stewardship, becomes merely logical. And here the trouble completes itself. The almost exclusive emphasis on production permits the way of working to be determined, not by the nature and character of the farm in its ecosystem and in its human community, but rather by the national or the global economy and the available or affordable technology. The farm and all concerns not immediately associated with production have in effect disappeared from sight. The farmer too in effect has vanished. He is no longer working as an independent and loyal agent of his place, his family, and his community, but instead as the agent of an economy that is fundamentally adverse to him and to all that he ought to stand for.

After mechanization it is certainly possible for a farmer to maintain a proper creaturely and stewardly awareness of the lives in her keeping. If you look, you can still find farmers who are farming well on mechanized farms. After mechanization, however, to maintain this kind of awareness requires a distinct effort of will. And if we ask what are the cultural resources that can inform and sustain such an effort of will, I believe that we will find them gathered under the heading of *husbandry*, and here my essay arrives finally at its subject.

The word *husbandry* is the name of a connection. In its original sense, it is the name of the work of a domestic man, a man who has accepted a bondage to the household. We have no cause here, I think, to raise the issue of "sexual roles." We need only to say that our earthly life requires both husbandry and housewifery, and that nobody, certainly no household, is excused from a proper attendance to both.

Husbandry pertains first to the household; it connects the farm to the household. It is an art wedded to the art of housewifery. To husband is to use with care, to keep, to save, to make last, to conserve. Old usage tells us that there is a husbandry also of the land, of the soil, of the domestic plants and animals—obviously because of the importance of these things to the household. And there have been times, one of which is now, when some people have tried to practice a proper human husbandry of the nondomestic creatures in recognition of the dependence of our households and domestic life upon the wild world. Husbandry is the name of all the practices that sustain life by connecting us conservingly to our places and our world; it is the art of keeping tied all the strands in the living network that sustains us.

And so it appears that most and perhaps all of industrial agriculture's manifest failures are the result of an attempt to make the land produce without husbandry. The attempt to remake agriculture as a science and an industry has excluded from it the age-old husbandry that was central and essential to it, and that denoted always the fundamental domestic connections and demanded a restorative care in the use of the land and its creatures.

This effort had its initial and probably its most radical success in separating farming from the economy of subsistence. Through World War II, farm life in my region (and, I think, nearly everywhere) rested solidly upon the garden, dairy, poultry flock, and meat animals that fed the farm's family. Especially in hard times these farm families, and their farms too, survived by means of their subsistence economy. This was the husbandry and the housewifery by which the farm lived. The industrial program, on the contrary, suggested that it was "uneconomic" for a farm family to produce its own food; the effort and the land would be better applied to commercial production. The result is utterly strange in human experience: farm families who buy everything they eat at the store.

An intention to replace husbandry with science was made explicit in the renaming of disciplines in the colleges of agriculture. "Soil husbandry" became "soil science," and "animal husbandry" became "animal science." This change is worth lingering over because of what it tells us about our susceptibility to poppycock. When any discipline is made or is called a science, it is thought by some to be much increased in preciseness, complexity, and prestige. When "husbandry" becomes "science," the lowly has been exalted and the rustic has become urbane. Purporting to increase the sophistication of the humble art of farming, this change in fact brutally oversimplifies it.

"Soil science," as practiced by soil scientists, and even more as it has been handed down to farmers, has tended to treat the soil as a lifeless matrix in which "soil chemistry" takes place and "nutrients" are "made available." And this, in turn, has made farming increasingly shallow—literally so—in its understanding of the soil. The modern farm is understood as a surface on which various mechanical operations are performed, and to which various chemicals are applied. The under-surface reality of organisms and roots is mostly ignored.

"Soil husbandry" is a different kind of study, involving a different kind of mind. Soil husbandry leads, in the words of Sir Albert Howard, to understanding "health in soil, plant, animal, and man as one great subject." We apply the word "health" only to living creatures, and to soil husbandry a healthy soil is a wilderness, mostly unstudied and unknown, but teemingly alive. The soil is at once a living community of creatures and their habitat. The farm's husband, its family, its crops and animals, all are members of the soil community; all belong to the character and identity of the place. To rate the farm family merely as "labor" and its domestic plants and animals merely as "production" is thus an oversimplification, both radical and destructive.

"Science" is too simple a word to name the complex of relationships and connections that compose a healthy farm—a farm that is a full membership of the soil community. If we propose, not the reductive science we generally have, but a science of complexity, that too will be inadequate, for any complexity that science can comprehend is going to be necessarily a human construct, and therefore too simple.

The husbandry of mere humans of course cannot be complex enough either. But husbandry always has understood that what is husbanded is ultimately a mystery. A farmer, as one of his farmer correspondents once wrote to Liberty Hyde Bailey, is "a dispenser of the 'Mysteries of God.'" The mothering instinct of animals, for example, is a mystery that husbandry must use and trust mostly without understanding. The husband, unlike the "manager" or the would-be objective scientist, belongs inherently to the complexity and the mystery that is to be husbanded, and so the husbanding mind is both careful and humble. Husbandry originates precautionary sayings like "Don't put all your eggs into one basket" and "Don't count your chickens before they hatch." It does not boast of technological feats that will "feed the world."

Husbandry, which is not replaceable by science, nevertheless uses science, and corrects it too. It is the more comprehensive discipline. To reduce husbandry to science, in practice, is to transform agricultural "wastes" into pollutants, and to subtract perennials and grazing animals from the rotation of crops. Without husbandry, the agriculture of science and industry has served too well the purpose of the industrial economy in reducing the number of landowners and the self-employed. It has transformed the United States from a country of many owners to a country of many employees.

Without husbandry, "soil science" too easily ignores the community of creatures that live in and from, that make and are made by, the soil. Similarly, "animal science" without husbandry forgets, almost as a requirement, the sympathy by which we recognize ourselves as fellow creatures of the animals. It forgets that animals are so called because we once believed them to be endowed with souls. Animal science has led us away from that belief or any such belief in the sanctity of animals. It has led us instead to the animal factory, which, like the concentration camp, is a vision of Hell. Animal husbandry, on the contrary, comes from and again leads to the psalmist's vision of good grass, good water, and the husbandry of God.

(It is only a little off my subject to notice also that the high and essential art of housewifery, later known as "home economics," has now become "family and consumer science." This presumably elevates the intellectual standing of the faculty by removing family life and consumption from the context—and the economy—of a home or household.)

Agriculture must mediate between nature and the human community, with ties and obligations in both directions. To farm well requires an elaborate courtesy toward all creatures, animate and inanimate. It is sympathy that most appropriately enlarges the context of human work. Contexts become wrong by being too small—too small, that is, to contain the scientist or the farmer or the farm family or the local ecosystem or the local community—and this is crucial. "Out of context," as Wes Jackson has said, "the best minds do the worst damage."

Looking for a way to give an exact sense of this necessary sympathy, the *feeling* of husbandry at work, I found it in a book entitled *Feed My Sheep* by Terry Cummins. Mr. Cummins is a man of about my age, who grew up farming with his grandfather in Pendleton County, Kentucky, in the 1940s and early 50s. In the following sentences he is remembering himself at the age of thirteen, in about 1947:

> When you see that you're making the other things feel good, it gives you a good feeling, too.
>
> The feeling inside sort of just happens, and you can't say this did it or that did it. It's the many little things. It doesn't seem that taking sweat-soaked harnesses off tired, hot horses would be something that would make you notice. Opening a barn door for the sheep standing out in a cold rain, or throwing a few grains of corn to the chickens are small things, but these little things begin

> to add up in you, and you can begin to understand that you're important. You may not be real important like people who do great things that you read about in the newspaper, but you begin to feel that you're important to all the life around you. Nobody else knows or cares too much about what you do, but if you get a good feeling inside about what you do, then it doesn't matter if nobody else knows. I do think about myself a lot when I'm alone way back on the place bringing in the cows or sitting on a mowing machine all day. But when I start thinking about how our animals and crops and fields and woods and gardens sort of all fit together, then I get that good feeling inside and don't worry much about what will happen to me.

This passage goes to the heart of what I am trying to say, because it goes to the heart of farming as I have known it. Mr. Cummins's sentences describe an experience regrettably and perhaps dangerously missing now from the childhood of most children. They also describe the communion between the farmer as husband and the well-husbanded farm. This communion is a cultural force that can exist only by becoming personal. To see it so described is to understand at once how necessary and how threatened it now is.

I have tried to say what husbandry is, how it works, and why it is necessary. Now I want to speak of two paramount accomplishments of husbandry to which I think we will have to pay more deliberate attention, in our present circumstances, than we ever have before. These are local adaptation and local coherence of form. It is strange that a science of agriculture founded on evolutionary biology, with its practical emphasis on survival, would exempt the human species from these concerns.

True husbandry, as its first strategy of survival, has always striven to fit the farming to the farm and to the field, to the needs and abilities of the farm's family, and to the local economy. Every wild creature is the product of such an adaptive process. The same process once was a dominant influence on agriculture, for the cost of ignoring it was hunger. One striking and well-known example of local adaptation in agriculture is the number and diversity of British sheep breeds, most of which are named for the localities in which they were developed. But local adaptation must be even more refined than this example suggests, for it involves consideration of the individuality of every farm and every field.

Our recent focus upon productivity, genetic and technological uniformity, and global trade—all supported by supposedly limitless supplies of fuel, water, and soil—has obscured the necessity for local adaptation. But our circumstances are changing rapidly now, and this requirement will be forced upon us again by terrorism and other kinds of political violence, by chemical pollution, by increasing energy costs, by depleted soils, aquifers, and streams, and by the spread of exotic weeds, pests, and diseases. We are going to have to return to the old questions about local nature, local carrying capacities, and local needs. And we are going to have to resume the breeding of plants and animals to fit the region and the farm.

The same obsessions and extravagances that have caused us to ignore the issue of local adaptation have at the same time caused us to ignore the issue of form. These two issues are so closely related that it is difficult to talk about one without talking about the other. During the half century and more of our neglect of local adaptation, we have subjected our farms to a radical oversimplification of form. The diversified and reasonably self-sufficient farms of my region and of many other regions have been conglomerated into larger farms with larger fields, increasingly specialized, and subjected increasingly to the strict, unnatural linearity of the production line.

But the first requirement of a form is that it must be comprehensive; it must not leave out something that essentially belongs within it. The farm that Terry Cummins remembers was remarkably comprehensive, and it was not any one of its several enterprises alone that made him feel good, but rather "how our animals and crops and fields and woods and gardens sort of all fit together."

The form of the farm must answer to the farmer's feeling for the place, its creatures, and its work. It is a never-ending effort of fitting together many diverse things. It must incorporate the life cycle and the fertility cycles of animals. It must bring crops and livestock into balance and mutual support. It must be a pattern on the ground and in the mind. It must be at once ecological, agricultural, economic, familial, and neighborly. It must be inclusive enough, complex enough, coherent, intelligible, and durable. It must have within its limits the completeness of an organism or an ecosystem, or of any other good work of art.

The making of a form begins in the recognition and acceptance of limits. The farm is limited by its topography, its climate, its ecosystem, its human neighborhood and local economy, and of course by the larger economies, and by the preferences and abilities of the farmer. The true husbandman shapes the farm within an assured sense of what it cannot be and what it should not be. And thus the problem of form returns us to that of local adaptation.

The task before us, now as always before, is to renew and husband the means, both natural and human, of agriculture. But to talk now about renewing husbandry is to talk about unsimplifying what is in reality an extremely complex subject. This will require us to accept again, and more competently than before, the health of the ecosystem, the farm, and the human community as the ultimate standard of agricultural performance.

Unsimplification is difficult, I imagine, in any circumstances; our present circumstances will make it especially so. Soon the majority of the world's people will be living in cities. We are now obliged to think of so many people demanding the means of life from the land, to which they will no longer have a practical connection, and of which they will have little knowledge. We are obliged also to think of the consequences of any attempt to meet this demand by large-scale, expensive, petroleum-dependent technological schemes that will ignore local conditions and local needs. The problem of renewing husbandry, and the need to promote a general awareness of everybody's agricultural responsibilities, thus becomes urgent.

How are we to do this? How can we restore a competent husbandry to the minds of the world's producers and consumers?

For a start of course we must recognize that this effort is already in progress on many farms and in many urban consumer groups scattered across our country and the world. But we must recognize too that this effort needs an authorizing focus and force that would grant it a new legitimacy, intellectual rigor, scientific respectability, and responsible teaching. There are many reasons to hope that this might be supplied by our colleges of agriculture, and there are some reasons to think that this hope is not fantastical.

With that hope in mind, I want to return to the precaution that I mentioned earlier. The effort of husbandry is partly scientific, but it is entirely cultural, and a cultural initiative can exist only by becoming personal. It will become increasingly clear, I believe, that agricultural scientists, and the rest of us as well, are going to have to be less specialized, or less isolated by our specialization. Agricultural scientists will need to work as indwelling members of agricultural communities or of consumer communities. Their scientific work will need to accept the limits and the influence of that membership. It is not irrational to propose that a significant number of these scientists should be farmers, and so subject their scientific work, and that of their colleagues, to the influence of a farmer's practical circumstances. Along with the rest of us, they will need to accept all the imperatives of husbandry as the context of their work. We cannot keep things from falling apart in our society if they do not cohere in our minds and in our lives.

The Burden of the Gospels

ANYBODY HALF AWAKE these days will be aware that there are many Christians who are exceedingly confident in their understanding of the Gospels, and who are exceedingly self-confident in their understanding of themselves in their faith. They appear to know precisely the purposes of God, and they appear to be perfectly assured that they are now doing, and in every circumstance will continue to do, precisely God's will as it applies specifically to themselves. They are confident, more-over, that God hates people whose faith differs from their own, and they are happy to concur in that hatred.

Having been invited to speak to a convocation of Christian seminarians, I at first felt that I should say nothing until I confessed that I do not have any such confidence. And then I understood that this would have to be my subject. I would have to speak of the meaning, as I understand it, of my lack of confidence, which I think is not at all the same as a lack of faith.

It is a fact that I have spent my life, for the most part willingly, under the influence of the Bible, particularly the Gospels, and of the Christian tradition in literature and the other arts. As a child, sometimes unwillingly, I learned many of the Bible's stories and teachings, and was affected more than I knew by the language of the King James Version, which is the translation I still prefer. For most of my adult life I have been an urgently interested and frequently uneasy reader of the Bible, particularly of the Gospels. At the same time I have tried to be a worthy reader of Dante, Milton, Herbert, Blake, Eliot, and other poets of the Christian tradition. As a result of this reading and of my experience, I am by principle and often spontaneously, as if by nature, a man of faith. But my reading of the Gospels, comforting and clarifying and instructive as they frequently are, deeply moving or exhilarating as they frequently are, has caused me to understand them also as a burden, sometimes raising the hardest of personal questions, sometimes bewildering, sometimes contradictory, sometimes apparently outrageous in their demands. This is the confession of an unconfident reader.

*

I will begin by dealing with the embarrassing questions that the Gospels impose, I imagine, upon any serious reader. There are two of these, and the first is this: If you had been living in Jesus's time and had heard Him teaching, would you have been one of His followers? To be an honest taker of this test, I think you have to try to forget that you have read the Gospels and that Jesus has been a "big name" for two thousand years. You have to imagine instead that you are walking past the local courthouse and you come upon a crowd listening to a man named Joe Green or Green Joe, depending on judgments whispered among the listeners on the fringe. You too stop to listen, and you soon realize that Joe Green is saying something utterly scandalous, utterly unexpectable from the premises of modern society. He is saying: "Don't resist evil. If somebody slaps your right cheek, let him slap your left cheek too. Love your enemies. When people curse you, you must bless them. When people hate you, you must treat them kindly. When people mistrust you, you must pray for them. This is the way you must act if you want to be children of God." Well, you know how happily *that* would be received, not only in the White House and the Capitol, but among most of your neighbors. And then suppose this Joe Green looks at you over the heads of the crowd, calls you by name, and says, "I want to come to dinner at your house."

I suppose that you, like me, hope very much that you would say, "Come ahead." But I suppose also that you, like me, had better not be too sure. You will remember that in Jesus's lifetime even His most intimate friends could hardly be described as overconfident.

The second question is this—it comes right after the verse in which Jesus says, "If ye love me, keep my commandments." Can you be sure that you would keep His commandments if it became excruciatingly painful to do so? And here I need to tell another story, this time one that actually happened.

In 1569, in Holland, a Mennonite named Dirk Willems, under threat of capital sentence as a heretic, was fleeing from arrest, pursued by a "thief-catcher." As they ran across a frozen body of water, the thief-catcher broke through the ice. Without help, he would have drowned. What did Dirk Willems do then?

Was the thief-catcher an enemy merely to be hated, or was he a neighbor to be loved as one loves oneself? Was he an enemy whom one must love in order to be a child of God? Was he "one of the least of these my brethren"?

What Dirk Willems did was turn back, put out his hands to his pursuer, and save his life. The thief-catcher, who then of course wanted to let his rescuer go, was forced to arrest him. Dirk Willems was brought to trial, sentenced, and burned to death by a "lingering fire."

I, and I suppose you, would like to be a child of God even at the cost of so much pain. But would we, in similar circumstances, turn back to offer the charity of Christ to an enemy? Again, I don't think we ought to be too sure. We should remember that "Christian" generals and heads of state have routinely thanked God for the deaths of their enemies, and that the persecutors of 1569 undoubtedly thanked God for the capture and death of the "heretic" Dirk Willems.

Those are peculiar questions. I don't think we can escape them, if we are honest. And if we are honest, I don't think we can answer them. We humans, as we well know, have repeatedly been surprised by what we will or won't do under pressure. A person may come to be, as many have been, heroically faithful in great adversity, but as long as that person is alive we can only say that he or she did well but remains under the requirement to *do* well. As long as we are alive, there is always a next time, and so the questions remain. These are questions we must live with, regarding them as unanswerable and yet profoundly influential.

The other burdening problems of the Gospels that I want to talk about are like those questions in that they are not solvable but can only be lived with as a sort of continuing education. These problems, however, are not so personal or dramatic but are merely issues of reading and making sense.

As a reader, I am unavoidably a writer. Many years of trying to write what I have perceived to be true have taught me that there are limits to what a human mind can know, and limits to what a human language can say. One may believe, as I do, in inspiration, but one must believe knowing that even the most inspired are limited in what they can tell of what they know.

We humans write and read, teach and learn, at the inevitable cost of falling short. The language that reveals also obscures. And these qualifications that bear on any writing must bear of course on the Gospels.

I need to say also that, as a reader, I am first of all a literalist, as I think every reader should be. This does not mean that I don't appreciate Jesus's occasional irony or sarcasm ("They have their reward"), or that I am against interpretation, or that I don't believe in "higher levels of meaning." It certainly does not mean that I think every word of the Bible is equally true, or that "literalist" is a synonym for "fundamentalist." I mean simply that I expect any writing to make literal sense before making sense of any other kind. Interpretation should not contradict or otherwise violate the literal meaning. To read the Gospels as a literalist is, to me, the way to take them as seriously as possible.

But to take the Gospels seriously, to assume that they say what they mean and mean what they say, is the beginning of troubles. Those would-be literalists who yet argue that the Bible is unerring and unquestionable have not dealt with its contradictions, which of course it does contain, and the Gospels are not exempt. Some of Jesus's instructions are burdensome, not because they involve contradiction, but merely because they are so demanding. The proposition that love, forgiveness, and peaceableness are the only neighborly relationships that are acceptable to God is difficult for us weak and violent humans, but it is plain enough for any literalist. We must either accept it as an absolute or absolutely reject it. The same for the proposition that we are not permitted to choose our neighbors ahead of time or to limit neighborhood, as is plain from the parable of the Samaritan. The same for the requirement that we must be perfect, like God, which seems as outrageous as the Buddhist vow to "save all sentient beings," and perhaps is meant to measure and instruct us in the same way. It is, to say the least, unambiguous.

But what, for example, are we to make of Luke 14:26: "If any man come to me, and hate not his father, and mother, and wife, and children, and brethren, and sisters, yea, and his own life also, he cannot be my disciple." This contradicts not only the fifth commandment but Jesus's own instruction to "Love thy neighbor as thyself." It contradicts His obedience to his mother at the marriage in Cana of Galilee. It contradicts the concern He shows for the relatives of his friends and followers. But the word in the King James Version is "hate." If you go to the New English Bible or the New Revised Standard Version, looking for relief, the word still is "hate." This clearly is the sort of thing that leads to "biblical exegesis." My own temptation is to become a literary critic, wag my head learnedly, and say, "Well, this obviously is a bit of hyperbole—the sort of exaggeration a teacher would use to shock his students awake." Maybe so, but it is not obviously so, and it comes perilously close to "He didn't really mean it"—always a risky assumption when reading, and especially dangerous when reading the Gospels. Another possibility, and I think a better one, is to accept our failure to understand, not as a misstatement or a textual flaw or as a problem to be solved, but as a question to live with and a burden to be borne.

We may say with some reason that such apparent difficulties might be resolved if we knew more, a further difficulty being that we *don't* know more. The Gospels, like all other written works, impose on their readers the burden of their incompleteness. However partial we may be to the doctrine of the true account or "realism," we must concede at last that reality is inconceivably great and any representation of it necessarily incomplete.

St. John at the end of his Gospel, remembering perhaps the third verse of his first chapter, makes a charming acknowledgement of this necessary incompleteness: "And there are also many other things which Jesus did, the which, if they should be written every one, I suppose that even the world itself could not contain the books that should be written." Our darkness, then, is not going to be completely lighted. Our ignorance finally is irremediable. We humans are never going to know everything, even assuming we have the capacity, because for reasons of the most insistent practicality we can't be told everything. We need to remember here Jesus's repeated admonitions to his disciples: You don't know; you don't understand; you've got it wrong.

The Gospels, then, stand at the opening of a mystery in which our lives are deeply, dangerously, and inescapably involved.

This is a mystery that the Gospels can only partially reveal, for it could be fully revealed only by more books than the world could contain. It is a mystery that we are condemned but also are highly privileged to live our way into, trusting properly that to our little knowledge greater knowledge may be revealed. It is this privilege that should make us wary of any attempt to reduce faith to a rigmarole of judgments and explanations, or to any sort of familiar talk about God. Reductive religion is just as objectionable as reductive science, and for the same reason: Reality is large, and our minds are small.

And so the issue of reality—What is the *scope* of reality? What is real?—emerges as the crisis of this discussion. Right at the heart of the religious impulse there seems to be a certain solicitude for reality: the fear of foreclosing it or of reducing it to some merely human estimate. Many of us are still refusing to trust Caesar, in any of his modern incarnations, with the power to define reality. Many of us are still refusing to entrust that power to science. As inhabitants of the modern world, we are religious now perhaps to the extent of our desire to crack open the coffin of materialism, and to give to reality a larger, freer definition than is allowed by the militant materialists of the corporate economy and their political servants, or by the mechanical paradigm of reductive science. Or perhaps I can make most plain what I'm trying to get at if I say that many of us are still withholding credence, just as properly and for the same reasons, from any person or institution claiming to have the definitive word on the purposes and the mind of God.

It seems to me that all the religions I know anything about emerge from an instinct to push against any merely human constraints on reality. In the Bible such constraints are conventionally attributed to "the world" in the pejorative sense of that term, which we may define as the world of the creation *reduced* by any of the purposes of selfishness. The contrary purpose, the purpose of freedom, is stated by Jesus in the fourth Gospel: "I am come that they might have life, and that they might have it more abundantly."

This astonishing statement can be thought about and understood endlessly, for it is endlessly meaningful, but I don't think it calls for much in the way of interpretation. It does call for a very strict and careful reading of the word "life."

To talk about or to desire more abundance of anything has probably always been dangerous, but it seems particularly dangerous now. In an age of materialist science, economics, art, and politics, we ought not to be much shocked by the appearance of materialist religion. We know we don't have to look far to find people who equate more abundant life with a bigger car, a bigger house, a bigger bank account, and a bigger church. They are wrong, of course. If Jesus meant only that we should have more possessions or even more "life expectancy," then John 10:10 is no more remarkable than an advertisement for any commodity whatever. Abundance, in this verse, cannot refer to an abundance of material possessions, for life does not require a material abundance; it requires only a material sufficiency. That sufficiency granted, life itself, which is a membership in the living world, is already an abundance.

But even life in this generous sense of membership in creation does not protect us, as we know, from the dangers of avarice, of selfishness, of the wrong kind of abundance. Those dangers can be overcome only by the realization that in speaking of more abundant life, Jesus is not proposing to free *us* by making us richer; he is proposing to set *life* free from precisely that sort of error. He is talking about life, which is only incidentally our life, as a limitless reality.

Now that I have come out against materialism, I fear that I will be expected to say something in favor of spirituality. But if I am going to go on in the direction of what Jesus meant by "life" and "more abundantly," then I have got to avoid that duality of matter and spirit at all costs.

As every reader knows, the Gospels are overwhelmingly concerned with the conduct of human life, of life in the human commonwealth. In the Sermon on the Mount and in other places Jesus is asking his followers to see that the way to more abundant life is the way of love. We are to love one another, and this love is to be more comprehensive than our love for family and friends and tribe and nation. We are to love our neighbors though they may be strangers to us. We are to love our enemies. And this is to be a practical love; it is to be practiced, here and now. Love evidently is not just a feeling but is indistinguishable from the willingness to help, to be useful to one another. The way of love is indistinguishable, moreover, from the way of freedom. We don't need much imagination to imagine that to be free of hatred, of enmity, of the endless and hopeless effort to oppose violence with violence, would be to have life more abundantly. To be free of indifference would be to have life more abundantly. To be free of the insane rationalizations for our desire to kill one another—that surely would be to have life more abundantly.

And where more than in the Gospels' teaching about love do we see that famously estranged pair, matter and spirit, melt and flow together? There was a Samaritan who came upon one of his enemies, a Jew, lying wounded beside the road. And the Samaritan had compassion on the Jew and bound up his wounds and took care of him. Was this help spiritual or material? Was the Samaritan's compassion earthly or heavenly? If those questions confuse us, that is only because we have for so long allowed ourselves to believe, as if to divide reality impartially between science and religion, that material life and spiritual life, earthly life and heavenly life, are two different things.

To get unconfused, let us go to a further and even more interesting question about the parable of the Samaritan: Why? Why did the Samaritan reach out in love to his enemy, a Jew, who happened also to be his neighbor? Why was the unbounding of this love so important to Jesus?

We might reasonably answer, remembering Genesis 1:27, that all humans, friends and enemies alike, have the same dignity, deserve the same respect, and are worthy of the same compassion because they are, all alike, made in God's image. That is enough of a mystery, and it implies enough obligation, to waylay us a while. It is certainly something we need to bear anxiously in mind. But it is also too human-centered, too potentially egotistical, to leave alone.

I think Jesus recommended the Samaritan's loving-kindness, what certain older writers called "holy living," simply as a matter of propriety, for the Samaritan was living in what Jesus understood to be a holy world. The foreground of the Gospels is occupied by human beings and the issues of their connection to one another and to God. But there is a background, and the background more often than not is the world in the best sense of the word, the world as made, approved, loved, sustained, and finally to be redeemed by God. Much of the action and the talk of the Gospels takes place outdoors: on mountainsides, lake shores, river banks, in fields and pastures, places populated not only by humans but by animals and plants, both domestic and wild. And these non-human creatures, sheep and lilies and birds, are always represented as worthy of, or as flourishing within, the love and the care of God.

To know what to make of this, we need to look back to the Old Testament, to Genesis, to the Psalms, to the preoccupation with the relation of the Israelites to their land that runs through the whole lineage of the prophets. Through all this, much is implied or taken for granted. In only two places that I remember is the always implicit relation—the practical or working relation—of God to the creation plainly stated. Psalm 104:30, addressing God and speaking of the creatures, says, "Thou sendest forth thy spirit, they are created . . ." And, as if in response, Elihu says to Job (34:14–15) that if God "gather unto himself his spirit and his breath; All flesh shall perish together . . ." I have cut Elihu's sentence a little short so as to leave the emphasis on the phrase "all flesh."

Those also are verses that don't require interpretation, but I want to stretch them out in paraphrase just to make as plain as possible my reason for quoting them. They are saying that not just humans but *all* creatures live by participating in the life of God, by partaking of His spirit and breathing His breath.* And so the Samaritan reaches out in love to help his enemy, breaking all the customary boundaries, because he has clearly seen in his enemy not only a neighbor, not only a fellow human or a fellow creature, but a fellow sharer in the life of God.

When Jesus speaks of having life more abundantly, this, I think, is the life He means: a life that is not reducible by division, category, or degree, but is one thing, heavenly and earthly, spiritual and material, divided only insofar as it is embodied

*We now know that this relationship is even more complex, more utterly inclusive and whole, than the biblical writers suspected. Some scientists would insist that the conventional priority given to living creatures over the nonliving is misleading. Try, for example, to separate life from the lifeless minerals on which it depends.

in distinct creatures. He is talking about a finite world that is infinitely holy, a world of time that is filled with life that is eternal. His offer of more abundant life, then, is not an invitation to declare ourselves as certified "Christians," but rather to become conscious, consenting, and responsible participants in the one great life, a fulfillment hardly institutional at all.

To be convinced of the sanctity of the world, and to be mindful of a human vocation to responsible membership in such a world, must always have been a burden. But it is a burden that falls with greatest weight on us humans of the industrial age who have been and are, by any measure, the humans most guilty of desecrating the world and of destroying creation. And we ought to be a little terrified to realize that, for the most part and at least for the time being, we are helplessly guilty. It seems as though industrial humanity has brought about phase two of original sin. We all are now complicit in the murder of creation. We certainly do know how to apply better measures to our conduct and our work. We know how to do far better than we are doing. But we don't know how to extricate ourselves from our complicity very surely or very soon. How could we live without degrading our soils, slaughtering our forests, polluting our streams, poisoning the air and the rain? How could we live without the ozone hole and the hypoxic zones? How could we live without endangering species, including our own? How could we live without the war economy and the holocaust of the fossil fuels? To the offer of more abundant life, we have chosen to respond with the economics of extinction.

If we take the Gospels seriously, we are left, in our dire predicament, facing an utterly humbling question: How must we live and work so as not to be estranged from God's presence in His work and in all His creatures? The answer, we may say, is given in Jesus's teaching about love. But that answer raises another question that plunges us into the abyss of our ignorance, which is both human and peculiarly modern: How are we to make of that love an economic practice?

That question calls for many answers, and we don't know most of them. It is a question that those humans who want to answer will be living and working with for a long time—if they are allowed a long time. Meanwhile, may Heaven guard us from those who think they already have the answers.

FROM

WHAT MATTERS?

(2010)

Money Versus Goods

MY ECONOMIC point of view is from ground-level. It is a point of view sometimes described as "agrarian." That means that in ordering the economy of a household or community or nation, I would put nature first, the economies of land use second, the manufacturing economy third, and the consumer economy fourth. The basis of such an economy would be broad, the successive layers narrowing in the order of their diminishing importance.

The first law of such an economy would be what the agriculturalist Sir Albert Howard called "the law of return." This law requires that what is taken from nature must be given back: The fertility cycle must be maintained in continuous rotation. The primary value in this economy would be the capacity of the natural and cultural systems to renew themselves. An authentic economy would be based upon renewable resources: land, water, ecological health. These resources, if they are to stay renewable in human use, will depend upon resources of culture that also must be kept renewable: accurate local memory, truthful accounting, continuous maintenance, un-wastefulness, and a democratic distribution of now-rare practical arts and skills. The economic virtues thus would be honesty, thrift, care, good work, generosity, and (since this is a creaturely and human, not a mechanical, economy) imagination, from which we have compassion. That primary value and these virtues are essential to what we have been calling "sustainability."

A properly ordered economy, putting nature first and consumption last, would start with the subsistence or household economy and proceed from that to the economy of markets. It would be the means by which people provide to themselves and to others the things necessary to support life: goods coming from nature and human work. It would distinguish between needs and mere wants, and it would grant a firm precedence to needs.

A proper economy, moreover, would designate certain things as priceless. This would not be, as now, the "pricelessness" of things that are extremely rare or expensive, but would refer to things of absolute value, beyond and above any price that could be set upon them by any market. The things of absolute value would be fertile land, clean water and air, ecological health, and the capacity of nature to renew herself in the economic landscapes. Our nearest cultural precedent for this assignment of absolute value is biblical, as in Psalm 24 ("The earth is the Lord's, and the fulness thereof . . .") and Leviticus 25:23 ("The land shall not be sold forever . . ."). But there are precedents in all societies and traditions that have understood the land or the world as sacred—or, speaking practically, as possessing a suprahuman value. The rule of pricelessness clearly imposes certain limits upon the idea of land ownership. Owners would enjoy certain customary privileges, necessarily, as the land would be entrusted to their intelligence and responsibility. But they would be expected to use the land as its servants and on behalf of all the living.

The present and now-failing economy is just about exactly opposite to the economy I have just described. Over a long time, and by means of a set of handy prevarications, our economy has become an anti-economy, a financial system without a sound economic basis and without economic virtues.

It has inverted the economic order that puts nature first. This economy is based upon consumption, which ultimately serves, not the ordinary consumers, but a tiny class of excessively wealthy people for whose further enrichment the economy is understood (by them) to exist. For the purpose of their further enrichment, these plutocrats and the great corporations that serve them have controlled the economy by the purchase of political power. The purchased governments do not act in the interest of the governed and their country; they act instead as agents for the corporations.

That this economy is, or was, consumption-based is revealed by the remedies now being proposed for its failure: stimulate, spend, create jobs. What is to be stimulated is spending. The government injects into the failing economy money to be spent, or to be loaned to be spent. If people have money to spend and are eager to spend it, demand for products will increase, creating jobs; industry will meet the demand with more products, which will be bought, thus increasing the amount of money in circulation; the greater amount of money in circulation will increase demand, which will increase spending, which will increase production—and so on until the old fantastical economy of limitless economic growth will have "recovered."

But spending is not an economic virtue. Miserliness is not an economic virtue either. Saving is. Not-wasting is. To encourage spending with no regard at all to what is being purchased may be pro-finance, but it is anti-economic. Finance, as opposed to economy, is always ready and eager to confuse wants with needs. From a financial point of view, it is good, even patriotic, to buy a new car whether you need one or not. From an economic point of view, however, it is wrong to buy anything you do not need. It is unpatriotic too: If you love your country, you don't want to burden or waste it by frivolous wants. Only in a financial system, an anti-economy, can it seem to make sense to talk about "what the economy needs." In an authentic economy, we would ask what the land and the people need. People do need jobs, obviously. But they need jobs that serve natural and human communities, not arbitrarily "created" jobs that serve only the economy.

From an economic point of view, a society in which every schoolchild "needs" a computer, and every sixteen-year-old "needs" an automobile, and every eighteen-year-old "needs" to go to college is already delusional and is well on its way to being broke.

In a so-called economy that is dependent on indiscriminate spending, "job creation" often implies an ability to "create" new "needs." Until lately this economy has been able to create jobs by creating needs. But this has involved much confusion and a kind of fraud, because it gives no priority to the meeting of needs, and cannot distinguish needs from wants. Our economy, having confused necessities with products or commodities that are merely marketable, deliberately reduces the indispensable service of providing needed goods to "selling" or "marketing" products, some of which have never been and will never be needed by anybody. The gullibility of the public thus becomes an economic resource.

The category of things sold that are not needed now includes even legally marketed foods and drugs. This involves the art (taught and learned in universities) of lying about products. A friend of mine remembers a teacher who said that advertising is "the manufacture of discontent." And so we have come to live in a world in which every brand of painkiller is better than every other brand, in which we have a "service economy" that does not serve and an "information economy" that does not distinguish good from bad or true from false.

The manufacturing sector of a financial system, which does not or cannot distinguish between needs and induced wants, will come willy-nilly into the service of wants, not needs. So it has happened with us. If in some state of emergency our manufacturers were suddenly called upon to supply us with certain necessities—shoes, for example—we would be out of luck. "Outsourcing" the manufacture of frivolities is at least partly frivolous; outsourcing the manufacture of necessities is entirely foolish.

As for the land economies, the academic and political economists seem mainly to ignore them. For years, as I have read articles on the economy, I have waited in vain for the author to "factor in" farming or ranching or forestry. The expert assumption appears to be that the products of the soil are not included in the economy until after they have been taken at the lowest possible cost from those who did the actual work of production, at which time they enter the economy as raw materials for the food, fiber, timber, and lately the fuel industries. The result is inevitable: The industrial system is disconnected from, is unconcerned about, and takes no responsibility for, its natural and human sources. The further result is that these sources are not maintained but merely used and thus are made as exhaustible as the fossil fuels.

As for nature herself, virtually nobody—not the "environmentalist," let alone the economist—regards nature as an economic resource. Nature, especially where she has troubled herself to be scenic, is understood to have a recreational and perhaps an aesthetic value that is to some extent economic. But for her accommodation of our needs to eat, drink, breathe, and be clothed and sheltered, our industrial and financial systems grant her no recognition, honor, or care.

Far from assigning an absolute value to those things we absolutely need, the financial system puts a price, though a highly variable price, on everything. We know from much experience that everything that is priced will sooner or later be sold. And from the accumulating statistics of soil loss, land loss, deforestation, overuse of water, various sorts of pollution, etc., we have reason to fear that everything that is sold will be ruined. When everything has a price, and the price is made endlessly variable by an economy without a stable relation to necessity or to real goods, then everything is disconnected from history, knowledge, respect, and affection—from anything at all that might preserve it—and so is implicitly eligible to be ruined.

What we have been pleased to call our economy does not acknowledge and apparently does not even recognize its continuing absolute dependence on the natural world, on the land economies, and on the work of farmers, ranchers, and foresters—all of which, given the use of available knowledge and precautions, would be self-renewing. At the same time, with a remarkable lack of foresight or even the sight to see what is presently obvious, this economy has made itself absolutely dependent on resources that are either exhaustible by nature or have been made exhaustible by our wastefulness and our refusal to husband and reuse: fossil fuels, metals, and other mined materials. By standards that are utterly absurd, it has long been "too expensive" to salvage perfectly good and usable materials from old buildings, which we knock down or blow up and haul to landfills, and so make even bricks and stones valueless and irrecoverable. Because of falsely cheap materials and energy, we have a "bubble" of houses too big to be heated efficiently or cheaply, or even to be paid for.

To use our agricultural land for the production of "biofuel," as some are now doing, is immediately to raise the question whether it can ever be right to replace food production by the production of a fuel to be burned. If this fuel is produced, like most of our food at present, without the close and loving care that the land requires, then the land becomes an exhaustible resource. Biofuel may be a product of the land and our world-changing technology, but it is just as much a product of ignorance and moral carelessness.

As commodities, the fossil fuels are in a category strictly their own. Unlike other minerals that (in a sensible economy) can be reused, and unlike waterpower that uses water and releases it to be used again, the fossil fuels can be made useful only by being destroyed. They are useful and therefore valuable only in the instant in which they are burning.

To be available for their brief usefulness, these fuels must be dug or pumped from the ground. Their extraction has nearly always damaged, often irreparably, the places and the human communities from which they are taken. For coal to feed the fires by which we live, whole landscapes are destroyed, forests and their soils and creatures are obliterated, streams are covered over, watersheds are degraded and polluted, poisonous residues are left behind, communities are degraded or flooded by toxic wastes or runoff from denuded watersheds, the people are exploited and endangered, their houses damaged, their drinking water poisoned, their complaints and needs ignored. When the fossil fuels, extracted at such a cost to people and nature, are burned, they pollute the atmosphere of all the world, with consequences that are fearful, infamous, and continuing.

In a consciously responsible economy, such abuses would be inconceivable. They could not happen. To damage or destroy an otherwise permanent resource for the sake of a temporary advantage would be readily perceived as senseless by every practical measure and, by the measure of human wholeness, as insane. To value human wants above all the natural and human resources that supply human needs, as the now-failing economy has done, is to run risks and defy paradoxes by which it was and is bound to fail. If we pursue limitless "growth" now, we impose ever-narrower limits on the future. If we put spending first, we put solvency last. If we put wants first, we put needs last. If we put consumption first, we put health last. If we put money first, we put food last. If for some spurious reason such as "economic growth" or "economic recovery," we put people and their comfort first, before nature and the land-based economies, then nature sooner or later will put people last.

But the fossil fuels, which involve destruction for the sake of production and again destruction as a consequence of production, are not the only typical products of our anti-economy. Also typical are products that replace, at high cost, goods that once were cheap or free. The genius of marketing and selling has given us, for example, bottled tapwater, for which we pay more than we pay for gasoline, because of our perfectly rational fear that our unbottled tapwater is polluted. The system of industry, finance, and "marketing" thus makes capital of its own viciousness and of the ignorance and gullibility of a supposedly educated public. By the influence of marketers and sellers, citizens and members are transformed into suckers. And so we have an alleged economy that is not only fire-dependent and consumption-dependent but also sucker-dependent.

For another example, consider the money-drenched entertainment industry. The human species, which has apparently outlived the name Homo sapiens, is said to be something like 200,000 years old. Except for the last seventy-five or so years of their life so far, and except for their decadent ruling classes, most humans have entertained themselves by remembering and telling stories, singing, dancing, playing games, and even by their work of providing themselves with necessities and things of beauty, which usually were the same things. All of this entertainment came free of charge, as a sort of overflow of human nature, local culture, and daily life. Even the beauty of good work and well-made things was a value added at no charge. The entertainment industry has improved upon this great freedom by providing at a high cost, in money but also in health and sanity, an egregiously overpaid corps of entertainers and athletes who tell or perform stories, sing, dance, and play games for us or sell games to us as we passively consume their often degrading productions. The wrong here may be at root only that of an inane and expensive redundancy. If you can read and have more imagination than a doorknob, what need do you have for a "movie version" of a novel?

This strange economy produces, typically and in the ordinary course of business, products that are destructive or fraudulent or unnecessary or useless, or all four at once. Another of its typical enterprises is remarkable for the production of what I suppose we will have to call no-product, or no product but money (to the extent that this works). The best-known or longest infamous example of a no-product financial industry is the practice of usury, which is to say the lending of money at exorbitant interest or (some have said) at any interest. In our cultural tradition, as opposed to financial precedent, the condemnation of usury seems to be unanimous.

The Hebrew Bible speaks emphatically against usury in ten of its chapters (by my count), calling it by name, but without much explanation, assuming apparently that its wrongfulness is obvious. From the context it is clear that usury is understood as an injustice and an offense against charity. It is a way for people of wealth to exploit the poor, whom they have been instructed to care for. Only the wealthy have a surplus of money to lend, and they should not use it to take advantage of the needs of others. Usury, moreover, cannot be consistent with the command (Leviticus 19:18) that "thou shalt love thy neighbor as thyself."

Aristotle in *The Nicomachean Ethics* also condemns usury and in language that is remarkably consistent with my description of our own economic malpractice. He classes usurers with pimps, as people who take "anything from any source" or who "take more than they ought and from wrong sources" (the Oxford edition, translated by Sir David Ross).

Dante is perfectly consistent with the Bible and Aristotle when he places the usurers in Hell (*Inferno* XI) with others who are guilty of violence against God. Virgil, explaining this fault to Dante, makes the case clearly and usefully. Usury is a violence against God because it is a violence against nature. Nature is the art of God, just as productive work, the making of useful things, is the art of humans. Humans prosper rightly when their goods come from nature by their good work. Usurers prosper, on the contrary, by making money grow from itself (by "making their money work for them," as we say), thus holding in contempt both nature and work, both divine art and human art.

Ezra Pound, a poet of our own time, was in Dante's tradition when he wrote the two versions of his eloquent poem against usury (*Cantos* XLV and LI). Pound who was (I hope) insane when at his worst, was perfectly sane when he wrote this:

> With usury has no man a good house
> made of stone, no paradise on his church wall
> With usury the stone cutter is kept from his stone
> the weaver is kept from his loom by usura

Wool does not come into market
the peasant does not eat his own grain
the girl's needle goes blunt in her hand
The looms are hushed one after another
. . . .
Usury kills the child in the womb
And breaks short the young man's courting
Usury brings age into youth; it lies between the bride
and the bridegroom
Usury is against Nature's increase.

The point—as I understand it, though I understand also that this poem offers far more than a point—is that when money is misused to grow from itself into heaps in the possession inevitably of fewer and fewer people, it cannot be rightly used for the production of goods or even to maintain the subsistence of the people. Workers will not be well paid for good work. The arts will not flourish, and neither will nature.

I need to say here that this issue of usury is far from simple, and that I am not capable even of giving usury a proper definition. The issue is simple only if usury is defined as the taking of *any* interest. It is so defined by Jesus in the Gospel of Luke (6:34–35):

> And if ye lend to them of whom ye hope to receive, what thank have ye? for sinners also lend to sinners, to receive as much again.
>
> But love ye your enemies, and do good, and lend, hoping for nothing again . . .

Such free lending would be possible among neighbors or in a small local economy, but in general we appear to be far from that, the churches along with the rest. In an extensive economy using money, banks appear to be necessary. If the poor, for instance, are to rise above their poverty, or if the young are to acquire houses or farms or small businesses, there probably needs to be an established means of lending them money, and that would be banking. If we are to have banks and banking, then we have to build and equip and maintain the necessary buildings, return a fair dividend to the necessary investors, and pay fair wages or salaries to the necessary employees. The needed funds would have to come to a considerable extent from interest on loans. If the money were to be loaned at no charge, there soon would be no lending institution and nobody to make loans.

And so we come to the uneasy question of what rate of interest would be neither too little nor too much. If too little, loans cannot be made. If too much, then lending becomes, not a service, but the exploitation and even the ruin of borrowers. I don't think a fair rate can be determined according to standards that are only financial. It would have to be determined by responsible bankers, acting also as community members, in the context of their community, local nature, and the local economy.

Such a determination, I believe, can take place only in a bank that is locally owned, conforming in scale to the size and needs of the local community, and by bankers who are aware that the prosperity of the bank is not and can never be separated from the prosperity of the community.

I know from my own experience and observation that a bank of community scale, owned principally by local investors, understanding its dependence on responsible service first of all to local customers—even in a fevered and delirious economy—can function usefully and considerately as a part of the community. Such a bank does not, because if it is to survive it cannot, adopt the lending practices that resulted in our recent housing bubble. In such a bank the loan officers understand necessarily that their responsibility is to the borrowers as much as to the bank. In a locally owned community bank, the lender is a neighbor of the borrower. You don't put your neighbors into trouble or into ruin by misleading them to assume debts they cannot pay—which ultimately, of course, would ruin the lender.

It is clear that if interest rates are not limited by a reasonable, workable concept of fairness, enforceable by law, then they will become exorbitant. Moreover, they are apt to become highly variable according to the whim of lenders inclined to "take more than they ought."

Among its other wrongs, usury destabilizes the relation of money to goods. So does inflation. So does the speculative trading in mortgages, "futures," and "commercial paper" that gives a monetary value to commodities having no present existence or no existence at all. To inflate or obscure the value of money in relation to goods is in effect to steal both from those who spend and from those who save. It is to subordinate real value to a value that is false.

By destabilizing the relation of money to goods, a financial system usurps an economy. Then, instead of the exchange of money for goods or goods for money, we have the conversion of goods into money, in the process often destroying the goods. Money, instead of a token signifying the value of goods, becomes a good in itself, which the wealthy can easily manipulate in their own favor. This is sometimes justified (by the favored) as freedom, as in "free trade" or "the free market," but such a freedom is calculated to reduce substantially the number of the free. The tendency of this freedom necessarily is toward monopoly—toward the one economic entity that will own or control everything. The undisguised aim of Monsanto, for example, is to control absolutely the economy of food. It would do so by setting its own price on its products sold to dependent purchasers who can set a price neither on what they buy nor on what they sell.

To permit so much wealth, power, influence, and ambition to one corporation is an egregious error in a polity supposedly democratic. From the point of view of nature and agriculture, it is an error even larger and more dangerous. For by this error agriculture is forced to subserve the rule of industrialism, which is in most respects antithetical to the healthful practice of agriculture and to the laws of nature, by which, and only by which, agriculture can be made sustainable.

The dominance of agriculture by agribusiness is made possible by the dominance of the economy by interests that are industrial or purely financial. Agribusiness is immensely more profitable monetarily than agriculture, which customarily for the last fifty or sixty years has been either barely profitable or unprofitable. Hence the drastic decline in the agricultural population. One cost of this error is economic injustice, characteristic of industrialism, to the people who do the work: ranchers, farmers, and farm workers. Another cost is first agricultural and then ecological: Under the rule of industrialism the land is forced to produce but is not maintained; the fertility cycle is broken; soil nutrients become water pollutants; toxic chemicals and fossil energy replace human work.

We have allowed, and even justified as "progress," a fundamental disconnection between money and food. And so we are led to the assumption, by ignorant leaders who apparently believe it, that if we have money we will have food, an assumption that is destructive of agriculture and food. It is a superstition just as wicked, and hardly different from, the notion that the world is conformable to our wants and we can be whatever we want to be.

Apparently it takes a lot of money, a lot of power, and even a lot of education to obscure the knowledge that food comes from the land and from the human ability to cause the land to produce and to remain productive. Under the rule of an economy perverted by industrial and financial presumptions, we are destroying both the land and the human means of using the land and caring for it.

We are destroying the land by exposing it to erosion, by infusing it year after year with toxic chemicals (which incidentally poison the water), by surface mining, and by so-called development. We are destroying the cultures and the communities of land use and land husbandry by deliberately slanting the economy of the food system against the primary producers.

We are losing and degrading our agricultural soils because we no longer have enough competent people available to take proper care of them. And we will not produce capable and stewardly farmers, ranchers, and foresters by what we are calling "job creation." The fate of the land is finally not separable from the fate of the people of the land (and the fate of country people is finally not different from the fate of city people). Industrial technology does not and cannot adequately replace human affection and care. Industrial and financial procedures cannot replace stable rural communities and their cultures of husbandry. One farmer, if that name applies, cannot farm thousands of acres of corn and soybeans in the Midwest without production costs that include erosion and toxicity, which is to say damages that are either long-term or permanent.

The farm population has now declined almost to nonexistence

because, since the middle of the last century, we have deliberately depressed farm income while allowing production costs to rise, for the sake of "cheap food" and to favor agribusiness. No wonder that farm-raised young people have been moving into the cities and suburbs by millions for two generations, leaving the farms without heirs or successors. The young people decide against too much investment and too much work for too little return. Even if they love farming or ranching enough to want to stay, paying the inevitable economic and personal penalties, they are more than likely to find that they cannot buy land and pay for it by using it. The one reason for this is the disequilibrium between the economy of money and the land economies. Professional people in the cities, who have done well financially, have been "investing" in farmland and rangeland and so lifting the market value of the land above the reach of farmers and ranchers who are *not* doing well economically. The result is that we have an enormous population of dependent people with the subservient mentality of industrial employees, helpless to feed themselves, who are being fed by the tiniest minority of exploited people and from land that is more cruelly exploited than the people.

If we are destroying both the productive land and the rural communities and cultures, how can we assume that money will somehow attract food to us whenever we need it? If, on the contrary, we should decide to right the economic balance by paying a just price to producers, then money could revert to its proper function of encouraging and supporting both food production and the proper husbanding of the land. This, if it could happen, would solve a number of problems. The right answer to urban sprawl, for example, is to make agriculture pay well enough that farmers and ranchers would want to keep the land in use, and their children would want to inherit it to use.

To a ground-level observer, it is obvious that the economic failures I have described involve moral issues of the gravest sort. An essentially immoral system of economy-as-finance, or an economy run by the sole standard of monetary profit, has been allowed to flourish to the point of catastrophe by a fairly general consent to the proposition that economy and morality are two professional specialties that either do not converge, or that can be made to converge by a simple moral manipulation, as follows.

In 1986 the "conservative" columnist William Safire wrote that "Greed is finally being recognized as a virtue . . . the best engine of betterment known to man." This was not, I think, the news that Mr. Safire thought it was, but was merely a repetition of a time-worn rationalization. What may have been new was the "professional" falsehood that greed is the exclusive motive in every choice—that, for example, the *only* way to have good teachers or good doctors is to pay them a lot of money.

Mr. Safire's error, and that of the people he spoke for, is in the idea that everybody can be greedy up to some limit—that, once you have made greed a virtue, it will not crowd out other virtues such as temperance or justice or charity. The virtuously greedy perhaps would agree to let one another be greedy, so long as one person's greed did not interfere with the greed of another person. This would be the Golden Rule of greedy persons, who no doubt would thank God for it.

But that rule appears to be honored entirely in the breach. There are still a good many people who choose or accept a vocation that will not make them rich—many teachers, for instance, and most writers. But for the greedy there appears to be no such concept as *greedy enough.* The greedy consume the poor, the moderately prosperous, and each other with the same relish and with an ever-growing appetite.

Part two of Mr. Safire's error is his assumption that we can restrict the honor of virtuehood to greed alone, leaving the other sins to pine away in customary disfavor: "I hold no brief," he said, "for Anger, Envy, Lust, Gluttony, Pride . . . or Sloth." But he was already too late. A glance at magazine advertising in 1986 would have suggested that these sins had been virtues of commerce long enough already to be taken for granted. As we have sometimes been told, the sins, like the virtues, are inclined to enjoy one another's company.

Mr. Safire's announcement was not a moral innovation, but rather a confession of the depravity of what in 1986 we were calling, and are still calling, "the economy"—a ramshackle, propped-up, greed-enforced anti-economy that is delusional, vicious, wasteful, destructive, hard-hearted, and so fundamentally dishonest as to have resorted finally to "trading" in various pure-nothings. Might it not have been better and safer to have assumed that there is no partition between economy and morality, that the test of both is practicality, and that morality is long-term practicality?

The problem with "the economy" is not only that it is anti-economic, destructive of the natural and human bases of any authentic economy, but that it has been out of control for a long time. At the root of our problem, we now need to suppose, is industrialism and the Industrial Revolution itself. As the original Luddites saw clearly and rightly, the purpose of industrialism from the first has been to replace human workers with machines. This has been justified and made unquestionable by the axiom that machines, according to standards strictly mechanical, work more efficiently and cheaply than people. They answer directly the perpetual need of the greedy to get more for less. This is yet another of our limitless "progressive" ideas: The industrial academics or academic industrialists who subserve the technological cutting edge are now nominating robots as substitutes for parents, nurses, and surgeons. Soon, surely, we will have robots that can worship and make love faster and cheaper than we mere humans, who have been encumbered in those activities by flesh and blood and our old-fashioned ways.

But to replace people by machines is to raise a difficult, and I would say an urgent, question: What are the replaced people to do? Or, since this is a question not all replaced people have been able to answer satisfactorily for themselves, What is to be done with or for them? This question has never received an honest answer from either liberals or conservatives, communists or capitalists. Replaced people have entered into a condition officially euphemized as "mobility." If you have left your farm or your country town and found a well-paying city job or entered a profession, then you are said to have been "upwardly mobile." If you have left the country for the city with visions of bright lights and more money, or if you have gone to the city because you have been replaced as a farm worker by machines and you have no other place to go and you end up homeless or living in a slum without a job, then I suppose you are downwardly mobile—but this is still "progress," for at least you have been relieved of "the idiocy of rural life" or the "mind-numbing work" of agriculture.

When replacement leads to "mobility" or displacement, and displacement leads to joblessness or homelessness, then we have a problem as characteristic of the industrialized world as land waste and pollution. To this problem the two political sides have produced nonsolutions that are hopeless and more cynical (I hope) than many of their advocates realize: versions of "Get a job," job training, job retraining, "better" education, job creation, and "safety nets" such as welfare, Social Security, varieties of insurance, retirement funds, etc. All of these "solutions," along with joblessness itself, serve the purposes of an economy of bubbling money. And every one of them fails to address the problem of "mobility," which is to say a whole society that is socially and economically unstable. In this state of perpetual mobility, even the most lucratively employed are likely to be homeless, if "home" means anything at all, for they are endlessly moving at the dictates of their careers or at the whims of their employers.

To escape the cynicism, heartlessness, and damage implicit in all this mobility, it is necessary to ask another question: Might it not have been that these replaced and displaced people were needed in the places from which they were displaced? I don't mean to suggest that this is a question easily answered, or that anybody should be required to stay put. I do mean that the question ought to be asked. It ought to be asked if only because it calls up another question that might lead to actual thinking: By what standard, or from what point of view, are we permitted to suppose that the displaced people were not needed in their original places? According to the industrial standard and point of view, persons are needed only when they perform a service valuable to an employer. When a machine can perform the same service, a person then is not needed.

Not-needed persons must graduate into mobility, which will take them elsewhere to a job newly vacated or "created," or to job training, or to some safety net, or to netlessness, joblessness, and homelessness. But this version of "not-needed" fits uncomfortably into the cultural pattern by which we define ourselves as civilized or humane or human. It grates achingly against the political and religious traditions that have affirmed for us the inherent worth and even preciousness of individual people. Our mobility, whether enforced or fashionable, has dismembered and scattered families and communities. Politicians and opinion dealers from far left to far right predictably and loudly regret these disintegrations, prescribing for them (in addition to the "solutions" already mentioned) year-round schools, day care, expert counseling, drugs, and prisons.

And so: Might it not be that the displaced persons were needed by their families and their neighbors, not only for their economic assistance to the home place and household, but for their love and understanding, for their help and comfort in times of trouble? Of the Americans known to me, only the Amish have dealt with such questions openly and conscientiously as families, neighbors, and communities. The Amish are Amish by choice. There is no requirement either to subscribe to the religion or to stay in the community. The Amish have their losses and their failures, as one would expect. Lately some of their communities have become involved in the failure of the larger economy. But their families and communities nevertheless have been held together by principle and by the deliberate rejection of economic and technological innovations that threaten them. With the Amish—as once with the rest of us—a family member or a neighbor is by definition needed, and is needed not according to any standard of usefulness or any ratio of cost and price, but according to the absolute standards of kindness, mutuality, and affection. Unlike the rest of us, the Amish have remembered that the best, most dependable, most kind safety net or social security or insurance is a coherent, neighborly, economically sound, local community.

To speak of the need for affection and loyalty and social stability is not at all to slight the need for life-supporting work. Of course people need to work. Everybody does. And in a money-using economy, people need to earn money by their work. Even so, to speak of "a job" as if it were the only economic need a person has, as if it doesn't matter what the job is or where a person must go in order to have it, is brutally reductive. To speak so is to leave out virtually everything that is humanly important: family and community ties, connection to a home place, the questions of vocation and good work. If you have "a job," presumably, you won't mind being a stranger among strangers in a strange place, doing work that is demeaning or unethical or work for which you are unsuited by talent or calling.

When people accept mobility as a condition of work, it means that they have accepted a kind of homelessness. It used to be a part of good manners to ask a person you had just met, "Where are you from?" That question has now become a social embarrassment, for it is too likely to be answered, "I'm not from anywhere." But to be not from anywhere is part of the definition of helplessness. Mobility is a condition in which you can do little or nothing to help yourself, and in which you live apart from family and old neighbors who would be the people most likely to help you.

Usury, for example, is "a job." But it happens to be a job that nobody ought to do. It is a violence against fellow humans who happen to be in need, a violence against work, or against good work, a violence against nature, and therefore (for those to whom it matters) a violence against God. It is a job also that estranges and isolates one from other people, who are perceived by the usurer, not as neighbors, but as potential victims.

To be mobile is not only to be in a new sense homeless. It is also to be in an old sense landless. If you have plenty of money to buy the necessities of life, and the stores are well-stocked with those necessities, then you may not see landlessness as a threat. But suppose you are a poor migrant, black or white, from the cotton or cane fields of the South or the Appalachian coal fields, and you wind up jobless in some "inner city." You have come from the country, and now, cooped up in a strange and unyielding place, without the mutual usefulness of a functioning neighborhood, you experience a helplessness that is new to you: the practical difficulty or impossibility of helping or being helped by somebody you know. A most significant part of that helplessness is the impossibility of helping yourself, and this is the condition of landlessness. I am not talking here about owning land, but merely of having access to it or the use of it. In your new circumstances of displacement, you have no place to grow a garden or keep a few chickens or gather firewood or hunt or fish. Maybe you were, by the official definition, poor where you came from, but there your abilities to do for yourself and others were given scope and efficacy by the landscape. You have come, in short, to the difference, defined by Paul Goodman a long time ago, between competent poverty and abject poverty. A home landscape enables personal subsistence but also generosity. It enables a community to exist and function.

When country people leave home to find work, even when "jobs" are available, they incur liabilities that cannot easily be discounted. The liabilities of homelessness and landlessness may not be noticeable in times of easy money and lots of stuff to buy. But in a time of economic failure and rising unemployment, as now, the liabilities once again rise undeniably into view.

Now the following sentences by Lowell H. Harrison and James C. Klotter, in *A New History of Kentucky*, make a different sense than they would have made to most readers a year ago:

> Yet [in the 1930s] the commonwealth weathered the drought and the floods and survived the depression better than many places. . . . [The] general absence of industry meant relatively little damage there, the overall lack of wealth left people only a little way to fall, and the rural nature of the commonwealth allowed families to live off the land. In fact, people returned to Kentucky, and the decade of the 1930s saw the state's population increase faster than the national average. . . . From distant places, those who had migrated in search of jobs that were now gone came home to crowd in with their families . . .

They "came home" because at home they still had families who were growing a garden, keeping a milk cow, raising chickens, fattening hogs, and gathering their cooking and heating fuel from the woods. Now, eighty years and much "progress" later, where will the jobless go? Not home, for there are no 1930s homes to go to.

Since the end of the Great Depression, and even more since the end of World War II, country people have crowded into the cities. They have come because they have attended colleges and been "overeducated" for country life. They have come for available jobs. They have come because television and the movies have taught them to be unhappy in their "provincial" or "backward" or "nowhere" circumstances. They have come because machines have displaced them from their work and their homes. Many who have come were already poor, and were entirely unprepared for a life away from home. Immense numbers of them have ended up in slums. Some live from some variety of "safety net." Some, the homeless or insane or addicted poor, sleep in doorways or under bridges. Some beg or steal.

In the long run, these surplus people, the not-needed, have overfilled the "labor pool" and so have made labor relatively cheap. If we run short of exploitable poor people in the United States, then we "outsource" our work to the exploitable poor of other countries. Industrial workers and labor unions are having a hard time, and so are farmers, ranchers, and farm workers. People who do the actual work of producing actual products must expect to work cheap, for they are not of the quality of the professionals who "deserve" to charge too much for their services or the financial nobility who sell worthless mortgages. As an exploitable underclass, those who perform actual work have raised a vexing question for their superiors, and they seem to have fallen somewhat short of the right answer: How could they get the cheapest work out of their workers and still pay them enough to afford the products they have made? Though mere workers may be crippled by debt for their houses or farms or their children's education, they must still be able with some frequency to buy a new car or pickup truck or television set or motorboat or tractor or combine. If they have such things along with an occasional stunt in Outer Space, then maybe they won't covet a financial noble's private jet and three or four "homes."

Decades of cheap labor, cheap energy, and cheap food (all more expensive than has been imagined) have allowed our society to incorporate itself in a material structure that will have to be seen as top-heavy. We have flooded the country, the roadsides and landfills, with shoddy "consumer goods." We have too many houses that are too big, too many public buildings that are gigantic, too much useless space enclosed in walls that are too high and under roofs that are too wide. We replaced an until-then-adequate system of railroads with an interstate highway system, expensive to build, disruptive of neighborhoods and local travel, increasingly expensive to maintain and use. We replaced an until-then-adequate system of local schools with consolidated schools, letting the old buildings tumble down, replacing them with bigger ones, breaking the old ties between neighborhoods and schools, and making education entirely dependent on the fossil fuels. Every rural school now runs a fleet of buses for the underaged and provides a large parking lot for those over sixteen who "need" a car to go to school. Education has been oversold, overbuilt, overelectrified, and overpriced. Colleges have grown into universities. Universities have become "research institutions" full of undertaught students and highly accredited "professionals" who are overpaid by the public to job-train the young and to invent cures and solutions for corporations to "market" for too much money to the public. And we have balanced this immense superstructure, immensely expensive to use and maintain, upon the frail stem of the land economy that we conventionally abuse and ignore.

There is no good reason, economic or otherwise, to wish for the "recovery" and continuation of the economy we have had. There is no reason, really, to expect it to recover and continue, for it has depended too much on fantasy. An economy cannot "grow" forever on limited resources. Energy and food cannot stay cheap forever. We cannot continue forever as a tax-dependent people who do not wish to pay taxes. Delusion and the future cannot serve forever as collateral. An untrustworthy economy dependent on trust cannot beguile the people's trust forever. The old props have been kicked away. The days when we could be safely crazy are over. Our airborne economy has turned into a deadfall, and we have got to jack it down. The problem is that all of us are under it, and so we have got to jack it down with the least possible suffering to our land and people. I don't know how this is to be done, and I am inclined to doubt that anybody does. You can't very confidently jack something down if you didn't know what you were doing when you jacked it up.

I do know that the human economy as a whole depends, as it always has, on nature and the land economy. The economy of land use is our link with nature. Though economic failure has not yet called any official attention to the land economy and its problems, those problems will have to be rightly solved if we are to solve rightly our other economic problems. Before we can make authentic solutions to the problems of credit and spending, we have got to begin by treating our land with the practical and effective love that alone deserves the name of patriotism. From now on, if we would like to continue here, our use of our land will have to be ruled by the principles of stewardship and thrift, using as the one indispensable measure, not monetary profit or industrial efficiency or professional success, but ecological health. And so I will venture to propose the following agenda of changes that would amount to a new, long-term agricultural policy:

1. There should be no further price supports or subsidies without production controls. This is because surplus production is an economic weapon, allowing corporations to reduce income to farmers while increasing their own income.
2. Return to 100 percent parity between agriculture and industry. Parity (fair) prices for agricultural products would make proposed payments for "ecological services" unnecessary, and would solve other problems as well.
3. Enforce anti-trust and anti-monopoly laws. Don't let *any* corporation get big, rich, or powerful enough to hold the nation for ransom. This applies with exceptional force to agribusiness and food corporations.
4. Help young farmers to own farms. In a sane economy such help would be unnecessary, but the departure of farm-raised young people from farming is now an agricultural crisis of the greatest urgency. And we don't have enough farm-raised young people. Others need to be drawn in. Here are some measures we should consider. We should set appropriate and reasonable acreage limits, according to region, for family-scale farms and ranches. Taxes should be heavy on holdings above those limits. Holdings within the prescribed limits should be

taxed at their agricultural value. There should be inheritance taxes on large holdings; none on small holdings. No-interest loans should be made available to young farmers and ranchers buying acreages under the limits. (These suggestions raise a lot of problems, and I flinch in making them. Acreage limits are hard to set appropriately, as we learned from the homestead laws. Also some of these measures would be unnecessary if land prices were not inflated above agricultural value, and if food prices were not deflated below their actual economic and ecological cost.)

5. Phase out toxic chemicals, which are inconsistent with the principles of good agriculture, and which are polluting the rivers and the oceans.
6. Phase out biofuels as quickly as possible. We have got to observe a strict distinction between fire and food, driving and eating. We can't "feed the hungry" and feed automobiles from the same land, using the same land-destroying technologies and methods, forever.
7. Phase in perennial plants—for pasture, winter forage, and grain crops—to replace annual crops requiring annual soil disturbance or annual applications of "no-till" chemicals. This would bring a substantial reduction of soil erosion and toxicity.
8. Set and enforce high standards of water quality.
9. High water quality standards (enforced) and a program to replace annual crops with perennials would tend strongly toward the elimination of animal factories. But let us be forthright on this issue. We should get rid of animal factories, those abominations, as quickly as we can. Get the farm animals, including hogs and chickens, back on grass. Put the animals where they belong, and their manure where it belongs.
10. Animal production should be returned to the scale of localities and communities. Do away with subsidies, incentives, and legislation favorable to gigantism in dairy, meat, and egg production.
11. Encourage the development of local food economies, which make more sense agriculturally and economically

than our present overspecialized, too-concentrated, long-distance food economy. Local food economies are desirable also from the standpoints of public health, "homeland security," and the energy economy. Provide economic incentives and supportive legislation for the establishment of local, small-scale food-processing plants, canneries, year-round farmers' markets, etc.

12. Local food economies, to be genuine, require local adaptation of domestic species and varieties of plants and animals. The universal evolutionary requirement of local adaptation has unaccountably been waived with respect to humans. But this waiver is potentially disastrous. We need ways of agriculture that are preservingly adapted to the ecological mosaic and even to individual farms and ranches. For the sake of local adaptation, and the genetic diversity that is necessary to it, we need to put an end to the U.S. Department of Agriculture's proposed National Animal Identification System, to the patenting of species, and to genetic engineering—all of which aim at a general agricultural uniformity and corporate control of agriculture and food. Central planning and its inevitable goal of uniformity cannot work in agriculture because of the requirement of local adaptation and the consequent need for local intelligence. Central planning and uniformity are effective only for the diminishment of genetic and biological diversity and the destruction of small producers.
13. Help and encourage small-scale forestry and owners of small woodlands. See that current market prices for sawlogs and other forest products are readily available everywhere. Tax fairly.
14. Study and teach sustainable forestry, using examples such as the Menominee Forest in Wisconsin and the Pioneer Forest in Missouri.
15. Promote the good use and care of farm woodlands as assets integral to the economy of farms.
16. Encourage the development, in forested regions, of local forest economies, providing economic incentives for local processing and value-adding, as for food.

Would such measures increase significantly the number of people at work in the land economy? Of course they would. This would be an authentic version, for a change, of "job creation." This work would help our economy, our people, and our country all at the same time. And that is the authentic test of practicality, for it makes complete economic sense.

Faustian Economics

THE GENERAL REACTION to the apparent end of the era of cheap fossil fuel, as to other readily foreseeable curtailments, has been to delay any sort of reckoning. The strategies of delay, so far, have been a sort of willed oblivion, or visions of large profits to the manufacturers of such "biofuels" as ethanol from corn or switchgrass, or the familiar unscientific faith that "science will find an answer." The dominant response, in short, is a dogged belief that what we call "the American way of life" will prove somehow indestructible. We will keep on consuming, spending, wasting, and driving, as before, at any cost to anything and everybody but ourselves.

This belief was always indefensible—the real names of global warming are "waste" and "greed"—and by now it is manifestly foolish. But foolishness on this scale looks disturbingly like a sort of national insanity. We seem to have come to a collective delusion of grandeur, insisting that all of us are "free" to be as conspicuously greedy and wasteful as the most corrupt of kings and queens. (Perhaps by devoting more and more of our already abused cropland to fuel production, we will at last cure ourselves of obesity and become fashionably skeletal, hungry but—Thank God!—still driving.)

The problem with us is not only prodigal extravagance, but also an assumed godly limitlessness. We have obscured the issue by refusing to see that limitlessness is a godly trait. We have insistently, and with relief, defined ourselves as animals or as "higher animals." But to define ourselves as animals, given our specifically human powers and desires, is to define ourselves as *limitless* animals—which of course is a contradiction in terms. Any definition is a limit, which is why the God of Exodus refuses to define Himself: "I am that I am."

Even so, that we have founded our present society upon delusional assumptions of limitlessness is easy enough to demonstrate. A recent "summit" in Louisville, Kentucky, was entitled "Unbridled Energy: The Industrialization of Kentucky's Energy Resources." Its subjects were "clean-coal generation, biofuels, and other cutting-edge applications," the conversion of coal to "liquid fuels," and the likelihood that all this will be "environmentally friendly." These hopes, which "can create jobs and boost the nation's security," are to be supported by government "loan guarantees . . . investment tax credits and other tax breaks." Such talk we recognize as completely conventional. It is, in fact, a tissue of clichés that is now the common tongue of promoters, politicians, and journalists. This language does not allow for any question about the *net* good of anything proposed. The entire contraption of "Unbridled Energy" is supported only by a rote optimism: "The United States has 250 billion tons of recoverable coal reserves —enough to last 100 years even at double the current rate of consumption." We humans have inhabited the earth for many thousands of years, and now we can look forward to surviving for another hundred by doubling our consumption of coal? *This* is national security? The world-ending fire of industrial fundamentalism may already be burning in our furnaces and engines, but if it will burn for a hundred more years, that will be fine. Surely it would be better to intend straightforwardly to contain the fire and eventually put it out? But once greed has been made an honorable motive, then you have an economy without limits, a contradiction in terms. This supposed economy has no plan for temperance or thrift or the ecological law of return. It will do anything. It is monstrous by definition.*

In keeping with our unrestrained consumptiveness, the commonly accepted basis of our present economy is the fantastical possibility of limitless growth, limitless wants, limitless wealth, limitless natural resources, limitless energy, and limitless debt. The idea of a limitless economy implies and requires a doctrine of general human limitlessness: *All* are entitled to pursue without limit whatever they conceive as desirable—a license that classifies the most exalted Christian capitalist with the lowliest pornographer.

This fantasy of limitlessness perhaps arose from the coincidence of the industrial revolution with the suddenly exploitable resources of the "new world." Or perhaps it comes from

*This is abundantly demonstrated by the suddenly ubiquitous rationalizations of "clean" and "safe" nuclear energy, ignoring the continuing problem of undisposable waste.

the contrary apprehension of the world's "smallness," made possible by modern astronomy and high-speed transportation. Fear of the smallness of our world and its life may lead to a kind of claustrophobia and thence, with apparent reasonableness, to a desire for the "freedom" of limitlessness. But this desire paradoxically reduces everything. The life of this world *is* small to those who think it is, and the desire to enlarge it makes it smaller, and can reduce it finally to nothing.

However it came about, this credo of limitlessness clearly implies a principled wish, not only for limitless possessions, but also for limitless knowledge, limitless science, limitless technology, and limitless progress. And necessarily it must lead to limitless violence, waste, war, and destruction. That it should finally produce a crowning cult of political limitlessness is only a matter of mad logic.

The normalization of the doctrine of limitlessness has produced a sort of moral minimalism: the desire to be "efficient" at any cost, to be unencumbered by complexity. The minimization of neighborliness, respect, reverence, responsibility, accountability, and self-subordination—this is the "culture" of which our present leaders and heroes are the spoiled children.

Our national faith so far has been "There's always more." Our true religion is a sort of autistic industrialism. People of intelligence and ability seem now to be genuinely embarrassed by any solution to any problem that does not involve high technology, a great expenditure of energy, or a big machine. Thus an X marked on a paper ballot no longer fulfills our idea of voting. One problem with this state of affairs is that the work now most needing to be done—that of neighborliness and caretaking—cannot be done by remote control with the greatest power on the largest scale. A second problem is that the economic fantasy of limitlessness in a limited world calls fearfully into question the value of our monetary wealth, which does not reliably stand for the real wealth of land, resources, and workmanship, but instead wastes and depletes it.

That human limitlessness is a fantasy means, obviously, that its life expectancy is limited. There is now a growing perception, and not just among a few experts, that we are entering a time of inescapable limits. We are not likely to be granted another world to plunder in compensation for our pillage of this one. Nor are we likely to believe much longer in our ability to outsmart, by means of science and technology, our economic stupidity. The hope that we can cure the ills of industrialism by the homeopathy of more technology seems at last to be losing status. We are, in short, coming under pressure to understand ourselves as limited creatures in a limited world.

This constraint, however, is not the condemnation it may seem. On the contrary, it returns us to our real condition and to our human heritage, from which our self-definition as limitless animals has for too long cut us off. Every cultural and religious tradition that I know about, while fully acknowledging our animal nature, defines us specifically as *humans*—that is, as animals (if the word still applies) capable of living, not only within natural limits, but also within cultural limits, self-imposed. As earthly creatures we live, because we must, within natural limits, which we may describe by such names as "earth" or "ecosystem" or "watershed" or "place" or "neighborhood." But as humans we may elect to respond to this necessary placement by the self-restraints implied in neighborliness, stewardship, thrift, temperance, generosity, care, kindness, friendship, loyalty, and love.

In our limitless selfishness, we have tried to define "freedom," for example, as an escape from all restraint. But, as my friend Bert Hornback has explained in his book *The Wisdom in Words*, "free" is etymologically related to "friend." These words come from the same Germanic and Sanskrit roots, which carry the sense of "dear" or "beloved." We set our friends free by our love for them, with the implied restraints of faithfulness or loyalty. This suggests that our "identity" is located not in the impulse of selfhood but in deliberately maintained connections.

Thinking of our predicament has sent me back again to Christopher Marlowe's *Tragical History of Doctor Faustus.* This is a play of the Renaissance: Faustus, a man of learning, longs to possess "all nature's treasury," to "Ransack the ocean . . . / And search all corners of the new-found world . . ." To assuage his thirst for knowledge and power, he deeds his soul to Lucifer, receiving in compensation for twenty-four years the services of the subdevil Mephistophilis, nominally his slave but in fact his master. Having the subject of limitlessness in mind, I was astonished on this reading to come upon Mephistophilis' description of hell. When Faustus asks, "How comes it then that thou art out of hell?" Mephistophilis replies, "Why, this is hell, nor am I out of it." A few pages later he explains:

> Hell hath no limits, nor is circumscribed
> In one self place, but where we [the damned] are is hell,
> And where hell is must we ever be.

For those who reject heaven, hell is everywhere, and thus is limitless. For them, even the thought of heaven is hell.

It is only appropriate, then, that Mephistophilis rejects any conventional limit: "Tut, Faustus, marriage is but a ceremonial toy. If thou lovest me, think no more of it." Continuing this theme, for Faustus' pleasure the devils present a sort of pageant of the seven deadly sins, three of which—Pride, Wrath, and Gluttony—describe themselves as orphans, disdaining the restraints of parental or filial love.

Seventy or so years later, and with the issue of the human definition more than ever in doubt, John Milton in Book VII of *Paradise Lost* returns again to a consideration of our urge to know. To Adam's request to be told the story of creation, the "affable Archangel" Raphael agrees "to answer thy desire / Of knowledge *within bounds* [my emphasis] . . . ," explaining that

> Knowledge is as food, and needs no less
> Her temperance over appetite, to know
> In measure what the mind may well contain;
> Oppresses else with surfeit, and soon turns
> Wisdom to folly, as nourishment to wind.

Raphael is saying, with angelic circumlocution, that knowledge without wisdom, limitless knowledge, is not worth a fart; he is not a humorless archangel. But he also is saying that knowledge without measure, knowledge that the human mind cannot appropriately use, is mortally dangerous.

I am well aware of what I risk in bringing this language of religion into what is normally a scientific discussion—if economics is in fact a science. I do so because I doubt that we can define our present problems adequately, let alone solve them, without some recourse to our cultural heritage. We are, after all, trying now to deal with the failure of scientists, technicians, and politicians to "think up" a version of human continuance that is economically probable and ecologically responsible, or perhaps even imaginable. If we go back into our tradition, we are going to find a concern with religion, which at a minimum shatters the selfish context of the individual life and thus forces a consideration of what human beings are and ought to be.

This concern persists at least as late as our Declaration of Independence, which holds as "self-evident, that all men are created equal; that they are endowed by their Creator with certain inalienable rights . . ." Thus among our political roots we have still our old preoccupation with our definition as humans, which in the Declaration is wisely assigned to our Creator; our rights and the rights of all humans are not granted by any human government but are innate, belonging to us by birth. This insistence comes, not from the fear of death or even extinction, but from the ancient fear, readily documentable in our cultural tradition, that in order to survive we might become inhuman or monstrous.

Our cultural tradition is in large part the record of our continuing effort to understand ourselves as beings specifically human—to say that, as humans, we must do certain things and we must not do certain things. We must have limits or we will cease to exist as humans; perhaps we will cease to exist, period. At times, for example, some of us humans have thought that human beings, properly so-called, did not make war against civilian populations, or hold prisoners without a fair trial, or use torture for any reason.

Some of us would-be humans have thought too that we should not be free at anybody else's expense. And yet in the phrase "free market," the word "free" has come to mean unlimited economic power for some, with the necessary consequence of economic powerlessness for others. Several years ago, after I had spoken at a meeting, two earnest and obviously troubled young veterinarians approached me with a question: How could they practice veterinary medicine without serious economic damage to the farmers who were their clients? Underlying their question was the fact that for a long time veterinary help for a sheep or a pig has been likely to cost more than the animal is worth. I had to answer that, in my opinion, so long as their practice relied heavily on selling patented drugs, they had no choice, since the market for medicinal drugs was entirely controlled by the drug companies, whereas most farmers had no control at all over the market for agricultural products. My questioners were asking in effect if a predatory economy can have a beneficent result. The answer usually is No. And that is because there is an absolute discontinuity between the economy of the seller of medicines and the economy of the buyer, as there is in the health industry as a whole. The drug industry is interested in the survival of patients, we have to suppose, because surviving patients will continue to consume drugs.

Now let us consider a contrary example. Recently at another meeting I talked for some time with an elderly, some would say old-fashioned, farmer from Nebraska. Unable to farm any longer himself, he had rented his land to a younger farmer on the basis of what he called "crop share" instead of a price paid or owed in advance. Thus, as the old farmer said of his renter, "If he has a good year, I have a good year. If he has a bad year, I have a bad one." This is what I would call community economy. It is a sharing of fate. It assures an economic continuity and a common interest between the two partners to the trade. This is as far as possible from the economy in which the young veterinarians were caught, in which the economically powerful are limitlessly "free" to trade to the disadvantage, and ultimately the ruin, of the powerless.

It is this economy of community destruction that, wittingly or unwittingly, most scientists and technicians have served for the last two hundred years. These scientists and technicians have justified themselves by the proposition that they are the vanguard of progress, enlarging human knowledge and power. Thus have they romanticized both themselves and the predatory enterprises that they have served.

As a consequence, our great need now is for sciences and technologies of limits, of domesticity, of what Wes Jackson of the Land Institute in Salina, Kansas, has called "homecoming." These would be specifically human sciences and technologies, working, as the best humans always have worked, within self-imposed limits. The limits would be, as they always have been, the accepted contexts of places, communities, and neighborhoods, both natural and human.

I know that the idea of such limitations will horrify some people, maybe most people, for we have long encouraged ourselves to feel at home on "the cutting edges" of knowledge and power or on some "frontier" of human experience. But I know too that we are talking now in the presence of much evidence that improvement by outward expansion may no longer be a good idea, if it ever was. It was not a good idea for the farmers who "leveraged" secure acreage to buy more during the 1970s. It has proved tragically to be a bad idea in a number of recent wars. If it is a good idea in the form of corporate gigantism, then we must ask, For whom? Faustus, who wants all knowledge and all the world for himself, is a man supremely lonely and finally doomed. I don't think Marlowe was kidding. I don't think Satan is kidding when he says in *Paradise Lost*, "myself am Hell."

If the idea of appropriate limitation seems unacceptable to us, that may be because, like Marlowe's Faustus and Milton's Satan, we confuse limits with confinement. But that, as I think Marlowe and Milton and others were trying to tell us, is a great and potentially a fatal mistake. Satan's fault, as Milton understood it and perhaps with some sympathy, was precisely that he could not tolerate his proper limitation; he could not subordinate himself to anything whatsoever. Faustus' error was his unwillingness to remain "Faustus, and a man." In our age of the world it is not rare to find writers, critics, and teachers of literature, as well as scientists and technicians, who regard Satan's and Faustus' defiance as salutary and heroic.

On the contrary, our human and earthly limits, properly understood, are not confinements, but rather are inducements to formal elaboration and elegance, to *fullness* of relationship and meaning. Perhaps our most serious cultural loss in recent centuries is the knowledge that some things, though limited, can be inexhaustible. For example, an ecosystem, even that of a working forest or farm, so long as it remains ecologically intact, is inexhaustible. A small place, as I know from my own experience, can provide opportunities of work and learning, and a fund of beauty, solace, and pleasure—in addition to its difficulties—that cannot be exhausted in a lifetime or in generations.

To recover from our disease of limitlessness, we will have to give up the idea that we have a right to be godlike animals, that we are at least potentially omniscient and omnipotent, ready to discover "the secret of the universe." We will have to start over, with a different and much older premise: the naturalness and, for creatures of limited intelligence, the necessity of limits. We must learn again to ask how we can make the most of what we are, what we have, what we have been given. If we always have a theoretically better substitute available from somebody or someplace else, we will never make the most of anything. It is hard enough to make the most of one life. If we each had two lives, we would not make much of either. One of my best teachers said of people in general: "They'll never be worth a damn as long as they've got two choices."

To deal with the problems, which after all are inescapable, of living with limited intelligence in a limited world, I suggest that we may have to remove some of the emphasis we have lately placed on science and technology and have a new look at the arts. For an art does not propose to enlarge itself by limitless extension, but rather to enrich itself within bounds that are accepted prior to the work.

It is the artists, not the scientists, who have dealt unremittingly with the problem of limits. A painting, however large, must finally be bounded by a frame or a wall. A composer or playwright must reckon, at a minimum, with the capacity of an audience to sit still and pay attention. A story, once begun, must end somewhere within the limits of the writer's and the reader's memory. And of course the arts characteristically impose limits that are artificial: the five acts of a play, or the fourteen lines of a sonnet. Within these limits artists achieve elaborations of pattern, of sustaining relationships of parts with one another and with the whole, that may be astonishingly complex. Probably most of us can name a painting, a piece of music, a poem or play or story that still grows in meaning and remains fresh after many years of familiarity.

We know by now that a natural ecosystem survives by the same sort of formal intricacy, ever-changing, inexhaustible, and perhaps finally unknowable. We know further that if we want to make our economic landscapes sustainably and abundantly productive, we must do so by maintaining in them a living formal complexity something like that of natural ecosystems. We can do this only by raising to the highest level our mastery of the arts of agriculture, animal husbandry, forestry, and, ultimately, the art of living.

It is true that insofar as scientific experiments must be conducted within carefully observed limits, scientists also are artists. But in science one experiment, whether it succeeds or fails, is logically followed by another in a theoretically infinite progression. According to the underlying myth of modern science, this progression is always replacing the smaller knowledge of the past with the larger knowledge of the present, which will be replaced by the yet larger knowledge of the future.

In the arts, by contrast, no limitless sequence of works is ever implied or looked for. No work of art is necessarily followed by a second work that is necessarily better. Given the methodologies of science, the law of gravity and the genome were bound to be discovered by somebody; the identity of the discoverer is incidental. But in the arts there are no second chances. We must assume that we had one chance each for *The Divine Comedy* and *King Lear*. If Dante and Shakespeare had died before they wrote those poems, nobody ever would have written them.

The same is true of our arts of land use, our economic arts, which are our arts of living. With these it is once-for-all. We will have no chance to redo our experiments with bad agriculture leading to soil loss. The Appalachian mountains and forests we have destroyed for coal are gone forever. It is now and forevermore too late to use thriftily the first half of the world's supply of petroleum. In the art of living we can only start again with what remains.

As we confront the phenomenon of "peak oil," we are really confronting the end of our customary delusion of "more." Whichever way we turn, from now on, we are going to find a limit beyond which there will be no more. To hit these limits at top speed is not a rational choice. To start slowing down, with the idea of avoiding catastrophe, *is* a rational choice, and a viable one if we can recover the necessary political sanity. Of course it makes sense to consider alternative energy sources, provided *they* make sense. But also we will have to reexamine the economic structures of our life, and conform them to the tolerances and limits of our earthly places. Where there is no more, our one choice is to make the most and the best of what we have.

FROM

IMAGINATION IN PLACE

(2010)

Imagination in Place

By an interworking of chance and choice, I have happened to live nearly all my life in a place I don't remember not knowing. Most of my forebears for the last two hundred years could have said the same thing. I was born to people who knew this place intimately, and I grew up knowing it intimately. For a long time the intimacy was not very conscious, but I certainly did not grow up here thinking of the place as "subject matter," and I have never thought of it in that way. I have not lived here, or worked with my neighbors and my family, or listened to the storytellers and the rememberers, in order to be a writer. The place is precedent to my work, especially my fiction, and is, as I shall try to show, inevitably different from it.

By the same interworking of chance and choice, though somewhat expectably, I have lived here as a farmer. Except for one great-grandfather, all of my family that I know about have been farming people, and I grew up under instruction, principally from my father but also from others, to learn farming, to know the difference between good farming and bad, to regard the land as of ultimate value, and to admire and respect those who farmed well. I never heard a farmer spoken of as "just a farmer" or a farm woman as "just a housewife." To my father and his father especially, the knowledge of land and of farming was paramount. They thought the difference between a good farmer and a bad one was just as critical as the difference between a good politician and a bad one.

In 1964, after several years of wandering about, my wife Tanya and I returned to Kentucky with our two children and bought the property known as Lanes Landing, on the Kentucky River, about a mile from the house where my mother was born and raised and about five miles from my father's home place. The next summer we fixed up the house and moved in. We have been here ever since. Or Tanya and I have; our children are farming nearby.

Before we moved here, I had known this place for thirty-one years, and we have now lived here for thirty-nine. We raised our children here. We have taken from this place most of our food, much of our fuel, and always, despite the difficulties and frustrations of a farming life, a sustaining pleasure. Also, nearly everything I have written has been written here. When I am asked how all this fits together, I have to say, "Awkwardly." Even so, this has been the place of my work and of my life.

This essay is most immediately obstructed by the difficulty of separating my work from my life, and the place from either. The place included in some of my work is also the place that has included me as a farmer and as a writer.

In the course of my life and of my work as a farmer, I have come to know familiarly two small country towns and about a dozen farms. That is, I have come to know them well enough at one time or another that I can shut my eyes and see them as they were, just as I can see them now as they are. The most intimate "world" of my life is thus a small one. The most intimate "world" of my fiction is even smaller: a town of about a hundred people, "Port William," and a few farms in its neighborhood. Between these two worlds, the experienced and the imagined, there is certainly a relationship. But it is a relationship obscure enough as it is, and easy to obscure further by oversimplification. Another difficulty of this essay is the temptation to oversimplify.

As a lot of writers must know, it is easy for one's family or neighbors to identify fictional characters with actual people. A lot of writers must know too that these identifications are sometimes astonishingly wrong, and are always at least a little wrong. The inevitability of this sort of error is explainable, and it is significant.

Some of my own fiction has seemed to me to be almost entirely imagined. Some of it has drawn maybe as close as possible to actual experience. The writing has sometimes grown out of a long effort to come to terms with an actual experience. But one must not be misled by the claims of "realism." There is, true enough, a kind of writing that has an obligation to tell the truth about actual experience, and therefore it is obliged to accept the limits of what is actually or provably known. But works of imagination come of an impulse to transcend the limits of experience or provable knowledge in order to make a thing that is whole. No human work can become whole by including everything, but it can become whole in another way: by accepting its formal limits and then answering within those limits all the questions it raises. Any reasonably literate reader can understand Homer without the benefit of archaeology, or Shakespeare without resort to his literary sources.

It seems to me that my effort to come to terms in writing with an actual experience has been, every time, an effort to imagine the experience, to see it clear and whole in the mind's eye. One might suppose, reasonably enough, that this could be accomplished by describing accurately what one actually knows from records of some sort or from memory. But this, I believe, is wrong. What one actually or provably knows about an actual experience is never complete; it cannot, within the limits of memory or factual records, be made whole. Imagination "completes the picture" by transcending the actual memories and provable facts. For this reason, I have often begun with an actual experience and in the end produced what I have had to call a fiction. In the effort to tell a whole story, to see it whole and clear, I have had to imagine more than I have known. "There's no use in telling a pretty good story when you can tell a really good one," my mother's father told me once. In saying so, he acknowledged both a human limit and a human power, as well as his considerable amusement at both.

I believe I can say properly that my fiction originates in part in actual experience of an actual place: its topography, weather, plants, and animals; its language, voices, and stories. The fiction I have written here, I suppose, must somehow belong here and must be different from any fiction I might have written in any other place. I am pleased to suppose so, but the issue of influence is complex and obscure, and the influence of this place alone cannot account for the fiction and the other work I have written here.

Both my writing and my involvement with this place have been in every way affected by my reading. My work would not exist as it is if the influence of this place were somehow subtracted from it. Just as certainly it would not exist as it is, if at all, without my literary mentors, exemplars, teachers, and guides. Lists are dangerous, but as a placed writer I have depended on the examples of Andrew Marvell at Appleton

House, Jane Austen in Hampshire, Thomas Hardy in Dorset, Mark Twain in Hannibal, Thoreau in Concord, Sarah Orne Jewett on the Maine coast, Yeats in the west of Ireland, Frost in New England, William Carlos Williams in Rutherford, William Faulkner and Eudora Welty in Mississippi, Wallace Stegner in the American West, and in Kentucky, James Still, Harlan Hubbard, and Harry Caudill—to name only some of the dead and no contemporaries. I have kept fairly constantly in my mind the Bible, Homer, Dante, Shakespeare, Herbert, Milton, and Blake. I have taken much consolation and encouragement from Paul Cézanne's devotion to his home landscapes in Provence and from Samuel Palmer's work at Shoreham. I have remembered often the man of Psalm 128 who shall eat the labor of his hands, and Virgil's (and Ronsard's) old Cilician of *Georgics* IV. Over the last twenty years or so, I have contracted a large debt to certain writers about religious and cultural tradition, principally Ananda Coomaraswamy, Titus Burckhardt, Kathleen Raine, and Philip Sherrard—again, to name only the dead. Now that I have listed these names, I am more aware than before how incomplete any such list necessarily must be, and how necessarily confusing must be the issue of influence.

I will allow the list to stand, not as an adequate explanation, but as a hint at the difficulty of locating the origins of a work of fiction by me (or, I assume, by anybody else). And I must add further to the difficulty by saying that I don't believe I am conscious of all the sources of my work. I dislike learned talk about "the unconscious," which always seems to imply that the very intelligent are able somehow to know what they don't know, but I mean only to acknowledge that much of what I have written has taken me by surprise. What I know does not yield a full or adequate accounting for what I have imagined. It seems to have been "given." My experience has taught me to believe in inspiration, about which I think nobody can speak with much authority.

My fiction, anyhow, has come into being within the contexts of local geography and local culture, of the personal culture of reading, listening, and looking, and also within the contexts of what is not known and of the originating power we call inspiration. But there is another context, that of agriculture, which I will need to deal with at more length.

I was brought up, as I have said, by agrarians and was conscientiously instructed in a set of assumptions and values that could be described only as agrarian. But I never saw that word in print or heard it pronounced until I was a sophomore at the University of Kentucky. At that time I was in a composition class whose instructor, Robert D. Jacobs, asked us to write an argument. I wrote, as I recall, a dialogue between two farmers on the condemnation of land for the construction of a highway or an airport. The gist of my argument was that the land was worth more than anything for which it might be destroyed. Dr. Jacobs didn't think much of my argument, but he did me a valuable service by identifying it as "agrarian" and referring me to a group of writers, the Southern Agrarians, who had written a book called *I'll Take My Stand.* I bought the book and read at least part of it about three years later, in 1956. It is a valuable book, in some ways a wonder, and I have returned to it many times since. My debt to it has increased.

I must have become a good deal interested in the Southern Agrarians during my last years at the university, for with my friend and fellow student Mac Coffman (Edward M. Coffman, the historian) I drove up to Kenyon College to talk with John Crowe Ransom on that subject. But it is hard now for me to tell how much I may have been influenced by the Southern Agrarians and their book at that time. (Ransom by then was disaffected from *I'll Take My Stand*, though his elegant introduction, "A Statement of Principles," is still the best summary of agrarian principles versus the principles of industrialism). And I think I encountered not much at the University of Kentucky that would have confirmed my native agrarianism. It seems to me now that my agrarian upbringing and my deepest loyalties were obscured by my formal education. Only after I returned to Kentucky in 1964 did I begin to reclaim what I had been taught at home as a growing boy. Once I was home again, the purpose and point of that teaching became clear to me as it had not before, and I became purposefully and eagerly an agrarian. Moreover, because I had settled here as a farmer, I knew that I was not a literary agrarian merely but also a practical one.

*

In 1970 I published in *The Southern Review* a small essay, "The Regional Motive," that I suppose was descended from, or at least a cousin to, the essays of *I'll Take My Stand*. But in my essay I said that "the withdrawal of the most gifted of [the Southern Agrarians] into . . . Northern colleges and universities invalidated their thinking, and reduced their effort to the level of an academic exercise." Whatever the amount of truth in that statement, and there is some, it is also a piece of smartassery.

I received in response a letter from Allen Tate. As I knew, Tate could be a combative man, and so I was moved, as I still am, by the kindness of his letter. He simply pointed out to me that I did not know the pressing reasons why he and his friends had moved to the North. And so when I reprinted my essay I added a footnote apologizing for my callowness and ignorance, but saying, even so—and, as I remember, with Tate's approval—that I might appropriately "warn that their departure should not be taken either as disproof of the validity of their [agrarian] principles, or as justification of absentee regionalism (agrarianism without agriculture)."

The parentheses around that concluding phrase suggest to me now that I was making a point I had not quite got. The phrase, which appears to have been only an afterthought thirty-two years ago, indicates what to me now seems the major fault of *I'll Take My Stand*: The agrarianism of most of the essays, like the regionalism of most of them, is abstract, too purely mental. The book is not impractical—none of its principles, I believe, is in conflict with practicality—but it is too often remote from the issues of practice. The legitimate aim (because it is the professed aim) of agrarianism is not some version of culture but good farming, though a culture complete enough may be implied in that aim. By 1970 I had begun to see the flaws and dangers of absentee regionalism, and especially of Southern absentee regionalism. Identifying with "The South," as if it were somehow all one and the same place, would not help you to write any more than it would help you to farm. As a regional book, *I'll Take My Stand* mostly ignores the difficulty and the discipline of locality. As an agrarian book, it mostly ignores also the difficulty and the discipline of farming, but this problem is more complicated, and dealing with it took me longer.

Of the twelve essayists, only Andrew Lytle and John Donald Wade appear to speak directly from actual knowledge of actual farming in an actual place. And a passage of Andrew Lytle's essay, "The Hind Tit," points the direction I now must take with this essay of mine. He has begun to write about "a type" of farmer who has two hundred acres of land, but he does so with a necessary precaution:

> This example is taken, of course, with the knowledge that the problem on any two hundred acres is never the same: the richness of the soil, its qualities, the neighborhood, the distance from market, the climate, water, and a thousand such things make the life on every farm distinctly individual.

Thus he sets forth the fundamental challenge, not only to all forms of industrial land use, but to all other approaches to land use, including agrarianism, that are abstract.

The most insistent and formidable concern of agriculture, wherever it is taken seriously, is the distinct individuality of every farm, every field on every farm, every farm family, and every creature on every farm. Farming becomes a high art when farmers know and respect in their work the distinct individuality of their place and the neighborhood of creatures that lives there. This has nothing to do with the set of personal excuses we call "individualism" but is akin to the holy charity of the Gospels and the political courtesy of the Declaration of Independence and the Bill of Rights. Such practical respect is the true discipline of farming, and the farmer must maintain it through the muddles, mistakes, disappointments, and frustrations, as well as the satisfactions and exultations, of every actual year on an actual farm.

And so it has mattered, undoubtedly it has mattered to my fiction, that I have lived in this place both as a farmer and as a writer. I am not going to pretend here to a judgment or criticism of the writing I have done. I mean only to say something about the pressures and conditions that have been imposed on my writing by my life here as a farmer. Rather than attempt to say what I have done, I will attempt to speak of farming as an influence.

Having settled even in so marginal a place as this, undertaking to live in it even by such marginal farming as I have done, one is abruptly and forcibly removed from easy access to the abstractions of regionalism, politics, economics, and the academic life. To farm is to be placed absolutely. To do the actual work of an actual farm, one must shed the clichés that constitute "The South" or "My Old Kentucky Home" and come to the ground.

One may begin as an agrarian, as some of us to our good fortune have done, but for a farmer agrarianism is not enough. Southern agrarianism is not enough, and neither is Kentucky agrarianism or Henry County agrarianism. None of those can be local enough or particular enough. To live as a farmer, one has to come into the local watershed and the local ecosystem, and deal well or poorly with them. One must encounter directly and feelingly the topography and the soils of one's particular farm, and treat them well or poorly.

If one wishes to farm well, and agrarianism inclines to that wish above all, then one must submit to the unending effort to change one's mind and ways to fit one's farm. This is a hard education, which lasts all one's life, never to be completed, and it almost certainly will involve mistakes. But one does not have to do this alone, or only with one's own small intelligence. Help is available, as one had better hope.

In my farming I have relied most directly on my family and my neighbors, who have helped me much and taught me much. And my thoughts about farming have been founded on a few wonderful books: *Farmers of Forty Centuries* by F. H. King, *An Agricultural Testament* and *The Soil and Health* by Sir Albert Howard, *Tree Crops* by J. Russell Smith, and *A Sand County Almanac* by Aldo Leopold. These writers bring the human economy face to face with ecology, the local landscape, and the farm itself. They teach us to think of the ecological problems and obligations of agriculture, and they do this by seeing in nature the inescapable standard and in natural processes the necessary pattern for any human use of the land. Their thinking has had its finest scientific result thus far in the Natural Systems Agriculture of the Land Institute in Salina,

Kansas. Natural Systems Agriculture returns to the classical conception of art as an imitation of nature. But whereas Hamlet saw art as holding a mirror up to nature, and thus in a sense taking its measure, these agricultural thinkers have developed the balancing concept of nature as the inevitable mirror and measure of art.

In addition to books specifically about agriculture and ecology, I have been steadily mindful, as a farmer, of the writers mentioned earlier as literary influences. And I have depended for many years on the writing and the conversation of my friends Gene Logsdon, Maurice Telleen, Wes Jackson, and David Kline. I have been helped immeasurably also by the examples of Amish agriculture, of the traditional farming of Tuscany as I saw it more than forty years ago, of the ancient agricultures of the Peruvian Andes and the deserts of the American Southwest, of the also ancient pastoral landscapes of Devonshire, and of the best farming here at home as I knew it in the 1940s and early '50s, before industrialization broke up the old pattern.

What I have learned as a farmer I have learned also as a writer, and vice versa. I have farmed as a writer and written as a farmer. For the sake of clarity, I wish that this were more divisible or analyzable or subject to generalization than it is. But I am talking about an experience that is resistant to any kind of simplification. It is an experience of what I will go ahead and call complexification. When I am called, as to my astonishment I sometimes am, a devotee of "simplicity" (since I live supposedly as a "simple farmer"), I am obliged to reply that I gave up the simple life when I left New York City in 1964 and came here. In New York, I lived as a passive consumer, supplying nearly all my needs by purchase, whereas here I supply many of my needs from this place by my work (and pleasure) and am responsible besides for the care of the place.

My point is that when one passes from any abstract order, whether that of the consumer economy or Ransom's "Statement of Principles" or a brochure from the Extension Service, to the daily life and work of one's own farm, one passes from a relative simplicity into a complexity that is irreducible except by disaster and ultimately is incomprehensible. It is the complexity of the life of a place uncompromisingly itself, which is at the same time the life of the world, of all Creation. One meets not only the weather and the wildness of the world, but also the limitations of one's knowledge, intelligence, character, and bodily strength. To do this, of course, is to accept the place as an influence.

My further point is that to do this, if one is a writer, is to accept the place and the farming of it as a literary influence. One accepts the place, that is, not just as a circumstance, but as a part of the informing ambience of one's mind and imagination. I don't dare to claim that I know how this "works," but I have no doubt at all that it is true. And I don't mind attempting some speculations on what might be the results.

To begin with, the work of a farmer, or of the sort of farmer I have been, is particularizing work. As farmers themselves never tire of repeating, you can't learn to farm by reading a book. You can't lay out a fence line or shape a plowland or fell a tree or break a colt merely by observing general principles. You can't deal with things merely according to category; you are continually required to consider the distinct individuality of an animal or a tree, or the uniqueness of a place or a situation, and to do so you draw upon a long accumulation of experience, your own and other people's. Moreover, you are always under pressure to explain to somebody (often yourself) exactly what needs to be done. All this calls for an exactly particularizing language. This is the right kind of language for a writer, a language developing, so to speak, from the ground up. It is the right kind of language for anybody, but a lot of our public language now seems to develop downward from a purpose. Usually, the purpose is to mislead, the particulars being selected or invented to suit the purpose; or the particulars dangle loosely and unregarded from the dislocated intellectuality of the universities. This is contrary to honesty and also to practicality.

The ability to speak exactly is intimately related to the ability to know exactly. In any practical work such as farming the penalties for error are sometimes promptly paid, and this is valuable instruction for a writer. A farmer who is a writer will at least call farming tools and creatures by their right names, will be right about the details of work, and may extend the same courtesy to other subjects.

A writer who is a farmer will in addition be apt actually to know some actual country people, and this is a significant advantage. Reading some fiction, and this applies especially to some Southern fiction, one cannot avoid the impression that the writers don't know any country people in particular and are in general afraid of them. They fill the blank, not with anybody they have imagined, but with the rhetorically conjured stereotype of the hick or hillbilly or redneck who is the utter opposite of the young woman with six arms in the picture by the late ("Alas") Emmeline Grangerford, and perhaps is her son. He comes slouching into the universe with his pistol in one hand, his penis in another, his Bible in another, his bottle in another, his grandpappy's cavalry sword in another, his plug of chewing tobacco in another. This does harm. If you wish to steal farm products or coal or timber from a rural region, you will find it much less troubling to do so if you can believe that the people are too stupid and violent to deserve the things you wish to steal from them. And so purveyors of rural stereotypes have served a predatory economy. Two of the Southern Agrarians, I should add, countered this sort of thing with knowledge. I am thinking of John Donald Wade's essay "The Life and Death of Cousin Lucius" in *I'll Take My Stand*, and *A Wake for the Living* by Andrew Lytle.

If you understand that what you do as a farmer will be measured inescapably by its effect on the place, and of course on the place's neighborhood of humans and other creatures, then if you are also a writer, you will have to wonder too what will be the effect of your writing on that place. Obviously this is going to be hard for anybody to know, and you yourself may not live long enough to know it, but in your own mind you are going to be using the health of the place as one of the indispensable standards of what you write, thus dissolving the university and "the literary world" as adequate contexts for literature. It also is going to skew your work away from the standard of realism. "How things really are" is one of your concerns, but by no means the only one. You have begun to ask also how things will be, how you want things to be, how things ought to be. You want to know what are the meanings, both temporal and eternal, of the condition of things in this world. "Realism," as Kathleen Raine said, "cannot show us what we are, but only our failure to become that to which the common man and the common woman inadequately, but continually, aspire and strive." If, in other words, you want to write a whole story about whole people—living souls, not "higher animals"—you must reach for a reality which is inaccessible merely to observation or perception but which in addition requires imagination, for imagination knows more than the eye sees, and also inspiration, which you can only hope and pray for. You will find, I think, that this effort involves even a sort of advocacy. Advocacy, as a lot of people will affirm, is dangerous to art, and you must beware the danger, but if you accept the health of the place as a standard, I think the advocacy is going to be present in your work. Hovering over nearly everything I have written is the question of how a human economy might be conducted with reverence, and therefore with due respect and kindness toward everything involved. This, if it ever happens, will be the maturation of American culture.

I have tried (clumsily, I see) to define the places, real and imagined, where I have taken my stand and done my work. I have made the imagined town of Port William, its neighborhood and membership, in an attempt to honor the actual place where I have lived. By means of the imagined place, over the last fifty years, I have learned to see my native landscape and neighborhood as a place unique in the world, a work of God, possessed of an inherent sanctity that mocks any human valuation that can be put upon it. If anything I have written in this place can be taken to countenance the misuse of it, or to excuse anybody for rating the land as "capital" or its human members as "labor" or "resources," my writing would have been better unwritten. And then to hell with any value anybody may find in it "as literature."

American Imagination and the Civil War

SOME SENTENCES of the Irish poet Patrick Kavanagh have been prominent in my thoughts for many years:

> Parochialism and provincialism are opposites. The provincial has no mind of his own; he does not trust what his eyes see until he has heard what the metropolis . . . has to say . . . The parochial mentality on the other hand is never in any doubt about the social and artistic validity of his parish.

In spite of necessary qualifications, which I will get to in a minute, Kavanagh's distinction has become indispensable to me in thinking about my native place and history. In Kentucky, a state famously characterized as barefooted, we might oversimplify Kavanagh by saying that those of us who are always admiring our shoes are provincial, whereas the unself-consciously bare or shod are parochial. Or we could more legitimately paraphrase him by saying that people who fear they are provincial are provincial.

I believe I can say truthfully that my particular part of Kentucky, at the time of my growing up in it, was in Kavanagh's terms more parochial than provincial. The parochial in any locality probably always is subject to qualification and inexact in geographical extent. I grew up in a county at that time almost exclusively preoccupied with farming: the county of Henry, a few miles south of the Ohio River. But the country truly native to my family and my experience is in the watersheds of Town Branch of Drennon Creek, Emily's Run, and Cane Run.

During my first twenty or so years, the "social validity" of that place at that time certainly was impaired by racial segregation. That phrase now has the currency of an abstraction, but segregation itself could be experienced only in particular. We were living in the history of segregation, but we were living in it in our place, with our neighbors, and as ourselves. In our small communities segregation involved the wicked prejudice on which it was based, but it also involved much familiarity and many exceptions. Racial inequality was a theory that performed its customary disservices and sometimes justified horrors, but that theory was inevitably qualified by the daily life in which the two races were separate only to an extent. In those places the history of segregation was lived out familiarly by black and white people who knew one another, told stories about one another to one another, helped or harmed one another, liked or disliked one another, and often worked together. Separate and different as the races were, it is impossible to imagine a white person of that place and time whose knowledge did not include the stories, songs, sayings, teachings, and characters of black persons. An honest accounting of the ancestry of my own mind would have to include prominently several black people. Despite segregation, the communities of my young life were, in function and in their consciousness of themselves, more intact than they are now.

As for the "artistic validity" of our place at that time, I have to be both careful and modest. We did have a local music that came to the fore at square dances, though not everybody granted much value to it, and it had begun to be supplanted by music from the radio and jukebox. Most of us were familiar with the Protestant hymns and the King James Bible. But we were not greatly concerned with issues of art, local or otherwise.

The arts that we took for granted, and that did gather us all together, were the arts of farming, gardening, cooking, and talking. Our economy was either agricultural or in service to agriculture. Vegetable gardens, grape arbors, and fruit trees were still commonplace. It was still ordinary to see poultry flocks, fattening hogs, or milkcows in the back yards or back lots of the towns. The grocery stores still bought surplus produce from the farms. Most of the food was homegrown, and excellent cooking was customary. Most of the cooking was done by women, but everybody talked about it.

Everybody, in fact, talked about everything. It seems to me that I grew up totally immersed in talk. Talk was a fifth element: talk in hayfields and tobacco patches, in tobacco barns and stripping rooms, in kitchens and living rooms, on porches and out in the yards. Sometimes, as we sat out in the yard or on the porch after a hot day, the dark would gradually disembody us, and we would become just voices going on until weariness re-embodied us and we would go into the house to bed. My best gift as a writer was that circumstance of talk. We had no cultivated art of conversation. Our talk was practical, local in reference, but was carried on also for pleasure or comfort. It was sometimes crude enough, but it was also articulate enough, humorous, precise, expressive, and sometimes beautifully so.

Though by then most of us had listened to the radio and seen at least a few movies, our talk as yet bore no hint of apology for the way we talked, or for our status as country or small-town people. We knew we were not Yankees, for we had heard Yankees talk and we knew we did not talk like them. We also had listened to people from "down South," and we knew we did not have what we called a "Southern accent." Maybe I can be excused for concluding, when I got old enough to read a map, that I spoke a perfectly average language, Standard American, since I could see that I lived at about the middle of the north–south axis. And maybe I can elicit a little sympathy for my surprise when, having clung to this notion all the way to some literary party in California, I delivered an undoubtedly sophisticated opinion to a literary young lady, whose eyes thereupon grew round with recognition. "Wayull!" she said in Yankee-Southern. "Wheah *you*'all frum, honey chile?"

And so I turned out to be a Southerner—legitimately so, as that term is used. I was born on the south side of the Ohio, was descended from slave-owners, and certainly did not talk like a Yankee.

The problem, as I am hardly the first to know, is that being a Southerner is less a condition than a job. The job, unendingly, is to distinguish between local life and the abstractions that we have allowed to obscure it. There is a huge difference between knowledge and classification. "South" and "Southerner" are not terms that are invariably useful. They belong sometimes to a taxonomy of clichés, stereotypes, and prejudices that have intruded between ourselves and our actual country. These shallow, powerful abstractions have worked to depreciate local knowledge and provincialize local life, and so have denied us the imaginative realizations that alone could have saved our country from the damage that has befallen it.

These old habits of mind and speech have continued in the babbled-to-nonsense polarity of "conservative" and "liberal," and of "red states" and "blue states." This oversimplified language of the media and politics is as far as possible from the best of the local speech I heard as a child, which was like no other in the world because it was of and about our place, which was like no other in the world. In it we were at least beginning to imagine ourselves somewhat as we actually were, and even somewhat as we should have been. Now, under the influence of media speech, we can only pretend and try to be like everybody else.

The problem is that there can be no general or official or sectional or national imagination. The chief instrument of economic and political power now is a commodified speech, wholly compatible with the old clichés, that can distinguish neither general from particular nor false from true. Local life is now a wren's egg brooded by an eagle or a buzzard. As Guy Davenport saw, nothing now exists that is so valuable as whatever theoretically might replace it. Every place must anticipate the approach of the bulldozer. No place is free of the threat implied in such phrases as "economic growth," "job creation," "natural resources," "human capital," "bringing in industry," even "bringing in culture"—as if every place is adequately identified as "the environment" and its people as readily replaceable parts of a machine. Devotion to any particular place now carries always the implication of heartbreak.

I suppose that human minds have always been threatened by the slur and blur of general bias, but it seems to me that this curse fell upon us Americans with a great fatefulness in the circumstances leading to the Civil War, and that the curse has persisted.

The Civil War and the rhetoric associated with it become penetrable by actual thought only when one asks why. Why could people of good sense on both sides not have treated slavery as a problem with a practical solution short of war? The answers, I suppose, are foolishness, fanaticism, sectional loyalty and pride, the wish to protect one's faults from correction by others, moral outrage, self-righteousness, the desire to punish sinners, and sectional hatred.

The Civil War was caused undoubtedly by disagreements over slavery and secession. It was contested so fiercely and so long by the Confederacy undoubtedly because of a truth that our federal government has never learned: People generally don't like to be invaded. But why was there no lenity?

Shakespeare's Henry V, incongruously in the midst of his invasion of France, gives "lenity" a pertinent definition:

> . . . we give express charge that in our marches through the country there be nothing compelled from the villages, nothing taken but paid for, none of the French upbraided or abused in disdainful language; for when lenity and cruelty play for a kingdom, the gentler gamester is the soonest winner.

The word occurs more credibly in Edmund Burke's *Speech On American Taxation*, in which he pleads desperately against the impositions that brought on the American Revolution:

> Yet now, even now, I should confide in the prevailing virtue and efficacious operation of lenity, though working in darkness and in chaos, in the midst of all this unnatural and turbid combination: I should hope it might produce order and beauty in the end.

The American Revolution may have been another "irrepressible conflict," but Burke, who saw it as a civil war, seems never to have doubted that there were two other possibilities: reconciliation on terms of justice or amicable separation.

Lenity can be understood as lenience or gentleness or mercy, and there was too little of it in Burke's England in 1774. There was too little in our North and South from 1861 to 1865, and before, and after. Failing lenity in any conceivable form, relishing its differences, savoring its animosities and divergent patriotisms, the nation divided and went to war. The two sides met in a series of great battles, and at last the strongest won in the name of emancipation and union.

That is the official version, and it is right enough as far as it goes. But to grant a just complexity to this history let us add a third side: that of the dead. Armies, by the necessity and purpose of military organization, are abstractions. We think of battles as convergences not of individuals but of "units."

Survivors, in their memoirs, speak as participants. Only in the aftermath of battle, on the nighttime battlefields horribly littered with the dead and the dying, do the individual soldiers begin to enter our imagination in their mere humanity. Imagination gives status in our consciousness and our hearts to a suffering that the statisticians would undoubtedly render in gallons of blood and gallons of tears. Maybe I am speaking only for myself, though I doubt it, when I say that to me the dead in Mathew Brady's photographs don't look like Unionists or Confederates; they look like dead boys, once uniquely themselves, undiminished by whichever half of the national quarrel they died for. In those photographs we meet war as a great maker of personal tragedies, not as a great enterprise of objectives.

Mathew Brady was by no means the first to show us this, nor was Shakespeare, but Shakespeare did show us, with a poignance unsurpassed in my reading, the tragedy specifically of civil war. In *Henry VI, Part III*, there is a battle scene in which first "a Son" and then "a Father," not identified as to side, enter separately, each bearing the body of a dead man whom he has killed and whom he now looks at. The Son says:

> Who's this? O God! it is my father's face . . .

And the Father says:

> But let me see: is this our foeman's face?
> Ah, no, no, no, it is mine only son!

Of our own civil war Walt Whitman saw clearly the pageantry and glamour and "all the old mad joy" of battle that Robert E. Lee acknowledged. But he saw also the personal tragedy much as Shakespeare saw it. With the same anonymity as to side, he speaks of coming at dawn upon three of the dead lying covered near a hospital tent:

> Curious I halt and silent stand,
> Then with light fingers I from the face of the nearest the
> first just lift the blanket;
> Who are you elderly man so gaunt and grim, with well-gray'd
> hair, and flesh all sunken about the eyes?
> Who are you my dear comrade?

Then to the second I step—and who are you my child and darling?
Who are you sweet boy with cheeks yet blooming?

Then to the third—a face nor child nor old, very calm, as of beautiful yellow-white ivory;
Young man I think I know you—I think this face is the face of the Christ himself,
Dead and divine and brother of all, and here again he lies.

Once dead, the dead in war are conscripted again into abstraction by political leaders and governments, and this is a great moral ugliness. The dead are made hostages of policy to sanctify the acts and intentions of their side: These have died in a holy cause; that they may not have died in vain, more must be killed. And to benefit the victors, there is always the calculation, frequently alluded to but never openly performed: At the cost of so many deaths, so much suffering, so much destruction, so much money or so much debt, we have got what we wanted, and at a fair price.

There is no doubt that wars may have moral purposes. Union and emancipation were moral purposes. So were secession and independence, however muddied by the immoral purpose of slavery. But battles don't have the same purposes as wars. The only purpose of a battle, once joined, is victory. And any price for victory is acceptable to the generals and politicians of the victorious side, who are under great pressure to say that it is acceptable. But the accounting is conventionally not attempted. Victors do not wish to evaluate their victory as a net gain, for fear that it will prove a net loss.

I doubt that such a calculation is possible, even if somebody were willing to try it. But that should not stop us from asking, if only to keep the question open, what we gained, as a people, by the North's expensive victory. My own impression is that the net gain was more modest and more questionable than is customarily said.

The Northern victory did preserve the Union. But despite our nationalist "mystique," our federation of states is a practical condition maintained only by the willing consent of the states and the people. And secession, today, is still not a dead issue. There is now, for instance, a vigorous and strictly principled secession movement in Vermont.

The other large Northern objective, the emancipation of the slaves, also was achieved. But this too appears in retrospect to be an achievement painfully limited. It does not seem unreasonable to say that emancipation was achieved and, almost by the same stroke, botched. The slaves were set free only to remain an exploited people for another hundred years. My guess is that, after the decision was taken to make slavery an issue of war, emancipation was inevitably botched. The North in effect abandoned the ex-slaves to the mercy of its embittered and still dissident former enemy, to whom they would be ever-present reminders, symbols virtually, of defeat.

Furthermore, we have remained a people in need of a racially designated underclass of menial laborers to do the work that the privileged (of whatever race) are too good, too well educated, and too ignorant to do for themselves. Our Stepanfetchits at present are Mexican immigrants, whom we fear for the familiar reasons that we exploit them and that we depend on them.

And so our Civil War raised questions that have been raised a number of times since: Can you force people to change their hearts and minds? Can you make them good by violence? Again and again human nature has replied no. Again and again, ignoring human nature and history, politicians have answered yes. And yet it seems true that Martin Luther King Jr. and his followers, by refusing to answer violence with violence, did more to alter racial attitudes in the South than was done by all the death and damage of the Civil War.

Is this reading of history too idealistic and unforgiving? Probably. Must we not say, pragmatically, that a botched emancipation is better than legal slavery? Well, I am a farmer, therefore a pragmatist: Half a crop beats none; a botched emancipation is better than none. But, as I am a farmer, I am also a critic, and I know the difference between a bad result and a good one. Of our history, though we cannot change it, we must still try for a true accounting. And to me it seems that the resort to violence is the death of imagination. Once the killing has started, lenity and the hope for order and beauty vanish along with causes and aims. Edmund Wilson's logic of the two sea slugs, the larger eating the smaller, then goes into effect: "not virtue but . . . the irrational instinct of an active power organism in the presence of another such organism . . ."

Once opponents become enemies, then the rhetoric of violence prevents them from imagining each other. Or it reduces imagination to powerlessness. Men such as Lincoln and Lee, from what I have read of them, seem not to have been destitute of imagination; of this I take as a sign their grief, their regret for the war even while they fought it. I see them as figures of tragedy, each an instrument of an immense violence which, once begun, was beyond their power to mitigate or stop, and which made of their imagination only a feckless suffering of the suffering of others. Once the violence has started, the outcome must be victory for one side, defeat for the other—with perhaps unending psychological and historical consequences.

When my thoughts circle about, trying to give my disturbance a location that is specific and familiar enough, they light sooner or later on "The Battle Hymn of the Republic." This song has a splendid tune, but the words are perfectly insane. Suppose, if you doubt me, that an adult member of your family said to you, without the music but with the same triumphal conviction, "Mine eyes have seen the glory of the coming of the Lord"—would you not, out of fear and compassion, try to find help? And yet this sectional hymn, by an alchemy obscure to me, seems finally to have given us all—North and South, East and West—a sort of official judgment of our history. It renders our ordeal of civil war into a truly terrifying simplemindedness, in which we can still identify Christ with military power and conflate "the American way of life" with the will of God.

I have made clear, I hope, my failure to perceive the glory of the coming of the Lord in the Civil War and its effects. The North was not uniformly abolitionist; the South was not uniformly proslavery or even prosecession. Theirs was not a conflict of pure good and pure evil. The Civil War was our first great industrial war, which was good for business, like every war since. The Civil War established violence against noncombatants as acceptable military policy. The Army of the United States, no longer the Northern army, proceeded from the liberation of the slaves to racist warfare against the native tribespeople of the West. Moreover, as the historian Donald Worster has said, the Civil War supplanted the "slave power" of the South with the "money power" of the North: "The fact of the matter is we have not even today figured out how to come to terms with the money power that replaced the slave power . . ." The great advantage of the aftermath went, certainly not to the ex-slaves or to the farmers and small tradesmen of either side, not to the people Wallace Stegner called "stickers," but rather to those he called "boomers": the speculators and exploiters, the main-chancers, the Manifest Destinarians, the railroads, the timber and mineral companies.

My purpose in reciting these problems is not to suggest that a Southern victory would have been better—which I doubt—but only to point out that the Northern victory set the tone of overconfidence, of self-righteousness and assumed privilege, that became the political tone of the whole nation.

The Civil War was followed, perhaps as a matter of course—and would have been followed, no matter who won—by the industrial exploitation of our land and people that still continues. While we have stood at our school desks or in our church pews asserting the divine prerogative of "The Battle Hymn," we have been destroying our country. This is not an impression. By measures empirical enough, we have wasted perhaps half of our country's topsoil; we are destroying by "development" thousands of acres every day; we have polluted the atmosphere and the water cycle; we have destroyed or damaged or brought under threat all of our natural ecosystems; in our agriculture and forestry we are treating renewable resources as carelessly as we have burned the fossil fuels; we have severely damaged all of our human communities. We have established unregarding violence as our means of choice in everything from international relations to land use to entertainment.

What are we to conclude? Only, I fear, that violence is its own way, which is entirely unlike the ways of thought or dialogue or work or art or any manner of caretaking. Once you have committed yourself to the way of violence, you can only suffer it through to exhaustion and accept the always unforeseen results.

I have been describing an enormous failure, and to me this appears to be a failure of imagination. Though we are now far advanced in the destruction of our country, we have only begun to imagine it. We are destroying it *because* of our failure to imagine it.

By "imagination" I do not mean the ability to make things up or to make a realistic copy. I mean the ability to make real to oneself the life of one's place or the life of one's enemy—and therein, I believe, is implied imagination in the highest sense. When I use this word I never forget its definitions by Coleridge and Blake, but for present purposes I am going to refer to the writings of William Carlos Williams, whose understanding of imagination, though compatible with that of his English predecessors, is peculiarly American in its urgency.

Three generations and more ago, Williams was fretting about the inclination of Americans to debase their land and, with it, themselves,

> as if the earth under our feet
> were
> an excrement of some sky
>
> and we degraded prisoners
> destined
> to hunger until we eat filth . . .

We were, he thought, "like a chicken with a broken neck, that aims where it cannot peck and pecks where it cannot aim, which a hog-plenty everywhere prevents from starving to death . . ."

Williams seems to have been one of the few so far who could see the vulnerability of a highly centralized economy. In a letter to James Laughlin on November 28, 1950, he tells of the disruptions of a recent storm, and then he writes:

> But witnessing what one small storm can do to a community in these parts I am awonder over the thought of what a single small atom bomb might not accomplish. Disruption of every service, now become more and more centralized, would starve us out in 3 days . . .

Against such craziness he set the "single force" of imagination: "To refine, to clarify, to intensify that eternal moment in which we alone live . . ." And imagination, in this sense, is not passively holding up a mirror to nature; it is a changing force. It does not produce illusions, or copies of reality, or "plagiarism after nature." And yet it does not produce artificiality. It does not lead away from reality but toward it. It can be used to show relationships. By it "the old facts of history" are "re-united in present passion." Thus I have pieced together Williams' thoughts from the prose fragments of *Spring and All.* Thirty or so years later, in "The Host," one of the devotional poems in *The Desert Music*, he lays it out more plainly, giving it, like Coleridge and Blake, a religious significance:

> There is nothing to eat,
>   seek it where you will,
>   but of the body of the Lord.
>
> The blessed plants
>   and the sea, yield it
>   to the imagination
> intact. And by that force
>   it becomes real . . .

If what we see and experience, if our country, does not become real in imagination, then it never can become real to us, and we are forever divided from it. And for Williams, as for Blake, imagination is a particularizing and a local force, native to the ground underfoot. If that ground is not in a great cultural center, but only in a New Jersey suburb, so be it. Imagination is as urgently necessary in Rutherford, New Jersey, or in Knott County, Kentucky, or in Point Coupée Parish, Louisiana, as it is in San Francisco or New York. As I am understanding it, imagination in this high sense shatters the frameworks of realism in the arts and empiricism in the sciences. It does so by placing the world and its creatures within a context of sanctity in which their worth is absolute and incalculable.

The particularizing force of imagination is a force of justice with obvious crucial correspondences in biology and in our legal system. Robert Ulanowicz says that "in ecosystems comprised of hundreds or thousands of distinguishable organisms, one must reckon not just with the occasional unique event, but with *legions* of them. Unique, singular events are occurring all the time, everywhere!" And, except for identical twins, every creature that comes into being by way of sexual reproduction is genetically unique. Recognition of the uniqueness of creatures and events is the reason for the standing we humans grant (when we do grant it) to one another before the law, and it is the reason we "return thanks" (when we do so) for food and other gifts that come to us from the living world. Without imagination there is no right appreciation of these rarities—no lenity, amity, or mercy. And, I think, there is no satisfaction either. Imagination, amply living in a place, brings what we want and what we have ever closer to being the same. It is the power that can save us from the prevailing insinuation that our place, our house, our spouse, and our automobile are not good enough.

Historians and scientists work toward generalizations from their knowledge, just as all of us do. We must do this, for generalization is a part of our means of making sense. But generalization alone, without the countervailing, particularizing power of imagination, is dehumanizing and destructive.

"The South," for example, as the name of an historical side, can have a reckonable and useful meaning. But as the name merely of a part of the country, it means less. If "region" means anything at all, then the South, like the North or the West, is a region of many regions. But so is Kentucky. My county has several distinct regions. My neighbors don't look like Southerners or Kentuckians to me. The better I know them, the more they look like *themselves.* The better I know my place, the less it looks like other places and the more it looks like itself. It is imagination, and only imagination, that can give standing to these distinctions.

If imagination is to have a real worth to us, it needs to have a practical, an *economic*, effect. It needs to establish us in our places with a practical respect for what is there besides ourselves. I think the highest earthly result of imagination is probably local adaptation. If we could learn to belong fully and truly where we live, then we would all finally be native Americans, and we would have an authentic multiculturalism.

And yet the problem I began with has never been resolved: How do we equilibrate or even negotiate between local identity and the abstractions of regional or national identity with the attendant clichés of "economic growth"? Obviously there can be no general answer to this question. If we see the need for an answer, then we must attempt it for ourselves in our communities. I believe that there is hope in the increasing uneasiness of people who see themselves as dispossessed or displaced and therefore as economically powerless. Growing out of this uneasiness, there is now a widespread effort toward local economy, local self-determination, and local adaptation. In this there is the potential of a new growth of imagination, and at last an authentic settlement of our country.

But we must not fool ourselves. This movement toward local adaptation necessarily is being led from the bottom. And it confronts a leadership from the top—in government, in the corporate economy, in the universities—that is utterly lacking in imagination, local loyalty, and local knowledge. Both conservatives and liberals, having accepted the ecological and social damages of industrialism as inevitable, even normal, have conceived the individual as subject alone either to the economy or to the government. In this official numbness, though it is clearly self-doomed, there is for the moment an almost overwhelming power.

Let me give you an example of the way a failure of imagination works against people and land. At present, in the eastern mountains of my state, the coal companies are blasting the tops off the mountains and pushing them into the valleys, covering the streams. They are doing this without concern for the land, the topsoil, the forest, the waterways and the water, or for the homes and lives of the people. This total, permanent destruction is not anomalous in our economy or without causes in our history. I need not delay you here by retelling the history of the corporate pillage of eastern Kentucky, which you will find well told in *Night Comes to the Cumberlands* and other books by Harry M. Caudill, or by describing the culmination of that malignant history in "mountaintop removal," which Erik Reece has accomplished fully in *Lost Mountain*. But if you want to know how this hardly credible or bearable waste could have happened, consider the chapter "Mountain Passes of the Cumberland" in *The Bluegrass Region of Kentucky and Other Kentucky Articles* by James Lane Allen, published in 1892.

Allen was a "genteel" writer of the Bluegrass and an outsider to the mountains, about which he had a curiously divided mind. On the one hand he regarded the then mostly unspoiled forests of that region with a sort of rapture. But he was equally rapturous about the economics and machinery of industrialization. Between these enthusiasms he saw no possible contradiction. The beautiful forest, of course, was invaluable in itself, and it would be nearly completely cut down by the middle of the next century. Beneath the forest lay enormously valuable deposits of coal. Allen accurately foresaw the industrial exploitation of the region, but he thought only good could come of it.

As for the local people, he characterized them in the person of "a faded, pinched, and meager mountain boy" driving a team of oxen and eating from a sack of candy:

> In one dirty claw-like hand he grasped a small paper bag, into the open mouth of which he had thrust the other hand . . . He had just bought . . . some sweetmeat of civilization which he was about for the first time to taste.

These people, according to Allen, needed to be civilized and Christianized—which was the official rationale of the federal campaigns against the Indians in the same era.

I don't think Allen was an evil man. Probably, like us, he was pretty good. But he also was ignorant and naive, as we too have been about our continuing bonfire of coal and oil. Being only a prophet, Allen had no doubt about the beneficence of industrialization; he thought it was the coming of the Lord: "You begin with coke and end with Christianity." Compare this bit of prophecy with James Still's *River of Earth*, written half a century later, and you will see what I mean by failure of imagination.

I would like to end by turning to the work of a Southern writer who, for several reasons, is exemplary and dear to me. Ernest J. Gaines inherited the harder side of the racial history that I inherited, and I believe I know some of the questions he faced as he made his way into his work. Could he imagine sympathetically a Southern white person? Could he imagine, so as to require us to imagine, an uneducated black farmhand as a person of dignity, wisdom, and eloquence? Yes, as we know, he could. He has imagined also the community of his people as a part of the life of their place and the hardships of that community. He has imagined the community's belonging to its place, the houses that had the names of people, the flower-planted doorvards, the church, the graveyard, the shared history and experience, the shared stories, the talk of old people on the galleries in the summer evenings and the young people listening. He has imagined also the loss of those things.

In a time when the provincial fear of provinciality has brought the local into suspicion, Ernest Gaines has been true to his place, his people, and their story. He has shown that the local, fully imagined, becomes universal. He has brought his place and his people to such a pitch of realization that again and again as I read him he seems to speak also for me and mine. He has done this in a language like no other, belonging to a place like no other.

His novel *In My Father's House* contains a passage that alarms and consoles me every time I read it. The reason for my alarm is obvious, for the passage is about somebody's damage that becomes everybody's danger. But I consent to the author's understanding of the damage, and I am consoled by the companionship of that.

The main character in the novel, Phillip Martin, gives a ride to a terribly angry young black man named Billy. Billy has been to Vietnam. He is now a would-be revolutionary. He would like to burn the whole country by setting fire to the gasoline in every filling station. In his fury, this Billy is thoroughly frightening. I am frightened of him and for him. But then Billy says:

> You see all them empty fields round here, mister . . . Go all over this place—empty fields, empty houses, empty roads. Where the people used to be—nothing. Machines. Every time they build another machine that takes work from the people, they hire another hundred cops to keep the people quiet.

And I am caught. I see that Billy and I are joined by a mutual sense of calamity and loss. From my well-wishing in the safety of my chair, I have been carried into the trouble itself that has so nearly consumed Billy. Suddenly, in the midst of his rant, he has spoken from the grief felt by many rural Americans, of whatever race, and certainly by me. I know well that it is possible for me, like Billy, to respond with anger and despair. But I know also that it is possible for me, as for Ernest Gaines, to respond with work, hope, and love.

Sweetness Preserved

WHAT I AM GOING TO DO is talk about some poems—lyric poems—as the products of a story. Most poems, whether or not they tell or contain stories, come out of stories, and often they bear reference to the stories they come out of. For a couple of generations now, critics and teachers have not thought it wise to approach poems by way of stories. They have thought that poems should be read "as poems" or "as texts," as words written or printed on a page, ignoring the story the poems come out of.

I am willing to suppose (for the sake of supposing) that such poems as I am going to talk about must finally shrug off their stories, and all else they do not explicitly contain, and stand before us on their own. I am aware that some poems stand before us on their own because they must; we do not know the stories they came from. But in reading poems that are perfectly anonymous, we still know that there is more to them than the "text." We think of a poem and in the same thought think of what it is about, if it is about anything. Literature involves more than literature, or we would not be grateful for it.

Suppose we know not only a body of poetry but also the story that the poetry came out of. How then are we to help knowing what we know? How are we to help knowing, for instance, that some poets are pilgrims, and that their poems are not just objets d'art, but records, reports, road signs, or trail markers? By what curious privilege are we allowed to ignore what we know?

But I would like to go a little further still, and honor the possibility that the stories that poems come out of are valuable in themselves, so far as they are known. Those who are living and writing at a given time are not isolated poetry dispensers more or less equivalent to soft-drink machines, awaiting the small change of critical approval. We are, figuratively at least, members of a community, joined together by our stories. We are inevitably collaborators. We are never in any simple sense the authors of our own work. The body of work we make for ourselves in our time is only remotely a matter of literary history. The work we make is the work we are living by, and not in the hope of making literary history, but in the hope of using, correcting so far as we are able, and passing on the art of human life, of human flourishing, which includes the arts of reading and writing poetry.

There is a danger of presumption or imposition in what I am about to do, for I am not an authority on the story I am going to talk about. The story of a person's life cannot be entirely known or told. However, I do know unavoidably and unforgettably that Jane Kenyon's poems came out of a story. I know as well that Donald Hall's did. And I know that the poems of both poets came to a considerable extent out of the same story—or perhaps out of the intersection or overlapping of two stories; I want my language to be accurate and courteous, and am not confident of my ability to make it so.

The story of Donald Hall and Jane Kenyon is not my story. And yet their story is not absolutely distinct from mine, for their story is one I have depended on, and have spent a good deal of time telling over to myself and thinking about. And I have been attentive to the poems that came out of it. Their story and their poems have affected, instructed, troubled, consoled, and clarified my understanding of my own story.

Because I am a storyteller and was from childhood a hearer and reader (and believer) of stories, I have always known that people live in stories. And so it has been a little shocking to me to realize also that it is possible for people to wander outside their stories. When Donald Hall and I first met, at a literary party in Manhattan in the winter of 1963, both of us were living outside our stories. I find it readily supposable that Don didn't know what to make of me, even if it is supposable that he tried to make anything of me at all. I certainly didn't know what to make of him, and the reason was that I didn't know what to make of anything. Later, when both of us were again living inside our stories, we would recognize each other and become friends. This happened, I think, because we both loved our grandparents and we both derived from childhood homeplaces that we did not like to forsake. We have corresponded in two ways.

At the time of the party in 1963, the two of us were in "exile." I give that word an emphasis because it was so important and applied so peculiarly to young writers in our generation. We came to our calling in the shadow (and the glamour) of eminent literary exiles: James, Pound, Eliot, Joyce, Stein, Hemingway, and others. Moreover, those in charge of our education tended to think that they were preparing us for careers, not for settling down someplace. The question before us seemed to be, not how we might fit ourselves and our book knowledge into our home landscapes, but how we would fit into our careers, which is to say our exile.

This is confirmed by Donald Hall's early poem entitled "Exile." Looking back at that poem now, I find nothing in it that surprises me. It is a good poem. It is also an inevitable product of the poet's era and education; it *had* to be written by somebody. It states the case beautifully:

> Imagining, by exile kept from fact,
> We build of distance mental rock and tree,
> And make of memory creative act . . .

This is an exact enough description of the poet's job of work in Don's "Elegy for Wesley Wells," also an early poem. It too is a good poem—I don't mean at all to be denigrating this work of "exile." In the elegy, the poet mourns and celebrates his grandfather Wesley Wells, a New Hampshire farmer, a good one apparently, but one belonging to an age that, at the time of the poem, is "bygone." The poet (in the way of a young elegist, which I was once myself) is hoping to grant a measure of immortality to his grandfather by means of his poem. It is, to me, an extraordinarily moving poem. I have never read it without being moved by it, though by now I have lived beyond the notion that immortality can be conferred by a poem, and though by now my reading of the poem is influenced by my knowledge of a story that the poem, so to speak, does not know. When I read "Elegy for Wesley Wells" now, I feel a humorousness and a sadness that the poem did not anticipate.

In immortalizing his grandfather Wells, Donald Hall the young elegist is also immortalizing a part of his own life which he now considers to be finished. That life, if it is to have a present life, must have the immortal life of art. Maybe you are outside your life when you think your past has ended. Maybe you are outside your life when you think you are outside it. I don't know what Donald Hall in later life would say. I know only what I in later life would say. I would say, partly from knowing the story I am talking about, that though you may get a new life, you can't get a new past. You don't get to leave your story. If you leave your story, then how you left your story *is* your story, and you had better not forget it.

Now I want to speak of another poem that is a landmark both in the story I am dealing with and in my own consciousness of poetry and of the world. This is the poem called "Maple Syrup," written about twenty-five years after "Elegy for Wesley Wells." The poem tells about an experience shared by a couple designated merely as "we." Since I am observing no critical conventions here, I will say that this "we" refers to the poet Donald Hall and the poet Jane Kenyon, who have returned to the house of Donald Hall's grandparents Kate and Wesley Wells. The two poets, married to each other, will live their life together in this house on this farm, relinquished and immortalized in the "Elegy" so many years before. In the poem "Maple Syrup" they go through the house together, to "the back chamber" full of artifacts and relics, and then down into "the root cellar," where they find a quart of maple syrup left there by Wesley Wells. And here I must let the poem speak:

Today
we take my grandfather's last
quart of syrup
upstairs, holding it gingerly,
and we wash off twenty-five years
of dirt, and we pull
and pry the lid up, cutting the stiff
dried rubber gasket, and dip our fingers
in, you and I both, and taste
the sweetness, you for the first time,
the sweetness preserved, of a dead man . . .

We (that is, now, the poet and his readers) have come a long way from "Elegy for Wesley Wells." We have left the immortality of art and have come, by way of a sort of mortalization, to a communion of lovers with one another and with the dead, which is to say that we have come to a marriage rite, joining two mortals to one another, to a place, and to other mortals, bringing them perhaps within imaginable reach of a more authentic idea of immortality. Donald Hall, who in the "Elegy" is maybe a sort of bard, has now become, in the full and mysterious sense, a love poet. Or we might say that, having started out to be a professional, he has become an amateur, working (like the best kind of professional) for love. The sign of this is that the memory of Wesley Wells, once elegized into a mental landscape of the finished past, has become a living faculty of the poet's mind and imagination. The sweetness of the dead man, now, is not preserved in an artifact but in the lives of those who taste it.

One more thing. Because this rite of marriage occurs in this story, it does not give new life just to the couple, who now enter into its "one flesh"; it gives new life also to the dead and to an old house. It matters that this is an old house that is familiar to the bridegroom. If the house had been sold to strangers, according to the common destiny of old houses in our day, Wesley Wells' quart of syrup, if found, would have been thrown away. It would have seemed fearfully old and fearfully anonymous. To Don, and to Jane, trusting Don, it was mortal and everlasting, old and new, and sweet.

Having set up (so to speak) these two landmarks, an elegy and a celebration of marriage, I am much more moved than I would be by either of the poems alone, for I know the story that joins them. The two poems are joined by this story because the story of Donald Hall had become also (to the degree that separate stories do converge) the story of Jane Kenyon.

What had happened was that these two stories had converged in one of the stages of Donald Hall's exile, teaching at the University of Michigan, and their convergence had made him free to return to the family house in New Hampshire. The agent of this freedom was Jane Kenyon, who said, according to her husband, "Why are we thinking of *here*, when there is New Hampshire?"

Not long after Don and Jane were settled at Eagle Pond, Don wrote to me, telling what they had done, and I wrote back some advice: Don't take on too much farming too quickly. Don has pointed out that the advice was wasted, since he did not intend to take on any farming at all—leaving me with the consolation that, anyhow, if he had needed it, it would have been good advice.

I am not sure when I met Jane, except that it was a good while ago, when she and Don were still heating their house with a very handsome wood-burning stove. I was on a speaking and reading trip in New England, and was able to stop by just for a short visit and lunch at a local eating place. I remember a tour of the house, but not much that was said. I remember being impressed by Jane's self-possession and dignity and quietness. These qualities continued to impress me after I knew her better. She was a writer, but she appeared to be watching "the literary world" without anxiety or great excitement.

Now the requirement of honesty is going to embarrass me, for I have to confess that I didn't read anything by Jane for a long time after I met her. For one reason, I felt a certain complicated sympathy for her—a poet who had set up shop smack in the middle of another poet's subject. The other poet's claim to this subject was well established; the other poet was her husband. It was easy to wish that she might have been, say, a painter. Another reason was that I liked her, and if she was a bad poet I did not want to know.

And then Bert Hornback invited Don and Galway Kinnell and Seamus Heaney and me to give a reading at the University of Michigan in January of 1986. For this there was a reason and a real reason. The reason was the public reading on Friday night. The real reason was that Bert wanted his students to have a late breakfast and conversation with the visiting poets on Saturday. In this age of careerist "research professors," Bert is a real teacher who thinks nothing of the trouble it takes to capture poets alive to talk with his students.

The visitors gathered at Bert's house for supper before the reading. When I came into the kitchen as the mingling and the talking began, Jane was standing by the refrigerator, watching the situation develop with the composure that I mentioned before. For the sake of political correctness I have been trying to avoid saying that Jane was beautiful, but of course she was, and of course I could see that she was. When we greeted each other, she said, "Wendell, I can't give you a hug. I have a bad cold." Baffled utterly by this generosity, I remember thinking that I had nothing better to do than catch a bad cold.

I have to go ahead and confess also that I do not greatly love literary occasions. The reading on Friday night was as readings are. The occasion beginning at breakfast on Saturday, however, was a literary occasion that surpassed itself. It was a *friendly* occasion, one of the loveliest that I have known. What I so liked about it was that everybody was talking for pleasure. There was no contention. Nobody defended a "position." There was much laughter. The students were hesitant to take part, but after a while they too entered into the conversation, and we had that additional pleasure.

Finally, late in the day, somebody—I don't remember who; it wasn't me—said, "Jane, why don't you read us a poem?"

Jane, who had been sitting almost outside the room, saying little, perhaps nothing at all, during the conversation, fished up from somewhere a page that she had brought with her and spread it open to read. For me, this was the only uncomfortable moment of that day. I don't remember what I thought, but it would have been like me to have started trying to think of some ambiguous compliment to make in case I thought the poem was bad—something like "Well, Jane, you certainly do write poetry." And then that quiet woman read beautifully her poem "Twilight: After Haying":

> Yes, long shadows go out
> from the bales; and yes, the soul
> must part from the body:
> what else could it do?
>
> The men sprawl near the baler,
> reluctant to leave the field.
> They talk and smoke,
> and the tips of their cigarettes
> blaze like small roses
> in the night air. (It arrived
> and settled among them
> before they were aware.)
>
> The moon comes
> to count the bales,
> and the dispossessed—
> *Whip-poor-will, Whip-poor-will*
> —sings from the dusty stubble.

These things happen . . . the soul's bliss
and suffering are bound together
like the grasses. . . .

The last, sweet exhalations
of timothy and vetch
go out with the song of the bird;
the ravaged field
grows wet with dew.

I hope I have adequately prepared you to imagine my relief. Now I must ask you to imagine something else. However many poets there may be who know from experience the subject of this poem by Jane Kenyon, I surely am one of them. I have lived countless times through that moment at the end of a day's work when its difficulty and heat and weariness take on a kind of sublimity and you know that you are alive both in the world and in something greater, when it is time to go and yet you stay on, charmed. I had never tried to write a poem about that moment, and on that day, when I had heard Jane read her poem, I knew that I would not need to write one; Jane had written better about it than I could. Sometimes I feel competitive or jealous when I *suspect* that somebody has written better than I can about something I know. When I am *certain* that somebody has done so, then I am relieved and I feel happy. "Twilight: After Haying" made me happy that day in 1986, and it has made me happy every time I have read it since.

Wittgenstein said that "In art [and, I assume, in writing about art] it is hard to say anything as good as: saying nothing." I believe and honor that, and I keep it in mind. But also we obviously need to speak from time to time of the things that have moved us. We need to wonder, for instance, why we remember some things and forget others. I have remembered Jane's reading of her poem that day, I think, because it was impossible to mistake the revelation of the event: Here was a poet present in her work with an authority virtually absolute. I don't mean that she is in the poem personally, but that all her gifts are in it: her quietness, gentleness, compassion, elegance, and clarity, her awareness of mystery, her almost severe good sense. This poem, like just about every one of her poems, is unconditional; it is poetry without qualification. It has no irony, no cynicism, no self-conscious reference to literary history, no anxiety about its place in literary history, no glance at the reader, no anticipation of the critic, no sensationalism, no self-apology or self-indulgence. How many poets of our time have been so unarmed as to say, "The moon comes / to count the bales . . ."? As she herself said (in the next poem of *The Boat of Quiet Hours*):

> These lines are written
> by an animal, an angel,
> a stranger sitting in my chair;
> by someone who already knows
> how to live without trouble
> among books, and pots and pans. . . .

—which is to say that she was authentically a poet of inspiration.

And this, to return to the story, seems to have settled pretty quickly the artistic problem of a shared life and subject. My wife, Tanya, has pointed out to me, from her knowledge of her own story, that Jane Kenyon had become, in fact, an exile in the very place that her husband had once felt himself to be exiled from. For a while after coming to Eagle Pond, she seems to have remembered "Ruth amid the alien corn":

> I'm the one who worries
> if I fit in with the furniture
> and the landscape.

And:

> Maybe
> I don't belong here.
> Nothing tells me that I don't.

But such lines as these testify to a radically different approach to the problem of exile. The modern American version of exile is a rootless and wandering life in foreign lands or (amounting to about the same thing) in American universities. Jane Kenyon, like Ruth of old, understood her exile as resettlement. Very few American exiles, and not many American settlers, have asked "if I fit in with . . . the landscape" or worried about belonging to a place. And already one is aware of her originality, as one continues always to be aware of it. I mean "originality" in what I take to be the best sense: not the enactment of a certain kind of literary intention or ambition, but the grace to submit to influence—the influence of places, passages of scripture, works of art; the influence of all her subjects—and the grace and patience to find within herself the means to respond. Her contribution to this story is hers distinctly.

When I read a disparagement of the book *Otherwise* in *The Hudson Review*, I was offended, but also puzzled. How could anybody able to read fail to see the quality of that book? But after a while, I believe, I figured it out. Jane Kenyon's work, in fact, makes an unnegotiable demand upon a reader. It doesn't demand great intellect or learning or even sympathy; it demands quiet. It demands that in this age of political, economic, educational, and recreational pandemonium, and a concomitant rattling in the literary world, one must somehow become quiet enough to listen. Her poems raise unequivocally the issue of the quality of the poet's ear.

A true poem, we know, forms itself within hearing. It must live in the ear before it can live in the mind or the heart. The ear tells the poet when and how to break the silence, and when enough has been said. If one has no ear, then one has no art and is no poet. There is no appeal from this. If one has no ear, it does not matter what or how one writes. Without an ear, the traditional forms will not produce Andrew Marvell, nor will "free verse" give us William Carlos Williams.

Jane Kenyon had a virtually faultless ear. She was an exquisite master of the art of poetry. Her voice always carries the tremor of feeling disciplined by art. This is what over and over again enabled her to take the risk of plainness, or of apparent plainness. Her ear controls rhythm and sound, and also tone. It is tone as much as anything that makes one able to say what is unusual or unexpected. It is because of her perfection of tone that Jane Kenyon is able to say, "The moon comes / to count the bales . . ."

It is her perfection of tone that makes her poems able to accommodate sudden declarations of spiritual knowledge or religious faith, and that gives to so many of her poems the quality of prayer. It rules in her poems and passages of humor. It is the enabling principle in the political stroke of the poem entitled "Fat," and of the affirmation always present in her poems of sorrow. I am suggesting what I suppose cannot be demonstrated: that there is a practical affinity between the life of her soul and the technique of her poems.

The poems assemble themselves with a seeming arbitrariness, which is perhaps a comment. The poet looks at her subjects and experiences as they come to her and sees that they are ordinary; they are the stuff of life in this world; they could have come to anybody, at any time, in any order. They are revelations of ordinary satisfactions, joys, sufferings, deliverances which, in being revealed, become somehow numinous and resonant—extraordinary. In seeing that the poems are revelations, you see that they are not arbitrary but inevitable; in the course of the poem, form has occurred.

Sometimes the poems are poems of suspense; everything waits for the final line, as in the poem called "Things":

> The hen flings a single pebble aside
> with her yellow, reptilian foot.
> Never in eternity the same sound—
> a small stone falling on a red leaf.
>
> The juncture of twig and branch,
> scarred with lichen, is a gate
> we might enter, singing.
>
> The mouse pulls batting
> from a hundred-year-old quilt.
> She chewed a hole in a blue star
> to get it, and now she thrives. . . .
> Now is her time to thrive.
>
> Things: simply lasting, then
> failing to last: water, a blue heron's
> eye, and the light passing
> between them: into light all things
> must fall, glad at last to have fallen.

The poem gathers itself as quietly as a snowy night, and then by the end a kind of dawn has come and everything is shining. That seems to be about all there is to say. This poem confirms for me as well as any I have read what I think is the fundamental fact of poetry: If you can explain it, it is something else.

Nor am I able to say much more about this story that I have undertaken to talk about. It is, I think, a good and valuable story. Two poets entered into it together, consenting to its foretold cost, lived it out, met its occasions, and made, separately and together, a life and a body of work that, for some of us, the world is now unimaginable without. They tasted a sweetness stored up by others; they stored up a sweetness to be tasted by others. And what are we friends and beneficiaries to say? Well, finally, maybe no more than "Thank you."

The Uses of Adversity

IT HAS BEEN useful to me to think of *As You Like It* and *King Lear* as versions of the same archetypal story, belonging to human experience both before and after the plays. This is the story: In the instituted life of a society "things fall apart" because the people of power have grown selfish, cruel, and dishonest. The effect of this is centrifugal; the powerless and the disempowered are sent flying from their settled domestic life into the wilderness or the world's wildness—the state of nature. Thus deprived of civil society and exposed to the harshness of the natural world and its weather, they suffer correction, and their suffering eventually leads to a restoration of civility and order.

The outline of this story is clearly apparent in *As You Like It*. In *King Lear* the story is subjected to nearly intolerable stresses, and yet the outline remains unbroken; it is the major source of the play's coherence and meaning. What I believe is the proper understanding of both plays depends on our ability to take seriously the assumptions of the archetypal story—on how we answer the following questions: Do all human societies have in them the seeds of their failure? Are those seeds likely to be the selfishness and dishonesty of the dominant people? Does failure typically reduce the society, or persons in it, to some version of the state of nature? And is there something possibly instructive and restorative in this reduction?

For most readers nowadays these questions will be an unwelcome dose. We have read some history, and we do not doubt that other societies have failed, but we are not much inclined to credit the possible failure of our own, even though we are less and less able to deny the implications of our propensity to waste or to mechanical violence, or of our entire dependence on cheap petroleum. We have pretty much made a virtue of selfishness as the mainstay of our economy, and we have provided an abundance of good excuses to dishonesty. Most of us give no thought to the state of nature as the context of our lives, because we conventionally disbelieve in natural limits.

Another problem is that there is a considerable overlap

between this archetypal story and the pastoral tradition. In the pastoral tradition, as Shakespeare was fully aware, there is a prominent strain of frivolity. What is frivolous is the sentimentalization of rural life, which is supposedly always pretty, pleasant, and free of care. The famous example is Christopher Marlowe's:

> Come live with me and be my love,
> And we will all the pleasures prove
> That valleys, groves, hills, and fields,
> Woods, or steepy mountains yields.

To this Sir Walter Raleigh justly and just as famously replied:

> The flowers do fade, and wanton fields
> To wayward winter reckoning yields . . .

What neither poet acknowledged is the possibility of a real need, as Robert Frost put it, "of being versed in country things."

Shakespeare knew of course the pastoral conventions represented by Marlowe's poem. But he was a countryman, and he knew the truth of Raleigh's admonition; he knew also the need of being versed in country things. He knew that "a true laborer" might have something to say to a courtier that the courtier might need to hear—because, for one reason, the courtier lives by eating country things.

Another obstacle between modern readers and the archetypal story underlying these plays is our popular, and uncritical, egalitarianism. To us, the order of the natural world is horizontal, and so, we would like to think, is the order of human society: Any creature is as important as any other; any citizen is as important as any other.

But to Shakespeare the order of the world, as of human society, is also vertical and hierarchical. The order of created things descends in a Chain of Being from God down to the simplest organisms. In human society, order descends downward from the monarch. Every creature and every human has a place in this hierarchy according to "degree." Ulysses' discourse on degree in the first act of *Troilus and Cressida* can serve as a clarifying prologue to a reading of *As You Like It* and *King Lear*:

O, when degree is shaked,
Which is the ladder of all high designs,
The enterprise is sick. How could communities,
Degrees in schools, and brotherhoods in cities,
Peaceful commerce from dividable shores,
The primogenity and due of birth,
Prerogative of age, crowns, sceptres, laurels,
But by degree, stand in authentic place?
. . . .
right and wrong,
Between whose endless jar justice resides,
Should lose their names, and so should justice too;
Then everything include itself in power,
Power into will, will into appetite.
And appetite, an universal wolf,
So doubly seconded with will and power,
Must make perforce an universal prey
And last eat up himself.

This speech, by which Ulysses calls the "tortive and errant" Greeks to order, tells us precisely how to understand Orlando's complaint at the beginning of *As You Like It*. Orlando's oldest brother, Oliver, charged by their father, now dead, with Orlando's education, has forsaken his duty. Orlando's "keeping," he says to his old servant Adam, "differs not from the stalling of an ox." As the younger brother, lacking the "primogenity . . . of birth," Orlando is a man of lower degree than Oliver. But he is, even so, a man, his father's son, and Oliver's brother; Oliver's mistreatment of him, as if he were no more than a beast, is an affront to order, both human and natural; it is a symptom of a sick enterprise.

The trouble, for Oliver as for the villains of *King Lear* and other Shakespearean villains, is that the human place in the order of things, between the angels and the animals, is precisely and narrowly delimited, and it is precarious. To fall from one's rightful place, to become less than human, is not to become an animal; it is to become monstrous. And so Oliver's mere dislike and neglect of Orlando decline fairly predictably to a plot to kill him, which forces Orlando into exile.

In scene iii of Act I, a parallel estrangement occurs. The scene is in the palace of Duke Frederick, who has usurped the place of his brother, the carelessly named Duke Senior. Duke Senior, as we have already learned, is in exile in the Forest of Arden where he and some "merry men," his followers, "live like the old Robin Hood of England." Duke Senior's daughter, Rosalind, has been permitted to remain in the palace as the companion of Celia, Duke Frederick's daughter. The two young women not only are cousins and companions but are dearest friends. The two, Celia says, are "coupled and inseparable." She says to Rosalind, "Thou and I am one." And it is Celia also who, in attempting to console Rosalind, states one of the main themes common to this play and *King Lear*, that of affection or gentleness or generosity versus force: "what he [her father] hath taken away from thy father perforce, I will render thee again in affection." This affection is soon tested by Duke Frederick's determination to send Rosalind into exile:

> Within these ten days if that thou beest found
> So near our public court as twenty miles,
> Thou diest for it.

His reason is that he does not trust her. His distrust originates of course in his knowledge that he himself is not trustworthy. His daughter and niece, by contrast, possess in full the trust and the trustworthiness now lacking in the court, and so they must leave. In proof both of their friendship and of Duke Frederick's failure to know them, they decide to disguise themselves as "Ganymede" and "Aliena" and run away to join Duke Senior in Arden.

Act II, I think, is the paramount act of the play and is one of the greatest acts in all of Shakespeare. Both its poetry and its drama are exceedingly fine. It is also the crisis of the play for its readers, who have to decide here whether or not to take the play seriously. From what I have read and seen, some readers and directors have found it easy to understand the play as a pastoral diversion, merely sentimental and "comic," which I think is an insult to the play and its poet.

The test comes immediately with Duke Senior's speech that begins the first scene of Act II. The speech develops a standard pastoral theme: the honesty of the pastoral or rural life in contrast to life at court; it is the same theme expounded by Meliboe in *The Faerie Queene*, VI. The duke asks rhetorically, "Are not these woods / More free from peril than the envious court?" And we know the answer as well as his fellow exiles: These woods are free from the envy, jealousy, hypocrisy, power-hunger, and fraud that imperil the court or any other center of power.

There is an editorial crux in line five that we have to settle before reading further. The duke says, "Here feel we not [or: but] the penalty of Adam . . ." I am quoting the new Pelican edition, in which the editor chooses "not." But that usage, if it stands, reduces the speech to nonsense, and the duke to a fool. The problem with this reading is that the duke is *not* a fool, and the exiles, according to the play, are still subject to the penalty of Adam—that is, to mortality, discord, and the need to earn their living. And so the line necessarily is "Here feel we *but* the penalty of Adam . . ." The intended contrast is not between Eden and the fallen world, but between the unadorned life of the forest and the "painted pomp" of the court.

That "the icy fang / And churlish chiding of the winter's wind" are not flatterers but "feelingly persuade me what I am" we may take without argument to be merely true. "Sweet are the uses of adversity" may oversweeten the point, and yet we know that adversity can be corrective, and is sometimes indispensably so.

For modern readers, the largest difficulty in this speech may come in the last three lines, in which the duke proclaims that

> this our life, exempt from public haunt,
> Finds tongues in trees, books in the running brooks,
> Sermons in stones, and good in everything.

To the modern ear, this is likely to sound naive—an instance of the "pathetic fallacy," an almost cartoonish sentimentalization of nature. And yet this is a play solidly biblical and Christian in its moral basis, and this is one of its passages that most insistently depends on our knowledge of scripture. The overarching concept is that of the "good in everything," and the authority for this is Genesis 1:31: "And God saw every thing that he had made, and, behold, it was very good." As for "tongues in trees, books in the running brooks, / Sermons in stones," Shakespeare may be paraphrasing Job 12:7–9:

> Aske now the beastes, and they shal teache thee, and the foules of the heaven, and they shal tel thee:
> Or speake to the earth, and it shal shewe thee: or the fishes of the sea, and they shal declare unto thee.
> Who is ignorant of all these, but that the hand of the Lord hathe made these?

And he could as well be alluding to the long tradition in which nature is seen as a second or supplementary revelation.

The third scene of Act II parallels thematically the third scene of Act I. In the earlier scene Rosalind is confronted by Duke Frederick, sentenced to exile, and she and Celia make their plan to escape together in disguise. In II, iii, Adam, a servant loyal to Orlando, warns his young master that he, like Rosalind, must go into exile, for his envious and vengeful brother is plotting to kill him.

This play is not an allegory, but some of its characters have a semiallegorical or representative function; they represent human qualities or kinds. Adam, for one, is "the old Adam," father of us all, the fallen humanity which we all share, but he is furthermore the old Adam redeemed by good and faithful service to his master. He was first the servant of Orlando's father, the good Sir Rowland de Boys. In this scene, out of loyalty to the father and love for the son, he makes an absolute gift of his service and his fortune to Orlando, trusting that in his old age he will be comforted by him "that doth the ravens feed" and "caters for the sparrow." Thus, as a true servant to good men, he understands himself as a true servant of God. Orlando reciprocates by saying, like Celia in Act I, that the two of them will join their fates: "we'll go along together" in the belief that, before they have spent all of Adam's savings, they will "light upon some settled low content." The idea of a "settled low content" is the moral baseline of the play. It is what human beings most authentically have a right to expect and to achieve. It is the possibility that adversity most usefully and sweetly reveals. A settled low content is what Thomas Jefferson wished for America's small farmers; it is what Henry Thoreau was seeking at Walden Pond.

In scene iv, having arrived in the Forest of Arden, Rosalind and Celia encounter two other representative figures: Silvius, the young shepherd, and the old shepherd Corin. Silvius, classically named, represents what is most artificial in literary pastorals. He is an "uncouth swain" stricken by love into utter silliness and uselessness; wherever his sheep are, he is not going to think of them during this play.

Corin, by contrast as Englishly named as Spenser's Colin Clout or Hardy's Hodge, is strongly drawn as an individual and at the same time as a representative countryman. He is an "ideal character" of the same honest family as Chaucer's Plowman, who was "A trewe swinkere and a good . . ." Another critical question that this play imposes on its readers and directors is what to make of Corin. Here I have to depart from the sequence of the action to quote Corin's characterization of himself to Touchstone in III, ii:

> Sir, I am a true laborer; I earn that I eat, get that I wear, owe no man hate, envy no man's happiness, glad of other men's good, content with my harm; and the greatest of my pride is to see my ewes graze and my lambs suck.

To many readers that last clause would seem fatally countrified; from them the best rating it could hope for would be "quaint." Many Americans now would see this speech unhesitatingly as the utterance of a "hick" or a "redneck," hopelessly "retro." Nevertheless, any husbander of livestock would recognize Corin as a good shepherd, and Thomas Jefferson would have appraised him highly. In his independence he is democratic, and in his charity, fortitude, and humility he is Christian. Shakespeare knew that the human world survives by the work and responsibility of such people, and Corin's character is one of the standards by which we are to measure the other persons of the play.

In II, iv, Touchstone, assuming the role of sophisticated urbanite, sees Corin on their first encounter as a hick and addresses him accordingly: "Holla, you clown!" But Rosalind, as Ganymede, displaying her extraordinary good sense, recognizes him immediately for what he is: "Good even to you, friend." And Corin replies with perfect courtesy: "And to you, gentle sir, and to you all." Corin, offering hospitality to the strangers, is obliged to reveal that he is poorly paid:

> I am shepherd to another man
> And do not shear the fleeces that I graze.
> My master is of a churlish disposition
> And little recks to find the way to heaven
> By doing deeds of hospitality.

This ungenerous master, moreover, is preparing to sell his flock and land. Rosalind and Celia arrange with Corin to buy "the cottage, pasture, and the flock," Celia promising, "we will mend thy wages." Receiving gratefully this offer of economic justice, Corin sounds again the play's theme of the good servant: "I will your very faithful feeder be . . ."

A fourth representative character is Jaques, whose dominant trait is self-indulgence. "The melancholy Jaques," as he is called in Act II, scene i, manages to be both sentimental and cynical. He is uselessly sensitive and intellectual, a dilettante of his own moods, a boastfully free-speaking critic who corrects nothing. In II, i, Duke Senior speaks of his proper regret at having to kill the deer of the forest for food. But Jaques, as his fellows report, sentimentalizes this regret, making the same equation between human beings and animals as some animal rights advocates of our own day. And, like them, he offers no practicable alternative.

In II, v, after Amiens has sung a song that closely paraphrases Duke Senior's speech in II, i, Jaques responds by supplying a verse of his own which suggests that the forest company are asses and fools. So far he has been a peripheral character, looking on and commenting from the margin as a sort of fecklessly disapproving "chorus." Presently he will serve the play much more vitally, though still passively.

The next scene is brief, containing only two speeches, but to fail to take it seriously enough is again to be seriously in error about the play. Old Adam, weakened by hunger, cannot go on: "Here lie I down and measure out my grave. Farewell, kind master." As a "fallen" man, Adam cannot save himself. Nor can he survive as the servant of Orlando. But *As You Like It* is a play of transformations, and this scene presents the first one.

Adam has completed his servanthood. As a servant, he knows, he is as good as dead. His life now depends upon a change in Orlando. And Orlando changes; he becomes his servant's servant—as Edgar in *King Lear*, his father being reduced to helplessness, becomes his father's parent. Shakespeare is relying again on our knowledge of scripture, and the reference here is to Matthew 20:25–27:

> Ye knowe that the lords of the Gentiles have domination over them, and they that are great, exercise autoritie over them.
>
> But it shall not be so among you: but whosoever wil be great among you, let him be your servant . . .

The apparent lightheartedness of Orlando's reply must be understood as tenderness: as his attempt to lighten the heart of old Adam and as his pledge of service. His words also recall the measure of a "settled low content":

> Live a little, comfort a little, cheer thyself a little . . . For my sake be comfortable; hold death awhile at the arm's end. I will here be with thee presently, and if I bring thee not something to eat, I will give thee leave to die . . .

Here as elsewhere, and despite his allegiance to "degree," there is a strong democratic impulse in Shakespeare. But he is a democrat, not in the fashion of Jefferson, but in the fashion of Christ. "The least of these my brethren" also have their place in the order of things and their entitlement to be loved and served.

What is the relevance of this to the archetypal story that is my interest? Let us remember, to start with, that this play begins after the old state of things, the old "power structure," has fallen. We don't know what the error or fault of Duke Senior might have been; we know only that he became so weakened —perhaps so misled by flattery—that he was driven into exile by his power-hungry brother. Also the good Sir Rowland de Boys has died, and has been replaced by his selfish eldest son, Oliver. There is nothing more disorderly and disordering in civilized life than the selfishness of people of power—that is, their failure to be servants either to God or to their subjects. ("Public servants," as they and we too often forget, are meant not to rule but to serve the people.) The corrective to this is begun in the exiles by their recognition of the need to serve. And, in exile, this need is insistently practical. Outcasts in the forest—or on the stormy heath—cannot survive by selfishness.

In the long seventh and final scene of Act II, the theme of the forest (or adversity) as the corrective of selfishness and misrule, the theme of the necessity of servanthood, and the theme of affection or gentleness versus force are all joined in the play's moral climax. In my opinion, this scene threatens also to be the play's dramatic climax—to be both more dramatic and more moving than anything in the three acts that follow. Shakespeare's problem (and I assume a director's also) is to make the rest of the play worthy in moral interest and drama of what he has done in the first two acts.

The seventh scene begins with a leisurely, bantering conversation at first about Jaques and then between Jaques and Duke Senior. Jaques, having encountered Touchstone in the forest, wishes that he too could be a fool, apparently without in the least suspecting that he already is one. If, he says, he were given the liberty of a fool—that is, if the duke should grant him an official tolerance, permitting him to speak the truth as he sees it—then he would prove himself so purgative a critic as to "Cleanse the foul body of th' infected world . . ." The duke says that as such a critic Jaques would necessarily be a hypocrite, "For thou thyself hast been a libertine . . ." Jaques thereupon discourses on the universality of sin and hypocrisy in a speech that prefigures a much better one by the maddened King Lear.

Jaques' speech is interrupted by the entrance of Orlando with his sword drawn, and the scene then gets serious. Dinner has been laid out in the camp of the exiles, and Orlando is desperately in need of food for Adam and for himself. His sword is drawn because, like Touchstone in his encounter with Corin, he is mistaken about the circumstances. He assumes, as he will presently say, that he is in a "savage" place, and therefore will have to take the food by force. In his own savagery, then, he finds himself comically and wonderfully reproved by the duke in the name of "good manners" and "civility." Having fled from the failed civility of civilization, he has come into the presence of a civility reconstituted in the "savage" forest. Instead of drawing his own sword to defend his dinner, the duke welcomes Orlando as a guest:

> What would you have? Your gentleness shall force
> More than your force move us to gentleness.
> . . . .
> Sit down and feed, and welcome to our table.

Orlando, surprised, acknowledges his error and apologizes. He and Duke Senior then speak an antiphonal celebration of their common tradition of charity. Orlando says:

> If ever you have looked on better days,
> If ever been where bells have knolled to church,
> If ever sat at any good man's feast,
> If ever from your eyelids wiped a tear
> And know what 'tis to pity and be pitied,
> Let gentleness my strong enforcement be;
> In the which hope I blush, and hide my sword.

And the duke replies:

> True is it that we have seen better days,
> And have with holy bell been knolled to church,
> And sat at good men's feasts, and wiped our eyes
> Of drops that sacred pity hath engendered;
> And therefore sit you down in gentleness,
> And take upon command what help we have
> That to your wanting may be ministered.

But Orlando is not yet ready to sit down. He remains true to his promise to Adam, and he asks the company to "forbear your food a little while . . ." When he speaks of Adam now his kindness is forthright: "like a doe, I go to find my fawn / And give it food." There could be no more tender expression of loving servanthood, and no more apt a simile.

While Orlando is away, Jaques, in response to no encouragement, delivers his famous speech on the seven ages of man. This is a dandy set piece, but it is also utterly cynical. It is the life history of a lone specimen, such as one might find in a modern zoology manual. The last age, which is described most heartlessly,

> Is second childishness and mere oblivion,
> Sans teeth, sans eyes, sans taste, sans everything.

What Shakespeare thought of this may be inferred from the stage direction that immediately follows: "Enter Orlando, with Adam." That Orlando enters carrying Adam in his arms we know from Duke Senior's next speech, which also seems a rebuke to Jaques: "Welcome. Set down your venerable burden / and let him feed." Far from "sans everything," old Adam has a young friend who is his faithful servant—and who moreover, seeing that Adam is in his "second childishness," treats him with a mother's tenderness.

The scene ends with Duke Senior's recognition of Orlando, in which he implicitly affirms love as the right bond between generations and the members of a community: "Be truly welcome hither. I am the duke / That loved your father."

After Act II, *As You Like It* becomes a play of lovers, and the comedy of it, I think, is brilliant enough to follow worthily the eminent scene I have just described. The theme of transformation is worked out in greatest detail and most delightfully in the courtship of Rosalind and Orlando. In this courtship, which is both farcical and serious, Rosalind in the guise of Ganymede assumes the role of "Rosalind," so that Orlando, in the guise of his love-maddened self, may practice as a lover and so be "cured." The premise of this masquerade is set forth by Rosalind in III, ii: "Love is merely a madness . . ." There is good sense in this. She and Orlando fell in love "at first sight" in Act I. Rosalind, who is as smart and resourceful as she needs to be, realizes that such a love requires testing. Lovers in the madness of new love are, as Albany says of Goneril in *King Lear*, "self-covered." Rosalind's "cure," as it turns out, is a trial for herself as well as for Orlando. It removes the "cover" of selfhood; it tests them and proves them worthy of each other and ready for marriage. It is important to notice that these lovers do not turn seriously toward each other and toward their marriage until each of them has explicitly rejected the company of the cynical and sentimental Jaques.

The issue, for Rosalind and for the play, is how to make a

civil thing of the wildness of sexual love. The forest is the right place for courtship, which puts lovers in the state of nature. By the same token it is the right place to transform "mad" lovers, if they wish, into grown-up lovers fully prepared for the marriage rite and the "blessed bond of board and bed" with which the play ends.

By the end of the play its "self-covered" villains also have been transformed: Oliver by becoming the conscious and grateful beneficiary of his brother's courage and forgiveness, and Duke Frederick by his encounter with "an old religious man" in "the skirts of this wild wood." Jaques even has resolved to go and learn from "these convertites." Duke Senior and his fellow exiles, as we know from Act II, will return from the forest to a domestic world far better than the one they fled, for they too have been changed, renewed in their specifically human nature, their civility and charity, by this time of adversity in the natural world.

Thus by the play's end all of its principal characters have been changed, and for the better, by their time in the forest. Shakespeare saw, and wants us to see, that the forest can be corrective and restorative to disordered human life. But he goes further. At once explicitly and indirectly he invests the forest with a mysterious and even a mystical transformative power. Partly this is accomplished by Touchstone, speaking with implication apparently beyond his intention. In III, ii, he says to Rosalind, "You have said; but whether wisely or no, let the forest judge." And, twitting Audrey in III, iii, he says, "here we have no temple but the wood . . ." Also Rosalind, in her masquerade with Orlando, alludes to "an old religious uncle of mine" and to "a magician" she has "conversed with." Orlando, in V, iv, conflates the two when he speaks of Ganymede's uncle,

> Whom he reports to be a great magician,
> Obscurèd in the circle of this forest.

Is there, then, a great magician in the forest? Is the forest a holy place of judgment and magical or miraculous transformation? We must ask, but we must not answer. The play must not answer. *As You Like It* is not the voice out of the whirlwind. Once upon a time several people fled from a disordered and murderous society into a forest, and there they were profoundly changed. That is all we know.

II.

In *King Lear*, both the Lear story and the Gloucester story grow out of corruption at the center of wealth and power, just as does the action of *As You Like It*. Initially, in *King Lear*, this is the corruption merely of selfishness: self-complacency, self-indulgence, self-ignorance, the lack of critical self-knowledge. From this selfishness grows, in turn, an infection of monstrous proportions that is described, though unwittingly, by Gloucester's bastard son, Edmund, in Act I, scene ii, as he works his deception upon Edgar, his legitimate elder brother:

> . . . unnaturalness between the child and the parent; death, dearth, dissolutions of ancient amities; divisions in state, menaces and maledictions against king and nobles; needless diffidences, banishment of friends, dissipation of cohorts, nuptial breaches . . .

Because Lear is king, his self-absorption becomes in effect state policy. Like any head of state he is able, temporarily, to invest his fantasies with power. His fantasy is a primitive instance of "early retirement." He believes that by dividing his kingdom among his daughters he can free himself of care and responsibility while retaining the initiative and the privileges of his kingship. This will prove to be almost limitlessly foolish. He is an "idle old man," as his daughter Goneril calls him, "That still would manage those authorities / That he hath given away." His daughter Regan also is right when she says of him that "he hath ever but slenderly known himself." Because he does not know himself, he cannot know others. He has failed disastrously to know Goneril and Regan. He fails to learn of them in the play's first scene what is obvious to everybody else: that they are eloquent, clever, heartless, false, and greedy.

Having apparently determined already the portions of land that he will give to his daughters, Lear involves them pointlessly and cruelly in a contest in which they are to compete for his "largest bounty" by declaring their love for him. Goneril and Regan, good poets and good actors, give him precisely the groveling flattery he has asked for. Only his third daughter, Cordelia, who lacks neither sense nor eloquence, and who in fact truly loves him, refuses to tell him more than the plain truth: She loves him as she ought. She loves him *completely* as she ought, and the play will reveal this, but her refusal to participate in the love contest is entirely proper. It is a refusal to falsify her love by indulging her father's frivolous abuse of his power, which she both disdains and fears.

Predictably infuriated, Lear disinherits Cordelia. In doing so, Martin Lings argues in *The Secret of Shakespeare*, Lear banishes "the Spirit," by which Lings means "the Holy Spirit" or "the pearl of great price." I am unwilling so to allegorize the play, but I think nevertheless that Lings has pointed us in the right direction. In disinheriting Cordelia, in making her "a stranger to my heart and me . . . forever," Lear has, in the face of great evil, estranged himself from goodness. He then deepens and ratifies this estrangement by exiling the Earl of Kent, who has dared to call folly by its right name.

Thus in *King Lear*, exactly as in *As You Like It*, corruption at the center of power sets loose a centrifugal force that ultimately will send the powerless and the defeated into the wildness of the natural world. But in the six or so years between the two plays Shakespeare evidently saw a need to raise the stakes. The archetypal story is, after all, not necessarily a comedy. Selfishness does not necessarily involve one in a limited evil. Evil people are not necessarily relenting or easily converted. The state of nature is not necessarily the relatively hospitable Forest of Arden. The uses of adversity are not necessarily "sweet." The terms and the affirmation of *As You Like It* now required a harsher test.

Shakespeare brought the earlier play to trial by imagining a set of villains who in the course of the play will reveal—and discover for themselves, to their cost—that they have limitlessly consigned themselves to evil. Lacking self-knowledge and too "self-covered" even to suspect that he does, Lear rids his court of love, goodness, and honesty, and thus in effect abandons himself to the purposes of Goneril, Regan, and Regan's husband, the Duke of Cornwall. These three, like Lear, are selfish, but there is a difference. Lear, in his selfishness, is self-deluded: He thinks he is a loving and generous father, as no doubt he wishes to be. Goneril, Regan, and Cornwall, by contrast, are selfish by policy; there is no inconsistency between what they are doing and what they think they are doing. By dividing his kingdom, by isolating himself from Cordelia and Kent, Lear places himself in a deadly trap. Escape will cost him everything he has, or everything he thinks he has in the opening scene. In his quarrel with Kent, he unknowingly foretells his fate: "So be my grave my peace . . ."

In outline, through II, i, the Gloucester story exactly parallels that of Lear. Gloucester has two sons, Edgar and Edmund. Edgar loves his father as Cordelia loves hers. Edmund, the illegitimate younger son, is as contemptuous of Gloucester as Goneril and Regan are of Lear. Edmund, like those daughters, is a good actor and flatterer. He wants to cheat Edgar out of their father's estate, and he succeeds in convincing Gloucester that Edgar is planning to kill him. Edgar is then forced to flee to save his own life.

By the beginning of scene iii of Act II, when the fugitive Edgar transforms himself into Tom o'Bedlam, the villains are successful and in control; their schemes are working and they have what they want. Love, goodness, honesty, and fidelity have been directly confronted by evil, and evil so far has won. But this working of evil, by its very successes, has instigated a countermovement, and in parallel to *As You Like It* this movement is the work of good and faithful servants. The proscriptions against Cordelia, Kent, and Edgar have set them free to serve Lear and Gloucester.

Disguise in this play is just as important as in *As You Like It*, and more portentously so. Goneril, Regan, and Cornwall are "self-covered." Their better selves have been utterly and finally renounced. Edmund is only barely, but significantly, less self-obscured than they. Gloucester, like Lear, is a naively selfish old man. Neither is sinful or evil beyond the measure of ordinary human behavior, but both are obscured, obscured most consequentially to themselves, by foolishness and complacency. They are deluded and self-deluded. They are deludable *because* they are self-deluded. Goneril's husband, the Duke of Albany, for the time being is disguised to himself because of his hesitancy in recognizing and denouncing the evil character of his wife. In order to serve Lear and Gloucester in their time of greatest need, Kent and Edgar must serve in disguise. Of all the major characters, Cordelia alone always appears, to us and to herself, only as she is. She is good, and her understanding of her goodness is constant, profound, and absolutely assured. Much of the drama and meaning of the play comes from the actions of the characters in relation to their disguises, and we understand those actions by the measure of Cordelia's transparency, clarity, and candor.

In II, i, the villains of the Lear plot, recognizing their own kind in Edmund, claim him as an ally. Cornwall tells him, with terrible import, "you shall be ours . . . You we first seize on." And so, early in the play, the party of evil, of power-lust and greed, recognizes superficially the usefulness of cooperation, and for a while they are a coherent force. By contrast the party of goodness—the party of Cordelia, Kent, the Fool, Edgar, and, finally, Albany—in its early defeat is widely scattered. As the play proceeds, however, the party of evil, because of the nature of evil, disintegrates while the members of the other party recognize one another and draw together.

Edmund's soliloquy in I, ii, introduces another set of contraries into the play as he subordinates his specifically human nature to nature:

> Thou, Nature, art my goddess; to thy law
> My services are bound. Wherefore should I
> Stand in the plague of custom, and permit
> The curiosity of nations to deprive me,
> For that I am some twelve or fourteen moonshines
> Lag of a brother? Why bastard? Wherefore base . . . ?
> . . . .
> Legitimate Edgar, I must have your land.

If we are fair-minded, we must see the justice of Edmund's indictment of the prejudice against bastards as "base born," just as we see the justice of Goneril and Regan's perception that their father is foolish and intemperate. But evil characteristically supports and disguises itself by such partial claims of justice. In his dire intention to deceive his father and his brother, putting both their lives at risk, Edmund offends against moral law, specifically the fifth, sixth, eighth, ninth, and tenth of the Ten Commandments, and against the order of "degrees" as set forth in Ulysses' speech in *Troilus and Cressida*. Edmund understands "nature" as exclusive self-interest, which he by implication ascribes to all "natural" creatures. As a person self-consciously "enlightened," later in the same scene he rejects his father's astrological determinism, and so accepts full moral responsibility for what he is doing—and again we are tempted to sympathize. But in rejecting his father's superstition, he defines himself as self-determined. By thus subordinating human nature to nature, he means that he accepts *no* subordination. By putting himself at the service of nature's law, he means, perhaps more absolutely than he intends, that he rejects all service to anybody but himself, and will honor *no* law. This speech of Edmund's is answered in IV, vi, by the "Gentleman" who says of Cordelia in an apostrophe to Lear, "Thou hast one daughter / Who redeems nature . . ."

And so the thrusting-out of Lear and Gloucester into the wild world is as profoundly and purposefully thematic in this play as is the forest exile of the sufferers in *As You Like It*. When Lear speaks of Goneril and Regan as "unnatural," he means that they have, like Edmund, subscribed to nature's supposed law of entire selfishness, as opposed to human nature's laws of filiality and love. These virtually opposite uses of the word "nature" may be confusing, but the word in fact has this duplicity in our language, and Shakespeare exploits it fully, to serious purpose, in both plays.

By the unnaturalness of his bad daughters Lear is driven out into nature. Nature now is not the Forest of Arden, but the open heath in the midst of a "pitiless storm." The pitilessness of the storm, which is set before us in its full extremity in the dialogue, is the measure of the pitilessness of Goneril, Regan, and Cornwall—though the pitilessness of the storm, unlike that of these familial villains, is not unkind, as Lear understands and says in III, ii. The heath and the storm belong to the moral landscape of the tragedy, just as the forest belongs to the moral landscape of the comedy. And Lear's dreadful exile upon the heath in the storm and the darkness forces almost immediately a change upon his character. Even as he announces to the Fool that he is going mad—"My wits begin to turn"—he speaks for the first time unselfishly, in compassion and concern for the Fool's suffering: "How dost, my boy? Art cold?" And so his wits are turning, we may say, not just to madness, but through his madness, which is the utter frustration and destruction of his sanity as of Act I, to a better sanity.

Lear's adversity is not "sweet" but it *is* useful: It has made him tender; it has feelingly persuaded him what he is; it has reduced him from a king to a mere human, sharing the lot of other humans. And in III, iv, he speaks in compassion, confession, and repentance, his words recalling both Duke Senior's speech on the uses of adversity and Rosalind and Celia's act of justice toward Corin. These two themes of *As You Like It* recur, with heightened urgency and purpose, in *King Lear*. The disguised Kent, the faithful servant, has led the old king and the Fool to no welcome in Arden, but to a "hovel" that will provide them some meager shelter from the storm. At the doorway Lear says:

> Poor naked wretches, wheresoe'er you are,
> That bide the pelting of this pitiless storm,
> How shall your houseless heads and unfed sides,
> Your looped and windowed raggedness, defend you
> From seasons such as these? O, I have ta'en
> Too little care of this! Take physic, pomp;
> Expose thyself to feel what wretches feel,
> That thou mayst shake the superflux to them
> And show the heavens more just.

Lear's admission, "O, I have ta'en / Too little care of this!" is the turning point of his story. He has heretofore "ta'en care" mainly of himself; that has now become his calamity, and he knows it. His reproof, "Take physic, pomp," recalls Duke Senior's denunciation of "painted pomp" in *As You Like It*, at the same time that it takes up with greater force the earlier play's concern for economic justice. Recognition of the suffering of "Poor naked wretches" leads directly here to the biblical imperative of charity to the poor, for as long as people are painfully in want there is an implicit cruelty in anybody's "superflux" of wealth. This theme is repeated in full by Gloucester in IV, i, after he has given his purse to "Poor Tom":

> Heavens, deal so still!
> Let the superfluous and lust-dieted man,
> That slaves your ordinance, that will not see
> Because he does not feel, feel your pow'r quickly;
> So distribution should undo excess,
> And each man have enough.

*

In revenge for his kindness and service to the king during the storm, Gloucester has been captured by Cornwall and Regan, who bind him and put out his eyes. He too is then thrust out, blind and (as his tormentors believe) alone, into the world and the weather—to "smell / His way to Dover," as Regan says in as cruel a speech as was ever written.

But immediately after the terrible scene of his blinding, we find that Gloucester is not after all alone. He is helped first by an elderly servant who, in the little he tells of himself, answers exactly to the description of the old Adam of *As You Like It*:

> O my good lord,
> I have been your tenant, and your father's tenant,
> These fourscore years.

And then he is helped by Edgar in the guise of Tom o'Bedlam or Poor Tom, who is in fact his father's faithful servant, guide, and teacher, and who at last "save[s] him from despair."

Cast out into the storm and the darkness, Lear too is accompanied—first by the Fool and then by Kent and then by Gloucester, at what cost we know, and then by Cordelia and Albany. These are the good and faithful servants of this play, and they continue here *As You Like It*'s theme of service and, with it, the earlier play's theme of affection and generosity as opposed to force. But good service in *King Lear* is more costly than in *As You Like It*, also less effective, and thus it emerges in what Shakespeare must have concluded is its true character. Kent is exiled because he understands faithful service, not only as loyalty, as faithful help, but also as truth-telling, requiring even opposition. In this Kent is contrasted with Oswald, who is a bad servant because he connives in the evil of his masters and does as he is told; and Oswald is again contrasted with Cornwall's "First Servant," who opposes his master's cruelty to Gloucester and dies for his insubordination.

By the time he wrote *King Lear*, Shakespeare clearly had begun to doubt that it is possible to consent to use by policy a little or a limited evil to serve some perceived "good," and then to stop before the evil has enlarged to some unforeseen perfection of itself. This perfection, as Shakespeare saw it, was the destruction of the evildoers along with whatever else might be destroyed by them. (This is the tragedy itself of *Macbeth*, which was written in the next year.)

Self-destruction, after selfishness has been accepted as a policy, is merely a matter of logic, as explained in Ulysses' speech. The key insight is given by Cornwall in III, vii, when, before the blinding of Gloucester, he speaks of "our wrath, which men / May blame, but not control"; and again by Goneril in V, iii: "the laws are mine . . ." If the laws belong to individual persons—if Goneril, as queen, does not rule "under the law" —then those persons are in effect lawless. The party of evil is by definition out of control from the start. Its members are out of control as individuals dedicated to self-interest. People who are united by the principle of unrestrained self-interest have inevitably a short-lived union. However large and however costly to their victims their successes may be, their failure is assured. But their espousal of evil as a deliberate policy assures also that they will be unrelenting while they last.

To this great force of relentless if self-doomed evil Shakespeare opposes the counterforce of good and faithful service. As Lear and Gloucester are made powerless, poor, and helpless, the theme of help manifests itself in the presence and the acts of people entirely dedicated to serving them. But Lear and Gloucester in their selfishness are too vulnerable, and the wickedness of their adversaries is too great, to permit to the good servants any considerable practical success. They can give no victory and achieve no restoration, as the world understands such things. Their virtues do not lead certainly or even probably to worldly success, as some bad teachers would have us believe. They stand by, suffering what they cannot help, as parents stand by a dying or disappointing child. This assures only the survival in this world of faithfulness, compassion, and love —which is no small thing.

But this play refuses to stop at what the world understands of service or success. For Lear and Gloucester worldly failure is fully assured; it is too late for worldly vindication. What the good servants can do, and this they succeed in doing, is restore those defeated old men to their true nature as human beings. They can waken them to love and save them from despair.

*

It is obvious by now that I have begun to argue against what we might call the "dark interpretation" of *King Lear*. The dark interpretation is well represented by Stephen Orgel, editor of the new Pelican edition of "the traditional conflated text," who sums the meaning of the play as follows:

> The world is an instrument of torture, and the only comfort is in the nothing, the never, of death. The heroic vision is of suffering, unredeemed and unmitigated.

It is impossible to see this nihilistic reading of the play as valid, and hard to see it as "heroic." There is a kind of modern mind that finds Hell more imaginable and believable than Heaven and nihilism more palatable than redemption. What is "heroic" to this mind is the "courage" to face the immitigable pointlessness of human experience. This is the same mind that, in default of any structure of meaning, finds all bad outcomes, political or economic or ecological, to be "inevitable."

Before even approaching the issues of this play's ending, one ought at least to consider the biblical context within which Shakespeare oriented his work. *King Lear* was written in reference to three passages in the Gospel of Matthew. Like *As You Like It*, it alludes repeatedly, and more insistently and sternly, to the call to service in Matthew 20 that I quoted earlier. And more forcibly than in the earlier play it extends the obligation of service to "the least of these my brethren" (Matthew 25:40), for who more than Lear and Gloucester in their injury and helplessness could be counted as "least"? And unlike the comedy, the tragic play broods constantly on the idea, in Matthew 10:39 and also in the other three Gospels, of losing one's life in order to find it. This theme is stated plainly in II, i, when the King of France says to Cordelia, "Thou losest here, a better where to find," and this strikes so nearly to the heart of the play as to be virtually its subject.

In I, iv, Kent, newly exiled and in disguise, his old life thus lost, says to himself:

> If thou canst serve where thou dost stand condemned,
> So may it come thy master whom thou lov'st
> Shall find thee full of labors.

This is a literal description of Kent's predicament in the play, if we read "thy master" as King Lear, and "condemned" as Kent's exile. But it is also, and just as literally, a description of the human predicament and consequent obligation, if we read "thy master" as Christ, and "where thou dost stand condemned" as the fallen world.

In III, iv, Edgar as Poor Tom, in his feigning madness, recites a number of pertinent biblical laws, four of them from the Ten Commandments.

In IV, ii, when Albany says to Goneril, "Wisdom and goodness to the vile seem vile . . . ," Shakespeare could be recalling Isaiah 5:20: "Wo unto them that speake good of evil, and evil of good . . ."

Cordelia's sentence, lines 23–24 in IV, iv, has the same ambiguity as Kent's speech cited earlier: "O dear father, / It is thy business I go about." This either is an apostrophe to Lear or it is a prayer recalling Luke 2:49, in which Jesus says to his parents, "knewe ye not that I must go about my fathers busines?"

In IV, vi, when Edgar, seeing Lear in his madness "bedecked with weeds," exclaims, "O thou side-piercing sight!" he is recalling John 19:34: "But one of the soldiers with a spear pierced his side . . ."

When Lear in his mad sermon in IV, vi, says, "See how yond justice rails upon yond simple thief. Hark in thine ear: change places and, handy-dandy, which is the justice, which is the thief?" his words paraphrase Romans 2:1: "Therefore thou art inexcusable, o man, whosoever thou art that judgest: for in that thou judgest another, thou condemnest thy self: for thou that judgest, doest the same things."

Gloucester's prayer in the same scene—

> You ever-gentle gods, take my breath from me;
> Let not my worser spirit tempt me again
> To die before you please

—is the daunting submission of the Lord's Prayer and of Christ's agony in Gethsemane: "Thy wil be done . . ."

Cordelia's lamentation over her father in IV, vii—". . . and wast thou fain, poor father, / To hovel thee with swine and rogues forlorn . . ."—recalls the story of the Prodigal Son. She wakens Lear a few lines later in what he perceives as a resurrection. And her forgiveness of his offenses against her—"No cause, no cause"—is Christ's: not a mere human excusing or overlooking of an error, but the cancellation of its cause in Lear's fallen nature; his wrong no longer exists to be forgiven.

The trumpets that sound in V, iii, must have sounded in Shakespeare's imagination, and to the ears of many in his audiences, like the trumpets of Revelation, for they are a summoning to judgment.

When Albany in V, iii, offers to "friends . . . / The wages of their virtue," the words evoke Romans 6:23: "For the wages of sinne is death; but the gifte of God is eternal life . . ."

And when Kent, in his final speech, says, "My master calls me; I must not say no," he is confirming earlier suggestions of his impending death. But here we have again that term "master," which in I, iv, we could take to be ambiguous, but which here we are bound to understand as referring to Christ. To assume that "master" refers to Lear is to assume that Kent thinks Lear will require his services in the hereafter, a sentimentality that puts Kent far out of character.

The play, furthermore, contains three references to the miraculous, always in circumstances of great misery, reminding us that Christ's miracles are almost always performed on behalf of those who are seemingly beyond help. Kent, in the stocks in II, ii, says, "Nothing almost sees miracles / But misery." And Edgar, after Gloucester's "suicide" in IV, vi, says to him in the playacting of his "cure," but in fact urging the realization on him: "Thy life's a miracle." A few lines further on, he says, "the clearest gods who make them honors / Of men's impossibilities have preserved thee."

The foregoing list of biblical references may be incomplete or otherwise at fault. But it is at least sufficient to show that Shakespeare thought of the action of this play as occurring in a context far larger than that of what we have come to mean by "realism."

Anybody looking for meaninglessness or nihilism in *King Lear* can find it in abundance, but nearly all of it is in the deeds and the implicit principles of the villains. There are, however, three statements in the play that are explicitly and pointedly nihilistic.

The first is in one of Lear's speeches in III, iv, when in the storm he is reduced, as he thinks, virtually to nothing, and in his madness he adopts a fierce reductionism of his own. "Is man no more than this?" he asks of the nearly naked Edgar. And then, addressing Edgar, he says, "Thou art the thing itself; unaccommodated man is no more but such a poor, bare, forked animal as thou art." This, now under the weight of tragedy, is Jaques' conclusion in his "seven ages" speech; "unaccommodated man" is a lone specimen "sans everything."

A considerable part of the purpose of this play is to answer such statements, and this one is answered simply by the circumstances in which it is uttered. Lear's despair at this point is over the failure of a mere man to be successfully selfish. He cannot secure for himself his own wishes, and he cannot, alone, save himself even from the weather. But as bad as his predicament is, as nearly hopeless as it is, he is not "unaccommodated." Like the old Adam of *As You Like It*, he is not alone. Kent, the Fool, and Edgar are with him. A little later Gloucester enters with a torch to offer what help he can. With one most consequential exception, the good people of the play are going to be with him, doing all they can for him, to the end. What they can do is not enough, but they stand nonetheless for all that is opposite to his trouble and his suffering, his rage and his despair. They stand for the faithfulness that is opposed to treachery and the gentleness that is opposed to force.

The second expression of utter despair needing some comment is blinded Gloucester's accusation against "th' gods" in IV, i. This phrase "th' gods" is in keeping with a parliamentary proscription of the use of the word "God" on the stage, which the Puritans thought to be blasphemous. And so Gloucester was reduced to blaming the Greek and Roman deities: "As flies to wanton boys are we to th' gods; / They kill us for their sport."

Gloucester says this just after he has met Edgar-as-Poor-Tom. From now until Gloucester's death, Edgar's ruling purpose is to save his father from despair. Gloucester's sentence, while avoiding the appearance of blasphemy so fearful to the Puritan politicians, is authentically blasphemous, as Edgar understands. It is blasphemous, desperate, and perfectly self-centered. It is self-pity in extremis, driving him to say what he can hardly bear to say and cannot know. To save him from despair is to save him from the death of "a poor, bare, forked animal" reduced to the self-indulgence of self-pity. And by the end of the Gloucester story Edgar has led his father to a proper care for his life ("Thy life's a miracle") and to the proper submission to divine will that I quoted earlier. Edgar's service to Gloucester is clearly to be understood as redemptive, and he is not being frivolous when he says that his father died "smilingly" between the two extremes of "joy and grief."

My final exhibit in this line of nihilism is from a speech of Edmund's. Near the end of the play he sends a "captain" to follow Lear and Cordelia to prison, with instructions to kill them. Here is his justification: "Know thou this, that men / Are as the time is." This is a crude, self-serving determinism, the counterpart of "It is inevitable." All the energy and passion of *King Lear* gathers to refute this speech. *Some* men are as the time is, some always are, and they have always said so in self-justification. But Cordelia is not as the time is, Kent is not, Edgar is not, Albany is not, even the Fool is not.

And so these three assertions of hopelessness and meaninglessness are answered with three resounding nos that are passionately affirmative: No, "unaccommodated man" is not the type specimen of humanity. No, you are not eligible to conclude that the gods kill us for their sport. No, all men are not "as the time is."

By the end of the play we can have no doubt that we have watched a deadly skirmish in the battle between good and evil. We have watched the passage of tormented souls and a human community through a profound disorder, in which they have been driven away from their comforts and customary assurances into the world's unaccommodating wildness. The consequences of this casting out are surely tragic and horrifying. The death of Cordelia, as Dr. Johnson and others have testified, is shocking; it is nearly unbearable. The survivors are clearly in shock themselves, barely able to speak. And so we now must ask if in fact *King Lear* conforms to the archetypal story I outlined at the beginning. Are Lear and Gloucester in any sense reformed or redeemed by their great suffering? Is there any promise of a return to civil order? Does the play (to quote J. A. Bryant Jr.) "satisfy the society's impulse to renewal"?

Well, before concluding with the proponents of darkness that the play merely demonstrates the meaninglessness of suffering, we need to deal patiently with certain facts. The first of these is that by the play's end every one of the villains is dead —and not one of them is dead by chance. The death of each has come as a logical consequence of the assumption that human nature can be satisfactorily subordinated to nature. This assumption has proved to be as uncontrollable as the storm on the heath. There is a right relation between nature and human nature, and to get it wrong is eventually to perish. Shakespeare does not present this as an issue of justice, for such wrongs may destroy the innocent as well as the guilty; he presents it as the natural result of unnatural (that is, inhuman) behavior. The conflict of the two natures is revealed in Edmund's dying effort to redeem himself: "Some good I mean to do, / Despite of mine own nature."

According to this view, it would be too much to expect a "comic" outcome, for great evil will victimize the good. But the good people, unlike the evil ones, are not inevitably destroyed. Another fact is that, as the play ends, Kent, Edgar, and Albany are still alive. Kent's life is in doubt, but Edgar and Albany are young and will live on. In them is a reasonable hope for the restoration of civil order. And not only have those three survived, but in the course of the play, together with Cordelia and the Fool, they have grown ever greater in our respect and love, as has Shakespeare himself for imagining such people.

Another fact hard to ignore is the work of forgiveness. Both Cordelia and Edgar freely forgive their erring fathers, and by this forgiveness those fathers are made more truly and fully human.

Now we must deal with the reconciliation of the two plots. This is not really a problem, except that it has been made so by bad reading. The "problem" for the dark interpreter is that the Lear and the Gloucester stories are parallel, each enlarging our understanding of the other by resonating with it. The "problem" is that the Gloucester story is explicitly redemptive, for Edgar intends, as he says, to save his father from despair, and he succeeds, whereas the Lear story, according to the dark interpretation, ends in despair: "suffering, unredeemed and unmitigated." How can a play thus have two plots and two meanings that absolutely contradict each other and still deserve our respect? Stephen Orgel solves this problem by asserting that "Gloucester is effectively abandoned by the play." But this only raises a worse problem: Why would Shakespeare have given so much of his play, and so much magnificent poetry, to a secondary plot that he later "abandons"? And why should we indulge or forgive his doing so?

Another, more sensible way to deal with the supposed problem is to ask if Lear's story actually ends in despair, and thus in contradiction of Gloucester's story. To answer, we must look with the greatest care at Lear's final speech. It is possible, I suppose, to read or speak those lines as an unpausing scramble of outrage, grief, and despair. But the speech in fact has five parts, involving four profound changes of mind and mood. It begins with a complaint:

> And my poor fool is hanged: no, no, no life?

This is tenderness, heartbreaking enough, but it bears still a taint of the old selfishness. Cordelia is "*my* poor fool." Her death here is perceived as Lear's loss, not hers. Then comes a natural outrage:

> Why should a dog, a horse, a rat, have life,
> And thou no breath at all?

All grief over the death of the young, especially the death of one's own child, must bear the burden of such a question. And only after that ineluctable and futile question, which also comes from his own loss, can Lear turn his thoughts fully to the dead girl in his arms, and, forgetting himself, speak to her of her death:

> Thou'lt come no more,
> Never, never, never, never, never.

And then, turning to one of the bystanders, he says:

> Pray you undo this button. Thank you, sir.

This is literally meant. His clothes are somehow binding; he asks and receives help with a button. But the button is symbolic as well; it, or this small discomfort, is the last thing holding him to the world. This is not the renunciation of Gloucester's "suicide," but rather a profound submission and relinquishment of his will. At this point all the emotions of the preceding lines, and of his tragedy, pass from him, with the result that at last he *sees* Cordelia:

> Do you see this? Look on her! Look, her lips,
> Look there, look there—

Martin Lings, in *The Secret of Shakespeare*, understands Lear's story as at once a descent into Hell and a Purgatory, and he thinks that when Lear speaks these final two lines he is seeing Cordelia again, and this time in truth, as "a soul in bliss." I have no doubt that the play can be read or presented as Dr. Lings suggests. And yet I hesitate. The difficulty is that Shakespeare, as it seems to me, was not a visionary Christian; he was not Dante. The redemption he saw as possible for Gloucester and Lear did not come by way of an intercession from Heaven. It was earned, or lived out, or suffered out, in an unrelenting confrontation both with the unregenerate self, the self-covered self, and with the deliberate evil of others. The straight way was lost to Gloucester and Lear as it was to Dante, but it was recoverable to them by a self-loss more painful, and nearer too late, than Dante's.

I am content to rest with the more literal understanding that Cordelia, the play's only wholly undisguised character, has been disguised to Lear until the end by his self-preoccupying pride, anger, outrage, guilt, grief, and despair; and that, when his vision clears at last and he can see her as she was and is, he is entirely filled with love and wonder. And so the play may be said to show us at last a miracle: that Lear, dying, is more alive than he has ever been until this moment.

God, Science, and Imagination

AMONG THE ODDEST bedevilers of our time are the eminent scientists who use their heaped-up credentials, achievements, and awards as pedestals from which to foretell the future and pronounce upon the ultimate questions of life and religion. One of the most recent of these is Steven Weinberg's essay "Without God" in *The New York Review of Books* for September 25, 2008.

The oddity of these ventures, of which Professor Weinberg's is fairly typical, is in their ready—and, it seems to me, their thoughtless—abandonment of scientific rigor and methodology. For example, despite his protest that he does not want "to try to talk anyone out of their [*sic*] religion," Prof. Weinberg sets forth an elaborate argument for the nonexistence of God, an argument obviously meant to be persuasive but one that is based entirely on opinion.

As a fundamentalist of science, like the fundamentalists of religion, he is clearly evangelizing, hoping to convert or at least to disturb those who disagree. And like the religious fundamentalists, he uses a language that presents belief as knowledge. But more troubling than the authority he grants to his own opinions is his claim to know what cannot be known. "As religious belief weakens," he writes, "more and more of us know that after death there is nothing." The only fact available here is that Prof. Weinberg and more and more of us do not, and will never, know any such thing. There is no proof of this "nothing," and there is no scientific or other procedure by which to attempt such a proof.

Prof. Weinberg is a physicist, and he says that he is "professionally concerned with finding out what is true." But as a mere person he evidently is concerned, like too many others, merely with investing his opinions with power. This is the concern of fundamentalists of all kinds: religious, atheistic, scientific, technological, economic, and political. They all seek power—they seek victory, in fact—by abandoning the proprieties that permit us to seek and to honor what is true while acknowledging the limits of our ability to know.

Not far into his essay, Prof. Weinberg says, with proper humility, "Of course, not everything has been explained, nor will it ever be." But, two paragraphs later, speaking of "religious conservatives," he abandons the careful and exacting speech of humility, and prognosticates with the absolute confidence and gleeful vengefulness of a religious conservative: "I can imagine how disturbed they will feel in the future, when at last scientists learn how to understand human behavior in terms of the chemistry and physics of the brain, and nothing is left that needs to be explained by our having an immaterial soul."

This is something else that he does not know. Nor does he hesitate over the apparent difficulty of a material proof of the nonexistence of something immaterial.

The argument about the existence of God necessarily must be conducted in the absence of evidence that would stand as proof in either a laboratory or a court of law. There is no objective or empirical or experimental evidence on either side. The argument, as such, is by definition hopeless—a piece of foolishness and a waste of time. Even so, it has long existed and no doubt it will long continue, but only for the paltry reason that it cannot be won. Chaucer defined the problem about six hundred years ago, and I doubt that it can be more clearly defined:

A thousand tymes have I herd men telle
That ther ys joy in hevene and peyne in helle,
And I acorde wel that it ys so;
But, natheles, yet wot I wel also
That ther nis noon dwellyng in this contree,
That eyther hath in hevene or helle ybe,
Ne may of hit noon other weyes witen,
But as he hath herd seyd, or founde it writen;
For by assay ther may no man it preve.
—*The Prologue to the Legend of Good Women*

People of religion, and not just fundamentalists, can speak with tiresome confidence of knowing what in fact they don't know but instead believe. None of us is immune to the temptation to do this. Modern science itself, ignoring its famous devotion to empirical proof and factuality, has pampered and marketed itself by beliefs that have proved to be empirically flimsy and unimaginably damaging. Chemistry, while helping us to "live better," has poisoned the whole world; the "elegant" science of nuclear physics, while making us "safe" from our enemies and offering us "cheap" and "peaceful" power, has littered the world with lethal messes that apparently are irremediable; and so on to genetic engineering and other giddy "miracles." These developments, at least in their origin, are scientific. But the science involved has not been comprehensive or humble or self-critical or neighborly or publicly responsible. Mere self-interest obliges us to doubt the scientific faith that facts alone can assure the proper or safe use of facts. Modern science, as we have known it and as it has represented itself to us, has encouraged a healthy skepticism of everything but itself. But surely it implies no disrespect for science if we regard it with the skepticism upon which it prides itself.

We human beings, because we are short-lived creatures of limited intelligence, are going to remain under the necessity of talking about things that we don't provably know. But respect for "what is true," for what we don't know, and for our neighbors and fellow creatures, requires us to know and to say when we *don't* know.

Prof. Weinberg understands religious belief as "the belief in facts about God or the afterlife." This is a mistake. If "fact" means what we have agreed it means, and if we respect the word, then we have to say with Chaucer that none of us knows any facts about God or the afterlife. If we did, there would be no issue of "belief." We know for sure that it is possible to speak of beliefs and opinions as facts, but that does not make the beliefs and opinions factual; it only makes them lies.

Most writers about religion, however, have not been scientists, or consciously subject to the methodological strictures of science. If they speak of knowledge, they may mean the things one knows from tradition or from unreplicable experience or "from the heart." Even so, in the Bible the language of belief often falls far short of the confidence of factual knowledge. It is most moving—and, to me, it seems most authentic—when it is honestly confronting its own imperfection, or the inadequacy or failure of knowledge. Far from the cocksureness of fundamentalism, the starting place of authentic belief or faith is not-knowing.

One of the primary characteristics of the biblical God is his irreducibility; he cannot be confined in any structure of human comprehension. And so in 1 Kings, having completed his temple, Solomon cries out, not with confident religiosity, but with despair; his mighty work has contradicted what apparently was its purpose:

> But will God indeed dwell on the earth? behold, the
> heaven and heaven of heavens cannot contain thee;
> how much less this house that I have builded? (8:27)

The supplication of Mark 9:24 is likewise authenticated by its honest unknowing, unconfidence, and sense of struggle: "Lord, I believe; help thou my unbelief." And Paul's letter to the Romans is precise and unrelenting in his definition of hope:

> For we are saved by hope: but hope that is seen is not
> hope: for what a man seeth, why doth he yet hope for?
> But if we hope for that we see not, then do we with
> patience wait for it. (8:24–25)

"Faith," at root, is related to "bide" and "abide." It has certainly the sense of belief, but also the sense of difficult belief —of waiting, of patience, of endurance, of hanging on and holding together.

And so Prof. Weinberg's definition of "belief" involves not only a misuse of the word "facts," but also an implicit misunderstanding of the word "faith."

I dislike very much the disciplinary provincialisms of the universities; therefore, as a literary person, I ought to be delighted that Prof. Weinberg finds literature as irresistible as religion. But I am obliged instead to regret that he speaks of it with complacent oversimplification and ineptitude. When he says, for example, that "nothing prevents those of us who have no religious belief from enjoying religious poetry," does he mean that such enjoyment is the same for believers and unbelievers? If so, how does he know this? Does he have a way of comparing objectively the degrees and kinds of enjoyment? I would gladly agree that enjoyment is a desirable, maybe even a necessary, result of any art; but is enjoyment the only or the highest effect of religious art? What is it about religious art that unbelievers enjoy? The underlying question here, and an important one, is this: How do you authenticate, and make credible to somebody else, your response to a work of art? Prof. Weinberg seems not to have suspected that this question exists, or that it implies a careful, difficult job of work.

He likewise suspects no danger in his assertion that "we see already that little English language poetry written in the past few decades owes anything to belief in God." But who is "we"? How many decades does he have in mind? What is signified by "little"? The existence of God is not a statistical issue to be proved or disproved by quantities of belief or numbers of believers. If God exists, then, like Prof. Weinberg, he exists independently of anybody's knowledge or anybody's belief or disbelief in his existence. If *nobody* believed in God, Prof. Weinberg would still have his case to make, and evidence would be required of him that he cannot produce.

He says further that "very great poetry can be written without religion," using Shakespeare as his example, apparently unaware that Shakespeare's religion is still a controversial issue among Shakespearean scholars. But you don't have to be an expert—you have only to read the Bible and the plays—to know of Shakespeare's frequent allusions to scripture and his concern with scriptural themes such as mercy and forgiveness and with scriptural characters such as the good and faithful servant. But if great poetry can be written without religion, what does that prove about religion? It proves only that great poetry can be written without it.

In the boring, pointless, and destructive quarrel between fundamentalist science and fundamentalist religion, it seems to me that both sides are wrong. The religious fundamentalists are wrong because their disrespect for the materiality of the world involves, as a matter of course, disrespect for material evidence. They are like a jury that sees no significance in a "smoking gun" because its members don't believe in guns or smoke. The fundamentalist scientists are wrong because they counter one absolutism with another. Against their own history and tradition, they assume the posture of absolute certainty and unquestionability. Both sides assume that they are right now and forever. Neither can say "I don't know" or "I wonder." Both are bigoted, unforgiving, and humorless.

In his troubled and consoling last book of poems, *Second Space*, Czeslaw Milosz includes a poem of exemplary generosity:

> If there is no God,
> Not everything is permitted to man.
> He is still his brother's keeper
> And he is not permitted to sadden his brother,
> By saying there is no God.

This instruction, as Milosz undoubtedly knew, is perfectly reversible: If there is a God, that does not justify condescension or insult to your atheist neighbor. Such differences, so far as I can see, become issues of justice only when one side attempts to abridge or deny the freedom of the other.

If in fact the fundamentalist scientists were as smart as they think they are, and if the religious fundamentalists were as secure in their belief as they claim to be, then they would (except for issues of justice) leave one another in peace. They keep pestering each other because they need each other. The sort of mind that is inclined to fundamentalism is not content within itself or within its own convictions or principles. It needs to humiliate its opponents. It needs the sustenance of converts. It is fundamentally insecure and ungenerous.

Worst of all, the fundamentalists of both science and religion do not adequately understand or respect imagination. Is imagination merely a talent, such as a good singing voice, the ability to "make things up" or "think things up" or "get ideas"? Or is it, like science, a way of knowing things that can be known in no other way? We have much reason to think that it is a way of knowing things not otherwise knowable. As the word itself suggests, it is the power to make us *see*, and to see, moreover, things that without it would be unseeable. In one of its aspects it is the power by which we sympathize. By its means we may see what it was to be Odysseus or Penelope, or David or Ruth, or what it is to be one's neighbor or one's enemy. By it, we may "see ourselves as others see us." It is also the power by which we see the place, the predicament, or the story we are in.

To use what is by now the most notorious example, the creation story in Genesis is neither science nor pseudo-science, neither history nor pseudo-history. Like other traditional creation stories, it welled up out of the oldest, deepest human imagination to help us, even now, to see what it is to have a wondrous world that had a beginning in time. It is not true by the corroboration of contemporary documents or physical evidence. It is the imagination, in the high sense given to it by the greatest poets, that assents to its truth, just as it assents to the story of King Lear or Blake's rendering of Jacob's vision. The following lines, from Hayden Carruth's *Toward the Distant Islands*, rightly ignoring the unwinnable contest of science versus religion, were written with a proper deference to mystery and a proper respect for imagination:

> The *Iliad*, the *Odyssey*, the *Book of Genesis*,
> These were acts of love, I mean deeply felt gestures, which
> continuously bestow upon us
> What we are.

As for the afterlife, it has been imagined by Homer, Virgil, the biblical writers, Dante, and others, with the result that at least some of us, their willing heirs, have imagined it also.

I don't see that scientists would suffer the loss of any skin from their noses by acknowledging the validity and the power of imaginative truths, which are harmless to the truths of science, even though imagination in the highest sense seems allied less to science than to religion. The first chapters of Genesis are imagined and imaginable, whereas the big bang theory is the result of calculation. If you have read Dante, you can imagine Hell, Purgatory, and Heaven, but reading Prof. Weinberg cannot help you to imagine "nothing."

Perhaps the most interesting thing that Prof. Weinberg says in his essay is this: "There are plenty of people without religious faith who live exemplary moral lives (as for example, me)." This of course is a joke, modeled on the shameless self-commendation of politicians, but it is a joke without a context sufficient to reveal how large and sad a joke it is. The large sad fact that gives the joke its magnitude and its cutting edge is that there is probably not one person now living in the United States who, by a strict accounting, could be said to be living an exemplary moral life.

We are still somewhere in the course of the most destructive centuries of human history. And, though I believe I know some pretty good people whom I love and admire, I don't know one who is not implicated, by direct participation and by proxies given to suppliers, in an economy, recently national and now global, that is the most destructive, predatory, and wasteful the world has ever seen. Our own country in only a few hundred years has suffered the loss of maybe half its arable topsoil, most of its original forest and prairie, much too much of its mineral wealth and underground water. Most of its surface water and all of its air are polluted. Its rural cultures—the cultures, at their best, of husbandry—have been almost annihilated. Many of its plants and animals, both wild and domestic, are extinct or in danger. It is littered with wastelands, landfills, and, most shameful and fearful of all, dumps, industrial sites, and whole landscapes made dangerous virtually forever by radioactive waste. An immense part of this damage has been done in the years after World War II, when the machinery and chemicals of industrial warfare were turned upon the land—to make production "efficient" by the most doubtful standards and to replace the people of the land economies. I have no doubt that the dualisms of body and soul, heaven and earth, too prominent among the religious, have been damaging both to people and to the world, for that division has made it easy to withhold the necessary protections from material things. But the materialists of the science-technology-industry complex, whose minds are not so divided, and who might have been expected to value highly the material world, have instead held it in contempt and damaged it more than anybody.

Scientists and scholars in the knowledge industry—corporate or academic, if there is a difference—are probably in the greatest moral jeopardy of anybody except political, military, and corporate leaders. All knowledge now is potentially a commodity, and there is no way for its originators to control, or even foresee, the uses to which it may be put.

*

I would like, in conclusion, to bring up a question of religion and politics that I think needs more attention from everybody, but maybe especially from atheists. Before going on, I had better say that I adhere absolutely to the First Amendment. Any form of religious coercion by religious organizations or by governments would be intolerable to me. The idea of the separation of church and state seems to me fairly clear when it is a matter simply of limiting the powers of institutions. But the world is not so simple as to allow a neat, clear separation of politics and religion or of politics and irreligion.

My question is about the origin and existence of human rights. How did we get them? How are they authenticated? In ancient traditional cultures such as those of surviving peasant or hunter-gatherer communities, the people may be said to possess certain rights by tradition and inheritance: Because they have possessed them immemorially, they possess them still. No modern government, younger and of shallower origin, could rightfully revoke or ignore them. A younger nation of recent immigrants such as the United States possesses no rights by immemorial tradition and inheritance. The founders understood this, and so they stated in our Declaration of Independence the principle that "all men" (which we now construe as "all people") "are endowed by their Creator with certain inalienable Rights."

I don't think it is adequately appreciated how essential, and what a stroke of political brilliance, that statement is. The purport of it is that, as humans, we have rights that precede the existence of any government. We therefore were not on our knees to the government of England then, and we are not kneeling to our own government now, beseeching a grant of rights. As a would-be free people we were, and we are, requiring any government whatever to recognize and honor the rights we have always possessed by divine gift. The difference between rights granted by a government and rights given by "our Creator" is critical, for it is the difference between rights that are absolute and rights that are contingent upon the will or the whim of those in power.

The possession of rights by divine endowment obviously is an article of faith, for it has no objective or empirical standing.

It would have no standing at all with a government in principle atheistic. (We had better hope, I think, that the separation of church and state implies the separation of institutional atheism and the state.) But vulnerable as this principle may be as an article merely of faith, I know of no other authorization of human rights that can adequately replace it.

It is easy to anticipate that some who will not allow any validity to divine rights will bring forward "natural rights" as an alternative. But this too would be an article of faith, and it forces upon us the probably unanswerable question of what, in the nature of nature, might bring forth and confer upon us rights specifically human.

I am not able to settle such questions, even to my own satisfaction. And so I am obliged to conclude by offering the possibility that we humans are by definition, and perhaps by nature, creatures of faith (which we are as likely to place in luck or science or the free market or our own intelligence as in some version of God); and that we are further defined by principles and cultural properties, not objectively verifiable, that we inherit.

FROM
IT ALL TURNS ON AFFECTION
(2012)

It All Turns on Affection

> "Because a thing is going strong now, it need not go strong for ever," [Margaret] said. "This craze for motion has only set in during the last hundred years. It may be followed by a civilization that won't be a movement, because it will rest upon the earth."
>
> E. M. Forster, *Howards End* (1910)

ONE NIGHT in the winter of 1907, at what we have always called "the home place" in Henry County, Kentucky, my father, then six years old, sat with his older brother and listened as their parents spoke of the uses they would have for the money from their 1906 tobacco crop. The crop was to be sold at auction in Louisville on the next day. They would have been sitting in the light of a kerosene lamp, close to the stove, warming themselves before bedtime. They were not wealthy people. I believe that the debt on their farm was not fully paid, there would have been interest to pay, there would have been other debts. The depression of the 1890s would have left them burdened. Perhaps, after the income from the crop had paid their obligations, there would be some money that they could spend as they chose. At around two o'clock the next morning, my father was wakened by a horse's shod hooves on the stones of the driveway. His father was leaving to catch the train to see the crop sold.

He came home that evening, as my father later would put it, "without a dime." After the crop had paid its transportation to market and the commission on its sale, there was nothing left. Thus began my father's lifelong advocacy, later my brother's and my own, and now my daughter's and my son's, for small farmers and for land-conserving economies.

The economic hardship of my family and of many others, a century ago, was caused by a monopoly, the American Tobacco Company, which had eliminated all competitors and thus was able to reduce as it pleased the prices it paid to farmers. The American Tobacco Company was the work of James B. Duke of Durham, North Carolina, and New York City, who, disregarding any other consideration, followed a capitalist logic to absolute control of his industry and, incidentally, of the economic fate of thousands of families such as my own.

My effort to make sense of this memory and its encompassing history has depended on a pair of terms used by my teacher, Wallace Stegner. He thought rightly that we Americans, by inclination at least, have been divided into two kinds: "boomers" and "stickers." Boomers, he said, are "those who pillage and run," who want "to make a killing and end up on Easy Street," whereas stickers are "those who settle, and love the life they have made and the place they have made it in." "Boomer" names a kind of person and ambition that is the major theme, so far, of the history of the European races in our country. "Sticker" names a kind of person and desire that is, so far, a minor theme of that history, but a theme persistent enough to remain significant and to offer, still, a significant hope.

The boomer is motivated by greed, the desire for money, property, and therefore power. James B. Duke was a boomer, if we can extend the definition to include pillage in absentia. He went, or sent, wherever the getting was good, and he got as much as he could take.

Stickers on the contrary are motivated by affection, by such love for a place and its life that they want to preserve it and remain in it. Of my grandfather I need to say only that he shared in the virtues and the faults of his kind and time, one of his virtues being that he was a sticker. He belonged to a family who had come to Kentucky from Virginia, and who intended to go no farther. He was the third in his paternal line to live in the neighborhood of our little town of Port Royal, and he was the second to own the farm where he was born in 1864 and where he died in 1946.

We have one memory of him that seems, more than any other, to identify him as a sticker. He owned his farm, having bought out the other heirs, for more than fifty years. About forty of those years were in hard times, and he lived almost continuously in the distress of debt. Whatever has happened in what economists call "the economy," it is generally true that the land economy has been discounted or ignored. My grandfather lived his life in an economic shadow. In an urbanizing and industrializing age, he was the wrong kind of man. In one of his difficult years he plowed a field on the lower part of a long slope and planted it in corn. While the soil was exposed, a heavy rain fell and the field was seriously eroded. This was heartbreak for my grandfather, and he devoted the rest of his life first to healing the scars and then to his obligation of care. In keeping with the sticker's commitment, he neither left behind the damage he had done nor forgot about it, but stayed to repair it, insofar as soil loss can be repaired. My father, I think, had his father's error in mind when he would speak of farmers attempting, always uselessly if not tragically, "to plow their way out of debt." From that time, my grandfather and my father were soil conservationists, a commitment that they handed on to my brother and to me.

It is not beside the point, or off my subject, to notice that these stories and their meanings have survived because of my family's continuing connection to its home place. Like my grandfather, my father grew up on that place and served as its caretaker. It has now belonged to my brother for many years, and he in turn has been its caretaker. He and I have lived as neighbors, allies, and friends. Our long conversation has often taken its themes from the two stories I have told, because we have been continually reminded of them by our home neighborhood and topography. If we had not lived there to be reminded and to remember, nobody would have remembered. If we, like most of our generation, had moved away, the place with its memories would have been lost to us and we to it.

Because I have never separated myself from my home neighborhood, I cannot identify myself to myself apart from it. I am fairly literally flesh of its flesh. It is present in me, and to me, wherever I go. This undoubtedly accounts for my sense of shock when, on my first visit to Duke University, and by surprise, I came face-to-face with James B. Duke in his dignity, his glory perhaps, as the founder of that university. He stands imperially in bronze in front of a Methodist chapel aspiring to be a cathedral. He holds between two fingers of his left hand a bronze cigar. On one side of his pedestal is the legend: INDUSTRIALIST. On the other side is another single word: PHILANTHROPIST. The man thus commemorated seemed to me terrifyingly ignorant, even terrifyingly innocent, of the connection between his industry and his philanthropy. But I did know the connection. I felt it instantly and physically. The connection was my grandparents and thousands of others more or less like them. If you can appropriate for little or nothing the work and hope of enough such farmers, then you may dispense the grand charity of "philanthropy."

After my encounter with the statue, the story of my grandfather's 1906 tobacco crop slowly took on a new dimension and clarity in my mind. I still remembered my grandfather as himself, of course, but I began to think of him also as a kind of man standing in thematic opposition to a man of an entirely different kind. And I could see finally that between these two kinds there was a failure of imagination that was ruinous, that belongs indelibly to our history, and that has continued, growing worse, into our own time.

The term "imagination" in what I take to be its truest sense refers to a mental faculty that some people have used and thought about with the utmost seriousness. The sense of the verb "to imagine" contains the full richness of the verb "to see." To imagine is to see most clearly, familiarly, and understandingly with the eyes, but also to see inwardly, with "the mind's eye." It is to see, not passively, but with a force of vision and even with visionary force. To take it seriously we must give up at once any notion that imagination is disconnected from reality or truth or knowledge. It has nothing to do either with clever imitation of appearances or with "dreaming up."

I will say, from my own belief and experience, that imagination thrives on contact, on tangible connection. For humans to have a responsible relationship to the world, they must imagine their places in it. To have a place, to live and belong in a place, to live from a place without destroying it, we must imagine it. By imagination we see it illuminated by its own unique character and by our love for it. By imagination we recognize with sympathy the fellow members, human and nonhuman, with whom we share our place. By that local experience we see the need to grant a sort of preemptive sympathy to all the fellow members, the neighbors, with whom we share the world. As imagination enables sympathy, sympathy enables affection.

And in affection we find the possibility of a neighborly, kind, and conserving economy.

Obviously there is some risk in making affection the pivot of an argument about economy. The charge will be made that affection is an emotion, merely "subjective," and therefore that all affections are more or less equal: People may have affection for their children and their automobiles, their neighbors and their weapons. But the risk, I think, is only that affection is personal. If it is not personal, it is nothing; we don't, at least, have to worry about governmental or corporate affection. And one of the endeavors of human cultures, from the beginning, has been to qualify and direct the influence of emotion. The word "affection" and the terms of value that cluster around it —love, care, sympathy, mercy, forbearance, respect, reverence —have histories and meanings that raise the issue of worth. We should, as our culture has warned us over and over again, give our affection to things that are true, just, and beautiful. When we give affection to things that are destructive, we are wrong. A large machine in a large, toxic, eroded cornfield is not, properly speaking, an object or a sign of affection.

My grandfather knew, urgently, the value of money, but only of such comparatively small sums as would have paid his debts and allowed to his farm and his family a decent prosperity. He certainly knew of the American Tobacco Company. He no doubt had read and heard of James B. Duke, but nothing in his experience could have enabled him to imagine the life of the man himself.

James B. Duke came from a rural family in the tobacco country of North Carolina. In his early life he would have known men such as my grandfather. But after he began his rise as an industrialist, the life of a small tobacco grower would have been to him a negligible detail incidental to an opportunity for large profits. In the minds of the "captains of industry," then and now, the people of the land economies have been reduced to statistical numerals. Power deals "efficiently" with quantities that affection cannot recognize.

It may seem plausible to suppose that the head of the American Tobacco Company would have imagined at least that a dependable supply of raw material to his industry would depend upon a stable, reasonably thriving population of farmers and upon the continuing fertility of their farms. But he imagined no such thing. In this he was like apparently all agribusiness executives. They don't imagine farms or farmers. They imagine perhaps nothing at all, their minds being filled to capacity by numbers leading to the bottom line. Though the corporations, by law, are counted as persons, they do not have personal minds, if they can be said to have minds. It is a great oddity that a corporation, which properly speaking has no self, is by definition selfish, responsible only to itself. This is an impersonal, abstract selfishness, limitlessly acquisitive, but unable to look so far ahead as to preserve its own sources and supplies. The selfishness of the fossil fuel industries by nature is self-annihilating; but so, always, has been the selfishness of the agribusiness corporations. Land, as Wes Jackson has said, has thus been made as exhaustible as oil or coal.

There is another difference between my grandfather and James B. Duke that may finally be more important than any other, and this was a difference of kinds of pleasure. We may assume that, as a boomer, moving from one chance of wealth to another, James B. Duke wanted only what he did not yet have. If it is true that he was in this way typical of his kind, then his great pleasure was only in prospect, which excludes affection as a motive.

My grandfather, on the contrary, and despite his life's persistent theme of hardship, took a great and present delight in the modest good that was at hand: in his place and his affection for it, in its pastures, animals, and crops.

He did not participate in the least in what we call "mobility." He died, after eighty-two years, in the same spot he was born in. He was probably in his sixties when he made the one longish trip of his life. He went with my father southward across Kentucky and into Tennessee. On their return, my father asked him what he thought of their journey. He replied: "Well, sir, I've looked with all the eyes I've got, and I wouldn't trade the field behind my barn for every inch I've seen."

In such modest joy in a modest holding is the promise of a stable, democratic society, a promise not to be found in "mobility": our forlorn modern progress toward something indefinitely, and often unrealizably, better. A principled dissatisfaction with whatever one has promises nothing or worse.

James B. Duke would not necessarily have thought so far of the small growers as even to hold them in contempt. The Duke trust exerted an oppression that was purely economic, involving a mechanical indifference, the indifference of a grinder to what it grinds. It was not, that is to say, a political oppression. It did not *intend* to victimize its victims. It simply followed its single purpose of the highest possible profit, and ignored the "side effects." Confronting that purpose, any small farmer is only one, and one lost, among a great multitude of others, whose work can be quickly transformed into a great multitude of dollars.

Corporate industrialism has tended to be, and as its technological and financial power has grown it has tended increasingly to be, indifferent to its sources in what Aldo Leopold called "the land-community": the land, all its features and "resources," and all its members, human and nonhuman, including of course the humans who do, for better or worse, the work of land use. Industrialists and industrial economists have assumed, with permission from the rest of us, that land and people can be divorced without harm. If farmers come under adversity from high costs and low prices, then they must either increase their demands upon the land and decrease their care for it, or they must sell out and move to town, and this is supposed to involve no ecological or economic or social cost. Or if there are such costs, then they are rated as "the price of progress" or "creative destruction."

But land abuse *cannot* brighten the human prospect. There is in fact no distinction between the fate of the land and the fate of the people. When one is abused, the other suffers. The penalties may come quickly to a farmer who destroys perennial cover on a sloping field. They *will* come sooner or later to a land-destroying civilization such as ours.

And so it has seemed to me less a choice than a necessity to oppose the boomer enterprise with its false standards and its incomplete accounting, and to espouse the cause of stable, restorative, locally adapted economies of mostly family-sized farms, ranches, shops, and trades. Naïve as it may sound now, within the context of our present faith in science, finance, and technology—the faith equally of "conservatives" and "liberals" —this cause nevertheless has an authentic source in the sticker's hope to abide in and to live from some chosen and cherished small place—which is the agrarian vision that Thomas Jefferson spoke for, a sometimes honored human theme, minor and even fugitive, but continuous from ancient times until now. Allegiance to it, however, is not a conclusion but the beginning of thought.

The problem that ought to concern us first is the fairly recent dismantling of our old understanding and acceptance of human limits. For a long time we knew that we were not, and could never be, "as gods." We knew, or retained the capacity to learn, that our intelligence could get us into trouble that it could not get us out of. We were intelligent enough to know that our intelligence, like our world, is limited. We seem to have known and feared the possibility of irreparable damage. But beginning in science and engineering, and continuing, by imitation, into other disciplines, we have progressed to the belief that humans are intelligent enough, or soon will be, to transcend all limits and to forestall or correct all bad results of the misuse of intelligence. Upon this belief rests the further belief that we can have "economic growth" without limit.

Economy in its original—and, I think, its proper—sense refers to household management. By extension, it refers to the husbanding of all the goods by which we live. An authentic economy, if we had one, would define and make, on the terms of thrift and affection, our connections to nature and to one another. Our present industrial system also makes those connections, but by pillage and indifference. Most economists think of this arrangement as "the economy." Their columns and articles rarely if ever mention the land-communities and land-use economies. They never ask, in their professional oblivion, why we are willing to do permanent ecological and cultural damage "to strengthen the economy."

In his essay, "Notes on Liberty and Property," Allen Tate gave us an indispensable anatomy of our problem. His essay begins by equating, not liberty and property, but liberty and *control* of one's property. He then makes the crucial distinction between ownership that is merely legal and what he calls "effective ownership." If a property, say a small farm, has one owner, then the one owner has an effective and assured, if limited, control over it—as long as he or she can afford to own it, and is free to sell it or use it, and (I will add) free to use it poorly or well. It is clear also that effective ownership of a small property is personal and therefore can, at least possibly, be intimate, familial, and affectionate. If, on the contrary, a person owns a small property of stock in a large corporation, then that person has surrendered control of the property to larger shareholders. The drastic mistake our people made, as Tate believed and I agree, was to be convinced "that there is *one* kind of property—just *property*, whether it be a thirty-acre farm in Kentucky or a stock certificate in the United States Steel Corporation." By means of this confusion, Tate said, "Small ownership . . . has been worsted by big, dispersed ownership—the giant corporation." (It is necessary to append to Tate's argument the further fact that by now, owing largely to corporate influence, land ownership implies the right to destroy the land-community entirely, as in surface mining, and to impose, as a consequence, the dangers of flooding, water pollution, and disease upon communities downstream.)

Tate's essay was written for the anthology *Who Owns America?*, the publication of which was utterly without effect. With other agrarian writings before and since, it took its place on the far margin of the national dialogue, dismissed as anachronistic, retrogressive, nostalgic, or reactionary in the face of the supposedly "inevitable" dominance of corporate industrialism. *Who Owns America?* was published in the Depression year of 1936. It is at least ironic that talk of "effective property" could have been lightly dismissed at a time when many rural people who had migrated to industrial cities were returning to their home farms to survive. In 1936, when to the dominant minds a thirty-acre farm in Kentucky was becoming laughable, Tate's essay would have seemed irrelevant as a matter of course. At that time, despite the Depression, faith in the standards and devices of industrial progress was nearly universal and could not be shaken.

But now, three-quarters of a century later, we are no longer talking about theoretical alternatives to corporate rule. We are talking with practical urgency about an obvious need. Now the two great aims of industrialism—replacement of people by technology and concentration of wealth into the hands of a small plutocracy—seem close to fulfillment. At the same time the *failures* of industrialism have become too great and too dangerous to deny. Corporate industrialism itself has exposed the falsehood that it ever was inevitable or that it ever has given precedence to the common good. It has failed to sustain the health and stability of human society. Among its characteristic signs are destroyed communities, neighborhoods, families, small businesses, and small farms. It has failed just as conspicuously and more dangerously to conserve the wealth and health of nature. No amount of fiddling with capitalism to regulate and humanize it, no pointless rhetoric on the virtues of capitalism or socialism, no billions or trillions spent on "defense" of the "American dream," can for long disguise this failure. The evidences of it are everywhere: eroded, wasted, or degraded soils; damaged or destroyed ecosystems; extinction of species; whole landscapes defaced, gouged, flooded, or blown up; pollution of the whole atmosphere and of the water cycle; "dead zones" in the coastal waters; thoughtless squandering of fossil fuels and fossil waters, of mineable minerals and ores; natural health and beauty replaced by a heartless and sickening ugliness. Perhaps its greatest success is an astounding increase in the destructiveness, and therefore the profitability, of war.

In 1936, moreover, only a handful of people were thinking about sustainability. Now, reasonably, many of us are thinking about it. The problem of sustainability is simple enough to state. It requires that the fertility cycle of birth, growth, maturity, death, and decay—what Albert Howard called "the Wheel of Life"—must turn continuously in place, so that the law of return is kept and nothing is wasted. For this to happen in the stewardship of humans, there must be a cultural cycle, in harmony with the fertility cycle, also continuously turning in place. The cultural cycle is an unending conversation between old people and young people, assuring the survival of local memory, which has, as long as it remains local, the greatest practical urgency and value. This is what is meant, and is all that can be meant, by "sustainability." The fertility cycle turns by the law of nature. The cultural cycle turns on affection.

That we live now in an economy that is not sustainable is not the fault only of a few mongers of power and heavy equipment.

We all are implicated. We all, in the course of our daily economic life, consent to it, whether or not we approve of it. This is because of the increasing abstraction and unconsciousness of our connection to our economic sources in the land, the land-communities, and the land-use economies. In my region and within my memory, for example, human life has become less creaturely and more engineered, less familiar and more remote from local places, resources, pleasures, and associations. Our knowledge, in short, has become increasingly statistical.

Statistical knowledge once was rare. It was a property of the minds of great rulers, conquerors, and generals, people who succeeded or failed by the manipulation of large quantities that remained, to them, unimagined because unimaginable: merely accountable quantities of land, treasure, people, soldiers, and workers. This is the sort of knowledge we now call "data" or "facts" or "information." By means of such knowledge a category assumes dominion over its parts or members. With the coming of industrialism, the great industrialists, like kings and conquerors, became exploiters of statistical knowledge. And finally virtually all of us, in order to participate and survive in their system, have had to agree to their substitution of statistical knowledge for personal knowledge. Virtually all of us now share with the most powerful industrialists their remoteness from actual experience of the actual world. Like them, we participate in an absentee economy, which makes us effectively absent even from our own dwelling places. Though most of us have little wealth and perhaps no power, we consumer-citizens are more like James B. Duke than we are like my grandfather. By economic proxies thoughtlessly given, by thoughtless consumption of goods ignorantly purchased, now we all are boomers.

The failure of imagination that divided the Duke monopoly and such farmers as my grandfather seems by now to be taken for granted. James B. Duke controlled remotely the economies of thousands of farm families. A hundred years later, "remote control" is an unquestioned fact, the realization of a technological ideal, and we have remote entertainment and remote war. Statistical knowledge is remote, and it isolates us in our remoteness. It is the stuff itself of unimagined life. We may, as we say, "know" statistical sums, but we cannot imagine them.

It is by imagination that knowledge is "carried to the heart" (to borrow again from Allen Tate). The faculties of the mind —reason, memory, feeling, intuition, imagination, and the rest —are not distinct from one another. Though some may be favored over others and some ignored, none functions alone. But the human mind, even in its wholeness, even in instances of greatest genius, is irremediably limited. Its several faculties, when we try to use them separately or specialize them, are even more limited.

The fact is that we humans are not much to be trusted with what I am calling statistical knowledge, and the larger the statistical quantities the less we are to be trusted. We don't learn much from big numbers. We don't understand them very well, and we are not much affected by them. The reality that is responsibly manageable by human intelligence is much nearer in scale to a small rural community or urban neighborhood than to the "globe."

When people succeed in profiting on a large scale, they succeed for themselves. When they fail, they fail for many others, sometimes for us all. Propriety of scale in all human undertakings is paramount, and we ignore it. We are now betting our lives on quantities that far exceed all our powers of comprehension. We believe that we have built a perhaps limitless power of comprehension into computers and other machines, but our minds remain as limited as ever. Our trust that machines can manipulate to humane effect quantities that are unintelligible and unimaginable to humans is incorrigibly strange.

As there is a limit only within which property ownership is effective, so is there a limit only within which the human mind is effective and at least possibly beneficent. We must assume that the limit would vary somewhat, though not greatly, with the abilities of persons. Beyond that limit the mind loses its wholeness, and its faculties begin to be employed separately or fragmented according to the specialties or professions for which it has been trained.

In my reading of the historian John Lukacs, I have been most instructed by his understanding that there is no knowledge but human knowledge, that we are therefore inescapably central to our own consciousness, and that this is "a statement not of arrogance but of humility. It is yet another recognition of the inevitable limitations of mankind." We are thus isolated within our uniquely human boundaries, which we certainly cannot transcend or escape by means of technological devices.

But as I understand this dilemma, we are not *completely* isolated. Though we cannot by our own powers escape our limits, we are subject to correction from, so to speak, the outside. I can hardly expect everybody to believe, as I do (with caution), that inspiration can come from the outside. But inspiration is not the only way the human enclosure can be penetrated. Nature too may break in upon us, sometimes to our delight, sometimes to our dismay.

As many hunters, farmers, ecologists, and poets have understood, Nature (and here we capitalize her name) is the impartial mother of all creatures, unpredictable, never entirely revealed, not my mother or your mother, but nonetheless our mother. If we are observant and respectful of her, she gives good instruction. As Albert Howard, Wes Jackson, and others have carefully understood, she can give us the right patterns and standards for agriculture. If we ignore or offend her, she enforces her will with punishment. She is always trying to tell us that we are not so superior or independent or alone or autonomous as we may think. She tells us in the voice of Edmund Spenser that she is of all creatures "the equall mother, / And knittest each to each, as brother unto brother." Nearly three and a half centuries later, we hear her saying about the same thing in the voice of Aldo Leopold: "In short, a land ethic changes the role of *Homo sapiens* from conqueror of the land-community to plain member and citizen of it."

We cannot know the whole truth, which belongs to God alone, but our task nevertheless is to seek to know what is true. And if we offend gravely enough against what we know to be true, as by failing badly enough to deal affectionately and responsibly with our land and our neighbors, truth will retaliate with ugliness, poverty, and disease. The crisis of this line of thought is the realization that we are at once limited and unendingly responsible for what we know and do.

The discrepancy between what modern humans presume to know and what they can imagine—given the background of pride and self-congratulation—is amusing and even funny. It becomes more serious as it raises issues of responsibility. It becomes fearfully serious when we start dealing with statistical measures of industrial destruction.

To hear of a thousand deaths in war is terrible, and we "know" that it is. But as it registers on our hearts, it is not more terrible than one death fully imagined. The economic hardship of one farm family, if they are our neighbors, affects us more painfully than pages of statistics on the decline of the farm population. I can be heartstruck by grief and a kind of compassion at the sight of one gulley (and by shame if I caused it myself), but, conservationist though I am, I am not nearly so upset by an accounting of the tons of plowland sediment borne by the Mississippi River. Wallace Stevens wrote that "Imagination applied to the whole world is vapid in comparison to imagination applied to a detail"—and that appears to have the force of truth.

It is a horrible fact that we can read in the daily paper, without interrupting our breakfast, numerical reckonings of death and destruction that ought to break our hearts or scare us out of our wits. This brings us to an entirely practical question: Can we—and, if we can, *how* can we—make actual in our minds the sometimes urgent things we say we know? This obviously cannot be accomplished by a technological breakthrough, nor can it be accomplished by a big thought. Perhaps it cannot be accomplished at all. Yet another not very stretchable human limit is in our ability to tolerate or adapt to change. Change of course is a constant of earthly life. You can't step twice into exactly the same river, nor can you live two successive moments in exactly the same place. And always in human history there have been costly or catastrophic sudden changes. But with relentless fanfare, at the cost of almost indescribable ecological and social disorder, and to the almost incalculable enrichment and empowerment of corporations, industrialists have substituted what they fairly accurately call "revolution" for the slower, kinder processes of adaptation or evolution. We have had in only about two centuries a steady and ever-quickening sequence of industrial revolutions in manufacturing, transportation, war, agriculture, education, entertainment, homemaking and family life, health care, and so-called communications.

Probably everything that can be said in favor of all this has been said, and it is true that these revolutions have brought some increase of convenience and comfort and some easing of pain. It is also true that the industrialization of everything has incurred liabilities and is running deficits that have not been adequately accounted. All of these changes have depended upon industrial technologies, processes, and products, which have depended upon the fossil fuels, the production and consumption of which have been, and are still, unimaginably damaging to land, water, air, plants, animals, and humans. And the cycle of obsolescence and innovation, goaded by crazes of fashion, has given the corporate economy a controlling share of everybody's income.

The cost of this has been paid also in a social condition that apologists call "mobility," implying it has been always "upward" to a "higher standard of living," but which in fact has been an ever-worsening unsettlement of our people, and the extinction or near-extinction of traditional and necessary communal structures.

For this also there is no technological or large-scale solution. Perhaps, as they believe, the most conscientiously up-to-date people can easily do without local workshops and stores, local journalism, a local newspaper, a local post office, all of which supposedly have been replaced by technologies. But what technology can replace personal privacy or the coherence of a family or a community? What technology can undo the collateral damages of an inhuman rate of technological change?

The losses and damages characteristic of our present economy certainly cannot be stopped, let alone restored, by "liberal" or "conservative" tweakings of corporate industrialism, against which the ancient imperatives of good care, homemaking, and frugality can have no standing. The possibility of authentic correction comes, I think, from two already-evident causes. The first is scarcity and other serious problems arising from industrial abuses of the land-community. The goods of nature so far have been taken for granted and, especially in America, assumed to be limitless, but their diminishment, sooner or later unignorable, will enforce change.

A positive cause, still little noticed by high officials and the media, is the by now well-established effort to build or re-build local economies, starting with economies of food. This effort to connect cities with their surrounding rural landscapes rests exactly upon the recognition of human limits and the necessity of human scale. Its purpose, to the extent possible, is to bring producers and consumers, causes and effects, back within the bounds of neighborhood, which is to say the effective reach of imagination, sympathy, affection, and all else—including enough food—that neighborhood implies. An economy genuinely local and neighborly offers to localities a measure of security they cannot derive from a national or a global economy controlled by people who, by principle, have no local commitment.

In this age so bewildered by technological magnifications of power, people who stray beyond the limits of their mental competence typically find no guide except for the supposed authority of market price. "The market" thus assumes the standing of ultimate reality. But market value is an illusion, as is proven by its frequent changes; it is determined solely by the buyer's ability and willingness to pay.

By now our immense destructiveness has made clear that the actual value of some things exceeds human ability to calculate, and therefore must be considered absolute. For the destruction of these things there is never, under any circumstances, any justification. Their absolute value is recognized by the mortal need of those who do not have them, and by affection. Land, to people who do not have it and who are thus without the means of life, is absolutely valuable. Ecological health, in a land dying of abuse, is not worth "something"; it is worth everything. And abused land relentlessly declines in value to its present and succeeding owners, whatever its market price.

But we need not wait, as we are doing, to be taught the absolute value of land and of land health by hunger and disease. Affection can teach us, and soon enough, if we grant appropriate standing to affection. For this we must look to the stickers, who "love the life they have made and the place they have made it in."

All thoughtful people have begun to feel our eligibility to be instructed by ecological disaster and mortal need. But we endangered ourselves first of all by dismissing affection as an honorable and necessary motive. Our decision in the middle of the last century to reduce the farm population, eliminating the allegedly "inefficient" small farmers, was enabled by the discounting of affection. As a result, we now have barely enough farmers to keep the land in production, with the help of increasingly expensive industrial technology and at an increasing ecological and social cost. Far from the plain citizens and members of the land-community, as Aldo Leopold wished them to be, farmers are now too likely to be merely the land's exploiters.

I don't hesitate to say that damage or destruction of the land-community is morally wrong, just as Leopold did not hesitate to say so when he was composing his essay "The Land Ethic" in 1947. But I do not believe, as I think Leopold did not, that morality, even religious morality, is an adequate motive for good care of the land-community. The *primary* motive for good care and good use is always going to be affection, because affection involves us entirely. And here Leopold himself set the example. In 1935 he bought an exhausted Wisconsin farm and, with his family, began its restoration. To do this was morally right, of course, but the motive was affection. Leopold was an ecologist. He felt, we may be sure, an informed sorrow for the place in its ruin. He imagined it as it had been, as it was, and as it might be. And a profound, delighted affection radiates from every sentence he wrote about it.

Without this informed, practical, and *practiced* affection, the nation and its economy will conquer and destroy the country.

In thinking about the importance of affection, and of its increasing importance, I have been guided most directly by E. M. Forster's novel *Howards End*, published in 1910. By then, Forster was aware of the implications of "rural decay," and in this novel he spoke, with some reason, of his fear that "the literature of the near future will probably ignore the country and seek inspiration from the town. . . . and those who care for the earth with sincerity may wait long ere the pendulum swings back to her again." Henry Wilcox, the novel's "plain man of business," speaks the customary rationalization, which has echoed through American bureaus and colleges of agriculture, almost without objection, for at least sixty years: "the days for small farms are over."

In *Howards End*, Forster saw the coming predominance of the machine and of mechanical thought, the consequent deracination and restlessness of populations, and the consequent ugliness. He saw an industrial ugliness, "a red rust," already creeping out from the cities into the countryside. He seems to have understood by then also that this ugliness was the result of the withdrawal of affection from places. To have beautiful buildings, for example, people obviously must want them to be beautiful and know how to make them beautiful, but evidently they also must love the places where the buildings are to be built. For a long time, in city and countryside, architecture has disregarded the nature and influence of places. Buildings have become as interchangeable from one place to another as automobiles. The outskirts of cities are virtually identical and as depressingly ugly as the corn-and-bean deserts of industrial agriculture.

What Forster could not have foreseen in 1910 was the *extent* of the ugliness to come. We still have not understood how far at fault has been the prevalent assumption that cities could be improved by pillage of the countryside. But urban life and rural life have now proved to be interdependent. As the countryside has become more toxic, more eroded, more ecologically degraded and more deserted, the cities have grown uglier, less sustainable, and less livable.

The argument of *Howards End* has its beginning in a manifesto against materialism:

> It is the vice of a vulgar mind to be thrilled by bigness, to think that a thousand square miles are a thousand times more wonderful than one square mile . . . That is not imagination. No, it kills it. . . . Your universities? Oh, yes, you have learned men who collect . . . facts, and facts, and empires of facts. But which of them will rekindle the light within?

"The light within," I think, means affection, affection as motive and guide. Knowledge without affection leads us astray. Affection leads, by way of good work, to authentic hope.

The climactic scene of Forster's novel is the confrontation between its heroine, Margaret Schlegel, and her husband, the self-described "plain man of business," Henry Wilcox. The issue is Henry's determination to deal, as he thinks, "realistically" with a situation that calls for imagination, for affection, and then forgiveness. Margaret feels at the start that she is "fighting for women against men." But she is not a feminist in the popular or political sense. What she opposes with all her might is Henry's hardness of mind and heart that is "realistic" only because it is expedient and because it subtracts from reality the life of imagination and affection, of living souls. She opposes his refusal to see the practicality of the life of the soul.

Margaret's premise, as she puts it to Henry, is the balance point of the book: "It all turns on affection now . . . Affection. Don't you see?"

In a speech delivered in 2006, "Revitalizing Rural Communities," Frederick Kirschenmann quoted his friend Constance Falk, an economist: "There is a new vision emerging demonstrating how we can solve problems and at the same time create a better world, and it all depends on collaboration, love, respect, beauty, and fairness."

Those two women, almost a century apart, speak for human wholeness against fragmentation, disorder, and heartbreak. The English philosopher and geometer Keith Critchlow brings his own light to the same point: "The human mind takes apart with its analytic habits of reasoning but the human heart puts things together because it loves them . . ."

The great reassurance of Forster's novel is the wholeheartedness of his language. It is to begin with a language not disturbed by mystery, by things unseen. But Forster's interest throughout is in soul-sustaining habitations: houses, households, earthly places where lives can be made and loved. In defense of such dwellings he uses, without irony or apology, the vocabulary that I have depended on in this talk: truth, nature, imagination, affection, love, hope, beauty, joy. Those words are hard to keep still within definitions; they make the dictionary hum like a beehive. But in such words, in their resonance within their histories and in their associations with one another, we find our indispensable humanity, without which we are lost and in danger.

No doubt there always will be some people willing to do anything at all that is financially or technologically possible, who look upon the world and its creatures without affection and therefore as exploitable without limit. Against that limitlessness, in which we foresee assuredly our ruin, we have only our ancient effort to define ourselves as human and humane. But this ages-long, imperfect, unendable attempt, with its magnificent record, we have virtually disowned by assigning it to the ever more subordinate set of school subjects we call "arts and humanities" or, for short, "culture." Culture, so isolated, is seen either as a dead-end academic profession or as a mainly useless acquisition to be displayed and appreciated "for its own sake." This definition of culture as "high culture" actually debases it, as it debases also the presumably low culture that is excluded: the arts, for example, of land use, life support, healing, housekeeping, homemaking.

I don't like to deal in categorical approvals, and certainly not of the arts. Even so, I do not concede that the "fine arts," in general, are useless or unnecessary or even impractical. I can testify that some works of art, by the usual classification "fine," have instructed, sustained, and comforted me for many years in my opposition to industrial pillage.

But I would insist that the economic arts are just as honorably and authentically refinable as the fine arts. And so I am nominating economy for an equal standing among the arts and humanities. I mean, not economics, but economy, the making of the human household upon the earth: the *arts* of adapting kindly the many human households to the earth's many ecosystems and human neighborhoods. This is the economy that the most public and influential economists never talk about, the economy that is the primary vocation and responsibility of every one of us.

My grandparents were fortunate. They survived their debts and kept their farm. But in the century and more since that hard year of 1907, millions of others have not been so fortunate. Owing largely to economic constraints, they have lost their hold on the land, and the land has lost its hold on them.

They have entered into the trial of displacement and scattering that we try to dignify as "mobility."

The land and the people nevertheless have suffered together, as invariably they must. Under the rule of industrial economics, the land, our country, has been pillaged for the enrichment, supposedly, of those humans who have claimed the right to own or exploit it without limit. Of the land-community much has been consumed, much has been wasted, almost nothing has flourished.

But this has not been inevitable. We do not have to live as if we are alone.

About Civil Disobedience

MATT ROTHSCHILD has asked me to write a short piece about civil disobedience. This is a request I would ignore, if I had my druthers. But Matt was representing Duty, perhaps, and several hours of hard thought have not afforded me an adequate excuse.

The reason for civil disobedience, as I understand it, is simple necessity. If you belong to a political side with a significant grievance, and if your side is not represented—is in effect not heard—in the halls of your nominally representative government, then you have two choices: You can passively submit, accepting your grievance as irremediable, or you can, perhaps uselessly, insist by some act of civil disobedience upon your right to have your grievance redressed by your government. I am speaking here of *nonviolent* civil disobedience, of course. I subscribe to the commandment to love our neighbors, even if they are our enemies. But I am also in favor of making sense. Answering violence with violence is understandable, but it is also nonsense.

Matt asked me to write this piece because he knows that last February, in protest against coal mining by "mountaintop removal," I committed myself to an act of civil disobedience in the office of Kentucky's governor. In fact, I have made that commitment three times. The first was on June 3, 1979, in opposition to a nuclear power plant then being built at Marble Hill on the Ohio River near Madison, Indiana. The second was in Washington, D.C., on March 2, 2009, in protest, with a host of others, generally against mountaintop removal and air pollution by the burning of fossil fuels, and immediately against the burning of coal by a power plant within a few blocks of the national capitol. The third was on the eleventh of last February: the aforementioned attempt to discover conscience in official Frankfort.

Only one of these adventures resulted in actual civil disobedience and arrest. After we crossed the fence at Marble Hill, we were arrested and booked—and turned loose. In Washington, the number of us offering to get arrested—two or three thousand, maybe—overwhelmed the police, who, thinking perhaps of the hours it would take to write down our names and addresses, declined the opportunity to know us better. Or so we thought. We then had to choose between climbing the fence, potentially a felony, or, after far too many speeches, dispersing. We dispersed. In Frankfort, the governor, somewhat delightfully, outsmarted us. Instead of calling the police, he invited us to camp in his waiting room, which we did, from Friday until Monday morning.

And so my career in civil disobedience, so far, has been an exercise in anticlimax. Also it has been, by any practical reckoning, pretty useless. Owing probably not much, if anything, to our civil disobedience, the power plant at Marble Hill finally was stopped. But nothing that my side has done has come anywhere near to stopping mountaintop removal.

At a time when virtuous behavior tends to be measured in degrees of misery, I had better confess that all three of these episodes were mostly pleasant. The police and other officials in Indiana were nice to us, and we were nice to them and to one another. The march in Washington, in spite of cold weather, was a social success, better by far than any cocktail party I ever attended. And our weekend in the Governor's office was, I think, for all of us, an extraordinarily happy time, even a joyful time. We were warned only that if we left the building we could not return; we were to that extent confined. We stayed put, we worked hard at getting our message out to the media, we told stories, we laughed a lot, we ate the good food sent in by allies, and slept well on bedding likewise sent in. The security people, the office people, and the police were kind to us, and we reciprocated. I am proud to say that we were model guests. We damaged nothing, and we cleaned up after ourselves.

It may seem odd to speak of pleasure as a result of trouble, but there is nothing wrong with decent pleasure, however it comes. It is a gift, and we should be grateful. The pleasures I have mentioned certainly do not reduce the seriousness of civil disobedience. I am sure that all who have undertaken it have felt intensely and complexly the seriousness of it. There are a number of considerations that come in a hurry and are inescapable. I will list them, not in the order of their importance, but as I have thought of them in my own efforts to decide.

Civil disobedience will likely be considered, first of all, as an inconvenience. It will, and not for a predictable length of time, interrupt one's life and one's work. I have always been suspicious of people who seem to devote their entire lives to forms of protest. We all ought to have better things to do. Ken Kesey once said that the reason not to resist evil is that such resistance is dependent on evil; it makes *you* dependent on evil. He was right. And Edward Abbey said that saving the world is a good hobby—though he worked hard to save at least parts of it. As for me, the older I get, the less happy I am to leave home. All the places I go seem to be getting farther away. Frankfort, Kentucky, now appears as far off as the planet Saturn, and I wish it more remote. Reluctance, then, may be a dependable enforcer of thoughtfulness. Protest becomes properly a part of a citizen's life and work after political and legal processes have failed, and other recourse is exhausted. Civil disobedience is properly the last resort.

It is also an unhappiness of citizenship. By it, you make yourself, publicly, an exception. It involves a kind of loneliness. I, at least, have felt no pleasure in opposing constitutional authority, however corrupt and irresponsible I have found it to be.

Civil disobedience is also plenty scary. At least to me it is. I have never felt one bit brave even in thinking about it. It involves a strange and risky paradox: You and your friends will be exploiting your obvious powerlessness to recover to your cause, and to your own citizenship, a just measure of power. But your acknowledged condition is powerlessness. Your commitment to nonviolence makes you vulnerable to violence. You can get hurt, or worse. It is fearful also to make yourself available to be treated with contempt. And you are, in effect, volunteering to go to jail.

During the Washington protest, some genius with a microphone asked me, "Do you want to go to jail?"

I said, "*Hell no!*"

There is a world of difference between wanting to go to jail and being willing to go. I thank God for every minute I am not in jail.

A final consideration, and the one you may ponder longest, is the possibility that your act of civil disobedience may be useless. In Kentucky, where the state government is owned outright by the coal industry and the "flagship university" is the coal industry's trophy wife, opposition to mountaintop removal faces the unhappiest of odds.

This means, I believe, that it is a mistake to make your opposition conditional upon winning. If you do that, you won't last. I am in this struggle with the firm intention of winning, but I don't forget that I first wrote against strip mining in 1965. If I had required even a reasonable expectation of victory, I would have given up long ago. It is unlikely that anybody now opposing mountaintop removal thinks that victory is in sight. But the opposition is now astonishingly well-populated in comparison to what it was forty years ago. Why do so many people go to so much trouble, learning the hard things they need to know, organizing, arguing with politicians, making speeches, marching, risking arrest, getting arrested, keeping on, making the same try again and again?

I think they do it, above all else, to keep alive the possibility of decency, and to refuse to accept as normal the indecency of public officials.

Forty-some years ago, maybe in 1966, I attended in Frankfort a hearing on coal company abuses. In those days some of the companies were getting the coal out by opening contour gashes along the sides of the mountains, pushing the "overburden" regardlessly downslope. They were doing this, at that time, under the infamous "broad form deeds," which, as construed by the courts, gave all rights to owners of the coal and none to the surface owners whose properties were being damaged or destroyed.

Among the audience at this hearing was a group of maybe fifteen people whose homes and lands were damaged or threatened by "contour stripping." There was, on that occasion, no "demonstrating," but those people were there nonetheless in protest. Sitting with them was a man in appearance somewhat different from them. He was wearing a well-fitting summer suit and a tie. He held a rather elegant straw hat on his lap.

I became curious about him, and at one of the recesses of the hearing, I approached him. We introduced ourselves. He was Courtney Wells, he said, of the town of Hazard.

I asked if he owned land that was affected by strip mining.

He said no, he had no land at risk. He was a lawyer.

"Oh," I said. "You're here to represent these people."

Again he said no. He represented only himself.

I said, "Well, then, why are you here?"

And he replied: "I want to be on the side of the right."

I have never forgotten him, for he gave me the one reason that will always be enough.

FROM
OUR ONLY WORLD
(2015)

Paragraphs from a Notebook

WE NEED TO acknowledge the formlessness inherent in the analytic science that divides creatures into organs, cells, and ever smaller parts or particles according to its technological capacities.

I recognize the possibility and existence of this knowledge, even its usefulness, but I also recognize the narrowness of its usefulness and the damage it does. I can see that in a sense it is true, but also that its truth is small and far from complete.

In and by all my thoughts and acts, I am opposed to any claim that such knowledge is adequate to the sustenance of human life or the health of the ecosphere.

Do even the professionals and experts believe in it, in the sense of acting on it in their daily lives? I doubt that they do.

To this science, the body is an assembly of parts provisionally joined, a "basket case" sure enough. A mountain is a heap of "resources" unfortunately mixed with substances that are not marketable.

There is an always significant difference between knowing and believing. We may know that the earth turns, but we believe, as we say, that the sun rises. We know by evidence, or by trust in people who have examined the evidence in a way that we trust is trustworthy. We may sometimes be persuaded to believe by reason, but within the welter of our experience reason is limited and weak. We believe always by coming, in some sense, to see. We believe in what is apparent, in what we can imagine or "picture" in our minds, in what we feel to be true, in what our hearts tell us, in experience, in stories—above all, perhaps, in stories.

We can, to be sure, see parts and so believe in them. But there has always been a higher seeing that informs us that parts, in themselves, are of no worth. Genesis is right: "It is not good that the man should be alone." The phrase "be alone" is a contradiction in terms. A brain alone is a dead brain. A man alone is a dead man.

We are thus as likely to be wrong in what we know as in what we believe.

We may know, or think we know, and often say, that humans are "only" animals, but we teach our children specifically human virtues—evidently because we believe that they are not "only" animals.

Another question of knowledge and belief that keeps returning to my mind is this: Are there not some things that cannot be known apart from belief? This question refers not just to matters of religion—as in Job 19:25: "I know that my redeemer liveth, and that he shall stand at the latter day upon the earth . . ."—but also to ordinary motives of family and community life such as love, compassion, and forgiveness. Do people who believe that such motives are genetically determined have the same knowledge as people who believe that they are the results of choice, culture, cultivation, and discipline? Or: Do people who believe in the sanctity or intrinsic worth of the world and its creatures have the same knowledge as people who recognize only market value? If there is no way to measure or prove such differences of knowledge, that does at least prove one of my points: There is more to us than some of us suppose.

We may know the anatomy of the body to the extent of the anatomy of atoms, and yet we love and instruct our children as whole persons. And we accept an obligation to help them to preserve their wholeness, which is to say their health. This is not an obligation that we can safely transfer to the subdivided and anatomizing medical industry, not even for the sake of cures. Cures, to industrial medicine, are marketable products extractable from bodies. To cure in this sense is not to heal. To heal is to make whole, not so ideologically definable or so technologically possible or so handily billable.

*

This applies as well to the industries of landscapes: agriculture, forestry, and mining. Once they have been industrialized, these enterprises no longer recognize landscapes as *wholes*, let alone as the homes of people and other creatures. They regard landscapes as sources of extractable products. They have "efficiently" shed any other interest or concern.

We have come to this by way of the disembodiment of thought —a mentalization, almost a puritanization, of thought— depriving us of the physical basis of a sympathy that might join us kindly to landscapes and their creatures, including their human creatures. This purity or sublimity of thought is hard to account for, for it has come about under the sponsorship of materialism. Perhaps it happened because materialists, instead of assigning ultimate value to materiality as would have been reasonable, have abstracted "material" to "mechanical," and thus have removed from it all bodily or creaturely attributes. Or perhaps the abstracting impulse branched in either of two directions: one toward the mechanical, the other toward the financial, which is to say toward the so-called economy of money as opposed to the actual economy (oikonomia or "house-keeping") of goods. Either way the result is the same: the scientific-industrial culture, founded nominally upon materialism, arrives at a sort of fundamentalist disdain for material reality. The living world is then treated as dead matter, the worth of which is determined exclusively by the market.

This highly credentialed, highly politicized disdain, now allied with the similar disdain of highly spiritualized religions, is limitlessly destructive. We cannot say that its destructiveness has been unnoticed as it has been happening, or that the dissolutions, and the dissoluteness, of mechanical thought have not been, by some, well understood. The poet William Butler Yeats prayed: "God guard me from the thoughts men think / In the mind alone . . ." ("A Prayer for Old Age"). He wrote in 1916: "We only believe in those thoughts which have been conceived not in the brain but in the whole body" (Introduction to *Certain Noble Plays of Japan*, by Ezra Pound and Ernest Fenollosa). In the same essay he spoke with foreboding of "a mechanical sequence of ideas."

*

As another example, more explicit, here is the critic and translator Philip Sherrard on the Greek poet Anghelos Sikelianos: "He saw [the western world of his time] as increasingly alienated from those principles which give life significance and beauty and as approaching the condition of a machine out of control and hastening towards destruction. . . . The organic sense of life was being shattered into countless unconnected fragments. . . . A system of learning which made extreme demands on the purely mechanical and sterile processes of memory had the effect of absorbing all the spontaneous movements of body and soul of the younger generations."

Or here is a passage, by the poet and critic John Crowe Ransom, pointed more directly at our specialist system, which he identified as a phase of the Puritanism that began in religion: "You may dissociate the elements of experience and exploit them separately. But then at the best you go on a schedule of small experiences, taking them in turn, and trusting that when the rotation is complete you will have missed nothing. And at the worst you will become so absorbed in some one small experience that you will forget to go on and complete the schedule; in that case you will have missed something. The theory that excellence lies in the perfection of the single functions, and that society should demand that its members be hard specialists, assumes that there is no particular harm in missing something."

A proper attention to our language, moreover, informs us that the Greek root of "anatomy" means "dissection," and that of "analysis" means "to undo." The two words have essentially the same meaning. Neither suggests a respect for formal integrity. I suppose that the nearest antonym to both is a word we borrow directly from Greek: *poiesis*, making or creation, which suggests that the work of the poet, the composer or maker, is the necessary opposite to that of the analyst and the anatomist. Some scientists, I think, are in this sense poets.

But we appear to be deficient in learning or teaching a competent concern for the way that parts are joined. We certainly are not learning or teaching adequately the arts of forming parts into wholes, or the arts of preserving the formal integrity of the things we receive as wholes already formed.

Without this concern and these arts, our efforts of conservation are probably futile. Without some sense of necessary connections and a competent awareness of human and natural limits, the issues of scale and form are not only pointless, but cannot even enter our consciousness.

My premise is that there is a scale of work at which our minds are as effective and even as harmless as they ought to be, at which we can be fully responsible for consequences and there are no catastrophic surprises. But such a possibility does not excite us.

What excites us is some sort of technological revolution: the fossil fuel revolution, the automotive revolution, the assembly line revolution, the antibiotic revolution, the sexual revolution, the computer revolution, the "green revolution," the genomic revolution, and so on. But these revolutions—all with something to sell that people or their government "must" buy—are mere episodes of the one truly revolutionary revolution perhaps in the history of the human race: the Industrial Revolution, which has proceeded from the beginning with only two purposes: to replace human workers with machines, and to market its products, regardless of their usefulness or their effects, at the highest possible profit—and so to concentrate wealth into ever fewer hands.

This revolution has, so far, fulfilled its purposes, with remarkably few checks or thwarts. I say "so far" because its great weakness obviously is its dependence on what it calls "natural resources," which it has used ignorantly and foolishly, and which it has progressively destroyed. Its weakness, in short, is that its days are numbered. Having squandered nature's "resources," it will finally yield to nature's correction, which in prospect grows ever harsher.

We have formed our present life, including our economic and intellectual life, upon specialization, professionalism, and competition. Certified smart people expect to solve all problems by analysis, dividing wholes into ever smaller parts. Science and industry do give room to synthesis, but by that they do not mean putting back together the things once together that they have taken apart; they mean making something "synthetic." They mean engineering the disassembled parts, by some manner of violence, into profitable new commodities. In such a state of things we don't see or, apparently, suspect the complexity of connections among ecology, agriculture, food, health, and medicine (if by "medicine" we mean healing). Nor can we see how this complexity is necessarily contained within, and at the mercy of, human culture, which in turn is necessarily contained within the not very expandable limits of human knowledge and human intelligence.

We have accumulated a massive collection of "information" to which we may have "access." But this information, by being accessible, does not become knowledge. We might find, if such a computation were possible, that the amount of human knowledge over many millennia has remained more or less constant —that is it has always filled the available mental capacity—and therefore that learning invariably involves forgetting. To have the Renaissance, we had to forget the Middle Ages. To the extent that we have learned about machines, we have forgotten about plants and animals. Every nail we drive in, as I believe C. S. Lewis said, drives another out.

The thing most overlooked by scientists, and by the enviers and emulators of science in the humanities, is the complicity of science in the Industrial Revolution, which science has served not by supplying the "scientific" checks of skepticism, doubt, criticism, and correction, but by developing of marketable products, from refined fuels to nuclear bombs to computers to poisons to pills.

It has been remarkable how often science has hired out to the ready-made markets of depravity, as when it has served the military-industrial complex, which is solidly founded upon the hopeless logic of revenge, or the medical and pharmaceutical industries, which are based somewhat on the relief of suffering but also on greed, on the vicious circles of hypochondria, and on the inducible fear of suffering yet to come. The commodification of genome-reading rides upon the same fears of the future that palmistry and phrenology rode upon.

We may say with some confidence that the most apparently beneficent products of science and industry should be held in suspicion if they are costly to consumers or bring power to governments or profits to corporations.

There are, we know, scientists who are properly scrupulous, responsible, and critical, who call attention to the dangers of oversold and under-tested products, and who are almost customarily ignored. They are often called "independent scientists," and the adjective is significant, for it implies not only certain moral virtues but also political weakness. The combination of expertise, prestige, wealth, and power, incapable of self-doubt or self-criticism, is hardly to be deterred by a few "independent scientists."

Scientists in general, like humanists and artists in general, have accepted the industrialists' habit, or principle, of ignoring the contexts of life, of place, of community, and even of economy.

The capitalization of fear, weakness, ignorance, bloodthirst, and disease is certainly financial, but it is not, properly speaking, economic.

Criticism of scientific-industrial "progress" need not be balked by the question of how we would like to do without anesthetics or immunizations or antibiotics. Of course there have been benefits. Of course there have been advantages—at least to the advantaged. But valid criticism does not deal in categorical approvals and condemnations. Valid criticism attempts a just description of our condition. It weighs advantages against disadvantages, gains against losses, using standards more general and reliable than corporate profit or economic growth. If criticism involves computation, then it aims at a full accounting and an honest net result, whether a net gain or a net loss. If we are to hope to live sensibly, correcting mistakes that need correcting, we need a valid general criticism.

*

Scared for health, afraid of death, bored, dissatisfied, vengeful, greedy, ignorant, and gullible—these are the qualities of the ideal consumer. Can we imagine a way of education that would turn passive consumers into active and informed critics, capable of using their own minds in their own defense? It will not be the purely technical education-for-employment now advocated by the most influential "educators" and "leaders."

We have good technical or specialized criticism: A given thing is either a good specimen of its kind or it is not. A valid *general* criticism would measure work against its context. The health of the context—the body, the community, the ecosystem—would reveal the health of the work.

The Commerce of Violence

ON THE DAY of the bombing in Boston, *The New York Times* printed an op-ed piece by a human being who had been imprisoned at Guantanamo for more than eleven years, uncharged and of course untried. The occurrence of these two events on the same day was a coincidence, but that does not mean that they are unrelated.

What connects them is our devaluation, and when convenient our disvaluation, of human life as well as the earthly life of which human life is a dependent part. This cheapening of life, and the violence that inevitably accompanies it, is surely the dominant theme of our time. The ease and quickness with which we resort to violence would be astounding if it were not conventional.

In the Appalachian coal fields we mine coal by destroying a mountain, its forest, its waterways, and its human community without counting the destruction as a cost. Our military technicians, our representatives, sit in armchairs and kill our enemies, and our enemies' children, by remote control. In the Guantanamo prison, guards force their fasting prisoners to live, and they do so as routinely as in other circumstances they would kill them.

And the Boston bombing? Like most people, I was not there and I don't know anybody who was, but I was grieved and frightened by the news. This fearful grief has grown familiar to me since I first felt it at the start of World War II, but at each of its returns it is worse. Each new resort to violence enlarges the argument against our species, and the task of hope becomes harder.

I am absolutely in sympathy with those who suffered the bombing in Boston and with their loved ones. They have been singled out by a violence that was general in its intent, not aimed particularly at anybody. The oddity, the mystery, of a particular hurt from a general violence—the necessity to ask, "Why me? Why my loved ones?"—must compound the suffering. What I am less and less in sympathy with is the rhetoric and the tone of official indignation. Public officials cry out for justice against the perpetrators. I too wish them caught and punished. But I am unwilling to have my wish spoken for me in a tone of surprise and outraged innocence. The event in Boston is not unique or rare or surprising or in any way new. It is only another transaction in the commerce of violence: the unending, the not foreseeably endable, exchange of an eye for an eye, with customary justifications on every side, in which we fully participate; and beyond that, it is our willingness to destroy anything, any place, or anybody standing between us and whatever we are "manifestly destined" to have.

We congratulate ourselves perpetually upon our Civil War by which the slaves were, in a manner of speaking, "freed." We forget, if we have ever learned, that the same army that "freed the slaves" established for us the "right" of military violence against a civilian population, and then acted upon that "right" by a war of extermination against the native people of the West. Nobody who knows our history, from the "Indian wars" to our contemporary foreign wars of "homeland defense," should find anything unusual in the massacre of civilians and their children.

It is not possible for us to reduce the value of life, including human life, to nothing *only* to suit our own convenience or our own perceived need. By making this reduction for ourselves, we make it for everybody and anybody, even for our enemies, even for the maniacs whose enemies are schoolchildren or spectators at a marathon.

We forget also that violence is so securely founded among us—in war, in forms of land use, in various methods of economic "growth" and "development"—because it is immensely profitable. People do not become wealthy by treating one another or the world kindly and with respect. Do we not need to remember this? Do we have a single eminent leader who would dare to remind us?

On the second day after the catastrophe in Boston, Thomas L. Friedman announced in the *Times* that "the right reaction is: Wash the sidewalk, wipe away the blood, and let whoever did it know that . . . they have left no trace on our society or way of life." We should, said Mr. Friedman, "let there be no reminder whatsoever." And he asserted, with a shocking indifference to evidence and his own language, that "the benefits

—living in an open society—always outweigh the costs." He is speaking to (among others) people whose loved ones have been killed and people who will never again stand on their own legs. How can he think that all the traces of any violence can be easily wiped away? How would he wipe away the traces of a bombed village or a strip mine or a gullied field or a wrecked forest? The dead in Boston no longer live in an open society. How have *they* benefitted?

Mr. Friedman, like other journalists, asks us, as he wrote, to "notice how many people came *running toward the blast* within seconds to help." And that is very well. To know that people would run to help, perhaps at the risk of their lives, is consoling and reassuring. But we have got to acknowledge that the help that comes after the violence has been done, though it undeniably helps, is not a solution to violence.

The solution, many times more complex and difficult, would be to go beyond our ideas, obviously insane, of war as the way to peace and of permanent damage to the ecosphere as the way to wealth. Actually to help our suffering of one man-made horror after another, we would have to revise radically our understanding of economic life, of community life, of work, and of pleasure. We employ thousands of scientists and spend billions of dollars to reduce matter to its smallest particles and to search for farther stars. How many scientists and how many dollars are devoted to harmony between economy and ecology, or to amity and lenity in the face of hatred and killing?

To learn to meet our needs without continuous violence against one another and our only world would require an immense intellectual and practical effort, requiring the help of every human being perhaps to the end of human time.

This would be work worthy of the name "human." It would be fascinating and lovely.

A Forest Conversation

AN ECOSYSTEM, the web of relationships by which a place and its creatures sustain a mutual life, ultimately is mysterious, like life itself. We can know enough, and probably only enough, to tell us how little we know and to make us careful. At present, too ignorant to know how ignorant we are, we believe that we are free to impose our will upon the land with the utmost power and speed to gain the largest profit in the shortest time, and we believe that there are no penalties for this.

Industrial farming and forestry use extreme methodologies that barter the long-term health and fertility, which is to say the long-term productivity, of local ecosystems for a short-term monetary gain. Like mining, they tend, though not so suddenly, toward the same totality of ruin. The issue of land *use* is not on the agenda of most conservation organizations, which have been primarily concerned throughout their history with the preservation of wilderness and wildlife habitat, even though most land is being used, and used badly, and though no wilderness or wildlife can survive the prolonged abuse of the economic landscapes. Governments typically are preoccupied with the politics and commerce related to land use rather than the ever more pressing issues of land use itself: soil erosion, toxicity, ecological degradation, the destruction of rural communities, and the substitution of the global consumer culture for local cultures of husbandry. The considerable force of the colleges of agriculture has been applied mostly to promotion of industrial technologies and the economics of agribusiness, ignoring the great teachers and advocates of good land use: Liberty Hyde Bailey, J. Russell Smith, Hugh Hammond Bennett, Albert Howard, Aldo Leopold, and others. So utterly dominant has industrial agriculture become that the officially recommended soil conservation technology, "no-till agriculture," undertakes to "save" the soil by poisoning it. By this method, the noncommercial plant cover is killed with an herbicide, and then the "undisturbed" soil is planted in corn or soybeans genetically engineered for "herbicide resistance." Among the results of continuous no-till cropping on sloping land is severe soil erosion. Everything human and natural is sacrificed for the sake of annual production.

On the mostly wooded valley sides where I live, near the lower end of the Kentucky River, the same bad trade is under way. The woodlands here, which for the last three quarters of a century or so have enjoyed the flimsy beneficence of neglect, are now often as heartlessly cropped as the fields.

The story of a bad job of logging is easy to imagine. "We need money," a forest-owning family decides. "We'll sell our trees." "The environment" being in fashion, the family takes comfort in the thought that after the marketable trees are gone, many smaller trees will remain as the "next crop." And so they sell their "standing timber" to a logging company, whose representative comes in and marks every tree that can be sold.

And then, all too often, a sawyer and a driver of a mechanical skidder, employees of the logging company, arrive with, inevitably, the single purpose of cutting and removing the marked trees as quickly, which is to say as cheaply, as possible. The logging company is in every sense an absentee, and of all the short-term economies in forestry, the absentee logging company's is the shortest. Any concern for the "next crop" cherished by the owners, who also have in effect made themselves absentees, has become forceless. Standing trees, if they are unmarked, are now regarded merely as obstacles. There is little concern for directional felling. Many of the smaller trees are broken off or permanently bent or in other ways damaged by the marked trees as they fall. As the tree-length logs are dragged out of the woods, sometimes straight up or down a steep slope, the trees of the "next crop" are damaged by the skidder or by the dragged logs.

The woods is left a shambles, for nobody thought of the forest rather than the trees. All along the way, the economic interest was shifted from the forest to the marketable "standing timber," though the *source* of the timber is not the trees but the forest. Finally there is no significant difference between forest ecology and a long-term forest economy. To forget this profound kinship is to abandon the forest to bad work.

The example I have given is a worst case scenario, but my point is that when land abuse is normal and not a public issue, and when local land economies have disappeared into a global financial system, the worst case can happen easily. I asked William Martin, a forest ecologist living in Lexington, "What percentage of Kentucky's forest is sustainably managed?" He replied, "I would say less than ten." Jim Finley, a forester at Penn State, would apply the same fraction to the forest of Pennsylvania, but he suggests that "healthy" is a better term than "sustainably managed."

The damage extends to the human community when, as often happens, the cut timber is transported out of the neighborhood and even out of the state without so much as a stop at a sawmill, thus yielding the least possible local benefit from a local resource.

The U.S. Department of Agriculture and the land grant colleges of agriculture seem to regard forestry as a kind, or division, of farming. This risks confusion and requires thought. For example, the use of "harvesting" as a synonym for "logging" seems at first to be merely euphemistic; people of delicacy wish to have their woodlands "harvested" as they wish to have their meat "processed." But "harvest" (Middle English *hervest*, meaning "autumn") was associated originally with annual grain crops that ripen in the fall. When we harvest a crop of corn, we take all that is of economic worth—which, with some special exceptions, is a bad way to treat a forest.

Like the grasses and forbs of farm pastures and hayfields, trees are perennials. But *unlike* the grasses and forbs, which mature and can be cut once, or more than once, every year, trees keep growing bigger year after year, sometimes for hundreds of years. Forests also are more complex in structure and diversity of species. You can't learn to manage a forest by managing a pasture.

Farming, moreover, must always be to some extent a compromise with the local ecosystem. Whereas the farmer requires from the farm, specifically for human use, several plants and animals that the ecosystem in its natural state would not produce, the best foresters ask the local forest ecosystem only to continue to produce what, according to its nature, it produces best: primarily trees of the native species. A properly diversified forest economy produces more than trees, of course. And a healthy forest contributes to the health of the soil, water, air, and all other constituents of the natural world. The forest submits to compromise only by allowing the foresters to take timber and other forest products for human use, if they can do so without destroying the integrity and the continuous productivity of the ecosystem.

Forest owners, however, may have a good deal to learn from livestock farmers—little as such farmers may know about forestry. No sane farmer would sell all her brood cows, keep their heifer calves, and wait for another calf crop until the heifers have become old enough to breed and calve. But forest owners do substantially that when they sell off every marketable tree—except that the forest owners (or their descendants) may have to wait for generations, not years, for another marketable "crop."

Nor would a sane cattle farmer "highgrade" his herd by selling his best cows and keeping and breeding the worst. But people who highgrade their woodlands do exactly the same thing, selling the best and keeping the worst—which, if not thoughtless, as it usually is, would be insane.

The best logging, however, we may rightly call culling. The foresters' name for it is "worst-first single tree selection." By this method, the trees in the area to be logged are looked at individually, evaluated according to standards of worth and health, and the worst are carefully removed, leaving only small openings in the canopy, and doing the least possible damage to the trees, young and old, that remain, and to the forest floor. The "worst" are trees that are diseased or dying, leaning trees that are more likely to fall as they grow more top-heavy, trees that branch too low or are otherwise inferior in conformation.

But I need to interrupt myself here to notice that my need to be clear has led me into oversimplification. A stand consisting only of the "best" trees would be a tree farm, not a forest. A healthy forest would necessarily include some less valuable or unmarketable species and some diseased and dying trees, so as to sustain the finally unimaginable diversity of creatures that belong to the natural forest. Good foresters know this and are charitable toward members of the forest community that by the standards merely of economics would be considered worthless.

The essential point is that by applying the standard of ecological health, which is the standard of long-term economy, the good forester will leave standing numerous marketable trees because they are healthy and flourishing, and by doing so will maintain the forest at its highest productivity. The explanation for this comes from elementary geometry: A quarter inch of annual growth on a tree two feet in diameter is far more than a quarter inch of annual growth on a tree six inches in diameter. A flourishing large tree makes more marketable wood per year than many small ones. A woodland logged in this way can be logged fairly frequently, at intervals of ten or fifteen or twenty years. And at each successive logging the quality and marketability of the trees that are cut will have increased.

Forestry of this kind, though it is still far too rare, is not new. That it is more than four hundred years old we know from Aldo Leopold's essay, "The Last Stand," in *The River of the Mother of God*:

> I know a hardwood forest called the Spessart, covering a mountain on the north flank of the Alps. Half of it has sustained cuttings since 1605, but was never slashed. The other half was slashed during the 1600s, but has been under intensive forestry during the last 150 years. Despite this rigid protection, the old slashing now produces only mediocre pine, while the unslashed portion grows the finest cabinet oak in the world; one of those oaks fetches a higher price than a whole acre of the old slashings. On the old slashings the litter accumulates without rotting, stumps and limbs disappear slowly, natural reproduction is slow. On the unslashed portion litter disappears as it falls, stumps and limbs rot at once, natural reproduction is automatic. Foresters attribute the inferior performance of the old slashing to its depleted microflora, meaning that underground community of bacteria, molds, fungi, insects, and burrowing mammals which constitute half the environment of a tree.

By "slashed" Leopold evidently meant something like what we now mean by "clear-cut." "Slashing" destroyed the integrity of the forest ecosystem.

The literature on what we are now calling "sustainable agriculture," laying out the ecological principles of good land use, is fairly substantial, but not nearly so much has been written on sustainable forestry. To learn about that, it is necessary to become familiar with one or more of the few exemplary forests where good practice is well established and for which there is some published accounting, or one may study the ways and the work of exemplary loggers and forest managers. Bad logging, like bad farming, comes from estrangement between the industrial economy and nature's ecosystems. Most land users, as they come more and more to adopt industrial technologies and attitudes, know nothing or think nothing of ecology. And few ecologists or people oriented to ecology are involved with, or familiar with or even interested in, the economies of land use. When you find a farmer or a forester who has united the inescapable economic concern with an equally compelling interest in ecology, that is when you had better stop and take notice.

I first met Jason Rutledge a good many years ago when both of us were visiting Washington and Lee University in Lexington, Virginia, which is in Jason's home country. He was talking then, and he is talking still, on the closely related themes of sustainable forestry and logging with horses. Since that first encounter, our paths have crossed fairly often in the woods and at various field days and conferences. "Worst-first single tree selection" is language I first heard from Jason, a logger who uses horses (*Suffolk* horses, he would want me to say) to skid the logs out of the woods. He is also a teacher whose alumni are scattered about in the woods, horse logging on their own and earning a living by doing so.

It was Jason Rutledge who introduced me to Troy Firth. Like Jason, Troy is an advocate for sustainable forestry and is worth listening to because his advocacy, like Jason's, rests upon practice. Troy is a forest owner whose holdings are extensive. He is a forester and logger who operates his own sawmill. He is also owner and manager of Firth Maple Products, the chief product of which is maple syrup. This year Troy and his crew tapped 17,000 sugar maple trees and made 6,100 gallons of syrup. Next year they plan to tap 22,000 trees.

Troy has lived and worked all his life in the neighborhood near Spartansburg, Pennsylvania, long occupied by his family; his house is on Firth Road. Troy's involvement in the woodland economy of his region is elaborate and of long standing. He is not a waster of words, and I think this is because he is confident both of his knowledge and of the limits of his knowledge. He has a reliable sense of humor and is generous with his knowledge, but he generally keeps quiet until he has something worthwhile to say. The result is that when he talks you listen, and you are apt to be surprised at how much you remember of what he has said.

From his student days, Troy has been inclined toward the woods. While he was in college he took two years off to work as a logger. He accumulated some money and in 1972 bought woodland, which at that time was cheap. He counted his purchase as an investment, but it was more than that. His interest in forestry had become a passion, as it still is.

In college he majored in history, not forestry, which he thinks was a good thing, for a person's education should be broad rather than "vocational." As he went on with his work in the woods, he realized that the conventional (and recommended) brand of forestry was wrong. It went against nature and was not sustainable. I asked him how long it took him to see this. He said, "Not long. It was obvious."

He had, he says, "no role model or mentor." Nor was he acquainted with examples to show him what to try for. This part of his education, the postgraduate part, was up to him. He simply rejected the conventional practices and tolerances that over time reduced the productivity of the forest, and from there he was on his own. The change of mind from forestry as an extractive industry to the study of sustainability must always involve a radical shift in perspective. One ceases to think of the source of sawlogs as trees, which can be cut according to wishes or needs or standards that are merely economic, and begins the understanding—far more complex and difficult, but also far more interesting—that the source is the forest ecosystem.

In 1985 Troy bought more woodland, and he has continued to do so at the rate of "a property or two a year." He and his wife, Lynn, got married in 1988. The wedding took place on a beautiful day, out in the snowy woods, in January. Such a wedding certainly suggests a strong liking for the woods, on the part of both bride and groom. And I would say that it suggests, more than most weddings, a strong mutual interest in getting married. Lynn and Troy have a daughter who, like Troy, is an only child.

Marriage and parenthood, for responsible people, lead to considerations of futurity and mortality, duties and limits. Troy and Lynn now had their daughter's future to think of, along with the future of their woodlands. This led them to the deliberations and arrangements of "estate planning." Like in fact a good many people, Troy thinks there ought to be a limit to a parent's bequest to a child. Too much makes things too easy. Eventually he asked his mother to change her will, removing him as her heir, and leaving her entire estate to his daughter. His mother's estate, he says, amounted to "something, not too much."

Providing for the future of the woodlands was a more difficult problem, for it involved more time, more unknowns, and evidently there were no precedents. Good management for the length of a human life is certainly a gift to the forest, but it is pathetically brief. A single tree may live several times as long as a human. A forest, kindly used, may outlive unimaginably any of its trees. Troy doesn't know whether he or his lawyer thought of it first, but the idea for the Firth Family Foundation emerged in 2004 from the need to provide the Firths' forest properties an inheritor that might keep them intact and well managed beyond the limits of a human lifetime. This is the same impulse that has led to the formation of land trusts to hold development rights or "conservation easements" in order to preserve farmland or open space beyond the lifetimes of present owners. But Troy's vision for his foundation is more complex, and it addresses more problems.

One of the most daunting problems in forestry, especially in forested lands that are privately owned, is fragmentation. Small holdings in farmland make sense, or can be made to make sense. But small holdings in a forested landscape—and the average privately owned forest acreage in Kentucky is just under 20 acres; in Pennsylvania it is 16.7—are too small both economically and ecologically. A small woodlot on a farm can contribute significantly to the economy and the life of the farm, but a scattering of small forest properties, even if well used, cannot sustain a forest ecosystem.

Another problem is short-term management of large tracts of forest by financiers or investors who, for an acceptable return in fifteen years, may highgrade their properties, or cut everything, and then subdivide the land for "second homes." The problem, predictably, is that "short term" land management leads almost inevitably to long-term exhaustion.

As the land resource declines, in forestry as in farming, so necessarily does the local land economy and the local human community. As the human community declines, there are always fewer actual or potential good husbanders of the land. As this cycle turns, whatever the nation and the national economy may be doing, the country and its native health and wealth spiral downward.

Troy understood that the good health and productivity of his own woodlands depends on good use sustained over a long time. But good health cannot be bounded and isolated like a property. The good health of part of a forest depends on the good health of the whole forest. It would not be enough for the foundation to serve merely as heir and steward of the Firths' personal holdings; the foundation could serve as an heir also to other forest owners, who might bequeath or donate their properties to it. Beyond that, the need is to exemplify and promote sustainable forestry for the sake of the forest ecosystem as a whole. The goal of the Firth Family Foundation was "to promote and practice long-term sustainable forestry while conserving working timberlands throughout Pennsylvania, New York and the surrounding states."

By 2010 the Firth Family Foundation had transformed itself into the Foundation for Sustainable Forests: A Land Trust. The name had become more general, but the regional boundary had been retained. Troy hopes that eventually there might be "numerous regional associations that could work cooperatively." The goal "is an abundance of intact forest ecosystems that provide for the widest range of native biodiversity possible, sustainable forest products, the economic viability of rural communities, and recreational opportunities."

What Troy had forming in his mind—what apparently is still forming in his mind—is the idea of a sustainable forest economy. Like any sustainable land economy, it would have to form locally, taking shape in response to local conditions, limits, and opportunities. The influence of its example and working principles would then extend to anybody who might be interested.

Troy's ultimate wish is "to do whatever can be done to protect land." An example of sustainable forestry adapted to one place cannot be applied like a stamp to a different place. What *can* be applied are right principles and standards, and some understanding of what is involved in the effort of adaptation.

But a forest economy, however well adapted, however concordant with its ecosystem, cannot stand or continue alone. For a local economy to become truly sustainable, it must function as a belonging, a support and an artifact, of a local culture. The great need we are talking about is to hold the local forest ecosystem together, but that leads to a second need, just as great, which is to hold the local forest economy and therefore the local human community together. The two needs could be answered only by a thriving, confident, stable local culture in which the young would learn from their elders. Change and innovation would naturally occur, but would not be imposed from the outside or for the benefit primarily of outsiders. Nor would the changes and innovations arrive at the breakneck speed of industrial innovation. The community would understand that it exists, in part, to cherish the forest and to preserve its knowledge of the forest.

There is a good deal of supposing in that paragraph. In America, or in the history of Old World races in America, we have no example of an economic enterprise ecologically and socially sustainable. To imagine such a thing, we have to consult scattered pieces. Any conversation at home between grandparents and grandchildren is potentially the beginning of a local culture, even of a sustaining local culture, however it might be cut short and wasted.

Troy's practice of forestry, even when he was working mainly alone, was already one of the scattered pieces of a sustainable economy, and Troy knew that. But as he also knew, it was not nearly enough. He needed to be talking about his work with somebody, a younger somebody, at work with him. And so in 2005 he looked about in the Spartansburg area and hired Guy Dunkle, a bright young man just out of college, who, like Troy, was locally born and raised. I had gathered from my conversations with Troy that he had hired Guy as an apprentice or understudy, supplying the younger half of the necessary conversation. But when I asked the name of Guy's "position," hoping for better understanding, Troy said, not to my surprise, "We're not much on titles." He spoke of Guy simply as "a forester for Firth Maple Products." In the most practical terms, Guy is a younger forester coming behind an older forester and learning from him. "Everything depends on that," Troy said.

It is more than passingly significant, I think, that this complex enterprise of forestry, sawmilling, "maple products," and an emergent foundation is "not much on titles," and that its executive staff consists of two "foresters" involved in the fate of the same woodlands. It is significant also that Guy's college degree is in environmental science, not forestry, and that his job interview was conducted in the course of a week's work in the woods with Troy, tapping maple trees.

Guy Dunkle enjoys describing his job as whatever Troy doesn't much want to do. And so it was Guy who drove to Erie to meet my plane on May 16, 2012, when I came to walk and talk in the woods with the two of them and to take part in a couple of public events promoting the foundation. I had met Guy in October 2005 at one of Jason Rutledge's logging sites in Virginia, but our drive from the Erie airport to supper with Troy and Lynn Firth at their house was our first opportunity actually to become acquainted.

Guy is a tall, lean, pleasant man, impressive in several ways, not least for his apparent innocence of the too-common wish to "make an impression." He is not as quiet as Troy, but he is quiet enough. He seems, right away, to be a man of settled character, and you suppose he has answered several of the important questions about his life and his vocation. He and his wife, Wilma, have two young sons. They are living on their own farm, which is partly wooded, partly planted in Christmas trees, and partly in pasture. We spent a good many miles of our drive in a conversation about raising sheep. A young man with a farm, and fully interested in it, may be assumed to have some permanent intentions.

When we arrived at the house on Firth Road, Troy had come in from his day's work. Lynn gave us a good supper, most welcome to me, for my noon fare at the Cincinnati airport had been both meager and bad. I had asked Troy to show me as much of his practice of forestry as I would be able to see in the time available, and he had laid out a crowded itinerary. Soon after we had eaten, he and Guy and I used up the rest of the day on the first three stops of my tour of places important to see.

First we made a fairly quick drive-through at Troy's sawmill, most memorable to me for the ricks of beautiful large black cherry logs. To be profitable, a mill of this relatively small scale must have a specialty, or "niche," which here is the lucrative trade in veneer-quality cherry. That happens to be a proper niche for this area, which is uniquely productive of the finest cherry wood. To maintain the necessary stock of good logs, since he can't supply all he needs from his own relatively limited acreage, Troy regularly logs timber sales in the Allegheny National Forest or the Susquehannock State Forest.

We next went to the headquarters of Firth Maple Products, where the sap of thousands of trees is collected, reduced by boiling to the proper consistency of maple syrup, and bottled. My own country has a lot of sugar maples, but we have no tradition of syrup-making. We do, or did, have something of a tradition of sorghum molasses-making, and so I understand in principle the boiling down of a thin, mildly sweet liquid to a viscous liquid that is intensely sweet. But the equipment of Troy's maple syrup factory far exceeded in both scale and complexity any that I had seen before.

Troy's woodlands are full of pipelines. The black plastic pipe, most of it an inch and a quarter in diameter, is tightly stretched, reinforced by high-tensile, galvanized steel wire, and secured at about knee-height by other wires guyed to trees on either side. Small transparent tubes run the sap from the trees to the black pipes. I was impressed by the sturdiness of the pipelines, which are installed to last for several years, and also by the care with which they had been attached to the trees. Every wire that went around a tree was buffered by blocks of scrapwood to prevent injury to the bark. The pipelines carry the sap to tanks at well-sited collection points, from which it is trucked to the boilers. Troy is alert for any way to reduce costs. The maple syrup operation requires large tanks of various sizes and kinds, and many of these have been bought cheap—used or damaged but still good. The trucks are mostly stainless steel milk tankers, also used but good. If you have limited or no control of the markets you buy and sell on, thrift becomes an essential virtue.

Our third stop on the evening of my arrival was at a woodland that Troy had logged, he thought, ten times in the last forty-five years. And here we came to our principal subject and the purpose of my trip. Clearly, no patch of forest could have been so steadily productive for so many years if it had not been knowingly and carefully logged. And in fact this was an excellent demonstration of the results, over time, of worst-first single tree selection. The last cutting had been recent, the tops of the stumps were still bright and unweathered, and yet you could see immediately that the forest in that place was ecologically whole. There were no too-large openings in the canopy. The remaining trees were of a diversity of sizes, from large to small. None of the remaining trees had been damaged by the felling or skidding. Because the skidding had been done with horses, the ground had been only slightly scarred. The skid trails would mostly disappear by the next year.

My visit with Troy and Guy, and then with Jason Rutledge and his son Jagger, who arrived on Thursday afternoon, was from Wednesday evening until mid-morning on Sunday. Subtracting the several hours when we merely sat and talked, we probably spent two longish days walking, looking, and talking in the woods.

In the bright early sunlight of Thursday morning we walked into an open woodland of tall trees, and under them a uniform stand, virtually a garden, of graceful hay-scented ferns. This was a pleasant, "scenic" place where one might like to have a picnic, the sort of "natural" place that many a nature or wilderness enthusiast would like to "preserve." But there was a problem. There was no understory to speak of: no seedling trees, saplings, shrubs, or the smaller plants that in a thriving forest would occupy the spaces between the ground and the lower branches of the tallest trees. The succession of the trees had been broken. If one should be felled or blown down, there would be none to replace it. The culprit was the beautiful hay-scented fern, whose rhizomes and roots grow so thickly in the top layer of the soil that they cannot be penetrated by the roots of seedlings, and whose fronds steal the light from other plants.

This domineering population of ferns is caused by an overpopulation of white-tailed deer who eat and destroy the plants that compete with the ferns. Once the fern carpet is established, it remains dominant. Reducing the deer population then has no effect, and another remedy must be found. Among the ferns a few blackberry briars were scattered about, and these may offer some hope. It may be that where the blackberry roots have pierced the fern mat, the roots of other species may follow and take hold. Or it may be that the blackberries, because of their earlier leafing in the spring, will shade out the ferns. The interaction of the two species is as yet unclear. Alternatively, the ferns may have to be set back by the use of an herbicide. The immediate problem here for the long-term commercial forester—this is the point—is an ecological problem: how to restore to this impoverished woodland the highly diverse "vertical structure" indicative of ecological health.

The predominant and most valuable trees of this region are black cherry and sugar maple. In addition, there are northern red oaks, some white oaks, white pines, and red or soft maples. Because of their yield of sap, the sugar maples have a special status in the Firth woodlands. They are dual-purpose trees. Tapping them for syrup degrades them somewhat as timber, but at appropriate times they too are logged.

But when you are with him in the woods, you can feel Troy's partiality for the tall, veneer-grade cherry trees with their dark, thick, flawless trunks. Looking with him at these trees, you see one of the primary ways in which good forest management increases yield. We were discussing a tall, thick-boled cherry tree with a fork at about half its height: two long, large prongs with many smaller branches. This woodland, Troy says, was highgraded perhaps ninety years ago. "Highgrading," he says, "takes the top off the woods." This reduces the competition for space and light, with the result that the trees branch or fork lower than they normally or naturally would have done. This particular cherry is a valuable tree that will yield a log of high quality, but the main log will be half to two-thirds as long as it would have been in a properly managed forest. The tree, moreover, is vulnerable in a way that a single-stemmed tree is not. Cherry is a brittle wood. As the two great prongs of the fork become more top-heavy, and their leverage increases, it becomes more likely that they will cause the tree to break or split. A tree like this one must be closely watched. At some point, a skilled climber will top the tree so that it can be safely felled.

We visited, by contrast, a single-stemmed cherry tree in a woodland that had not been high-graded. This tree was perhaps ninety years old, about twenty inches in diameter at breast height, still flourishing, still making wood, and not nearly so old as, with right management, it may become. It was maybe a hundred and twenty feet tall, eighty feet to the first branch. Some day it will yield five sixteen-foot logs, the first two of veneer quality. It was by any measure a beautiful tree. We spent a good while looking up at it and complimenting it.

The great event of Friday morning was our encounter in the woods with two young Amish teamsters driving their two-horse teams from the high seats of the elegantly engineered carts known as "logging arches," which are used for skidding logs out of the forest. The carts are so constructed that, when the horses tighten against a log, the log's own weight causes its fore-end to lift off the ground, thus reducing friction, lightening the draft, and minimizing damage to the woods floor.

More remarkable were the two teams of horses that provide the traction power. These horses do the hardest sort of work daylong every weekday, drawing heavy loads over rough ground, and yet they had the appearance of show horses. They were sound, bright-eyed, in excellent flesh, their coats unflawed and shining. Their condition spoke well of everything involved, but principally of the intelligence, care, and skill of the two teamsters. These were friendly, articulate men in their late twenties or early thirties. In conversing with us, and particularly with Jason, who is himself an excellent horseman and teamster, they showed themselves to be constantly attentive to every point of contact between horse and harness. This comes from a fine kind of sympathy: A good horseman wants the horses to work as comfortably in their harness as he himself works in his clothes. Sometimes in horse logging one sawyer, felling the marked trees, will keep two or three teamsters busy skidding the logs out. Here, the teamsters themselves do the felling, and this gives an obvious advantage to the horses, who rest while the men work.

The use of horses in the woods is fundamental to Troy's way of logging. It may finally prove to be necessary to anybody's version of sustainable forestry. Horses are kinder to the forest than mechanical skidders. Their skid trails are narrower and require no digging or grading of the forest floor. Using less force, they are less violent. A horse's hooves may slip now and then, scarring the surface of the ground, but the wheels of a heavy skidder compact the soil and, in the roughest places, tend to dig. Horses, moreover, can work on wetter ground than skidders.

Horses and horse equipment also cost much less and work much cheaper than skidders, and this can yield a large advantage to the forest. A new skidder will cost in the neighborhood of $150,000, a used one something like $50,000. A teamster's outfit—horses, logging arch, chains, saw, etc.—might cost as little as $5,000. (The greatest "cost" of using horses is the knowledge of how to care for, breed, raise, train, and work them, which is now in short supply. But knowledge, once you know it, is free.) This is a significant and influential difference. The more expensive the logging equipment, the higher the operating costs, the greater must be the pressure to increase the daily volume of board feet, increasing in turn recklessness in the use of the forest and the likelihood of damage. The urge toward the highest possible volume works against selectivity, judgment, and forbearance. In forestry as in farming, low production costs can increase the quality of work and so of care for the land. The logger who is free of financial anxiety can stop and think.

Moreover, the use of horses in logging increases human employment. One person operating a skidder may replace three teamsters. We are taught to see that as an advantage, and it is, but only to the company that makes skidders, and to the suppliers of fuel. The advantages of horse logging, much to the contrary, accrue to the forest and the human community. The Amish horsemen we talked with earn a good living, working as "independent contractors." In a time officially obsessed with "job creation," perhaps we don't need to argue about the worth of three people in effect self-employed as opposed to one employee driving somebody else's machine. Horse logging, in addition, is principally maintained by the local economy, which in turn it helps to maintain, whereas the purchase of large machines and the fuel to run them siphons money out of the local economy to enrich the shareholders of remote corporations.

In keeping with his commitment to sustainability, Troy does what he can to favor continuity of employment. At present he has six teams at work in the woods. One of the teamsters, who happens to be one of Jason Rutledge's alumni, has worked for Troy for fifteen years, another for eight, two others for six. He tries to keep them busy all the year round. Except for times in the spring when the ground is too wet—and, when using horses, these times are not long—the logging is continuous. When it is too wet, Troy finds other work for the teamsters.

The connections and interactions among all the creatures in a thriving forest ecosystem are complex beyond the human ability to think. This is the starting point, the primary axiom. And so humility is the primary virtue of good forestry. One must get the scale right, so as not to put too much at risk. One must not use too much power, or be in too much of a hurry. Troy accordingly does not talk like an expert. Because the forest is complex, it requires a due complexity of knowledge and of thought. Ultimately it requires work that honors, not only the known complexity, but also the unknown, the mystery of the nature of any place. By now Troy has a lot of experience. He has thought and worked with care. But how does he know when he has worked well?

His summary statement to me was that he finds out from the songbirds. If one of his woodlands as a result of his care and management is ecologically healthy, if there is enough diversity of the species and ages of the plants from the lowliest flowers and shrubs to the tallest trees, if the "vertical structure" is right, then you will hear a lot of birds singing. This is not a Disney sentimentalism. The birds are not singing to compliment Troy. They are singing because they have found a diversity of places in which to nest and feed, as the nature of the forest requires. Their songs indicate that all, or at least enough, of the pieces are in place. They are singing specifically to Troy only in the sense that he knows at least this much of what they mean.

Another virtue essential to good forestry is generosity. You don't have to see much bad logging before you become unable to imagine that a selfish or greedy person could be a good forester. A good forester thinks first of the forest. I remember Troy saying at a forestry meeting several years ago, "A bad logger goes to the woods thinking of what he can take out. A good logger goes to the woods thinking of what he should leave." It is this generosity toward the forest that enables the necessary thought. Generosity involves the forester in the history of the forest, its past and its future.

Troy's knowledge of his woodlands is historical, extensive, highly particular, and intimate. If you follow him through the woods for several hours, you realize that you are passing through a succession of distinct places, each different from any other, in its community of plants, slope, exposure, soil quality, history, problems, and so on. About each place Troy is likely to know a more or less adequate history: from visible evidence, from local or personal memory, from dated aerial photographs. He will know that a given place, until a certain year, was cleared and farmed. Or he will know that another woodland was high-graded eighty years ago, or that another was carelessly logged fifty years ago. He is aware, of course, of general principles, ideas, and theories of forestry, some better than others, but he says, with emphasis, "Forestry is mainly observational, rather than theoretical." By "observation" Troy means walking and looking, paying attention, season after season, for many years. Eventually, a profound familiarity will grow between an observant forester and the places of the forest. Such knowledge is what we mean, maybe, by "sympathy" or "sixth sense" or "intuition." It is the knowledge that tells one, in a given situation, where to look or what to expect or how much is enough. It tells what to take and what to leave.

Though this certainly is knowledge, and though it certainly comes by a kind of education, it cannot be conveyed in courses or curricula or majors. This education is "observational" and it takes many years. To know competently a tract of forest, Troy says, "is going to take decades. That's all there is to it."

You can't learn one woodland by studying another, and you can load only a limited amount of competent or workable knowledge into one mind in one lifetime. This is why it is important for good foresters both to stay put and to have local successors. The United States Forest Service makes a practice of moving people around. We recognize this as an industrial ideal: The supposedly easy mobility of human populations is an exploitable asset. Troy's idea, on the contrary, is to stay in place himself, and to hire local people for life. There is a sound economic reason for this, in addition to its obvious ecological value: Local knowledge of the local forest, like the forest itself, is an asset.

A local or personal economy, no matter how intelligently and kindly managed, cannot be made entirely immune to changes in the larger economy, no matter how false or fantastical the larger economy may be. And so in the "recession" that began in 2008, Troy says, "I lost a bundle." His inventory of standing timber "fell off a cliff." Now things are improving, and he recently hired two more skidding teams. In general the Firth economy, within a larger economy never favorable to the economies of land use, has done remarkably well. Probably the most interesting sentence in the circular issued by the Firth Family Foundation is this: "A 2006 study by LandVest showed that the Firths averaged 11 percent annual return on their timberlands for the previous 20 years."

Troy Firth's years of work in the forest have carried him far beyond the usual definitions of land ownership, of land use, and even of forestry. Rather than requiring the forest to submit to his economic demand, he has learned painstakingly to fit his economy into the forest. He has taken his living from the forest, not in defiance of the nature of the forest ecosystem, but to the best of his ability cooperating with it. Under his management, the production of timber has not reduced the productivity of the forest. On the contrary, his management has increased productivity—it can, he says, almost double the production of poorly managed woodlands—at the same time that it increases the number of human workers. It would be hard to overestimate the importance of this, as an accomplishment and as an example: several human livelihoods taken from the forest, to the forest's benefit, which becomes in turn and complexly a further human benefit.

Troy is an exemplary user of the land. At a time when such exemplars are desperately needed, he is somebody to turn to for confirmation of certain good possibilities. But the value of this, as he well knows, is strictly limited. One or two good foresters here and there are not enough. Nor even would be some widespread brand of "sustainable forestry," given an official label in the fashion of "organic agriculture." To develop and promote a forest ethic would do some good, but it would not do much good in the absence of working and paying examples of good practice. Governments may provide some checks against bad forestry by laws, regulations, restraints, "incentives," etc., but governments do not have the means to bring about good forestry. In a note of June 25, 1942, Aldo Leopold recorded his sense of this governmental limit: "I am skeptical about government timber production as a sole remedy for the apathy of private timber owners."

A sustainable local forest economy would supply as many as possible of its own needs, and it would perform most of the value-adding to most of its products. A country-wide project of sustainable forestry would require for many years an increasing number of such economies, which would in turn depend on the coincidence of available decent livelihoods from the forest with the passion for forestry that came to Troy in his student days and has continued with him ever since. If you are not squeamish about the word, you might say the "love" of forestry. To say that the good care of the forest, as of all the world's places, depends upon love is, sure enough, to define a difficulty. But not an impossibility. The impossibility is that humans would ever take good care of anything that they don't love. And we can take courage from the knowledge that millions of Americans once loved their vegetable gardens, cared well for them, and kept them dependably productive—and that a good many still do.

The Foundation for Sustainable Forests, at present, is the work of a few people. It has so far acquired in its own name only 740 acres. It is assured eventually of having, by bequest from Troy and Lynn, more than 6,000 additional acres. What seems most hopeful about it is its solid basis in practicality. Unlike most foundations, this one will be a viable commercial enterprise. It can use donated money to purchase woodlands, but these properties will become self-sustaining.

Like The Land Institute, Tilth, the Quivira Coalition, the Southern Agriculture Working Group, the Land Stewardship Project, and other organizations concerned with sustainable agriculture, the Foundation for Sustainable Forests will occupy the intersection of ecology and economy, which ought to be occupied by everybody but at present has a surplus of elbow room. These organizations promoting sustainable land economies are identifiable by their endless agendas of work, by their virtual invisibility to the worlds of policy and the news, and by the swarms of questions that hover about them—questions asked and waiting for answers, questions as yet unasked.

For example: If sustainable forestry depends upon sustainable local forest economies, how will we develop the necessary small-scale value-adding industries? I don't know the answer, and I doubt that anybody does. But then I remember hearing that small Amish furniture factories did more than a hundred million dollars worth of business last year, and I draw another breath. The necessary pieces may already exist, though widely scattered and perhaps lost from one another. No one person will be capable of putting them all together. Troy Firth knows this, but he also knows a further truth: One person can begin. "And can begin better with the help of others."

Local Economies to Save the Land and the People

As often before, my thoughts begin with the modern history of rural Kentucky, which in all of its regions has been deplorable. In my county, for example, as recently as the middle of the last century, every town was a thriving economic and social center. Now all of them are either dying or dead. If there is any concern about this in any of the state's institutions, I have yet to hear about it. The people in these towns and their tributary landscapes once were supported by their usefulness to one another. Now that mutual usefulness has been removed, and the people relate to one another increasingly as random particles.

To help in understanding this, I want to quote a few sentences of a letter written on June 22, 2013, by Anne Caudill. Anne is the widow of Harry Caudill. For many years she was involved in Harry's study of conditions in Eastern Kentucky and in his advocacy for that region. Since Harry's death, she has maintained on her own the long interest and devotion she once shared with Harry, and she is always worth listening to. She wrote:

> The Lexington Herald Leader last Sunday . . . published a major piece on the effects of the current downturn in the coal industry . . . Perhaps the most telling statement quoted came from Karin Slone of Knott County whose husband lost his job in the mines . . . finally found a job in Alabama and the family had to leave their home. Karin said, "There should have been greater efforts to diversify the economy earlier."
>
> [Fifty] years ago and more Harry tried . . . everything he could think of to encourage diversity. My heart goes out to those families who yet again are being battered by a major slump in available jobs. . . . Again they are not being exploited, but discarded.

This is a concise and useful description of what Anne rightly calls a tragedy, and "tragedy" rightly applies, not just to the present condition of Eastern Kentucky, but to the present condition of just about every part of rural Kentucky. The tragedy of Eastern Kentucky is the most dramatic and obvious because that region was so extensively and rapidly industrialized so early. The industrialization of other regions (mine, for example) began with the accelerated industrialization of agriculture after World War II, and it has accelerated increasingly ever since. The story of industrialization is the same story everywhere, and everywhere the result is ruin. Though it has developed at different rates of speed in different areas, that story is now pretty fully developed in all parts of our state.

To know clearly what industrialization is and means, we need to consider carefully some of the language of Anne Caudill's letter. We see first of all that she is speaking of a region whose economy is dependent upon "jobs." This word, as we now use it in political clichés such as "job creation," entirely dissociates the idea of work from any idea of calling or vocation or vocational choice. A "job" exists without reference to anybody in particular or any place in particular. If a person loses a "job" in Eastern Kentucky and finds a "job" in Alabama, then he has ceased to be "unemployed" and has become "employed," it does not matter who the person is or what or where the "job" is. "Employment" in a "job" completely satisfies the social aim of the industrial economy and its industrial government.

Perhaps there have always been "jobs" and "employees" to fill them. The point here is that the story of industrialization radically enlarges the number of both. It also enlarges the number of the unemployed and the unemployable. I can tell you confidently that the many owners of small farms, shops, and stores, and the self-employed craftspeople who were thriving in my county in 1945, did not think of their work as "a job." Most of those people, along with most skilled employees who worked in their home county or home town, have now been replaced by a few people working in large chain stores and by a few people using large machines and other human-replacing industrial technologies. Local economies, local communities, even local families, in which people lived and worked as members, have been broken. The people who once were members of mutually supportive memberships are now "human resources" in the "labor force," whose fate (to return to the language of Anne Caudill's letter) is either to be "exploited" by an employer or "discarded" by an employer when the economy falters or as soon as a machine or a chemical can perform their "job." The key word in Anne's letter is "discarded," which denotes exactly the meaning and the sorrow of our tragedy.

How can it be that the people of rural Kentucky can first become dependent upon officially favored industries, the "job-creating industries" that their politicians are always talking of "bringing in," and then by those industries be discarded? To answer that question, I need to refer again to Eastern Kentucky and something I learned there—or began consciously to learn there—nearly fifty years ago.

In the summer of 1965 I paid a visit of several days to my friend Gurney Norman, who was then a reporter for the *Hazard Herald*. At that time a formidable old man, Dan Gibson, armed with a .22 rifle, stopped a strip miner's bulldozer. The land Mr. Gibson was defending belonged to his stepson, who was serving with the Marines in Vietnam. Mr. Gibson's defiance and his arrest caused a considerable disturbance, and a crowd of troubled people gathered on a Friday night in the courthouse in Hindman. Gurney and I attended the meeting. That night Harry Caudill made a speech that recalled certain meetings in Philadelphia in the summer of 1776, for he spoke against the domestic successors of the British colonialists: "the mindless oafs who are destroying the world and the gleeful yahoos who abet them."

I am indebted to another speech of the same night. That speech was made by Leroy Martin, chairman of the Appalachian Group to Save the Land and the People. Mr. Martin bore witness to the significance of Dan Gibson's act, his loyalty, and his courage. He spoke impressively also of the forest that stood on the mountainside that Mr. Gibson had defended. He spoke the names of the trees. He reminded his hearers, many of whom were local people, that they knew the character and the value of such woodlands.

Three lines of thought have stayed with me pretty constantly from that time until now.

The first concerns the impossibility of measuring, understanding, or expressing either the ecological cost or the human heartbreak of the permanent destruction of any part of our only world.

The second consists of repeated returns to the impossibility, at least so far, of permanently stopping this permanent damage by confronting either actual machines or political machines. Dan Gibson's unlawful weapon was answered by the lawful weapons of thirteen state police, a sheriff, and two deputies. Our many attempts to confront the political machine that authorizes the industrial machinery have really not been answered at all. If money is speech, as our dominant politicians believe, then we may say that all our little speeches have been effectively answered by big money, which speaks powerfully though in whispers.

The third line of thought, the one I want to follow now, has to do with the hopefulness, and the correction, implied in the name of the Appalachian Group to Save the Land *and* the People. The name of that organization—and, if I have remembered it correctly, Leroy Martin's speech—assumed that we must not speak or think of the land alone or of the people alone, but always and only of both together. If we want to save the land, we must save the people who belong to the land. If we want to save the people, we must save the land the people belong to.

To understand the absolute rightness of that assumption, I believe, is to understand the work that we must do. The connection is necessary of course because it is inescapable. All of us who are living owe our lives directly to our connection to the land. I am not talking about the connection that is implied by such a term as "environmentalism." I am talking about the connection that we make economically, by work, by living, by making a living. This connection, as we see every day, is going to be either familiar, affectionate, and saving, or distant, uncaring, and destructive.

The loss of a saving connection between the land and the people begins and continues with the destruction of locally based household economies. This happens, whether in the United States after World War II or in present day China, by policies more or less forcibly moving people off the land. It happens also when the people remaining on the land are convinced by government or academic experts that they "can't afford" to produce anything for themselves, but must employ all their land and all their effort in making money with which to buy the things they need or can be persuaded to want. Leaders of industry, industrial politics, and industrial education decide, for example, that there are "too many farmers," and that the surplus would be "better off" working at urban "jobs." The movement of people off the land and into industry, away from local subsistence and into the economy of jobs and consumption, was one of our national projects after World War II, and it has succeeded.

This division between the land and the people has happened in all the regions of rural Kentucky, just as it has happened or is happening in rural places all over the world. The problem, invisible equally to liberals and conservatives, is that the forces that destroy the possibility of a saving connection between the land and the people destroy at the same time essential values and practices. The conversion of an enormous number of somewhat independent producers into entirely dependent consumers is a radical change that in many ways is immediately catastrophic. Without a saving connection to the land, people become useless to themselves and to one another except by the intervention of money. Everything they need must be bought. Things they cannot buy they do not have.

This great change is the subject of Harriette Arnow's novel *The Dollmaker*. In the early pages of this book we recognize its heroine, Gertie Nevels, as an entirely competent woman. Her competence does not come from any "success," political or social or economic. She is powerful because, within the circumstances of her agrarian life in the mountain community of Ballew, Kentucky, she is eminently practical. Among the varied resources of her native place, she is resourceful. She has, from her own strength and willingness and from her heritage of local knowledge, the means of doing whatever needs to be done. These are the means, for her, of being content in Ballew where she is at home. Her husband, Clovis, is not content or at home in Ballew. He is an off-and-on mechanic and coal hauler whose aspiration and frustration are embodied in a decrepit truck. This is during World War II. The world is changing, and people are being changed. Physically unfit for the draft, attracted to modern life and "big money," Clovis goes to Detroit and finds a job as a "machine repair man." Gertie and their children follow him to the city where, to Gertie, the cars seem to be "driving themselves through a world not meant for people." They find that Clovis has rented a disheartening, small, thin-walled apartment, and is already in debt for a used car, a radio, and other things that he has bought on credit.

In these circumstances, Gertie's practical good sense is depreciated nearly to nothing, except for the meaning it gives to her grief. Back home, she had dreamed of buying, and had almost bought, a small farm that would have given greater efficacy to her abilities and greater scope to her will. As her drastically narrowed life in Detroit closes upon her, she thinks: "Free will, free will: only your own place on your own land brought free will." (And now we should notice that those who have lived in the saving way preferred by Gertie Nevels—and some have done so—are solvent still, and Detroit is bankrupt.)

It is a small logical step from understanding that self-determination for an individual depends on "your own place on your own land" to understanding that self-determination for a community depends on the same thing: its home ground, and a reasonable measure of local initiative in the use of it. This gives us a standard for evaluating the influence of an "outside interest" upon a region or a community. It gives us a standard for evaluating the policy of "bringing in industry" and any industry that is brought in. Outside interests do not come in to a place to help the local people or to make common cause with the local community or to care responsibly for the local countryside. There is nothing at all to keep a brought-in industry in place when the place has become less inviting, less exploitable, or less profitable than another place.

We may not want to oppose any and all bringing in or coming in of industry, but localities and communities should insist upon dealing for themselves with any outside interest that proposes to come in. They should not permit themselves merely to be dealt *for* by state government or any other official body. This of course would require effective, unofficial local organizing, and I believe we are developing the ability to do that.

But the most effective means of local self-determination would be a well-developed local economy based upon the use and protection of local resources, including local human intelligence and skills. Local resources have little local value when they are industrially produced or extracted and shipped out. They become far more valuable when they are developed, produced, processed, and marketed by, and first of all to, the local people—when, that is, they support, and are supported by, a local economy. And here we realize that a local economy, supplying local needs so far as possible from local fields and woodlands, is necessarily diverse.

As things now stand, the land and people of rural Kentucky are not going to be saved by the state and the federal governments or any of their agencies and institutions. All of those great official forces are dedicated primarily to the perpetuation of the corporate economy, not to new life and livelihood in small Kentucky communities. We must not make of that a reason to give up our efforts for better politics, better policy, better representation, better official understanding of our problems and needs. But to quit *expecting* the help we need from government bureaus, university administrations, and the like will give us an increase of clarity and freedom. It will give us back the use of our own minds.

For the fact is that if the land and the people are ever to be saved, they will be saved by local people enacting together a proper respect for themselves and their places. They can do this only in ways that are neighborly, convivial, and generous, but also, and in the smallest details, practical and economic. How might they do this? I will offer a few suggestions:

1. We must reject the idea—promoted by politicians, commentators, and various experts—that the ultimate reality is political, and therefore that the ultimate solutions are political. If our project is to save the land and the people, the real work will have to be done locally. Obviously we could use political help, if we had it. Mostly, we don't have it. There is, even so, a lot that can be done without waiting on the politicians. It seems likely that politics will improve after the people have improved, not before. The "leaders" will have to be led.

2. We should accept help from the centers of power, wealth, and advice only *if*, by our standards, it is actually helpful. The aim of the corporations and their political and academic

disciples is large, standardized industrial solutions to be applied everywhere. Our aim, to borrow language from John Todd, must be "elegant solutions predicated on the uniqueness of [every] place."

3. The ruling ideas of our present national or international economy are competition, consumption, globalism, corporate profitability, mechanical efficiency, technological change, upward mobility—and in all of them there is the implication of acceptable violence against the land and the people. We, on the contrary, must think again of reverence, humility, affection, familiarity, neighborliness, cooperation, thrift, appropriateness, local loyalty. These terms return us to the best of our heritage. They bring us home.

4. Though many of our worst problems are big, they do not necessarily have big solutions. Many of the needed changes will have to be made in individual lives, in families and households, and in local communities. And so we must understand the importance of scale, and learn to determine the scale that is right for our places and needs. Brought-in industries are likely to overwhelm small communities and local ecosystems because both the brought-in and the bringers-in ignore the issue of scale.

5. We must understand and reaffirm the importance of subsistence economies for families and communities.

6. For the sake of cultural continuity and community survival, we must reconsider the purpose, the worth, and the cost of education—especially of higher education, which too often leads away from home, and too often graduates its customers into unemployment or debt or both. When young people leave their college or university too much in debt to afford to come home, we need to think again. There can never be too much knowledge, but there certainly can be too much school.

7. Every community needs to learn how much of the local land is locally owned, and how much is available for local needs and uses.

8. Every community and region needs to know as exactly as possible the local need for local products.

9. There must be a local conversation about how best to meet that need, once it is known.

10. The high costs of industrial land-using technology encourage and often enforce land abuse. This technology is advertised as "labor-saving," but in fact it is people-replacing. The people, then, are gone or unemployed, the products of the land are taken by violence and exported, the land is wasted, and the streams are poisoned. For the sake of our home places and our own survival, we need many more skilled and careful people in the land-using economies. The problems of achieving this will be difficult, and probably they will have to be solved by unofficial people working at home. We can't expect a good land-based economy from people who wish above all to continue a land-destroying economy.

11. The people who do the actual work and take the most immediate risks in the land economies have almost always been the last to be considered and the poorest paid. And so we must do everything we can to develop associations of land owners and land users for the purpose of land use planning, but also of supply management and the maintenance of just prices. The nearest, most familiar model here in Kentucky is the federal tobacco program, which gave the same economic support to the small as to the large producers.

12. If we are interested in saving the land and the people of rural Kentucky, we will have to confront the issue of prejudice. Too many rural Kentuckians are prejudiced against themselves. They have been told and have believed that they are provincial, backward, ignorant, ugly, and thus not worthy to "stand in the way of progress," even when "progress" will destroy their land and their homes. It is hard to doubt that good places have been destroyed (as in the coal fields) or appropriated by hostile taking (as in Land Between the Lakes) because, in official judgment, nobody lived there but "hicks" or "hillbillies." But prejudice against other disfavored groups still is alive and well in rural Kentucky. This is isolating, weakening, and distracting. It reduces the supply of love to our needs and our work.

To end, I want to say how grateful I am to have this audience for this speech. I remember when there was no organization called (or *like*) Kentuckians for the Commonwealth, and so I know its worth. I am proud to be one of you. In speaking to you, I've felt that I could reach, beyond several false assumptions, toward our actual neighborhoods and the actual ground under our feet. If we keep faithful to our land and our people, both together, never apart, then we will always find the right work to do, and our long, necessary, difficult, happy effort will continue.

Caught in the Middle

IN THE PRESENT political atmosphere it is assumed that everybody must be on one of only two sides, liberal or conservative. It doesn't matter that neither of these labels signifies much in the way of intellectual responsibility or that both are paralyzed in the face of the overpowering issue of our time: the destruction of land and people, of life itself, by means either economic or military. What does matter is that a person should choose one side or the other, accept the "thinking" and the "positions" of that side and its institutions and be so identified forevermore. How you vote is who you are.

We appear thus to have evolved into a sort of teenage culture of wishful thinking, of contending "positions," oversimplified and absolute, requiring no knowledge and no thought, no loss, no tragedy, no strenuous effort, no bewilderment, no hard choices.

Depending on the issues, I am often in disagreement with both of the current political sides. I am especially in disagreement with them when they invoke the power and authority of government to enforce the moral responsibilities of persons. The appeal to government is made, whether or not it is defensible, when families and communities fail to meet their prescribed moral responsibilities. Between the two moralities now contending for political dominance, the middle ground is so shaken as to be almost no ground at all. The middle ground is the ground once occupied by communities and families whose coherence and authority have now been destroyed, with the connivance of both sides, by the economic determinism of the corporate industrialists. The fault of both sides is that, after accepting and abetting the dissolution of the necessary structures of family and community as an acceptable "price of progress," they turn to government to fill the vacancy, or they allow government to be sucked into the vacuum. This, I think, explains both Prohibition and the War on Drugs, to name two failed government remedies. Another may be government-prescribed compulsory education.

To believe, as I do, that families and communities are necessary despite their present decrepitude is to be in the middle and to be most uncomfortable there. My stand nevertheless is practical. I do not think a government should be asked or expected to do what a government cannot do. A government cannot effectively exercise familial authority, nor can it effectively enforce communal or personal standards of moral conduct.

The collapse of families and communities—so far, more or less disguisable as "mobility" or "growth" or "progress" or "liberation"—comes from or with the collapse of personal character and is a social catastrophe. It leaves individuals subject to no requirements or restraints except those imposed by government. The liberal individual desires freedom from restraints upon personal choices and acts, which often has extended to freedom from familial and communal responsibilities. The conservative individual desires freedom from restraints upon economic choices and acts, which often extends to freedom from social, ecological and even economic responsibilities. Preoccupied with these degraded freedoms, both sides have refused to look straight at the dangers and the failures of government-by-corporations.

The Christian or social conservatives who wish for government protection of their version of family values have been seduced by the conservatives of corporate finance who wish for government protection of their semireligion of personal wealth earned in contempt for families. The liberals, calling for too few restraints upon incorporated wealth, wish for government enlargement of their semireligion of personal rights and liberties. One side espouses family values pertaining to temporary homes that are empty all day, every day. The other promotes liberation that vouchsafes little actual freedom and no particular responsibility. And so we are talking about a populace in which nearly everybody is needy, greedy, envious, angry, and alone. We are talking therefore about a politics of mutual estrangement, in which the two sides go at each other with the fervor of extreme righteousness in defense of rickety absolutes that are indefensible and therefore cannot be compromised.

Nowhere has this callow politics asserted itself more thoughtlessly and noisily than on the so-called rights of abortion and homosexual marriage. The real issue here is the politicization of personal or private life, and inevitably, given the absence of authentic political discourse or dialogue, the reduction of the issues to two absolute positions. In addition to distracting from interests authentically public and political, the politicization of personal life, involving as it must the publicization of privacy, is inhumane and inherently tyrannical.

After Boris Pasternak's *Doctor Zhivago* was published in the West and Pasternak received the Nobel Prize in 1958, thus earning the Soviet government's reprisal, Thomas Merton wrote:

> Communism is not at home with nonpolitical categories, and it cannot deal with a phenomenon which is not in some way political. It is characteristic of the singular logic of Stalinist-Marxism that when it incorrectly diagnoses some phenomenon as "political," it corrects the error by forcing the thing to *become* political.

Now, after many decades of anticommunism, Merton's sentences have come remarkably to be descriptive of our own politics. Maybe people who focus their minds for a long time upon enmity finally begin to resemble their enemies. This has happened before. It is deeply embedded in the logic of warfare.

Whatever the cause, we seem to have become as adept as the old Soviet Union at politicizing the nonpolitical. Most notably, by the connivances of both political sides, we have invented a sexual politics, which, by the standards of our own political tradition, is a contradiction in terms. Or it is if there is to be a continuing political distinction between public life and private life. This distinction, after all, is the basis of the freedoms protected by the First Amendment, which holds essentially that people's thoughts and beliefs are of no legitimate interest to the government. The government is not in charge of our personal lives, our private affections, our prayers, or our political opinions. It is not in charge of our souls. Those who formed our government also limited it, forbidding it any freehold in our homes or in our minds.

I am as ready as any so-called conservative to worry about big government, though I would remember that government has gotten big in the much-needed effort to regulate big corporations and to help their victims. To my fear of big government I add my at least equal fear of unlimited government, which is to say total government. It is not entirely surprising that after our long, costly resistance to communist dictatorship, we should now see the rise of passions and excuses tending toward capitalist dictatorship. The most insidious of these passions tends toward state religion and government regulation of private behavior.

Sexual politics has to do with public disagreements about rights that, however valid, are newly proclaimed, obscure in origin, extremely controversial, and productive of conflicts that probably are not politically resolvable—the prime example being the apparently unendable conflict over abortion.

Not so long ago abortion was illegal in the United States. It was illegal, one must suppose, because of an innate aversion to a woman's destruction of her own child. And then the Supreme Court ruled in *Roe v. Wade* (1973) that abortion, within certain limits, was legal. The ruling is based on the right to privacy under the 14th Amendment and, more remarkably and controversially, on the proposition that a human fetus is not legally a person and therefore is not eligible for the protections guaranteed by that amendment. This distinction between a fetus and a person is, to common sense, arbitrary and therefore inevitably a source of trouble. *Fetus*, to begin with, is a technical term which once was rarely used by pregnant women, who had conventionally and naturally referred to the creature forming in their wombs as a *baby*, which is to say a human being, a person. The abortion debate involves endless, unendable disagreement about such issues as when a fetus becomes a human or a person, when life begins, when or whether abortion should be legal, whether we should call it "killing" or "termination." Some enlightened people hold in derision the idea that life begins at conception. But if life does not begin at conception, then we are at the beginning of a kind of sophistry: an argument about when life may be *said* to begin.

The right to have an abortion has been popularly justified as a woman's right to control her own body. Such a right seems to be implied by a number of other rights, but only recently has it been stated in this way. So stated, it is somewhat confusing, for many of our laws, legal and moral, *require* one to control one's body—to restrain it, for instance, from killing the body of another person, except of course when ordered to do so by the government. To say when and why a requirement may become a right, and when and why the requirement or the right should be suspended or opposed, needs a lot of spelling out—if such a spelling out is possible.

The facts remain, on one side, that abortions are still proscribed by some religious traditions and the old aversion is still felt by many people, and, on the other side, that the legalization of abortion answers a need desperately felt for real and pressing reasons by many women, and legal abortion would at least put an end to illegal abortions badly performed in bad circumstances by incompetent and disreputable people.

Also involved are questions of ultimate seriousness and importance: questions of life and death that exceed the competence of human intelligence and are forever veiled in mystery. The trains of causation run quickly out of sight. I know a man who said, plausibly, "Life begins with erection." Elders used to refer young people to a time "when you were just a look in your mother's eyes." But when I asked the geneticist Wes Jackson, "Does life begin at conception?" he replied, "Life *continues* at conception." This, I felt, was at last a statement sufficiently serious.

In making any choice, we choose for the future, and so all our choices involve us in mystery and in a kind of tragedy. To choose to have a baby, to abort a fetus, to save a life, to destroy a life is to make a whole change on the basis of partial knowledge. One chooses in light of what one knows now about the past and thus changes the future inevitably and forever. What would have been, had the choice been different, will never be known.

To reduce this complexity and mystery to a public contest between two absolutes seems to wrong everything involved. Some equivocation seems natural and appropriate because one is attending to two possibilities, both unknown. Saints, heroes, great artists and scientists began as fetuses. So did tyrants, torturers, and mass murderers. Choices do not invariably cut cleanly between good and evil. Sometimes we poor humans must choose between two competing goods, sometimes between two evils. Responsibility or circumstances will require us to choose. But we cannot choose to be unbewildered or not to grieve.

The theologian William E. Hull, worrying over the destructive animosities that divide religious organizations, asked, "How can we avoid the wrangling that breeds hostility?" And he answered: "By seeking clarity rather than victory." This sounds exactly right to me, and I find little clarity in the public argument about abortion. I know that both sides are made up of individual humans whose thoughts and feelings may differ in significant ways from the public positions of their sides. But the problem with those positions as they are generalized and vented into the political atmosphere is that they substitute simplicity for clarity. By separating the statistical facts of abortion from the lived experience—from the mystery, bewilderment, and suffering that attend it—the simplicity becomes obscure and heartless. To the proabortion side, abortion is simply a right, the creature to be aborted is a fetus, the act itself is termination of a pregnancy by a forthright medical procedure. To the antiabortion side, abortion is simply a wrong to be refused or opposed in obedience to a moral or religious law that ought to be the law of the land. The two sides seem about equally to disregard both the truth of human suffering and the possibility of human compassion. Sexual politics is overflowing with principles and abstractions, but otherwise seems deserted. No actual woman wanting or needing or refusing an abortion is present.

The issue, I think, can be clarified only by imagining a woman to whom an abortion is one of two heartbreaking alternatives, one of which she alone must choose, and between which, however she chooses, she will remain emotionally divided perhaps for the rest of her life. This woman, troubling as she is to the political atmosphere of opposed absolutes, cannot be admitted by either side into the public argument. But her example is starkly clarifying. Her absence from the argument stupefies both sides.

I am unsure of the whereabouts or even the possibility of truth in the abortion strife, but I, with perhaps a good many others, am somewhere in the middle, where I see no chance of a public reconciliation. In fairness, we have to acknowledge that within the experience and history of abortion there must be many shades and mixtures of right and wrong. As in the human condition generally, we are not dealing with a choice between a shadowless light and utter darkness.

I have said several times that I am opposed to abortion except when it is necessary to save the mother's life. I stick to that, for I still feel strongly the old aversion. Unlike the proabortion side, I think that abortion is killing. What else could it be? And I think that the creature killed is a human being, for it can be a being of no other kind, and it is not a nonbeing. But I feel just as strongly an aversion to our life-destroying economy and way of life, and every day increases our need to cherish life in all its forms. I oppose the official killings that bear the names of justice and defense and also the killing that is a cost or by-product of certain industrial enterprises. I do, however, recognize the cruelty that is inherent and inescapable in the life of this world, in which no creature lives but at the expense of other creatures, as I recognize right and wrong ways of exacting and recompensing such costs.

But when I have spoken of my opposition to abortion, I hope I have never neglected to say that I can imagine circumstances in which I would willingly aid and comfort a girl or a woman getting an abortion. And here I arrive at what is for me the moral difficulty, even the moral obscurity, of this problem: Though I can say that, in some circumstances, I would willingly help somebody get an abortion, I cannot say that I would willingly aid and abet a murder.

Whatever one may think of a woman's right to control her own body, the inexpressibly intimate involvement of her own body in a woman's decision to have an abortion is a real and urgent consideration, and for a man it is a special one. That it does not involve, and could never have involved, my body does not invalidate my belief that abortion is wrong, but it does require me to be carefully aware of the bodily difference. Whereas a person's demonstrated willingness to kill another person already born requires us to look upon that killer as a public menace, a woman's decision to kill the baby in her womb does not require us to look upon her as a menace to anybody else. In fact we *don't* look upon her in that way.

So far as I can see, there are four possible legislative choices in relation to the abortion controversy:

1. Abortion could be forbidden absolutely, with no exceptions.
2. It could be forbidden, with specified exceptions.
3. It could be permitted, with specified exceptions.
4. We could permit it without exception, which to me means that we would have no law related specifically to abortion.

The first of these would outrage the proabortion side, it would settle the controversy merely by ignoring it, it is immitigably harsh, and it makes little sense. Absolute forbidding would choose the life of the unborn child over any and all other considerations, including that of the mother's life. The government thus would abandon any obligation to protect the mother's life in order to protect the life of the child. If, for want of an abortion, mother and child *both* should die, then the state would accomplish no good at all except for the pacification of fanatics.

Any law forbidding abortion would be ineffective, and it could easily do harm. To forbid an established practice for which the demand is widespread and the supply dispersed and readily available would be virtually to license an illicit and lucrative economy that would reward the greed and enterprise of the worst people. It would repeat the futility of the War on Drugs and Prohibition. Such a situation undermines government authority and brings law enforcement into disrespect.

The two middle solutions, as opposed to an outright ban, would require niggling official regulation of the conduct of individual persons, conduct at least semiprivate. This would require an increase in police power that would be expensive and also a danger to everybody's freedom. We could, for example, make a law forbidding abortion except to save the mother's life, but what would we mean by "the mother's life"? Would it be denoted only by her vital signs or, more reasonably, by her ability to live and thrive in the world—in which case the definition of her life would include her economic life, the life of her family (if she has one), even the life of her community (if she has one). For another example, we could make a law permitting abortion except during the third trimester. But this would require a lot of official watching. And who is to say exactly when the third trimester begins? Such legislating can only strand everybody, including the government, in permanently painful and dangerous confusion.

The problem in forbidding or permitting with exceptions is that the exceptions apparently cannot be decided upon by precise determinants, but rather by "approximate" or "appropriate" judgments by experts. The language of *Roe v. Wade*, as the ruling implicitly acknowledges, is vague and uneasy. What exactly is meant, with respect to abortion, by *life*, *conception*, *viability*, *privacy*, and *person*? *Roe v. Wade* does not, to my mind, settle the meaning of any of those words. The legal definition of a person evaporated when the Supreme Court defined a corporation as a person. If a corporation is a person, contrary to all previous usage and to common sense, then personhood can be conferred upon virtually anything merely by decree. Issues are thus quickly carried not just into vagueness but beyond the bounds of language.

I am going to take the risk, therefore, of saying that there should be *no* law either for or against abortion. Like certain other wrongs—various addictions, let us say—this one is more personal than public and would be best dealt with by the persons immediately involved: the pregnant girl or woman, her family or her friends, her doctor.

This is my attempt to make a statement on abortion that is reasonably complete—and that, in result, may be necessarily incomplete. I should add that I may find further reasons that will require me to revise. To have a mind, I think, depends upon one's willingness to change it.

Regarding homosexual marriage, the fault that I again must acknowledge is that what I have said before has been incomplete. As far as I remember, I have made only two public statements about this issue. My argument, much abbreviated both times, was that sexual practices of consenting adults ought not to be subjected to the government's approval or disapproval and that domestic partnerships, in which people who live together and devote their lives to one another, ought to receive the spousal rights, protections, and privileges that the government allows to heterosexual couples.

In those two statements I was considering homosexual marriage as an issue of law—with reference to the contention from both sides that marriage is a right to be granted or withheld by the government. This puts me again in the middle but this time with more certainty of my whereabouts and with good reasons to object.

First, this "right to marriage" is still birth-wet. It exists only to be selectively withheld. Apart from its momentary political expediency there is no reason for it. Whatever one may think of all that is presently implied and entailed by the legalization of marriage, surely nobody can claim that marriage is either the government's invention or that the government has an inherent right to determine who may marry. A government that can forbid two women or two men to marry might with better reason forbid two bigots to marry.

Second, this right depends upon a curious agreement between liberals and conservatives that human rights originate in government, to be dispensed to the people according to their pleading and at the government's pleasure, implying inescapably that any right, being so expediently the government's gift, can just as expediently be withheld or withdrawn. This flatly contradicts the founding principle of American democracy that human rights are precedent to the government's existence, that the government is established to protect them, and that the government must be restrained from violating them.

Third, it cannot be allowable, under the above principles, for the government, on the pleading of *some* of the people, to establish a right solely for the purpose of withholding it from some other people. If this were to happen, it would amount to a punishment imposed on a disfavored group for no crime except their existence. I don't need to point out that this has happened before.

That the liberals, who so often have been rightly anxious about the protection of rights and liberties, should define those rights and liberties as the gifts of a generous and parental government is absurd.

The conservative program on this issue, promoting as it does a constitutional apportionment of rights according to sexual category, in implicit violation of the Fifteenth and Nineteenth Amendments, is more darkly absurd. The theory that accreditation of the sexual practices of individuals is a function proper to a "small" and noninterfering government is comical.

That homosexuals have been denied the right to marry, supposing for the moment that such a right can exist, surely violates the Fourteenth Amendment, which forbids the state to "deprive any person of life, liberty, or property, without due process of law; [or] to deny to any person within its jurisdiction the equal protection of the laws." There is no need for homosexuals to be granted a right to marry that is at all different from the right of heterosexuals to do so. There is no good reason for the government to treat homosexuals as a special category of persons.

To support their strategy of outlawing homosexual marriage, Christians of a certain disposition have found several ways to categorize homosexuals as a different kind from themselves, who are in the category of heterosexuals and therefore normal and therefore good. They are mindful that the Bible looks upon homosexual acts as sinful or perverse. But it is not clear to me why perversion should have been specifically assigned to homosexuality. The Bible has a lot more to say against fornication and adultery than against homosexuality. If one accepts the 24th and 104th Psalms as scriptural norms, then surface mining and other forms of earth destruction clearly are perversions. If we take the Gospels seriously, how can we not see industrial warfare and its unavoidable massacre of innocents as a most shocking perversion? By the standard of all scripture, neglect of the poor, of widows and orphans, of the sick, the homeless, the insane, is an abominable perversion. Jesus taught that hating your neighbor is tantamount to hating God, and yet some Christians hate their neighbors as a matter of policy and are busy hunting biblical justifications for doing so. Are they not perverts in the fullest and fairest sense of that term? And yet none of these offenses, not all of them together, has made as much political-religious noise as homosexual marriage.

Another way to categorize and isolate homosexuals from the general citizenry and the prerogatives of citizenship is to define homosexuality as a disease having a cause that can be discovered and removed or cured by some sort of therapy. This seems most promising as long-term job security for scientists and doctors. Ken Kesey once saw an inscription in a men's room: "My mother made me a homosexual." Under it somebody else had written: "If I gave her the yarn would she make me one?" My own speculation is that we will never do much better than that. We will discover that, like all the rest of us, homosexuals are made what they are by their mothers, their fathers, their genes, their germs, their upbringing and their education, by their friends and neighbors, their dwelling places, their time and its culture, by their economic and social status, their personal history, and by history.

Yet another such argument is that homosexuality is unnatural. If the nature in question is merely biological—the realm of the ape and the naked ape—that may prove too roomy and accommodating to be of much help. By the standard of that nature, monogamy is unnatural, an artifact of *some* cultures. If it is argued that homosexual marriage cannot be reproductive, is therefore unnatural and should be forbidden, must we not then argue that any childless marriage is unnatural and should be annulled?

Specifically *human* nature, by contrast, has always had a definition more complex and demanding than that of a naked ape. William Blake thought we are made human by being made in the image of God:

For mercy, pity, peace, and love
Is God our father dear;
And mercy, pity, peace, and love
Is man, his child and care.

Are homosexuals capable of mercy, pity, peace, and love? Some certainly are, as some heterosexuals certainly are. To deny that distinction to homosexuals is to deny categorically that they are human, which is hardly a proper employment for mercy, pity, peace, and love. Oversimplified moral certainties—always requiring hostility, always potentially violent—isolate us from mercy, pity, peace, and love and leave us lonely and dangerous. The only perfect laws are absolute, but perfect laws are only approximately fitted to imperfect humans. That is why we have needed to think of mercy, and of the spirit, as opposed to the letter, of the law.

One may find the sexual practices of homosexuals to be unattractive or displeasing and therefore unnatural. But anything that can be done in that line by homosexuals can be done, and is done, by heterosexuals. Do we need a political remedy for this? Would conservative Christians like a small government bureau to inspect, approve, and certify their sexual behavior? Would they like a colorful tattoo, verifying government approval, on the rumps of lawfully copulating persons? We have the technology, after all, to monitor everybody's sexual behavior, but so eager an interest in other people's most private intimacy is both prurient and totalitarian.

The oddest of the strategies to condemn and isolate homosexuals is to propose that homosexual marriage is opposed to and a threat to heterosexual marriage—as if the marriage market is about to be cornered and monopolized by homosexuals. If this is not industrial-capitalist paranoia, it at least follows the pattern of industrial-capitalist competitiveness, according to which you *must* destroy the competition. If somebody else wants what you've got—from money to marriage—you must not hesitate to use the government (small, of course) to keep them from getting it.

But if heterosexual marriage is so satisfying to heterosexual couples, why can they not just reside in their satisfaction? So-called traditional marriage, now mostly divested of a traditional household and traditional bonds to a community, is for sure suffering a statistical failure, but this is not the result of a homosexual plot. Heterosexual marriage does not need defending. It only needs to be practiced, which is pretty hard to do just now. But the difficulty is rooted mainly in the values and priorities of our industrial-capitalist system, in which every one of us is complicit.

It certainly is possible for a government to withhold the legal perquisites of marriage from any group that it may be persuaded to designate in our present civil cold war. That is mainly to say that a government can forbid its officers to license weddings for people in the designated group.

But a wedding is not a marriage. A wedding is traditionally an exchange of vows of fidelity and love in all circumstances until death. In some circumstances, for some people, a wedding may be a sacrament. But however complicated and costly the preparations, the costumes, the photography, and the reception, a wedding is over and done with in a few minutes.

A marriage, by contrast, is the *making* of marriage, by daily effort to live out the vows until death. The vows may be taken seriously or not, broken or not, but there is no way of withholding them from homosexuals. You cannot copyright the vows, which a homosexual couple is perfectly free to make. The government cannot forbid them to do so, nor can any church.

Conjugal love, Kierkegaard wrote,

> is faithful, constant, humble, patient, long-suffering, indulgent, sincere, contented, vigilant, willing, joyful. All these virtues have the characteristic that they are inward qualifications of the individual. The individual is not fighting with external foes but fights with himself. . . . Conjugal love does not come with any outward sign . . . with whizzing and bluster, but it is the imperishable nature of a quiet spirit.

No church can *make* a homosexual marriage, because it cannot make any marriage, nor can it withhold any degree of blessedness or sanctity from any pledged couple striving day by day to be at one. If I were one of a homosexual couple, the same as I am one of a heterosexual couple, I would place my faith and hope in the mercy of Christ, not in the judgment of Christians.

Condemnation by category is the lowest form of hatred, for it is cold-hearted and abstract, lacking the heat and even the courage of a personal hatred. Categorical hatred is the hatred of the mob, which makes cowards brave. And there is nothing more fearful than a religious mob overflowing with righteousness, as at the crucifixion, and before, and since. This sort of violence can happen only after we have made a categorical refusal of kindness to heretics, foreigners, enemies, or any other group different from ourselves.

Kindness is not a word much at home in current political and religious speech, but it is a rich word and a necessary one. There is good reason to think that we cannot live without it.

Kind is obviously related to *kin*, but also to *race* and to *nature*. In the Middle Ages *kind* and *nature* were synonyms. *Equal*, in the famous phrase of the Declaration of Independence, could be well translated by these terms: All men are created kin, or of a kind, or of the same race or nature.

Jesus saves the life of the woman taken in adultery by removing her from the category in which her accusers (another mob) have placed her and placing her within kindness, his own kindness first and then that of her accusers: "He that is without sin among you, let him first cast a stone at her."

The accusers take this kindness as a defeat, as we all are too likely to do, and they depart without another word. The brief dialogue that follows is wonderfully animated by Jesus's sense of humor:

> "Woman, where are those thine accusers? hath no man condemned thee?"
>
> "No man, Lord."
>
> "Neither do I condemn thee: go, and sin no more."

Good advice—but can we suppose he could have given it without smiling, knowing as he did the vast repertory of sins and the endless human susceptibility?

Within the larger story of the Gospels this story is not exceptional. It does show us Jesus's way of dealing with one of the biblically denominated sins, but he simply reaches out to the woman in her great need as he did many times to many others. In the Gospels the sinfulness of all humans is assumed. It is neediness that is exceptional, and in Jesus's ministry need clearly takes a certain precedence over sin. His kindness is best exemplified by his feedings and healings with no imputation at all of deserving.

But the wealth of this idea of kindness is not exhausted by kindnesses to humans. It is far more encompassing. From some Christians as far back as the twelfth century, certainly from farther back in so-called primitive cultures, and from some ecologists of our own time, we have the idea of a great kindness including and binding together all beings: the living and the nonliving, the plants and animals, the water, the air, the stones. All, ultimately, are of a kind, belonging together, interdependently, in this world. From the point of view of Genesis 1 or of the 104th Psalm, we would say that all are of one kind, one kinship, one nature, because all are *creatures.*

Much happiness, much joy, can come to us from our membership in a kindness so comprehensive and original. It is a shame, as I know from long acquaintance with myself, to be divided from it by the autoerotic pleasure of despising other members.

Our Deserted Country

IF WE ARE to understand the history of our landscapes, which mostly are economic—farms, ranches, working forests, mines—we will have to begin by understanding the impetus and motive of the Industrial Revolution. Many people, still, will regard this suggestion as odd or unthinkable, though even some economists have begun to acknowledge what has always been obvious: One result of replacing human workers by industrial technologies is "joblessness."

But joblessness has been exactly the aim of industrialization from its beginning, as the so-called Luddites understood immediately. The purpose of industrial technology has always been to cheapen work by displacing human workers, thus increasing the flow of wealth from the less wealthy to the more wealthy. We have dealt with the violence always implicit in these substitutions by disregarding it, or disguising it by an official, quasi-religious litany of synonyms: *labor-saving*, *efficiency*, *progress*, *convenience*, *speed*, *comfort*, even *creative destruction*. Maybe we have to grant the possibility of some degree of altruism. Painless dentistry is often invoked as a justification of technological progress, and surely we must concede that painlessness is preferable to pain, just as comfort is preferable to discomfort. But what might be the costs of the "God-wrought" painlessness and comfort that have come with industrialization? "There is no such thing as a free lunch," the hard-headed realists love to advise us, implying that they have completely "done the numbers." But actually in all the industrial world the least popular mathematical operation is subtraction. We habitually imply that all the gains of technological progress are entirely net. *Our* painlessness and comfort are *everybody's* painlessness and comfort. Nobody pays, nobody loses.

But obsolescent human workers, characteristically, have been both replaced and displaced. The costs of progress have routinely been borne by discarded workers, and often the costs have been exorbitant. Of the outmoded coal miners of eastern Kentucky, Harry Caudill wrote:

> Many voices have decried [mechanization], contending . . . that when machines relegate men to idleness their condition will be worse than before. There is a certain logic to this reasoning, but it is a logic hard to swallow when one watches a . . . bucket-excavator . . . digging a ditch at a rate scores of men could not maintain. But what is the result when the machines sweep through an industry in a short time, replacing . . . fathers, sons and even grandsons . . . so that they simultaneously lose their livelihoods and all claim to status and standing? What happens to entire communities leveled by such traumas? What befalls the psyche and spirit of people so afflicted . . . ?

Such questions have never burdened the consciences of coal companies, and have seldom disturbed the sympathy of politicians. A society dominantly industrial has no effective means either of democratizing the gains or of ameliorating the human costs of industrialization. The fate of workers is abandoned to "the market" and "the labor market."

Of the eastern Kentucky coal miners one needs to add that when they moved into the "coal camps" they left not only their homes but also the families, neighborhoods, and land-based household economies that had supported them at home. They thus became entirely dependent on the money economy and their wages. Once progress had taken away their wages, they became entirely dependent on "welfare," or they migrated in search of jobs to the cities, where again they were wage-dependent and now in circumstances for which their rural experience had in no way prepared them.

Similarly, when industrial machinery and chemicals came into the sugarcane and cotton lands of the South, the mostly black field hands and their families became immediately obsolete, useless and wageless, with no recourse but to take their chances in urban ghettos. They too had to move from rural homes, communities, and a land-based subsistence—in which, however poor, they had lived with a long-established competence—into situations alien to their experience and abilities.

Those who have been in this way uprooted, replaced and

displaced, and who by their sufferings pay far more than their share of "the price of progress," are then subjected to versions of political regard dependent on the same social disconnection, and about equally irresponsible. The "conservatives," eagerly serving both God and Mammon, have concluded much to their convenience that the poor are poor or "jobless" because they are lazy, requiring only the incentive of starvation, or the starvation of their children, to become "productive members of society." The "liberals," serving an abstracted and oblique political compassion, assume that poverty and joblessness can somehow be corrected by the tiding over or new start supposedly granted by public charity and government "programs."

These attitudes issue from the great blank in the political-industrial mind that has forgotten, if it ever knew, the public and political value of securing for all citizens a reasonable permanence of dwelling place and vocation, which depends upon the widespread ownership of small farms and other economically supportive small properties. Lacking that fundamental connection, individuals, families, and neighborhoods can originate nothing in their own support, but become helplessly dependent on the money economy and the government, as now in fact we all are. Lacking that connection, the public economy becomes little more than a financial system irrelevant to economic life and to need, as ours now has done.

The following sentences speak as well as any I have ever read of the essentially democratic connection between people and land. The writer praises

> the spirit of independence which is generated in countries where the free cultivators of the soil constitute the major part of the population. It can scarcely be imagined how proudly man feels, however small his property may be, when he has a spot of arable land and pasture . . . [H]is independence being founded on permanent property, he has an interest in the welfare of the state, by supporting which he renders his own property more secure, and, although the value of the property may not be great, it is every day in his view . . . Those who wish to see only the two castes of capitalists and day-labourers, may smile at this union of independence and poverty.

That statement is a part of Alexander Mackenzie's assessment of the "nineteenth-century clearances" of the small holders of the Scottish Highlands. It is a statement that Thomas Jefferson would have recognized and honored. So would Virgil and several of the authors of the Bible. So would have my father and many other rural Kentuckians whose minds were formed a generation before World War II. They and many others spoke for an authentic, ancient human need to have and to belong to a piece of land, however small or poor, to live on and from, and to care for, a place offering a significant measure of life-support to themselves and their families and, as needed, to their neighbors. This is in no way like the land-greed of "them that join house to house, that lay field to field . . . that they may be placed alone in the midst of the earth!"

The Scottish Highlanders who were "cleared" from their small holdings, to make way for landlords' sheep pastures and game preserves, were abandoned mostly to the choice between emigration and starvation. As I have already suggested, their fate was not unique. Such regardless harm was the deliberate result also of England's Enclosure Acts. The same dispossession of country people took place in Stalin's Russia, and is taking place now in China. The resemblance of these developments to our own clearances of the American Indians is plain enough. The modern American version, in addition to the casual substitution of machines for field hands, and also following World War II, was the semi-official agricultural policy of "too many farmers" or "Get big or get out." It was semi-official because no act of Congress gave it a legal basis. Its basis was the "expert" opinion of corporate, academic, and political leaders. The relatively self-sufficient producers on small farms, according to this opinion, needed to become members of the industrial "labor force" and consumers of industrial commodities. Reducing the number of farmers and farms became a devastatingly successful political goal. The "free market" was allowed to have its way, which meant, among much else, that the primary producers in the industrial food economy would buy dear and sell cheap. By now nearly all of the land-using population have left their farms and home places to be industrially or professionally employed, or unemployed, and to be entirely dependent on the ways and the products of industrialism. Or they have remained, as "farmers," to pilot enormous machines over thousands of acres continuously in annual row crops such as soybeans and corn.

From earliest times we have known, if we were willing to know, having learned by experience and example, that when people are disconnected from their land they suffer. But that is only half of the truth. The other half is that when its rightful people, the people who rightfully care for it, are absent from it, the land suffers. It is the mutual, indivisible suffering of land and people that sets in right perspective the suffering of either.

The first problem of a drastic reduction of the land-using population is to keep the land producing in the absence of the people. The expert solution followed unsurprisingly the expert doctrine of "too many farmers." At the end of World War II, the war industries conveniently could "gear up" to adapt the mechanical and chemical technologies of war to the "needs of agriculture." The departing farm families would be replaced by the re-rigged war technologies, depending upon a seemingly limitless bounty of "natural resources," mainly ores and fuels. Agriculture would become an industry. Farms would become factories, like other factories ever more automated and remotely controlled. Industrial land use thus has become a front in a war against the living world. For the remaining fewer and fewer farmers, this has required a shift of interest from husbanding the fertility of the land to the various means of consuming the fossil fuels—with consequences perhaps foreseeable even by the experts, had their eyes been open.

But there was a further problem that the experts did not recognize then, and have not recognized yet: Agricultural production without land maintenance can lead only to exhaustion. Land that is in use, if the use is to continue, must be used with care. And care is not, it can never be, an industrial product or an industrial result. It cannot be prescribed or enforced by a market, free or unfree. Care can come only from what we used to understand as the human heart, so called because it is central to human concerns and to human being. The human heart is informed by the history of care and by the need for it, also by the heritage of the skills of caring and of caretaking.

The replacement of our displaced farm families by technologies derived from warfare has involved, beyond appeal, a supposedly acceptable, and generally accepted, violence against land and people. By it we have established an analogy between land use and war that has remained remarkably consistent to our present wars with their transferable "precision" technologies of remote observation and control. The common theme is a terrible pragmatism that grants an absolute predominance of the end over the means, in oblivion or defiance of any natural or moral law that may stand in the way. In the industrial land economies, from agriculture to mining, anything coming from the land that cannot be sold is treated as an enemy, and this includes natural and human communities.

The quickest way is the best way: This is the industrial version of efficiency. The industrial economy thrives on the rapidest possible changes of technology and fashion. Industrial land use thrives, or its suppliers thrive, by ever-swifter passages of machines over the land. Anything obstructing or reducing speed must be cleared away. To realize this highest aim of industrial agriculture, everything must give way to the rule of the widest expanse and the straightest line. Every surviving woodland, every tree, every fencerow must be removed. So must the animals, their pastures and pens. So must the surplus people and their buildings. Streams must be straightened and ponds filled. Every acre that will support a tractor must be cropped. Such use of the land is determined entirely by the market, and is limited entirely by the capabilities of the available technology. Questions relating to ecological and human health, or to the health of the local economy, are easily ignored because there are no industrial answers to such questions.

The rhetoric and indignation of conservationists often leads to stereotyping and condemnation by category. I need now to be careful to avoid that. From the prevalence of villainous ways of land use, it does not necessarily follow that all industrial farmers and foresters are villains. For the sake of fairness and in the interest of valid remedies, we need to understand how economic systems and constraints, plus the availability of shortcutting technologies, incline or attract or force land users toward abuse. We need to understand how abuse is favored or rewarded in the absence of sound and conserving land-use policies, and of any informed public concern and discussion that might lead to such policies.

Moreover, though a large portion of our remnant of farmers are now fully industrialized and their "operations" more extensive than ever before, I believe that they have inherited the economic standing of farmers in general and nearly always. For reasons usually of scarcity or of crop failure elsewhere, some of them sometimes experience a good year or a few good years, but their future is never reasonably secure, and in the industrial era their children are less and less likely to become farmers. Though the corporations that supply "industrial inputs" to farmers and the corporations that purchase "farm products" may prosper exceedingly, the farmers themselves, the people who bear the primary financial burdens and who do the actual work, will be the last considered, the least respected, and the lowest paid.

And so however severe may be the current abuses of the land, and however urgent the need for conservation, we have got to bear in mind that the land will not be well used, because it cannot be, by people who cannot afford to use it well. I recently attended a meeting on agriculture and water quality, at which a thoughtful farmer made a point shockingly obvious. What he said ought, in reason, to have ended the meeting, though of course it did not. He said that keeping animal manure out of the streams makes sense, it makes both agricultural and economic sense, "but it is hard to think of your environmental responsibilities when you're wondering who will be the next family to live in your house."

To make as much sense as I can of our predicament, I must turn now to Wes Jackson's perception that for any parcel of land in human use there is an "eyes-to-acres ratio" that is right and necessary to save it from destruction. By "eyes" Wes means a competent watchfulness, aware of the nature and the history of the place, constantly present, always alert for signs of harm and signs of health. The necessary ratio of eyes to acres is not constant from one place to another, nor is it scientifically predictable or computable for any place, because from place to place there are too many natural and human variables. The need for the right eyes-to-acres ratio appears nonetheless to have the force of law.

Economic landscapes, in short, require the most careful watching. And "careful" is the right adjective here: People who don't care, or know enough to care, or care enough to know, don't watch. And here I need to add an indispensable comment from the biologist Robert B. Weeden, who read an earlier attempt at this essay:

> [T]he essential eye is both attentive and loving. I think you use the term caring, which can have the same meaning as loving. However, one may have a variety of motives for caring, including looking after an investment. The attentiveness needed is a whole and balanced thing, not allowed to skew over into mere analysis. Narrow focus is necessary but not sufficient. As to loving, it must never be left alone, at least in your context of relation to place. Romantics of the eighteenth and nineteenth centuries loved rural places, but as ideas and ideals—as projections of perfection. Someone has to check whether the lowing herd is chopping the life out of the saturated lea!

What is necessary and attractive here is the introduction of the idea of a practical and practicing love. Reading Bob Weeden's letter caused me to think again of something I have, from my own experience, begun to know: how intimately related, how nearly synonymous, are the terms "love" and "know," how likely impossible it is to know authentically or well what one does not love, and how certainly impossible it is to love what one does not know.

To anybody who takes seriously the eyes-to-acres ratio, and who is carefully and lovingly watching, it is apparent that the working landscapes of our country are now virtually deserted. In the vast, relatively flat acreages of the Midwest now given over exclusively to the production of corn and soybeans, the number of farmers is lower than it has ever been. I don't know what the average number of acres per farmer now is, but I do know that you often can drive for hours through those corn-and-bean deserts without seeing a human being beyond the road ditches, or any green plant other than corn and soybeans.

Any people you may see at work, if you see any at work anywhere, almost certainly will be inside the temperature-controlled cabs of large tractors, the connection between the human organism and the soil organism perfectly interrupted by the machine. Thus we have transposed our culture, our cultural goal, of sedentary, indoor work to the fields. Some of the "field work" is now done by airplanes.

This contact, such as it is, between land and people is now brief and infrequent, occurring mainly at the times of planting and harvest. The speed and scale of this work have increased until it is impossible to give close attention to anything beyond the performance of the equipment. The condition of the crop of course is of concern and is observed, but not the condition of the land. And so the technological focus of industrial agriculture, by which species diversity has been reduced to one or two crops, is reducing human participation ever nearer to zero. Under the preponderant rule of "labor-saving," the worker's attention to the work *place* has been effectively nullified even when the worker is present. The "farming" of corn-and-bean farmers—and of others as fully industrialized—has been brought down from the complex arts of tending or husbanding the land to the application of "purchased inputs" according to the instructions conveyed by labels and operators' manuals.

Almost suddenly in the last three years the Midwestern version of corn-and-bean farming has invaded the highly vulnerable sloping fields of my part of Kentucky—where the officially recommended and encouraged "no-till" farming, dependent upon herbicides, does not, as claimed, prevent erosion, and especially not under continuous cropping. (Until recently, weed control in crops was accomplished by plowing or other mechanical means of stirring and loosening the soil. The no-till method controls weeds, instead, by poisoning them with herbicides. Loosening the soil, of course, makes it vulnerable to erosion, the vulnerability increasing with the steepness of the land. The selling point that no-till stops erosion thus seems to justify planting on land that is too steep.) This expansion of the "cutting edge" has been caused by high grain prices, caused in turn by the officially recommended and encouraged production of "biofuels," supposedly sustainable but not so by ecological standards, and doubtfully so even economically.

We can suppose that the eyes-to-acres ratio is approximately correct when a place is thriving in human use and care. The sign of its thriving would be the evident good health and diversity, not just of its crops and livestock but also of its population of native and noncommercial creatures, including the community of creatures living in the soil. Equally indicative and necessary would be the signs of a thriving local and locally adapted human economy.

The great and characteristic problem of industrial agriculture is that it does not distinguish one place from another. In effect, it blinds its practitioners to where they are. It cannot, by definition, be adapted to local ecosystems, topographies, soils, economies, problems, and needs.

The sightlessness and thoughtlessness of the imposition of the corn-and-bean industry upon the sloping or rolling countryside hereabouts is made vividly objectionable to me by my memory of the remarkably careful farming that was commonly practiced in these central Kentucky counties in the 1940s and 1950s—though, even then, amid much regardlessness and damage. The best farming here was then highly diversified in both plants and animals. Its basis was understood to be grass and grazing animals; cattle, sheep, hogs, and, during the 1940s, the workstock, all were pastured. Grain crops typically were raised to be fed; the farmers would say, "The grain raised here must *walk* off." And so in any year only a small fraction of the land would be plowed. I knew an excellent farmer of that older kind who thought that only about 5 percent of most upland farms could be safely broken for row crops (best by rotating from sod and back to sod) in any year. This was farming fitted to the land, as J. Russell Smith said it should be. And the commercial economy of the farms was augmented and supported by the elaborate subsistence economies of the households. "I may be sold out or run out," the farmers would say, "but I'll not be *starved* out."

My brother recently reminded me how carefully our father thought about the nature of our home countryside. He had witnessed the ultimate futility—the high costs to both farmer and farm—of raising corn for cash during the hard times of the 1920s and 1930s. He concluded, rightly, that the only crop that could be raised here both abundantly and profitably in the long run was grass. That was because we did not have large acreages that could safely be used for growing grain, but our land was aboundingly productive of grass, which moreover it produced more cheaply than any other crop. And the grass sod, which was perennial, covered and preserved the soil the year round.

A further indication of the quality of the farming here in the 1940s and 1950s is that the Soil Conservation Service was more successful during those years than it would or could be again in the promotion of plowing and terracing on the contour to control soil erosion. Those measures at that time were permitted by the right scale of the farming and of the equipment then in use. Anybody familiar with topographic maps will know that contour lines, remaining strictly horizontal, over the irregularities of the land's surfaces, cannot be regularly spaced. This variability presents no significant problem to a farmer using one- or two-row equipment in relatively small lands or fields. And so for a while contour farming became an established practice on many farms, and to good effect. It was defeated primarily by the enlargement of fields and the introduction of larger equipment. Eventually many farmers simply ignored their terraces, plowing over them, the planted rows sometimes running straight downhill. Earlier a good many farmers had taken readily to the idea of soil conservation. A farmer in a neighboring county said, "I want the water to *walk* off my land, not run." But beyond a certain scale, the farming begins to conform to the demands of the machines, not to the nature of the land.

I should pause here to notice that within three paragraphs I have twice quoted farmers who used "walk" as an approving figure of speech: Grain leaving a farm hereabouts should *walk* off; and the rainwater fallen upon a farm should *walk*, not run. This is not accidental. The gait most congenial to agrarian thought and sensibility is walking. It is the gait best suited to paying attention, most conservative of land and equipment, and most permissive of stopping to look or think. Machines, companies, and politicians "run." Farmers studying their fields travel at a walk.

Farms that are highly diversified and rightly scaled tend, by their character and structure, toward conservation of the land, the human community, and the local economy. Such farms are both work places and homes to the families who inhabit them and who are intimately involved in the daily life of land and household. Without such involvement, farmers cease to be country people and become in effect city people, industrial workers and consumers, living in the country.

To understand the complex and demanding requirements of good agriculture, and to know the vast acreage now given over to bad agriculture (leaving aside for now the vast acreage consigned to bad forestry), is to recognize the utter futility of the notion, apparently still prevalent among conservation groups, that the health of the natural world, revealingly called "the environment," can be preserved in parks and "wilderness areas." This drastic abbreviation of land stewardship permits no competent concern for the effects of the lowing herd upon the lea or of corn and beans upon the slopes. It holds that the gated communities of "the wild" will somehow preserve the natural health of "the planet."

Such conservationists, one imagines, might take instruction from scientists about the need for ecological health in the food-producing landscapes. But such scientists are rare and scattered. The predominant agricultural science of the universities, the corporations, and the government is still almost unanimously promoting industrial agriculture despite the by now overwhelming evidence of its failure: soil erosion, salinization, aquifer depletion, nutrient depletion, dependence on fossil fuels and toxic chemicals, pollution of streams and rivers, loss of genetic and ecological diversity, destruction of rural communities and the cultures of husbandry. The agricultural scientists and experts go doggedly on in their "cutting edge" rut because they either are employed by agribusiness or because their universities are now helplessly dependent on grants from agribusiness.

Urban conservationists, university scientists and intellectuals, journalists, powerful officials and politicians moreover are unlikely to live in the economic landscapes. In general they don't like the "boondocks" and the "nowheres" of rural America, and they don't know anything about them. Most of them know nothing of the issues of land use, and they think them unimportant. The sentimentalized ignorance of the romantic wilderness lovers, and the institutionalized, fear-enforced ignorance of agricultural scientists, are thus in turn permitted and supported by the ignorance of the general public, most of whom see the economic landscapes only through the windows of their speeding automobiles. If, by some rare chance, they should get out of their cars and walk in the fields, pastures, and woodlands, they most likely would take the present look of things to be "normal." Knowing no history of places, having no memories of them, they could not distinguish the country as it is from the country as it ever was. They would not recognize the signs of deterioration, or the numerous alien species that have come in as a side effect of global trade.

Nowhere is this ignorance more poignantly manifest than in the common use of words such as "land" and "ecosystem" simply as ideas or metaphors. I recently read *Ill Fares the Land*, a mostly admirable book, by Tony Judt, an admirable man. The book's title comes from a line in "The Deserted Village," Oliver Goldsmith's poem of protest against the Enclosure Acts. As Goldsmith used the word, "land" meant land. But Tony Judt's book never mentions the land. It speaks of "environmental well-being," of climate change and "environmental effects," it quotes John Maynard Keynes on "the beauty of the countryside," but its author clearly had not thought of the land itself, the land-use economies, or the natural world. To him "the land" is merely a figure of speech denoting "the nation" or "the national economy."

In the absence of a widely practiced and capable attention to our use of the land, to the land-use economies, and to the natural sources of our life, we have a national, or global, economy consisting entirely of capital (rated at monetary value), minimal labor ("jobs," merely numbered, and the numbers always liable to reduction by technology), information (infinite perhaps, but never sufficient), marketing (seduction of the gullible), and consumption (conversion of goods into waste or poison). And so we have lost patriotism in the old sense of love for one's country, and have replaced it with an ignorant, hard-hearted military-industrial nationalism that devours the country.

Under such a dominance it is understandable that land use should be reduced to the application, at the greatest possible speed and with the least possible labor, of technology and information. Since no limit is implied in the economic assumptions of such use, its technology or its methods, its destructiveness is unlimited, far exceeding the reach of any envisionable public regulation or supervision, as in the mountaintop removal method of coal mining, which destroys absolutely and beyond remedy the original, invaluable forest ecosystem of the Appalachian coal fields. An economy operating on the basis merely of quantities runs oddly toward both infinity and nothing, limitless desire and final exhaustion. No billionaire, evidently, can be satisfied with one billion or with any conceivable number of billions. But the effect of so much wealth, uncontrollable because unaccountable and unknowable by any human mind, is to use up the world's real wealth, which is its ability to live and to renew its life.

I don't think we can set the terms of a restorative and conserving land-use economy simply by juggling and somehow fixing the technologies, methods, resources, assumptions, and regulations of the present economy. We seem to have exhausted the capacity of our present system to improve much of anything. Or it may be better to say that we improve things by means of costs that we never count or subtract from the supposed benefits, and so we relinquish any notion of net improvement, not to speak of net loss.

The possibility of actual improvement in our economic life, which is to say our way of living from our land, seems to lie, not just in stabilizing our occupation of our country on the basis of knowledgeable attachment to its localities, but also in studying the economic value of such intangible goods as knowledge, memory, familiarity, imagination, affection, sympathy, neighborliness, and so on. One might greatly lengthen this list simply by allowing the several goods to call forth their kindred, each of which, like the ones I have listed, would impose certain conditions and certain limits. An economy, of course, must deal in quantities, but an economy answerable to terms such as I have listed would be opposite, in its effects, to an economy merely of quantities. I resist the inclination to call such terms qualities or ideas, just as I resist the inclination to call them feelings. They certainly are informed by qualities, ideas, and feelings, but they also are mental powers capable of enforcing care in our treatment of persons, places, and things.

In thinking about the economy, not of agriculture or farming, but of a farm, we come quickly to see the value of knowledge. The knowledge of how to farm, how and when to do the work, has an obvious, and fairly reckonable, economic worth. But the knowledge that is limited to one's own particular farm —knowledge of its nature, character, history, limitations, and right use, gathered out of years of experience—also has an economic value that is far more specific and not so readily accounted. Its value can only be suggested by pointing out that it clearly is an asset to the farmer who has it, that it cannot be sold or adequately conveyed by instruction to a new owner, whose want of it will be an economic disadvantage requiring much time, and perhaps some costly mistakes, to remedy. The value of such knowledge may be further suggested by seeing that it is depreciated virtually to nothing by the continuous monocultures I have described.

The force and effect of such intangibles is nowhere better exemplified than in the better communities of the Amish, which, so far as my own observation goes, are the only communities in the United States that are successful by every appropriate standard. Some would argue that the fundamental power or principle of these communities is their religious faith, and some would argue, further, that no community can be successful without religious faith as a fundamental principle. I am inclined to accept their conclusion, though I doubt that it can be adequately demonstrated, let alone proved. And so I will deal here only with issues of economy—though, among the Amish, the economy is formed or informed at every point by religious faith.

I think it is reasonable to say that the fundamental principle of Amish economy is rightness of scale: rightness of the size of their farms, and the appropriateness of that size to the nature of the local landscape, to the community's way of farming, and to the scale of its work. Rightness of scale gives scope and efficacy to the powers I have listed, and to others. It is the principle that allows diversity, flexibility, and local adaptation of farm enterprises.

Amish economy is an economy dependent upon limits strictly understood and observed. And each of the limits produces an economic advantage. Traction for field work, for example, is limited to the use of horses or mules. And that limit implies a limit of farm size; it implies diversity, which implies a structure of interdependent and mutually supportive plant and animal enterprises; it makes the farm the source of most of its operating energy and fertility—all of which, together, lower the cost of production.

Another limit, of almost incalculable significance, is neighborliness. This means that you would rather have a neighbor than to have your neighbor's farm, and here is another limitation on the size of farms. If their farms are of the right size, neighbors can help one another in times of trouble or when the work calls for many hands. If one has a neighbor—in the sense derived from the Gospels, as it is here—then one has help, and help is an economic advantage. If neighbors exchange work, they don't have to hire help and pay in cash.

The Amish famously, or infamously, limit formal schooling to eight grades. (This is not at all to say that they limit learning. Some Amishmen, for example, have gone on to learn mechanical engineering.) They limit schooling in order to keep their children in the community. This makes sense if you *want* to keep your children in the community, and if you have understood that the purpose of mainstream education is to prepare children, and especially country children, to *leave* the community. If you contrive in general to keep the community's children in the community, there are two desirable results: 1) The children, from earliest childhood, learn the community's work, by observing it and, as they become able, by doing it; and 2) If you keep all or most of the community's children in the community, then as a matter of course you keep the brightest and most talented ones. To keep the community's own brightest and most talented members in the community, in the community's local circumstances, doing the community's work, is to have the best intelligence and talent applied to that work, producing good examples that can be followed by neighbors.

This is an economic advantage. Moreover, the community may thus gain the competence to deal with tools, to quote again my friend Robert B. Weeden, "that take more skill not to use, than to use."

If on a farm of a conscientiously limited size you balance row crops and forages with an appropriate number of animals, then as a matter of course you become familiar with those animals. You can know them individually, and so sympathy can enter into your association with them. The best-known example of this is Psalm 23, a shepherd's psalm in which the shepherd, identifying himself as one of God's sheep, identifies profoundly with his own sheep: "He maketh *me* to lie down in green pastures: he leadeth *me* beside the still waters" [my emphases].

Just so, if by the same order of limitation you use a team of horses for field work, you know by sympathy, by imagining yourself "in their place," when they are overheating or overtired or when you are asking too much. From there it is not far to sympathy for the field you are working in, which also can be too much demanded upon, can be overworked and in need of rest. Imagination leads to sympathy, sympathy leads to good care, and all three convey economic advantages.

Once this pattern of interdepending limits and advantages has been understood, it can be described perhaps endlessly. But I have said enough to show why, if you drive through one of the good Amish communities, you will see a lot of people outdoors and busy. You will see that they have honored their places with the visible signs of good work lovingly done. And you will think again, with sharpened sadness, of the nearly always deserted "farms" of thousands of acres of corn and beans.

II

I have spoken so far of the decline of country work, but the decline of country pleasures is at least equally significant. If the people who live and work in the country don't also enjoy the country, a valuable and necessary part of life is missing. And for families on farms of a size permitting them to be intimately lived on and from, the economic life of the place is itself the primary country pleasure. As one would expect, not every day or every task can be a pleasure, but for farmers who love their livestock there is pleasure in watching the animals graze and in winter feeding. There is pleasure in the work of maintenance, the redemption of things worn or broken, that must go on almost continuously. There is pleasure in the growing, preserving, cooking, and eating of the good food that the family's own land provides. But around this core of the life and work of the farm are clustered other pleasures, in their way also life-sustaining, and most of which are cheap or free.

I live in a country that would be accurately described as small-featured. There are no monumental land forms, no peaks or cliffs or high waterfalls, no wide or distant vistas. Though it is by nature a land of considerable beauty, there is little here that would attract vacationing wilderness lovers. It is blessed by a shortage of picturesque scenery and mineable minerals. The topography, except in the valley bottoms, is rolling or sloping. Along the sides of the valleys, the slopes are steep. It is divided by many hollows and streams, and it has always been at least partly wooded.

Because of the brokenness and diversity of the landscape, there was never until lately a clean separation here between the pursuits of farming and those of hunting and gathering. On many farms the agricultural income, including the homegrown and homemade subsistence of the households, would be supplemented by hunting or fishing or trapping or gathering provender from the woods and berry patches—perhaps by all of these. And beyond their economic contribution, these activities were forms of pleasure. Many farmers kept hounds or bird dogs. The gear and skills of hunting and fishing belonged to ordinary daily and seasonal life. More ordinary was the rambling about and looking that kept people aware of the condition of the ground, the crops, the pastures, and the livestock, of the state of things in the house yard and the garden, in the woods, and along the sides of the streams.

My own community, centered upon the small village of Port Royal, is along the Kentucky River and in the watersheds of local tributaries. Its old life, before the industrialization of much of the farmland and the urbanization of the people, was under the influence of the river, as other country communities of that time were under the influence of the railroads. In the neighborhood of Port Royal practically every man and boy, some girls and women too, fished from time to time in the Kentucky River. Some of the men fished "all the time" or "way too much." Until about a generation ago, there was some commercial fishing. And I can remember when hardly a summer day would pass when, from the house where eventually I would live, you could not hear the shouts of boys swimming in the river, often flying out into the water from the end of a swinging rope. I remember when I was one of them. My mother, whose native place this was, loved her girlhood memories of swimming parties and picnics at the river. In hot weather she and her friends would walk the mile from Port Royal down to the river for a cooling swim, and then would make the hot walk back up the hill to town.

Now all of that belongs to the past. The last of the habituated fishermen of the local waters are dead. They have been replaced by fishermen using expensive "bassboats," almost as fast as automobiles. This sport is less describable as "fishing" than as "using equipment." In the last year only one man, comparatively a newcomer, has come to the old landing where I live to fish with trotlines—and, because of the lack of competition, he has caught several outsize catfish. Some local people, and a good many outsiders, hunt turkeys and deer. There is still a fair amount of squirrel hunting. The bobwhite, the legendary gamebird of this region, is almost extinct here, and the bird hunters with them. A rare few still hunt with hounds.

Most remarkable is the disappearance of nearly all children and teenagers from the countryside, and in general from the out-of-doors. The technologies of large-scale industrial agriculture are too complicated and too dangerous to allow the participation of children. For most families around here, the time is long gone when children learned to do farmwork by playing at it, and then taking part in it, in the company of their parents. It seems that most children now don't play much in their house yards, let alone in the woods and along the creeks. Many now descend from their school buses at the ends of lanes and driveways to be carried the rest of the way to their houses in parental automobiles. Most teenagers apparently divide their out-of-school time between indoor entertainment and travel in motor vehicles. The big boys no longer fish or swim or hunt or camp out. Or work. The town boys, who used to hire themselves out for seasonal or part-time work on the farms, no longer find such work available, or they don't wish to do the work that is available.

Not so long ago, talking with one of my contemporaries about the time before school consolidation, when there were five high schools in Henry County, I said, "When we were boys, there were five basketball teams in this county."

He said, "I'll tell you something else. In my senior year, any of the five could have beaten this one we've got now."

"Well, you all were in shape," I said.

"Yes," he said. "We *worked*."

More recently I heard a woman complaining that her husband could not hire anybody to help him with farmwork, "not even young people."

Somebody asked, "What *are* the young people doing?"

She was indignant: "Nothing!"

They aren't doing nothing, of course, but nothing might be preferable to what they are likely to be doing. For young people accustomed (by habit and indoctrination) to indoor life, the dominant attractions are drugs (including alcohol), sex, and various digital devices. The drug use is merely predictable in a society in which drugs are everywhere recommended for every imaginable discomfort and "need." Sex (arranged by telecommunication, using various digital devices) is promoted everywhere as another cure-all, as an incentive to participate in the economy of spending and consuming, and no doubt as another of our new crop of "rights." The digital devices are recommended or required in order to prepare the young for "the world of the future." The cost of this expensive preparation is virtual exile from the present world that is available at no cost outside their front doors. And so they spend their liveliest years mostly sitting and looking at screens.

Local people who regularly hunted or fished or foraged or walked or played in the local countryside served the local economy and stewardship as inspectors, rememberers, and story tellers. They gave their own kind of service to the eyes-to-acres ratio. Now most of those people are gone or absent, along with most of the farming people who used to be at work here.

With them have gone the local stories and songs. When

people begin to replace stories from local memory with stories from television screens, another vital part of life is lost. I have my own memories of the survival in a small rural community of its own stories. By telling and retelling those stories, people told themselves who they were, where they were, and what they had done. They thus maintained in ordinary conversation their own living history. And I have from my neighbor, John Harrod, a thorough student of Kentucky's traditional fiddle music, his testimony that every rural community once heard, sang, and danced to at least a few tunes that were uniquely its own. What is the economic value of stories and songs? What is the economic value of the lived and living life of a community? My argument here is directed by my belief that the art and the life of settled rural communities are necessary to the sustainability of a life-supporting economy. But their value is incalculable. It can only be acknowledged and respected, and our present economy is incapable, and cannot on its own terms be made capable, of such acknowledgement and respect.

Meanwhile, the farmlands and woodlands of this neighborhood are being hurt worse and faster by bad farming and bad logging than at any other time in my memory. The signs of this abuse are often visible even from the roads, but nobody is looking. Or to people who are looking, but seeing from no perspective of memory or knowledge, the country simply looks "normal." Outsiders who come visiting almost always speak of it as "beautiful." But along this river, the Kentucky, which I have known all my life and have lived beside for half a century, there is a large and regrettable recent change, clearly apparent to me, and to me indicative of a drastic change in water quality, but perfectly invisible to nearly everybody else.

I don't remember what year it was when I first noticed the disappearance of the native black willows from the low-water line of this river. Their absence was sufficiently noticeable, for the willows were both visually prominent and vital to the good health of the river. Wherever the banks were broken by "slips" or the uprooting of large trees, and so exposed to sunlight, the willows would come in quickly to stabilize the banks. Their bushy growth and pretty foliage gave the shores of the river a distinctive grace, now gone and much missed by the few who remember. Like most people, I don't welcome bad news, and so I said to myself that perhaps the willows were absent only from the stretch of the river that I see from my house and work places. But in 2002 for the first time in many years I had the use of a motorboat, and I examined carefully the shores of the twenty-seven-mile pool between locks one and two. I saw a few old willows at the tops of the high banks, but none at or near the low-water line, and no young ones anywhere.

The willows still live as usual along other streams in the area, and they thrive along the shore of the Ohio River just above the mouth of the Kentucky at Carrollton. The necessary conclusion is that their absence from the Kentucky River must be attributable to something seriously wrong with the water. And so, since 2002, I have asked everybody I met who might be supposed to know: "Why have the black willows disappeared from the Kentucky River?" I have put this question to conservationists, to conservation organizations specifically concerned with the Kentucky River, to water-quality officials and to university biologists. And I have found nobody who could tell me why. Except for a few old fishermen, I have found nobody who knew they were gone.

This may seem astonishing. At least, for a while, it astonished me. I thought that in a state in which water pollution is a permanent issue, people interested in water quality surely would be alert to the disappearance of a prominent member of the riparian community of a major river. But finally I saw that such ignorance is more understandable than I had thought.

A generation or so ago, when fishing and the condition of the river were primary topics of conversation in Port Royal, the disappearance of the willows certainly would have been noticed. Fishermen used to tie their trotlines to the willows. That time, as I've said, is past, and I was seeking local knowledge from conservationists and experts and expert conservationists. But most conservationists, like most people now, are city people. They "escape" their urban circumstances and preoccupations by going on vacations. They thus go into the countryside only occasionally, and their vacations are unlikely to take them into the economic landscapes. They want to go to parks, wilderness areas, or other famous "destinations." Government and university scientists often have economic concerns or responsibilities, and some of them do venture into farmland or working forests or onto streams and rivers that are not "wild." But it seems they are not likely to have a particular or personal or long-term interest in such places, or to go back to them repeatedly and often over a long time, or to maintain an economic or recreational connection to them. Such scientists affect the eyes-to-acres ratio probably less than the industrial farmers.

It seems to me significant, and not a bit surprising, that among the many conservationists I have encountered in my home state, the most competent witness by far is Barth Johnson, a retired game warden who is a dedicated trapper, hunter, and fisherman, as he has been all his life. Barth has devoted much of his life to conservation. Like most conservationists he is informed about issues and problems. Unlike most, he is exceptionally alert to what is happening in the actual countryside that needs to be conserved. This is because he is connected to the neighboring fields and woods and waters by bonds of economy and pleasure, both at once. Because those bonds are long-sustained and continued from year to year, he knows those places well. He has, as a matter of course, a year-round interest in the health—which is, in the best sense, the productivity—of the countryside. Moreover, he has lived for thirty years in the same place at the lower end of the Licking River. This greatly increases the value of his knowledge, for he can speak of changes *over time*. People who stay put and remain attentive know that the countryside changes, as it must, and for better or worse.

In his grasp of ecological principles, Barth is completely up with the times, and in some ways maybe ahead of the times, but he also is genuinely a countryman of a kind that is old and now rare. As opposed to writers, the best story *tellers* are people who spend a lot of time outdoors. Barth is well supplied with stories, most of them funny, all of them interesting, but some of them could be better classed as "reports."

He tells a story about Harris Creek, a small stream along which he had trapped for many years. It was richly productive, and Barth was careful never to ask too much of it. But in 2007, confident that it would be as it always had been, he went there with his traps and discovered that the stream was dead. He could not find a live minnow or crawfish. There were no animal tracks. So far as he could tell, there could be only one reason for this: In the spring of that year, the bottomland along the creek had been herbicided in preparation for a seeding of alfalfa. In 2008, the stream was still dead. In 2009, there was "a little coon activity." Finally, in 2013, the stream was "close to normal."

I have also learned from Barth that upstream as far as he has looked, to a point two and a half miles above the small town of Boston, the black willows are gone from the Licking River. And in October 2013, he wrote me that the river had turned a brownish "brine" color that he had never seen before.

What happened to the willows? Two young biologists at Northern Kentucky University are now at work on the question, and perhaps they will find the answer. But other scientists have led me to consider the possibility that such questions may never be answered. It may be extremely difficult or impossible to attach a specific effect to a specific cause in a large volume of flowing water.

What killed Harris Creek? Barth's evidence is "anecdotal," without scientifically respectable proof. I have read scientific papers establishing that the herbicide glyphosate and its "degradation products" are present in high concentrations in some Mississippi River tributaries, but the papers say nothing about the effects. I have called up scientists working on water quality, including one of the authors of one of the papers on glyphosate. What about the *effects*? Good question. Nobody knows the answer. It seems that the research projects and the researchers are widely scattered, making such work somewhat incoherent. And besides there is always the difficulty of pinning a specific cause to a specific effect. To two of these completely friendly and obliging people I told Barth's story of Harris Creek: Does that surprise you? One said it did not surprise him. The other said it was possible but unlikely that the stream was killed by an herbicide. Was an insecticide also involved?

What caused the strange discoloration of the Licking River? Since the discoloration was visible until obscured by mud in the water when the river rose, I suppose that, if it happens again, the odd color could be traced upstream to a source.*

*As of July 2014, it has not happened again.

Will somebody do that? I don't know. Is any scientist from any official body monitoring the chemical runoff from croplands and other likely sources? I have been asking that question too, and so far I have asked nobody who could answer.

In my search for answers, it may be that I have been making a characteristic modern mistake of relying on experts, which has revealed a characteristic modern failure: Experts often don't know and sometimes can never know. Beneficiaries of higher education, of whom I am one, often give too much credit to credentials.

By now, my interest has necessarily shifted from an attempt to find answers to an attempt to understand the implications, not just of my failure so far to find answers, but of my failure so far to find any reason to think that answers are likely to be found. And so I must back up to what I do actually know and start again.

I know that the willows have disappeared, apparently because of some toxic chemical in the water. And I know that this is the result of a scientific success in developing a chemical that serves some industrial purpose, but with water pollution as a side effect. It is also the result of a scientific failure to notice—or to care, if noticed—that this chemical has "accidentally" polluted the river.

The countryside and its waterways are now being significantly, recognizably, and sometimes measurably damaged by industrialized science. Because of scientific success and scientific failure, toxic chemicals are on the loose, the effects of which the available science does not know and therefore cannot hope to remedy. This obviously is beyond the competence of people who are not scientists. If anybody now is going to maintain an effective vigilance over the runoff of toxic chemicals from crop monocultures or surface mines or other industrial sites, that vigilance will have to be maintained by scientists (such as they are). But there is no indication that there is enough money anywhere to hire enough scientists to do this. And how could enough scientists be found to watch carefully enough over such an extent of country and so many miles of creeks and rivers?

To be fair, there are a lot of amateurs who voluntarily watch over some of our waterways, and some of these people become competent at testing for the presence of pollutants. But such vigilance, to be effective, must be constant and sustained over a long time, and I am unsure of the extent to which this has been, or can be, achieved. The problem is the "mobility" of our people, which obviously limits their ability to pay attention to places.

In the middle of the last century we had within easy reach the possibility of preserving a right ratio of eyes to acres. From people and ways then in place, we could have cultivated, educated, and encouraged a population of vigilant land stewards attached to our economic landscapes by bonds of economy, pleasure, affection, and long memory, possessing the cultural means and imperatives of good care. We not only abandoned that possibility but condemned it as of no worth, substituting a program of industrial innovation, scientific responsibility, and governmental regulation that appears definitively to have failed. Of the "spill" of 4-methylcyclohexane methanol that contaminated the water supply of 300,000 West Virginians, a *New York Times* editorial noticed that this was "the third major chemical accident in the region in five years," adding that "the EPA has tested just 200 of the roughly 85,000 chemicals in use today." The *Times* editors called for "meaningful reform." But by what conceivable reform could any agency locate, in all their inevitable wanderings, transformations, and combinations, and make harmless 85,000 chemicals already in use?

As another example, the Water Quality Branch of the Kentucky state government's Division of Water has nine biologists to monitor the state's 92,000 surface miles of streams. That is a biologist-to-miles ratio of 1 to more than 10,000. What, one has to ask, can be the adequacy of one biologist per 10,000 miles of streams? Or: What would be the value of one biologist per 10,000 miles compared to the value of one vigilant and stewardly farmer or trapper every 10 or even 100 miles—with, let us say, a staff of public biologists to help when there is a question or a problem?

Confronting industrial agriculture in particular, we are requiring ourselves to substitute science for citizenship, community membership, and land stewardship. But science fails at all of these. Science as it now predominantly is, by definition and on its own terms, does not make itself accountable for unintended effects. The intended effect of chemical nitrogen fertilizer, for example, is to grow corn, whereas its known effect on the Mississippi River and the Gulf of Mexico is a catastrophic accident. Moreover science of this kind is invariably limited and controlled by the corporations that pay for it.

In the economic landscapes, now so desperately in want of proper stewards and so little watched over, the industrial economy is free to do anything it can do, according to its wishes. This is a triumph of laissez-faire economics riding upon a triumph of its laissez-faire science—a science free to invent causes, and free of worry about effects. This triumph, highly complicated, highly profitable to a few corporations, intelligible (somewhat) only to experts, is nevertheless limited by its dependence on exhaustible materials, including, as now used, the soil. The bad ecological effects, even more complicated, and as now "regulated," may be limitable only by disease and death. This is the industrial version of the human predicament.

To this the only rational response is that it has been a mistake to allow industrialism, from the beginning, to measure its conduct exclusively by its own success, using such standards as mechanical efficiency or monetary profit, and ignoring all else. This great project of continuous technical innovation and obsolescence, substituting technologies for human workers, was intentionally wrong from the start in its evasion of the long-established moral requirement of neighborliness. Its ecological, and therefore its long-term economic, wrongs may at first have been to some degree innocent, for the earliest industrial abuses of nature were comparatively slight. No doubt the availability of long-distance transportation from "remote" areas encouraged the assumption that the world's supply of raw materials was all but infinite. And the availability in North America of "new land" to the west for three hundred years encouraged, and more or less subsidized, land abuses in the east. Now, though it certainly is possible to know better, and many people do, most industrialists still have not learned better. The frontier superstitions of the inexhaustibility of natural supplies and of the adequacy of human concern still prevail in the face of the overwhelming evidence against them.

Our original and continuing mistake has been to ignore the probability, even the inevitability, of a formal misfitting between the human economy and the economy of nature, or between economy and ecology. This misfitting has been dangerous and damaging at least since the beginnings of agriculture. The reason for this is the limited competence of the human mind, which will never fully comprehend the forms and functions of the natural world. With the development of industrialism, this misfitting has become increasingly a contradiction or opposition between industrial technologies and the creatures of nature, tending always toward the destruction of creatures, creaturely habitat, and creaturely life. We can respond rationally to this predicament only by honest worry, unrelenting caution, and propriety of scale. We must not put too much, let alone everything, at risk. We must never tolerate permanent damage to the ecosphere or to any of its parts. We must not, because we cannot for very long, tolerate compromises with soil erosion and agricultural poisons.

It is anyhow clear that if we want to do better, we will have to recognize the old mistake as a mistake: no more euphemisms such as "creative destruction," no more "sacrificing" of a present good for "greater good in the future." We will have to repudiate the too-simple industrial standards and replace them with the comprehensive standard of ecological health, realizing that this standard involves necessarily the humane obligation of neighborliness both to other humans and to other creatures. This means that all our uses of the natural world must be governed by our willingness to learn the nature of every place, and to submit to nature's limits and requirements for the use of every place. In short, agriculture and forestry must finally submit to ecology. Mining of course is not natural by any stretch of the term. Unlike agriculture and forestry, it cannot be made analogous to natural processes, because by method and purpose it exhausts its sources. But, at least, it must no longer be allowed to extract wealth from under the ground by destroying the invariably greater, self-renewing wealth that is on the surface. The human economy must operate, to the always extendable limits of its ability, as a good neighbor in both the natural and the human communities, because in the long run the health of one is the health of the other.

*

The history of industrialism has been an ever-ramifying series of substitutions. It began and has continued by substitutions of two kinds: the substitution of hotter to hottest fuels for the cooler energies of gravity, sunlight, wind, and food; and the substitution of ever more "sophisticated" technologies for human work and care.

The side effects of our use of fossil and nuclear fuels are too infamous to require much notice here. Their extraction has been increasingly, and often irreparably, damaging to the ecosphere, as well as to human communities and workers. Their use produces enormous volumes of earthborne and airborne poisons that human intelligence has never learned what to do with, probably because the safe disposal of permanently dangerous substances cannot be learned. The farce of such "disposability" is a major theme of the history of industrialism.

Another major theme is the disposability of people, and this is not a farce. It is one of the versions of "creative destruction," which is to say the theme of heartlessness, heartbreak, and permanent damage to people and their communities, endlessly repeated from the beginning, and with no proposed or theoretical end. We now use "Luddite" as a term of contempt, and this usage, often by people who consider themselves compassionate and humane, implies a sort of progressivist etiquette by which, in the interest of the future (of the more fortunate), we are to submit passively to our obsolescence, disemployment, displacement, and (likely enough) impoverishment. We smear this over with talk of social mobility, upward mobility, and retraining, but this is as false and cynical as the association of "safe" with the extraction, transportation, and use of fossil and nuclear fuels. Customarily we ignore the possibility that people's knowledge, intelligence, and skills may be needed in the places where their minds were formed, also the likelihood that such assets will become worthless as soon as the people leave home. We seem to have overlooked as well the possibility of using technologies to ameliorate the working conditions and the lives of workers while keeping them in place. Coal companies, for example, readily have used the newest technology to speed and cheapen the extraction of coal, but they often have used it reluctantly or too late to provide for the health and safety of their employees.

Replacement of workers by machines becomes more serious when it is enabled by the degradation of work. We have ignored the limits of compatibility between labor-saving and good work. And we have degraded almost all work by reducing the generously qualitative idea of "vocation" or "calling" to the merely quantitative integer, "a job." The purpose of education now is to make everybody eligible for "a job." A primary function of politics is "job creation." Persons deprived of work that they have loved and enjoyed and performed with pride are to consider their loss well-remedied by some form of "welfare" or "another job."

The idea of vocation attaches to work a cluster of other ideas, including devotion, skill, pride, pleasure, the good stewardship of means and materials. Here we have returned to intangibles of economic value. When they are subtracted, what remains is "a job," always implying that work is something good only to escape: "Thank God it's Friday." "A job" pretty much equals bad work, which can be performed as well or better by a machine. Once the scale and speed of farmwork have overridden any care for the health of the land community and any pride in the beauty of the farm, then we can talk, as we now are talking, of farming by remote control.

If such substitutions appear to work, we must consider the likelihood that they work only temporarily or according to criteria that are too simple or false. And we must acknowledge that some do not work at all: A "service economy" is immediately a falsehood when it is staffed by phone-answering robots. The computerization and robotification of the United States Postal "Service," far from improving the service, has impaired its ability to transport and deliver the mail on time or at any time. Somewhere along the line of industrial substitutions, it appears certain that we will find ourselves again confined within, and sharing the fate of, the natural, naturally-limited, industrially-depleted world that we have thought to transcend by more and more engineering. And then we will know, needing no experts to tell us, that our world, like our bodies, cannot survive unstanched bleeding and repeated doses of poison.

*

To replace industrial standards by the standard of ecological health would undo a failed substitution. It would not undo the history and the legacy of industrialism. We are not going to have the privilege of a clean slate or a new start. A change of standards would certainly change our ways of living and working, but those changes, even as they are chosen or forced upon us, cannot come with the speed of our various technological revolutions. They will require more patience than haste. (It may be that we can keep without harm some industrial comforts: warm baths in wintertime maybe, maybe painless dentistry.)

Though a clean slate is impossible, as it has always been, we are not destitute of instructions and examples. Though our present anxieties incline us toward theories and illustrations of the natural rapaciousness of humans, not all humans and not all human communities have been so. I don't think the present bunch of living humans can be allowed to make the (very restful) claim that there is nothing they can do, pleading the incorrigibility of their nature or their circumstances. And so I will end this essay with an inventory of the resources we have at hand that will support us in our effort to do better. This is *an* inventory, not *the* inventory. My obligation here is only to show that we do have resources, probably enough, if we would pay attention to them.

1. First and fundamental are the examples of nature's own ways of land care in the native ecosystems that precede the human economy, also in some humanly modified ecosystems that preceded the industrial economy. These ways have been carefully studied by agricultural scientists from Albert Howard, Aldo Leopold, and others before the middle of the last century to Wes Jackson and his colleagues at The Land Institute in Kansas right now. One of our most urgent needs is for ecologists, trained in the native forests and prairies, to apply their knowledge in the economic landscapes. We humans, because of our limited intelligence, are never going to understand perfectly the nature of any place, or fit our local economies perfectly into local ecosystems. But the right instructions, or the right "model," for our use of a place can come only from the nature of the place.

*

2. We have from all over the world, from F. H. King's *Farmers of Forty Centuries*, 1911, to the ongoing work of Vandana Shiva, examples of long-enduring traditional or peasant agricultures. These seem to have come from respectful and competent observation of the nature of places, and they have embodied natural processes or analogues thereof.

3. We have Thomas Jefferson's great principle: "The small landholders are the most precious part of a state." If we understand that the state and the state's economy depend upon the land, and if we understand the numerousness, diversity, uniqueness, and vulnerability of the land's small places, and so their dependence upon human care, we cannot doubt the truth of Jefferson's rule.

4. Scattered about the country, and concentrated in some Amish communities, we have, still surviving, farmers and foresters who have held to established good practices and tried for better. It is not altogether a secret that in many a rural community you will find farmers or memories of farmers who have survived, even prospered, in hard times by limiting their farms to a manageable size, by taking care of them, by avoiding or minimizing debt, by limiting or avoiding purchases of new equipment, by subsisting so far as possible from their own land, and in general by saving rather than spending.

5. Over the last two or three decades, there has been a growing national and international movement toward local economies, starting with economies of local food. This has been little noticed and poorly understood by the news media, mostly ignored in the state capitals and in Washington, D.C. Some city and county governments, however, have taken notice of this movement and understood its importance. For example, the local food effort is now well established as a part of the economic development plan of Louisville, Kentucky. To bring a local demand and a local supply into existence more or less simultaneously and on the principle of cooperation is a task obviously difficult, complicated, and long. But it has begun (there are now many examples in many places), and from nothing only a few years ago it has come a remarkable distance, though

the distance ahead is much longer. This project involves food production *in* cities as well as around them, and it is fostering a necessary urban agrarianism among gardeners and consumers.

6. We have a fairly long history of organic farming and gardening. We need to say, I think, that organic is as organic does. The term has often been too negatively defined (You *don't* use chemicals), and it can be too loosely defined, but it has always implied the standard of good health, and it seems by now to have escaped its old shadow of kookiness.

7. For half a dozen years we have had the 50-Year Farm Bill, which addresses the specifically agricultural problems of soil erosion, toxicity, loss of diversity, and the destruction of rural communities. It proposes to invert the ratio of 80 percent annual and 20 percent perennial crops at present to 20 percent annual and 80 percent perennial crops in fifty years. This bill certainly comes from beyond the present margin of farm policy, and of course the habitual objectors are happy to point out that there is at present no chance of its passage. But such a farsighted bill could not have been written by people foolish enough to believe in its immediate passage.

8. The 50-Year Farm Bill was published in 2009 by The Land Institute in Salina, Kansas, with the concurrence of other farm and conservation groups all over the country. That it came from The Land Institute should be no surprise, for that organization's great project for nearly forty years has been the perennialization of agriculture, arriving finally at perennial grainfields of mixed species, in analogy to the native prairie. This project of perennial grain crops has now spread from Kansas to other states and other countries. The first of the necessary medley of perennial grains, a domesticated intermediate wheatgrass, is already well developed and may be ready for distribution to farmers in eight to ten years. That in itself is a remarkable achievement, though much remains to be done. But I want to suggest that the kind of science practiced at The Land Institute is itself a great and necessary resource. It is by definition a local science, carried on conscientiously in the contexts of the local ecosystem and the local human community. Whatever

is developed in that place will require local adaptation, and the careful employment of many minds in other places. This science, moreover, is carried on with respect for local nature and local humanity. It is not going to produce a poison or an explosive.

9. This is somewhat speculative, since I am not an economist or an accountant, but I believe that our present systems of economy and accounting can be greatly and usefully improved simply by becoming more inclusive and more honest. My impression is that the prominent economists who advise and influence the government pay little attention to the so-called economy's basis in nature or to the use and produce of the land. The ruling assumption seems to be that a thriving economy can be sustained by abuse of the land and of the people who do the land's work. I remember, of course, my argument that some things of real economic worth cannot be quantified and accounted. But there should be a full and fair accounting of things that are accountable. Many of the costs of water pollution, for example, can be determined, and those costs should be charged to water-polluting industries and their customers. We need to be less interested in the "growth" of our gross income, and more interested in the subtraction of real costs and in net benefits.

10. Finally, and most necessarily, we have the ancient and long-enduring cultural imperative of neighborly love and work. This becomes ever more important as hardly imaginable suffering is imposed upon all creatures by industrial tools and industrial weapons. If we are to continue, in our only world, with any hope of thriving in it, we will have to expect neighborly behavior of sciences, of industries, and of governments, just as we expect it of our citizens in their neighborhoods.

On Being Asked for "A Narrative for the Future"

So far as I can see, the future has no narrative. The future does not exist until it has become the past. To a very limited extent, prediction has worked. The sun, so far, has set and risen as we have expected it to do. And the world, I suppose, will predictably end, but all of its predicted deadlines, so far, have been wrong.

The End of Something—history, the novel, Christianity, the human race, the world—has long been an irresistible subject. Many of the things predicted to end have so far continued, evidently to the embarrassment of none of the predictors. The future has been equally, and relatedly, an irresistible subject. How can so many people of certified intelligence have written so many pages on a subject about which nobody knows anything? Perhaps we need a book—in case we don't already have one—on the end of the future.

None of us knows the future. Fairly predictably, we are going to be surprised by it. That is why "Take . . . no thought for the morrow . . ." is such excellent advice. Taking thought for the morrow is, fairly predictably, a waste of time.

I have noticed, for example, that most of the bad possibilities I have worried about have never happened. And so I have taken care to worry about all the bad possibilities I could think of, in order to keep them from happening. Some of my scientific friends will call this a superstition, but if I did not forestall all those calamities, who did? However, after so much good work, even I must concede that by taking thought for the morrow we have invested, and wasted, a lot of effort in preparing for morrows that never came. Also by taking thought for the morrow we repeatedly burden today with undoing the damage and waste of false expectations—and so delaying our confrontation with the actuality that today has brought.

The question, of course, will come: If we take no thought for the morrow, how will we be *prepared* for the morrow?

I am not an accredited interpreter of Scripture, but taking thought for the morrow is a waste of time, I believe, because all we can do to prepare rightly for tomorrow is to do the right things today.

The passage continues: "for the morrow shall take thought for the things of itself. Sufficient unto the day is the evil thereof." The evil of the day, as we know, enters into it from the past. And so the first right thing we must do today is to take thought of our history. We must act daily as critics of history so as to prevent, so far as we can, the evils of yesterday from infecting today.

Another right thing we must do today is to appreciate the day itself and all that is good in it. This also is sound biblical advice, but good sense and good manners tell us the same. To fail to enjoy the good things that are enjoyable is impoverishing and ungrateful.

The one other right thing we must do today is to provide against want. Here the difference between "prediction" and "provision" is crucial. To predict is to foretell, as if we know what is going to happen. Prediction often applies to unprecedented events: human-caused climate change, the end of the world, etc. Prediction is "futurology." To provide, literally, is to see ahead. But in common usage it is to look ahead. Our ordinary, daily understanding seems to have accepted long ago that our capacity to see ahead is feeble. The sense of "provision" and "providing" comes from the past, and is informed by precedent.

Provision informs us that on a critical day—St. Patrick's Day, or in a certain phase of the moon, or when the time has come and the ground is ready—the right thing to do is plant potatoes. We don't do this because we have predicted a bountiful harvest; history warns us against that. We plant potatoes because history informs us that hunger is possible, and we must do what we can to provide against it. We know from the past only that, if we plant potatoes today, the harvest *might* be bountiful, but we can't be sure, and so provision requires us to think today also of a diversity of food crops.

What we must *not* do in our efforts of provision is to waste or permanently destroy anything of value. History informs us that the things we waste or destroy today may be needed on the morrow. This obviously prohibits the "creative destruction" of the industrialists and industrial economists, who think that evil is permissible today for the sake of greater good tomorrow. There is no rational argument for compromise with soil erosion or toxic pollution.

For me—and most people are like me in this respect—"climate change" is an issue of faith; I must either trust or distrust the scientific experts who predict the future of the climate. I know from my experience, from the memories of my elders, from certain features of my home landscape, from reading history, that over the last 150 years or so the *weather* has changed and is changing. I know without doubt that to change is the nature of weather.

Just so, I know from as many reasons that the alleged causes of climate change—waste and pollution—are wrong. The right thing to do today, as always, is to stop, or start stopping, our habit of wasting and poisoning the good and beautiful things of the world, which once were called "divine gifts" and now are called "natural resources." I always suppose that experts may be wrong. But even if they are wrong about the alleged human causes of climate change, we have nothing to lose, and much to gain, by trusting these particular experts.

Even so, we are not dummies, and we can see that for all of us to stop, or start stopping, our waste and destruction today would be difficult. And so we chase our thoughts off into the morrow where we can resign ourselves to "the end of life as we know it" and come to rest, or start devising heroic methods and technologies for coping with a changed climate. The technologies will help, if not us, then the corporations that will sell them to us at a profit.

I have let the preceding paragraph rest for two days to see if I think it is fair. I think it is fair. As evidence, I will mention only that, while the theme of climate change grows ever more famous and fearful, land abuse is growing worse, noticed by almost nobody.

A steady stream of poisons is flowing from our croplands into the air and water. The land itself continues to flow or blow away, and in some places erosion is getting worse. High grain prices are now pushing soybeans and corn onto more and more sloping land, and "no-till" technology does not prevent erosion on continuously cropped grainfields. Industrial agriculture, moreover, is entirely dependent on burning the fossil fuels, the most notorious of the alleged causes of climate change.

Climate change, supposedly, is recent. It is apocalyptic, "big news," and the certified smart people all are talking about it, thinking about it, getting ready to deal with it in the future.

Land abuse, by contrast, is ancient as well as contemporary. There is nothing futurological about it. It has been happening a long time, it is still happening, and it is getting worse. Most people have not heard of it. Most people would not know it if they saw it.

The laws for conservation of land in use were set forth by Sir Albert Howard in the middle of the last century. They were nature's laws, he said, and he was right. Those laws are the basis of the 50-Year Farm Bill, which outlines a program of work that can be started now, which would help with climate change, but which needs to be done anyhow. Millions of environmentalists and wilderness preservers are dependably worried about climate change. But they are not conversant with nature's laws, they know and care nothing about land use, and they have never heard of Albert Howard or the 50-Year Farm Bill.

II

If we understand that Nature can be an economic asset, a help and ally, to those who obey her laws, then we can see that she can help us now. There is work to do now that will make us her friends, and we will worry less about the future. We can begin backing out of the future into the present, where we are alive, where we belong. To the extent that we have moved out of the future, we also have moved out of "the environment" into the actual places where we actually are living.

If, on the contrary, we have our minds set in the future, where we are sure that climate change is going to play hell with the environment, we have entered into a convergence of abstractions that makes it difficult to think or do anything in particular. If we think the future damage of climate change to the environment is a big problem only solvable by a big solution, then thinking or doing something in particular becomes more difficult, perhaps impossible.

It is true that changes in governmental policy, if the changes were made according to the right principles, would have to be rated as big solutions. Such big solutions surely would help, and a number of times I have tramped the streets to promote them, but just as surely they would fail if not accompanied by small solutions. And here we come to the reassuring difference between changes in policy and changes in principle. The needed policy changes, though addressed to present evils, wait upon the future, and so are presently nonexistent. But changes in principle can be made now, by so few as just one of us. Changes in principle, carried into practice, are necessarily small changes made at home by one of us or a few of us. Innumerable small solutions emerge as the changed principles are adapted to unique lives in unique small places. Such small solutions do not wait upon the future. Insofar as they are possible now, exist now, are actual and exemplary now, they give hope. Hope, I concede, is for the future. Our nature seems to require us to hope that our life and the world's life will continue into the future. Even so, the future offers no validation of this hope. That validation is to be found only in the knowledge, the history, the good work, and the good examples that are now at hand.

There is in fact much at hand and in reach that is good, useful, encouraging, and full of promise, although we seem less and less inclined to attend to or value what is at hand. We are always ready to set aside our present life, even our present happiness, to peruse the menu of future exterminations. If the future is threatened by the present, which it undoubtedly is, then the present is more threatened, and often is annihilated, by the future. "Oh, oh, oh," cry the funerary experts, looking ahead through their black veils. "Life as we know it soon will end. If the governments don't stop us, we're going to destroy the world. The time is coming when we will have to do something to save the world. The time is coming when it will be too late to save the world. Oh, oh, oh." If that is the way our minds are afflicted, we and our world are dead already. The present is going by and we are not in it. Maybe when the present is past, we will enjoy sitting in dark rooms and looking at pictures of it, even as the present keeps arriving in our absence.

Or maybe we could give up saving the world and start to live savingly in it. If using less energy would be a good idea for the future, that is because it is a good idea. The government could enforce such a saving by rationing fuels, citing the many good reasons, as it did during World War II. If the government should do something so sensible, I would respect it much more than I do. But to wish for good sense from the government only displaces good sense into the future, where it is of no use to anybody and is soon overcome by prophesies of doom. On the contrary, so few as just one of us can save energy right now by self-control, careful thought, and remembering the lost virtue of frugality. Spending less, burning less, traveling less may be a relief. A cooler, slower life may make us happier, more present to ourselves, and to others who need us to be present. Because of such rewards, a large problem may be effectively addressed by the many small solutions that, after all, are necessary, no matter what the government might do. The government might even do the right thing at last by imitating the people.

In this essay and elsewhere, I have advocated for the 50-Year Farm Bill, another big solution I am doing my best to promote, but not because it will be good in or for the future. I am for it because it is good now, according to present understanding of present needs. I know that it is good now because its principles are now satisfactorily practiced by many (though not nearly enough) farmers. Only the present good is good. It is the presence of good—good work, good thoughts, good acts, good places—by which we know that the present does not have to be a nightmare of the future. "The kingdom of heaven is at hand" because, if not at hand, it is nowhere.

FROM

THE ART OF LOADING BRUSH

(2017)

The Thought of Limits in a Prodigal Age

Is there, at bottom, any real distinction between esthetics and economics?

—Aldo Leopold, *A Sand County Almanac & Other Writings on Ecology and Conservation*

I WANT TO SAY something about the decline, the virtual ruin, of rural life, and about the influence and effect of agricultural surpluses, which I believe are accountable for more destruction of land and people than any other economic "factor." This is a task that ought to be taken up by an economist, which I am not. But economists, even agricultural economists, farm-raised as many of them have been, do not live in rural communities, as I do, and they appear not to care, as I do, that rural communities like mine all over the country are either dying or dead. And so, only partly qualified as I am, I will undertake this writing in the hope that I am contributing to a conversation that will attract others better qualified.

I have at hand an article from the *Wall Street Journal* of February 22, 2016, entitled "The U.S. Economy Is in Good Shape." The article is by Martin Feldstein, "chairman of the Council of Economic Advisors under President Ronald Reagan . . . a professor at Harvard and a member of the Journal's board of contributors." Among economists Prof. Feldstein appears to be somewhere near the top of the pile. And yet his economic optimism is founded entirely upon current measures of "incomes," "unemployment," and "industrial production," all abstractions narrowly focused. Nowhere in his analysis does he mention the natural world, or the economies of land use by which the wealth of nature is made available to the "American economy." Mr. Feldstein believes that "the big uncertainties that now hang over our economy are political."

But from what I see here at home in the watershed of the Kentucky River, and from what I have seen and learned of other places, I know that industrial agriculture is in serious failure, which is to say that it is not sustainable. Projecting from the damages of the comparatively brief American histories of states such as Kentucky and Iowa, one must conclude that the present use of the farmland cannot be sustained for another hundred years: The rates of soil erosion are too high, the run-off is too toxic, the ecological impoverishment is too great, the surviving farmers are too few and too old. To anybody who knows these things, by witness of sight or by numerical measures, they would appear to qualify significantly the "good shape" of the economy. I conclude that Prof. Feldstein does not know these things, but is conventionally ignorant of them. Like other people of privilege for thousands of years, far more numerous now than ever before, he appears to take for granted the bounty of nature and the work that provides it to the human economy.

In remarkable contrast to the optimism of Prof. Feldstein, the *New York Times* of March 10, 2016, printed an article, "Who's Killing Global Growth?" by Steven Rattner, "a Wall Street executive and a contributing opinion writer." Mr. Rattner's downhearted assessment, like Prof. Feldstein's upbeat one, is based upon measures that are entirely economic or monetary, quantitative, and abstract: "financial markets," "projections for future growth," "wages," "consumer spending," "rising supply," "disappointing demand," etc. The "global growth" Mr. Rattner has in mind is purely financial and is without reference to the effect of such "growth" upon the health and the welfare of the globe's actual people and other creatures. Like Prof. Feldstein, Mr. Rattner appears to suppose, consciously or not, that the natural world and human workers will continue to supply their necessary goods without limit and to the allure simply of money.

I have at hand also a sentence from the *New York Review of Books*, September 24, 2015, by James Surowiecki, another highly credentialed economist. Mr. Surowiecki is reviewing among others a book by Joseph E. Stiglitz and Bruce C. Greenwald, *Creating a Learning Society: A New Approach to Growth, Development, and Social Progress*. This book, the reviewer says, "is dedicated to showing how developing countries can use government policy to become high-growth, knowledge-intensive economies, rather than remaining low-cost producers of commodities." I have kept this sentence in mind because of the problems it raises, all relating to my concern about the damages imposed by national and global "economies" upon land and people. Mr. Surowiecki's sentence seems to be highly condensed and allusive, a sort of formula for increasing economic growth—or, as it actually says, for turning countries into economies. The sentence no doubt is clear to economists, but it has put me to some trouble. My interest is not in the analyses and theories of these economists, since they seem mainly to ignore the natural world and the human communities that are my concern. I am interested here in their public language, by which they reveal what they accept, and expect most others to accept, as axioms—what one might call their lore or more accurately their faith.

I assume, then, that by "low-cost producers of commodities" Mr. Surowiecki means "poorly paid producers of cheap commodities," that these commodities are material goods or raw materials produced from the land, that "knowledge-intensive economies" are based upon the abilities to exploit, trade, add value to, and market the cheaply produced commodities. Apparently it is taken for granted that this improving formula applies to *all* developing countries, their people, their land, and their natural resources, without regard to differences or distinctions among them. Such disregard of local and personal differences is a major article of this faith. It takes for granted furthermore that a knowledge-intensive economy, by causing growth, development, and social progress, will change a developing country into a developed country, and that this will be an all-around improvement. From the standpoint of industrial economists and their clients, this apparently is self-evident and unquestionable. It becomes immediately and urgently questionable from the standpoint of a dweller in a rural countryside who is bound to the land and the community by ties of history, family, and affection.

Here we arrive at a fundamental division of interest and allegiance, as probably also at the difference between two kinds of mind. The attention of these economists and others like them is directed as a matter of course to the monetary economy and to what, according to their abstract measurements, is good or bad for it. The attention of settled dwellers, at home in their chosen or hereditary places, is directed partially to the monetary economy, of course, and often in fear or sufferance, but their attention is directed also, out of natural affection and solicitude, to their places, the particular, unique, and irreplaceable patches of ground under their feet. Another difference involved here, if the settled dwellers are farmers, is that between people whose livelihoods are primarily dependent upon salaries and people whose livelihoods are primarily dependent upon the weather.

The settled dwellers, then, in their natural desire to remain settled, and facing the "promises" of development, certainly are going to have questions for the developers, and the first would be this: What is the net good that industrial economists, their employers, and their clients appear customarily to credit to growth, development, and "social progress"? In the United States, since at least the Civil War, and ever more rapidly after World War II, we have achieved industrial versions of all of those goals. But almost nobody is asking what is the worth of that achievement after we deduct its ecological and human costs. We have, in fact, been turning our country into an economy as fast as possible, and we have been doing so by an unaccounted squandering of its actual, its natural and its cultural, wealth.

As a second question, we should ask why commodities, the material goods that support our life, and the work of producing them, should be "low cost" or significantly cheaper than the goods and services of a "knowledge-intensive economy." There is no reason to believe that the present market values of technological (developmental) knowledge and of commodities are absolute or in any way permanent. Nor is there reason to believe that such issues of value are, or can be, reliably settled by the free market of our present economy, or by any market. The good health of the land economies is a value that a market as such cannot consider and cannot protect. Moreover, agribusiness in all of its aspects is a knowledge-intensive system, which uses knowledge ruthlessly to control and exploit land and people.

Apparently it is assumed that a country's economy of commodity-production, which could be as diverse as the country's climate and soils permit, can safely be replaced or further depreciated by an economy of knowledge only. And so, as a third question, we must ask how secure and how beneficent is a one-product economy. Is the market for knowledge infinite in its demand, or can it be over-supplied and depressed, as the one-product economy of coal in the Appalachian coal fields has often been? And it hardly needs to be said that in the Appalachian coal fields the benefits of the coal economy to a rich and distant few has never adequately been measured against its impoverishment of the local people and their land.

Perhaps no outsider—no visiting expert, no dispassionate observer, certainly no outside investor—will notice the inherent weakness and cruelty of a one-product economy in a region or a country. But the adverse effects will certainly be visible and acutely feelable to the resident insiders. Those who live and must make their lives within the boundary of such an economy experience daily the readiness of their political leaders to endorse and excuse the destructiveness of "the economy," as well as the public unwillingness to remedy or compensate the damages to the land and the people. The Appalachian coal economy has not only inflicted immeasurable and immeasurably lasting ecological and social damage to its region, but it has also distracted attention and care from the region's other assets: its forests, soils, streams, and the (too often exported) talents of its people.

And so a fourth question: How, even in a knowledge-intensive economy, even unendingly "growing" and wealthy, are the people's needs for food, clothing, and shelter to be met? Does the development of a highly lucrative knowledge economy entirely eliminate the need for the fundamental economies of subsistence? Do people eat and wear knowledge? Do they sleep warm in it? I know very well what the far-seeing economists will answer: People earning large salaries from "high-growth, knowledge-intensive economies" will *buy* their material subsistence from "low-cost producers of commodities" at home or, if not at home, then elsewhere. It is assumed that where there is a demand, and enough money, there will be unfailingly a supply, and this is another article of the industrial economic faith: The land and its "resources" will be always with us, and so will the poor who will dig, hack, and whittle an everlasting supply of low-cost commodities until they can be replaced by knowledge-intensive machines that will dig, hack, and whittle, no doubt on solar energy, faster and at a lower cost.

And so we come to question five: Do the economists of development ever attempt a fair assessment, or any assessment, of the value to a knowledge-intensive economy of a dependable local supply of life-supporting commodities? The answer, so far as I have learned, is that the developmental economists do no such thing. Their dream of human progress calls simply for the *replacement* of the commodity economy by the knowledge economy, and that is that. As evidence, let us consider a review of our economic past and future, "Moving On from Farm and Factory," by Eduardo Porter, in the *New York Times* of April 27, 2016.

Mr. Porter's premise is that the economies of farming and manufacturing, as a fixed and final consequence of historical trends, are now obsolete, or nearly so, and his statistics are sufficiently precipitous:

> Over the course of the 20th century, farm employment in the United States dropped to 2 percent of the work force from 41 percent, even as output soared. Since 1950, manufacturing's share has shrunk to 8.5 percent of non-farm jobs, from 24 percent.

To this state of things Mr. Porter grants something like half an approval. Whereas nearly all of the "work force" once employed in farming have been "liberated . . . from their chains," he thinks that "The current transition, from manufacturing to services, is more problematic." Though for workers in the United States there are "options: health care, education and clean energy, just to name a few," these options "present big economic and political challenges." The principal challenges will be to get the politicians to abandon their promises of an increase of employment in manufacturing, and then to provide the government help necessary to make "the current transition from manufacturing to services" without too much rebellion by workers "against the changing tide." Mr. Porter's conclusion, despite these challenges, is optimistic:

> Yet just as the federal government once provided a critical push to move the economy from its agricultural past into its industrial future, so, too, could it help build a postindustrial tomorrow.

Mr. Porter's article, which clearly assumes the agreement or consent of a large number of his fellow economists and fellow citizens, rests upon the kind of assumptions that I have been calling articles of faith. Though it is certain that a lot of people, economists and others, are putting their faith in these assumptions, they are nonetheless entirely groundless. The assumptions, so far as I can trace them out, are as follows:

1. The economy of a country or a nation needs only to provide employment, it does not matter at what. And so of course no particular value can be assigned to the production of commodities.
2. So long as there is enough employment at work of some kind, a country or a nation can safely dispense with employment in sustainable farming and manufacturing, which is to say a sustainable dependence upon natural resources and the natural world.
3. Farming has little economic worth. It is of the past, and better so. Farm work involves no significant responsibilities, and requires no appreciable intelligence, knowledge, skill, or character. It is, as is often said, "mind-numbing," a servile condition from the "chains" of which all workers, even owners who work on their own farms, need to be "liberated."
4. The "output" of industrial agriculture will continue to "soar" without limit as ever more farmers are "liberated from their chains" by technology, and as technologies are continuously succeeded by more advanced technologies.
5. There is no economic or intrinsic difference between agriculture and industry: A farm is no more than a factory, a plant or an animal is no more than a machine or a substance.

6. The technological advances that have disemployed so many people from farming and manufacturing will never take away the jobs of service or postindustrial workers.
7. History, including economic history, is a forward motion, a progress, made up of irreversible changes. These changes can be established absolutely and forever in so little time as a century or two. Thus the great technological progress since perhaps the steam engine—a progress enabled by the fossil fuels, war, internal combustion, external combustion, and a sequence of poisons—will carry us right on into the (climactic? everlasting?) "postindustrial tomorrow."
8. The need for "health care, education and clean energy, just to name a few," will go securely on and on, supplying without limit the need for jobs, whereas the need for food, clothing, shelter, and manufactured goods will be supplied by *what*?

Now I must tell why, as a comparatively prosperous and settled resident of my home country in the United States, I should be as troubled as I am by the faith or superstition or future-fantasy of the economists of so-called development. My family and I live, as we know and fear, in what the orthodox economists consider a backward, under-developed, and to-be-developed country. This is "rural America," the great domestic colony that we have made of our actual country, as opposed to the nation, the government, and the economy. This particular fragment of it is called "The Golden Triangle," a wedge of country bounded by the three interstate highways connecting Louisville, Lexington, and Cincinnati. The three are connected also by rail and by air. The Triangle is bounded on its northwest side also by the Ohio River. Because it is so fortunately located with respect to transportation and markets, this area is thought (by some) to be "Golden," which is to say eminently suited to (future) development.

The landscapes within the Triangle are topographically diverse —rolling uplands, steep valley sides, fairly level bottomlands —all, though varyingly, fragile and vulnerable to various established abuses. The soils are fertile, productive, responsive to good treatment, but much diminished by erosion and misuse in the years of "settlement," severely eroded in some places, still eroding in others. The native forest is predominantly hardwood, much diminished and fragmented, suffering from diseases and invasive species, largely undervalued, neglected or ill-used, but potentially of great economic worth if well used and cared for. The watercourses are numerous, often degraded, mostly polluted by silt or chemicals or both. There is, in most years, abundant rainfall.

The three cities seem generally to be prospering and expanding, but are expectably troubled by social disintegration, drugs, poverty, traffic congestion, and violence. The towns, including county seats, are in decay or dead, preyed upon by the cities and chain stores, diseased by urban and media culture, cheap energy, family disintegration, drugs, and the various electronic screens.

Especially during the early decades of the tobacco program, the farming here was highly diversified and, at its best, exemplary in its husbanding of the land. Because of the program, tobacco was the basis of a local agrarian culture that was both economically and socially stabilizing. The farms were mostly small, farmed by their resident families and neighborly exchanges of work. In addition to tobacco and provender for the households, they produced (collectively and often individually) corn, small grains, hogs, chickens and other poultry, eggs, cream, milk, and an abundance of pasture for herds of beef cattle and flocks of sheep.

The tobacco program with its benefits ended in 2004. Though it served growers of a crop that after the Surgeon General's report of 1965 could not be defended, the program itself was exemplary. Both the people and the land benefitted from it. By the combination of price supports with production control, limiting supply to anticipated demand, the program maintained the livelihoods of the small farms, and so maintained the livelihoods of shops and stores in the towns. It gave the same protection on the market to the small producer as to the large. By limiting the acreage of a high-paying crop, it provided a significant measure of soil conservation. Most important, it supported the traditional family and social structure of the region and its culture of husbandry.

For once and for a while, then, the farmers of this region

stood together, stood up for themselves, and secured for themselves prices reasonably fair for one of their products. The tobacco program, once and for a while, gave them an asking price, with results in every way good. Before and after the program, which was *their* program, they have had simply to accept whatever the buyers have been pleased to offer. When producers of commodities have no asking price, the result is plunder of both land and people, as in any colony. By "asking price" I mean a fair price, as determined for example by "parity," which would enable farmers to prosper "on a par with" their urban counterparts; a fair price, then, supported by bargaining power.

After the demise of the tobacco program, and with it the economy and way of life it had preserved and stood for, this so-favored Triangle and its region have declined economically, agriculturally, and socially. The tobacco that is still grown here is grown mainly in large acreages under contracts written by the tobacco companies, and primarily with migrant labor. Most of the farms that are still working are mainly or exclusively producing beef cattle—which is good, insofar as it gives much of our vulnerable countryside the year-round protection of perennial pastures and hay crops, but it is a far cry from the old diversity of crops and livestock. Of much greater concern is the continuous planting of large acreages of soy beans and corn, a way of farming unsuited to our sloping land (or, in fact, to any land), erosive, toxic, requiring large expenditures of money for uncertain returns. For such cropping the fences are removed, making the land useless for grazing. Farms are being subdivided and "developed," or cash-rented for corn and beans.

The land is no longer divided and owned in the long succession, by inheritance or purchase, of farmer after farmer. It has now become "real estate," ruled by the land market, owned increasingly by urban investors, or by urban escapees seeking the (typically short-lived) consolation and relaxation of "a place in the country." The government now subsidizes land purchases by some young farmers, "helping" them by involving them in large long-term debts and in ways of farming that degrade the land they may, late in their lives, finally own. For many young people whose vocation once would have been farming, farming is no longer possible. You have to be too rich to farm before you can afford a farm in my county.

Only a few years ago, I received a letter from a man extraordinarily thoughtful, who described himself as an ex-addict whose early years were spent under the teaching and influence of a family elder, in the tobacco patches of a neighborhood of small farms. Caught up by the centrifugal force of a disintegrating community and way of life, he drifted into addiction, from which, with help, he got free. He wrote to me, I think, believing that I should know his story. People, he said, were wondering what comes after the tobacco program. He answered: drug addiction. He was right. Or he was partly right. His answer would have been complete if he had added screen addiction to drug addiction.

As long as the diverse economy of our small farms lasted, our communities were filled with people who needed one another and knew that they did. They needed one another's help in their work, and from that they needed one another's companionship. Most essentially, the grownups and the elders needed the help of the children, who thus learned the family's and the community's work and the entailed duties, pleasures, and loyalties. When that work disappears, when parents leave farm and household for town jobs, when the upbringing of the young is left largely to the schools, then the children, like their parents, live as individuals, particles, loved perhaps but not needed for any usefulness they may have or any help they might give. As the local influences weaken, the outside influences grow stronger.

And so the drugs and the screens are with us. The day is long past when most school-age children benefitted from work and instruction that gave them in turn a practical assurance of their worth. They have now mostly disappeared from the countryside and from the streets and houseyards of the towns. In this new absence and silence of the children, parents, teachers, church people, and public officials hold meetings to wonder what to do about the drug problem. The screen problem receives less attention, but it may be the worse of the two because it wears the aura of technological progress and social approval.

The old complex life, at once economic and social, was fairly coherent and self-sustaining because each community was focused upon its own local countryside and upon its own people, their needs, and their work. That life is now almost entirely gone. It has been replaced by the dispersed lives of dispersed individuals, commuting and consuming, scattering in every direction every morning, returning at night only to their screens and carryout meals. Meanwhile, in a country everywhere distressed and taxed by homelessness, once-used good farm buildings, built by local thrift and skill, rot to the ground. Good houses, that once sheltered respectable lives, stare out through sashless windows or have disappeared.

I have described briefly and I am sure inadequately my home country, a place dauntingly complex both in its natural history and in its human history, offering much that is good, much good also that is unappreciated or unrecognized. Outsiders passing through, unaware of its problems, are apt to think it very beautiful, which partly it still is. To me, and to others known to me, it is also a very needy place. When I am wishing, as I often do, I wish its children might be taught thoroughly and honestly its own history, and its history as a part of American history. I wish every one of its schools had enough biologists and ecologists to lead the students outdoors, to show them where they are in relation to drainages, soils, plants, and animals. I wish we had an economy wisely kind to the land and the people.

A good many years ago somebody, or several somebodies, named this parcel of land "The Golden Triangle." Like I assume most people here, I don't know who the somebodies were. I don't know how or what they were thinking or what their vision was. I know that the name "The Golden Triangle" is allied to other phrases or ideas, equally vague and doubtful, that have been hovering over us: the need for "job creation," the need to "bring in industry," the obligation (of apparently everybody) to "compete in the global economy," the need (of apparently everybody) for "a college education," the need for or the promises of "the service economy," and "the knowledge economy." None of these by now weary foretellings has anything in particular to do with anything that is presently here. They and the thinking they represent all gesture somewhat heroically toward "the Future," another phrase, obsessively repeated by the people out front in politics and education, signifying not much. Perhaps the most influential "future" right now is that of "the knowledge economy," as yet not here but surely expected. This means that in order to get jobs and to compete in the global economy, our eligible young people need to major in courses of science, technology, engineering, and mathematics while they are still in high school. This is the so-called STEM curriculum, dear to the hearts of our several too expensive, overadministered, underfunded, and ravenous state universities. And STEM is promoted by slurs, coming from the highest offices of state government, against such studies as literature and history.

The advantage of the STEM-emphasis to the education industry is fairly obvious. And if the great corporations of the global knowledge economy settle in The Golden Triangle or somewhere nearby, they surely will be glad to have a highly trained workforce readily available. But no supreme incarnation of the knowledge economy has yet arrived. If such an arrival is imminent or expected, that has not been announced to the natives. No doubt for that reason, the authorities have not predicted how many STEM graduates the future is going to need (and, as predicted, pay well). The possibility that the schools may turn out too many expensively educated, overspecialized STEM graduates evidently is not being considered. Nor evidently is the possibility that a surplus of such graduates, like their farming ancestors, will have no asking price, and so will come cheap to whomever may hire them. Maybe someday the people living here will have a fine, affluent Scientific, Technological, Engineered, and Mathematized Future to live in. Or, of course, maybe not.

That, anyhow, is development as we know it in The Golden Triangle. Meanwhile, our land is going to the devil, and too many of our people are addicted to drugs or screens or to mere distraction.

For a person living here, it is possible to imagine an economic project that would be locally appropriate and might actually help us. This likely is a project that could not be accomplished by economists only, but economists surely will be needed. The project would be to define a local or regional economy that, within the given limits, would be diverse, coherent, and lasting. If they were not so fad-ridden, economists might see that a knowledge economy, or any other single economy, cannot and should not occupy a whole region or a whole future. They might consider the possibility of a balance or parity of necessary occupations.

I am assuming a need for any locality, region, or nation to provide itself so far as possible with food, clothing, and shelter. Such fundamental economic provision, one would think, should be considered normal or fitting to human inhabitation of the earth. In addition to the economic benefits to local people of local supplies, a future-oriented society such as ours ought to consider the possibility that any locality might become stranded by lasting interruptions of long-distance transportation. Since for many years I have been trying to think as a pacifist, I feel a little strange in addressing issues of military strategy. But it seems preposterous to me that we should maintain an enormous, enormously expensive armory of weapons, including nuclear bombs, ready at every moment to defend a country in which most people live far from the sources of their food, clothing, and other necessities. Arguing from our leaders' own premises, then, the need for balanced local economies is obvious. From the recognition of local needs, both visible and supposable, the people of this or any region might reasonably proceed to a set of questions needing to be answered. Eventually, I think, there would be many such questions. I am sure that I don't know them all, but it is easy to foresee some of them:

1. After so long a history of diminishment and loss, what remains here, in the land and people of this place, that is valuable and worth keeping? Or: What that is here do the local people need for their own use and sustenance, and then, the local needs met, to market elsewhere?
2. What is the present use or value of the local land and its products to the local people?
3. How might we earn a sustainable income from the local land and its products? This would require adding value locally to the commodities—the goods!—coming from our farms and woodlands, but how might that be done?
4. What kinds of work are necessary to preserve and to live from the productivity of our land and people?

5. What do our people need to know, or learn and keep in mind, in order to accomplish the necessary work? The STEM courses might help, might be indispensable, but what else is needed? We are talking of course about education for livelihood, but also for responsible membership, citizenship, and stewardship.
6. What economic balances are necessary to reward adequately, and so to maintain indefinitely, the necessary work?

To answer those questions, close and patient study will be required of economists and others. The difficulty here is that, within the terms and conditions of the dominant economy over the last century and a half, the communities and economies of land use have been increasingly vulnerable. The effort to make them something like sustainable would have to begin with attention to the difference between the industrial economy of inert materials and monetary abstractions and an authentic land economy that must include the kindly husbandry of living creatures. This is the critical issue. As for many years, we are still hearing that almost any new technology will "transform farming." This implies an almost-general approval of the so far unrestrainable industrial prerogative to treat living creatures as comprising a sort of ore, and the food industry as a sort of foundry. If farming is no more than an industry to be unendingly transformed by technologies, as is still happening, then farmers can be replaced by engineers, and engineers finally by robots, in the progress toward our evident goal of human uselessness. If, on the contrary, because of the uniqueness and fragility of each one of the world's myriad of small places, the land economies must involve a creaturely affection and care, then we must look back fifty or sixty years and think again. If, as even some scientists have recognized, there are natural and human limits beyond which farming (and forestry) cannot be industrialized, then we need a more complex and particularizing language than the economists so far use.

The six questions I have proposed for my or any region do not derive from a wished-for or a predicted future. They have to do with what I would call "provision," which depends upon being attentively and responsibly present in the present. We do not, for example, love our children because of their potential to become well-trained workers in a future knowledge economy. We love them because we are alive to them in this present moment, which is the only time when we and they are alive. This love implicates us in a present need to *provide*: to be living a responsible life, which is to say a responsible economic life.

Provision, I think, is never more than caring properly for the good that you have, including your own life. As it relates to the future, provision does only what our oldest, longest experience tells us to do. We must continuously attend to our need for food, clothing, and shelter. We must care for the land, care for the forest, plant trees, plant gardens and crops, see that the brood animals are bred, keep the house and the household intact. We must teach the children. But provision does not foresee, predict, project, or theorize the future. Provision instructs us to renew the roof of our house, not to shelter us when we are old—we may die or the world may end before we are old—but so we may live under a sound roof now. Provision merely accepts the chances we must take with the weather, mortality, fallibility. Perhaps the wisest of the old sayings is "Don't count your chickens before they hatch." Provision accepts, next, the importance of diversity. Perhaps the next-wisest old saying is "Don't put all your eggs in one basket." When the bad, worse, or worst possibility presents itself, provision only continues to take the best possible care of what we have, or of what we have left.

The answers to my questions of course will affect the future. They might even bring about the "better future for our children" so famous with some politicians. But the answers will not come from the future. We must study what exists: what we know of the past, what we know now, what we can see now, if we look. It is likely that, if we look, we will see a need for the STEM disciplines, for we know already their capacity to serve some good purposes. But we will see that the need for them is limited by, for one thing, the need for other disciplines. And we will see a need also not to allow the value of highly technical knowledge to depress the value of the equally necessary and respectable knowledge of land use and land husbandry.

From its beginning, industrialism has depended on a general

willingness to ignore everything that does not serve the cheapest possible production of merchandise and, therefore, the highest possible profit. And so to look back and think again, we must acknowledge real needs that have continued through the years to be real, though unacknowledged: the need to see and respect the inescapable dependence even of our present economy, as of our lives, upon nature and the natural world; and upon the need, just as important, to see and respect our inescapable dependence upon the economies—of farming, ranching, forestry, fishing, and mining—by which the goods of nature are made serviceable to human good.

And now, because it seems to be somewhat conventionally assumed that we are "moving on from farm and factory," we need to recognize again our inescapable dependence upon manufacturing. This does not imply that we must be dependent always and for every product upon large corporations and a global economy. If manufacturing as we have known it is in decline, then that gives room to the thought of a genuinely domestic and conserving economy of provision. This would be a national economy made up of local economies, which, to an extent naturally and reasonably possible, would be complete, self-sustaining, and local in scale. For example, in a town not far from where I am writing, we have recently gained a small, clean, well-equipped, federally inspected slaughtering plant, which completes locally the connection between local pastures and local kitchens, while providing work to local people. There is no reason for this connection and provision to be more extensive. To make the connection between pasture and kitchen by way of the industrial food system is to siphon livelihoods and life itself out of the rural communities.

We also have woodlands here that could even now produce a sustainable yield of valuable hardwoods. But trees cut here at present leave here as raw lumber or saw logs, at the most minimal benefit to the community. Other places and other people may prosper on the bounty of our forest, but not our place and, except minimally for the sellers and a few workers, not our people. It is not hard, considering this, to imagine a local forest economy, made up of small enterprises that would be, within the given limits, complete and coherent, yielding local livelihoods from the good use and care of the living forest to the production of lumber for buildings to finished cabinetry. The thought of such economies is of the nature of provision, not of projection, prediction, or contingency planning. The land and the people are here now. The *present* economic questions are about the work by which land and people might thrive mutually in the best health for the longest time, starting now.

To think well of such enterprises, and of the possibility of combining them in a diverse and coherent local economy, is to think of the need for sustaining all of the necessary occupations. Because a local, a *placed*, economy would be built in sequence from the ground up, from primary production to manufacturing to marketing, a variety of occupations would be necessary. Because all occupations would be necessary, all would be equally necessary. Because of the need to keep them all adequately staffed, it would be ruinous to prefer one above another by price, custom, or social prejudice. There must be a sustained economic parity among them.

In such an economic structure the land-using occupations are primary. We must be mindful of what is, or should be, the fundamental difference between agriculture or forestry and mining, but until the farmers, ranchers, foresters, and miners have done their work, nothing else that we count as economic can happen. And unless the land users do their work *well*—which is to say without depleting the fertility of the earth's surface—nothing we count as economic can happen for very long.

The land-using occupations, then, are of primary importance, but they are also the most vulnerable. We must notice, to begin with, that almost nobody in the supposedly "higher" occupational and social strata has ever recognized the estimable care, intelligence, knowledge, and artistry required to use the land without degrading or destroying it. It is as customary now as it was in the Middle Ages to regard farmers as churls—"mind-numbed," backward, laughable, and dispensable. Farmers may be the last minority that even liberals freely stereotype and insult. If farmers live and work in an economic squeeze between inflated purchases and depressed sales, if their earnings are severely depressed by surplus production, if they are priced out of the land market, it is assumed that they deserve no better: They need only to be "liberated from their chains."

*

The problem to be dealt with here is that the primary producers in agriculture and forestry do not work well inevitably. On the contrary, in our present economy there are constraints and even incentives that favor bad work, the result of which is waste of fertility and of the land itself. Good work in the use of the land is work that goes beyond production to maintenance. Production must not reduce productivity. Every mine eventually will be exhausted. But where the laws of nature are obeyed in use—as we know they can be, given sufficient care and skill—a farm, a ranch, or a forest will remain fertile and productive as long as nature lasts. Good work also is informed by traditional, locally adapted ways that must be passed down, taught and learned, generation after generation. The standard of such work, as the lineages of good farmers and of agrarian scientists have demonstrated, cannot be established only by "the market." The standard must be partly economic, for people have to live, but it must be equally ecological in order to sustain the possibility of life, and if it is to be ecological it must be cultural. The economies of agriculture and forestry are vulnerable also because they are exceptional, in this way, to the rule of industry.

To obtain the best work in the economies of land use, those who use the land must be enabled to afford the time and patience necessary to do the best work. They must know how, and must desire, to do it well. Owners and workers in the land economy who grow their own food will not likely be starved into mistreating their land. But they can be taxed and priced into mistreating it. And so the parity of necessary occupations must be supported by parity of income.

Parity in this sense is not a new thought, although new thinking may be required in applying it to the variety of crops and commodities produced in a variety of regions. But we do fortunately have some precedence for such thinking. The Agricultural Adjustment Act defines parity as "that gross income from agriculture which will provide the farm operator and his family with a standard of living equivalent to those afforded persons dependent upon other gainful occupation." Perhaps the idea of parity does not need much explanation or defense. If, as now and always, a sufficient staff of land-users is necessary to the health of the land and therefore to the lives of all of us, then they should be assured a decent livelihood. And this the so-called free market cannot provide except by accident.

The concept of parity, as fair-minded as it is necessary, addresses one of the problems of farming and farmers in the industrial economy. Another such problem, more fundamental and most in need of understanding, is that of overproduction. "Other gainful employment" in the cities escapes this problem because the large industrial corporations have not characteristically overproduced. Overproduction moreover is not a problem of subsistence farming, or of those enterprises of any farm that are devoted to the subsistence of the farm family. The aim of the traditional economy of the farm household—a garden, poultry, family milk cow, meat animals, vines, fruit trees—was *plenty*, enough for the family to eat in season and to preserve, plus some to share or to sell. Surpluses and scraps were fed to the dogs or the livestock. There were no leftovers.

Surplus production is a risk native to commercial agriculture. This is because farmers individually and collectively do not know, and cannot learn ahead of time, the extent either of public need or of market demand. Given the right weather and the "progressive" application of technologies, their failure to control production, even in their own interest, is thus inevitable. This is not so much because they won't, but because, on their own, they can't. Either because the market is good and they are encouraged, or because the market is bad and they are desperate, farmers tend to produce as much as they can. They tend logically, and almost by nature, toward overproduction. In the absence of imposed limits, overproduction will fairly predictably occur in agriculture as long as farmers and the land remain productive. It has only to be allowed by a political indifference prescribed by evangels of the "free market." For the corporate purchasers the low price attendant upon overproduction is the greatest benefit, as for farmers it is the singular cruelty, of the current agricultural economy. Farm subsidies without production controls further encourage overproduction. In times of high costs and low prices, such subsidies are paid ultimately, and quickly, to the corporations.

This version of a farm economy pushes farmers off their farms. By increasing the wealth of urban investors and shoppers for "country places," it increases the price of farmland, making it impossible especially for small farmers, or would-be small farmers, to compete on the land market. The free market lays down the rule: Good land for investors and escapists, poor land or none for farmers. Young people wishing to farm are crowded to the economic margins and to the poorest land, or to no land at all. Meanwhile overproduction of farm commodities always implies overuse and abuse of the land.

The traditional home economies of subsistence, while they lasted, gave farmers some hope of surviving their hard times. This was true especially when the chief energy source was the sun, and the dependence on purchased supplies was minimal. As farming became less and less subsistent and more and more commercial, it was exposed ever more nakedly to the vagaries and the predation of an economy fundamentally alien to it. When farming is large in scale, is highly specialized, and all needs and supplies are purchased, the farmer's exposure to "the economy" is total.

It ought to be obvious that an economy that works against its sources will finally undercut the law of supply and demand in the most fatal way, that is, by destroying the supply. A food economy staffed by producers who are always fewer and older, whose increasing dependence on industrial technologies puts them and their land at ever greater risk, obviously cannot feed without limit an increasing population. But the reality of such an increasing scarcity is unaccounted for by the doctrine of the free market as applied to agriculture. Even less can this version of freedom comprehend the need for strict limits upon land use in order to preserve for an unlimited time the land's ability to produce. In a natural ecosystem, even on a conservatively managed farm, the fertility cycle may turn from life to death to life again to no foreseeable limit. By opposing to this cycle the delusion of a limitlessness exclusively economic and industrial, the supposedly free market overthrows the limits of nature and the land, thus imposing a mortal danger upon the land's capacity to produce.

When agricultural production is not controlled by a marketing cooperative such as the tobacco program, the market becomes, from the standpoint of the farmers, a sort of limitless commons, the inevitable tragedy of which is inherent in its limitlessness. In the absence of any imposed limit that they collectively agree to and abide by, all producers may have as large a share of the market as they want or can take. Only in this sense is the market, to them, "free." To limit production as a way of assuring an equitable return to producers is assuredly an abridgement of freedom. But freedom for what? For producers, it is the freedom to produce themselves into bankruptcy—to fail, that is, by succeeding. For the purchaser, it is the freedom to destroy the producers as a normal and acceptable expense. The only solution to the tragedy of the limitless market is for the producers to divide their side, the selling side, of the market into limited fair shares by limiting production, which is exactly what the tobacco program accomplished here in my region. By preventing the farmers' overuse of the market, it prevented as well the overuse of the land.

Agribusiness corporations of course don't openly advocate overproduction. They don't have to assume visibly the moral burden of their bad motive. All they have to do is stand by, praising American agriculture's record-breaking harvests, while either hope or despair drives the remaining farmers to produce as much as they can. The agribusinesses then are glad to sell the very expensive surplus of seeds, chemicals, and machines needed for surplus production.

The agricultural tragedy of the market is in part political. And how was the by-now entirely dominant political position on the agricultural free market defined? In the middle of the twentieth century, think tanks containing corporate and academic experts laid down the decree that there were too many farmers. They decreed further that the excess should be removed as rapidly as possible, and that the instrument of this removal should be the free market, with all price supports and production controls eliminated. The assumption evidently was that the removed farmers would be replaced by industrial technologies, recommended by the land-grant universities, and supplied by the corporations. The surplus farmers would increase the industrial labor force, and they and their families would enlarge the population of consumers of industrial products. It was proposed of course that all of society, including the displaced farm families and farm workers, would benefit from this. There would be no costs, social or agricultural, no problems, no debits, nothing at all to subtract from the accrued economic and social assets. This would institute an evolutionary process that would unerringly eliminate "the least efficient producers." Only the fittest would survive.

In short, by granting a limitless permission and scope to the free market and technological progress—which is assumed to work invariably for the best—politicians, by doing merely nothing, could rid themselves of any concern for farmers or farmland. The representatives of the people and the guardians of the common good were thus able to "free" the market to promote the (allegedly) inefficient farmers to (supposedly) the suburbs while subjecting the countryside to limitless progress and modernization. Against this heartless determinism, it is useful to remember that it was the aim of the program for burley tobacco in my region to include and help every farmer, even the smallest, who wanted to grow the crop. The difference was in the minds of the people whose work during four decades at last shaped an effective program. Those people, unlike the experts of the midcentury think tanks, were thoughtful of the needs of farming and farmers as opposed to the needs of the corporate free market known as the economy. The doctrine of "too many farmers" has never been revoked. No limit to the attrition has been proposed.

As evidence of the persistence of this doctrine, here is a passage from a letter of October 3, 2016, from John Logan Brent, Judge Executive of Henry County, Kentucky:

> I have taken a couple of afternoons to work on the accounting for farming cattle under the current terms. Enclosed you will find that product based upon a real example, which is our 100 acre farm . . . and its approximately 25 cow herd. . . . The good news is that for a young man wishing to earn a middle, to slightly below middle class annual salary of $45,000 farming cattle full-time, he only has to have $3,281,000 in capital to get started. If he can find 780 acres to rent, he only has to have $551,000 for used cows and equipment. I say this is the good news, because the reality is that this was based on a weaned calf price of $850 from June of this year. According to today's sales reports, that same calf is now $650 at best.

That alone, forgetting other adverse agricultural markets, would be an excellent recipe for the elimination of farmers. And conservationists should take note, as mostly they have not done, that in the absence of the eliminated farmers and with the consequent increase of agricultural dependence on the fossil fuels and toxic chemicals, there will be more pollution of water and air.

The related problems of low prices and overproduction of a single but significant crop were solved for about sixty years, in my part of the country and in others, in the only way they could be solved: by a combination of price supports and production controls. This was the purpose and the work of the tobacco program. I want now to look more closely at the Burley Tobacco Growers Co-operative Association, not this time as the brightest public occurrence in the history of my home countryside, but in terms of the suitability of its economic strategy to farming everywhere.

Here I must acknowledge that this organization and, more important, its economic principles have had the allegiance and the service of members of my family for three generations. Beginning in the winter of 1941, when the "Burley Association" renewed its work under the New Deal, my father, John M. Berry, Sr., served as vice president for sixteen and as president for eighteen years, retiring in 1975 but serving as an advisor until a few years before his death in 1991. My brother, John M. Berry, Jr., served as president from 1987 to 1993. My daughter, Mary Berry, started the Berry Center in New Castle, Kentucky, in 2011 for the purpose mainly of remembering, advocating, and applying the Association's proven economic strategy and its purpose of assuring a decent livelihood for small farmers. My son serves on the Berry Center board.

Under this program, support prices for the various grades of tobacco were set according to a formula for assuring a fair return on the cost of production. Production was controlled by allotting to each farm, according to its history of production, at first an acreage, and later a poundage, that would be eligible for price supports under the program. The total of the allotments for each year was determined by the supply, worldwide, that was available for manufacture. The rule was that the supply on hand should be sufficient for 2.4 years. If the supply was less than the predicted demand for 2.4 years, allotments would be increased; if more, allotments would be reduced. I don't know why the factor was set at 2.4. Its significance, however, is that production was limited according to an established measure of expected demand.

To buy a crop or a portion of a crop protected by the program, a purchaser had to bid a penny a pound above the support price. The government's assistance to the program consisted of a loan, made annually "against the crop," which permitted the program to purchase, store, and resell the portion of any year's crop that did not earn the extra penny a pound—which, thanks to the loan, would be purchased by the Association and the grower paid at the warehouse. The cost to the government was only administrative until, in response to protests, this cost was charged to the farmers, and the program then operated on the basis of "no net cost" to the government. This program succeeded remarkably well in doing what it was designed to do, and a part of its success is that it still provides a pattern for the thought and hope of those who are working for the survival of land and people.

The tobacco program is an example of a necessary service that government can provide to people who cannot provide it to themselves. The point most needing to be made now is that parity of pricing under the tobacco program was in no sense a subsidy. It did not involve a grant of money, a government giveaway, or a public charity. The concept of parity was used, by intention, to *prevent* government subsidation. Its purpose was to achieve fair prices, fairly determined, and with minimal help from the government. My father defended parity as an appropriate incentive: "It accords with our way of life, and it gives real and tangible meaning to the philosophy of 'equal opportunity.'" He thought of "direct subsidy payments" as virtually opposite to parity and an "abominable form of regimentation."

The tobacco program in all of its versions was finally defeated and destroyed in 2004 by the political free marketers who had always opposed it, and who had resented it in proportion to its success. During the six decades of its life, the Burley Tobacco Growers Co-operative Association helped keep farm families on their farms and gainfully employed in Kentucky, Missouri, Indiana, Ohio, and West Virginia. One measure of its success was the decrease of farm tenancy among the growers from 33 percent in 1940 to 9 percent in 1970. During those years some of the population of tenant farmers undoubtedly died, and some left farming, but most of them ceased to be tenant farmers by becoming owners of farms. This was a defining event in the lives of a considerable number of worthy people whom I knew. The farmer-members of the Association overwhelmingly renewed their support in referendum after referendum.

The Burley Association was thus truly a commons and a common good, based not only upon correct political and economic principles, but also upon the common history and culture, and thus upon the understanding consent of its sharers. So complete was the understanding of the members that in 1955, because of an oversupply of tobacco in storage, they voted for a 25 percent reduction of their allotments. On April 8, 2016, my neighbor Thomas Grissom, by far the best historian of the Association, wrote in a personal letter to me:

> After years of research, I have concluded that the most distinctive characteristic of the Kentucky [Burley] Tobacco Program is its design and application of an industrial agriculture commodity program to the cultivation and production of an agrarian crop indigenous to an agrarian society.

I think that Tom's perception is exactly right and that he found the right and necessary terms to describe it.

Burley tobacco, despite the dire health problems that it was found to cause and the consequent disfavor, was very much an agrarian crop. It was characteristically and mainly the product of small family farms, produced mainly by family labor and exchanges of work among neighbors. It was for a long time the staple crop in a highly diversified way of farming on landscapes that typically required considerate and affectionate care. As long as the market paid highly for high quality (which it finally ceased to do), the production of burley tobacco demanded, and from its many highly competent producers it received, both conscientious land husbandry and a fine artistry.

Industrialism and agrarianism are almost exactly opposite and opposed. Industrialism regards mechanical or technical functions as ideal. It rates its accomplishments by quantitative measures. Though it values the prestige of public charity, it is motivated necessarily by the antisocial traits that assure success in competition. Agrarianism, by contrast, arises from the primal wish for a home land or home place—the wish, in the terms of our tradition, for the freedom and independence that come with dependence on a parcel of land, however small, that one owns and is owned by or has at least the use of. Agrarianism grants its highest practical value to the good husbandry of the land. It is motivated, to an extent effective and significant, by neighborliness, family loyalty, and devotion to the coherence and longevity of communities.

As long as it has a sufficiency of "natural resources" and remains free of imposed political or economic restraint, an industrial economy will dominate and destroy an agrarian economy —no matter that the agrarian economy is indispensable for a continuing supply of resources. This defines precisely the need for the "design and application of an industrial agricultural commodity program to the cultivation and production of an agrarian crop indigenous to an agrarian society." For a while the Burley Tobacco Growers Co-operative Association—never mind the deserved infamy of tobacco—did preserve a sort of balance between the interests of industrialism and agrarianism, which prevented their inherent difference and opposition from becoming absolute, and thus absolutely destructive of the agrarian society. This balance was fair enough to the industry and it permitted the growers to prosper. The program worked in fact to the best interest of both economies.

From the perspective of this balance during the decades when it worked as it should have, it is possible to see that a step too far toward industrialism was probably taken by the Burley Association itself when, in 1971, it permitted the "lease and transfer" of production quotas away from the farms to which they had been assigned. This change, made under pressure from industrializing members, permitted the accumulation of allotments finally into very large acreages dependent upon more extensive technology and migrant labor. The program then was obliged to "balance" a reduced agrarianism against an increased industrialism.

With the demise of the program in 2004, the region's indigenous agrarianism could survive only as a history, a memory, and a set of vital principles that someday may be revived and reincarnated in reaction against the damages of industrialism.

*

For the past six decades, except for such a remnant of the New Deal, the government has done nothing for farmers except to quiet them down by subsidizing uncontrolled production, which really is worse than nothing. But this "policy," in the minds of the dominant politicians, signified that they were "doing something for agriculture" and so relieved them of thinking or knowing about agriculture's actual requirements. For example, the Democratic platform preceding President Clinton's first term initially contained no agricultural plank. My brother, John M. Berry, Jr., who was on the platform committee, was dismayed by this innovation, and he said so. He was then told that a plank was being drafted. When he saw the result, he laughed. He asked if he might draft a more meaningful plank. After much resistance, he was allowed to do so. He then "spent the next six hours redrafting the amendment so as to satisfy the Clinton staff." I am quoting his letter of June 29, 1992, to Dr. Grady Stumbo, chair of the Kentucky Democratic Party. The letter goes on to say that Clinton's staff refused to permit any reference to

> "supply management," "price support" or any government guarantee of a fair price for farmers. They also refused to permit any reference to agribusiness control of farm policy or the level of agribusiness profits. They also refused to permit any language that could be construed as a commitment . . . to anything specific for agriculture or the rural community. . . .
>
> I had already been advised that Chairman Ron Brown had formed an agriculture task force and sold seats to its members for $15,000* contributions to the Democratic National Committee.
>
> Those seats went to representatives of agribusiness and other interests that have traditionally written farm policy for the Republicans.

The doctrine of "too many farmers" thus had become the established orthodoxy of the leaders of both parties. My brother was then president of the Burley Tobacco Growers

*My brother told me not long before his death in 2016 that the "contributions" actually were $30,000.

Co-operative Association, which was still a major life support of our state's small farmers. By 1992 tobacco had become indefensible as a product, and it bore too great a public stigma to be touchable by a national candidate. My brother understood that, and he did not expect approval specifically of the tobacco program. But he knew that the working principles of that program would protect farmers who produced commodities other than tobacco everywhere in the country—and would also protect our own farmers when they no longer produced tobacco. He knew that Mr. Clinton, if he wanted to, could endorse the program's principles without endorsing its product. The agricultural plank of the 1992 Democratic platform, as published, gave a general approval to "family farmers receiving a fair price," to "a sufficient and sustainable agricultural economy . . . achieved through fiscally responsible programs," and to "the private–public partnership to ensure that family farmers get a fair return for their labor and investment." And of course it condemned "Republican farm policy." It committed Mr. Clinton and his party to do nothing. And nothing was what they did.

In 1995 President Clinton spoke to an audience of farmers and farm leaders in Billings, Montana. He acknowledged that the farm population by then was "dramatically lower . . . than it was a generation ago." But, he said, "that was inevitable because of the increasing productivity of agriculture." Nevertheless, he wanted to save the family farm, which he held to be "alive and well" in Montana. He believed we had "bottomed out in the shrinking of the farm sector." He said he wanted to help young farmers. He spoke of the need to make American agriculture "competitive with people around the world." And so on.

He could not have meant what he said, because he was speaking without benefit of thought. And why should he have thought when he was not expected to do so? He was speaking forty or fifty years after politicians and their consulting experts had abandoned any effort to think about agriculture. "Inevitable" is a word much favored by people in positions of authority who do not wish to think about problems. When and why did Mr. Clinton in 1995 think that the inevitable "shrinking of the farm sector" had ceased? In fact, "the farm sector" had not bottomed out in 1995; there is no good reason to think that it has bottomed out, at less than 1 percent of the population, in 2016. And how could he have helped young farmers except by giving them the protections against the free market that my brother had recommended three years before? Mr. Clinton was talking nonsense in Billings in 1995 because he did not have, and could not have had from his advisers, the means to think about what he thought he was talking about. The means of actual thought about the use and care of the land had been intentionally discounted and forgotten by people such as themselves.

It appears to be widely assumed by politicians, executives, academics, public intellectuals, industrial economists, and the like that they have a competent understanding of agriculture because their grandparents were farmers, or they have met some farmers, or they worked on a farm when they were young. But they invoke their understanding, which they do not have, only to excuse themselves from actual thought about actual issues of agriculture. These people have found "inevitability" a sufficient explanation for the deplorable history of industrial agriculture. They see the reason for the present discontent of "blue-collar" voters as low or "stagnant" wages. They don't see, in back of that, the dispossession that made many of them wage-workers in the first place. The loss everywhere of small farms and small towns and the respectable livelihoods that they provided was ruled "inevitable" and thus easily explained and forgotten. In their perceived worthlessness and dispensability, at least, the people of the farms and small towns were in effect racially equal. If, for instance, black small farmers were helped to prosper, as some liberals would have liked, then white small farmers would have had to be helped to prosper, which would have pleased neither liberals nor conservatives.

It was, then, "inevitable" that the independent livelihoods in the old economies of the countryside and the small towns should be replaced by the mainly subservient livelihoods in industry, or by unemployment. But if the "working class" counted for nothing and were dispensable as small farmers or farmhands or as small independent keepers of shops and stores or as independent tradespeople and craftspeople, why then should they count for something and be more than dispensable as "blue-collar workers"? In the corporate and urban economy the blue-collar workers were just as "inevitably" replaceable by technologies as they had been before. They were then notified that they were losing out because they were "uneducated." They needed "a college education," in default of which they were offered "retraining" and "job creation." But these were only political baits, which left the blue-collared ones to their "inevitable" fate of low or stagnant wages or unemployment.

This doctrine of inevitability, also known as technological progress, is in fact a poor excuse for an economic and technological determinism, as heartless as it is ignorant, which has belonged about equally to the political establishment of both parties. Realizing that they were the broken eggs of an omelet that others would eat, the blue-collar workers became angry. Their anger turned them to Donald Trump, who at least recognized their existence and the political usefulness of their anger.

In the pre-Trump version of the history of progress, determinism and inevitability overruled any need for actual knowledge and actual thought. But with the ascendancy of Mr. Trump, at least some of the determinists seem to be reverting to free will. While the conservatives, who have strained at a gnat and swallowed a camel, endeavor to digest their dinner, the liberals talk of "connecting" with the blue-collar workers of rural America, to whom they have given not a substantive thought since Ezra Taft Benson, Eisenhower's secretary of agriculture, pronounced to their grandparents the political death sentence, "Get big or get out."

Let us remember also the workers, white and black, who in their thousands became simply obsolete at the instant when "efficient" machines were brought into the coal mines, the factories, and the fields of sugarcane and cotton. I thought of them when I read in a column by Roger Cohen in the *New York Times* of November 19, 2016, that "the very essence of the modern world" is "the movement of people and ever greater interconnectedness, driven by technology." Mr. Cohen approves of this "essence" and is afraid that Mr. Trump will stop it. What he has in mind surely must be the *voluntary* movement of people. The movements of people *actually* "driven by technology" are outside Mr. Cohen's field of vision, surely only because of his political panic. Millions of people, as we know, have been driven away from their homes in the modern world by the similarly imperative technologies of industrial production and industrial war.

I am uncertain what value Mr. Cohen assigns to "interconnectedness," but he cannot be referring to the interconnectedness of families in their home places, or of neighbors in their neighborhoods. How the loss of those things can be compensated by movement, driven or not, is far from clear. The same obscurity clouds over any massive "movement of peoples," as over the arguments by which these movements are excused or justified. It does not require a great refinement of intellect to see the harm that is in all of them.

The experts who decided in the middle of the twentieth century that there were too many farmers had in fact no agricultural knowledge or competence upon which to base such a judgment. They and their successors certainly had not the competence to assume any responsibility for, or in any way to mitigate, the totalitarian displacement of about twenty million farmers.

Farming is one of the major enactments of the connection between the human economy and the natural world. In the industrial age farming also enacts the connection, far more complicated and perilous than industrialists admit, between industrial technologies and living creatures. Some science certainly needs to be involved, also more and better accounting. But good farming is first and last an art, a way of doing and making that involves human histories, cultures, minds, hearts, and souls. It is not the application by dullards of methods and technologies under the direction of a corporate-academic intelligentsia.

If we should want to revive, or begin, in a public way the actual thinking about agriculture that has actually taken place in some cultures, that is still taking place in some small organizations and on some farms, what would we have to do? We would have to begin, I think, by giving the most careful attention to issues of carrying capacity, scale, and form, to issues of production, of course, but also and just as necessarily to issues of maintenance or conservation. The indispensable issue of conservation would apply, not just to the farm's agricultural "resources," but also to the ecosystem that includes the farm and to the waterways that drain it. I think, moreover, that this attention to issues must be paid always outdoors in the presence of examples. The thing of greatest importance is to think about the land with the land's people in the presence of the land. Every theory, calculation, graph, diagram, idea, study, model, method, scheme, plan, and hope must be caught firmly by the ear and led out into the weather, onto the ground.

It is obvious that this effort of thinking has to confront everywhere the limits both of nature and of human nature, limits imposed by the ecosphere and ecosystems, limits of human intelligence, human cultures, and the capacities of human persons. Such thought is authenticated by its compatibility with limits, its willingness to accept limits and to limit itself. This will not be easy in a time overridden by fantasies of limitlessness. A market limitlessly usable by sellers and limitlessly exploitable by buyers is merely normal in such a time. And limitlessness is the common denominator of the dominant political sides, both of which tend to refer to limitlessness as "freedom."

We have the liberal freedom of unrestrained personal behavior, and the conservative freedom of unrestrained economic behavior. These two freedoms are more alike, more allied, and more collaborative than either side would like to admit. Opposition to the industrial economy's ravaging of the landscapes of farming and forestry now comes from a small and scattered alliance of agrarians, not from liberals or conservatives.

Conservatives and liberals disagree passionately about climate change, for example, yet liberal protests against climate change far exceed protests against the waste and pollution that occur locally in industrial agriculture and are its reputed causes. And neither the conservatives who esteem the fossil fuels nor the liberals who deplore them have advocated rationing their use, either to make them last or to reduce their harm. For these people the old ideals of *enough* and *plenty* have been overruled by the ideals of *all you want* and *all you can get.* They cannot imagine that for farmers a limitless market share, like a limitless appetite, can lead only to the related diseases of *too much* and *too little.*

Science, apart from moral limits in scientists, seems to be limitless, for it has produced nuclear and chemical abominations that humans, with their very limited intelligence, can neither limit nor safely live with. "Anything goes" and "Stop at nothing" are the moral principles that some scientists have borrowed apparently from the greediest of conservatives and the most libertine of liberals. The faith that limitless technological progress will finally solve the problems of limitless contamination seems to depend upon some sort of neo-religion.

The good care of land and people, on the contrary, depends primarily upon arts, ways of making and doing. One cannot be, above all, a good neighbor without such ways. And the arts, all of them, are limited. Apart from limits they cannot exist. The making of any good work of art depends, first, upon limits of purpose and attention, and then upon limits specific to the kind of art and its means.

It is a formidable paradox that in order to achieve the sort of limitlessness we have begun to call "sustainability," whether in human life or the life of the ecosphere, strict limits must be observed. Enduring structures of household and family life, or the life of a community or the life of a country, cannot be formed except within limits. We must not outdistance local knowledge and affection, or the capacities of local persons to pay attention to details, to the "minute particulars" *only* by which, William Blake thought, we can do good to one another. Within limits, we can think of rightness of scale. When the scale is right, we can imagine completeness of form.

The first limit to be encountered in making a farm—or a regional or national economy—is carrying capacity: How much can we ask of *this* land, this field or this pasture or this woodland, without diminishing the land's response? And then we come to other limits, perhaps many of them, each one addressing directly our imagination, sympathy, affection, forbearance, knowledge, and skill. And now I must call to mind Aldo Leopold, who, unlike most conservationists since John Muir, could think beyond wilderness conservation to conservation of the country's economic landscapes of farming and forestry. His conception of humanity's relation to the natural world was eminently practical, and this must have come from his experience as a hunter and fisherman, his study of game management, and his and his family's restoration of their once-exhausted Sand County farm. He knew that land-destruction is easy, for it requires only ignorance and violence. But the obligation to restore the land and conserve it requires humanity in its highest, completest sense. The Leopold family renewed the fertility and health of their land by their work, their pleasure, and their love for their place and for one another.

Aldo Leopold thought carefully about farming and forestry because he knew that far more land would be put to those uses than ever could be safeguarded in wilderness preserves. In an essay of 1945, "The Outlook for Farm Wildlife," he laid side by side "two opposing philosophies of farm life" (the italics being his):

1. *The farm is a food-factory*, and the criterion of its success is salable products.
2. *The farm is a place to live.* The criterion of success is a harmonious balance between plants, animals, and people; between the domestic and the wild; between utility and beauty.

This is a statement about form, contrasting a form that is too simple and too exclusive with a form that may be complex enough to accommodate the interest of what is actually involved. Under the rule of the first form, "the trend of the landscape is toward a monotype." This form can be adequately described as the straightest, shortest line between input and income. All else is left out or denied. Such a form concedes nothing to its whereabouts. It is placed upon whatever landscape merely by imposition, as a cookie cutter is imposed upon dough. In its simplicity and rigidity, such a form is bad art, but also, as Leopold knew and as we now know better than he could have, it is bad science.

The second form is described as "a harmonious balance" among a diversity of interests. On such a farm, made whole by the high artistry of farming, every part is both limited and enabled by the others. This harmonious balance, I should not need to say, cannot be prefabricated. It can be realized only uniquely within the boundary of any given farm, according to the natures and demands of its indwelling plants and animals, and according to the abilities, needs, and wishes of its resident human family. Wherever this is fully accomplished, it is a grand masterpiece to behold.

The Presence of Nature in the Natural World: A Long Conversation

THE GREAT TROUBLE of our age, involving the whole human economy from agriculture to warfare, is in our relationship to the natural world—to what we call "nature" or even, still, "Nature" or "Mother Nature." The old usage persists even seriously, among at least some humans, no matter how "objectivity" weighs upon us. "Of all the pantheon," C. S. Lewis wrote, "Great Mother Nature has . . . been the hardest to kill." With Nature we have, properly speaking, a relationship, for the responses go both ways: Nature is fully as capable of responding to us as we are of responding to her. In the age of industrialism, this relationship has been radically brought down to a pair of hopeless assumptions: that the natural world is passively subject either to unlimited pillage as a "natural resource," or to partial and selective protection as "the environment."

We seem to have forgotten that there might be, or that there ever were, mutually sustaining relationships between resident humans and their home places in the world of Nature. We seem to have no idea that the absence of such relationships, almost everywhere in our country and the world, might be the cause of our trouble. Our trouble nonetheless exists, is severe, and is getting worse. Instead of settled husbanders of cherished home places, we have become the willing parasites of any and every place, destroying the source and substance of our lives, as parasites invariably do.

This critical state of things has not always been explicitly the subject of my writing, but it has been constantly the circumstance in which I have written, for I have had constantly and consciously in sight the progressive decline of my home countryside and community. I have been perforce aware that this is a local manifestation of a decline that is now worldwide, affecting not only every place but also the oceans and the air, and I have of course felt a need to understand, and to oppose so far as I have been able, this downslope of all creation. In this effort of thought, I have been always in need of teachers, friends, and allies among the living and the dead. The mercy or the generosity accompanying this effort has been that I have found perhaps not all but many of the teachers, friends, and allies I have needed, often when I have needed them most.

By that I do not mean to suggest that my looking for help has been easy or carefree. To begin with, I never trusted, and after a while I rejected, the hope that many people have invested in what we might call the industrial formula: Science + Technology + Political Will = The Solution. This assumes that the science is adequate or soon will be, that the technology is adequate or soon will be, and therefore that the only essential task is to increase political pressure favoring the right science and the right technology.

An outstanding example of the industrial formula in action is the present campaign against global warming. If the scientific calculations and predictions about global warming are correct (which I am willing to assume that they are, though I have no science of my own by which to know), then its causes are waste and pollution. But "global warming" is a ravenously oversimplifying phrase that gobbles up and obscures all the myriad local instances of waste and pollution, for which local solutions will have to be worked out if waste and pollution are ever to be stopped. The phrase "global warming" suggests no such thing. Global warming, the language insists, is a *global* emergency: a global problem requiring a global solution. To solve it we have the science, we have the technology, and now we have only to prevail upon the world's politicians to enforce the recommended solution: burning less fossil fuel.

But from where I live and watch, I see the countryside and the country communities being wrecked by industrial violences: a heartless gigantism of scale and power, massive and irreparable soil erosion, pollution by toxic chemicals, ecological and social disintegration—and of course an immense burning of fossil fuels, making in turn an immense contribution, as alleged by scientists, to global warming.

I have no doubt at all that even if the global climate were getting better, our abuses of the land would still be the disaster most seriously threatening to the survival of humans and other creatures. Land abuse, I know, is pretty much a global phenomenon. But it is not happening in the whole world as climate change happens in the whole sky. It is happening, because it can happen, only locally, in small places, where the people who commit the abuses also live. And so my question has been, and continues to be, What can cause people to destroy the places where they live, the humans and other creatures who are their neighbors, and ultimately themselves? How can humans willingly turn against the earth, of which they are made, from which they live? To treat that as a scientific and technological or political question is not enough, is even misleading. The question immediately and at least is economic: What is wrong with the way we are keeping house, the way we make our living, the way we live? (What is wrong with our minds?) And to take the economic question seriously enough is right away to ask another that is also but not only economic: What is happening to our souls?

There is no industrial answer to such questions. Industrialism has never provided a standard by which such questions can be answered. I long ago hatched out of the egg in which I could believe that industrialism is capable of competent judgments of its effects, let alone competent solutions to the problems it has caused. Its "solutions," on the contrary, tend to increase the problems, as in the desperate example of industrial war, in which more never produces less; or the example, equally desperate, of agricultural pesticides, which must become more toxic and diverse as immunities develop among the pests. Since industry has no language with which to speak to us as living souls and children of Nature, but only as interchangeable employees, customers, or victims, by what language can we, in the fullness of our being, speak back to industrialism?

Seeing that my need for help was defined by that question, I have faced another difficulty: I am inescapably a product of "Western" culture, first as I was born and grew up in it, and then as I, by my work, have made myself able to know it and more responsibly to inherit it. The difficulty was that Western culture, especially when it is understood as Christian culture, however decayed, has been for many years in disfavor among writers and intellectuals, some of whom have seen it as the very origin of our unkindness toward "the environment." At its best, this disfavor has produced useful criticism, for professedly Christian institutions, nations, and armies have much cruelty and violence to answer for. At its worst, it is fashionably attitudinal and dismissive, so that we now have "Shakespeare scholars" who cannot recognize Shakespeare's frequent references to the Gospels.

For some, it has been possible and useful to turn away from the Western or Christian inheritance to find instruction and sustenance in, mainly, oriental or tribal cultures. I am glad to acknowledge my own considerable indebtedness to the little I have managed to learn of oriental, American Indian, and other cultures, which have been often confirming and sometimes clarifying of the cultural lineage and faith that I consider my own. From Gary Snyder's Buddhism, for instance, I understood more clearly than from any other source that the practice of a religion is necessarily economic: how we live on and from the earth.

But I am too completely involved in Western culture by the history of my mind, my people, and my place to be capable of a new start in another tradition. My need to make as much sense as I could of my history and experience, as I began fifty years ago to think of my task, clearly depended on my willingness to do so, not only as a native of a small patch of country in Henry County, Kentucky, but also as an heir and inevitably a legator of Western culture. If I hoped to make sense, my culture would have to be always, at least implicitly, my subject, and I would have to be its critic. If I was troubled by our epidemic mistreatment of the natural or given world, especially in the economic landscapes that we live from, then I would have to search among the artifacts and records of my culture, in the time I had and so far as I was able, to find probable causes, appropriate standards, and possible corrections. This search began, as I would later realize, in my school years, before I could have explained, even to myself, my need and purpose.

Eventually, provoked by attacks on Western culture as founded upon a supposedly biblical permission to humans to use the earth and its creatures in any way they might please, I read through the Bible to see how far it might support any such permission, or if, on the contrary, it might impose on humans the obligation to take good care of a world both given to them as a dwelling place, as to the other creatures, and to which they had been given as caretakers.

I found of course Genesis 1:28 ("Be fruitful and multiply, and replenish the earth, and subdue it: and have dominion over the fish of the sea, and over the fowl of the air, and over every living thing that moveth upon the earth"), which is the verse the detractors have found, and is pretty much the sum of their finding. But I also found that the Bible as a whole is a context that very sternly qualifies even the "dominion" given in Genesis 1:28. And I found many verses and passages that, even out of context, require humans to take the best possible care of the earth and its creatures. To speak to Genesis 1:28, for instance, there is the first verse of Psalm 24: "The earth is the Lord's, and the fullness thereof: the world and they that dwell therein." To that verse, singly and without context, if one takes it seriously, there are only two sane responses: (1) fear and trembling, and (2) a human economy that would conserve and revere the natural world on which it obviously depends. But it is possible for materialist environmentalists (as for many professing Christians) to read and not see, even to memorize a passage of language and not know what it says. It is easy, even intellectually permissible, to quote from a book that one has not read. And now most people are too distracted automotively and electronically to know what world they are in, let alone what the Bible might say about it.

In a world allegedly endangered by biblical religion, it is dangerous to be ignorant of the Bible. I must say, now, that I am not an uncritical reader of the Bible. There are parts of it that I dislike, but there are parts of it also in which I place my faith. From those parts, or some of them, I gather an old belief that God is not merely above this world, or behind it, but is *in* it, and it lives by sharing His life and breathing His breath. "If he gather unto himself his spirit and his breath," Elihu says to Job, "All flesh shall perish together, and man shall turn again unto dust" (Job 34:14–15). Like Psalm 24, these fearful verses clearly imply a mandate for the good care of all creation, and this becomes explicit in the Gospels' paramount moral commandment (Matthew 22:39) that we must love our neighbors as ourselves, even when our neighbors happen to be our enemies. This neighborly love cannot be a merely human transaction, for you cannot love your neighbor while you destroy the earth and its community of creatures on which you and your neighbor mutually depend.

The Bible, then, is not defined by misreading, or Western culture by economic violence, any more than a body is defined by a disease.

Over the years, for my soul's sake and for the sake of my work, I have returned many times to the Bible, testing it and myself by such understanding, never enough, as I have been able to bring to it.

I have also returned repeatedly to the pages of several English poets, who lived and worked of course under the influence of the Bible, and whose interest in the natural world and the magisterial figure of Nature has been instructive and sustaining to me.

Both Chaucer and Spenser testify that "this noble goddesse Nature" or "great dame Nature" comes to them from a Latin allegory written in about the seventh decade of the twelfth century by Alanus De Insulis, or Alan of Lille. This book, *De Planctu Naturae* (*Plaint of Nature*), is composed of alternating chapters of verse and prose. It tells the story of Alan's dream-vision in which he converses with Nature, whom he recognizes both as his "kinswoman" and as the Vicar of God. The Latin original apparently is extremely hard to translate. According to James J. Sheridan, whose translation I have used,

> The author revels in every device of rhetoric. . . . He so interweaves the ordinary, etymological and technical signification of words that, when one extracts the meaning of many a section, one despairs of approximating a satisfactory translation.

Despite its intricacy even in English, and the strangeness in our time of Alan's leisurely pleasure in rhetorical devices—he elaborates, for example, a sexual symbolism of grammar—this is a book extraordinarily useful to a reader interested in the history of our thought about the natural world as well as the history of conservation.

At the beginning of his book, Alan's happiness has been overcome by grief because of humanity's abandonment of Nature's laws. His own immediate complaint is against homosexuality, which seems to have been widespread and fairly openly practiced and acknowledged in his time. This he understands as a consequence that comes "when Venus wars with Venus." For my own sake in my own time, I am sorry that he begins this way, for I am much troubled by our current politics of private life. Some of us now may reasonably object to Alan's view of homosexuality as unnatural, or anti-natural, but it would be a mistake to dismiss him on that account at the end of his first sentence.

As Alan continues the story of his vision, his objection to homosexuality becomes incidental to his, and Nature's, objection to lust, which is one of the whole set of sins that are opposed both to the integrity of Nature and to the integrity of human nature. The two Venuses that are at war are the Venus of lust or mere instinct and the Venus of responsible human love. This duality is analogous to, and dependent upon, the double nature of humankind. In the order of things, as in the Chain of Being, humans are placed between the angels and the animals. Strongly attracted in two conflicting directions, they have the hardship of a moral obligation to keep themselves in their rightful place.

Nature, then, as "faithful Vicar of heaven's prince," makes two requirements of us. She requires us to be natural in the general sense that, like other animals, we are born into physical embodiment, we reproduce, and we die. But she most sternly conditions her first requirement by requiring us also to be natural in the sense of being true to our specifically human nature. This means that we must practice the virtues—chastity, temperance, generosity, and humility, as they are named, their procession led by Hymenaeus,* in the *Plaint*—that keep us in our rightful place in the order of things. We must not be prompted by lust, avarice, arrogance, pride, envy, and the other sins, prodigality being the sum of them all, either to claim the prerogatives of gods or to lapse into bestiality and monstrosity.

*Hymenaeus was once, until forsaken by her, the husband of Venus. The son of that marriage is Desire. In Alan's mythology, Hymenaeus is "closely related" to Nature and has "the honor of her right hand." He represents the sanctity, and the bad history, of marriage.

This is Nature's requirement because her own integrity and survival depend upon it, but she herself is unable to enforce it. She very properly acknowledges the limit of her power when she explains to Alan that although "man is born by my work, he is reborn by the power of God; through me he is called from non-being into being, through Him he is led from being into higher being."

To put it another way: Humanity as an animal species—"the only extant species of the primate family Hominidae," says *The American Heritage Dictionary*—is a limitless category in which anything we may do can be explained, even justified, as "natural." But the necessity of our communal life as fellow humans, as well as our shared and interdepending life as a species among species, imposes limits, defining, for the sake of our survival and that of the living world, a higher or moral human nature, specified by Nature herself in her conversation with Alan. The nature of every species is uniquely specific to itself. Human nature in its fullest sense is the most complex of all, for it involves standards and choices.

To speak of this from the perspective of our own time and the fashionable environmentalist prejudice against "anthropocentrism," it is our mandated human nature that allows us, by understanding the legitimacy of our own self-interest, or self-love as in Matthew 22:39, to understand the self-interest of other humans and other creatures. A human-centered and even a self-centered point of view is inevitable—What other point of view can a human have?—but by imagination, sympathy, and charity *only* are we able to recognize the actuality and necessity of other points of view.

It is impossible to speak of such things in a positivist or "objective" language, pruned of its upper branches and presided over by professional or specialist consciences embarrassed by faith, hope, and charity. And so it is a relief to resume the freedom and completeness of the language in which we can say that Nature, the Vicar of God, is the maker or materializer of a good world, as affirmed by Genesis 1:31, entrusted to the keeping of humans who must prove worthy of it by keeping true to their own, specifically human, nature, which is defined by that worthiness, which they must choose.

According to Alan's great instructor, the integrity of the

natural world depends upon the maintenance by humans of *their* integrity by the practice of the virtues. The two integrities are interdependent. They cannot be separated and they must not be separately thought about. This is the moral framework of *Plaint of Nature*, and I would argue that it persists in a lineage of English poets from Chaucer to Pope, whether or not they can be shown to have read Alan. The same set of traditional assumptions can be shown also to support the work of a lineage of agricultural writers and scientists of about the last hundred years, as I will later demonstrate. So far as my reading, observation, and experience have informed me, I believe that Nature's imperative as set forth in the *Plaint* is correct, not just for the sake of morality or for the sake of Heaven, but in its conformity to the practical terms and demands of human life in the natural world. The high standards of Nature and of specifically human nature obviously will not be comforting to humans of the industrial age, among whom I certainly must include myself, and they are not in fashion, but that hardly proves them false.

As Alan grieves over the degeneracy of humankind, Nature appears to him. She is "a woman" whose hair shines with "a native luster surpassing the natural." He proceeds to describe in patient detail her hair, her face, and her neck. He devotes seven fairly substantial paragraphs, in Sheridan's translation, to her crown, the jewels of which are elaborately allusive and symbolic. He describes with the same attention to detail and reference her dress, on which is depicted "a packed convention" of all the birds; her mantle, which displays "in pictures . . . an account of the nature of aquatic animals," mainly fish; and her embroidered tunic, a part of which has been damaged by man's abuse of his reason, but on the remainder of which "a kind of magic picture makes the land animals come alive." He discreetly refrains from looking at her shoetops and her undergarments, but is "inclined to think that a smiling picture made merry there in the realms of herbs and trees"—which he then describes.

The presence of Nature is so numinous and exalted, of such "starlike beauty," that Alan faints and falls facedown. Though my interest is fully invested in this book, my own reaction here is much less excited, and I doubt that any reader now could find much excitement in this exhaustive portraiture, though I agree with C. S. Lewis that "the decorations do not completely obscure the note of delight." The problem is that Alan's description of Nature is too distracted by his rhetorical extravagances to give us anything like a picture of her.

We get a better sense of her, I think, in his later invocation, in which he describes Nature's character rather than her appearance:

> O child of God, mother of creation, bond of the universe and its stable link . . . you, who by your reins guide the universe, unite all things in a stable and harmonious bond and wed heaven to earth in a union of peace; who, working on the pure ideas of [Divine Wisdom], mould the species of all created things, clothing matter with form . . .

It is no doubt wrong, or at least pointless, to ask Alan for a "realistic" portrait, as if he were dealing merely with a personification. His effort to describe Nature's appearance, exaggerated as it is, failing as it does, speaks nonetheless of his need to realize her, not merely as an allegorical person, but as a presence actually felt or known.

The really useful question, it seems to me, is not how Alan's Nature functions in a medieval allegory, but what she means historically as a figure very old and long-lasting, of whom people have needed, of whom some people need even now, to speak. How might one explain this need? My reading is sufficient only to assure me that Alan's vision of Nature draws upon a variety of sources, classical and Christian. She came to him as a figure of attested standing and power, he did imagine her, she did appear to him in something like a vision.

In Alan's *Plaint of Nature* and in Christian poems much later, the classical deities continue to appear, still needed by imagination, no longer perhaps as vital gods and goddesses, but at least as allegorical figures standing for their qualities or powers. That Nature takes her place prominently among them in the work of later poets appears to be owing largely to Alan. Sheridan says, "It is in Alan of Lille that she reaches full stature." And according to Sheridan it was Alan who "theorized that God first created the world and then appointed Natura as His substitute and vice-regent . . . in particular to ensure that by like producing like all living creatures should increase and multiply . . ."

And so Nature comes into the plot of Alan's Christian allegory reasonably enough, to keep company with Venus, Cupid, Genius, and a cast of personified vices and virtues. But if allegory tends toward simplification and a kind of shallowness —Truth is going to be purely true, and Greed purely greedy —Alan's representation of Nature tends in virtually the opposite direction: She is complex, vulnerable, mysterious, somewhat ghostly, less a personification than the presence, felt or intuited, of the natural world's artificer. She comes from an intuition of order and harmony in creation that is old and independent of empirical proofs. That we have personified her for so long testifies that we know her as an active, purposeful, and demanding force. Her domain, in the *Plaint* and elsewhere, is both natural and supernatural. As Vicar of God, she joins Heaven and Earth, resolving the duality of spirit and matter.

To say, as I just did, that we know her is obviously to raise the question of what we mean, and what we ever have meant, by saying that we "know." In our time an ideological tide has been carrying us toward a sort of apex at which none of us will claim to know anything that has not been proved and certified by scientists. But to read Alan of Lille is to realize that there was a time, a *long* time, when such knowledge was neither available nor wanted, but when many necessary things were nonetheless known. Thomas Carlyle, writing from his own perch in the modern world about the twelfth-century monk Jocelin of Brakelond, had to confront this very question: "Does it never give thee pause . . . that men then had a *soul*, not by hearsay alone, and as a figure of speech; but as a truth that they *knew* . . ."

A century and a half later in the same darkening age, I asked my friend Maurice Telleen, who had much experience of livestock shows: "Does a good judge measure every individual by the breed standards, or does he go by intuition and use the breed standards to check himself?" In reply Maury said something remarkable: "He goes by intuition! A slow judge is always wrong." I thought his reply remarkable, of course, because it confirmed my belief that, among people still interested in the qualities of things, intuition still maintains its place and its standing as a way to know.

Intuition tells us, and has told us maybe as long as we have been human, that the nature of the world is a great being, the one being in which all other beings, living and not-living, are joined. And for a long time, in our tradition, we have called this being "Natura" or "Kind" or "Nature." And if we forget, our language remembers for us the relation of "natural" (by way of "kind") to "kindness" and "kin," and to "natal," "native," "nativity," and "nation." Moreover, as understood by Alan of Lille and the poets who descend from him, the being and the name of "Nature" also implicates the history of human responsibility toward the being of all things, and Nature's continuing requirement of that responsibility.

Contrast "Nature," then, with the merely clever poeticism "Gaia" used by some scientists to name the idea of the unity of earthly life, and perhaps to warm up and make congenial the term "biosphere." "Gaia" is the name of the Greek goddess of the earth, in whom no modern humans, let alone modern scientists, have even pretended to believe. Her name was used, no doubt, because, unlike "Nature," it had no familiar or traditional use, could attract no intuitive belief, or appeal at all to imagination.

By name Nature familiarly belongs to us, and she has so belonged to us for many hundreds of years, but considered from a viewpoint strictly biblical and doctrinal, she still may seem an intruder. Like the other classical deities, she joins the biblical tradition somewhat brazenly. The writers needed her perhaps because she has persisted in oral tradition, and because there remained not a Christian only but a general human need for her. That she usurps or threatens the roles of all three persons of the Trinity suggests that there was also a felt need for her to some degree in spite or because of them. Speaking only for myself, I will say that for me it has always been easy to be of two minds about the Trinity. The Father, the Son, and the Holy Spirit, as they appear in their places, as "characters" so to speak, in the Gospels, I have found simply recognizable or imaginable as such, but I think of them as members of "the Trinity" only deliberately and without so much interest. As an idea, the Trinity, the three-in-one, the three-part godhead, seems to me austerely abstract, complicated, and cold. The more it is explained, the less believable it becomes.

Perhaps it is this aloofness of the Trinity that calls into being and causes us to need this other, more lowly presence, holy yet familiar, matronly, practical, concerned, and eager to teach: Nature, the mother of creatures as the Virgin is the mother of God. And here I remember that Thomas Merton, in his prose poem "Hagia Sophia," sees both Nature and the Virgin as sharing in the identity or person of Holy Wisdom: "There is in all visible things an invisible fecundity, a dimmed light, a meek namelessness, a hidden wholeness. This mysterious Unity and Integrity is Wisdom, the Mother of all, *Natura naturans*." And later: "*Natura* in Mary becomes pure Mother." In Merton's treatment these three—Natura, Mary, Sophia—seem to fade into one another, shadowing forth an always evasive reality. And this, I think, is what we must expect when human thought reaches toward the mystery of the world's existence: an almost visible, almost palpable presence of a reality more accessible to poetry than to experiment, never fully to be revealed by any medium of human knowledge. And yet it is to be, it can be, learned and known, not by peering through a lens or by assemblies of data, but perhaps only by being quietly observant for a long time. Merton's vision does not explain or clarify Alan of Lille's, but between the two there is a sort of recognition and a mutual verification.

I am not entirely confident of my grasp of Alan's *Plaint of Nature*, partly because of my very reasonable doubt of my scholarship and understanding, but also because I am looking back at Alan from my knowledge of later writers who seem to be his successors, and from my own long preoccupation with the issues he addresses. For I do not wish, and I have not tried, to read Alan as a writer of historical or literary interest only. Despite the always considerable differences of times and languages, I have thought of him as a fellow writer with immediately useful intelligence of the world that is both his and mine.

His concept of the integrity of the natural world, and of the dependence of the world's integrity upon the integrity of human nature, leads by the most direct and simple logic to our own more scientific recognition of the integrity of ecosystems, the integrity ultimately of the ecosphere, and to the recognition (by not enough of us) of the necessity of an ecological ethic.

But a large part of the value of the *Plaint* is that, though I suppose it is as full as it could be of the biology and taxonomy of its time, it cannot be reduced to the sort of knowledge that we call scientific. It cannot, for that matter, be reduced to the sort of knowledge that we call poetic. It belongs instead to the great Western family of writings that warn us against what we now call reductionism, but which traditionally we have called the deadly sin of pride or hubris: the wish to be "as gods," or the assumption that our small competence in dealing with small things implies or is equal to a great competence in dealing with great things. In our time we have ceased to feel the traditional fear of that equation, and we have a world of waste, pollution, and violence to show for it.

Plaint of Nature cannot be comprehended within the bounds of any of our specialties, which exclude themselves by definition from dealing with one another. But once we acknowledge, once we permit our language to acknowledge, the immense miracle of the existence of this living world, in place of nothing, then we confront again that world and our existence in it, forever more mysterious than known. And then the air swarms with questions that are scientific, artistic, religious, and all of them insistently economic. Some of the questions are answerable, some are not. The summary questions are: What are our responsibilities? and What must we do? The connection of all questions to the human economy is finally not escapable. For our economy (how we live) cannot leave the world or any of its parts alone, as the ideal of the wilderness preserve seems to hope. We have only one choice: We must either care properly for all of it or continue our lethal damage to all of it.

That this is true we may be unable to know until we have understood how, and how severely, we have been penalized by the academic and professional divorces among the sciences and the arts. Division into specialties as a necessity or convenience of thought and work may be as old as civilization, but industrialism certainly has exaggerated it. We could hardly find a better illustration of this tendency than Shelley's "Defence of Poetry," in which he makes, with characteristic passion, a division between imagination and reason, assigning things eternal and spiritual to imagination, and things temporal and material to reason. *Plaint of Nature*, as if in answer, resolves exactly that duality in the person of Nature herself, who joins Heaven and Earth, and whose discourse is of the vices that break her laws and her world, and of the virtues by which her laws would be obeyed, her rule restored, and her work made whole.

Nearly all of the *Plaint* occurs in Alan's dream-vision, which is to say in his mind. It is an intensely mental work, if only because he worked so hard to render the style of it, and it becomes rather stuffy. Going from Alan to Chaucer is like stepping outdoors. Nature, the "noble emperesse" herself, is presented twice in Chaucer's work, once with acknowledgement of "Aleyn" and "the Pleynt of Kynde." But in Chaucer, as never in Alan, the natural world itself is present also.

It has been said that in Chaucer nature is mainly idealized in dreams or paradisal visions. And I suppose it could be said that in his treatment especially of birds, he is a mere fabulist or a protocartoonist. But to me, his attitude toward the outdoor world seems in general to be familiar, affectionate, sympathetic, and, as only he could be, humorous. This we would know if he had written only the first twelve lines of *The Canterbury Tales.* A teacher required me to memorize those lines, and I did, more or less, sixty-some years ago. Since then I have found no other writing that conveys so immediately the *presence* (the freshness of the sight, feel, smell, and sound) of an early spring morning, as well as the writer's excitement and happiness at the thought of it.

Chaucer was a man of the city and the court, but I think he was not a "city person," as we now mean that phrase. London in Chaucer's time was comparatively a small city. Inundated as we are by the commotions of internal combustion, I doubt that we can easily imagine the quietness of fourteenth-century London, from any part of which, above or beyond the street noises, Chaucer probably could have heard the birds singing.

Travel on horseback through the countryside was then an ordinary thing. This involved familiar relationships with horses, and on horseback at a comfortable gait anybody at all observant participates in the life of the roadside and the countryside with an intimacy impossible for a traveler in a motor vehicle. City people then would have had country knowledge as a matter of course. If they dreamed or imagined idealized landscapes, that may have been a "natural" result of their close knowledge of real ones.

And so in *The Parliament of Fowls* we have an ideal or visionary garden with real trees. And Chaucer's list of its trees is not just an inventory. He clearly enjoys sounding their names, but he also introduces them, so to speak, by their personalities: "The byldere ok, and ek the hardy asshe, / The piler elm," and so on. A little later he identifies the birds in the same way. *The Knight's Tale* also has its list of trees, but this one names the membership of a grove that has been cut down, "disinheriting" its further membership of gods, beasts, and birds, and leaving the ground "agast" at the novelty of the sunlight falling upon it. Ronsard too felt this shock, this sympathy and dismay, at the exposure and vulnerability of the ground after the "butchery" of the Forest of Gastine (Elegy XXIV), and some of us are feeling it still at the sight of our own clear-cut forests. There is nothing fabulous or visionary about the Knight's small elegy for the fallen grove. It is informed by observation of such events, and by a real regret.

That the ground under the fallen grove was "agast" (aghast, shocked, terrified) at its new nakedness to the sunlight is not something that the Knight or Chaucer knew by any of the "objective facts" that industrial foresters would use to deny that such a thing is possible. The much older, long-enduring knowledge came from sympathy and compassion.

If Chaucer heard the birds singing in his own English, that was owing to his sympathy for them. Birds and animals use human speech by convention in literary fables, but that usage came into literature, I am sure, from the conversation of country people who lived, as some of them live still, in close daily association with the birds and animals of farmsteads and with those of the natural world. People who are in the habit of speaking to their nonhuman neighbors and collaborators are likely also to have the habit of translating into their own speech the languages and thoughts of those other creatures. There are practical reasons for this, and obviously it can be amusing, but it comes from sympathy, and in turn it increases sympathy.

Some years ago I wrote the introduction for a new printing of Theodora Stanwell-Fletcher's *Driftwood Valley*, a book I had loved since I was in the seventh or eighth grade. When I sent my typescript to the author, her strongest response was to my anticipation of the objection of some readers "to Mrs. Stanwell-Fletcher's anthropomorphizing of animal thoughts and feelings." One of her degrees was in animal ecology; she was "scientific" enough to know that I was right. But her marginal note, as I remember, said: "How else could we understand them?" She spoke with authority, for she and her husband had spent three years in a remote part of British Columbia, where their nearest neighbors were the native plants and animals they had come to study.

From *The Nun's Priest's Tale* we know that Chaucer was a perfect master of the literary fable, but that tale signifies also that he was a close observer of the manners of household poultry, and he no doubt had listened with care to the conversation of country people—several of whom, after all, were in his company of Canterbury pilgrims. That his attention to them and their kind had been deliberate and lively we know from two of his portraits in *The Canterbury Tales*: those of the Plowman in the *General Prologue* and the "poor widow" of *The Nun's Priest's Tale*. The descriptions of those two characters give us a sort of compendium of agrarian values—that are, by no accident, agreeable to Nature's laws.

The Plowman is said to be one of Chaucer's "ideal characters," and I suppose he is, but his idealization does not make him in any way simple. His description, though brief, is ethically complex. The Plowman is an exemplary man and countryman because he is an exemplary Christian. He loves God best, and then he loves his neighbor as himself. He is a hard worker, good and true, and a man of peace. And those virtues enable his charity, for he will work without pay for those who need his help. If I am not mistaken, Chaucer's sense of humor is at work here, pointed at those who think of goodness as a sacrifice paid here for admission to Heaven. The Plowman's virtues are understood as solidly practical and economic. He is a good neighbor, and good neighbors are likely to *have* good neighbors. The payoff, to complete the joke, may be Heavenly, but it is also earthly: A good neighborhood is an economic asset to all of its members.

The portrait of the elderly widow in *The Nun's Priest's Tale* attends to other practical and necessary virtues: patience, frugality, and good husbandry. These are congenial with the virtues of the Plowman and complete them, as his virtues complete hers. That the whole set of virtues is divided between the two characters is a matter only of appropriateness. The widow lives with her two daughters in a small cottage. She owns (God has "sente" her) three large sows, three cows, a sheep named Malle (Mollie?), seven hens, and a rooster. She works "out" as a "dairymaid," but it is clear that her economy is most securely founded upon her own small holding and her few head of livestock, also upon needing little. She needs no "poignant sauce" for her food because the food is good and hunger seasons it well. Her charity is to need no charity, another recognized way of being a good neighbor in a country community. This widow is a first cousin to the old Corycian farmer in Virgil's fourth Georgic. She practices exactly the "cottage economy" later praised and advocated by William Cobbett (and others).

My approach to Chaucer's first representation of Nature has been backward, from late work to early, as a way of knowing how he understood her. If we read *The Parliament of Fowls* simply as an exemplary "dream-vision" by a master poet and courtier, then perhaps we are free to regard the figure of Nature as a pleasing, even a beautiful, "picture" borrowed from Alan's *Plaint* and requiring little more of us than to place the poem historically and to appreciate it critically. But if we would like to take seriously this appearance or apparition of Nature, then we have to ask how seriously Chaucer took it, and that is not so easy. It helps, I think, if we conceive of Chaucer not only as the great poet and sophisticated "man of the world" that he certainly was, but also, and on the evidence of his writing, a man on easy speaking terms with the countryside and all of its inhabitants.

The poem is a dream-vision, a lighthearted fantasy, above all a comedy. It also takes place in a landscape in which the poet is well acquainted with the trees, the flowers, and the birds. The birds speak English, sometimes at length, but they speak also in their own tongues. We hear from the goose, the cuckoo, and the duck all together and all in the same line: "Kek kek! kokkow! quek quek!" When they speak English, they use images that might have been used at court or by a farmer's hearth, but they certainly came from people who had spent time outdoors at night: "There are more stars, God knows, than a pair" and "You fare by love as owls do by light."

Faced with the possibility of copying Alan's vision of Nature in her extremely elaborate finery, Chaucer politely declines by referring us to the *Plaint*, where we can find her in "such array." This is a part of the comedy, but it leaves him free to describe her in his own way, which he does by a single graceful compliment which serves to acknowledge her sanctity and her standing: She is so far fairer "than any creature" as the summer sun's light surpasses that of any star. And then, it seems, he draws aside a curtain:

She is sitting in a forest glade on a flowery hill. She is "set off" by halls and bowers of branches—evergreen branches, as Chaucer's readers must always have assumed—that have been "wrought" according to her design. This is Nature in her English vicarage. She sits before us as a distinctly hieratic figure, yet made familiar by the homely setting that combines art and nature. This is clearly akin to the scene in which, two and a half centuries later, Robert Herrick's Corinna will go a-Maying:

Come, my Corinna, come; and comming, marke
How each field turns a street; each street a Parke
Made green, and trimm'd with trees: see how
Devotion gives each House a Bough,
Or Branch: Each Porch, each doore, ere this, An arke a
  Tabernacle is
Made up of white-thorn neatly enterwove . . .

Chaucer's "noble emperesse, ful of grace" seems perfectly to belong here, in the center of such a mating ritual as he must have witnessed many times. She is there, not in May but on Saint Valentine's day, as in every year, to see to the matchmaking of all the birds, according to her judgment. To see that they in their great numbers find mates and reproduce after their kinds is her high office. Following Alan's doctrine, Chaucer has her presiding as

> the vicaire of the almyghty Lord,
> That hot, cold, hevy, light, moyst, and dreye
> Hath knyt by evene noumbres of acord . . .

Her work is to reconcile the world's opposites and contentions into a lasting, self-renewing composure.

From Nature enthroned our attention is drawn to the assembly of all the birds, to the clamorous low "common sense" of the goose, the cuckoo, and the duck, and finally to the elegant "roundel" that the mated birds all sing together at the end. These incongruities take nothing away from Nature's dignity and they don't need to be justified or explained. They notify us of the range of Chaucer's art and his knowledge, authenticated by the world itself, which often puts high seriousness and low comedy into the same event or the same instant.

But let us go ahead and ask the modern question: Did Chaucer "believe in" this Nature? Did he "know" her? Did he "actually see" her seated in her glade "upon an hil of floures"? Well, he seems to have seen her there, he seems to have invited me to see her, and I too seem to have seen her. Where are the proofs? There are of course no proofs, no photograph, no second witness. But we are talking here about the imaginative life of a country—not a nation, a *country*—which, in its apprehension of the natural world and its "invisible fecundity," its "hidden wholeness," always must outreach its proofs, its sciences, its mechanic arts, its political economy, its market. If it fails in that, as it has with us, then we get probably what we have got: a country, mainly unknown to its occupying humans, rapidly melting into a toxic slurry and flowing away through its rivers.

That Alan's Nature appears again, years later, in *The Canterbury Tales* suggests that Chaucer thought her important enough to keep her in mind to the end of his life. At the beginning of his tale, the Physician "quotes" what he imagines she would say of the singular beauty of Virginia, his chaste heroine. Though the Physician is a man whose honesty needs watching, we need not doubt his characterization of Nature. Following his professional habit of learned speech, he has Nature describe herself in keeping with Alan's description, and with Chaucer's in *The Parliament of Fowls*. She is God's "vicar general," in charge of the sublunar creation, whose "forming and painting" of the creatures is work done "to the worship of my lord." What may be unique here is her insistence that her work cannot be "counterfeited." In this, she takes the side of William Carlos Williams* and others in our own time, against Hamlet et al. who have argued that the "end" of art "was and is to hold . . . the mirror up to Nature . . ."

This disagreement is of interest now because it clearly defines the problem confronted by scientists of the last hundred or so years who have, against the industrial declivity, taken an ecological approach to our use of the land, primarily in agriculture. To those scientists, whose work I will discuss later in this essay, two truths have been obvious: first, that we humans cannot live in unaltered natural settings, for we passed up our chance to be "rational apes" too long ago; and, second, though we cannot make mirror images of natural places, even if we could do so, even if we could live in them if we could make them, we nonetheless are obliged to obey Nature's laws, which are imposed absolutely and will never change.

From Alan's vision of Nature, and then from Chaucer's, came Spenser's (as he testifies) in the fragmentary seventh book of *The Faerie Queene*, and of these three Spenser's is the finest. It is the most fully developed and detailed and the richest in meaning. The story he places Nature in, as he tells it, is the most dramatic. And the problem he sets for her is one that was urgent for him then, and that seems still more urgent for us.

Book VII of *The Faerie Queene* contains only the sixth and seventh cantos and two stanzas of an eighth. These are known as "the Mutability cantos," for they tell how "the Titanesse," Mutability, rebels against the classical deities and attempts to establish her sovereignty over them, and over all creation. She disdains the rule of Jove, and refuses to accept his judgment, appealing instead to "the highest him . . . Father of Gods and men . . . the God of Nature." And so great Nature herself, as God's deputy, comes to preside over a trial to determine the justice of Mutability's claim. For this all the creatures are assembled on Arlo hill, where they are "well disposed" by Order, who is "Natures Sergeant."

*In especially some of the most lucid prose passages of *Spring and All.*

And then Nature enters with the great processional dignity that Spenser seems to have learned from the twenty-fourth Psalm,* as maybe of all poets only he could have learned from it:

Then forth issewed (great goddesse) great dame *Nature*,
With goodly port and gracious Maiesty;
Being far greater and more tall of stature
Then any of the gods or Powers on hie:
Yet certes by her face and physnomy,
Whether she man or woman inly were,
That could not any creature well descry:
For, with a veile that wimpled euery where,
Her head and face was hid, that mote to none appeare.

That some doe say was so by skill deuized,
To hide the terror of her vncouth hew,
From mortall eyes that should be sore agrized;
That eye of wight could not indure to view:
But others tell that it so beautious was,
And round about such beames of splendor threw,
That it the Sunne a thousand times did pass,
Ne could be seene, but like an image in a glass.

That well may seemen true: for, well I weene
That this same day, when she on *Arlo* sat,
Her garment was so bright and wondrous sheene,
That my fraile wit cannot deuize to what
It to compare, nor finde like stuffe to that,
As those three sacred *Saints*, though else most wise,
Yet on mount *Thabor* quite their wits forgat,
When they their glorious Lord in strange disguise
Transfigur'd sawe; his garments so did daze their eyes.

. . . .

This great Grandmother of all creatures bred
Great *Nature*, euer young yet full of eld,
Still moouing, yet vnmoued from her sted;
Vnseene of any, yet of all beheld . . .

*This assumption is fairly obvious if one is thinking also of the twelfth stanza of his "Epithalamion." Only a poet of the greatest skill, and confidence, would have attempted this.

This representation of Nature clearly derives from those of Alan of Lille and Chaucer, but the resemblance, though unmistakable, is distant. Here there is nothing at all of Alan's relentless accumulation of details. And in our own time, when poets are supposed or expected to disown their forebears, it is a relief to come upon Spenser's filial devotion to Chaucer, but nothing here reminds us of the at-home conviviality of Nature's precedence over the Parliament of Fowls.

Spenser's Nature is altogether hieratic and luminous. In a way that recalls Dante's frustration in the *Paradiso*, Spenser describes this "great goddesse" mostly by describing his inability to describe her. Though she is heavily veiled like perhaps a nun and he refers to her always by feminine pronouns, she may be "inly" either a man or a woman, and her face is either unendurably terrible or so radiantly beautiful that it could not be seen except "through a glass darkly." And in describing her, Spenser, unlike Alan and Chaucer, recalls the Gospels, for her garment was "so bright and wondrous sheene" as to recall the garments of Christ at the Transfiguration. And here Spenser must be remembering also the Gospel of John 1:3: "All things were made by him; and without him was not any thing made that was made."

That she is veiled we may take as Spenser's admonishment to the scientists, from Francis Bacon to the highest-funded prophets and oracles of our time, who have proposed, by their merely human wits and devices, to wring from Nature her ultimate Truth, and thus to mate every problem with its ultimate solution. There is nothing so "progressive" in Spenser's view of Nature—or, hence, of nature. Knowing as he knew, he would have recognized Bacon's arrogance as old under the sun. He could not have expected Science, our own allegorical giant, to quiz Nature face to face.

Furthermore, Mutability, in pleading her case before Nature, addresses her as "greatest goddesse, only great"

Who Right to all dost deale indifferently,
Damning all Wrong and tortuous Iniurie,
Which any of thy creatures doe to other
(Oppressing them with power, vnequally)
Sith of them all thou art the equall mother,
And knittest each to each, as brother vnto brother.

*

Any twenty-first-century reader familiar with the formal principle of interdependence, as it operates in ecosystems, will recognize this "knitting" for what it is and means, as they will recognize also Nature's "equal" motherhood of "all" the creatures. Her "indifference" is not apathetic; she is merely impartial, preferring no single species over any other, just as the realists of present-day biology know her to be. And this supposedly modern perception is much older than Spenser, for he took it from the *Plaint* in which, four hundred years earlier, Nature had warned Alan that "my bounteous power does not shine forth in you alone individually but also universally in all things." We have our lives by no right of our own, but instead by the privilege of sharing in the life that sustains all creatures. This great convocation is the work of Nature, its "equall mother," which makes her not only, as Alan saw, our teacher, but also, as Spenser was first to see, our judge.

And so, submitting to Nature as the supreme worldly authority, Mutability presents her argument, and summons a procession of witnesses: the earth, the four elements, the seasons, the months, day and night, and finally life and death, all of whom support her argument. Jove then argues that worldly change is ruled by the "heavenly" gods. But Mutability charges and proves the changingness of the planetary deities, of the sun, which is sometimes eclipsed, and of Jove himself, who once was not and then was born. She is an unbluffable, brilliant lawyer, and her case is nearly perfect, as is the poetry that Spenser gives to it.

Finally, after a long considering pause, Nature gives her verdict:

I well consider all that ye haue sayd,
And find that all things stedfastnes doe hate
And changed be: yet being rightly wayd
They are not changed from their first estate;
But by their change their being doe dilate:
And turning to themselues at length againe,
Doe worke their owne perfection so by fate:
Then ouer them Change doth not rule and raigne;
But they raigne ouer change, and doe their states maintaine.

C. S. Lewis speaks, with justice, of the "deep obscurity" of those lines. But we may clarify them somewhat by placing beside them these earlier lines in which Mutability argues that

For, all that from [Earth] springs, and is ybredde,
How-euer fayre it flourish for a time,
Yet see we soone decay; and, being dead,
To turne again vnto their earthly slime:
Yet, out of their decay and mortall crime,
We daily see new creatures to arize;
And of their Winter spring another Prime,
Vnlike in forme, and chang'd by strange disguise . . .

Mutability here seems to argue, rather craftily, that change is absolute, leading invariably to something "unlike" and entirely new. Nature, if at the end she is remembering those lines, sees them as an imperfect description of a natural cycle capable of endlessly repeating itself—with, we would now say, occasional variations or "mutations," and depending on appropriate human cooperation. In her verdict, Nature readily acknowledges the ceaselessness of change, but she confirms, if not quite clearly, its cyclicality as greater, and as a form, or *the* form, of stability. She would have been confirmed in this by the *Plaint* where, in Alan's vision of her, Nature says (more clearly) that

> it was God's will that by a mutually related circle of birth and death, transitory things should be given stability by instability, endlessness by endings, eternity by temporariness, and that the series of things should ever be knit by successive renewals of birth.

This will be better understood about three and a half centuries later by scientists aware of the biology, and the supreme economic importance, of the fertility cycle.

At the end of Canto VII, Nature, having completed her work for the time being, "did vanish, whither no man wist." *The Faerie Queene*, as we have it, ends two stanzas later with Spenser's prayer for the "Sabaoths sight"

Of that same time when no more *Change* shall be,
But stedfast rest of all things firmely stayd
Vpon the pillours of Eternity . . .

Nature's standing in the order of things, as Spenser understood it, is exalted, well above that of humanity, and she has about her the nimbus of sanctity. Her equitable motherhood of all the creatures and her judgeship over them impose upon humans a responsibility that is both worshipful and relentlessly practical. But the order in which she is placed is firmly Christian, and her jurisdiction is limited to the incarnate world. Reassuring as may be her verdict against Mutability, it offers little comfort to individual humans in their suffering of their own mortality and that of their loved ones. And so *The Faerie Queene* as we have it, though incomplete, ends appropriately by invoking our so far undying hope for a "time," beyond Nature's world and all of its stories, "when no more change shall be."

Within the compass of my reading, Spenser's vision of Nature is the highest and fullest, the most responsibly imagined, the most complete, and the most instructive. And this, I think, is because it is the most thoughtful. In the Mutability cantos, Spenser confronts a question serious enough to have no definitive or final answer: On what terms are we to live with the perpetual changingness of this world? And he answers with an argument meticulously constructed. That his stanzas on Nature's appearance are so complex and beautiful must be partly the result of his thoughtfulness. I do not mean that he used his poetry as a vehicle to express or communicate his finished thought, but rather that his poetry was the vital means by which his thinking was done. Strange as it may seem to say this after the division of the mental functions into departments, it is clear that some poets have recognized that poetry in its way, like prose in its way, can be serviceable to thought, and when they have needed to do so they have used it as a way to think.

After Spenser, so far as I have read or remember, no other English poet acknowledged the influence of Alan of Lille or thought so carefully about Nature—probably because, after Spenser, no poet needed to think so carefully about her. But there certainly have been other English poets who appear to have been influenced by the earlier visions of Nature, and who have contributed to a line of thought about the proper human use of the natural world.

Passages in the eleventh and twelfth parts of *Piers Plowman* suggest that William Langland, for one, had read *Plaint of Nature*. C. S. Lewis assumed that he had. A remarkable difference is that, to Langland, Nature or "Kynde," rather than the Vicar of God, is God Himself. Langland, anyhow, was a better naturalist than either Chaucer or Spenser. Kynde instructs him in a dream to study the creatures, "the wonders of this world," to gain understanding and to learn to love his creator. And he observes carefully the variety of the creatures and their ways of mating, the skills of the nesting birds, the woodland flowers and their colors. His intimate knowledge of these things authenticates his wonder at them and his sense of the miraculousness of their existence. His wonder involves a tenderness unlike any other that I know: Of the flowers, the stones, and the stars only Kynde "himself" knows the causes; He is the magpie's patron and tells her, "putteth it in hir ere," to build her nest where the thorns are thickest.

Langland and Chaucer both died in 1400. After them, and after Spenser, I have in mind poems by Milton and Pope, in which they seem to have remembered Alan's Nature or Chaucer's or Spenser's and called her, so to speak, by name.

John Milton's *Comus* is a masque, an elegant play in verse, presented at Ludlow Castle in 1634 when the poet was twenty-five years old. It tells the story of a temptation, remembering the story of Satan's temptation of Christ (Matthew 4:1–11), and anticipating *Paradise Regained*, first published in 1671. As Spenser in the Mutability cantos had asked on what terms we are to live with the perpetual changing of this world, a question that had become urgent for him in the latter part of his life, so Milton in *Comus* was asking a question no doubt urgent in his youth but approximately parallel to Spenser's: On what terms are we to live with the material abundance of this world? Human nature, by any honest measure, is limited strictly and narrowly—we don't live very long, and we don't know very much—whereas the nature of the world at large by comparison seems a limitless plenitude. The two poets, then, were asking how to make human sense, a *little* sense, of an immensity.

In *Comus* a young maiden, identified simply as "The Lady," becomes lost in the woods. Alone and vulnerable, she meets Comus, perhaps the Tempter himself, disguised as a shepherd, who offers to guide her to safety. Instead, he takes her to "a stately palace, set out with all manner of deliciousness: soft music, tables spread with all dainties." He then proves himself the masterful and eloquent seducer he really is. He makes the conventional argument of *carpe diem*, "seize the day," which Jonson, Waller, Herrick, and Marvell also made in famous poems of elegant wit, giving to their dire and perfect logic a characteristic lightness of heart. The poet, as would-be lover, reminds his lady, as Marvell would put it, that "The grave's a fine and private place, / But none, I think, do there embrace."

This is exactly Comus's argument, but he makes it with a philosophic impudence and gravity that greatly enlarges its bearing. Milton's poem is sometimes described as a defense of "the sage / And serious doctrine of Virginity." It is that, but also far more than that. The poem's great question, as Comus himself raises it, is about the proper use and care of natural gifts:

> Wherefore did Nature pour her bounties forth
> With such a full and unwithdrawing hand,
> Covering the earth with odors, fruits, and flocks,
> Thronging the seas with spawn innumerable,
> But all to please and sate the curious taste?
>
> . . . .
>
> If all the world
> Should in a pet of temperance feed on pulse,
> Drink the clear stream, and nothing wear but frieze,
> The All-giver would be unthanked, would be unpraised,
> Not half his riches known, and yet despised . . .

He eventually takes up the conventional *carpe diem* theme of the transience of mortal beauty and roses that wither, but he is arguing, as the Lady quickly understands, not for using the world but for using it up. His ideology goes beyond mere personal gluttony and lust to a modern avarice and utilitarianism: the assumption, laid bare in our own time, that all of the natural world that we humans do not consume either is worthless or is wasted.

The Lady, threatened by Comus's passionate intensity and his power of enchantment, is protected only by her inner light —even Comus can see that "something holy lodges in that breast"—and her hope of rescue, which eventually comes, but she easily overmatches Comus as a debater, arguing from a better premise and with sufficient courage. Nature, she says,

> Means her provision only to the good,
> That live according to her sober laws
> And holy dictate of spare Temperance.
> If every just man that now pines with want
> Had but a moderate and beseeming share
> Of that which lewdly-pampered luxury
> Now heaps upon some few with vast excess,
> Nature's full blessings would be well dispensed
> In unsuperfluous even proportion . . .

Comus and the Lady are too allegorical in character to allow for much in the way of drama, but I don't think Milton can be accused of rigging their debate. He seems to have taken care to make Comus's argument as attractive as a vital man of twenty-five would have known it to be. The Lady's argument, attractive in a way perfectly opposite, more soundly appreciative of Nature's abundance, and approving temperance as the only safeguard of abundance, is Christian and democratic. (The "Christian conservatives" of our day would call it socialism.) Her argument serves my own by what I take to be its completion of the poets' long-evolving characterization of Nature. The Lady enlarges the import of Nature's demand upon humanity by making it, at last, explicitly economic. It remained for Milton to perceive clearly that Nature requires of us a *practical* reverence. Temperance in the use of natural gifts is certainly a religious obligation, but it is also an economic virtue. *Comus* requires us to think of the right use of gifts. To be in one's right mind is to know the right use of gifts. The Lady reacts so fiercely to Comus's proposition, not just because it assaults her personal virtue, but because it disdains and destroys the idea of economy. The word "economy," taken literally, as I am taking it, does not designate a financial system, but rather the management and care of the given means of life.

Alexander Pope was a poet in many ways unlike John Milton, and yet from the Lady's rebuke to Comus it is only a step to certain poems and passages of Pope. In spite of his physical debilities, Pope had more fun than Milton, but he certainly would have recognized his kinship to the Lady in *Comus.* As the Lady puts it, the human obligation to Nature is defined by obedience "to her sober laws" and the "holy dictate" of temperance. To Pope, that obligation is defined by "Sense" or "Good Sense," which in his use of those terms is pretty much the same. As he understood her, Nature requires of us certain proprieties, not only of manners but also of work. To him, everything depended on a proper sense of scale. We must act and work with the awareness always of the magnitude of Nature's work and of our own comparative smallness as individuals and as a species. And so Pope was another enemy of prodigality, of ostentation and the utter silliness of every kind of extravagance or waste.

In the "Epistle to Burlington," his satire is against the owners of country estates who surround themselves with houses and gardens magnificent to the point of ugliness and discomfort, far exceeding in dimensions and cost any use or pleasure: "huge heaps of littleness" built to display the wealth of the owner, who by comparison is "A puny insect, shiv'ring at a breeze." This is contrary to "Good Sense, which only is the gift of Heaven" and is worth as much as all the sciences. Against the Prodigals who waste their wealth on expensive, fashionable things they don't even like, and especially against their grotesque extravagances in what we now call "landscaping," Pope lays down two rules—"In all, let Nature never be forgot" and "Consult the Genius of the Place in all"—that my friend Wes Jackson and I have quoted back and forth for years in confirmation of our efforts for good husbandry of the land.

I need to confess, however, that I have often wondered how seriously I ought to take, not Pope's rules, which I think are sound, but Pope himself as a critic of land use. In the "Epistle to Burlington," after all, he is talking about the country houses and pleasure gardens of wealthy gentlemen, not working farms and forests. He goes so far as to say that even the worst examples, which he curses for "lavish cost, and little skill," are pardonable at least for giving employment to the poor—too much as polluting industries now are justified by "job creation." I forgive him for that because of his prophecy, clearly hopeful and immediately following, that these show places of extravagant littleness in "another age" will be wheat fields. Moreover, his understanding of the relation of art and nature is authentically complex and practical, appropriate to land use of any kind. Good sense leads to a proper mindfulness of Nature, which leads to collaboration between Nature and the gardener. The gardener's intention or design is completed by Nature's gathering of the parts into wholeness. By her gift, moreover, the land is made useful:

> 'Tis Use alone that sanctifies expense,
> And Splendor borrows all her rays from Sense.

The lawns of the estate should not be ashamed to be grazed by livestock, or the beautiful forests to yield timber. Though with less passion and not so explicitly, Pope thus consents to Milton's argument in *Comus*, that the human economy should be appropriate to the human dependence on Nature.

Pope is the last of the English poets to be mindful of Nature as mother, maker, teacher, giver of patterns and standards, and judge—so far as I know. I have repeatedly acknowledged the limits of my knowledge, first as a duty, but also with the hope that my deficiencies will be supplied by better scholars. What I am sure of is that we have lost the old apprehension of Nature as a being accessible to imagination, linking Heaven and Earth, making and informing the incarnate creation, and requiring of humanity an obedience at once worshipful, ethical, and economic. Her stern instruction, never disproved, that we humans have a rightful but responsible place in the order of things, has disappeared, and has been absent a long time from our working consciousness and our formal schooling.

Nature, as she appeared to Chaucer, Spenser, Milton, and Pope, does not appear in the "nature poetry" of the Romantic poets, and she is absent from the history of their influence upon both poetry and the conservation movement. By the time we come to Wordsworth, who often wrote about the natural world and often was on foot in it, there is already a powerful sense of being alienated from it, with a concomitant longing to escape into it from "the din / Of towns and cities," "the heavy and the weary weight / Of all this unintelligible world," "the fretful stir / Unprofitable, and the fever of the world."

I am quoting now from the poem familiarly called "Tintern Abbey," which displays pretty fully our modern love for nature, our often-lamented distance from it, and the vacationer's sensibility and economy that bring us occasionably "close" to it again, allowing us to feel more or less a religious sense of beauty and peace.

In "Tintern Abbey," addressed to his sister, Dorothy, who accompanied him, the poet has returned after a five-year absence to the Wye valley, "a wild secluded scene" of "beautiful forms," where the weight of the busy world is lightened and

> with an eye made quiet by the power
> Of harmony, and the deep power of joy,
> We see into the life of things.

In the quiet and beauty of that "wild" place he feels or perhaps recalls

> A presence that disturbs me with the joy
> Of elevated thoughts; a sense sublime
> Of something far more deeply interfused,
> Whose dwelling is the light of setting suns . . .

There can be no doubt of the strength of his emotion here, or of the loftiness of his language. Perhaps the presence he feels is that of Nature as the older poets imagined her, but these lines, however intense, are vague by comparison, and his thought is entirely dissipated by his resort to "something." His claim, soon following, that he recognizes

> In nature and the language of the sense
> The anchor of my purest thoughts, the nurse,
> The guide, the guardian of my heart, and soul
> Of all my moral being

is devoid of any particular thought or any implication of a practical responsibility. His version of "nature" thus lacks altogether the intelligence and moral energy of Nature as she appeared to the older poets. Of Wordsworth she seems to have required nothing at all in particular, except perhaps his admiration.

In line 122 the poet personifies Nature by capitalizing her name, but he then also sentimentalizes her:

> Nature never did betray
> The heart that loved her; 'tis her privilege,
> Through all the years of this our life, to lead
> From joy to joy . . .

and thus she becomes the poet's guardian against human evils and "The dreary intercourse of daily life . . ." To perceive in Nature this favoritism is clearly more self-indulgent and less true than Spenser's characterization of her as the "equall mother" of all creatures, which conforms exactly to Jesus's reminder that God "maketh his sun to rise on the evil and on the good, and sendeth rain on the just and on the unjust." Since Nature is so exceptionally kind to Wordsworth and his sister, it is in a manner logical that he declares himself a "worshipper of Nature." Here he departs maybe as far as possible from the Nature of his predecessors, to whom she was God's Vicar and thus forever subordinate to Him. The older poets were, as C. S. Lewis said of Spenser, Christians, not pantheists.

In comparison to the imaginative force and complexity of the earlier poets, this poem looks simpleminded and slack. Nature is understood merely as the purveyor of a sort of consolation or what we now call "mental health." Nobody could take from this version of Nature any sense of our economic dependence upon her, much less of her dependence upon our virtue. The wonder is that this poem contains in lines 11–17 a fine and moving description of an economic landscape—"pastoral farms / Green to the very door," "plots of cottage ground," orchards, and hedge rows—but he makes no approach to the economic life of that place or to its farmers, who certainly could have enlightened him about Nature's special preferences and favors. In this poem, the farms rate only as scenery, as they do for nature lovers of our own time.

A further wonder is that two years later Wordsworth wrote "Michael," a poem that penetrates the scenery. In that poem, the poet imagines in plenitude of detail the lives of a "pastoral" family of the Lake District: Michael, an elderly shepherd, Isabel, his wife, younger by twenty years, and their late-born only son, Luke. They live in difficult country, in weather that can be harsh. Like the many generations of their forebears, they live by endless work, from daylight to dusk and on into the night, the aging parents

> neither gay perhaps,
> Nor cheerful, yet with objects and with hopes,
> Living a life of eager industry.

There is something of the quality of legend in the telling, for their life was old in the poet's memory, and old beyond memory. But I think there is little if any idealization, no "romanticizing." Above all, Wordsworth passed unregarding the temptation to present these people as "clowns" or, as we would say, "hicks." Michael's mind "was keen, / intense, and frugal." As a shepherd, he is "prompt / And watchful." He knows "the meaning of all winds."

The poem tells of Luke's upbringing, during which he learned his people's world and their work by accompanying his father almost from infancy, by being "Something between a hindrance and a help" at the age of five in helping his father to manage their sheep flock, and by becoming his father's "companion" in the work by the age of ten.

Theirs is a "world" in itself almost complete and everlasting. But when Luke is eighteen trouble comes from the outside, as trouble is apt to come from the outside in, for instance, the novels of Thomas Hardy. Michael has mortgaged his land "in surety for his brother's son," to whom misfortune has come, and Michael is called upon to pay the debt, which amounts in value to about half his property. To preserve the land undivided as Luke's inheritance, his old parents decide to send him away, to work for another kinsman, "A prosperous man, / Thriving in trade." Working for this man, Luke would earn enough to lift the debt from the land. As Michael says to Isabel,

> He quickly will repair this loss, and then
> He may return to us. If here he stay,
> What can be done? Where everyone is poor,
> What can be gained?

For a time, Luke does well, the kinsman is pleased with him, he writes "loving letters" home. And then doom falls with terrible swiftness upon the family and their long history:

> Luke began
> To slacken in his duty; and, at length,
> He in the dissolute city gave himself
> To evil courses: ignominy and shame
> Fell on him, so that he was driven at last
> To seek a hiding-place beyond the seas.

Thus an ancient story receives almost abruptly its modern version: The Prodigal, the far-wandering son, does not now finally make his way home to continue the family lineage in its home place. Now he is gone forever. The forsaken parents live on alone, and die, and the land, at such great cost held to, is sold "into a stranger's hand."

In "Tintern Abbey," without of course intending to do so, Wordsworth laid out pretty fully the model of industrial-age conservation, which reduces too readily to the effort to preserve "wilderness" and "the wild," in certain favored places, as if to compensate or forget the ongoing industrial devastation of the other landscapes. This version of conservation, industrial and romantic, orthodox and dominant for at least a century, simplifies and sentimentalizes nature as friendly, wild, virgin, spectacular or scenic, picturesque or photogenic, distant or remote from work or workplaces, ever-pleasing, consoling, restorative of a kind of norm of human sanity. Conservationists of this order have thus established and ratified a division, even a hostility between nature and our economic life that is both utterly false and limitlessly destructive of the world that they are intent upon "saving." Such conservationists are no threat at all to the economy of industry, science, and technology, of recreational equipment and vacations, which threatens everything those conservationists think they are defending, including "wilderness" and "the wild." Meanwhile the absolute dependence, even of our present so-called economy, even of our lives, upon the natural world is ignored. In my many years of advocacy for better care of farms and working forests, the silence of conservationists and their organizations has been conspicuous. They oppose sensational abuses such as global warming or fracking or (sometimes) surface mining, but they don't oppose bad farming. Most of them would not recognize bad farming if they saw it, and they see plenty of it even from the highways as they drive toward the virgin forests and the snowcapped mountains. It seems never to have occurred to them that soil erosion and stream pollution in agricultural lands threaten all of this natural world, even "the wild," and that such abuses are caused by an economy that ruins farmers and farms by policy.

"Tintern Abbey" is an archetypal poem, for it gives us the taste, tone, and "spiritual" justification of the escapist nature-love of the many romantic nature poems that descended from it, and of the still-prevailing mentality of conservation. "Michael" also is an archetypal poem, but in a sense nearly opposite. Wordsworth understood it as no more than a family tragedy. Two centuries later, we must see it as uncannily predictive of millions of versions of its story all over the world: the great and consequential tragedy, little acknowledged or understood, of the broken succession of farm families, farm communities, and the cultures of husbandry. Generation after generation the children of farm couples have moved away, leaving the land's human memory, often of great ecological and economic worth, to die with their parents. Whether they have gone away to fail or succeed by the measures of their time, they still have gone away, and their absence is a permanent and enormous loss. And so "Michael" is great both because of its achievement and stature as a work of art and because of its rarity and significance as a cultural landmark. It stands high among the poems that have meant most to me. I first read it more than sixty years ago.

The second tragedy of "Michael" has to do with the history of the poem itself. Unlike "Tintern Abbey," it has had, so far as I am aware, no influence on our thinking about the natural world and our use of it, and little or no influence on the subsequent history of poetry. It stands almost alone. The only poem I know that I think worthy of its company is "Marshall Washer," by my friend Hayden Carruth. Marshall Washer was for many years Hayden's neighbor, a dairy farmer near Johnson, Vermont. Like Michael, Marshall lived a life of hard work on a small farm in a demanding place and climate. As, like Michael, the master and artist of his circumstances, Marshall earned Hayden's love and respect, and from his companionship and example Hayden learned many things that he valued. Hayden, moreover, saw Marshall in perspective of the hard history of such farmers before and after the time of Wordsworth. In the poem, Marshall has lived and worked and aged into a loneliness known to millions of his kind and time. His wife has died, and his sons have

> departed, caring little for the farm because
> he had educated them—he who left school
> in 1931 to work by his father's side
> on an impoverished farm in an impoverished time.

Beyond Marshall's life, the life of the farm had become unknowable and unimaginable, and all such farms were coming under the influence of "development." Land prices and taxes were increasing. It was becoming less possible for a small farmer to own a small farm. And Hayden imagined Marshall's sorrow:

> farming is an obsolete vocation—
> while half the world goes hungry. Marshall walks
> his fields and woods, knowing every useful thing
> about them, and knowing his knowledge useless.

This describes the breaking of what Hayden called "the link of the manure"—"manure" in the senses both of fertilizing and caring for, hand-working, the land—which cannot be ignorantly maintained. The "link of the manure" is the fundamental economic link between humans and the natural world. No matter the plain necessity of this link, it is breaking, or is broken, everywhere.

Ignorance certainly will break it. But so also will forces imposed upon it by what we falsely and too readily think of as the "larger" economy. There obviously can be no economy larger than its own natural sources and supports. Less obvious, farming being "an obsolete vocation" as far as most of us are concerned, is the impossibility that any economy can be larger or more important or more valuable than the economies of land use that connect us practically to the natural sources.

Nevertheless, we have this small contrivance we call "the economy," utterly detached from our households and our need for food, clothing, and shelter, in which people "put their money to work for them" and sit down to await the increase, in which money interbreeding with money enlarges itself to monstrosity, glutting on the world's goods. This small economy, centralized and concentrated in the larger cities, imposes in its great equation of ignorance and power a determining and limitlessly destructive influence upon the economies of land use, of farming and forestry, which are large, dispersed, and weak.

Those of us who would like to understand this could do worse than consult with poets, or with the too few of them who have taken an interest and paid attention. Those two widely separated and lonely poems, "Michael" and "Marshall Washer," are tragedies of the modern world. The stories they tell become tragic because the interest of the land, the human investment of interest and affection in the land, becomes subordinated to the interest of a "larger" economy that removes the human interest native to a place and replaces it with its own interest in itself.

After the story of the westward migration, the dominant American story so far is that of the young people who have departed from their rural birthplaces, "humble" or "small" or "backward" or "poor," to find success or failure in the big city. The story of the loneliness of the elders left behind, though surely as common, is rarely told. The implication of its rarity is that it does not matter, but is only a small sorrow incidental to the quest for a greater happiness. But sometimes a backward look will occur, bringing a recognition of loss and suffering that matter more than expected. The example I have in mind is a Carter Family song, "The Homestead on the Farm":

> I wonder how the old folks are at home,
> I wonder if they miss me when I'm gone,
> I wonder if they pray for the boy who went away
> And left his dear old parents all alone.

The song, which I like, carries the story only as far as nostalgia: The departed boy looks back with fondness for his old home, which his memory no doubt has improved, and with some regret, but he clearly has no plans to return. Maybe we can do no better than this, having as yet no common standards or a common language to deal with social disintegration, much less with diminishments of culture and the loss of local memory, all of which will enter into an accounting indivisibly cultural, ecological, and economic.

I don't know if Ezra Pound ever actually knew such people as Michael or Marshall Washer. But his poems on usury, "sin against nature," cast on the stories of these men what has been for me an indispensable light. Maybe nobody ever gave more passionate attention than Pound to the ability of a monetary system by means of usury to drive the cost of land and its products beyond any human measure of their worth, and thus to prey upon and degrade the work and the health that sustain us. And nobody has ever instructed us about this with more economy or grace or beauty:

> With usury has no man a good house
> made of stone, no paradise on his church wall
> With usury the stonecutter is kept from his stone
> the weaver is kept from his loom by usura
> Wool does not come into market
> the peasant does not eat his own grain
> The girl's needle goes blunt in her hand
> The looms are hushed one after another
>
> . . . .
>
> Usury kills the child in the womb
> And breaks short the young man's courting
> Usury brings age into youth; it lies between the bride
> and the bridegroom
> Usury is against Nature's increase.

Pound had the misfortune, self-induced, of becoming more notorious for his prejudices and political mistakes than famous for sanity. But he did have in him a broad streak of good sense. He could see all the way to the ground, not invariably a talent among poets, and he had moments of incisive agrarianism. He wrote in these lines a clarifying history of modern land husbandry:

Dress 'em in folderols
                                        and feed 'em with dainties,
In the end they will sell out the homestead.

He praised with perfect soundness "Chao-Kong the surveyor," who "Gave each man land for his labour"—wages being always vulnerable, whereas the value of land, like the value of a life, is unreckonable and absolute. And he wrote these lines,

                                        Pull down thy vanity, I say pull down.
Learn of the green world what can be thy place
In scaled invention or true artistry . . .

in which he stood before Nature much in the posture of Alan of Lille eight hundred years before.

Whoever would think at the same time of the home ecosystem on the one hand, and on the other of the home community (ecosystem plus humans), is all but forced to think of the local economy—and its tributary local economies elsewhere. Very few poets—very few people—have thought of both at once, because, I suppose, of the intensity of the stretch. It is something like standing with one foot on shore and the other in a loosely floating small boat. It requires a big heart and a strong crotch.

Of the poets I know of my own time, Gary Snyder is the one whose thought and work (in poetry and prose) have most insistently inclined toward the daily and practical issues of our economic life, which is to say our life: the possibility of bringing our getting and spending into concordance with terrestrial reality. This has been, for Gary, a lifelong effort, involving the events and materials of his life, much travel, much reading, much study and thought of his home geography. At a loss as to how to represent this effort in its complexity, I will say only that I think it has begun and renewed itself again and again in realizations of the profound, indissoluble link, the virtual identity, between the world's life and the lives of creatures. How is it with us who live our ever-changing lives as parts or members of the ever-changing world? Or to use language Gary has borrowed from Dogen: How is it with us who are walking on mountains that are walking? This question is not comfortable, and we sometimes would like to ignore or wish against it. But it means at least that the world is always with us, new and fresh:

> Clear running stream
> clear running stream
>
> Your water is light
> to my mouth
>
> And a light to my dry body
>
> your flowing
> Music,
> in my ears. free,
>
> Flowing free!
> With you
> in me.

It is not hard to imagine thoughts of economy and of "the economy" starting from the tenderness, intimacy, and inherent delight of this kindness, this kinship. But the poem also is a thanksgiving, the proper conclusion to such thoughts.

And I ask myself, Can it be that, by way of Buddhism and American Indian anthropology, Gary Snyder is another who has come face to face with the great goddess, great dame Nature, and he makes her this cordial greeting? Maybe.

It must be clear enough by now that I am a reader who reads for instruction. I have always read, even literature, even poetry, for instruction. I am a poet, and so I have read other poets as a poet, to learn about poetry, and, just as important, to learn from poetry. I am also in a small way a farmer devoted to farming, as my father and his father made me, and I have read poetry, and everything else, also as a farmer. As a farmer I have lived daily with the inherent hardships and pleasures, but also with an ever-fascinating and utterly intimidating question: What should I be doing to care properly for this (as it happens) very difficult and demanding place? And a further one: How can I make myself a man capable of seeing what is the right thing to do, and of doing it?

Such questions now are typically addressed to experts, and are answered by a letter or a pamphlet replete with statistics, graphs, diagrams, and instructions, all presumed to be applicable to all persons and to any and every place within a designated region or zone. This may in fact prove helpful, but it is not enough. It is not enough because it includes no knowledge of the particular and unique place where the expertise is to be applied. Moreover, expertise which is credentialed and much dependent on the demeanor and language of authority is typically unable to acknowledge either its inner ignorance or the immense mystery that surrounds equally the asker and the answerer. And so askers are in effect left alone with the expert answer on their singular small places within the mostly uninformative universe.

Supplements and alternatives are available. One can talk with one's neighbors and elders, who never are oracles but are usually worth listening to. One can consult with other experts. I have always liked to consult older or earlier experts, whose knowledge may be seasoned with affection and humility. Or one can read, following one's nose as good hounds and readers do, always on watch for what people in other times and places have known or learned, what penalties they paid for ignorance, what satisfactions they gained by knowledge. These benefits can come from books of every kind.

Such broad and even random conversation is necessary because the present industrial world is not an isolated empire fortified against history or incomparable by its improvements with other times and ways. As I have conversed and read for the last fifty years, I have been reminding myself that many new ways of doing work have been adopted not because they were better for land and people, but simply because the old ways were technologically outmoded. Industrial agriculture, as probably the paramount example, is better than pre-industrial agriculture by the standard mainly of corporate profitability, excluding by a sort of conventional or fashionable blindness the paramount standard of mental, bodily, and ecological health. One can learn this by the study of rural landscapes, but conversation and reading also are necessary. To think responsibly about land use, the whole known spectrum of means and ways must be available to one's thoughts. If, for example, one cannot compare a tractor (and its attendant economy and ecological effects) with a team of horses (and its attendant economy and ecological effects), then one has the use of less than half of one's mind.

Do I, then, think that farmers, or persons of other vocations, can find actual help in reading Chaucer or Milton or Spenser or Pope? Yes, that is what I think. To know Chaucer's plowman is not to know something merely of historical interest. It is to know, to recognize immediately, something that one needs to know. It would be too much to say that *all* farmers, foresters, economists, ecologists, and conservationists should know what Chaucer, Spenser, Milton, and Pope wrote about Nature, but it is a pity and a danger that none or only a few of them do.

Having argued that it is possible to learn valuable, even useful and necessary things from poetry, I now have to answer two further questions. First: Is it necessary for a poem to be instructive in order to be good? I hope not, and I don't think so. There is nothing very instructive, for example, in hearing that "the cow jumped over the moon," but who is not delighted by that poem's exuberant indifference to the possibility of making sense? It is a masterpiece. Even so, I am happy to know that some poems delight *and* instruct, which is a richer possibility.

Second: How does an instructive poem instruct? The answer seems obvious—by containing something worth knowing—but there is one condition: It must teach without intending to do so. In support of this I offer a sentence by Jacques Maritain, who said of the cathedral builders: "Their achievement revealed God's truth, but without *doing it on purpose*, and because it was not done on purpose." The point, I believe, is that what the cathedral builders were doing on purpose was building a cathedral. Any other purpose would have distracted them from the thing they were making and spoiled their work. Teaching as a purpose, as such, is difficult to prescribe or talk about because the thing it is proposing to make is usually something so vague as "understanding." My own best teachers, as I remember, sometimes undertook deliberately to teach me something: "You handed me that board wrong, and made my work harder. Now I'm going to hand it back to you *right*. Now you hand it back to me *right*." But this was rare. When I was most learning from them, I think, they were attempting something besides teaching, if only to make a sentence that made sense. They taught me best by example, unaware that they were teaching or that I was learning. Just so, an honest poet who is making a poem is doing neither more nor less than making a poem, undistracted by the thought even that it will be read. Poets, or some poets, bear witness as faithfully as possible to what they have experienced or observed, suffered or enjoyed, and this inevitably is instructive to anybody able to be instructed. But the instruction is secondary. It must be embodied in the work.

One remarkable thing I have noticed in my reading is that agrarianism, readily identifiable by certain themes (the importance of the small holding, the relationship to Nature, etc.), appears throughout the written record from as far back as Homer and the Bible. A second remarkable thing is that these appearances are intermittent, sometimes widely separated. When this vital strand of human thought and concern has disappeared from writing—as it does, in poetry, from Wordsworth's "Michael" to Hayden Carruth's "Marshall Washer"—it has continued in the conversation of farmers. I was (I am still) Hayden Carruth's friend. I knew him from much conversation and many letters. In Hayden's company I met and talked at some length with Marshall Washer. Though Hayden was far better read than I am, I am confident that the agrarianism of Hayden's poem did not come from his reading. It came straight from Marshall himself, who had it from his father, who had it from a lineage of living voices going back and back, more or less parallel to, but rarely, maybe never, intersecting with the lineage of writings. I know this also from my own long participation in such conversation. The appearances of agrarian thought in the written record are like stepping stones, separate and irregularly spaced, but resting on a firm and (so far) a continuous bottom.

Who were, and to a much-reduced extent still are, the people who have passed this tradition by living word from generation to generation for so long? The answer, obviously, is the country people who have done the work of field and forest, generation after generation. They are the people once known as peasants, serfs, or churls, and now, still doing the same work, as farmers or country people, when "farmer" and "country" are still as readily used as terms of abuse as "peasant" and "churl" ever were. My friend Gene Logsdon says that when he was a schoolboy in Wyandot County, Ohio, "*Dumb farmer* was one word." Here I need to remember two sentences from G. G. Coulton's *Medieval Panorama*. The first is this: "Four-fifths, at least, of the medieval population had grown their own food in their own fields; had spun and woven wool from their own sheep or linen from their own toft; and very often it was they themselves who made it into clothes." I should add that they also kept themselves housed, and they fed and sheltered their animals. Anybody who has done any part of such work knows that it involves knowledge of the highest order and value, for the human species has survived by it, and it is neither simple nor easy to learn.

But how were these people with their elaborate and indispensable knowledge valued in the Middle Ages? Here is the second sentence from *Medieval Panorama*: "Froissart shows us plainly enough how, after the bloody capture of a town or castle, the gentles [knights and nobles] were spared, but the common soldiers [churls] were massacred without protest from their more exalted comrades in arms." They were thought to be, in a word, dispensable.

I don't believe that this thought has changed very substantially from then until now. We, whose humane instincts have so famously evolved and improved, are firmly opposed to outmoded forms of massacre, but we still regard farmers as dispensable, and by various economic constraints and social fashions we have dispensed with many millions of them over the last sixty or so years. If the twenty-first-century American farmer is farming several thousand acres and employing a million dollars' worth of equipment, he still is a member of a disparaged, and therefore a vanishing, population. He does not control any part of his economy. He has no more influence over the markets on which he buys and sells than he has over the weather. No prominent politician, economist, or intellectual is thoughtful either of him or of the condition of the land he works, or of the health of the ecosystem that includes his land. In the entire food industry he will be dependably the most at risk, the least valued, and the lowest paid, except, of course, for the migrant laborers he may at times be constrained to hire. If he miscalculates or has a bad year and is ruined, that will register among the professional onlookers as a minor instance of "creative destruction."

In the course of this progress of industrialization and depopulation, farming itself has become so radically simplified as to be unworth the name. Most damaging has been the division between the field crop industry and the meat-animal industry. To remove the farm animals from farming is to remove more than half of the need for knowledge, skill, and intelligence, and nearly all of the need for sympathy. To crowd the animals into the tightest possible confinement to live and function exclusively as meat-makers is to do away with sympathy as a precondition, to reduce mindfulness to routine, and to replace all the free helps of Nature and natural health with purchased machinery and medicines.

But when plants and animals, croplands and pastures, are gathered into the care of a single farmer, this calls for a mind versatile and accomplished, competent to deal with the fairly stable natures of the kinds of plants and animals, as well as the variable and sometimes surprising circumstances of mortality, the weather, and the economy. Among the good farmers I have known, I have found also a formal intelligence by which they ordered the spatial arrangement of fences, fields, and buildings to be most conserving of the land and most usable, and by which they formed also the temporal structures of their crop rotations and the days, seasons, and years of their work.

Good farmers, whose minds are comparable to the minds of artists of the "fine arts," have been instructed so often for so long that they are "dumbfarmers," that they only half believe how smart and capable they actually are. I have heard too many of them describe themselves as "just a farmer." It is, at any rate, impossible for highly credentialed professionals and academics to appraise justly the intelligence of a good farmer. They are too ignorant for that. You might as well send a bird dog to judge the competence of a neurologist.

About such things there are no "objective" measures. I can offer only my testimony. In my by-now long life, I have known well and observed closely the work of a good many farmers whom I have respected. I have thought carefully about them for a long time, and some of them I have written about. In general: I have found them alert, observant, interested, interesting, thoughtful of their experience, conversant with the experience of others. By midlife or sooner they have come to know many things so completely as to be unconscious of knowing them. They have had, in general, the humorous intelligence that recognizes natural limits and their own limits. They have had also a commonplace sense of tragedy that permitted them to accept their helplessness in the face of loss and suffering—as Marshall Washer faced his burning barn with thirteen heifers trapped inside, and kept on, still working, by the testimony of his friend Hayden Carruth, within weeks of his death at eighty-eight. That is not indicative either of "dullness" or "stoicism," nor does it mean that farmers in general are better or more virtuous than other persons. It means simply that farmers live and work in circumstances unremittingly practical, and for this certain strengths of mind are necessary.

In general, the good farmers I have known have had no taste for self-display, did not parade their knowledge, and would say little where their knowledge might not be respected. But if you had a mind and ear for their conversation, and some ground of friendship or common knowledge, you were likely to hear sound sense that came from their experience of the natural world, and of farming and its local history. As late as my own generation, because the young then were still working with and listening to their elders, the talk of farmers was still carrying the traditional themes and attitudes. Many of the things I heard from them I have kept always in my mind, their words and their voices.

Probably in the summer of 1965, after my family and I had moved here to the "twelve acres more or less" where we still live, I was not equipped for some mowing that needed to be done. I hired a neighbor of my parents' generation, a man I had known from my childhood, liked, and respected. When he had finished the work, he turned off the tractor engine and we talked a while, he much in the spirit of welcoming me to my new place. He had been renewing his own acquaintance with the place as he worked. We spoke of the ways it had been used (not always kindly), and of the uses I might make of it.

When he was ready to go, he started the tractor and, to end the conversation, said cheerfully, "Well, try a little of everything, you'll hit on something."

He knew, and he knew I knew, how little of "everything" the old place was actually capable of producing on its steep slopes and its one pretty good "garden spot" by the river. But he had said a good thing, and we both knew that. He had stated one of Nature's laws: the Law of Diversity, vital both to ecological and to economic health. One of my father's father's rules was "Sell something every week," a different version of the Law of Diversity.

Another of my older friends, a fine farmer whose friendship I inherited from my father, once told me a crucial part of the story of his beginning on his own farm. This, I think, would have been in 1940 or a little earlier. He and his wife had gone into debt to buy their place, and they had no money. My friend went early in the year to a grocery store in town and asked the owner if he would "carry" him until he sold his crop. The owner knew him, trusted him, and agreed. When my friend sold his crop at the year's end and went to settle up at the store, his bill came to eleven dollars and some cents. He and his wife had been thrifty and careful, living so far as possible from their place, thus obeying another of Nature's laws, the Law of Frugality: Don't be prodigal.

A man of about the same age, a dear friend who was conscientiously my teacher for thirty years, once told me with an emphasis amounting to passion: "If you've got grass and room to keep a milk cow [for family use] and you don't do it, you've *lost* one milk cow." The absent cow he saw, not as a neutral matter of mere preference, but as a negative economic force, a subtraction from thrift and thriving. This was in respect for the Law of Diversity and the Law of Frugality.

A steadying influence on my mind for most of my life has been the metaphor by which my father referred to a patch of abused land that had been healed and grassed as "haired over." He thus perceived a wound to the earth as a wound to a living creature, perhaps a collar gall on the shoulder of a workhorse. We could say that this phrase observes another of Nature's laws—Keep the ground covered—but much else of value is concentrated in it. By it my father wakened my mind to the thought of the system of kinships by which the world survives, and to the thought of the whole significance of the smallest healing. I don't know whether or not this language originated with my father, but I knew *his* father and so I know that it was not the thought of one mind. Such a thought is as far as possible unlike the thought of the earth as an inert material mass to be shoved about, poisoned, and blasted at human discretion.

None of those sayings comes from a note I made at the time or from any record I kept, but only from memory, and to me this signifies their importance and value. I have remembered them and many others, I believe, because I recognized them not quite consciously as parts of a whole, a kind of mind which, like my mentors, I inherited, which probably is as old as farming and necessary to it. This mind, which I think has never been fully conscious or coherent in any one person, never perfect, always in some manner failing, almost never having the incentive of public appreciation or adequate economic support, has nonetheless cohered, coming to consciousness as needed, and expressing itself in an articulate local speech that takes its heft and inflection from the reconciliations between people and their circumstances, their work and their feelings about it. This mind, much to the credit of its inherent good sense, has survived all its adversities until now, living on, talking to itself, so to say, in the conversation of its local memberships. It still speaks, where it survives, of the importance of the well-doing of small tasks that the dominant culture has always considered degrading but which are nonetheless essential and worthy of care: "Don't think of the dollar," my friend and teacher said to me. "Think of the *job*." And it still speaks of our dependence on the particular natures of creatures and on the natural world, and, more practically, of a necessary respect and deference toward Nature. In this speech, Nature customarily is personified: the great dame herself, who knows best and will have her way. Always there is the implication that Nature helps when you work *with* her, with knowledge of her ways, their value, and their ultimate dominance, and that she does not help but works against you when you work against her.

Recently my son brought me an issue of a conventional industrial farming magazine, containing an article, by a "semi-retired" rancher, Walt Davis, about the need to manage a cow herd so as "to be in sync with nature." Though some of the vocabulary ("in sync") is new, this advice is old. I know nothing that suggests it is wrong. The point, as often, is economic: To be in sync with Nature is to make full use of her helps that are free or cheap, as opposed to the use of industrial substitutes that are expensive. By observance of Nature's laws, the land survives and even thrives in human use. By the *same* laws, farmers can hope at least to survive in the almost conventional adversity of the farm economy.

Almost typical of young farmers is the impulse to intervene in natural processes with some "latest" expertise, requiring typically some "latest" merchandise in order to increase production or fend off some perceived threat. Older farmers are more likely to be suspicious of anything that costs money, and to rely instead on the gifts of their own intelligence and the nature of creatures. Young shepherds, in their eager and self-regarding sympathy, may try to help a newborn lamb to stand and suck—may even succeed, or seem to. Older shepherds will know that such "help" is most likely a waste of time. They know when to walk away and "leave it to Nature." They are likely to know also that Nature bids them to get rid of a ewe whose lamb can survive its birth only with human help.

Going on forty years ago, because I knew his grand reputation as a breeder of Southdown sheep, I had the honor of becoming a little acquainted with Henry Besuden of Vinewood Farm in Clark County, Kentucky. Mr. Besuden was seventy-six years old then and not in good health; it had been several years since he had owned a sheep. But when I visited him, which I did at least twice, and he showed me his place and we talked at length, I found that his accomplishments as a farmer excelled in fact and in interest his accomplishments as a sheepman.

In 1927, at the age of twenty-three, he inherited a farm of 632 acres, a large holding for that country, but it came to him as something less than a privilege. The land had been ruined by the constant row-cropping of renters: "Corned to death." Gullies were everywhere, some of them deep enough to hide a standing man. Between Mr. Besuden then and an inheritance of nothing was the stark need virtually to remake his farm.

To do this he had to return the land from its history of human carelessness to the care of Nature: Every gully, through which the land was flowing away, had to be transformed into a grassed swale that would check and retain the runoff. This required a lot of work and a long time, but by 1950 all of the scars at last were grassed over.

To this effort the sheep were not optional but necessary. They would thrive on the then-inferior, weedy and briary pastures, and, rightly managed, they were "land builders." He became a sheepman in order to become the farmer Nature required for his land. The constant theme of his work was "a way of farming compatible with nature" or, as Pope would have put it, with "the Genius of the place." I have not known a farmer in whose mind the traditional agrarianism was more complete or more articulate.

He wrote in one of his several articles entitled "Sheep Sense": "It is good to have Nature working for you. She works for a minimum wage." Soil conservation, he wrote, "also involves the heart of the man managing the land. If he loves his soil he will save it." He wrote again, obedient to Nature's guidance, of the need "to study the possibilities of grass fattening." He wrote of the importance of "little things done on time." He told me about a farmer who would wait until he came to a spot bare of grass to scrape the manure off his shoes: "That's what I mean. You have to keep it in your mind."

In any economy that becomes exploitive of land or people or (as usual) both, it appears that some of the people working within it will recognize that its special standards of judgment are inadequate. They will see the need for a more comprehending standard by which the constrictive system can be judged, so to speak, from the outside. At present, for example, some doctors and others who work in the medical profession have the uneasy awareness that their industrial standards are failing, and their thoughts are going, as always in such instances, toward Nature's laws, which is to say, toward the health that is at once bodily, familial, communal, economic, and ecological. The outside standard invariably will turn out to be health, and perceptive people, looking beyond industrial medicine, see that health is not the painlessness of a body part, or the comfort of parts of a society. They see that the idea of a healthy individual in an unhealthy community in an unhealthy place is an absurdity that, by the standards of industry, can only become more absurd.

In agricultural science the same realization was occurring at the beginning of the twentieth century. The first prominent sign of this, so far as I know, was a book, *Farmers of Forty Centuries*, in which F. H. King, who had been professor of agricultural physics at the University of Wisconsin, recounted his travels among the small farms of China, Korea, and Japan. As a student of soil physics and soil fertility, King must be counted an expert witness, but he was also a sympathetic one. He was farm-raised, and his interest in agriculture seems to have come from a lifelong affection for farming, farmers, and the details of their work.

The impetus for his Asian journey seems to have been his recognition of the critical deficiencies of American agriculture, the chief one being the relentless exploitation of soil and soil fertility by practices that were supportable only by the importation to the farms of "cargoes of feeding stuffs and mineral fertilizers." This, he knew, could not last. It was, we would say now, unsustainable, as it still is. It was possible in the United States because the United States was still a thinly populated country.

Such extravagance had never been possible in China, Korea, and Japan, where "the people . . . are toiling in fields tilled more than three thousand years and who have scarcely more than two acres per capita, more than one-half of which is uncultivable mountain land." This long-enduring agriculture was made possible by keeping the fertility cycle intact and in place. The "plant food materials" that we were wasting, "through our modern systems of sewage disposal and other faulty practices," the Asian farmers "held . . . sacred to agriculture, applying them to their fields."

As a sample of the economic achievement of the peasant farms he visited, King says in his introduction that "in the Shantung province [of China] we talked with a farmer having 12 in his family and who kept one donkey, one cow, both exclusively laboring animals, and two pigs on 2.5 acres of cultivated land where he grew wheat, millet, sweet potatoes and beans." The introduction ends with this summary:

> Almost every foot of land is made to contribute material for food, fuel or fabric. Everything which can be made edible serves as food for man or domestic animals. Whatever cannot be eaten or worn is used for fuel. The wastes of the body, of fuel and of fabric worn beyond other use are taken back to the field . . .

The chapters that follow contain finely detailed descriptions of one tiny farm after another, every one of them exemplifying what I have called Nature's Law of Frugality, or what Sir Albert Howard would later call her "law of return." Fertility, we might say, was understood as borrowed from Nature on condition of repayment in full. "Nothing," King wrote, "jars on the nerves of these people more than incurring of needless expense, extravagance in any form, or poor judgment in making purchases." No debt was to be charged to the land and left unpaid, as we were doing in King's time and are doing still.

Farmers of Forty Centuries was published in 1911. In 1929, J. Russell Smith, then professor of economic geography at Columbia University, published *Tree Crops*, another necessary book, another of my stepping stones. I don't know whether or not Smith had read King's book, but that does not matter, for Smith's book was motivated by the same trouble: our senseless waste of fertility and the prospect of land exhaustion.

What Smith saw was not only our fracture of the fertility cycle, but also another, an opposite, cycle: "Forest—field—plow—desert—that is the cycle of the hills under most plow agricultures . . ." Smith was worried about soil erosion. He had seen a part of China that King had not visited:

> The slope below the Great Wall was cut with gullies, some of which were fifty feet deep. As far as the eye could see were gullies, gullies, gullies—a gashed and gutted countryside.

And Smith knew that there were places similarly ruined in the "new" country of the United States. The mistake, everywhere, was obvious: "Man has carried to the hills the agriculture of the flat plain." The problem, more specifically, was the agriculture of annual plants. "As plants," he wrote, "the cereals are weaklings." They cannot protect the ground that their cultivation exposes to the weather.

As Henry Besuden had realized on his own farm two years before the publication of Smith's book, the reaction against the damage of annual cropping could only go in the direction of perennials: Slopes that were not wooded needed to be permanently grassed. So Mr. Besuden rightly thought. J. Russell Smith took the same idea one step further: to what he called "two-story agriculture," which would work anywhere, but which he saw as a necessity, both natural and human, for hillsides. The lower story would be grass; the upper story would consist of "tree crops" producing nuts and fruits either as forage for livestock or as food for humans.

The greater part of Smith's book is devoted to species and varieties of trees suitable for such use, and to examples (with photographic plates and explanatory captions) of two-story agriculture in various parts of the world. He was, he acknowledged, a visionary: "I see a million hills green with crop-yielding trees and a million neat farm homes snuggled in the hills." But he was careful to show, by his many examples, that two-story agriculture *could* work because in some places it was working, and in some places it had been working for hundreds of years. He was promoting a proven possibility, not a theory.

Like Alan of Lille, Smith understood Nature's condemnation of prodigality, and he accepted her requirement that we should save what we have been given. The passion that informs his book comes from his realization that Nature's most precious gift was given only once. The Prodigal Son could squander his inheritance of money, repent, and be forgiven and restored to his father's favor. But Nature, "the equall mother" of all the creatures, has no bias in favor of humans. She gave us, along with all her other children, the great gift of life in a rich world, the wealth of which reduces finally to its thin layer of fertile soil. When we have squandered that, no matter how we may repent, it is simply and finally gone.

Smith was above all a practical man, and he stated one of Nature's laws in terms exquisitely practical: "*farming should fit the land.*" The italics are his. He said that this amounted to "*a new point of view*," and again his italics indicate his sense of the urgency of what he was saying. He was undoubtedly right in assuming that this point of view was new to his countrymen, or to most of them, but of course it was also very old. It was implicit in Nature's complaint to Alan of Lille, it was explicit in Virgil's first Georgic, more than a thousand years before Alan, and it certainly informed Pope's instructions to consult "the Genius of the Place." The point is that, in using land, you cannot know what you are doing unless you know well the place where you are doing it.

In the long lineage that I am discussing, the fundamental assumption appears to be that Nature is the perfect—and, for our purposes, the exemplary—proprietor and user of any of her places. In our agricultural uses of her land, we are not required to imitate her work, because, as Chaucer's physician says, she is inimitable, and in order to live we are obliged to interpose our own interest between her and her property. We are required instead to do, not *as* she does, but *what* she does to protect the land and preserve its health. For our farming in our own interest, she sets the pattern and provides the measure. We learn to farm properly only under the instruction of Nature: "What are the main principles underlying Nature's agriculture? These can most easily be seen in operation in our woods and forests."

Those are the key sentences of *An Agricultural Testament*, by Sir Albert Howard, published in 1940. Howard had read *Farmers of Forty Centuries*, and his testimony begins with the worry that had sent King to Asia:

> Since the Industrial Revolution the processes of growth have been speeded up to produce the food and raw materials needed by the population and the factory. Nothing effective has been done to replace the loss of fertility involved in this vast increase in crop and animal production. The consequences have been disastrous. Agriculture has become unbalanced: the land is in revolt: diseases of all kinds are on the increase: in many parts of the world Nature is removing the worn-out soil by means of erosion.

From this perception of human error and failure he made the turn to Nature, that we can now recognize, across a span of many centuries, as characteristic and continuing in a lineage of some poets, some intelligent farmers and farm cultures, and some scientists. Howard's study of agriculture rests solidly upon his study of the nature of the places where farming was done. If the land in use was originally forested, as much of it was and is, then to learn to farm well, the farmer should study the forest.

Industrial agriculture, far from consulting the Genius of the Place or fitting the farming to the land or remembering at all the ecological mandate for local adaptation, has instead and from the beginning forced the land to submit to the capabilities and the limitations of the available technology. From the ruinous and ugly consequences, now visible and obvious everywhere the land can be farmed, one turns with relief to the great good sense, the mere sanity, the cheerful confidence of Howard's advice to farmers: Go to the woods and see what Nature would be doing on your land if you were not farming it, for you are asking her, not just for her "resources," but to accept you as her student and collaborator. And Howard summarizes the inevitable findings in the following remarkable paragraph:

> The main characteristic of Nature's farming can therefore be summed up in a few words. Mother earth never attempts to farm without livestock; she always raises mixed crops; great pains are taken to preserve the soil and to prevent erosion; the mixed vegetable and animal wastes are converted into humus; there is no waste; the processes of growth and the processes of decay balance one another; ample provision is made to maintain large reserves of fertility; the greatest care is taken to store the rainfall; both plants and animals are left to protect themselves against disease.

Howard was familiar with the work of F. H. King and other scientific predecessors. *An Agricultural Testament*, he says, was founded upon his own "work and experience of forty years, mainly devoted to agricultural research in the West Indies, India, and Great Britain." He had the humility and the good sense to learn from the peasant farmers his work was meant to serve. He came from a Shropshire farm family "of high reputation locally," and so I assume he had grown up hearing the talk of farmers. I don't know what, if anything, he had learned from the poets I have learned from, but of course I am delighted that he called Nature personally by name as Chaucer, Spenser, Milton, and Pope had done.

As Milton in *Comus* had enlarged the earlier characterizations of Nature by recognizing the economic significance of her religious and moral demands, making it explicit and practical, so the work of Sir Albert Howard completed their ecological significance. The agricultural economy, and the economies of farms, as determined by the economies and technologies of industrialism, would by logical necessity run to exhaustion. Agriculture in general, and any farm in particular, could survive only by recognizing, respecting, and incorporating the integrity of ecosystems. And this, as Howard showed by his work, could be understood practically and put into practice.

Spenser's problem of stability within change, moreover, we may think of as waiting upon Howard for its proper solution, which he gives in *The Soil and Health*, published in 1947:

> It needs a more refined perception to recognize throughout this stupendous wealth of varying shapes and forms the principle of stability. Yet this principle dominates. It dominates by means of an ever-recurring cycle, a cycle which, repeating itself silently and ceaselessly, ensures the continuation of living matter. This cycle is constituted of the successive and repeated processes of birth, growth, maturity, death, and decay. An eastern religion calls this cycle the Wheel of Life and no better name could be given to it. The revolutions of this Wheel never falter and are perfect. Death supersedes life and life rises again from what is dead and decayed.

As Howard of course knew, religion would construe that last sentence analogically or symbolically, but he does not diminish it or make it less miraculous by treating it as fact. The Wheel of Life, or the fertility cycle, is not an instance of stability as opposed to change, as Spenser may have hoped, but rather an instance, much more interesting and wonderful, of stability dependent upon change. The Wheel of Life is a religious principle of which Howard saw the scientific validity, or it is a principle of Nature, eventually of science, which has long been "natural" to religion. This suggests that beyond the experimental or empirical proofs of science, there may be other ways of determining the truth of a solution, a very prominent one being versatility. Is the solution valid economically as well as ecologically? Does it serve the interest both of the land and of the land's people? The Wheel of Life certainly has that double validity. But we can go on: Does it fit both the farm and the local ecosystem? Does it satisfy the needs of both biology and religion? Is it imaginatively as well as factually true? Can it reconcile utility and beauty? Is it compatible with the practice of the virtues? As a solution to the problem of change and stability, the Wheel of Life answers affirmatively every one of those questions. It is complexly, and joyously, true.

Such a solution could not have originated, and cannot be accommodated, under the rule of industrial or technological progress, which does not cycle, which recycles only under compulsion, and makes no returns. Waste is one of its definitive products, and waste is profitable so long as you are not liable for the cost or replacement of what you have wasted.

Waste of the gifts of Nature from the beginning of industrialism has subsidized our system of corporate profits and "economic growth." Howard, judging by the unforgiving standard of natural health, described this "economy" unconditionally: "The using up of fertility is a transfer of past capital and of future possibilities to enrich a dishonest present: it is banditry pure and simple."

The falsehood of the industrial economy has been disguised for generations by the departmentalization and over-specialization of the essential academic and intellectual disciplines. In the absence of any common or unifying standard of judgment, the professionals of the schools have been free to measure their work by professional standards exclusively, submitting only to the "peer review" of fellow professionals. As in part a professional himself, Howard understood the ethical blindness of this system. Against it he raised the standard of the integrity of the Wheel of Life, which is comprehensive and universal, granting no exemptions.

Whether because of his "instinctive awareness of the importance of natural principle" or because of respect for his own intelligence or because of an evident affection for Nature's living world and the life of farming, Howard kept the way always open between his necessarily specialized work as a scientist and his responsibility to the Wheel of Life. His personal integrity was to honor the integrity of what he called the "one great subject": "the whole problem of health in soil, plant, animal, and man." The practical effect of this upon his work was his insistence upon thinking of everything in relation to its context. In his "revolt against 'fragmentation' of knowledge," his investigation of a crop would include its "whole existence": "the plant itself in relation to the soil in which it grows, to the conditions of village agriculture under which it is cultivated, and with reference to the economic uses of the product." As a researcher and teacher, he would not offer an innovation to his farmer clients that they could not afford, or that they lacked the traction power to use.

There clearly is no break or barrier between Howard's principles and his practice and the "land ethic" of his contemporary, Aldo Leopold, who wrote:

> Health is the capacity of the land for self-renewal. Conservation is our effort to understand and preserve this capacity.

And:

> A thing is right when it tends to preserve the integrity, stability, and beauty of the biotic community. It is wrong when it tends otherwise.

Moreover, there is no discontinuity between the ethics of Howard and Leopold and Nature's principle, given to Alan of Lille, that her good, the good of the natural world, depends upon human goodness, which is to say the human practice of the human virtues. And here I will say again that the needs of the land and the needs of the people tend always to be the same. There is always the convergence of what Nature requires for the survival of the land with what economic demand or economic adversity requires for the survival of the farmer.

From Alan's *Plaint of Nature*, and surely from the conversation of farmers and other country people before and after, it is possible to trace a living tradition of deference, respect and responsibility to Nature and her laws, carried forward by many voices speaking in agreement and in mutual help and amplification. They speak in fact with more agreement and continuity than I expected when I began this essay. And they speak, as also I now see, without so much as a glance toward moral relativism.

For many years I have known that the books by King, Smith, Howard, and Leopold, supported by other intellectually respectable books that I have not mentioned, constitute a coherent, sound, and proven argument against industrial agriculture and for an agricultural ecology and economy that, had it prevailed, would have preserved both the economic landscapes and their human communities. It is significant that all of those books, and others allied with them, were in print by the middle of the last century. Together, they might have provided a basis for the reformation of agriculture according to principles congenial to it. If that had happened, an immeasurable, and so far an illimitable, damage to the land and the people could have been prevented.

The opposite happened. All specifically agricultural and ecological standards were replaced by the specifically industrial standards of productivity, mechanical efficiency, and profitability (to agri-industrial and other corporations). Meanwhile, the criticisms and the recommendations of King, Smith, Howard, Leopold, et al., have never been addressed or answered, let alone disproved.

That they happen to have been right, by any appropriate standard, simply has not mattered to the academic and official forces of agriculture. The writings of the agrarian scientists, the validity of their science notwithstanding, have been easily ignored and overridden by wealth and power. The perfection of the industrial orthodoxy becomes clear when we remember that King, Smith, and Howard traveled the world to search out and study examples of sustainable agriculture, whereas agri-industrial scientists have traveled the world as evangelists for an agriculture unsustainable by every measure.

The truth of agrarian scientists, and their long cultural tradition, casts nevertheless a bright and exposing light upon the silliness and superficiality of the industrial economy, the industrial politicians and economists, and the industrial conservationists. Industrial politicians and economists ignore everything that can be ignored, mainly the whole outdoors. Industrial conservationists ignore everything but wilderness preservation ("Give us the 'wild' land; do as you please with the rest") and the most sensational and fashionable "environmental disasters."

In opposition to industrial and all other sorts and ways of subhuman consumption of the living world, the tradition of cooperation with Nature has persisted for many centuries, through many changes, sustained by right principles, the proofs of experience and eventually of science, and good sense. Its diminishment in our age has been tragic. If ever it should be lost entirely, that would be a greater "environmental disaster" than global warming. It is reassuring therefore to know that it has lasted until now, in the practice of some farmers, and in the work of some scientists. I now have in mind particularly the scientists of The Land Institute, which to me has been a source of instruction and hope for nearly half my life, and which certainly belongs in, and is in good part explained by, the cultural lineage I have been tracking.

The Land Institute was started forty years ago. It is amusing, amazing, and most surely confirmative of my argument, that when Wes Jackson laid out in his mind the order of thought that formed The Land Institute, he had never heard of Sir Albert Howard. The confirming gist of this is that Wes's ignorance of the writings of Sir Albert Howard at that time did not matter. What matters, and matters incalculably, is that Wes's formative thought developed exactly according to the pattern followed by the thinking of King, Smith, and Howard: perception of the waste of fertility and of soil, recognition of the failure of the current standards, and the turn to Nature for a better standard.

Wes had read *Tree Crops*, which had given him the fundamental principle and pattern, which he must more and more consciously have needed. He had in himself the alertness and the sense to look about and see where he was. And like King and Howard he had grown up farming—in, as I know, an agrarian family of extraordinary intelligence and attentiveness. He grew up loving to farm. He knew, and probably long before he knew he knew, the essential truth spoken by Henry Besuden: that soil conservation "involves the heart of the man managing the land. If he loves his soil he will save it." The entire culture of husbandry is implicit in that last sentence. The practical result was, again exactly, Albert Howard's comparison of human farming to Nature's farming—or, as Wes has phrased it in many a talk, a comparison of "human cleverness" to "Nature's wisdom." Whereas Howard had looked to the woods for this, Wes, a Kansan, looked to the native prairie, a possibility that Howard himself had foreseen and approved. If you want to know what is wrong with a Kansas wheat field, study the prairie. And then, taking his own advice, Wes conceived the long-term and ongoing project of The Land Institute: to replace the monocultures of annual grains with polycultures of grain-bearing perennials, which would have to be developed by a long endeavor of plant-breeding.

Among those who are interested, this project is well known. I want to say only two things about it: First, it is radical, for it goes to the roots of the problem and to the roots of the plants. Second, there is nothing sensational about it. To Wes, having thought so far, it was obviously the next thing to be done. To any agricultural scientist whose mind had been formed as Wes's had been formed—and Wes, unique as he is, could not have been *that* unique—it ought to have been obvious. If one of only two possibilities has failed, the alternative is not far to seek. That the need for an agriculture of perennials was obvious to nobody but Wes, and that it is still by principle unobvious to many agricultural scientists, suggests that the purpose of higher education has been to ignore or obscure the obvious.

The work of The Land Institute is as dedicated to the best interests of the land and the people as if it, rather than the present universities, had sprung from the land-grant legislation. And so the science of The Land Institute is significantly different from the industrial sciences of product development and product addiction—also from the "pure" science that seeks, with the aid of extremely costly and violent technology, the ultimate truth of the universe.

I am talking now of a science subordinate and limited, dedicated to the service of things greater than itself, as every science and art ought to be. There are some things it won't do, some dangers it won't risk. It will not, I think, commit "creative destructions," for the sake of some future good or higher truth. It is a science founded upon the traditional respect for Nature, the natural world, the farmland, and those farmers whose use of the land enacts this respect.

This old respect, really one of the highest forms of human self-respect, provides an indispensable basis, unifying and congenial, for work of all kinds. It might unify a university. In evidence, I offer the friendship that for nearly forty years has kept Wes Jackson and me talking with each other. According to the departmentalism of thought and the other fragmentations that beset us, a useful friendship or a conversation inexhaustibly interesting ought to be impossible between a scientist and a poet. That such a conversation is possible is in fact one of its most interesting subjects. Different as we necessarily are, we have much in common: an interest in the natural world, an interest in farming, an inherited agrarianism, an unresting concern about the problems of farming, of land use in general, and of rural life. Making a common interest of these subjects depends of course on speaking common English, which we can do. But that is not all.

I have heard Wes say many times that "the boundaries of causation always exceed the boundaries of consideration." The more I have thought about that statement, the more interesting it has become. The key word is "always." Mystery, the unknown, our ignorance, always will be with us, to be dealt with. The farther we extend the radius of knowledge, the larger becomes the circumference of mystery. There is, in other words, a boundary that may move somewhat, but can never be removed, between what we know and what we don't, between our human minds and the mind of Nature or the mind of God. To ignore or defy that division, wishing to be as gods, believing that the human mind is so capacious as to contain the whole universe and its whole truth, is characteristic of a kind of science that is at once romantic and industrial, ever in search of new worlds to conquer. From its work, I fear, we can expect only a continuing spillover of violence, to the world and to ourselves.

But a scientist who knows that the boundary exists and accepts, even welcomes, its existence, who knows that the boundary has a human side and elects to stay on it, is a scientist of a kind opposed to the would-be masters of the universe. A poet, too, can choose the human side of that boundary. Any artist, any scientist can so choose. Having so chosen, they can speak to one another congenially, in good faith and friendship. When we know securely the smallness of our minds relative to the immensity of our ignorance, then a certain poise and grace may become possible for us, and we can think responsibly of the circumstances in which we work, of the issues of limits and of the proprieties of scale. If we can talk of limits and of scale, we can slack off our obsession with quantities and immensities and take up the study of form: of the forms of Nature's work, of the forms by which our work might be adapted to hers. We may then become capable of the hope, that Wes and I caught years ago from our friend John Todd, for "Elegant solutions predicated upon the uniqueness of place."

And now the long journey of my essay is ended. I have written it in order to encounter once again, and more coherently than before, the writings and the thoughts among which my own writings have formed themselves. The newest to me of the predecessors I have discussed is the oldest: Alan of Lille's *Plaint of Nature*. My reading and re-reading of that book over the last four or five years helped me to see the thread that connects the others: the books and the voices—to borrow Alan's happiest metaphor—that through many years have been in conversation in "the little town of my mind."

CHRONOLOGY

NOTE ON THE TEXTS

NOTES

INDEX

Chronology

1934 Born Wendell Erdman Berry on August 5 in Henry County, Kentucky, the first child of John Marshall and Virginia Erdman Perry Berry. Father, born November 8, 1900, near Lacie, Kentucky, earned a bachelor's degree at Georgetown College in central Kentucky in 1922, and then worked as secretary to Congressman Virgil Chapman while earning a law degree from The George Washington University in Washington, D.C. He graduated in 1927 and later that year opened a private legal practice in Henry County. Mother born November 6, 1907, in Port Royal, in Henry County. She attended Randolph-Macon Woman's College in Lynchburg, Virginia, a Methodist women's college tied to the all-male Randolph-Macon College in Ashland, Virginia, graduating with a degree in English in 1930. Both families can trace roots back to early years of Henry County's history; both families also owned slaves before the Civil War (at time of Berry's birth, there are many children of both slaves and slaveholders in the community). Parents married on July 14, 1933.

1935 Brother John Marshall Jr. born October 13.

1936 Family moves to New Castle, Kentucky, where father practices law and helps with paternal grandfather's farm, the "Home Place."

1938 Sister Mary Jo born March 18.

1939 Sister Martha Frances born July 22.

1940 Enters New Castle Elementary School in Henry County. Over the next few years, mother introduces Berry to books about King Arthur's knights and Robin Hood, as well as *The Swiss Family Robinson*, *Treasure Island*, and *The Yearling*. Later will read Mary O'Hara's *My Friend Flicka*, *Thunderhead*, and *Green Grass of Wyoming*.

1941 The Burley Tobacco Growers Cooperative Association, representing burley tobacco farmers in Kentucky, Missouri, Indiana, Ohio, and West Virginia with an average farm size of 150 acres, which had been founded in 1921, is revived under the New Deal, establishing parity prices and production control as protections for farmers. Father is one of the principal authors of the program and serves as counsel and vice-president and later president.

1942 Uncle Morgan Perry serves in the navy medical corps during World War II, and because of his knowledge of agriculture, is placed in charge of the garden of the naval hospital at Pearl Harbor.

1944 On July 3, uncle Wendell Holmes Berry, with whom Berry was extremely close, is shot at a defunct lead mine property, where he had been working with a crew salvaging lumber and roofing, after a quarrel with Floyd Martin, one of the owners of the property, and dies the same day.

1948 Finishes the eighth grade at the New Castle School. Enters Millersburg Military Institute in Millersburg, Kentucky, at the same time as his brother, John, Jr. Berry will remember, "It was very confining and military, of course. I never liked the military part of it. I did have some good teachers there, and I began there to read more seriously than I had before. I began to have a favorite subject: literature." Father speaks before Congress on behalf of the Tobacco Program. In high school, begins tentatively to write poems.

1952 Graduates from Millersburg Military Institute. In the fall, enters University of Kentucky, in Lexington, Kentucky. Co-edits freshman magazine, *Green Pen*.

1953 Takes "Introduction to Literature" from Thomas Stroup. Begins bringing Stroup poems for criticism. Later takes classes from Hollis Summers and Robert Hazel. Reads T. S. Eliot and Ezra Pound. In May, publishes first essay, "The Wings of the Future," in *Green Pen*, about the survival of society in the light of two world wars and the entrance of the U.S. into the Korean conflict; he writes that it is the "American debt to tomorrow" to ensure that a democratic form of government succeeds over Soviet-style communism. During a composition class, the instructor Robert D. Jacobs introduces Berry to the term "agrarian" and to the 1930s manifesto of the twelve Southern Agrarian writers (who include Allen Tate, John Crowe Ransom, and Robert Penn Warren), *I'll Take My Stand*, a work that will be significant (both in influence and in reaction) to the development of Berry's thought.

1954 Publishes first poem, "Spring," and first short story, "Summer Crop," in the spring issue of *Stylus*, the university literary magazine. Becomes editor of *Stylus*. At a writers' conference at Morehead State College in Kentucky meets writer James Still, whose work will become an influence (Berry will later write, "His short stories, I think, are as near to perfect as any writing I know"). In fall, meets James Baker Hall, who will become a close friend, in a creative writing class taught by Hollis Summers, for which Berry writes story "The Brothers" (later part of *Nathan Coulter*).

1955 Wins *Stylus*'s Dantzler Award for short story "The Brothers." In fall, meets Tanya Amyx (born April 30, 1936, in Berkeley, California), a French and music major and daughter of University of Kentucky art professor Clifford Amyx and textile artist Dee Amyx. "The first time I ever saw Tanya," Berry would later remark, "she was standing by [a] wooden newel post in Miller Hall at the University of Kentucky. Years later, they started to remodel the place. I went over and said, 'Look. When you tear that post out, I want it.'" (The post is now in the Berrys' home.) With fellow student Edward M. Coffman, travels to Kenyon College to meet John Crowe Ransom.

1956 Wins *Stylus*'s Dantzler Award for short story "The Chestnut Stud." Meets fellow student Gurney Norman. On May 16, submits term paper on *I'll Take My Stand*: "The Regional Context: A Consideration of the Southern Agrarians as Southerners." Graduates from the University of Kentucky with a bachelor's degree in English. In summer, Berry and James Baker Hall join literature seminars at Indiana University School of Letters; Berry takes course in Joyce and Yeats from Richard Ellman, and course in modern poetry from Karl Shapiro. "The Brothers" wins first prize in the *Carolina Quarterly*'s 6th Annual Fiction Contest and is published in its summer issue. In the fall, enters the master's program in English at the University of Kentucky.

1957 In February, poem "Rain Crow" is published in *Poetry* magazine; in the next five years, his work will be published in *Poetry* five times. Meets writer Ed McClanahan during a graduate course. Completes MA in English. "Elegy" wins the Farquhar Award for Poetry from the University of Kentucky. Two short stories, "Whippoorwills" and "Apples," are published in *Coraddi: Arts*

Forum 1957. Marries Tanya Amyx, May 29. In summer, they live in a cabin built by Berry's great-uncle Curran Mathews, called "the Camp," on the Kentucky River near Port Royal, which was used as a family retreat from the 1920s. In the fall, they move to an apartment in Georgetown, Kentucky; Berry begins teaching freshman composition and sophomore literature at Georgetown College, father's alma mater, while Tanya continues to study at the University of Kentucky.

1958 Daughter Mary Dee born May 10 in Lexington. Awarded a Wallace Stegner Fellowship at Stanford University; in August, family moves to Mill Valley, California, to live while he studies creative writing at Stanford University under Stegner (about whom Berry will later write, "My debt to him is probably greater than I know") and Richard Scowcroft; fellow members of the fiction seminars include Ernest J. Gaines (also a Stegner fellow), Ken Kesey, and Nancy Packer. Lives with family in small house owned by Tanya's aunt and uncle, Ann and Dick O'Hanlon, which would later become part of the O'Hanlon Center. Works on novel *Nathan Coulter*: "I had about half of it written when I went out there, and I just continued to work on it." Introduced by Holman Hamilton, one of Berry's history professors at the University of Kentucky, to Craig Wylie of Houghton Mifflin, to whom Berry submits the manuscript of *Nathan Coulter.* Houghton Mifflin purchases an option on the book for $250.

1959 Accepts appointment as Edward H. Jones Lecturer in Creative Writing at Stanford University for one year.

1960 In January, begins writing second novel, *A Place on Earth.* First novel, *Nathan Coulter*, published in April by Houghton Mifflin. In spring, family moves back to Kentucky, living on the "Home Place" in Henry County, and also farming with Berry's friend and neighbor Owen Flood.

1961 Receives a Guggenheim Foundation Fellowship. Will later write of Henry County that summer: "I began to understand that so long as I did not know the place fully, or even adequately, I belonged to it only partially. That summer I began to see, however dimly, that one of my ambitions, perhaps my governing ambition, was to belong fully to this place, to belong as the thrushes and the herons and the muskrats belonged, to be altogether at home here." In August, leaves with family for a year in Tuscany and southern France to study and write: "the land in Tuscany has had about 2,000 years of good care. And it looked like it when I was there. . . . The sight of that changed my mind about what was possible in land use."

1962 Meets critic and translator Wallace Fowlie when both stay at La Napoule Arts Foundation near Cannes. Awarded Vachel Lindsay Prize by *Poetry* magazine. Returns to Kentucky from Europe in July. Son Pryor Clifford (Den) born August 19 in Lexington. In the fall, moves with family to New York to take up post as assistant professor of English and director of freshman writing at New York University's University Heights campus in the Bronx. Rents apartment in New Rochelle, New York. Meets Bobbie Ann Mason.

1963 Family moves to loft at 277 Greenwich Street, New York City, downstairs from poet Denise Levertov (whom Berry had met the previous fall, although they had corresponded since 1958, when Levertov wrote Berry about a poem he had published in *Poetry*) and writer Mitchell Goodman. Tanya arranges playdates for daughter Mary and a fellow student at St. Luke's School, the daughter of James Parks Morton, later dean of St. John the Divine. Spends summer in Kentucky. Dismantles the Camp, which has flooded several times over the years, and rebuilds it higher on the bank of the river. Works on novel *A Place on Earth*. Reads Harry Caudill's *Night Comes to the Cumberlands*, an important influence ("It showed me what it might mean to be a responsible Kentucky writer living in Kentucky, and it affected me deeply"). Returns to New York for fall semester. Meets poet Donald Hall, who becomes friend and correspondent, at a cocktail party on Riverside Drive in Manhattan. Offered and accepts position as professor of English at the University of Kentucky to begin in the fall of 1964, where he will teach creative writing and other courses.

1964 In May, visits Levertov and Goodman's Maine farm, where he reads in typescript Hayden Carruth's poetry collection *North Winter* (Berry will later begin a correspondence with Carruth that will last until Carruth's death in 2008). Leaves New York for Kentucky soon after. A poem about Kennedy's assassination, *November Twenty Six Nineteen Hundred Sixty Three* (with drawings by Ben Shahn) published by George Braziller in May. Begins commuting weekly from Lexington (where family lives) to the Camp to write. Meets Kentucky artist Harlan Hubbard and his wife Anna by chance while on a canoe trip on the Ohio River; later writes that their homestead "differed from Thoreau's economy radically in some respects, and also advanced and improved upon it. The main differences were that, whereas Thoreau's was a bachelor's economy, the Hubbards' was that of a married couple." First poetry collection *The Broken Ground* published by Harcourt, Brace & World in September (a *New York Times* review will begin: "The quiet but sure and melodious voice of Wendell Berry makes 'The Broken Ground' an immediate pleasure"). In November, purchases the Lanes Landing property as a weekend place, a twelve-acre hillside farm adjacent to the Camp and bordering his maternal grandfather's farm on the west side of the Kentucky River. Berry's editor at Harcourt, Dan Wickenden, introduces him to the work of organic farming pioneer Sir Albert Howard.

1965 Awarded a Rockefeller Foundation Fellowship. On July 4, moves with family to Lanes Landing Farm, where they will continue to live and farm year round. "I remember a moment—in 1965, or a little after—when I realized that I didn't have to be a writer; there were other kinds of work also that required artistry and offered satisfaction. From here, looking back, I can see what a defining moment that was. I had, in effect, decided not to be a 'professional' writer, but instead, in the literal sense, an amateur: I would work for love." Over time, expands the farm until it consists of about 117 acres in two tracts; at first, raises only food for the family. In July, sees first strip mine above Hardburly, Kentucky, while visiting writer Gurney Norman; Norman introduces him to Kentucky lawyer, writer, and environmentalist Harry Caudill and Caudill's wife Anne, who will become lifelong friends, at a meeting of the Appalachian Group to Save the Land and People. Will one day write, "Gurney has been my Virgil on the upper end (the mountain end) of the Kentucky River watershed, where he is at home. I'm at home on the lower end."

1967 In summer, makes first notes for a work that will become *The Unsettling of America* after reading about the report of President Johnson's "special commission on federal food and fiber policies" that defined the nation's greatest agriculture problem as a surplus of farmers. Novel *A Place on Earth* published by Harcourt, Brace & World in September. Receives Bess Hoken Prize from *Poetry* magazine.

Becomes president of the Cumberland Chapter of the Sierra Club in Kentucky. Begins work with farmers and landowners to oppose a dam across the Red River Gorge that would destroy the gorge's unique topography and ecosystem. In December, visits Trappist monk Thomas Merton at Abbey of Our Lady of Gethsemani near Bardstown, Kentucky, with Ralph Eugene Meatyard and his wife Madelyn, Tanya, and Denise Levertov.

1968 On February 10, delivers speech, "A Statement Against the War in Vietnam," during the Kentucky Conference on War and the Draft in Lexington, Kentucky: "I have come to the realization that I can no longer imagine a war that I would believe to be either useful or necessary. I would be against any war." In fall, takes leave from University of Kentucky to become visiting professor of creative writing at Stanford University during fall 1968 and winter 1969 quarters; lives with family in Menlo Park. Colleagues at Stanford include Wallace Stegner, Ed McClanahan, and Ken Fields. Meets writer John Haines when he gives a reading at Stanford. During the Christmas holiday, writes *The Hidden Wound*, an extended meditation on race and memories of two older black people who were his friends during his childhood. Poetry collection *Openings* published by Harcourt, Brace & World in October; *The New York Times* reviewer calls it a book "to win the respect of anyone who cares about contemporary verse."

1969 Returns with family to Kentucky. Essay collection *The Long-Legged House*, of which the title essay is about the history of the Camp, published by Harcourt, Brace & World in April; other essays concern place and belonging, agriculture, community, small farming, the Vietnam War, and consumerism. The same month, poetry collection *Findings* published by The Prairie Press. Wins first prize in the Borestone Mountain Poetry Awards. Receives grant from the National Endowment for the Arts.

1970 On March 7, speaks at the march at the state capitol in Frankfort, Kentucky, to protest the war in Vietnam; his speech notes government disdain for the will of the people and the huge expense of the war: "With the very earth dying under our feet, with the air full of pollution; we are spending billions of dollars and thousands of lives to assure the success of tyranny in Vietnam." At the first Earth Day celebration at the University of Kentucky, delivers speech "Think Little." Publishes essay "The Regional Motive" in the autumn issue of *The Southern Review*, a response to the Southern Agrarians; in it, he is critical of those Agrarians who had moved to northern universities, saying that it "invalidated their thinking, and reduced their effort to the level of an academic exercise." (Receives letter from Allen Tate in response, and when the essay is included in *A Continuous Harmony* in 1972, Berry appends an apologetic footnote; he later writes that "whatever the amount of truth in that statement, and there is some, it is also a piece of smartassery.") Poetry collection *Farming: A Hand Book* published by Harcourt Brace Jovanovich in September. Meets writer and farmer Gene Logsdon, who becomes a friend and ally, after Logsdon reads *Farming: A Hand Book*. *The Hidden Wound* published by Houghton Mifflin in September.

1971 *The Unforeseen Wilderness: An Essay on Kentucky's Red River Gorge* published by University Press of Kentucky in April, with photographs by Ralph Eugene Meatyard; it is a response to the plan to dam the Red River, which Berry had opposed since 1967. (Congress will withdraw funds for the project in 1976. In 1993 the river is designated by Congress in the National Wild and Scenic Rivers System Act.) Elected Distinguished Professor of the Year by the University of Kentucky. Receives the Arts and Letters Literary Award from the American Academy of Arts and Letters.

1972 With his brother John Berry Jr. and the Save Our Land Committee, helps in the successful opposition to the Jefferson County Air Board's plan for an international jetport in Henry and Shelby Counties. In the fall, takes a one-year sabbatical from the University of Kentucky; his courses are covered by Ed McClanahan. Essay collection *A Continuous Harmony: Essays Cultural and Agricultural* published by Harcourt Brace Jovanovich in September. Meets bookseller and future publisher Jack Shoemaker when Shoemaker visits Port Royal after reading *The Long-Legged House*.

1973 *The Country of Marriage* published by Harcourt Brace Jovanovich in February, his fifth poetry collection, with thirty-five poems, including the title poem, "Prayer After Eating," and "Manifesto: The Mad Farmer Liberation

Front." Buys neighboring forty-acre farm that had been sold to a developer, bulldozed, and graveled; begins process of repair and restoration. Invites writer and photographer James Baker Hall to take photographs during a tobacco harvest (they will be published with an essay by Berry in 2004). Begins correspondence with poet and essayist Gary Snyder, and after a few months Snyder travels to Port Royal to visit Lanes Landing. In November, brother elected to the Kentucky senate, in which he will serve until 1981.

1974 Serves as Elliston Poetry Lecturer at the University of Cincinnati for winter 1974. Novel *The Memory of Old Jack* published by Harcourt Brace Jovanovich in February (*The New York Times Book Review* calls it "a slab of rich Americana" and *Library Journal* says the novel is "worthy of a place among the best pieces of prose written by American writers of this century"). Meets Maurice Telleen at a draft horse sale in Waverly, Iowa, whose *Draft Horse Journal* will publish many of Berry's stories. Gives speech on July 1 at "Agriculture for a Small Planet Symposium" in Spokane, Washington, which will form the basis for the first chapters of *The Unsettling of America*: "I was asked to talk about 'Labor Intensive Micro-Systems Agriculture.' That's not my language, and it's not the sort of language I wish to use because it's the way people speak when they don't want to be understood by most people. I'm not sure what to make of these particular phrases, but they seem to suggest a very methodological or technological approach to agriculture. Part of my purpose here is to suggest that any such approach will necessarily be too simple." Mentions critically Secretary of Agriculture Earl Butz's "adapt or die" policy. Speech helps encourage the Tilth movement, a regional network of organic farmers in the Northwest (still active today). Poetry collection *An Eastward Look* published by Sand Dollar Press in the fall, Berry's first publication with editor Jack Shoemaker. Wins the Emily Clark Balch Prize from *The Virginia Quarterly Review*.

1975 *The Memory of Old Jack* wins the Friends of American Writers Award. A poetry pamphlet, *Horses*, published by Larkspur Press in Kentucky in April; another, *To What Listens*, published by Best Cellar Press. Poetry collection *Sayings and Doings* published by Gnomon Press in December.

1977 Serves as writer-in-residence for the winter quarter at Centre College in Danville, Kentucky. Poetry collection *Clearing* published by Harcourt Brace Jovanovich in March. *The Unsettling of America: Culture and Agriculture* published by Sierra Club Books in August and dedicated to Maurice Telleen, the editor of *Draft Horse Journal*. Donald Hall reviews both volumes for *The New York Times*, writing, "Berry is a prophet of our healing, a utopian poet-legislator like William Blake." Resigns from the University of Kentucky to become contributing editor at Rodale Press, including its magazines *Organic Gardening* and *The New Farm*.

1978 With Tanya, buys first flock (six ewes and one buck) of Border Cheviot sheep, a Scottish breed, which they will raise for the next four decades. In November, debates Secretary of Agriculture Earl Butz on the agricultural crisis at Manchester University, North Manchester, Indiana: "As I see it, the farmer standing in his field is not simply a component of a production machine. He stands where lots of cultural lines cross. The traditional farmer, that is the farmer who first fed himself off his farm and then fed other people, who farmed with his family, who passed the land on down to people who knew it and had the best reasons to take care of it—that farmer stood at the convergence of traditional values, our values: independence, thrift, stewardship, private property, political liberties, family, marriage, parenthood, neighborhood—values that decline as that farmer is replaced by a technologist whose only standard is efficiency."

1979 On June 3, opposes and with eighty-nine others is arrested during nonviolent protests against the construction of the Marble Hill nuclear power plant on the Ohio River near Madison, Indiana. The company eventually abandons the effort to finish it. Poetry collection *The Gift of Gravity* published by Deerfield Press.

1980 Fired from Rodale Press: "I think this was because I was more for small farmers than I was for organic farmers. Also, I don't think I've ever been a very good employee." Ernest J. Gaines visits the Berrys in Kentucky. Begins working with editor Jack Shoemaker, who had left bookselling to cofound North Point Press in Berkeley, California, in 1979, with William Turnbull. (Most of Berry's books will henceforward be published with Shoemaker.) Poetry collection

A Part published in October. Joins with brother and other residents of Henry County along with the group Kentuckians for the Commonwealth in a successful effort to oppose the building of a hazardous chemical waste incinerator in the county. Publishes poetry pamphlet *The Salad*. On November 11, writes to Wes Jackson, founder of The Land Institute, after reading Jackson's book, *New Roots for Agriculture*. Visits Jackson in December to write an article about The Land Institute for *The New Farm* magazine. The two begin a long correspondence and friendship.

1981 *Recollected Essays: 1965–1980* published in August. Essay collection *The Gift of Good Land: Further Essays Cultural and Agricultural*, dedicated to Gene Logsdon, published in November. Both volumes are reviewed in *The Washington Post* by Larry Woiwode, who calls them "reference works of the body and soul." Granddaughter Katie Jean Smith born, December 16.

1982 Poetry collection *The Wheel* published in October.

1983 Speaking at Oberlin College in February, meets David and Elsie Kline, an Amish couple from Holmes County who had read *The Unsettling of America* and came to hear him speak. Kline (with Meatyard, Telleen, and Jackson) is one of the only four agrarian friends Berry has from outside of Henry County for the next decade. Essay collection *Standing by Words* published in October, which *The Christian Science Monitor* review calls "nothing short of splendid." Publishes substantially revised and shortened version of 1967 novel *A Place on Earth* in March; in a new introduction, writes that the original book was "clumsy, overwritten, wasteful."

1985 Co-edits with Wes Jackson and Bruce Coleman *Meeting the Expectations of the Land: Essays in Sustainable Agriculture and Stewardship*, published by North Point in February; Berry's contribution is the essay "Whose Head is the Farmer Using? Whose Head is Using the Farmer?" Granddaughter Virginia Dee Smith born, March 6. *Collected Poems 1957–1982* published in May; *The New York Times* reviewer writes that Berry "can be said to have returned American poetry to a Wordsworthian clarity of purpose." Publishes substantially revised version of *Nathan Coulter* in May. Helps found Community Farm Alliance, a group of Kentucky farmers, bankers, businessmen, and clergy members organized to help small farmers shift from tobacco to other products. Begins correspondence with writer John Haines.

1986 *The Wild Birds: Six Stories of the Port William Membership* published in March. Receives honorary doctorate from the University of Kentucky. In November, delivers lecture "Preserving Wildness" at the first Temenos Academy Conference in Devon, England.

1987 Returns to the English Department of the University of Kentucky, now teaching courses for "future teachers and farmers or anyone preparing to work in practical and comprehensive ways with young minds or with nature"; courses include "Composition for Teachers," aimed at public school English teachers; "Readings in Agriculture," which assigns Spenser, Milton, Shakespeare, Pope, and Wordsworth, as well as Sir Albert Howard, J. Russell Smith, Wes Jackson, and Gene Logsdon; and "The Pastoral." Publishes *Home Economics: Fourteen Essays* in June; the *Christian Science Monitor*, in reviewing it, calls Berry "*the* prophetic American voice of our day." Serves as writer-in-residence for the interim term at Bucknell University. Receives the American Academy of Arts and Letters Jean Stein Award and the Kentucky Governor's Milner Award in the Arts. In the fall, "Why I Am Not Going to Buy a Computer" published in the *New England Review and Bread Loaf Quarterly* and reprinted in *Harper's*; the essay inspires many critical responses. *Sabbaths: Poems* published in September, the first of what will become a sustained project of poems on the themes of work and rest, fields and woods. *The Landscape of Harmony: Two Essays on Wildness and Community*, including the lecture "Preserving Wildness," published by Five Seasons Press, United Kingdom, in September. Poetry collection *Some Differences* published by Confluence Press in December.

1988 Novel *Remembering* published in October and reviewed favorably in the *Los Angeles Times*.

1989 Receives Lannan Foundation Award for nonfiction. Poetry collection *Traveling at Home* published in November. Delivers the Blazer Lecture on the life and work of Kentucky artist and author Harlan Hubbard at the University of Kentucky.

1990 Granddaughter Tanya Christine Smith born February 19.

Essay collection *What Are People For?*, dedicated to Gurney Norman, published in March; it includes the oft-repeated line, "Eating is an agricultural act." (Berry later calls this quote, as repeated without context, "an oversimplification he is now damned sorry to have written.") Bill McKibben publishes a major profile of Berry in *The New York Review of Books*, writing that "wherever we live, however we do so, we desperately need a prophet of responsibility; and although the days of the prophets seem past to many of us, Berry may be the closest to one we have. But, fortunately, he is also a poet of responsibility. He makes one believe that the good life may not only be harder than we're used to but sweeter as well." Inducted into the Fellowship of Southern Writers. Berry's Blazer lecture, *Harlan Hubbard: Life and Work*, published by University Press of Kentucky in November.

1991 *Standing on Earth: Selected Essays* published in the United Kingdom by Brian Keeble's Golgonooza Press in April. Publishes poetry collection *Sabbaths 1987* with Larkspur Press in October. Father dies October 31. North Point Press closes, and Berry follows Shoemaker to Pantheon.

1992 Receives the Victory of Spirit Ethics Award from the University of Louisville and the Louisville Community Foundation. From late August to late September, teaches a course at Schumacher College in South Devon, England, on "Nature as Teacher: The Lineage of Writings Which Link Culture and Agriculture"; writers assigned include Shakespeare, Sir Albert Howard, and Wes Jackson. On September 8, meets Charles, Prince of Wales. "It was a much easier meeting than I expected, for he is intelligent, considerate, and talks and listens well. He is an ally, is deeply concerned about what is happening to rural life and to civilization," Berry wrote in a letter to Wes Jackson. *Fidelity: Five Stories* published in October to favorable reviews, including in *The New York Times* and *The New Criterion*.

1993 Granddaughter Emily Rose Berry born, May 2. *Sex, Economy, Freedom and Community: Eight Essays* published in October; *The New York Times* writes that the book "in its eight essays distills the author's radically conservative views (in the good senses of both words)." Receives The Orion Society's John Hay Award for nature writing. Quits his job at the University of Kentucky.

1994 Receives T. S. Eliot Award for creative writing from the

Ingersoll Foundation. Poetry collection *Entries* published in May. *Watch With Me and Six Other Stories of the Yet-Remembered Ptolemy Proudfoot and His Wife, Miss Minnie, Née Quinch* published in August. Berry follows his editor to Counterpoint Press, which Shoemaker cofounds in Washington, D.C., with Frank H. Pearl.

1995 Grandson Marshall Amyx Berry born June 22. In October, long poem *The Farm* (one of the *Sabbath* series) published by Larkspur Press and the essay collection *Another Turn of the Crank* published by Counterpoint.

1996 Co-authors *Three on Community* with Gary Snyder and Carole Koda, published by Limberlost Press in April. Receives the Harry M. Caudill Conservationist Award from the Cumberland Chapter of the Sierra Club. Novel *A World Lost* published in October.

1997 Mother dies, January 3. Father-in-law Clifford Amyx dies, July 30. *Two More Stories of the Port William Membership* published by Gnomon Press. Receives the Lyndhurst Prize for individuals making a significant contribution to the arts. Speaks at a benefit for The Garden Project in San Francisco, November 10.

1998 Gives keynote address, "In Distrust of Movements," at the Northeast Organic Farming Association conference. *A Timbered Choir: The Sabbath Poems 1979–1997* published in April. *The Selected Poems of Wendell Berry* published in October.

1999 Receives the Thomas Merton Award from the Thomas Merton Center for Peace and Social Justice in Pittsburgh. On November 10, joins Gary Snyder and Jack Shoemaker in reading and conversation at a Lannan Foundation event in Santa Fe, New Mexico.

2000 Wins Poets' Prize for *The Selected Poems of Wendell Berry.* In May publishes *Life Is a Miracle: An Essay Against Modern Superstition*; Bill McKibben, reviewing it for *The Washington Monthly*, writes, "it's hard to imagine that the millennium has seen a more important book than this slim volume from our finest essayist." Novel *Jayber Crow* published in September; *The New York Times* reviewer writes that "by the end this melancholy barber has won both our attention and our hearts." In October, teaches one-week course at Schumacher College in England on "Community, Sustainability and Globalisation."

2001 In response to the events following 9/11, writes "Thoughts in the Presence of Fear," which is published on October 30 by *The Land Report*; in December it is joined by two other essays, "The Idea of a Local Economy" and "In Distrust of Movements," and published by The Orion Society in paperback as *In the Presence of Fear: Three Essays for a Changed World*. Publishes *Sonata at Payne Hollow: A Play* with Larkspur Press.

2002 *The Art of the Commonplace: The Agrarian Essays of Wendell Berry*, edited and introduced by Norman Wirzba, published in April. Berry follows his editor to Shoemaker & Hoard, which Shoemaker cofounds with Trish Hoard.

2003 On February 9, essay "A Citizen's Response to the National Security Strategy of the United States," critical of the Bush administration's post-9/11 strategy, published as a full-page advertisement in *The New York Times*. Essay collection *Citizenship Papers* published in August. *Citizen's Dissent: Security, Morality, and Leadership in an Age of Terror: Essays*, co-authored with David James Duncan, published by The Orion Society. Receives Lifetime Achievement Award from the Cathedral Heritage Foundation in Louisville, Kentucky.

2004 *That Distant Land: The Collected Stories* published in February. Receives the Eli M. Oboler Memorial Award in Orlando in June, shared by David James Duncan, awarded by the American Library Association Intellectual Freedom Round Table, for *Citizen's Dissent*. Mother-in-law Dee Rice Amyx dies, July 3. Awarded the Charity Randall Citation from the International Poetry Forum, as well as the Soil and Water Conservation Society Honor Award from the Kentucky Bluegrass Chapter for dedicated efforts toward the preservation of unique rural and cultural resources. Writes essay for *Tobacco Harvest: An Elegy*, to accompany photographs by James Baker Hall; published by University Press of Kentucky in September. Novel *Hannah Coulter* published in September. Poetry collection *Sabbaths 2002* published by Larkspur Press.

2005 Receives O. Henry Prize for short story "The Hurt Man." *Given: Poems* published in May. Inducted into the University of Kentucky Arts and Sciences Hall of Fame. *Blessed Are the Peacemakers: Christ's Teachings about Love, Compassion & Forgiveness*, selections of the gospels compiled and introduced by Berry, published in October. *The Way of Ignorance and Other Essays* published in October. Essay "Not a Vision of our Future, But of Ourselves," about the earth's destruction due to coal mining, included in *Missing Mountains: We Went to the Mountaintop but It Wasn't There*, published by Wind Publications in October. Receives Lifetime Achievement Award from the Conference on Christianity and Literature, December 29.

2006 In August, speaks at "One Thing to Do About Food: A Forum" with Alice Waters, Eric Schlosser, Marion Nestle, Peter Singer, and others. In October, gives the keynote address for the thirtieth anniversary of The Land Institute in Salina, Kansas. Novel *Andy Catlett: Early Travels* published in November.

2007 Visits Ernest J. Gaines in Louisiana. Shoemaker and Charlie Winton acquire both Counterpoint Press and Soft Skull Press; merge both with Shoemaker & Hoard to re-form Counterpoint.

2008 On February 14, at the Kentuckians for the Commonwealth's annual I Love Mountains Day, delivers speech at the Kentucky state capitol in Frankfort calling for nonviolent resistance and civil disobedience to protest mountaintop removal in Kentucky. In May, awarded honorary doctorate by Duke University. *The Mad Farmer Poems* published by Counterpoint in November. Children's book *Whitefoot: A Story from the Center of the World* published in December. Receives the Cynthia Pratt Laughlin Medal from the Garden Club of America.

2009 On January 4, publishes an op-ed in *The New York Times* with Wes Jackson, "A 50-Year Farm Bill," calling for legislation that would address soil loss, pollution, fossil fuels, and the preservation of rural communities. On January 23, releases public statement against the death penalty through the Kentucky Coalition to Abolish the Death Penalty: "As I am made deeply uncomfortable by the taking of a human life before birth, I am also made deeply uncomfortable by the taking of a human life after birth." On March 2, joins protests in Washington, D.C., against mountaintop removal and burning of fossil fuels. On April 3, awarded the Fellowship of Southern Writers' Cleanth Brooks Medal for Excellence in Southern Letters. Essay collection *Bringing It to the Table: On Farming and Food* with an introduction by Michael Pollan, published in August. Poetry collection *Leavings* published in October. In November, joins public letter signed by forty Kentucky writers to Kentucky governor and attorney general asking for a moratorium on the death penalty. On December 20, removes personal papers on loan to the University of Kentucky Archives as a result of the University's too close relationship with the coal industry. (In August 2012, those papers will be donated to the Kentucky Historical Society in Frankfort.)

2010 Receives O. Henry Prize for short story "Stand By Me." Essay collections *Imagination in Place* (including reflections on Wallace Stegner, Gurney Norman, Hayden Carruth, Donald Hall, Jane Kenyon, John Haines, James Still, Gary Snyder, and Kathleen Raine) published in January, and *What Matters? Economics for a Renewed Commonwealth*, with a foreword by Herman E. Daly, published in May.

2011 In mid-February, joins nineteen other activists protesting mountaintop removal in eastern Kentucky by occupying the office of Governor Steve Beshear in Frankfort for a weekend. The governor agrees only to tour coal communities to see the damage. *The Poetry of William Carlos Williams of Rutherford*, a personal tribute to the poet, published in February and is reviewed in *The New York Review of Books*. On March 2, awarded National Humanities Medal by President Barack Obama at the White House. On May 4, speaks at the Future of Food conference at Georgetown University: "the future of food is not distinguishable from the future of the land, which is indistinguishable, in turn, from the future of human care." Daughter Mary Berry establishes The Berry Center in New Castle, Kentucky, to continue the work of Berry and his father and brother toward healthy and sustainable agriculture.

2012 Great-granddaughter Charlcye Anne Johnson born February 7. Receives O. Henry Prize for short story "Nothing Living Lives Alone." *New Collected Poems* published in March; *Christian Science Monitor* hails it as "a welcome complement to the rest of his work." Receives the Steward of God's Creation Award at the National Cathedral in Washington, D.C., April 22. Delivers the 41st Jefferson Lecture, the highest honor the federal government confers for distinguished intellectual achievement in the humanities, at the John F. Kennedy Center for the Performing Arts on April 23: "We cannot know the whole truth, which belongs to God alone, but our task nevertheless is to seek to know what is true. And if we offend gravely enough against what we know to be true, as by failing badly enough to deal affectionately and responsibly with our land and our neighbors, truth will retaliate with ugliness, poverty, and disease." *It All Turns on Affection: The Jefferson Lecture and Other Essays* published in September. *A Place in Time: Twenty Stories of the Port William Membership* published in October. On October 17, receives the James Beard Foundation Leadership Award for efforts toward a healthier, safer, and more sustainable food system, in New York City. On October 30, receives the inaugural Green Cross Award, given by the Bishop of California, for increasing cultural engagement in environmental issues.

2013 In April, elected into the American Academy of Arts and Sciences. The Berry Center and St. Catharine College in Springfield, Kentucky, establish The Berry Farming Program, an undergraduate, multidisciplinary degree inspired by Wes Jackson and Berry's philosophy and work, and designed for students from generational farm families. The program accepts its first students in the fall (St. Catharine College will close in 2016, and the following year the Berry Farming Program will move to partnership with Sterling College in Vermont). *This Day: Sabbath Poems Collected & New 1979–2013* published in October. On October 17, awarded the Freedom Medal from the Roosevelt Institute's Four Freedoms Awards in New York City. Daughter Mary marries Steve Smith, October 26.

2014 *Distant Neighbors: The Selected Letters of Wendell Berry and Gary Snyder*, edited by Chad Wriglesworth, published in June with reviews in the *Los Angeles Review of Books*, *Commonweal*, and other publications. *Terrapin and Other Poems*, with illustrations by Tom Pohrt, published in November. Poetry collection *Roots to the Earth*, with woodcuts by Wesley Bates, published by Larkspur Press.

2015 On January 28, inducted into the Kentucky Writers Hall of Fame, the first living writer to be so honored. *Our Only World: Ten Essays* published in February. Poetry collection *Sabbaths 2013* published by Larkspur Press.

2016 Receives O. Henry Prize for short story "Dismemberment." On March 17, receives the Ivan Sandrof Lifetime Achievement Award from the National Book Critics Circle in New York City, presented by Nick Offerman. *A Small Porch: Sabbath Poems 2014 and 2015* together with *"The Presence of Nature in the Natural World: A Long Conversation"* published in June. On September 25, delivers the Strachan Donnelley Lecture on Conservation and Restoration at the 2016 Prairie Festival in Salina, Kansas, celebrating the fortieth anniversary of The Land Institute and the tenure and retirement of Wes Jackson as president. Brother John M. Berry Jr. dies, October 27. Speaks in conversation with Wes Jackson, and moderated by Mary Berry, at the 36th Annual Schumacher Lectures, October 22. Great-granddaughter Wendy Jean Johnson born, October 29. In December, speaks with Eric Schlosser at the twentieth anniversary of the Center for a Livable Future at Johns Hopkins University.

2017 *The World Ending Fire: The Essential Wendell Berry*, selected by Paul Kingsnorth, published by Allen Lane in the United Kingdom in January. Contributes to *Letters to a Young Farmer: On Food, Farming, and Our Future* by Stone Barns Center for Food and Agriculture, published by Princeton Architectural Press. *The Art of Loading Brush* published by Counterpoint in October.

Note on the Texts

This volume—the second of a two-volume set, *What I Stand On: The Collected Essays of Wendell Berry 1969–2017*—gathers essays from the latter half of Berry's career. It contains the whole of *Life Is a Miracle* (2000) and forty-two essays culled from nine other books by Berry published from 1993 to 2017: *Sex, Economy, Freedom & Community* (1993), *Another Turn of the Crank* (1995), *Citizenship Papers* (2003), *The Way of Ignorance* (2005), *What Matters?* (2010), *Imagination in Place* (2010), *It All Turns on Affection* (2012), *Our Only World* (2015), and *The Art of Loading Brush* (2017). The contents were chosen by the author and his longtime editor and publisher, Jack Shoemaker, and are arranged in chronological order by first book publication. In accord with Shoemaker and Berry's wishes, and as described below, this volume prints the texts of the essays from the most recent Counterpoint Press printings, which reflect corrections made by the author. The companion volume presents selections from books published from 1969 to 1990.

After the publication of *The Unsettling of America* (1977) by Sierra Club Books, Berry began to look for a new publisher. In 1980 he embarked on his long collaboration with Jack Shoemaker, then a thirty-four-year-old editor who, with William Turnbull, had recently cofounded North Point Press in Berkeley, California. From that point forward most of Berry's books would be edited by Shoemaker. Berry followed his editor first to Pantheon when North Point closed in 1991; then to Counterpoint Press, which Shoemaker cofounded with Frank H. Pearl in Washington, D.C., in 1994; then to Shoemaker & Hoard in 2002, which Shoemaker cofounded with Trish Hoard; and finally back to Counterpoint, which merged with Shoemaker & Hoard when Shoemaker and Charlie Winton purchased both Counterpoint Press and Soft Skull Press in 2007. None of the books mentioned above appeared in British editions.

Sex, Economy, Freedom & Community was published by Pantheon Books on October 19, 1993; the most recent Counterpoint Press paperback printing, the source for the texts printed here, was published in 2018. The first publication of the essays selected from *Sex, Economy, Freedom & Community* is given below.

Conservation and Local Economy. Lecture delivered at the Southern Baptist Theological Seminary in Louisville, Kentucky, during a conference sponsored by the Clarence Jordan

Center for Christian Ethical Concerns, on March 23, 1992, published in *Sex, Economy, Freedom & Community* (New York and San Francisco: Pantheon, 1993).

Conservation Is Good Work. Lecture delivered at the Southern Baptist Theological Seminary in Louisville, Kentucky, on March 25, 1992, published in *Wild Earth*, Spring 1992, and in *The Amicus Journal*, Winter 1992.

Christianity and the Survival of Creation. Lecture delivered at the Southern Baptist Theological Seminary in Louisville, Kentucky, on March 24, 1992, published in *CrossCurrents*, Summer 1993.

Sex, Economy, Freedom, and Community. Lecture given at the University of Louisville School of Business, on April 8, 1992, Parts I and II printed as a pamphlet under the title "Sex, Economy, and Community" by the Louisville Community Foundation, 1992. The essay also appeared prior to its book publication in *Pantheon Previews*, Fall 1993.

Another Turn of the Crank was published by Counterpoint Press on October 18, 1995; the most recent Counterpoint paperback printing, the source for the texts printed here, was published in 2011. The first publication of the essays selected from *Another Turn of the Crank* is given below.

Farming and the Global Economy. *Another Turn of the Crank* (Washington, D.C.: Counterpoint Press, 1995).

Conserving Forest Communities. Speech from the Kentucky Forest Summit and Governor's Conference on the Environment, September 29–31, 1994, published by the State of Kentucky as a proceedings paper in 1994 and excerpted in *The Land Report*, Summer 1995.

Health Is Membership. Speech at a conference, "Spirituality and Healing," at Louisville, Kentucky, on October 17, 1994, published in *Utne Reader*, September/October 1995.

Life Is a Miracle was published by Counterpoint Press on May 16, 2000. The text printed in this volume is that of the Counterpoint paperback printing, published in 2001.

Citizenship Papers was published by Shoemaker & Hoard on August 8, 2003; the most recent Counterpoint Press paperback printing, the source for the texts printed here, was published in 2014. The first publication of the essays selected from *Citizenship Papers* is given below.

A Citizen's Response to the National Security Strategy of the United States. Originally published in the United States on Sunday, February 9, 2003, in *The New York Times* in a full-page ad taken out by *Orion Magazine* and on www.Orion Online.org; and in the United Kingdom in *Irish Pages*, Autumn–Winter, 2002/2003. It appeared as the cover article in the March/April 2003 issue of *Orion Magazine*.

Thoughts in the Presence of Fear. Originally published October 30, 2001, on www.OrionOnline.org, in a feature entitled "Thoughts on America: Writers Respond to Crisis."

The Failure of War. Originally published online at www.yesmagazine.org, November 5, 2001.

In Distrust of Movements. *Orion Magazine*, Summer 1999, adapted from a keynote address delivered at the 1998 Northeast Organic Farmers' Association Conference.

The Total Economy. *Orion Magazine*, Winter 2001.

Two Minds. *Citizenship Papers* (Washington, D.C.: Shoemaker & Hoard, 2003).

The Whole Horse. Originally published in a shorter version in *The Amicus Journal*, Winter 1998.

The Agrarian Standard. *Orion Magazine*, Summer 2002.

Conservationist and Agrarian. Originally published as "For Love of the Land" in *Sierra*, May–June 2002.

The Way of Ignorance was published by Shoemaker & Hoard on October 21, 2005; the Counterpoint Press paperback printing, the source for the texts printed here, was published in 2006. The first publication of the essays selected from *The Way of Ignorance* is given below.

Secrecy vs. Rights. *The Way of Ignorance* (Washington, D.C.: Shoemaker & Hoard, 2005).

Contempt for Small Places. *Farming*, Spring 2004.

Compromise, Hell! Speech for the annual Earth Day celebration of the Kentucky Environmental Quality Commission at Frankfort, Kentucky, April 22, 2004, published on May 3, 2004, in the *Lexington Herald-Leader* as "People Can't Survive If the Land Is Dead" and, under its present title, in *Orion Magazine*, November/December 2004.

Charlie Fisher. Originally published in the spring 1996 issue of *The Draft Horse Journal* as "Trees for My Son and Grandson to Harvest."

The Way of Ignorance. *New Letters*, Summer 2005.

Quantity vs. Form. Originally published in the Fall 2005 issue of *Southern Arts Journal* as "Quantity and Form."

Renewing Husbandry. *Crop Science: A Journal Serving the International Community of Crop Scientists*, May–June 2005.

The Burden of the Gospels. Speech for the first joint convocation of the Lexington Theological Seminary and the Baptist Seminary of Kentucky, Lexington, Kentucky, August 30, 2005, printed in *The Christian Century*, September 20, 2005.

What Matters? was published by Counterpoint Press on May 10, 2010, as a paperback original. The texts of the essays selected from *What Matters?* are taken from that edition. The first publication of the essays chosen for inclusion is given below.

Money Versus Goods. Originally published in two parts in *The Progressive*: "Inverting the Economic Order" (September 2009) and "The Cost of Displacement" (December 2009/ January 2010).

Faustian Economics. *Harper's Magazine*, May 2008.

Imagination in Place was published by Counterpoint Press on January 19, 2010; the Counterpoint paperback printing, the source for the texts printed here, was published in 2011. The first publication of the essays selected from *Imagination in Place* is given below.

Imagination in Place. H. L. Weatherby and George Core, eds., *Place in American Fiction: Excursions and Explorations* (Columbia: University of Missouri Press, 2004).

American Imagination and the Civil War. *The Sewanee Review*, Fall 2007.

Sweetness Preserved. Bert G. Hornback and Peter Lang, eds., *"Bright Unequivocal Eye": Poems, Papers, and Remembrances from the First Jane Kenyon Conference* (New York: Peter Lang Publishing Group, 2000).

The Uses of Adversity. *The Sewanee Review*, Spring 2007.

God, Science, and Imagination. *The Sewanee Review*, Winter 2010.

It All Turns on Affection was published by Counterpoint Press on September 11, 2012, as a paperback original. The texts of the essay selected from *It All Turns on Affection* are taken from that edition. Their first publication is given below.

It All Turns on Affection. Jefferson Lecture delivered April 23, 2012, published in *It All Turns on Affection* (Berkeley, CA: Counterpoint Press, 2012).

About Civil Disobedience. *The Progressive*, January 2012.

Our Only World was published by Counterpoint Press on February 10, 2015; the Counterpoint paperback printing, the source for the texts printed here, was published in 2016. The first publication of the essays selected from *Our Only World* is given below.

Paragraphs from a Notebook. *Our Only World* (Berkeley, CA: Counterpoint Press, 2015).

The Commerce of Violence. *The Progressive*, June 2013.

A Forest Conversation. *Our Only World* (Berkeley, CA: Counterpoint Press, 2015).

Local Economies to Save the Land and the People. *Our Only World* (Berkeley, CA: Counterpoint Press, 2015), based on a speech to Kentuckians for the Commonwealth in Carrollton, Kentucky, on August 16, 2013.

Caught in the Middle. *Christian Century*, April 2013.

Our Deserted Country. *Our Only World* (Berkeley, CA: Counterpoint Press, 2015).

On Being Asked for "A Narrative for the Future." *Our Only World* (Berkeley, CA: Counterpoint Press, 2015).

The Art of Loading Brush was published by Counterpoint Press on October 31, 2017; the Counterpoint paperback printing, the source for the texts printed here, was published in 2019. The first publication of the essays selected from *The Art of Loading Brush* is given below.

The Thought of Limits in a Prodigal Age. *The Art of Loading Brush* (Berkeley, CA: Counterpoint Press, 2017).

The Presence of Nature in the Natural World. *A Small Porch: Sabbath poems 2014 and 2015 together with The Presence of Nature in the Natural World* (Berkeley, CA: Counterpoint Press, 2016).

This volume presents the texts of the printings chosen for inclusion here, but it does not attempt to reproduce features of their typographic design. Original citations have been altered to conform to Library of America practices; however, original footnotes that take the form of personal observation or argument have been preserved verbatim. In a few instances quotations have been corrected with the author's approval. Original in-text citations have been removed where both the author and title of the work under discussion are evident. The texts are presented without change, except for the correction of typographical errors and minor changes necessitated by the formatting of notes. Spelling, punctuation, and capitalization are often expressive features, and they are not altered, even when inconsistent or irregular. The following is a list of typographical errors corrected, cited by page and line number: 6.8, urbanindustrial; 8.7, L'il; 83.18, dreary ,; 121.19, book,; 133.16–17, "Whenever; 133.17, physiochemical; 172.34, agriculturalist,; 180.33, The *ABC*; 181.26, generation; 217.20, the Land; 227.34, entomologist,; 265.25, 1930's; 311.37, September,; 344.32, becomes; 383.11, discreet; 384.2, them to do; 447.11, cost.; 499.17, wife Tanya; 505.22, brother Oliver,; 506.10, says are; 545.7 (and *passim*), *Howard's*; 553.21, anthology,; 561.14, essays,; 561.15, Ethic,"; 561.31, novel,; 563.5, geometer,; 576.25, something.; 577.35, life upon; 596.4, Tory; 626.27, care."; 639.15, agriculture by; 640.28, famer; 641.10, 1950s, is; 670.6, prophesies; 675.3, Surowieki's; 683.25, worst; 712.18, western; 723.3, Defense; 731.8, presidence; 737.1, breast–; 742.13, keen, intense,; 752.15, sometime.

Notes

In the notes below, the reference numbers denote page and line of this volume (the line count includes headings). No note is made for material included in standard desk-reference books. Quotations from Shakespeare are keyed to *The Riverside Shakespeare*, ed. G. Blakemore Evans (Boston: Houghton Mifflin, 1974). Biblical quotations are keyed to the King James Version. Original citations have been altered to conform to Library of America practices; however, original footnotes that take the form of personal observation or argument have been preserved verbatim. For references to other studies and further information than is included in the Chronology, see *Wendell Berry: Life and Work*, ed. Jason Peters (Lexington: The University Press of Kentucky, 2007); and *Conversations with Wendell Berry*, ed. Morris Allen Grubbs (Jackson: University Press of Mississippi, 2007).

From SEX, ECONOMY, FREEDOM & COMMUNITY

7.13–20 "the county . . . earned it."] From an article by editor William Allen White (1868–1944) in the *Emporia Gazette* (1912), a small-town paper with a national reputation.

7.39–40 the General Agreement on Tariffs and Trade] Also known as GATT, an international trade agreement signed by multiple nations in Geneva on October 30, 1947, that was designed to reduce trade barriers. Negotiations between GATT members in the years 1986 to 1994 would lead to trade liberalization in agricultural products and the establishment of the World Trade Organization (WTO). In the early 1990s, Berry argued against proposed changes to GATT and has written subsequently on how trade liberalization has been damaging to American farmers.

8.7 Snuffy Smith or Li'l Abner] Snuffy Smith, a gun-toting, moonshine-swilling hillbilly in the still-running American comic strip *Barney Google and Snuffy Smith*, which debuted in 1919; Li'l Abner Yokum, a gullible hillbilly who lives in Dogpatch, USA, in the American comic strip *Li'l Abner*, which ran from 1934 to 1977.

8.9 "The business of America is business,"] Frequently attributed to President Calvin Coolidge (1872–1933), who in a speech of January 17, 1925, before the American Society of Newspaper Editors, Washington, D.C., remarked, "After all, the chief business of the American people is business."

10.26–27 what Edward Abbey called . . . cancer cell"] Edward Abbey (1927–1989), American essayist, novelist, ecologist, and a dedicatee of Berry's *Sex, Economy, Freedom & Community* (1993). See Abbey's essay "Arizona: How Big Is Enough?," reprinted in his collection *One Life at a Time* (1987): "Growth for the sake of growth is the ideology of the cancer cell."

15.18–20 Wilderness preserves . . . marketable timber.] Dave Foreman, *Confessions of an Eco-Warrior* (1991).

15.21 Wes Jackson] American plant geneticist, advocate for sustainable practices, and cofounder of The Land Institute in Salina, Kansas (b. 1936). Jackson was one of the four agrarian allies of Berry for several decades (outside his family), along with Gene Logsdon, Maurice Telleen, and David Kline.

17.4–5 Love Canal, Bhopal, Chernobyl, the burning oil fields of Kuwait.] Love Canal, near Niagara Falls, New York, the site of a chemical waste dump in the 1940s and 1950s that forced the evacuation and resettlement of more than a thousand families in the Love Canal neighborhood, after President Jimmy Carter (twice) declared a state of emergency in the area, in 1978 and 1981; Bhopal, the capital city in Madhya Pradesh, India, which gained international attention in 1984 when a pesticide-manufacturing plant leaked toxic gas, killing thousands; Chernobyl, an uninhabitable city in northern Ukraine, synonymous with the nuclear disaster of 1986, when the Chernobyl nuclear power station emitted large amounts of radiation into the atmosphere; the burning fields of Kuwait, more than 600 oil wells set aflame by the Iraqi military army as it withdrew from Kuwait in 1991, in the Persian Gulf War.

17.17 the Sacramento River is now dead] On July 14, 1991, a Southern Pacific tanker car spilled 19,000 gallons of metam-sodium (a soil fumigant and weed killer) into the Upper Sacramento River.

28.1 *Christianity and the Survival of Creation*] Originally delivered as a lecture by Berry at the Southern Baptist Theological Seminary in Louisville, Kentucky, in 1992.

30.13–14 "The earth . . . dwell therein."] Psalms 24:1.

30.25–26 "The land . . . with me."] Leviticus 25:23.

30.29–30 "All things . . . was made."] John 1:3.

31.2 John 3:16] "For God so loved the world, that he gave his only begotten Son, that whosoever believeth in him should not perish, but have everlasting life."

31.13–14 "gather unto . . . perish together."] Job 34:14–15.

31.17–18 "Creation is . . . hidden Being."] Philip Sherrard, *Human Image: World Image* (1992).

31.22–23 Thou art . . . and all.] George Herbert, "Providence," stanza 11, first published in *The Temple* (1633).

31.29–30 "despising Nature . . . against God.] Dante, *Inferno*, XI.46–48, Charles S. Singleton translation (1970). All subsequent quotations of *The Divine Comedy* are from the Singleton translation.

31.36–37 "condemns Nature . . . hope elsewhere."] Dante, *Inferno*, XI.109–111.

31.39 "everything that lives is holy."] The concluding line of Willian Blake's *The Marriage of Heaven and Hell* (1790–93).

32.1–2 "the sense . . . human norm."] See *Golgonooza: City of Imagination* (1991), by Kathleen Raine (1908–2003), English poet, essayist, and scholar whose expertise was William Blake and W. B. Yeats.

33.24–26 "Behold, the heaven . . . have builded?"] 1 Kings 8:27.

33.29–33 God that made . . . have said.] Acts 17:24 and 28.

34.10–11 "where two . . . my name."] Matthew 18:20.

34.30–39 Your new moons . . . the widow.] Isaiah 1:13–17.

35.4 "hypaethral book,"] "Hypaethral" means roofless, open to the sky. Writing in his *Journal* on June 29, 1851, Henry David Thoreau (1817–1862) characterized *A Week on the Concord and Merrimack Rivers* (1849) as "a hypaethral or unroofed book, lying open under the ether and permeated by it, open to all weathers, not easy to be kept on a shelf."

35.36–38 "The invisible things . . . are made."] Romans 1:20.

36.30–39 And of Joseph . . . the bush.] Deuteronomy 33:13–16.

37.32–33 "For what . . . own soul?"] Matthew 18:20.

38.17 "bruised to pleasure soul."] W. B. Yeats, "Among School Children," stanza VIII, published in his collection *The Tower* (1928).

39.33–34 "It is a tale . . . Signifying nothing,"] *Macbeth*, V.v.26–28.

39.35 "supp'd . . . the sun."] *Macbeth*, V.v.13, 48.

40.22–25 "the normal view . . . of artist."] Ananda K. Coomaraswamy, "What Is the Use of Art, Anyway?," collected in his book of essays *Christian and Oriental Philosophy of Art* (1956). Ananda Kentish Coomaraswamy (1877–1947) was a Ceylonese historian of Indian art and culture, art theorist, and scholar of comparative religious thought who was instrumental in bringing ancient Indian art and culture to the attention of the English-speaking world.

40.33–34 "the plowman . . . cosmic function."] Walter Shewring, *Artist and Tradesman* (1984).

41.9–11 Mother Ann Lee's famous instruction . . . die tomorrow."] In his original note Berry cites as his source for Lee's instruction June Sprigg's *By Shaker Hands* (1990). Ann Lee (1736–1784), also known as Mother Ann Lee, was the spiritual leader and founder of the Shaker movement in America.

41.12–26 "the perfection . . . reformation must begin."] Coomaraswamy, *Selected Papers*, vol. 1 (1977).

42.9–11 "Genius inhabits . . . has neighbors."] Coomaraswamy, "What Is the Use of Art, Anyway?," collected in *Christian and Oriental Philosophy of Art*.

42.23–25 "far worse . . . the art."] Dante, *Paradiso*, XIII.121–123.

43.34 "I made . . . try for it."] See the beginning of *The Adventures of Huckleberry Finn* (1884) by Mark Twain (1835–1910).

44.31–33 "Love your enemies . . . persecute you."] Matthew 5:44.

46.5–6 the Clarence Thomas hearing] In 1991, during Clarence Thomas's U.S. Supreme Court confirmation hearings, Thomas's former subordinate Anita Hill testified that Thomas had sexually harassed her at the Department of Education and the EEOC. Thomas denied any wrongdoing, and the Senate confirmed Thomas on October 15, 1991.

52.27–53.18 The . . . reason for . . . afterwards absorbed.] From *Hawkshead (The Northernmost Parish of Lancashire): Its History, Archaeology, Industries, Dialect, Etc.* (1899) by English antiquary Henry Swainson Cowper (1865–1941).

53.21–23 The industrialization . . . pattern exactly.] Berry, in his original note, indicates his indebtedness to Harry M. Caudill's *Theirs Be the Power* (1983) for this observation. Caudill (1922–1990), from Letcher County, Kentucky, was an author, legislator, opponent of strip-mining in the Appalachian coalfields, and professor of Appalachian Studies at the University of Kentucky. He is a dedicatee of Berry's *Sex, Economy, Freedom & Community*.

53.37–54.13 In the traditional culture . . . its value.] Helena Norberg-Hodge, *Ancient Futures: Learning from Ladakh* (1991).

55.12–18 Increasingly, people . . . longer afford.] Norberg-Hodge, *Ancient Futures*.

55.25 Luddites] Disaffected nineteenth-century English artisans—primarily textile workers and weavers—who destroyed new labor-saving machinery that was replacing them, in a widespread rebellion that lasted from 1811 to 1817.

55.33–56.8 By the adoption . . . his hire.] In his original note Berry cites as his source for Byron's speech Humphrey Jennings's *Pandaemonium, 1660–1886: The Coming of the Machine as Seen by Contemporary Observers* (1985).

56.9–16 The Luddites did . . . hangings and transportations."] Berry's source is *Encyclopedia Britannica*, 14th ed., s.v., "Luddite."

57.25–26 the Community Farm Alliance] Founded in 1985, an alliance of Kentucky farmers dedicated to helping independent family farmers across Kentucky and to promoting sustainable farming methods and traditional practices.

58.23–24 "fearfully and wonderfully made."] Psalms 139:14.

59.32–33 If sex . . . wished-for words.] Wallace Stevens, "Le Monocle de Mon Oncle" (1918), stanza XI, reprinted in Stevens's first collection of poems, *Harmonium* (1923).

61.10–12 "Of a surety . . . would return."] Dante, *La Vita Nuova*, Chapter XIV, translation by D. G. Rossetti (1828–1882), first published in *The Early Italian Poets* (1861).

62.22–23 "Losing kindness . . . to justness."] *The Way of Life According to Lao-tzu*, translation by Witter Bynner (1962).

67.26–27 "Ye shall know . . . you free,"] John 8:32.

68.15 "no ethical . . . an indictable one."] Mary McGrory, "Ed Meese on the Stand," the Louisville *Courier-Journal* (March 31, 1989).

70.11–13 I am indebted to Judith Weissman . . . the individual.] Berry, in his original note, directs readers "especially" to the Introduction and Chapter 1 of Judith Weissman's *Half Savage and Hardy and Free: Women and Rural Radicalism in the Nineteenth-Century Novel* (1987).

70.33–38 The Liberation . . . feminist writers. . . .] Weissman, *Half Savage and Hardy and Free.*

73.10–16 I am immodestly . . . American life."] William Mootz, "Arthur Kopit Plans to Offend Almost Everyone," the Louisville *Courier-Journal* (February 26, 1989).

77.26–27 "the purpose . . . *to move*"] Coomaraswamy, *Selected Papers*, vol. 2 (1977).

81.37–38 "certainly it . . . will die"] Dante, *La Vita Nuova*, Chapter XXIII, translation by D. G. Rossetti.

83.17–23 People were bathing . . . like water-lilies.] D. H. Lawrence, "The Gods! The Gods!," published in the posthumous collection *Last Poems* (1932).

From ANOTHER TURN OF THE CRANK

94.32 the Stamp Act] To recoup expenses incurred as a result of the French and Indian War, in 1765 the British parliament imposed a tax on printed matter in the American colonies. After protests and colonial boycotts, the Stamp Act was repealed the following year.

96.1 *Conserving Forest Communities*] Delivered as a speech at the Kentucky Forest Summit (in conjunction with the Nineteenth Governor's Conference on the Environment) at Louisville, Kentucky (September 29, 1994).

98.13–14 the Burley Tobacco Growers Co-operative Association] Founded in 1921, the Association resumed its work under the New Deal's Federal Tobacco Program starting in 1941. By a combination of production controls and price supports, it gave vital help to small farmers in Kentucky and other burley-growing states for six decades.

98.18–20 "is owned . . . 26 acres."] From "Kentucky's Forest Resources," an unpublished paper by William H. Martin, Mark Matuszewski, Robert N. Muller, and Bradley E. Powell (1994). In his original note Berry says, "I have taken statistics and other information on Kentucky forests from this paper and also from William H. Martin, 'Sustainable Forestry in Kentucky,' *In Context* (Center for Economic Development at Eastern Kentucky University, winter 1993) . . . and William H. Martin, 'Characteristics of Old-Growth Mixed Mesophytic Forests,' *Natural Areas Journal* 12, no. 3 (July 1992)."

103.10 an article in *California Forestry Notes*] Dave McNamara, "Horse Logging at Latour," *California Forestry Notes* (September 1983).

106.39 the Menominee Indians] Algonquian-speaking tribe of Native American people who originally lived along the Menominee River on the present-day border between Wisconsin and Michigan; the Menominee Indian Reservation in Wisconsin is approximately 60 miles west of the mouth of the Menominee. Berry visited the Menominee tribal forest in June 1994.

108.21–23 "they could . . . secondary processing."] Scott Landis, "Seventh-Generation Forestry," *Harrowsmith Country Life* (November/December 1992).

110.1 *Health Is Membership*] Delivered as a speech at a conference entitled "Spirituality and Healing," in Louisville, Kentucky, on October 17, 1994.

112.17 Sir Albert Howard . . . *The Soil and the Health*] Albert Howard (1873–1947), British botanist and agriculturist. Howard's publications drew on extensive practical research in India, where, working as an agricultural advisor, he came to see the wisdom of traditional Indian farming practices. *The Soil and Health: A Study of Organic Agriculture* was originally published under the title *Farming and Gardening for Health or Disease* (1945) before its 1947 reissue.

121.19–22 "Machines . . . calculate with."] Neil Postman, *Technopoly: The Surrender of Culture to Technology* (1992).

122.25–27 "Are not . . . before God?"] Luke 12:6.

122.28–31 "If a clod . . . diminishes me."] John Donne, *Devotions upon Emergent Occasions* (1623), Meditation XVII.

LIFE IS A MIRACLE

128.2 *Lionel Basney (1946–1999)*] American poet, essayist, environmental activist, and professor of English whose interests included the history of the Luddite movement and the writings of Wendell Berry; a personal friend of Berry.

131.34 "I stumbled when I saw."] *King Lear*, IV.i.19.

132.5–6 "'Twixt two . . . joy and grief. . . ."] *King Lear*, V.iii.199.

132.12 "Falls forward and swoons."] *King Lear*, IV.vi.41 s.d. Berry quotes the stage directions in the Pelican Shakespeare edition of the play.

132.18 "Away, and let me die."] *King Lear*, IV.vi.48.

132.22 "Thy life's . . . yet again."] *King Lear*, IV.vi.55.

133.16–18 "Wherever one . . . as such."] See *Physics and Philosophy: The Revolution in Modern Science* (1958) by German theoretical physicist Werner Heisenberg (1901–1976), a pioneer of quantum mechanics.

133.26–28 What seems . . . seems to Be . . .] William Blake, *Jerusalem* (1804–20), plate 32 [36], lines 51–53.

134.36–37 Kathleen Raine was correct . . . being experienced.] See Kathleen Raine, *The Inner Journey of the Poet* (1982).

135.17–19 O you mighty . . . affliction off.] *King Lear*, IV.vi.34–36.

135.27–29 You ever-gentle gods . . . you please] *King Lear*, IV.vi.217–219.

135.33 "Well pray you, father."] *King Lear*, IV.vi.219.

140.26–27 Dante saw it from a higher level of accomplishment] See Dante, *Paradiso*, XXII.133–154.

143.4–5 C. P. Snow spoke . . . "automatic corrective"] In *The Two Cultures and the Scientific Revolution* (1959), based on his Rede Lecture of the same year. Charles Percy Snow (1905–1980), English novelist and chemist.

145.13–14 *Consilience* by Edward O. Wilson] E. O. Wilson (b. 1929), American biologist, sometimes called the "father of sociobiology." *Consilience* was published in 1998 by Alfred A. Knopf.

152.17 so-called Y2K problem] Also known as the Y2K bug, the Millennium bug, the Year 2000 problem, or simply Y2K, the Y2K problem arose because many computer programs used two digits to signify the calendar year, raising the possibility of catastrophic consequences on January 1, 2000, if computers were unable to distinguish between 2000 and 1900. Estimates vary, but hundreds of billions of dollars were spent around the globe to make computers and application programs Y2K-compliant.

153.21–22 "We murder to dissect."] William Wordsworth, "The Tables Turned," published in Wordsworth and Coleridge's *Lyrical Ballads* (1798).

161.20–21 "Life," wrote Erwin Chargaff . . . the inexplicable."] Erwin Chargaff, *Heraclitean Fire: Sketches from a Life Before Nature* (1978).

162.36–37 Tarzan theory of the mind . . . Edgar Rice Burroughs.] Burroughs (1875–1950), American author of more than sixty works, including *Tarzan of the Apes* (1912) and other books in the Tarzan series. In *Tarzan of the Apes*, Tarzan as a young boy teaches himself to read and write English by puzzling over alphabet books and primers he finds in the ruins of his parents' home.

169.8–9 Cortés . . . Montezuma] Spanish conquistador Hernán Cortés (1485–1547) held Aztec emperor Montezuma II (1466–1520) prisoner after

Montezuma welcomed him to the Aztec capital Tenochtitlán; Montezuma was killed in fighting for control of the city, largely destroyed by the Spanish and their native allies.

169.20–27 The intellect . . . the night's remorse.] W. B. Yeats, "The Choice," published in *The Winding Stair and Other Poems* (1933).

170.1 Dante . . . found this Ulysses in Hell] Dante encounters Ulysses in the eighth trench of the eighth circle of hell, a place reserved for evil counsellors. See the *Inferno*, XXVI.

179.30–32 "all great scientific discoveries . . . to lose."] Chargaff, *Heraclitean Fire.*

179.35–39 "The Nazi experiment . . . modern science."] Chargaff, *Heraclitean Fire.*

181.7–12 "There can be . . . ideals or opinions."] Coomaraswamy, "The Christian and Oriental, or True Philosophy of Art," collected in his book of essays *Christian and Oriental Philosophy of Art* (1956).

181.13–15 "when there . . . new or old."] Coomaraswamy, *The Transformation of Nature in Art* (1936), Chapter 1.

182.12 "dark Satanic mills"] William Blake, *Milton* (1810), Preface.

182.13 "we murder to dissect"] See note 153.21–22.

182.20 Hitler's troops marched into Prague] March 15, 1939.

182.20–21 the Scottish poet Edwin Muir] Edwin Muir (1887–1959) and his wife, Willa Muir (1890–1970), translated the works of the German-speaking Jewish Prague writer Franz Kafka into English.

182.23–30 "So many goodly citties . . . oh base conquest."] Montaigne, "Of Coaches," *The Essayes of Michael Lord of Montaigne*, John Florio translation (1603).

182.30–41 "Think of all . . . merely new."] Edwin Muir, *The Story and the Fable* (1940), revised and reissued in 1954 as *An Autobiography.*

183.2–7 "Dreams of . . . for progress."] C. S. Lewis, *That Hideous Strength* (1945), Chapter 9.

183.9–21 "It is a severe question . . . photographing the corpse . . ."] Albert Howard, *The Soil and Health* (1947), Chapter 6. See also note 112.17.

183.25–28 "The wonderful . . . be recalled."] Chargaff, *Heraclitean Fire.*

183.38–39 "deep-set repugnances"] C. S. Lewis, *That Hideous Strength*, Chapter 9.

184.8–9 "We ought to stay out of the nuclei."] Wes Jackson in conversation with Berry.

187.6–8 a panel discussion . . . published in the *Authors Guild Bulletin.*] "Whose Life Is It, Anyway?," the *Authors Guide Bulletin* (Winter 1999).

198.8 "Thou'lt come . . . never never."] *King Lear*, V.iii.308–309.

198.13–14 General Lee was overheard saying . . . too bad!"] Berry's source is *Lee* (1961), Richard Harwell's abridgment of Douglas Southall Freeman's four-volume biography *R. E. Lee* (1934).

198.24–27 "I know . . . shall I see God. . . ."] Job 19:25–26.

201.25–27 "I could be bounded . . . bad dreams."] *Hamlet*, II.ii.254–255.

202.16–17 All flesh . . . breath of God.] Cf. Job 34:14–15.

202.17 "live, and move, and have our being"] Acts 17:28.

202.20 the fall of every sparrow] Cf. Matthew 10:29.

202.21 "the very hairs of your head"] Matthew 10:30.

202.22 "Thy life's a miracle,"] *King Lear*, IV.vi.55.

202.23–24 "everything that lives is holy."] See note 31.39.

202.25–27 "want us . . . simple things as well."] See *Revelations of Divine Love* (1395), Chapter 32, by English anchoress and Christian mystic Julian of Norwich (1342–c. 1416).

202.28–29 "*A Sand County Almanac* . . . distinct beings. . . ."] Stephanie Mills, *In Service of the Wild: Restoring and Reinhabiting Damaged Land* (1995).

203.12–14 "The gods' presence . . . the gods."] See Guy Davenport's translation of the fragments of pre-Socratic Greek philosopher Herakleitos (c. 535–c. 475 B.C.) in *Herakleitos and Diogenes* (1979).

206.30–31 "assert Eternal Providence . . . to men."] Milton, *Paradise Lost*, I.25–26.

211.6–7 To see a World . . . Wild Flower.] The opening lines of William Blake's "Auguries of Innocence" (1866).

213.10 Gerard David's *Annunciation*] Oil on wood painting, by Netherlandish painter Gerard David (c. 1460–1523), of the announcement of the Incarnation by the angel Gabriel to Mary, commissioned in 1506.

213.20 Luke 1:28–35] The Annunciation is narrated in Luke 1:28–35.

217.19–20 Wes Jackson . . . co-founder of The Land Institute] Wes Jackson cofounded The Land Institute in 1976 with Dana Jackson. The Land Institute is dedicated to the development of robust locally adapted, perennial grain crops, to reducing or eliminating soil erosion and the use of herbicides, and to the overall reduction of dependence on industrial agriculture. See also note 15.21–23.

221.24 Wallace Stegner] American novelist, short story writer, historian, environmentalist, teacher, and founder of Stanford University's creative writing program (1909–1993). Stegner won the Pulitzer Prize for his novel *Angle of Repose* (1971) and the National Book Award for the novel that followed, *The Spectator Bird* (1977).

221.27–30 "boomers . . . made it in."] Wallace Stegner, *Where the Bluebird Sings to the Lemonade Springs: Living and Writing in the West* (1992), Introduction.

226.19–20 "the tones given off by the heart."] From *The Great Digest*, translation by Ezra Pound (1928), reprinted in his anthology *Confucius* (1951).

227.35 French entomologist Jean Henri Fabre] Largely self-taught, Fabre (1823–1915) conducted important research on bees and wasps, beetles, and grasshoppers.

227.38 "a patch of pebbles . . . four walls."] Jean Henri Fabre, "The Laboratory of Open Fields," in *The Insect World of J. Henri Fabre* (1949), translated by Alexander Teixeira de Mattos, with an introduction and comments by Edwin Way Teale.

228.20 *The Wheelwright's Shop*] Autobiography published in 1923 by English writer George Sturt (1863–1927), who after his father's death managed the family wheelwright shop in the town of Farhnam, in Surrey, where he spent the remainder of his life.

230.5–11 Richard Strohman has made clear . . . predictability] In *The Daily Californian* (April 1, 1999), and *The Wild Duck Review* (Summer 1999).

235.1–2 I learned from the theologian Philip Sherrard to ask this question] In an embedded citation in the original text, Berry points readers to Sherrard, *Human Image: World Image* (1992).

From CITIZENSHIP PAPERS

239.20 The new "National Security Strategy"] Issued by the executive branch of the U.S. government, the "National Security Strategy" is a report to Congress that articulates national security concerns and how an administration plans to address them. The "new" report to which Berry refers is the one published on September 20, 2002, by the George W. Bush administration.

241.8 the events of September 11, 2001] Coordinated attacks by al-Qaeda on U.S. targets using four hijacked commercial planes; 2,996 people including the 19 hijackers died at the World Trade Center, the Pentagon, and in a field near Shanksville, Pennsylvania, where one of the planes crashed before reaching its intended target (believed to be either the U.S. Capitol or the White House).

243.22 "There is none . . . God,"] Matthew 19:17.

243.23–24 "He that is . . . at her."] John 8:7.

243.25–27 "for nothing . . . superintendence."] Thomas Jefferson, letter to Mann Page, August 30, 1795.

245.21–24 "Laws are commanded . . . to uphold."] Edmund Burke, *Reflections of the Revolution in France* (1790).

246.32 Saddam Hussein] President of Iraq from 1979 to 2003, when Hussein's Baathist regime was overthrown by U.S. and coalition forces in the invasion of Iraq. Found guilty of crimes against humanity by an Iraqi court, Hussein was hanged on December 30, 2006.

247.4 America Whose Business Is Business] See note 8.9.

249.3–4 the Alien and Sedition Laws of 1798] Four laws passed by U.S. Congress in 1798 amid growing fears of war with France. Aimed at French and Irish immigrants, the laws extended naturalization from five to fourteen years, authorized the imprisonment and expulsion of any noncitizens determined to be dangerous, and restricted speech critical of the U.S. government. Negative public reaction to the laws contributed to the victory of Thomas Jefferson over the incumbent, John Adams.

251.24 Monsanto] On June 7, 2018, Bayer AG purchased the Monsanto Company for $66 billion; shortly before the purchase was completed Bayer announced its intention to drop the Monsanto name.

254.21–22 the Gulf War of 1991] Also known as the First Gulf War, the First Iraq War, the Kuwait War, and the Persian Gulf War, a war conducted by coalition forces led by the United States against Iraq in response to Saddam Hussein's invasion of Kuwait.

259.7–8 The modern doctrine of such warfare . . . General William Tecumseh Sherman] During the Civil War, Sherman (1820–1891) conducted a scorched earth campaign in his march through Georgia and the Carolinas.

264.38–265.2 "Both . . . answered fully."] Lincoln's Second Inaugural Address, March 4, 1865.

269.24–29 "Love your enemies . . . the unjust."] Matthew 5:44–45.

275.40–276.1 in Mary Austin's phrase "summer and winter with the land."] See the Preface to *The Land of Little Rain* (1903) by Mary Hunter Austin (1868–1934), a writer of the American Southwest.

283.14 "Love thy neighbor as thyself"] Matthew 22:39.

283.15 "Sentient beings . . . to save them."] One of the Four Bodhisattva Vows in Mahayana Buddhism, an expression of commitment to the enlightenment of all sentient beings.

284.39–40 the Committee for Economic Development] Public policy organization created in 1942 by leaders of American corporations, in anticipation of World War II's conclusion and transition to a peace-time economy. The CED championed American economic internationalism.

286.38–39 the World Trade Organization] Under the Marrakesh Agreement of April 15, 1994, signed by 124 member nations, the WTO replaced the General Agreement on Tariffs and Trade (GATT). See also note 7.39–40.

290.7–10 "the principle . . . fitted by nature. . . ."] Ananda K. Coomaraswamy, "What Is Civilization?" (1946), in *What Is Civilization and Other Essays* (1989).

290.12 "the mammon of injustice,"] Cf. Luke 16:9.

290.15 "to work is to pray"] Benedictine motto (*laborare est orare*).

293.22–25 "Whenever the timber . . . as possible."] See Albert Schweitzer's *Out of My Life and Thought: An Autobiography* [*Aus meinem Leben und Denken*], first published in German in 1931. Berry quotes from the English translation by Antje Bultmann Lemke (1990). An Alsatian-German theologian, philosopher, organist, and philanthropist, Schweitzer (1875–1965) was awarded the 1952 Nobel Prize in part for his medical missionary work in west-central Africa.

293.27–29 "These people . . . own needs."] Schweitzer, *Out of My Life and Thought.*

295.12–13 Thomas Kuhn described this process] In place of the gradual accumulation of scientific knowledge, American physicist Kuhn (1922–1996) saw alternating phases of "normal science" and revolutionary invention that brought "paradigm shifts."

301.33–34 "the instinctive integrations . . . for living."] Wallace Stevens, "Imagination as Value," *The Necessary Angel: Essays on Reality and Imagination* (1951).

303.7–12 "If a man . . . not astray."] Matthew 18:12.

308.17 Bertie Wooster] Character and narrator in the Jeeves fictions by English writer P. G. Wodehouse (1881–1975); Wooster, an English gentleman, relies on his valet, Jeeves, to assist him in the simplest tasks and to rescue him from more complicated situations.

309.1 James Laughlin's poem "Above the City,"] James Laughlin (1914–1997), American poet and the founder of the publishing house New Directions. "Above the City" first appeared in *New Directions in Prose & Poetry 9* (1946) and was reprinted in the collection *A Small Book of Poems* (1948).

311.33–35 "let us build . . . the whole earth."] Genesis 11:4.

320.3–4 "I think . . . with animals"] Walt Whitman, "Song of Myself," section 32, initially published without sections in the 1855 edition of *Leaves of Grass.*

323.12–13 J. Russell Smith, Liberty Hyde Bailey, Albert Howard, Wes Jackson, John Todd] Joseph Russell Smith (1874–1966), American economic geographer, conservationist, and writer whose books include *Tree Crops: A Permanent Agriculture* (1929); Liberty Hyde Bailey (1858–1954), American botanist, horticulturalist, and cofounder of the American Society for Horticultural Science; Albert Howard, see note 112.17; Wes Jackson, see note 15.21–23; John Todd (b. 1939), Canadian ecological designer whose design innovations address challenges of food production, waste treatment, and environmental repair through ecosystem technologies that emulate nature.

323.16–17 "let the forest judge"] *As You Like It*, III.ii.121.

323.17 "consult the Genius of the Place"] Alexander Pope, "Epistle to Burlington" (1731), line 57.

323.29–33 "the eternal task . . . lived in."] John Haines, "On a Certain Attention to the World," published in *Fables and Distances: New and Selected Essays* (1996). An American poet and essayist, Haines (1924–2011) was poet laureate of Alaska from 1969 to 1973.

324.29–30 the Twelve Southerners of *I'll Take My Stand*] The Twelve Southerners, also known as the Southern Agrarians, were a circle of Southern writers and academics formed in the 1920s and united in the cause of preserving an agrarian way of life against the forces of industrialism. Among the group's most prominent members were John Crowe Ransom, Robert Penn Warren, Donald Davidson, and Allen Tate. Each of the Twelve Southerners contributed an essay to the landmark anthology *I'll Take My Stand: The South and Agrarian Tradition* (1930).

332.21–23 The purpose . . . "food dictatorship."] Vandana Shiva, "Food Democracy v. Food Dictatorship: The Politics of Genetically Modified Food," *Z Magazine* (April 2003).

336.35–337.6 I saw . . . unbought delicacies.] Virgil, *Georgics*, IV.132. Berry quotes from L. P. Wilkinson's translation (1982).

338.13–16 "it is not . . . a state. . . ."] Thomas Jefferson, letter to James Madison Fontainebleau, October 28, 1785.

339.19–20 "the ideology of the cancer cell"] See note 10.26–27.

341.32 "a false nostalgia . . . never existed."] Richard Lewontin, "Genes in the Food!," *The New York Review of Books* (June 21, 2001).

348.10–11 "fit the farming to the land"] See J. Russell Smith, *Tree Crops: A Permanent Agriculture* (1929), which describes the traditional use of tree crops and how trees such as the carob, honey locust, mulberry, chestnut, and black walnut can be an important source for food, soil conservation, and sustainable agriculture.

From THE WAY OF IGNORANCE

362.4–10 The contradictions . . . to live."] James R. Carroll, "Mountaintop Mining at Center of Controversy," the Louisville *Courier-Journal* (June 14, 2003).

363.1 *Compromise, Hell!*] Written originally as a speech for the annual Earth Day celebration of the Kentucky Environmental Quality Commission at Frankfort, Kentucky, April 22, 2004.

368.33 Joe Begley] Joseph Taylor Begley (1919–2000), storekeeper, longtime resident of Blackey, Kentucky, and activist who was instrumental in mounting opposition to strip-mining in eastern Kentucky.

377.1 *The Way of Ignorance*] Written as a preliminary paper for a Land Institute conference of the same title at Matfield Green, Kansas, June 3–5, 2004.

377.9–10 Richard Dawkins's assertion, in an open letter to the Prince of Wales] Dated May 21, 2000.

378.13–17 "we cannot imagine . . . not possess."] Kathleen Raine, "On the Symbol," published in her essay collection *Defending Ancient Springs* (1967).

385.39 "man as artist . . . after form."] David Jones, "Art in Relation to War," published in his *The Dying Gaul and Other Writings* (1978).

387.32–38 But Eliot . . . one part.] Kathleen Raine, "The Poet of Our Time," in *T. S. Eliot: A Symposium* (1948), edited by Tambimuttu and Richard March.

388.19–20 "As to diseases . . . no harm."] Hippocrates, *Of the Epidemics*, Book I.

389.21–22 "the way of ignorance,"] T. S. Eliot, "East Coker," first published in *New English Weekly* (March 1940) for its "Easter Number" and reprinted as the second poem in *Four Quartets* (1943).

391.1 *Quantity vs. Form*] Written originally for the conference "Evidence-Based, Opinion-Based, and Real World Agriculture and Medicine," convened by Charlie Sing at Emigrant, Montana, October 10–15, 2004.

392.11–12 Tiresias' foretelling of the death of Odysseus] Recounted in *The Odyssey*, Book XI.

392.23–24 Robert Southey's . . . biography of Lord Nelson.] *The Life of Horatio, Lord Nelson* (1813).

394.14–15 I wrote in a poem . . . infinite form,"] Wendell Berry, "From the Crest," first published in *The Virginia Quarterly Review* (Summer 1974). See Berry's *New Collected Poems* (2012).

395.4–6 "our common responsibility . . . can recognize."] Jeffery L. Amestoy, "Uncommon Humanity: Reflections on Judging in a Post-Human

Era," *New York Law Review* (November 2003), first delivered as a speech on February 6, 2003, at New York University School of Law for the annual Justice William J. Brennan, Jr. Lecture on State Courts and Social Justice.

399.3–4 McCormick High Gear No. 9] An animal-powered mower manufactured by International Harvester from 1939 to 1951.

399.5 Farmall A] A compact row-crop tractor manufactured by International Harvester from 1939 to 1947. In 1948, Harvester introduced the Super A, the model purchased by Berry's father.

405.33–34 "health in soil . . . one great subject."] Albert Howard, *The Soil and Health* (1947), Introduction.

406.16 "a dispenser of the 'Mysteries of God.'"] Liberty Hyde Bailey, *The Holy Earth* (1915).

412.1 *The Burden of the Gospels*] Written originally as a speech for the first joint convocation of the Lexington Theological Seminary and the Baptist Seminary of Kentucky, Lexington, Kentucky, August 30, 2005.

413.15–20 "Don't resist . . . children of God."] Cf. Matthew 5:39, 5:44–45.

413.31 "If ye love . . . my commandments."] John 14:15.

414.4 "one of the least of these my brethren"] Matthew 25:40.

415.7–8 "They have their reward"] Matthew 6:2.

415.31 parable of the Samaritan] Told in Luke 10:25–37.

415.33 Buddhist vow to "save all sentient beings,"] See note 283.15.

415.40–416.1 "Love thy neighbor as thyself."] Matthew 22:39.

416.28–31 "And there are . . . be written."] John 21:25.

417.35–37 "I am come . . . more abundantly."] John 10:10.

419.27 remembering Genesis 1:27] "So God created man in his own image, in the image of God created he him; male and female created he them."

From WHAT MATTERS?

432.16–18 He classes usurers . . . wrong sources."] Aristotle, *Nicomachean Ethics*, Book IV. Berry quotes from the English translation of 1908 by Sir William David Ross.

438.4–5 "Greed is . . . known to man."] William Safire, "Ode to Greed," *The New York Times* (January 5, 1986).

438.29–31 "I hold no brief . . . or Sloth."] Safire, "Ode to Greed."

439.11 Luddites] See note 55.25.

442.2–3 a stranger . . . a strange place] Cf. Exodus 2:22.

443.2–4 the difference, defined by Paul Goodman . . . between competent poverty and abject poverty] American psychologist, author, and filmmaker Paul Goodman (1911–1972) advocated "decent poverty," as opposed to misery, as a way of life for those interested in pursuits not concerned primarily with making money. Speaking of himself, Goodman claimed not to see a significant difference between being decently poor and modestly rich.

448.33–34 Menominee Forest . . . the Pioneer Forest] Berry's visit in June 1994 to the Menominee Indians of northern Wisconsin to learn about sustainable forest management is described in "Conserving Forest Communities," on pages 96 to 109 in this volume; another, similar information-gathering trip to Missouri's Pioneer Forest, made by Berry in December 2007, is described in his op-ed piece, "Better Ways to Manage, Study Robinson Forest," the *Lexington Herald-Leader* (December 28, 2007), which also appeared as "Sustainable Forestry: Lessons for Kentucky," Louisville *Courier-Journal* (December 28, 2007).

450.31 "I am that I am."] Exodus 3:14.

451.10–13 "The United States . . . rate of consumption."] Bill Wolfe, "Ky. Urged to Be Power Player," the Louisville *Courier-Journal* (October 21, 2006).

453.35–36 "all nature's treasury . . . new-found world . . ."] Christopher Marlowe, *The Tragical History of Doctor Faustus* (1604), I.i.77, 85–86. Citations of Marlowe's play correspond to those in *English Renaissance Drama: A Norton Anthology* (2002), edited by David Bevington et al.

454.3–5 "How comes . . . out of it." *Faustus*, I.iii.77.

454.6–8 Hell hath . . . ever be.] *Faustus*, II.i.122–124.

454.12–13 "Tut, Faustus . . . more of it."] *Faustus*, II.i.152–153.

454.22–23 "to answer . . . *within bounds*] Milton, *Paradise Lost*, VII.119–120.

454.25–29 Knowledge is . . . to wind.] Milton, *Paradise Lost*, VII.126–30.

457.19–20 "myself am Hell."] Milton, *Paradise Lost*, IV.75.

457.29 "Faustus, and a man."] *Faustus*, I.i.23.

From IMAGINATION IN PLACE

466.6–7 James Still, Harlan Hubbard, and Harry Caudill] James Still (1906–2001), of Knott County, Kentucky, a poet, novelist, and story writer who spent most of his writing life in a log house near the Dead Mare Branch of Little Carr Creek; Harlan Hubbard (1900–1988), of Trimble County, Kentucky, a painter and writer who with his wife, Anna, lived a life of subsistence in remote Payne

Hollow on the Ohio River (Berry's portrait of Hubbard, *Harlan Hubbard: Life and Work*, was published in 1990); Harry Caudill, see note 53.21–23.

466.15 old Cilician of *Georgics*] Virgil's aged gardener, who works a few acres of land on otherwise unproductive soil near Taranto in Apulia.

466.17–18 Ananda Coomaraswamy, Titus Burckhardt, Kathleen Raine, and Philip Sherrard] Ananda Coomaraswamy, see note 40.22–25; Titus Burckhardt (1908–1984), Swiss historian of sacred art, cultural anthropologist, and publisher; Kathleen Raine, see note 32.1–2; Philip Owen Arnould Sherrard (1922–1995), English poet, translator, theologian, and interpreter of Orthodox Christianity.

467.15 the Southern Agrarians] See note 324.29–30.

467.23–24 John Crowe Ransom] American poet, critic, one of the founders of the magazine *The Fugitive*, and a prominent member of the Southern Agrarians (1888–1974). His name remains closely associated with the New Critics, who argued for the autonomy of the literary work.

468.11 Allen Tate] American poet, critic, novelist, and biographer of Stonewall Jackson and Jefferson Davis who was a member of the Southern Agrarians (1899–1979). Like Ransom, Tate was a New Critic.

469.3–5 only Andrew Lytle and John Donald Wade . . . knowledge of actual farming] Novelist, essayist, and teacher Andrew Lytle (1902–1995) grew up on the family farm near Guntersville, Alabama; noted biographer, essayist, scholar, and editor John Donald Wade (1892–1963) lived in the family home in Macon County, Georgia, where farming remains an important part of the economy.

470.40 The Land Institute] See note 217.19–20.

471.11–12 Gene Logsdon, Maurice Telleen, Wes Jackson, and David Kline.] Gene Logsdon (1931–2016), American farmer and author of many books including the posthumously published *Letter to a Young Farmer: How to Live Richly Without Wealth on the New Garden Farm* (2017), with a foreword by Wendell Berry; Maurice Telleen (1928–2011), American farmer, writer, and founder of *The Draft Horse Journal*, to whom Berry dedicated *The Unsettling of America*; Wes Jackson, see note 15.21–23; David Kline (b. 1945), American author, farmer, editor of *Farming Magazine*, and a member of the Old Order Amish community of Fredericksburg, Ohio.

473.9–10 woman with six arms in the picture by . . . Emmeline Grangerford] Emmeline Grangerford, Buck Grangerford's deceased sister in *The Adventures of Huckleberry Finn* (1884), a juvenile artist whose crayon drawings include an unfinished "picture of a young woman in a long white gown, standing on the rail of a bridge all ready to jump off, with her hair all down her back, and looking up to the moon, with the tears running down her face, and she had two arms folded across her breast, and two arms stretched out in front, and two more reaching up towards the moon—and the idea was to see which pair would look best, and then scratch out all the other arms; but, as I was saying, she died before she got her mind made up, and now they kept this picture over the head of the bed in her room, and every time her birthday come they hung flowers on it."

474.1–4 "Realism . . . aspire and strive."] Kathleen Raine, "The Use of the Beautiful," published in her collection of essays *Defending Ancient Springs* (1967).

475.4–9 Parochialism and provincialism . . . his parish.] Patrick Kavanagh, "The Parish and the Universe," *Collected Pruse* (1967).

478.17–19 As Guy Davenport saw . . . replace it.] Throughout his writing career, American essayist and fiction writer Guy Davenport (1927–2005) was a fierce critic of money-valued society. Davenport began teaching at the University of Kentucky in 1964, the same year Berry joined the faculty as a creative writing instructor.

479.8–13 . . . we give . . . soonest winner.] *Henry V*, III.vi.108–113.

479.14–15 Edmund Burke's *Speech on American Taxation*] Delivered in the British House of Commons on April 19, 1774, in the debate on the repeal of tea-duty.

480.9 Mathew Brady's photographs] American photographer Mathew Brady (1823–1896) and his associates documented camp life, war preparations, and the grisly aftermath of battle during the Civil War.

480.22–25 Who's this? . . . only son!] *Henry VI*, Part III, II.v.61, 82–83.

480.27 "all the mad old joy"] Walt Whitman, "The Artilleryman's Vision," originally entitled "The Veteran's Vision," *Drum-Taps* (1865).

480.32–481.8 Curious I halt . . . he lies.] Walt Whitman, "A Sight in Camp in the Daybreak Gray and Dim," *Drum-Taps* (1865).

483.1–3 "not virtue . . . such organism . . ."] Edmund Wilson, *Patriotic Gore* (1962), Introduction.

483.18 "The Battle Hymn of the Republic."] Pro-Union, antislavery lyric written by American social reformer Julia Ward Howe (1819–1910) for the song "John Brown's Body," published in *The Atlantic* (1862).

484.1–6 Donald Worster has said . . . slave power . . ."] Comments made in the Chautauqua discussion at the Prairie Festival, The Land Institute, Salina, Kansas (1999).

485.7–9 imagination . . . definitions by Coleridge and Blake] See Coleridge's *Biographia Literaria* (1817), Chapter XIII, where he differentiates between fancy (the lower, associative faculty) and imagination (the higher, creative faculty, which shapes and unifies). For Blake, imagination is more than an artistic faculty: "Imagination is the real and eternal world of which this vegetable universe is but a faint shadow."

485.16–21 as if the earth . . . eat filth . . .] William Carlos Williams, "To Elsie," *Spring and All* (1923).

485.22–25 "like a chicken . . . to death . . ."] William Carlos Williams, *In the American Grain* (1925).

486.33–37 Robert Ulanowicz says . . . everywhere!"] Ulanowicz's comments were made in a public lecture, "Naturalism and/or Immanent Divine Action?," in Santa Barbara, California (January 26, 2006).

489.25–26 James Still's *River of the Earth*, written half a century later] Still's novel, set in the 1930s coalfields of eastern Kentucky, appeared in 1940.

489.29–30 Ernest J. Gaines inherited the harder . . . racial history] Gaines (b. 1933), an African American writer, grew up on the River Lake Plantation in Pointe Coupee Parish, Louisiana.

492.10–13 Jane Kenyon's poems . . . Donald Hall's . . . out of the same story] American poet Jane Kenyon (1947–1995), who grew up in the Midwest, moved with her husband, the American poet and essayist Donald Hall (1928–2018), her former teacher at the University of Michigan, to Eagle Pond Farm in Wilmot, New Hampshire, Hall's grandparents' former home. For Berry's account of his own return home to Henry County, Kentucky, see "The Long-Legged House" and "A Native Hill."

496.23 Bert Hornback] Bert G. Hornback (b. 1935), a member of the English faculty at the University of Michigan from 1964 to 1992.

498.26–28 "In art . . . saying nothing."] See the selection of Ludwig Wittgenstein's notebooks entitled *Culture and Value* [Vermischte Bemerkungen], published in German in 1977 and in an English translation by Peter Winch in 1980. Berry quotes from the Winch translation.

499.33 like Ruth of old] Moabite woman of the Old Testament and Hebrew Bible's Book of Ruth who after her husband's death moved with her mother-in-law, Naomi, to Judah, where she remarried and began a new life.

500.7–8 a disparagement . . . in the *Hudson Review*] Robert Phillips, "Poems, Mostly Personal, Some Historical, Many Unnecessary," *The Hudson Review* (Winter 1997).

503.5 "things fall apart"] W. B. Yeats, "The Second Coming," published in *The Dial* (November 1920) and first collected in *Michael Robartes and the Dancer* (1921).

504.7–10 Come live . . . mountains yields.] Christopher Marlowe, "The Passionate Shepherd to His Love" (1599).

504.12–13 The flowers . . . reckoning yields . . .] Sir Walter Raleigh, "The Nymph's Reply to the Shepherd" (1600).

504.15–16 "of being versed in country things"] Robert Frost, "The Need of Being Versed in Country Things," published in *Harper's Magazine* (December 1920) and first collected in *New Hampshire* (1923).

505.1–18 O, when degree . . . eat up himself.] *Troilus and Cressida*, I.iii.101–108, 116–124.

507.3 Meliboe in *The Faerie Queene*] An old shepherd. A pastoral interlude in Book VI of *The Faerie Queene* brings Sir Calidore to the house of old Meliboe and his foster daughter, Pastorella.

507.33 "pathetic fallacy,"] A term coined by English art critic John Ruskin (1819–1900), which he defined as the "extraordinary, or false appearances" assumed by objects "when under the influence of emotion, or contemplative fancy."

509.8–9 Spenser's Colin Clout or Hardy's Hodge] Colin Clout, a pastoral persona making appearances in several of Edmund Spenser's works, including *The Shepheardes Calendar* (1579), *Colin Clouts Come Home Againe* (1595), and *The Faerie Queene* (1590–96); Hodge, the subject of Thomas Hardy's poem "Drummer Hodge or the Dead Drummer" (1899), first collected in *Poems of Past and Present* (1901).

509.12 "A trewe swinkere and a good . . ."] Chaucer, *The Canterbury Tales*, Prologue, line 531.

511.25 "The least of these my brethren"] Matthew 25:40.

525.35 Christ's agony in Gethsemane] Described in Matthew 26:36–46.

525.38 story of the Prodigal Son] Recounted in Luke 15:11–32.

528.40 "satisfy the society's impulse to renewal."] J. A. Bryant, Jr., *Shakespeare and the Uses of Comedy* (1986).

533.23–31 A thousand tymes . . . it preve.] See the opening lines of the Prologue to Chaucer's *The Legend of Good Women*.

535.9–11 But will God . . . I have builded?] 1 Kings 8:27.

535.16–19 For we . . . wait for it.] Romans 8:24–25.

538.18–21 The *Iliad* . . . What we are.] Hayden Carruth, "The Impossible Indispensability of the Ars Poetica," published in *Toward the Distant Islands: New and Selected Poems* (2006).

From IT ALL TURNS ON AFFECTION

545.1 *It All Turns on Affection*] Delivered as the 41st Annual Jefferson Lecture of the National Endowment for the Humanities at the John F. Kennedy Center for the Performing Arts, Washington, D.C. (April 23, 2012).

545.18 The depression of the 1890s] One of the worst economic crises in American history, with unemployment exceeding 10 percent for five or six consecutive years. A declining economy combined with surplus production caused numerous foreclosures on mortgaged farms.

545.35 American Tobacco Company] Incorporated in 1890 after the takeover and merger of the nation's leading tobacco manufacturers, J. B. Duke's American Tobacco Company controlled four-fifths of the U.S. tobacco industry (excluding cigars) by 1906.

546.6–7 my teacher, Wallace Stegner] Berry was a Stegner Fellow in the Stanford Creative Writing Program in the school year 1958–59.

546.8–12 "boomers" and . . . made it in."] Wallace Stegner, *Where the Bluebird Sings to the Lemonade Springs: Living and Writing in the West* (1992), Introduction.

550.15–16 Land, as Wes Jackson has said . . . as exhaustible as oil or coal.] A common theme of Jackson in his and Berry's conversations.

551.17 "the land-community"] Aldo Leopold, *A Sand County Almanac and Sketches Here and There* (1949), Part 3.

556.1 "carried to the heart"] Allen Tate, "Ode to the Confederate Dead," published in *Mr. Pope and Other Poems* (1928) and reprinted in *Collected Poems, 1919–1976*.

556.39–557.2 "a statement . . . limitations of mankind."] John Lukacs, *Last Rites* (2009).

557.24–25 "the equall mother . . . unto brother."] *The Faerie Queene*, VII, canto vii, stanza 14.

557.27–29 "In short . . . citizen of it."] Leopold, *A Sand County Almanac*, Part 3.

558.14–16 "Imagination implied . . . a detail"] One of Wallace Stevens's aphorisms from his "Adagia" (1934–40), published in *Opus Posthumous* (1957).

563.17–18 a speech . . . "Revitalizing Rural Communities,"] Delivered by Frederick Kirschenmann at the forty-third annual meeting of the Northern Plains Conference of the United Church of Christ, Bismarck, North Dakota (June 10, 2006), reprinted in Kirschenmann's *Cultivating an Ecological Conscience: Essays from a Farmer Philosopher* (2010).

563.26–28 "The human mind . . . loves them . . ."] Keith Critchlow, *The Hidden Geometry of Flowers: Living Rhythms, Form, and Number* (2011).

566.2 Matt Rothschild] Senior editor at *The Progressive* from 1994 to 2014.

566.22 I committed . . . an act of civil disobedience] On the weekend of February 12, 2011, Berry and nineteen other protestors occupied Kentucky governor Steve Beshear's outer office to protest surface mining for coal in the Kentucky mountains.

568.5–7 Ken Kesey once said . . . *you* dependent on evil.] Kesey in conversation with Berry.

568.8–9 Edward Abbey once said . . . good hobby] According to Berry, a response by Abbey to an invitation to protest. Abbey makes a similar remark in *Postcards from Ed* (2006): "Saving the world is only a hobby. Most of the time I do nothing."

From OUR ONLY WORLD

574.1–2 "It is not good . . . be alone."] Genesis 2:18.

575.32–33 "God guard me . . . mind alone. . . ."] W. B. Yeats, "A Prayer for Old Age," published in *Parnell's Funeral and Other Poems* (1935).

575.34–35 "We only believe . . . whole body"] See Yeats's Introduction to *Certain Noble Plays of Japan* (1916), edited by Ernest Fenollosa and Ezra Pound.

576.4–12 "He saw . . . younger generations."] Philip Sherrard, *The Wound of Greece: Studies in Neo-Hellenism* (1978).

576.16–26 "You may dissociate . . . missing something."] John Crowe Ransom, "Poets Without Laurels," published in his collection of essays *The World's Body* (1938).

581.2 the day of the bombing in Boston] April 15, 2003. Two bombs were detonated on Boylston Street near the finish line of the Boston Marathon, killing three people and injuring 264 others.

582.34–35 Thomas L. Friedman announced in the *Times*] In an op-ed piece by Friedman entitled "Bring on the Next Marathon," April 17, 2003. All quotes attributed to Friedman come from this piece.

584.29–30 Liberty Hyde Bailey, J. Russell Smith, Hugh Hammond Bennett, Albert Howard, Aldo Leopold] Liberty Hyde Bailey, see note 323.12–13; Joseph Russell Smith, see note 323.12–13; Hugh Hammond Bennett (1881–1960), American soil conservationist and founder of the Soil Conservation Service (now the Natural Resources Conservation Service); Albert Howard, see note 112.17; Aldo Leopold (1887–1948), American conservationist, ecologist, forester, and writer best known for his book *A Sand County Almanac* (1949).

589.22 *Suffolk* horses] Heavy chestnut-colored breed of English draft horses developed specifically for the purpose of farmwork.

603.37–39 The Land Institute, Tilth, the Quivira Coalition, the Southern Agriculture Working Group, the Land Stewardship Project] The Land Institute, see note 217.19–20; the Tilth Association (now the Tilth Alliance), incorporated in 1977, a sustainable agricultural movement and organization in the Pacific Northwest inspired by Berry's speech at the "Agriculture for a Small Planet" symposium in Spokane, Washington, July 1, 1974; the Quivira Coalition, an organization dedicated to restoring ecological and social health to the working landscapes of the American West and finding a "middle way" between ranchers and environmentalists; the Southern Agriculture Working Group, operating across thirteen southern states, was formed in 1991 "to foster a movement towards a more sustainable farming and food system"; the Land Stewardship Project, founded in 1982, a Minnesota-based organization dedicated to fostering "an ethic of stewardship for farmland, to promote sustainable agriculture and to develop sustainable communities."

605.1–2 *Local Economies to Save the Land and the People*] Originally delivered as a speech for Kentuckians for the Commonwealth, Carrollton, Kentucky (August 16, 2013).

605.16 Harry Caudill] See note 53.21–23.

607.29–30 the Appalachian Group to Save the Land and the People.] An early grassroots effort to put an end to strip-mining in eastern Kentucky, organized in Hindman, Kentucky, in 1965, before Dan Gibson's standoff with strip miners at Clear Creek.

612.2–4 to borrow language from John Todd, must be "elegant solutions . . . place."] John Todd, "Tomorrow Is Our Permanent Address," published in *The Book of the New Alchemists*, edited by Nancy Jack Todd (1977).

617.10–15 Communism is not . . . *become* political.] Thomas Merton, "The Pasternak Affair," published in his collection of essays *Disputed Questions* (1960).

620.6–7 "How can we . . . than victory"] William E. Hull, *Beyond the Barriers* (1981).

625.22 24th and 104th Psalms as scriptural norms] Psalms 24:1: "The earth is the Lord's, and the fulness thereof; the world, and they that dwell therein." Psalms 104:31: "The glory of the Lord shall endure for ever: the Lord shall rejoice in his works."

626.1–3 Ken Kesey once saw . . . me one?"] According to Berry, Kesey "told me that in conversation just before or just after I told him of a warning I had seen on a condom machine in a men's room: 'Don't buy this gum. It tastes like rubber.'"

626.24–27 For mercy, pity, peace . . . and care.] William Blake, "The Divine Image," published in *Songs of Innocence* (1789).

628.12–20 Conjugal love . . . a quiet spirit.] See Volume II of *Either/Or* (1843), section on "Aesthetic Validity of Marriage," by Danish philosopher Søren Kierkegaard (1813–1855). Berry quotes from the 1958 revised English translation of *Enten-Eller* by Walter Lowrie.

629.15–18 "Woman, where . . . no more."] John 8:7, John 8:10–11.

632.1–13 Many voices . . . so afflicted . . . ?] Harry M. Caudill, *The Watches of the Night* (1979).

633.28–39 the spirit of independence . . . and poverty.] From *The History of the Highland Clearances* (1883) by Alexander Mackenzie (1838–1898), Scottish ploughman, editor, and historian.

634.12–14 "them that join . . . the earth!"] Isaiah 5:8.

634.20 England's Enclosure Acts] A series of acts passed by Parliament—many of them coming from 1750 to 1860—that empowered the "enclosure" of open fields and common land in England and Wales, stripping the rights of local people to land they had worked for generations. Evicted rural laborers and poor farmers often migrated to cities to find work.

634.20–21 The same dispossession of country people . . . Stalin's Russia] Under collectivization, carried out by the Soviet government from 1928 to 1940, peasant farmers were forced to give up their land and join larger collective farms.

634.21–22 taking place now in China.] Collectivization of agriculture in China began in 1952 and was complete by 1956; after Mao Zedong's death in 1976, farming communes began to be dismantled, as part of the Four Modernizations campaign. Berry refers to China's more recent efforts to convert agricultural land to industrial sites and the resulting displacement of farmers.

634.26 semi-official agricultural policy] Beginning in the 1950s, the Committee for Economic Development championed the idea that there were too many farmers. Eisenhower's secretary of agriculture, Ezra Taft Benson (1899–1994), admonished farmers "to get big or get out," which would be echoed in the 1970s by Secretary of Agriculture Earl Butz (1909–2008) under Nixon and Ford.

637.31–32 Wes Jackson's . . . "eyes-to-acres ratio"] A formulation by Jackson and often quoted by Berry to indicate the proper scale of farm management.

638.10–21 [T]he essential eye . . . saturated lea!] Robert B. Weeden, letter to Wendell Berry (January 23, 2013). Weeden (b. 1933) is an ecologist and writer who worked in Alaska.

640.31–32 farming fitted to the land, as J. Russell Smith said it should be.] J. Russell Smith, *Tree Crops: A Permanent Agriculture* (1929), Chapter XXIII. See also note 323.12–13.

643.21 Oliver Goldsmith's poem] Oliver Goldsmith (1728–1774), Irish poet, dramatist, and novelist who emigrated to England. "The Deserted Village" appeared in 1770 as the English countryside was being transformed by enclosure.

643.25 John Maynard Keynes] British economist (1883–1946) whose refutation of laissez-faire economics proved influential in the practices of Western democracies.

647.3–4 "that take . . . to use."] From an unpublished paper by Robert B. Weeden entitled "Small Forays into Big Spaces."

656.18 "spill" of 4-methylcyclohexane methanol] On January 9, 2014, 10,000 gallons of crude MCHM, a foaming agent used to wash coal, was released into West Virginia's Elk River.

656.19–23 a *New York Times* editorial noticed . . . "meaningful reform."] "Contaminated Water in West Virginia," *New York Times* (January 17, 2014).

658.21 "creative destruction,"] A term coined by Austrian political economist Joseph Schumpeter (1883–1950) to refer to the way in which "industrial mutation" in capitalist economies creates new ways of doing things even as it destroys the old ways. According to Berry, "It is often quoted still to justify any ecological or human cost of industrial development such as fracking."

659.23 "Luddite"] See note 55.25.

661.28–30 Albert Howard, Aldo Leopold . . . Wes Jackson] Albert Howard, see note 112.17; Aldo Leopold, see note 584.29–30; Wes Jackson, see note 15.21–23.

662.1–2 F. H. King's *Farmers of Forty Centuries*] Franklin Hiram King (1848–1911), American agricultural scientist and author of *Farmers of Forty Centuries: Or Permanent Agriculture in China, Korea and Japan* (1911), drawn from observations made during his extensive travels to Asia.

662.2–3 Vandana Shiva] Indian physicist, ecofeminist, and food sovereignty advocate (b. 1959). Her many books include *Biopiracy: The Plunder of Nature and Knowledge* (1997), *Stolen Harvest: The Hijacking of the Global Food Supply* (1999), and *Earth Democracy: Justice, Sustainability, and Peace* (2005).

662.7–8 "The small landholders . . . a state."] Thomas Jefferson, letter to James Madison, October 28, 1785.

663.10 the 50-Year Farm Bill] Proposed by Wes Jackson and The Land Institute, fruitlessly advocated by Jackson and Berry in a *New York Times* op-ed piece of January 2, 2009. Later that year, accompanied by Fred Kirschenmann of Iowa State University's Leopold Center for Sustainable Agriculture, the two men traveled to Washington, D.C., to lobby for the proposed bill.

665.19–20 "Take . . . no thought for the morrow . . ."] Matthew 6:34.

670.26–27 "The kingdom of heaven is at hand"] Matthew 4:17.

From THE ART OF LOADING BRUSH

691.34–35 Agricultural Adjustment Act] The nation's first omnibus farm bill, signed into law on February 16, 1938.

707.13–14 an essay of 1945, "The Outlook for Farm Wildlife,"] Contributed by Aldo Leopold for the 10th North American Wildlife Conference, scheduled to be held in New York, February 26–28, 1945, but cancelled that year due to government bans on all conventions.

709.8–10 "Of all the pantheon . . . hardest to kill."] C. S. Lewis, *Studies in Words* (1960), Chapter 2.

713.16–17 fear and trembling] Cf. Psalms 55:5 and Philippians 2:12.

714.16–17 "this noble goddesse Nature"] Chaucer, *The Parliament of Fowls*, line 303.

714.17 "great dame Nature"] Edmund Spenser, *The Faerie Queene*, VII, canto vii, stanza 5.

714.19 Alanus De Insulis, or Alan of Lille] French theologian and poet (1128–1202/3). All quotations of Alan of Lille are from his allegorical prose-and-verse work *De Planctu Naturae* (1168–70), translation and commentary by James J. Sheridan: *The Plaint of Nature* (1980).

716.24 self-love as in Matthew 22:39] "Thou shalt love thy neighbour as thyself."

716.36 Genesis 1:31] "And God saw every thing that he had made, and, behold, it was very good. And the evening and the morning were the sixth day."

718.3–4 "the decorations . . . note of delight."] C. S. Lewis, *The Allegory of Love* (1936), Chapter 2.

718.38–719.5 Sheridan says . . . increase and multiply . . ."] James J. Sheridan, in his commentary on Alan's *The Plaint of Nature* (1980).

719.31–34 "Does it never . . . they *knew* . . ."] Thomas Carlyle, *Past and Present* (1843), Book II, Chapter 2.

719.36 Maurice Telleen] See note 471.11–12.

723.17–18 once with acknowledgement of . . . "the Pleynt of Kynde."] Chaucer, *The Parliament of Fowls*, line 316.

724.14–15 "The byldere ok . . . piler elm,"] *The Parliament of Fowls*, lines 176–177.

724.20 Ronsard] Pierre de Ronsard, French poet of noble birth (1524–1585). Berry published a partial translation of Ronsard's Elegy XXIV (1584) in *The Hudson Review* (Spring 1979), under the title "Ronsard's Lament for the Cutting of the Forest of Gastine."

725.4–5 I wrote the introduction for a new printing of . . . Stanwell-Fletcher's *Driftwood Valley*] In his Introduction (1989), Berry says that as a teenage boy it was "the only book I read for a year or two, the end serving only to permit a new return to the beginning." Originally published in 1946, *Driftwood Valley* recounts the life and work of Theodora Stanwell-Fletcher (1906–2000), an American naturalist, in remote northern British Columbia before World War II.

725.8–10 "to Mrs. Stanwell-Fletcher's . . . feelings."] From Berry's Introduction to *Driftwood Valley*.

726.20 old Corycian farmer] See note 466.15.

726.21–22 "cottage economy" . . . advocated by William Cobbett] William Cobbett (1763–1835), an English journalist, farmer, and political reformer, championed traditional rural English practices versus many changes brought by industrialization. His book *Cottage Economy* (1821), first published as a series of pamphlets, is a practical guide to economic self-sufficiency, offering instructions on "the brewing of beer, making of bread, keeping of cows, pigs, bees, ewes, goats, poultry, and rabbits, and . . . [on] other matters deemed useful in the conducting of the affairs of a labourer's family."

727.4–5 "Kek kek! . . . quek quek!"] Chaucer, *The Parliament of Fowls*, line 499.

727.8–9 "There are more stars . . . by light."] *The Parliament of Fowls*, lines 595, 599.

727.12–13 "such array"] *The Parliament of Fowls*, line 318.

727.16 "than any creature"] *The Parliament of Fowls*, line 301.

727.22 "wrought" according to her design.] *The Parliament of Fowls*, line 305.

727.27–33 Come, my Corinna . . . neatly enterwove . . .] See "Corinna's Going A-Maying" by English poet and cleric Robert Herrick (1591–1674), published in his magnum opus, *Hesperides* (1648).

727.34 "noble emperesse . . . grace"] *The Parliament of Fowls*, line 319.

728.3–5 the vicaire . . . of accord . . .] *The Parliament of Fowls*, lines 379–381.

728.40–729.4 "vicar general," in charge . . . "counterfeited."] See the first two stanzas of Chaucer's "The Physician's Tale."

729.6–7 "was and is . . . up to Nature . . ."] *Hamlet*, III.ii.21–22.

729.33–34 "the highest him . . . of Nature."] Spenser, *The Faerie Queene*, VII, canto vi, stanza 35.

729.37–38 "well disposed" . . . "Natures Sergeant."] *The Faerie Queene*, VII, canto vii, stanza 4.

730.4–34 Then forth issewed . . . all beheld . . ."] *The Faerie Queene*, VII, canto vii, stanzas 5, 6, 7, and 13.

731.16 "through a glass darkly."] 1 Corinthians 13:12.

731.23 Francis Bacon] Early English empiricist philosopher (1561–1626).

731.29–30 old under the sun] Cf. Ecclesiastes 1:9.

731.34–39 Who Right . . . vnto brother.] *The Faerie Queene*, VII, canto vii, stanza 14.

732.31–39 I well consider . . . states maintaine.] *The Faerie Queene*, VII, canto vii, stanza 3.

733.1–2 C. S. Lewis speaks . . . those lines.] See Lewis's *The Allegory of Love*, Chapter 7, in which he discusses Spenser's poem.

733.4–11 For, all that . . . strange disguise . . .] *The Faerie Queene*, VII, canto vii, stanza 18.

735.1–2 Langland . . . read *Plaint of Nature.* C.S. Lewis assumed he had.] William Langland, presumed author of *Piers Plowman* (c. 1332–c. 1386). Lewis makes this observation in *The Allegory of Love*, in a footnote in his chapter on Chaucer.

735.13–14 Of the flowers . . . the causes] Langland, *The Vision of Piers the Plowman: A Critical Edition of the B-Text*, edited by A.V.C. Schmidt (1978), passus XI, lines 320–366.

735.15 "putteth it in hir ere,"] Langland, *The Vision of Piers the Plowman: A Critical Edition of the B-Text*, edited by A.V.C. Schmidt (1978), passus XII, lines 218–228.

735.40–736.2 "a stately palace . . . all dainties."] John Milton, *Comus*, stage direction after line 658.

736.8–9 "The grave's . . . there embrace."] Andrew Marvell, "To His Coy Mistress," lines 31–32, written from 1650 to 1652 but not published until 1681, three years after the poet's death.

736.13 "the sage . . . Virginity."] Milton, *Comus*, lines 786–787.

736.17–27 Wherefore did Nature . . . yet despised . . .] *Comus*, lines 710–14, 720–724.

736.38–737.1 "something holy lodges in that breast."] *Comus*, line 246.

737.4–12 Means her provision . . . even proportion . . .] *Comus*, lines 765–773.

738.19–21 "huge heaps . . . a breeze."] Alexander Pope, "Epistle to Burlington" (1731), lines 109, 108. See also note 323.17.

738.21–22 "Good Sense . . . Heaven."] Pope, "Epistle to Burlington," line 43.

738.26–27 "In all, let Nature . . . Place in all."] "Epistle to Burlington," lines 50 and 57.

738.36 "lavish cost, and little skill,"] "Epistle to Burlington," line 167.

739.10–11 'Tis Use alone . . . from Sense.] "Epistle to Burlington," lines 179–180.

739.37–740.2 "the din . . . the world."] William Wordsworth, "Lines Composed a Few Miles Above Tintern Abbey," lines 25–26, 39–40, 52–53, the concluding poem in Wordsworth and Coleridge's *Lyrical Ballads* (1798).

740.13–15 with an eye . . . of things.] Wordsworth, "Tintern Abbey," lines 47–49.

740.18–21 A presence . . . setting suns . . .] "Tintern Abbey," lines 94–97.

740.28–31 In nature . . . moral being] "Tintern Abbey," lines 108–111.

741.3–6 Nature never . . . joy to joy . . .] "Tintern Abbey," lines 122–125.

741.8 "The dreary intercourse of daily life."] "Tintern Abbey," line 131.

741.12–13 "maketh his sun . . . the unjust."] Matthew 5:45.

741.15–16 "worshipper of Nature."] "Tintern Abbey," line 152.

742.5–7 neither gay . . . eager industry.] Wordsworth, "Michael," lines 120–122, first published in the 1800 edition of *Lyrical Ballads.*

742.13–15 "was keen . . . all winds."] "Michael," lines 44–48.

742.18–21 "Something between . . . age of ten.] "Michael," lines 189, 194–198.

742.25–26 "in surety . . . brother's son,"] "Michael," line 211.

742.30–31 "A prosperous man . . . in trade."] "Michael," lines 249–250.

742.34–37 He quickly . . . be gained?] "Michael," lines 252–255.

743.2 "loving letters"] "Michael," line 433.

743.4–9 Luke began . . . the seas.] "Michael," lines 442–447.

743.15 "into a stranger's hand."] "Michael," line 475.

745.10–13 departed, caring little . . . impoverished time.] Hayden Carruth, "Marshall Washer," section 2, published in *Brothers, I Loved You All* (1978) and reprinted in *Collected Shorter Poems, 1946–1991* (1992).

745.20–23 farming is . . . knowledge useless.] Hayden Carruth, "Marshall Washer," section 6.

746.31 a Carter family song, "The Homestead on the Farm"] Written by Herbert S. Lambert and Harry J. Lincoln, "The Homestead on the Farm" was recorded by the Carter Family on November 22, 1929, and released as a single by Victor Records in February 1930.

747.17–30 With Usury . . . Nature's increase.] Ezra Pound, *The Cantos,* Canto LI.

748.1–3 Dress 'em . . . the homestead.] Pound, Canto XCIX.

748.4–5 "Chao-Kong . . . his labour"] Pound, Canto LIII.

748.8–10 Pull down . . . true artistry . . .] Pound, Canto LXXXI.

748.36 Dogen] Dōgen Zenji (1200–1253), Japanese Buddhist monk who founded Soto Zen in Japan, based on the practice of sitting meditation.

749.3–13 Clear running stream . . . in me.] Gary Snyder, "Running Water Music II," published in *The Last Wildlands* (1978) and reprinted in *No Nature: New and Selected Poems* (1992). Snyder (b. 1930) is an American poet of the Pacific Rim, essayist, and environmental activist, whose many works include the 1974 Pulitzer Prize–winning poetry collection *Turtle Island*. *Distant Neighbors* (2015), edited by Chad Wriglesworth, is a selection of Berry and Snyder's correspondence.

751.25–27 "Their achievement . . . on purpose."] Jacques Maritain, *Art and Scholasticism with Other Essays* (1947), translated by J. F. Scanlan.

753.1 Gene Logsdon] See note 471.11–12.

753.5–8 "Four-fifths . . . into clothes."] G. C. Coulton, *Medieval Panorama: The English Scene from the Conquest to the Reformation* (1938), Chapter IX.

753.16–20 "Froissart shows . . . in arms."] G. C. Coulton, *Medieval Panorama*, Chapter XX.

754.2 "creative destruction."] See note 658.21.

757.39–40 an article, by a "semi-retired" rancher] Walt Davis, "Consider the Cow, and Her Purpose on the Ranch," *Beef Producer* (October 2015).

758.25–26 Henry Besuden of Vinewood Farm] See Berry's essay "A Talent for Necessity," where he tells Besuden's story at greater length. Several of Besuden's quoted remarks come from that essay.

760.9 F. H. King] See note 662.1–2.

760.21–22 "cargoes of . . . mineral fertilizers."] See the Introduction to F. H. King's *Farmers of Forty Centuries* (1911).

760.27–30 "the people . . . mountain land."] King, *Farmers of Forty Centuries*, Introduction.

760.33–35 "through our modern . . . their fields."] *Farmers of Forty Centuries*, Introduction.

761.14–17 "Nothing," King wrote . . . making purchases."] *Farmers of Forty Centuries*, Chapter X.

761.27–29 "Forest—field . . . plow agricultures . . ."] J. Russell Smith's *Tree Crops: A Permanent Agriculture* (1929), Chapter I.

761.31–34 The slope . . . gutted countryside.] *Tree Crops*, Chapter I.

761.37–762.3 "Man has . . . cereals are weaklings."] *Tree Crops*, Chapter II.

762.11 "two-story agriculture,"] *Tree Crops*, Chapter II.

762.20–22 "I see a million . . . the hills."] *Tree Crops*, Chapter XXIII.

762.39–763.1 "*farming should fit . . . point of view.*"] *Tree Crops*, Chapter XXIII.

763.21–23 "What are . . . woods and forests."] Albert Howard, *An Agricultural Testament* (1940), Introduction.

763.28–36 Since the Industrial Revolution . . . of erosion.] Howard, *An Agricultural Testament*, Preface.

764.22–32 The main characteristic . . . against disease.] *An Agricultural Testament*, Introduction.

764.35–37 "work and experience . . . Great Britain."] *An Agricultural Testament*, Preface.

764.39–765.1 "of high reputation locally,"] See *Sir Albert Howard in India* (1953) by Louise E. Howard (1880–1969), a classics scholar and second wife of Albert Howard.

765.20–31 It needs . . . dead and decayed.] Albert Howard, *The Soil and Health* (1947), Chapter 2.

766.26–28 "The using up . . . pure and simple."] *The Soil and Health*, Chapter 15.

767.1–2 "instinctive awareness . . . natural principle"] L. E. Howard, *Sir Albert Howard in India.*

767.7–9 "one great subject . . . and man."] *The Soil and Health*, Introduction.

767.11–15 "revolt against . . . the product."] L. E. Howard, *Sir Albert Howard in India.*

767.22–28 Health is the capacity . . . tends otherwise.] Aldo Leopold, *A Sand County Almanac* (1949), Part IV.

769.21 The Land Institute] See note 217.19–20.

772.20–21 "Elegant solutions . . . uniqueness of place."] See note 612.2–4.

Index

This book is set in 10 point ITC Galliard, a face designed for digital composition by Matthew Carter and based on the sixteenth-century face Granjon. The paper is acid-free lightweight opaque that will not turn yellow or brittle with age. The binding is sewn, which allows the book to open easily and lie flat. The binding board is covered in Brillianta, a woven rayon cloth made by Van Heek–Scholco Textielfabrieken, Holland. Composition by Dedicated Book Services. Printing and binding by LSC Communications. Designed by Bruce Campbell.

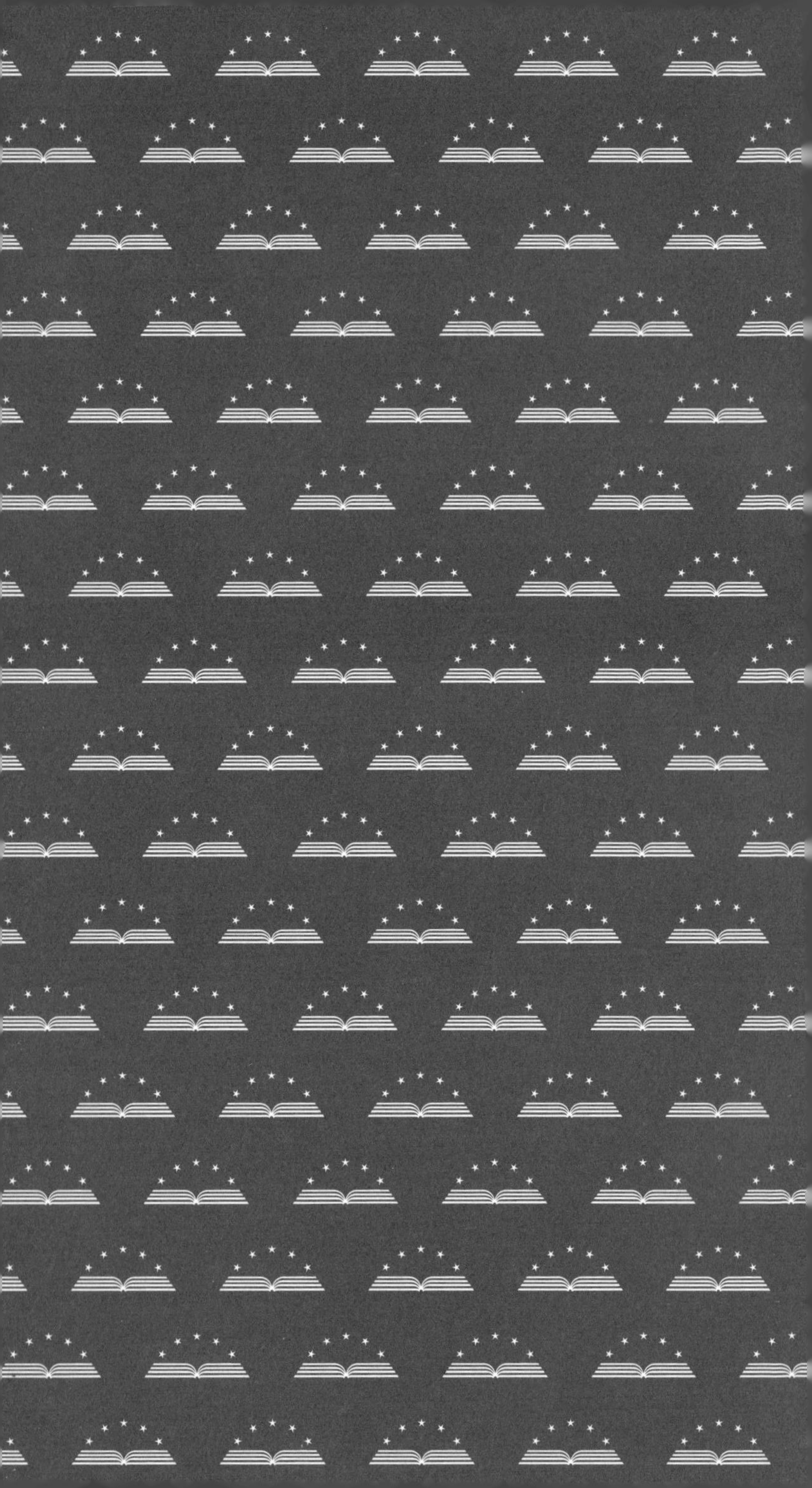

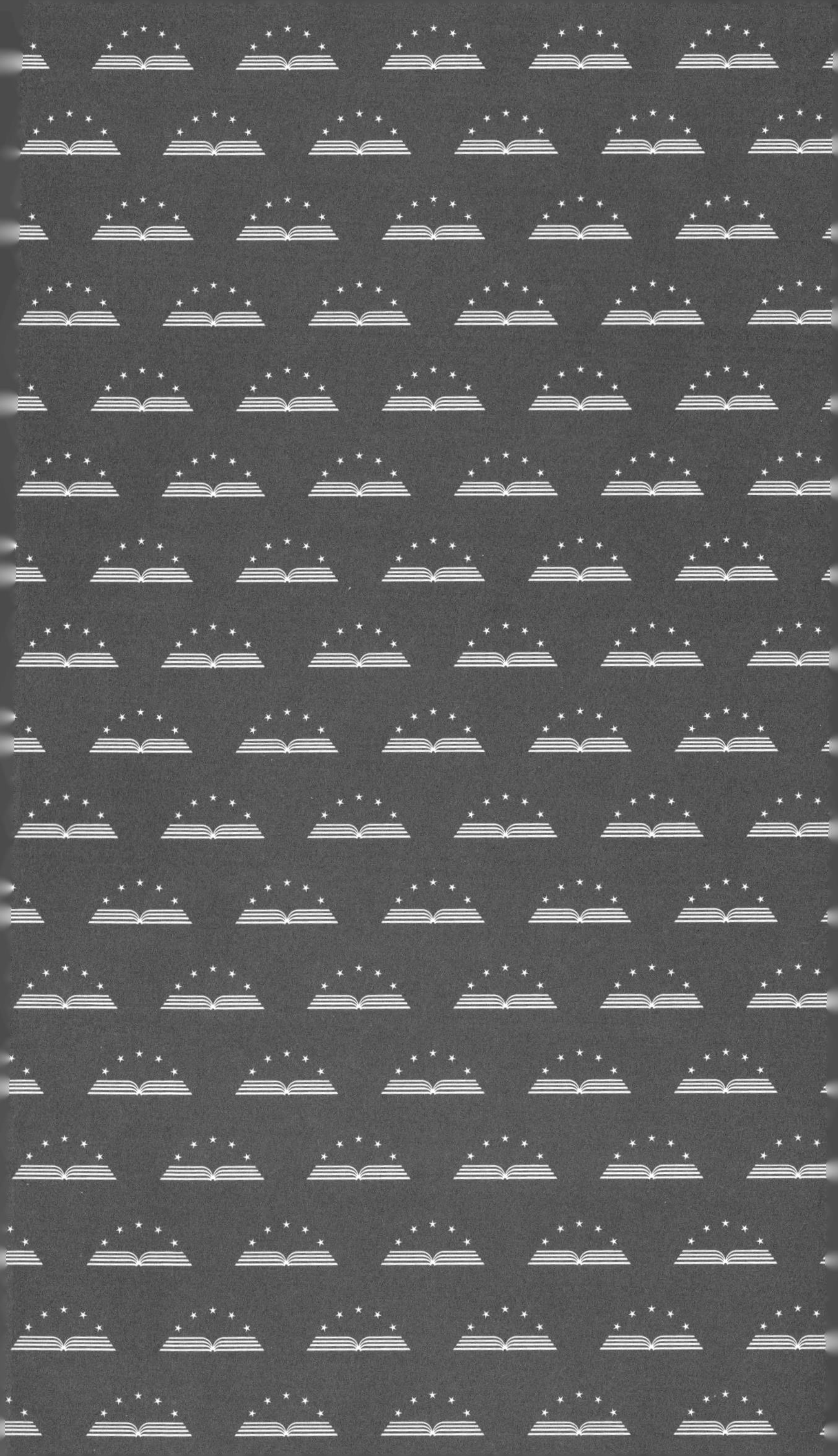

WENDELL BERRY

Wendell Berry

ESSAYS 1969–1990

INCLUDING

The Unsettling of America

AND SELECTIONS FROM

The Long-Legged House

The Hidden Wound

A Continuous Harmony

Recollected Essays

The Gift of Good Land

Standing by Words

Home Economics

What Are People For?

Jack Shoemaker, *editor*

THE LIBRARY OF AMERICA

Published in the United States by Library of America.
Visit our website at www.loa.org.

All texts reprinted by arrangement with Counterpoint Press.

This paper exceeds the requirements of
ANSI/NISO Z39.48–1992 (Permanence of Paper).

Distributed to the trade in the United States
by Penguin Random House Inc.
and in Canada by Penguin Random House Canada Ltd.

Library of Congress Control Number: 2018952818
ISBN 978-1-59853-606-5

First Printing
The Library of America—316

Manufactured in the United States of America

Wendell Berry: Essays 1969–1990
is kept in print with a gift from

WALTER E. ROBB
with the deepest appreciation and gratitude
for the words, life, and inspiration of Wendell Berry

to the Guardians of American Letters Fund,
established by Library of America
to ensure that every volume in the series
will be permanently available.

Wendell Berry: Essays 1969–1990
is published with support from

THE GOULD FAMILY FOUNDATION

WALTER E. ROBB

To Wes Jackson and David Kline

and to the memory of Gene Logsdon and Maury Telleen

Contents

FROM
THE LONG-LEGGED HOUSE
(1969)

The Rise

WE PUT THE canoe in about six miles up the Kentucky River from my house. There, at the mouth of Drennon Creek, is a little colony of summer camps. We knew we could get down to the water there with some ease. And it proved easier than we expected. The river was up maybe twenty feet, and we found a path slating down the grassy slope in front of one of the cabins. It went right into the water, as perfect for launching the canoe and getting in as if it had been worn there by canoeists.

To me that, more than anything else, is the excitement of a rise: the unexpectedness, always, of the change it makes. What was difficult becomes easy. What was easy becomes difficult. By water, what was distant becomes near. By land, what was near becomes distant. At the water line, when a rise is on, the world is changing. There is an irresistible sense of adventure in the difference. Once the river is out of its banks, a vertical few inches of rise may widen the surface by many feet over the bottomland. A sizable lagoon will appear in the middle of a cornfield. A drain in a pasture will become a canal. Stands of beech and oak will take on the look of a cypress swamp. There is something Venetian about it. There is a strange excitement in going in a boat where one would ordinarily go on foot—or where, ordinarily, birds would be flying. And so the first excitement of our trip was that little path; where it might go in a time of low water was unimaginable. Now it went down to the river.

Because of the offset in the shore at the creek mouth, there was a large eddy turning in the river where we put in, and we began our drift downstream by drifting upstream. We went up inside the row of shore trees, whose tops now waved in the current, until we found an opening among the branches, and then turned out along the channel. The current took us. We were still settling ourselves as if in preparation, but our starting place was already diminishing behind us.

There is something ominously like life in that. One would always like to settle oneself, get braced, say "Now I am going to begin"—and then begin. But as the necessary quiet seems about to descend, a hand is felt at one's back, shoving. And that is the way with the river when a current is running: once the connection with the shore is broken, the journey has begun.

We were, of course, already at work with the paddles. But we were ahead of ourselves. I think that no matter how deliberately one moved from the shore into the sudden fluid violence of a river on the rise, there would be bound to be several uneasy minutes of transition. It is another world, which means that one's senses and reflexes must begin to live another kind of life. Sounds and movements that from the standpoint of the shore might have come to seem even familiar now make a new urgent demand on the attention. There is everything to get used to, from a wholly new perspective. And from the outset one has the currents to deal with.

It is easy to think, before one has ever tried it, that nothing could be easier than to drift down the river in a canoe on a strong current. That is because when one thinks of a river one is apt to think of *one* thing—a great singular flowing that one puts one's boat into and lets go. But it is not like that at all, not after the water is up and the current swift. It is not one current, but a braiding together of several, some going at different speeds, some even in different directions. Of course, one *could* just let go, let the boat be taken into the continuous mat of drift—leaves, logs, whole trees, cornstalks, cans, bottles, and such—in the channel, and turn and twist in the eddies there. But one does not have to do that long in order to sense the helplessness of a light canoe when it is sideways to the current. It is out of control then, and endangered. Stuck in the mat of drift, it can't be maneuvered. It would turn over easily; one senses that by a sort of ache in the nerves, the way bad footing is sensed. And so we stayed busy, keeping the canoe between the line of half-submerged shore trees and the line of drift that marked the channel. We weren't trying to hurry—the currents were carrying us as fast as we wanted to go—but it took considerable labor just to keep straight. It was like riding a spirited horse not fully bridle-wise: We kept our direction *by intention*; there could be no dependence on habit or inertia; when our minds wandered the river took over and turned us according to inclinations of its own. It bore us like a consciousness, acutely wakeful, filling perfectly the lapses in our own.

But we did grow used to it, and accepted our being on it as one of the probabilities, and began to take the mechanics of it for granted. The necessary sixth sense had come to us, and we began to notice more than we had to.

There is an exhilaration in being *accustomed* to a boat on dangerous water. It is as though into one's consciousness of the dark violence of the depths at one's feet there rises the idea of the boat, the buoyancy of it. It is always with a sort of triumph that the boat is realized—that it goes *on top of the water*, between breathing and drowning. It is an ancient-feeling triumph; it must have been one of the first ecstasies. The analogy of riding a spirited horse is fairly satisfactory; it is mastery over something resistant—a buoyancy that is not natural and inert like that of a log, but desired and vital and to one's credit. Once the boat has fully entered the consciousness it becomes an intimate extension of the self; one feels as competently amphibious as a duck, whose feet are paddles. And once we felt accustomed and secure in the boat, the day and the river began to come clear to us.

It was a gray, cold Sunday in the middle of December. In the woods on the north slopes above us we could see the black trunks and branches just faintly traced with snow, which gave them a silvery, delicate look—the look of impossibly fine handwork that nature sometimes has. And they looked cold. The wind was coming straight up the river into our faces. But we were dressed warmly, and the wind didn't matter much, at least not yet. The force that mattered, that surrounded us, and inundated us with its sounds, and pulled at or shook or carried everything around us, was the river.

To one standing on the bank, floodwater will seem to be flowing at a terrific rate. People who are not used to it will commonly believe it is going three or four times as fast as it really is. It is so all of a piece, and so continuous. To one drifting along in a boat this exaggerated impression of speed does not occur; one is going the same speed as the river then and is not fooled. In the Kentucky when the water is high a current of four or five miles an hour is about usual, I would say, and there are times in a canoe that make that seem plenty fast.

What the canoeist gets, instead of an impression of the river's speed, is an impression of its power. Or, more exactly, an impression of the *voluminousness* of its power. The sense of the volume alone has come to me when, swimming in the summertime, I have submerged mouth and nose so that the plane of the water spread away from the lower eyelid; the awareness of its bigness that comes then is almost intolerable; one feels how falsely assuring it is to look down on the river, as we usually do. The sense of the power of it came to me one day in my boyhood when I attempted to swim ashore in a swift current, pulling an overturned rowboat. To check the downstream course of the boat I tried grabbing hold of the partly submerged willows along the shore with my free hand, and was repeatedly pulled under as the willows bent, and then torn loose. My arms stretched between the boat and the willow branch might have been sewing threads for all the holding they could do against that current. It was the first time I realized that there could be circumstances in which my life would count for nothing, absolutely nothing—and I have never needed to learn that again.

Sitting in a canoe, riding the back of the flooding river as it flows down into a bend, and turns, the currents racing and crashing among the trees along the outside shore, and flows on, one senses the volume and the power all together. The sophistications of our age do not mitigate the impression. To some degree it remains unimaginable, as is suggested by the memory's recurrent failure to hold on to it. It can never be remembered as wild as it is, and so each new experience of it bears some of the shock of surprise. It would take the mind of a god to watch it as it changes and not be surprised.

These long views that one gets coming down it show it to move majestically. It is stately. It has something of the stylized grandeur and awesomeness of royalty in a Sophoclean tragedy. But as one watches, there emanates from it, too, an insinuation of darkness, implacability, horror. And the nearer look tends to confirm this. Contained and borne in the singular large movement are hundreds of smaller ones: eddies and whirlpools, turnings this way and that, cross-currents rushing out from the shores into the channel. One must simplify it in order to speak of it. One probably simplifies it in some way in order to look at it.

There is something deeply horrifying about it, roused. Not, I think, because it is inhuman, alien to us; some of us at least must feel a kinship with it, or we would not loiter around it for pleasure. The horror must come from our sense that, so long as it remains what it is, it is not subject. To say that it is indifferent would be wrong. That would imply a malevolence, as if it could be aware of us if only it wanted to. It is more remote from our concerns than indifference. It is serenely and silently not subject—to us or to anything else except the other natural forces that are also beyond our control. And it is apt to stand for and represent to us all in nature and in the universe that is not subject. That is its horror. We can make use of it. We can ride on its back in boats. But it won't stop to let us get on and off. It is not a passenger train. And if we make a mistake, or risk ourselves too far to it, why then it will suffer a little wrinkle on its surface, and go on as before.

That horror is never fully revealed, but only sensed piecemeal in events, all different, all shaking, yet all together falling short of the full revelation. The next will be as unexpected as the last.

A man I knew in my boyhood capsized his motorboat several miles upriver from here. It was winter. The river was high and swift. It was already nightfall. The river carried him a long way before he drowned. Farmers sitting in their houses in the bottoms heard his cries passing down in the darkness, and failed to know what to make of them. It is hard to imagine what they could have done if they had known.

I can't believe that anyone who has heard that story will ever forget it. Over the years it has been as immediate to me as if I had seen it all—almost as if I had *known* it all: the capsized man aching and then numb in the cold water, clinging to some drift log in the channel, and calling, seeing the house lights appear far off across the bottoms and dwindle behind him, the awful power of the flood and his hopelessness in it finally dawning on him—it is amazingly real; it is happening to him. And the families in their lighted warm kitchens, eating supper maybe, when the tiny desperate outcry comes to them out of the darkness, and they look up at the window, and then at each other.

"Shh! Listen! What was that?"

"By God, it sounded like somebody hollering out there on the river."

"But it *can't* be."

But it makes them uneasy. Whether or not there *is* somebody out there, the possibility that there *may* be reminds them of their lot; they never know what may be going by them in the darkness. And they think of the river, so dark and cold.

The history of these marginal places is in part the history of drownings—of fishermen, swimmers, men fallen from boats. And there is the talk, the memory, the inescapable *feeling* of dragging for the bodies—that terrible fishing for dead men lost deep in the currents, carried downstream sometimes for miles.

Common to river mentality, too, are the imaginings: step-offs, undertows, divers tangled in sunken treetops, fishermen hooked on their own lines.

And yet it fascinates. Sometimes it draws the most fearful to it. Men must test themselves against it. Its mystery must be forever tampered with. There is a story told here of a strong big boy who tried unsuccessfully to cross the river by walking on the bottom, carrying an iron kettle over his head for a diving bell. And another story tells of a young man who, instead of walking under it, thought he would walk *on* it, with the help of a gallon jug tied to each foot. The miracle failing, of course, the jugs held his feet up, and his head under, until somebody obliged him by pulling him out. His pride, like Icarus', was transformed neatly into his fall—the work of a river god surely, *hybris* being as dangerous in Henry County as anywhere else.

To sense fully the power and the mystery of it, the eye must be close to it, near to level with the surface. I think that is the revelation of George Caleb Bingham's painting of trappers on the Missouri. The painter's eye, there, is very near the water, and so he sees the river as the trappers see it from their dugout —all the space coming down to that vast level. One feels the force, the aliveness, of the water under the boat, close under the feet of the men. And there they are, isolated in the midst of it, with their box of cargo and their pet fox—men and boat and box and animal all so strangely and poignantly coherent on the wild plain of the water, a sort of island.

But impressive as the sights may be, the river's wildness is most awesomely announced to the ear. Along the channel, the area of the most concentrated and the freest energy, there is silence. It is at the shore line, where obstructions are, that the currents find their voices. The water divides around the trunks of the trees, and sucks and slurs as it closes together again.

Trunks and branches are ridden down to the surface, or suddenly caught by the rising water, and the current pours over them in a waterfall. And the weaker trees throb and vibrate in the flow, their naked branches clashing and rattling. It is a storm of sound, changing as the shores change, increasing and diminishing, but never ceasing. And between these two storm lines of commotion there is that silence of the middle, as though the quiet of the deep flowing rises into the air. Once it is recognized, listened to, that silence has the force of a voice.

After we had come down a mile or two we passed the house of a fisherman. His children were standing on top of the bank, high at that place, waiting for him to come in off the river. And on down we met the fisherman himself, working his way home among the nets he had placed in the quieter water inside the rows of shore trees. We spoke and passed, and were soon out of sight of each other. But seeing him there changed the aspect of the river for us, as meeting an Arab on a camel might change the aspect of the desert. Problematic and strange as it seemed to us, here was a man who made a daily thing of it, and went to it as another man would go to an office. That race of violent water, which would hang flowing among the treetops only three or four days, had become familiar country to him, and he sunk his nets in it with more assurance than men sink wells in the earth. And so the flood bore a pattern of his making, and he went his set way on it.

And he was not the only creature who had made an unexpected familiarity with the risen water. Where the drift had matted in the shore eddies, or caught against trees in the current, the cardinals and chickadees and titmice foraged as confidently as on dry land. The rise was an opportunity for them, turning up edibles they would have found with more difficulty otherwise. The cardinals were more irresistibly brilliant than ever, kindling in the black-wet drift in the cold wind. The sight of them warmed us.

The Kentucky is a river of steep high banks, nearly everywhere thickly grown with willows and water maples and elms and sycamores. Boating on it in the summer, one is enclosed in a river-world, moving as though deep inside the country. One sees only the river, the high walls of foliage along the banks, the hilltops that rise over the trees first on one side and then the other. And that is one of the delights of this river. But one of the delights of being out on a winter rise is in seeing the country, and in seeing it from a vantage point that one does not usually see it from. The rise, that Sunday, had lifted us to the bank tops and higher, and through the naked trees we could look out across the bottoms. It was maybe like boating on a canal in Holland, though we had never done that. We could see the picked cornfields, their blanched yellow seeming even on that cloudy day to give off a light. We could see the winter grain spiking green over the summer's tobacco patches, the thickly wooded hollows and slews, the backs of houses and farm buildings usually seen only by the people who live there.

Once, before the man-made floods of modern times, and before the automobile, all the river country turned toward the river. In those days our trip would probably have had more witnesses than it did. We might have been waved to from house windows, and from barn doors. But now the country has turned toward the roads, and we had what has come to be the back view of it. We went by mostly in secret. Only one of the fine old river houses is left on this side of the river in the six miles of our trip, and it is abandoned and weathering out; the floods have been in it too many times in the last thirty-five years, and it is too hard to get back to from the road. We went by its blank windows as the last settlers going west passed the hollow eyes of the skulls of their predecessors' oxen.

The living houses are all out along the edges of the valley floor, where the roads are. And now that all the crops had been gathered out of the bottoms, men's attention had mostly turned away. The land along the river had taken on a wildness that in the summer it would not have. We saw a pair of red-tailed hawks circling low and unafraid, more surprised to see us than we were to see them.

Where the river was over the banks a stretch of comparatively quiet water lay between the trees on the bank top and the new shore line. After a while, weary of the currents, we turned into one of these. As we made our way past the treetops and approached the shore we flushed a bobwhite out of a brush pile near the water and saw it fly off downstream. It seemed strange to see only one. But we didn't have to wait long for an explanation, for presently we saw the dogs, and then the hunters coming over the horizon with their guns. We knew where their bird had gone, but we didn't wait to tell them.

These men come out from the cities now that the hunting season is open. They walk in these foreign places, unknown to them for most of the year, looking for something to kill. They wear and carry many dollars' worth of equipment, and go to a great deal of trouble, in order to kill some small creatures that they would never trouble to know alive, and that means little to them once they have killed it. If those we saw had killed the bobwhite they would no doubt have felt all their expense and effort justified, and would have thought themselves more manly than before. It reminds one of the extraordinary trouble and expense governments go to in order to kill men—and consider it justified or not, according to the "kill ratio." The diggers among our artifacts will find us to have been honorable lovers of death, having been willing to pay exorbitantly for it. How much better, we thought, to have come upon the *life* of the bird as we did, moving peaceably among the lives of the country that showed themselves to us because we were peaceable, than to have tramped fixedly, half oblivious, for miles in order to come at its death.

We left the hunters behind and went down past a green grainfield where cattle were grazing and drinking at the waterside. They were not disturbed that the river had come up over part of their pasture, no more troubled by the height of today's shore line than they were by the height of yesterday's. To them, no matter how high it was, so long as the ground was higher it was as ordinary as a summer pond. Surely the creatures of the fifth day of Creation accepted those of the sixth with equanimity, as though they had always been there. Eternity is always present in the animal mind; only men deal in beginnings and ends. It is probably lucky for man that he was created last. He would have got too excited and upset over all the change.

Two mallards flew up ahead of us and turned downriver into the wind. They had been feeding in the flooded corn rows, reminding us what a godsend the high water must be for ducks. The valley is suddenly full of little coves and havens, so that they can scatter out and feed safer and more hidden, and more abundantly too, than they usually can, never having to leave the river for such delicacies as the shattered corn left by the pickers. A picked cornfield under a few inches of water must be the duck Utopia—Utopia being, I assume, more often achieved by ducks than by men.

If one imagines the shore line exactly enough as the division between water and land, and imagines it rising—it comes up too slowly for the eye usually, so one *must* imagine it—there is a sort of magic about it. As it moves upward it makes a vast change, far more than the eye sees. It makes a new geography, altering the boundaries of worlds. Above it, it widens the freehold of the birds; below it, that of the fish. The land creatures are driven back and higher up. It is a line between boating and walking, gill and lung, standing still and flowing. Along it, suddenly and continuously, all that will float is picked up and carried away: leaves, logs, seeds, little straws, bits of dead grass.

And also empty cans and bottles and all sorts of buoyant trash left behind by fishermen and hunters and picnickers, or dumped over creek banks by householders who sometimes drive miles to do it. We passed behind a house built on one of the higher banks whose backyard was simply an avalanche of kitchen trash going down to the river. Those people, for all I know, may be champion homebodies, but their garbage is well-traveled, having departed for the Gulf of Mexico on every winter rise for years.

It is illuminating and suitably humbling to a man to recognize the great power of the river. But after he has recognized its power he is next called upon to recognize its limits. It can neither swallow up nor carry off all the trash that people convenience themselves by dumping into it. It can't carry off harmlessly all the sewage and pesticides and industrial contaminants that we are putting into it now, much less all that we will be capable of putting into it in a few years. We haven't accepted—we can't really believe—that the most characteristic product of our age of scientific miracles is junk, but that is so. And we still think and behave as though we face an unspoiled continent, with thousands of acres of living space for every man. We still sing "America the Beautiful" as though we had not created in it, by strenuous effort, at great expense, and with dauntless self-praise, an unprecedented ugliness.

The last couple of miles of our trip we could hear off in the bottoms alongside us the cries of pileated woodpeckers, and we welcomed the news of them. These belong to the big trees and the big woods, and more than any other birds along this river they speak out of our past. Their voices are loud and wild, the cries building strongly and then trailing off arrhythmically and hesitantly as though reluctant to end. Though they never seemed very near, we could hear them clearly over the commotion of the water. There were probably only a pair or two of them, but their voices kept coming to us a long time, creating beyond the present wildness of the river, muddy from the ruin of mountainsides and farmlands, the intimation of another wildness that will not overflow again in *our* history.

The wind had finally made its way into our clothes, and our feet and hands and faces were beginning to stiffen a little with the cold. And so when home came back in sight we thought it wasn't too soon. We began to slant across the currents toward the shore. The river didn't stop to let us off. We ran the bow out onto the path that goes up to my house, and the current rippled on past the stern as though it were no more than the end of a stranded log. We were out of it, wobbling stiff-legged along the midrib on our way to the high ground.

With the uproar of the water still in our ears, we had as we entered the house the sense of having been utterly outside the lives we live as usual. My warm living room was a place we seemed to have been away from a long way. It needed getting used to.

The Long-Legged House

SOMETIME IN THE TWENTIES, nobody knows exactly when any more, Curran Mathews built a cabin of two small rooms near Port Royal between the Kentucky River and the road, on a narrow strip of land belonging to my mother's father. Curran Mathews was my grandmother's bachelor brother. He died of cancer when I was a child and I have come to know him better since his death than I did while he was alive. At the time he died I knew him mostly as a sort of wanderer visiting in the family households, an inspired tinkerer with broken gadgetry and furniture, a man of small disciplines and solutions without either a home or a profession, and a teller of wonderful bedtime stories.

At that time the family still owned the building in Port Royal in which James Mathews, my great-grandfather, an immigrant from Ireland, had made shoes and run an undertaking business. There was a little partitioned-off space at the back of this building that Curran used as a workshop. He was good with tools, and a thorough workman—though I don't believe he ever made any money in that way, or ever intended to. I loved that little shop, and the smells of it, and spent some happy hours there. It smelled of varnish, and was filled with tools and objects of mysterious use. It was tucked away there, as if secretly, out of the way of the main coming and going of the town, looking out across the back lots into a pasture. And everything about it partook of the excitement I felt in this man, home from his wanderings, with his precise ways and peculiarly concentrated silences. I sat in the open doorway there one afternoon, a rich plot of sunlight on the floor around me, Curran quietly at work in the room at my back; I looked up at the ridge beyond the town, the open, still sunlit country of the summer afternoon, and felt a happiness I will never forget.

As for the bedtime stories, they were inexhaustible and always lasted a fine long while past bedtime. He was a reader of adventure stories, and a deeply restless man who spent a large part of his life in distant places. And his stories fed on his reading and on his travels. They could keep going night after night for weeks, and then he would go off someplace. By the time he got back he would have "forgot" the story, and we would have to tell it to him up to where he left off, and then after grunting a little with effort he would produce the next installment. Some of these were straight out of books—I discovered later that he had told us, almost page by page, the first of the *Tarzan* stories; and some were adapted from Zane Grey. But others are harder to account for. I remember that the best one was a sort of Swiftian yarn about a boy who was kidnaped by the fairies, and escaped, and was captured and then adopted by a tribe of Indians, and escaped, and so on and on through a summer in which bedtime was, for a change, looked forward to.

He had a family side, and another side. The family side I knew, I suppose, well enough. The other side I still pick up the hearsay of in conversations here and there. The other side is wild, extravagant, funny in the telling, and for me always more than a little troubling and sad. There was something he needed that he never found. That other side of his life he lived alone. Though some of his doings are legendary among people who remember him, I believe he kept quiet about them himself, and I think I will do the same.

He built the cabin on the river because he thought it would improve his health to live more in the open. His health had never been good. He had permanently injured his leg in a buggy accident in his youth—he disguised his limp so well that he was taken into the army during World War I. And there seems reason to think that the disease that finally killed him had begun years back in his life.

But health must have been only the ostensible reason; the best reason for the cabin he built here must be that it was in his nature to have a house in the woods and to return now and again to live in it. For there was something deep about him, something quiet-loving and solitary and kin to the river and the woods. My mother remembers summer evenings in her girlhood when they would drive his old car five or six miles out in the country to a railroad section pond, where as it grew dark they would sit and listen for a certain bullfrog to croak—and when the frog croaked at last they would drive back to Port Royal. He did that, of course, partly for the entertainment of his young niece; but partly, too, he was entertaining himself.

Given the same niece and the same need for entertainment, another man would not have thought of that. It must have been meaningful to him—a sort of ritual observance. Hard as it may be to know the needs and feelings of a man who said little about himself, dead now more than twenty years, that jaunting to hear the bullfrog does suggest how far his life escaped the categories. It was a life a man could hardly have carried in his wallet, or joined comfortably to the usual organizations.

And so his building of the cabin on the riverbank had a certain logic. The state road had not yet been built up the river on our side—that would have to wait for the Depression and the W.P.A. There was only a sort of rocked wagon-track going up through the bottoms, winding around the heads of slew-hollows, fording the creeks; farm fences crossed it, and one could not travel it without opening and closing gates. Only a few hundred yards down the river from the site of the cabin was Lane's Landing, the port of Port Royal, with its coal tipple and general store, but even so Curran Mathews' wooded stretch of riverbank was far more remote and isolated then than it is now. The voices of the frogs were here—there must have been summer nights when he sat awake a long time, listening to them. And there were other wild voices, too, that I know he heard, because I have heard them here myself. Through a notch in the far bank a little stream poured into the river, a clear moist sound, a part of the constancy of the place. And up in the thickets on the hillside there would be a cowbell, pastoral music, *tinkle-tink tink tink tinkle*, a sort of rhythmless rhythm late into the night sometimes, and nearby, when all else would be still. It was a place where a man, staying by himself, could become deeply quiet. It would have been a quiet that grew deeper and wider as the days passed, and would have come to include many things, both familiar and unexpected. If the stories about him are correct, and they doubtless are, Curran Mathews was not always alone here, and not always quiet. But there were times when he was both. He is said to have owned an old fiddle that he played at these times.

He did most of the building himself. And I imagine that those were days of high excitement. Except for the floors and the one partition, the lumber came out of an old log house that was built up on the hilltop by my grandfather's great-grandfather,

Ben Perry. It surely must have been one of the first houses built in this part of the country. In my childhood the main part of it was still standing, and was known then as the Aunt Molly Perry House, after the last member of the family to live in it. It was in bad shape when I knew it, and hogs were being fed in it. Now there is nothing left but the stones of the chimneys piled up, and the well. The lot where it stood is still known, though, as the Old House Lot, and it is a lovely place, looking out over the woods into the river valley. Some of the old locusts that stood in the yard are still there, and in the spring the little white starry flowers of old-time dooryards still bloom in the grass. It is a place I like, and like to go to and sit down. I am not oppressed by it as I would be by an ancient and venerated family seat, full of old records and traditions and memories. I figure a ponderous amount of my historical inheritance took place there—it is one of the main routes on my way here—but none of it was written down, and most of it has been forgotten, and the house itself is gone. Last summer for a while I pastured a mare and colt up there in the Old House Lot, and where my forebears had sat down to meals three times a day for generations, the horses grazed and thought nothing of it, it being only daylight and grass to them. I sat sometimes on the piled stones of the old chimneys, watching them graze, and it seemed to me that my line had issued out of the ground there like a spring, as regardless of itself, in the historical sense anyhow, and as little able to memorialize itself, as water—and had trickled off into oblivion, as I said to myself that I would too, leaving the hill there looking into the valley, and the horses grazing in the sunlight. And I was more grateful for the silence of their departure than I would be for the lineage of a king. Not knowing who had planted the flowers there, I did not have to weep over them, or grow reminiscent, but could enjoy them as they were. I had them free as wildflowers.

But the house—or, anyhow, part of it—had a very pleasing resurrection. I think of Curran Mathews coming there one summer morning, full of the excitement of his vision of a cabin on the riverbank, to begin work. I can see him carefully prizing off the boards of the kitchen L—broad poplar and walnut boards cut to irregular thicknesses in some crude sawmill; and fine handmade tongue-and-groove paneling out of, I think, the dining room. The nails would probably have rusted tight in their holes, and it would have been slow, nagging work, getting boards off without splitting them, but Curran was a man capable of great persistence and patience when he wanted to be, and he got it done, and drew out the old hand-forged square-headed nails, and loaded his lumber, and hauled it down the hill to the riverbank where he intended to settle a new place. Figuring backward from my own experiences here later, I think I partly know what was on his mind then. He was striking out beyond the bounds of the accepted, beyond the ordinary and common things that most people respect—as any man does who sets a willing foot into the woods—and there would have been an exhilaration in that. As soon as he marked out the dimensions of his house on the ground the place would have begun to look different to him, would have begun to have an intimacy for him that it could never have had before. Earlier, any place he stood was more or less equal to any other place he stood; he would move on to another place. But once those boundaries were marked on the ground, there would have begun to be a permanent allegiance. Here was the tree that would stand by the door. This limb would reach across the porch. Looking out here would give a fine view of the water. Here was where the steppingstones must come down the slope. When he began clearing the ground he was an eager man, and he felt an eagerness that is felt only by men who are doing what they want passionately to do, and who are not justifying it in an account book. All this I have experienced myself, much of it in the same place and in the same way that he experienced it, and I know it must be so. But there is a great deal, too, that I don't know, and never will. What did he think in the nights he spent here alone? What desires and dreams came to him in his solitudes? Had he read Thoreau, I wonder—or any of the other writers and poets who have so shaped my life? I doubt it, and feel in that a large difference between us.

He built a cabin of two rooms, a bedroom and a kitchen. The bedroom opened onto a small screened porch, and an open porch went the rest of the way across the front and along the upriver side, railed with locust poles. He painted it with a green paint that in my time took on a bluish cast, weathered and natural-looking, a color unlike any I ever saw. It has somehow come to be associated in my mind with the lichens that grow here on the trunks of the trees, but it was more vivid. He planted a few shrubs around it, some of which are still living: two spiraeas and a bush honeysuckle whose translucent red berries I remember wanting badly to eat when I was a child. Though I could eat them now if I wanted to, I never have, certain they would taste worse than they look.

A few years ago I was given an old dim photograph of Curran from back in the days when he lived actively here on the river. It shows him at the top of a flight of steps he had just finished digging into the steep bank below the cabin. He is bent slightly forward, holding a shovel in his hands, looking up the slope. He is wearing an old limber-brimmed felt hat. You can barely see his face. Behind him his new steps go down to the river, where, under the lacy top of an overhanging willow, a boat rests lightly on the water.

I have no memory of the cabin as it stood on its original site. Before I was three years old, the flood of 1937 picked it up from where Curran had built it and carried it several feet downriver and up the bank, where it lodged against some trees, and was then anchored and given new underpinnings. But I knew it early, and my first memories of it, trivial as they are, involve a delight that I associate with no other place. The cabin had come to be known simply as the Camp—or, when one wanted to be very specific, Curran's Camp. Throughout my childhood "going down to the Camp" was to be going to the most exciting place. In my very early years it was often used as a family gathering place in the summers. We children had an Uncle Herbert who got to be known as Uncle River because we hardly ever saw him except at the Camp. And I had another uncle, Uncle Doc, with whom it was a ritual for me to go down the long flight of stone steps to the river to watch for boats—which usually never appeared, but whose mere possibility was charming enough. And I remember being put to bed there for a nap on the promise that when I woke up I could have a dish of chocolate pudding. I needed to be bribed, of course, for I knew even then that those were indispensable times. And Clio, the bitter Muse, has had her way even with that, for I remember the nap but have forgot the pudding.

Throughout my childhood there were trips to the Camp

—some memorable, some forgotten. Those visits put the place deeply into my mind. It was a place I often thought about. I located a lot of my imaginings in it. Very early, I think, I began to be bound to the place in a relation so rich and profound as to seem almost mystical, as though I knew it before birth and was born for it. It remained so attractive to me, for one reason, because I had no bad associations with it. It was the family's wilderness place, and lay beyond the claims and disciplines and obligations that motivated my grownups. From the first I must have associated it with freedom. And I associated it with Curran, who must have associated it with freedom. It always had for me something of the charming strangeness of that man's life, as though he had carried it with him in his travels out in the West—as indeed I know he had, in his mind. Like Curran's life, the Camp seemed open to experiences not comprehended in the regularities of the other grown people. That is only to suggest the intensity and the nature of the bond; such feelings, coming from so far back in childhood, lie deeper than the reasons that are thought of afterward. But given the bond, it was no doubt inevitable that I would sooner or later turn to the place on my own, without the company of the grownups.

The first such visit I remember was during the flood of March, 1945. With my brother and a friend, I slipped away one bright Saturday and hitchhiked to Port Royal. We walked down to the Camp from there. It is one of the clearest days I have in my memory. The water was up over the floor of the Camp, and we spent a good while there, straddling out along the porch railings over the water—a dangerous business, of course, but we were in danger the rest of that day. There is something about a flooded house that is endlessly fascinating. Mystery again. One world being supplanted by another. But to a boy on a bright spring day it is without horror. I remember well the fascination I felt that day: the river pouring by, bright in the sun, laced with the complex shadows of the still naked trees, its currents passing the porch posts in little whirlpools the size of half dollars. The world suddenly looked profoundly alive and full of new interest. I had a transfiguring sense of adventure. The world lay before me, and Saturday lay before the world.

We left the Camp after a while and walked on up the river to where several johnboats were tied, and helped ourselves to the biggest one, a white one, built long and wide and heavy, as a boat must be when it is used for net fishing. We unfastened its mooring chain and pushed off. For oars we had an oak board about four feet long and an old broom worn off up to the bindings, but that seemed the paltriest of facts. We were free on the big water. Once we broke with the shore and disappeared into the bushes along the backwater, we were farther beyond the reach of our parents than we had ever been, and we knew it—and if our parents had known it, what a dark day it would have been for them! But they didn't. They were twelve miles away in New Castle, their minds on other things, assuming maybe that we had just gone to somebody's house to play basketball. And we had the world to ourselves—it would be years before we would realize how far we were from knowing what to do with it. That clumsy boat and our paddles seemed as miraculous as if we were cave men paddling the first dugout.

We paddled up to the Cane Run bridge and found the water over the road. Another wonder! And since it was unthinkable to refuse such an opportunity, we boated over the top of the road and entered the creek valley. We worked our way up through the treetops for some way, and broke out into a beautiful lake of the backwater, the sun on it and all around it the flooded woods. It was deeply quiet—nobody around anywhere, and nobody likely to be. The world expanded yet again. It would have taken an airplane to find us. From being mere boys—nine, ten, and eleven years old, I believe—we were enlarged beyond our dreams.

We went up the long backwater and into the woods. The woods is still there and I often go there these days to walk or to sit and look. It is a large stand of trees, and there is some big timber in it. I know it well now, but then I had only been there a few times to gather ferns with the rest of the family, and it was still new to me. The woods stands on a series of narrow low ridges divided by deep hollows. On the day of our adventure the backwater was in all these hollows, and we could go far into the woods without ever setting foot on the ground. That was another wondrous opportunity, and the hollows were numerous and long, allowing for endless variation and elaboration. The first wildflowers were blooming among the dead leaves above the water line, and I remember how green and suggestive the pads of moss looked in the warm water light, that time of the year. I gathered some and wrapped it in my handkerchief and put it in the boat, as if to take that place and the day with me when I left. We found a groundhog basking on a point of land, and got out and gave chase, thinking we had him cornered, only to see him take to the water and swim away. It never had occurred to me that groundhogs could swim, and I felt like a naturalist because of that discovery.

But it had to end, and it ended hard. When we started back the wind had risen, and drove strong against our boat. Our crude paddles lost their gleeful aura of makeshift, and turned against us. We moved like the hands of a clock, and only by the greatest effort. I don't think it occurred to any of us to worry about getting out, or even about being late, but being thwarted that way, pitted so against the wind and the water, had terror in it. Before we got safely back it had become a considerable ordeal—which, as soon as we *were* safely back, began to look like the best adventure of all. We tied the boat and walked up to the road, feeling like Conquerors, weary as Conquerors.

And there my grandparents were, waiting anxiously for us. We had been spied upon, it turned out. We had been seen taking the boat. Phone calls had been made. And my grandparents had driven down—in great fear, I'm sure—to see what was to be done. When we finally came off the river they were ready to give us up for drowned. They told us that. And suddenly we felt a good deal relieved to know we weren't.

That was the second bad flood to hit the Camp. These were modern floods, and man-made. Too many of the mountain slopes along the headwater streams had been denuded by thoughtless lumbering and thoughtless farming. Too little humus remained in the soil to hold the rains. After this second flood more of the old houses built near the river in earlier times would be abandoned. And in that spring of 1945 the war was about to end. Before long the country, as never before, would be full of people with money to spend. Men's demands upon nature were about to begin an amazing increase that would continue until now. The era when Curran Mathews conceived and built the Camp was coming to an end.

And two springs later Curran lay dying in the town of Carrollton at the mouth of the river. He had gone to spend his last days in the home of two of my aunts. From his bedroom in Carrollton the windows looked out across back yards and gardens toward the bluff over the valley of the Ohio. He couldn't see the river, but as always in a river town it would have been present to his mind, part of his sense of the place, the condition of everything else. The valley of the river that had been home to him all his life, and where his cabin stood in its woods patch on the bank, lay across town, behind him. He knew he wouldn't see it again.

That spring he watched the lengthening of the buds and the leafing out of a young beech near his window. And after the leaves had come he continued to watch it—the lights and moods and movements of it. He spoke often to my aunts of its beauty. I learned of this only a year or so ago. It has made his illness and death real to me, as they never were before. It has become one of the most vivid links in my kinship with him. How joyous the young tree must have been to him, who could have had so little joy then in himself. That he watched it and did not turn away from it must mean that he found joy in it, affirmed the life in it, even though his own life was painfully going out.

In those years of the war and of Curran's illness the Camp fell into disuse and neglect. The screens rotted; leaves piled up on the porch, and the boards decayed under them; the weeds and bushes grew up around it; window glasses were broken out by rock-throwers; drunks and lovers slipped in and out through the windows. And then one evening in early summer —it must have been the year after Curran's death, when I was nearly fourteen—I came down by myself just before dark, and cooked a supper on the riverbank below the house, and spent the night in a sleeping bag on the porch. I don't remember much about that night, and don't remember at all what reason I had for coming. But that night began a conscious relation between me and the Camp, and it has been in my mind and figured in my plans ever since.

I must have come, that first time, intending to stay for a while, though I can't remember how long. The next morning as I was finishing breakfast a car pulled in off the road and an old man got out, saying he had come to fish. Did I mind? Maybe I felt flattered that my permission had been asked. No, I didn't mind, I said. In fact, I'd fish with him. He was a friendly white-haired old man, very fat, smiling, full of talk. Would he like to fish from a boat? I asked him. I'd go get one. He said that would be fine, and I went down along the bank to the landing and got one—again, I think, by the forthright method of "borrowing"—and brought it back. It was a good big boat, but I had managed to find only a broken oar to paddle with. He got in with his equipment and we set off. We fished the whole day together. Not much was said. I don't remember that we ever introduced ourselves. But it was a friendly day. I was slow in moving us between fishing spots with my broken oar and I remember how he sat in the middle seat and hunched us along like a kid in a toy car. It helped a great deal. We fished until the late afternoon and caught—or, rather, *he* caught—a sizable string of little perch and catfish. When he left he gave me all the catch. He fished, he said, for the love of it. I don't yet know who he was. I never saw him again. In the evening, after I had taken the boat home and had started cleaning the fish on the porch, my friend Pete stopped by with his father. They had been working on their farm up the river that day, and were on their way home. Pete liked the look of things and decided to stay on with me. And that began a partnership that lasted a long time. The next morning we began scraping out the dried mud that the last high water had deposited on the floors—and suddenly the place entered our imagination. We quit being campers and became settlers. Twenty or so years before, the same metamorphosis must have taken place in the mind of Curran Mathews. A dozen or two strokes, flaking up the sediment from the floor—and we had got an idea, and been transformed by it: we'd clean her out and fix her up.

And that is what we did. We gave it a cleaning, and then scrounged enough furnishings and food from our families to set up in free and independent bachelorhood on the river. Pete's family had a well-stocked frozen-food locker that we acquired a deep respect for. The idea was to be well enough supplied that we wouldn't need to go back home for a while. That way, if somebody wanted to impose work or church on us they would have to come get us. That was another realization the Camp had suddenly lit up for us: We were against civilization, and wanted as little to do with it as possible. On the river we came aware of a most inviting silence: the absence of somebody telling us what to do and what not to do. We swam and fished and ate and slept. We could leave our clothes in the cabin and run naked down the bank and into the water. We could buy cigars and lie around and smoke. We could go out on the porch in the bright damp shadowy mornings and pee over the railing. We made a ritual of dispensing with nonessentials; one of us would wash the dishes with his shirttail, and the other would dry them with his. We were proud of that, and thought it had style.

For several years we came to the Camp every chance we got, and made good memories for ourselves. We will never live days like those again, and will never live happier ones. I suppose it was growing up that put an end to them—college, jobs, marriages and families of our own. We began to have a stake in civilization, and could no longer just turn our backs on it and go off. Back in our free days we used to tell people—a good deal to their delight, for we were both sons of respected lawyers—that we intended to be bootleggers. I think the foreknowledge of our fate lay in that. For to contrive long to be as free and careless as we were in those days we would have had to become bootleggers, or something of the kind. Ambition and responsibility would take us a different way.

During those years when I would come down to the Camp with Pete, I would also fairly often come alone. Pete, who was older than I was, would be having a date. Or he would have to stay at home and work at a time when I could get free. And at times I came alone because I wanted to. I was melancholic and rebellious, and these moods would often send me off to the river.

Those times were quiet and lonely, troubled often by the vague uneasiness and dissatisfaction of growing up. The years I spent between childhood and manhood seem as strange to me now as they did then. Clumsy in body and mind, I knew no place I could go to and feel certain I ought to be there. I had no very good understanding of what I was rebelling against: I was going mostly by my feelings, and so I was rarely calm. And I didn't know with enough certainty what I wanted to be purposeful about getting it. The Camp offered no escape from these troubles, but it did allow them the dignity of solitude. And there were days there, as in no other place, when as if by accident, beyond any reason I might have had, I was deeply at peace, and happy. And those days that gave me peace suggested to me the possibility of a greater, more substantial peace—a decent, open, generous relation between a man's life and the world—that I have never achieved; but it must have begun to be then, and it has come more and more consciously to be, the hope and the ruling idea of my life.

I remember one afternoon when I tied my boat to a snag in the middle of the river outside the mouth of Cane Run, and began fishing. It was a warm sunny afternoon, quiet all around, and I had the river to myself. There is some nearly mystical charm about a boat that I have always been keenly aware of, and tied there that afternoon in midstream, a sort of island, it made me intensely alive to the charm of it. It seemed so intact and dry in its boatness, and I so coherent and satisfied in my humanness. I fished and was happy for some time, until I became *conscious* of what a fine thing I was doing. It came to me that this was one of the grand possibilities of my life. And suddenly I became deeply uneasy, even distressed. What I had been at ease with, in fact and without thinking, had become, as a possibility, too large. I hadn't the thoughts for it. I hadn't the background for it. My cultural inheritance had prepared me to exert myself, work, move, "get someplace." To be idle, simply to live there in the sunlight in the middle of the river, was something I was not prepared to do deliberately. I tried to stay on, forcing myself to do what I now *thought* I ought to be doing, but the spell was broken. That I had nothing to do but what I wanted to do, and what I was in fact doing, had become utterly impotent as an idea. I had to leave. I would have to live to twice that age before I could do consciously what I wanted so much then to do. And even now I can do it only occasionally.

I read a good deal during those stays I made at the river. I read *Walden* here then—though I can't remember whether it was for the first or second time. And I was beginning to read poetry with some awareness that it interested me and was important to me—an awareness I had not yet come by in any classroom. I had a paperback anthology of English and American poems, and I would lie in bed at night and read it by the light of a kerosene lamp. One night I lay there late and read for the first time and was deeply stirred by Gray's "Elegy Written in a Country Churchyard." It seemed to me to be a fine thing, and I thought I understood it and knew why it was fine. That was a revealing experience. While I read a dog was howling somewhere down the river, and an owl was calling over in the Owen County bottom, and my father's old nickel-plated Smith and Wesson revolver lay in the bedding under my head. I loved this place and had begun to understand it a little—but I didn't love it and understand it well enough yet to be able to trust it in the dark. That revolver belonged on top of the bookcase at home, but nobody ever thought of it, and I could borrow it and bring it back without fear that it would be missed.

The Camp was not the only place that was important to me. But during the winters of those years, starting when I was fourteen, when I was away at school, it lay in my mind with the other places that were important, part of what I dreamed of coming home to. During the years of high school, particularly, I believe that the thought of it was indispensable to me. The school I attended was a military school. There military correctness and regularity were always the aim—thwarted constantly, to be sure, by the natural high spirits of the students and by the natural mediocrity of most of the teachers—but when thwarted always exacting vengeance on somebody. Sympathy and intelligence were in everything replaced by rules, and by a long ago outworn—hence, threatened and fanatical—moral dogmatism. The highest aim of the school was to produce a perfectly obedient, militarist, puritanical moron who could play football. That aim, or course, inspired a regime that was wonderfully vindictive against anything that threatened to be exceptional. And having a lively and independent mind, I became a natural enemy of the regime. Take a simpleton and give him power and confront him with intelligence—and you have a tyrant. I was once struck by one of the teachers for using a dictionary (not an authorized textbook) during a study hall, and another time was openly chastised for reading a story by Balzac entitled "A Passion in the Desert," the only passion authorized by the regime being a passionate servility. The "discipline" exercised by the student officers was often equally stupid, and often more violent. I waged four years there in sustained rebellion against everything the place stood for, paying the cost both necessarily and willingly. I was not, during all those years, well equipped for such a struggle—though I was a conscious student of resistance, and got pretty good at it toward the end. I had, maybe because of the prolonged awkwardness of my adolescence, an enormous craving for personal dignity—and in the military school dignity simply was not possible for one who was not an athlete and who could not regard mechanical obedience as the summit of virtue. I don't think I could have survived that struggle intact if I hadn't had a history that taught me that there was dignity of another kind, and more desirable. I had known from the beginning a few men who accepted and required of themselves as men with a great simplicity of pride, who could be lonely in their virtues and excellences if they had to be, and who could move in their lives without either crawling or marching—and the thought of those men was before me. But also I had lived days of my own, perhaps mainly at the Camp, when my life had seemed to come to me naturally, with an ease and rightness, as life must come to the kingfishers on the river. I knew that these were my best days, and knew they had not come to me on the orders of anybody or because of anybody's opinion of them or because somebody had allowed me to have them. I knew they were my holy days, my sabbaths, and they had come to me freely because I was free. Knowing that, I knew that men were most admirable singly, and that standing in lines under the command of other men is the least becoming thing they do. Though I stood in line many times each day, my privilege was that my mind was *not* in line. It didn't *have* to be. It had better places to go. Now that I think of it, I had, by that time, a superbly furnished mind. It was not yet much furnished with books, for I had not read many books, or many good ones, then; it was furnished with the knowledge of a few good men and good places and good days that had come to me in my life. I was not servile because I knew what it was *not* to be servile. It is bound to be in some sense true that a man is not a slave if his mind is free. It is the man who can think of no alternative to his enslavement who is truly a slave. Though I found the life at the military school to be often painful and interminably hateful, I was coherent and steadfast in my rebellion against it because I knew, I *must* have known, that I was the creature of another place, and that my life was already given to another way.

What I remember best from those years are the days in early summer when I would first come down to the Camp to clean the place up. By the time school would be out the weeds there on the damp riverbank would have already grown waist-high. I would come down with ax and scythe, preparatory to some stay I intended to make, and drive back to the wilderness. I would mow the wild grass and horseweeds and nettles and elderberries, and chop down the tree sprouts, and trim obstructive branches off the nearby trees. It would be hard to describe the satisfaction this opening up would give me. I would sit down now and again to rest and dry the sweat a little, and look at what I had done. I would meditate on the difference I had made, and my mind would be full of delight. It was some instinctive love of wildness that would always bring me back here, but it was by the instincts of a farmer that I established myself. The Camp itself was not imaginable until the weeds were cut around it; until that was done I could hardly bring myself to enter it. A house is not simply a building, it is also an enactment. That is the first law of domesticity, even the most meager. The mere fact must somehow be turned into meaning. Necessity must be made a little ceremonious. To ever arrive at what one would call home even for a few days, a decent, thoughtful approach must be made, a clarity, an opening.

Only after I had mowed and trimmed around it, so that it stood clear of its surroundings, or clearly within them, could I turn to the house. The winter's leaves would have to be swept off the porch, the doors opened, the shutters opened, daylight filling the rooms for the first time in months, the sashes drawn back in their slots, the walls and floors swept. And then, finished, having earned the right to be there again, I would go out on the porch and sit down. Tired and sweaty, the dust of my sweeping still flavoring the air, I would have a wonderful sense of order and freedom. The old recognitions would come back, the familiar sights and sounds slowly returning to their places in my mind. It would seem inexpressibly fine to be living, a joy to breathe.

During the last of my college years and the year I spent as a graduate student, the Camp went through another period of neglect. There were distractions. I had begun to be preoccupied with the girl I was going to marry. In a blundering, half-aware fashion I was becoming a writer. And, as I think of it now, school itself was a distraction. Although I have become, among other things, a teacher, I am skeptical of education. It seems to me a most doubtful process, and I think the good of it is taken too much for granted. It is a matter that is overtheorized and overvalued and always approached with too much confidence. It is, as we skeptics are always discovering to our delight, no substitute for experience or life or virtue or devotion. As it is handed out by the schools, it is only theoretically useful, like a randomly mixed handful of seeds carried in one's pocket. When one carries them back to one's own place in the world and plants them, some will prove unfit for the climate or the ground, some are sterile, some are not seeds at all but little clods and bits of gravel. Surprisingly few of them come to anything. There is an incredible waste and clumsiness in most efforts to prepare the young. For me, as a student and as a teacher, there has always been a pressing anxiety between the classroom and the world: how can you get from one to the other except by a blind jump? School is not so pleasant or valuable an experience as it is made out to be in the theorizing and reminiscing of elders. In a sense, it is not an experience at all, but a hiatus in experience.

My student career was over in the spring of 1957, and I was glad enough to be done with it. My wife and I were married in May of that year. In the fall I was to take my first teaching job. We decided to stay through the summer at the Camp. For me, that was a happy return. For Tanya, who was hardly a country girl, it was a new kind of place, confronting her with hardships she could not have expected. We were starting a long way from the all-electric marriage that the average modern American girl supposedly takes for granted. If Tanya had been the average modern American girl, she would probably have returned me to bachelorhood within a week—but then, of course, she would have had no interest in such a life, or in such a marriage, in the first place. As it was, she came as a stranger into the country where I had spent my life, and made me feel more free and comfortable in it than I had ever felt before. That seems to me the most graceful generosity that I know.

For weeks before the wedding I spent every spare minute at the Camp, getting it ready to live in. I mowed around it, and cleaned it out, and patched the roof. I replaced the broken windowpanes, and put on new screens, and white-washed the walls, and scrounged furniture out of various family attics and back rooms. As a special wedding gift to Tanya I built a new privy—which never aspired so high as to have a door, but did sport a real toilet seat.

All this, I think, was more meaningful and proper than I knew at the time. To a greater extent than is now common, or even possible for most men, I had by my own doing prepared the house I was to bring my wife to, and in preparing the house I prepared myself. This was the place that was more my own than any other in the world. In it, I had made of loneliness a good thing. I had lived days and days of solitary happiness there. And now I changed it, to make it the place of my marriage. A complex love went into those preparations —for Tanya, and for the place too. Working through those bright May days, the foliage fresh and full around me, the river running swift and high after rain, was an act of realization: as I worked getting ready for the time when Tanya would come to live there with me, I understood more and more what the possible meanings were. If it had gone differently—if it had followed, say, the prescription of caution: first "enough" money, and then the "right" sort of house in the "right" sort of place —I think I would have been a poorer husband. And my life, I know, would have been poorer. It wasn't, to be sure, a permanent place that I had prepared; we were going to be there only for the one summer. It was, maybe one ought to say, no more than a ritual. But it was a meaningful and useful ritual.

That is more than I am able to say of our wedding. Our wedding reminds me a little of the Kentucky Derby; the main event, which lasted only a couple of minutes, required days of frantic prologue. During this commotion one must discontinue one's own life and attempt to emulate everybody. Nobody can be still until convinced that this marriage will be like everybody else's. The men insinuate. The women gloat. The church is resurrected and permitted to interfere. As at funerals, the principals cannot be decently let alone, but must be overhauled, upgraded, and messed with until they are not recognizable. During the week up to and including the wedding I am sure I was at the center of more absurdity than I hope ever to be at the center of again. One of my unforgettable moments, as they are called, was a quaint session with the minister who having met me for the first time that moment, and being scarcely better acquainted with Tanya, undertook to instruct us in the marital intimacies. A display of preacherish cant and presumption unusual even for a preacher, and all carried out with a slogging joyless dutifulness. As we were leaving he handed us a book on marriage. In a country less abject before "expert advice," the effrontery of it would be incredible. Well, be damned to him and his book, too. We thrive in spite of him, and in defiance of some of his rules. We are, I like to think, his Waterloo—though I know that, like most of his kind, he had come to his Waterloo before that, and survived by his inability to recognize it.

In any sense that is meaningful, our wedding was made in our marriage. It did not begin until the ceremony was over. It began, it seems to me, the next morning when we went together to the Camp for the first time since I started work on it. I hesitate to try to represent here the pleasure Tanya may have felt on this first arrival at our house, or the pride that I felt—those feelings were innocent enough, and probably had no more foundation than innocence needs. The point is that, for us, these feelings were substantiated by the Camp; they had its atmosphere and flavor, and partook of its history. That morning when Tanya first came to it as my wife, its long involvement in my life was transformed, given a richness and significance it had not had before. It had come to a suddenly illuminating promise. A new life had been added to it, as a new life had been added to my life. The ramshackle old house and my renewal of it particularized a good deal more for us than we could have realized then. We began there. It began that morning to have as profound a significance in our marriage as it had already had in my life.

We carried in what we had brought in the way of baggage, and then went out and bought the kitchen utensils and groceries we needed. In that way we began our marriage. And in that way—which may be only to say the same thing—we out-distanced what the sterile formalities of the wedding had expected of us or prescribed to us. We escaped the dead hands of the conventions and the institutions into a life that was distinctly our own. Our marriage became then, and has remained, the center of our life. And it is peculiarly true that the Camp is at the center of our marriage, both as actuality and as symbol. The memories of that first summer are strong and clear, and they stay in our minds. After we left the Camp at the summer's end, we continued to think of it and to talk about it and to make plans for it. We lived in other places in Kentucky, in California, in Europe, in New York—and now we have come back to live in sight and in calling distance of the place where we began. It is our source and our emblem, and it keeps its hold.

It would be a mistake to imply that two lives can unite and make a life between them without discord and pain. Marriage is a perilous and fearful effort, it seems to me. There can't be enough knowledge at the beginning. It must endure the blundering of ignorance. It is both the cause and the effect of what happens to it. It creates pain that it is the only cure for. It is the only comfort for its hardships. In a time when divorce is as accepted and conventionalized as marriage, a marriage that lasts must look a little like a miracle. That ours lasts—and in its own right and its own way, not in pathetic and hopeless parody of some "expert" notion—is largely, I believe, owing to the way it began, to the Camp and what it meant and came to mean. In coming there, we avoided either suspending ourselves in some honeymoon resort or sinking ourselves into the stampede for "success." In the life we lived that summer we represented to ourselves what we wanted—and it was *not* the headlong pilgrimage after money and comfort and prestige. We were spared that stress from the beginning. And there at the Camp we had around us the elemental world of water and light and earth and air. We felt the presences of the wild creatures, the river, the trees, the stars. Though we had our troubles, we had them in a true perspective. The universe, as we could see any night, is unimaginably large, and mostly empty, and mostly dark. We knew we needed to be together more than we needed to be apart.

There were physical hardships, or what pass these days for

physical hardships, that scandalized certain interested onlookers. How, they wondered, could I think of bringing a girl like Tanya into a place like that? The question ought to have been: How could a girl like Tanya think of it? They will never know. We had no electricity, no plumbing, no new furniture. Our house would, no doubt, have been completely invisible to the average American bride and groom of that year, and when it rained hard enough the roof leaked. I think our marriage is better for it. By these so-called "hardships"—millions of people put up with much worse as a matter of course and endlessly—we freed our marriage of things. Like Thoreau at Walden, we found out what the essentials are. Our life will never be distorted by the feeling that there are luxuries we cannot do without. We will not have the anxiety of an abject dependence on gadgets and corporations. We are, we taught ourselves by our beginning, the dependents of each other, not of the local electric company.

That summer has no story; it has not simplified itself enough in my memory to have the consistency of a story and maybe it never will; the memories are too numerous and too diverse, and too deeply rooted in my life.

One of the first things I did after we got settled was to put some trotlines in the river—an early outbreak of male behavior that Tanya, I think, found both mystifying and depraved. On a dark rainy night in early June we had stayed up until nearly midnight, making strawberry preserves, and I decided on an impulse to go and raise my lines. Working my way along the line in pitch dark a few minutes later, I pulled out of the middle of the river a catfish that weighed twenty-seven pounds. Tanya had already gone to bed, and had to get up again to hear me congratulate myself in the presence of the captive. And so, indelibly associated with the early days of my marriage is a big catfish. Perhaps it is for the best.

The night of the Fourth of July of that year there came one of the worst storms this part of the country ever knew. For hours the rain spouted down on our tin roof in a wild crashing that did not let up. The Camp had no inner walls or ceilings; it was like trying to sleep inside a drum. The lightning strokes overlapped, so that it would seem to stay light for minutes at a stretch, and the thunder kept up a great knocking at the walls.

After a while it began to seem unbelievable that the rain did not break through the roof. It was an apocalyptic night. The next morning we went out in bright sunshine to find the river risen to the top of its banks. There was a lasting astonishment in looking at it, and a sort of speculative fear; if that storm had reached much farther upstream we would have had to swim out of bed. Upstream, we could see several large trees that Cane Run had torn out by the roots and hurled clear across the river to lodge against the Owen County bank. The marks of that rain are still visible here.

And I remember a quiet night of a full moon when we rowed the boat up into the bend above the mouth of Cane Run, and let the slow current bring us down again. I sat on the rower's seat in the middle of the boat, and Tanya sat facing me in the stern. We stayed quiet, aware of a deep quietness in the country around us, the sky and the water and the Owen County hills all still in the white stare of the moon. The wooded hill above the Camp stood dark over us. As it bore us, the current turned us as though in the slow spiral of a dance.

Summer evenings here on the river have a quietness and a feeling of completion about them that I have never known in any other place, and I have kept in mind the evenings of that summer. The wind dies about sundown, and the surface of the river grows smooth. The reflections of the trees lie inverted and perfect on it. Occasionally a fish will jump, or a kingfisher hurry, skreaking, along the fringe of willows. In the clearing around the house the phoebes and pewees call from their lookout perches, circling out and back in their hunting flights as long as the light lasts. Out over the water the swallows silently pass and return, dipping and looping, climbing and dipping and looping, sometimes skimming the surface to drink or bathe as they fly. The air seems to come alive with the weaving of their paths. As I sat there watching from the porch those evenings, sometimes a profound peacefulness would come to me, as it had at other times, but now it came of an awareness not only of the place, but of my marriage, a completeness I had not felt before. I was there not only because I wanted to be, as always before, but now because Tanya was there too.

But most of all I like to remember the mornings. We would get up early, and I would go out on the river to raise my lines the first thing. There is no light like that on the river on a clear early morning. It is fresh and damp and full of glitters. The intense linear reflections off the wind waves wobble up the tree trunks and under the leaves. It was fishing that paid well, though not always in fish. When I came back to the Camp Tanya would have a big breakfast waiting.

After we ate I would carry a card table out into a corner of the little screened porch, and sit down to write. I would put in the morning there, conscious always as I worked of the life of the river. Fish would jump. A kingfisher would swing out over the water, blue and sudden in the water light, making his harsh ratcheting boast to startle the world. The green herons would pass intently up and down, low to the water, just outside the willows, like busy traffic. Or one would stop to fish from a snag or a low-bending willow, a little nucleus of stillness; sitting at my own work on the other side of the river I would feel an emanation of his intent silence; he was an example to me. And in the trees around the Camp all the smaller birds would be deep in their affairs. It might be that a towboat—the *Kentuckian* or the old *John R. Kelly*, that summer—would come up, pushing two bargeloads of sand to Frankfort. Or there would be somebody fishing from the other bank, or from a boat. And the river itself was as intricately and vigorously alive as anything on it or in it, always shifting its lights and its moods.

That confirmed me in one of my needs. I have never been able to work with any pleasure facing a wall, or in any other way fenced off from things. I need to be in the presence of the world. I need a window or a porch, or even the open outdoors. I have always had a lively sympathy for Thoreau's idea of a hypaethral book, a roofless book. Why should I shut myself up to write? Why not write and live at the same time?

There on the porch of the Camp that summer I wrote the first poetry that I still feel represented by—a long poem rather ostentatiously titled "Diagon," about a river—and did some of the most important reading I have ever done.

In the spring of that year I had read attentively through the poems of Andrew Marvell, and had felt a strong kinship with him. The poem of his that interested me most was not one of the familiar short poems, but the strange, imperfect long one entitled "Upon Appleton House, to my Lord *Fairfax*." This is a complimentary piece, evidently very deliberately undertaken, and in long stretches it is amply boring. But in his description of the countryside Marvell's imagination seems abruptly to break out of the limitations of subject and genre, and he wrote some stunning poetry. It has remained for me one of the most exciting poems I know—not just in spite of its faults, but to a considerable extent because of them. I need to quote extensively from it, both to show the quality of the best work in the poem and to show what I think is remarkable about it. To begin with, all this is most *particularly* observed—as rarely happens in English nature poetry; the scene is of interest in itself, not just as the manifestation of something transcendent or subjective. The grass is grass, and one feels the real rankness and tallness of it. Equally, the killed bird is real; its blood is on the scythe's edge, and we feel the mower's regret of the useless death. What natural things manifest, if observed closely enough, is their nature, and their nature is to change. Marvell's landscape is in constant metamorphosis, and so metaphor is peculiarly necessary to its poetry—it is continuously being carried beyond what it was. The comparative image is not imposed from without by the poet, but is seen by the poet to be implicit in the nature of the thing: it is in the nature of a meadow to be like a sea.

> And now to the Abbyss I pass
> Of that unfathomable Grass,
> Where Men like Grashoppers appear,
> But Grashoppers are Gyants there:
> They, in there squeking Laugh, contemn
> Us as we walk more low then them:
> And, from the Precipices tall
> Of the green spir's, to us do call.
>
> To see Men through this Meadow Dive,
> We wonder how they rise alive.
> As, under Water, none does know
> Whether he fall through it or go.
> But, as the Marriners that sound,
> And show upon their Lead the Ground,
> They bring up Flow'rs so to be seen,
> And prove they've at the Bottom been.

No Scene that turns with Engines strange
Does oftner then these Meadows change.
For when the Sun the Grass hath vext,
The tawny Mowers enter next;
Who seem like *Israalites* to be,
Walking on foot through a green Sea.
To them the Grassy Deeps divide,
And crowd a Lane to either Side.

With whistling Sithe, and Elbow strong,
These Massacre the Grass along:
While one, unknowing, carves the *Rail*,
Whose yet unfeather'd Quils her fail.
The Edge all bloody from its Breast
He draws, and does his stroke detest . . .

The Mower now commands the Field;
In whose new Traverse seemeth wrougth
A Camp of Battail newly fought:
Where, as the Meads with Hay, the Plain
Lyes quilted ore with Bodies slain . . .

When after this 'tis pil'd in Cocks,
Like a calm Sea it shews the Rocks:
We wondring in the River near
How Boats among them safely steer.

This *Scene* again withdrawing brings
A new and empty Face of things;
A levell'd space, as smooth and plain,
As Clothes for *Lilly* stretched to stain.
The World when first created sure
Was such a Table rase and pure.

For to this naked equal Flat,
Which *Levellers* take Pattern at,
The Villagers in common chase
Their Cattle, which it closer rase . . .

They feed so wide, so slowly move,
As *Constellations* do above.
Then, to conclude these pleasant Acts,
Denton sets ope its *Cataracts*;

And makes the Meadow truly be
(What it but seem'd before) a Sea.

The River in it self is drown'd,
And Isl's th' astonished Cattle round.
Let others tell the *Paradox*,
How Eels now bellow in the Ox;
How Horses at their Tails do kick,
Turn'd as they hang to Leeches quick;
How Boats can over Bridges sail;
And Fishes do the Stables scale.
How *Salmons* trespassing are found;
And Pikes are taken in the Pound.

As I worked that summer I had these lines of Marvell very fresh in my mind, and they were having a deeper influence on me than I knew. My problem as a writer, though I didn't clearly know it yet, was that I had inherited a region that had as a literary tradition only the corrupt and crippling local colorism of the "Kentucky" writers. This was both a mythologizing chauvinism and a sort of literary imperialism, tirelessly exploiting the clichés of rural landscape, picking and singing and drinking and fighting lazy hillbillies, and Bluegrass Colonels. That is a blinding and tongue-tying inheritance for a young writer. And one doesn't even have to read the books to get it; it is so thoroughly established and accredited that it is propagated by schoolteachers, politicians, official bulletins, postcards, and the public at large. Kentucky is a sunny, beautiful land, full of happy country folks, whose very failures are quaint and delightful and to be found only here. It is surely no accident that along with this tradition of literary falsification there has been a tradition, equally well-established, and one could almost say equally respected, of political and industrial exploitation that has defaced and destroyed more of the state's beauty and wealth than there is left. The truth was too hard to tell; the language of the state's writers was dead in their mouths and they could not tell it. And when one finally told the clean truth, as Harry Caudill did in *Night Comes to the Cumberlands*, how few could hear it! And for the same reason. The people's ears are stuffed with the dead language of their literature. All my life as a writer I have had this rag bag of chauvinistic clichés to struggle with.

It is not difficult to see how serviceable and clarifying I found those lines from "Upon Appleton House." They showed me the poet's vision breaking out of its confines into the presence of its subject. I feel yet the exhilaration and release when Marvell turns from his elaborate overextended compliment to the noble family, and takes up a matter that really interested him; the full powers of his imagination and intelligence become suddenly useful to him, and necessary:

And now to the Abbyss I pass
Of that unfathomable Grass. . . .

He is talking about a river valley of farms and woodlands such as I had known all my life, and now had before me as I wrote and read through that summer. As a child I had even believed, on what I then considered the best advice, in the metamorphosis of horsehairs that Marvell alludes to in the last stanza of his description of the meadows. I would put hairs from tails and manes into the watering trough at night, confident that by morning they would be turned to snakes. But I was a poor scientist, and in the mornings always forgot to look —and so kept the faith.

With Marvell's work in my mind, I began that summer of my marriage the surprisingly long and difficult labor of *seeing* the country I had been born in and had lived my life in until then. I think that this was peculiarly important and necessary to me; for whereas most American writers—and even most Americans —of my time are displaced persons, I am a placed person. For longer than they remember, both sides of my family have lived within five or six miles of this riverbank where the old Camp stood and where I sit writing now. And so my connection with this place comes not only from the intimate familiarity that began in babyhood, but also from the even more profound and mysterious knowledge that is inherited, handed down in memories and names and gestures and feelings, and in tones and inflections of voice. I never, for reasons that could perhaps be explained, lost affection for this place, as American writers have almost traditionally lost affection for their rural birthplaces. I have loved this country from the beginning, and I believe I was grown before I ever really confronted the possibility that I could live in another place. As a writer, then, I have had this place as my fate. For me, it was never a question of *finding* a subject, but rather of learning what to do with the subject I had had from the beginning and could not escape. Whereas most of this country's young writers seem able to relate to no place at all, or to several, I am related absolutely to one.

And this place I am related to not only shared the state's noxious literary inheritance, but had, itself, never had a writer of any kind. It was, from a writer's point of view, undiscovered country. I have found this to be both an opportunity and a disadvantage. The opportunity is obvious. The disadvantage is that of solitude. Everything is to be done. No beginnings are ready-made. One has no proof that the place can be written about, no confidence that it can produce such a poet as one suspects one might be, and there is a hesitance about local names and places and histories because they are so naked of associations and assigned values—none of which difficulties would bother a poet beginning in Concord, say, or the Lake District. But here I either had to struggle with these problems or not write. I was so intricately dependent on this place that I did not begin in any meaningful sense to be a writer until I began to see the place clearly and for what it was. For me, the two have been the same.

That summer I was only beginning, and my poem "Diagon" came out of the excitement of that first seeing, and the first inklings that there might be viable meanings in what I knew. I was seeing consciously the lights and colors and forms of my own world for the first time:

> The sun sets vision afloat,
> Its hard glare down
> All the reaches of the river,
> Light on the wind waves
> Running to shore. Under the light
> River and hill divide. Two dead
> White trees stand in the water,
> The shimmering river casts
> A net of light around them,
> Their snagged shapes break through.

*

It is a descriptive poem mostly, and I have worried at times because in my work I have been so often preoccupied with description. But I have begun to think of that as necessary. I had to observe closely—be disciplined by the look and shape and feel of things and places—if I wanted to escape the blindness that would have made my work sound like an imitation of some Kentucky politician's imitation of the Romantic poets.

Sustaining and elaborating the effect of Marvell were the poems of William Carlos Williams, whose work I had known before but read extensively and studiously during that summer of 1957. I had two books of his, the *Collected Earlier Poems*, and his newest one, *Journey to Love*. I saw how his poems had grown out of his life in his native city in New Jersey, and his books set me free in my own life and my own place as no other books could have. I'll not forget the delight and hopefulness I felt in reading them. They relieved some of the pressure in the solitude I mentioned earlier. Reading them, I felt I had a predecessor, if not in Kentucky then in New Jersey, who confirmed and contemporized for me the experience of Thoreau in Concord.

Another book that deeply affected me that summer was Kenneth Rexroth's *100 Poems from the Chinese*, which immediately influenced my work and introduced me to Oriental poetry, not to mention the happy reading it made. I still think it is one of the loveliest books I know.

All these—my new marriage, the Camp, the river, the reading and writing—are intimately associated in my mind. It would be impossible to do more than imply the connections. Those were probably the three most important months in my life, as well as the happiest. When the summer was over it was a sharp sorrow to have to go. I remember us loading our borrowed household things into and onto our old Jeep station wagon and driving off up the river road on a brilliant day, the fields in the bottoms all yellow with the fall flowers. And I remember the troubling sense that what we were going to would be more ordinary than what we were leaving behind. And it was.

II

It has been almost exactly a year since I began this history. My work was interrupted by the spring weather, when gardening and other outside concerns took me away from writing. But now it is deep winter again. Yesterday snow fell all day and covered the ground. This morning, though the sun came up clear, the thermometer read four above—a good morning to sit in the Camp in the warmth of the stove and the brisk snow light from the big window over the table. It is a morning for books and notebooks and the inviting blank pages of writing paper.

For people who live in the country there is a charming freedom in such days. One is free of obligations to the ground. There is no outside work that one ought to do, simply because, with the ground frozen deep and covered with snow, no such work is possible. Growth has stopped; there is plenty of hay and grain in the barn; the present has abated its urgencies. And the mind may again turn freely to the past and look back on the way it came. This morning has been bearing down out of the future toward this bit of riverbank forever. And for perhaps as long, in a sense, my life has been approaching from the opposite direction. The approach of a man's life out of the past is history, and the approach of time out of the future is mystery. Their meeting is the present, and it is consciousness, the only time life is alive. The endless wonder of this meeting is what causes the mind, in its inward liberty of a frozen morning, to turn back and question and remember. The world is full of places. Why is it that I am here?

What has interested me in telling the history of the Camp is the possibility of showing how a place and a person can come to belong to each other—or, rather, how a person can come to belong to a place, for places really belong to nobody. There is a startling reversal of our ordinary sense of things in the recognition that we are the belongings of the world, not its owners. The social convention of ownership must be qualified by this stern fact, and by the humility it implies, if we are not to be blinded altogether to where we are. We may deeply affect a place we own for good or ill, but our lives are nevertheless included in its life; it will survive us, bearing the results. Each of us is a part of a succession. I have come here following Curran Mathews. Who was here or what was done before he came, I do not know. I know that he had predecessors. It is certain that at some time the virgin timber that once stood here was cut down, and no doubt somebody then planted corn among the stumps, and so wore out the ground and allowed the trees to return. Before the white men were the Indians, who generation after generation bequeathed the country to their children, whole, as they received it. The history is largely conjecture. The future is mystery altogether: I do not know who will follow me. These realizations are both aesthetic and moral; they clear the eyes and prescribe an obligation.

At the point when my story was interrupted, my life no longer seemed to be bearing toward this place, but away from it. In the early fall of 1957 Tanya and I left the Camp, and through the following year I taught at Georgetown College. In the spring of 1958 our daughter Mary was born, and in the fall we left Kentucky for the West Coast. The Camp was closed and shuttered. Even though it had taken a new and lasting hold on my mind, it had entered another time of neglect. Three years would pass before I would come back to it. Toward the end of that summer of 1958 my friend Ed McClanahan and I made a canoe trip down the river and spent a night in the Camp, sleeping on the floor. For me, that night had the sadness of a parting. I was about to leave the state; the past was concluded, and the future, not yet begun, was hardly imaginable. The Camp was empty, dark, full of finished memories, already falling back into the decay of human things that humans have abandoned. I was glad when the morning came, and we loafed on down along the shady margin of the river, watching the muskrats and the wood ducks.

I did not go back to the Camp again until the May of 1961. After two years on the West Coast, we had spent another year in Kentucky, this time on the farm, and again we had a departure ahead of us; late in the summer we would be going to Europe for a year. In order to prepare myself for this experience I began spending some mornings and rainy days at the Camp. My intention at first was to do some reading that would help me to understand the life of the places I would be seeing in Europe. But as I might have expected it was not Europe that most held my attention on those days, but the Camp and the riverbank and the river. It soon became clear that I was not so much preparing for an important experience as *having* an important one. I had been changed by what had happened to me and what I had learned during the last three years, and I was no sooner back at the Camp, with the familiar trees around me and the river in front of me again, than I began to see it differently and in some ways more clearly than I had before. Through that summer I wrote a sort of journal, keeping account of what I saw.

I first went back to the Camp that year on a rainy Monday, May 8. Heavy rains had begun the Saturday before, and the river was in flood. The water was under the house, within about a foot and a half of the floor. I had come to read, but mostly I sat and looked. How can one read history when the water is rising? The presence of the present had become insistent, undeniable, and I could not look away; the past had grown still, and could be observed at leisure in a less pressing time. The current was driving drift logs against the legs of the cabin beneath my chair. The river flexed and throbbed against the underpinnings like a great muscle, its vibrations too set in my nerves to permit thinking of anything else.

The river had become a lake, but a lake *flowing*, a continuous island of drift going down the channel, moving swiftly and steadily but forever twisting and eddying within itself; and along the edges the water was picking up little sticks and leaves and bits of grass as it rose. The house's perspective on the river had become the same as that of a boat. I kept an uneasy sense of its nearness, knowing that it was coming nearer; in an hour and a quarter it rose five inches. And there was a sort of permanent astonishment at its massiveness and flatness and oblivious implacable movement.

That day a new awareness of the Camp came to me, an awareness that has become a part of my understanding of all houses. It was a boat—a futile, ill-constructed, doomed boat—a boat such as a child might make on a hilltop. It had been built to stand there on the bank according to the rules of building on solid ground. But now the ground was under water, and the water was rising. The house would have the river to contend with. It would be called on to be a boat, as it had been called on before—as in the 1937 flood it had been called on, and had made its short voyage downstream and up the bank until it lodged among the trees. It was a boat by necessity, but not by nature, which is a recipe for failure. It was built with kinder hopes, to fare in a gentler element than water. All houses are not failed boats, but all are the failed, or failing, vehicles of some alien element; of wind, or fire, or time. When I left that night the river was only a foot beneath the floor, and still rising.

It rose two feet into the Camp, cresting sometime Wednesday. It was out of the house again by Thursday afternoon, and I opened the doors to dry the mud. On Sunday I was at work at a table on the screened porch, sitting in a chair where I could have sat in a boat a few days before, looking down into a landscape that still bore everywhere the marks of overflowing. As far as I could see, up and down across the valley, there was the horizon of the flood, a level in the air below which everything was stained in the dull gray-brown of silt, and the tree branches were hung with tatters of drift, as though the flood was still there in ghostly presence. Above that horizon the spring went on uninterrupted, the new clear green of the leaves unfolding. It was as though I sat with my feet in one world and my head in another. And so with the mud still drying on the floors, I resumed my connection with the Camp.

Other creatures had worse luck getting started that spring than I did. On the Monday of the rise I watched a pair of prothonotary warblers hovering and fluttering around their nest hole near the top of a box elder snag. The snag stood on the bank directly below the porch of the Camp, its top about level with the floor. And so that night when the river crept into the Camp it had already filled the warblers' nest. When I came back after the water went down the birds were back. They nested again in the same hole. And then sometime around the middle of June the snag blew over. Since then I believe that no pair has nested near the Camp, though they nest around the slew across the river and I often see them feeding here.

That spring a pair of phoebes nested under the eaves in front of the house just above the door to the screened porch. A dead elm branch reaching over the porch made them a handy place to sit and watch for insects. I would often pause in my writing or reading to watch them fly out, pick an insect out of the air, and return to the branch. Sometimes I could hear their beaks snap when they made a catch.

A pair of starlings was nesting in a woodpecker hole in a maple down the bank in front of my writing table and a pair of crested flycatchers in another hollow maple a few yards upstream. Titmice were in a woodpecker hole in the dead locust near the kitchen door. High up in an elm, in the fork of a branch hanging over the driveway, pewees built their neat cup of a nest and covered the outside with lichens, so that it looked strangely permanent and ancient, like a small rock.

Wood thrushes lived in the thicker woods upriver. They never came near the house while I was there, but their music did, as though their feeling toward me was both timid and generous. One of the unforgettable voices of this place is their exultant fluting rising out of the morning shadows.

On two separate days, while I sat at work, a hummingbird came in through a hole in the rotted porch screen to collect spiderwebs for his nest. He would stand in the air, deliberate as a harvester, and gather the web in his beak with a sort of winding motion.

Later, in July, I would often watch a red-bellied woodpecker who hunted along the tree trunks on this side of the river to feed his young in a hollow snag on the far side. He would work his way slowly up the trunks of the sycamores, turning his head to the side, putting his eye close to the trunk to search under the loose bark scales. And then he would fly to the far side of the river, where his snag jutted up over the top of a big willow. With the binoculars I could see him perch at his hole and feed his nestlings.

For the first time in all my staying at the Camp I had a pair of binoculars, and perhaps more than anything else during that time, they enlarged and intensified my awareness of the place. With their help I began to know the warblers. At a distance these little birds usually look drab, and the species are hardly distinguishable, but the binoculars show them to be beautifully colored and marked, and wonderfully various in their kinds. There is always something deeply enticing and pleasing to me in the sight of them. Perhaps because I was only dimly aware of them for so long, I always see them at first with a certain unexpectedness, and with the sense of gratitude that one feels for any goodness unearned and almost missed. In their secretive worlds of treetop and undergrowth, they seem among the most remote of the wild creatures. They see little of us, and we see even less of them. I think of them as being aloof somehow from common life. Certain of the most beautiful of them, I am sure, have lived and died for generations in some of our woods without being recognized by a human being.

But the binoculars not only give access to knowledge of lives that are usually elusive and distant; they make possible a peculiar imaginative association with those lives. While opening and clarifying the remote, they block out the immediate. Where one is is no longer apparent. It is as though one stood at the window of a darkened room, lifted into a world that cannot be reached except by flying. The treetops are no longer a ceiling, but a spacious airy zone full of perching places and nervously living lights and shadows. One sees not just the bird, but something of how it is to *be* the bird. One's imagination begins to reach and explore into the sense of how it would be to be without barriers, to fly over the river, to perch at the frailest, most outward branchings of the trees.

In those days I began the long difficult realization of the complexity of the life of this place. Until then—at the level of consciousness, at least—I had thoughtlessly accepted the common assumption of my countrymen that the world is merely an inert surface that man lives on and uses. I don't believe that I had yet read anything on the subject of ecology. But I had read Thoreau and Gilbert White and a little of Fabre, and from seeing natural history displays I knew the concept of the habitat group. And that summer, I remember, I began to think of myself as living within rather than upon the life of the place. I began to think of my life as one among many, and one kind among many kinds. I began to see how little of the beauty and the richness of the world is of human origin, and how superficial and crude and destructive—even self-destructive—is man's conception of himself as the owner of the land and the master of nature and the center of the universe. The Camp with its strip of riverbank woods, like all other places of the earth, stood under its own widening column of infinity, in the neighborhood of the stars, lighted a little, with them, within the element of darkness. It was more unknown than known. It was populated by creatures whose ancestors were here long before my ancestors came, and who had been more faithful to it than I had been, and who would live as well the day after my death as the day before.

Seen as belonging there with other native things, my own nativeness began a renewal of meaning. The sense of belonging began to turn around. I saw that if I belonged here, which I felt I did, it was not because anything here belonged to me. A man might own a whole county and be a stranger in it. If I belonged *in* this place it was because I belonged *to* it. And I began to understand that so long as I did not know the place fully, or even adequately, I belonged to it only partially. That summer I began to see, however dimly, that one of my ambitions, perhaps my governing ambition, was to belong fully to this place, to belong as the thrushes and the herons and the muskrats belonged, to be altogether at home here. That is still my ambition. I have made myself willing to be entirely governed by it. But now I have come to see that it proposes an enormous labor. It is a spiritual ambition, like goodness. The wild creatures belong to the place by nature, but as a man I can belong to it only by understanding and by virtue. It is an ambition I cannot hope to succeed in wholly, but I have come to believe that it is the most worthy of all.

Whenever I could during that summer I would come to the Camp—usually on Sundays, and on days when it was too wet for farm work. I remember vividly the fine feeling I would have, starting out after breakfast with a day at the river ahead of me. The roadsides would be deep in the fresh clear blue of chicory flowers that in the early morning sun appeared to give off a light of their own. And then I would go down into the fog that lay deep in the valley, and begin work. Slowly the sun would burn through the fog, and brighten the wet foliage along the riverbanks in a kind of second dawn. When I got tired of reading or writing I would cut weeds around the house, or trim the trees to open up the view of the river. Or I would take a walk into the woods.

And as before, as always, there was the persistent consciousness of the river, the sense of sitting at the edge of a great opening passing through the country from the Appalachians to the Gulf of Mexico. The river is the ruling presence of this place. Here one is always under its influence. The mind, no matter how it concentrates on other things, is never quite free of it, is always tempted and tugged at by the nearness of the water and the clear space over it, ever widening and deepening into the continent. Its life, in the long warm days of summer when the water is low, is as leisurely and self-preoccupied as the life of a street in a country town. Fish and fishermen pass along it, and so do the kingfishers and the herons. Rabbits and squirrels and groundhogs come down in the late afternoon to drink. The birds crisscross between shores from morning to night. The muskrats graze the weed patches or browse the overhanging willows or carry whole stalks of green corn down out of the fields to the water's edge. Molting wood ducks skulk along the banks, hiding under the willows and behind the screens of grapevines. But of all the creatures, except the fish, I think the swallows enjoy the river most. Whole flocks of barn swallows and rough wings will spend hours in the afternoon and evening circling and dipping over the water, feeding, bathing, drinking—and rejoicing, too, as I steadfastly believe, for I cannot imagine that anything could fly as splendidly as the swallows and not enjoy it.

One afternoon as I was sitting at my worktable on the porch a towboat came up with two barges loaded with sand. A man and a boy sat on the edge of the bow of the head barge, their feet hanging over the water. They were absorbed in their talk, remote from observance, the river world wholly surrounding and containing them, like the boatmen in the paintings of George Caleb Bingham. For the moment they belonged to movement and I to stillness; they were bound in kinship with the river, which is always passing, and I in kinship with the trees, which stand still. I watched them out of sight, intensely aware of them and of their unawareness of me. It is an eloquent memory, full of the meaning of this river.

And on those days, as on all the days I have spent here, I was often accompanied by the thought of Curran Mathews. When I am here, I am always near the thought of him, whether I think about him or not. For me, his memory will always be here, as indigenous and congenial as the sycamores. As long as I am here, I think, he will not be entirely gone. Shortly after the May flood of that year of 1961 I was walking in the weeds on the slope above the original site of the Camp, and I came unexpectingly on a patch of flowers that he had planted there thirty years before. They were trillium, lily of the valley, woodland phlox, Jacob's-ladder, and some ferns. Except for the lily of the valley, these flowers are native to the woods of this area —though they apparently had not returned to this stretch of the riverbank since it was cleared and plowed. And so in digging them up and bringing them here, Curran was assisting amiably in the natural order. They have remained through all the years because they belong here; it is their nature to live in such a place as this. But his pleasure in bringing them here is an addition to them that does not hamper them at all. All their lives they go free of him. Because the river had carried the Camp some distance from the old site, leaving the flowers out of the way of our usual comings and goings, I had never seen them before. They appeared at my feet like some good news of Curran, fresh as if he had spoken it to me, tidings of a day when all was well with him.

Another insistent presence that summer was that of time. The Camp was rapidly aging and wearing out. It had suffered too much abandonment, had been forgotten too much, and the river had flowed into it too many times. Its floors were warped and tilted. The roof leaked where a falling elm branch had punched through the tin. Some of the boards of the walls had begun to rot where the wet weeds leaned against them. What was purposive in it had begun to be overtaken by the necessary accidents of time and weather. Decay revealed its kinship with the earth, and it seemed more than ever to belong to the riverbank. The more the illusion of permanence fell away from it, the easier it fit into the flux of things, as though it entered the fellowship of birds' nests and of burrows. But as a house, it was a failed boat. As a place to sit and work, it was a flimsy, slowly tilting shelf. As a shelter, it was like a tree.

And the day was coming when I would leave again. And again I did not know when I would be back. But this time I did consciously intend to come back. Tanya and I had even begun to talk of building a house someday there on the riverbank, although the possibility seemed a long way off, and the plan was more to comfort ourselves with than to act on. But the plan, because it represented so deep a desire, was vivid to us and we believed in it. Near the original site of the Camp were two fine sycamores, and we thought of a house standing on the slope above them, looking down between them at the river. As a sort of farewell gesture, and as a pledge, I cut down some box elders and elms whose branches had begun to grow obstructively into the crowns of the sycamores. And so, before leaving, I made the beginning of a future that I hoped for and dimly foresaw.

On the twelfth of March the following spring, a letter came to us in Florence, Italy, saying that the river was in the Camp again. Far away as I was, the letter made me strangely restless and sad. I could clearly imagine the look of it. The thought of inundation filled me: the river claiming its valley, making it over again and again—the Camp, my land creature, inhabited by water. The thought, at that distance, that what I knew might be changed filled me and held me in abeyance, as the river filled and held the Camp.

After the year in Europe we lived in New York, and I taught at New York University. We had a second child by then, a boy. In the winter of 1963, I accepted a teaching job at the University of Kentucky to begin in the fall of the following year. That suited us. Our hopes and plans had already turned us back toward Kentucky. We had already spent several years living in other places, and after a second year in New York we would be ready to think of settling down at home.

We returned to Kentucky when school was out that year to spend the summer. And encouraged by the prospect that my relation to the place might soon be permanent, I planned to rebuild the Camp. For one reason, I would be needing a place set aside to do my work in. Another reason, and the main one, was that I needed to preserve the Camp as an idea and a possibility here where it had always been. So many of the good days of my life had been lived here that I could not willingly separate myself from it.

At first I considered repairing the Camp as it stood. But as I looked it over it appeared to be too far gone to be worth the effort and money I would have to spend on it. Besides, it was too near the river; as the watershed deteriorated more and more through misuse, the spring rises had begun to come over the floors too often. Much as I still valued it, the old house had become a relic, and there were no more arguments in its favor. And so on the sixth of June, 1963, I began the work of tearing it down and clearing a place farther up the bank to build it back again. That afternoon I took out the partition between the two rooms, and then cut down the elm that stood where the new building was to be.

The next afternoon I cleared the weeds and bushes off the building site, and with that my sadness at parting with the old house began to give way to the idea of the new. I was going to build the new house several feet higher up the slope than the old one, and to place it so that it would look out between the two big sycamores. Unlike a wild place, a human place gone wild can be strangely forbidding and even depressing. But that afternoon's work made me feel at home here again. My plans suddenly took hold of me, and I began to visualize the new house as I needed it to be and as I thought it ought to look. My work had made the place inhabitable, had set my imagination free in it. I began again to belong to it.

During the next several days I worked back and forth between the old and the new, tearing down and preparing to build. The tearing down was slow work, for I wanted to save and reuse as much of the old lumber as I could, and the floods had rusted all the nails tight in their holes. The sediment of the flood of 1937 still lay on the tops of the rafters. And on the sheeting under the tin of the roof I found the wallpaper the boards had worn when they stood in the walls of the old family house up on the top of the hill. I squared the outlines of the new house on the slope—measuring off a single room twelve by sixteen feet, and a porch eight by eight feet—and dug the holes for the posts it would stand on.

By June 18 I was ready to begin building, and that day two carpenters, friends of mine I had hired to help me, came early in the morning, and we began work. By the night of the twenty-fifth the new house was up, the roof and siding were on, and from that day I continued the work by myself. The old Camp provided the roof, the floor, and three walls of the new, as well as the two doors and some windows. This dependence on the old materials determined to a considerable extent the shape of the new house, for we would shorten or lengthen the dimensions as we went along in accordance with the lengths of the old boards. And so the new house was a true descendant of the old, as the old in its time was the descendant of one still older.

That summer I was deep into the writing on a long book called *A Place on Earth*. And as soon as the heavier work on the house was done and I no longer needed the carpenters, I returned to work on the book, writing in the mornings and continuing work on the house in the afternoons. I nailed battens over the cracks between the boards, braced the underpinnings, made screens and screen doors and shutters and steps, painted the roof and the outside walls. At the end of the summer I had a satisfactory nutshell of a house, green-roofed and brown-walled, that seemed to fit well enough into the place. Standing on its long legs, it had a peering, aerial look, as though built under the influence of trees. But it was heron-like, too, and made for wading at the edge of the water.

The most expensive member of the new house was a big window, six feet by four and a half feet, with forty panes. This was the eye of the house, and I put it in the wall facing the river and built a long worktable under it. In addition, three window sashes from the old Camp were set into each of the end walls. And so the house became a container of shifting light, the sunlight entering by the little windows in the east wall in the morning, and by those in the west wall in the afternoon, and the steadier light from the northward-staring big window over the worktable.

When I began tearing down the old Camp a pair of phoebes was nesting as usual under one of the eaves. Four eggs were already in the nest. I took an old shutter and fixed a little shelf and a sort of porch roof on it, and nailed it to a locust tree nearby, and set the nest with its eggs carefully on it. As I feared, the old birds would have nothing to do with it. The nest stayed where I had put it for another year or so, a symbol of the ended possibilities of the old Camp, and then it blew away.

As I wrote through the mornings of the rest of that summer, a green heron would often be fishing opposite me on the far bank. An old willow had leaned down there until it floated on the water, reaching maybe twenty feet out from the bank. The tree was still living, nearly the whole length of it covered with leafy sprouts. It was dead only at the outer end, which bent up a few inches above the surface of the river. And it was there that the heron fished. He stooped a little, leaning a little forward, his eyes stalking the river as it flowed down and passed beneath him. His attention would be wonderfully concentrated for minutes at a stretch, he would stand still as a dead branch on the trunk of his willow, and then he would itch and have to scratch among his feathers with his beak. When prey swam within his reach, he would crouch, tilt, pick it deftly out of the water, sit back, swallow. Once I saw him plunge headlong into the water and flounder out with his minnow—as if the awkward flogging body had been literally yanked off its perch by the accurate hunger of the beak. When a boat passed he did not fly, but walked calmly back into the shadows among the sprouts of the willow and stood still.

Another bird I was much aware of that summer was the sycamore warbler. Nearly every day I would see one feeding in the tall sycamores in front of the house, or when I failed to see the bird I would hear its song. This is a bird of the tall trees, and he lives mostly in their highest branches. He loves the sycamores. He moves through their crowns, feeding, and singing his peculiar quaking seven-note song, a voice passing overhead like the sun. I am sure I had spent many days of my life with this bird going about his business high over my head, and I had never been aware of him before. This always amazes me; it has happened to me over and over; for years I will go in ignorance of some creature that will later become important to me, as though we are slowly drifting toward each other in time, and then it will suddenly become as visible to me as a star. It is at once almost a habit. After the first sighting I see it often. I become dependent on it, and am uneasy when I do not see it. In the years since I first saw it here and heard its song, the sycamore warbler has come to hold this sort of power over me. I never hear it approaching through the white branches of its trees that I don't stop and listen; or I get the binoculars and watch him as he makes his way from one sycamore to another along the riverbank, a tiny gray bird with a yellow throat, singing from white branch to white branch, among the leaf lights and shadows. When I hear his song for the first time in the spring, I am deeply touched and reassured. It has come to be the most characteristic voice of this place. He is the Camp's emblem bird, as the sycamore is its emblem tree.

From the old Camp the new Camp inherited the fate of a river house. High up as I had built it, I hadn't been able to move it beyond the reach of the river; there was not room enough here below the road to do that. Like the old house, the new was doomed to make its way in the water—to be a failed boat, and survive by luck. The first spring after it was built the river rose more than two feet over the floor. When the water went down, my uncle hosed out the silt and made the place clean again, and when I came back from New York later in the spring it looked the way it had before. But the idea of it had changed. If it had been built in the hope that the river would never rise into it and in the fear that it would, it now lived in the fact that the river had and in the likelihood that it would again. Like all river houses, it had become a stoical house. Sitting in it, I never forget that I am within the reach of an awesome power. It is a truthful house, not indulging the illusion of the permanence of human things. To be here always is not its hope. Long-legged as it is, it is responsive to the natural vibrations. When the dog scratches under the table there is a tremor in the rafters.

Our return from New York in early June of 1964 changed our lives. We were coming back to Kentucky this time with the intention of staying and making a home. Our plans were still unsettled, but our direction was clear. For the first time, we were beginning to have a foreseeable future. From then on, my relation to my native country here might be interrupted occasionally, but it would not be broken. For the summer, we would stay on the farm and I would spend the days working at the Camp, as before.

The previous summer, when I had moved the phoebes' nest from under the eaves of the old Camp before tearing it down, I had said to myself that it would be a good omen if the birds should nest the next year at the new Camp. And they did, building their mossy nest under the roof of the porch. I felt honored by this, as though my work had been received into the natural order. The phoebes had added to its meaning. Later in the summer a pair of Carolina wrens also nested under the porch roof. Instead of one room, I had begun to have a house of apartments where several different kinds of life went on together. And who is to say that one kind is more possible or natural here than another? My writing and the family life of the phoebes go along here together, in a kind of equality.

In that summer of 1964 one of my first jobs was to insulate the Camp, and to wall and ceil it on the inside with six-inch tongue and groove. Once that was done, I resumed my mornings of writing. The afternoons I often spent working around the Camp, or reading, or walking in the woods. As fall approached I had a bottle-gas heater and a two-burner cooking stove installed, which made the Camp ready for cold weather use and for what is known as batching.

We still were unsure what shape our life in Kentucky was going to take, and so we had rented a furnished house in Lexington for the winter. Since I was still at work on my book, the plan was that I would do my teaching at the university on Tuesdays, Wednesdays, and Thursdays, and then drive down to the Camp to write during the other four days. Difficult as this was, it seemed the best way of assuring the quiet and the concentration that I needed for my work. But as a sort of by-product, it also made for the most intense and prolonged experience of the Camp and the river that I had ever known. In many ways that was to be a most critical time for me. Before it was over it would make a deep change in my sense of myself, and in my sense of the country I was born in.

From the beginning of September, when school started, until the first of May, when it was over, I would leave Lexington after supper Thursday night and stay at the Camp through Monday afternoon. Occasionally I would stay for some meeting or another at the university on into Thursday night, and those late drives are the ones I remember best. I would leave the university and drive across Lexington and the suburbs, and then the sixty miles through the country and down along the river to the Camp. As I went, the roads would grow less traveled, the night quieter and lonelier, the darkness broken by fewer lights. I would reach the Camp in the middle of the night, the country quiet and dark all around when I turned off the car engine and the headlights. I would hear an owl calling, or the sound of the small stream on the far bank tumbling into the river. I would go in, light the lamp and the stove, read until the room warmed, and then sleep. The next morning I would be at work on my book at the table under the window. It was always a journey from the sound of public voices to the sound of a private quiet voice rising falteringly out of the roots of my mind, that I listened carefully in the silence to hear. It was a journey from the abstract collective life of the university and the city into the intimate country of my own life. It is only in a country that is well known, full of familiar names and places, full of life that is always changing, that the mind goes free of abstractions, and renews itself in the presence of the creation, that so persistently eludes human comprehension and human law. It is only in the place that one belongs to, intimate and familiar, long watched over, that the details rise up out of the whole and become visible: the hawk stoops into the clearing before one's eyes; the wood drake, aloof and serene in his glorious plumage, swims out of his hiding place.

One clear morning as the fall was coming on I saw a chipmunk sunning on a log, as though filling himself with light and warmth in preparation for his winter sleep. He was wholly preoccupied with the sun, for though I watched him from only a few steps away he did not move. And while he mused or dozed rusty-golden sycamore leaves bigger than he was were falling all around him.

With the approach of winter the country opened. Around the Camp the limits of seeing drew back like the eyelids of a great eye. The foliage that since spring had enclosed it slowly fell away, and the outlook from its windows came to include the neighboring houses. It was as though on every frosty night the distances stole up nearer.

On the last morning of October, waking, I looked out the window and saw a fisherman in a red jacket fishing alone in his boat tied against the far bank. He sat deeply quiet and still, unmoved as a tree by my rising and the other events that went on around him. There was something heron-like in his intent waiting upon what the day might bring him out of the dark. In his quietness and patience he might have been the incarnation of some river god, at home among all things, awake while I had slept.

One bright warm day in November it was so quiet that I could hear the fallen leaves ticking, like a light rain, as they dried and contracted, scraping their points and edges against each other. That day I saw the first sapsuckers, which are here only in the winter.

Another day I woke to see the trees below the house full of birds: chickadees, titmice, juncos, bluebirds, jays. They had found a red screech owl asleep in a hollow in one of the water maples. It was a most noisy event, and it lasted a long time. The bluebirds would hover, fluttering like sparrow hawks, over the owl's hole, looking in. The titmice would perch on the very lip of the hole and scold and then, as if in fear of their own bravery, suddenly startle away. They all flew away and came back again two or three times. Everybody seemed to have a great backlog of invective to hurl down upon the head of the owl, who apparently paid no attention at all—at least when I climbed up and looked in he paid no attention to me.

I believe that the owl soon changed his sleeping place; when I next looked in he was gone. But for days afterward the birds of the neighborhood pretended he was still there, and would stop in passing to enact a sort of ritual of outrage and fright. The titmice seemed especially susceptible to the fascination of that hole. They would lean over it, yell down into it, and then spring away in a spasm of fear. They seemed to be scaring themselves for fun, like children playing around a deserted house. And yet for both birds and children there must be a seriousness in such play; they mimic fear in order to be prepared for it when it is real.

While I was eating breakfast the morning after the birds' discovery of the owl, I heard several times a voice that I knew was strange to this place. I was reading as I ate, and at first I paid little attention. But the voice persisted, and when I put my mind to it I thought it must be that of a goose. I went out with the binoculars, and saw two blue geese, the young of that year, on the water near the far bank. Though the morning was clear and the sun well up, there was a light fog blowing over the river, thickening and thinning out and thickening again, making it difficult to see, but I made out their markings clearly enough. While I watched they waded out on shore at a place where the bank had slipped, preened their feathers, and drank, and after about fifteen minutes flew away.

That afternoon I found them again in the same slip where I had seen them in the morning. I paddled the canoe within twenty feet of them, and then they only flew out onto the water a short distance away. I thought that since these birds nest to the north of Hudson Bay and often fly enormous distances in migration, I might have been the first man these geese had ever seen. Not wanting to call attention to them and so get them shot, I went on across the river and walked into the bottom on the far side. I spent some time there, looking at the wood ducks and some green-winged teal on the slews, and when I returned to the canoe a little before five the geese were gone. But my encounter with them cast a new charm on my sense of the place. They made me realize that the geography of this patch of riverbank takes in much of the geography of the world. It is under the influence of the Arctic, where the winter birds go in summer, and of the tropics, where the summer birds go in winter. It is under the influence of forests and of croplands and of strip mines in the Appalachians, and it feels the pull of the Gulf of Mexico. How many nights have the migrants loosened from their guide stars and descended here to rest or to stay for a season or to die and enter this earth? The geography of this place is airy and starry as well as earthy and watery. It has been arrived at from a thousand other places, some as far away as the poles. I have come here from great distances myself, and am resigned to the knowledge that I cannot go without leaving it better or worse. Here as well as any place I can look out my window and see the world. There are lights that arrive here from deep in the universe. A man can be provincial only by being blind and deaf to his province.

In December the winter cold began. Early in the mornings when it would be clear and cold the drift logs going down the channel would be white with frost, not having moved except as the current moved all night, as firmly embedded in the current, almost, as in the ground. A sight that has always fascinated me, when the river is up and the water swift, is to see the birds walking about, calmly feeding, on the floating logs and the mats of drift as they pass downward, slowly spinning in the currents. The ducks, too, like to feed among the drift in the channel at these times. I have seen mallards drift down, feeding among the uprooted tree trunks and the cornstalks and the rafted brush, and then fly back upstream to drift down again.

A voice I came to love and listen for on the clear cold mornings was that of the Carolina wren. He would be quick and busy, on the move, singing as he went. Unlike the calls of the other birds, whose songs, if they sang at all, would be faltering and halfhearted in the cold, the wren's song would come big and clear, filling the air of the whole neighborhood with energy, as though he could not bear to live except in the atmosphere of his own music.

Toward the end of that December a gray squirrel began building a nest in a hollow sycamore near the house. He seemed unable or unwilling to climb the trunk of the sycamore; in all the time I watched him he never attempted to do so. He always followed the same complicated route to his nest: up a grapevine, through the top of an elm and then, by a long leap, into the sycamore. On his way down this route offered him no difficulties, but the return trip, when he carried a load of sycamore leaves in his mouth, seemed fairly risky. The big leaves gave him a good deal of trouble; he frequently stopped and worked with his forepaws to make the load more compact. But because he persisted in carrying as big loads as he could, his forward vision seemed usually to be blocked altogether. I believe that his very exacting leaps were made blind, by memory, after a bit of nervous calculation beforehand. When he moved with a load of leaves, apparently because of the obstruction of his vision, he was always extremely wary, stopping often to listen and look around.

I remember one night of snow, so cold that the snow squeaked under my feet when I went out. The valley was full of moonlight; the fields were dazzling white, the woods deep black, the shadows of the trees printed heavily on the snow. And it was quiet everywhere. As long as I stood still there was not a sound.

That winter a pair of flickers drilled into the attic and slept there. Sometimes at night, after I went to bed, I could hear them stirring. But contrary to my sense of economy—though, I suppose, in keeping with theirs—they did not make do with a single hole, but bored one in one gable and one in the other. I shared my roof with them until the cold weather ended, and then evicted them. Later I put up a nesting box for them, which was promptly taken over by starlings.

One sunny morning of high water in April while I sat at work, keeping an eye on the window as usual, there were nineteen coot, a pair of wood ducks, and two pairs of blue-winged teal feeding together near the opposite bank. They fed facing upstream, working against the current, now and then allowing themselves to be carried downstream a little way, then working upstream again. After a while the four teal climbed out onto a drift log caught in the bushes near the shore. For some time they sat sunning and preening there. And then the log broke loose and began to drift again. The four birds never moved. They rode it down the river and out of sight. They accepted this accident of the river as much as a matter of fact as if it had been a purpose of theirs. Able both to swim and to fly, they made a felicity of traveling by drift log, as if serendipity were merely a way of life with them.

On the tenth of April, I woke at about six o'clock, and the first sound I heard was the song of the sycamore warbler, returned from the South. With that my thoughts entered spring. I went into the woods and found the bloodroot in bloom. Curran's flowers were coming up on the slope beside the house. In the warm evening I noticed other spring music: the calling of doves, and the slamming of newly hung screen doors.

During that winter I had spent many days and nights of watchfulness and silence here. I had learned the power of silence in a place—silence that is the imitation of absence, that permits one to be present as if absent, so that the life of the place goes its way undisturbed. It proposes an ideal of harmlessness. A man should be in the world as though he were not in it, so that it will be no worse because of his life. His obligation may not be to make "a better world," but the world certainly requires of him that he make it no worse. That, at least, was man's moral circumstance before he began his ruinous attempt to "improve" on the creation; now, perhaps, he is under an obligation to leave it better than he found it, by undoing some of the effects of his meddling and restoring its old initiatives—by making his absence the model of his presence.

But there was not only the power of silence; there was the power of attentiveness, of permanence of interest. By coming back to Kentucky and renewing my devotion to the Camp and the river valley, I had, in a sense, made a marriage with the place. I had established a trust, and within the assurance of the trust the place had begun to reveal its life to me in moments of deep intimacy and beauty. I had to come here unequivocally, accepting the place as my fate and privilege, before I could see it with clarity. I had become worthy to see what I had inherited by being born here. I had been a native; now I was beginning to belong. There is no word—certainly not *native* or *citizen*—to suggest the state I mean, that of belonging willingly and gladly and with some fullness of knowledge to a place. I had ceased to be a native as men usually are, merely by chance and legality, and had begun to be native in the fashion of the birds and animals; I had begun to be born here in mind and spirit as well as in body.

For some months after our return to Kentucky we assumed that we would settle more or less permanently in Lexington, near the university, and perhaps have a place here in the country to come to for the summers. We had thought of building a couple of more rooms onto the Camp for that purpose, and we had thought of buying a piece of woods out of reach of the river and building there. But in November of that first winter the Lane's Landing property, adjoining the Camp on the downriver side, came up for sale, and I was able to buy it. We would, we told ourselves, fix it up a little, use it in the summers, and perhaps settle there permanently some day. The previous owners moved out in February, and I began spending some of my afternoons there, working to get ready for the carpenters who would work there later on. But on the weekends Tanya and the children would often come down from Lexington, and we would walk over the place, and through the rooms of the house, talking and looking and measuring and planning. And soon we began to see possibilities there that we could not resist. Our life began to offer itself to us in a new way, in the terms of that place, and we could not escape it or satisfy it by anything partial or temporary. We made up our minds to live there. By the morning in April when I first heard the sycamore warbler, we had begun a full-scale overhaul of the house, and I had planted two dozen fruit trees. Early in July, with work completed on only three of the rooms, we moved in. After eight years, our lives enlarged in idea and in concern, our marriage enlarged into a family, we had come back to where we had begun. The Camp, always symbolically the center of our life, had fastened us here at last.

It has been several years now since I first consciously undertook to learn about the natural history of this place. My desire to do this grew out of the sense that the human life of the country is only part of its life, and that in spite of the extreme *effects* of modern man's presence on the land, his relation to it is largely superficial. In spite of all his technical prowess, nothing we have built or done has the permanence, or the congeniality with the earth, of the nesting instincts of the birds.

As soon as I felt a necessity to learn about the nonhuman world, I wished to learn about it in a hurry. And then I began to learn perhaps the most important lesson that nature had to teach me: that I could not learn about her in a hurry. The most important learning, that of experience, can be neither summoned nor sought out. The most worthy knowledge cannot be acquired by what is known as study—though that is necessary, and has its use. It comes in its own good time and in its own way to the man who will go where it lives, and wait, and be ready, and watch. Hurry is beside the point, useless, an obstruction. The thing is to be attentively present. To sit and wait is as important as to move. Patience is as valuable as industry. What is to be known is always there. When it reveals itself to you, or when you come upon it, it is by chance. The only condition is your being there and being watchful.

Though it has come slowly and a little at a time, by bits and fragments sometimes weeks apart, I realize after so many years of just being here that my knowledge of the life of this place is rich, my own life part of its richness. And at that I have only made a beginning. Eternal mysteries are here, and temporal ones too. I expect to learn many things before my life is over, and yet to die ignorant. My most inspiring thought is that this place, if I am to live well in it, requires and deserves a lifetime of the most careful attention. And the day that will finally enlighten me, if it ever comes, will come as the successor of many days spent here unenlightened or benighted entirely. "It requires more than a day's devotion," Thoreau says, "to know and to possess the wealth of a day."

At the same time my days here have taught me the futility of living for the future. Men who drudge all their lives in order to retire happily are the victims of a cheap spiritual fashion invented for their enslavement. It is no more possible to live in the future than it is to live in the past. If life is not now, it is never. It is impossible to imagine "how it will be," and to linger over that task is to prepare a disappointment. The tomorrow I hope for may very well be worse than today. There is a great waste and destructiveness in our people's desire to "get somewhere." I myself have traveled several thousand miles to arrive at Lane's Landing, five miles from where I was born, and the knowledge I gained by my travels was mainly that I was born into the same world as everybody else.

Days come to me here when I rest in spirit, and am involuntarily glad. I sense the adequacy of the world, and believe that everything I need is here. I do not strain after ambition or heaven. I feel no dependence on tomorrow. I do not long to travel to Italy or Japan, but only across the river or up the hill into the woods.

And somewhere back of all this, in a relation too intricate and profound to trace out, is the life that Curran Mathews lived here before me. Perhaps he, too, experienced holy days here. Perhaps he only sensed their possibility. But if he had not come here and made a firm allegiance with this place, it is likely that I never would have. I am his follower and his heir. "For an inheritance to be really great," René Char says, "the hand of the deceased must not be seen." The Camp is my inheritance from Curran Mathews, and though certain of his meanings continue in it, his hand is not on it. As an inheritance, he touched it only as a good man touches the earth—to cherish and augment it. Where his hand went to the ground one forgotten day the flowers rise up spring after spring.

Now it is getting on toward the end of March. Just as the grass had started to grow and the jonquils were ready to bloom, we had a foot of snow and more cold. Today it is clear and thawing, but the ground is still white. Though the redbird sings his mating song, it is still winter, and my thoughts keep their winter habits. But soon there will come a day when, without expecting to, I will hear the clear seven-note song of the sycamore warbler passing over the Camp roof. Something will close and open in my mind like a page turning. It will be another spring.

A Native Hill

Pull down thy vanity, it is not man
Made courage, or made order, or made grace,
Pull down thy vanity, I say pull down.
Learn of the green world what can be thy place. . . .
—Ezra Pound, Canto LXXXI

THE HILL IS NOT a hill in the usual sense. It has no "other side." It is an arm of Kentucky's central upland known as The Bluegrass; one can think of it as a ridge reaching out from that center, progressively cut and divided, made ever narrower by the valleys of the creeks that drain it. The town of Port Royal in Henry County stands on one of the last heights of this upland, the valleys of two creeks, Gullion's Branch and Cane Run, opening on either side of it into the valley of the Kentucky River. My house backs against the hill's foot where it descends from the town to the river. The river, whose waters have carved the hill and so descended from it, lies within a hundred steps of my door.

Within about four miles of Port Royal, on the upland and in the bottoms upriver, all my grandparents and great-grandparents lived and left such memories as their descendants have bothered to keep. Little enough has been remembered. The family's life here goes back to my mother's great-great-grandfather and to my father's great-grandfather, but of those earliest ones there are only a few vague word-of-mouth recollections. The only place antecedent to this place that has any immediacy to any of us is the town of Cashel in County Tipperary, Ireland, which one of my great-grandfathers left as a boy to spend the rest of his life in Port Royal. His name was James Mathews, and he was a shoemaker. So well did he fit his life into this place that he is remembered, even in the family, as having belonged here. The family's only real memories of Cashel are my own, coming from a short visit I made there five years ago.

And so such history as my family has is the history of its life here. All that any of us may know of ourselves is to be known in relation to this place. And since I did most of my growing up here, and have had most of my most meaningful experiences here, the place and the history, for me, have been inseparable, and there is a sense in which my own life is inseparable from the history and the place. It is a complex inheritance, and I have been both enriched and bewildered by it.

I began my life as the old times and the last of the old-time people were dying out. The Depression and World War II delayed the mechanization of the farms here, and one of the first disciplines imposed on me was that of a teamster. Perhaps I first stood in the role of student before my father's father, who, halting a team in front of me, would demand to know which mule had the best head, which the best shoulder or rump, which was the lead mule, were they hitched right. And there came a time when I knew, and took a considerable pride in knowing. Having a boy's usual desire to play at what he sees men working at, I learned to harness and hitch and work a team. I felt distinguished by that, and took the same pride in it that other boys my age took in their knowledge of automobiles. I seem to have been born with an aptitude for a way of life that was doomed, although I did not understand that at the time. Free of any intuition of its doom, I delighted in it, and learned all I could about it.

That knowledge, and the men who gave it to me, influenced me deeply. It entered my imagination, and gave its substance and tone to my mind. It fashioned in me possibilities and limits, desires and frustrations, that I do not expect to live to the end of. And it is strange to think how barely in the nick of time it came to me. If I had been born five years later I would have begun in a different world, and would no doubt have become a different man.

Those five years made a critical difference in my life, and it is a historical difference. One of the results is that in my generation I am something of an anachronism. I am less a child of my time than the people of my age who grew up in the cities, or than the people who grew up here in my own place five years after I did. In my acceptance of twentieth-century realities there has had to be a certain deliberateness, whereas most of my contemporaries had them simply by being born to them.

In my teens, when I was away at school, I could comfort myself by recalling in intricate detail the fields I had worked and played in, and hunted over, and ridden through on horseback—and that were richly associated in my mind with people and with stories. I could recall even the casual locations of certain small rocks. I could recall the look of a hundred different kinds of daylight on all those places, the look of animals grazing over them, the postures and attitudes and movements of the men who worked in them, the quality of the grass and the crops that had grown on them. I had come to be aware of it as one is aware of one's body; it was present to me whether I thought of it or not.

I believe that this has made for a high degree of particularity in my mental processes. When I have thought of the welfare of the earth, the problems of its health and preservation, the care of its life, I have had this place before me, the part representing the whole more vividly and accurately, making clearer and more pressing demands, than any *idea* of the whole. When I have thought of kindness or cruelty, weariness or exuberance, devotion or betrayal, carelessness or care, doggedness or awkwardness or grace, I have had in my mind's eye the men and women of this place, their faces and gestures and movements. It seems to me that because of this I have a more immediate feeling for abstract principles than many of my contemporaries; the values of principles are more vivid to me.

I have pondered a great deal over a conversation I took part in a number of years ago in one of the offices of New York University. I had lived away from Kentucky for several years—in California, in Europe, in New York City. And now I had decided to go back and take a teaching job at the University of Kentucky, giving up the position I then held on the New York University faculty. That day I had been summoned by one of my superiors at the university, whose intention, I had already learned, was to persuade me to stay on in New York "for my own good."

The decision to leave had cost me considerable difficulty and doubt and hard thought—for hadn't I achieved what had become one of the almost traditional goals of American writers? I had reached the greatest city in the nation; I had a good job; I was meeting other writers and talking to them and learning from them; I had reason to hope that I might take a still larger part in the literary life of that place. On the other hand, I knew I had not escaped Kentucky, and had never really wanted to. I was still writing about it, and had recognized that I would probably need to write about it for the rest of my life. Kentucky was my fate—not an altogether pleasant fate, though it had much that was pleasing in it, but one that I could not leave behind simply by going to another place, and that I therefore felt more and more obligated to meet directly and to understand. Perhaps even more important, I still had a deep love for the place I had been born in, and liked the idea of going back to be part of it again. And that, too, I felt obligated to try to understand. Why should I love one place so much more than any other? What could be the meaning or use of such love?

The elder of the faculty began the conversation by alluding to Thomas Wolfe, who once taught at the same institution. "Young man," he said, "don't you know you can't go home again?" And he went on to speak of the advantages, for a young writer, of living in New York among the writers and the editors and the publishers.

The conversation that followed was a persistence of politeness in the face of impossibility. I knew as well as Wolfe that there is a certain *metaphorical* sense in which you can't go home again—that is, the past is lost to the extent that it cannot be lived in again. I knew perfectly well that I could not return home and be a child, or recover the secure pleasures of childhood. But I knew also that as the sentence was spoken to me it bore a self-dramatizing sentimentality that was absurd. Home—the place, the countryside—was still there, still pretty much as I left it, and there was no reason I could not go back to it if I wanted to.

As for the literary world, I had ventured some distance into that, and liked it well enough. I knew that because I was a writer the literary world would always have an importance for me and would always attract my interest. But I never doubted that the world was more important to me than the literary world; and the world would always be most fully and clearly present to me in the place I was fated by birth to know better than any other.

And so I had already chosen according to the most intimate and necessary inclinations of my own life. But what keeps me thinking of that conversation is the feeling that it was a confrontation of two radically different minds, and that it was a confrontation with significant historical overtones.

I do not pretend to know all about the other man's mind, but it was clear that he wished to speak to me as a representative of the literary world—the world he assumed that I aspired to above all others. His argument was based on the belief that once one had attained the metropolis, the literary capital, the worth of one's origins was canceled out; there simply could be nothing *worth* going back to. What lay behind one had ceased to be a part of life, and had become "subject matter." And there was the belief, long honored among American intellectuals and artists and writers, that a place such as I came from could be returned to only at the price of intellectual death; cut off from the cultural springs of the metropolis, the American countryside is Circe and Mammon. Finally, there was the assumption that the life of the metropolis is *the* experience, the *modern* experience, and that the life of the rural towns, the farms, the wilderness places is not only irrelevant to our time, but archaic as well because unknown or unconsidered by the people who really matter—that is, the urban intellectuals.

I was to realize during the next few years how false and destructive and silly those ideas are. But even then I was aware that life outside the literary world was not without honorable precedent: if there was Wolfe, there was also Faulkner; if there was James, there was also Thoreau. But what I had in my mind that made the greatest difference was the knowledge of the few square miles in Kentucky that were mine by inheritance and by birth and by the intimacy the mind makes with the place it awakens in.

I can ask now why it should be so automatically supposed that a man is wrong when he turns to face what he cannot escape, and permits himself to accept what he has desired. And I know that the answer is that in so turning back I turned directly against the current of intellectual fashion, which for so long in America has paralleled the movement of the population from country to city. At times I have found the going difficult. Though at the time of my return I felt that I was acting in faithfulness to my nature and possibilities, and now am certain that I was, there were many months when virtually none of my friends agreed with me. During the first year I was back in Kentucky, though my work apparently gave no evidence of decline, I received letters warning me against the Village Virus and the attitudes of Main Street, counseling me to remain broad-minded and intellectually aware, admonishing that I should be on the lookout for signs of decay in my work and in my mind. It was feared that I would grow paunchy and join the Farm Bureau.

It is of interest to me that, at the start I was not at all immune to this sort of thing. I felt an obligation to listen, and to take these warnings to heart. I kept an eye on myself, as though the promised signs of decay might appear at any moment. Even at the first I never really thought they would, but I continued, as though honor-bound, to expect them even after I had become sure that they wouldn't. Which is to say that, although I had maintained a vital connection with my origins, I had been enough influenced by the cultural fashion to have become compulsively suspicious both of my origins and of myself for being unwilling to divide myself from them.

I have come finally to see a very regrettable irony in what happened. At a time when originality is more emphasized in the arts, maybe, than ever before, I undertook something truly original—I returned to my origins—and it was generally thought by my literary friends that I had worked my ruin. As far as I can tell, this was simply because *my* originality, my faith in my own origins, had not been anticipated or allowed for by the *fashion* of originality.

What finally freed me from these doubts and suspicions was the insistence in what was happening to me that, far from being bored and diminished and obscured to myself by my life here, I had grown more alive and more conscious than I had ever been.

I had made a significant change in my relation to the place: before, it had been mine by coincidence or accident; now it was mine by choice. My return, which at first had been hesitant and tentative, grew wholehearted and sure. I had come back to stay. I hoped to live here the rest of my life. And once that was settled I began to *see* the place with a new clarity and a new understanding and a new seriousness. Before coming back I had been willing to allow the possibility—which one of my friends insisted on—that I already knew this place as well as I ever would. But now I began to see the real abundance and richness of it. It is, I saw, inexhaustible in its history, in the details of its life, in its possibilities. I walked over it, looking, listening, smelling, touching, alive to it as never before. I listened to the talk of my kinsmen and neighbors as I never had done, alert to their knowledge of the place, and to the qualities and energies of their speech. I began more seriously than ever to learn the names of things—the wild plants and animals, the natural processes, the local places—and to articulate my observations and memories. My language increased and strengthened, and sent my mind into the place like a live root system. And so what has become the usual order of things reversed itself with me; my mind became the root of my life rather than its sublimation. I came to see myself as growing out of the earth like the other native animals and plants. I saw my body and my daily motions as brief coherences and articulations of the energy of the place, which would fall back into the earth like leaves in the autumn.

In this awakening there has been a good deal of pain. When I lived in other places I looked on their evils with the curious eye of a traveler; I was not responsible for them; it cost me nothing to be a critic, for I had not been there long, and I did not feel that I would stay. But here, now that I am both native and citizen, there is no immunity to what is wrong. It is impossible to escape the sense that I am involved in history. What I am has been to a considerable extent determined by what my forefathers were, by how they chose to treat this place while they lived in it; the lives of most of them diminished it, and limited its possibilities, and narrowed its future. And every day I am confronted by the question of what inheritance I will leave. What do I have that I am using up? For it has been our history that each generation in this place has been less welcome to it than the last. There has been less here for them. At each arrival there has been less fertility in the soil, and a larger inheritance of destructive precedent and shameful history.

I am forever being crept up on and newly startled by the realization that my people established themselves here by killing or driving out the original possessors, by the awareness that people were once bought and sold here by my people, by the sense of the violence they have done to their own kind and to each other and to the earth, by the evidence of their persistent failure to serve either the place or their own community in it. I am forced, against all my hopes and inclinations, to regard the history of my people here as the progress of the doom of what I value most in the world: the life and health of the earth, the peacefulness of human communities and households.

And so here, in the place I love more than any other and where I have chosen among all other places to live my life, I am more painfully divided within myself than I could be in any other place.

I know of no better key to what is adverse in our heritage in this place than the account of "The Battle of the Fire-Brands," quoted in Collins' *History of Kentucky* "from the autobiography of Rev. Jacob Young, a Methodist minister." The "Newcastle" referred to is the present-day New Castle, the county seat of Henry County. I give the quote in full:

> The costume of the Kentuckians was a hunting shirt, buckskin pantaloons, a leathern belt around their middle, a scabbard, and a big knife fastened to their belt; some of them wore hats and some caps. Their feet were covered with moccasins, made of dressed deer skins. They did not think themselves dressed without their powder-horn and shot-pouch, or the gun and the tomahawk. They were ready, then, for all alarms. They knew but little. They could clear ground, raise corn, and kill turkeys, deer, bears, and buffalo; and, when it became necessary, they understood the art of fighting the Indians as well as any men in the United States.
>
> Shortly after we had taken up our residence, I was called upon to assist in opening a road from the place where Newcastle now stands, to the mouth of the Kentucky

river. That country, then, was an unbroken forest; there was nothing but an Indian trail passing the wilderness. I met the company early in the morning, with my axe, three days' provisions, and my knapsack. Here I found a captain, with about 100 men, all prepared to labor; about as jovial a company as I ever saw, all good-natured and civil. This was about the last of November, 1797. The day was cold and clear. The country through which the company passed was delightful; it was not a flat country, but, what the Kentuckians called, rolling ground—was quite well stored with lofty timber, and the undergrowth was very pretty. The beautiful canebrakes gave it a peculiar charm. What rendered it most interesting was the great abundance of wild turkeys, deer, bears, and other wild animals. The company worked hard all day, in quiet, and every man obeyed the captain's orders punctually.

About sundown, the captain, after a short address, told us the night was going to be very cold, and we must make very large fires. We felled the hickory trees in great abundance; made great log-heaps, mixing the dry wood with the green hickory; and, laying down a kind of sleepers under the pile, elevated the heap and caused it to burn rapidly. Every man had a water vessel in his knapsack; we searched for and found a stream of water. By this time, the fires were showing to great advantage; so we warmed our cold victuals, ate our suppers, and spent the evening in hearing the hunter's stories relative to the bloody scenes of the Indian war. We then heard some pretty fine singing, considering the circumstances.

Thus far, well; but a change began to take place. They became very rude, and raised the war-whoop. Their shrill shrieks made me tremble. They chose two captains, divided the men into two companies, and commenced fighting with the fire-brands—the log heaps having burned down. The only law for their government was, that no man should throw a brand without fire on it—so that they might know how to dodge. They fought, for two or three hours, in perfect good nature; till brands became scarce, and they began to violate the law. Some were severely wounded, blood began to flow freely, and

> they were in a fair way of commencing a fight in earnest. At this moment, the loud voice of the captain rang out above the din, ordering every man to retire to rest. They dropped their weapons of warfare, rekindled the fires, and laid down to sleep. We finished our road according to directions, and returned home in health and peace.

The significance of this bit of history is in its utter violence. The work of clearing the road was itself violent. And from the orderly violence of that labor, these men turned for amusement to disorderly violence. They were men whose element was violence; the only alternatives they were aware of were those within the comprehension of main strength. And let us acknowledge that these were the truly influential men in the history of Kentucky, as well as in the history of most of the rest of America. In comparison to the fatherhood of such as these, the so-called "founding fathers" who established our political ideals are but distant cousins. It is not John Adams or Thomas Jefferson whom we see night after night in the magic mirror of the television set; we see these builders of the road from New Castle to the mouth of the Kentucky River. Their reckless violence has glamorized all our trivialities and evils. Their aggressions have simplified our complexities and problems. They have cut all our Gordian knots. They have appeared in all our disguises and costumes. They have worn all our uniforms. Their war whoop has sanctified our inhumanity and ratified our blunders of policy.

To testify to the persistence of their influence, it is only necessary for me to confess that I read the Reverend Young's account of them with delight; I yield a considerable admiration to the exuberance and extravagance of their fight with the firebrands; I take a certain pride in belonging to the same history and the same place that they belong to—though I know that they represent the worst that is in us, and in me, and that their presence in our history has been ruinous, and that their survival among us promises ruin.

"They knew but little," the observant Reverend says of them, and this is the most suggestive thing he says. It is surely understandable and pardonable, under the circumstances, that these men were ignorant by the standards of formal schooling.

But one immediately reflects that the American Indian, who was ignorant by the same standards, nevertheless knew how to live in the country without making violence the invariable mode of his relation to it; in fact, from the ecologist's or the conservationist's point of view, he did it *no* violence. This is because he had, in place of what we would call education, a fully integrated culture, the content of which was a highly complex sense of his dependence on the earth. The same, I believe was generally true of the peasants of certain old agricultural societies, particularly in the Orient. They belonged by an intricate awareness to the earth they lived on and by, which meant that they respected it, which meant that they practiced strict economies in the use of it.

The abilities of those Kentucky road builders of 1797 were far more primitive and rudimentary than those of the Stone Age people they had driven out. They could clear the ground, grow corn, kill game, and make war. In the minds and hands of men who "know but little"—or little else—all of these abilities are certain to be destructive, even of those values and benefits their use may be intended to serve.

On such a night as the Reverend Young describes, an Indian would have made do with a small shelter and a small fire. But these road builders, veterans of the Indian War, "felled the hickory trees in great abundance; made great log-heaps . . . and caused [them] to burn rapidly." Far from making a small shelter that could be adequately heated by a small fire, their way was to make no shelter at all, and heat instead a sizable area of the landscape. The idea was that when faced with abundance one should consume abundantly—an idea that has survived to become the basis of our present economy. It is neither natural nor civilized, and even from a "practical" point of view it is to the last degree brutalizing and stupid.

I think that the comparison of these road builders with the Indians, on the one hand, and with Old World peasants on the other, is a most suggestive one. The Indians and the peasants were people who belonged deeply and intricately to their places. Their ways of life had evolved slowly in accordance with their knowledge of their land, of its needs, of their own relation of dependence and responsibility to it. The road builders, on the contrary, were *placeless* people. That is why they "knew but little." Having left Europe far behind, they had not yet in any meaningful sense arrived in America, not yet having *devoted* themselves to any part of it in a way that would produce the intricate knowledge of it necessary to live in it without destroying it. Because they belonged to no place, it was almost inevitable that they should behave violently toward the places they came to. We *still* have not, in any meaningful way, arrived in America. And in spite of our great reservoir of facts and methods, in comparison to the deep earthly wisdom of established peoples we still know but little.

But my understanding of this curiously parabolic fragment of history will not be complete until I have considered more directly that the occasion of this particular violence was the building of a road. It is obvious that one who values the idea of community cannot speak against roads without risking all sorts of absurdity. It must be noticed, nevertheless, that the predecessor to this first road was "nothing but an Indian trail passing the wilderness"—a path. The Indians, then, who had the wisdom and the grace to live in this country for perhaps ten thousand years without destroying or damaging any of it, needed for their travels no more than a footpath; but their successors, who in a century and a half plundered the area of at least half its topsoil and virtually all of its forest, felt immediately that they had to have a road. My interest is not in the question of whether or not they *needed* the road, but in the fact that the road was then, and is now, the most characteristic form of their relation to the country.

The difference between a path and a road is not only the obvious one. A path is little more than a habit that comes with knowledge of a place. It is a sort of ritual of familiarity. As a form, it is a form of contact with a known landscape. It is not destructive. It is the perfect adaptation, through experience and familiarity, of movement to place; it obeys the natural contours; such obstacles as it meets it goes around. A road, on the other hand, even the most primitive road, embodies a resistance against the landscape. Its reason is not simply the necessity for movement, but haste. Its wish is to *avoid* contact with the landscape; it seeks so far as possible to go over the country, rather than through it; its aspiration, as we see clearly in the example of our modern freeways, is to be a bridge; its tendency is to translate place into space in order to traverse it with the least effort. It is destructive, seeking to remove or destroy all obstacles in its way. The primitive road advanced by the destruction of the forest; modern roads advance by the destruction of topography.

That first road from the site of New Castle to the mouth of the Kentucky River—lost now either by obsolescence or metamorphosis—is now being crossed and to some extent replaced by its modern descendant known as I-71, and I have no wish to disturb the question of whether or not *this* road was needed. I only want to observe that it bears no relation whatever to the country it passes through. It is a pure abstraction, built to serve the two abstractions that are the poles of our national life: commerce and expensive pleasure. It was built, not according to the lay of the land, but according to a blueprint. Such homes and farmlands and woodlands as happened to be in its way are now buried under it. A part of a hill near here that would have caused it to turn aside was simply cut down and disposed of as thoughtlessly as the pioneer road builders would have disposed of a tree. Its form is the form of speed, dissatisfaction, and anxiety. It represents the ultimate in engineering sophistication, but the crudest possible valuation of life in this world. It is as adequate a symbol of our relation to our country now as that first road was of our relation to it in 1797.

But the sense of the past also gives a deep richness and resonance to nearly everything I see here. It is partly the sense that what I now see other men that I have known once saw, and partly that this knowledge provides an imaginative access to what I do not know. I think of the country as a kind of palimpsest scrawled over with the comings and goings of people, the erasure of time already in process even as the marks of passage are put down. There are the ritual marks of neighborhood—roads, paths between houses. There are the domestic paths from house to barns and outbuildings and gardens, farm roads threading the pasture gates. There are the wanderings of hunters and searchers after lost stock, and the speculative or meditative or inquisitive "walking around" of farmers on wet days and Sundays. There is the spiraling geometry of the rounds of implements in fields, and the passing and returning scratches of plows across croplands. Often these have filled an interval, an opening, between the retreat of the forest from the virgin ground and the forest's return to ground that has been worn out and given up. In the woods here one often finds cairns of stones picked up out of furrows, gullies left by bad farming, forgotten roads, stone chimneys of houses long rotted away or burned.

Occasionally one stumbles into a coincidence that, like an unexpected alignment of windows, momentarily cancels out the sense of historical whereabouts, giving with an overwhelming immediacy an awareness of the reality of the past.

The possibility of this awareness is always immanent in old homesites. It may suddenly bear in upon one at the sight of old orchard trees standing in the dooryard of a house now filled with baled hay. It came to me when I looked out the attic window of a disintegrating log house and saw a far view of the cleared ridges with wooded hollows in between, and nothing in sight to reveal the date. Who was I, leaning to the window? When?

It broke upon me one afternoon when, walking in the woods on one of my family places, I came upon a gap in a fence, wired shut, but with deep-cut wagon tracks still passing through it under the weed growth and the fallen leaves. Where that thicket stands there was crop ground, maybe as late as my own time. I knew some of the men who tended it; their names and faces were instantly alive in my mind. I knew how it had been with them—how they would harness their mule teams in the early mornings in my grandfather's big barn and come to the woods-rimmed tobacco patches, the mules' feet wet with the dew. And in the solitude and silence that came upon them they would set to work, their water jugs left in the shade of bushes in the fencerows.

As a child I learned the early mornings in these places for myself, riding out in the wagons with the tobacco-cutting crews to those steep fields in the dew-wet shadow of the woods. As the day went on the shadow would draw back under the feet of the trees, and it would get hot. Little whirlwinds would cross the opening, picking up the dust and the dry "ground leaves" of the tobacco. We made a game of running with my grandfather to stand, shoulders scrunched and eyes squinched, in their middles.

Having such memories, I can acknowledge only with reluctance and sorrow that those slopes should never have been broken. Rich as they were, they were too steep. The humus stood dark and heavy over them once; the plow was its doom.

Early one February morning in thick fog and spattering rain I stood on the riverbank and listened to a towboat working its way downstream. Its engines were idling, nudging cautiously through the fog into the Cane Run bend. The end of the head barge emerged finally like a shadow, and then the second barge appeared, and then the towboat itself. They made the bend, increased power, and went thumping off out of sight into the fog again.

Because the valley was so enclosed in fog, the boat with its tow appearing and disappearing again into the muffling whiteness within two hundred yards, the moment had a curious ambiguity. It was as though I was not necessarily myself at all. I could have been my grandfather, in his time, standing there watching, as I knew he had.

II

I start down from one of the heights of the upland, the town of Port Royal at my back. It is a winter day, overcast and still, and the town is closed in itself, humming and muttering a little, like a winter beehive.

The dog runs ahead, prancing and looking back, knowing the way we are about to go. This is a walk well established with us—a route in our minds as well as on the ground. There is a sort of mystery in the establishment of these ways. Any time one crosses a given stretch of country with some frequency, no matter how wanderingly one begins, the tendency is always toward habit. By the third or fourth trip, without realizing it, one is following a fixed path, going the way one went before. After that, one may still wander, but only by deliberation, and when there is reason to hurry, or when the mind wanders rather than the feet, one returns to the old route. Familiarity has begun. One has made a relationship with the landscape, and the form and the symbol and the enactment of the relationship is the path. These paths of mine are seldom worn on the ground. They are habits of mind, directions and turns. They are as personal as old shoes. My feet are comfortable in them.

From the height I can see far out over the country, the long open ridges of the farmland, the wooded notches of the streams, the valley of the river opening beyond, and then more ridges and hollows of the same kind.

Underlying this country, nine hundred feet below the highest ridgetops, more than four hundred feet below the surface of the river, is sea level. We seldom think of it here; we are a long way from the coast, and the sea is alien to us. And yet the attraction of sea level dwells in this country as an ideal dwells in a man's mind. All our rains go in search of it and, departing, they have carved the land in a shape that is fluent and falling. The streams branch like vines, and between the branches the land rises steeply and then rounds and gentles into the long narrowing fingers of ridgeland. Near the heads of the streams even the steepest land was not too long ago farmed and kept cleared. But now it has been given up and the woods is returning. The wild is flowing back like a tide. The arable ridgetops reach out above the gathered trees like headlands into the sea, bearing their human burdens of fences and houses and barns, crops and roads.

Looking out over the country, one gets a sense of the whole of it: the ridges and hollows, the clustered buildings of the farms, the open fields, the woods, the stock ponds set like coins into the slopes. But this is a surface sense, an exterior sense, such as you get from looking down on the roof of a house. The height is a threshold from which to step down into the wooded folds of the land, the interior, under the trees and along the branching streams.

I pass through a pasture gate on a deep-worn path that grows shallow a little way beyond, and then disappears altogether into the grass. The gate has gathered thousands of passings to and fro that have divided like the slats of a fan on either side of it. It is like a fist holding together the strands of a net.

Beyond the gate the land leans always more steeply toward the branch. I follow it down, and then bear left along the crease at the bottom of the slope. I have entered the downflow of the land. The way I am going is the way the water goes. There is something comfortable and fit-feeling in this, something free in this yielding to gravity and taking the shortest way down. The mind moves through the watershed as the water moves.

As the hollow deepens into the hill, before it has yet entered the woods, the grassy crease becomes a raw gully, and along the steepening slopes on either side I can see the old scars of erosion, places where the earth is gone clear to the rock. My people's errors have become the features of my country.

It occurs to me that it is no longer possible to imagine how this country looked in the beginning, before the white people drove their plows into it. It is not possible to know what was the shape of the land here in this hollow when it was first cleared. Too much of it is gone, loosened by the plows and washed away by the rain. I am walking the route of the departure of the virgin soil of the hill. I am not looking at the same land the firstcomers saw. The original surface of the hill is as extinct as the passenger pigeon. The pristine America that the first white man saw is a lost continent, sunk like Atlantis in the sea. The thought of what was here once and is gone forever will not leave me as long as I live. It is as though I walk knee-deep in its absence.

The slopes along the hollow steepen still more, and I go in under the trees, I pass beneath the surface. I am enclosed, and my sense, my interior sense, of the country becomes intricate. There is no longer the possibility of seeing very far. The distances are closed off by the trees and the steepening walls of the hollow. One cannot grow familiar here by sitting and looking as one can up in the open on the ridge. Here the eyes become dependent on the feet. To see the woods from the inside one must look and move and look again. It is inexhaustible in its standpoints. A lifetime will not be enough to experience it all. Not far from the beginning of the woods, and set deep in the earth in the bottom of the hollow, is a rock-walled pool not a lot bigger than a bathtub. The wall is still nearly as straight and tight as when it was built. It makes a neatly turned narrow horseshoe, the open end downstream. This is a historical ruin, dug here either to catch and hold the water of the little branch, or to collect the water of a spring whose vein broke to the surface here—it is probably no longer possible to know which. The pool is filled with earth now, and grass grows in it. And the branch bends around it, cut down to the bare rock, a torrent after heavy rain, other times bone dry. All that is certain is that when the pool was dug and walled there was deep topsoil on the hill to gather and hold the water. And this high up, at least, the bottom of the hollow, instead of the present raw notch of the stream bed, wore the same mantle of soil as the slopes, and the stream was a steady seep or trickle, running most or all of the year. This tiny pool no doubt once furnished water for a considerable number of stock through the hot summers. And now it is only a lost souvenir, archaic and useless, except for the bitter intelligence there is in it. It is one of the monuments to what is lost.

Wherever one goes along the streams of this part of the country, one is apt to come upon this old stonework. There are walled springs and pools. There are the walls built in the steeper hollows where the fences cross or used to cross; the streams have drifted dirt in behind them, so that now where they are still intact they make waterfalls that have scooped out small pools at their feet. And there used to be miles of stone fences, now mostly scattered and sifted back into the ground.

Considering these, one senses a historical patience, now also extinct in the country. These walls were built by men working long days for little wages, or by slaves. It was work that could not be hurried at, a meticulous finding and fitting together, as though reconstructing a previous wall that had been broken up and scattered like puzzle pieces along the stream beds. The wall would advance only a few yards a day. The pace of it could not be borne by most modern men, even if the wages could be afforded. Those men had to move in closer accord with their own rhythms, and nature's, than we do. They had no machines. Their capacities were only those of flesh and blood. They talked as they worked. They joked and laughed. They sang. The work was exacting and heavy and hard and slow. No opportunity for pleasure was missed or slighted. The days and the years were long. The work was long. At the end of this job the next would begin. Therefore, be patient. Such pleasure as there is, is here, now. Take pleasure as it comes. Take work as it comes. The end may never come, or when it does it may be the wrong end.

Now the men who built the walls and the men who had them built have long gone underground to be, along with the buried ledges and the roots and the burrowing animals, a part of the nature of the place in the minds of the ones who come after them. I think of them lying still in their graves, as level as the sills and thresholds of their lives, as though resisting to the last the slant of the ground. And their old walls, too, re-enter nature, collecting lichens and mosses with a patience their builders never conceived of.

Like the pasture gates, the streams are great collectors of comings and goings. The streams go down, and paths always go down beside the streams. For a while I walk along an old wagon road that is buried in leaves—a fragment, beginningless and endless as the middle of a sentence on some scrap of papyrus. There is a cedar whose branches reach over this road, and under the branches I find the leavings of two kills of some bird of prey. The most recent is a pile of blue jay feathers. The other has been rained on and is not identifiable. How little we know. How little of this was intended or expected by any man. The road that has become the grave of men's passages has led to the life of the woods.

And I say to myself: Here is your road
without beginning or end, appearing
out of the earth and ending in it, bearing
no load but the hawk's kill, and the leaves
building earth on it, something more
to be borne. Tracks fill with earth
and return to absence. The road was worn
by men bearing earth along it. They have come
to endlessness. In their passing
they could not stay in, trees have risen
and stand still. It is leading to the dark,
to mornings where you are not. Here
is your road, beginningless and endless as God.

Now I have come down within the sound of the water. The winter has been rainy, and the hill is full of dark seeps and trickles, gathering finally, along these creases, into flowing streams. The sound of them is one of the elements, and defines a zone.

When their voices return to the hill after their absence during summer and autumn, it is a better place to be. A thirst in the mind is quenched.

I have already passed the place where water began to flow in the little stream bed I am following. It broke into the light from beneath a rock ledge, a thin glittering stream. It lies beside me as I walk, overtaking me and going by, yet not moving, a thread of light and sound. And now from below comes the steady tumble and rush of the water of Camp Branch—whose nameless camp was it named for?—and gradually as I descend the sound of the smaller stream is lost in the sound of the larger.

The two hollows join, the line of the meeting of the two spaces obscured even in winter by the trees. But the two streams meet precisely as two roads. That is, the stream *beds* do; the one ends in the other. As for the meeting of the waters, there is no looking at that. The one flow does not end in the other, but continues in it, one with it, two clarities merged without a shadow.

All waters are one. This is a reach of the sea, flung like a net over the hill, and now drawn back to the sea. And as the sea is never raised in the earthly nets of fishermen, so the hill is never caught and pulled down by the watery net of the sea. But always a little of it is. Each of the gathering strands of the net carries back some of the hill melted in it. Sometimes, as now, it carries so little that the water seems to flow clear; sometimes it carries a lot and is brown and heavy with it. Whenever greedy or thoughtless men have lived on it, the hill has literally flowed out of their tracks into the bottom of the sea.

There appears to be a law that when creatures have reached the level of consciousness, as men have, they must become conscious of the creation; they must learn how they fit into it and what its needs are and what it requires of them, or else pay a terrible penalty: the spirit of the creation will go out of them, and they will become destructive; the very earth will depart from them and go where they cannot follow.

My mind is never empty or idle at the joinings of the streams. Here is the work of the world going on. The creation is felt, alive and intent on its materials, in such places. In the angle of the meeting of the two streams stands the steep wooded point of the ridge, like the prow of an upturned boat—finished, as it was a thousand years ago, as it will be in a thousand years. Its becoming is only incidental to its being. It will be because it is. It has no aim or end except to be. By being, it is growing and wearing into what it will be. The fork of the stream lies at the foot of a finished sculpture. But the stream is no dead tool; it is alive, it is still at its work. Put your hand to it to learn the health of this part of the world. It is the wrist of the hill.

Perhaps it is to prepare to hear some day the music of the spheres that I am always turning my ears to the music of streams. There is indeed a music in streams, but it is not for the hurried. It has to be loitered by and imagined. Or imagined *toward*, for it is hardly for men at all. Nature has a patient ear. To her the slowest funeral march sounds like a jig. She is satisfied to have the notes drawn out to the lengths of days or weeks or months. Small variations are acceptable to her, modulations as leisurely as the opening of a flower.

The stream is full of stops and gates. Here it has piled up rocks in its path, and pours over them into a tiny pool it has scooped at the foot of its fall. Here it has been dammed by a mat of leaves caught behind a fallen limb. Here it must force a narrow passage, here a wider one. Tomorrow the flow may increase or slacken, and the tone will shift. In an hour or a week that rock may give way, and the composition will advance by another note. Some idea of it may be got by walking slowly along and noting the changes as one passes from one little fall or rapid to another. But this is a highly simplified and diluted version of the real thing, which is too complex and widespread ever to be actually heard by us. The ear must imagine an impossible patience in order to grasp even the unimaginableness of such music.

But the creation is musical, and this is a part of its music, as bird song is, or the words of poets. The music of the streams is the music of the shaping of the earth, by which the rocks are pushed and shifted downward toward the level of the sea.

And now I find lying in the path an empty beer can. This is the track of the ubiquitous man Friday of all our woods. In my walks I never fail to discover some sign that he has preceded me. I find his empty shotgun shells, his empty cans and bottles, his sandwich wrappings. In wooded places along roadsides one is apt to find, as well, his overtraveled bedsprings, his outcast refrigerator, and heaps of the imperishable refuse of his modern kitchen. A year ago, almost in this same place where I have found this beer can, I found a possum that he had shot dead and left lying, in celebration of his manhood. He is the true American pioneer, perfectly at rest in his assumption that he is the first and the last whose inheritance and fate this place will ever be. Going forth, as he may think, to sow, he only broadcasts his effects.

As I go on down the path alongside Camp Branch, I walk by the edge of croplands abandoned only within my own lifetime. On my left are the south slopes where the woods is old, long undisturbed. On my right, the more fertile north slopes are covered with patches of briars and sumacs and a lot of young walnut trees. Tobacco of an extraordinary quality was once grown here, and then the soil wore thin, and these places were given up for the more accessible ridges that were not steep, where row cropping made better sense anyway. But now, under the thicket growth, a mat of bluegrass has grown to testify to the good nature of this ground. It was fine dirt that lay here once, and I am far from being able to say that I could have resisted the temptation to plow it. My understanding of what is best for it is the tragic understanding of hindsight, the awareness that I have been taught what was here to be lost by the loss of it.

We have lived by the assumption that what was good for us would be good for the world. And this has been based on the even flimsier assumption that we could know with any certainty what was good even for us. We have fulfilled the danger of this by making our personal pride and greed the standard of our behavior toward the world—to the incalculable disadvantage of the world and every living thing in it. And now, perhaps very close to too late, our great error has become clear. It is not only our own creativity—our own capacity for life—that is stifled by our arrogant assumption; the creation itself is stifled.

We have been wrong. We must change our lives, so that it will be possible to live by the contrary assumption that what is good for the world will be good for us. And that requires that we make the effort to *know* the world and to learn what is good for it. We must learn to co-operate in its processes, and to yield to its limits. But even more important, we must learn to acknowledge that the creation is full of mystery; we will never entirely understand it. We must abandon arrogance and stand in awe. We must recover the sense of the majesty of creation, and the ability to be worshipful in its presence. For I do not doubt that it is only on the condition of humility and reverence before the world that our species will be able to remain in it.

Standing in the presence of these worn and abandoned fields, where the creation has begun its healing without the hindrance or the help of man, with the voice of the stream in the air and the woods standing in silence on all the slopes around me, I am deep in the interior not only of my place in the world, but of my own life, its sources and searches and concerns. I first came into these places following the men to work when I was a child. I knew the men who took their lives from such fields as these, and their lives to a considerable extent made my life what it is. In what came to me from them there was both wealth and poverty, and I have been a long time discovering which was which.

It was in the woods here along Camp Branch that Bill White, my grandfather's Negro hired hand, taught me to hunt squirrels. Bill lived in a little tin-roofed house on up nearer the head of the hollow. And this was, I suppose more than any other place, his hunting ground. It was the place of his freedom, where he could move without subservience, without considering who he was or who anybody else was. On late summer mornings, when it was too wet to work, I would follow him into the woods. As soon as we stepped in under the trees he would become silent and absolutely attentive to the life of the place. He was a good teacher and an exacting one. The rule seemed to be that if I wanted to stay with him, I had to make it possible for him to forget I was there. I was to make no noise. If I did he would look back and make a downward emphatic gesture with his hand, as explicit as writing: Be quiet, or go home. He would see a squirrel crouched in a fork or lying along the top of a branch, and indicate with a grin and a small jerk of his head where I should look; and then wait, while I, conscious of being watched and demanded upon, searched it out for myself. He taught me to look and to listen and to be quiet. I wonder if he knew the value of such teaching or the rarity of such a teacher.

In the years that followed I hunted often here alone. And later in these same woods I experienced my first obscure dissatisfactions with hunting. Though I could not have put it into words then, the sense had come to me that hunting as I knew it—the eagerness to kill something I did not need to eat—was an artificial relation to the place, when what I was beginning to need, just as inarticulately then, was a relation that would be deeply natural and meaningful. That was a time of great uneasiness and restlessness for me. It would be the fall of the year, the leaves would be turning, and ahead of me would be another year of school. There would be confusions about girls and ambitions, the wordless hurried feeling that time and events and my own nature were pushing me toward what I was going to be—and I had no notion what it was, or how to prepare.

And then there were years when I did not come here at all—when these places and their history were in my mind, and part of me, in places thousands of miles away. And now I am here again, changed from what I was, and still changing. The future is no more certain to me now than it ever was, though its risks are clearer, and so are my own desires: I am the father of two young children whose lives are hostages given to the future. Because of them and because of events in the world, life seems more fearful and difficult to me now than ever before. But it is also more inviting, and I am constantly aware of its nearness to joy. Much of the interest and excitement that I have in my life now has come from the deepening, in the years since my return here, of my relation to this countryside that is my native place. For in spite of all that has happened to me in other places, the great change and the great possibility of change in my life has been in my sense of this place. The major difference is perhaps only that I have grown able to be wholeheartedly present here. I am able to sit and be quiet at the foot of some tree here in this woods along Camp Branch, and feel a deep peace, both in the place and in my awareness of it, that not too long ago I was not conscious of the possibility of. This peace is partly in being free of the suspicion that pursued me for most of my life, no matter where I was, that there was perhaps another place I *should* be, or would be happier or better in; it is partly in the increasingly articulate consciousness of being here, and of the significance and importance of being here.

After more than thirty years I have at last arrived at the candor necessary to stand on this part of the earth that is so full of my own history and so much damaged by it, and ask: What *is* this place? What is in it? What is its nature? How should men live in it? What must I do?

I have not found the answers, though I believe that in partial and fragmentary ways they have begun to come to me. But the questions are more important than their answers. In the final sense they *have* no answers. They are like the questions—they are perhaps the same questions—that were the discipline of Job. They are a part of the necessary enactment of humility, teaching a man what his importance is, what his responsibility is, and what his place is, both on the earth and in the order of things. And though the answers must always come obscurely and in fragments, the questions must be persistently asked. They are fertile questions. In their implications and effects, they are moral and aesthetic and, in the best and fullest sense, practical. They promise a relationship to the world that is decent and preserving.

They are also, both in origin and effect, religious. I am uneasy with the term, for such religion as has been openly practiced in this part of the world has promoted and fed upon a destructive schism between body and soul, heaven and earth. It has encouraged people to believe that the world is of no importance, and that their only obligation in it is to submit to certain churchly formulas in order to get to heaven. And so the people who might have been expected to care most selflessly for the world have had their minds turned elsewhere—to a pursuit of "salvation" that was really only another form of gluttony and self-love, the desire to perpetuate their own small lives beyond the life of the world. The heaven-bent have abused the earth thoughtlessly, by inattention, and their negligence has permitted and encouraged others to abuse it deliberately. Once the creator was removed from the creation, divinity became only a remote abstraction, a social weapon in the hands of the religious institutions. This split in public values produced or was accompanied by, as it was bound to be, an equally artificial and ugly division in people's lives, so that a man, while pursuing heaven with the sublime appetite he thought of as his soul, could turn his heart against his neighbors and his hands against the world. For these reasons, though I know that my questions *are* religious, I dislike having to *say* that they are.

But when I ask them my aim is not primarily to get to heaven. Though heaven is certainly more important than the earth if all they say about it is true, it is still morally incidental to it and dependent on it, and I can only imagine it and desire it in terms of what I know of the earth. And so my questions do not aspire beyond the earth. They aspire *toward* it and *into* it. Perhaps they aspire *through* it. They are religious because they are asked at the limit of what I know; they acknowledge mystery and honor its presence in the creation; they are spoken in reverence for the order and grace that I see, and that I trust beyond my power to see.

The stream has led me down to an old barn built deep in the hollow to house the tobacco once grown on those abandoned fields. Now it is surrounded by the trees that have come back on every side—a relic, a fragment of another time, strayed out of its meaning. This is the last of my historical landmarks. To here, my walk has had insistent overtones of memory and history. It has been a movement of consciousness through knowledge, eroding and shaping, adding and wearing away. I have descended like the water of the stream through what I know of myself, and now that I have there is a little more to know. But here at the barn, the old roads and the cow paths—the formal connections with civilization—come to an end.

I stoop between the strands of a barbed-wire fence, and in that movement I go out of time into timelessness. I come into a wild place. I walk along the foot of a slope that was once cut bare of trees, like all the slopes of this part of the country—but long ago; and now the woods is established again, the ground healed, the trees grown big, their trunks rising clean, free of undergrowth. The place has a serenity and dignity that one feels immediately; the creation is whole in it and unobstructed. It is free of the strivings and dissatisfactions, and partialities and imperfections of places under the mechanical dominance of men. Here, what to a housekeeper's eye might seem disorderly is nonetheless orderly and within order; what might seem arbitrary or accidental is included in the design of the whole as if by intention; what might seem evil or violent is a comfortable member of the household. Where the creation is whole nothing is extraneous. The presence of the creation here makes this a holy place, and it is as a pilgrim that I have come—to give the homage of awe and love, to submit to mystification. It is the creation that has attracted me, its perfect interfusion of life and design. I have made myself its follower and its apprentice.

One early morning last spring, I came and found the woods floor strewn with bluebells. In the cool sunlight and the lacy shadows of the spring woods the blueness of those flowers, their elegant shape, their delicate fresh scent kept me standing and looking. I found a rich delight in them that I cannot describe and that I will never forget. Though I had been familiar for years with most of the spring woods flowers, I had never seen these and had not known they grew here. Looking at them, I felt a strange feeling of loss and sorrow that I had never seen them before. But I was also exultant that I saw them now—that they were here.

For me, in the thought of them will always be the sense of the joyful surprise with which I found them—the sense that came suddenly to me then that the world is blessed beyond my understanding, more abundantly than I will ever know. What lives are still ahead of me here to be discovered and exulted in, tomorrow, or in twenty years? What wonder will be found here on the morning after my death? Though as a man I inherit great evils and the possibility of great loss and suffering, I know that my life is blessed and graced by the yearly flowering of the bluebells. How perfect they are! In their presence I am humble and joyful. If I were given all the learning and all the methods of my race I could not make one of them, or even imagine one. Solomon in all his glory was not arrayed like one of these. It is the privilege and the labor of the apprentice of creation to come with his imagination into the unimaginable, and with his speech into the unspeakable.

III

On weekends the Air Force reservists practice war. Their jet fighters come suddenly over the crest of the hill into the opening of the valley, very low, screeching insistently into our minds—perfecting deadliness. We can see the rockets nestling under their wings. They do not represent anything I understand as my own or that I identify with. Unable to imagine myself flying one or employing one for any of my purposes, I am left with the alternative of imagining myself their target. They strike fear into me. I fear them, I think, with the same fear—though not, certainly, with the same intensity—with which they are feared in the villages and the rice paddies of Vietnam. I am afraid that nothing I value can withstand them. I am unable to believe that what I most hope for can be served by them.

On weekends, too, the hunters come out from the cities to kill for pleasure such game as has been able to survive the constraints and destructions of the human economy, the highway traffic, the poison sprays. They come weighted down with expensive clothes and equipment, all purchased for the sake of a few small deaths in a country they neither know nor care about. We hear their guns, and see their cars parked at lane ends and on the roadsides.

Sometimes I can no longer think in the house or in the garden or in the cleared fields. They bear too much resemblance to our failed human history—failed, because it has led to this human present that is such a bitterness and a trial. And so I go to the woods. As I enter in under the trees, dependably, almost at once, and by nothing I do, things fall into place. I enter an order that does not exist outside, in the human spaces. I feel my life take its place among the lives—the trees, the annual plants, the animals and birds, the living of all these and the dead—that go and have gone to make the life of the earth. I am less important than I thought, the human race is less important than I thought. I rejoice in that. My mind loses its urgings, senses its nature, and is free. The forest grew here in its own time, and so I will live, suffer and rejoice, and die in my own time. There is nothing that I may decently hope for that I cannot reach by patience as well as by anxiety. The hill, which is a part of America, has killed no one in the service of the American government. Then why should I, who am a fragment of the hill? I wish to be as peaceable as my land, which does no violence, though it has been the scene of violence and has had violence done to it.

How, having a consciousness, an intelligence, a human spirit —all the vaunted equipment of my race—can I humble myself before a mere piece of the earth and speak of myself as its fragment? Because my mind transcends the hill only to be filled with it, to comprehend it a little, to know that it lives on the hill in time as well as in place, to recognize itself as the hill's fragment.

The false and truly belittling transcendence is ownership. The hill has had more owners than its owners have had years —they are grist for its mill. It has had few friends. But I wish to be its friend, for I think it serves its friends well. It tells them they are fragments of its life. In its life they transcend their years.

The most exemplary nature is that of the topsoil. It is very Christ-like in its passivity and beneficence, and in the penetrating energy that issues out of its peaceableness. It increases by experience, by the passage of seasons over it, growth rising out of it and returning to it, not by ambition or aggressiveness. It is enriched by all things that die and enter into it. It keeps the past, not as history or as memory, but as richness, new possibility. Its fertility is always building up out of death into promise. Death is the bridge or the tunnel by which its past enters its future.

Ownership presumes that only man lives in time, and that the hill is merely present, merely a place. But the poet who lives, heeded or not, within the owner knows that the hill is also taking place. It is as alive as its owner. The two lives go side by side in time together, and ultimately they are the same.

To walk in the woods, mindful only of the *physical* extent of it, is to go perhaps as owner, or as knower, confident of one's own history and of one's own importance. But to go there, mindful as well of its temporal extent, of the age of it, and of all that led up to the present life of it, and of all that will probably follow it, is to feel oneself a flea in the pelt of a great living thing, the discrepancy between its life and one's own so great that it cannot be imagined. One has come into the presence of mystery.

After all the trouble one has taken to be a modern man, one has come back under the spell of a primitive awe, wordless and humble.

In the centuries before its settlement by white men, among the most characteristic and pleasing features of the floor of this valley, and of the stream banks on its slopes, were the forests and the groves of great beech trees. With their silver bark and their light graceful foliage, turning gold in the fall, they were surely as lovely as any forests that ever grew on earth. I think so because I have seen their diminished descendants, which have returned to stand in the wasted places that we have so quickly misused and given up. But those old forests are all gone. We will never know them as they were. We have driven them beyond the reach of our minds, only a vague hint of their presence returning to haunt us, as though in dreams—a fugitive rumor of the nobility and beauty and abundance of the squandered maidenhood of our world—so that, do what we will, we will never quite be satisfied ever again to be here.

The country, as we have made it by the pretense that we can do without it as soon as we have completed its metamorphosis into cash, no longer holds even the possibility of such forests, for the topsoil that they made and stood upon, like children piling up and trampling underfoot the fallen leaves, is no longer here.

It is thought that the beech forests of the Midwest, in the time before the white invasion, produced an annual nut crop of a billion bushels. Suppose that early in our history we had learned, like the Indians, to make use of this bounty, and of the rest of the natural produce of the country. We would have been unimaginably different, and would have been less in need of forgiveness. Instead of asking what was already here that might be of use to us, we hastened to impose on the face of the new country, like the scraps and patches of a collage, the fields and the crop rows and the fences of Europe. We destroyed the abundance that lay before us simply by being unwilling or unable to acknowledge that it was there. We did not know where we were, and to avoid the humility and the labor of our ignorance, we pretended to be where we had come from. And so there is a sense in which we are still not here. Because we have ignored the place we have come to, our presence here remains curiously accidental, as though we came by misapprehension or mistake, like birds driven out to sea by a storm.

We still want to be Europeans. We have abandoned Europe only reluctantly and only by necessity. We tried for two hundred years to grow the European grape here, without success, before turning to the development of the native vines that abounded in variety and in health in all parts of the country.

There is an ominous—perhaps a fatal—presumptuousness, like the *hybris* of the ancient kings, in living in a place by the *imposition* on it of one's preconceived or inherited ideas and wishes. And that is the way we white people have lived in America throughout our history, and it is the way our history now teaches us to live here.

Surely there could be a more indigenous life than we have. There could be a consciousness that would establish itself on a place by understanding its nature and learning what is potential in it. A man ought to study the wilderness of a place before applying to it the ways he learned in another place. Thousands of acres of hill land, here and in the rest of the country, were wasted by a system of agriculture that was fundamentally alien to it. For more than a century, here, the steepest hillsides were farmed, by my forefathers and their neighbors, as if they were flat, and as if this was not a country of heavy rains. And that symbolizes well enough how alien we remain, in our behavior and in our thoughts, to our country. We haven't yet, in any meaningful sense, arrived in these places that we declare we own. We undertook the privilege of the virgin abundance of this land without any awareness at all that we undertook at the same time a responsibility toward it. That responsibility has never yet impressed itself upon our character; its absence in us is signified on the land by scars.

Until we understand what the land is, we are at odds with everything we touch. And to come to that understanding it is necessary, even now, to leave the regions of our conquest—the cleared fields, the towns and cities, the highways—and re-enter the woods. For only there can a man encounter the silence and the darkness of his own absence. Only in this silence and darkness can he recover the sense of the world's longevity, of its ability to thrive without him, of his inferiority to it and his dependence on it. Perhaps then, having heard that silence and seen that darkness, he will grow humble before the place and begin to take it in—to learn *from it* what it is. As its sounds come into his hearing, and its lights and colors come into his vision, and its odors come into his nostrils, then he may come into *its* presence as he never has before, and he will arrive in his place and will want to remain. His life will grow out of the ground like the other lives of the place, and take its place among them. He will be *with* them—neither ignorant of them, nor indifferent to them, nor against them—and so at last he will grow to be native-born. That is, he must re-enter the silence and the darkness, and be born again.

One winter night nearly twenty years ago I was in the woods with the coon hunters, and we were walking toward the dogs, who had moved out to the point of the bluff where the valley of Cane Run enters the valley of the river. The footing was difficult, and one of the hunters was having trouble with his lantern. The flame would "run up" and smoke the globe, so that the light it gave obscured more than it illuminated, an obstacle between his eyes and the path. At last he cursed it and flung it down into a hollow. Its little light went looping down through the trees and disappeared, and there was a distant tinkle of glass as the globe shattered. After that he saw better and went along the bluff easier than before, and lighter, too.

Not long ago, walking up there, I came across his old lantern lying rusted in the crease of the hill, half buried already in the siftings of the slope, and I let it lie. But I've kept the memory that it renewed. I have made it one of my myths of the hill. It has come to be truer to me now than it was then.

For I have turned aside from much that I knew, and have given up much that went before. What will not bring me, more certainly than before, to where I am is of no use to me. I have stepped out of the clearing into the woods. I have thrown away my lantern, and I can see the dark.

In order to know the hill it is necessary to slow the mind down, imaginatively at least, to the hill's pace. For the hill, like Valéry's sycamore, is a voyager standing still. Never moving a step, it travels through years, seasons, weathers, days and nights. These are the measures of its time, and they alter it, marking their passage on it as on a man's face. The hill has never observed a Christmas or an Easter or a Fourth of July. It has nothing to do with a dial or a calendar. Time is told in it mutely and immediately, with perfect accuracy, as it is told by the heart in the body. Its time is the birth and the flourishing and the death of the many lives that are its life.

The hill is like an old woman, all her human obligations met, who sits at work day after day, in a kind of rapt leisure, at an intricate embroidery. She has time for all things. Because she does not expect ever to be finished, she is endlessly patient with details. She perfects flower and leaf, feather and song, adorning the briefest life in great beauty as though it were meant to last forever.

In the early spring I climb up through the woods to an east-facing bluff where the bloodroot bloom in scattered colonies around the foot of the rotting monument of a tree trunk. The sunlight is slanting, clear, through the leafless branches. The flowers are white and perfect, delicate as though shaped in air and water. There is a fragility about them that communicates how short a time they will last. There is some subtle bond between them and the dwindling great trunk of the dead tree. There comes on me a pressing wish to imitate them and so preserve them. But I know that what draws me to them would not pass over into anything I can *do*. They will be lost. In a few days none will be here. Suddenly I want to be, I don't know which, a worshiper or a god.

Coming upon a mushroom growing out of a pad of green moss between the thick roots of an oak, the sun and the dew still there together, I have felt my mind irresistibly become small, to inhabit that place, leaving me standing vacant and bewildered, like a boy whose captured field mouse has just leaped out of his hand.

As I slowly fill with the knowledge of this place, and sink into it, I come to the sense that my life here is inexhaustible, that its possibilities lie rich behind and ahead of me, that when I am dead it will not be used up.

Too much that we do is done at the expense of something else, or somebody else. There is some intransigent destructiveness in us. My days, though I think I know better, are filled with a thousand irritations, worries, regrets for what has happened and fears for what may, trivial duties, meaningless torments—as destructive of my life as if I wanted to be dead. Take today for what it is, I counsel myself. Let it be enough.

And I dare not, for fear that if I do, yesterday will infect tomorrow. We are in the habit of contention—against the world, against each other, against ourselves.

It is not from ourselves that we will learn to be better than we are.

In spite of all the talk about the law of tooth and fang and the struggle for survival, there is in the lives of the animals and birds a great peacefulness. It is not all fear and flight, pursuit and killing. That is part of it, certainly; and there is cold and hunger; there is the likelihood that death, when it comes, will be violent. But there is peace, too, and I think that the intervals of peace are frequent and prolonged. These are the times when the creature rests, communes with himself or with his kind, takes pleasure in being alive.

This morning while I wrote I was aware of a fox squirrel hunched in the sunlight on a high elm branch beyond my window. The night had been frosty, and now the warmth returned. He stayed there a long time, warming and grooming himself. Was he not at peace? Was his life not pleasant to him then?

I have seen the same peacefulness in a flock of wood ducks perched above the water in the branches of a fallen beech, preening and dozing in the sunlight of an autumn afternoon. Even while they dozed they had about them the exquisite alertness of wild things. If I had shown myself they would have been instantly in the air. But for the time there was no alarm among them, and no fear. The moment was whole in itself, deeply satisfying both to them and to me.

Or the sense of it may come with watching a flock of cedar waxwings eating wild grapes in the top of the woods on a

November afternoon. Everything they do is leisurely. They pick the grapes with a curious deliberation, comb their feathers, converse in high windy whistles. Now and then one will fly out and back in a sort of dancing flight full of whimsical flutters and turns. They are like farmers loafing in their own fields on Sunday. Though they have no Sundays, their days are full of sabbaths.

One clear fine morning in early May, when the river was flooded, my friend and I came upon four rough-winged swallows circling over the water, which was still covered with frail wisps and threads of mist from the cool night. They were bathing, dipping down to the water until they touched the still surface with a little splash. They wound their flight over the water like the graceful falling loops of a fine cord. Later they perched on a dead willow, low to the water, to dry and groom themselves, the four together. We paddled the canoe almost within reach of them before they flew. They were neat, beautiful, gentle birds. Sitting there preening in the sun after their cold bath, they communicated a sense of domestic integrity, the serenity of living within order. We didn't belong within the order of the events and needs of their day, and so they didn't notice us until they had to.

But there is not only peacefulness, there is joy. And the joy, less deniable in its evidence than the peacefulness, is the confirmation of it. I sat one summer evening and watched a great blue heron make his descent from the top of the hill in to the valley. He came down at a measured deliberate pace, stately as always, like a dignitary going down a stair. And then, at a point I judged to be midway over the river, without at all varying his wingbeat he did a backward turn in the air, a loop-the-loop. It could only have been a gesture of pure exuberance, of joy—a speaking of his sense of the evening, the day's fulfillment, his descent homeward. He made just the one slow turn, and then flew on out of sight in the direction of a slew farther down in the bottom. The movement was incredibly beautiful, at once exultant and stately, a benediction on the evening and on the river and on me. It seemed so perfectly to confirm the presence of a free nonhuman joy in the world—a joy I feel a great need to believe in—that I had the skeptic's impulse to doubt that I had seen it. If I had, I thought, it would be a sign of the presence of something heavenly in the earth. And then, one evening a year later, I saw it again.

Every man is followed by a shadow which is his death—dark, featureless, and mute. And for every man there is a place where his shadow is clarified and is made his reflection, where his face is mirrored in the ground. He sees his source and his destiny, and they are acceptable to him. He becomes the follower of what pursued him. What hounded his track becomes his companion.

That is the myth of my search and my return.

I have been walking in the woods, and have lain down on the ground to rest. It is the middle of October, and around me, all through the woods, the leaves are quietly sifting down. The newly fallen leaves make a dry, comfortable bed, and I lie easy, coming to rest within myself as I seem to do nowadays only when I am in the woods.

And now a leaf, spiraling down in wild flight, lands on my shirt front at about the third button below the collar. At first I am bemused and mystified by the coincidence—that the leaf should have been so hung, weighted and shaped, so ready to fall, so nudged loose and slanted by the breeze, as to fall where I, by the same delicacy of circumstance, happened to be lying. The event, among all its ramifying causes and considerations, and finally its mysteries, begins to take on the magnitude of history. Portent begins to dwell on it.

And suddenly I apprehend in it the dark proposal of the ground. Under the fallen leaf my breastbone burns with imminent decay. Other leaves fall. My body begins its long shudder into humus. I feel my substance escape me, carried into the mold by beetles and worms. Days, winds, seasons pass over me as I sink under the leaves. For a time only sight is left to me, a passive awareness of the sky overhead, birds crossing, the mazed interreaching of the treetops, the leaves falling—and then that, too, sinks away. It is acceptable to me, and I am at peace.

When I move to go, it is as though I rise up out of the world.

FROM
THE HIDDEN WOUND
(1970)

Chapter 4

When I was three years old Nick Watkins, a black man, came to work for my Grandfather Berry. I don't remember when he came, which is to say that I don't remember not knowing him. When I was older and Nick and I would reminisce about the beginnings of our friendship, he used to laugh and tell me that when he first came I would follow him around calling him Tommy. Tommy was the hand who had lived there just before Nick. It was one of those conversations that are repeated ritually between friends. I would ask Nick to tell how it had been when he first came, and he would always tell about me calling him Tommy, and he would laugh. When I was eight or nine the story was important to me because it meant that Nick and I had known each other since way back, and were old buddies.

I have no idea of Nick's age when I first knew him. He must have been in his late fifties, and he worked for us until his death in, I believe, 1945—a period of about eight years. During that time one of my two or three chief ambitions was to be with him. With my brother or by myself, I dogged his steps. So faithful a follower, and so young and self-important and venturesome as I was, I must have been a trial to him. But he never ran out of patience.

From something philosophical and serene in that patience, and from a few things he said to me, I know that Nick had worked hard ever since his childhood. He told me that when he was a small boy he had worked for a harsh white woman, a widow or a spinster. When he milked, the cow would often kick the bucket over, and he would have to carry it back to the house empty, and the white woman would whip him. He had worked for hard bosses. Like thousands of others of his race he had lived from childhood with the knowledge that his fate was to do the hardest of work for the smallest of wages, and that there was no hope of living any other way.

White people thought of Nick as "a good nigger," and within the terms of that designation he had lived his life. But in my memory of him, and I think in fact, he was possessed of a considerable dignity. I think this was because there was a very conscious peace and faithfulness that he had made between himself and his lot. When there was work to be done, he did it dependably and steadily and well, and thus escaped the indignity of being bossed. I do not remember seeing him servile or obsequious. My grandfather, within the bounds of the racial bias, thought highly of him. He admired him particularly, I remember, as a teamster, and was always pointing him out to me as an example: “Look a yonder how old Nick sets up to drive his mules. Look how he takes hold of the lines. Remember that, and you’ll know something.”

In the eight or so years that Nick lived on the place, he and my grandfather spent hundreds and hundreds of work days together. When Nick first came there my grandfather was already in his seventies. Beyond puttering around and “seeing to things,” which he did compulsively as long as he could stand up, he had come to the end of his working time. But despite the fact that my father had quietly begun to make many of the decisions in the running of the farm, and had assumed perhaps most of the real worries of running it, the old man still thought of himself as the sovereign ruler there, and it could be a costly mistake to attempt to deal with him on the assumption that he was not. He still got up at four o’clock as he always had, and when Nick and the other men on the place went to work he would be with them, on horseback, following the mule teams to the field. He rode a big bay mare named Rose; he would continue to ride her past the time when he could get into the saddle by himself. Through the long summer days he would stay with Nick, sitting and watching and talking, reminiscing, or riding behind him as he drove the rounds of a pasture on a mowing machine. When there was work that he could do, he would be into it until he tired out, and then he would invent an errand so he could get away with dignity.

Given Nick’s steadiness at work, I don’t think my grandfather stayed with him to boss him. I think he stayed so close because he couldn’t stand not to be near what was going on, and because he needed the company of men of his own kind, working men. I have the clearest memory of the two of them passing again and again in the slowly shortening rounds of a big pasture, Nick driving a team of good black mules hitched to a mowing machine, my grandfather on the mare always only two or three steps behind the cutter bar. I don't know where I am in the memory, perhaps watching from the shade of some bush in a fencerow. In the bright hot sun of the summer day they pass out of sight and the whole landscape falls quiet. And then I hear the chuckling of the machine again, and then I see the mules' ears and my grandfather's hat appear over the top of the ridge, and they all come back into sight and pass around again. Within the steady monotonous racket of the machine, they keep a long silence, rich, it seems to me, with the deep camaraderie of men who have known hard work all their lives. Though their long labor in barns and fields had been spent in radically different states of mind, with radically different expectations, it was a common ground and a bond between them—never by men of their different colors, in that time and place, to be openly acknowledged or spoken of. Nick drives on and on into the day, deep in his silence, erect, alert and solemn faced with the patience that has kept with him through thousands of such days before, the elemental reassurance that dinnertime will come, and then quitting time, supper and rest. Behind him as the day lengthens, my grandfather dozes on the mare; when he sways in the saddle the mare steps under him, keeping him upright. Nick would claim that the mare did this out of a conscious sense of responsibility, and maybe she did.

On those days I know that Nick lived in constant fear that the mare too would doze and step over the cutter bar, and would be cut and would throw her rider before the mules could be stopped. Despite my grandfather's unshakable devotion to the idea that he was still in charge of things, it was clearly Nick who bore the great responsibility of those days. Because of childishness or whatever, the old man absolutely refused to accept the limits of age. He was fiercely headstrong in everything, and so was constantly on the verge of doing some damage to himself. I can see Nick working along, pretending not to watch him, but watching him all the same out of the corner of his eye, and then hustling anxiously to the rescue: "Whoa, boss. Whoa. Wait, boss." When he had my brother and me, and maybe another boy or two, to look after as well, Nick must have been driven well nigh out of his mind, but he never showed it.

When they were in the mowing or other such work, Nick

and my grandfather were hard to associate with. Of course we could get on horseback ourselves and ride along behind the old man's mare, but it was impossible to talk and was consequently boring. But there was other work, such as fencing or the handwork in the crops, that allowed the possibility of conversation, and whenever we could we got into that—in everybody's way, whether we played or tried to help, often getting scolded, often aware when we were not being scolded that we were being stoically put up with, but occasionally getting the delicious sense that we were being kindly indulged and catered to for all our sakes, or more rarely that we were being of use.

I remember one fine day we spent with Nick and our grandfather, cutting a young sassafras thicket that had grown up on the back of the place. Nick would fell the little trees with his ax, cutting them off about waist high, so that when they sprouted the cattle would browse off the foliage and so finally kill them. We would pile the trees high on the sled, my brother and I would lie on the mass of springy branches, in the spiced sweetness of that foliage, among the pretty leaves and berries, and Nick would drive us down the hill to unload the sled in a wash that our grandfather was trying to heal.

That was quiet slow work, good for talk. At such times the four of us would often go through a conversation about taking care of Nick when he got old. I don't remember how this conversation would start. Perhaps Nick would bring the subject up out of some anxiety he had about it. But our grandfather would say, "Don't you worry, Nick. These boys'll take care of you."

And one of us would say, "Yessir, Nick, we sure will."

And our grandfather would shake his head in sober emphasis and say, "By God, they'll do it!"

Usually, then, there would follow an elaborately detailed fantasy in which Nick would live through a long carefree old age, with good foxhounds and time to hunt, looked after by my brother and me who by then would have grown up to be lawyers or farmers.

Another place we used to talk was in the barn. Usually this would be on a rainy day, or in the late evening after work. Nick and the old man would sit in the big doorway on upturned buckets, gazing out into the lot. They would talk about old times. Or we would all talk about horses, and our grandfather would go through his plan to buy six good colts for my brother and me to break and train. Or we would go through the plan for Nick's old age.

Or our grandfather would get into a recurrent plan all his own about buying a machine, which was his word for automobile. According to the plan, he would buy a good new machine, and Nick would drive it, and they would go to town and to "Louis-ville" and maybe other places. The intriguing thing about this plan was that it was based on the old man's reasoning that since Nick was a fine teamster he was therefore a fine automobile driver. Which Nick wasn't; he couldn't drive an automobile at all. But as long as Nick lived, our grandfather clung to that dream of buying a good machine. Under the spell of his own talk about it, he always believed that he was right on the verge of doing it.

We also talked about the war that was being fought "across the waters." The two men were deeply impressed with the magnitude of the war and with the ominous new weapons that were being used in it. I remember sitting there in the barn door one day and hearing our grandfather say to Nick: "They got cannons now that'll shoot clean across the water. Good God Amighty!" I suppose he meant the English Channel, but I thought then that he meant the ocean. It was one of the ways the war and modern times became immediate to my imagination.

A place I especially liked to be with Nick was at the woodpile. At his house and at my grandparents' the cooking was still done on wood ranges, and Nick had to keep both kitchens supplied with stove wood. The poles would be laid up in a sawbuck and sawed to the proper lengths with a crosscut saw, and then the sawed lengths would be split on the chopping block with an ax. It was a daily thing throughout the year, but more wood was needed of course in the winter than in the summer. When I was around I would often help Nick with the sawing, and then sit up on the sawbuck to watch while he did the splitting, and then I would help him carry the wood in to the woodbox in the kitchen. Those times I would carry on long conversations, mostly by myself; Nick, who needed his breath for the work, would reply in grunts and monosyllables.

Summer and winter he wore two pairs of pants, usually an old pair of dress pants with a belt under a pair of bib overalls, swearing they kept him cool in hot weather and warm in cold. Like my grandfather, he often wore an old pair of leather puttees, or he would have his pants legs tied snugly above his shoe tops with a piece of twine. This, he liked to tell us, was to keep the snakes and mice from running up his britches legs. He wore old felt hats that were stained and sweaty and shaped to his character. He had a sober open dignified face and gentle manners, was quick to smile and to laugh. His teeth were ambeer-stained from chewing tobacco. His hands were as hard as leather; one of my hopes was someday to have hands as hard as his. He seemed instinctively to be a capable handler of stock. He could talk untiringly of good saddle horses and good work teams that he had known. He was an incurable fox hunter and was never without a hound or two. I think he found it easy to be solitary and quiet.

I heard my grandfather say to him one day: "Nick, you're the first darkie I ever saw who didn't sing while he worked."

But there were times, *I* knew, when Nick did sing. It was only one little snatch of a song that he sang. When the two of us would go on horseback to the store or to see about some stock—Nick on my grandfather's mare, I on a pony—and we had finished our errand and started home, Nick would often sing: "Get along home, home, Cindy, get along home!" And he would laugh.

"Sing it again," I would say.

And he would sing it again.

Chapter 5

BUT before I can tell any more about Nick, I will have to tell about Aunt Georgie. She came to live in the little two-room house with Nick in perhaps the third year he worked for us. Why did we call her Aunt when we never called Nick Uncle? I suppose it was because Nick was informal and she was formal. I remember my grandparents insisting that my brother and I should say Uncle Nick, but we would never do it. Our friendship was somehow too democratic for that. They might as well have insisted that we boys call each other mister. But we used the title Aunt Georgie from the first. She was a woman of a rather stiff dignity and a certain aloofness, and the term of respect was clearly in order.

She was short and squat, bowlegged, bent, her hands crooked with arthritis, her two or three snaggling front teeth stained with snuff. I suppose she was ugly—though I don't believe I ever made that judgment in those days. She looked like Aunt Georgie, who looked like nobody else.

Her arrival at Nick's house suddenly made that one of the most intriguing places that my brother and I had ever known. We began to spend a lot of our time there just sitting and talking. She would always greet us and make us at home with a most gracious display of pleasure in seeing us. Though she could cackle with delight like an old child, though she would periodically interrupt her conversation to spit ambeer into a coffee can she kept beside her chair, there was always a reserve about her, an almost haughty mannerliness that gave a peculiar sense of *occasion* to these visits. When she would invite us to eat supper, as she sometimes did, her manner would impose a curious self-consciousness on us—not a *racial* self-consciousness, but the demanding self-consciousness of a child who has been made, in the fullest sense, the guest of an adult, and of whom therefore a certain dignity is expected. Her house was one of the few places I visited as a child where I am certain I always behaved well. There was something in her presence that kept you always conscious of how you were acting; in response to her you became capable of social delicacy.

Thinking of her now, in spite of my rich experience of her, I realize how little I really *know* about her. The same is true of Nick. I knew them as a child knows people, as they revealed themselves to a child. What they were in themselves, as they spoke to each other, or thought in solitude, I can only surmise from that child's knowledge. I will never know.

Aunt Georgie had lived in a small rural community thirty or forty miles away. She had also lived in Louisville, where I believe she had relatives. She was a great reader of the Bible, and I assume of tracts of various sorts, and I don't know what else.

The knowledge that came out in her talk—fantastical, superstitious, occult, theological, Biblical, autobiographical, medical, historical—was amazing in variety and extent. She was one of the most intricate and powerful characters I have ever known. Perhaps not twenty-five percent of her knowledge was subject to any kind of proof—a lot of it was the stuff of unwritten fairy tales and holy books, a lot I think she had made up herself—but all of it, every last occult or imaginary scrap of it, was caught up and held together in the force of her personality. Nothing was odd to her, nothing she knew stood aside from her unassimilated into the restless wayward omnivorous force of her mind. I write of her with fear that I will misrepresent or underestimate her. I believe she had great intelligence, which had been forced to grow and form itself on the strange straggling wildly heterogeneous bits of information that sifted down to her through various leaks in the stratification of white society.

The character of Aunt Fanny in my book *A Place on Earth* is to some extent modeled on Aunt Georgie: "She's an accomplished seamstress, and the room is filled with her work: quilts, crocheted doilies, a linen wall-hanging with the Lord's Prayer embroidered on it in threads of many colors. In the house she's nearly always occupied with her needle, always complaining of her dim eyesight and arthritic hands. Her dark hands, though painfully crooked and drawn by the disease, are still . . . dexterous and capable. She's extremely attentive to them, always anointing them with salves and ointments of her own making. The fingers wear rings made of copper wire, which she believes to have the power of prevention and healing. She's an excellent persistent canny gardener. The garden beside the house is her work. She makes of its small space an amplitude . . . rows of vegetables and flowers—and herbs, for which she knows the recipes and the uses." If that is her daylight aspect, there is also an aspect of darkness: ". . . the intricate quilts which are always prominent in the room, on the bed or the quilting frame . . . always seem threatened, like earthquake country, by an ominous nearness of darkness in the character of their creator. Aunt Fanny has seen the Devil, not once but often, especially in her youth, and she calls him familiarly by his name: Red Sam. Her obsessions are Hell and Africa, and she has the darkest, most fire-lit notions of both. Her idea of Africa is a hair-raising blend of lore and hearsay and imagination. She thinks of it with nostalgia and longing—a kind of earthly Other Shore, Eden and Heaven—and yet she fears it because of its presumed darkness, its endless jungles, its stock of deliberately malevolent serpents and man-eating beasts. And by the thought of Hell she's held as endlessly fascinated as if her dearest ambition is to go there. She can talk at any length about it, cataloguing its tortures and labyrinths in almost loving detail. . . . Stooped in the light of a coal-oil lamp at night, following her finger down some threatening page of the Bible, her glasses opaquely reflecting the yellow of the lamp, her pigtails sticking out like compass points around her head, she looks like a black Witch of Endor.

"She possesses a nearly inexhaustible lore of snakes and deaths, and bottomless caves and pools, and mysteries and ghosts and wonders."

But the purpose of Aunt Fanny in that book is not to represent Aunt Georgie, and she comes off as a much simpler character. Aunt Fanny had spent all her life in the country, but Aunt Georgie had lived for some time in the city, and her mind, in its way of rambling and sampling, had become curiously cosmopolitan. Like Aunt Fanny, she had an obsession with Africa, but I think it must have started with her under the influence of the Back to Africa movement. She must at some time or other have heard speakers involved in this movement, for I remember her quoting someone who said, "Don't let them tell you they won't know you when you go back. They know their own people, and they'll *welcome* you." There was much more of this talk, but I had no context in which to place it and understand it, and so I have lost the memory. She used to tell a lot of stories about Africa, and I remember only one of them: a story of a woman and her small child who somehow happened to be passing through the jungle alone. A lion was following them, and the woman was terrified. They ran on until they were exhausted and could go no farther. Not knowing what else to do, the woman resigned herself and sat down to wait for death. And the lion came up to them. But instead of attacking, he walked calmly up and laid his paw in the woman's lap. She saw that the paw had a thorn in it that had made a very sore wound. She removed the thorn and treated the wound, and the lion became her protector, driving away the other large animals that threatened them in the jungle, hunting and providing for them, and in all ways taking care of them. I suppose I remember that story among all the others she told because I visualized it very clearly as she told it; it has stayed in my mind all these years with the straightforwardness and innocence of a Rousseau painting, but darker, the foliage ranker and more blurred in detail. But also I have remembered it, I think, because of the sense that began then, and that remains poignant in the memory, that the lion had become the woman's husband.

It was from Aunt Georgie, sitting and listening to her by the hour when I was seven and eight and nine and ten years old, that I first heard talk of the question of civil rights for Negroes. Again, I could supply no context for what I heard, and so have forgotten most of it. But one phrase has stuck in my mind along with her manner of saying it. She said that many times the white people had promised the Negro people "a right to the flag," and they never had given it to them. The old woman was capable of a moving eloquence, and I was deeply disturbed by what she said. I remember that, and I remember the indignation of my white elders when I would try to check the point with them. After Aunt Georgie moved away it was probably ten years before I paid attention to any more talk about civil rights, and it was longer than that before I felt again anything like the same disturbed sense of personal responsibility that she made me feel.

She used to tell a story about the end of the First World War, of people dancing wildly through the streets, carrying the Kaiser's head impaled on the end of a fence rail. I keep a clear image of that scene, too. Was it some celebration she had seen in Louisville, at which the Kaiser had been mutilated in effigy? Was it something she had read or imagined? I don't know. I assumed then that it was really the Kaiser's head, that she had seen that barbaric celebration herself, and that it was one of the central events of the history of the world.

Germany was also a sort of obsession with her, I suppose because of the two world wars, and she had nothing good to say for it. If to her Africa was a darkish jungly place of marvels that enticed the deepest roots of her imagination, Germany was a medieval torture chamber, a place of dire purposes and devices, pieces of diabolical machinery. In her sense of it the guillotine had come to be its emblem. It was a sort of earthly hell where people were sent to be punished for such crimes as public drunkenness. According to her, if you were caught drunk once you were *warned*, if you were caught twice you were put in jail, but the *third* time you were put on the train and sent to Germany where the Kaiser would cut your head off. She told of a man and his son in the town of Finchville, Kentucky, who had been caught drunk for the third time. The people of the town went down to the railroad station to witness their condemned neighbors' departure for Germany. Through the coach windows the doomed father and son could be seen finding an empty seat and sitting down together to begin their awesome journey. Over the condemned men and over the watching crowd there hung a great heaviness of finality and fate and horror. She told it as one who had been there and had seen it. And I remember it as if *I* had seen it; it is as vivid in my mind as anything that ever happened to *me*. The first time I heard her tell that story I believe I must have spent an hour or two cross-examining her, trying to find some small mitigation of the implacable finality of it, and she would not yield me the tiniest possibility of hope. Thinking about it now I feel the same contraction of despair that I felt then.

She told about awful sicknesses, and acts of violence, and terrible deaths. There was once a lion tamer who used to put his head in a lion's mouth, and one morning he nicked himself shaving and when he put his head into the lion's mouth that day, the lion tasted his blood and bit his head off. She talked about burials, bodies lying in the ground, dug-up bodies, ghosts.

And snakes. She believed that woman should bruise the head of the serpent with her heel. She would snick them up into little pieces with her sharp hoe. But they fascinated her, too, and they lived in her mind with the incandescence that her imagination gave to everything it touched. She was full of the lore of snakes—little and big snakes, deadly poisonous snakes, infernal snakes, *blue* snakes, snakes with yellow stripes and bright red heads, snakes that sucked cows, snakes that swallowed large animals such as people, hoop snakes. Once there was a woman who was walking in the woods, and her feet got tired and she sat down on a log to rest, and she took off her shoe and beat it on the log to dislodge a rock that was in it, and slowly that log began to *move*. My goodness gracious sakes alive!

The most formidable snake of all was the hoop snake. The hoop snake traveled by the law of gravity—a concept as staggering to the mind as atomic energy or perpetual motion. The hoop snake could travel on his belly like any other snake, but when he found himself on the top of a hill he was apt to whip himself up into the shape of a wheel, and go flying off down the slope at a dazzling speed. Once he had started his descent, of course, he had no sense of direction, and no way to stop or to turn aside and he had a deadly poisonous sharp point on the end of his tail that killed whatever it touched. Trees, horses and cattle, cats, dogs, men, women, and even *boys* were all brought down, like grass before the breath of the Lord, in the furious free-wheeling descent of the hoop snake, quicker than the eye. How could you know when he was coming? You couldn't. How could you get out of his way? You couldn't.

It has occurred to me to wonder if there wasn't a degree of conscious delight that Aunt Georgie took in scaring the wits out of two gullible little white boys. I expect there was. There was undoubtedly some impulse of racial vengeance in it; and it was bound to have given her a sense of power. But there was no cynicism in it; she believed what she said. I think, in fact, that she was motivated somewhat by a spirit of evangelism; she was instructing us, warning us against human evils, alerting us to the presence of ominous powers. Also she must have been lonely; we were company for her, and these things that she had on her mind were what she naturally talked to us about. And probably beyond any other reasons was that, like all naturally talkative people, she loved to listen to herself and to provide herself with occasions for eloquence.

But she was also a tireless loving gardener, a rambler in the fields, a gatherer of herbs and mushrooms, a raiser of chickens. And she talked well about all these things. She was a marvelous teller of what went on around the farm: How the mules ran the goat and the goat jumped up in the loading chute, safe, and looked back at the mules and wagged his head as if to say, "Uh *huh*!" How the hawk got after her chickens, and how she got her old self out of the house just in time and said, "Whoooeee! Hi! Get *out* of here, you *devil* you, and don't you come *back*!"

She knew about healing herbs and tonics and poultices and ointments. In those years I was always worrying about being skinny. I knew from the comic books that a ninety-eight-pound weakling was one of the worst things you could grow up to be; I foresaw my fate, and dreaded it. I confided my worry to Aunt Georgie, as I did to nearly everybody who seemed willing to listen, and she instantly prescribed a concoction known as Dr. Bell's Pine Tar Honey. By propounding the cure she ratified the ill; no one would ever persuade me again that I was not disastrously skinny. But Dr. Bell's Pine Tar Honey was nowhere to be found. No drugstore had it. The more I failed to find it, the more I believed in its power. It could turn a skinny boy into a *normal* boy. It was the elusive philosopher's stone of my childhood. And Aunt Georgie stood by her truth. She never quit recommending it. She would recommend no substitute.

She and Nick never married, though I believe she never abandoned the hope that they would. She always referred to Nick, flirtatiously and a little wistfully, as Nickum-Nackum. But Nick, with what I believe to have been a very pointed discretion, always called her *Miss* Georgie.

Chapter 6

If Aunt Georgie was formal and austere, Nick was casual and familiar. If the bent of her mind was often otherworldly, Nick's belonged incorrigibly to this world; he was a man of the fields and the barns and the long nighttime courses of fox hunts. When he spoke of the Lord he called him, as my grandfather did, the Old Marster, by which they meant a god of mystery, the maker of weather and seasons, of abundance and dearth, of growth and death—a god far more remote, and far less talkative, than the god of the churches.

At night after I had finished supper in my grandmother's kitchen I would often walk down across the field to where Nick's house sat at the corner of the woods. He would come out, lighting his pipe, and we would sit down on the stones of the doorstep, and Nick would smoke and we would talk while it got dark and the stars came out. Up the hill we could see the lighted windows of my grandparents' house. In front of us the sloping pasture joined the woods. The woods would become deep and softly massive in its darkness. At dusk a toad who lived under the doorstep would come out, and always we would notice him and Nick would comment that he lived there and that he came out every night. Often as we sat there, comfortable with all the outdoors and the night before us, Aunt Georgie would be sitting by the lamp in the house behind us, reading aloud from the Bible, or trying to lecture to Nick on the imminence of eternity, urging him to think of the salvation of his soul. Nick's reluctance to get disturbed over such matters was always a worry to her. Occasionally she would call out the door: "Nickum-Nackum? Are you listening to me?" and Nick would serenely interrupt whatever he was telling me—"Yessum, Miss Georgie"—and go on as before.

What we would often be talking about was a fine foxhound named Waxy that Nick had owned a long time ago. He would tell of the old fox hunts, saying who the hunters had been and what kind of dogs they'd had. He would tell over the whole course of some hunt. By the time the race was over his Waxy would always have far outdistanced all the other dogs, and everybody would have exclaimed over what a great, fleet, wise hound she was, and perhaps somebody would have tried to buy her from Nick at a high price, which Nick would never take. Thinking about it since, I have had the feeling that those dogs so far outhunted and outrun by Nick's Waxy were white men's dogs. But I don't know for sure. In addition to telling how well Waxy had performed in some fox race or other, Nick would always tell how she looked, how she was marked and made. And he would frequently comment that Waxy was a fine name for a foxhound. He had thought a lot about how things ought to be named. In his mind he had lists of the best names for milk cows and horses and dogs. Blanche, I remember, he thought to be the prettiest name for a certain kind of light-coated Jersey cow. As long as he was with us I don't think he ever had the luck to give that name to a cow. Among others whose names I have forgotten, I remember that he milked one he called Mrs. Williams.

There were certain worldly ideals that always accompanied him, as hauntingly, I think, as Aunt Georgie was accompanied by devils and angels and snakes and ghosts. There was the ideal foxhound and the ideal team of mules and the ideal saddle horse, and he could always name some animal he had known that had not been quite perfect but had come close. Someday he would like to own a foxhound like Waxy but just a little better. Someday he would like to work a team of gray mare mules each just a little better than the one named Fanny that my grandfather owned.

We talked a great deal about the ideal saddle horse, because I persistently believed that I was going to pick out and buy one of my own. In the absence of any particular horse that I intended to buy, the conversational possibilities of this subject were nearly without limit. Listing points of color, conformation, breeding, disposition, size, and gait, we would arrive within a glimpse of the elusive outlines of the ideal. And then, supposing some divergence from one of the characteristics we had named, we might prefigure a horse of another kind. We were like those experts who from a track or a single bone can reconstruct an extinct animal: give us a color or a trait of conformation or character, and we could produce a horse to go with it. And once we had the horse, we had to settle on a name and on the way it should be broke and fed and kept.

But our *great* plan, the epic of our conversation, was to go camping and hunting in the mountains. I don't think that either Nick or my brother and I knew very much about the mountains. None of us had been there. My own idea seems to have blended what I had read of the virgin Kentucky woods of Boone's time with Aunt Georgie's version of Africa. My brother and I believed there was a great tract of wilderness in the mountains, thickly populated with deer and bears and black panthers and mountain lions. We thought we would go up there and kill some of those beasts and eat their flesh and dress in their skins and make ornaments and weapons out of their long teeth, and live in the untouched maidenhood of history.

I don't know what idea Nick may have had of the mountains; he always seemed more or less to go along with our idea of them; whether or not he contributed to the notions we had, I don't remember. Though he was an ignorant man, he was knowing and skilled in the realities that had been available to him, and so he was bound to have known that we would never make any such trip. But he elaborated this plan with us year after year; it was, in fact, in many ways more his creature than ours. It was generous—one of the most generous things anybody ever did for us. And yet it was more than generous, for I think Nick believed in that trip as a novelist believes in his novel: his imagination was touched by it; he couldn't resist it.

We were going to get new red tassels to hang on the mules' harness, and polish the brass knobs on their hames. And then one night we were going to load the wagon—we had told over many times what all we were going to put in it. And the next morning early, way before daylight so my grandfather wouldn't catch us, we would slip off and go to the mountains. We had a highly evolved sense of the grandeur of the spectacle we would make as we went through New Castle, where everybody who saw us would know where we were going and what a fine adventure it was. Here we would come through the town just as everybody was getting up and out into the street. Nick would be driving, my brother and I sitting on either side of him, his ready henchmen. We would have the mules in a spanking trot, the harness jangling, the red tassels swinging, and everything fine—right on through town past the last house. From there to the tall wilderness where we would make our camp we had no plans. The trip didn't interest us after people we knew quit watching.

When we got to the mountains we were going to build a log cabin, and we talked a lot about the proper design and construction of that. And there were any number of other matters, such as cooking, and sewing the clothes we would make out of the skins of animals. Sometimes, talking of the more domestic aspects of our trip, we would get into a bind and have to plan to take Aunt Georgie. She never could work up much enthusiasm for the project; it was too mundane for her.

There were several times when we actually named the day of our departure, and my brother and I went to bed expecting to leave in the morning. We had agreed that to wake us up Nick would rattle a few tin cans in the milk bucket. But we never woke up until too late. I remember going out one morning and finding the very tin cans that Nick had rattled floating in the rain barrel. Later I realized that of course Nick had never rattled those cans or any others. But in the years since I have often imagined him taking the time to put those cans there for me to see, and then slipping off to his day's work—conspiring with himself like Santa Claus to deceive and please me.

In the present time, when the working man in the old sense of the phrase is so rapidly disappearing from the scene, his conversation smothered in the noise of machinery, it may be necessary to point out that such talk as we carried on with Nick was by no means uncommon among the farming people of that place, and I assume of other places as well. It was one of the natural products and pleasures of the life there. With the hands occupied in the work, the mind was set free; and since there was often considerable misery in the work, the mind turned naturally to what would be desirable or pleasant. When one was at work with people with whom one had worked many days or years before, one naturally spoke of the good things that were on one's mind. And once the conversation was started, it began to be a source of pleasure in itself, producing and requiring an energy that propelled it beyond the bounds of reality or practicality. As long as one was talking about what would be good, why not go ahead and talk about what would be *really* good? My childhood was surrounded by a communal daydreaming, the richest sort of imaginative talk, that began in this way—in work, in the misery of work, to make the work bearable and even pleasant. Such talk ranged all the way from a kind of sensuous realism to utter fantasy, but because the bounds of possibility were almost always ignored I would say that the impetus was always that of fantasy. I have heard crews of men, weary and hungry and hot near the end of a day's work, construct long elaborate conversations on the subject of what would be good to eat and drink, dwelling at length and with subtlety on the taste and the hotness or the coldness of various dishes and beverages, and on combinations of dishes and beverages, the menu lengthening far beyond the capacity of any living stomach. I knew one man who every year got himself through the ordeal of the tobacco harvest by elaborating from one day to the next the fantasy of an epical picnic and celebration which was always to occur as soon as the work was done—and which never did occur except, richly, marvelously, in the minds of his listeners. Such talk could be about anything: where you would like to go, what you would like to do when you got there, what you would like to be able to buy, what would be a good kind of a farm, what science was likely to do, and so on.

Nick often had fox hunting on his mind, and he always knew where there was a fox, and he was always going to hunt as soon as he had time. Occasionally he *would* have time—a wet afternoon or a night when he didn't feel too tired—and he would go, and would sometimes take one of us. I like the thought of Nick, alone as he often was, out in the woods and the fields at night with his dogs. I think he transcended his lot then, and was free in the countryside and in himself—beyond anybody's knowledge of where he was and anybody's notion of what he ought to do and anybody's claim on his time, his dogs mouthing out there in the dark, and all his senses with them, discovering the way the fox had gone. I don't remember much about any hunt I ever made with him, but I do remember the sense of accompanying him outside the boundaries of his life as a servant, into that part of his life that clearly belonged to him, in which he was competent and knowing. And I remember how it felt to be riding sleepily along on the pony late at night, the country all dark around us, not talking, not sure where we were going. Nick would be ahead of me on old Rose, and I could hear her footsteps soft on the grass, or hard on the rocks when we crossed a creek. I knew that Nick knew where we were, and I felt comfortable and familiar with the dark because of that.

When Joe Louis fought we would eat supper and then walk across the fields to the house of the man who was raising my grandfather's tobacco; we would have been planning for days to go there to hear the fight on the radio. My grandparents had a radio, too, but Nick would never have thought of going there. That he felt comfortable in going to another household of white people and sitting down by the radio with them was always a little embarrassing to me, and it made me a little jealous. And yet I understood it, and never questioned it. It was one of those complex operations of race consciousness that a small child will comprehend with his feelings, and yet perhaps never live long enough to unravel with his mind. The prowess of Joe Louis, anyhow, was something Nick liked to talk about; it obviously meant a great deal to him. And so the championship matches of those days were occasions when I felt called on to manifest my allegiance to Nick. Though I could have listened to the fight in my grandparents' living room, I would go with Nick, and the more my grandmother declared she didn't know why I had to do that the more fierce and clear my loyalty to Nick became.

The people of that house had a boy a little older than I was, and so I also had my own reasons for going there. We boys would listen to the fight and then, while the grownups talked, go out to play, or climb the sugar pear trees and sit up in the branches, eating. And then holding Nick's hand, I would walk with him back home across the dark fields.

One winter there suddenly appeared in the neighborhood a gaunt possessed-looking white man who was, as people said, "crazy on religion" or "religious crazy." He was, I believe, working on a nearby farm. Somehow he discovered Aunt Georgie, and found that she could quote the Bible and talk religion as long as he could. He called Nick Mix, and so in derision and reprisal Nick and my brother and I called *him* Mix, and that was the only name we ever knew him by. We would see him coming over the hill, his ragged clothes flapping in the cold wind, and see him go into the little house. And then after a while we would go down, too, and listen. The tone of their conversations was very pious and oratorical; they more or less preached to each other, and the sessions would go on for hours.

Of it all, I remember only a little fragment: Mix, sitting in a chair facing Aunt Georgie, sticks his right hand out into the air between them, dramatically, and asks her to put hers out there beside it. She does, and then he raises his left hand like a Roman orator and, glaring into her eyes, says: "There they are, Auntie! *There* they are! Look at 'em! One of 'em's white, and one of 'em's black! But *inside*, they both white!"

What impresses me now, as I remember those times, is that

while Mix was never anything but crazy and absurd, ranting mad, Aunt Georgie always seemed perfectly dignified and sane. It seems to me that this testifies very convincingly to the strength of her intelligence: she had *comprehended* the outlandishness of her own mind, and so she could keep a sort of grace before the unbridled extravagance of his. In her way she had faced the worst. The raging of Lear himself would not have perturbed her; she would have told him stories worse than his, and referred him to the appropriate Scripture. To Nick, who took a very detached and secular view of most things, these sessions were a source of ironic amusement. He liked to look over at us through the thick of the uproar and wink.

Chapter 7

I SUPPOSE that it is usually the aim of a writer to produce a definitive statement, one that will prove him to be the final authority on what he has said. But though my aim here is to tell the truth as nearly as I am able, I am aware that the truth I am telling may be a very personal one, the truth, that is, as distorted and qualified by my own heritage and personality. I am, after all, writing about people of another race and a radically different heritage, whom I knew only as a child, and whose lives parted from mine nearly a quarter of a century ago. As I write I can hardly help but think of the possibility that if Nick and Aunt Georgie were alive to read this, they might not recognize themselves. And in the face of the extreme racial sensitivity of the present time, I can hardly ignore the possibility that my black contemporaries may find some of my assumptions highly objectionable. And so I write with the feeling that the truth I may tell will not be definitive or objective or even demonstrable, but in the strictest sense subjective, relative to the peculiar self-consciousness of a diseased man struggling toward a cure. I am trying to establish the outlines of an understanding of myself in regard to what was fated to be the continuing crisis of my life, the crisis of racial awareness—the sense of being doomed by my history to be, if not always a racist, then a man always limited by the inheritance of racism, condemned to be always conscious of the necessity *not* to be a racist, to be always dealing deliberately with the reflexes of racism that are embedded in my mind as deeply at least as the language I speak.

(In the Western tradition of individualism there is the assumption that art can grow out of a personal or a cultural disease, and triumph over it. I no longer believe that. It is related to the idea that a man can achieve personal immortality in a work of art, which I also no longer believe. Though I believe that the liveliest art is suffused with the energy of the creation, and in that sense participates in immortality, I do not believe that any one work of art is immortal any more than I believe that a grove of trees or a nation is immortal. A man cannot be immortal except by saving his soul, and he cannot save his soul except by freeing his body and mind from the destructive forces in his history. A work of art that grows out of a diseased culture has not only the limits of art but the limits of the disease—if it is not an affirmation of the disease, it is a reaction against it. The art of a man divided within himself and against his neighbors, no matter how sophisticated its techniques or how beautiful its forms and textures, will never have the communal *power* of the simplest tribal song.)

There is an inescapable tentativeness in writing of one's own formative experiences: as long as the memory of them stays alive they *remain* formative; the power of growth and change remains in them, and they never become quite predictable in their influence. If memory is in a way the ancestor of consciousness, it yet remains dynamic within consciousness. And so, though I can write about Nick and Aunt Georgie as two of the significant ancestors of my mind, I must also deal with their memory as a live resource, a *power* that will live and change in me as long as I live. To fictionalize them, as I did Aunt Georgie to some extent in *A Place on Earth*, would be to give them an imaginative stability at the cost of oversimplifying them. To attempt to tell the "truth" about them as they really were is to resign oneself to enacting a small fragment of an endless process. Their truth is inexhaustible both in their lives as they were, and in my life as I think they were.

The peculiar power of my memory of them comes, I think, from the fact that all my association with them occurred within a tension between the candor and openness of a child's view of things, and the racial contrivances of the society we lived in. As I have already suggested, there were times when I was inescapably aware of the conflict between what I felt about them, in response to what I knew of them, and the feelings that were prescribed to me by the society's general prejudice against their race. Being with them, it was hard to escape for very long from the sense of racial difference prepared both in their minds and in my own.

The word *nigger* might be thought of as rattling around, with devastating noise and impact, within the silence, that black-manshaped hollow, inside our language. When the word was spoken abstractly, I believe that it seemed as innocuous and casual to me as any other word. I used it that way myself, in the *absence* of black people, without any consciousness that I was participating in a judgment and a condemnation; certainly I used it without any feeling that my use of it manifested anything that was wrong with *me*. But when it was used with particular reference to a person one cared for, as a child cares, it took on a tremendous force; its power reached ominously over one's sense of things. I remember the shock and confusion I felt one night when, saying good-bye to Nick, I impulsively kissed his hand, which I had held as we brought the milk to the house, and then came into the kitchen to hear: "Lord, child, how can you kiss that old nigger's hand?"

And I remember sitting down at Nick's house one day when he and Aunt Georgie had company. There was a little boy, four or five years old, who was rambunctious and continually on the verge of mischief. To keep him in hand Nick would lean over and rattle his old leather leggings that he had taken off and put behind the drum stove, and he would say to the boy: "John's going to get you. If you don't quit that, John's going to get you." All this—the rattling of the leggings, the threat, the emphatic result—impressed me a good deal, and I wondered about it. Was this John-the-Leggings somebody Nick had thought up on the spur of the moment? Was he one of the many associates of Red Sam? And then it hit me. That John that my friend Nick was speaking of in so formidable a tone was my father! The force of a realization like that is hardly to be measured; it's the sort of thing that can initiate a whole epoch of the development of a mind, and yet remain on as a force. It gave me the strongest sort of a hint of the existence of something large and implacable and rigid that I had been born into, and lived in—something I have been trying to get out of ever since. In justice to my father, I must say that I don't believe his name was used in this way because of anything he had *done*, but because of his place in the system. In spite of my grandfather's insistences to the contrary, Nick knew as I did that my father had become the man really in charge of things. Thus he had entered the formidable role of "boss man": whoever he was, whatever he did, he had the power and the austerity of that role; the society assigned it to him, as it assigned to Nick the role of "nigger."

At times in my childhood I was made to feel the estranging power of this role myself—though I believe always by white men. When I would be playing where the men were working in the fields—I suppose in response to some lapse of tact on my part, some unconscious display of self-importance or arrogance—one of them would sing or quote:

> He ain't the boss, he's the boss's son,
> But he'll be the boss when the boss is done.

A complementary rhyme, which might have been made up as a companion to that, goes this way:

> A naught's a naught and a figger's a figger.
> All for the white man, none for the nigger.

Between them these two little rhymes make a surprisingly complete definition of the values and the class structure of the place and the time of my childhood, and they define its most erosive psychic tensions.

That sense of difference, given the candor and the affection and the high spirits of children, could only beget in us its opposite: a very strong and fierce sense of allegiance. Whenever there was a conflict of interest between Nick and Aunt Georgie and our family, neither my brother nor I ever hesitated to take the side of Nick and Aunt Georgie. We had an uncle who would make a game of insulting Nick in our presence because it tickled him the way we would come to the rescue; we would attack him and fight him as readily as if he had been another child. And there were winter Saturdays that we spent with our goat hitched to a little sled we had made, hauling coal from our grandfather's pile down to Nick's. That was stealing, and we knew it; it had the moral flair and the dangerous loyalty of the adventures of Robin Hood.

But the clearest of all my own acts of taking sides happened at a birthday party my grandmother gave for me when I must have been nine or ten years old. As I remember, she invited all the family and perhaps some of the neighbors. I issued one invitation of my own, to Nick. I believe that in my eagerness to have him come, and assuming that as my friend he ought to be there, I foresaw none of the social awkwardness that I created. But I had, in fact, surrounded us all with the worst sort of discomfort. Nick, trying to compromise between his wish to be kind to me and his embarrassment at my social misconception, quit work at the time of the party and came and sat on the cellar wall behind the house. By that time even I had begun to sense the uneasiness I had created: I had done a thing more powerful than I could have imagined at the time; I had scratched the wound of racism, and all of us, our heads beclouded in the social dream that all was well, were feeling the pain. It was suddenly evident to me that Nick neither would nor could come into the house and be a member of the party. My grandmother, to her credit, allowed me to follow my instincts in dealing with the situation, and I did. I went out and spent the time of the party sitting on the cellar wall with Nick.

It was obviously the only decent thing I could have done; if I had thought of it in moral terms I would have had to see it as my duty. But I didn't. I didn't think of it in moral terms at all. I did simply what I *preferred* to do. If Nick had no place at my party, then I would have no place there either; my place would be where he was. The cellar wall became the place of a definitive enactment of our friendship, in which by the grace of a child's honesty and a man's simple-hearted generosity, we transcended our appointed roles. I like the thought of the two of us sitting out there in the sunny afternoon, eating ice cream and cake, with all my family and my presents in there in the house without me. I was full of a sense of loyalty and love that clarified me to myself as nothing ever had before. It was a time I would like to live again.

Chapter 8

ONE day when we were all sitting in the barn door, resting and talking, my grandfather sat gazing out a while in silence, and then he pointed to a far corner of the lot and said: "When I'm dead, Nick, I want you to bury me there."

And Nick laughed and said, "Boss, you got to go farther away from here than that."

My grandfather's words bore the forlorn longing to remain in the place, in the midst of the work and the stirring, that had interested him all his life. It was also a characteristic piece of contrariness: if he didn't want to go, then by God he *wouldn't.* Nick's rejoinder admitted how tough it would be on everybody to have such a stubborn old boss staying on there forever, and it told a plain truth: there wasn't to be any staying; good as it might be to be there, we were all going to have to leave.

Those words stand in my memory in high relief; strangely powerful then (it was an exchange that we younger ones quoted over and over in other conversations) they have assumed a power in retrospect that is even greater. That whole scene—the two old men, two or three younger ones, two boys, all looking out that big doorway at the world, and beyond the world—seems to me now not to belong to my childhood at all, but to stand like an illuminated capital at the head of the next chapter. It presages a series of deaths and departures and historical changes that would put an end both to my childhood and to the time and the way of life I knew as a child. Within two or three years of my tenth summer both Nick and my grandfather died, and after Nick's death Aunt Georgie moved away. Also the war ended, and our part of the country moved rapidly into the era of mechanized farming. People, especially those who had worked as hired hands in the fields, began to move to the cities, and the machines moved from the cities out into the fields. Soon nearly everybody had a tractor; the few horses and mules still left in the country were being kept only for old time's sake.

It was a school day in the winter and I was at home in New Castle eating breakfast, when my grandmother telephoned from the farm to tell us that Nick had had a stroke. He had eaten breakfast and was starting to go up to the barn when he fell down in the doorway. He had opened the door before he fell and Aunt Georgie was unable to move him, and so he had lain in the cold for some time before help came. Until he died a few days later he never moved or spoke. I wasn't allowed to go to see him, but afterward my grandmother told me how he had been. He lay there on his back, she said, unable to move, looking up at the ceiling. When she would come in he would recognize her; she could tell that by his eyes. And I am left with that image of him: lying there still as death, his life showing only in his eyes. As if I had been there, I am aware of the intelligence and the gentleness and the sorrow of his eyes. He had fallen completely into the silence that had so nearly surrounded him all his life. It was not the silence of death, which men may speak of with words, and so know each other. It was the racial silence in the speech of white men, the wound of their history, formed three hundred years before my birth to stand between him and me, so that when I think of him now, as important as his memory is to me, it must be partly to wonder if I knew him.

One night not long after Nick's death I went with my father down to the little house to see Aunt Georgie. She had a single lamp lighted in the bedroom, and we went in and sat down. For a time, while they talked, I sat there without hearing, saddened and bewildered by the heavy immanence of change I felt in the room and by Nick's absence from it. And then I began to listen to what was being said. Aunt Georgie's voice was without the readiness of laughter I had always known in her. She spoke quietly and deliberately, avoiding betrayal of her grief. My father was talking, I remember, very gently and generously to her, helping her to get her affairs in order, discussing an insurance policy that Nick had, determining what her needs were. And all at once I realized that Aunt Georgie was going to leave. I hadn't expected it; she was deeply involved in my history and my affections; she had been there what seemed to me a long long time. Perhaps because of those conversations we used to have about taking care of Nick when he got old, I had supposed that such permanent arrangements were made as a matter of course. I interrupted and said something to the effect that I thought Aunt Georgie ought to stay, I didn't want her to go. And they turned to me and explained that it was not possible. She had relatives in Louisville; she would go to live with them. Their few words made it clear to me that there were great and demanding realities that I had never considered: realities of allegiance, realities of economics. I waited in silence then until their talk was finished, and we went out into the winter night and walked back up the hill.

I don't know if I saw Aunt Georgie again before she left. That is the last memory I have of her: strangely detached and remote from the house as though she already no longer lived there, an old woman standing in the yellow lamplight among the shadowy furnishings of the room, stooped and slow as if newly aware of the heaviness of her body.

And so it ended.

FROM

A CONTINUOUS HARMONY

(1972)

Think Little

FIRST THERE WAS Civil Rights, and then there was the War, and now it is the Environment. The first two of this sequence of causes have already risen to the top of the nation's consciousness and declined somewhat in a remarkably short time. I mention this in order to begin with what I believe to be a justifiable skepticism. For it seems to me that the Civil Rights Movement and the Peace Movement, as popular causes in the electronic age, have partaken far too much of the nature of fads. Not for all, certainly, but for too many they have been the fashionable politics of the moment. As causes they have been undertaken too much in ignorance; they have been too much simplified; they have been powered too much by impatience and guilt of conscience and short-term enthusiasm, and too little by an authentic social vision and long-term conviction and deliberation. For most people those causes have remained almost entirely abstract; there has been too little personal involvement, and too much involvement in organizations that were insisting that *other* organizations should do what was right.

There is considerable danger that the Environmental Movement will have the same nature: that it will be a public cause, served by organizations that will self-righteously criticize and condemn other organizations, inflated for a while by a lot of public talk in the media, only to be replaced in its turn by another fashionable crisis. I hope that will not happen, and I believe that there are ways to keep it from happening, but I know that if this effort is carried on solely as a public cause, if millions of people cannot or will not undertake it as a *private* cause as well, then it is *sure* to happen. In five years the energy of our present concern will have petered out in a series of public gestures—and no doubt in a series of empty laws—and a great, and perhaps the last, human opportunity will have been lost.

It need not be that way. A better possibility is that the movement to preserve the environment will be seen to be, as I think it has to be, not a digression from the civil rights and peace movements, but the logical culmination of those movements.

For I believe that the separation of these three problems is artificial. They have the same cause, and that is the mentality of greed and exploitation. The mentality that exploits and destroys the natural environment is the same that abuses racial and economic minorities, that imposes on young men the tyranny of the military draft, that makes war against peasants and women and children with the indifference of technology. The mentality that destroys a watershed and then panics at the threat of flood is the same mentality that gives institutionalized insult to black people and then panics at the prospect of race riots. It is the same mentality that can mount deliberate warfare against a civilian population and then express moral shock at the logical consequence of such warfare at My Lai. We would be fools to believe that we could solve any one of these problems without solving the others.

To me, one of the most important aspects of the Environmental Movement is that it brings us not just to another public crisis, but to a crisis of the protest movement itself. For the environmental crisis should make it dramatically clear, as perhaps it has not always been before, that there is no public crisis that is not also private. To most advocates of civil rights, racism has seemed mostly the fault of someone else. For most advocates of peace, the war has been a remote reality, and the burden of the blame has seemed to rest mostly on the government. I am certain that these crises have been more private, and that we have each suffered more from them and been more responsible for them, than has been readily apparent, but the connections have been difficult to see. Racism and militarism have been institutionalized among us for too long for our personal involvement in those evils to be easily apparent to us. Think, for example, of all the Northerners who assumed—until black people attempted to move into *their* neighborhoods—that racism was a Southern phenomenon. And think how quickly—one might almost say how naturally—among some of its members the Peace Movement has spawned policies of deliberate provocation and violence.

But the environmental crisis rises closer to home. Every time we draw a breath, every time we drink a glass of water, we are suffering from it. And more important, every time we indulge in, or depend on, the wastefulness of our economy—and our economy's first principle is waste—we are *causing* the crisis. Nearly every one of us, nearly every day of his life, is contributing *directly* to the ruin of this planet. A protest meeting on the issue of environmental abuse is not a convocation of accusers, it is a convocation of the guilty. That realization ought to clear the smog of self-righteousness that has almost conventionally hovered over these occasions and let us see the work that is to be done.

In this crisis it is certain that every one of us has a public responsibility. We must not cease to bother the government and the other institutions to see that they never become comfortable with easy promises. For myself, I want to say that I hope never again to go to Frankfort to present a petition to the governor on an issue so vital as that of strip mining, only to be dealt with by some ignorant functionary—as several of us were not so long ago, the governor himself being "too busy" to receive us. Next time I will go prepared to wait as long as necessary to see that the petitioners' complaints and their arguments are heard *fully*—and by the governor. And then I will hope to find ways to keep those complaints and arguments from being forgotten until something is done to relieve them. The time is past when it was enough merely to elect our officials. We will have to elect them and then go and *watch* them and keep our hands on them, the way the coal companies do. We have made a tradition in Kentucky of putting self-servers, and worse, in charge of our vital interests. I am sick of it. And I think that one way to change it is to make Frankfort a less comfortable place. I believe in American political principles, and I will not sit idly by and see those principles destroyed by sorry practice. I am ashamed that American government should have become the chief cause of disillusionment with American principles.

And so when the government in Frankfort again proves too stupid or too blind or too corrupt to see the plain truth and to act with simple decency, I intend to be there, and I trust that I won't be alone. I hope, moreover, to be there, not with a sign or a slogan or a button, but with the facts and the arguments. A crowd whose discontent has risen no higher than the level of slogans is *only* a crowd. But a crowd that understands the reasons for its discontent and knows the remedies is a vital community, and it will have to be reckoned with. I would rather go before the government with two men who have a competent understanding of an issue, and who therefore deserve a hearing, than with two thousand who are vaguely dissatisfied.

But even the most articulate public protest is not enough. We don't live in the government or in institutions or in our public utterances and acts, and the environmental crisis has its roots in our *lives.* By the same token, environmental health will also be rooted in our lives. That is, I take it, simply a fact, and in the light of it we can see how superficial and foolish we would be to think that we could correct what is wrong merely by tinkering with the institutional machinery. The changes that are required are fundamental changes in the way we are living.

What we are up against in this country, in any attempt to invoke private responsibility, is that we have nearly destroyed private life. Our people have given up their independence in return for the cheap seductions and the shoddy merchandise of so-called "affluence." We have delegated all our vital functions and responsibilities to salesmen and agents and bureaus and experts of all sorts. We cannot feed or clothe ourselves, or entertain ourselves, or communicate with each other, or be charitable or neighborly or loving, or even respect ourselves, without recourse to a merchant or a corporation or a public service organization or an agency of the government or a stylesetter or an expert. Most of us cannot think of dissenting from the opinions or the actions of one organization without first forming a new organization. Individualism is going around these days in uniform, handing out the party line on individualism. Dissenters want to publish their personal opinions over a thousand signatures.

The Confucian *Great Digest* says that the "chief way for the production of wealth" (and it is talking about real goods, not money) is "that the producers be many and that the mere consumers be few. . . ." But even in the much-publicized rebellion of the young against the materialism of the affluent society, the consumer mentality is too often still intact: The standards of behavior are still those of kind and quantity, the security sought is still the security of numbers, and the chief motive is still the consumer's anxiety that he is missing out on what is "in." In this state of total consumerism—which is to say a state of helpless dependence on things and services and ideas and motives that we have forgotten how to provide ourselves—all meaningful contact between ourselves and the earth is broken. We do not understand the earth in terms either of what it offers us or of what it requires of us, and I think it is the rule that people inevitably destroy what they do not understand. Most of us are not directly responsible for strip mining and extractive agriculture and other forms of environmental abuse. But we are guilty nevertheless, for we connive in them by our ignorance. We are ignorantly dependent on them. We do not know enough about them; we do not have a particular enough sense of their danger. Most of us, for example, not only do not know how to produce the best food in the best way—we don't know how to produce any kind in any way. Our model citizen is a sophisticate who before puberty understands how to produce a baby, but who at the age of thirty will not know how to produce a potato. And for this condition we have elaborate rationalizations, instructing us that dependence for everything on somebody else is efficient and economical and a scientific miracle. I say, instead, that it is madness, mass produced. A man who understands the weather only in terms of golf is participating in a public insanity that either he or his descendants will be bound to realize as suffering. I believe that the death of the world is breeding in such minds much more certainly and much faster than in any political capital or atomic arsenal.

For an index of our loss of contact with the earth we need only look at the condition of the American farmer—who must enact our society's dependence on the land. In an age of unparalleled affluence and leisure, the American farmer is harder pressed and harder worked than ever before; his margin of profit is small, his hours are long; his outlays for land and equipment and the expenses of maintenance and operation are growing rapidly greater; he cannot compete with industry for labor; he is being forced more and more to depend on the use of destructive chemicals and on the wasteful methods of haste. As a class, farmers are one of the despised minorities. So far as I can see, farming is considered marginal or incidental to the economy of the country, and farmers, when they are thought of at all, are thought of as hicks and yokels, whose lives do not fit into the modern scene. The average American farmer is now an old man whose children have moved away to the cities. His knowledge, and his intimate connection with the land, are about to be lost. The small independent farmer is going the way of the small independent craftsmen and storekeepers. He is being forced off the land into the cities, his place taken by absentee owners, corporations, and machines. Some would justify all this in the name of efficiency. As I see it, it is an enormous social and economic and cultural blunder. For the small farmers who lived on their farms *cared* about their land. And given their established connection to their land—which was often hereditary and traditional as well as economic—they could have been encouraged to care for it more competently than they have so far. The corporations and machines that replace them will never be bound to the land by the sense of birthright and continuity, or by the love that enforces care. They will be bound by the rule of efficiency, which takes thought only of the volume of the year's produce, and takes no thought of the life of the land, not measurable in pounds or dollars, which will assure the livelihood and the health of the coming generations.

If we are to hope to correct our abuses of each other and of other races and of our land, and if our effort to correct these abuses is to be more than a political fad that will in the long run be only another form of abuse, then we are going to have to go far beyond public protest and political action. We are going to have to rebuild the substance and the integrity of private life in this country. We are going to have to gather up the fragments of knowledge and responsibility that we have parceled out to the bureaus and the corporations and the specialists, and put those fragments back together in our own minds and in our families and households and neighborhoods. We need better government, no doubt about it. But we also need better minds, better friendships, better marriages, better communities. We need persons and households that do not have to wait upon organizations, but can make necessary changes in themselves, on their own.

For most of the history of this country our motto, implied or spoken, has been Think Big. A better motto, and an essential one now, is Think Little. That implies the necessary change of thinking and feeling, and suggests the necessary work. Thinking Big has led us to the two biggest and cheapest political dodges of our time: plan-making and law-making. The lotus-eaters of this era are in Washington, D.C., Thinking Big. Somebody perceives a problem, and somebody in the government comes up with a plan or a law. The result, mostly, has been the persistence of the problem, and the enlargement and enrichment of the government.

But the discipline of thought is not generalization; it is detail, and it is personal behavior. While the government is "studying" and funding and organizing its Big Thought, nothing is being done. But the citizen who is willing to Think Little, and, accepting the discipline of that, to go ahead on his own, is already solving the problem. A man who is trying to live as a neighbor to his neighbors will have a lively and practical understanding of the work of peace and brotherhood, and let there be no mistake about it—he is *doing* that work. A couple who make a good marriage, and raise healthy, morally competent children, are serving the world's future more directly and surely than any political leader, though they never utter a public word. A good farmer who is dealing with the problem of soil erosion on an acre of ground has a sounder grasp of that problem and *cares* more about it and is probably doing more to solve it than any bureaucrat who is talking about it in general. A man who is willing to undertake the discipline and the difficulty of mending his own ways is worth more to the conservation movement than a hundred who are insisting merely that the government and the industries mend *their* ways.

If you are concerned about the proliferation of trash, then by all means start an organization in your community to do something about it. But before—*and while*—you organize, pick up some cans and bottles yourself. That way, at least, you will assure yourself and others that you mean what you say. If you are concerned about air pollution, help push for government controls, but drive your car less, use less fuel in your home. If you are worried about the damming of wilderness rivers, join the Sierra Club, write to the government, but turn off the lights you're not using, don't install an air conditioner, don't be a sucker for electrical gadgets, don't waste water. In other words, if you are fearful of the destruction of the environment, then learn to quit being an environmental parasite. We all are, in one way or another, and the remedies are not always obvious, though they certainly will always be difficult. They require a new kind of life—harder, more laborious, poorer in luxuries and gadgets, but also, I am certain, richer in meaning and more abundant in real pleasure. To have a healthy environment we will all have to give up things we like; we may even have to give up things we have come to think of as necessities. But to be fearful of the disease and yet unwilling to pay for the cure is not just to be hypocritical; it is to be doomed. If you talk a good line without being changed by what you say, then you are not just hypocritical and doomed; you have become an agent of the disease. Consider, for an example, President Nixon, who advertises his grave concern about the destruction of the environment, and who turns up the air conditioner to make it cool enough to build a fire.

Odd as I am sure it will appear to some, I can think of no better form of personal involvement in the cure of the environment than that of gardening. A person who is growing a garden, if he is growing it organically, is improving a piece of the world. He is producing something to eat, which makes him somewhat independent of the grocery business, but he is also enlarging, for himself, the meaning of food and the pleasure of eating. The food he grows will be fresher, more nutritious, less contaminated by poisons and preservatives and dyes than what he can buy at a store. He is reducing the trash problem; a garden is not a disposable container, and it will digest and reuse its own wastes. If he enjoys working in his garden, then he is less dependent on an automobile or a merchant for his pleasure. He is involving himself directly in the work of feeding people.

If you think I'm wandering off the subject, let me remind you that most of the vegetables necessary for a family of four can be grown on a plot of forty by sixty feet. I think we might see in this an economic potential of considerable importance, since we now appear to be facing the possibility of widespread famine. How much food could be grown in the dooryards of cities and suburbs? How much could be grown along the extravagant right-of-ways of the interstate system? Or how much could be grown, by the intensive practices and economics of the garden or small farm, on so-called marginal lands? Louis Bromfield liked to point out that the people of France survived crisis after crisis because they were a nation of gardeners, who in times of want turned with great skill to their own small plots of ground. And F. H. King, an agriculture professor who traveled extensively in the Orient in 1907, talked to a Chinese farmer who supported a family of twelve, "one donkey, one cow . . . and two pigs on 2.5 acres of cultivated land"—and who did this, moreover, by agricultural methods that were sound enough to have maintained his land in prime fertility through several thousand years of such use. These are possibilities that are readily apparent and attractive to minds that are prepared to Think Little. To Big Thinkers—the bureaucrats and businessmen of agriculture—they are invisible. But intensive, organic agriculture kept the farms of the Orient thriving for thousands of years, whereas extensive—which is to say, exploitive or extractive—agriculture has critically reduced the fertility of American farmlands in a few centuries or even a few decades.

A person who undertakes to grow a garden at home, by practices that will preserve rather than exploit the economy of the soil, has set his mind decisively against what is wrong with us. He is helping himself in a way that dignifies him and that is rich in meaning and pleasure. But he is doing something else that is more important: He is making vital contact with the soil and the weather on which his life depends. He will no longer look upon rain as a traffic impediment, or upon the sun as a holiday decoration. And his sense of humanity's dependence on the world will have grown precise enough, one would hope, to be politically clarifying and useful.

What I am saying is that if we apply our minds directly and competently to the needs of the earth, then we will have begun to make fundamental and necessary changes in our minds. We will begin to understand and to mistrust *and to change* our wasteful economy, which markets not just the produce of the earth, but also the earth's ability to produce. We will see that beauty and utility are alike dependent upon the health of the world. But we will also see through the fads and the fashions of protest. We will see that war and oppression and pollution are not separate issues, but are aspects of the same issue. Amid the outcries for the liberation of this group or that, we will know that no person is free except in the freedom of other persons, and that our only real freedom is to know and faithfully occupy our place—a much humbler place than we have been taught to think—in the order of creation.

But the change of mind I am talking about involves not just a change of knowledge, but also a change of attitude toward our essential ignorance, a change in our bearing in the face of mystery. The principles of ecology, if we will take them to heart, should keep us aware that our lives depend upon other lives and upon processes and energies in an interlocking system that, though we can destroy it, we can neither fully understand nor fully control. And our great dangerousness is that, locked in our selfish and myopic economy, we have been willing to change or destroy far beyond our power to understand. We are not humble enough or reverent enough.

Some time ago, I heard a representative of a paper company refer to conservation as a "no-return investment." This man's thinking was exclusively oriented to the annual profit of his industry. Circumscribed by the demand that the profit be great, he simply could not be answerable to any other demand—not even to the obvious needs of his own children.

Consider, in contrast, the profound ecological intelligence of Black Elk, "a holy man of the Oglala Sioux," who in telling his story said that it was not his own life that was important to him, but what he had shared with all life: "It is the story of all life that is holy and it is good to tell, and of us two-leggeds sharing in it with the four-leggeds and the wings of the air and all green things. . . ." And of the great vision that came to him when he was a child he said: "I saw that the sacred hoop of my people was one of many hoops that made one circle, wide as daylight and as starlight, and in the center grew one mighty flowering tree to shelter all the children of one mother and father. And I saw that it was holy."

Discipline and Hope

I: INTRODUCTORY NOTE

I BEGIN WITH what I believe is a safe premise, at least in the sense that most of the various sides of the current public argument would agree, though for different reasons: that we have been for some time in a state of general cultural disorder, and that this disorder has now become critical. My interest here has been to examine to what extent this disorder is a failure of discipline—specifically, a failure of those disciplines, both private and public, by which desired ends might be reached, or by which the proper means to a desired end might be determined, or by which it might be perceived that one apparently desirable end may contradict or forestall another more desirable. Thus we have not asked how the "quality of life," as we phrase it, may be fostered by social and technological means that are sensitive only to quantitative measures; we have not really questioned the universal premise of power politics that peace (among the living) is the natural result of war; we have hardly begun to deal with the fact that an economy of waste is not compatible with a healthy environment.

I realize that I have been rather severely critical of several, perhaps all, sides of the present disagreement, having in effect made this a refusal to be a partisan of *any* side. If critical intelligence has a use it is to prevent the coagulation of opinion in social or political cliques. My purpose has been to invoke the use of principle, rather than partisanship, as a standard of behavior, and to clear the rhetoric of the various sides from what ought to be the ground of personal experience and common sense.

It appears to me that the governing middle, or the government, which supposedly represents the middle, has allowed the extremes of left and right to force it into an extremism of its own. These three extremes of left, right, and middle, egged on by and helplessly subservient to each other's rhetoric, have now become so self-righteous and self-defensive as to have no social use. So large a ground of sanity and good sense and decency has been abandoned by these extremes that it becomes possible now to think of a New Middle made up of people conscious and knowledgeable enough to despise the blandishments and oversimplifications of the extremes—and roomy and diverse enough to permit a renewal of intelligent cultural dialogue. That is what I hope for: a chance to live and speak as a person, not as a function of some political bunch.

2: THE POLITICS OF KINGDOM COME

Times of great social stress and change, when realities become difficult to face and to cope with, give occasion to forms of absolutism, demanding perfection. We are in such a time now, and it is producing the characteristic symptoms. It has suddenly become clear to us that practices and ambitions that we have been taught from the cradle to respect have made us the heirs apparent of a variety of dooms; some of the promised solutions, on which we have been taught to depend, are not working, are probably not going to work. As a result the country is burdened with political or cultural perfectionists of several sorts, demanding that the government or the people create *right now* one or another version of the ideal state. The air is full of dire prophecies, warnings, and threats of what will happen if the Kingdom of Heaven is not precipitately landed at the nearest airport.

It is important that we recognize the childishness of this. Its ancestor is the kicking fit of childhood, a sort of behavioral false rhetoric that offers the world two absolute alternatives: "If I can't have it, I'll tear it up." Its cultural model is the fundamentalist preacher, for whom there are no degrees of behavior, who cannot tell the difference between a shot glass and a barrel. The public *demand* for perfection, as opposed to private striving for it, is almost always productive of violence, and is itself a form of violence. It is totalitarian in impulse, and often in result.

The extremes of public conviction are always based upon rhetorical extremes, which is to say that their words—and their actions—have departed from facts, causes, and arguments, and have begun to follow the false logic of a feud in which nobody remembers the cause but only what was last said or done by the other side. Language and behavior become purely negative in function. The opponents no longer speak in support of their vision or their arguments or their purposes, but only in opposition to each other. Language ceases to bind head to heart, action to principle, and becomes a weapon in a contention deadly as war, shallow as a game.

But it would be an oversimplification to suggest that the present contention of political extremes involves only the left and the right. These contend not so much against each other as against the middle—the administration in power—which each accuses of *being* the other. In defending itself, the middle characteristically adopts the tactics of the extremes, corrupting its language by a self-congratulatory rhetoric bearing no more kinship to the truth and to honest argument than expediency demands, and thus it becomes an extreme middle. Whereas the extreme left and right see in each other the imminence of Universal Wrong, the extreme middle appears to sense in itself the imminence of the Best of All Possible Worlds, and therefore looks upon all critics as traitors. The rhetoric of the extreme middle equates the government with the country, loyalty to the government with patriotism, the will of the chief executive with the will of the people. It props itself with the tone of divine good will and infallibility, demanding an automatic unquestioning faith in its actions, upholding its falsehoods and errors with the same unblinking piety with which it obscures its truths and its accomplishments.

Because the extreme middle is characteristically in power, its characteristic medium is the one that is most popular—television. How earnestly and how well this middle has molded itself to the demands of television is apparent when one considers how much of its attention is given to image making, or remaking, and to public relations. It has given up almost altogether the disciplines of political discourse (considerations of fact and of principle and of human and historical limits and possibilities), and has taken up the cynical showmanship of those who have cheap goods for sale. Its catch phrases do not rise from any viable political tradition; their next of kin are the TV jingles of soup and soap. It is a politics of illusion, and its characteristic medium is pre-eminently suited—as it is almost exclusively limited—to the propagation of illusion.

Of all the illusions of television, that of its much-touted "educational value" is probably the first. Because of its utter transience as a medium and the complete passivity of its audience, television is doomed to have its effect within the limits of the most narrow and shallow definition of entertainment—that is, entertainment as diversion. The watcher sees the program at the expense of no effort at all; he is inert. All the live connections are broken. More important, a TV program can be seen only once; it cannot be re-examined or judged upon the basis of study, as even a movie can be; a momentous event or a serious drama slips away from us among the ordinary furniture of our lives, as transient and fading as the most commonplace happenings of every day. For these reasons a political speech on television has to be first and last a show, simply because it has no chance to become anything else. The great sin of the medium is not that it presents fiction as truth, as undoubtedly it sometimes does, but that it cannot help presenting the truth as fiction, and that of the most negligible sort—a way to keep awake until bedtime.

In depending so much upon a medium that will not permit scrutiny, the extreme middle has perhaps naturally come to speak a language that will not *bear* scrutiny. It has thus abased its own part in the so-called political dialogue to about the level of the slogans and chants and the oversimplified invective of the extreme left and right—a fact that in itself might sufficiently explain the obsession with image-making. Hearing the televised pronouncements of the political leaders of our time upon the great questions of human liberty, community obligation, war and peace, poverty and wealth, one might easily forget that such as John Adams and Thomas Jefferson once spoke here upon those questions. Indeed, our contemporary men of power have produced in their wake an industry of journalistic commentary and interpretation, because it is so difficult to determine *what* they have said and whether or not they meant it. Thus one sees the essential contradiction in the expedient doctrine that the end may justify the means. Corrupt or false means must inevitably corrupt or falsify the end. There is an important sense in which the end *is* the means.

What is disturbing, then, about these three "sides" of our present political life is not their differences but their similarities.

They have all abandoned discourse as a means of clarifying and explaining and defending and implementing their ideas. They have taken almost exclusively to the use of the rhetoric of ad-writers: catch phrases, slogans, clichés, euphemisms, flatteries, falsehoods, and various forms of cheap wit. This has led them —as such rhetoric must—to the use of power and the use of violence against each other. But however their ideological differences might be graphed, they are, in effect, all on the same side. They are on the side of their quarrel, and are against all other, and all better, possibilities. There is a political and social despair in this that is the greatest peril a country can come to, short of the inevitable results of such despair should it continue very long. "We are fatalists," Edward Dahlberg wrote, "only when we cease telling the truth, but, so long as we communicate the truth, we move ourselves, life, history, men. There is no other way."

Our present political rhetoric is the desperation of argument. It is like a weapon in its inflexibility, in its insensitivity to circumstance, and in its natural inclination toward violence. It is by this recourse to loose talk—this willingness to say whatever will be easiest to say and most willingly heard—that the left permits its methods to contradict its avowed aims, as when it contemplates violence as a means of peace or permits arguments to shrink into slogans. By this process the right turns from its supposed aim of conserving the best of the past and undertakes the defense of economic privilege and the deification of symbols. And by this process the middle abandons its obligation to lead and enlighten the majority it claims as its constituency, and takes to the devices of a sterile showmanship, by which it hopes to elude criticism and obscure its failures.

The political condition in this country now is one in which the means or the disciplines necessary to the achievement of professed ends have been devalued or corrupted or abandoned altogether. We are offered peace without forbearance or tolerance or love, security without effort and without standards, freedom without risk or responsibility, abundance without thrift. We are asked repeatedly by our elected officials to console ourselves with that most degenerate of political arguments: Though we are not doing as well as we might, we could do worse, and we are doing better than some.

3: THE KINGDOM OF EFFICIENCY AND SPECIALIZATION

But this political indiscipline is exemplary of a condition that is widespread and deeply rooted in almost all aspects of our life. Nearly all the old standards, which implied and required rigorous disciplines, have now been replaced by a new standard of efficiency, which requires not discipline, not a mastery of means, but rather a carelessness of means, a relentless subjection of means to immediate ends. The standard of efficiency displaces and destroys the standards of quality because, by definition, it cannot even consider them. Instead of asking a man what he can do well, it asks him what he can do fast and cheap. Instead of asking the farmer to practice the best husbandry, to be a good steward and trustee of his land and his art, it puts irresistible pressures on him to produce more and more food and fiber more and more cheaply, thereby destroying the health of the land, the best traditions of husbandry, and the farm population itself. And so when we examine the principle of efficiency as we now practice it, we see that it is not really efficient at all. As we use the word, efficiency means no such thing, or it means short-term or temporary efficiency, which is a contradiction in terms. It means cheapness at any price. It means hurrying to nowhere. It means the profligate waste of humanity and of nature. It means the greatest profit to the greatest liar. What we have called efficiency has produced among us, and to our incalculable cost, such unprecedented monuments of destructiveness and waste as the strip-mining industry, the Pentagon, the federal bureaucracy, and the family car.

Real efficiency is something entirely different. It is neither cheap (in terms of skill and labor) nor fast. Real efficiency is long-term efficiency. It is to be found in means that are in keeping with and preserving of their ends, in methods of production that preserve the sources of production, in workmanship that is durable and of high quality. In this age of planned obsolescence, frivolous horsepower and surplus manpower, those salesmen and politicians who talk about efficiency are talking, in reality, about spiritual and biological death.

Specialization, a result of our nearly exclusive concern with the form of exploitation that we call efficiency, has in its turn become a destructive force. Carried to the extent to which we have carried it, it is both socially and ecologically destructive. That specialization has vastly increased our knowledge, as its defenders claim, cannot be disputed. But I think that one might reasonably dispute the underlying assumption that knowledge per se, undisciplined knowledge, is good. For while specialization has increased knowledge, it has fragmented it. And this fragmentation of knowledge has been accompanied by a fragmentation of discipline. That is, specialization has tended to draw the specialist toward the discipline that will lead to the discovery of new facts or processes within a narrowly defined area, and it has tended to lead him away from or distract him from those disciplines by which he might consider the *effects* of his discovery upon human society or upon the world.

Nowhere are these tendencies more apparent than in agriculture. For years now the agricultural specialists have tended to think and work in terms of piecemeal solutions and in terms of annual production, rather than in terms of a whole and coherent system that would maintain the fertility and the ecological health of the land over a period of centuries. Focused nearly exclusively upon so-called efficiency with respect to production, as if the only discipline pertinent to agriculture were that of economics, they have eagerly abetted a rapid industrialization of agriculture that is potentially catastrophic, both in the ecological deterioration of farm areas and in the dispossession and displacement of the rural population.

Ignoring the ample evidence that a healthy, ecologically sound agriculture is highly diversified, using the greatest possible variety of animals and plants, and that it returns all organic wastes to the soil, the specialists of the laboratories have promoted the specialization of the farms, encouraging one-crop agriculture and the replacement of humus by chemicals. And as the pressures of urban populations upon the land have grown, the specialists have turned more and more not to the land, but to the laboratory.

Ignoring the considerable historical evidence that to have a productive agriculture over a long time it is necessary to have a stable, prosperous rural population closely bound in sympathy and association to the land, the specialists have either connived in the dispossession of small farmers by machinery and technology, or have actively encouraged their migration into the cities.

The result of the short-term vision of these experts is a whole series of difficulties that together amount to a rapidly building ecological and social disaster, which there is little disposition at present to regret, much less to correct. The organic wastes of our society, for which our land is starved and which in a sound agricultural economy would be returned to the land, are flushed out through the sewers to pollute the streams and rivers and, finally, the oceans; or they are burned and the smoke pollutes the air; or they are wasted in other ways. Similarly, the small farmers who in a healthy society ought to be the mainstay of the country—whose allegiance to their land, continuing and deepening in association from one generation to another, would be the motive and guarantee of good care—are forced out by the homicidal economics of efficiency, to become emigrants and dependents in the already overcrowded cities. In both instances, by the abuse of knowledge in the name of efficiency, assets have been converted into staggering problems.

The metaphor governing these horrendous distortions has been that of the laboratory. The working assumption has been that nature and society, like laboratory experiments, can be manipulated by processes that are for the most part comprehensible toward ends that are for the most part foreseeable. But the analogy, as any farmer knows, is too simple, for both nature and humanity are unpredictable and ultimately mysterious. Sir Albert Howard was speaking to this problem in *An Agricultural Testament*: "Instead of breaking up the subject into fragments and studying agriculture in piecemeal fashion by the analytical methods of science, appropriate only to the discovery of new facts, we must adopt a synthetic approach and look at the wheel of life as one great subject and not as if it were a patchwork of unrelated things." A much more appropriate model for the agriculturist, scientist, or farmer is the forest, for the forest, as Howard pointed out, "manures itself" and is therefore self-renewing; it has achieved that "correct relation between the processes of growth and the processes of decay that is the first principle of successful agriculture." A healthy agriculture can take place only within nature, and in cooperation with its processes, not in spite of it and not by "conquering" it. Nature, Howard points out, in elaboration of his metaphor, "never attempts to farm without livestock; she always raises mixed crops; great pains are taken to preserve the soil and to prevent erosion; the mixed vegetable and animal wastes are converted into humus; *there is no waste* [my emphasis]; the processes of growth and the processes of decay balance one another; ample provision is made to maintain large reserves of fertility; the greatest care is taken to store the rainfall; both plants and animals are left to protect themselves against disease."

The fact is that farming is not a laboratory science, but a science of practice. It would be, I think, a good deal more accurate to call it an art, for it grows not only out of factual knowledge but out of cultural tradition; it is learned not only by precept but by example, by apprenticeship; and it requires not merely a competent knowledge of its facts and processes, but also a complex set of attitudes, a certain culturally evolved stance, in the face of the unexpected and the unknown. That is to say, it requires *style* in the highest and richest sense of that term.

One of the most often repeated tenets of contemporary optimism asserts that "a nation that can put men on the moon certainly should be able to solve the problem of hunger." This proposition seems to me to have three important flaws, which I think may be taken as typical of our official view of ourselves:

1—It construes the flight to the moon as an historical event of a complete and coherent significance, when in fact it is a fragmentary event of very uncertain significance. Americans have gone to the moon as they came to the frontiers of the New World: with their minds very much upon getting there, very little upon what might be involved in *staying* there. I mean that, because of our history of waste and destruction here, we have no assurance that we can survive in America, much less on the moon. And until we can bring into balance the processes of growth and decay, the white man's settlement of this continent will remain an incomplete event. When a Japanese peasant went to the fields of his tiny farm in the preindustrial age, he worked in the governance of an agricultural tradition that had sustained the land in prime fertility for thousands of years, in spite of the pressures of a population that in 1907 had reached a density, according to F. H. King's *Farmers of Forty Centuries*, of "more than three people to each acre." Such a farmer might look upon his crop year as a complete and coherent historical event, suffused and illuminated with a meaning and mystery that were both its own and the world's, because in his mind and work agricultural process had come into an enduring and preserving harmony with natural process. To him the past confidently promised a future. What are we to say, by contrast, of a society that places no value at all upon such a tradition or such a man, that instead works the destruction of such imperfect agricultural tradition as it has, that replaces the farm people with machines, that values the techniques of production far above the techniques of land maintenance, and that has espoused as an ideal a depopulated countryside farmed by a few technicians for the supposedly greater benefit of hundreds of millions crowded into cities and helpless to produce food or any other essential for themselves?

2—The agricultural optimism that bases itself upon the moon landings assumes that there is an equation between agriculture and technology, or that agriculture is a kind of technology. This grows out of the much-popularized false assumptions of the agricultural specialists, who have gone about their work as if agriculture was answerable only to the demands of economics, not to those of ecology or human culture, just as most urban consumers conceive eating to be an activity associated with economics but not with agriculture. The ground of agricultural thinking is so narrowly circumscribed, one imagines, to fit the demands of laboratory science, as well as the popular prejudice that prefers false certainties to honest doubts. The discipline proper to eating, of course, is not economics but agriculture. The discipline proper to agriculture, which survives not just by production but also by the return of wastes to the ground, is not economics but ecology. And ecology may well find its proper disciplines in the arts, whose function is to refine and enliven *perception*, for ecological principle, however publicly approved, can be enacted only upon the basis of each man's perception of his relation to the world.

Under the governance of the laboratory analogy, the *device*, which is simple and apparently simplifying, becomes the focal point and the standard rather than the human need, which is complex. Thus an agricultural specialist, prescribing the best conditions for the use of a harvesting machine, thinks only of the machine, not its cultural or ecological effects. And because of the fixation on optimum conditions, big-farm technology has come to be highly developed, whereas the technology of the family farm, which must still involve methods and economies that are "old-fashioned," has been neglected. For this reason, and others perhaps more pressing, small-farm technology is rapidly passing from sight, along with the small farmers. As a result we have an increasing acreage of supposedly "marginal" but potentially productive land for the use of which we have neither methods nor people—an alarming condition in view of the likelihood that someday we will desperately need to farm these lands again.

The drastic and incalculably dangerous assumption is that farming can be considered apart from farmers, that the land may be conceptually divided in its use from human need and human care. The assumption is that moving a farmer into a factory is as simple a cultural act as moving a worker from one factory to another. It is inconceivably more complicated, and more final. American agricultural tradition has been for the most part inadequate from the beginning, and we have an abundance of diminished land to show for it. But American farmers are nevertheless an agricultural population of long standing. Most settlers who farmed in America farmed in Europe. The farm population in this country therefore embodies a knowledge and a set of attitudes and interests that have been thousands of years in the making. This mentality is, or was, a great resource upon which we might have built a truly indigenous agriculture, fully adequate to the needs and demands of American regions. Ancient as it is, it is destroyed in a generation in every family that is forced off the farm into the city—or in less than a generation, for the farm mentality can survive only in sustained vital contact with the land.

A truer agricultural vision would look upon farming not as a function of the economy or even of the society, but as a function of the land; and it would look upon the farm population as an indispensable and inalienable part of the ecological system. Among the Incas, according to John Collier (*Indians of the Americas*), the basic social and economic unit was the tribe, or *ayllu*, but "the *ayllu* was not merely its people, and not merely the land, but people and land wedded through a mystical bond." The union of the land and the people was indissoluble, like marriage or grace. Chief Rekayi of the Tangwena tribe of Rhodesia, in refusing to leave his ancestral home, which had been claimed by the whites, is reported in recent newspaper accounts to have said: "I am married to this land. I was put here by God . . . and if I am to leave, I must be removed by God who put me here." This altogether natural and noble sentiment was said by the Internal Affairs Minister to have been "Communist inspired."

3—The notion that the moon voyages provide us assurance of enough to eat exposes the shallowness of our intellectual confidence, for it is based upon our growing inability to distinguish between training and education. The fact is that a man can be made an astronaut much more quickly than he can be made a good farmer, for the astronaut is produced by training and the farmer by education. Training is a process of conditioning, an orderly and highly efficient procedure by which a man learns a prescribed pattern of facts and functions. Education, on the other hand, is an obscure process by which a person's experience is brought into contact with his place and his history. A college can train a person in four years; it can barely begin his education in that time. A person's education begins before his birth in the making of the disciplines, traditions, and attitudes of mind that he will inherit, and it continues until his death under the slow, expensive, uneasy tutelage of his experience. The process that produces astronauts may produce soldiers and factory workers and clerks; it will never produce good farmers or good artists or good citizens or good parents.

White American tradition, so far as I know, contains only one coherent social vision that takes such matters into consideration, and that is Thomas Jefferson's. Jefferson's public reputation seems to have dwindled to that of Founding Father and advocate of liberty, author of several documents and actions that have been enshrined and forgotten. But in his thinking, democracy was not an ideal that stood alone. He saw that it would have to be secured by vigorous disciplines or its public offices would become merely the hunting grounds of mediocrity and venality. And so those who associate his name only with his political utterances miss both the breadth and depth of his wisdom. As Jefferson saw it, two disciplines were indispensable to democracy: on the one hand, education, which was to produce a class of qualified leaders, an aristocracy of "virtue and talents" drawn from all economic classes; and on the other hand, widespread land ownership, which would assure stable communities, a tangible connection to the country, and a permanent interest in its welfare. In language that recalls Collier's description of the *ayllu* of the Incas, and the language of Chief Rekayi of the Tangwenans, Jefferson wrote that farmers "are tied to their country, and wedded to its liberty and interests, by the most lasting bonds." And: ". . . legislators cannot invent too many devices for subdividing property. . . ." And: ". . . it is not too soon to provide by every possible means that as few as possible shall be without a little portion of land. The small landholders are the most precious part of a state." For the discipline of education of the broad and humane sort that Jefferson had in mind, to produce a "natural aristocracy . . . for the instruction, the trusts, and government of society," we have tended more and more to substitute the specialized training that will most readily secure the careerist in his career. For the ownership of "a little portion of land" we have, and we apparently wish, to substitute the barbarous abstraction of nationalism, which puts our minds within the control of whatever demagogue can soonest rouse us to self-righteousness.

On September 10, 1814, Jefferson wrote to Dr. Thomas Cooper of the "condition of society" as he saw it at that time: ". . . we have no paupers, the old and crippled among us, who possess nothing and have no families to take care of them, being too few to merit notice as a separate section of society. . . . The great mass of our population is of laborers; our rich . . . being few, and of moderate wealth. Most of the laboring class possess property, cultivate their own lands . . . and from the demand for their labor are enabled . . . to be fed abundantly, clothed above mere decency, to labor moderately. . . . The wealthy . . . know nothing of what the Europeans call luxury." This has an obvious kinship with the Confucian formula: ". . . that the producers be many and that the mere consumers be few; that the artisan mass be energetic and the consumers temperate. . . ."

In the loss of that vision, or of such a vision, and in the abandonment of that possibility, we have created a society characterized by degrading urban poverty and an equally degrading affluence—a society of undisciplined abundance, which is to say a society of waste.

4: THE KINGDOM OF CONSUMPTION

The results have become too drastic to be concealed by our politicians' assurances that we have built a "great society" or that we are doing better than India. Official pretense has begun to break down under the weight of the obvious. In the last decade we have become unable to condition our children's minds to approve or accept our errors. Our history has created in the minds of our young people a bitter division between official pretense and social fact, and we have aggravated this division by asking many of them to fight and die in support of official pretense. In this way we have produced a generation whose dissidence and alienation are without precedent in our national experience.

The first thing to be said about this rebelliousness is that it is understandable, and that it deserves considerate attention. Many of this generation have rejected values and practices that they believe to be destructive, and they should do so. Many of them have begun to search for better values and forms of life, and they should do so. But the second thing to be said is that this generation is as subject as any other to intelligent scrutiny and judgment, and as deserving of honest criticism. It has received much approbation and condemnation, very little criticism.

One of its problems is that it has been isolated in its youthfulness, cut off from the experience and the counsel of older people, as probably no other generation has ever been. It is true that the dissident young have had their champions among the older people, but it is also true, I think, that these older people have been remarkably uncritical of the young, and so have abdicated their major responsibility to them. Some appear to have *joined* the younger generation, buying their way in by conniving in the notion that idealistic youth can do no wrong —or that one may reasonably hope to live without difficulty or effort or tragedy, or that surfing is "a life." The uncritical approval of a band of senior youth freaks is every bit as isolating and every bit as destructive as the uncritical condemnation of those who have made hair length the foremost social issue of the time.

And so a number of the problems of the young people now are problems that have always attended youthfulness, but which isolation has tended to aggravate in the present generation: impetuousness, a haste to undertake work that one is not yet prepared for; a tendency to underestimate difficulty and overestimate possibility, which is apt, through disillusionment, to lead to the overestimation of difficulty and the underestimation of possibility; oversimplification, as when rejection takes the place of evaluation; and, finally, naïve prejudice, as when people who rightly condemn the use of such terms as "nigger" or "greaser" readily use such terms as "pig" and "redneck."

Another of its problems, and a much larger one, is that the propaganda both of the "youth culture" and of those opposed to it has inculcated in many minds, both young and old, the illusion that this is a wholly new generation, a generation free of history. The proposition is dangerously silly. The present younger generation is, as much as any other, a product of the past; it would not be as it is if earlier generations had not been as they were. Like every other young generation, this one bears the precious human burden of new possibility and new hope, the opportunity to put its inheritance to better use. And like every other, it also bears the germ of historical error and failure and weakness—which it rarely forgives in its predecessors, and seldom recognizes in itself. In the minds of those who do not know it well, and who have not mastered the disciplines of self-criticism, historical error is a subtle virus indeed. It is of the greatest importance that we recognize in the youth culture the persistence in new forms of the mentality of waste, certain old forms of which many of the young have rightly repudiated.

Though it has forsworn many of the fashions and ostentations of the "affluent society," the youth culture still supports its own forms of consumerism, the venerable American doctrine which holds that if enough is good, too much is better. As an example, consider the present role of such drugs as marijuana and the various hallucinogens. To deal sensibly with this subject, it is necessary to say at the outset that the very concept of drug abuse implies the possibility of drug use that is *not* abusive. And it is, in fact, possible to produce examples of civilizations that have employed drugs in disciplines and ceremonies that have made them culturally useful and prevented their abuse.

Tobacco, for instance, is a drug that we have used so massively and thoughtlessly that we have, in typical fashion, come to be endangered by it. This is the pattern of the consumer economy and it applies not just to drugs, but to such commodities as the automobile and electrical power. But American Indians attached to tobacco a significance that made it more valuable to them than it has ever been to us, and at the same time kept them from misusing it as we have. In the Winnebago Origin Myth tobacco had a ceremonial and theological role. According to this myth, Paul Radin wrote in his introduction to *The Road of Life and Death*, man "is not to save himself or receive the wherewithal of life through the accidental benefactions of culture-heroes. On the contrary, he is to be in dire straits and saved. Earthmaker is represented as withholding tobacco from the spirits in order to present it to man and as endowing these same spirits with an overpowering craving for it. In short, it is to be the mechanism for an exchange between man and the deities. He will give them tobacco; they will give him powers to meet life and overcome obstacles."

The use of alcohol has had, I believe, a similar history: a decline and expansion from ceremonial use to use as a commodity and extravagance, from cultural usefulness to cultural liability.

The hallucinogenic drugs have now also run this course of cultural diminishment, and at the hands not of the salesmen of the corporations and the advertising agencies, but of the self-proclaimed enemies of those salesmen. Most of these drugs have been used by various cultures in association with appropriate disciplines and ceremonies. Anyone who reads an account of the Peyote Meeting of the Native American Church will see that it resembles very much the high ritual and art of other cultures but very little indeed the usual account of the contemporary "dope scene."

A very detailed and well-understood account of the disciplined use of such drugs is in Carlos Castaneda's *The Teachings of Don Juan: A Yaqui Way of Knowledge*. Don Juan, a medicine man and sorcerer, a Yaqui Indian from Sonora, Mexico, undertakes to teach Castaneda the uses of jimson weed, peyote, and the psilocybe mushrooms. The book contains some remarkable accounts of the author's visions under the influence of these drugs, but equally remarkable is the rigor of the disciplines and rituals by which his mentor prepared him for their use. At one point early in their association the old Indian said to him: "A man goes to knowledge as he goes to war, wide awake, with fear, with respect, and with absolute assurance. Going to knowledge or going to war in any other manner is a mistake, and whoever makes it will live to regret his steps."

The cultural role of both hallucinogens and intoxicants, in societies that have effectively disciplined their use, has been strictly limited. They have been used either for the apprehension of religious or visionary truth or, a related function, to induce in conditions prescribed by ceremony and festivity a state of self-abandon in which one may go free for a limited time of the obscuring and distorting preoccupation with one's own being. Other cultures have used other means—music or dance or poetry, for examples—to produce these same ends, and although the substance of Don Juan's teaching may be somewhat alien to the mainstream of our tradition, the terms of its discipline are not.

By contrast, the youth culture tends to use drugs in a way very similar to the way its parent culture uses alcohol: at random, as a social symbol and crutch, and with the emphasis upon the fact and quantity of use rather than the quality and the content of the experience. It would be false to say that these drugs have come into contemporary use without any of their earlier cultural associations. Indeed, a good deal of importance has been assigned to the "religious" aspect of the drug experience. But too often, it seems to me, the tendency has been to make a commodity of religion, as if in emulation of some churches. Don Juan looked upon drugs as a way to knowledge, difficult and fearful as other wise men have conceived other ways to knowledge, and therefore to be rigorously prepared for and faithfully followed; the youth culture, on the other hand, has tended to look upon drugs as a sort of instant Holy Truth, of which one need not become worthy. When they are inadequately prepared for the use of drugs, that is to say, people "consume" and waste them.

The way out of this wastefulness obviously cannot lie simply in a shift from one fashionable commodity to another commodity equally fashionable. The way out lies only in a change of mind by which we will learn not to think of ourselves as "consumers" in any sense. A consumer is one who uses things up, a concept that is alien to the creation, as are the concepts of waste and disposability. A more realistic and accurate vision of ourselves would teach us that our ecological obligations are to use, not use up; to use by the standard of real need, not of fashion or whim; and then to relinquish what we have used in a way that returns it to the common ecological fund from which it came.

The key to such a change of mind is the realization that the first and final order of creation is not such an order as men can impose on it, but an order in the creation itself by which its various parts and processes sustain each other, and which is only to some extent understandable. It is, moreover, an order in which things find their places and their values not according to their inert quantities or substances but according to their energies, their powers, by which they cooperate or affect and influence each other. The order of the creation, that is to say, is closer to that of drama than to that of a market.

This relation of power and order is another of the major concerns of the Winnebago Creation Myth. "Having created order within himself and established it for the stage on which man is to play," Paul Radin says, "Earthmaker proceeds to create the first beings who are to people the Universe, the spirits and deities. To each one he assigns a fixed and specific amount of power, to some more, to some less. . . . This principle of gradation and subordination is part of the order that Earthmaker is represented as introducing into the Universe. . . . The instant it is changed there is danger and the threat of disruption." The principle is dramatized, according to Radin, in the legend of Morning Star: ". . . one of the eight great Winnebago deities, Morning Star, has been decapitated by his enemy, a water-spirit. The body of the hero still remains alive and is being taken care of by his sister. The water-spirit, by keeping the head of Morning Star, has added the latter's power to his own. So formidable is this combination of powers that none of the deities [is] a match for him now. In fact only Earthmaker is his equal. Here is a threat of the first magnitude to the order ordained by Earthmaker and it must be met lest destruction overtake the world."

The point is obvious: To take and keep, to consume the power of another creature is an act profoundly disordering, contrary to the nature of the creation. And equally obvious is its applicability to our own society, which sees its chief function in such accumulations of power. A man grows rich by strip mining, adding the power of a mountain to himself in such a way that he cannot give it back. As a nation, we have so far grown rich by adding the power of the continent to ourselves in such a way that we cannot give it back. "Here is a threat of the first magnitude to the order ordained by Earthmaker and it must be met lest destruction overtake the world."

Though we generally concede that a man may have more of the world's goods than he deserves, I think that we have never felt that a man may have more light than he deserves. But an interesting implication of the Winnebago doctrine of power and order is that a man must not only become worthy of enlightenment, but has also an ethical obligation to make himself worthy of the world's goods. He can make himself worthy of them only by using them carefully, preserving them, relinquishing them in good order when he has had their use. That a man shall find his life by losing it is an ethical concept that applies to the body as well as the spirit.

An aspect of the consumer mentality that has cropped up with particular virulence in the youth culture is an obsessive fashionableness. The uniformity of dress, hair style, mannerism, and speech is plain enough. But more serious, because less conscious and more pretentious, is an intellectual fashionableness pinned up on such shibboleths as "the people" (the most procrustean of categories), "relevance" (the most reactionary and totalitarian of educational doctrines), and "life style."

This last phrase furnishes a particularly clear example of the way poor language can obscure both a problem and the possibility of a solution. Compounded as "alternate life style," the phrase becomes a part of the very problem it aspires to solve. There are, to begin with, two radically different, even opposed meanings of style: style as fashion, an imposed appearance, a gloss upon superficiality; and style as the signature of mastery, the efflorescence of long discipline. It is obvious that the style of mastery can never become the style of fashion, simply because every master of a discipline is different from every other; his mastery is suffused with his own character and his own materials. Cézanne's paintings could not have been produced by a fad, for the simple reason that they could not have been produced by any other person. As a popular phrase, "life style" necessarily has to do only with what is imitable in another person's life, its superficial appearances and trappings; it cannot touch its substances, disciplines, or devotions. More important is the likelihood that a person who has identified his interest in another person as an interest in his "life style" will be *aware* of nothing but appearances. The phrase "alternate life style" attempts to recognize our great need to change to a kind of life that is not wasteful and destructive, but stifles the attempt, in the same breath, by infecting it with that superficial concept of style. An essential recognition is thus obscured at birth by the old lie of advertising and public relations: that you can alter substance by altering appearance. "Alternate life style" suggests, much in the manner of the fashion magazines, that one can change one's life by changing one's clothes.

Another trait of consumption that thrives in the youth culture is that antipathy to so-called "drudgery" that has made us, with the help of salesmen and advertisers, a nation of suckers. This is the pseudoaristocratic notion, early popularized in America, that one is too good for the fundamental and recurring tasks of domestic order and biological necessity; to dirty one's hands in the soil or to submerge them for very long in soapy water is degrading and brutalizing. With one's hands thus occupied, the theory goes, one is unlikely to reach those elusive havens of "self-discovery" and "self-fulfillment"; but if one can escape such drudgery, one then has a fair chance of showing the world that one is *really* better than all previous evidence would have indicated. In every drudge there is an artist or a tycoon yearning to breathe free.

The entire social vision, as I understand it, goes something like this: Man is born into a fallen world, doomed to eat bread in the sweat of his face. But there is an economic redemption. He should go to college and get an education—that is, he should acquire the "right" certificates and meet the "right" people. An education of this sort should enable him to get a "good" job—that is, short hours of work that is either easy or prestigious for a lot of money. Thus he is saved from the damnation of drudgery, and is presumably well on the way to proving the accuracy of his early suspicion that he is *really* a superior person.

Or, in a different version of the same story, the farmer at his plow or the housewife at her stove dreams of the neat outlines and the carefree boundaries of a factory worker's eight-hour day and forty-hour week, and his fat, unworried paycheck. They will leave their present drudgery to take the bait, in this case, of leisure, time, and money to enjoy the "good things of life."

In reality, this despised drudgery is one of the constants of life, like water only changing its form in response to changes of atmosphere. Our aversion to the necessary work that we call drudgery and our strenuous efforts to avoid it have not diminished it at all, but only degraded its forms. The so-called drudgery has to be done. If one is "too good" to do it for oneself, then it must be done by a servant, or by a machine manufactured by servants. If it is not done at home, then it must be done in a factory, which degrades both the conditions of work and the quality of the product. If it is not done well by the hands of one person, then it must be done poorly by the hands of many. But somewhere the hands of someone must be soiled with the work. Our aversion to this was once satisfied by slavery, or by the abuse of a laboring class; now it is satisfied by the assembly line, or by similar redundancy in bureaus and offices. For decades now our people have streamed into the cities to escape the drudgery of farm and household. Where do they go to escape the drudgery of the city? Only home at night, I am afraid, to the spiritual drudgery of factory-made suppers and TV. On weekends many of them continue these forms of urban drudgery in the country.

The youth culture has accepted, for the most part uncritically, the conviction that all recurring and necessary work is drudgery, even adding to it a uniquely gullible acquiescence in the promoters' myth that the purpose of technology is to free mankind for spiritual and cultural pursuits. But to the older idea of economic redemption from drudgery, the affluent young have added the even more simple-minded idea of redemption by spontaneity. Do what you feel like, they say—as if every day one could "feel like" doing what is necessary. Any farmer or mother knows the absurdity of this. Human nature is such that if we waited to do anything until we felt like it, we would do very little at the start, even of those things that give us pleasure, and would do less and less as time went on. One of the common experiences of people who regularly do hard work that they enjoy is to find that they begin to "feel like it" only after the task is begun. And one of the chief uses of discipline is to assure that the necessary work gets done even when the worker *doesn't* feel like it.

Because of the prevalence of the economics and the philosophy of laborsaving, it has become almost a heresy to speak of hard work, especially manual work, as an inescapable human necessity. To speak of such work as good and ennobling, a source of pleasure and joy, is almost to declare oneself a pervert. Such work, and any aptitude or taste for it, are supposedly mere relics of our rural and primitive past—a past from which it is the business of modern science and technology to save us.

Before one can hope to use any intelligence in this matter, it is necessary to resurrect a distinction that was probably not necessary before the modern era, and that has so far been made only by a few eccentrics and renegades. It is a distinction not made in business and government, and very little made in the universities. I am talking about the distinction between work that is necessary and therefore meaningful, and work that is unnecessary or devoid of meaning. There is no intelligent defense of what Thoreau called "the police of meaningless labor." The unnecessary work of producing notions or trinkets or machines intended to be soon worn out, or necessary work the meaning of which has been destroyed by mechanical process, is as degrading as slavery. And the purpose of such slavery, according to the laborsaving philosophy, is to set men free from work. Freed from work, men will presumably take to more "worthy" pursuits such as "culture." Noting that there have always been some people who, when they had leisure, studied literature and painting and music, the prophets of the technological paradise have always assured us that once we have turned all our work over to machines we will become a nation of artists or, at worst, a nation of art critics. This notion seems to me highly questionable on grounds both of fact and of principle.

In fact, we already know by experience what the "leisure" of most factory and office workers usually is, and we may reasonably predict that what it is it is likely to continue to be. Their leisure is a frantic involvement with salesmen, illusions, and machines. It is an expensive imitation of their work—anxious, hurried, unsatisfying. As their work offers no satisfactions in terms of work but must always be holding before itself the will-o'-the-wisp of freedom from work, so their leisure has no leisurely goals but must always be seeking its satisfaction outside itself, in some activity or some thing typically to be provided by a salesman. A man doing wholesome and meaningful work that he is pleased to do well is three times more at rest than the average factory or office worker on vacation. A man who does meaningless work does not have his meaning at hand. He must go anxiously in search of it—and thus fail to find it. The farmer's Sunday afternoon of sitting at home in the shade of a tree has been replaced by the "long weekend" of a thousand miles. The difference is that the farmer was where he wanted to be, understood the value of being there, and therefore when he had no work to do could sit still. How much have we spent to obscure so simple and obvious a possibility? The point is that there is an indissoluble connection and dependence between work and leisure. The freedom from work must produce not leisure, but an ever more frantic search for something to do.

The principle was stated by Thoreau in his *Journal*: "Hard and steady and engrossing labor with the hands, especially out of doors, is invaluable to the literary man and serves him directly. Here I have been for six days surveying in the woods, and yet when I get home at evening, somewhat weary at last . . . I find myself more susceptible than usual to the finest influences, as music and poetry." That is, certainly, the testimony of an exceptional man, a man of the rarest genius, and it will be asked if such work could produce such satisfaction in an ordinary man. My answer is that we do not have to look far or long for evidence that all the fundamental tasks of feeding and clothing and housing—farming, gardening, cooking, spinning, weaving, sewing, shoemaking, carpentry, cabinetwork, stonemasonry—were once done with consummate skill by ordinary people, and as that skill indisputably involved a high measure of pride, it can confidently be said to have produced a high measure of satisfaction.

We are being saved from work, then, for what? The answer can only be that we are being saved from work that is meaningful and ennobling and comely in order to be put to work that is unmeaning and degrading and ugly. In 1930, the Twelve Southerners of *I'll Take My Stand* issued as an introduction to their book "A Statement of Principles," in which they declared for the agrarian way of life as opposed to the industrial. The book, I believe, was never popular. At the time, and during the three decades that followed, it might have been almost routinely dismissed by the dominant cultural factions as an act of sentimental allegiance to a lost cause. But now it has begun to be possible to say that the cause for which the Twelve Southerners spoke in their introduction was not a lost but a threatened cause: the cause of human culture. "The regular act of applied science," they said, "is to introduce into labor a labor-saving device or a machine. Whether this is a benefit depends on how far it is advisable to save the labor. The philosophy of applied science is generally quite sure that the saving of labor is a pure gain, and that the more of it the better. This is to assume that labor is an evil, that only the end of labor or the material product is good. On this assumption labor becomes mercenary and servile. . . . The act of labor as one of the happy functions of human life has been in effect abandoned. . . .

"Turning to consumption, as the grand end which justifies the evil of modern labor, we find that we have been deceived. We have more time in which to consume, and many more products to be consumed. But the tempo of our labors communicates itself to our satisfactions, and these also become brutal and hurried. The constitution of the natural man probably does not permit him to shorten his labor-time and enlarge his consuming-time indefinitely. He has to pay the penalty in satiety and aimlessness."

The outcry in the face of such obvious truths is always that if they were implemented they would ruin the economy. The peculiarity of our condition would appear to be that the implementation of *any* truth would ruin the economy. If the Golden Rule were generally observed among us, the economy would not last a week. We have made our false economy a false god, and it has made blasphemy of the truth. So I have met the economy in the road, and am expected to yield it right of way. But I will not get over. My reason is that I am a man, and have a better right to the ground than the economy. The economy is no god for me, for I have had too close a look at its wheels. I have seen it at work in the strip mines and coal camps of Kentucky, and I know that it has no moral limits. It has emptied the country of the independent and the proud, and has crowded the cities with the dependent and the abject. It has always sacrificed the small to the large, the personal to the impersonal, the good to the cheap. It has ridden to its questionable triumphs over the bodies of small farmers and tradesmen and craftsmen. I see it, still, driving my neighbors off their farms into the factories. I see it teaching my students to give themselves a price before they can give themselves a value. Its principle is to waste and destroy the living substance of the world and the birthright of posterity for a monetary profit that is the most flimsy and useless of human artifacts.

Though I can see no way to defend the economy, I recognize the need to be concerned for the suffering that would be produced by its failure. But I ask if it is necessary for it to fail in order to change; I am assuming that if it does not change it must sooner or later fail, and that a great deal that is more valuable will fail with it. As a deity the economy is a sort of egotistical French monarch, for it apparently can see no alternative to itself except chaos, and perhaps that is its chief weakness. For, of course, chaos is not the only alternative to it. A better alternative is a better economy. But we will not conceive the possibility of a better economy, and therefore will not begin to change, until we quit deifying the present one.

A better economy, to my way of thinking, would be one that would place its emphasis not upon the *quantity* of notions and luxuries but upon the *quality* of necessities. Such an economy would, for example, produce an automobile that would last at least as long, and be at least as easy to maintain, as a horse.* It would encourage workmanship to be as durable as its materials; thus a piece of furniture would have the durability not of glue but of wood. It would substitute for the pleasure of frivolity a pleasure in the high quality of essential work, in the use of good tools, in a healthful and productive countryside. It would encourage a migration from the cities back to the farms, to assure a work force that would be sufficient not only to the production of the necessary quantities of food, but to the production of food of the best *quality* and to the maintenance of the land at the highest fertility—work that would require a great deal more personal attention and care and hand labor than the present technological agriculture that is focused so exclusively upon production. Such a change in the economy would not involve large-scale unemployment, but rather large-scale changes and shifts of employment.

"You are tilting at windmills," I will be told. "It is a hard world, hostile to the values that you stand for. You will never enlist enough people to bring about such a change." People who talk that way are eager to despair, knowing how easy despair is. The change I am talking about appeals to me precisely because it need not wait upon "other people." Anybody who wants to do so can begin it in himself and in his household as soon as he is ready—by becoming answerable to at least some of his own needs, by acquiring skills and tools, by learning what his real needs are, by refusing the glamorous and the frivolous. When a person learns to *act* on his best hopes he enfranchises and validates them as no government or public policy ever will. And by his action the possibility that other people will do the same is made a likelihood.

But I must concede that there is also a sense in which I *am* tilting at windmills. While we have been preoccupied by various ideological menaces, we have been invaded and nearly overrun by windmills. They are drawing the nourishment from our soil and the lifeblood out of our veins. Let us tilt against the windmills. Though we have not conquered them, if we do not keep going at them they will surely conquer us.

*If automobiles are not more durable and economic than horses, then obviously a better economy would replace them with horses. It would be progressive to do so.

5: THE KINGDOM OF ABSTRACTION AND ORGANIZATION

I do not wish to discount the usefulness of either abstraction or organization, but rather to point out that we have given them such an extravagant emphasis and such prodigal subsidies that their *functioning* has come to overbear and obscure and even nullify their usefulness. Their ascendancy no doubt comes naturally enough out of the need to deal with the massive populations of an urban society. But their disproportionate, their almost exclusive, importance among us can only be explained as a disease of the specialist mentality that has found a haven in the government bureaus and the universities.

The bureaucrat who has formulated a plan, the specialist who has discovered a new fact or process, and the student who has espoused a social vision or ideal, all are of a kind in the sense that they all tend to think that they are at the end of a complete disciplinary process when in fact they have little more than reached the beginning of one. And this is their weakness: that they conceive abstractions to be complete in themselves, and therefore have only the simplest and most mechanical notions of the larger processes within which the abstractions will have their effect—processes that are apt, ultimately, to be obscure or mysterious in their workings and are therefore alien to the specialist mentality in the first place.

Having produced or espoused an abstraction, they next seek to put it to use by means of another abstraction—that is, an organization. But there is a sense in which organization is not a means of implementation, but rather a way of clinging to the clear premises and the neat logic of abstraction. The specialist mentality, unable by the terms of its narrow discipline to relinquish the secure order of abstraction, is prevented by a sort of Zeno's law from ever reaching the real ground of proof in the human community or in the world; it never *meets* the need it purports to answer. Demanding that each step toward the world be a predictable one, the specialist is by that very token not moving in the direction of the world at all, but on a course parallel to it. He can reach the world, not by any organizational process, but only by a reverse leap of faith from the ideal realm of the laboratory or theoretical argument onto the obscure and clumsifying ground of experience, where other and larger disciplines are required.

The man who must actually put the specialist's abstraction to use and live with its effects is never a part of the specialist organization. The organization can only deliver the abstraction to him and, of necessity, largely turn him loose with it. The farmer is not a part of the college of agriculture and the extension service; he is, rather, their object. The impoverished family is not a part of the welfare structure, but its object. The abstraction handed to these object-people is either true only in theory or it has been tried only under ideal (laboratory) conditions. For the bureaucrat, social planning replaces social behavior; for the agricultural scientist, chemistry and economics replace culture and ecology; for the political specialist (student or politician), theory replaces life, or tries to. Thus we institutionalize an impasse between the theoretical or ideal and the real, between the abstract and the particular; the specialist maintains a sort of esthetic distance between himself and the ground of proof and responsibility; and we delude ourselves that precept can have life and useful force without example.

Abstractions move toward completion only in the particularity of enactment or of use. Their completion is only in that mysterious whole that Sir Albert Howard and others have called the wheel of life. A vision or a principle or a discovery or a plan is therefore only *half* a discipline, and, practically speaking, it is the least important half. Black Elk, the holy man of the Sioux, said in his autobiography, *Black Elk Speaks*: "I think I have told you, but if I have not, you must have understood, that a man who has a vision is not able to use the power of it until after he has performed the vision on earth for the people to see." And only a few years later another American, William Carlos Williams, said much the same: "No ideas but in things." The difference of which both men spoke is that between knowledge and the *use* of knowledge. Similarly, one may speak of the difference between the production of an idea or a thing and its use. The disciplines of production are always small and specialized. The disciplines of use and continuity are both different and large. A man who produces a fact or an idea has not completed his responsibility to it until he sees that it is well used in the world. A man may grow potatoes as a specialist of sorts, but he falsifies himself and his potatoes too if he eats them and fails to live as a man.

If the culture fails to provide highly articulate connections between the abstract and the particular, the organizational and the personal, knowledge and behavior, production and use, the ideal and the world—that is, if it fails to bring the small disciplines of each man's work within the purview of those larger disciplines implied by the conditions of our life in the world—then the result is a profound disorder in which men release into their community and dwelling place powerful forces the consequences of which are unconsidered or unknown. New knowledge, political ideas, technological innovations, all are injected into society merely on the ground that to the specialists who produce them they appear to be good in themselves. A "labor-saving" device that does the work it was intended to do is thought by its developers to be a success: In terms of their discipline and point of view it *works.* That, in working, it considerably lowers the quality of a product and makes obsolete a considerable number of human beings is, to the specialists, merely an opportunity for other specialists.

If this attitude were restricted to the elite of government and university it would be bad enough; but it has been so popularized by their propaganda and example that the general public is willing to attribute to declarations, promises, mere words, the force of behavior. We have allowed and even encouraged a radical disconnection between our words and our deeds. Our speech has drifted out of the world into a realm of fantasy in which whatever we say is true. The President of the republic* openly admits that there is no connection between what he says and what he does—this in spite of his evident wish to be re-elected on the strength of what he says. We find it not extraordinary that lovers of America are strip mining in Appalachia, that lovers of peace are bombing villages in Southeast Asia, that lovers of freedom are underwriting dictatorships. If we *say* we are lovers of America and peace and freedom, then this must be what lovers of America and peace and freedom *do.* Having no need to account for anything they have done, our politicians do not find it necessary to trouble

*Nixon.

us with either evidence or argument, or to confess their errors, or to subtract their losses from their gains; they speak like the gods of Olympus, assured that if they *say* they are our servants anything they do in their own interest is right. Our public discourse has been reduced to the manipulation of uprooted symbols: good words, bad words, the names of gods and devils, emblems, slogans, flags. For some the flag no longer stands for the country, it *is* the country; they plant their crops and bury their dead in it.

There is no better example of this deterioration of language than in the current use of the word "freedom." Across the whole range of current politics this word is now being mouthed as if its devotees cannot decide whether it should be kissed or eaten, and this adoration has nothing to do with its meaning. The government is protecting the freedom of people by killing them or hiding microphones in their houses. The government's opponents, left and right, wish to set people free by telling them exactly what to do. All this is for the sake of the political power the word has come to have. The up-to-date politician no longer pumps the hand of a prospective constituent; he offers to set him—or her—free. And yet it seems to me that the word has no political meaning at all; the government cannot serve freedom except negatively—"by the alacrity," in Thoreau's phrase, "with which it [gets] out of its way."*

The going assumption seems to be that freedom can be granted only by an institution, that it is the gift of the government to its people. I think it is the other way around. Free men are not set free by their government; they have set their government free of themselves; they have made it unnecessary. Freedom is not accomplished by a declaration. A declaration of freedom is either a futile and empty gesture, or it is the statement of a finished fact. Freedom is a personal matter; though we may be enslaved as a group, we can be free only as persons. We can set each other free only as persons. It is a matter of discipline. A person can free himself of a bondage that has been imposed on him only by accepting another bondage that he

*And—still negatively—by keeping the selfish or vicious intentions of people out of its way.

has chosen. A man who would not be the slave of other men must be the master of himself—that is the real meaning of self-government. If we all behaved as honorably and honestly and as industriously as we expect our representatives to behave, we would soon put the government out of work.

A person dependent on somebody else for everything from potatoes to opinions may declare that he is a free man, and his government may issue a certificate granting him his freedom, but he will not be free. He is that variety of specialist known as a consumer, which means that he is the abject dependent of producers. How can he be free if he can do nothing for himself? What is the First Amendment to him whose mouth is stuck to the tit of the "affluent society"? Men are free precisely to the extent that they are equal to their own needs. The most able are the most free.

6: DISCIPLINE AND HOPE, MEANS AS ENDS

The various problems that I have so far discussed can best be understood, I think, as failures of discipline caused by a profound confusion as to the functions and the relative values of means and ends. I do not suggest simply that we fall with the ease of familiarity into the moral expedient of justifying means by ends, but that we have also come to attribute to ends a moral importance that far outweighs that which we attribute to means. We expect ends not only to justify means, but to rectify them as well. Once we have reached the desired end, we think, we will turn back to purify and consecrate the means. Once the war that we are fighting for the sake of peace is won, then the generals will become saints, the burned children will proclaim in heaven that their suffering is well repaid, the poisoned forests and fields will turn green again. Once we have peace, we say, or abundance or justice or truth or comfort, everything will be all right. It is an old dream.

It is a vicious illusion. For the discipline of ends is no discipline at all. The end is preserved in the means; a desirable end may perish forever in the wrong means. Hope lives in the means, not the end. Art does not survive in its revelations, or agriculture in its products, or craftsmanship in its artifacts, or civilization in its monuments, or faith in its relics.

That good ends are destroyed by bad means is one of the dominant themes of human wisdom. The *I Ching* says: "If evil is branded it thinks of weapons, and if we do it the favor of fighting against it blow for blow, we lose in the end because thus we ourselves get entangled in hatred and passion. Therefore it is important to begin at home, to be on guard in our own persons against the faults we have branded. . . . For the same reasons we should not combat our own faults directly. . . . As long as we wrestle with them, they continue victorious. Finally, the best way to fight evil is to make energetic progress in the good." Confucius said of riches that "if not obtained in the right way, they do not last." In the Sermon on the Mount, Jesus said: "Ye have heard that it hath been said, An eye for an eye, and a tooth for a tooth: But I say unto you, That ye resist not evil. . . ." And for that text Ken Kesey supplies the modern exegesis: "As soon as you resist evil, as soon as it's gone, you fold, because it's what you're based on." In 1931, Judge Lusk of the Chattanooga criminal court handed down a decision in which he wrote: "The best way, in my judgment, to combat Communism, or any other movement inimical to our institutions, is to show, if we can, that the injustices which they charge against us are, in fact, nonexistent." And a friend of mine, a graduate of the University of Emily's Run, was once faced with the argument that he could "make money" by marketing some inferior lambs; he refused, saying that his purpose was the production of *good* lambs, and he would sell no other kind. He meant that his disciplines had to be those of a farmer, and that he would be diminished as a farmer by adopting the disciplines of a money-changer. It is a tragedy of our society that it neither pays nor honors a man for this sort of integrity —though it depends on him for it.

It is by now a truism that the great emphasis of our present culture is upon things, things as things, things in quantity without respect to quality; and that our predominant techniques and attitudes have to do with production and acquisition. We persist in the belief—against our religious tradition, and in the face of much evidence to the contrary—that if we leave our children wealthy we will assure their happiness. A corollary of this is the notion, rising out of the work of the geneticists, that we can assure a brighter future for the world by *breeding* a more intelligent race of humans—even though the present problems of the world are the result, not of human stupidity, but of human intelligence without adequate cultural controls. Both ideas are typical of the materialist assumption that human destiny can be improved by being constantly tinkered at, as if it were a sort of balky engine. But we can do nothing for the human future that we will not do for the human present. For the amelioration of the future condition of our kind we must look, not to the wealth or the genius of the coming generations, but to the quality of the disciplines and attitudes that we are preparing now for their use.

We are being virtually buried by the evidence that those disciplines by which we manipulate *things* are inadequate disciplines. Our cities have become almost unlivable because they have been built to be factories and vending machines rather than communities. They are conceptions of the desires for wealth, excitement, and ease—all illegitimate motives from the standpoint of community, as is proved by the fact that without the community disciplines that make for a stable, neighborly population, the cities have become scenes of poverty, boredom, and dis-ease.

The rural community—that is, the land and the people—is being degraded in complementary fashion by the specialists' tendency to regard the land as a factory and the people as spare parts. Or, to put it another way, the rural community is being degraded by the fashionable premise that the exclusive function of the farmer is production and that his major discipline is economics. On the contrary, both the function and the discipline of the farmer have to do with provision: He must provide, he must look ahead. He must look ahead, however, not in the economic-mechanistic sense of anticipating a need and fulfilling it, but the sense of using methods that preserve the source. In his work sound economics becomes identical with sound ecology. The farmer is not a factory worker, he is the trustee of the life of the topsoil, the keeper of the rural community. In precisely the same way, the dweller in a healthy city is not an office or a factory worker, but part and preserver of the urban community. It is in thinking of the whole citizenry as factory workers—as readily interchangeable parts of an entirely mechanistic and economic order—that we have reduced our people to the most abject and aimless of nomads, and displaced and fragmented our communities.

An index of the health of a rural community—and, of course, of the urban community, its blood kin—might be found in the relative acreages of field crops and tree crops. By tree crops I mean not just those orchard trees of comparatively early bearing and short life, but also the fruit and nut and timber trees that bear late and live long. It is characteristic of an unsettled and anxious farm population—a population that feels itself, because of economic threat or the degradation of cultural value, to be ephemeral—that it farms almost exclusively with field crops, within economic and biological cycles that are complete in one year. This has been the dominant pattern of American agriculture. Stable, settled populations, assured both of an economic sufficiency in return for their work and of the cultural value of their work, tend to have methods and attitudes of a much longer range. Though they have generally also farmed with field crops, established farm populations have always been planters of trees. In parts of Europe, according to J. Russell Smith's important book, *Tree Crops*, steep hillsides were covered with orchards of chestnut trees, which were kept and maintained with great care by the farmers. Many of the trees were ancient, and when one began to show signs of dying, a seedling would be planted beside it to replace it. Here is an agricultural discipline that could develop only among farmers who felt secure—as individuals and also as families and communities—in their connection to their land. Such a discipline depends not just on the younger men in the prime of their workdays but also on the old men, the keepers of tradition. The model figure of this agriculture is an old man planting a young tree that will live longer than a man, that he himself may not live to see in its first bearing. And he is planting, moreover, a tree whose worth lies beyond any conceivable market prediction. He is planting it because the good sense of doing so has been clear to men of his place and kind for generations. The practice has been continued because it is ecologically and agriculturally sound; the economic soundness of it must be assumed. While the planting of a field crop, then, may be looked upon as a "short-term investment," the planting of a chestnut tree is a covenant of faith.

An urban discipline that in good health is closely analogous to healthy agriculture is teaching. Like a good farmer, a good teacher is the trustee of a vital and delicate organism: the life of the mind in his community. The standard of his discipline is his community's health and intelligence and coherence and endurance. This is a high calling, deserving of a life's work. We have allowed it to degenerate into careerism and specialization. In education as in agriculture we have discarded the large and enlarging disciplines of community and place, and taken up in their stead the narrow and shallow discipline of economics. Good teaching is an investment in the minds of the young, as obscure in result, as remote from immediate proof as planting a chestnut seedling. But we have come to prefer ends that are entirely foreseeable, even though that requires us to shorten our vision. Education is coming to be not a long-term investment in young minds and in the life of the community, but a short-term investment in the economy. We want to be able to tell how many dollars an education is worth and how soon it will begin to pay.

To accommodate these frivolous desires, education becomes training and specialization, which is to say, it institutionalizes and justifies ignorance and intellectual irresponsibility. It produces a race of learned mincers, whose propriety and pride it is to keep their minds inside their "fields," as if human thoughts were a kind of livestock to be kept out of the woods and off the roads. Because of the obsession with short-term results that may be contained within the terms and demands of a single life, the interest of community is displaced by the interest of career. The careerist teacher judges himself, and is judged by his colleagues, not by the influence he is having upon his students or the community, but by the number of his publications, the size of his salary and the status of the place to which his career has taken him thus far. And in ambition he is where he is only temporarily; he is on his way to a more lucrative and prestigious place. Because so few stay to be aware of the *effects* of their work, teachers are not judged by their teaching, but by the short-term incidentals of publication and "service." That teaching is a long-term service, that a teacher's best work may be published in the children or grandchildren of his students, cannot be considered, for the modern educator, like his "practical" brethren in business and industry, will honor nothing that he cannot see. That is not to say that books do not have their progeny in the community, or that a legitimate product of a teacher's life may not be a book. It *is* to say that if *good* books are to be written, they will be written out of the same resources of talent and discipline and character and delight as always, and not by institutional coercion.

It is not from the standpoint of the university itself that we will see its faults, but from the standpoint of the whole community. Looking only at the university, one might perhaps believe that its first obligation is to become a better exemplar of its species: a *bigger* university, with more prestigious professors publishing more books and articles. But look at the state of Kentucky—whose land is being vandalized and whose people are being impoverished by the absentee owners of coal; whose dispossessed are refugees in the industrial cities to the north; whose farm population and economy are under the heaviest threat of their history; whose environment is generally deteriorating; whose public schools have become legendary for their poor quality; whose public offices are routinely filled by the morally incompetent. Look at the *state* of Kentucky, and it is clear that, more than any publication of books and articles, or any research, we need an annual increment of several hundred competently literate *graduates* who have some critical awareness of their inheritance and a sense of their obligation to it, and who know the use of books.

That, and that only, is the disciplining idea of education, and the methods must be derived accordingly. It has nothing to do with number or size. It would be impossible to value economically; it is the antithesis of that false economy which thrives upon the exploitation of stupidity. It stands forever opposed to the assumption that you can produce a good citizen by subjecting a moral simpleton to specialized training or expert advice.

It is the obsession with immediate ends that is degrading, that destroys our disciplines, and that drives us to our inflexible concentration upon number and price and size. I believe that the closer we come to correct discipline, the less concerned we are with ends, and with questions of futurity in general.

Correct discipline brings us into alignment with natural process, which has no explicit or deliberate concern for the future. We do not eat, for instance, because we want to live until tomorrow, but because we are hungry today and it *satisfies* us to eat. Similarly, a good farmer plants, not because of the abstractions of demand or market or his financial condition, but because it is planting time and the ground is ready—that is, he plants in response to his discipline and to his place. And the real teacher does not teach with reference to the prospective job market or some program or plan for the society's future; he teaches because he has something to teach and because he has students. A poet could not write a poem in order to earn a place in literary history. His place in literary history is another subject, and as such a distraction. He writes because he has a poem to write, he knows how, the work pleases him, and he has forgotten all else. "Take therefore no thought for the morrow: for the morrow shall take thought for the things of itself." This passage rests, of course, on the fact that we do not know what tomorrow will be, and are therefore strictly limited in our ability to take thought for it. But it also rests upon the assumption of correct disciplines. The man who works and behaves well today *need* take no thought for the morrow; he has discharged today's only obligation to the morrow.

7: THE ROAD AND THE WHEEL

There are, I believe, two fundamentally opposed views of the nature of human life and experience in the world: one holds that though natural processes may be cyclic, there is within nature a human domain the processes of which are linear; the other, much older, holds that human life is subject to the same cyclic patterns as all other life. If the two are contradictory that is not so much because one is wrong and the other right as because one is partial and the other complete. The linear idea, of course, is the doctrine of progress, which represents man as having moved across the oceans and the continents and into space on a course that is ultimately logical and that will finally bring him to a man-made paradise. It also sees him as moving through time in this way, discarding old experience as he encounters new. The cyclic vision, on the other hand, sees our life ultimately not as a cross-country journey or a voyage of discovery, but as a circular dance in which certain basic *and necessary* patterns are repeated endlessly. This is the religious and ethical basis of the narrative of Black Elk: "Everything the Power of the World does is done in a circle. The sky is round, and I have heard that the earth is round like a ball, and so are all the stars. The wind, in its greatest power, whirls. Birds make their nests in circles, for theirs is the same religion as ours. The sun comes forth and goes down again in a circle. The moon does the same, and both are round. Even the seasons form a great circle in their changing, and always come back again to where they were. The life of a man is a circle from childhood to childhood, and so it is in everything where power moves. Our tepees were round like the nests of birds, and these were always set in a circle, the nation's hoop, a nest of many nests, where the Great Spirit meant for us to hatch our children." The doctrine of progress suggests that the fluctuations of human fortune are a series of ups and downs in a road tending generally upward toward the earthly paradise. To Black Elk earthly blessedness did not lie ahead or behind; it was the result of harmony within the circle of the people and between the people and the world. A man was happy or sad, he thought, in proportion as he moved toward or away from "the sacred hoop of [his] people [which] was one of many hoops that made one circle, wide as daylight and as starlight. . . ."

Characteristic of the linear vision is the idea that anything is justifiable only insofar as it is immediately and obviously good for something else. The linear vision tends to look upon everything as a cause, and to require that it proceed directly and immediately and obviously to its effect. What is it good for? we ask. And only if it proves immediately to be good *for* something are we ready to raise the question of value: How much is it worth? But we mean how much money, for if it can only be good for something else then obviously it can only be *worth* something else. Education becomes training as soon as we demand, in this spirit, that it serve some immediate purpose and that it be worth a predetermined amount. Once we accept so specific a notion of utility, all life becomes subservient to its use; its value is drained into its use. That is one reason why these are such hard times for students and old people: They are living either before or after the time of their social utility. It is also the reason why so many non-human species are threatened with extinction. Any organism that is not contributing obviously and directly to the workings of the economy is now endangered—which means, as the ecologists are showing, that human society is to the same extent endangered. The cyclic vision is more accepting of mystery and more humble. Black Elk *assumes* that all things have a use—that is the condition of his respect for all things—but he does not know what all their uses are. Because he does not value them for their uses, he is free to value them for their own sake: "The Six Grandfathers have placed in this world many things, all of which should be happy. Every little thing is sent for something, and in that thing there should be happiness and the power to make happy." It should be emphasized that this is ecologically sound. The ecologists recognize that the creation is a great union of interlocking lives and processes and substances, all dependent on each other; and because they cannot discover the whole pattern of interdependence they recognize the need for the greatest possible care in the use of the world. Black Elk and his people, however, were further advanced, for they possessed the cultural means for the enactment of a ceremonious respect for and delight in the lives with which they shared the world, and that respect and delight afforded those other lives an effective protection.

The linear vision looks fixedly straight ahead. It never looks back, for its premise is that there is no return. The doctrine of possession is complemented by no doctrine of relinquishment. Our shallow concept of use does not imply good use or preservation; thus quantity depresses quality, and we arrive at the concepts of waste and disposability. Similarly, life is lived without regard or respect for death. Death thus becomes accidental, the chance interruption of a process that might otherwise go on forever—therefore, always a surprise and always feared. Dr. Leon R. Kass, of the National Academy of Sciences, recently said that "medicine seems to be sharpening its tools to do battle with death as though death were just one more disease." The cyclic vision, at once more realistic and more generous, recognizes in the creation the essential principle of return: What is here will leave to come again; if there is to be having there must also be giving up. And it sees death as an integral and indispensable part of life. In one of the medicine rites of the Winnebago, according to Paul Radin, an old woman says: "Our father has ordained that my body shall fall to pieces. I am the earth. Our father ordained that there should be death, lest otherwise there be too many people and not enough food for them." Because death is inescapable, a biological and ecological necessity, its acceptance becomes a spiritual obligation, the only means of making life whole: "Whosoever shall seek to save his life shall lose it; and whosoever shall lose his life shall preserve it."

The opposing characteristics of the linear and cyclic visions might, then, be graphed something like this:

Linear	*Cyclic*
Progress. The conquest of nature.	Atonement with the creation.
The Promised Land motif in the Westward Movement.	Black Elk's sacred hoop, the community of creation.
Heavenly aspiration without earthly reconciliation or stewardship. The creation as commodity.	Reconciliation of heaven and earth in aspiration toward responsible life. The creation as source *and end*.
Training. Programming.	Education. Cultural process.
Possession.	Usufruct, relinquishment.
Quantity.	Quality.
Newness. The unique and "original."	Renewal. The recurring.
Life.	Life and death.

The linear vision flourishes in ignorance or contempt of the processes on which it depends. In the face of these processes our concepts and mechanisms are so unrealistic, so *impractical*, as to have the nature of fantasy. The processes are invariably cyclic, rising and falling, taking and giving back, living and dying.

But the linear vision places its emphasis entirely on the rising phase of the cycle—on production, possession, life. It provides for no returns. Waste, for instance, is a concept that could have been derived only from the linear vision. According to the scheme of our present thinking, every human activity produces waste. This implies a profound contempt for correct discipline; it proposes, in the giddy faith of prodigals, that there can be production without fertility, abundance without thrift. We take and do not give back, and this causes waste. It is a hideous concept, and it is making the world hideous. It is consumption, a wasting disease. And this disease of our material economy becomes also the disease of our spiritual economy, and we have made a shoddy merchandise of our souls. We want the truth to be easy and spectacular, and so we waste our verities; we are always hastening from the essential to the novel. We want to have love without a return of devotion or fidelity; to us, Aphrodite is a peeping statistician, the seismographer of orgasms. We want a faith that demands no return of good work. And art—we want it to be instantaneous and effortless; we want it to involve no apprenticeship to a tradition or a discipline or a master, no devotion to an ideal of workmanship. We want our art to support the illusion that high achievement is within easy reach, for we want to believe that, though we are demeaned by our work and driven half crazy by our pleasures, we are all mute inglorious Miltons.

To take up again my theme of agriculture, it is obvious that the modern practice concentrates almost exclusively on the productive phase of the natural cycle. The means of production become more elaborate all the time, but the means of return —the building of health and fertility in the soil—are reduced more and more to the shorthand of chemicals. According to the industrial vision of it, the life of the farm does not rise and fall in the turning cycle of the year; it goes on in a straight line from one harvest to another. This, in the long run, may well be more productive of waste than of anything else. It wastes the soil. It wastes the animal manures and other organic residues that industrialized agriculture frequently fails to return to the soil. And what may be our largest agricultural waste is not usually recognized as such, but is thought to be both an urban product and an urban problem: the tons of garbage and sewage that are burned or buried or flushed into the rivers. This, like all waste, is the abuse of a resource. It was ecological stupidity of exactly this kind that destroyed Rome. The chemist Justus von Liebig wrote that "the sewers of the immense metropolis engulfed in the course of centuries the prosperity of Roman peasants. The Roman Campagna would no longer yield the means of feeding her population; these same sewers devoured the wealth of Sicily, Sardinia and the fertile lands of the coast of Africa."

To recognize the magnitude and the destructiveness of our "urban waste" is to recognize the shallowness of the notion that agriculture is only another form of technology to be turned over to a few specialists. The sewage and garbage problem of our cities suggests, rather, that a healthy agriculture is a cultural organism, not merely a universal necessity but a universal obligation as well. It suggests that, just as the cities exist within the environment, they also exist within agriculture. It suggests that, like farmers, city-dwellers have agricultural responsibilities: to use no more than necessary, to waste nothing, to return organic residues to the soil.

Our ecological or agricultural responsibilities, then, call for a corresponding set of disciplines that would be a part of the cultural common ground, and that each person would have an obligation to preserve in his behavior. Seeking his own ends by the correct means, he transcends selfishness and makes a just return to the ecological source; by his correct behavior, both the source and the means for its proper use are preserved. This is equally true of other cultural areas: it is the discipline, not the desire, that is the common ground. In politics, for example, it is only the personal career that can be advanced by "image making." *Politics* cannot be advanced except by honest, informed, open discourse—by determining and telling and implementing the truth, by assuming the truth's heavy responsibilities and great risks. The political careerist, by serving his "image" rather than the truth, becomes a consumer of the political disciplines. Similarly, in art, the common ground is workmanship, the artistic means, the technical possibility of art—not the insights or visions of particular artists. A person who practices an art without mastering its disciplines becomes his art's consumer; he obscures the means, and encumbers his successors. The art lives *for* its insights and visions, but it cannot live *upon* them. An art is inherited and handed down in its workmanly aspects. Workmanship is one of the means by which the artist prepares for—becomes worthy of, earns—his vision and his insights. "Art," A. R. Ammons says, "is the conscious preparation for the unconscious event. . . ."

Learning the correct and complete disciplines—the disciplines that take account of death as well as life, decay as well as growth, return as well as production—is an indispensable form of cultural generosity. It is the one effective way a person has of acknowledging and acting upon the fact of mortality: He will die, others will live after him.

One reason, then, for the disciplinary weakness of the linear vision is that it is incomplete. Another is that it sees history as always leading not to renewal but to the new: the road may climb hills and descend into valleys, but it is always going ahead; it never turns back on itself. "We have constructed a fate . . . that never turns aside," Thoreau said of the railroad; "its orbit does not look like a returning curve. . . ." But when the new is assumed to be a constant, discipline fails, for discipline is preparation, and the new cannot be prepared for; it cannot, in any very meaningful way, be expected. Here again we come upon one of the reasons for the generational disconnections that afflict us: all times, we assume, are different; we therefore have nothing to learn from our elders, nothing to teach our children. Civilization is thus reduced to a sequence of last-minute improvisations, desperately building today out of the wreckage of yesterday. There are two genres of writing that seem to me to be characteristic (or symptomatic) of the linear vision. The first is, not prophecy, though it is sometimes called that, but the most mundane and inquisitive taking of thought for the morrow. What will tomorrow be like? (We mean, what new machines or ideas will be invented by then?) What will the world be like in ten or fifty or a hundred years? Our preoccupation with these questions, besides being useless, is morbid and scared; mistaking appearance for substance, it assumes a condition of *absolute* change: The future will be *entirely* different from the past and the present, we think, because our vision of history and experience has not taught us to imagine persistence or recurrence or renewal. We disregard the necessary persistence of ancient needs and obligations, patterns and cycles, and assume that the human condition is entirely determined by human *devices.*

The other genre—complementary, obviously—is that of the death sentence. Because we see the human situation as perpetually changing, the new bearing down with annihilating force, we appear to ourselves to be living always at the end of possibilities. For the new to happen, the old must be destroyed. Our own lives, which are pleasant to us at least insofar as they represent a *kind* of life that we recognize, seem always on the verge of being replaced by a kind of life that is unrecognizable, or by a death that is equally so. Thus a common theme for the writers of feature articles and critical essays is the death or the impending death of something: of the fashions of dress and appearance, of the novel, of printing, of freedom, of Christianity, of Western civilization, of the human race, of the world, of God.

This genre is difficult to criticize, because there is always a certain justice, or likelihood of justice, in it. There is no denying that we fear the end of things because our way of life has brought so many things to an end. The sunlit road of progress never escapes a subterranean dread that threatens to undermine the pavement. Thoreau knew that the railroad was built upon the bodies of Irishmen, and not one of us but secretly wonders when *he* will be called upon to lie down and become a sleeper beneath the roadbed of progress. And so some of this death-sentence literature is faithfully reporting the destructiveness of the linear vision; it is the chorus of accusation and dread and mourning that accompanies Creon's defiance of the gods.

But at times it is no more than one of the sillier manifestations of the linear vision itself: the failure to see any pattern in experience, the failure to transform experience into useful memory. There is a sort of journalistic greenness in us that is continuously surprised by the seasons and the weather, as if these were no more than the inventions (or mistakes) of the meteorologists. History, likewise, is always a surprise to us; we read its recurring disasters as if they were the result merely of miscalculations of our intelligence—as if they could not have been foreseen in the flaws of our character. And the heralds of the push-button Eden of the future would be much put off to consider that those button-pushers will still have to deal with problems of food and sewage, with the picking up of scraps and the disposal of garbage, with building and maintenance and reclamation—with, that is, the fundamental work, much of it handwork, that is necessary to life; they would be even more put off to consider that the "quality of life" will not depend nearly so much on the distribution of push-buttons as on the manner and the quality of that fundamental and endlessly necessary work.

Some cycles revolve frequently enough to be well known in a man's lifetime. Some are complete only in the memory of several generations. And others are so vast that their motion can only be assumed: Like our galaxy, they appear to us to be remote and exalted, a Milky Way, when in fact they are near at hand, and we and all the humble motions of our days are their belongings. We are kept in touch with these cycles, not by technology or politics or any other strictly human device, but by our necessary biological relation to the world. It is only in the processes of the natural world, and in analogous and related processes of human culture, that the new may grow usefully old, and the old be made new.

The ameliorations of technology are largely illusory. They are always accompanied by penalties that are equal and opposite. Like the weather reports, they suggest the possibility of better solutions than they can provide; and by this suggestiveness—this glib and shallow optimism of gimcrackery—they have too often replaced older skills that were more serviceable to life in a mysterious universe. The farmer whose weather eye has been usurped by the radio has become less observant, has lost his old judicious fatalism with respect to the elements—and he is no more certain of the weather. It is by now obvious that these so-called blessings have not made us better or wiser. And their expense is growing rapidly out of proportion to their use. How many plastic innards can the human race afford? How many mountains can the world afford, to the strip miners, to light the whole outdoors and overheat our rooms? The limits, as Thoreau knew, have been in sight from the first: "To make a railroad round the world available to all mankind," he said in *Walden*, "is equivalent to grading the whole surface of the planet. Men have an indistinct notion that if they keep up this activity of joint stocks and spades long enough all will at length ride . . . but though a crowd rushes to the depot . . . when the smoke is blown away and the vapor condensed, it will be perceived that a few are riding, but the rest are run over. . . ."

We cannot look for happiness to any technological paradise or to any New Earth of outer space, but only to the world as it is, and as we have made it. The only life we may hope to live is here. It seems likely that if we are to reach the earthly paradise at all, we will reach it only when we have ceased to strive and hurry so to get there. There is no "there." We can only wait here, where we are, in the world, obedient to its processes, patient in its taking away, faithful to its returns. And as much as we may know, and all that we deserve, of earthly paradise will come to us.

8: KINDS OF DISCIPLINE

The disciplines we most readily think of are technical; they are the means by which we define and enact our relationship to the creation, and destroy or preserve the commonwealth of the living. There is no disputing the importance of these, and I have already said a good deal in their support. But alone, without the larger disciplines of community and faith, they fail of their meaning and their aim; they cannot even continue for very long. People who practice the disciplines of workmanship for their own sake—that is, as specializations—begin a process of degeneracy in those disciplines, for they remove from them the ultimate sense of use or effect by which their vitality and integrity are preserved. Without a proper sense of use a discipline declines from community responsibility to personal eccentricity; cut off from the common ground of experience and need, vision escapes into wishful or self-justifying fantasy, or into greed. Artists lose the awareness of an audience, craftsmen and merchants lose the awareness of customers and users.

Community disciplines, which can rise only out of the cyclic vision, are of two kinds. First there is the discipline of principle, the essence of the experience of the historical community. And second there is the discipline of fidelity to the living community, the community of family and neighbors and friends. As our society has become increasingly rootless and nomadic, it has become increasingly fashionable among the rhetoricians of dissatisfaction to advocate, or to seem to advocate, a strict and solitary adherence to principle in simple defiance of other people. "I don't care what they think" has become public currency with us; saying it, we always mean to imply that we are persons solemnly devoted to high principle—"rugged individuals" in the somewhat fictional sense Americans usually give to that term. In fact, this ready defiance of the opinions of others is a rhetorical fossil from our frontier experience. Once it meant that if our neighbors' opinions were repugnant to us, we were prepared either to kill our neighbors or to move west. Now it doesn't mean anything; it is adolescent bluster. For when there is no frontier to retreat to, the demands of one's community will be felt, and ways must be found to deal with them. The great moral labor of any age is probably not in the conflict of opposing principles, but in the tension between a living community and those principles that are the distillation of its experience. Thus the present anxiety and anguish in this country have very little to do with the much-heralded struggle between capitalism and communism, or left and right, and very much indeed to do with the rapidly building discord and tension between American principles and American behavior.

This sort of discord is the subject of tragedy. It is tragic because—outside the possibility of a renewal of harmony, which may depend upon the catharsis of tragedy—there are no possible resolutions that are not damaging. To choose community over principle is to accept in consequence a diminishment of the community's moral inheritance; it is to accept the great dangers and damages of life without principle. To choose principle over community is even worse, it seems to me, for that is to accept as the condition of being "right" a loneliness in which the right is ultimately meaningless; it is to destroy the only ground upon which principle can be enacted and renewed; it is to raise an ephemeral hope upon the ground of final despair.

Facing exactly this choice between principle and community, on April 20, 1861, Robert E. Lee resigned his commission in the army of the United States. Lee had clearly understood the evil of slavery. He disapproved and dreaded secession; almost alone among the Virginians, he foresaw the horrors that would follow. And yet he chose to go with his people. Having sent in his resignation, he wrote his sister: ". . . though I recognize no necessity for this state of things, and would have forborne and pleaded to the end for a redress of grievances, real or supposed, yet in my own person I had to meet the question whether I should take part against my native state. With all my devotion to the Union and the feeling of loyalty and duty of an American citizen, I have not been able to make up my mind to raise my hand against my relatives, my children, my home."

He was right. As a highly principled man, he could not bring himself to renounce the very ground of his principles. And devoted to that ground as he was, he held in himself much of his region's hope of the renewal of principle. His seems to me to have been an exemplary American choice, one that placed the precise Jeffersonian vision of a rooted devotion to community and homeland above the abstract "feeling of loyalty and duty of an American citizen." It was a tragic choice on the theme of Williams's maxim: "No ideas but in things."

If the profession of warfare has so declined in respectability since 1861 as to obscure my point, then change the terms. Say that a leader of our own time, in spite of his patriotism and his dependence on "the economy," nevertheless held his people and his place among them in such devotion that he would not lie to them or sell them shoddy merchandise or corrupt their language or degrade their environment. Say, in other words, that he would refuse to turn his abilities against his people. That is what Lee did, and there have been few public acts of as much integrity since.*

It is the intent of community disciplines, of course, to prevent such radical conflicts. If these disciplines are practiced at large among the members of the community, then the community holds together upon a basis of principle that is immediately clarified in feeling and behavior; and then destructive divisions, and the moral agonies of exceptional men, are averted.

*If loyalty to home and community allied one with a manifest evil, then it would be necessary to stand alone on principle. But it would be foolish to expect a choice so clean-cut.

In the Sermon on the Mount a major concern is with the community disciplines. The objective of this concern is a social ideal: ". . . all things whatsoever that men should do to you, do ye even so to them. . . ." But that everyone would do as he would be done by is hardly a realistic hope, and Jesus was speaking out of a moral tradition that was eminently realistic and tough-minded. It was a tradition that, in spite of its spiritual aspirations, was very worldly in its expectations: It would have spared Sodom for the sake of ten righteous men. And so the focus of the Sermon is not on the utopian social ideal of the Golden Rule, but on the *personal* ideal of nonresistance to evil: ". . . whosoever shall smite thee on thy right cheek, turn to him the other also. . . . Love your enemies, bless them that curse you, do good to them that hate you, and pray for them which despitefully use you, and persecute you. . . ." The point, I think, is that the anger of one man need not destroy the community, if it is contained by the peaceableness and long-suffering of another, but in the anger of *two* men, in anger repaid—"An eye for an eye, and a tooth for a tooth"—it is destroyed altogether.

Community, as a discipline, extends and enlarges the technical disciplines by looking at them within the perspective of their uses or effects. Community discipline imposes upon our personal behavior an ecological question: What is the effect, on our neighbors and on our place in the world, of what we do? It is aware that *all* behavior is social. It is aware, as the ecologists are aware, that there is a unity in the creation, and that the behavior and the fate of one creature must therefore affect the whole, though the exact relationships may not be known.

But essential as are the disciplines of technique and community, they are not sufficient in themselves. All such disciplines reach their limit of comprehensibility and at that point enter mystery. Thus an essential part of a discipline is that relinquishment or abandonment by which we acknowledge and accept its limits. We do not finally know what will be the result of our actions, however correct and excellent they may be. The good work we do today may be undone by some mere accident tomorrow. Our neighborly acts may be misunderstood and repaid with anger. With respect to what is to come, our real condition is that of abandonment; one of the primary functions of religion is to provide the ceremonial means of acknowledging this: We are in the hands of powers that we do not know.

The ultimate discipline, then, is faith: faith, if in nothing else, in the propriety of one's disciplines. We have obscured the question of faith by pretending that it is synonymous with the question of "belief," which is personal and not subject to scrutiny. But if one's faith is to have any public validity or force, then obviously it must meet some visible test. The test of faith is consistency—not the fanatic consistency by which one repudiates the influence of knowledge, but rather a consistency between principle and behavior. A man's behavior should be the creature of his principles, not the creature of his circumstances. The point has great practical bearing, because belief and the principles believed in, and whatever hope and promise are implied in them, are destroyed in contradictory behavior; hypocrisy salvages nothing but the hypocrite. If we put our faith in the truth, then we risk everything—the truth included—by telling lies. If we put our faith in peace, then we must see that violence makes us infidels. When we institute repressions to protect democracy from enemies abroad, we have already damaged it at home. The demands of faith are absolute: We must put all our eggs in one basket; we must burn our bridges.

An exemplary man of faith was Gideon, who reduced his army from thirty-two thousand to three hundred in earnest of his trust, and marched that remnant against the host of the Midianites, armed, not with weapons, but with "a trumpet in every man's hand, with empty pitchers, and lamps within the pitchers."

Beside this figure of Gideon, the hero as man of faith, let us place our own "defender," the Pentagon, which has faith in nothing except its own power. That, as the story of Gideon makes clear, is a dangerous faith for mere men; it places them in the most dangerous moral circumstance, that of *hubris*, in which one boasts that "mine own hand hath saved me." To be sure, the Pentagon is supposedly founded upon the best intentions and the highest principles, and there is a plea that justifies it in the names of Christianity, peace, liberty and democracy. But the Pentagon is an institution, not a person; and unless constrained by the moral vision of persons in them, institutions move in the direction of power and self-preservation, not high principle. Established, allegedly, in defense of "the free world," the Pentagon subsists complacently upon the involuntary servitude of millions of young men whose birthright, allegedly, is freedom. To wall our enemies out, it is walling us in.

Because its faith rests entirely in its own power, its mode of dealing with the rest of the world is not faith but suspicion. It recognizes no friends, for it knows that the face of friendship is the best disguise of an enemy. It has only enemies, or prospective enemies. It must therefore be prepared for *every possible* eventuality. It sees the future as a dark woods with a gunman behind every tree. It is passing through the valley of the shadow of death without a shepherd, and thus is never still. But as long as it can keep the public infected with its own state of mind, this spiritual dis-ease, it can survive without justification, and grow huge. Whereas the man of faith may go armed with only a trumpet and an empty pitcher and a lamp, the institution of suspicion arms with the death of the world; trusting nobody, it must stand ready to kill everybody.

The moral is that those who have no faith are apt to be much encumbered by their equipment, and overborne by their precautions. For the institution of suspicion there is no end of toiling and spinning. The Pentagon exists continually, not only on the brink of war, but on the brink of the exhaustion of its moral and material means. But the man of faith, even in the night, in the camp of his enemies, is at rest in the assurance of his trust and the correctness of his ways.* He has become the lily of the field.

9: THE LIKENESSES OF ATONEMENT (AT-ONE-MENT)

Living in our speech, though no longer in our consciousness, is an ancient system of analogies that clarifies a series of mutually defining and sustaining unities: of farmer and field, of husband and wife, of the world and God. The language both of our literature and of our everyday speech is full of references and

*That is not to suggest that faith may not be extremely dangerous, especially to the faithful.

allusions to this expansive metaphor of farming and marriage and worship. A man planting a crop is like a man making love to his wife, and vice versa: He is a husband or a husbandman. A man praying is like a lover, or he is like a plant in a field waiting for rain. As husbandman, a man is both the steward and the likeness of God, the greater husbandman. God is the lover of the world and its faithful husband. Jesus is a bridegroom. And he is a planter; his words are seeds. God is a shepherd and we are his sheep. And so on.

All the essential relationships are comprehended in this metaphor. A farmer's relation to his land is the basic and central connection in the relation of humanity to the creation; the agricultural relation *stands for* the larger relation. Similarly, marriage is the basic and central community tie; it begins and stands for the relation we have to family and to the larger circles of human association. And these relationships to the creation and to the human community are in turn basic to, and may stand for, our relationship to God—or to the sustaining mysteries and powers of the creation.

(These three relationships are dependent—and even intent—upon renewals of various sorts: of season, of fertility, of sexual energy, of love, of faith. And these concepts of renewal are always accompanied by concepts of loss or death; in order for the renewal to take place, the old must be not forgotten but relinquished; in order to become what we may be, we must cease to be as we are; in order to have life we must lose it. Our language bears abundant testimony to these deaths; the year's death that precedes spring; the burial of the seed before germination; sexual death, as in the Elizabethan metaphor; death as the definitive term of marriage; the spiritual death that must precede rebirth; the death of the body that must precede resurrection.)

As the metaphor comprehends all the essential relationships, so too it comprehends all the essential moralities. The moralities are ultimately emulative. For the metaphor does not merely perceive the likeness of these relationships. It perceives also that they are understandable only in terms of each other. They are the closed system of our experience; no instructions come from outside. A man finally cannot act upon the basis of absolute law, for the law is more fragmentary than his own experience; finally, he must emulate in one relationship what he knows of another. Thus, if the metaphor of atonement is alive in his consciousness, he will see that he should love and care for his land as for his wife, that his relation to his place in the world is as solemn and demanding, and as blessed, as marriage; and he will see that he should respect his marriage as he respects the mysteries and transcendent powers—that is, as a sacrament. Or—to move in the opposite direction through the changes of the metaphor—in order to care properly for his land he will see that he must emulate the Creator: to learn to use and preserve the open fields, as Sir Albert Howard said, he must look into the woods; he must study and follow natural process; he must understand the *husbanding* that, in nature, always accompanies providing.

Like any interlinking system, this one fails in the failure of any one of its parts. When we obscure or corrupt our understanding of any one of the basic unities, we begin to misunderstand all of them. The vital knowledge dies out of our consciousness and becomes fossilized in our speech and our culture. This is our condition now. We have severed the vital links of the atonement metaphor, and we did this initially, I think, by degrading and obscuring our connection to the land, by looking upon the land as merchandise and ourselves as its traveling salesmen.

I do not know how exact a case might be made, but it seems to me that there is an historical parallel, in white American history, between the treatment of the land and the treatment of women. The frontier, for instance, was notoriously exploitive of both, and I believe for largely the same reasons. Many of the early farmers seem to have worn out farms and wives with equal regardlessness, interested in both mainly for what they would produce, crops and dollars, labor and sons; they clambered upon their fields and upon their wives, struggling for an economic foothold, the having and holding that cannot come until both fields and wives are properly cherished. And today there seems to me a distinct connection between our nomadism (our "social mobility") and the nearly universal disintegration of marriages and families.

The prevalent assumption appears to be that marriage problems are problems strictly of "human relations": If the husband and wife will only assent to a number of truisms about "respect for the other person," "giving and taking," et cetera, and if they will only "understand" each other, then it is believed that their problems will be solved. The difficulty is that marriage is only partly a matter of "human relations," and only partly a circumstance of the emotions. It is also, and as much as anything, a practical circumstance. It is very much under the influence of things and people outside itself; that is, it must make a household, it must make a place for itself in the world and in the community. But with us, getting someplace always involves going somewhere. Every professional advance leads to a new place, a new house, a new neighborhood. Our marriages are always being cut off from what they have made; their substance is always disappearing into the thin air of human relations.

I think there is a limit to the portability of human relationships. Tribal nomads, when they moved, moved as a tribe; their personal and cultural identity—their household and community—accompanied them as they went. But our modern urban nomads are always moving away from the particulars by which they know themselves, and moving into abstraction (a house, a neighborhood, a job) in which they can only feel threatened by new particulars. The marriage becomes a sort of assembly-line product, made partly here and partly there, the whole of it never quite coming into view. Provided they stay married (which is unlikely) until the children leave (which is usually soon), the nomadic husband and wife who look to see what their marriage has been—that is to say, what it *is*—are apt to see only the lines in each other's face.

The carelessness of place that must accompany our sort of nomadism makes a vagueness in marriage that is its antithesis. And vagueness in marriage, the most sacred human bond and perhaps the basic metaphor of our moral and religious tradition, cannot help but produce a diminishment of reverence, and of the care for the earth that must accompany reverence.

When the metaphor of atonement ceases to live in our consciousness, we lose the means of relationship. We become isolated in ourselves, and our behavior becomes the erratic behavior of people who have no bonds and no limits.

10: THE PRACTICALITY OF MORALS

What I have been preparing at such length to say is that there is only one value: the life and health of the world. If there is only one value, it follows that conflicts of value are illusory, based upon perceptual error. Moral, practical, spiritual, esthetic, economic, and ecological values are all concerned ultimately with the same question of life and health. To the virtuous man, for example, practical and spiritual values are identical; it is only corruption that can see a difference. Esthetic value is always associated with sound values of other kinds. "Beauty is truth, truth beauty," Keats said, and I think we may take him at his word.* Or to say the same thing in a different way: Beauty is wholeness; it is health in the ecological sense of amplitude and balance. And ecology is long-term economics. If these identities are not apparent immediately, they are apparent *in time.* Time is the merciless, infallible critic of the specialized disciplines. In the ledgers that justify waste the ink is turning red.

Moral value, as should be obvious, is not separable from other values. An adequate morality would be ecologically sound; it would be esthetically pleasing. But the point I want to stress here is that it would be *practical.* Morality is long-term practicality.

Of all specialists the moralists are the worst, and the processes of disintegration and specialization that have characterized us for generations have made moralists of us all. We have obscured and weakened morality, first, by advocating it for its own sake—that is, by deifying it, as esthetes have deified art—and then, as our capacity for reverence has diminished, by allowing it to become merely decorative, a matter of etiquette.

What we have forgotten is the origin of morality in fact and circumstance; we have forgotten that the nature of morality is essentially practical. Moderation and restraint, for example, are necessary, not because of any religious commandment or any

*I now consider this reference to Keats to be misleading. The ideal beauty of the "Ode on a Grecian Urn" is very different from the common, earthly, mortal beauty I had in mind.

creed or code, but because they are among the assurances of good health and a sufficiency of goods. Likewise, discipline is necessary if the necessary work is to be done; also if we are to know transport, transcendence, joy. Loyalty, devotion, faith, self-denial are not ethereal virtues, but the concrete terms upon which the possibility of love is kept alive in this world. Morality is neither ethereal nor arbitrary; it is the definition of what is humanly possible, and it is the definition of the penalties for violating human possibility. A person who violates human limits is punished or he prepares a punishment for his successors, not necessarily because of any divine or human law, but because he has transgressed the order of things.* A live and adequate morality is an accurate perception of the order of things, and of humanity's place in it. By clarifying the human limits, morality tells us what we risk when we forsake the human to behave like false gods or like animals.

One would not wish to say—indeed, it is precisely my point that one *should* not say—that social *forms* will not change with changing conditions. They probably *will* do so, wholly regardless of whether or not they *should*. But I believe that it is erroneous to assume that a change of form implies a change of discipline. Under the influence of the rapid changes of modern life, it is persistently assumed that we are moving toward a justifiable relaxation of disciplines. This is wishful thinking, and it invites calamity, for the human place in the order of things, the human limits, the human tragedy remain the same. It seems altogether possible, as a final example, that for various reasons the forms of marriage will change.† But this does not promise a new age of benefit without obligation—which, I am afraid, is what many people mean by freedom. Though the forms of marriage may change, if it continues to exist in any form it will continue to rest upon the same sustaining disciplines, and to incorporate the same tragic awareness: that it is made "for better for worse, for richer for poorer, in sickness and in health, to love and to cherish, till death. . . ."

*The order of things, of course, *is* a law—and not a human one.

†But to change them quickly is simply to destroy them. It is, I think, foolish to think that ancient community forms can be satisfactorily altered or safely discarded by the intention of individuals.

II: THE SPRING OF HOPE

The most destructive of ideas is that extraordinary times justify extraordinary measures. This is the ultimate relativism, and we are hearing it from all sides. The young, the poor, the minority races, the Constitution, the nation, traditional values, sexual morality, religious faith, Western civilization, the economy, the environment, the world are all now threatened with destruction—so the arguments run—therefore let us deal with our enemies by whatever means are handiest and most direct; in view of our high aims history will justify and forgive. Thus the violent have always rationalized their violence.

But as wiser men have always known, all times are extraordinary in precisely this sense. In the condition of mortality all things are always threatened with destruction. The invention of atomic holocaust and the other man-made dooms renews for us the immediacy of the worldly circumstance as the religions have always defined it: We know "neither the day nor the hour. . . ."

Our bewilderment is not the time but our character. We have come to expect too much from outside ourselves. If we are in despair or unhappy or uncomfortable, our first impulse is to assume that this cannot be our fault; our second is to assume that some institution is not doing its duty. We are in the curious position of expecting from others what we can only supply ourselves. One of the Confucian ideals is that the "archer, when he misses the bullseye, turns and seeks the cause of the error in himself."

Goodness, wisdom, happiness, even physical comfort, are not institutional conditions. The real sources of hope are personal and spiritual, not public and political. A man is not happy by the dispensation of his government or by the fortune of his age. He is happy only in doing well what is in his power, and in being reconciled to what is not in his power. Thoreau, who knew such happiness, wrote in "Life Without Principle": "Of what consequence, though our planet explode, if there is no character involved in the explosion? In health we have not the least curiosity about such events. We do not live for idle amusement. I would not run round a corner to see the world blow up."

Asked why the Shakers, who expected the end of the world at any moment, were nevertheless consummate farmers and craftsmen, Thomas Merton replied: "When you expect the world to end at any moment, you know there is no need to hurry. You take your time, you do your work well."

In Defense of Literacy

In a country in which everybody goes to school, it may seem absurd to offer a defense of literacy, and yet I believe that such a defense is in order, and that the absurdity lies not in the defense, but in the necessity for it. The published illiteracies of the certified educated are on the increase. And the universities seem bent upon ratifying this state of things by declaring the acceptability in their graduates of adequate—that is to say, of mediocre—writing skills.

The schools, then, are following the general subservience to the "practical," as that term has been defined for us according to the benefit of corporations. By "practicality" most users of the term now mean whatever will most predictably and most quickly make a profit. Teachers of English and literature have either submitted, or are expected to submit, along with teachers of the more "practical" disciplines, to the doctrine that the purpose of education is the mass production of producers and consumers. This has forced our profession into a predicament that we will finally have to recognize as a perversion. As if awed by the ascendency of the "practical" in our society, many of us secretly fear, and some of us are apparently ready to say, that if a student is not going to become a teacher of his language, he has no need to master it.

In other words, to keep pace with the specialization—and the dignity accorded to specialization—in other disciplines, we have begun to look upon and to teach our language and literature as specialties. But whereas specialization is of the nature of the applied sciences, it is a perversion of the disciplines of language and literature. When we understand and teach these as specialties, we submit willy-nilly to the assumption of the "practical men" of business, and also apparently of education, that literacy is no more than an ornament: When one has become an efficient integer of the economy, *then* it is permissible, even desirable, to be able to talk about the latest novels. After all, the disciples of "practicality" may someday find themselves stuck in conversation with an English teacher.

I may have oversimplified that line of thinking, but not

much. There are two flaws in it. One is that, among the self-styled "practical men," the practical is synonymous with the immediate. The long-term effects of their values and their acts lie outside the boundaries of their interest. For such people a strip mine ceases to exist as soon as the coal has been extracted. Short-term practicality is long-term idiocy.

The other flaw is that language and literature are always *about* something else, and we have no way to predict or control what they may be about. They are about the world. We will understand the world, and preserve ourselves and our values in it, only insofar as we have a language that is alert and responsive to it, and careful of it. I mean that literally. When we give our plows such brand names as "Sod Blaster," we are imposing on their use conceptual limits that raise the likelihood that they will be used destructively. When we speak of man's "war against nature," or of a "peace offensive," we are accepting the limitations of a metaphor that suggests, and even proposes, violent solutions. When students ask for the right of "participatory input" at the meetings of a faculty organization, they are thinking of democratic process, but they are *speaking* of a convocation of robots, and are thus devaluing the very traditions that they invoke.

Ignorance of books and the lack of a critical consciousness of language were safe enough in primitive societies with coherent oral traditions. In our society, which exists in an atmosphere of prepared, public language—language that is either written or being read—illiteracy is both a personal and a public danger. Think how constantly "the average American" is surrounded by premeditated language, in newspapers and magazines, on signs and billboards, on TV and radio. He is forever being asked to buy or believe somebody else's line of goods. The line of goods is being sold, moreover, by men who are trained to make him buy it or believe it, whether or not he needs it or understands it or knows its value or wants it. This sort of selling is an honored profession among us. Parents who grow hysterical at the thought that their son might not cut his hair are *glad* to have him taught, and later employed, to lie about the quality of an automobile or the ability of a candidate.

What is our defense against this sort of language—this language-as-weapon? There is only one. We must know a better language. We must speak, and teach our children to speak, a language precise and articulate and lively enough to tell the truth about the world as we know it. And to do this we must know something of the roots and resources of our language; we must know its literature. The only defense against the worst is a knowledge of the best. By their ignorance people enfranchise their exploiters.

But to appreciate fully the necessity for the best sort of literacy we must consider not just the environment of prepared language in which most of us now pass most of our lives, but also the utter transience of most of this language, which is meant to be merely glanced at, or heard only once, or read once and thrown away. Such language is by definition, and often by calculation, not memorable; it is language meant to be replaced by what will immediately follow it, like that of shallow conversation between strangers. It cannot be pondered or effectively criticized. For those reasons an unmixed diet of it is destructive of the informed, resilient, critical intelligence that the best of our traditions have sought to create and to maintain —an intelligence that Jefferson held to be indispensable to the health and longevity of freedom. Such intelligence does not grow by bloating upon the ephemeral information and misinformation of the public media. It grows by returning again and again to the landmarks of its cultural birthright, the works that have proved worthy of devoted attention.

"Read not the Times. Read the Eternities," Thoreau said. Ezra Pound wrote that "literature is news that STAYS news." In his lovely poem, "The Island," Edwin Muir spoke of man's inescapable cultural boundaries and of his consequent responsibility for his own sources and renewals:

> Men are made of what is made,
> The meat, the drink, the life, the corn,
> Laid up by them, in them reborn.
> And self-begotten cycles close
> About our way; indigenous art
> And simple spells make unafraid
> The haunted labyrinths of the heart . . .

These men spoke of a truth that no society can afford to shirk for long: We are dependent, for understanding, and for consolation and hope, upon what we learn of ourselves from songs and stories. This has always been so, and it will not change.

I am saying, then, that literacy—the mastery of language and the knowledge of books—is not an ornament, but a necessity. It is impractical only by the standards of quick profit and easy power. Longer perspective will show that it alone can preserve in us the possibility of an accurate judgment of ourselves, and the possibilities of correction and renewal. Without it, we are adrift in the present, in the wreckage of yesterday, in the nightmare of tomorrow.

FROM

RECOLLECTED ESSAYS: 1965–1980 (1981)

The Making of a Marginal Farm

ONE DAY in the summer of 1956, leaving home for school, I stopped on the side of the road directly above the house where I now live. From there you could see a mile or so across the Kentucky River Valley, and perhaps six miles along the length of it. The valley was a green trough full of sunlight, blue in its distances. I often stopped here in my comings and goings, just to look, for it was all familiar to me from before the time my memory began: woodlands and pastures on the hillsides; fields and croplands, wooded slew-edges and hollows in the bottoms; and through the midst of it the tree-lined river passing down from its headwaters near the Virginia line toward its mouth at Carrollton on the Ohio.

Standing there, I was looking at land where one of my great-great-great-grandfathers settled in 1803, and at the scene of some of the happiest times of my own life, where in my growing-up years I camped, hunted, fished, boated, swam, and wandered—where, in short, I did whatever escaping I felt called upon to do. It was a place where I had happily been, and where I always wanted to be. And I remember gesturing toward the valley that day and saying to the friend who was with me: "That's all I need."

I meant it. It was an honest enough response to my recognition of its beauty, the abundance of its lives and possibilities, and of my own love for it and interest in it. And in the sense that I continue to recognize all that, and feel that what I most need is here, I can still say the same thing.

And yet I am aware that I must necessarily mean differently —or at least a great deal more—when I say it now. Then I was speaking mostly from affection, and did not know, by half, what I was talking about. I was speaking of a place that in some ways I knew and in some ways cared for, but did not live in. The differences between knowing a place and living in it, between cherishing a place and living responsibly in it, had not begun to occur to me. But they are critical differences, and understanding them has been perhaps the chief necessity of my experience since then.

I married in the following summer, and in the next seven years lived in a number of distant places. But, largely because I continued to feel that what I needed was here, I could never bring myself to want to live in any other place. And so we returned to live in Kentucky in the summer of 1964, and that autumn bought the house whose roof my friend and I had looked down on eight years before, and with it "twelve acres more or less." Thus I began a profound change in my life. Before, I had lived according to expectation rooted in ambition. Now I began to live according to a kind of destiny rooted in my origins and in my life. One should not speak too confidently of one's "destiny"; I use the word to refer to causes that lie deeper in history and character than mere intention or desire. In buying the little place known as Lanes Landing, it seems to me, I began to obey the deeper causes.

We had returned so that I could take a job at the University of Kentucky in Lexington. And we expected to live pretty much the usual academic life: I would teach and write; my "subject matter" would be, as it had been, the few square miles in Henry County where I grew up. We bought the tiny farm at Lanes Landing, thinking that we would use it as a "summer place," and on that understanding I began, with the help of two carpenter friends, to make some necessary repairs on the house. I no longer remember exactly how it was decided, but that work had hardly begun when it became a full-scale overhaul.

By so little our minds had been changed: this was not going to be a house to visit, but a house to live in. It was as though, having put our hand to the plow, we not only did not look back, but could not. We renewed the old house, equipped it with plumbing, bathroom, and oil furnace, and moved in on July 4, 1965.

Once the house was whole again, we came under the influence of the "twelve acres more or less." This acreage included a steep hillside pasture, two small pastures by the river, and a "garden spot" of less than half an acre. We had, besides the house, a small barn in bad shape, a good large building that once had been a general store, and a small garage also in usable condition. This was hardly a farm by modern standards, but it was land that could be used, and it was unthinkable that we would not use it. The land was not good enough to afford the possibility of a cash income, but it would allow us to grow our food—or most of it. And that is what we set out to do.

In the early spring of 1965 I had planted a small orchard; the next spring we planted our first garden. Within the following six or seven years we reclaimed the pastures, converted the garage into a henhouse, rebuilt the barn, greatly improved the garden soil, planted berry bushes, acquired a milk cow—and were producing, except for hay and grain for our animals, nearly everything that we ate: fruit, vegetables, eggs, meat, milk, cream, and butter. We built an outbuilding with a meat room and a food-storage cellar. Because we did not want to pollute our land and water with sewage, and in the process waste nutrients that should be returned to the soil, we built a composting privy. And so we began to attempt a life that, in addition to whatever else it was, would be responsibly agricultural. We used no chemical fertilizers. Except for a little rotenone, we used no insecticides. As our land and our food became healthier, so did we. And our food was of better quality than any that we could have bought.

We were not, of course, living an idyll. What we had done could not have been accomplished without difficulty and a great deal of work. And we had made some mistakes and false starts. But there was great satisfaction, too, in restoring the neglected land, and in feeding ourselves from it.

Meanwhile, the forty-acre place adjoining ours on the downriver side had been sold to a "developer," who planned to divide it into lots for "second homes." This project was probably doomed by the steepness of the ground and the difficulty of access, but a lot of bulldozing—and a lot of damage—was done before it was given up. In the fall of 1972, the place was offered for sale and we were able to buy it.

We now began to deal with larger agricultural problems. Some of this new land was usable; some would have to be left in trees. There were perhaps fifteen acres of hillside that could be reclaimed for pasture, and about two and a half acres of excellent bottomland on which we would grow alfalfa for hay. But it was a mess, all of it badly neglected, and a considerable portion of it badly abused by the developer's bulldozers. The hillsides were covered with thicket growth; the bottom was shoulder high in weeds; the diversion ditches had to be restored; a bulldozed gash meant for "building sites" had to be mended; the barn needed a new foundation, and the cistern a new top; there were no fences. What we had bought was less a farm than a reclamation project—which has now, with a later purchase, grown to seventy-five acres.

While we had only the small place, I had got along very well with a Gravely "walking tractor" that I owned, and an old Farmall A that I occasionally borrowed from my Uncle Jimmy. But now that we had increased our acreage, it was clear that I could not continue to depend on a borrowed tractor. For a while I assumed that I would buy a tractor of my own. But because our land was steep, and there was already talk of a fuel shortage—and because I liked the idea—I finally decided to buy a team of horses instead. By the spring of 1973, after a lot of inquiring and looking, I had found and bought a team of five-year-old sorrel mares. And—again by the generosity of my Uncle Jimmy, who has never thrown any good thing away—I had enough equipment to make a start.

Though I had worked horses and mules during the time I was growing up, I had never worked over ground so steep and problematical as this, and it had been twenty years since I had worked a team over ground of any kind. Getting started again, I anticipated every new task with uneasiness, and sometimes with dread. But to my relief and delight, the team and I did all that needed to be done that year, getting better as we went along. And over the years since then, with that team and others, my son and I have carried on our farming the way it was carried on in my boyhood, doing everything with our horses except baling the hay. And we have done work in places and in weather in which a tractor would have been useless. Experience has shown us—or re-shown us—that horses are not only a satisfactory and economical means of power, especially on such small places as ours, but are probably *necessary* to the most conservative use of steep land. Our farm, in fact, is surrounded by potentially excellent hillsides that were maintained in pasture until tractors replaced the teams.

Another change in our economy (and our lives) was accomplished in the fall of 1973 with the purchase of our first wood-burning stove. Again the petroleum shortage was on our minds, but we also knew that from the pasture-clearing we had ahead of us we would have an abundance of wood that otherwise would go to waste—and when that was gone we would still have our permanent wood lots. We thus expanded our subsistence income to include heating fuel, and since then have used our furnace only as a "backup system" in the coldest weather and in our absences from home. The horses also contribute significantly to the work of fuel-gathering; they will go easily into difficult places and over soft ground or snow where a truck or a tractor could not move.

As we have continued to live on and from our place, we have slowly begun its restoration and healing. Most of the scars have now been mended and grassed over, most of the washes stopped, most of the buildings made sound; many loads of rocks have been hauled out of the fields and used to pave entrances or fill hollows; we have done perhaps half of the necessary fencing. A great deal of work is still left to do, and some of it—the rebuilding of fertility in the depleted hillsides—will take longer than we will live. But in doing these things we have begun a restoration and a healing in ourselves.

I should say plainly that this has not been a "paying proposition." As a reclamation project, it has been costly both in money and in effort. It seems at least possible that, in any other place, I might have had little interest in doing any such thing. The reason I have been interested in doing it here, I think, is that I have felt implicated in the history, the uses, and the attitudes that have depleted such places as ours and made them "marginal."

I had not worked long on our "twelve acres more or less" before I saw that such places were explained almost as much by their human history as by their nature. I saw that they were not "marginal" because they ever were unfit for human use, but because in both culture and character *we* had been unfit to use them. Originally, even such steep slopes as these along the lower Kentucky River Valley were deep-soiled and abundantly fertile; "jumper" plows and generations of carelessness impoverished them. Where yellow clay is at the surface now, five feet of good soil may be gone. I once wrote that on some of the nearby uplands one walks as if "knee-deep" in the absence of the original soil. On these steeper slopes, I now know, that absence is shoulder-deep.

That is a loss that is horrifying as soon as it is imagined. It happened easily, by ignorance, indifference, "a little folding of the hands to sleep." It cannot be remedied in human time; to build five feet of soil takes perhaps fifty or sixty thousand years. This loss, once imagined, is potent with despair. If a people in adding a hundred and fifty years to itself subtracts fifty thousand from its land, what is there to hope?

And so our reclamation project has been, for me, less a matter of idealism or morality than a kind of self-preservation. A destructive history, once it is understood as such, is a nearly insupportable burden. Understanding it is a disease of understanding, depleting the sense of efficacy and paralyzing effort, unless it finds healing work. For me that work has been partly of the mind, in what I have written, but that seems to have depended inescapably on work of the body and of the ground. In order to affirm the values most native and necessary to me —indeed, to affirm my own life as a thing decent in possibility —I needed to know in my own experience that this place did not have to be abused in the past, and that it can be kindly and conservingly used now.

With certain reservations that must be strictly borne in mind, our work here has begun to offer some of the needed proofs.

Bountiful as the vanished original soil of the hillsides may have been, what remains is good. It responds well—sometimes astonishingly well—to good treatment. It never should have been plowed (some of it never should have been cleared), and it never should be plowed again. But it can be put in pasture without plowing, and it will support an excellent grass sod that will in turn protect it from erosion, if properly managed and not overgrazed.

Land so steep as this cannot be preserved in row crop cultivation. To subject it to such an expectation is simply to ruin it, as its history shows. Our rule, generally, has been to plow no steep ground, to maintain in pasture only such slopes as can be safely mowed with a horse-drawn mower, and to leave the rest in trees. We have increased the numbers of livestock on our pastures gradually, and have carefully rotated the animals from field to field, in order to avoid overgrazing. Under this use and care, our hillsides have mended and they produce more and better pasturage every year.

As a child I always intended to be a farmer. As a young man, I gave up that intention, assuming that I could not farm and do the other things I wanted to do. And then I became a farmer almost unintentionally and by a kind of necessity. That wayward and necessary becoming—along with my marriage, which has been intimately a part of it—is the major event of my life. It has changed me profoundly from the man and the writer I would otherwise have been.

There was a time, after I had left home and before I came back, when this place was my "subject matter." I meant that too, I think, on the day in 1956 when I told my friend, "That's all I need." I was regarding it, in a way too easy for a writer, as a mirror in which I saw myself. There was obviously a sort of narcissism in that—and an inevitable superficiality, for only the surface can reflect.

In coming home and settling on this place, I began to *live* in my subject, and to learn that living in one's subject is not at all the same as "having" a subject. To live in the place that is one's subject is to pass through the surface. The simplifications of distance and mere observation are thus destroyed. The obsessively regarded reflection is broken and dissolved. One sees that the mirror was a blinder; one can now begin to see where one is. One's relation to one's subject ceases to be merely emotional or aesthetical, or even merely critical, and becomes problematical, practical, and responsible as well. Because it must. It is like marrying your sweetheart.

Though our farm has not been an economic success, as such success is usually reckoned, it is nevertheless beginning to make a kind of economic sense that is consoling and hopeful. Now that the largest expenses of purchase and repair are behind us, our income from the place is beginning to run ahead of expenses. As income I am counting the value of shelter, subsistence, heating fuel, and money earned by the sale of livestock. As expenses I am counting maintenance, newly purchased equipment, extra livestock feed, newly purchased animals, reclamation work, fencing materials, taxes, and insurance.

If our land had been in better shape when we bought it, our expenses would obviously be much smaller. As it is, once we have completed its restoration, our farm will provide us a home, produce our subsistence, keep us warm in winter, and earn a modest cash income. The significance of this becomes apparent when one considers that most of this land is "unfarmable" by the standards of conventional agriculture, and that most of it was producing nothing at the time we bought it.

And so, contrary to some people's opinion, it *is* possible for a family to live on such "marginal" land, to take a bountiful subsistence and some cash income from it, and, in doing so, to improve both the land and themselves. (I believe, however, that, at least in the present economy, this should not be attempted without a source of income other than the farm. It is now extremely difficult to pay for the best of farmland by farming it, and even "marginal" land has become unreasonably expensive. To attempt to make a living from such land is to impose a severe strain on land and people alike.)

I said earlier that the success of our work here is subject to reservations. There are only two of these, but both are serious.

The first is that land like ours—and there are many acres of such land in this country—can be conserved in use only by competent knowledge, by a great deal more work than is required by leveler land, by a devotion more particular and disciplined than patriotism, and by ceaseless watchfulness and care. All these are cultural values and resources, never sufficiently abundant in this country, and now almost obliterated by the contrary values of the so-called "affluent society."

One of my own mistakes will suggest the difficulty. In 1974 I dug a small pond on a wooded hillside that I wanted to pasture occasionally. The excavation for that pond—as I should have anticipated, for I had better reason than I used—caused the hillside to slump both above and below. After six years the slope has not stabilized, and more expense and trouble will be required to stabilize it. A small hillside farm will not survive many mistakes of that order. Nor will a modest income.

The true remedy for mistakes is to keep from making them. It is not in the piecemeal technological solutions that our society now offers, but in a change of cultural (and economic) values that will encourage in the whole population the necessary respect, restraint, and care. Even more important, it is in the possibility of settled families and local communities, in which the knowledge of proper means and methods, proper moderations and restraints, can be handed down, and so accumulate in place and stay alive; the experience of one generation is not adequate to inform and control its actions. Such possibilities are not now in sight in this country.

The second reservation is that we live at the lower end of the Kentucky River watershed, which has long been intensively used, and is increasingly abused. Strip mining, logging, extractive farming, and the digging, draining, roofing, and paving that go with industrial and urban "development," all have seriously depleted the capacity of the watershed to retain water. This means not only that floods are higher and more frequent than they would be if the watershed were healthy, but that the floods subside too quickly, the watershed being far less a sponge, now, than it is a roof. The floodwater drops suddenly out of the river, leaving the steep banks soggy, heavy, and soft. As a result, great strips and blocks of land crack loose and slump, or they give way entirely and disappear into the river in what people here call "slips."

The flood of December 1978, which was unusually high, also went down extremely fast, falling from banktop almost to pool stage within a couple of days. In the aftermath of this rapid "drawdown," we lost a block of bottomland an acre square. This slip, which is still crumbling, severely damaged our place, and may eventually undermine two buildings. The same flood started a slip in another place, which threatens a third building. We have yet another building situated on a huge (but, so far, very gradual) slide that starts at the river and, aggravated by two state highway cuts, goes almost to the hilltop. And we have serious river bank erosion the whole length of our place.

What this means is that, no matter how successfully we may control erosion on our hillsides, our land remains susceptible to a more serious cause of erosion that we cannot control. Our river bank stands literally at the cutting edge of our nation's consumptive economy. This, I think, is true of many "marginal" places—it is true, in fact, of many places that are not marginal. In its consciousness, ours is an upland society; the ruin of watersheds, and what that involves and means, is little considered. And so the land is heavily taxed to subsidize an "affluence" that consists, in reality, of health and goods stolen from the unborn.

Living at the lower end of the Kentucky River watershed is what is now known as "an educational experience"—and not an easy one. A lot of information comes with it that is severely damaging to the reputation of our people and our time. From where I live and work, I never have to look far to see that the earth does indeed pass away. But however that is taught, and however bitterly learned, it is something that should be known, and there is a certain good strength in knowing it. To spend one's life farming a piece of the earth so passing is, as many would say, a hard lot. But it is, in an ancient sense, the human lot. What saves it is to love the farming.

THE UNSETTLING OF AMERICA (1977)

For Maurice Telleen

Contents

Preface

This book was meant to be a criticism of what I have called modern or orthodox agriculture. As I now realize, it is more a review than a criticism. Criticism requires a subject that is "finished." When agriculture is "finished," no would-be critic will be available. I am therefore constrained to accept my demotion as a privilege.

Nevertheless, there is a difficulty in writing a book on so inherently topical a subject as agricultural policy, and this difficulty is time: events that were the immediate cause of the book may be "finished" before the writing is. No reader of this book can fail to observe that it deals at length with the assumptions and policies of former Secretary of Agriculture Earl L. Butz, though Mr. Butz and the administration he served are now out of office.

I can only insist that my book is not for that reason out-of-date. Secretary Butz's tenure in the Department of Agriculture, and even his influence, are matters far more transient than the power and the values of those whose interests he represented. Moreover, the cultural issues that I attempt to deal with have been with us since our history began, and, barring miracle or catastrophe, they will be with us for a long time to come.

As a matter of fact, this book's origins go back farther than the secretaryship of Mr. Butz. The first notes I made for it were incited by a news story in the summer of 1967 on the report of President Johnson's "special commission on federal food and fiber policies."

The commission said, according to an article in the *Louisville Courier-Journal*, that the country's biggest farm problem was a surplus of farmers: ". . . the technological advances in agriculture have so greatly reduced the need for manpower that too many people are trying to live on a national farm income wholly inadequate for them." The proposed solutions were to find "better opportunities for the farm people," "a more comprehensive national employment policy," "retraining programs," "improved general educational facilities," etc.

Both the commission and the writer of the article had obviously *taken for granted* that the lives and communities of small farmers then still on the farm—and those of the 25 million who had left the farm since 1940—were of less value than "technological advances in agriculture." There seemed also to be no official doubt that adequate solutions were to be found in government-supplied "opportunities," facilities, and programs. Reading that article, I realized that my values were not only out of fashion, but under powerful attack. I saw that I was a member of a threatened minority. That is what set me off.

W.B.

Preface to the Second Edition

When I was working on this book—from 1974 to 1977—the long agricultural decline that it deals with was momentarily disguised as a "boom." The big farmers were getting bigger with the help of inflated land prices and borrowed money, and the foreign demand for American farm products was strong, so from the official point of view the situation looked good. The big were *supposed* to get bigger. Foreigners were *supposed* to be in need of our products. The official point of view, foreshortened as usual by statistics, superstitious theory, and wishful prediction, was utterly complacent. Then Secretary of Agriculture Earl L. Butz issued the most optimistic, the most widely obeyed, and the worst advice ever given to farmers: that they should plow "fencerow to fencerow."

That the situation was *not* good—for farms or farmers or rural communities or nature or the general public—was even then evident to any experienced observer who would turn aside from the preconceptions of "agribusiness" and look at the marks of deterioration that were plainly visible. And now, almost a decade later, it is evident to everyone that, at least for farmers and rural communities, the situation is catastrophic: Farmers are losing their farms, some are killing themselves, some in the madness of despair are killing other people, and rural economy and rural life are gravely stricken. The agricultural economists chart the "liquidations of assets," the "shake-outs," and the "downturns," apparently amazed that now even the large "progressive" and "efficient" farmers are in trouble.

But this is not just a financial crisis for country people. Critical questions are being asked of our whole society: Are we, or are we not, going to take proper care of our land, our country? And do we, or do we not, believe in a democratic distribution of usable property? At present, these questions are being answered in the negative. Our soil erosion rates are worse now than during the years of the Dust Bowl. In the arid lands of the West, we are overusing and wasting the supplies of water. Toxic pollution from agricultural chemicals is a growing problem. We are closer every day to the final destruction of private ownership not only of small family farms, but of small usable properties of all kinds. Every problem I dealt with in this book, in fact, has grown worse since the book was written.

The one improvement has been in public concern about the problems. Among farmers there is growing distrust of the "agribusiness" line of talk and growing interest in agricultural health and sanity. Among city people there is a growing awareness that sane and healthy agriculture requires an informed urban constituency. There is hope in these developments and in the continued existence of a remnant of excellent small farms and farmers.

Some prominent agricultural economists are still finding it possible to pretend that the only issues involved are economic, but that possibility is diminishing. I recently attended a meeting at which an agricultural economist argued that there is no essential difference between owning and renting a farm. A farmer stood up in the audience and replied: "Professor, I don't think our ancestors came to America in order to *rent* a farm."

'Nough said.

W.B.
March 1986

Acknowledgments

Anything that I will ever have to say on the subject of agriculture can be little more than a continuation of talk begun in childhood with my father and with my late friend Owen Flood. Their conversation, first listened to and then joined, was my first and longest and finest instruction. From them, before I knew I was being taught, I learned to think of the meanings, the responsibilities, and the pleasures of farming.

But this book's greatest immediate debt is acknowledged in the dedication. Maury Telleen has been an indefatigable friend both to my book and to me. He has arranged indispensable meetings, written me letters, sent me clippings and books, talked to me on the phone, read my manuscript, borne up through hours of fervent conversation in his house, in my house, and over many hundreds of miles of happy agricultural travels. To his wife, Jeannine, I am indebted for understanding and for hospitality.

I am, of course, hopelessly in debt to my own wife, Tanya, for keeping farm and household together during my absences, and for enduring both my travels and my travails. But she has also participated in the thinking-out of this book, has been its critic, and, finally, its typist.

My neighbor, Tom Grissom, gave me invaluable help with the preliminary reading, and in many talks helped me to clarify my ideas.

For reading the manuscript and other kindnesses, I am grateful to Robert Rodale, Jerry Goldstein, Cia and Ed McClanahan, James Baker Hall, Joan Hall, Gary Snyder, Jim and Barbara Foote, and my father.

Stephen B. Brush, Clarence Van Sant, Gurney Norman, Ben Webb, and David Budbill generously gave me permission to quote from their letters, and Stephen Brush allowed me to make use of his excellent paper on Andean agriculture.

For hospitality and other kindnesses on my travels, I thank Tommy Shoup, Everett Hildebrandt, Roger Blobaum, Mr. and Mrs. Ilo Kusserow, Mr. and Mrs. Mike Jessen, Mr. and Mrs.

Monroe J. Miller, Arnold Hockett, Clarence Van Sant, Mr. and Mrs. John Heinrich, Patty Kaminsky.

This list of my debts will make clear that much of the good of my book has been the gift of other people. Its flaws, on the other hand, have been furnished exclusively by me.

Who so hath his minde on taking,
hath it no more on what he hath taken.

MONTAIGNE, III. VI

So many goodly citties ransacked and razed; so many nations destroyed and made desolate; so infinite millions of harmelesse people of all sexes, states and ages, massacred, ravaged and put to the sword; and the richest, the fairest and the best part of the world topsiturvied, ruined and defaced for the traffick of Pearles and Pepper: Oh mechanicall victories, oh base conquest.

MONTAIGNE

CHAPTER ONE

The Unsettling of America

One of the peculiarities of the white race's presence in America is how little intention has been applied to it. As a people, wherever we have been, we have never really intended to be. The continent is said to have been discovered by an Italian who was on his way to India. The earliest explorers were looking for gold, which was, after an early streak of luck in Mexico, always somewhere farther on. Conquests and foundings were incidental to this search—which did not, and could not, end until the continent was finally laid open in an orgy of goldseeking in the middle of the 19th century. Once the unknown of geography was mapped, the industrial marketplace became the new frontier, and we continued, with largely the same motives and with increasing haste and anxiety, to displace ourselves—no longer with unity of direction, like a migrant flock, but like the refugees from a broken ant hill. In our own time we have invaded foreign lands and the moon with the high-toned patriotism of the conquistadors, and with the same mixture of fantasy and avarice.

That is too simply put. It is substantially true, however, as a description of the dominant tendency in American history. The temptation, once that has been said, is to ascend altogether into rhetoric and inveigh equally against all our forebears and all present holders of office. To be just, however, it is necessary to remember that there has been another tendency: the tendency to stay put, to say, "No farther. This is the place." So far, this has been the weaker tendency, less glamorous, certainly less successful. It is also the older of these tendencies, having been the dominant one among the Indians.

The Indians did, of course, experience movements of population, but in general their relation to place was based upon old usage and association, upon inherited memory, tradition, veneration. The land was their homeland. The first and greatest American revolution, which has never been superseded, was the coming of people who did *not* look upon the land as a homeland. But there were always those among the newcomers who saw that they had come to a good place and who saw its domestic possibilities. Very early, for instance, there were men who wished to establish agricultural settlements rather than quest for gold or exploit the Indian trade. Later, we know that every advance of the frontier left behind families and communities who intended to remain and prosper where they were.

But we know also that these intentions have been almost systematically overthrown. Generation after generation, those who intended to remain and prosper where they were have been dispossessed and driven out, or subverted and exploited where they were, by those who were carrying out some version of the search for El Dorado. Time after time, in place after place, these conquerors have fragmented and demolished traditional communities, the beginnings of domestic cultures. They have always said that what they destroyed was outdated, provincial, and contemptible. And with alarming frequency they have been believed and trusted by their victims, especially when their victims were other white people.

If there is any law that has been consistently operative in American history, it is that the members of any *established* people or group or community sooner or later become "redskins"—that is, they become the designated victims of an utterly ruthless, officially sanctioned and subsidized exploitation. The colonists who drove off the Indians came to be intolerably exploited by their imperial governments. And that alien imperialism was thrown off only to be succeeded by a domestic version of the same thing; the class of independent small farmers who fought the war of independence has been exploited by, and recruited into, the industrial society until by now it is almost extinct. Today, the most numerous heirs of the farmers of Lexington and Concord are the little groups scattered all over the country whose names begin with "Save": Save Our Land, Save the Valley, Save Our Mountains, Save Our Streams, Save Our Farmland. As so often before, these are *designated* victims—people without official sanction, often without official friends, who are struggling to preserve their places, their values, and their lives as they know them and prefer to live them against the agencies of their own government which are using their own tax moneys against them.

The only escape from this destiny of victimization has been to "succeed"—that is, to "make it" into the class of exploiters, and then to remain so specialized and so "mobile" as to be unconscious of the effects of one's livelihood. This escape is, of course, illusory, for one man's producer is another's consumer, and even the richest and most mobile will soon find it hard to escape the noxious effluents and fumes of the various public services.

Let me emphasize that I am not talking about an evil that is merely contemporary or "modern," but one that is as old in America as the white man's presence here. It is an intention that was *organized* here almost from the start. "The New World," Bernard DeVoto wrote in *The Course of Empire*, "was a constantly expanding market . . . Its value in gold was enormous but it had still greater value in that it expanded and integrated the industrial systems of Europe."

And he continues: "The first belt-knife given by a European to an Indian was a portent as great as the cloud that mushroomed over Hiroshima . . . Instantly the man of 6000 B.C. was bound fast to a way of life that had developed seven and a half millennia beyond his own. He began to live better and he began to die."

The principal European trade goods were tools, cloth, weapons, ornaments, novelties, and alcohol. The sudden availability of these things produced a revolution that "affected every aspect of Indian life. The struggle for existence . . . became easier. Immemorial handicrafts grew obsolescent, then obsolete. Methods of hunting were transformed. So were methods—and the purposes—of war. As war became deadlier in purpose and armament a surplus of women developed, so that marriage customs changed and polygamy became common. The increased usefulness of women in the preparation of pelts worked to the same end . . . Standards of wealth, prestige, and honor changed. The Indians acquired commercial values and developed business cults. They became more mobile. . . .

"In the sum it was cataclysmic. A culture was forced to change much faster than change could be adjusted to. All corruptions of culture produce breakdowns of morale, of communal integrity, and of personality, and this force was as strong as any other in the white man's subjugation of the red man."

I have quoted these sentences from DeVoto because, the obvious differences aside, he is so clearly describing a revolution that did not stop with the subjugation of the Indians, but went on to impose substantially the same catastrophe upon the small farms and the farm communities, upon the shops of small local tradesmen of all sorts, upon the workshops of independent craftsmen, and upon the households of citizens. It is a revolution that is still going on. The economy is still substantially that of the fur trade, still based on the same general kinds of commercial items: technology, weapons, ornaments, novelties, and drugs. The one great difference is that by now the revolution has deprived the mass of consumers of any independent access to the staples of life: clothing, shelter, food, even water. Air remains the only necessity that the average user can still get for himself, and the revolution has imposed a heavy tax on that by way of pollution. Commercial conquest is far more thorough and final than military defeat. The Indian became a redskin, not by loss in battle, but by accepting a dependence on traders that made *necessities* of industrial goods. This is not merely history. It is a parable.

DeVoto makes it clear that the imperial powers, having made themselves willing to impose this exploitive industrial economy upon the Indians, could not then keep it from contaminating their own best intentions: "More than four-fifths of the wealth of New France was furs, the rest was fish, and it had no agricultural wealth. One trouble was that whereas the crown's imperial policy required it to develop the country's agriculture, the crown's economy required the colony's furs, an adverse interest." And La Salle's dream of developing Louisiana (agriculturally and otherwise) was frustrated because "The interest of the court in Louisiana colonization was to secure a bridgehead for an attack on the silver mines of northern Mexico. . . ."

One cannot help but see the similarity between this foreign colonialism and the domestic colonialism that, by policy, converts productive farm, forest, and grazing lands into strip mines. Now, as then, we see the abstract values of an industrial economy preying upon the native productivity of land and people. The fur trade was only the first establishment on this continent of a mentality whose triumph is its catastrophe.

My purposes in beginning with this survey of history are (1) to show how deeply rooted in our past is the mentality of exploitation; (2) to show how fundamentally revolutionary it is; and (3) to show how crucial to our history—hence, to our own minds—is the question of how we will relate to our land. This question, now that the corporate revolution has so determinedly invaded the farmland, returns us to our oldest crisis.

We can understand a great deal of our history—from Cortés' destruction of Tenochtitlán in 1521 to the bulldozer attack on the coalfields four-and-a-half centuries later—by thinking of ourselves as divided into conquerors and victims. In order to understand our own time and predicament and the work that is to be done, we would do well to shift the terms and say that we are divided between exploitation and nurture. The first set of terms is too simple for the purpose because, in any given situation, it proposes to divide people into two mutually exclusive groups; it becomes complicated only when we are dealing with situations in succession—as when a colonist who persecuted the Indians then resisted persecution by the crown. The terms exploitation and nurture, on the other hand, describe a division not only between persons but also within persons. We are all to some extent the products of an exploitive society, and it would be foolish and self-defeating to pretend that we do not bear its stamp.

Let me outline as briefly as I can what seem to me the characteristics of these opposite kinds of mind. I conceive a strip-miner to be a model exploiter, and as a model nurturer I take the old-fashioned idea or ideal of a farmer. The exploiter is a specialist, an expert; the nurturer is not. The standard of the exploiter is efficiency; the standard of the nurturer is care. The exploiter's goal is money, profit; the nurturer's goal is health—his land's health, his own, his family's, his community's, his country's. Whereas the exploiter asks of a piece of land only how much and how quickly it can be made to produce, the nurturer asks a question that is much more complex and difficult: What is its carrying capacity? (That is: How much can be taken from it without diminishing it? What can it produce *dependably* for an indefinite time?) The exploiter wishes to earn as much as possible by as little work as possible; the nurturer expects, certainly, to have a decent living from his work, but his characteristic wish is to work *as well* as possible. The competence of the exploiter is in organization; that of the nurturer is in order—a human order, that is, that accommodates itself both to other order and to mystery. The exploiter typically serves an institution or organization; the nurturer serves land, household, community, place. The exploiter thinks in terms of numbers, quantities, "hard facts"; the nurturer in terms of character, condition, quality, kind.

It seems likely that all the "movements" of recent years have been representing various claims that nurture has to make against exploitation. The women's movement, for example, when its energies are most accurately placed, is arguing the cause of nurture; other times it is arguing the right of women to be exploiters—which men have no *right* to be. The exploiter is clearly the prototype of the "masculine" man—the wheeler-dealer whose "practical" goals require the sacrifice of flesh, feeling, and principle. The nurturer, on the other hand, has always passed with ease across the boundaries of the so-called sexual roles. Of necessity and without apology, the preserver of seed, the planter, becomes midwife and nurse. Breeder is always metamorphosing into brooder and back again. Over and over again, spring after spring, the questing mind, idealist and visionary, must pass through the planting to become nurturer of the real. The farmer, sometimes known as husbandman, is by definition half mother; the only question is how good a mother he or she is. And the land itself is not mother or father only, but both. Depending on crop and season, it is at one time receiver of seed, bearer and nurturer of young; at another, raiser of seed-stalk, bearer and shedder of seed. And in response to these changes, the farmer crosses back and forth from one zone of spousehood to another, first as planter and then as gatherer. Farmer and land are thus involved in a sort of dance in which the partners are always at opposite sexual poles, and the lead keeps changing; the farmer, as seed-bearer, causes growth; the land, as seed-bearer, causes the harvest.

The exploitive always involves the abuse or the perversion of nurture and ultimately its destruction. Thus, we saw how far the exploitive revolution had penetrated the official character when our recent secretary of agriculture remarked that "Food is a weapon." This was given a fearful symmetry indeed when, in discussing the possible use of nuclear weapons, a secretary of defense spoke of "palatable" levels of devastation. Consider the associations that have since ancient times clustered around the idea of food—associations of mutual care, generosity, neighborliness, festivity, communal joy, religious ceremony—and you will see that these two secretaries represent a cultural catastrophe. The concerns of farming and those of war, once thought to be diametrically opposed, have become identical. Here we have an example of men who have been made vicious, not presumably by nature or circumstance, but by their *values.*

Food is *not* a weapon. To use it as such—to foster a mentality willing to use it as such—is to prepare, in the human character and community, the destruction of the sources of food. The first casualties of the exploitive revolution are character and community. When those fundamental integrities are devalued and broken, then perhaps it is inevitable that food will be looked upon as a weapon, just as it is inevitable that the earth will be looked upon as fuel and people as numbers or machines. But character and community—that is, culture in the broadest, richest sense—constitute, just as much as nature, the source of food. Neither nature nor people alone can produce human sustenance, but only the two together, culturally wedded. The poet Edwin Muir said it unforgettably:

> Men are made of what is made,
> The meat, the drink, the life, the corn,
> Laid up by them, in them reborn.
> And self-begotten cycles close
> About our way; indigenous art
> And simple spells make unafraid
> The haunted labyrinths of the heart
> And with our wild succession braid
> The resurrection of the rose.

To think of food as a weapon, or of a weapon as food, may give an illusory security and wealth to a few, but it strikes directly at the life of all.

The concept of food-as-weapon is not surprisingly the doctrine of a Department of Agriculture that is being used as an instrument of foreign political and economic speculation. This militarizing of food is the greatest threat so far raised against the farmland and the farm communities of this country. If present attitudes continue, we may expect government policies that will encourage the destruction, by overuse, of farmland. This, of course, has already begun. To answer the official call for more production—evidently to be used to bait or bribe foreign countries—farmers are plowing their waterways and permanent pastures; lands that ought to remain in grass are being planted in row crops. Contour plowing, crop rotation, and other conservation measures seem to have gone out of favor or fashion in official circles and are practiced less and less on the farm. This exclusive emphasis on production will accelerate the mechanization and chemicalization of farming, increase the price of land, increase overhead and operating costs, and thereby further diminish the farm population. Thus the tendency, if not the intention, of Mr. Butz's confusion of farming and war, is to complete the deliverance of American agriculture into the hands of corporations.

The cost of this corporate totalitarianism in energy, land, and social disruption will be enormous. It will lead to the exhaustion of farmland and farm culture. Husbandry will become an extractive industry; because maintenance will entirely give way to production, the fertility of the soil will become a limited, unrenewable resource like coal or oil.

This may not happen. It *need* not happen. But it is necessary to recognize that it *can* happen. That it can happen is made evident not only by the words of such men as Mr. Butz, but more clearly by the large-scale industrial destruction of farmland already in progress. If it does happen, we are familiar enough with the nature of American salesmanship to know that it will be done in the name of the starving millions, in the name of liberty, justice, democracy, and brotherhood, and to free the world from communism. We must, I think, be prepared to see, and to stand by, the truth: that the land should not be destroyed for *any* reason, not even for any apparently good reason. We must be prepared to say that enough food, year after year, is possible only for a limited number of people, and that this possibility can be preserved only by the steadfast, knowledgeable *care* of those people. Such "crash programs" as apparently have been contemplated by the Department of Agriculture in recent years will, in the long run, cause more starvation than they can remedy.

Meanwhile, the dust clouds rise again over Texas and Oklahoma. "Snirt" is falling in Kansas. Snow drifts in Iowa and the Dakotas are black with blown soil. The fields lose their humus and porosity, become less retentive of water, depend more on pesticides, herbicides, chemical fertilizers. Bigger tractors become necessary because the compacted soils are harder to work—and their greater weight further compacts the soil. More and bigger machines, more chemical and methodological shortcuts are needed because of the shortage of manpower on the farm—and the problems of overcrowding and unemployment increase in the cities. It is estimated that it now costs (by erosion) two bushels of Iowa topsoil to grow one bushel of corn. It is variously estimated that from five to twelve calories of fossil fuel energy are required to produce one calorie of hybrid corn energy. An official of the National Farmers Union says that "a farmer who earns $10,000 to $12,000 a year typically leaves an estate valued at about $320,000"—which means that when that farm is financed again, either by a purchaser or by an heir (to pay the inheritance taxes), it simply cannot support its new owner and pay for itself. And the *Progressive Farmer* predicts the disappearance of 200,000 to 400,000 farms each year during the next twenty years if the present trend continues.

The first principle of the exploitive mind is to divide and conquer. And surely there has never been a people more ominously and painfully divided than we are—both against each other and within ourselves. Once the revolution of exploitation is under way, statesmanship and craftsmanship are gradually replaced by salesmanship.* Its stock in trade in politics is to sell despotism and avarice as freedom and democracy. In business it sells sham and frustration as luxury and satisfaction. The "constantly expanding market" first opened in the New World by the fur traders is still expanding—no longer so much by expansions of territory or population, but by the calculated outdating, outmoding, and degradation of goods and by the hysterical self-dissatisfaction of consumers that is indigenous to an exploitive economy.

*The *craft* of persuading people to buy what they do not need, and do not want, for more than it is worth.

This gluttonous enterprise of ugliness, waste, and fraud thrives in the disastrous breach it has helped to make between our bodies and our souls. As a people, we have lost sight of the profound communion—even the union—of the inner with the outer life. Confucius said: "If a man have not order within him / He can not spread order about him. . . ." Surrounded as we are by evidence of the disorders of our souls and our world, we feel the strong truth in those words as well as the possibility of healing that is in them. We see the likelihood that our surroundings, from our clothes to our countryside, are the products of our inward life—our spirit, our vision—as much as they are products of nature and work. If this is true, then we cannot live as we do and be as we would like to be. There is nothing more absurd, to give an example that is only apparently trivial, than the millions who wish to live in luxury and idleness and yet be slender and good-looking. We have millions, too, whose livelihoods, amusements, and comforts are all destructive, who nevertheless wish to live in a healthy environment; they want to run their recreational engines in clean, fresh air. There is now, in fact, no "benefit" that is not associated with disaster. That is because power can be disposed morally or harmlessly only by thoroughly unified characters and communities.

What caused these divisions? There are no doubt many causes, complex both in themselves and in their interaction. But pertinent to all of them, I think, is our attitude toward work. The growth of the exploiters' revolution on this continent has been accompanied by the growth of the idea that work is beneath human dignity, particularly any form of hand work. We have made it our overriding ambition to escape work, and as a consequence have debased work until it is only fit to escape from. We have debased the products of work and have been, in turn, debased by them. Out of this contempt for work arose the idea of a nigger: at first some person, and later some thing, to be used to relieve us of the burden of work. If we began by making niggers of people, we have ended by making a nigger of the world. We have taken the irreplaceable energies and materials of the world and turned them into jimcrack "labor-saving devices." We have made of the rivers and oceans and winds niggers to carry away our refuse, which we think we are too good to dispose of decently ourselves. And in doing this to the world that is our common heritage and bond, we have returned to making niggers of people: we have become each other's niggers.

But is work something that we have a right to escape? And can we escape it with impunity? We are probably the first entire people ever to think so. All the ancient wisdom that has come down to us counsels otherwise. It tells us that work is necessary to us, as much a part of our condition as mortality; that good work is our salvation and our joy; that shoddy or dishonest or self-serving work is our curse and our doom. We have tried to escape the sweat and sorrow promised in Genesis—only to find that, in order to do so, we must forswear love and excellence, health and joy.

Thus we can see growing out of our history a condition that is physically dangerous, morally repugnant, ugly. Contrary to the blandishments of the salesmen, it is not particularly comfortable or happy. It is not even affluent in any meaningful sense, because its abundance is dependent on sources that are being rapidly exhausted by its methods. To see these things is to come up against the question: Then what *is* desirable?

One possibility is just to tag along with the fantasists in government and industry who would have us believe that we can pursue our ideals of affluence, comfort, mobility, and leisure indefinitely. This curious faith is predicated on the notion that we will soon develop unlimited new sources of energy: domestic oil fields, shale oil, gasified coal, nuclear power, solar energy, and so on. This is fantastical because the basic cause of the energy crisis is not scarcity; it is moral ignorance and weakness of character. We don't know *how* to use energy, or what to use it *for*. And we cannot restrain ourselves. Our time is characterized as much by the abuse and waste of human energy as it is by the abuse and waste of fossil fuel energy. Nuclear power, if we are to believe its advocates, is presumably going to be well used by the same mentality that has egregiously devalued and misapplied man- and womanpower. If we had an unlimited supply of solar or wind power, we would use that destructively, too, for the same reasons.

Perhaps all of those sources of energy are going to be developed. Perhaps all of them can sooner or later be developed without threatening our survival. But not all of them together can guarantee our survival, and they cannot define what is desirable. We will not find those answers in Washington, D.C., or in the laboratories of oil companies. In order to find them, we will have to look closer to ourselves.

I believe that the answers are to be found in our history: in its until now subordinate tendency of settlement, of domestic permanence. This was the ambition of thousands of immigrants; it is formulated eloquently in some of the letters of Thomas Jefferson; it was the dream of the freed slaves; it was written into law in the Homestead Act of 1862. There are few of us whose families have not at some time been moved to see its vision and to attempt to enact its possibility. I am talking about the idea that as many as possible should share in the ownership of the land and thus be bound to it by economic interest, by the investment of love and work, by family loyalty, by memory and tradition. How much land this should be is a question, and the answer will vary with geography. The Homestead Act said 160 acres. The freedmen of the 1860s hoped for forty. We know that, particularly in other countries, families have lived decently on far fewer acres than that.

The old idea is still full of promise. It is potent with healing and with health. It has the power to turn each person away from the big-time promising and planning of the government, to confront in himself, in the immediacy of his own circumstances and whereabouts, the question of what methods and ways are best. It proposes an economy of necessities rather than an economy based upon anxiety, fantasy, luxury, and idle wishing. It proposes the independent, free-standing citizenry that Jefferson thought to be the surest safeguard of democratic liberty. And perhaps most important of all, it proposes an agriculture based upon intensive work, local energies, care, and long-living communities—that is, to state the matter from a consumer's point of view: a dependable, long-term food supply.

This is a possibility that is obviously imperiled—by antipathy in high places, by adverse public fashions and attitudes, by the deterioration of our present farm communities and traditions, by the flawed education and the inexperience of our young people. Yet it alone can promise us the continuity of attention and devotion without which the human life of the earth is impossible.

Sixty years ago, in another time of crisis, Thomas Hardy wrote these stanzas:

Only a man harrowing clods
  In a slow silent walk
With an old horse that stumbles and nods
  Half asleep as they stalk.

Only thin smoke without flame
  From the heaps of couch-grass;
Yet this will go onward the same
  Though Dynasties pass.

Today most of our people are so conditioned that they do not wish to harrow clods either with an old horse or with a new tractor. Yet Hardy's vision has come to be more urgently true than ever. The great difference these sixty years have made is that, though we feel that this work must go onward, we are not so certain that it will. But the care of the earth is our most ancient and most worthy and, after all, our most pleasing responsibility. To cherish what remains of it, and to foster its renewal, is our only legitimate hope.

. . . wanting good government in their states, they first established order in their own families; wanting order in the home, they first disciplined themselves . . .

CONFUCIUS, *The Great Digest*

CHAPTER TWO

The Ecological Crisis as a Crisis of Character

IN JULY OF 1975 it was revealed by William Rood in the *Los Angeles Times* that some of our largest and most respected conservation organizations owned stock in the very corporations and industries that have been notorious for their destructiveness and for their indifference to the concerns of conservationists. The Sierra Club, for example, had owned stocks and bonds in Exxon, General Motors, Tenneco, steel companies "having the worst pollution records in the industry," Public Service Company of Colorado, "strip-mining firms with 53 leases covering nearly 180,000 acres and pulp-mill operators cited by environmentalists for their poor water pollution controls."

These investments proved deeply embarrassing once they were made public, but the Club's officers responded as quickly as possible by making appropriate changes in its investment policy. And so if it were only a question of policy, these investments could easily be forgotten, dismissed as aberrations of the sort that inevitably turn up now and again in the workings of organizations. The difficulty is that, although the investments were absurd, they were *not* aberrant; they were perfectly representative of the modern character. These conservation groups were behaving with a very ordinary consistency; they were only doing as organizations what many of their members were, and are, doing as individuals. They were making convenience of enterprises that they knew to be morally, and even practically, indefensible.

We are dealing, then, with an absurdity that is not a quirk or an accident, but is fundamental to our character as a people. The split between what we think and what we do is profound. It is not just possible, it is altogether to be expected, that our society would produce conservationists who invest in strip-mining companies, just as it must inevitably produce asthmatic executives whose industries pollute the air and vice-presidents of pesticide corporations whose children are dying of cancer.

And these people will tell you that this is the way the "real world" works. They will pride themselves on their sacrifices for "our standard of living." They will call themselves "practical men" and "hardheaded realists." And they will have their justifications in abundance from intellectuals, college professors, clergymen, politicians. The viciousness of a mentality that can look complacently upon disease as "part of the cost" would be obvious to any child. But this is the "realism" of millions of modern adults.

There is no use pretending that the contradiction between what we think or say and what we do is a limited phenomenon. There is no group of the extra-intelligent or extra-concerned or extra-virtuous that is exempt. I cannot think of any American whom I know or have heard of, who is not contributing in some way to destruction. The reason is simple: to live undestructively in an economy that is overwhelmingly destructive would require of any one of us, or of any small group of us, a great deal more work than we have yet been able to do. How could we divorce ourselves completely and yet responsibly from the technologies and powers that are destroying our planet? The answer is not yet thinkable, and it will not be thinkable for some time—even though there are now groups and families and persons everywhere in the country who have begun the labor of thinking it.

And so we are by no means divided, or readily divisible, into environmental saints and sinners. But there *are* legitimate distinctions that need to be made. These are distinctions of degree and of consciousness. Some people are less destructive than others, and some are more conscious of their destructiveness than others. For some, their involvement in pollution, soil depletion, strip-mining, deforestation, industrial and commercial waste is simply a "practical" compromise, a necessary "reality," the price of modern comfort and convenience. For others, this list of involvements is an agenda for thought and work that will produce remedies.

People who thus set their lives against destruction have necessarily confronted in themselves the absurdity that they have recognized in their society. They have first observed the tendency of modern organizations to perform in opposition to their stated purposes. They have seen governments that exploit and oppress the people they are sworn to serve and protect, medical procedures that produce ill health, schools that preserve ignorance, methods of transportation that, as Ivan Illich says, have "created more distances than they . . . bridge." And they have seen that these public absurdities are, and can be, no more than the aggregate result of private absurdities; the corruption of community has its source in the corruption of character. This realization has become the typical moral crisis of our time. Once our personal connection to what is wrong becomes clear, then we have to choose: we can go on as before, recognizing our dishonesty and living with it the best we can, or we can begin the effort to change the way we think and live.

The disease of the modern character is specialization. Looked at from the standpoint of the social *system*, the aim of specialization may seem desirable enough. The aim is to see that the responsibilities of government, law, medicine, engineering, agriculture, education, etc., are given into the hands of the most skilled, best prepared people. The difficulties do not appear until we look at specialization from the opposite standpoint—that of individual persons. We then begin to see the grotesquery—indeed, the impossibility—of an idea of community wholeness that divorces itself from any idea of personal wholeness.

The first, and best known, hazard of the specialist system is that it produces specialists—people who are elaborately and expensively trained *to do one thing*. We get into absurdity very quickly here. There are, for instance, educators who have nothing to teach, communicators who have nothing to say, medical doctors skilled at expensive cures for diseases that they have no skill, and no interest, in preventing. More common, and more damaging, are the inventors, manufacturers, and salesmen of devices who have no concern for the possible effects of those devices. Specialization is thus seen to be a way of institutionalizing, justifying, and paying highly for a calamitous disintegration and scattering-out of the various functions of character: workmanship, care, conscience, responsibility.

Even worse, a system of specialization requires the abdication to specialists of various competences and responsibilities that were once personal and universal. Thus, the average—one is tempted to say, the ideal—American citizen now consigns the problem of food production to agriculturists and "agribusinessmen," the problems of health to doctors and sanitation experts, the problems of education to school teachers and educators, the problems of conservation to conservationists, and so on. This supposedly fortunate citizen is therefore left with only two concerns: making money and entertaining himself. He earns money, typically, as a specialist, working an eight-hour day at a job for the quality or consequences of which somebody else—or, perhaps more typically, nobody else—will be responsible. And not surprisingly, since he can do so little else for himself, he is even unable to entertain himself, for there exists an enormous industry of exorbitantly expensive specialists whose purpose is to entertain him.

The beneficiary of this regime of specialists ought to be the happiest of mortals—or so we are expected to believe. *All* of his vital concerns are in the hands of certified experts. He is a certified expert himself and as such he earns more money in a year than all his great-grandparents put together. Between stints at his job he has nothing to do but mow his lawn with a sit-down lawn mower, or watch other certified experts on television. At suppertime he may eat a tray of ready-prepared food, which he and his wife (also a certified expert) procure at the cost only of money, transportation, and the pushing of a button. For a few minutes between supper and sleep he may catch a glimpse of his children, who since breakfast have been in the care of education experts, basketball or marching-band experts, or perhaps legal experts.

The fact is, however, that this is probably the most unhappy average citizen in the history of the world. He has not the power to provide himself with anything but money, and his money is inflating like a balloon and drifting away, subject to historical circumstances and the power of other people. From morning to night he does not touch anything that he has produced himself, in which he can take pride. For all his leisure and recreation, he feels bad, he looks bad, he is overweight, his health is poor. His air, water, and food are all known to contain poisons. There is a fair chance that he will die of suffocation. He suspects that his love life is not as fulfilling as other people's. He wishes that he had been born sooner, or later. He does not know why his children are the way they are. He does not understand what they say. He does not care much and does not know why he does not care. He does not know what his wife wants or what he wants. Certain advertisements and pictures in magazines make him suspect that he is basically unattractive. He feels that all his possessions are under threat of pillage. He does not know what he would do if he lost his job, if the economy failed, if the utility companies failed, if the police went on strike, if the truckers went on strike, if his wife left him, if his children ran away, if he should be found to be incurably ill. And for these anxieties, of course, he consults certified experts, who in turn consult certified experts about *their* anxieties.

It is rarely considered that this average citizen is anxious because he *ought* to be—because he still has some gumption that he has not yet given up in deference to the experts. He ought to be anxious, because he is helpless. That he is dependent upon so many specialists, the beneficiary of so much expert help, can only mean that he is a captive, a potential victim. If he lives by the competence of so many other people, then he lives also by their indulgence; his own will and his own reasons to live are made subordinate to the mere tolerance of everybody else. He has *one* chance to live what he conceives to be his life: his own small specialty within a delicate, tense, everywhere-strained system of specialties.

From a public point of view, the specialist system is a failure because, though everything is done by an expert, very little is done well. Our typical industrial or professional product is both ingenious and shoddy. The specialist system fails from a personal point of view because a person who can do only one thing can do virtually nothing for himself. In living in the world by his own will and skill, the stupidest peasant or tribesman is more competent than the most intelligent worker or technician or intellectual in a society of specialists.

What happens under the rule of specialization is that, though society becomes more and more intricate, it has less and less structure. It becomes more and more organized, but less and less orderly. The community disintegrates because it loses the necessary understandings, forms, and enactments of the relations among materials and processes, principles and actions, ideals and realities, past and present, present and future, men and women, body and spirit, city and country, civilization and wilderness, growth and decay, life and death—just as the individual character loses the sense of a responsible involvement in these relations. No longer does human life rise from the earth like a pyramid, broadly and considerately founded upon its sources. Now it scatters itself out in a reckless horizontal sprawl, like a disorderly city whose suburbs and pavements destroy the fields.

The concept of country, homeland, dwelling place becomes simplified as "the environment"—that is, what surrounds us. Once we see our place, our part of the world, as *surrounding* us, we have already made a profound division between it and ourselves. We have given up the understanding—dropped it out of our language and so out of our thought—that we and our country create one another, depend on one another, are literally part of one another; that our land passes in and out of our bodies just as our bodies pass in and out of our land; that as we and our land are part of one another, so all who are living as neighbors here, human and plant and animal, are part of one another, and so cannot possibly flourish alone; that, therefore, our culture must be our response to our place, our culture and our place are images of each other and inseparable from each other, and so neither can be better than the other.

Because by definition they lack any such sense of mutuality or wholeness, our specializations subsist on conflict with one another. The rule is never to cooperate, but rather to follow one's own interest as far as possible. Checks and balances are all applied externally, by opposition, never by self-restraint. Labor, management, the military, the government, etc., never forbear until their excesses arouse enough opposition to *force* them to do so. The good of the whole of Creation, the world and all its creatures together, is never a consideration because it is never thought of; our culture now simply lacks the means for thinking of it.

It is for this reason that none of our basic problems is ever solved. Indeed, it is for this reason that our basic problems are getting worse. The specialists are profiting too well from the symptoms, evidently, to be concerned about cures—just as the myth of imminent cure (by some "breakthrough" of science or technology) is so lucrative and all-justifying as to foreclose any possibility of an interest in prevention. The problems thus become the stock in trade of specialists. The so-called professions survive by endlessly "processing" and talking about problems that they have neither the will nor the competence to solve. The doctor who is interested in disease but not in health is clearly in the same category with the conservationist who invests in the destruction of what he otherwise intends to preserve. They both have the comfort of "job security," but at the cost of ultimate futility.

One of the most troubling characteristics of the specialist mentality is its use of money as a kind of proxy, its willingness to transmute the powers and functions of life into money. "Time is money" is one of its axioms and the source of many evils—among them the waste of both time and money. Akin to the idea that time is money is the concept, less spoken but as commonly assumed, that we may be adequately represented by money. The giving of money has thus become our characteristic virtue.

But to give is not to do. The money is given *in lieu* of action, thought, care, time. And it is no remedy for the fragmentation of character and consciousness that is the consequence of specialization. At the simplest, most practical level, it would be difficult for most of us to give enough in donations to good causes to compensate for, much less remedy, the damage done by the money that is taken from us and used destructively by various agencies of the government and by the corporations that hold us in captive dependence on their products. More important, even if we *could* give enough to overbalance the official and corporate misuse of our money, we would still not solve the problem: the willingness to be represented by money involves a submission to the modern divisions of character and community. The remedy safeguards the disease.

This has become, to some extent at least, an argument against institutional solutions. Such solutions necessarily fail to solve the problems to which they are addressed because, by definition, they cannot consider the real causes. The only real, practical, hope-giving way to remedy the fragmentation that is the disease of the modern spirit is a small and humble way—a way that a government or agency or organization or institution will never think of, though a *person* may think of it: one must begin in one's own life the private solutions that can only *in turn* become public solutions.

If, for instance, one is aware of the abuses and extortions to which one is subjected as a modern consumer, then one may join an organization of consumers to lobby for consumer-protection legislation. But in joining a consumer organization, one defines oneself as a consumer *merely*, and a mere consumer is by definition a dependent, at the mercy of the manufacturer and the salesman. If the organization secures the desired legislation, then the consumer becomes the dependent not only of the manufacturer and salesman, but of the agency that enforces the law, and is at its mercy as well. The law enacted may be a good one, and the enforcers all honest and effective; even so, the consumer will understand that one result of his effort has been to increase the number of people of whom he must beware.

The consumer may proceed to organization and even to legislation by considering only his "rights." And most of the recent talk about consumer protection has had to do with the consumer's rights. Very little indeed has been said about the consumer's responsibilities. It may be that whereas one's rights may be advocated and even "served" by an organization, one's responsibilities cannot. It may be that when one hands one's responsibilities to an organization, one becomes by that divestiture irresponsible. It may be that responsibility is intransigently a personal matter—that a responsibility can be fulfilled or failed, but cannot be got rid of.

If a consumer begins to think and act in consideration of his responsibilities, then he vastly increases his capacities as a person. And he begins to be effective in a different way—a way that is smaller perhaps, and certainly less dramatic, but sounder, and able sooner or later to assume the force of example.

A responsible consumer would be a critical consumer, would refuse to purchase the less good. And he would be a moderate consumer; he would know his needs and would not purchase what he did not need; he would sort among his needs and study to reduce them. These things, of course, have been often said, though in our time they have not been said very loudly and have not been much heeded. In our time the rule among consumers has been to spend money recklessly. People whose governing habit is the relinquishment of power, competence, and responsibility, and whose characteristic suffering is the anxiety of futility, make excellent spenders. They are the ideal consumers. By inducing in them little panics of boredom, powerlessness, sexual failure, mortality, paranoia, they can be made to buy (or vote for) virtually anything that is "attractively packaged." The advertising industry is founded upon this principle.

What has not been often said, because it did not need to be said until fairly recent times, is that the responsible consumer must also be in some way a producer. Out of his own resources and skills, he must be equal to some of his own needs. The household that prepares its own meals in its own kitchen with some intelligent regard for nutritional value, and thus depends on the grocer only for selected raw materials, exercises an influence on the food industry that reaches from the store all the way back to the seedsman. The household that produces some or all of its own food will have a proportionately greater influence. The household that can provide some of its own pleasures will not be helplessly dependent on the entertainment industry, will influence it by not being helplessly dependent on it, and will not support it thoughtlessly out of boredom.

The responsible consumer thus escapes the limits of his own dissatisfaction. He can choose, and exert the influence of his choosing, because he has given himself choices. He is not confined to the negativity of his complaint. He influences the market by his freedom. This is no specialized act, but an act that is substantial and complex, both practically and morally. By making himself responsibly free, a person changes both his life and his surroundings.

It is possible, then, to perceive a critical difference between responsible consumers and consumers who are merely organized. The responsible consumer slips out of the consumer category altogether. He is a responsible consumer incidentally, almost inadvertently; he is a responsible consumer because he lives a responsible life.

The same distinction is to be perceived between organized conservationists and responsible conservationists. (A responsible consumer *is*, of course, a responsible conservationist.) The conservationists who are merely organized function as specialists who have lost sight of basic connections. Conservation organizations hold stock in exploitive industries because they have no clear perception of, and therefore fail to be responsible for, the connections between what they say and what they do, what they desire and how they live.

The Sierra Club, for instance, defines itself by a slogan which it prints on the flaps of its envelopes. Its aim, according to the slogan, is ". . . to explore, enjoy, and protect the nation's scenic resources . . ." To some extent, the Club's current concerns and attitudes belie this slogan. But there is also a sense in which the slogan defines the limits of organized conservation—some that have been self-imposed, others that are implicit in the nature of organization.

The key word in the slogan is "scenic." As used here, the word is a fossil. It is left over from a time when our comforts and luxuries were accepted simply as the rewards of progress to an ingenious, forward-looking people, when no threat was perceived in urbanization and industrialization, and when conservation was therefore an activity oriented toward vacations. It was "good to get out of the city" for a few weeks or weekends a year, and there was understandable concern that there should remain pleasant places to go. Some of the more adventurous vacationers were even aware of places of unique beauty that would be defaced if they were not set aside and protected. These people were effective in their way and within their limits, and they started the era of wilderness conservation. The results will give us abundant reasons for gratitude as long as we have sense enough to preserve them. But wilderness conservation did little to prepare us either to understand or to oppose the general mayhem of the all-outdoors that the industrial revolution has finally imposed upon us.

Wilderness conservation, we can now see, is specialized conservation. Its specialization is memorialized, in the Sierra Club's slogan, in the word "scenic." A scene is a place "as seen by a viewer." It is a "view." The appreciator of a place perceived as scenic is merely its observer, by implication both different and distant or detached from it. The connoisseur of the scenic has thus placed strict limitations both upon the sort of place he is interested in and upon his relation to it.

But even if the slogan were made to read ". . . to explore, enjoy, and protect the nation's resources . . . ," the most critical concern would still be left out. For while conservationists are exploring, enjoying, and protecting the nation's resources, they are also *using* them. They are drawing their lives from the nation's resources, scenic and unscenic. If the resolve to explore, enjoy, and protect does not create a moral energy that will define and enforce responsible use, then organized conservation will prove ultimately futile. And this, again, will be a failure of character.

Although responsible use may be defined, advocated, and to some extent required by organizations, it cannot be implemented or enacted by them. It cannot be effectively enforced by them. The use of the world is finally a personal matter, and the world can be preserved in health only by the forbearance and care of a multitude of persons. That is, the possibility of the world's health will have to be defined in the characters of persons as clearly and as urgently as the possibility of personal "success" is now so defined. Organizations may promote this sort of forbearance and care, but they cannot provide it.

CHAPTER THREE

The Ecological Crisis as a Crisis of Agriculture

ONE REASON THAT an organization cannot properly enact our relationship to the world is that an organization cannot define that relationship except in general terms, and no matter how general may be a person's attitude toward the world, his impact upon it must become specific and tangible at some point. Sooner or later in his behalf—whether he approves or understands or not—a strip-miner's bulldozer tears into a mountainside, a stand of trees is clear-cut, a gully washes through a cornfield.

The conservation movement has never resolved this dilemma. It has never faced it. Until very recently—until pollution and strip-mining became critical issues—conservationists divided the country into that land which they wished to preserve and enjoy (the wilderness areas) and that which they consigned to use by *other* people. With the increase of pollution and mining, their interest has become two-branched, to include, along with the pristine, the critically abused. At present the issue of *use* is still in its beginning.

Because of this, the mentality of conservation is divided, and disaster is implicit in its division. It is divided between its intentional protection of some places and some aspects of "the environment" and its inadvertent destruction of others. It is variously either vacation-oriented or crisis-oriented. For the most part, it is not yet sensitive to the impact of daily living upon the sources of daily life. The typical present-day conservationist will fight to preserve what he enjoys; he will fight whatever directly threatens his health; he will oppose any ecological violence large or dramatic enough to attract his attention. But he has not yet worried much about the impact of his own livelihood, habits, pleasures, or appetites. He has not, in short, addressed himself to the problem of use. He does not have a definition of his relationship to the world that is sufficiently elaborate and exact.

The problem is well defined in a letter I received from David Budbill of Wolcott, Vermont:

"What I've noticed around here with the militant ecology people (don't get me wrong, I, like you, consider myself one of them) is a syndrome I call the Terrarium View of the World: nature always at a distance, under glass.

"Down-country people come up here, buy a 30-acre meadow, then when you ask them what they plan to *do* with it, they look at you like you're some kind of war criminal and say, 'Why, nothing! We want to leave it *just* the way it is!' They think they're protecting the environment, even though they've forgotten, or never knew, that nature abhors a vacuum . . . and in a couple of years their meadow is full of hardhack and berries and young gray birch and red maple. Pretty soon they can't even walk through the brush it's so thick. They treat the land like any other possession, object, they own, set it aside, watch it, passively, not wanting to, nay! thinking it abhorrent to engage in a living relationship with it. . . .

"Another thing folks like this do is buy land and immediately post it (to protect the animals, or their investment, I guess) then go back home. . . . The old guy or the young guy who has always hunted deer on that piece is mad. The excuse for posting (protection) is a thinly disguised cover for the real notion which has to do with the possessive, capitalist ideas about property. I'm not opposed to private property, like it even, but the folks I'm talking about, in their posting, violate . . . a strong local tradition of free trespass. There are disadvantages to free trespass, abuses, we've suffered them, but what's good about it is it understands something about use and sharing. The upper-class eco-folks lack this understanding . . .

". . . we always, with our neighbor, pick apples in the fall off trees on a down-country owner's land. There is a feeling we have the *right* to do that, a feeling that the sin is not trespass, the sin is letting the apples go to waste.

"What I'm trying to get at is that in the environmental movement there are some ugly, elitist, class-struggle type things operating. The best example of this around here is the controversy over trailers. The Audubon types (I'm a member of Audubon) are fighting . . . terribly hard to zone trailers out of areas like this, put them in trailer parks or eliminate them altogether. Well, a trailer is the only living space a working man around here can afford. And if he, say, inherits 3 acres from a parent and wants to put a trailer on it, the eco-folks would like to say no, which is a dandy way to ghettoize the poor. There are so many elements of class struggle lying under the attitudes of a lot of environmentalists; it's scary. . . . Their view of the natural world is so delicate and precious, terrarium-like, picture-windowish. I know nature is precious and delicate. I also know it is incredibly tough and resilient, has unbelievable power to respond to and flourish with kindly use.

". . . I don't care about the landscape if I am to be excluded from it. Why should I? In Audubon magazine almost always the beautiful pictures are without man; the ugly ones with him. Such self hatred! I keep wanting to write to them and say, 'Look! my name is David Budbill and I belong to the chain of being too, as a participant not an observer (nature is not television!) and the question isn't to use or not to use but rather *how* to use.'"

The conservationist congratulates himself, on the one hand, for his awareness of the severity of human influence on the natural world. On the other hand, in his own contact with that world, he can think of nothing but to efface himself—to leave it *just* the way it is.

This is an important issue, and I want to be careful not to oversimplify it. What has to be acknowledged at the outset is that wilderness conservation is important and that it has its place in any conservation program, just as the wilderness has its place in human memory and culture. It seems likely to me that the concern for wilderness must stand at the apex of the conservation effort, just as it probably must stand at the apex of consciousness in any decent culture. There are several reasons for this:

1. Our biological roots as well as our cultural roots are in nature. We began in a world that was pristine, undiminished by anything we had done, and at various times in our history the unspoiled wilderness has again imposed itself, its charming and forbidding *invitation*, upon our consciousness. It is important that we should preserve this memory. We need places in reach of every community where children can imagine the prehistoric and the beginning of history: the unknown, the trackless, the first comers.

2. If we are to be properly humble in our use of the world, we need places that we do not use at all. We need the experience of leaving something alone. We need places that we forbear to change, or influence by our presence, or impose on even by our understanding; places that we accept as influences upon us, not the other way around, that we enter with the sense, the pleasure, of having nothing to do there; places that we must enter in a kind of cultural nakedness, without comforts or tools, to submit rather than to conquer. We need what other ages would have called sacred groves. We need groves, anyhow, that we would treat as if they were sacred—in order, perhaps, to perceive their sanctity.

3. We need wilderness as a standard of civilization and as a cultural model. Only by preserving areas where nature's processes are undisturbed can we preserve an accurate sense of the impact of civilization upon its natural sources. Only if we know how the land *was* can we tell how it *is*. Records, figures, statistics will not suffice; to know, in the true sense, is to see. We must see the difference—in rates of erosion, for instance, or in soil structure or fertility—in order to keep it as small as possible. As a cultural model, the wilderness is probably indispensable. Sir Albert Howard suggests that it is when he says that farmers should pattern the maintenance of their fields after the forest floor, for the forces of growth and the forces of decay are in balance there.

But we cannot hope—for reasons practical and humane, we cannot even wish—to preserve more than a small portion of the land in wilderness. Most of it we will have to use. The conservation mentality swings from self-righteous outrage to self-deprecation because it has neglected this issue. Its self-contradictions can only be reconciled—and the conservation impulse made to function as ubiquitously and variously as it needs to—by understanding, imagining, and living out the possibility of "kindly use." Only that can dissolve the boundaries that divide people from the land and its care, which together are the source of human life. There are many kinds of land use, but the one that is most widespread and in need of consideration is that of agriculture.

For us, the possibility of kindly use is weighted with problems. In the first place, this is not ultimately an organizational or institutional solution. Institutional solutions tend to narrow and simplify as they approach action. A large number of people can act together only by defining the point or the line on which their various interests converge. Organizations tend to move toward single objectives—a ruling, a vote, a law—and they find it relatively simple to cohere under acronyms and slogans.

But kindly use is a concept that of necessity broadens, becoming more complex and diverse, as it approaches action. The land is too various in its kinds, climates, conditions, declivities, aspects, and histories to conform to any generalized understanding or to prosper under generalized treatment. The use of land cannot be both general and kindly—just as the forms of good manners, generally applied (applied, that is, without consideration of differences), are experienced as indifference, bad manners. To treat every field, or every part of every field, with the same consideration is not farming but industry. Kindly use depends upon intimate knowledge, the most sensitive responsiveness and responsibility. As knowledge (hence, use) is generalized, essential values are destroyed. As the householder evolves into a consumer, the farm evolves into a factory—with results that are potentially calamitous for both.

The understanding of kindly use in agriculture must encompass both farm and household, for the mutuality of influence between them is profound. Once, of course, the idea of a farm included the idea of a household: an integral and major part of a farm's economy was the economy of its own household; the family that owned and worked the farm lived from it. But the farm also helped to feed other households in towns and cities. These households were dependent on the farms, but not passively so, for their dependence was limited in two ways. For one thing, the town or city household was itself often a producer of food: at one time town and city lots routinely included garden space and often included pens and buildings to accommodate milk cows, fattening hogs, and flocks of poultry. For another thing, the urban household carefully selected and prepared the food that it bought; the neighborhood shops were suppliers of kitchen raw materials to local households, of whose needs and tastes the shopkeepers had personal knowledge. The shopkeepers were under the direct influence and discipline of their customers' wants, which they had to supply honestly if they hoped to prosper. The household was therefore not merely a unit in the economy of food production; its members practiced essential productive skills. The consumers of food were also producers or processors of food, or both.

This collaboration of household and farm was never, in America, sufficiently thrifty or sufficiently careful of soil fertility. It is tempting to suppose that, given certain critical historical and cultural differences, they might have developed sufficient thrift and care. As it happened, however, the development went in the opposite direction. The collaborators purified their roles—the household became simply a house or residence, purely consumptive in its function; the farm ceased to be a place to live and a way of life and became a unit of production—and their once collaborative relationship became competitive. Between them the merchant, who had been only a supplier of raw materials, began to usurp the previous functions of both household and farm, becoming increasingly both a processor and producer. And so an enterprise that once had some susceptibility to qualitative standards—standards of personal taste and preference at one end and of good husbandry at the other—has come more and more under the influence of standards that are merely economic or quantitative. The consumer wants food to be as cheap as possible. The producer wants it to be as expensive as possible. Both want it to involve as little labor as possible. And so the standards of cheapness and convenience, which are irresistibly simplifying and therefore inevitably exploitive, have been substituted for the standard of health (of both people and land), which would enforce consideration of essential complexities.

Social fashion, delusion, and propaganda have combined to persuade the public that our agriculture is for the best of reasons the envy of the Modern World. American citizens are now ready to believe without question that it is entirely good, a grand accomplishment, that each American farmer now "feeds himself and 56 others." They are willing to hear that "96 percent of America's manpower is freed from food production" —without asking what it may have been "freed" *for*, or how many as a consequence have been "freed" from employment of any kind. The "climate of opinion" is now such that a recent assistant secretary of agriculture could condemn the principle of crop rotation without even an acknowledgment of the probable costs in soil depletion and erosion, and former Secretary of Agriculture Butz could say with approval that in 1974 "only 4 percent of all U.S. farms . . . produced almost 50 percent of all farm goods," without acknowledging the human—and, indeed, the agricultural—penalties.

What these men were praising—what such men have been praising for so long that the praise can be uttered without thought—is a disaster that is both agricultural and cultural: the generalization of the relationship between people and land. That one American farmer can now feed himself and fifty-six other people may be, within the narrow view of the specialist, a triumph of technology; by no stretch of reason can it be considered a triumph of agriculture or of culture. It has been made possible by the substitution of energy for knowledge, of methodology for care, of technology for morality. This "accomplishment" is not primarily the work of farmers—who have been, by and large, its victims—but of a collaboration of corporations, university specialists, and government agencies. It is therefore an agricultural development not motivated by agricultural aims or disciplines, but by the ambitions of merchants, industrialists, bureaucrats, and academic careerists. We should not be surprised to find that its effect on both the farmland and the farm people has been ruinous. It has divided all land into two kinds—that which permits the use of large equipment and that which does not. And it has divided all farmers into two kinds—those who have sufficient "business sense" and managerial ability to handle the large acreages necessary to finance large machines and those who do not.

Those lands that are too steep or stony or small-featured to be farmed with big equipment are increasingly not farmed at all, but are abandoned to weeds and bushes, often with the gullies of previous bad use unrepaired. That these lands can often be made highly productive with kindly use is simply of no interest; we now have neither the small technology nor the small economics nor the available work force necessary to make use of them. What might be the importance of these "marginal" lands, and of an agricultural technology and economy appropriate to them, in light of population growth is a question that the agriculture experts apparently would be embarrassed to consider, so entranced are they by the glamor of bigness.

As for the farm families who cannot "get bigger" and therefore have to "get out," they are apparently written off as a reasonable, quite ordinary, and altogether bearable expense. Former Secretary Butz could praise the business acumen of the new big-time American farmer ("In all likelihood he knows as much about financing and business accountability as his banker"), evidently without wondering what may be the *agricultural* import or effect of such knowledge, or if somewhere there might not be an excellent farmer who is *not* more acute, in a business way, than his banker. But this is the catch in our almost religious dependence on experts: Mr. Butz is a farm expert, and a farm expert is by definition not a farmer; he has changed sides. I have at hand fifteen speeches by Mr. Butz and his assistant secretaries, all of which praise the productivity—that is, the business success—of the American farmer, and none of which mentions any problem of land maintenance or any problem of the small farmer.

A sampling of quotations from one of these speeches—one made by former Assistant Secretary Richard E. Bell—will give the gist and the manner of official agricultural thinking:

". . . true agripower . . . generates agridollars through agricultural exports."

"True agripower is the capacity of less than 5 percent of America's population to feed itself and the remaining 95 percent with enough food left over to meet market demands of other nations and still provide food assistance for poor people throughout the world."

"Agripower should not be a political tool. Feeding people . . . is too serious a matter to be left to political manipulation."

"Once again growth in U.S. farm productivity . . . is on the rise. . . . We no longer have the acreage limitations which for so many years served to restrict grain and cotton production. . . ."

". . . the real measure [of agricultural strength] is productivity, combined with processing and marketing efficiency."

"Years ago, farm operations were highly diversified, but today, farmers are concentrating on fewer and much larger crop or livestock enterprises. Now, many one- or two-enterprise farms exist where there were formerly three to five enterprises.

"And with the spread of sophisticated machinery, farm sizes have expanded as their numbers have declined—stretching from an average 195 acres in the 1940s to about 390 in the 1970s.

"Specialization and growth are aided by the ready availability of purchased inputs and custom services."

"With additional income earned from exports, U.S. farmers are able to purchase more household appliances, farm equipment, building supplies, and other capital and consumer goods."

"Agridollars have gone a long way toward offsetting our petro-dollar drain."

"Less than 5 percent . . . of all grain moving between countries goes for food assistance."

"It is evident that U.S. agripower is a major force in the world's exchange of goods and services. Agripower is, unquestionably, an even greater force than petropower in man's survival in the future. Man can and has survived without petroleum, but he cannot live without food."

And that was the official line on agriculture during the Butz years. There is nothing in it that was not representative: the self-congratulation, the confusions of purpose, the complacency, the jargon, the sprains and ruptures of sense, the ignorance or ignoring of consequence, the social and economic prejudices ritualized in progressivist clichés. And nowhere that I have seen was the official line more complicated than this, more aware of costs or inequities or conflicts or problems.

We would do well to examine these statements in more detail, for they are not just the political policies of ex-officials. They represent very well the prevalent assumptions of agricultural bureaucrats, academicians, and businessmen.

"Agripower," it will be noted, is not measured by the fertility or health of the soil, or the health, wisdom, thrift, or stewardship of the farming community. It is measured by its ability to produce a marketable surplus, which "generates agridollars." It is to be measured by "productivity, combined with processing and marketing efficiency." The income from this increased production, we are told, is spent by farmers not for soil maintenance or improvement, water conservation, or erosion control, but for "purchased inputs": "household appliances, farm equipment, building supplies, and other capital and consumer goods." I do not mean that we should necessarily begrudge the farmer these purchases; I am only noticing that, to Mr. Bell, the farmer does not prosper to become a better farmer, but to become a bigger spender. The assistant secretary was applying to farming a standard of judgment that is economic, not agricultural. Farming is defined here purely to suit the purposes of a businessman.

Mr. Bell makes the benign assertion that this "agripower" feeds people, including the poor of the world, and is therefore too important to be put to political *use*. But when this subject is reverted to at the end of the speech, we find that "U.S. agripower" is a major force in world trade, a force intended to offset the "petropower" of other countries. And we have the assurance that, after all, "Less than 5 percent . . . of all grain moving between countries goes for food assistance" to the poor. (And, of course, all of this must be weighed against former Secretary Butz's avowal that "Food is a weapon.")*

Next we hear the routine self-congratulation of the department on the increase of productivity following the removal of production controls (the only agricultural problem acknowledged in any of these speeches). Our agriculture policy is now based on the principle of "full production"—an obscure notion that former Secretary Butz and his colleagues paraded

*A friend has pointed out the "incredible cheek" of calling food "agripower" and then warning against its use as "a political tool."

before their audiences like the True Cross. As businessmen and politicians, perhaps they did not know how strenuously agricultural production must be qualified by the restraints and disciplines of soil maintenance and conservation. Perhaps they did not know what "full production" means in present practice—present technology, methods, and economic urgencies having replaced those restraints and disciplines. In practice, however, "full production" means that on farm after farm fence rows, windbreaks, and waterways have been plowed, steep slopes put under cultivation, and soil stewardship generally neglected. It means that production is being paid for, not just with labor, money, and fuel, *but with land.*

But the most remarkable and significant part of Mr. Bell's speech is the one in which he applauds the most degenerative, dangerous, costly, and socially disruptive "achievements" of American agriculture: (1) "economy of size," which means the gathering of farmland into the ownership of fewer and fewer people—not farmers necessarily but an "agribusiness elite"—and the consequent dispossession of millions of small farmers and farm families; and (2) specialization, which means the abandonment of the ancient, proven principle of agricultural diversity—agricultural stability through diversity—with its attendant principles of mixed husbandry of plants and animals and crop rotation. It is now, for the first time, deemed provident and wise to put all the eggs in one basket.

The giveaway is in the curiously pleased-sounding statement that "specialization and growth are aided by the ready availability of purchased inputs. . . ." This betrays, for one thing, how far we have abandoned the old ideal that the farm should aim at economic independence; that is, it should be far more productive than consumptive, more a source than a consumer of material goods. This old ideal sought to preserve the farmer on the farm; that was of necessity its first objective. But it also sought to keep the source of food independent of any but agricultural means—an aim that ought to recommend itself, it would seem, to a fairly ordinary intelligence. Its desirability becomes altogether clear when one considers that a farm—given the *appropriate* technology, the recovery and return of organic wastes to the soil, an economy that is not exploitive, and a sufficient human work force—can achieve a high measure of economic independence.

None of this was clear to the intellectuals of the Department of Agriculture, and no doubt they were thereby saved a good deal of worry. For one of the "purchased inputs," on the "ready availability" of which our agriculture now absolutely depends, is petroleum—for which we are not only dependent on non-agricultural sources, but on other nations. That we should have an agriculture based as much on petroleum as on the soil—that we need petroleum exactly as much as we need food and must have it *before* we can eat—may seem absurd. It *is* absurd. It is nevertheless true. And it exposes the hollowness of Mr. Bell's contention that "Agripower is, unquestionably, an even greater force than petropower in man's survival in the future. Man can and has survived without petroleum, but he cannot live without food." The two powers are now clearly the same. That the two are not only interdependent, but competitive as well, suggests more forcibly than Mr. Butz's words that "Food is a weapon."

And so, far from the concerns of "kindly use" that alone can assure a permanent agriculture and a permanent food supply, the Department of Agriculture is lost in the paper clouds of "agribusiness," propagating statistical proofs of visibly ruinous agricultural practices. One can imagine the average American nodding over these "expert" reports and projections. Whether he is nodding because he agrees or because he is asleep does not matter; there is no difference.

Thus the estrangement of consumer and producer, their evolution from collaborators in food production to competitors in the food market, involves a process of oversimplification on both sides. The consumer withdraws from the problems of food production, hence becomes ignorant of them and often scornful of them; the producer no longer sees himself as intermediary between people and land—the people's representative on the land—and becomes interested only in production. The consumer eats worse, and the producer farms worse. And, in their estrangement, waste is institutionalized. Without regret, with less and less interest in the disciplines of thrift and conservation, with, in fact, the assumption that this is the way of the world, our present agriculture wastes topsoil, water, fossil fuel, and human energy—to name only the most noticeable things. Consumers participate "innocently" or ignorantly in all these farm wastes and add to them wastes that are urban or consumptive in nature: mainly all the materials and energy that go into unnecessary processing and packaging, as well as tons of organic matter (highly valuable—and certainly, in the long run, necessary—as fertilizer) that they flush down their drains or throw out as garbage.

What this means for conservationists is that, as consumers, they may be using—and abusing—more land by proxy than they are conserving by the intervention of their organizations. We now have more people using the land (that is, living from it) and fewer thinking about it than ever before. We are eating thoughtlessly, as no other entire society ever has been able to do. We are eating—drawing our lives out of our land—thoughtlessly. If we study carefully the implications of that, we will see that the agricultural crisis is not merely a matter of supply and demand to be remedied by some change of government policy or some technological "breakthrough." It is a crisis of culture.

CHAPTER FOUR

The Agricultural Crisis as a Crisis of Culture

IN MY BOYHOOD, Henry County, Kentucky, was not just a rural county, as it still is—it was a *farming* county. The farms were generally small. They were farmed by families who lived not only upon them, but within and *from* them. These families grew gardens. They produced their own meat, milk, and eggs. The farms were highly diversified. The main money crop was tobacco. But the farmers also grew corn, wheat, barley, oats, hay, and sorghum. Cattle, hogs, and sheep were all characteristically raised on the same farms. There were small dairies, the milking more often than not done by hand. Those were the farm products that might have been considered major. But there were also minor products, and one of the most important characteristics of that old economy was the existence of markets for minor products. In those days a farm family could easily market its surplus cream, eggs, old hens, and frying chickens. The power for field work was still furnished mainly by horses and mules. There was still a prevalent pride in workmanship, and thrift was still a forceful social ideal. The pride of most people was still in their homes, and their homes looked like it.

This was by no means a perfect society. Its people had often been violent and wasteful in their use of the land and of each other. Its present ills had already taken root in it. But I have spoken of its agricultural economy of a generation ago to suggest that there were also good qualities indigenous to it that might have been cultivated and built upon.

That they were not cultivated and built upon—that they were repudiated as the stuff of a hopelessly outmoded, unscientific way of life—is a tragic error on the part of the people themselves; and it is a work of monstrous ignorance and irresponsibility on the part of the experts and politicians, who have prescribed, encouraged, and applauded the disintegration of such farming communities all over the country.

In the decades since World War II the farms of Henry County have become increasingly mechanized. Though they are still comparatively diversified, they are less diversified than they used to be. The holdings are larger, the owners are fewer. The land is falling more and more into the hands of speculators and professional people from the cities, who—in spite of all the scientific agricultural miracles—still have much more money than farmers. Because of big technology and big economics, there is more abandoned land in the county than ever before. Many of the better farms are visibly deteriorating, for want of manpower and time and money to maintain them properly. The number of part-time farmers and ex-farmers increases every year. Our harvests depend more and more on the labor of old people and young children. The farm people live less and less from their own produce, more and more from what they buy. The best of them are more worried about money and more overworked than ever before. Among the people as a whole, the focus of interest has largely shifted from the household to the automobile; the ideals of workmanship and thrift have been replaced by the goals of leisure, comfort, and entertainment. For Henry County plays its full part in what Maurice Telleen calls "the world's first broad-based hedonism." The young people expect to leave as soon as they finish high school, and so they are without permanent interest; they are generally not interested in anything that cannot be reached by automobile on a good road. Few of the farmers' children will be able to afford to stay on the farm—perhaps even fewer will wish to do so, for it will cost too much, require too much work and worry, and it is hardly a fashionable ambition.

And nowhere now is there a market for minor produce: a bucket of cream, a hen, a few dozen eggs. One cannot sell milk from a few cows anymore; the law-required equipment is too expensive. Those markets were done away with in the name of sanitation—but, of course, to the enrichment of the large producers. We have always had to have "a good reason" for doing away with small operators, and in modern times the good reason has often been sanitation, for which there is apparently no small or cheap technology. Future historians will no doubt remark upon the inevitable association, with us, between sanitation and filthy lucre. And it is one of the miracles of science and hygiene that the germs that used to be in our food have been replaced by poisons.

In all this, few people whose testimony would have mattered have seen the connection between the "modernization" of agricultural techniques and the disintegration of the culture and the communities of farming—and the consequent disintegration of the structures of urban life. What we have called agricultural progress has, in fact, involved the forcible displacement of millions of people.

I remember, during the fifties, the outrage with which our political leaders spoke of the forced removal of the populations of villages in communist countries. I also remember that at the same time, in Washington, the word on farming was "Get big or get out"—a policy which is still in effect and which has taken an enormous toll. The only difference is that of method: the force used by the communists was military; with us, it has been economic—a "free market" in which the freest were the richest. The attitudes are equally cruel, and I believe that the results will prove equally damaging, not just to the concerns and values of the human spirit, but to the practicalities of survival.

And so those who could not get big have got out—not just in my community, but in farm communities all over the country. But as a social or economic goal, bigness is totalitarian; it establishes an inevitable tendency toward the *one* that will be the biggest of all. Many who got big to stay in are now being driven out by those who got bigger. The aim of bigness implies not one aim that is not socially and culturally destructive.

And this community-killing agriculture, with its monomania of bigness, is not primarily the work of farmers, though it has burgeoned on their weaknesses. It is the work of the institutions of agriculture: the university experts, the bureaucrats, and the "agribusinessmen," who have promoted so-called efficiency at the expense of community (and of real efficiency), and quantity at the expense of quality.

In 1973, 1,000 Kentucky dairies went out of business. They were the victims of policies by which we imported dairy products to compete with our own and exported so much grain as to cause a drastic rise in the price of feed. And, typically, an agriculture expert at the University of Kentucky, Dr. John Nicolai, was optimistic about this failure of 1,000 dairymen, whose cause he is supposedly being paid—partly with *their* tax money—to serve. They were inefficient producers, he said, and they needed to be eliminated.

He did not say—indeed, there was no indication that he had ever considered—what might be the limits of his criterion or his logic. Did he propose to applaud this process year after year until "biggest" and "most efficient" become synonymous with "only"? Did these dairymen have any value not subsumed under the heading of "efficiency"? And who benefited by their failure? Assuming that the benefit reached beyond the more "efficient" (that is, the bigger) producers to lower the cost of milk to consumers, do we then have a formula by which to determine how many consumer dollars are equal to the livelihood of one dairyman? Or is *any* degree of "efficiency" worth any cost? I do not think that this expert knows the answers. I do not think that he is under any pressure—scholarly, professional, moral, or otherwise—to ask the questions. This sort of regardlessness is invariably justified by pointing to the enormous productivity of American agriculture. But any abundance, in any amount, is illusory if it does not safeguard its producers, and in American agriculture it is now virtually the accepted rule that abundance will destroy its producers.

And along with the rest of society, the established agriculture has shifted its emphasis, and its interest, from quality to quantity, having failed to see that in the long run the two ideas are inseparable. To pursue quantity alone is to destroy those disciplines in the producer that are the only assurance of quantity. What is the effect on quantity of persuading a producer to produce an inferior product? What, in other words, is the relation of pride or craftsmanship to abundance? That is another question the "agribusinessmen" and their academic collaborators do not ask. They do not ask it because they are afraid of the answer: The preserver of abundance is excellence.

My point is that food is a cultural product; it cannot be produced by technology alone. Those agriculturists who think of the problems of food production solely in terms of technological innovation are oversimplifying both the practicalities of production and the network of meanings and values necessary to define, nurture, and preserve the practical motivations. That the discipline of agriculture should have been so divorced from other disciplines has its immediate cause in the compartmental structure of the universities, in which complementary, mutually sustaining and enriching disciplines are divided, according to "professions," into fragmented, one-eyed specialties. It is suggested, both by the organization of the universities and by the kind of thinking they foster, that farming shall be the responsibility only of the college of agriculture, that law shall be in the sole charge of the professors of law, that morality shall be taken care of by the philosophy department, reading by the English department, and so on. The same, of course, is true of government, which has become another way of institutionalizing the same fragmentation.

However, if we conceive of a culture as one body, which it is, we see that all of its disciplines are everybody's business, and that the proper university product is therefore not the whittled-down, isolated mentality of expertise, but a mind competent in all its concerns. To such a mind it would be clear that there are agricultural disciplines that have nothing to do with crop production, just as there are agricultural obligations that belong to people who are not farmers.

A culture is not a collection of relics or ornaments, but a practical necessity, and its corruption invokes calamity. A healthy culture is a communal order of memory, insight, value, work, conviviality, reverence, aspiration. It reveals the human necessities and the human limits. It clarifies our inescapable bonds to the earth and to each other. It assures that the necessary restraints are observed, that the necessary work is done, and that it is done well. A healthy *farm* culture can be based only upon familiarity and can grow only among a people soundly established upon the land; it nourishes and safe-guards a human intelligence of the earth that no amount of technology can satisfactorily replace. The growth of such a culture was once a strong possibility in the farm communities of this country. We now have only the sad remnants of those communities. If we allow another generation to pass without doing what is necessary to enhance and embolden the possibility now perishing with them, we will lose it altogether. And then we will not only invoke calamity—we will deserve it.

Several years ago I argued with a friend of mine that we might make money by marketing some inferior lambs. My friend thought for a minute and then he said, "I'm in the business of producing *good* lambs, and I'm not going to sell any other kind." He also said that he kept the weeds out of his crops for the same reason that he washed his face. The human race has survived by that attitude. It can survive *only* by that attitude—though the farmers who have it have not been much acknowledged or much rewarded.

Such an attitude does not come from technique or technology. It does not come from education; in more than two decades in universities I have rarely seen it. It does not come even from principle. It comes from a passion that is culturally prepared—a passion for excellence and order that is handed down to young people by older people whom they respect and love. When we destroy the possibility of that succession, we will have gone far toward destroying ourselves.

It is by the measure of culture, rather than economics or technology, that we can begin to reckon the nature and the cost of the country-to-city migration that has left our farmland in the hands of only five percent of the people. From a cultural point of view, the movement from the farm to the city involves a radical simplification of mind and of character.

A competent farmer is his own boss. He has learned the disciplines necessary to go ahead on his own, as required by economic obligation, loyalty to his place, pride in his work. His workdays require the use of long experience and practiced judgment, for the failures of which he knows that he will suffer. His days do not begin and end by rule, but in response to necessity, interest, and obligation. They are not measured by the clock, but by the task and his endurance; they last as long as necessary or as long as he can work. He has mastered intricate formal patterns in ordering his work within the overlapping cycles—human and natural, controllable and uncontrollable—of the life of a farm.

Such a man, upon moving to the city and taking a job in industry, becomes a specialized subordinate, dependent upon the authority and judgment of other people. His disciplines are no longer implicit in his own experience, assumptions, and values, but are imposed on him from the outside. For a complex responsibility he has substituted a simple dutifulness. The strict competences of independence, the formal mastery, the complexities of attitude and know-how necessary to life on the farm, which have been in the making in the race of farmers since before history, all are replaced by the knowledge of some fragmentary task that may be learned by rote in a little while.

Such a simplification of mind is easy. Given the pressure of economics and social fashion that has been behind it and the decline of values that has accompanied it, it may be said to have been gravity-powered. The reverse movement—a reverse movement *is* necessary, and some have undertaken it—is uphill, and it is difficult. It cannot be fully accomplished in a generation. It will probably require several generations—enough to establish complex local cultures with strong communal memories and traditions of care.

There seems to be a rule that we can simplify our minds and our culture only at the cost of an oppressive social and mechanical complexity. We can simplify our society—that is, make ourselves free—only by undertaking tasks of great mental and cultural complexity. Farming, the *best* farming, is a task that calls for this sort of complexity, both in the character of the farmer and in his culture. To simplify either one is to destroy it.

That is because the best farming requires a farmer—a husbandman, a nurturer—not a technician or businessman. A technician or a businessman, given the necessary abilities and ambitions, can be made in a little while, by training. A good farmer, on the other hand, is a cultural product; he is made by a sort of training, certainly, in what his time imposes or demands, but he is also made by generations of experience. This essential experience can only be accumulated, tested, preserved, handed down in settled households, friendships, and communities that are deliberately and carefully native to their own ground, in which the past has prepared the present and the present safeguards the future.

The concentration of the farmland into larger and larger holdings and fewer and fewer hands—with the consequent increase of overhead, debt, and dependence on machines—is thus a matter of complex significance, and its agricultural significance cannot be disentangled from its cultural significance. It *forces* a profound revolution in the farmer's mind: once his investment in land and machines is large enough, he must forsake the values of husbandry and assume those of finance and technology. Thenceforth his thinking is not determined by agricultural responsibility, but by financial accountability and the capacities of his machines. Where his money comes from becomes less important to him than where it is going. He is caught up in the drift of energy and interest away from the land. Production begins to override maintenance. The economy of money has infiltrated and subverted the economies of nature, energy, and the human spirit. The man himself has become a consumptive machine.

For some time now ecologists have been documenting the principle that "you can't do one thing"—which means that in a natural system whatever affects one thing ultimately affects everything. Everything in the Creation is related to everything else and dependent on everything else. The Creation is one. It is a uni-verse, a whole, the parts of which are all "turned into one."

A good agricultural system, which is to say a durable one, is similarly unified. In the 1940s, the great British agricultural scientist, Sir Albert Howard, published *An Agricultural Testament* and *The Soil and Health*, in which he argued against the influence in agriculture of "the laboratory hermit" who had substituted "that dreary precept [official organization] for the soul-shaking principle of that essential freedom needed by the seeker after truth." And Howard goes on to speak of the disruptiveness of official organization: "The natural universe, which is one, has been halved, quartered, fractioned. . . . Real organization always involves real responsibility: the official organization of research tries to retain power and avoid responsibility by sheltering behind groups of experts." Howard himself began as a laboratory hermit: "I could not take my own advice before offering it to other people." But he saw the significance of the "wide chasm between science in the laboratory and practice in the field." He devoted his life to bridging that chasm. His is the story of a fragmentary intelligence seeking both its own wholeness and that of the world. The aim that he finally realized in his books was to prepare the way "for treating the whole problem of health in soil, plant, animal, and man as one great subject." He unspecialized his vision, in other words, so as to see the necessary unity of the concerns of agriculture, as well as the convergence of these concerns with concerns of other kinds: biological, historical, medical, moral, and so on. He sought to establish agriculture upon the same unifying cycle that preserves health, fertility, and renewal in nature: the Wheel of Life (as he called it, borrowing the term from religion), by which "Death supersedes life and life rises again from what is dead and decayed."

It remains only to say what has often been said before—that the best human cultures also have this unity. Their concerns and enterprises are not fragmented, scattered out, at variance or in contention with one another. The people and their work and their country are members of each other and of the culture. If a culture is to hope for any considerable longevity, then the relationships within it must, in recognition of their interdependence, be predominantly cooperative rather than competitive. A people cannot live long at each other's expense or at the expense of their cultural birthright—just as an agriculture cannot live long at the expense of its soil or its work force, and just as in a natural system the competitions among species must be limited if all are to survive.

In any of these systems, cultural or agricultural or natural, when a species or group exceeds the principle of usufruct (literally, the "use of the fruit"), it puts itself in danger. Then, to use an economic metaphor, it is living off the principal rather than the interest. It has broken out of the system of nurture and has become exploitive; it is destroying what gave it life and what it depends upon to live. In all of these systems a fundamental principle must be the protection of the source: the seed, the food species, the soil, the breeding stock, the old and the wise, the keepers of memories, the records.

And just as competition must be strictly curbed within these systems, it must be strictly curbed *among* them. An agriculture cannot survive long at the expense of the natural systems that support it and that provide it with models. A culture cannot survive long at the expense either of its agricultural or of its natural sources. To live at the expense of the source of life is obviously suicidal. Though we have no choice but to live at the expense of other life, it is necessary to recognize the limits and dangers involved: past a certain point in a unified system, "other life" is our own.

The definitive relationships in the universe are thus not

competitive but interdependent. And from a human point of view they are analogical. We can build one system only within another. We can have agriculture only within nature, and culture only within agriculture. At certain critical points these systems have to conform with one another or destroy one another.

Under the discipline of unity, knowledge and morality come together. No longer can we have that paltry "objective" knowledge so prized by the academic specialists. To know anything at all becomes a moral predicament. Aware that there is no such thing as a specialized—or even an entirely limitable or controllable—effect, one becomes responsible for judgments as well as facts. Aware that as an agricultural scientist he had "one great subject," Sir Albert Howard could no longer ask, What can I do with what I know? without at the same time asking, How can I be responsible for what I know?

And it is within unity that we see the hideousness and destructiveness of the fragmentary—the kind of mind, for example, that can introduce a production machine to increase "efficiency" without troubling about its effect on workers, on the product, and on consumers; that can accept and even applaud the "obsolescence" of the small farm and not hesitate over the possible political and cultural effects; that can recommend continuous tillage of huge monocultures, with massive use of chemicals and no animal manure or humus, and worry not at all about the deterioration or loss of soil. For cultural patterns of responsible cooperation we have substituted this moral ignorance, which is the etiquette of agricultural "progress."

Dreams of the far future destiny of man were dragging up from its shallow and unquiet grave the old dream of Man as God. The very experience of the dissecting room and the pathological laboratory were breeding a conviction that the stifling of all deep-set repugnances was the first essential for progress.

C. S. LEWIS, *That Hideous Strength*

CHAPTER FIVE

Living in the Future: The "Modern" Agricultural Ideal

THE DOMESTICATION OF ABSENCE

IT is impossible to divorce the question of what we do from the question of where we are—or, rather, where we think we are. That no sane creature befouls its own nest is accepted as generally true. What we conceive to be our nest, and where we think it is, are therefore questions of the greatest importance. Do we, for instance, carry on our work in our nest or do we only reside and get our mail there? Is our nest a place of consumption only or is it also a place of production? Is it the source of necessary goods, energies, and "services," or only their destination?

I have already spoken of the highly simplified role of the modern household with respect to the production and preparation of food: it has set itself increasingly aside from production and preparation and become more and more a place for the consumption of food produced and prepared elsewhere. But this setting aside of the nest or residence from the sources of life is more general and even more serious than that would indicate. The modern home, even more than the government and universities, has institutionalized the divisions and fragmentations of modern life.

With its array of gadgets and machines, all powered by energies that are destructive of land or air or water, and connected to work, market, school, recreation, etc., by gasoline engines, the modern home is a veritable factory of waste and destruction. It is the mainstay of the economy of money. But within the economies of energy and nature, it is a catastrophe. It takes in the world's goods and converts them into garbage, sewage, and noxious fumes—for none of which we have found a use.

And the modern household's direct destructiveness of the world bears a profound relation—as cause or effect or both—to the fundamental moral disconnections for which it also stands. It divorces us from the sources of our bodily life; as a people, we no longer know the earth we come from, have no respect for it, keep no responsibilities to it. And few who are acquainted with the young can doubt that the modern home has also failed as a place of instruction and that the schools are failing under the burden of that deeper failure.

But nowhere is the destructive influence of the modern home so great as in its remoteness from work. When people do not live where they work, they do not feel the effects of what they do. The people who make wars do not fight them. The people responsible for strip-mining, clear-cutting of forests, and other ruinations do not live where their senses will be offended or their homes or livelihoods or lives immediately threatened by the consequences. The people responsible for the various depredations of "agribusiness" do not live on farms. They—like many others of less wealth and power—live in ghettos of their own kind in homes full of "conveniences" which signify that all is well. In an automated kitchen, in a gleaming, odorless bathroom, in year-round air-conditioning, in color TV, in an easy chair, the world is redeemed. If what God made can be made by humans into *this*, then what can be wrong?

The modern home is so destructive, I think, because it is a generalization, a product of factory and fashion, an everyplace or a noplace. Modern houses, like airports, are extensions of each other; they do not vary much from one place to another. A person standing in a modern room anywhere might imagine himself anywhere else—much as he could if he shut his eyes. The modern house is not a response to its place, but rather to the affluence and social status of its owner. It is the first means by which the modern mentality imposes itself upon the world. The industrial conquistador, seated in his living room in the evening in front of his TV set, many miles from his work, can easily forget where he is and what he has done. He is everywhere or nowhere. Everything around him, everything on TV, tells him of his success: *his* comfort is the redemption of the world. His home is the emblem of his status, but it is not the center of his interest or of his consciousness. The history of our time has been to a considerable extent the movement of the center of consciousness away from home.

Once, some farmers, particularly in Europe, lived in their barns—and so were both at work and at home. Work and rest, work and pleasure, were continuous with each other, often not distinct from each other at all. Once, shopkeepers lived in, above, or behind their shops. Once, many people lived by "cottage industries"—home production. Once, households were producers and processors of food, centers of their own maintenance, adornment, and repair, places of instruction and amusement. People were born in these houses, and lived and worked and died in them. Such houses were not generalizations. Similar to each other in materials and design as they might have been, they nevertheless looked and felt and smelled different from each other because they were articulations of particular responses to their places and circumstances.

THE VAGRANT SOVEREIGN

The modern specialist and/or industrialist in his modern house can probably have no very clear sense of where he is. His sense of his whereabouts is abstract: he is in a certain "line" as signified by his profession, in a certain "bracket" as signified by his income, and in a certain "crowd" as signified by his house and his amusements. Where he is matters only in proportion to the number of other people's effects he has to put up with. Geography is defined for him by his house, his office, his commuting route, and the interiors of shopping centers, restaurants, and places of amusement—which is to say that his geography is artificial; he could be anywhere, and he usually is.

This generalized sense of worldly whereabouts is a reflection of another kind of bewilderment: this modern person does not know where he is morally either. He assumes, as he has clearly been taught to assume, that as a member of the human race he is sovereign in the universe. He assumes that there is nothing that he *can* do that he should not do, nothing that he *can* use that he should not use. His "success"—which at present is indisputable—is that he has escaped any order that might imply restraints or impose limits. He has, like the heroes of fantasy, left home—left behind all domestic ties and restraints—and gone out into the world to seek his fortune.

This mentality has been long in the making, and its rise evidently parallels the exploitation of the New World. Carl Sauer wrote: "The Modern Age began with the extension of royal absolution overseas. The crowns gave patents to individuals to discover, take possession, and govern islands or mainland, inhabited or uninhabited. The crown took to itself the title to land and people, first claimed for it by formal act. Thus Columbus planted the flag as he landed, the natives being bemused spectators. Thus Cabot without having sight of a native. Thus Juan de la Cosa entered on his map the flags of three nations. *The course of colonial empire began with disregard of native rights and persons* [my emphasis]. The Portuguese loaded the first cargo of black slaves when they reached the Bay of Arguin, and they did the same with Indians in New Foundland. Columbus estimated the prospects of slave trade when he landed in the West Indies. The Colonial idea as it took shape in the fifteenth century was untroubled by any concern other than to establish priority over other European nations."

Economic exploitation and competition as we now know them were thus established at the beginning of American history. Or perhaps it would be truer to say that they were established by the beginning of American history—for they do not seem to have risen so much out of theory or vision or desire or decree as out of newly opened distance and space. The new reaches of oceanic navigation, the discovery of new lands across what shortly before had been inconceivable distances, seem to have forced the European mind out of its old moral order. Those first discoverers carried the patents of their sovereigns, but they carried them into places altogether new to them, beyond what had been imagined, much less what had been culturally ordered. And so no matter the flags and pronouncements and the other trappings of fealty—the sovereignty that crossed the surf onto the shore of the New World was a new sovereignty of the human mind. What appeared to the eyes of the discoverers was not one of the orders of Creation that required respect or deference for its own sake. What they saw was a great concentration of "natural resources"—to be used according to purposes exterior to them. That some of those resources were human beings mattered not at all.

And so at the same time that they "discovered" America, these men invented the modern condition of being away from home. On the new shores the old orders of domesticity, respect, deference, humility fell away from them; they arrived contemptuous of whatever existed before their own coming, disdainful beyond contempt of native creatures or values or orders, ravenous for their own success. They began the era of absolute human sovereignty—which is to say the era of absolute human presumption. They invented us: the flag of Ferdinand and Isabella in the hand of Columbus on the shores of the Indies becomes Old Glory in the hand of Neil Armstrong on the moon. An infinitely greedy sovereign is afoot in the universe, staking his claims.

THE MANUFACTURED PARADISE

But our experience of sovereignty suggests that it becomes dangerous when it defines itself exclusively in terms of what is inferior to it, neglecting or ignoring what is superior to it. That is to say that sovereignty is a safe concept only when its place is symmetrically defined. Thus, once, the place of humans was thought to be above the animals and below the angels—between the natural and the divine. Then, by understanding and accepting that human place in the order of things, people could see that their privileges were limited and safeguarded by certain responsibilities. They could see, moreover, that only evil could be the result of the transgression of these limits: one could not escape the human condition except sinfully, by pride or by degradation.

The growth of what is called the Modern World has been, by turns, both the cause and the effect of the destruction of that old sense of universal order. The most characteristically modern behavior, or misbehavior, was made possible by a redefinition of humanity which allowed it to claim, not the sovereignty of its place, neither godly nor beastly, in the order of things, but rather an absolute sovereignty, placing the human will in charge of itself and of the universe.

And having thus usurped the whole Chain of Being, conceiving itself, in effect, both creature and creator, humanity set itself a goal that in those circumstances was fairly predictable: it would make an Earthly Paradise. This projected Paradise was no longer that of legend: the lost garden that might be rediscovered by some explorer or navigator. This new Paradise was to be invented and built by human intelligence and industry.

And by machines. For the agent of our escape from our place in the order of Creation, and of our godlike ambition to make a Paradise, was the machine—not only as instrument, but even more powerfully as metaphor. Once, the governing human metaphor was pastoral or agricultural, and it clarified, and so preserved in human care, the natural cycles of birth, growth, death, and decay. But modern humanity's governing metaphor is that of the machine. Having placed ourselves in charge of Creation, we began to mechanize both the Creation itself and our conception of it. We began to see the whole Creation merely as raw material, to be transformed by machines into a manufactured Paradise.

And so the machine did away with mystery on the one hand and multiplicity on the other. The Modern World would respect the Creation only insofar as it could be *used* by humans. Henceforth, by definition, by principle, we would be unable to leave anything as it was. The usable would be used; the useless would be sacrificed in the use of something else. By means of the machine metaphor we have eliminated any fear or awe or reverence or humility or delight or joy that might have restrained us in our use of the world. We have indeed learned to act as if our sovereignty were unlimited and as if our intelligence were equal to the universe. Our "success" is a catastrophic demonstration of our failure. The industrial Paradise is a fantasy in the minds of the privileged and the powerful; the reality is a shambles.

THE COLONIZATION OF THE FUTURE

The generalization of vital connections and the assumption of unlimited human sovereignty go a long way toward explaining the displacement of the modern mind. But they do not explain *how* it happened. It can be said that the motive has often been greed, but even that does not satisfy, for greed has always existed. It is necessary to account for a new intensity of greed—a greed newly empowered, under no constraint to see itself as evil, allied (so it believes) with a manifest destiny and the way of the world. There must have been, not just a shift of basic assumptions, not just a motive, but also some kind of vision or dream or psychic lure.

It has been, I think, the future. What has drawn the Modern World into being is a strange, almost occult yearning for the future. The modern mind longs for the future as the medieval mind longed for Heaven. The great aim of modern life has been to improve the future—or even just to *reach* the future, assuming that the future will inevitably be "better." One of the oddest terms of praise in our language is "futuristic." "Far out," as a term of universal approbation, is perhaps a lineal descendant. Such terms are used to identify the signs and landmarks that confirm that we are indeed on the right road to the future, that we are getting there, that at any moment we may at last arrive. And this is no elitist obsession; it is commonplace. Politicians understand very well the power of the promise to build a better or more prosperous or more secure future. Parents characteristically strive and sacrifice to make a better or more secure future for their children. Workers work toward a secure future in which they will retire and enjoy themselves. Our obsession with security is a measure of the power we have granted the future to hold over us.*

The future has been envisioned, dreamed, projected, painted for us by prophets of every kind: scientists, comic-book writers, novelists, philosophers, politicians, industrialists, professors. And, of course, by ourselves; the cult of the future has turned us all into prophets. The future is the time when science will have solved all our problems, gratified all our desires; when we will all live in perfect ease in an air-conditioned, fully automated womb; when all the work will be done by machines so sophisticated that they will not only clothe, house, and feed us, but think for us, play our games, paint our pictures, write our poems. It is the Earthly Paradise, the Other Shore, where all will be well. And if we are living for the future, then history is on our side—or so we are at liberty to think, for the needed proofs are never at hand. That there has for some time been growing a cult of dread of the future testifies not only to the

*The following sentences are from a recent oil company advertisement:

"We have always been a nation more interested in the promise of the future than in the events of the past.

"Here at Atlantic Richfield we see the future as an exciting time. The best of times."

innate silliness and frivolity of this vision, but to its power. The adoration of the future may be beginning to falter, but it is still dominant, still available and useful to the exploitive mind.

There is no aspect of our life as a people that is not now under the dominance of this industrial dream of the future-as-Paradise. All our implements—automobiles, tractors, kitchen utensils, etc.—have always been conceived by the modern mind as in a kind of progress or pilgrimage toward their future forms. The automobile-of-the-future, the kitchen-of-the-future, the classroom-of-the-future have long figured more actively in our imaginations, plans, and desires than whatever versions of these things we may currently have. We long ago gave up the wish to have things that were adequate or even excellent; we have preferred instead to have things that were up-to-date. But to be up-to-date is an ambition with built-in panic: our possessions cannot be up-to-date more than momentarily unless we can stop time—or somehow get ahead of it. The only possibility of satisfaction is to be driving *now* in one's *future* automobile.

It is no doubt impossible to live without thought of the future; hope and vision can live nowhere else. But the only possible guarantee of the future is responsible behavior in the present. When supposed future needs are used to justify misbehavior in the present, as is the tendency with us, then we are both perverting the present and diminishing the future. But the most prolific source of justifications for exploitive behavior has been the future. The exploitive mind characteristically puts itself in charge of the future. The future is a time that cannot conceivably be reached except by industrial progress and economic growth. The future, so full of material blessings, is nevertheless threatened with dire shortages of food, energy, and security unless we exploit the earth even more "freely," with greater speed and less caution. The obvious paradoxes involved in this—that we are using up future necessities in order to make a more abundant future; that final loss has been made a calculated strategy of annual gain—have so far been understood to no great effect. The great convenience of the future as a context of behavior is that nobody knows anything about it. No rational person can *see* how using up the topsoil or the fossil fuels as quickly as possible can provide greater security for the future; but if enough wealth and power can conjure up the audacity to *say* that it can, then sheer fantasy is given the force of truth; the future becomes reckonable as even the past has never been. It is as if the future is a newly discovered continent which the corporations are colonizing. They have made "redskins" of our descendants, holding them subject to alien values, while their land is plundered of anything that can be shipped home and sold.

Nowhere is the cult of the future stronger than in agriculture. One reason for this is that farming has been harder to industrialize than manufacturing, and when industrialization has come, it has not brought shorter hours or greater ease or less worry. A great deal of the strain of the industrial revolution has been borne by farmers, and so it has been fairly easy to secure their allegiance to the future, when more industrialization will supposedly bring a better farm economy. The industrialization of farming as we now have it is not something that farmers would have bought all in a piece; as a group they have been too traditional or conservative for that. Instead, it has been sold to them in stages, one implement at a time. The reduction of available manpower by each new machine created the need for a better machine or a different one. In the practical circumstances of the modern farm, the popular yearning for the future is directly felt as a yearning for relief from weariness and worry.

Another reason for the dominance of the future over agriculture is that projected rates of population growth have become the all-purpose threat and justifier of the apologists of the agricultural establishment. Millions are threatened with starvation —so the argument runs—therefore we must continue to farm in larger monocultures on larger holdings with fewer farmers, larger and more expensive machines, more chemicals. The hunger of these future millions is now the foundation of policy in the Department of Agriculture. Hunger supports the department's charitable rhetoric ("Feeding people . . . is too serious a matter to be left to political manipulation"), its realpolitik ("Food is a weapon"), and its self-justification ("true agripower . . . generates agridollars through agricultural exports"). How the future might be served by careless and destructive practices in the present is a question that is simply overridden by the brazen glibness of official optimism. If there is a food crisis, then, according to specialist logic, we must produce more food more carelessly than ever before. The energy crisis has been used, by the same logic, to justify the squandering of fuels.

LET THEM EAT THE FUTURE

As a sampler both of prevalent agricultural trends and official attitudes, as well as of the popular gullibility with which they have been received, one could not do better than an article entitled "The Revolution in American Agriculture" in the *National Geographic* of February 1970.

We should remember that in 1970 revolution was a controversial subject in America. We had spent the past half century in various stages of panic over the fact or the alleged possibility of communist revolution. And, during the decade just past, a good many of our people, mostly young, had begun to think of themselves as revolutionaries; some of them had even begun to *act* like revolutionaries. That the *National Geographic* could speak at such a time of an agricultural revolution could only indicate that a revolution of this kind, as opposed to a political revolution, was entirely acceptable to most Americans; it was simply part of the industrial revolution, which, after all, had become their way of life. That the industrial revolution, and the agricultural revolution along with it, had been real revolutions, surely the most powerful ever experienced, with real consequences, some of them political, and by no means all good—none of that mattered. The agricultural revolution, so far as the *National Geographic* and its readers were concerned, was a "good" revolution.

The author of the article, Mr. Jules B. Billard, is identified as a member of the magazine's senior editorial staff. But nowhere does he display the independence of judgment that one would expect either of a geographer or an editor. During most of the article he is in the grip of the ignorant awe, the greenhorn's ecstasy, that has been as necessary to this revolution as the ball bearing. During his encounters with the various manifestations of agricultural progress, Mr. Billard "marveled" twice; he was "staggered," "fascinated," "astounded," and "jolted" once each; he experienced two "jolting awakenings," the second more jolting than the first; and once his "mind churned."

The following is an inventory of Mr. Billard's revolutionary wonders:

"You can have strawberries in January, fresh oranges and lettuce the year round."

"Of the 6,000 to 8,000 items in the typical supermarket, 40 percent were not there a dozen years ago."

". . . in a single lifetime United States agriculture has advanced more than in all the preceding millenniums of man's labor on the land."

". . . I watched a factory-on-wheels move down celery rows . . . doing the work of forty men."

"I handled tomatoes bred for machine harvesting."

"I learned about heating cables buried underground to warm the soil so asparagus can grow in December . . ."

"Because only one person in 43 is needed to produce food, others can become doctors, teachers, shoemakers, janitors . . ." [This is a quote from former Agriculture Secretary Clifford Hardin.]

"Today 90 percent of the [California] tomato crop is picked mechanically."

". . . an incredible parade of machines are at work today on U.S. farms: acre-eaters . . . self-propelled combines that permit a man to ride in an air-conditioned cab to harvest a crop of corn that used to take a crew of 80 hands. Monster road-building machinery to level terraces or shape rice fields. Helicopters to spray cucumber fields. In all such a host of devices that today U.S. farmers are investing eight times as much capital as they did thirty years ago."

"Automated feeders, waterers, ventilators, and other labor savers make it possible for one man to take care of 100,000 broilers . . ."

"Block-long buildings, each housing 90,000 White Leghorns, cooped five birds to a 16-by-18 inch cage . . ."

". . . meatless dishes tasting like chicken, beef, or ham."

These accomplishments sort themselves readily into two categories: the frivolous and the problematic. The frivolity of strawberries in January, asparagus in December, and wheat or soybean products that taste like chicken is simply never acknowledged. Nor are the implications of the enormous increase of "items" in the supermarkets. By the values of gee-whiz journalism *any* increase is marvelous.

Nor is there any acknowledgment of the influence of "monster" technology ("acre-eaters") on the soil, the produce, the farm communities, and the lives and characters of farmers.

It is harder to ignore the enormous increase of indebtedness and overhead that has accompanied the enlargement of farm technology. Mr. Billard quotes an Iowa banker: "In 1920 . . . $5,000 was a big loan, and people hesitated to borrow. Now a $40,000 loan is commonplace, and having mortgage after mortgage is an accepted thing. I occasionally wonder whether the average farmer will ever get out of debt." The article gives examples of the enormous acreages and costs involved in several up-to-date operations. But these figures are simply left lying; in Mr. Billard's mind they evidently stand for nothing except the bigness of modern agriculture—which he approves of, so far as one can tell, because it amazes him. The Iowa banker's statement, doubtful as it may seem out of context, is made *in praise* of credit. Nowhere is there a question of the advisability of basing so large an enterprise on credit, or of the influence of routine indebtedness on a people's character. Nowhere is there a suspicion that there might be any worth in the old rural virtues of solvency and thrift.

The economic and moral uncertainty of living on credit is evidently—and typically—thought to be compensated by an improved standard of living: "Today [the farm wife is] as likely to be mini-skirted as her city sister, and as likely to own a dishwasher or self-cleaning oven or color television set. And her husband, who drives a tractor with an automatic transmission and uses power tools to eliminate back-straining labor, is as likely to have gone to college as his town cousin." That this standard of living is entirely material and entirely urban is characteristic of the prejudices that underlie the article.

The industrialization of animal husbandry is likewise seriously oversimplified. In addition to the ethical questions involved, the use of animals as machines—penning them in feed lots and cages—creates an enormous pollution problem. Mr. Billard acknowledges that this problem exists. He even cites a dubious solution: spreading the manure of 20,000 cattle on the pastures of a 320-acre farm which also contains the feed lots, a drainage pond, and a feed mill. But he also notes that in 1968 American farmers spread "nearly forty million tons" of chemical fertilizers, or "260 pounds for each acre under cultivation." The manure problem is separated from these figures on fertilizer consumption by fourteen pages. The dependence of our farmers on chemical fertilizers is not seen as a problem, and so the connection is missed. Mr. Billard forgot, or he never knew, that once plants and animals were raised together on the same farms—which therefore neither produced unmanageable surpluses of manure, to be wasted and to pollute the water supply, nor depended on such quantities of commercial fertilizer. The genius of American farm experts is very well demonstrated here: they can take a solution and divide it neatly into two problems.

That the agricultural revolution has displaced large numbers of people and put large numbers out of work is also acknowledged by Mr. Billard. But like the society as a whole, he has no trouble accepting this as part of the inevitable cost of progress. Lest anyone should become concerned about it, he includes early in his article a formula that makes it all right: "'Machines do replace labor,' G. E. VandenBerg told me . . . in his office at the USDA's Agricultural Research Center in Beltsville, Maryland. 'However, it is the scarcity of labor that really spurs adoption of machines.'"

Nevertheless, twenty-four pages later, Mr. Billard is saying:

"Squeezed between higher operating costs and what he gets for his produce, the man on the farm must become more efficient or give up." So apparently there is a problem after all. But another "agribusiness" formula is immediately invoked to assuage the moral discomfort: when all else fails to disguise the indifference of official agriculture to all human concerns, one can always fall back on efficiency. And so it appears that the failure of so many small farmers over so many years is really a kind of justice: it is their own fault; they ought to have been more efficient; if they had to get bigger in order to be more efficient, then they ought to have got bigger.

But suppose there is no room to get bigger unless somebody is driven out. In that case, one must have recourse to the law of compensation. This is the favorite law of the exploiter. It holds that for every loss there is a gain that is opposite and at least equal. This law is good fortune itself, for it means that you can do no wrong. Mr. Billard is an ardent observer of the law of compensation. "How many have given up," he writes, "can be seen in such figures as these: In 1910 our farm population accounted for a third of the U.S. total. By 1969 it was a mere twentieth. People leave rural areas at an average rate of 650,000 a year; many drift into cities where they join past migrants in the ghettos—to become added tinder for the riots that can be labeled one of the social consequences of the agricultural revolution." And he goes on: "When people leave the farm, rural communities . . . likewise wither away."

Here surely is cause for mourning: a forced migration of people greater than any in history, the foretelling of riots in the cities and the failure of human community in the country. But no. On the contrary: "Not all small towns are dying. The smog and the traffic and the social unrest of megalopolis prompts a second look at advantages of living in smaller communities. Industry, freed by jet planes and superhighways from dependence on nearby markets, shifts its plants away from cities. Employees are drawn by such appeals as being able, ten minutes after leaving work, to be out on the golf course, or roaming the woods with gun and dog, or watching kids and crops grow in a handful of acres a man can call his own."

Thus, if country people are forced to move into the city, that is made up for, according to Mr. Billard, by the movement of city people, and the city itself, into the country. But that only *looks* like a balanced equation. The people who move into the city and those who move out into the country are hardly the same people. The country community (of "inefficient" and therefore socially negligible people) is broken up, to be replaced by an influx of urban people who (however "efficient") have no economic or cultural ties to the land and are not a community. In this exchange we lose country people, we lose community, and we lose land. And we also lose the "inner city," which is abandoned to those who cannot perform "efficiently" either in the city or in the country.

But probably the most interesting feature of Mr. Billard's account of this exchange is the importance he attaches to "watching kids and crops grow on a handful of acres a man can call his own." Why, one wonders, does this feeling assert itself when the handful of acres is owned by an urban migrant, but not when they are owned by a farmer? How, rationally, can one hold the small farm in contempt as the living of a farm family and then sentimentalize over it as the "country place" or hobby of an executive? It cannot be done unless it is assumed that an executive is more deserving of a small farm because, as an urban or a professional person, he is superior to a farmer.

The callousness and smugness of this attitude is fully displayed in the caption to two pictures, one showing several members of a black family in their house and the other showing a modern cotton-picking machine at work. The caption is headlined dramatically: "When machines displace people"—and it reads: "Through the years Ruth Anderson's husband had worked the sweltering cotton fields around Isola, Mississippi. In late spring Ed Anderson chopped cotton . . . Summers he picked the cotton at $2.50 a hundred pounds. Between having her nine children, four of whom she tends above in the family's one-room shanty, Mrs. Anderson worked beside her husband. During picking season they brought home as much as $10 a day, and they got by.

"Then onto the fields rolled machines . . . that harvested as much in a day as could 80 men. Picking jobs vanished. Herbicides came on the market to kill weeds; they killed the chopping, too.

"Lacking a skill for steady work, the Andersons joined the

hapless millions of rural refugees who, uprooted by mechanized farming, often drift to big cities seeking jobs.

"To help stem this flow, civil-rights groups, foundations, and the National Council of Churches support a self-help community called Freedom City . . ."

So much for the Andersons. We are evidently expected to assume that their plight, and the plight of millions like them, is exactly offset by Freedom City, which is trying "to help stem this flow"—as if the flow can be stemmed until every last "inefficient" field worker has entered a ghetto and gone on welfare. And then we are asked to turn away and marvel at the big machine that can do the work of eighty men, whose working conditions were, after all, "sweltering," and whose getting-by economy was out of fashion, if not slightly contemptible.

We are shown another farm family—three generations of them—at the dinner table on their 130-acre Long Island farm, part of which they have owned since 1737. These people, too, "stand at a crossroads: Either they mechanize and expand, or rising costs, high taxes, and big farm competition will drive them from the land." There is not a word or implication of so much as a doubt about the economic conditions that constitute this "crossroads," not the smallest curiosity as to what may be the cost. Nearly two-and-a-half centuries of family history on the same farm amounts simply to nothing if it can't pay the taxes. The fate of this family is offered as merely interesting, a kind of journal-fodder.

What excuses this human waste, this destruction of preserving traditions and associations, this moral indifference? It is the future—the future as both threat and lure—the secular Hell and Heaven of the enraptured booster.

Early in his article Mr. Billard refers to a possibility that is certainly grave, certainly to be taken seriously, but which has nevertheless become the routine curtain-raiser of "agribusiness" propagandists, who use it not for the rigorous self-evaluation that it requires, but shamelessly and tirelessly to justify themselves. They are fanatical believers in themselves, and like all fanatics they need an apocalypse—some ultimate bugaboo to shove into the face of doubt. What we have here is everybody's worry, but the farm experts and agribusinessmen would like us to leave it to them: "Earth's numbers now stand at 3.6 billion, and could double in 35 years. This . . . raises the specter of a famine more catastrophic than the world has ever seen."

Of course it does. And that means that we should be at work overhauling all our assumptions about ourselves and what we have done and what we are capable of doing, all our attitudes toward life and its complex sources, all our resources of technique and technology. If we are heading toward apocalypse, then obviously we must undertake an ordeal of preparation. We must cleanse ourselves of slovenliness, laziness, and waste. We must learn to discipline ourselves, to restrain ourselves, to need less, to care more for the needs of others. We must understand what the health of the earth requires, and we must put that before all other needs. If a catastrophic famine is possible, then let us undertake the labors of wisdom and make the necessary sacrifices of luxury and comfort.

But, according to Mr. Billard, this is not for ordinary people to worry about. The agriculture experts, industrialists, and scientists are going to take care of it: "The spread of modern agriculture can help assure the underdeveloped two-thirds of the world the freedom from hunger it gives the economically advanced one-third."

And by the end of his article Mr. Billard has entered into the glory of the true future-rapture. He talked to "Dr. Irving, of the Department of Agriculture," who said of the future: "Agriculture will be highly specialized. . . . Farms in one area will concentrate on growing oranges, those in another area tomatoes, in another potatoes—capitalizing on the competitive advantage soil or climate gives for a particular crop.

"Fields will be larger, with fewer trees, hedges, and roadways. Machines will be bigger and more powerful. . . . They'll be automated, even radio-controlled, with closed circuit TV to let an operator sitting on a front porch monitor what is going on. . . .

"Weather control may tame hailstorm and tornado dangers. . . . Atomic energy may supply power to level hills or provide irrigation water from the sea."

It was at this point that Mr. Billard's "mind churned with the implications of such developments building on the progress of the past." Gone are the fears of famine. Gone are any thoughts of displaced small farmers and farm workers, or of the threat of riots in the cities. Mr. Billard has risen right over apocalypse into Heaven itself. He ends by quoting triumphantly a remark by a Brazilian official: "'We are concerned about the future of agriculture in Brazil. . . . In your country, you *are* in the future.'"

And so, of course, are the Andersons, with their nine children, their "one-room shanty," and no job—which ought to be reassuring to the people in the "underdeveloped two-thirds of the world," who are still trapped back there in the present.

The final two pages of Mr. Billard's article carry an "artist's conception" of the agricultural future, which is a veritable paradigm of the agribusiness ambition. The caption reads as follows:

"Farm of the future: Grainfields stretch like fairways and cattle pens resemble high-rise apartments in a farm of the early 21st century, as portrayed by artist David Meltzer *with the guidance of U.S. Department of Agriculture specialists* [my emphasis].

"Attached to a modernistic farm house, a bubble-topped control tower hums with a computer, weather reports, and a farm-price ticker tape. A remote-controlled tiller-combine glides across a 10-mile-long wheat field on tracks that keep the heavy machine from compacting the soil. Threshed grain, funneled into a pneumatic tube beside the field, flows into storage elevators rising close to a distant city. The same machine that cuts the grain prepares the land for another crop. A similar device waters neighboring strips of soybeans as a jet-powered helicopter sprays insecticides.

"Across a service road, conical mills blend feed for beef cattle, fattening in multilevel pens that conserve ground space. Tubes carry the feed to be mechanically distributed. A central elevator transports the cattle up and down, while a tubular side drain flushes wastes to be broken down for fertilizer. Beside the farther pen, a processing plant packs beef into cylinders for shipment to market by helicopter and monorail. Illuminated plastic domes provide controlled environments for growing high-value crops such as strawberries, tomatoes, and celery. Near a distant lake and recreation area, a pumping plant supplies water for the vast operation."

THE ORGANIZATION OF DISORDER

The cooperation of Department of Agriculture specialists in this visualization of a completely industrialized agriculture-of-the-future makes it as official, it would seem, as any vision of the future could be. And that this sort of thing is not an isolated aberration of overexcited journalism, but a confirmed habit—even the theoretical context—of agricultural expertise, is suggested by an article in the October 1974 issue of the *American Farmer* (voice of the American Farm Bureau Federation). This article is about a "dream farm" of 2076 A.D.—a model constructed by a group of South Dakota State University agricultural engineering students. This farm of the future is described as follows:

"The farm of 9 square miles will use only about 1,800 acres, less than one-fourth of which is for production.* The remainder will be a buffer or 'relaxed' zone for recreation, wildlife, and living under the 'blending with human values' aspect of the overall planning.

"Livestock will be housed (and products processed) in a 15-story, 150′ × 200′ building. It will also contain power facilities, administrative headquarters, veterinary facilities, repair shops, refrigeration and packaging units, storage, research labs, water and waste treatment facilities. At capacity, the high-rise building will house 2,500 feeder cattle, 600 cow-calf units, 500 dairy cattle, 2,500 sheep, 6,750 finishing hogs, space for 150 sows and litters, 1,000 turkeys, and 15,000 chickens.

"Crops will be grown year around under plastic covers that provide precise climate control in three circular fields each a mile in diameter. At any given time, regardless of weather, one field or crop will be in the planting stage, another in the growing stage, and the third in the harvesting stage. Exceptionally high yields mean that only a fourth of the total 5,760-acre farm area would be needed for agricultural production.

"Only a half-inch of water will be needed for each crop. That's because evapotranspiration from growing plants would be recycled under massive, permanent plastic enclosures . . .

*This sentence is hard to understand. The acreage to be used for production is evidently one-fourth of the nine square miles, not of the 1800 acres. But 1800 acres is *more* than one-fourth of nine square miles, not less.

"Underground magnetic patterns, arranged to fit crop or machine, will attract specially-treated seed blasted from overhead tubes in the enclosures.

"If tillage is needed, it will be done by electromagnetic waves. Air-supported, remotely controlled machines will harvest entire plants because by 2076 A.D. the students believe multiple uses will be needed and found for most crops.

"'Trickle' irrigation is to be electronically monitored to provide subsurface moisture automatically whenever needed.

"Recycling human, animal and crop wastes will be a key to the operation of the farm. Carbon dioxide from the respiration of livestock is to be piped into the circular enclosures for use by crops in exchange for the oxygen transpired by crops for use by livestock.

"Weed control is not anticipated as a problem because weeds would be eradicated under the field covers."

Soon after reading this article I wrote to Dr. Milo A. Hellickson, Associate Professor in Agricultural Engineering at South Dakota State University. My letter asked the following questions:

"1. Was any attention given to the possible social and economic effects of the projected innovations? Was it envisioned that this sort of farm would entirely replace the relatively small owner-operated farm? What would be its effect upon population patterns? Would it make food more or less expensive? What would be the energy requirements of such an operation, and what would be the sources of the required energy?

"2. What political consequences were anticipated? What, for instance, would be the impact . . . upon the doctrines of personal liberty and private property?

"3. What would be the effect upon the consumer? Would there be more or less choice of variety and quality?

"4. What would be the effect on the environment? For instance, roofing so large an acreage would present an unprecedented drainage problem. What did your students propose to do with the runoff? Would such a farm be built only in a desert area, or would it be feasible in an area of abundant rain fall?"

I received a prompt and very cordial reply from Dr. Hellickson, who responded to my questions as follows:

"Attention was given to the social and economic effect of the innovations. Essentially we feel that these developments would most likely fit individually into various farming operations and would not necessarily be all concentrated into one farmstead. As a matter of convenience in construction and so as not to alienate any particular phase of the agricultural industry, the model is constructed incorporating all the areas. Therefore, it would be equally possible in the future to maintain the smaller owner-operated farm and this then would cause little change in the distribution of the population. Specifically, we are thinking of energy captured from the sun, solar energy, as the sole energy source. I wouldn't even attempt to make a guess concerning expense, since this is such an area of dynamic change.

"Hopefully the above paragraph also answers question two. We are in no way advocating the elimination of the free enterprise system or the reduction of privately owned land.

"As per question three, I would see little change in the variety or quality of products available. . . . If anything, quality might be improved through the reduction or elimination of disease and through better handling systems.

"Hopefully this system would improve the environment by eliminating air pollution from the livestock building and also eliminating erosion from the cropped area. Runoff from the roof areas is proposed to be used as the water source for the irrigation system and for live-stock and humans. Naturally, adequate facilities must be included to handle unusually large rainfalls."

There is no quarreling with the professed aim of either of these farms-of-the-future, which is an abundance of food. And they have other aspects that are praiseworthy: the conversion of wastes into fertilizer and the reliance of the South Dakota model on solar energy. But we are still left with the question of what will be the costs, not just of construction and materials, which would be passed on to consumers in the price of food, but costs of other kinds: social, cultural, political, nutritional, etc. And we still must ask if there may not be less costly ways to achieve the same ends.

The issue that is raised most directly by these farms-of-the-future is that of control. The ambition underlying these model farms is that of total control—a totally controlled agricultural environment. Nowhere is the essential totalitarianism and the essential weakness of the specialist mind more clearly displayed than in this ambition. Confronted with the living substance of farming—the complexly, even mysteriously interrelated lives on which it depends, from the microorganisms in the soil to the human consumers—the agriculture specialist can think only of subjecting it to total control, of turning it into a machine.

But total human control is just as impossible now as it ever was—or so the available evidence constrains one to believe. Nothing, for instance, could be more organized than one of our large cities, with its geometric streets, its numbered houses, its numbered citizens, its charted routes and zones, its great numbers of police and other functionaries charged to keep order—and yet nothing could be more chaotic than one of these same cities during rush hour or after dark or during a riot or a garbage collectors' strike. In the modern city unprecedented organization and unprecedented disorder exist side by side; one could argue that they have a symbiotic relationship, that they feed and thrive upon each other. It is not difficult to think of any number of such examples in government, education, industry, medicine, agriculture—wherever the specialist has come with his controls.

The reason would seem to be that the specialist and the idea of total control also have a symbiotic relationship, that neither can exist without the other. The specialist puts himself in charge of *one* possibility. By leaving out all other possibilities, he enfranchises his little fiction of total control. Leaving out all the "non-functional" or otherwise undesirable possibilities, he makes a rigid, exclusive boundary within which absolute control becomes, if not possible, at least conceivable.

But what the specialist never considers is that such a boundary is, in itself, profoundly disruptive. Its first disruption is in his mind, for having enclosed the possibility of control that is within his competence to imagine and desire, he becomes the enemy of all other possibilities. And, secondly, having chosen the possibility of total control within a small and highly simplified enclosure, he simply abandons the rest, leaves it totally *out* of control; that is, he forsakes or even repudiates the complex, partly mysterious patterns of interdependence and cooperation, controllable only within limits, by which human culture joins itself to its sources in the natural world.

This attempt at total control is an invitation to disorder. And the rule seems to be that the more rigid and exclusive is the specialist's boundary, and the stricter the control within it, the more disorder rages around it. One can make a greenhouse and grow summer vegetables in the wintertime, but in doing so one creates a vulnerability to the weather and a possibility of failure where none existed before. The control by which a tomato plant lives through January is much more problematic than the natural order by which an oak tree or a titmouse lives through January. The patterns of cooperation are safer than the mechanisms of exclusion, even though they lack the illusory safety of "control."

Because of his dependence on boundaries and controls, the genre or mode of the specialist is the "model." The necessary context of a model is the future. The qualifications of the present, of *living*, do not affect it, nor do the non-functional or the undesirable. It is remote even from probable difficulties of the future. Thus the language of the article in the *American Farmer*, having to do with the inward workings of the South Dakota State model, is confident and for the most part it is exact. But Dr. Hellickson's responses to my questions about its *influence* are tentative, conjectural, and hopeful. The model perfectly empowers the machine metaphor: only the "working parts" need be admitted. Therefore, if one is going to make a "model farm," one must give it a boundary, if possible a roof, that will keep out whatever does not "work." Weeds, insects, diseases do not work; leave them out. The weather works only sometimes, or on the average; leave the weather out. The work can be done by machines; leave the people out. But chemicals and drugs, no matter how dangerous, *do* work; they are part of the boundary, so they can be let in.

It may be a bit startling at this point to realize that what has been left out of this enclosure is health. As soon as pests, parasites, diseases, climatic fluctuations and extremes are left out, resistance to these things is also left out; and this resistance, in the soil and in the lives that come from the soil, is what we call health. And so for total control we have given up health—which is also a kind of control, safer by far than a plastic roof, but never total.

The model is an ideal and is surely meant to function as an ideal. But it is a *mechanical* ideal, and an exclusive one. Furthermore, its connections with the past and the present are severed; always implicit in a model is the idea of *replacing* what has survived of the past, what exists in the present. These characteristics divide the model radically from ideals of the more usual sort. Such ideals as honesty or generosity or gentleness or symmetry do indeed have an influence on the future, but we recognize them from what we have known of them in the past and from what they require of us in the present. Like health, they are required to survive among us in the presence of what they must resist; they survive in culture, in community, and in the characters of people. They are known and valued not because they have been modeled, but because they have been *exemplified*. The specialist, on the other hand, is interested only in the model, never in the example. He is interested in the future of farming, not in its history.

That is why the influence of his work does not interest him; if he puts a machine into the field to "save labor," he does not ask the fate of the replaced people.* He is working "in the future," which puts him at liberty simply to leave out whatever is displaced or whatever does not work. That is why there are no more people in these scenes of future farms than in the landscape photographs in conservation magazines; neither the agriculture specialist nor the conservation specialist has any idea where people belong in the order of things. Neither can conceive of a domesticated or a humane landscape. People are complex, contradictory, unpredictable; they are perceived by the specialist as a kind of litter, pollutants of pure nature on the one hand and of pure technology, total control, on the other.

WHERE ARE THE PEOPLE?

By the power of a model, the specialist turns the future into a greenhouse of fantasies. But the model also empowers the fantasies with influence over present life—and, of course, over future life. And so, considering these model farms, one asks,

*This is the flaw in the doctrine of labor-saving. Labor-saving machines are *supposed* to make jobs easier. In fact, they destroy jobs. Instead of ameliorating work, they replace workers. What makes work easier and more pleasant without reducing employment is collaboration, neighbors helping each other. "Many hands make light work."

Where are the people? out of self-interest and with some trepidation. The *National Geographic* model shows, as far as I can tell, only one "farmer." He is standing in the "bubble-topped control tower," presumably operating the whole farm by remote control. The article on the South Dakota State model mentions that "The hired hand gets the imposing new title of 'manager.'" Allowing for shifts, vacations, etc., these model farms evidently require a staff of only half a dozen or so to do the actual "work." Most of the jobs for people would evidently be non-agricultural: jobs of construction, maintenance, transport, etc.

And where are the *other* people—the ones who are not doing the computer-work of future farming? Well, the *National Geographic* picture shows some highway traffic that may or may not be remote-controlled. It shows "a distant city" and "a distant lake and recreation area." The South Dakota State model includes a zone for "recreation, wildlife, and living."

The agriculture prophets evidently think that they have left people pretty much to their own devices: they will have places to live and places to work and places for recreation, and, thanks to the completely controlled farms-of-the-future, they will have plenty to eat.

The specialists who conceived these models are American citizens. They undoubtedly believe in the doctrines of personal liberty and dignity, equality of opportunity, etc. If asked, they would undoubtedly say that the people outside the boundaries of these farms would benefit from them in every way: they would not only have areas especially allotted to them for living, working, and recreation, they would also have more freedom, dignity, and equality of opportunity than ever before.

But one must ask if they would not say these things thoughtlessly—because they are the right things to say in a democracy, or the most persuasive things to say, or because they are in the habit of saying them. It is clear, at least, that official policies—and these model farms represent official policy—have come to be *routinely* justified in this country on the grounds that they will uphold freedom, dignity, and equality of opportunity. There is no official depredation that one can think of that has not been initially so justified. The skids are greased with unctions of democracy.

But these assurances are always incidental, outside the boundary of whatever allegedly benign (and profitable) innovation is at hand. People are not going to be free or dignified or even well fed just because some specialist *says* that they will be. Or says that they will be *allowed to be, in certain areas*—for that is what these "agribusiness" visionaries are in fact saying. People will be *allowed* to be free to do *certain* things in *certain* places prescribed by *other* people. They will be free to work in the places set aside for work, free to play or relax in places set aside for recreation, free to live (whatever that may mean) in places set aside for living.

Thus there are several things that people will *not* be free to do in the nation-of-the-future that will be fed by these farms-of-the-future. They will not live where they work or work where they live. They will not work where they play. And they will not, above all, play where they work. There will be no singing in those fields. There will be no crews of workers or neighbors laughing and joking, telling stories, or competing at tests of speed or strength or skill. There will be no holiday walks or picnics in those fields because, in the first place, the fields will be ugly, all graces of nature having been ruled out, and, in the second place, they will be dangerous.

Very few people, more likely none of them, will own those farms. Very few will work on them. Most of them, more even than the ninety-five percent that *now* live in urban situations, will live remote from the farmland, divided from it by distance, by "buffer zones," by economics, by official structure. They will have nothing to say about how the land is used or the kind or quality of its produce. For these farms are obviously designed for the ownership and management of huge "agribusiness" corporations that will control them "privately" and control the market as well. The people will eat what the corporations decide for them to eat. They will be detached and remote from the sources of their life, joined to them only by corporate tolerance. They will have become consumers purely —consumptive machines—which is to say, the slaves of producers. What these model farms very powerfully suggest, then, is that the concept of total control may be impossible to confine within the boundaries of the specialist enterprise—that it is impossible to mechanize production without mechanizing consumption, impossible to make machines of soil, plants, and animals without making machines also of people.

It is important to recognize that in the minds both of the agribusiness specialists and of their believers and supporters among the public these representations of technological totalitarianism rest side by side with conventional good intentions. Mr. Billard, in his *National Geographic* article, is careful to write a paragraph of reassurance about the future of the family farm and the welfare of consumers: ". . . farms grossing more than $10,000 a year expand in number, with those in the more-than-$40,000 category increasing rapidly. The family farm figures largest in this growth. It accounts for 95 percent of all farms and 64 percent of total marketings. Corporate behemoths play no greater role today than 20 years ago; the specter of their progressively gobbling up all the farmland and in the end holding consumers at their mercy seems farfetched."

That, of course, depends on how you define "family farm" and "corporate behemoth." And beside the Department of Agriculture figures quoted earlier, and the testimony of his own article, Mr. Billard's reassurance is a mere hopeful assertion, not very reassuring.

And Dr. Hellickson, replying to my question about the possible influence of his students' work on personal liberty and private property, said: "We are in no way advocating the elimination of the free enterprise system or the reduction of privately owned land."

Sometimes I ask myself if it may not be that these reassurances are given cynically by people who know very well that they are turned against what they wish to appear to uphold. Though I leave open the possibility that this occasionally may be so, I have concluded so far that most often it is not. I believe that both Mr. Billard and Dr. Hellickson are sincere in their belief that the innovations they praise or advocate will not adversely affect traditional values, the supply or the quality of food, or the life of farming. I believe this of Dr. Hellickson in spite of his substitution of "the free enterprise system" for my phrase "personal liberty."

It is nevertheless a part of the significance of the statements of both men that they embody a large, if unconscious, moral contradiction. In this, it seems to me, they represent very accurately the flawed consciousness of our society, which is everywhere eagerly conniving in the destruction of what it says, and thinks, it wants to preserve.

For no matter what these gentlemen say, the private ownership of farmland and public concern for the health of farming are both diminishing at an alarming rate, and they are diminishing because of the big economics and big technology represented by these visions of future agriculture.

They are diminishing because as a society we have abandoned any interest in the survival of anything small. We seem to have adopted a moral rule of thumb according to which anything big is better than anything small. As a result, the agricultural establishment has simply looked away from the possibility of an economics and a technology suited to the needs and aims of the small farmer.

Some time ago I took part in a conference on agriculture, at which one of the speakers was an executive of Deere and Company. This man was asked by someone in the audience if he and his company were interested in small-farm technology. He replied that indeed they were. But as I remember that was all he said; he spoke of none of the particulars of such technology, in which he evidently had at least no personal interest. What did interest him, as I learned later in conversation, was the impending development of a 600-horsepower tractor, which eventually would be operated by remote control. In the face of such an interest, empowered as it is by official sanction, tax-supported research, and vast sums of money, the small farmer is not so much condemned as written off as a necessary expenditure. A price is put on his way of life which he is less and less able to meet.

DESERTS OF VAST TECHNOLOGY

As specialists, the agricultural scientists and "agribusinessmen" find it easy to talk as if the influence of big agricultural technology can be confined neatly to the "field" of agriculture. That it cannot be is already proven by the powerful *urban* influence that such technology has already had. And to the big-thinking, non-agricultural mind, food is merely a resource, like energy and raw materials, and so agricultural technology is not different from any other. About grain, fuel, and ore the only questions are: How much? and How fast?

Because big technology is so simplifying, the future looks, not bright, but absolutely perfect to F. M. Esfandiary, who teaches "long-range planning" in New York City's New School for Social Research. In an article entitled "Homo Sapiens, the Manna Maker," Mr. Esfandiary sees the future as an earthly Heaven in which, by the miracles of technology, humans will usurp the role of God—who, it may be recalled, was once thought to be the only maker of manna. The following quotations will give the gist of his argument:

"The world is moving toward an age of limitless abundance —abundant energy, food, raw materials."

"Solar power, nuclear fusion, geothermal energy, recycled energy, wind energy, hydrogen fuel—these sources will soon provide cheap, nonpolluting, limitless energy, enough to last for millions of years."

"Agriculture is undergoing an epochal revolution. We are evolving from feudal and industrial agriculture to cybernated food production. Computers, remote control cultivators, television monitors, sensors, data banks can now automatically run thousands of acres of cultivated land. A couple of telefarm operators can feed a million people."

"We now have the capability to extract limitless raw materials from recycled wastes, rocks, the earth's interiors, the ocean floors, space."

That is the "objective" part of the argument. There follows a series of paragraphs that must hold the world record for rhetorical passion. It appears that Mr. Esfandiary is mad at us because we do not duck our heads and hurry right on into the future.

"How absurd the American panic over scarcity when we are entering an age of abundance. How absurd to focus on 'finiteness' at the period in evolution when our world is transcending finiteness, opening up the infinite resources of an infinite universe.

"How outrageous that after centuries of privation and sacrifice leaders can come up with nothing more than yet more sacrifice. How short-sighted the exhortation to no-growth at precisely the time when we urgently need more and more growth—growth not *within* but *beyond* industrialism.

"How retrogressive the preachings to lower living standards of the relatively rich to raise conditions of the poor, at a time when we can raise *everybody's* living conditions by vigorously developing and spreading abundance, not sharing scarcity."

The common assumption is that mechanization involves the giving over of certain tasks or functions to machines. In these paragraphs by Mr. Esfandiary a very different assumption, that may always have been implicit in the advocacy of industrial revolution, comes to the surface: he is proposing that we give over *everything* to machines. He is berating us, with the fervor of an evangelist, because we do not abandon ourselves to machines as people of faith abandon themselves to God. He is berating us, in fact, for not *being* gods or at least acting as if we were gods.

The crucial concept here is that of "limitless" or "infinite" quantity. By "limitless" and "infinite" Mr. Esfandiary undoubtedly means only "inconceivable." At any rate, people who have desired material *quantities* on such a scale have always been recognized as evil, and their stories have always involved a sort of ecological justice: godly appetite very quickly led beyond human competence, invariably with disastrous consequences. Mr. Esfandiary's unlimited, if theoretical, gluttony is licensed and given an illusory respectability because of its claim to be "scientific"—godly appetite may be within the competence of a computer—and because, as a "long-range planner," he does his theorizing in the future, where it cannot very handily be called to account.

It is nevertheless clear that Mr. Esfandiary's "future" calls for unprecedented violence. It would require the sacrifice of every value that is not quantitative. The technology of infinity (however that might be defined) would be vast and exclusive. It would be completely totalitarian, whether "publicly" or "privately" owned. It would overthrow the whole issue of control, for it would *be* the control. Since everyone would be totally dependent upon it, it would necessarily be everyone's first consideration. It might at first seem that enormous power would lie in the hands of the "couple of telefarm operators" who would be feeding a million people; but it seems more likely that they, too, would be the absolute slaves of their machinery, no less dependent on it than the million. The machine would become an anti-god—if not infinite, at least absolute. To have even the illusion of infinite quantity, we would have to debase both the finite and the infinite; we would have to sacrifice both flesh and spirit. It is an old story. Evil is offering us the world: "All these things will I give thee, if thou wilt fall down and worship me." And we have only the old paradox for an answer: If we accept all on that condition, we lose all.

What is new is the *guise* of the evil: a limitless technology, dependent upon a limitless morality, which is to say upon no morality at all. How did such a possibility become thinkable? It seems to me that it is implicit in the modern separation of life and work. It is implicit in the assumption that we can live entirely apart from our way of making a living. It is implicit in the idea of the agricultural engineering students at South Dakota State University that their farm-of-the-future would require "blending with human values." To propose to blend such a farm with human values is simply to acknowledge that it *has* no human values, that human values have been removed from it. (The analogy is not accidental, I think, between this "blending with human values" and the "enrichment" of bread after the nutrients have been removed from the wheat.) If human values are removed from production, how can they be preserved in consumption? How can we value our lives if we devalue them in making a living?

If we do not live where we work, and when we work, we are wasting our lives, and our work too.

CHAPTER SIX

The Use of Energy

"ENERGY," SAID William Blake, "is Eternal Delight." And the scientific prognosticators of our time have begun to speak of the eventual opening, for human use, of "infinite" sources of energy. In speaking of the use of energy, then, we are speaking of an issue of religion, whether we like it or not.

Religion, in the root sense of the word, is what binds us back to the source of life. Blake also said that "Energy is the only life . . ." And it is superhuman in the sense that humans cannot create it. They can only refine or convert it. And they are bound to it by one of the paradoxes of religion: they cannot have it except by losing it; they cannot use it except by destroying it. The lives that feed us have to be killed before they enter our mouths; we can only use the fossil fuels by burning them up. We speak of electrical energy as "current": it exists only while it runs away; we use it only by delaying its escape. To receive energy is at once to live and to die.

Perhaps from an "objective" point of view it is incorrect to say that we can destroy energy; we can only change it. Or we can destroy it only in its current form. But from a human point of view, we can destroy it also by wasting it—that is, by changing it into a form in which we cannot use it again. As users, we can preserve energy in cycles of use, passing it again and again through the same series of forms; or we can waste it by using it once in a way that makes it irrecoverable. The human pattern of cyclic use is exemplified in the small Oriental peasant farms described in F. H. King's *Farmers of Forty Centuries*, in which all organic residues, plant and animal and human, were returned to the soil, thus keeping intact the natural cycle of "birth, growth, maturity, death, and decay" that Sir Albert Howard identified as the "Wheel of Life." The pattern of wasteful use is exemplified in the modern sewage system and the internal combustion engine. With us, the wastes that escape use typically become pollutants. This kind of use turns an asset into a liability.

We have two means of bringing energy to use: by living things (plants, animals, our own bodies) and by tools (machines, energy-harnesses). For the use of these we have skills or techniques. All three together comprise our technology. Technology joins us to energy, to life. It is not, as many technologists would have us believe, a simple connection. Our technology is the practical aspect of our culture. By it we enact our religion, or our lack of it.

I began thinking about this by trying to make a clear distinction between the living organisms and skills of technology and its mechanisms, and to say that the living aspect was better than the mechanical. I found it impossible to make such a distinction. I thought of going back through history to a point at which such a distinction would become possible, but found that the farther back I went the less possible it became. When people had no machines other than throwing stones and clubs, their technology was all of a piece. It stayed that way through their development of more sophisticated tools, their mastery of fire, their domestication of plants and animals. Lives, skills, and tools were culturally indivisible.

The question at issue, then, is not of distinction but of balance. The ideal seems to be that the living part of our technology should not be devalued or overpowered by the mechanical. Because the biological limits are probably narrower than the mechanical, this calls for restraint on the proliferation of machines.

At some point in history the balance between life and machinery was overthrown. I think this began to happen when people began to desire long-term stores or supplies of energy—that is, when they began to think of energy as volume as well as force—and when machines ceased to enhance or elaborate skill and began to replace it.

Though it seems impossible to distinguish between the living and the mechanical aspects of technology, it is possible to distinguish between two kinds of energy: that which is made available by living things and that which is made available by machines.

The energy that comes from living things is produced by combining the four elements of medieval science: earth, air, fire (sunlight), and water. This is current energy. Though it is possible to speak of a *reserve* of such energy, as Sir Albert Howard does, in the sense of a surplus of fertility, it is impossible to conceive of a *reservoir* of it. It is not available in long-term supplies in any form in which it can be preserved, as in humus, in the flesh of living animals, in cans or freezers or grain elevators, it still perishes fairly quickly in comparison, say, to coal or plutonium. It lasts over a long term only in the living cycle of birth, growth, maturity, death, and decay. The technology appropriate to the use of this energy, therefore, preserves its cycles. It is a technology that never escapes into its own logic but remains bound in analogy to natural law.

The energy that is made available, and consumed, by machines is typically energy that can be accumulated in stockpiles or reservoirs. Energy from wind and water obviously does not fit this category, but it suggests the possibility of bigger and better storage batteries, which one must assume will sooner or later be produced. And, of course, we already store water power behind hydroelectric dams. This mechanically derived energy is supposed to have set people free from work and other difficulties once considered native to the human condition. Whether or not it has done so in any meaningful sense is questionable—in my opinion, it is highly questionable. But there is no doubt that this sort of energy has freed machinery from the natural restraints that apply to the use of organic energy. We now have a purely mechanical technology that is very nearly a law unto itself.

And yet, in the long term, this liberation of the machine is illusory. Mechanical technology is based on quantities of materials and fuels that are finite. If the prophets of science foresee "limitless abundance" and "infinite resources," one must assume that they are speaking figuratively, meaning simply that they cannot comprehend how much there may be. In that sense, they are right: there are sources of energy that, given the necessary machinery, are inexhaustible *as far as we can see.*

The great difficulty, which these cheerful prophets do not acknowledge at all, is that we are trustworthy only so far as we can see. The length of our vision is our moral boundary. Even if these foreseen supplies *are* limitless, we can use them only within limits. We can bring the infinite to bear only within the finite bounds of our biological circumstance and our understanding. It is already certain that our planet alone—not to mention potential sources in space—can provide us with more energy and materials than we can use safely or well. By our abuse of our finite sources, our lives and all life are already in danger. What might we bring into danger by the abuse of "infinite" sources?

The difficulty with mechanically extractable energy is that so far we have been unable to make it available without serious geological and ecological damage, or to effectively restrain its use, or to use or even neutralize its wastes. From birth, right now, we are carrying the physical and the moral poisons produced by our crude and ignorant use of this sort of energy. And the more abundant the energy of this sort that we use, the more abounding must be the consequences.

It is typical of the mentality of our age that we cannot conceive of infinity except as an enormous quantity. We cannot conceive of it as orderly process, as pattern or cycle, as shapeliness. We conceive of it as inconceivable quantity—that is, as the immeasurable. Any quantity that we cannot measure we assume must be infinite. That is about as sophisticated as saying that the world is flat because it *looks* flat. The talk about "infinite" resources is thus a kind of scientific-sounding foolishness. And it involves some quaint paradoxes. If we think, for instance, of infinite energy as immeasurable fuel, we are committed in the same thought to its destruction, for fuel must be destroyed to be used. We thus arrive at the curious idea of a destructible infinity. Furthermore, we have become guilty not only of the demonstrably silly assumption that we know what to do with infinite energy, but also of the monstrous pride of thinking ourselves somehow entitled to undertake infinite destruction.

This mechanically rendered infinitude of energy is an ambition surrounded by terrific problems. Such energy cannot be used constructively without at the same time being used destructively. And which way the balance will finally fall is a question that baffles the best minds. Nobody knows what will be the ultimate consequences of our present use of fossil fuel, much less those of our future use of atomic fuel. The sun may prove an "infinite" source of energy—at least one that may last several billion years. But who will control the use of that energy? How and for what purposes will it be used?

How much can be used without overthrowing ecological or social or political balances? Nobody knows.

The energy that is made available to us by living things, on the other hand, is made available not as an inconceivable quantity, but as a conceivable pattern. And for the mastery of this pattern—that is, the ability to see its absolute importance and to preserve it in use—one does not need a Ph.D. or a laboratory or a computer. One can master it in this sense, in fact, without having any analytic or scientific understanding of it at all. It was mastered, better than our scientific experts have mastered it, by "primitive" peasants and tribesmen thousands of years before modern science. It is conceivable not so much to the analytic intelligence, to which it may always remain in part mysterious, as to the imagination, by which we perceive, value, and imitate order beyond our understanding.

We cannot create biological energy any more than we can create atomic or fossil fuel energy. But we *can* preserve it in *use*; we can probably even augment it in use, in the sense that, by proper care, we can "build" soil. We cannot do that with machine-derived energy. This is an extremely important difference, with respect both to the energy economy itself and to the moral order that is undoubtedly determined by, as much as it determines, the value we put on energy.

The moral order by which we use machine-derived energy is comparatively simple. Whatever uses this sort of energy works simply as a conduit that carries it beyond use: the energy goes in as "fuel" and comes out as "waste." This principle sustains a highly simplified economy having only two functions: production and consumption.

The moral order appropriate to the use of biological energy, on the other hand, requires the addition of a third term: production, consumption, *and return*. It is the principle of return that complicates matters, for it requires responsibility, care, of a different and higher order than that required by production and consumption alone, and it calls for methods and economies of a different kind. In an energy economy appropriate to the use of biological energy, all bodies, plant and animal and human, are joined in a kind of energy community. They are not divided from each other by greedy, "individualistic" efforts to produce and consume large quantities of energy, much less to store large quantities of it. They are indissolubly linked in complex patterns of energy exchange. They die into each other's life, live into each other's death. They do not consume in the sense of using up. They do not produce waste. What they take in they change, but they change it always into a form necessary for its use by a living body of another kind. And this exchange goes on and on, round and round, the Wheel of Life rising out of the soil, descending into it, through the bodies of creatures.

The soil is the great connector of lives, the source and destination of all. It is the healer and restorer and resurrector, by which disease passes into health, age into youth, death into life. Without proper care for it we can have no community, because without proper care for it we can have no life.

It is alive itself. It is a grave, too, of course. Or a healthy soil is. It is full of dead animals and plants, bodies that have passed through other bodies. For except for some humans—with their sealed coffins and vaults, their pathological fear of the earth—the only way into the soil is through other bodies. But no matter how finely the dead are broken down, or how many times they are eaten, they yet give into other life. If a healthy soil is full of death it is also full of life: worms, fungi, microorganisms of all kinds, for which, as for us humans, the dead bodies of the once living are a feast. Eventually this dead matter becomes soluble, available as food for plants, and life begins to rise up again, out of the soil into the light. Given only the health of the soil, nothing that dies is dead for very long. Within this powerful economy, it seems that death occurs only for the good of life. And having followed the cycle around, we see that we have not only a description of the fundamental biological process, but also a metaphor of great beauty and power. It is impossible to contemplate the life of the soil for very long without seeing it as analogous to the life of the spirit. No less than the faithful of religion is the good farmer mindful of the persistence of life through death, the passage of energy through changing forms.

And this living topsoil—living in both the biological sense and in the cultural sense, as metaphor—is the basic element in the technology of farming.

It is the nature of the soil to be highly complex and variable, to conform very inexactly to human conclusions and rules. It is itself a pattern of inexhaustible intricacy, and so it is easily damaged by the imposition of alien patterns. Out of the random grammar and lexicon of possibilities—geological, topographical, climatological, biological—the soil of any one place makes its own peculiar and inevitable sense. It makes an order, a pattern of forms, kinds, and processes, that includes any number of offsets and variables. By its permeability and absorbency, for example, the healthy soil corrects the irregularities of rainfall; by the diversity of its vegetation it protects against both disease and erosion. Most farms, even most fields, are made up of different kinds of soil patterns or soil sense. Good farmers have always known this and have used the land accordingly; they have been careful students of the natural vegetation, soil depth and structure, slope and drainage. They are not appliers of generalizations, theoretical or methodological or mechanical. Nor are they the active agents of their own economic will, working their way upon an inert and passive mass. They are responsive partners in an intimate and mutual relationship.

Because the soil is alive, various, intricate, and because its processes yield more readily to imitation than to analysis, more readily to care than to coercion, agriculture can never be an exact science. There is an inescapable kinship between farming and art, for farming depends as much on character, devotion, imagination, and the sense of structure, as on knowledge. It is a practical art.

But it is also a practical religion, a practice of religion, a rite. By farming we enact our fundamental connection with energy and matter, light and darkness. In the cycles of farming, which carry the elemental energy again and again through the seasons and the bodies of living things, we recognize the only infinitude within reach of the imagination. How long this cycling of energy will continue we do not know; it will have to end, at least here on this planet, sometime within the remaining life of the sun. But by aligning ourselves with it here, in our little time within the unimaginable time of the sun's burning, we touch infinity; we align ourselves with the universal law that brought the cycles into being and that will survive them.

The word *agriculture*, after all, does not mean "agriscience," much less "agribusiness." It means "cultivation of land." And *cultivation is* at the root of the sense both of *culture* and of *cult*. The ideas of tillage and worship are thus joined in *culture*. And these words all come from an Indo-European root meaning both "to revolve" and "to dwell." To live, to survive on the earth, to care for the soil, and to worship, all are bound at the root to the idea of a cycle. It is only by understanding the cultural complexity and largeness of the concept of agriculture that we can see the threatening diminishments implied by the term "agribusiness."

That agriculture is in so complex a sense a cultural endeavor —and that food is therefore a cultural product—would be regarded as heresy by most of the agencies, institutions, and publications of modern farming. The spokesmen of the official reckoning would doubtless respond that they are not cultural but scientific, that they are specialists of "agriscience." If agriculture is acknowledged to have anything to do with culture, then its study has to include people. But the agriculture experts ruled people out when they made their discipline a specialty —or, rather, when they sorted it into a collection of specialties —and moved it into its own "college" in the university. This specialty collection is interested in soils (in the limited sense of soil chemistry), in plants and animals, and in machines and chemicals. It is not interested in people.

But what respect is one to give to a science that parcels a unified discipline into discrete fragments, that has no interest in its effects if they are not immediately measurable in a laboratory, and that is founded upon the waste of topsoil, energy, and manpower, and upon the dissolution of communities? Not much. And it has been my experience that, with respect to this science, farmers are divided into two kinds: those who endanger their solvency, and often their sanity, by trusting it and those who hold it in contempt.

In the view of the experts, then, agriculture is not only not a concern of culture, but not even a concern of science, for they have abandoned interest in the health of the farming communities on the one hand and in the health of the land on the other. They appear to have concluded that agriculture is purely a commercial concern; its purpose is to provide as much food as quickly and cheaply and with as few man-hours as possible and to be a market for machines and chemicals. It is, after all, "agribusiness"—not the land or the farming people—that now benefits most from agricultural research and that can promote humble academicians to highly remunerative and powerful positions in corporations and in government. Former Secretary Earl Butz's career exemplifies the predominant direction of interest of the agriculture specialist. According to Lauren Soth, writing in the *Nation*, "Butz is the perfect example of the agribusiness, commercial-farming, agricultural-education establishment man. When dean of agriculture at Purdue University, he also sat on the boards of directors of the Ralston-Purina Co., the J. I. Case Co., International Minerals and Chemicals Corp., Stokely-Van Camp Co. and Standard Life Insurance Co. of Indiana." By such men and such careers the land-grant college system, originally meant to enhance the small-farm possibility, has been captured for the corporations.

The discipline of agriculture—the "great subject," as Sir Albert Howard called it, "of health in soil, plant, animal, and man"—has been reduced to fit first the views of a piecemeal "science" and then the purposes of corporate commerce. I can see no possibility of a doubt that this is true, though I cannot explain exactly how it happened. But it seems to me that the way was prepared when the specialized shapers or makers of agricultural thought simplified their understanding of energy and began to treat current, living, biological energy as if it were a *store* of energy extractable by machinery. At that point the living part of technology began to be overpowered by the mechanical. The machine was on its own, to follow its own logic of elaboration and growth apart from life, the standard that had previously defined its purposes and hence its limits. Let loose from any moral standard or limit, the machine was also let loose in another way: it replaced the Wheel of Life as the governing cultural metaphor. Life came to be seen as a road, to be traveled as fast as possible, never to return. Or, to put it another way, the Wheel of Life became an industrial metaphor; rather than turning in place, revolving in order to dwell, it began to roll on the "highway of progress" toward an ever-receding horizon. The idea, the responsibility, of return weakened and disappeared from agricultural discipline. Henceforth, any resource would be regarded as an ore.

If agriculture is founded upon life, upon the use of living energy to serve human life, and if its primary purpose must therefore be to preserve the integrity of the life cycle, then agricultural technology must be bound under the rule of life. It must conform to natural processes and limits rather than to mechanical or economic models. The culture that sustains agriculture and that it sustains must form its consciousness and its aspiration upon the correct metaphor of the Wheel of Life. The appropriate agricultural technology would therefore be diverse; it would aspire to diversity; it would enable the diversification of economies, methods, and species to conform to the diverse kinds of land. It would always use plants and animals together. It would be as attentive to decay as to growth, to maintenance as to production. It would return all wastes to the soil, control erosion, and conserve water. To enable care and devotion and to safeguard the local communities and cultures of agriculture, it would use the land in small holdings. It would aspire to make each farm so far as possible the source of its own operating energy, by the use of human energy, work animals, methane, wind or water or solar power.

The mechanical aspect of the technology would serve to harness or enhance the energy available on the farm. It would not be permitted to replace such energies with imported fuels, to replace people, or to replace or reduce human skills.

The damages of our present agriculture all come from the determination to use the life of the soil as if it were an extractable resource like coal, to use living things as if they were machines, to impose scientific (that is, laboratory) exactitude upon living complexities that are ultimately mysterious.

If animals are regarded as machines, they are confined in pens remote from the source of their food, where their excrement becomes, instead of a fertilizer, first a "waste" and then a pollutant. Furthermore, because confinement feeding depends so largely on grains, grass is removed from the rotation of crops and more land is exposed to erosion.

If plants are regarded as machines, we wind up with huge monocultures, productive of elaborate ecological mischiefs, which are in turn productive of agricultural mischief: monocultures are much more susceptible to pests and diseases than mixed cultures and are therefore more dependent on chemicals.

If the soil is regarded as a machine, then its life, its involvement in living systems and cycles, must perforce be ignored. It must be treated as a dead, inert chemical mass. If its life is ignored, then so must be the natural sources of its fertility —and not only ignored, but scorned. Alfalfa and the clovers, according to some of the most up-to-date practitioners, are "weeds"; the only legitimate source of nitrogen is the fertilizer manufacturer. And animal manures are "wastes"; "efficiency" cannot use them. Not long ago I found that the manure from a saddle-horse barn belonging to the University of Kentucky was simply being dumped. When I asked why it was not used somewhere on the farm, I was told that it would interfere with the College of Agriculture's experiments. The result is absurd: our agriculture, potentially capable of a large measure of independence, is absolutely dependent on petroleum, on the oil companies, and on the vagaries of politics.

If people are regarded as machines, they must be regarded as replaceable by other machines. They are regarded, in other words, as dispensable. Their place on the farm is safe only as long as they are mechanically necessary.

In modern agriculture, then, the machine metaphor is allowed to usurp and wipe from consideration not merely *some* values, but the very *issue* of value. Once the expert's interest is focused on the question of "what will work" within the exclusive confines of his theoretical model, values are no longer of any concern whatever. The confines of his specialty enable him to impose a biological totalitarianism on—he thinks, since he is an agricultural expert—the farm. When he leaves his office or laboratory he will, he assumes, go "home" to value.

But then it must be asked if we can remove cultural value from one part of our lives without destroying it also in the other parts. Can we justify secrecy, lying, and burglary in our so-called intelligence organizations and yet preserve openness, honesty, and devotion to principle in the rest of our government? Can we subsidize mayhem in the military establishment and yet have peace, order, and respect for human life in the city streets? Can we degrade all forms of essential work and yet expect arts and graces to flourish on weekends? And can we ignore all questions of value on the farm and yet have them answered affirmatively in the grocery store and the household?

The answer is that, though such distinctions can be made theoretically, they cannot be preserved in practice. Values may be corrupted or abolished in only one discipline at the start, but the damage must sooner or later spread to all; it can no more be confined than air pollution. If we corrupt agriculture we corrupt culture, for in nature and within certain invariable social necessities we are one body, and what afflicts the hand will afflict the brain.

The effective knowledge of this unity must reside not so much in doctrine as in skill. Skill, in the best sense, is the enactment or the acknowledgment or the signature of responsibility to other lives; it is the practical understanding of value. Its opposite is not merely unskillfulness, but ignorance of sources, dependences, relationships.

Skill is the connection between life and tools, or life and machines. Once, skill was defined ultimately in qualitative terms: How *well* did a person work; how good, durable, and pleasing were his products? But as machines have grown larger and more complex, and as our awe of them and our desire for labor-saving have grown, we have tended more and more to define skill quantitatively: How speedily and cheaply can a person work? We have increasingly wanted a *measurable* skill. And the more quantifiable skills became, the easier they were to replace with machines. As machines replace skill, they disconnect themselves from life; they come between us and life. They begin to enact our ignorance of value—of essential sources, dependences, and relationships.

The catch is that we cannot live in machines. We can only live in the world, in life. To live, our contact with the sources of life must remain direct: we must eat, drink, breathe, move, mate, etc. When we let machines and machine skills obscure the values that represent these fundamental dependences, then we inevitably damage the world; we diminish life. We begin to "prosper" at the cost of a fundamental degradation.

The digging stick, for example, brought in a profound technological revolution: it made agriculture possible. Its use required skill. But its *effect* also required skill, and this kind of skill was higher and more complex than the first, for it involved restraint and responsibility. The digging stick made it possible to grow food; that was one thing. It also made it possible, and necessary, to disturb the earth; and that was another thing. The first skill required others that were its moral elaboration: the skill used in disturbing the earth called directly for other skills that would preserve the earth and restore its fertility.

Until fairly recently, as agricultural tools became more efficient or powerful or both, they required an increase of both kinds of skill. One could do more with stone implements than with sticks, and more with metal implements than with stone implements; the skilled use of these tools enabled one to disturb more ground and so called for further elaboration of the skills of responsibility.

This remained true after the beginning of the use of draft animals. The skills of use had to become much greater, for the human mind had to relate to the animal mind in a new way: not by the magic and cunning of the hunt, but in the practical intricacies of collaboration. And the skills of responsibility had to increase proportionately. More ground could now be disturbed, and so the technology of preservation had to become much larger. Also, the investment of life in work greatly increased; people had to take responsibility not only for their own appetites and excrements but for those of their animals as well.

It was only with the introduction of self-powering machines, and of machine-extracted energy, into the fields that something really new happened to agricultural skills: they began a radical diminishment.

In the first place, it requires more skill to use a team of horses or mules or oxen than to use a tractor. It is more difficult to learn to manage an animal than a machine; it takes longer. Two minds and two wills are involved. A relationship between a person and a work animal is analogous to a relationship between two people. Success depends upon the animal's willingness and upon its health; certain moral imperatives and restraints are therefore pragmatically essential. No such relationship is either necessary or possible with a machine. Within the range of the possible, a machine is directly responsive to human will; it neither starts nor stops because it wants to. A machine has no life, and for this reason it cannot of itself impose any restraint or any moral limit on behavior.

In the second place, the substitution of machines for work animals is justified mainly by their ability to increase the volume of work per man—that is, by their greater speed. But as speed increases, care declines. And so, necessarily, do the skills of responsibility. If this were not so, we would not restrict the speed of traffic in residential areas. We know that there is a limit to the capacity of attention, and that the faster we go the less we see. This law applies with equal force to work; the faster we work the less attention we can pay to its details, and the less skill we can apply to it.

This is true of *any* productive work, and it has great cultural importance; at present we are all suffering, in various ways, from dependence on goods that are poorly made. But its importance in agricultural production is probably more critical than elsewhere. In any biological system the first principle is restraint—that is, the natural or moral checks that maintain a balance between use and continuity. The life of one year must not be allowed to diminish the life of the next; nothing must live at the expense of the source. Thus, in nature, the food species is dependent on its predator, and pests and diseases are agents of health; so populations are controlled and balanced. In agriculture these natural checks are removed and therefore must be replaced by the skills of responsibility, which have to do with the prevention of erosion, the diversification and rotation of plant and animal species, the return of wastes to the soil, and all the other provisionings of the source. When productive power—that is, speed—in machines replaces the productive skills of people, there is a consequent narrowing of attention. The machines are expensive and they run on purchased fuels; they feed upon money. The work of production is immediately profitable, whereas the work of responsibility is not. Once the machine is in the field it creates an economic pressure that enforces haste; the machine concentrates all the energy of the farm and hurries it toward the marketplace. The demands of immediate use eclipse the demands of continuity. As the skills of production decline, the skills of responsibility perish.

To argue for a balance between people and their tools, between life and machinery, between biological and machine-produced energy, is to argue for restraint upon the use of machines. The arguments that rise out of the machine metaphor—arguments for cheapness, efficiency, labor-saving, economic growth, etc.—all point to infinite industrial growth and infinite energy consumption. The moral argument points to restraint; it is a conclusion that may be in some sense tragic, but there is no escaping it. Much as we long for infinities of power and duration, we have no evidence that these lie within our reach, much less within our responsibility. It is more likely that we will have either to live within our limits, within the human definition, or not live at all. And certainly the knowledge of these limits and of how to live within them is the most comely and graceful knowledge that we have, the most healing and the most whole.

The knowledge that purports to be leading us to transcendence of our limits has been with us a long time. It thrives by offering material means of fulfilling a spiritual, and therefore materially unappeasable, craving: we would all very much like to be immortal, infallible, free of doubt, at rest. It is because this need is so large, and so different in kind from all material means, that the knowledge of transcendence—our entire history of scientific "miracles"—is so tentative, fragmentary, and grotesque. Though there are undoubtedly mechanical limits, because there are human limits, there is no mechanical restraint. The only logic of the machine is to get bigger and more elaborate. In the absence of *moral* restraint—and we have never imposed adequate moral restraint upon our use of machines—the machine is out of control by definition. From the beginning of the history of machine-developed energy, we have been able to harness more power than we could use responsibly. From the beginning, these machines have created effects that society could absorb only at the cost of suffering and disorder.

And so the issue is not of supply but of use. The energy crisis is not a crisis of technology but of morality. We already have available more power than we have so far dared to use. If, like the strip-miners and the "agribusinessmen," we look on all the world as fuel or as extractable energy, we can do nothing but destroy it. The issue is restraint. The energy crisis reduces to a single question: Can we forbear to do anything that we are able to do? Or to put the question in the words of Ivan Illich: Can we, believing in "the effectiveness of power," see "the disproportionately greater effectiveness of abstaining from its use"?

The only people among us that I know of who have answered this question convincingly in the affirmative are the Amish. They alone, as a community, have carefully restricted their use of machine-developed energy, and so have become the only true masters of technology. They are mostly farmers, and they do most of their farm work by hand and by the use of horses and mules. They are pacifists, they operate their own local schools, and in other ways hold themselves aloof from the ambitions of a machine-based society. And by doing so they have maintained the integrity of their families, their community, their religion, and their way of life. They have escaped the mainstream American life of distraction, haste, aimlessness, violence, and disintegration. Their life is not idly wasteful, or destructive. The Amish no doubt have their problems; I do not wish to imply that they are perfect. But it cannot be denied that they have mastered one of the fundamental paradoxes of our condition: we can make ourselves whole only by accepting our partiality, by living within our limits, by being human—not by trying to be gods. By restraint they make themselves whole.

But just stop for a minute and think about what it means to live in a land where 95 percent of the people can be freed from the drudgery of preparing their own food.

JAMES E. BOSTIC, JR., FORMER DEPUTY ASSISTANT SECRETARY OF AGRICULTURE FOR RURAL DEVELOPMENT

Find the shortest, simplest way between the earth, the hands and the mouth.

LANZA DEL VASTO

CHAPTER SEVEN

The Body and the Earth

ON THE CLIFF

THE question of human limits, of the proper definition and place of human beings within the order of Creation, finally rests upon our attitude toward our biological existence, the life of the body in this world. What value and respect do we give to our bodies? What uses do we have for them? What relation do we see, if any, between body and mind, or body and soul? What connections or responsibilities do we maintain between our bodies and the earth? These are religious questions, obviously, for our bodies are part of the Creation, and they involve us in all the issues of mystery. But the questions are also agricultural, for no matter how urban our life, our bodies live by farming; we come from the earth and return to it, and so we live in agriculture as we live in flesh. While we live our bodies are moving particles of the earth, joined inextricably both to the soil and to the bodies of other living creatures. It is hardly surprising, then, that there should be some profound resemblances between our treatment of our bodies and our treatment of the earth.

That humans are small within the Creation is an ancient perception, represented often enough in art that it must be supposed to have an elemental importance. On one of the painted walls of the Lascaux cave (20,000–15,000 B.C.), surrounded by the exquisitely shaped, shaded, and colored bodies of animals, there is the childish stick figure of a man, a huntsman who, having cast his spear into the guts of a bison, is now weaponless and vulnerable, poignantly frail, exposed, and incomplete. The message seems essentially that of the voice out of the whirlwind in the Book of Job: the Creation is bounteous and mysterious, and humanity is only a part of it—not its equal, much less its master.

Old Chinese landscape paintings reveal, among towering mountains, the frail outline of a roof or a tiny human figure passing along a road on foot or horseback. These landscapes are almost always populated. There is no implication of a dehumanized interest in nature "for its own sake." What is represented is a world in which humans belong, but which does not belong to humans in any tidy economic sense; the Creation provides a place for humans, but it is greater than humanity and within it even great men are small. Such humility is the consequence of an accurate insight, ecological in its bearing, not a pious deference to "spiritual" value.

Closer to us is a passage from the fourth act of *King Lear*, describing the outlook from one of the Dover cliffs:

> The crows and choughs that wing the midway air
> Show scarce so gross as beetles. Halfway down
> Hangs one that gathers samphire, dreadful trade!
> Methinks he seems no bigger than his head.
> The fishermen that walk upon the beach
> Appear like mice, and yond tall anchoring bark
> Diminished to her cock—her cock, a buoy
> Almost too small for sight.

And this is no mere description of a scenic "view." It is part of a play-within-a-play, a sort of ritual of healing. In it Shakespeare is concerned with the curative power of the perception we are dealing with: by understanding accurately his proper place in Creation, a man may be made whole.

In the lines quoted, Edgar, disguised as a lunatic, a Bedlamite, is speaking to his father, the Earl of Gloucester. Gloucester, having been blinded by the treachery of his false son, Edmund, has despaired and has asked the supposed madman to lead him to the cliff's edge, where he intends to destroy himself. But Edgar's description is from memory; the two are not standing on any such dizzy verge. What we are witnessing is the working out of Edgar's strategy to save his father from false feeling—both the pride, the smug credulity, that led to his suffering and the despair that is its result. These emotions are perceived as madness; Gloucester's blindness is literally the result of the moral blindness of his pride, and it is symbolic of the spiritual blindness of his despair.

Thinking himself on the edge of a cliff, he renounces this world and throws himself down. Though he falls only to the level of his own feet, he is momentarily stunned. Edgar remains with him, but now represents himself as an innocent bystander at the foot of what Gloucester will continue to think is a tall cliff. As the old man recovers his senses, Edgar persuades him that the madman who led him to the cliff's edge was in reality a "fiend." And Gloucester repents his self-destructiveness, which he now recognizes as another kind of pride; a human has no right to destroy what he did not create:

> You ever-gentle gods, take my breath from me.
> Let not my worser spirit tempt me again
> To die before you please.

What Gloucester has passed through, then, is a rite of death and rebirth. In his new awakening he is finally able to recognize his true son. He escapes the unhuman conditions of godly pride and fiendish despair and dies "smilingly" in the truly human estate "'Twixt two extremes of passion, joy and grief . . ."

Until modern times, we focused a great deal of the best of our thought upon such rituals of return to the human condition. Seeking enlightenment or the Promised Land or the way home, a man would go or be forced to go into the wilderness, measure himself against the Creation, recognize finally his true place within it, and thus be saved both from pride and from despair. Seeing himself as a tiny member of a world he cannot comprehend or master or in any final sense possess, he cannot possibly think of himself as a god. And by the same token, since he shares in, depends upon, and is graced by all of which he is a part, neither can he become a fiend; he cannot descend into the final despair of destructiveness. Returning from the wilderness, he becomes a restorer of order, a preserver. He sees the truth, recognizes his true heir, honors his forebears and his heritage, and gives his blessing to his successors. He embodies the passing of human time, living and dying within the human limits of grief and joy.

ON THE TOWER

Apparently with the rise of industry, we began to romanticize the wilderness—which is to say we began to institutionalize it within the concept of the "scenic." Because of railroads and improved highways, the wilderness was no longer an arduous passage for the traveler, but something to be looked at as grand or beautiful from the high vantages of the roadside. We became viewers of "views." And because we no longer traveled in the wilderness as a matter of course, we forgot that wilderness still circumscribed civilization and persisted in domesticity. We forgot, indeed, that the civilized and the domestic continued to *depend* upon wilderness—that is, upon natural forces within the climate and within the soil that have never in any meaningful sense been controlled or conquered. Modern civilization has been built largely in this forgetfulness.

And as we transformed the wilderness into scenery, we began to feel in the presence of "nature" an awe that was increasingly statistical. We would not become appreciators of the Creation until we had taken its measure. Once we had climbed or driven to the mountain top, we were awed by the view, but it was an awe that we felt compelled to validate or prove by the knowledge of how high we stood and how far we saw. We are invited to "see seven states from atop Lookout Mountain," as if our political boundaries had been drawn in red on the third morning of Creation.

We became less and less capable of sensing ourselves as small within Creation, partly because we thought we could comprehend it statistically, but also because we were becoming creators, ourselves, of a mechanical creation by which we felt ourselves greatly magnified. We built bridges that stood imposingly in titanic settings, towers that stood around us like geologic presences, single machines that could do the work of hundreds of people. Why, after all, should one get excited about a mountain when one can see almost as far from the top of a building, much farther from an airplane, farther still from a space capsule? We have learned to be fascinated by the statistics of magnitude and power. There is apparently no limit in sight, no end, and so it is no wonder that our minds, dizzy with numbers, take refuge in a yearning for infinitudes of energy and materials.

And yet these works that so magnify us also dwarf us, reduce us to insignificance. They magnify us because we are capable of them. They diminish us because, say what we will, once we build beyond a human scale, once we conceive ourselves as

Titans or as gods, we are lost in magnitude; we cannot control or limit what we do. The statistics of magnitude call out like Sirens to the statistics of destruction. If we have built towering cities, we have raised even higher the cloud of megadeath. If people are as grass before God, they are as nothing before their machines.

If we are fascinated by the statistics of magnitude, we are no less fascinated by the statistics of our insignificance. We never tire of repeating the commonizing figures of population and population growth. We are entranced to think of ourselves as specks on the pages of our own overwhelming history. I remember that my high-school biology text dealt with the human body by listing its constituent elements, measuring their quantities, and giving their monetary worth—at that time a little less than a dollar. That was a bit of the typical fodder of the modern mind, at once sensational and belittling—no accidental product of the age of Dachau and Hiroshima.

In our time Shakespeare's cliff has become the tower of a bridge—not the scene of a wakening rite of symbolic death and rebirth, but of the real and final death of suicide. Hart Crane wrote its paradigm, as if against his will, in *The Bridge*:

> Out of some subway scuttle, cell or loft
> A bedlamite speeds to thy parapets,
> Tilting there momentarily, shrill shirt ballooning,
> A jest falls from the speechless caravan.

In Shakespeare, the real Bedlamite or madman is the desperate and suicidal Gloucester. The supposed Bedlamite is in reality his true son, and together they enact an eloquent ritual in which Edgar gives his father a vision of Creation. Gloucester abandons himself to this vision, literally casting himself into it, and is renewed; he finds his life by losing it. Gloucester is saved by a renewal of his sense of the world and of his proper place in it. And this is brought about by an enactment that is communal, both in the sense that he is accompanied in it by his son, who for the time being has assumed the disguise of a madman but the role of a priest, and in the sense that it is deeply traditional in its symbols and meanings. In Crane, on the other hand, the Bedlamite is alone, surrounded by speechlessness, cut off within the crowd from any saving or renewing vision.

The height, which in Shakespeare is the traditional place of vision, has become in Crane a place of blindness; the bridge, which Crane intended as a unifying symbol, has become the symbol of a final estrangement.

HEALTH

After I had begun to think about these things, I received a letter containing an account of a more recent suicide. The following sentences from that letter seem both to corroborate Crane's lines and to clarify them:

"My friend ——— jumped off the Golden Gate Bridge two months ago. . . . She had been terribly depressed for years. There was no help for her. None that she could find that was sufficient. She was trying to get from one phase of her life to another, and couldn't make it. She had been terribly wounded as a child. . . . Her wound could not be healed. She destroyed herself."

The letter had already asked, "How does a human pass through youth to maturity without 'breaking down'?" And it had answered: "help from tradition, through ceremonies and rituals, rites of passage at the most difficult stages."

My correspondent went on to say: "Healing, it seems to me, is a necessary and useful word when we talk about agriculture." And a few paragraphs later he wrote: "The theme of suicide belongs in a book about agriculture . . ."

I agree. But I am also aware that many people will find it exceedingly strange that these themes should enter so forcibly into this book. It will be thought that I am off the subject. And so I want to take pains to show that I am on the subject —and on it, moreover, in the only way most people have of getting on it: by way of the issue of their own health. Indeed, it is when one approaches agriculture from any *other* issue than that of health that one may be said to be off the subject.

The difficulty probably lies in our narrowed understanding of the word *health*. That there is some connection between how we feel and what we eat, between our bodies and the earth, is acknowledged when we say that we must "eat right to keep fit" or that we should eat "a balanced diet." But by health we mean little more than how we feel. We are healthy, we think, if we do not feel any pain or too much pain, and if we are strong enough to do our work. If we become unhealthy, then we go to a doctor who we hope will "cure" us and restore us to health. By health, in other words, we mean merely the absence of disease. Our health professionals are interested almost exclusively in preventing disease (mainly by destroying germs) and in curing disease (mainly by surgery and by destroying germs).

But the concept of health is rooted in the concept of wholeness. To be healthy is to be whole. The word *health* belongs to a family of words, a listing of which will suggest how far the consideration of health must carry us: *heal, whole, wholesome, hale, hallow, holy.* And so it is possible to give a definition to health that is positive and far more elaborate than that given to it by most medical doctors and the officers of public health.

If the body is healthy, then it is whole. But how can it be whole and yet be dependent, as it obviously is, upon other bodies and upon the earth, upon all the rest of Creation, in fact? It becomes clear that the health or wholeness of the body is a vast subject, and that to preserve it calls for a vast enterprise. Blake said that "Man has no Body distinct from his Soul . . ." and thus acknowledged the convergence of health and holiness. In that, all the convergences and dependences of Creation are surely implied. Our bodies are also not distinct from the bodies of other people, on which they depend in a complexity of ways from biological to spiritual. They are not distinct from the bodies of plants and animals, with which we are involved in the cycles of feeding and in the intricate companionships of ecological systems and of the spirit. They are not distinct from the earth, the sun and moon, and the other heavenly bodies.

It is therefore absurd to approach the subject of health piecemeal with a departmentalized band of specialists. A medical doctor uninterested in nutrition, in agriculture, in the wholesomeness of mind and spirit is as absurd as a farmer who is uninterested in health. Our fragmentation of this subject cannot be our cure, because it is our disease. The body cannot be whole alone. Persons cannot be whole alone. It is wrong to think that bodily health is compatible with spiritual confusion or cultural disorder, or with polluted air and water or impoverished soil. Intellectually, we know that these patterns of interdependence exist; we understand them better now perhaps than we ever have before; yet modern social and cultural patterns contradict them and make it difficult or impossible to honor them in practice.

To try to heal the body alone is to collaborate in the destruction of the body. Healing is impossible in loneliness; it is the opposite of loneliness. Conviviality is healing. To be healed we must come with all the other creatures to the feast of Creation. Together, the above two descriptions of suicides suggest this very powerfully. The setting of both is urban, amid the gigantic works of modern humanity. The fatal sickness is despair, a wound that cannot be healed because it is encapsulated in loneliness, surrounded by speechlessness. Past the scale of the human, our works do not liberate us—they confine us. They cut off access to the wilderness of Creation where we must go to be reborn—to receive the awareness, at once humbling and exhilarating, grievous and joyful, that we are a part of Creation, one with all that we live from and all that, in turn, lives from us. They destroy the communal rites of passage that turn us toward the wilderness and bring us home again.

THE ISOLATION OF THE BODY

Perhaps the fundamental damage of the specialist system—the damage from which all other damages issue—has been the isolation of the body. At some point we began to assume that the life of the body would be the business of grocers and medical doctors, who need take no interest in the spirit, whereas the life of the spirit would be the business of churches, which would have at best only a negative interest in the body. In the same way we began to see nothing wrong with putting the body—most often somebody else's body, but frequently our own—to a task that insulted the mind and demeaned the spirit. And we began to find it easier than ever to prefer our own bodies to the bodies of other creatures and to abuse, exploit, and otherwise hold in contempt those other bodies for the greater good or comfort of our own.

The isolation of the body sets it into direct conflict with

everything else in Creation. It gives it a value that is destructive of every other value. That this has happened is paradoxical, for the body was set apart from the soul in order that the soul should triumph over the body. The aim is stated in Shakespeare's Sonnet 146 as plainly as anywhere:

> Poor soul, the center of my sinful earth,
> Lord of these rebel powers that thee array,
> Why dost thou pine within and suffer dearth,
> Painting thy outward walls so costly gay?
> Why so large cost, having so short a lease,
> Dost thou upon thy fading mansion spend?
> Shall worms, inheritors of this excess,
> Eat up thy charge? Is this thy body's end?
> Then, soul, live thou upon thy servant's loss,
> And let that pine to aggravate thy store;
> Buy terms divine in selling hours of dross;
> Within be fed, without be rich no more.
> So shalt thou feed on death, that feeds on men,
> And death once dead, there's no more dying then.

The soul is thus set against the body, to thrive at the body's expense. And so a spiritual economy is devised within which the only law is competition. If the soul is to live in this world only by denying the body, then its relation to worldly life becomes extremely simple and superficial. Too simple and superficial, in fact, to cope in any meaningful or useful way with the world. Spiritual value ceases to have any worldly purpose or force. To fail to employ the body in this world at once for its own good and the good of the soul is to issue an invitation to disorder of the most serious kind.

What was not foreseen in this simple-minded economics of religion was that it is not possible to devalue the body and value the soul. The body, cast loose from the soul, is on its own. Devalued and cast out of the temple, the body does not skulk off like a sick dog to die in the bushes. It sets up a counterpart economy of its own, based also on the law of competition, in which it devalues and exploits the spirit. These two economies maintain themselves at each other's expense, living upon each other's loss, collaborating without cease in mutual futility and absurdity.

You cannot devalue the body and value the soul—or value anything else. The prototypical act issuing from this division was to make a person a slave and then instruct him in religion —a "charity" more damaging to the master than to the slave. Contempt for the body is invariably manifested in contempt for other bodies—the bodies of slaves, laborers, women, animals, plants, the earth itself. Relationships with all other creatures become competitive and exploitive rather than collaborative and convivial. The world is seen and dealt with, not as an ecological community, but as a stock exchange, the ethics of which are based on the tragically misnamed "law of the jungle." This "jungle" law is a basic fallacy of modern culture. The body is degraded and saddened by being set in conflict against the Creation itself, of which all bodies are members, therefore members of each other. The body is thus sent to war against itself.

Divided, set against each other, body and soul drive each other to extremes of misapprehension and folly. Nothing could be more absurd than to despise the body and yet yearn for its resurrection. In reaction to this supposedly religious attitude, we get, not reverence or respect for the body, but another kind of contempt: the desire to comfort and indulge the body with equal disregard for its health. The "dialogue of body and soul" in our time is being carried on between those who despise the body for the sake of its resurrection and those, diseased by bodily extravagance and lack of exercise, who nevertheless desire longevity above all things. These think that they oppose each other, and yet they could not exist apart. They are locked in a conflict that is really their collaboration in the destruction of soul and body both.

What this conflict has done, among other things, is to make it extremely difficult to set a proper value on the life of the body in this world—to believe that it is good, howbeit short and imperfect. Until we are able to say this and know what we mean by it, we will not be able to live our lives in the human estate of grief and joy, but repeatedly will be cast outside in violent swings between pride and despair. Desires that cannot be fulfilled in health will keep us hopelessly restless and unsatisfied.

COMPETITION

By dividing body and soul, we divide both from all else. We thus condemn ourselves to a loneliness for which the only compensation is violence—against other creatures, against the earth, against ourselves. For no matter the distinctions we draw between body and soul, body and earth, ourselves and others—the connections, the dependences, the identities remain. And so we fail to contain or control our violence. It gets loose. Though there are categories of violence, or so we think, there are no categories of victims. Violence against one is ultimately violence against all. The willingness to abuse other bodies is the willingness to abuse one's own. To damage the earth is to damage your children. To despise the ground is to despise its fruit; to despise the fruit is to despise its eaters. The wholeness of health is broken by despite.

If competition is the correct relation of creatures to one another and to the earth, then we must ask why exploitation is not more successful than it is. Why, having lived so long at the expense of other creatures and the earth, are we not healthier and happier than we are? Why does modern society exist under constant threat of the same suffering, deprivation, spite, contempt, and obliteration that it has imposed on other people and other creatures? Why do the health of the body and the health of the earth decline together? And why, in consideration of this decline of our worldly flesh and household, our "sinful earth," are we not healthier in spirit?

It is not necessary to have recourse to statistics to see that the human estate is declining with the estate of nature, and that the corruption of the body is the corruption of the soul. I know that the country is full of "leaders" and experts of various sorts who are using statistics to prove the opposite: that we have more cars, more super-highways, more TV sets, motorboats, prepared foods, etc., than any people ever had before—and are therefore better off than any people ever were before. I can see the burgeoning of this "consumer economy" and can appreciate some of its attractions and comforts. But that economy has an inside and an outside; from the outside there are other things to be seen.

I am writing this in the north-central part of Kentucky on a morning near the end of June. We have had rain for two days, hard rain during the last several hours. From where I sit I can see the Kentucky River swiftening and rising, the water already yellow with mud. I know that inside this city-oriented consumer economy there are many people who will never see this muddy rise and many who will see it without knowing what it means. I know also that there are many who will see it, and know what it means, and not care. If it lasts until the weekend there will be people who will find it as good as clear water for motorboating and waterskiing.

In the past several days I have seen some of the worst-eroded corn fields that I have seen in this country in my life. This erosion is occurring on the cash-rented farms of farmers' widows and city farmers, absentee owners, the doctors and businessmen who buy a farm for the tax breaks or to have "a quiet place in the country" for the weekends. It is the direct result of economic and agricultural policy; it might be said to *be* an economic and agricultural policy. The signs of the "agridollar," big-business fantasy of the Butz mentality are all present: the absenteeism, the temporary and shallow interest of the land-renter, the row-cropping of slopes, the lack of rotation, the plowed-out waterways, the rows running up and down the hills. Looked at from the field's edge, this is ruin, criminal folly, moral idiocy. Looked at from Washington, D.C., from inside the "economy," it is called "free enterprise" and "full production."

And around me here, as everywhere else I have been in this country—in Nebraska, Iowa, Indiana, New York, New England, Tennessee—the farmland is in general decline: fields and whole farms abandoned, given up with their scars unmended, washing away under the weeds and bushes; fine land put to row crops year after year, without rest or rotation; buildings and fences going down; good houses standing empty, unpainted, their windows broken.

And it is clear to anyone who looks carefully at any crowd that we are wasting our bodies exactly as we are wasting our land. Our bodies are fat, weak, joyless, sickly, ugly, the virtual prey of the manufacturers of medicine and cosmetics. Our bodies have become marginal; they are growing useless like our "marginal" land because we have less and less use for them. After the games and idle flourishes of modern youth, we use them only as shipping cartons to transport our brains and our few employable muscles back and forth to work.

As for our spirits, they seem more and more to comfort themselves by buying things. No longer in need of the exalted drama of grief and joy, they feed now on little shocks of greed, scandal, and violence. For many of the churchly, the life of the spirit is reduced to a dull preoccupation with getting to Heaven. At best, the world is no more than an embarrassment and a trial to the spirit, which is otherwise radically separated from it. The true lover of God must not be burdened with any care or respect for His works. While the body goes about its business of destroying the earth, the soul is supposed to lie back and wait for Sunday, keeping itself free of earthly contaminants. While the body exploits other bodies, the soul stands aloof, free from sin, crying to the gawking bystanders: "I am not enjoying it!" As far as this sort of "religion" is concerned, the body is no more than the lusterless container of the soul, a mere "package," that will nevertheless light up in eternity, forever cool and shiny as a neon cross. This separation of the soul from the body and from the world is no disease of the fringe, no aberration, but a fracture that runs through the mentality of institutional religion like a geologic fault. And this rift in the mentality of religion continues to characterize the modern mind, no matter how secular or worldly it becomes.

But I have not stated my point exactly enough. This rift is not *like* a geologic fault; it *is* a geologic fault. It is a flaw in the mind that runs inevitably into the earth. Thought affects or afflicts substance neither by intention nor by accident, but because, occurring in the Creation that is unified and whole, it must; there is no help for it.

The soul, in its loneliness, hopes only for "salvation." And yet what is the burden of the Bible if not a sense of the mutuality of influence, rising out of an essential unity, among soul and body and community and world? These are all the works of God, and it is therefore the work of virtue to make or restore harmony among them. The world is certainly thought of as a place of spiritual trial, but it is also the confluence of soul and body, word and flesh, where thoughts must become deeds, where goodness is to be enacted. This is the great meeting place, the narrow passage where spirit and flesh, word and world, pass into each other. The Bible's aim, as I read it, is not the freeing of the spirit from the world. It is the handbook of their interaction. It says that they cannot be divided; that their mutuality, their unity, is inescapable; that they are not reconciled in division, but in harmony. What else can be meant by the resurrection of the body? The body should be "filled with light," perfected in understanding. And so everywhere there is the sense of consequence, fear and desire, grief and joy. What is desirable is repeatedly defined in the tensions of the sense of consequence. False prophets are to be known "by their fruits." We are to treat others as we would be treated; thought is thus barred from any easy escape into aspiration or ideal, is turned around and forced into action. The following verses from Proverbs are not very likely the original work of a philosopher-king; they are overheard from generations of agrarian grandparents whose experience taught them that spiritual qualities become earthly events:

> I went by the field of the slothful, and by the vineyard of the man void of understanding;
>
> And, lo, it was all grown over with thorns, and nettles had covered the face thereof, and the stone wall thereof was broken down.
>
> Then I saw, and considered it well. I looked upon it, and received instruction.
>
> Yet a little sleep, a little slumber, a little folding of the hands to sleep:
>
> So shall thy poverty come as one that traveleth; and thy want as an armed man.

CONNECTIONS

I do not want to speak of unity misleadingly or too simply. Obvious distinctions can be made between body and soul, one body and other bodies, body and world, etc. But these things that appear to be distinct are nevertheless caught in a network of mutual dependence and influence that is the substantiation of their unity. Body, soul (or mind or spirit), community, and world are all susceptible to each other's influence, and they are all conductors of each other's influence. The body is damaged by the bewilderment of the spirit, and it conducts the influence of that bewilderment into the earth, the earth conducts it into the community, and so on. If a farmer fails to understand what health is, his farm becomes unhealthy; it produces unhealthy food, which damages the health of the community. But this is a network, a spherical network, by which each part is connected to every other part. The farmer is a part of the community, and so it is as impossible to say exactly where the trouble began as to say where it will end. The influences go backward and forward, up and down, round and round, compounding and branching as they go. All that is certain is that an error introduced anywhere in the network ramifies beyond the scope of prediction; consequences occur all over the place, and each consequence breeds further consequences. But it seems unlikely that an error can ramify endlessly. It spreads by way of the connections in the network, but sooner or later it must also begin to break them. We are talking, obviously, about a circulatory system, and a disease of a circulatory system tends first to impair circulation and then to stop it altogether.

Healing, on the other hand, complicates the system by opening and restoring connections among the various parts—in this way restoring the ultimate simplicity of their union. When all the parts of the body are working together, are under each other's influence, we say that it is whole; it is healthy. The same is true of the world, of which our bodies are parts. The parts are healthy insofar as they are joined harmoniously to the whole.

What the specialization of our age suggests, in one example after another, is not only that fragmentation is a disease, but that the diseases of the disconnected parts are similar or analogous to one another. Thus they memorialize their lost unity, their relation persisting in their disconnection. Any severance produces two wounds that are, among other things, the record of how the severed parts once fitted together.

The so-called identity crisis, for instance, is a disease that seems to have become prevalent after the disconnection of body and soul and the other piecemealings of the modern period. One's "identity" is apparently the immaterial part of one's being—also known as psyche, soul, spirit, self, mind, etc. The dividing of this principle from the body and from any particular worldly locality would seem reason enough for a crisis. Treatment, it might be thought, would logically consist in the restoration of these connections: the lost identity would find itself by recognizing physical landmarks, by connecting itself responsibly to practical circumstances; it would learn to stay put in the body to which it belongs and in the place to which preference or history or accident has brought it; it would, in short, find itself in finding its work. But "finding yourself," the pseudo-ritual by which the identity crisis is supposed to be resolved, makes use of no such immediate references. Leaving aside the obvious, and ancient, realities of doubt and self-doubt, as well as the authentic madness that is often the result of cultural disintegration, it seems likely that the identity crisis is a conventional illusion, one of the genres of self-indulgence. It can be an excuse for irresponsibility or a fashionable mode of self-dramatization. It is the easiest form of self-flattery—a way to construe procrastination as a virtue—based on the romantic assumption that "who I really am" is better in some fundamental way than the available evidence would suggest.

The fashionable cure for this condition, if I understand the lore of it correctly, has nothing to do with the assumption of responsibilities or the renewal of connections. The cure is "autonomy," another illusory condition, suggesting that the self can be self-determining and independent without regard for any determining circumstance or any of the obvious dependences. This seems little more than a jargon term for indifference to the opinions and feelings of other people. There is, in practice, no such thing as autonomy. Practically, there is only a distinction between responsible and irresponsible dependence. Inevitably failing this impossible standard of autonomy, the modern self-seeker becomes a tourist of cures, submitting his quest to the guidance of one guru after another. The "cure" thus preserves the disease.

It is not surprising that this strange disease of the spirit—the self's loss of self—should have its counterpart in an anguish of the body. One of the commonplaces of modern experience is dissatisfaction with the body—not as one has allowed it to become, but as it naturally is. The hardship is perhaps greater here because the body, unlike the self, is substantial and cannot be supposed to be inherently better than it was born to be. It can only be thought inherently worse than it *ought* to be. For the appropriate standard for the body—that is, health—has been replaced, not even by another standard, but by very exclusive physical *models.* The concept of "model" here conforms very closely to the model of the scientists and planners: it is an exclusive, narrowly defined ideal which affects destructively whatever it does not include.

Thus our young people are offered the ideal of health only by what they know to be lip service. What they are made to feel forcibly, and to measure themselves by, is the exclusive desirability of a certain physical model. Girls are taught to want to be leggy, slender, large-breasted, curly-haired, unimposingly beautiful. Boys are instructed to be "athletic" in build, tall but not too tall, broad-shouldered, deep-chested, narrow-hipped, square-jawed, straight-nosed, not bald, unimposingly handsome. Both sexes should look what passes for "sexy" in a bathing suit. Neither, above all, should look old.

Though many people, in health, are beautiful, very few resemble these models. The result is widespread suffering that does immeasurable damage both to individual persons and to the society as a whole. The result is another absurd pseudo-ritual, "accepting one's body," which may take years or may be the distraction of a lifetime. Woe to the man who is short or skinny or bald. Woe to the man with a big nose. Woe, above all, to the woman with small breasts or a muscular body or strong features; Homer and Solomon might have thought her beautiful, but she will see her own beauty only by a difficult rebellion. And like the crisis of identity, this crisis of the body brings a helpless dependence on cures. One spends one's life dressing and "making up" to compensate for one's supposed deficiencies. Again, the cure preserves the disease. And the putative healer is the guru of style and beauty aid. The sufferer is by definition a customer.

SEXUAL DIVISION

To divide body and soul, or body and mind, is to inaugurate an expanding series of divisions—not, however, an *infinitely* expanding series, because it is apparently the nature of division sooner or later to destroy what is divided; the principle of durability is unity. The divisions issuing from the division of body and soul are first sexual and then ecological. Many other divisions branch out from those, but those are the most important because they have to do with the fundamental relationships —with each other and with the earth—that we all have in common.

To think of the body as separate from the soul or as soulless, either to subvert its appetites or to "free" them, is to make an object of it. As a thing, the body is denied any dimension or rightful presence or claim in the mind. The concerns of the body—all that is comprehended in the term *nurture*—are thus degraded, denied any respected place among the "higher things" and even among the more exigent practicalities.

The first sexual division comes about when nurture is made the exclusive concern of women. This cannot happen until a society becomes industrial; in hunting and gathering and in agricultural societies, men are of necessity also involved in nurture. In those societies there usually have been differences between the work of men and that of women. But the necessity here is to distinguish between sexual difference and sexual division.

In an industrial society, following the division of body and soul, we have at the "upper" or professional level a division between "culture" (in the specialized sense of religion, philosophy, art, the humanities, etc.) and "practicality," and both of these become increasingly abstract. Thinkers do not act. And the "practical" men do not work with their hands, but manipulate the abstract quantities and values that come from the work of "workers." Workers are simplified or specialized into machine parts to do the wage-work of the body, which they were initially permitted to think of as "manly" because for the most part women did not do it.

Women traditionally have performed the most confining—though not necessarily the least dignified—tasks of nurture: housekeeping, the care of young children, food preparation. In the urban-industrial situation the confinement of these traditional tasks divided women more and more from the "important" activities of the new economy. Furthermore, in this situation the traditional nurturing role of men—that of provisioning the household, which in an agricultural society had become as constant and as complex as the women's role —became completely abstract; the man's duty to the household came to be simply to provide money. The only remaining *task* of provisioning—purchasing food—was turned over to women. This determination that nurturing should become *exclusively* a concern of women served to signify to both sexes that neither nurture nor womanhood was very important.

But the assignment to women of a kind of work that was thought both onerous and trivial was only the beginning of their exploitation. As the persons exclusively in charge of the tasks of nurture, women often came into sole charge of the household budget; they became family purchasing agents. The time of the household barterer was past. Kitchens were now run on a cash economy. Women had become customers, a fact not long wasted on the salesmen, who saw that in these women they had customers of a new and most promising kind. The modern housewife was isolated from her husband, from her school-age children, and from other women. She was saddled with work from which much of the skill, hence much of the dignity, had been withdrawn, and which she herself was less and less able to consider important. She did not know what her husband did at work, or after work, and she knew that her life was passing in his regardlessness and in his absence. Such a woman was ripe for a sales talk: this was the great commercial insight of modern times. Such a woman must be told—or subtly made to understand—that she must not be a drudge, that she must not let her work affect her looks, that she must not become "unattractive," that she must always be fresh, cheerful, young, shapely, and pretty. All her sexual and mortal fears would thus be given voice, and she would be made to reach for money. What was implied was always the question that a certain bank finally asked outright in a billboard advertisement: "Is your husband losing interest?"

Motivated no longer by practical needs, but by loneliness and fear, women began to identify themselves by what they bought rather than by what they did. They bought labor-saving devices which worked, as most modern machines have tended to work, to devalue or replace the skills of those who used them. They bought manufactured foods, which did likewise.

They bought any product that offered to lighten the burdens of housework, to be "kind to hands," or to endear one to one's husband. And they furnished their houses, as they made up their faces and selected their clothes, neither by custom nor invention, but by the suggestion of articles and advertisements in "women's magazines." Thus housewifery, once a complex discipline acknowledged to be one of the bases of culture and economy, was reduced to the exercise of purchasing power.* The housewife's only remaining productive capacity was that of reproduction. But even as a mother she remained a consumer, subjecting herself to an all-presuming doctor and again to written instructions calculated to result in the purchase of merchandise. Breast-feeding of babies became unfashionable, one suspects, because it was the last form of home production; no way could be found to persuade a woman to purchase her own milk. All these "improvements" involved a radical simplification of mind that was bound to have complicated, and ironic, results. As housekeeping became simpler and easier, it also became more boring. A woman's work became less accomplished and less satisfying. It became easier for her to believe that what she did was not important. And this heightened her anxiety and made her even more avid and even less discriminating as a consumer. The cure not only preserved the disease, it compounded it.

There was, of course, a complementary development in the minds of men, but there is less to say about it. The man's mind was not simplified by a degenerative process, but by a kind of coup: as soon as he separated working and living and began to work away from home, the practical considerations of the household were excerpted from his mind all at once.

In modern marriage, then, what was once a difference of work became a division of work. And in this division the household was destroyed as a practical bond between husband and

*She did continue to do "housework," of course. But we must ask what this had come to mean. The industrial economy had changed the criterion of housekeeping from thrift to convenience. Thrift was a complex standard, requiring skill, intelligence, and moral character, and private thrift was rightly considered a public value. Once thrift was destroyed as a value, housekeeping became simply a corrupt function of a corrupt economy: its public "value" lay in the wearing out or using up of commodities.

wife. It was no longer a condition, but only a place. It was no longer a circumstance that required, dignified, and rewarded the enactment of mutual dependence, but the site of mutual estrangement. Home became a place for the husband to go when he was not working or amusing himself. It was the place where the wife was held in servitude.

A sexual difference is not a wound, or it need not be; a sexual division is. And it is important to recognize that this division—this destroyed household that now stands between the sexes—is a wound that is suffered inescapably by both men and women. Sometimes it is assumed that the estrangement of women in their circumscribed "women's world" can only be for the benefit of men. But that interpretation seems to be based on the law of competition that is modeled in the exploitive industrial economy. This law holds that for everything that is exploited or oppressed there must be something else that is proportionately improved; thus, men must be as happy as women are unhappy.

There is no doubt that women have been deformed by the degenerate housewifery that is now called their "role"—but not, I think, for any man's benefit. If women are deformed by their role, then, insofar as the roles are divided, men are deformed by theirs. Degenerate housewifery is indivisible from degenerate husbandry. There is no escape. This is the justice that we are learning from the ecologists: you cannot damage what you are dependent upon without damaging yourself. The suffering of women is noticed now, is noticeable now, because it is not given any considerable status or compensation. If we removed the status and compensation from the destructive exploits we classify as "manly," men would be found to be suffering as much as women. They would be found to be suffering for the same reason: they are in exile from the communion of men and women, which is their deepest connection with the communion of all creatures.

For example: a man who is in the traditional sense a *good* farmer is husbandman and husband, the begetter and conserver of the earth's bounty, but he is also midwife and motherer. He is a nurturer of life. His work is domestic; he is bound to the household. But let "progress" take such a man and transform him into a technologist of production (that is, sever his bonds to the household, make useless or pointless or "uneconomical" his impulse to conserve and to nurture), and it will have made of him a creature as deformed, and as pained, as it has notoriously made of his wife.

THE DISMEMBERMENT OF THE HOUSEHOLD

We are familiar with the concept of the disintegral life of our time as a dismembered cathedral, the various concerns of culture no longer existing in reference to each other or within the discipline of any understanding of their unity. It may also be conceived, and its strains more immediately felt, as a dismembered household. Without the household—not just as a unifying ideal, but as a practical circumstance of mutual dependence and obligation, requiring skill, moral discipline, and work—husband and wife find it less and less possible to imagine and enact their marriage. Without much in particular that they can *do* for each other, they have a scarcity of practical reasons to be together. They may "like each other's company," but that is a reason for friendship, not for marriage. Aside from affection for any children they may have and their abstract legal and economic obligations to each other, their union has to be empowered by sexual energy alone.

Perhaps the most dangerous, certainly the most immediately painful, consequence of the disintegration of the household is this isolation of sexuality. The division of sexual energy from the functions of household and community that it ought both to empower and to grace is analogous to that other modern division between hunger and the earth. When it is no longer allied by proximity and analogy to the nurturing disciplines that bound the household to the cycles of fertility and the seasons, life and death, then sexual love loses its symbolic or ritualistic force, its deepest solemnity and its highest joy. It loses its sense of consequence and responsibility. It becomes "autonomous," to be valued only for its own sake, therefore frivolous, therefore destructive—even of itself. Those who speak of sex as "recreation," thinking to claim for it "a new place," only acknowledge its displacement from Creation.

The isolation of sexuality makes it subject to two influences that dangerously oversimplify it: the lore of sexual romance and capitalist economics. By "sexual romance" I mean the sentimentalization of sexual love that for generations has been the work of popular songs and stories. By means of them, young people have been taught a series of extremely dangerous falsehoods:

1. That people in love ought to conform to the fashionable models of physical beauty, and that to be unbeautiful by these standards is to be unlovable.
2. That people in love are, or ought to be, young—even though love is said to last "forever."
3. That marriage is a solution—whereas the most misleading thing a love story can do is to end "happily" with a marriage, not because there is no such thing as a happy marriage, but because marriage cannot be happy except by being *made* happy.
4. That love, alone, regardless of circumstances, can make harmony and resolve serious differences.
5. That "love will find a way" and so finally triumph over any kind of practical difficulty.
6. That the "right" partners are "made for each other," or that "marriages are made in Heaven."
7. That lovers are "each other's all" or "all the world to each other."
8. That monogamous marriage is therefore logical and natural, and "forsaking all others" involves no difficulty.

Believing these things, a young couple could not be more cruelly exposed to the abrasions of experience—or better prepared to experience marriage as another of those grim and ironic modern competitions in which the victory of one is the defeat of both.

As experience frets away gullibility, the exclusiveness of the sentimental ideal gives way to the possessiveness of sexual capitalism. Failing, as they cannot help but fail, to be each other's all, the husband and wife become each other's only. The sacrament of sexual union, which in the time of the household was a communion of workmates, and afterward tried to be a lovers' paradise, has now become a kind of marketplace in which husband and wife represent each other as sexual property. Competitiveness and jealousy, imperfectly sweetened and disguised by the illusions of courtship, now become governing principles, and they work to isolate the couple inside their marriage. Marriage becomes a capsule of sexual fate. The man must look on other men, and the woman on other women, as threats. This seems to have become particularly damaging to women; because of the progressive degeneration and isolation of their "role," their worldly stock in trade has increasingly had to be "their" men. In the isolation of the resulting sexual "privacy," the disintegration of the community begins. The energy that is the most convivial and unifying loses its communal forms and becomes divisive. This dispersal was nowhere more poignantly exemplified than in the replacement of the old ring dances, in which all couples danced together, by the so-called ballroom dancing, in which each couple dances alone. A significant part of the etiquette of ballroom dancing is, or was, that the exchange of partners was accomplished by a "trade." It is no accident that this capitalization of love and marriage was followed by a divorce epidemic—and by fashions of dancing in which each one of the dancers moves alone.

The disintegration of marriage, which completes the disintegration of community, came about because the encapsulation of sexuality, meant to preserve marriage from competition, inevitably *enclosed* competition. The principle that fenced everyone else out fenced the couple in; it became a sexual cul-de-sac. The model of economic competition proved as false to marriage as to farming. As with other capsules, the narrowness of the selective principle proved destructive of what it excluded, and what it excluded was essential to the life of what it enclosed: the nature of sexuality itself. Sexual romance cannot bear to acknowledge the generality of instinct, whereas sexual capitalism cannot acknowledge its particularity. But sexuality appears to be *both* general and particular. One cannot love a particular woman, for instance, unless one loves womankind—if not all women, at least other women. The capsule of sexual romance leaves out this generality, this generosity of instinct; it excludes Aphrodite and Dionysus. And it fails for that reason. Though sexual love can endure between the same two people for a long time, it cannot do so on the basis of this pretense of the exclusiveness of affection. The sexual capitalist—that is, the disillusioned sexual romantic—in reaction to disillusion makes the opposite oversimplification; one acknowledges one's spouse as one of a general, necessarily troublesome kind or category.

Both these attitudes look on sexual love as ownership. The sexual romantic croons, "You be-long to me." The sexual capitalist believes the same thing but has stopped crooning. Each holds that a person's sexual property shall be sufficient unto him or unto her, and that the morality of that sufficiency is to be forever on guard against expropriation. Within the capsule of marriage, as in that of economics, one intends to exploit one's property and to protect it. Once the idea of property becomes abstract or economic, both these motives begin to rule over it. They are, of course, contradictory; all that one can really protect is one's "right" or intention to exploit. The proprieties and privacies used to encapsulate marriage may have come from the tacit recognition that exploitive sex, like exploitive economics, is a very dirty business. One makes a secret of the sexuality of one's marriage for the same reason that one posts "Keep Out/Private Property" on one's strip mine. The tragedy, more often felt than acknowledged, is that what is exploited becomes undesirable.

The protective capsule becomes a prison. It becomes a household of the living dead, each body a piece of incriminating evidence. Or a greenhouse excluding the neighbors and the weather for the sake of some alien and unnatural growth. The marriage shrinks to a dull vigil of duty and legality. Husband and wife become competitors necessarily, for their only freedom is to exploit each other or to escape.

It is possible to imagine a more generous enclosure—a household welcoming to neighbors and friends; a garden open to the weather, between the woods and the road. It is possible to imagine a marriage bond that would bind a woman and a man not only to each other, but to the community of marriage, the amorous communion at which all couples sit: the sexual feast and celebration that joins them to all living things and to the fertility of the earth, and the sexual responsibility that joins them to the human past and the human future. It is possible to imagine marriage as a grievous, joyous human bond, endlessly renewable and renewing, again and again rejoining memory and passion and hope.

But it is extremely difficult, now, to imagine marriage in terms of such dignity and generosity, and this difficulty is explained by the failure of these possessive and competitive forms of sexual love that have been in use for so long. This failure raises unavoidably the issue of fidelity: What is it, and what does it mean—in marriage, and also, since marriage is a fundamental relationship and metaphor, in other relationships?

No one can be glad to have this issue so starkly raised, for any consideration of it now must necessarily involve one's own bewilderment. We are apparently near the end of a degenerative phase of an evolutionary process—a long way from any large-scale regeneration. For that reason it is necessary to be hesitant and cautious, respectful of the complexity and importance of the problem. Marriage is not going to change because somebody thinks about it and recommends an "answer"; it can change only as its necessities are felt and as its circumstances change.

The idea of fidelity is perverted beyond redemption by understanding it as a grim, literal duty enforced only by willpower. This is the "religious" insanity of making a victim of the body as a victory of the soul. Self-restraint that is so purely negative is self-hatred. And one cannot be good, anyhow, just by not being bad. To be faithful merely out of duty is to be blinded to the possibility of a better faithfulness for better reasons.

It is reasonable to suppose, if fidelity is a virtue, that it is a virtue with a purpose. A purposeless virtue is a contradiction in terms. Virtue, like harmony, cannot exist alone; a virtue must lead to harmony between one creature and another. To be good for nothing is just that. If a virtue has been thought a virtue long enough, it must be assumed to have practical justification—though the very longevity that proves its practicality may obscure it. That seems to be what happened with the idea of fidelity. We heard the words "forsaking all others" repeated over and over again for so long that we lost the sense of their practical justification. They assumed the force of superstition: people came to be faithful in marriage not out of any understanding of the meaning of faith or of marriage, but out of the same fear of obscure retribution that made one careful not to break a mirror or spill the salt. Like other superstitions, this one was weakened by the scientific, positivist intellectuality of modern times and by the popular "sophistication" that came with it. Our age could be characterized as a manifold experiment in faithlessness, and if it has as yet produced no effective understanding of the practicalities of faith, it has certainly produced massive evidence of the damage and disorder of its absence.

It is possible to open this issue of the practicality of fidelity by considering that the modern age was made possible by the freeing, and concurrently by the cheapening, of energy. It can be said, of course, that the modern age was made possible by technologies that *control* energy and thus make it usable at an unprecedented rate. But such control is at best extremely limited: the devices by which industrial and military energies are used control them only momentarily; their moment of usefulness sets them loose in the world as social, ecological, and geological *forces*. We can use these energies only as explosives; we can control the rate, intensity, and time of combustion, but our effective control ends with the use of the small amount of the released energy that we are able to harness. Past that, the effects are on their own, to compound themselves as they will. In modern times we have never been able to subject our use of energy to a sense of responsibility anywhere near complex enough to be equal to its effects.

It may be that the principle of sexual fidelity, once it is again fully understood, will provide us with as good an example as we can find of the responsible use of energy. Sexuality is, after all, a form of energy, one of the most powerful. If we see sexuality as energy, then it becomes impossible to see sexual fidelity as merely a "duty," a virtue for the sake of virtue, or a superstition. If we made a superstition of fidelity, and thereby weakened it, by thinking of it as purely a moral or spiritual virtue, then perhaps we can restore its strength by recovering an awareness of its practicality.

At the root of culture must be the realization that uncontrolled energy is disorderly—that in nature all energies move in forms; that, therefore, in a human order energies must be *given* forms. It must have been plain at the beginning, as cultural degeneracy has made it plain again and again, that one can be indiscriminately sexual but not indiscriminately responsible, and that irresponsible sexuality would undermine any possibility of culture since it implies a hierarchy based purely upon brute strength, cunning, regardlessness of value and of consequence. Fidelity can thus be seen as the necessary discipline of sexuality, the practical definition of sexual responsibility, or the definition of the moral limits within which such responsibility can be conceived and enacted. The forsaking of all others is a keeping of faith, not just with the chosen one, but with the ones forsaken. The marriage vow unites not just a woman and a man with each other; it unites each of them with the community in a vow of sexual responsibility toward all others. The whole community is married, realizes its essential unity, in each of its marriages.

Another use of fidelity is to preserve the possibility of devotion against the distractions of novelty. What marriage offers —and what fidelity is meant to protect—is the possibility of moments when what we have chosen and what we desire are the same. Such a convergence obviously cannot be continuous. No relationship can continue very long at its highest emotional pitch. But fidelity prepares us for the return of these moments, which give us the highest joy we can know: that of union, communion, atonement (in the root sense of at-one-ment). The principle is stated in these lines by William Butler Yeats (by "the world" he means the world after the Fall):

> Maybe the bride-bed brings despair,
> For each an imagined image brings
> And finds a real image there;
> Yet the world ends when these two things,
> Though several, are a single light . . .

To forsake all others does not mean—because it *cannot* mean —to ignore or neglect all others, to hide or be hidden from all others, or to desire or love no others. To live in marriage is a responsible way to live in sexuality, as to live in a household is a responsible way to live in the world. One cannot enact or fulfill one's love for womankind or mankind, or even for all the women or men to whom one is attracted. If one is to have the power and delight of one's sexuality, then the generality of instinct must be resolved in a responsible relationship to a particular person. Similarly, one cannot live in the world; that is, one cannot become, in the easy, generalizing sense with which the phrase is commonly used, a "world citizen." There can be no such thing as a "global village." No matter how much one may love the world as a whole, one can live fully in it only by living responsibly in some small part of it. Where we live and who we live there with define the terms of our relationship to the world and to humanity. We thus come again to the paradox that one can become whole only by the responsible acceptance of one's partiality.

But to encapsulate these partial relationships is to entrap and condemn them in their partiality; it is to endanger them and to make them dangerous. They are enlivened and given the possibility of renewal by the double sense of particularity and generality: one lives in marriage *and* in sexuality, at home *and* in the world. It is impossible, for instance, to conceive that a man could despise women and yet love his wife, or love his own place in the world and yet deal destructively with other places.

HOME LAND AND HOUSE HOLD

What I have been trying to do is to define a pattern of disintegration that is at once cultural and agricultural. I have been groping for connections—that I think are indissoluble, though obscured by modern ambitions—between the spirit and the body, the body and other bodies, the body and the earth. If these connections do necessarily exist, as I believe they do, then it is impossible for material order to exist side by side with spiritual disorder, or vice versa, and impossible for one to thrive long at the expense of the other; it is impossible, ultimately, to preserve ourselves apart from our willingness to preserve other creatures, or to respect and care for ourselves except as we respect and care for other creatures; and, most to the point of this book, it is impossible to care for each other more or differently than we care for the earth.

This last statement becomes obvious enough when it is considered that the earth is what we all have in common, that it is what we are made of and what we live from, and that we therefore cannot damage it without damaging those with whom we share it. But I believe it goes farther and deeper than that. There is an uncanny *resemblance* between our behavior toward each other and our behavior toward the earth. Between our relation to our own sexuality and our relation to the reproductivity of the earth, for instance, the resemblance is plain and strong and apparently inescapable. By some connection that we do not recognize, the willingness to exploit one becomes the willingness to exploit the other. The conditions and the means of exploitation are likewise similar.

The modern failure of marriage that has so estranged the sexes from each other seems analogous to the "social mobility" that has estranged us from our land, and the two are historically parallel. It may even be argued that these two estrangements are very close to being one, both of them having been caused by the disintegration of the household, which was the formal bond between marriage and the earth, between human sexuality and its sources in the sexuality of Creation. The importance of this practical bond has not been often or very openly recognized in our tradition; in modern times it has almost disappeared under the burden of adverse fashion and economics. It is necessary to go far back to find it clearly exemplified.

To my mind, one of the best examples that we have is in Homer's *Odyssey*. Nowhere else that I know are the connections between marriage and household and the earth so fully and so carefully understood.

At the opening of the story Odysseus, after a twenty-year absence, is about to begin the last leg of his homeward journey. The sole survivor of all his company of warriors, having lived through terrible trials and losses, Odysseus is now a castaway on the island of the goddess Kalypso. He is Kalypso's lover but also virtually her prisoner. At night he sleeps with Kalypso in her cave; by day he looks across the sea toward Ithaka, his home, and weeps. Homer does not stint either feeling—the delights of Kalypso's cave, where the lovers "revel and rest softly, side by side," or the grief and longing of exile.

But now Zeus commands Kalypso to allow Odysseus to depart; she comes to tell him that he is free to go. And yet it is a tragic choice that she offers him: he must choose between her and Penélopê, his wife. If he chooses Kalypso, he will be immortal, but remain in exile; if he chooses Penélopê, he will return home at last, but will die in his time like other men:

> If you could see it all, before you go—
> all the adversity you face at sea—
> you would stay here, and guard this house, and be
> immortal—though you wanted her forever,
> that bride for whom you pine each day.
> Can I be less desirable than she is?
> Less interesting? Less beautiful? Can mortals
> compare with goddesses in grace and form?

And Odysseus answers:

> My quiet Penélopê—how well I know—
> would seem a shade before your majesty,
> death and old age being unknown to you,
> while she must die. Yet, it is true, each day
> I long for home . . .

This is, in effect, a wedding ritual much like our own, in which Odysseus forsakes all others, in renouncing the immortal womanhood of the goddess, and renews his pledge to the mortal terms of his marriage. But unlike our ritual, this one involves an explicit loyalty to a home. Odysseus' far-wandering through the wilderness of the sea is not merely the return of a husband; it is a journey home. And a great deal of the power as well as the moral complexity of *The Odyssey* rises out of the richness of its sense of home.

By the end of Book XXIII, it is clear that the action of the narrative, Odysseus' journey from the cave of Kalypso to the bed of Penélopê, has revealed a structure that is at once geographical and moral. This structure may be graphed as a series of diminishing circles centered on one of the posts of the marriage bed. Odysseus makes his way from the periphery toward that center.

All around, this structure verges on the sea, which is the wilderness, ruled by the forces of nature and by the gods. In spite of the excellence of his ship and crew and his skill in navigation, a man is alien there. Only when he steps ashore does he enter a human order. From the shoreline of his island of

Ithaka, Odysseus makes his way across a succession of boundaries, enclosed and enclosing, with the concentricity of a blossom around its pistil, a human pattern resembling a pattern of nature. He comes to his island, to his own lands, to his town, to his household and house, to his bedroom, to his bed.

As he moves toward this center he moves also through a series of recognitions, tests of identity and devotion. By these, his homecoming becomes at the same time a restoration of order. At first, having been for a while uncertain of his whereabouts, he recognizes his homeland by the conformation of the countryside and by a certain olive tree. He then becomes the guest of his swineherd, Eumaios, and tests his loyalty, though Eumaios will not be permitted to recognize his master until the story approaches its crisis. In the house of Eumaios, Odysseus meets and makes himself known to his son, Telémakhos. As he comes, disguised as a beggar, into his own house, he is recognized by Argus, his old hunting dog. That night, as the guest of Penélopê, who does not yet know who he is, he is recognized by his aged nurse, Eurýkleia, who sees a well-remembered scar on his thigh as she is bathing his feet.

He is scorned and abused as a vagabond by the band of suitors who, believing him dead, have been courting his wife, consuming his meat and wine, desecrating his household, and plotting the murder of his son. Penelope proposes a trial by which the suitors will compete for her: she will become the bride of whichever one can string the bow of her supposedly dead husband and shoot an arrow through the aligned helve-sockets of twelve axe heads. The suitors fail. Odysseus performs the feat easily and is thereby recognized as "the great husband" himself. And then, with the help of the swineherd, the cowherd, and Telémakhos, he proceeds to trap the suitors and slaughter them all without mercy. To so distinguished a commentator as Richmond Lattimore, their punishment "seems excessive." But granting the acceptability of violent means to a warrior such as Odysseus, this outcome seems to me appropriate to the moral terms of the poem. It is made clear that the punishment is not merely the caprice of a human passion: Odysseus enacts the will of the gods; he is the agent of a divine judgment. The suitors' sin is their utter contempt for the domestic order that the poem affirms. They do not respect or honor the meaning of the household, and in *The Odyssey* this meaning is paramount.

It is therefore the recognition of Odysseus by Penélopê that is the most interesting and the most crucial. By the time Odysseus' vengeance and his purification of the house are complete, Penélopê is the only one in the household who has not acknowledged him. It is only reasonable that she should delay this until she is absolutely certain. After all, she has waited twenty years; it is not to be expected that she would be less than cautious now. Her faith has been equal and more than equal to his, and now she proves his equal also in cunning. She tells Eurýkleia to move their bed outside their bedroom and to make it up for Odysseus there. Odysseus' rage at hearing that identifies him beyond doubt, for she knew that only Odysseus would know—it is their "pact and pledge" and "secret sign" —that the bed could not be moved without destroying it. He built their bedroom with his own hands, and an old olive tree, as he says,

> grew like a pillar on the building plot,
> and I laid out our bedroom round that tree . . .
> . . . I lopped off the silvery leaves and branches,
> hewed and shaped that stump from the roots up
> into a bedpost . . .

She acknowledges him then, and only then does she give herself to his embrace.

> Now from his breast into his eyes the ache
> of longing mounted, and he wept at last,
> his dear wife, clear and faithful in his arms,
> longed for
> as the sunwarmed earth is longed for by a swimmer
> spent in rough water where his ship went down . . .

And so in the renewal of his marriage, the return of Odysseus and the restoration of order are complete. The order of the kingdom is centered on the marriage bed of the king and queen, and that bed is rooted in the earth. The figure last quoted makes explicit at last the long-hinted analogy between

Odysseus' fidelity to his wife and his fidelity to his homeland. In Penélopê's welcoming embrace his two fidelities become one.

For Odysseus, then, marriage was not merely a legal bond, nor even merely a sacred bond, between himself and Penélopê. It was part of a complex practical circumstance involving, in addition to husband and wife, their family of both descendants and forebears, their household, their community, and the sources of all these lives in memory and tradition, in the countryside, and in the earth. These things, wedded together in his marriage, he thought of as his home, and it held his love and faith so strongly that sleeping with a goddess could not divert or console him in his exile.

In Odysseus' return, then, we see a complete marriage and a complete fidelity. To reduce marriage, as we have done, to a mere contract of sexual exclusiveness is at once to degrade it and to make it impossible. That is to take away its dignity and its potency of joy, and to make it only a pitiful little duty—not a union, but a division and a solitude.

The Odyssey's understanding of marriage as the vital link which joins the human community and the earth is obviously full of political implication. In this it will remind us of the Confucian principle that "The government of the state is rooted in family order." But *The Odyssey* goes further than the Confucian texts, it seems to me, in its understanding of agricultural value as the foundation of domestic order and peace.

I have considered the poem so far as describing a journey from the non-human order of the sea wilderness to the human order of the cleansed and reunited household. But it is also a journey between two kinds of human value; it moves from the battlefield of Troy to the terraced fields of Ithaka, which, through all the years and great deeds of Odysseus' absence, the peasants have not ceased to farm.

The Odyssey begins in the world of *The Iliad*, a world which, like our own, is war-obsessed, preoccupied with "manly" deeds of exploitation, anger, aggression, pillage, and the disorder, uprootedness, and vagabondage that are their result. At the end of the poem, Odysseus moves away from the values of that world toward the values of domesticity and peace. He restores order to his household by an awesome violence, it is true. But that finished and the house purified, he re-enters his marriage, the bedchamber and the marriage bed rooted in the earth. From there he goes into the fields.

The final recognition scene occurs between Odysseus and his old father, Laërtês:

> Odysseus found his father in solitude
> spading the earth around a young fruit tree.
>
> He wore a tunic, patched and soiled, and leggings—
> oxhide patches, bound below his knees
> against the brambles . . .

The point is not stated—the story is moving so evenly now toward its conclusion that it will not trouble to remind us that the man thus dressed is a *king*—but it is clear that Laërtês has survived his son's absence and the consequent grief and disorder *as a peasant*. Although Odysseus jokes about his father's appearance, the appropriateness of what he is doing is never questioned. In a time of disorder he has returned to the care of the earth, the foundation of life and hope. And Odysseus finds him in an act emblematic of the best and most responsible kind of agriculture: an old man caring for a young tree.

But the homecoming of Odysseus is still not complete. During his wanderings, he was instructed by the ghost of the seer Teirêsias to perform what is apparently to be a ritual of atonement. As the poem ends he still has this before him. Carrying an oar on his shoulder, he must walk inland until he comes to a place where men have no knowledge of the sea or ships, where a passerby will mistake his oar for a winnowing fan. There he must "plant" his oar in the ground and make a sacrifice to the sea god, Poseidon. Home again, he must sacrifice to all the gods. Like those people of the Biblical prophecy who will "beat their swords into plowshares, and their spears into pruning hooks" and not "learn war any more," Odysseus will not know rest until he has carried the instrument of his sea wanderings inland and planted it like a tree, until he has seen the symbol of his warrior life as a farming tool. But after his atonement has been made, a gentle death will come to him when he is weary with age, his countrymen around him "in blessed peace."

The Odyssey, then, is in a sense an anti-*Iliad*, posing against the warrior values of the other epic—the glories of battle and foreign adventuring—an affirmation of the values of domesticity and farming. But at the same time *The Odyssey* is too bountiful and wise to set these two kinds of value against each other in any purity or exclusiveness of opposition. Even less does it set into such opposition the two kinds of experience. The point seems to be that these apparently opposed experiences are linked together. The higher value may be given to domesticity, but this cannot be valued or understood alone. Odysseus' fidelity and his homecoming are as moving and instructive as they are precisely because they are the result of *choice*. We know—as Odysseus undoubtedly does also—the extent of his love for Penélopê because he can return to her only by choosing her, at the price of death, over Kalypso. We feel and understand, with Odysseus, the value of Ithaka as a homeland, because bound inextricably to the experience of his return is the memory of his absence, of his long wandering at sea, and even of the excitement of his adventures. The prophecy of the peaceful death that is to come to him is so deeply touching because the poem has so fully realized the experiences of discord and violent death. The farm life of the island seems so sweet and orderly because we know the dark wilderness of natural force and mystery within which the fields are cleared and lighted.

THE NECESSITY OF WILDNESS

Domestic order is obviously threatened by the margin of wilderness that surrounds it. Marriage may be destroyed by instinctive sexuality; the husband may choose to remain with Kalypso or the wife may run away with godlike Paris. And the forest is always waiting to overrun the fields. These are real possibilities. They must be considered, respected, even feared.

And yet I think that no culture that hopes to endure can afford to destroy them or to set up absolute safeguards against them. Invariably the failure of organized religions, by which they cut themselves off from mystery and therefore from sanctity, lies in the attempt to impose an absolute division between faith and doubt, to make belief perform as knowledge; when they forbid their prophets to go into the wilderness, they lose the possibility of renewal. And the most dangerous tendency in modern society, now rapidly emerging as a scientific-industrial ambition, is the tendency toward encapsulation of human order—the severance, once and for all, of the umbilical cord fastening us to the wilderness or the Creation. The threat is not only in the totalitarian desire for absolute control. It lies in the willingness to ignore an essential paradox: the natural forces that so threaten us are the same forces that preserve and renew us.

An enduring agriculture must never cease to consider and respect and preserve wildness. The farm can exist only within the wilderness of mystery and natural force. And if the farm is to last and remain in health, the wilderness must survive within the farm. That is what agricultural fertility *is*: the survival of natural process in the human order. To learn to preserve the fertility of the farm, Sir Albert Howard wrote, we must study the forest.

Similarly, the instinctive sexuality within which marriage exists must somehow be made to thrive within marriage. To divide one from the other is to degrade both and ultimately to destroy marriage.

Fidelity to human order, then, if it is fully responsible, implies fidelity also to natural order. Fidelity to human order makes devotion possible. Fidelity to natural order preserves the possibility of choice, the possibility of the renewal of devotion. Where there is no possibility of choice, there is no possibility of faith. One who returns home—to one's marriage and household and place in the world—desiring anew what was previously chosen, is neither the world's stranger nor its prisoner, but is at once in place and free.

The relation between these two fidelities, inasmuch as they sometimes appear to contradict one another, cannot help but be complex and tricky. In our present stage of cultural evolution, it cannot help but be baffling as well. And yet it is only the double faith that is adequate to our need. If we are to have a culture as resilient and competent in the face of necessity as it needs to be, then it must somehow involve within itself a ceremonious generosity toward the wilderness of natural force and instinct. The farm must yield a place to the forest, not as a wood lot, or even as a necessary agricultural principle, but as a sacred grove—a place where the Creation is let alone, to serve as instruction, example, refuge; a place for people to go, free of work and presumption, to let themselves alone. And marriage must recognize that it survives because of, as well as in spite of, Kalypso and Paris and the generosity of instinct that they represent. It must give some ceremonially acknowledged place to the sexual energies that now thrive outside all established forms, in the destructive freedom of moral ignorance or disregard. Without these accommodations we will remain divided: some of us will continue to destroy the world for purely human ends, while others, for the sake of nature, will abandon the task of human order.

What forms or revisions of forms may be adequate to this double faith, I do not know. Cultural solutions are organisms, not machines, and they cannot be invented deliberately or imposed by prescription. Perhaps all that one can do is to clarify as well as possible the needs and pressures that bear upon the process of cultural evolution. I am certain, however, that no satisfactory solution can come from considering marriage alone or agriculture alone. These are our basic connections to each other and to the earth, and they tend to relate analogically and to be reciprocally defining: our demands upon the earth are determined by our ways of living with one another; our regard for one another is brought to light in our ways of using the earth. And I am certain that neither can be changed for the better in the experimental, prescriptive ways we have been using. Ways of life change only in living. To live by expert advice is to abandon one's life.

"FREEDOM" FROM FERTILITY

The household is the bond of marriage that is most native to it, that grows with it and gives it substantial being in the world. It is the practical condition within which husband and wife can enact devotion and loyalty to each other. The motive power of sexual love is thus joined directly to constructive work and is given communal and ecological value. Without the particular demands and satisfactions of the making and keeping of a household, the sanctity and legality of marriage remain abstract, in effect theoretical, and its sexuality becomes a danger. Work is the health of love. To last, love must enflesh itself in the materiality of the world—produce food, shelter, warmth or shade, surround itself with careful acts, well-made things. This, I think, is what Millen Brand means in *Local Lives* when he speaks of the "threat" of love—"so that perhaps acres of earth and its stones are needed and drawn-out work and monotony/ to balance that danger. . . ."

Marriage and the care of the earth are each other's disciplines. Each makes possible the enactment of fidelity toward the other. As the household has become increasingly generalized as a function of the economy and, as a consequence, has become increasingly "mobile" and temporary, these vital connections have been weakened and finally broken. And whatever has been thus disconnected has become a ground of exploitation for some breed of salesman, specialist, or expert.

A direct result of the disintegration of the household is the division of sexuality from fertility and their virtual takeover by specialists. The specialists of human sexuality are the sexual clinicians and the pornographers, both of whom subsist on the increasing possibility of sex between people who neither know nor care about each other. The specialists of human fertility are the evangelists, technicians, and salesmen of birth control, who subsist upon our failure to see any purpose or virtue in sexual discipline. In this, as in our use of every other kind of energy, our inability to contemplate any measure of restraint or forbearance has been ruinous. Here the impulse is characteristically that of the laboratory scientist: to encapsulate sexuality by separating it absolutely from the problems of fertility.

This division occurs, it seems to me, in a profound cultural failure: the loss of any sense that sexuality and fertility might exist together compatibly in this world. We have lost this possibility because we do not understand, because we cannot bear to consider the meaning of restraint.* The sort of restraint I am talking about is illustrated in a recent *National Geographic*

*At the root of this failure is probably another sexual division: the assignment to women of virtually all responsibility for sexual discipline.

article about the people of Hunza in northern Pakistan. The author is a woman, Sabrina Michaud, and she is talking with a Hunza woman in her kitchen:

"'What have you done to have only one child?' she asks me. Her own children range from 12 to 30 years of age, and seem evenly spaced, four to five years apart. 'We leave our husband's bed until each child is weaned,' she explains simply. But this natural means of birth control has declined, and population has soared."

The woman's remark is thus passed over and not returned to; but if I understand the significance of this paragraph, it is of great importance. The decline of "this natural means of birth control" seems to have been contemporaneous with the coming of roads and "progress" and the opening up of a previously isolated country. What is of interest is that in their isolation in arid, narrow valleys surrounded by the stone and ice of the Karakoram Mountains, these people had practiced sexual restraint as a form of birth control. They had neither our statistical expertise nor our doom-prophets of population growth; it just happened that, placed geographically as they were, they lived always in sight of their agricultural or ecological limits, and they made a competent response.

We have been unable to see the difference between this kind of restraint—a cultural response to an understood practical limit—and the obscure, self-hating, self-congratulating Victorian self-restraint, of which our attitudes and technologies of sexual "freedom" are merely the equally obscure other side. This so-called freedom fragments us and turns us more vehemently and violently than before against our own bodies and against the bodies of other people.

For the care or control of fertility, both that of the earth and that of our bodies, we have allowed a technology of chemicals and devices to replace entirely the cultural means of ceremonial forms, disciplines, and restraints. We have gathered up the immense questions that surround the coming of life into the world and reduced them to simple problems for which we have manufactured and marketed simple solutions. An infertile woman and an infertile field both receive a dose of chemicals, at the calculated risk of undesirable consequences, and are thus equally reduced to the status of productive machines.

And for unwanted life—sperm, ova, embryos, weeds, insects, etc.—we have the same sort of ready remedies, for sale, of course, and characteristically popularized by advertisements that speak much of advantages but little of problems.

The result is that we are bringing up a generation of young people who feel that they are "free from worry" about fertility. The pharmacist or the doctor will look after the fertility of the body, and the farming experts and agribusinessmen will look after the fertility of the earth. This is to short-circuit human culture at its source. It is, in effect, to remove from consciousness the two fundamental issues of human life. It permits two great powers to be regarded and used as if they were unimportant.

More serious is the resort to "authorized" modes of direct violence. In land use, this is the permanent diminishment or destruction of fertility as an allowable cost of production, as in strip-mining or in the sort of agriculture that good farmers have long referred to also as "mining." This use of technological means cuts across all issues of health and culture for the sake of an annual quota of production.

The human analogue is in the "harmless" and "simple" surgeries of permanent sterilization, which are now being promoted by a propaganda of extreme oversimplification. The publicity on this subject is typically evangelical in tone and simplistically moral; the operations are recommended like commercial products by advertisings complete with exuberant testimonials of satisfied customers and appeals to the prospective customer's maturity, sexual pride, and desire for "freedom"; and the possible physical and psychological complications are played down, misrepresented, not mentioned at all, or simply not known. It is altogether possible that the operations will be performed by doctors as perfunctory, simplistic, presumptuous, and uninforming as the public literature.

I am fully aware of the problem of overpopulation, and I do not mean to say that birth control is unnecessary. What I do mean to say is that any means of birth control is a serious matter, both culturally and biologically, and that sterilization is the most serious of all: to give up fertility is a major change, as important as birth, puberty, marriage, or death.

The great changes having to do with a woman's fertility—puberty, childbirth, and menopause—have, like sexual desire, the unarguable sanction of biological determinism. They belong to a kind of natural tradition. As a result, they are not only endurable, but they belong to a process—the life process or the Wheel of Life—that we have learned to affirm with some understanding. We know, among other things, that this process includes tragedy and survives it, even triumphs over it. The same applies to the occasions of a man's fertility, although not so formidably, a man being less involved, physically, in the *predicament* of fertility and consciously involved in it only if he wants to be. Nevertheless, he comes to fertility and, if he is a moral person, to the same issues of responsibility that it poses for women.

One of the fundamental interests of human culture is to impose this responsibility, to subject fertility to moral will. Culture articulates needs and forms for sexual restraint and involves issues of value in the process of mating. It is possible to imagine that the resulting tension creates a distinctly human form of energy, highly productive of works of the hands and the mind. But until recently there was no division between sexuality and fertility, because none was possible.

This division was made possible by modern technology, which subjected human fertility, like the fertility of the earth, to a new kind of will: the technological will, which may not *necessarily* oppose the moral will, but which has not only tended to do so, but has tended to replace it. Simply because it became possible—and simultaneously profitable—we have cut the cultural ties between sexuality and fertility, just as we have cut those between eating and farming. By "freeing" food and sex from worry, we have also set them apart from thought, responsibility, and the issue of quality. The introduction of "chemical additives" has tended to do away with the issue of taste or preference; the specialist of sex, like the specialist of food, is dealing with a commodity, which he can measure but cannot value.

What is horrifying is not only that we are relying so exclusively on a technology of birth control that is still experimental, but that we are using it *casually*, in utter cultural nakedness, unceremoniously, without sufficient understanding, and as a substitute for cultural solutions—exactly as we now employ the technology of land use. And to promote these means without cultural and ecological insight, as merely a way to divorce sexuality from fertility, pleasure from responsibility —or to *sell* them that way for ulterior "moral" motives—is to try to cure a disease by another disease. That is only a new battle in the old war between body and soul—as if we were living in front of a chorus of the most literal fanatics chanting: "If thy right eye offend thee, pluck it out! If thy right hand offend thee, cut it off!"

The technologists of fertility exercise the powers of gods and the social function of priests without community ties or cultural responsibilities. The clinicians of sex change the lives of people —as the clinicians of agriculture change the lives of places and communities—to whom they are strangers and whom they do not know. These specialists thrive in a profound cultural rift, and they are always accompanied by the exploiters who mine that rift for gold. The pornographer exploits sexual division. And working the similar division between us and our land we have the "agribusinessmen," the pornographers of agriculture.

FERTILITY AS WASTE

But there is yet another and more direct way in which the isolation of the body has serious agricultural effects. That is in our society's extreme oversimplification of the relation between the body and its food. By regarding it as merely a consumer of food, we reduce the function of the body to that of a conduit which channels the nutrients of the earth from the supermarket to the sewer. Or we make it a little factory which transforms fertility into pollution—to the enormous profit of "agribusiness" and to the impoverishment of the earth. This is another technological and economic interruption of the cycle of fertility.

Much has already been said here about the division between the body and its food in the productive phase of the cycle. It is the alleged wonder of the Modern World that so many people take energy from food in which they have invested no energy, or very little. Ninety-five percent of our people, boasted the former deputy assistant secretary of agriculture, are now free of the "drudgery" of food production. The meanings of that division, as I have been trying to show, are intricate and degenerative. But that is only half of it. Ninety-five percent (at least) of our people are also free of any involvement or interest in the maintenance phase of the cycle. As their bodies take in and use the nutrients of the soil, those nutrients are transformed into what we are pleased to regard as "wastes"—and are duly wasted.

This waste also has its cause in the old "religious" division between body and soul, by which the body and its products are judged offensive. Once, living with this offensiveness was considered a condemnation, and that was bad enough. But modern technology "saved" us with the flush toilet and the waterborne sewage system. These devices deal with the "wastes" of our bodies by simply removing them from consideration. The irony is that this technological purification of the body requires the pollution of the rivers and the starvation of the fields. It makes the alleged offensiveness of the body truly and inescapably offensive and blinds an entire society to the knowledge that these "offensive wastes" are readily purified in the topsoil—that, indeed, from an ecological point of view, these are not wastes and are not offensive, but are valuable agricultural products essential both to the health of the land and to that of the "consumers."

Our system of agriculture, by modeling itself on economics rather than biology, thus removes food from the *cycle* of its production and puts it into a finite, linear process that in effect destroys it by trans-forming it into waste. That is, it transforms food into fuel, a form of energy that is usable only once, and in doing so it transforms the body into a consumptive machine.

It is strange, but only apparently so, that this system of agriculture is institutionalized, not in any form of rural life or culture, but in what we call our "urban civilization." The cities subsist in competition with the country; they live upon a one-way movement of energies out of the countryside—food and fuel, manufacturing materials, human labor, intelligence, and talent. Very little of this energy is ever returned. Instead of gathering these energies up into coherence, a cultural consummation that would not only return to the countryside what belongs to it, but also give back generosities of learning and art, conviviality and order, the modern city dissipates and wastes them. Along with its glittering "consumer goods," the modern city produces an equally characteristic outpouring of garbage and pollution—just as it produces and/or collects unemployed, unemployable, and otherwise wasted people.

Once again it must be asked, if competition is the appropriate relationship, then why, after generations of this inpouring of rural wealth, materials, and humanity into the cities, are the cities and the countryside in equal states of disintegration and disrepair? Why have the rural and urban communities *both* fallen to pieces?

HEALTH AND WORK

The modern urban-industrial society is based on a series of radical disconnections between body and soul, husband and wife, marriage and community, community and the earth. At each of these points of disconnection the collaboration of corporation, government, and expert sets up a profit-making enterprise that results in the further dismemberment and impoverishment of the Creation.

Together, these disconnections add up to a condition of critical ill health, which we suffer in common—not just with each other, but with all other creatures. Our economy is based upon this disease. Its aim is to separate us as far as possible from the sources of life (material, social, and spiritual), to put these sources under the control of corporations and specialized professionals, and to sell them to us at the highest profit. It fragments the Creation and sets the fragments into conflict with one another. For the relief of the suffering that comes of this fragmentation and conflict, our economy proposes, not health, but vast "cures" that further centralize power and increase profits: wars, wars on crime, wars on poverty, national schemes of medical aid, insurance, immunization, further industrial and economic "growth," etc.; and these, of course, are followed by more regulatory laws and agencies to see that our health is protected, our freedom preserved, and our money well spent. Although there may be some "good intention" in this, there is little honesty and no hope.

Only by restoring the broken connections can we be healed. Connection *is* health. And what our society does its best to disguise from us is how ordinary, how commonly attainable, health is. We lose our health—and create profitable diseases and dependences—by failing to see the direct connections between living and eating, eating and working, working and loving. In gardening, for instance, one works with the body to feed the body. The work, if it is knowledgeable, makes for excellent food. And it makes one hungry. The work thus makes eating both nourishing and joyful, not consumptive, and keeps the eater from getting fat and weak. This is health, wholeness, a source of delight. And such a solution, unlike the typical industrial solution, does not cause new problems.

The "drudgery" of growing one's own food, then, is not drudgery at all. (If we make the growing of food a drudgery, which is what "agribusiness" does make of it, then we also make a drudgery of eating and of living.) It is—in addition to being the appropriate fulfillment of a practical need—a sacrament, as eating is also, by which we enact and understand our oneness with the Creation, the conviviality of one body with all bodies. This is what we learn from the hunting and farming rituals of tribal cultures.

As the connections have been broken by the fragmentation and isolation of work, they can be restored by restoring the wholeness of work. There is work that is isolating, harsh, destructive, specialized or trivialized into meaninglessness. And there is work that is restorative, convivial, dignified and dignifying, and pleasing. Good work is not just the maintenance of connections—as one is now said to work "for a living" or "to support a family"—but the *enactment* of connections. It *is* living, and a way of living; it is not support for a family in the sense of an exterior brace or prop, but is one of the forms and acts of love.

To boast that now "95 percent of the people can be freed from the drudgery of preparing their own food" is possible only to one who cannot distinguish between these kinds of work. The former deputy assistant secretary cannot see work as a vital connection; he can see it only as a trade of time for money, and so of course he believes in doing as little of it as possible, especially if it involves the use of the body. His ideal is apparently the same as that of a real-estate agency which promotes a rural subdivision by advertising "A homelife of endless vacation." But the society that is so glad to be free of the drudgery of growing and preparing food also boasts a thriving medical industry to which it is paying $500 per person per year. And that is only the down payment.

We embrace this curious freedom and pay its exorbitant cost because of our hatred of bodily labor. We do not want to work "like a dog" or "like an ox" or "like a horse"—that is, we do not want to use ourselves as beasts. This as much as anything is the cause of our disrespect for farming and our abandonment of it to businessmen and experts. We remember, as we should, that there have been agricultural economies that used people as beasts. But that cannot be remedied, as we have attempted to do, by using people as machines, or by not using them at all.

Perhaps the trouble began when we started using animals disrespectfully: as "beasts"—that is, as if they had no more feeling than a machine. Perhaps the destructiveness of our use of machines was prepared in our willingness to abuse animals. That it was never necessary to abuse animals in order to use them is suggested by a passage in *The Horse in the Furrow*, by George Ewart Evans. He is speaking of how the medieval ox teams were worked at the plow: ". . . the ploughman at the handles, the team of oxen—yoked in pairs or four abreast—and the driver who walked alongside with his goad." And then he says: "It is also worth noting that in the Welsh organization . . . the counterpart of the driver was termed *y geilwad* or the *caller*. He walked *backwards* in front of the oxen singing to them as they worked. Songs were specially composed to suit the rhythm of the oxen's work . . ."

That seems to me to differ radically from our present customary use of any living thing. The oxen were not used as beasts or machines, but as fellow creatures. It may be presumed that this work used people the same way. It is possible, then, to believe that there is a kind of work that does not require abuse or misuse, that does not use anything as a substitute for anything else. We are working well when we use ourselves as the fellow creatures of the plants, animals, materials, and other people we are working with. Such work is unifying, healing. It brings us home from pride and from despair, and places us responsibly within the human estate. It defines us as we are: not too good to work with our bodies, but too good to work poorly or joylessly or selfishly or alone.

Instead of sending the experimenter into the fields and meadows to question the farmer and the land worker so as to understand how important quality is, and above all to take up a piece of land himself, the new authoritarian doctrine demands that he shut himself up in a study . . .

SIR ALBERT HOWARD, *The Soil and Health*

. . . his education had had the curious effect of making things that he read and wrote more real to him than things he saw. Statistics about agricultural labourers were the substance; any real ditcher, ploughman, or farmer's boy, was the shadow. Though he had never noticed it himself, he had a great reluctance, in his work, ever to use such words as "man" or "woman." He preferred to write about "vocational groups," "elements," "classes" and "populations": for, in his own way, he believed as firmly as any mystic in the superior reality of the things that are not seen.

C. S. LEWIS, *That Hideous Strength*

CHAPTER EIGHT

Jefferson, Morrill, and the Upper Crust

THE CONVICTION OF THOMAS JEFFERSON

IN the mind of Thomas Jefferson, farming, education, and democratic liberty were indissolubly linked. The great conviction of his life, which he staked his life upon and celebrated in a final letter two weeks before his death, was "that the mass of mankind has not been born with saddles on their backs, nor a favored few booted and spurred, ready to ride them legitimately, by the grace of God." But if liberty was in that sense a right, it was nevertheless also a privilege to be earned, deserved, and strenuously kept; to keep themselves free, he thought, a people must be stable, economically independent, and virtuous. He believed—on the basis, it should be remembered, of extensive experience both in this country and abroad—that these qualities were most dependably found in the farming people: "Cultivators of the earth are the most valuable citizens. They are the most vigorous, the most independent, the most virtuous, and they are tied to their country, and wedded to its liberty and interests by the most lasting bonds." These bonds were not merely those of economics and property, but those, at once more feeling and more practical, that come from the investment in a place and a community of work, devotion, knowledge, memory, and association.

By contrast, Jefferson wrote: "I consider the class of artificers as the panders of vice, and the instruments by which the liberties of a country are generally overturned." By "artificers" he meant manufacturers, and he made no distinction between "management" and "labor." The last-quoted sentence is followed by no explanation, but its juxtaposition with the one first quoted suggests that he held manufacturers in suspicion because their values were already becoming abstract, enabling them to be "socially mobile" and therefore subject preeminently to the motives of self-interest.

To foster the strengths and virtues necessary to citizenship in a democracy, public education was obviously necessary, and

Jefferson never ceased to be thoughtful of that necessity: ". . . I do most anxiously wish to see the highest degrees of education given to the higher degrees of genius, and to all degrees of it, so much as may enable them to read and understand what is going on in the world, and to keep their part of it going on right for nothing can keep it right but their own vigilant and distrustful superintendence."

And all these statements must be read in the light of Jefferson's apprehension of the disarray of agriculture and of agricultural communities in his time: ". . . the long succession of years of stunted crops, of reduced prices, the general prostration of the farming business, under levies for the support of manufacturers, etc., with the calamitous fluctuations of value in our paper medium, have kept agriculture in a state of abject depression, which has peopled the Western States by silently breaking up those on the Atlantic . . ."

JUSTIN MORRILL AND THE LAND-GRANT COLLEGE ACTS

On July 2, 1862, two days less than thirty-six years after the death of Jefferson, the first of the land-grant college acts became law. This was the Morrill Act, which granted "an amount of public land, to be apportioned to each State a quantity equal to thirty thousand acres for each Senator and Representative in Congress . . ." The interest on the money from the sale of these lands was to be applied by each state "to the endowment, support, and maintenance of at least one college where the leading object shall be . . . to teach such branches of learning as are related to agriculture and the mechanic arts . . . in order to promote the liberal and practical education of the industrial classes in the several pursuits and professions in life."

In 1887 Congress passed the Hatch Act, which created the state agricultural experiment stations, with the purpose, among others, of promoting "a sound and prosperous agriculture and rural life as indispensable to the maintenance of maximum employment and national prosperity and security." This act states that "It is also the intent of Congress to assure agriculture a position in research *equal to that of industry*, which will aid in maintaining an equitable balance between agriculture and other segments of the economy." (Emphasis mine—to call attention to the distinction made between agriculture and industry.) The act declares, further, that "It shall be the object and duty of the State agricultural experiment stations . . . to conduct . . . researches, investigations, and experiments bearing directly on and contributing to the establishment and maintenance of a permanent and effective agricultural industry . . . including . . . such investigations as have for their purpose the development and improvement of the rural home and rural life . . ."

And in 1914 the Smith-Lever Act created the cooperative extension service "In order to aid in diffusing among the people . . . useful and practical information on subjects relating to agriculture and home economics, and to encourage the application of the same . . ."

Together, these acts provide for what is known as the land-grant college complex. They fulfill the intention of Justin Smith Morrill, representative and later senator from Vermont. In clarification of the historical pertinence and the aims of the language of the several bills, it is useful to have Morrill's statement of his intentions in a memoir written "apparently in 1874."

Morrill was aware, as Jefferson had been, of an agricultural disorder manifested both by soil depletion and by the unsettlement of population: ". . . the very cheapness of our public lands, and the facility of purchase and transfer, tended to a system of bad-farming or strip and waste of the soil, by encouraging short occupancy and a speedy search for new homes, entailing upon the first and older settlements a rapid deterioration of the soil, which would not be likely to be arrested except by more thorough and scientific knowledge of agriculture and by a higher education of those who were devoted to its pursuit."

But Morrill, unlike Jefferson, had personal reason to be generously concerned for "the class of artificers": ". . . being myself the son of a hard-handed blacksmith . . . who felt his own deprivation of schools . . . I could not overlook mechanics in any measure intended to aid the industrial classes in the procurement of an education that might exalt their usefulness."

And he wished to break what seemed to him "a monopoly of education": ". . . most of the existing collegiate institutions and their feeders were based upon the classic plan of teaching those only destined to pursue the so-called learned professions, leaving farmers and mechanics and all those who must win their bread by labor, to the haphazard of being self-taught or not scientifically taught at all, and restricting the number of those who might be supposed to be qualified to fill places of higher consideration in private or public employments to the limited number of the graduates of the literary institutions."

THE LAND-GRANT COLLEGES

To understand what eventually became of the land-grant college complex, it will be worthwhile to consider certain significant differences between the thinking of Jefferson and that of Morrill. The most important of these is the apparent absence from Morrill's mind of Jefferson's complex sense of the dependence of democratic citizenship upon education. For Jefferson, the ideals and aims of education appear to have been defined directly by the requirements of political liberty. He envisioned a local system of education with a double purpose: to foster in the general population the critical alertness necessary to good citizenship and to seek out and prepare a "natural aristocracy" of "virtue and talents" for the duties and trusts of leadership. His plan of education for Virginia did not include any form of specialized or vocational training. He apparently assumed that if communities could be stabilized and preserved by the virtues of citizenship and leadership, then the "practical arts" would be improved as a matter of course by local example, reading, etc. Morrill, on the other hand, looked at education from a strictly practical or utilitarian viewpoint. He believed that the primary aims of education were to correct the work of farmers and mechanics and "exalt their usefulness." His wish to break the educational monopoly of the professional class was Jeffersonian only in a very limited sense: he wished to open the professional class to the children of laborers. In distinguishing among the levels of education, he did not distinguish, as Jefferson did, among "degrees of genius."

Again, whereas Jefferson regarded farmers as "the most valuable citizens," Morrill looked upon the professions as "places of higher consideration." We are thus faced with a difficulty in understanding Morrill's wish to "exalt the usefulness" of "those who must win their bread by labor." Would education exalt their usefulness by raising the quality of their work or by making them eligible for promotion to "places of higher consideration"?

Those differences and difficulties notwithstanding, the apparent intention in regard to agriculture remains the same from Jefferson to Morrill to the land-grant college acts. That intention was to promote the stabilization of farming populations and communities and to establish in that way a "permanent" agriculture, enabled by better education to preserve both the land and the people.

The failure of this intention, and the promotion by the land-grant colleges of an *impermanent* agriculture destructive of land and people, was caused in part by the lowering of the educational standard from Jefferson's ideal of public or community responsibility to the utilitarianism of Morrill, insofar as this difference in the aims of the two men represented a shift of public value. The land-grant colleges have, in fact, been very little—and have been less and less—concerned "to promote the liberal and practical education of the industrial classes" or of any other classes. Their history has been largely that of the whittling down of this aim—from education in the broad, "liberal" sense to "practical" preparation for earning a living to various "programs" for certification. They first reduced "liberal and practical" to "practical," and then for "practical" they substituted "specialized." And the standard of their purpose has shifted from usefulness to careerism. And if this has not been caused by, it has certainly accompanied a degeneration of faculty standards, by which professors and teachers of disciplines become first upholders of "professional standards" and then careerists in pursuit of power, money, and prestige.

The land-grant college legislation obviously calls for a system of local institutions responding to local needs and local problems. What we have instead is a system of institutions which more and more resemble one another, like airports and motels, made increasingly uniform by the transience or rootlessness of their career-oriented faculties and the consequent inability to respond to local conditions. The professor lives in his career, in a ghetto of career-oriented fellow professors. Where he may be geographically is of little interest to him. One's career is a vehicle, not a dwelling; one is concerned less for where it is than for where it will go.

The careerist professor is by definition a specialist professor. Utterly dependent upon his institution, he blunts his critical intelligence and blurs his language so as to exist "harmoniously" within it—and so serves his school with an emasculated and fragmentary intelligence, deferring "realistically" to the redundant procedures and meaningless demands of an inflated administrative bureaucracy whose educational purpose is written on its paychecks.

But just as he is dependent on his institution, the specialist professor is also dependent on his students. In order to earn a living, he must teach; in order to teach, he must have students. And so the tendency is to make a commodity of education: to package it attractively, reduce requirements, reduce homework, inflate grades, lower standards, and deal expensively in "public relations."

As self-interest, laziness, and lack of conviction augment the general confusion about what an education is or ought to be, and as standards of excellence are replaced by sliding scales of adequacy, these schools begin to depend upon, and so to institutionalize, the local problems that they were founded to solve. They begin to need, and so to promote, the mobility, careerism, and moral confusion that are victimizing the local population and destroying the local communities. The stock in trade of the "man of learning" comes to be ignorance.

The colleges of agriculture are focused somewhat more upon their whereabouts than, say, the colleges of arts and sciences because of the local exigencies of climate, soils, and crop varieties; but like the rest they tend to orient themselves within the university rather than within the communities they were intended to serve. The impression is unavoidable that the academic specialists of agriculture tend to validate their work experimentally rather than practically, that they would rather be professionally reputable than locally effective, and that they pay little attention, if any, to the social, cultural, and political consequences of their work. Indeed, it sometimes appears that they pay very little attention to its economic consequences.

There is nothing more characteristic of modern agricultural research than its divorcement from the sense of consequence and from all issues of value.

This is facilitated on the one hand by the academic ideal of "objectivity" and on the other by a strange doctrine of the "inevitability" of undisciplined technological growth and change. "Objectivity" has come to be simply the academic uniform of moral cowardice: one who is "objective" never takes a stand. And in the fashionable "realism" of technological determinism, one is shed of the embarrassment of moral and intellectual standards and of any need to define what is excellent or desirable. Education is relieved of its concern for truth in order to prepare students to live in "a changing world." As soon as educational standards begin to be dictated by "a changing world" (changing, of course, to a tune called by the governmental-military-academic-industrial complex), then one is justified in teaching virtually anything in any way—for, after all, one never knows for sure what "a changing world" is going to become. The way is thus opened to run a university as a business, the main purpose of which is to sell diplomas—after a complicated but undemanding four-year ritual—and thereby give employment to professors.

COLLEGES OF "AGRIBUSINESS" AND UNSETTLEMENT

That the land-grant college complex has fulfilled its obligation "to assure agriculture a position in research equal to that of industry" simply by failing to distinguish between the two is acknowledged in the term "agribusiness." The word does not denote any real identity either of function or interest, but only an expedient confusion by which the interests of industry have subjugated those of agriculture. This confusion of agriculture with industry has utterly perverted the intent of the land-grant college acts. The case has been persuasively documented by a task force of the Agribusiness Accountability Project. In the following paragraphs, Jim Hightower and Susan DeMarco give the task force's central argument:

"Who is helped and who is hurt by this research?

"It is the largest-scale growers, the farm machinery and chemicals input companies and the processors who are the primary beneficiaries. Machinery companies such as John Deere, International Harvester, Massey-Ferguson, Allis-Chalmers and J. I. Case almost continually engage in cooperative research efforts at land grant colleges. These corporations contribute money and some of their own research personnel to help land grant scientists develop machinery. In return, they are able to incorporate technological advances in their own products. In some cases they actually receive exclusive licences to manufacture and sell the products of tax-paid research.

"If mechanization has been a boon to agribusiness, it has been a bane to millions of rural Americans. Farmworkers have been the earliest victims. There were 4.3 million hired farmworkers in 1950. Twenty years later that number had fallen to 3.5 million . . .

"Farmworkers have not been compensated for jobs lost to mechanized research. They were not consulted when that work was designed, and their needs were not a part of the research that resulted. They simply were left to fend on their own—no re-training, no unemployment compensation, no research to help them adjust to the changes that came out of the land grant colleges.

"Independent family farmers also have been largely ignored by the land grant colleges. Mechanization research by land grant colleges is either irrelevant or only incidentally adaptable to the needs of 87 to 99 percent of America's farmers. The public subsidy for mechanization actually has weakened the competitive position of the family farmer. Taxpayers, through the land grant college complex, have given corporate producers a technological arsenal specifically suited to their scale of operation and designed to increase their efficiency and profits. The independent family farmer is left to strain his private resources to the breaking point in a desperate effort to clamber aboard the technological treadmill."

The task force also raised the issue of academic featherbedding —irrelevant or frivolous research or instruction carried on by colleges of agriculture, experiment stations, and extension services. Evidently, people in many states may expect to be "served" by such studies as one at Cornell that discovered that "employed home-makers have less time for housekeeping tasks than non-employed homemakers." An article in the *Louisville*

Courier-Journal lately revealed, for example, that "a 20-year-old waitress . . . recently attended a class where she learned 'how to set a real good table.'

"She got some tips on how to save steps and give faster service by 'carrying quite a few things' on the same tray. And she learned most of the highway numbers in the area, so she could give better directions to confused tourists.

"She learned all of that from the University of Kentucky College of Agriculture. Specialists in restaurant management left the Lexington campus to give the training to waitresses . . .

"The UK College of Agriculture promotes tourism.

"The college also helps to plan highways, housing projects, sewer systems and industrial developments throughout the state.

"It offers training in babysitting, 'family living' . . ."

This sort of "agricultural" service is justified under the Smith-Lever Act, Section 347a, inserted by amendment in 1955, and by Representative Lever's "charge" to the Extension Service in 1913. Both contain language that requires some looking at.

Section 347a is based mainly upon the following congressional insight: that "in certain agricultural areas," "there is concentration of farm families on farms either too small or too unproductive or both . . ." For these "disadvantaged farms" the following remedies were provided: "(1) Intensive on-the-farm educational assistance to the farm family in appraising and resolving its problems; (2) assistance and counseling to local groups in appraising resources for capability of improvement in agriculture or introduction of industry designed to supplement farm income; (3) cooperation with other agencies and groups in furnishing all possible information as to existing employment opportunities, particularly to farm families having underemployed workers; and (4) in cases where the farm family, after analysis of its opportunities and existing resources, finds it advisable to seek a new farming venture, the providing of information, advice, and counsel in connection with making such change."

The pertinent language of Representative Lever's "charge," which is apparently regarded as having the force of law, at least by the University of Kentucky Cooperative Extension Service, places upon extension agents the responsibility "to assume leadership in every movement, whatever it may be, the aim of which is better farming, better living, more happiness, more education and better citizenship."

If Section 347a is an example—as it certainly is—of special-interest legislation, its special interest is only ostensibly and vaguely in the welfare of small ("disadvantaged") farmers. To begin with, it introduces into law and into land-grant philosophy the startling concept that a farm can be "too small" or "too unproductive." The only standard for this judgment is implied in the clauses that follow it: the farmers of such farms "are unable to make adjustments and investments required to establish profitable operations"; such a farm "does not permit profitable employment of available labor"; and—most revealing—"many of these farm families are not able to make full use of current extension programs . . ."

The first two of these definitions of a "too small" or "too unproductive" farm are not agricultural but economic: the farm must provide, not a living, but a profit. And it must be profitable, moreover, in an economy that—in 1955, as now—favors "agribusiness." (Section 347a is a product of the era in which then Assistant Secretaries of Agriculture John Davis and Earl Butz were advocating "corporate control to 'rationalize' agriculture production"; in which Mr. Davis himself invented the term "agribusiness"; in which then Secretary of Agriculture Ezra Taft Benson told farmers to "Get big or get out.") Profitability may be a standard of a sort, but a most relative sort and by no means sufficient. It leaves out of consideration, for instance, the possibility that a family might farm a small acreage, take excellent care of it, make a decent, honorable, and independent living from it, and yet fail to make what the authors of Section 347a would consider a profit.

But the third definition is, if possible, even more insidious: a farm is "too small" or "too unproductive" if it cannot "make full use of current extension programs." The farm is not to be the measure of the service; the service is to be the measure of the farm.

It will be argued that Section 347a was passed in response to real conditions of economic hardship on the farm and that the aim of the law was to permit the development of *new* extension programs as remedies. But that is at best only half true. There certainly were economic hardships on the farm in 1955; we have proof of that in the drastic decline in the number of farms and farmers since then. But there was plenty of land-grant legislation at that time to permit the extension service to devise any program necessary to deal with agricultural problems *as such*. What is remarkable about Section 347a is that it permitted the land-grant colleges to abandon these problems as such, to accept the "agribusiness" revolution as inevitable, and to undertake non-agricultural solutions to agricultural problems. And the assistances provided for in Section 347a are so general and vague as to allow the colleges to be most inventive. After 1955, the agricultural academicians would have a vested interest, not in the welfare of farmers, but in virtually anything at all that might happen to ex-farmers, their families, and their descendants forevermore. They have, in other words, a vested interest in their own failure—foolproof job security.

But it is hard to see how the language of Section 347a, loose as it is, justifies the teaching of highway numbers to waitresses, the promotion of tourism, and the planning of industrial developments, sewer systems, and housing projects. For justification of these programs we apparently must look to the language of Representative Lever's "charge," which in effect tells the extension agents to do anything they can think of.

These new "services" seem little more than desperate maneuvers on the part of the land-grant colleges to deal with the drastic reduction in the last thirty years of their lawful clientele—a reduction for which the colleges themselves are in large part responsible because of their eager collaboration with "agribusiness." As the conversion of farming into agribusiness has depopulated the farmland, it has become necessary for the agriculture specialists to develop "programs" with which to follow their erstwhile beneficiaries into the cities—either that or lose their meal ticket in the colleges. If the colleges of agriculture have so assiduously promoted the industrialization of farming and the urbanization of farmers that now "96 percent of America's manpower is freed from food production," then the necessary trick of survival is to become colleges of industrialization and urbanization—that is, colleges of "agribusiness"

—which, in fact, is what they have been for a long time. Their success has been stupendous: as the number of farmers has decreased, the colleges of agriculture have grown larger.

The bad faith of the program-mongering under Section 347a may be suggested by several questions:

Why did land-grant colleges not address themselves to the *agricultural* problems of small or "disadvantaged" farmers?

Why did they not undertake the development of small-scale technologies and methods appropriate to the small farm?

Why have they assumed that the turn to "agribusiness" and big technology was "inevitable"?

Why, if they can promote tourism and plan sewer systems, have they not promoted cooperatives to give small farmers some measure of protection against corporate suppliers and purchasers?

Why have they watched in silence the destruction of the markets of the small producers of poultry, eggs, butter, cream, and milk—once the mainstays of the small-farm economy?

Why have they never studied or questioned the necessity or the justice of the sanitation laws that have been used to destroy such markets?

Why have they not tried to calculate the real (urban and rural) costs of the migration from farm to city?

Why have they raised no questions of social, political, or cultural value?

That the colleges of agriculture should have become colleges of "agribusiness"—working, in effect, *against* the interests of the small farmers, the farm communities, and the farmland—can only be explained by the isolation of specialization.

First we have the division of the study of agriculture into specialties. And then, within the structure of the university, we have the separation of these specialties from specialties of other kinds. This problem is outlined with forceful insight by Andre Mayer and Jean Mayer in an article entitled "Agriculture, the Island Empire," published in the summer 1974 issue of *Daedalus*. Like other academic professions, agriculture has gone its separate way and aggrandized itself in its own fashion: "As it developed into an intellectual discipline in the nineteenth century, it did so in academic divisions which were isolated from the liberal arts center of the university . . ." It "produced ancillary disciplines parallel to those in the arts and sciences . . ." And it "developed its own scientific organizations; its own professional, trade, and social organizations; its own technical and popular magazines; and its own public. It even has a separate political system . . ."

The founding fathers, these authors point out, "placed agriculture at the center of an Enlightenment concept of science broad enough to include society, politics, and sometimes even theology." But the modern academic structure has alienated agriculture from such concerns. The result is an absurd "independence" which has produced genetic research "without attention to nutritional values," which has undertaken the so-called Green Revolution without concern for its genetic oversimplification or its social, political, and cultural dangers, and which keeps agriculture in a separate "field" from ecology.

A BETRAYAL OF TRUST

The educational *ideal* that concerns us here was held clearly in the mind of Thomas Jefferson, was somewhat diminished or obscured in the mind of Justin Morrill, but survived indisputably in the original language of the land-grant college acts. We see it in the intention that education should be "liberal" as well as "practical," in the wish to foster "a sound and prosperous agriculture and rural life," in the distinction between agriculture and industry, in the purpose of establishing and maintaining a "permanent" agriculture, in the implied perception that this permanence would depend on the stability of "the rural home and rural life." This ideal is simply that farmers should be educated, liberally and practically, *as farmers*; education should be given and acquired with the understanding that those so educated would return to their home communities, not merely to be farmers, corrected and improved by their learning, but also to assume the trusts and obligations of community leadership, the highest form of that "vigilant and distrustful superintendence" without which the communities could not preserve themselves. This leadership, moreover, would tend to safeguard agriculture's distinction from and competitiveness with industry. Conceivably, had it existed, this leadership might have resulted in community-imposed restraints upon technology, such as those practiced by the Amish.

Having stated the ideal, it becomes possible not merely to perceive the degeneracy and incoherence of the land-grant colleges within themselves, but to understand their degenerative influence on the farming communities. It becomes possible to see that their failure goes beyond the disintegration of intellectual and educational standards; it is the betrayal of a trust.

The land-grant acts gave to the colleges not just government funds and a commission to teach and to do research, but also a purpose which may be generally stated as the preservation of agriculture and rural life. That this purpose is a practical one is obvious from the language of the acts; no one, I dare say, would deny that this is so. It is equally clear, though far less acknowledged, that the purpose is also moral, insofar as it raises issues of value and of feeling. It may be that pure practicality can deal with agriculture so long as agriculture is defined as a set of problems that are purely technological (though such a definition is in itself a gross falsification), but it inevitably falters at the meanings of "liberal," "sound and prosperous," "permanent and effective," "development and improvement"; and it fails altogether to address the concepts of "the rural home and rural life." When the Hatch Act, for instance, imposed upon the colleges the goals of "a permanent and effective agricultural industry" and "the development and improvement of the rural home and rural life," it implicitly required of them an allegiance to the agrarian values that have constituted one of the dominant themes of American history and thought.

The tragedy of the land-grant acts is that their moral imperative came finally to have nowhere to rest except on the careers of specialists whose standards and operating procedures were amoral: the "objective" practitioners of the "science" of agriculture, whose minds have no direction other than that laid out by career necessity and the logic of experimentation. They have no apparent moral allegiances or bearings or limits. Their work thus inevitably serves whatever power is greatest. That power at present is the industrial economy, of which "agribusiness" is a part. Lacking any moral force or vision of its own, the "objective" expertise of the agriculture specialist points like a compass needle toward the greater good of the "agribusiness" corporations. The objectivity of the laboratory functions in the world as indifference; knowledge without responsibility is merchandise, and greed provides its applications. Far from developing and improving the rural home and rural life, the land-grant colleges have blindly followed the drift of virtually the whole population away from home, blindly documenting or "serving" the consequent disorder and blindly rationalizing this disorder as "progress" or "miraculous development."

At this point one can begin to understand the violence that has been done to the Morrill Act's provision for a "liberal and practical education." One imagines that Jefferson might have objected to the inclusion of the phrase "and practical," and indeed in retrospect the danger in it is clearly visible. Nevertheless, the law evidently sees "liberal and practical" as a description of *one* education, not two. And as long as the two terms are thus associated, the combination remains thinkable: the "liberal" side, for instance, might offer necessary restraints of value to the "practical"; the "practical" interest might direct the "liberal" to crucial issues of use and effect.

In practice, however, the Morrill Act's formula has been neatly bisected and carried out as if it read "a liberal or a practical education." But though these two kinds of education may theoretically be divided and given equal importance, in fact they are no sooner divided than they are opposed. They enter into competition with one another, and by a kind of educational Gresham's Law the practical curriculum drives out the liberal.

This happens because the *standards* of the two kinds of education are fundamentally different and fundamentally opposed. The standard of liberal education is based upon definitions of excellence in the various disciplines. These definitions are in turn based upon example. One learns to order one's thoughts and to speak and write coherently by studying exemplary thinkers, speakers, and writers of the past.

One studies *The Divine Comedy* and the Pythagorean theorem not to acquire something to be exchanged for something else, but to understand the orders and the kinds of thought and to furnish the mind with subjects and examples. Because the standards are rooted in examples, they do not change.

The standard of practical education, on the other hand, is based upon the question of what will work, and because the practical is by definition of the curriculum set aside from issues of value, the question tends to be resolved in the most shallow and immediate fashion: what is practical is what makes money; what is most practical is what makes the most money. Practical education is an "investment," something acquired to be exchanged for something else—a "good" job, money, prestige. It is oriented entirely toward the future, toward what *will* work in the "changing world" in which the student is supposedly being prepared to "compete." The standard of practicality, as used, is inherently a degenerative standard. There is nothing to correct it except suppositions about what the world will be like and what the student will therefore need to know. Because the future is by definition unknown, one person's supposition about the future tends to be as good, or as forceful, as another's. And so the standard of practicality tends to revise itself downward to meet, not the needs, but the desires of the student who, for instance, does not want to learn a science because he *intends* to pursue a career in which he does not *think* a knowledge of science will be necessary.

It could be said that a liberal education has the nature of a bequest, in that it looks upon the student as the potential heir of a cultural birthright, whereas a practical education has the nature of a commodity to be exchanged for position, status, wealth, etc., *in the future*. A liberal education rests on the assumption that nature and human nature do not change very much or very fast and that one therefore needs to understand the past. The practical educators assume that human society itself is the only significant context, that change is therefore fundamental, constant, and necessary, that the future will be wholly unlike the past, that the past is outmoded, irrelevant, and an encumbrance upon the future—the present being only a time for dividing past from future, for getting ready.

But these definitions, based on division and opposition, are too simple. It is easy, accepting the viewpoint of either side, to find fault with the other. But the wrong is on neither side; it is in their division. One of the purposes of this book is to show how the practical, divorced from the discipline of value, tends to be defined by the immediate interests of the practitioner, and so becomes destructive of value, practical and otherwise. But it must not be forgotten that, divorced from the practical, the liberal disciplines lose their sense of use and influence and become attenuated and aimless. The purity of "pure" science is then ritualized as a highly competitive intellectual game without awareness of use, responsibility, or consequence, such as that described in *The Double Helix*, James D. Watson's book about the discovery of the structure of DNA. And the so-called humanities become a world of their own, a collection of "professional" sub-languages, complicated circuitries of abstruse interpretation, feckless exercises of sensibility. Without the balance of historic value, practical education gives us that most absurd of standards: "relevance," based upon the suppositional needs of a theoretical future. But liberal education, divorced from practicality, gives something no less absurd: the specialist professor of one or another of the liberal arts, the custodian of an inheritance he has learned much about, but nothing from.

And in the face of competition from the practical curriculum, the liberal has found it impossible to maintain its own standards and so has become practical—that is, career-oriented—also. It is now widely assumed that the only good reason to study literature or philosophy is to become a teacher of literature or philosophy—in order, that is, to get an income from it. I recently received in the mail a textbook of rhetoric in which the author stated that "there is no need for anyone except a professional linguist to be able to explain language operations specifically and accurately." Maybe so, but how does one escape the implicit absurdity that linguists should study the language only to teach aspiring linguists?

The education of the student of agriculture is almost as absurd, and it is more dangerous: he is taught a course of practical knowledge and procedures for which uses do indeed exist, but these uses lie outside the purview and interest of the school. The colleges of agriculture produce agriculture specialists and "agribusinessmen" as readily as farmers, and they are producing far more of them. Public funds originally voted to provide for "the liberal and practical education" of farmers thus become, by moral default, an educational subsidy given to the farmers' competitors.

THE VAGRANT ARISTOCRACY

But in order to complete an understanding of the modern disconnection between work and value, it is necessary to see how certain "aristocratic" ideas of status and leisure have been institutionalized in this system of education. This is one of the liabilities of the social and political origins not only of our own nation, but of most of the "advanced" nations of the world. Democracy has involved more than the enfranchisement of the lower classes; it has meant also the popularization of the more superficial upper-class values: leisure, etiquette (as opposed to good manners), fashion, everyday dressing up, and a kind of dietary persnicketiness. We have given a highly inflated value to "days off" and to the wearing of a necktie; we pay an exorbitant price for the *looks* of our automobiles; we pay dearly, in both money and health, for our predilection for white bread. We attach much the same values to kinds of profession and levels of income that were once attached to hereditary classes.

It is extremely difficult to exalt the usefulness of any productive discipline *as such* in a society that is at once highly stratified and highly mobile. Both the stratification and the mobility are based upon notions of prestige, which are in turn based upon these reliquary social fashions. Thus doctors are given higher status than farmers, not because they are more necessary, more useful, more able, more talented, or more virtuous, but because they are *thought* to be "better"—one assumes because they talk a learned jargon, wear good clothes all the time, and make a lot of money. And this is true generally of "office people" as opposed to those who work with their hands. Thus an industrial worker does not aspire to become a master craftsman, but rather a foreman or manager. Thus a farmer's son does not usually think to "better" himself by becoming a better farmer than his father, but by becoming, professionally, a better *kind* of man than his father.

It is characteristic of our present society that one does not think to improve oneself by becoming better at what one is doing or by assuming some measure of public responsibility in order to improve local conditions; one thinks to improve oneself by becoming different, by "moving up" to a "place of higher consideration." Thinkable changes, in other words, tend to be quantitative rather than qualitative, and they tend to involve movement that is both social and geographic. The unsettlement at once of population and of values is virtually required by the only generally acceptable forms of aspiration. The typical American "success story" moves from a modest rural beginning to urban affluence, from manual labor to office work.

We must ask, then, what must be the educational effect, the influence, of a farmer's son who believes, with the absolute authorization of his society, that he has mightily improved himself by becoming a professor of agriculture. Has he not improved himself by an "upward" motivation which by its nature avoids the issue of quality—which assumes simply that an agriculture specialist is better than a farmer? And does he not exemplify to his students the proposition that "the way up" leads away from home? How could he, who has "succeeded" by earning a Ph.D. and a nice place in town, advise his best students to go home and farm, or even assume that they might find good reasons for doing so?

I am suggesting that our university-based structures of success, as they have come to be formed upon quantitative measures, virtually require the degeneration of qualitative measures and the disintegration of culture. The university accumulates information at a rate that is literally inconceivable, yet its structure and its self-esteem institutionalize the likelihood that not much of this information will ever be taken *home*. We do not work where we live, and if we are to hold up our heads in the presence of our teachers and classmates, we must not live where we come from.

THE STATUS QUO

So far, in tracing the changes of an American educational ambition, this chapter has necessarily been to some extent conjectural. As elsewhere in this book, I have been writing what my experience has made it possible for me to say—with the understanding that it must then await confirmation, amplification, or contradiction from the experience of other people. I have intentionally placed experience ahead of "proof," feeling that the ordinary visibility of the deterioration of rural life ought to take precedence over statistics and expert testimony.

Nevertheless, the testimony of experts must be taken into account. It seems appropriate that I should conclude this chapter by examining in some detail a prominent expert's justification of the agricultural status quo. The article, "The Agriculture of the U.S.," comes from the September 1976 issue of *Scientific American*. Its author is Earl O. Heady, Curtiss Distinguished Professor at Iowa State University and director of that university's Center for Agriculture and Rural Development. Professor Heady "was born and raised on a farm in Nebraska" and received his degrees from the University of Nebraska and from Iowa State. He is author or co-author of "17 books and more than 725 journal articles, research bulletins and monographs." He is vice-president of the American Association of Agricultural Economists, vice-president of the Canadian Agricultural Economics Association, and permanent chairman of the East-West Seminars for Agricultural Economists. His biographical note quotes him as follows: "I do a lot of work in developing countries, consulting with planners, evaluating policies for economic and agricultural development and analyzing development in general."

Professor Heady begins his account with this statement: "Over the past 200 years the U.S. has had the best, the most logical and the most successful program of agricultural development anywhere in the world. Other countries would do well to copy it." The occurrence of such an absolute assertion at the *beginning* of a scientific article by an objective scientist can only strike one as remarkable. And a little consideration makes it even more so. Has he forgotten, or did he ever know, for instance, that in 1907, F. H. King, also an American professor of agriculture and chief of Division of Soil Management, United States Department of Agriculture, was traveling in China, Korea, and Japan, studying the ancient agricultural practices of those countries and finding them exemplary? Does Professor Heady know, for that matter, of the work of any critic of his assumptions? And who is he trying to convince? Surely not the readers of *Scientific American*, most of whom will at least wish to see his evidence. But, in fact, for the supremacy of American agriculture over that of all other countries, Professor Heady's article offers no evidence whatsoever. And the evidence he does supply leaves the logic and success of the American program very much in doubt.

"At the beginning of the nation's agricultural development," Professor Heady writes, "land was abundant and labor was cheap. Capital inputs such as farm machinery, fertilizer and food for the farmer's family were relatively modest, and most of them were produced on the farm. Farmers created their own power in the form of the physical work of family members and of animals raised on the farm. They also harnessed energy from the sun for that work in the form of crops grown on the farm and eaten by the people or the animals. The farmers generated their own fertilizer by rotating crops and by utilizing the wastes from the animals. The rotation of crops also controlled insects to some extent."

That description is not critical enough. In its general outline it describes the agriculture of many parts of this country as late as World War II. The greatest weakness of that agriculture was undoubtedly its wastefulness of the soil itself, but there were other weaknesses also. It was the knowledge of these weaknesses that sent F. H. King to the Orient, and his discoveries there, had they taken root here, might have made our farmers more solvent and productive, and much kinder in their use of the soil. But Professor Heady's description may be allowed to stand; it does represent accurately enough the possibility of a thrifty, independent, diversified, farm-based agriculture that remained easily within our reach until a generation ago.

That possibility and a virgin continent were the endowment that we started with. In the rest of his article Professor Heady tells what we have done with, and to, that endowment.

In the nineteenth century, he tells us, after the United States had expanded to its westward limits and the public land grants had all been taken up, the government's agricultural policy shifted its emphasis from expansion to productivity. The land-grant college system was created "to encourage research and to extend new technical knowledge to farmers." Science and technology became "an effective substitute for land." As a result, production "approximately doubled" in the period from 1910 to 1970, and "by 1970 the nation was producing its food on considerably fewer acres than it had been in 1910." Rapidly put into use, the new technology "became an effective substitute not only for land but also for labor. The result was that between 1950 and 1955 more than a million workers migrated out of the agricultural sector into other sectors of the economy."

We are asked to accept that our agricultural policy-makers displayed profound wisdom in shifting their emphasis from expansion to productivity—as if, after the possibility of expansion had ended, the choice was difficult. And we are asked to accept productivity as a sufficient criterion; nothing is said, here or elsewhere in Professor Heady's article, about the issues of restoration and maintenance. The displacement of a million workers in five years is cited merely as evidence of the efficacy of technology. One wonders what may have been the social and economic costs of that "migration." Into what "sectors of the economy" did those workers move? And it may not be impertinent in a democracy to ask, Did they want to go?

Next Professor Heady focuses on the period from 1950 to 1970: "Farms became larger and more specialized, handling either crops or livestock instead of both. Farms growing crops greatly increased their utilization of fertilizers, pesticides, farm machinery and other capital items . . . the use of fertilizer increased by 276 percent. . . . The use of powered machines increased by only 30 percent, but in 1972 there were substantially fewer farms than there were in 1950. The result was that farm labor declined by 54 percent over that period as labor productivity quadrupled and total farm output increased by 55 percent."

Again, highly problematic changes are cited solely as evidence of the advance of technology, which we are evidently expected to regard as simply good. And again a massive displacement of "labor" is treated as if people were merely underpowered, slow machines, now happily replaced by machines of a better make.

In 1974 and 1975, Professor Heady tells us, American farmers produced "record" yields, which brought them a "record" income. Records, as we know, are made by champions and are good beyond question. But Professor Heady goes on: "The rapid upward movement in income has put farmers in a highly favorable position with regard to capital assets. Although some farmers took advantage of the opportunity to repay their mortgage before it came due, the majority put their higher earnings into acquiring new farm equipment, upgrading their living facilities and enlarging their farms by buying more land. As a result farm real estate values more than doubled between 1970 and 1973."

This is the second time Professor Heady's article has spoken of the recent increases in the value of farmland to "record levels," as if this is some kind of grand agricultural achievement. But is this increase entirely due to competition among farmers for the land, or do inflation, urban development, and speculation have something to do with it? And are there dangers in these high prices? Although the fact of inflation is rather casually mentioned later in the article, the first question is really neither answered nor asked. The second question is answered later on, but the dangers are not admitted.

Meanwhile, Professor Heady acknowledges the existence of certain other problems: "The change in the very nature of farming, with its higher productivity and greater degree of mechanization, has severely affected rural communities. . . . With the decline in the farm population the demand for the goods and services of businesses in the country has been eroded. Employment and income opportunities in typical rural communities have therefore declined markedly. As people migrated out of the rural communities, there were fewer people left to participate in the services of schools, medical facilities and other institutions. With the lessened demand such services retreated in quantity and quality and advanced in cost.

"Nonfarm groups in the rural communities took large capital losses. . . ."

Professor Heady further acknowledges that "Rapid agricultural development . . . has also had a heavy impact on the environment." The larger and more specialized farms are "depleting the soil of certain specific nutrients and thus requiring larger amounts of fertilizer." This increase in the use of fertilizer has been accompanied by an increased use of pesticides and more intensive (that is, more continuous) cultivation. "The burden placed on streams and lakes by the runoff of silt and farm chemicals has therefore increased."

"On the other hand," he says, "the development of American

agriculture has fostered the growth of an entire agricultural industry—'agribusiness'—of which farming is only a small part."

Anyone who cares at all for the welfare of the rural home and rural life and for the good health of the farmland will see the arrogance of that phrase "on the other hand." It is the balancing point of a monstrous equation. Professor Heady has just described a serious impairment of rural life that is social, economic, and ecological, and he has said that it is justified and compensated by the growth of "agribusiness." The sacrifice of many and of much for the enrichment of a few is thus justified as if the Declaration of Independence had never been written.

The "industry" of modern agriculture, according to Professor Heady, has "three major components": "the input-processing industry," "the farm itself," and "the food-processing industry." I will quote, nearly complete, Professor Heady's description of the first and last of these, asking the reader to bear in mind the professor's earlier description of the kind of farming we had at the beginning of our "agricultural development."

"The input-processing industry now supplies many things that were once produced on the farm. Today tractors substitute for draft animals, fossil fuels for animal feeds, chemical fertilizers for manure and nitrogen-fixing crops. Such developments not only have shifted a greater proportion of the agricultural work force from the farms into the input-processing sector but also have increased the cash cost of farming. . . . The greater proportion of cash cost has made farm profits much more vulnerable to price fluctuations than they used to be.

"The food-processing sector has in recent years come to represent a larger proportion of the total agricultural industry than farming itself. In 1975, 42 cents of each consumer dollar spent for food at retail prices went to the farmer and 58 cents went to the food processor. Even the typical commercial farm family now buys frozen, packaged and ready-to-serve foods from the supermarket rather than consuming products raised and prepared on the farm."

So much for the ideal—and the practical values—of independence. If the farmer sells his foodstuff to "agribusiness" at a narrow profit, if any, and buys it back ready-to-serve from "agribusiness" to its great profit, then the cash flow has at that point deftly inserted its tail into its mouth, a wonder of sorts has been accomplished, and a reverent "Golly!" is heard from certain agricultural economists.

And now, sufficiently far from the question, Professor Heady gives us an answer as to the dangers of high land prices: "The change in the nature of agriculture has greatly enhanced the financial position of established farmers with large holdings. . . . The situation is not as favorable for farmers who are starting from scratch. . . . One can therefore expect to see an increasing trend toward more large commercial farms and fewer small ones."

Professor Heady's "therefore" is nearly as irresponsible as his "on the other hand." By various inequities, abuses, and misconceptions, a condition has come to exist in which big farmers thrive by the ruin of smaller ones. And Professor Heady enjoins this condition upon the future by a simple "therefore."

Aside from the urgent social and political questions that are obviously raised by Professor Heady's observation, it raises, in fact, some agricultural and economic questions that are also extremely serious. I shall mention two.

First, if hunger and malnutrition are now in prospect for many of the world's people, as hardly anyone (including Professor Heady) denies, and if productivity is therefore the major issue, can we afford this trend toward bigger and bigger farms? The question rises from the awareness, now shared by many experts, that large farms do not produce as abundantly or efficiently as small ones. Sterling Wortman, for instance, writing in the same issue of *Scientific American*, says that "mechanized agriculture is very productive in terms of output per man-year, but it is not as productive per unit of land as the highly intensive systems are." Why, then, does it not make sense to advocate a return to smaller, family-type farms, on which human and animal labor can be effectively substituted for machines?

Second, if the size of farms continues to increase, and the farm population proportionately decreases, will not that population become at the same time more vulnerable, less surely able to reproduce itself? According to Professor Heady, it is one of the grand achievements of American agriculture that it now employs "only 4.4 percent of the nation's population." But at what level does a population—especially one in precipitous decline—become threatened with extinction? I assume, as perhaps Professor Heady does not, that in order to run our farms productively we will have to have farmers, that a knowledge of farming and of land stewardship are of direct value to those who farm, and that the most obvious and economical way to get farmers with this knowledge is to raise them. By this accounting, the knowledge and interest of the many young farmers who are now being priced off the land amounts to a sizable loss.

According to Professor Heady, American agriculture still has plenty of room to expand: if necessary, in order to increase production, we can plow and plant some hundreds of millions of acres of fallows, pastures, forests, range lands, and wetlands. Land, then, so far as he is concerned, is not an agricultural problem. And he evidently has no doubt that we will continue to have plenty of farmers. His worries come from another direction: "The future of American agriculture will depend on a number of factors in addition to its productive capacity. The two most important factors will be the extent to which recent international conditions continue to prevail and the presence or absence of Government policies affecting output either through future supply-control programs or environmental limits on fertilizers, pesticides and soil erosion." In other words, American agriculture will continue to prosper so long as hunger remains an international threat, so long as "agribusiness" is not restrained, and so long as "established farmers with large holdings" are left free to continue the pollution and soil erosion that are the inevitable by-products of industrial agriculture.

By this "most logical" of developments, then, we have passed from a farm-based, family-based, independent agriculture to an agriculture abjectly dependent upon many kinds of industrial "inputs" and firmly based upon several kinds of disaster. We are producing, at an incalculable waste of topsoil and of human life and energy, and at the cost of destroying communities and poisoning the land and the streams, food to be used against the hungry as a weapon.*

*At this point one thinks with some solicitude of the "developing countries" in which Professor Heady does "a lot of work." They are apparently in danger of taking the advice of an agricultural consultant the success of whose policy requires them to get hungry.

EXPERIENCE AND EXPERIMENT

Having for some years attentively read and listened to the statements of agriculture experts, I cannot have the comfort of looking upon Professor Heady as an anomaly. I am constrained to regard him as representative of that academic upper crust that has provided a species of agricultural vandalism with the prestige of its professorships and the justifications of a bogus intellectuality, incomprehensible to any order of thought, but decked out in statistics, charts, and graphs to silence unspecialized skepticism and astonish gullibility.

In spite of his eagerness to defend what he calls a "logical" program, there is no logic in Professor Heady's defense. His defense is deduction *without* logic, a kind of disordered scholasticism that proceeds merely by flinging statistics at a premise. That his premise is called into serious question—if not disproved—by his "proof" does not cause Professor Heady to hesitate.

If Professor Heady and his kind had not so much power, they would deserve far less attention. But because they do have power, because they belong to that association of industrial conquistadors who would claim our future as their colony, it is important to understand how, and how poorly, they think.

Like most of that association, Professor Heady is a specialist. Within the enclosure of his specialty he is no doubt capable of order and sense of a very formidable kind. But when he tries to justify these in terms of value and to say that they and the assumptions on which they rest are "good," then he produces disorder and nonsense, because the order of his specialty does not comprehend a ground large enough to permit such a justification. The calculations that prove the efficacy of technology as a "substitute not only for land but also for labor" can do so convincingly only by ignoring the human and ecological contexts of the substitution. It would be possible to calculate the probable monetary cost of the unemployment, community and family breakdown, crime, vandalism, pollution, and soil loss that are the results of overwhelming "inputs" of technology—but apparently an agricultural economist is not expected to look either so widely around or so far ahead. Nor is any other agriculture expert. They are free to argue with the blind determination of fanatics from the premises that they prefer to the conclusions that they desire. It is an irony that would be amusing, were it not so frightening, that the prestigious "positions" that have relieved them of the necessity to use their hands have cost them the use of their heads.

No wonder they look forward so eagerly to the future. *We*, with our awkwardly divergent and valued lives, our bothersome rights and meanings, are not yet there. Only posterity is native there, and they have as yet produced nothing; they have no claim recognizable by an expert. The future is already surveyed and ribboned according to the claims of these people and their clients, the corporate industrialists and big businessmen. It is their New World, and they are its self-elected ruling class.

The expert knowledge of agriculture developed in the universities, like other such knowledges, is typical of the alien order imposed on a conquered land. We can never produce a native economy, much less a native culture, with this knowledge. It can only make us the imperialist invaders of our own country.

The reason is that this knowledge has no cultural depth or complexity whatever. It is concerned only with the most immediate practical (that is, economic and *sometimes* political) results. It has, for instance, never mastered the crucial distinction between experiment and experience. Experience, which is the basis of culture, tends always toward wholeness because it is interested in the *meaning* of what has happened; it is necessarily as interested in what does not work as in what does. It cannot hope or desire without remembering. Its approach to possibility is always conditioned by its remembrance of failure. It is therefore not "objective," but is at once personal and communal. The experimental intelligence, on the other hand, is only interested in what works; what doesn't work is ruled out of consideration. This sort of intelligence tends to be shallow in that it tends to impose upon experience the metaphor of experiment. It invariably sees innovation, not as adding to, but as replacing what existed or was used before. Thus machine technology is seen as a *substitute* for human or animal labor, requiring the "old way" to be looked upon henceforth with contempt. In technology, as in genetics, the experimental intelligence tends toward radical oversimplification, reducing the number of possibilities. Whereas the voice of experience, of culture, counsels, “Don’t put all your eggs in one basket,” the experimental intelligence, which behaves strangely like the intelligence of imperialists and religious fanatics, says, “This is the *only* true way.”

And this intelligence protects itself from the disruptive memories and questions of experience by building around itself the compartmental structure of the modern university, in which effects and causes need never meet. The experimental intelligence is a tyrant that is saved from the necessity of killing bearers of bad news because it lives at the center of a maze in which the bearers of bad news are lost before they can arrive.

But it is imperative to understand that this sort of intelligence *is* tyrannical. It is at least potentially totalitarian. To think or act without cultural value, and the restraints invariably implicit in cultural value, is simply to wait upon force. This sort of behavior is founded in the cultural disintegration and despair which are also the foundation of political totalitarianism. Whether recognized or not, there is in the workings of agricultural specialization an implicit waiting for the total state power that will permit experimentally derived, technologically pure solutions to be imposed by force.

Woe to those who add house to house
and join field to field
until everywhere belongs to them
and they are the sole inhabitants of the land.

ISAIAH 5:8

. . . it is not too soon to provide by every possible means that as few as possible shall be without a little portion of land. The small landholders are the most precious part of a state. . . .

THOMAS JEFFERSON,
LETTER TO REVEREND JAMES MADISON,
OCTOBER 28, 1785

CHAPTER NINE

Margins

"AGRIBUSINESS" AS ORTHODOXY

NOT all agricultural economists are blind to the human and ecological consequences of "agribusiness" economics. On March 1, 1972, Professor Philip M. Raup, of the University of Minnesota at St. Paul, testified as follows before the Subcommittee on Monopoly of the United States Senate Small Business Committee:

"Only in the past decade has serious attention been given to the fact that the large agricultural firm is . . . able to achieve benefits by externalizing certain costs. The disadvantages of large scale operation fall largely outside the decision-making framework of the large farm firm. Problems of waste disposal, pollution control, added burdens on public services, deterioration of rural social structures, impairment of the tax base, and the political consequences of a concentration of economic power have typically not been considered as costs of large scale, by the firm. They are unquestionably costs to the larger community.

"In theory, large-scale operation should enable the firm to bring a wide range of both benefits and costs within its internal decision-making framework. In practice, the economic and political power that accompanies large size provides a constant temptation to the large firm to take the benefits and pass on the costs.

"The rural community receives the immediate impact of this ability of large farm firms to practice selective internalization of benefits and externalization of costs. One of the most pervasive consequences is that the occupational composition of the population changes. Instead of a large number of small entrepreneurs, combining the functions of manager and laborer, the occupational structure includes a small number of managers and a large number of workers. In rural communities dominated by very large firms, the settlement and housing patterns reflect the increasingly transient nature of the labor force.

The symbol of the large corporate farm becomes the trailer house. Community institutions suffer from lack of leadership, and from the lack of a sense of commitment on the part of the labor force to long-run community welfare. Those institutions that survive take on a dependent character, reflecting the paternalistic role of the dominant firms. Income levels may stabilize, but at the expense of a decline in local capacity for risk-taking, decision-making, and investment of family labor in farms and local businesses."

And later in his testimony, Professor Raup spoke of the most ironic of these "externalized" costs: "Farmers who have succeeded in increasing their farm size to a scale that will enable them to achieve almost all of the economics of size in production now find that their capital structure is so large that their sons cannot finance a takeover of the family farm."

Professor Raup's distinction between internal and external accounting is of great usefulness in understanding our problem. It is by internal accounting that the modern American agricultural program may be thought "the best, the most logical and the most successful." External accounting brings us to a very different conclusion. External accounting pushes us back into our moral tradition, which asks us to consider that we are members of the human community and are therefore bound to help or harm it by our behavior. This sort of accounting involves much more than economics. It is broad and difficult, and it eludes quantification.

Modern American agriculture has made itself a "science" and has preserved itself within its grandiose and destructive assumptions by cutting itself off from the moral tradition (as it has done also from the agricultural tradition) and confining its vision and its thought within the bounds of internal accounting. Agriculture experts and "agribusinessmen" are free to believe that their system works because they have accepted a convention which makes "external," and therefore irrelevant, all evidence that it does not work. "External" questions are not asked or heard, much less answered.

But these people are human beings who inherit a community awareness, to whatever extent it may be suppressed, distorted, or ignored. Many, if not most, of them come from family farms, for which they feel some nostalgia, if no loyalty. And so it must be assumed that the claims of external accounting are still obscurely felt in the backs or the depths of their minds. Some of them may occasionally overhear their critics with a tremor of recognition; from time to time some of them may even come face to face with bad external results of internal purposes and recognize them as such. Internal accounting, then, must cohere under some pressure from the external. This obviously defines the necessary condition for a fierce and self-protective orthodoxy—a science-as-superstition, by which one clings to the assumption of the goodness of one kind of knowledge out of fear of knowledge of another kind. This fear makes the specialist scientist not merely willing to define a possibility, but *desperate* to define the *only* possibility. Only this desperation can explain the venomous contempt with which agricultural establishmentarians dismiss suggestions of other possibilities, old or new. These "objective" scientists exhibit an intense craving to be *right*—a craving hardly diminished by the profitability of their faith.

ORTHODOXY, MARGINS, AND CHANGE

Our history forbids us to be surprised that an orthodoxy of thought should become narrow, rigid, mercenary, morally corrupt, and vengeful against dissenters. This has happened over and over again. It might be thought the maturity of orthodoxy; it is what finally happens to a mind once it has consented to be orthodox. But one may be permitted a little amusement, if not surprise, that this should have befallen a modern science, which was set up, as it never tires of advertising, to *pursue* truth, not *protect* it.

But since what we now have in agriculture—as in several other "objective" disciplines—is a modern scientific orthodoxy as purblind, self-righteous, cocksure, and ill-humored as Cotton Mather's, our history also forbids us to expect it to change from within itself. Like many another orthodoxy, it would rather die than change, and may change only by dying.* This determination is enforced both from within and

*Orthodox agriculture is part of the larger orthodoxy of industrial progress and economic growth, which argues the necessity of pollution, unemployment, war, land spoliation, the exploitation of space, etc. And so the question is: Must we all die with it in order for it to change?

from without. It is enforced from within simply by prosperity: the professors, experts, and executives of the agrifaith do not want agricultural policy to change because they are eating very well off of it as it is. From without, it is enforced by the mistaken conviction of millions of believers that it is the only true way, that they have no choice but to accept the agribusiness philosophy or starve. But it is also enforced by the very nature of orthodoxy: one who presumes to *know* the truth does not *look* for it.

If change is to come, then, it will have to come from the outside. It will have to come from the margins. As an orthodoxy loses its standards, becomes unable to measure itself by what it ought to be, it comes to be measured by what it is not. The margins begin to close in on it, to break down the confidence that supports it, to set up standards clarified by a broadened sense of purpose and necessity, and to demonstrate better possibilities. Though it does not necessarily or always work for the better—though indeed this swing from the center to the margins and back again may be in itself a condemnation—this sort of change is a dominant theme of our tradition, whose "central" figures have often worked their way inward from the margins. It was the desert, not the temple, that gave us the prophets; the colonies, not the motherland, that gave us Adams and Jefferson.

The pattern of orthodoxy in religion, because it is well known, gives us a useful paradigm. The encrusted religious structure is not changed by its institutional dependents—they are part of the crust. It is changed by one who goes alone to the wilderness, where he fasts and prays, and returns with cleansed vision. In going alone, he goes independent of institutions, forswearing orthodoxy ("right opinion"). In going to the wilderness he goes to the margin, where he is surrounded by the possibilities—by no means all good—that orthodoxy has excluded. By fasting he disengages his thoughts from the immediate issues of livelihood; his willing hunger takes his mind off the payroll, so to speak. And by praying he acknowledges ignorance; the orthodox presume to *know*, whereas the marginal person is trying to find out. He returns to the community, not necessarily with new truth, but with a new vision of the truth; he sees it more whole than before.

In applying this pattern to agriculture, one is startled to realize that this is the first time it has been necessary, or possible, to do so. Not until recently have we had a widespread orthodoxy of agriculture in the same sense that we have had widespread orthodoxies of religion—an agriculture, that is to say, which is nearly uniform in technology and in its general assumptions and ambitions over a whole continent, and which, like many religions, aspires to become "universal" by means of a sort of evangelism, proclaiming that "Other countries would do well to copy it."

In agriculture there have always been prevalent patterns of technology, practice, and attitude that may have had the customary force of orthodoxy. But these patterns were local; they varied in response to local conditions. And, unlike orthodoxies, they were not imposed by external authority, but grew as part of a complex relationship between the human community and natural conditions. Until the triumph of the industrial values of the "agribusiness" vision, agriculture was very much a regional affair, a response at once to human need and to regional possibilities and limits, and it was successful and long-lasting in proportion to the sensitivity of that response.

A PRE-INDUSTRIAL EXAMPLE

By looking at an example of a sound pre-industrial agriculture we can get a sense of its ecological and cultural coherence and its geographical responsiveness and also a sense of its careful relationship to its margins. Perhaps no more vivid example exists in our time than that of the native agriculture of the Peruvian Andes. I take the following summary from an unpublished paper by Professor Stephen B. Brush, of the Department of Anthropology at the College of William and Mary.

Professor Brush's study focuses on the village of Uchucmarca in a valley in northern Peru. This valley has "one of the steepest environmental gradients in the world." Like other Andean farmers, the people of Uchucmarca farm in four different climatic zones, requiring four different kinds of agriculture:

"1. a tropical zone . . . which produces fruit (such as oranges and bananas), tropical root plants (such as manioc), chile peppers, and perhaps most importantly *coca* . . .

"2. a middle level mountain zone . . . where maize and wheat are grown . . .

"3. a relatively high mountain zone . . . where the staple of the Andean diet, potatoes, and other Andean tubers are grown . . .

"4. a high mountain zone . . . where llamas, alpacas, sheep, horses, and cattle are grazed on natural pasture . . ."

Within a distance of forty miles the valley rises from "roughly 3,200 feet" to "over 14,700 feet"—thus including a diversity of climates as great as that from west Texas to Alaska. The natives of the valley "recognize and name seven different production zones . . . which are variations on the four major zones."

The agriculture of the valley is based upon a highly evolved awareness of the nature of each of these climatic zones and of the differences among them. It involves a careful balance between the use and the maintenance of productivity. Professor Brush says that "The village economy may be understood as a set of subsistence strategies designed to provision each household with adequate food. . . . One of the most important features of the local economy is that it is able to function as a largely nonmonetized economy. The average family in Uchucmarca needs less than $100 yearly. . . ." It is significant that the verb in that last sentence is "needs," where "agribusiness" assumptions would require "has only." The governing concept of the agriculture of these Andean peasants, then, is *enough*, a long-term sufficiency, whereas the governing concept of ours is *profit* or *affluence*, without regard for long-term needs.*

Like most farmers, those of Uchucmarca must cope with the hazards of erosion, frost, too much or too little rain, pests, and diseases. They do this very effectively and without recourse to the industrial technology of machines and chemicals.

"The danger of erosion is avoided in three ways. First, fields are kept small—usually less than one acre. . . . A typical family

*In a letter to me, dated February 15, 1977, Professor Brush wrote as follows: ". . . I calculate that with their 'primitive' agriculture, the farmers of Uchucmarca produce 2700 calories and 80 grams of protein (vegetable) per capita per day. A very good diet and a well fed population. The worst malnutrition occurs in cities where people must depend on 'modern' agriculture."

of four to five persons cultivates between three and four acres, spread among as many fields. The small size of individual plots retards run-off and erosion. Second, each field is surrounded by a hedgerow constructed of rocks, brush, and living plants. Ostensibly built to keep out destructive livestock, these hedgerows effectively limit erosion. Their roots hold the soil, and horizontal plowing behind them tends to build up soil at the lower side of the field. This creates a quasi-terrace or lynchet. . . . Third, field rotation is practiced in the highest and steepest part of the valley where rainfall is heaviest and erosion most likely. Potatoes are cultivated under a regime of shifting cultivation in which fields are only planted for two or three years before being returned to a long fallow of five years or more. By using this method, the amount of soil washed off of fields is limited, and organic material is allowed to reaccumulate."

The people of Uchucmarca cope with climatic variations by "the exploitation of multiple zones and crops," so that if one crop fails they may rely on one of a different kind. "Another way is to plant several different fields of the same crop, hoping that if one field is destroyed, the other will survive. These means are reinforced by systems of economic reciprocity and mutual dependence which rely primarily on the kinship system." Within families, "individual households protect themselves from privation by exchanging land, labor, and goods."

Against insects and diseases, the main weapon of the Andean peasants is genetic diversity: "Botanists estimate that there are well over 2,000 potato varieties in Peru alone. In single villages like Uchucmarca people identify some fifty varieties . . ." And here we arrive at the greatest complexity, versatility, and responsiveness of this agriculture, as well as its most intense sensitivity to place. For these varieties are not used at random, but are delicately fitted into their appropriate ecological niches. "In Uchucmarca, a common practice is to plant fast growing varieties during the drier part of the year so as to avoid late blight which increases during the months of heavy rain. Another practice is to cultivate certain varieties believed to be somewhat frost resistant in flat, bottom areas of the high valley where frost but not late blight is common. Other varieties are chosen for hillside cultivation where late blight but not frost is common." Varieties are also chosen according to how well they do at certain altitudes or according to whether or not they need a soil that drains well. Of course, this description gives only a rough idea of the intricacy of possible adjustments among so many varieties and so many kinds of ground.

Professor Brush's work makes it plain that nearly all the methods of the Andean farmers are based upon the one principle of diversity. In their understanding and use of this principle, they have developed an agriculture much more sophisticated, efficient, and conservative of the soil than our own—and one that is also much more likely to survive a crisis. How finely this agriculture is attuned to the needs and circumstances of the community becomes apparent when Professor Brush describes recent attempts to change it by the introduction of industrial technology and "improved" potato varieties. Such change involves a gross simplification of the agriculture itself as well as a drastic complication of the economy. It requires a cash economy and credit, favors the larger producers, and threatens to destroy both the human community and the ecological viability of a farming system that is "the result of thousands of years of natural and human selection."

But the sophistication and durability of Andean agriculture cannot be fully appreciated until one has understood the way it utilizes—indeed, depends upon—its margins. The fifty potato varieties used in Uchucmarca are not a stable quantity, but rather a sort of genetic vocabulary in a state of continuous revision. Professor Brush says that "new varieties are constantly being created through cross-pollination between cultivated, wild and semidomesticated (weedy) species. . . . These wild and semidomesticated species thrive in the hedgerows around fields, and birds and insects living there assist cross-pollination." Thus, if an Andean farmer loses a crop because of an extremity of the weather or an infestation of insects or disease, he may find a plant of a new variety that has survived the calamity and produced in spite of it. If he finds such a plant, he may add it to his collection of domesticated varieties or substitute it for the one that has failed.

This Andean agriculture, then, does not push its margins back to land unsuitable for farming, as ours does, but incorporates them into the very structure of its farms. The hedgerows are marginal areas, little thoroughfares of wilderness closely crisscrossing the farmland, and in them agriculture is constantly renewing itself in direct response to what threatens it. This network of wilderness threading through the fields serves the Andean farmer as a college of agriculture and experiment station. And in at least one respect it serves him better: whatever is discovered there has already been tested in the circumstances of the farm itself, and its worth or worthlessness proven. The farmer, in whose mind culture and agriculture are wedded, acts as both teacher-researcher and student, both extension agent and client. Set thus in the light of a truly healthy agriculture, our land-grant college complex may be seen less as a symbol of our agricultural success than as a symptom of our failure.

And this integration of Andean farming with its margins may serve us in another way. It offers an example of a sort of reconciliation by which we might escape the endless swinging between center and margins, rigidity and revolt, that has dominated our culture for so long. The remedy is to accommodate the margin within the form, to allow the wilderness or nature to thrive in domesticity, to accommodate diversity within unity. It is surely by this means—this graceful, practical generosity toward the possible and the unexpected, toward time and history—that Andean agriculture has survived for so long, cohering even through the severe disturbances of the Spanish Conquest. By responding competently to whatever has threatened it, and by doing so in the most local and immediate fashion, it has kept its hold on the world, much as life itself has kept its hold. Having understood this reconciliation or integration of the human community with its natural margins, we may see how crude and dangerous are our absolute divisions between city and farmland, farmland and wilderness, by which we seek to exclude from our domestic enclosures everything for which we have foreseen no use or market.

This principle of accommodating the margins, of diversity within unity, underlies our Constitution and Bill of Rights. But we live by this principle only negatively and grudgingly: we *permit* or *tolerate* dissent and divergence because the law requires us to. And the law can do no more than that. To put dissent and divergence to use, to turn a curious eye to the margins, eager to see what may have been tried and proven there, we will need a sounder, saner culture than we have.

MARGINS AND HEALTH

By narrowing itself so fanatically, orthodox agriculture has, in one sense, left its margins extremely wide. For motive power, it has made itself almost exclusively dependent on the internal combustion engine, and its ambition is to become completely so—leaving out of use or consideration the large variety of tools and techniques for the employment of human and animal power. Its earlier dependence on wind and water power—for pumping, milling, etc.—has now been shifted to electricity. It has little interest in on-the-farm collection and use of methane gas or solar energy.

It has greatly reduced regional differences in technology, methods of tillage, soil husbandry, etc. At the same time, it has reduced the variety of production within regions. This is, as Maurice Telleen says, "the regional specialization, that inevitably flows from individual specialization." And, just as dangerously, it has reduced the genetic diversity of both field crops and animals.

It has drawn an ever straighter, stricter line between the domestic and the wild, crowding nature itself into the margins. For the complex biological wilderness of a healthy topsoil it has substituted a simple chemistry. It has plowed up fence rows and roadsides and waterways, bulldozed woodlands, drained and plowed marshes. It has made itself not only inhospitable but dangerous to wild animals, birds, and harmless or beneficial insects.

It has made a margin even of the agricultural past, which is no longer regarded as a resource, a fund of experience, or a lexicon of proven possibilities and understood mistakes, but only as an amusement for the idly curious or, in advertisements, a measure of "how far we have come." Farm-equipment corporations are fond of printing old photographs to show the "drawbacks" of the agricultural past in comparison to the shiny "sophistication" of modern times. But as a working principle, whatever has been displaced or outmoded is simply ignored. About anything "old-fashioned," whatever its worth, the invariable comment is that "You can't go back."

For the principle of diversity, in nature and in earlier agriculture, and for the principle of unity that includes and depends upon diversity, orthodox agriculture has substituted a dull, tight uniformity, not only ignorant of other possibilities, but scared of them, and vengeful in its ignorance.

People who remove their minds from this shadowy twilight of agribigotry find that they are surrounded by an abundance of divergent possibilities—from our own past, from the history and present practice of other peoples, from new technology, from new understandings of biology and ecology. But they soon become aware, especially if their interest in agriculture is personal and practical, that this wide margin is only a margin in the mind, seriously beset by speculations, questions, and doubts. The possibilities obviously do exist as possibilities, but where do they exist in proof? Where are they being enacted by a living farmer on a living farm? Having arrived at these questions, one realizes that as the margin of divergent possibility has widened around orthodox agriculture, the margins of geography and practice have been drastically narrowed. Who are the people who know how to farm in these other—and, one believes, better—ways? And where are they? They are few, as the saying goes, and they are far between.

In the last few years, I have made an effort to do a little traveling along the agricultural margins, to visit farms where unorthodox ways are working, to see for myself what these dissident farmers are doing, and to listen to what they have to say. In telling about them, I wish to respect their privacy, and so I will not give their names or say very specifically where they live.

Nor, except for some merely descriptive figures, will I deal very much in statistics. I have chosen instead to rely on the evidence that I have seen, and that other people can see, too, if they will look carefully. One need not be a specialist to understand the difference between good and bad farming. There is nothing mysterious or abstruse about it. It only requires enough acquaintance with land and people to have some sense of what a prospering farm and a prospering farm community ought to look like and the same acquaintance with the signs of greed, hopelessness, neglect, and abandonment.

The health of a farm is as apparent to the eye as the health of a person. To look at a farm in full health gives the same complex pleasure as looking at a fully healthy person or animal. It will give the same impression of abounding life. What grows on it will be thriving. It will seem to belong where it is; the form of it will be a considerate response to the nature of its place; it will not have the look of an abstract idea of a farm imposed upon an area somewhere or other. It will look cared for—groomed, so to speak—like a healthy person or animal; it will look lived in by people who care where they live. It will show no gullies or galls or other signs of erosion. The waterways and field edges and areas around buildings will be grassed, something that becomes more necessary the steeper the ground is.

The place will look well maintained. Buildings, fences, equipment, etc., will have been kept in good repair, carefully used, protected from the weather. One of the commonest sights associated with orthodox farming is a lot of huge, expensive farm machinery left sitting out in the weather, having, like the economy that produced it, outgrown the possibility of care. Like the land itself, the equipment is used but not protected. This is one of the first and most ironic results of the high costs of industrial agriculture. The farmer is forced to protect his investment at the expense of what he has invested in. He plows out his waterways, abusing the land to get the maximum use of it and his machinery, and then allows the machinery to rust to save the cost of the necessary buildings. Such an economy will make a difference very quickly in the looks of a farm, as it will make a difference in the looks of a person or a nation.

A healthy farm will have trees on it—woodlands, where forest trees are native, but also fruit and nut trees, trees for shade and for windbreaks. Trees will be there for their usefulness: for food, lumber, fence posts, firewood, shade, and shelter. But they will also be there for comfort and pleasure, for the wildlife that they will harbor, and for their beauty. The woodlands bespeak the willingness to let live that keeps wildness flourishing in the settled place. A part of the health of a farm is the farmer's wish to remain there. His long-term good intention toward the place is signified by the presence of trees. A family is married to a farm more by their planting and protecting of trees than by their memories or their knowledge, for the trees stand for their fidelity and kindness to what they do not know. The most revealing sign of the ill health of industrial agriculture—its greed, its short-term ambitions—is its inclination to see trees as obstructions and to strip the land bare of them.

Woodlands, orchards, and shade trees are part of the diversity of life that is another of the prime characteristics of a healthy farm. And this principle will extend to cropland and pasture. The aim of a healthy farm will be to produce as many kinds of plants and animals as it sensibly can. This will be an *ordered* diversity, the various species moving in rotation over the fields. The land will be fenced for livestock, and its aspect will change from field to field.

Related to the principle of diversity is that of carrying capacity: the various crops and animals will be sensibly proportionate to one another; the farm will strive as far as possible toward the balance, the symmetry, of an ecological system; there will not be too much of anything. The fields will not be overcropped; the pastures will not be overgrazed. It will be understood that the plants growing on a farm are not just its produce, but also its protection, and so a row crop will be followed by a cover crop, the cover crop by a sod of grass and clover.

And a healthy farm not only will have the right proportion of plants and animals; it will have the right proportion of people. There will not be so many as to impoverish themselves and the farm, but there will be enough to care for it fully and properly without overwork. On a healthy farm there will be the right proportion between work and rest. Outside the Amish communities I do not know where in American agriculture one can find people and land in healthy balance. As far as I know, the Amish are the only American community to have formed deliberate strategies to keep enough people on the farms. All the non-Amish, full-time working farms that I have seen in this country have showed the need of more human hands.

Finally, a healthy farm will be so far as possible independent and self-sustaining. It is necessary to say "so far as possible," for we are by no means talking here about a "closed system." Simply by selling produce, a farm involves itself with other places both economically and biologically. And unless it encapsulates itself under a glass roof—which is really to become less independent—a farm cannot produce its own weather. Many farms cannot provide their own water. The wild plants, animals, birds, and insects upon which a farm's health depends will not respect its boundaries any more than the rain. And, of course, the people of a farm will belong complexly to a larger human community. Nevertheless, a certain kind and a certain measure of independence is a practicable ambition for a farm, and it is a necessity of agricultural health and longevity.

For one thing, fertility, the major capital of any farm, can be largely renewed and maintained from sources on the farm itself—assuming that all else is in balance. By proper tillage, rotation, the use of legumes, and the return of manure and other organic wastes to the soil, the fields can be kept productive with minimal recourse to fertilizers from outside sources. If the organic or decayable wastes of the cities, which have their source on the farm, could be returned to the farm, that would greatly increase both the health of the land and the independence, if not of the individual farm, at least of agriculture.

Equally important, by the good use of human power, animal power, solar, wind, and water power, methane gas, firewood from its own woodlands, etc., a farm can produce by far the major part of its own energy. This, of course, calls for a revitalization of local skills. But given the skills, these sources of power are possible. They come from the past and/or from new technology.

As a farm measures up in these various ways to the standard of health, its troubles from pests and diseases will radically diminish, and so consequently will its dependence on chemicals. A healthy farm will have no more need for these expensive remedies than a healthy person has for medicine.

Health, then, does not "come from" independence or "lead to" it. Health *is* independence. The healthy farm sustains itself the same way that a healthy tree does: by belonging where it is, by maintaining a proper relationship to the ground. It is by this standard of health or independence that one recognizes the absurdity of a farm absolutely dependent upon a complex of industrial corporations, which are in turn dependent upon the actions of foreign governments and politicians whom the farmer did not vote for or against and cannot influence.

The ultimate good health of a farm is in its ability to produce independently of the ups and downs of the Dow Jones industrial averages or the vagaries of politics. (When I visit a farm I always look to see how many trademarks and brand names are in sight. The orthodox industrial farm is, among other things, an advertising space for any number of corporations.) Those who pride themselves on the "science" that has made agriculture an industry have found this sort of independence simply beneath their notice. But I have watched, in Tuscany, a plowman driving a team of white cattle to a wooden plow, and realized that I was seeing the continuance of a motion and a way and a preoccupation begun before the rise of Rome. It is not nostalgia or sentimentality or wishful thinking to say that that man and his plow and team on the hand-built terrace under the olive trees represented a value, perhaps an immeasurable value, that modern agriculture has superseded but has by no means replaced.

A MARGINAL PLACE

But one's travels should begin at home. Before speaking of my travels on the margins in other places, I would like to say something about the margins I live among. Perhaps that will give an idea of what I have had in my mind as a sort of basis, and of the meanings and possibilities I have been looking for.

Not far from my house there is a hillside whose soil, declivity, and history are fairly representative of much of the hillside land in my part of the country. At one time this hillside was covered with a fine hardwood forest, which was no doubt cut soon after the establishment of the white people's tenure. The logs may have been sawed into lumber, but more likely they were burned simply to rid the land of them. The land was used agriculturally, for both row crops and pasture, with results that will remain visible for many more generations than the land was in use. Around the time of the Second World War, when machines began to replace the horse and mule teams as well as the people, the hillside began to "go back to the bushes." The thicket growth that follows agriculture began to take it over.

It is still "in bushes." In some places the better forest hardwoods have begun to establish themselves again among the weed trees. In other places there are still tangles of briars, cedars, thorns, sumac, box elder, elm. Under the trees are the slowly healing scoops and gullies of old erosion—part of the "investment" in a way of farming unsuited to the place, which no generation's income will ever redeem.

Walking along the contour of the slope, one crosses at intervals a series of natural waterways cut to the rock and running straight down the hill. The plows stopped short of these places by somewhat more than the length of a horse. The land here is whole; one supposes that it is virgin. The trees here are larger, and species grow here that do not grow in the abandoned fields. Beside one of these hollows, high up the slope, there is a big tulip poplar, a loam-loving tree rarely found in the uplands. It is two feet thick at the butt, and its trunk rises thirty or forty feet to the first branch. It is comparatively young, not by many years a survivor of the original forest, but in its proportions and its great health it is a reminder of that forest. It stands there on the edge of the hollow, not just because it has been spared, but because it is growing in excellent soil.

And so on the one hillside you are aware of crossing agricultural margins of two radically different kinds: one that farming damaged and has virtually abandoned and one that farming never came to. The second is not only the indispensable measure of the first, telling us by how much our history here has failed, but it shows us just as exactly what we must aspire to. It is an indispensable example, a little border of health along the edge of bewilderment and defeat.

But what of the abandoned fields, hidden with their scars under the bushes? What are we to think of them? Many people would say that we should not think of them at all, that they are fit only for growing bushes, as they are doing. But I disagree. If we are to have a respectable agriculture we will have to think competently and kindly of lands of all sorts, even the apparently useless. But, in fact, this hillside is not even apparently useless. Its soil, even where badly eroded, is fertile and readily responsive to good treatment. Such hillsides can be made to produce excellent pasture. I know that this is so because I have seen it done and I have done it myself.

And pasture is not all that such slopes are good for. They might, with care, be made to support a kind of mixed or "two-story" agriculture of both pasture (with selectively located hay crops) and trees. Natural stands of walnut trees are already established and thriving on many of these overgrown hillsides. These stands can be managed for their yield of nuts, for timber, or for both. And they could be augmented by planting grafted varieties of walnuts and perhaps of other native nut and fruit trees. If these plantings were done on the contour, perhaps along the backslopes of terraces, they would be perfectly compatible with the use of the land for pasture and hay.

The fertility of these slopes is by no means unknown to local farmers. But at present the use of them is problematic. The almost invariable method of clearing them nowadays is to bulldoze all the forest or thicket growth off the entire hillside at once, occasionally leaving an exceptional walnut or shade tree, and then either pile the brush and burn it or shove it off into the hollows, where much of the topsoil that comes in with it may be washed away and wasted. The cleared land is usually sowed in fescue or a mixture of fescue and clover. Rarely can a farmer afford the time and expense of sowing rye or another quick-growing crop along with the grass. Until a sod is established, the slope is seriously vulnerable to erosion, and soil loss is frequently added to the other expenses of the job. Some farmers mow these cleared fields once a year with tractors and rotary mowers, and so keep the bushes from returning. You can find some hill pastures kept in good shape in this way. But mowing them with a tractor is both dangerous and expensive, and far more time-consuming than the same work on leveler land. And the use of a tractor tends to work against any hospitality the farmer may feel toward trees; a tractor driver on a steep slope will look at a tree as at best an obstacle and at worst a hazard.

Another common practice, used to save time and expense, is simply to bulldoze the trees and bushes off the land, sow it to pasture, and then stock it heavily with cattle. As the pasture becomes stale and overgrazed, the cattle turn to browsing on the sprouts that come up from the old tree roots, and so for a while the bushes are controlled. But the cost of this practice is high, for the hillside suffers serious erosion from the combination of overgrazing and heavy trampling in wet weather and is finally grown over by the thorns and other rough trees that the cattle refuse to eat.

The good use of such land (use that is at once full, efficient, and careful) requires something altogether different and is probably unthinkable in terms of our present agricultural economy and cultural values. Good use calls, first, for great care in clearing, minimal groundbreaking, minimal bulldozing. Clearing probably should be done in narrow strips on the contour, working from the top of the slope downward in successive years. Terracing should be considered, wherever feasible; it seems to me that slowing and retaining the run-off behind terraces might make excellent sense in combination with the planting of tree crops. In some situations, when there is time and when earth does not have to be moved to repair washes, the overgrowth may be taken off by sawing; the best thrift would salvage a great quantity of fence posts and firewood from this sort of clearing. The steeper slopes, of course, should not be cleared at all, but should be left in trees to be selectively logged for posts, firewood, or lumber. The rule would be to clear only what can safely be kept clear.

Second, after clearing, the land should be sowed as quickly as possible. This sowing should include as great a diversity of clovers and grasses as makes sense for the location. (I have lately been using both bluegrass and fescue, as well as a clover mixture consisting of red, ladino, and sweet clovers, and Korean lespedeza.) A quick-growing "shelter crop" should be sowed with the pasture mixture to hold the ground until the grass and clover can get established. The seed can be sowed right onto the disturbed ground, which then ought to be passed over with a light harrow to cover the seed a little and to smooth the surface.

Third, such land needs to be managed intensively and in small fields. Steep land requires close attention, thorough understanding, and selfless care. It must be mowed at least once a year to control weeds and bushes, to stimulate new growth, and to encourage uniform grazing. Stock should be rotated from field to field, both to keep enough growth on the ground to protect it and to prevent the wearing of paths. Grazing such land too closely endangers it, and paths can be disastrous, especially if they run up and down the hill. For these reasons, large numbers of animals are incompatible with the good use of hill land; a big herd can do severe damage to a slope when the animals all must converge daily on the same watering or feeding places, gates, or milking barns. A good hill farm, if it is located where climate and soil permit intensive use, is almost by definition a small farm; and, insofar as it benefits from long-standing knowledge and devoted care, it is almost by definition a family farm. Nothing could be more alien to healthy agriculture than a large, production- or profit-oriented hill farm whose owner or owners do not live on it. In such a situation the balance between use and care is overthrown, and waste is the result. The small differences may be the most important. A family farmer, for instance, will walk his fields out of interest, the industrial farmer or manager only out of necessity.

And, finally, the good use of hill land requires a technology appropriate to it in scale and cost. Here we approach what most of the agriculture specialists and all of the "agribusinessmen" would be quick to describe as nostalgia or fantasy or craziness. They would do this to protect themselves and their assumptions and to disguise their most serious error. For the true measure of agriculture is not the sophistication of its equipment, the size of its income, or even the statistics of its productivity, but the good health of the land. And we are talking here about seriously damaged but potentially useful land, where American agriculture has so far failed. One must assume that if these hills *could* be farmed well with big, expensive, "modern" technology, they *would* be. That they are not suggests both that the technology is ill-suited to the terrain and that the cost cannot be afforded.

What sort of technology might make sense on such land? It will be at least strongly suggestive at this point to quote Thomas P. Cooper, one-time dean and director of the Extension Division of the University of Kentucky College of Agriculture:

"In Kentucky and many other states there are farming areas where work animals are indispensable. Small farms, hillside fields, rolling land and poorly drained areas can be successfully and economically farmed only by the use of horses and mules. The economic advantage of the use of workstock instead of power machinery for farm use is that horses and mules can be raised on farm-produced grasses and grains and maintained, while at work, in the same way. Farmers who own rolling or infertile farms, or who are farming on a subsistence basis are unable to purchase tractors or other power machinery. Such farmers are able to raise horses or mules and to use them to do farm work with the outlay of but little money. . . .

"There is much work on large as well as on small farms that can be done successfully and economically by horses or mules."

That circular is dated November 1937. I would not argue that we ought to "go back" to 1937. That, I am sure, *would* be nostalgic, fantastical, and crazy. But I am not so sure that what was considered "indispensable" in 1937 can be simply dismissed as "out-of-date" in 1977. My doubt is strengthened by the fact that in the intervening forty years, on thousands of acres of such land as I have just described, Dean Cooper's successors have produced, not a better agriculture or even a different one, but virtually none at all. That is, they have removed from consideration a way of farming suited to certain kinds of land and have replaced it merely with neglect and waste. It is notable that Dean Cooper's approach is to look at both the land and the farmer and then to suggest a suitable technology, whereas the approach of his successors has been to focus on the most "up-to-date" technology and expect the land and the farmer to conform to it. They seem to have answered Dean Cooper's argument that horses and mules were indispensable for certain farmers on certain lands by declaring that those farmers and those lands were themselves dispensable. I suggest that in light of the staggering losses of both farmers and land since 1937, and in light of the social problems and food needs of 1977, this assumption may be seen to be what it has always been: an extremely serious error of judgment.

A MARGINAL PERSON

The hillside that I have described, then, represents both a marginal place and a marginal possibility. As such, it is a measure both of local agricultural history and of the capacities and limits of prevailing agricultural technology and practice. But the full force of the necessary judgment will not be felt until I have also described a marginal person.

Some years ago I frequently used to drive past a farm in a creek valley of narrow, scarce bottomlands and hillsides rougher and less fertile than the one I have been talking about. The farm was small, mostly hillside, with a few narrow ridges and a creek bottom that could not have been larger than an acre and a half. In an area of semi-abandoned land, this farm was outstanding, not because of its "improvements," which were old and few, but because it was clearly both well used and well cared for. It was farmed by an old man and woman and a team of Percheron horses.

Everything about the place was neatly kept. House and yard and barn always showed a resident pride. There was an orderly, abundant vegetable garden beside the house. The pastures were mowed every summer. The tiny bottomland where the old man grew his tobacco crop was cut into three or four pieces by waterways which were grassed and bridged. More than anything else, those little timber bridges bespoke the old man's care; the usual thing would have been to drive regardlessly across such shallow drains and so wear the banks away. In addition to the team of horses, the pastures were stocked with a little herd of excellent beef cows.

This place interested me because it was a *good* marginal farm and because it was obviously a relic, the lone survivor within hundreds of square miles of a kind of farm that had been commonplace only thirty or thirty-five years ago. And finally it, too, went the way of the rest of them.

As I watched the old man's farm, driving by it at intervals, I saw it suddenly begin to change. The yard began to look unkept. Disorder began to spread around the house. The team of horses disappeared. I learned a little of the story. The old man had died. His wife had moved to town to live with her children. The house had been rented to people who, though they had technically become its residents, clearly did not *live* there. The farm also had begun to be used by someone who did not belong to it.

I had stopped once and talked a while with the old man. He was busy fixing a fence at the time, and though he received me courteously enough, he did not permit himself to be much interrupted. I told him that I admired his farm. He thanked me, but without enthusiasm, obviously having spent little time yearning to be complimented by strangers. I said his team of horses looked like a good one. He said that they did very well.

One morning after I had learned of his death, I stopped at the farm again—in his honor, maybe, or in honor of my own sense of loss. It was a gray, wintery day. The place looked and felt forgotten. It had gone out of mind. Absence was in it like a force. The barn was closed, empty, the doors tied shut by someone who did not intend to come back very soon. Peering in through a crack, I found that I was looking into a milking room with homemade wooden stanchions, unused for years. I knew why: it had become impossible to be a *small* dairyman. I spent some time looking at the old man's horse-drawn equipment. Some antique collector had taken the metal seats off several of the machines; these had become bar stools, perhaps, in somebody's suburban ranch house. For the rest apparently nobody now had a use. Examining the pieces of equipment, I saw that they were nearly completely worn out, patched and wired together like the fences and buildings, made to do—the forlorn tools of a man who had heirs, but no successors.

By the standards of orthodox agriculture, as well as by those of the present economy and culture, this old man and his farm were merely anachronisms, leftovers. The possibility of their existence would seem contemptible, not just to the majority of agriculture experts, but to the majority of influential people of other kinds. And yet we must ask *why*. And we must be careful not to accept too hasty or easy an answer. For no matter what may be said by the current standards of economics or technology or cultural fashion about this old man's life, there is still no legitimate way of withholding respect from him. In a time when millions of people, including very able and expensively educated young people, are finding it easy to accept a dependence on welfare or unemployment, and when millions more are dependent on social security and other public means of support, here was a man who worked until he died, taking care of himself and of his part of the earth.

The curious thing is that many agriculture specialists and "agribusinessmen" see themselves as conservatives. They look with contempt upon governmental "indulgence" of those who have no more "moral fiber" than to accept "handouts" from the public treasury—but they look with equal contempt upon the most traditional and appropriate means of independence. What do such conservatives wish to conserve? Evidently nothing less than the great corporate blocks of wealth and power, in whose every interest is implied the moral degeneracy and economic dependence of the people. They do not esteem the possibility of a prospering, independent class of small owners because they are, in fact, not conservatives at all, but the most doctrinaire and disruptive of revolutionaries.

Nevertheless, the old man and his farm together made a sort of cultural unit, recognized and valued in this country from colonial times. And it is still a perfectly respectable human possibility. All it requires is the proper humanity.

TRADITION AND EXPERIENCE

It is of great importance to understand that the marginal possibility, the marginal place, and the marginal humanity that I have been describing are reinforced by a marginal way of thinking—until now a sort of counter-theme in our history, so far always subordinate to the theme of exploitation, but unbroken and still alive. This is the theme of settlement, of kindness to the ground, of nurture.

To exhibit this theme, in both its articulateness and its commonness, I offer the following quotations from the *Farmers Home Journal*, a regional farm magazine once published in Louisville, Kentucky. The quotations are taken from the issue of January 2, 1892.

One correspondent writes "to urge every man in Kentucky to set out nut-bearing trees." And this purpose is urged upon the writer by his sense of the necessity of *settling* on the land: "The first thing a young man should do is to get him a home; the next thing get him a wife, and next set him an orchard, but do not think an orchard complete till you have set a few nut-bearing trees."

Another writes that "No man . . . should spend his labor and time over so large an acreage as to fail in making a first class garden." (The reader should be reminded here that the agricultural orthodoxy boasts that farm families have become patrons of the supermarkets.)

Even as early as 1892, we meet industrial arrogance, already fully inflated: "That farmers do not apply more commercial manures to their gardens is mainly because they do not think."

But we also have an example of such not-thinking in a letter from "W. C." of Rural Neck, Kentucky, a place no longer on the map. W. C.'s letter is an exuberant essay on the economy of the soil, and he makes a direct connection between that economy and the economy of money. He recommends the use, as fertilizer, of manure fresh from the barn, and also of scrapings from the barn lot, rotten straw from last year's threshing, old piles of chips and ashes, anything that will rot. "Yes, rot is the word. Rot means death, and without death and rot there can be no new life." He says that one can even use bone dust or superphosphate. "But it won't do for a farmer to go in debt for special or commercial fertilizers, as a rule. You can more safely go in debt for a good stable manure. . . . Nature never loses anything: she preserves and protects herself. It is only a fool man who squanders his substance and makes himself poor, and everybody around him, and the land that he lives on too." He follows this with an attack on soil erosion and praise for manure, industry, and brains. And he concludes: "When people learn to preserve the richness of the land that God has given them, and the rights to enjoy the fruits of their own labors, then will be the time when all shall have meat in the smokehouse, corn in the crib and time to go to the election."

It is a remarkable letter. W. C.'s argument is the one we get—howbeit with greatly increased scientific authority—from Sir Albert Howard, but W. C. is stating it plainly enough fifty years before Howard's books were published. "Rot means death, and without death and rot there can be no new life." This is a principle as new and common as biology, as old and exalted as the Bible: "Except a corn of wheat fall into the ground and die, it abideth alone: but if it die, it bringeth forth much fruit." And W. C.'s voice is seamlessly joined to those of his fellow correspondents who were insisting on the importance of home, household and family, orchard and garden.

What are we to make of these undistinguished men from out-of-the-way places, who pled their cause with the eloquence of good sense and the exuberance of conviction? Jefferson spoke for them in politics. Albert Howard would speak for them, later on, in science. But they speak out of a much more particular engagement with the life of farming than Jefferson ever did. And Howard was still half a century ahead of them. We have to conclude, I think, that they were speaking out of tradition (the yeoman's or the agrarian tradition, which grew out of a peasant tradition still older) and out of experience —out of tradition proved and upheld by experience. This association of tradition and experience in the intelligence of a living person is humanly broad and deep. It is biologically, agriculturally, economically, politically, and culturally sound. It is deeply founded, solid enough to build a civilization upon, whereas the orthodox agriculture can support nothing but the shallow expansion of a bookkeeper's economy.

ORGANIC FARMS

The attitudes and values of traditional agriculture still survive in our time and are supported by the experience of our time. Their survival is marginal and is mostly ignored both by the colleges of agriculture and by the agricultural press, which, if they acknowledge it at all, do so in order to treat it with contempt. But survivors do exist. They are connected by a sort of network that one travels by hearsay and friendship. By now I have encountered a good many of them, and have been impressed as often by the excellence of their characters as by the excellence of their farms. They are people of principle, both stubborn and adventurous, independent enough to trust their own experience and strong enough to hold in considerable isolation to truths not officially or popularly favored. Their farms stand for their principles and prove them; one has only to notice their example, or their examples, to understand that the orthodox agriculture has founded its "scientific proofs" upon shallow assumptions.

In spite of some public notice in the last year or two, it probably still is not generally known that there are a number of large-scale, highly mechanized farms that do not use chemical fertilizers or pesticides. When I first began my search for examples of healthy agriculture, I did not realize how compatible organic soil management could be with a large scale of operation. And then in the spring of 1974, I visited a 900-acre organic farm in Iowa. This farm made extensive use of a commercial organic fertilizer. But that seemed to me probably the least important element of the farming there. More important, I thought, were a careful plan of crop rotation (corn for a year or two, oats, soybeans, and then two years of pasture), the use of animal manure on corn ground every year, and the use of a chisel plow rather than the conventional turning plow for the preparation of crop ground.

This system was said to have the following advantages: within the first year or two of its use, earthworms and other forms of life had again become abundant in the soil; the ground had become darker, looser, easier to work each year; crops could be planted earlier because the increased humus in the soil permitted it to dry more quickly; stock feed went farther every year, because as it became more palatable and nutritious it took less to satisfy the cattle; the farm had no insecticide program at all, either for crops or stock.

In its machinery, buildings, etc., this farm was as "modern" as any other of comparable size. Even though it was far more diversified than most large farms of these times, and did not use chemical shortcuts, it required only four full-time workers. Late in 1975 I visited another highly mechanized organic farm —this one a 700-acre farm in Nebraska—another extremely impressive example of organic farming on a large scale. The existence of such farms as these, on which crops, animals, and the farmers themselves are obviously thriving, invalidates out-of-hand the contempt of orthodox agriculturists and suggests strongly that their contempt must rest on ignorance or fright.

If all the farms in the country were managed organically, both our people and our land would undoubtedly be healthier and there would be a considerable ramification of the benefits. And yet the 700- or 800-acre organic farm equipped with up-to-the-minute machine technology cannot be considered the solution to all of our agricultural problems, or to the problems that grow out of our agricultural problems. If we accept this as a solution, we forswear, for one thing, any further discussion of the cultural and political importance of the small landowner.

Much more suggestive, in this light, was another Iowa farm that I visited, this one a family-size holding of 175 acres. Of this, 50 acres were in permanent hillside pasture for twenty-eight Charolais cows. On the remaining 125 acres, the farmer grew corn, oats, wheat, soybeans, and hay. In addition to his cow herd, he kept twelve brood sows and a laying flock of 200 hens.

This farm had been under a completely organic system of soil management for eleven years at the time I saw it in 1974. Here again some commercial organic fertilizer was used to supplement a careful plan of soil husbandry. The cycle of crop rotation was as follows: oats and/or wheat, legumes for hay, soybeans, corn. The application of animal manure was estimated (conservatively) at two tons per acre, and this was put on the bean ground before planting it in corn. The expenditure for commercial fertilizer, which was used only on the corn ground, came to twelve dollars per acre. Every three years or so the pastures were dressed with 300 pounds per acre of a natural mineral fertilizer.

The farmer here was a man of impressive intelligence and judgment and impressively independent in both. Prescribed measures had been altered as he felt necessary to fit his place and his needs. The secretary of agriculture had called for all-out production that year, and on many farms the plowlands had begun to edge out dangerously into waterways and hillside pastures. I asked this farmer if the secretary's recommendation had affected his program. He answered that it had not done so in the least.

On this farm I first had a chance to watch a chisel plow at work and to see the ground it had prepared. This is a favorite tool of many mechanized organic farmers, who give it enthusiastic praise. I could see why. To begin with, it does all the work of seed-bed preparation, replacing both turning plow and harrow. But its great advantage is that it leaves the top layer of soil on top, which is where it belongs. Loosely stirred into this top layer, animal manures and plant residues decompose aerobically. The resulting high content of organic matter causes the surface of the field to act as a sponge, readily absorbing and retaining water and also allowing it to percolate downward into the lower layers. Another advantage of this plow is that it does not cause a hardpan; it does not interfere with the downward course of water through the pores of the soil, worm holes, and old root channels deep into the subsoil. The result is that the soil becomes at once less drouthy and less subject to erosion. It also becomes looser, easier, and cheaper to work, and so operating money goes farther and machinery lasts longer.

On this farm, sod ground to be broken is plowed once with the straight chisels in the fall and is then plowed twice again with sixteen-inch sweeps in the spring before planting. These sweeps are very good for destroying deep-rooted weeds.

This farmer used no herbicides. The reason he gave was that he did not want to contaminate the streams. But he also appeared to have no great need for such chemicals. He tried to plant in the latter part of the planting season so as to allow more weeds to germinate and be killed in the preparation of the ground. He cultivated his row crops to remove large-stemmed weeds, and he found that taking three cuttings a year from his hay fields helped considerably to control weeds in the row crops that followed.

As for crop yields on this farm, I quote the following from a letter that the farmer wrote to me several months after my visit: "I would say our soy beans average 40 or more bu. per acre in an average year. The state average is 30 to 33 bu. Our corn has been yielding 90 to 100 bu. the past 5 years. Neighbors' yields are about the same for the same soil type and lay of land. Our wheat yielded over 25 bu. per acre which we feel is very good for this area. . . . Our oats have been 60 bu. on the average."

The cow herd on this farm was given a balanced mineral mixture as a supplement but was wintered on hay alone, without grain. No insecticides were used on the cattle. The farmer wrote that although his cattle have flies on them in the summertime, they do not have pinkeye or other eye problems usually associated with flies. He attributed this to their extraordinary good health. In December of 1974, he wrote me that the twenty-three March and April calves off this herd weighed variously from 400 to 700 pounds per head. These calves were in robust health, without pinkeye or any other disease.

Another mechanized organic farm is the new experimental farm belonging to Rodale Press, publisher of *Organic Gardening and Farming*. In 1972, 290 acres of this farm were rented to an excellent Mennonite farmer, who agreed to operate it according to strict organic principles. A five-year rotation cycle was set up (corn to rye to barley to wheat to timothy and clover), with twelve tons of manure per acre to be applied to the corn ground. The crops were planted in strips on the contour. Before 1972 this farm was cropped in the orthodox fashion, using heavy applications of chemicals, and the following corn-production figures are especially interesting for that reason. (Figures are available only for corn.) In the first year the yield was 40 bushels per acre; in the second year, 60; in the third year, 80; in the fourth year, 140. In that fourth year the top yield in the same county was 157 bushels per acre—obtained with an application of 190 pounds of nitrogen, 230 pounds of phosphorus, and 673 pounds of potassium.

DR. COMMONER'S ARGUMENT

There is, then, no way to deny that crops and animals can be produced in respectable yields by the methods generally designated as "organic." These methods work on large farms and on small ones. Available evidence indicates that they work at least as well as orthodox methods within the economy of the individual farm, and they will undoubtedly work better as the costs of chemical fertilizers, pesticides, and herbicides rise with the cost of petroleum. But perhaps the greatest benefits from the widespread adoption of organic methods of soil management would go to the general public—in greatly reduced soil and water pollution, in reduced public expenditures for pollution control, in better health, and at least eventually in cheaper food.

The abounding good health of the farms I have described is dramatically evident to an experienced observer. I believe that it would be just as evident to an inexperienced observer who would spend a few hours looking closely and comparing. But in support of the visual impression we now have some evidence from the Center for the Biology of Natural Systems at Washington University—a report published in 1975 and entitled *A Comparison of the Production, Economic Returns, and Energy Intensiveness of Corn Belt Farms That Do and Do Not Use Inorganic Fertilizers and Pesticides.* In *The Poverty of Power*, Barry Commoner makes this study the fulcrum of a powerful argument for organic soil management. I am going to make extensive reference to Dr. Commoner's argument both because it supports and completes my own and because I want to take issue, a little further on, with one of his assumptions.

Dr. Commoner begins by going over some ground often

traversed by the specialists and apologists of orthodox agriculture, but he goes further and sees much more clearly. From 1950 to 1970, he acknowledges, American agriculture made some impressive increases in productivity: corn production per acre tripled; "a broiler chicken gained nearly 50 percent more weight from its feed"; egg production increased by twenty-five percent; overall farm production "increased by 40 percent." *But* during that period the real farm income "*decreased* from about $18 billion in 1950 to $13 billion in 1971. . . . Because the number of farms also decreased by 50 percent, the income per farm rose by 46 percent. . . . However . . . the average increase in the family income of *all* U. S. families in that period [was] 76 percent. Meanwhile, the total mortgage debt of U. S. farms rose from about $8 billion in 1950 to $24 billion in 1971." During this time there was also a massive shift from diversified farming to monoculture, which reduced the time that the farmland was covered with plant growth, which in turn reduced the amount of solar energy put to use on the farm. The removal of animals from farms growing crops in monocultures reduced the amount of organic waste returned to the fields. And there was a shift from the use of nitrogen-fixing legumes to the use of commercial nitrogen fertilizers. From 1959 to 1973 there was a sixty-percent decrease in the production of legume seed. By these and other changes, "The farm's link to the sun has been weakened, replaced by a new and . . . dangerous liaison with industry." And this dependence on sources of energy off the farm explains the decline of farm income. The net farm income decreased from 1950 to 1970, not in spite of, but *because of* the new technology of machines and chemicals. Dr. Commoner concludes his analysis of the effects of this technology with the following indictment:

"One can almost admire the enterprise and clever salesmanship of the petrochemical industry. Somehow it has managed to convince the farmer that he should give up the free solar energy that drives the natural cycles and, instead, buy the needed energy—in the form of fertilizer and fuel—from the petrochemical industry. Not content with that commercial coup, these industrial giants have completed their conquest of the farmer by going into competition with what the farm produces. They have introduced into the market a series of competing synthetics: synthetic fiber, which competes with cotton and wool; detergents, which compete with soap made of natural oils and fat; plastics, which compete with wood; and pesticides that compete with birds and ladybugs, which used to be free.

"The giant corporations have made a colony out of rural America."

Dr. Commoner then turns to the organic farmers studied by the Washington University research group, of which he was a member. He sees in the methods of these farmers the obvious solution to the problem:

"The group analyzed the production of these farms for the 1974 season. The market value of the crops produced by the conventional farms was an average of $179 per acre, while the average value for the organic farm was $165 per acre. However, the operating costs of the conventional farms averaged $47 per acre, and those of the organic farms $31 per acre (the difference is largely due to the cost of the nitrogen fertilizer and pesticides used by the conventional farmers). As a result, the net income per acre of crop for the two types of farms is essentially the same . . . The yields of different crops obtained by the two groups of farms are about equal, except for a small excess (12 percent) of corn yields on conventional farms as compared with organic farms.

"The organic farms used only 6,800 BTU of energy to produce a dollar of output, while the conventional farms used 18,400 BTU. Thus, organic farms appear to yield about the same economic returns as the conventional ones, but do so by using about one-third as much energy."

THE USE OF DRAFT ANIMALS

Because "U.S. agriculture now consumes only about 4 percent of the total national energy budget," Dr. Commoner correctly perceives that the overriding issue here is not that of energy conservation, or even that of pollution resulting from farm use of fossil fuel energy. The overriding issue is economic: the colonization of the farmland by the petrochemical industry. But it seems to me that this perception is not carried far enough. Speaking of the adverse energy economy of the conventional farm, Dr. Commoner says that "when a farmer uses commercial nitrogen fertilizer, the amount of thermodynamic work expended to produce it is seven times greater than the minimum amount of work that is needed to accomplish the same result by planting vetch. But the external energy required to grow vetch *could* after all be reduced to essentially zero (for example, by using a horse fed on farm grown corn). On this albeit impractical standard, the fertilizer's thermodynamic efficiency is zero."

It is the qualifier in that last sentence that concerns me. Dr. Commoner is saying that he is willing to advocate only half the remedy that is called for by his argument. That is, he wishes to do away with agriculture's dependence on petroleum-derived fertilizers, pesticides, and herbicides, but he will not contemplate the reduction of its dependence on petroleum fuels. In the midst of an argument everywhere else incisively intelligent, he suddenly makes this perfunctory bow before the golden calf of "agribusiness"—this spurious standard of "practicality" by which any unorthodox technology may be loftily waved away. To suggest that anything besides a tractor could be used for motive power on the farm is like setting fire to the church—the righteous not only do not *do* it, they do not *think* about it.

But Dr. Commoner's routine refusal to defile the sanctuary is mild indeed in comparison to the official reaction to the same idea. In August of 1975, the *Farm Index*, a publication of the United States Department of Agriculture, carried an article entitled "Wanted[:] 61,000,000 Horses & Mules[,] 31,000,000 Farm Workers." This by now widely circulated article is "based on" a speech delivered by Earle E. Gavett of the National Resource Economics Division.

Mr. Gavett's purpose is to confound "some critics of today's farming practices" who, the article says, have advocated "an anti-technological revolution" involving an immediate and complete return to the use of horse and mule teams on American farms. This alleged proposal, the article is relieved to note, has "some serious—if not insurmountable—drawbacks" in that it requires sixty-one million horses and mules, of which there were only three million in the United States in 1975, and thirty-one million farm workers, of whom only four million were available. These figures were derived in the following way:

"The 1967 index was the yardstick. The 1918 crop had an index of 48—that is, 48 percent as large as the 1967 crop—compared with 109 in 1974. Thus 1974 production was about 2¼ times greater.

"As a peak year of nonmechanized farming, 1918 is an ideal choice in the comparison.

"A straight projection of 1918 resources to meet 1974 production can be made by simply multiplying the 26.7 million mules and horses and the 13½ million farm workers carrying on farming in 1918 by the 2.27 times larger output in 1974."

The article concedes that this is "only a guideline projection. Obviously, nonmechanical and nonchemical technology improvements . . . since 1918, such as hybrid seeds, would lessen the manpower and horsepower requirements by allowing greater yields for less work.

"But agricultural economists quickly emphasize that the point of the projection is valid: a complete abandonment of mechanized technology is a biologically impossible and sociologically impractical idea."

The necessary animals could not be produced, the article continues, before 1992 or 1993. To grow feed for these animals would require "180 million acres of prime farmland." And there would be "questions over feeding so many horses in this country while people abroad are starving."* Moreover, the necessary people are also in short supply, and "A movement of 26 million workers from city to farm would provide mind-boggling problems."

That gives the main line of the argument, which gets considerably more elaborate without ever becoming more intelligent. Like many another, this document would merit no more attention than it merits respect if it were not for its influence. It happens, however, that this argument was given the status of official policy of the United States Department of Agriculture in a speech by no other than former Secretary Butz himself.

*It is fascinating to observe the agriculture specialists' flexible mindfulness of the hungry, who are, according to the argument at hand, either to be compassionately fed or starved into compliance. Either way, their fate is directly bound to the ambitions of the "agribusiness" corporations, who have thus added an enlightened versatility to the originally narrow and primitive Christian concept of charity.

And so we have before us one of the characteristic political necessities of our time: to take seriously what we cannot respect.

The chief objection to this argument is that there was never a reason or an occasion for it. There are simply no serious critics of conventional agriculture who have advocated "a complete abandonment of mechanized technology." As the *Draft Horse Journal* noted in an editorial, "Most of the critics of today's agriculture . . . don't talk about any such complete anything, but rather a picking and choosing of techniques and tools to get the job done in the most energy conserving way possible." The key phrase here is "picking and choosing." There are indeed critics who believe that a much larger range of technological choices and alternatives ought to be available, that we will have neither a healthy agriculture nor a dependable food supply until such choices and alternatives are available—that, in short, the strength of agriculture is in diversity, of technology as of other things, and that the present agricultural orthodoxy ignores the principle of diversity altogether. A few of these critics have published articles in such magazines as *Organic Gardening and Farming*, *Mother Earth News*, and the *Draft Horse Journal*, in which they have pointed out that there are presently places in agriculture and forestry that can be competently and economically filled by horses or mules. Some have said that, given our difficult economies of both energy and money, much wider use might reasonably be made of draft animals in the future. No one, as far as I know, has *ever* proposed that such a change could, or should, be either complete or rapid.

That this small advocacy of a small diversity should have drawn a full-scale attack from the Department of Agriculture bespeaks both the totalitarianism and the paranoia of the "agribusiness" mentality. What can be the excuse for all this carrying on? If these critics are right, then as scientists, the agriculture experts might be expected simply to agree. If the critics are wrong, then it appears that they might safely be ignored, for orthodox farming is far too widely accepted to be seriously threatened by a bad idea. The truth is that these critics have offended, not by being either right or wrong, but by being *different*.

Even if we could grant that we are indeed threatened with "an anti-technological revolution," the competence of Mr.

Gavett's argument is still in question. As the *Draft Horse Journal* pointed out, the arithmetic of his "projection" is far too simple, assuming, as it does, "that it takes three times as many horses and mules to cultivate corn yielding 120 to 150 bushels to the acre as corn yielding 40 to 60." And of his assertion that it would require three acres of "prime farmland" to feed one horse, it can only be said that he does not know what he is talking about. By my figures, using the rations recommended in the twentieth edition of Morrison's *Feeds and Feeding*, a ton horse doing medium to heavy work every day of the year would require 104 bushels of ear corn and 7300 pounds (about 209 thirty-five-pound bales) of hay. Assuming that the hay is of grass and alfalfa, this much feed could be produced on less than two acres at today's yields. But not all draft horses would or should weigh a ton—1300 to 1800 pounds would be a realistic range. And very few indeed would work every day the year around. The above figures do not consider the horse's off-time subsistence on pasture alone or on a maintenance ration mostly of hay, and they do not consider his ability to utilize roughages such as cornstalks, now seldom used as feed. A more realistic accounting might be that of the *Draft Horse Journal*'s editorial, which states that a horse eats "the energy equivalent of 70 bushels" of corn.

One also notes this article's easy assumption that *all* of the thirty-one million needed people would be "workers" and not farmers. And that the mind that is "boggled" by the problems of "a movement of 26 million workers from city to farm" is apparently not boggled at all by the continuing and appalling problems of the recent movement of many more than that from farm to city. If there were any suspicion that such a reverse migration might be profitable to "agribusiness," we may be sure that there would be an unhesitating effort to bring it about. This is a kind of mind that is boggled only at its convenience.

The same is true of the "questions over feeding so many horses in this country while people abroad are starving." This serviceable charity is not at all troubled by work now being done at the University of Nebraska on the possibility of using grain alcohol as a motor fuel. It is morally questionable to feed grain to a work horse; but if the grain is to be consumed by engines to the profit of energy corporations and the machinery and automobile manufacturers, then the starving are forgotten. Nor do the people who attack the use of horses for farm work ever say a word against their use for racing, show competition, and other frivolous purposes.

"Horses," the *Draft Horse Journal* said, ". . . are no more anti-technological than legs on humans." They are simply a technological possibility that we have almost ceased to consider. We must learn to consider it again, for until we do we cannot complete the logic of Dr. Commoner's argument, nor can we answer the questions raised by the existence of, and the potential need for, such "marginal" lands as I described earlier. There are certain problems for which the use of horses is the appropriate solution—or for which we have so far found no more appropriate solution. There is also the possibility that a revival of the lapsed technology of horse-powered agriculture is necessary to complete our agricultural intelligence and judgment—to give us the diversity of choices required for the subsistence of intelligence and judgment.

The issue of economics merges finally into the much larger issue of health, just as the issue of the health of any one creature merges into that of the health of Creation. In the context of that issue, Dr. Commoner's "impractical standard" of near-perfect thermodynamic efficiency becomes not just thinkable but indispensable. It is no more impractical than the standard of perfect health, which we all apply to our bodies. We desire —our bodies desire—to be perfectly healthy. That is what we hope and strive for. And it is the way we understand our effort; without the ideal of perfect health, we could not know how healthy we are. To be three-quarters or seven-eighths healthy is not an ambition that ever occurs to us.

There is no point in saying that perfection of health, as of all else, is not attainable by humans. The point is that we must have the vision of perfection, we must strive for it, we must sense the possibility of approaching it, or we cannot live. Jesus enjoined his followers to be perfect—not, I think, because they could hope for perfection, but because perfection is the necessary standard. People cannot understand themselves, or live fully and humanly, without it. To reconcile ourselves to imperfection, to place great practical barriers in our own way, is brutish. It condemns us immediately to great suffering of the spirit and undoubtedly, in the long run, of the body as well. Common sense alone requires us to consider well any technology that might bring us nearer the vision of perfect health. To repudiate such technology on the ground that it is "old-fashioned" is madness.

It is hard to overestimate the importance of applying the correct standard to agricultural performance. I do not see how a stable, abundant, long-term agriculture can be built up and maintained by any standard less comprehensive than that of the perfect health of individual human bodies, of the community, and of the community's sources and supports in the natural world—whereas the standards of orthodox agriculture tend to be extremely simple and exclusive: productivity (as determined by "records" and by the equation between the number of eaters and the amount of food) and the financial prosperity of "agribusiness."

It is easy to say, as former Secretary Butz said in his own fatuous attack on the "anti-technological revolution," that "To return to the 'good old days' in agriculture, or indeed just to cling stubbornly to the farming methods of today, would be to condemn hundreds of millions of people to a lingering death by malnutrition and starvation in the years ahead." But that is simply the oldest—and the most profitable—cliché of the industrial revolution, supported only by a thoughtless obeisance to "progress." We must look beyond that to what is assumed. So far as I can make out, Mr. Butz's statement rests upon two main assumptions, both suspect: that the health of humans may be safely distinguished from the health of the rest of Creation and that there is no distinction between affluence and survival.

People who argue for ways of farming that are ecologically sound, says Mr. Butz, are "placing the needs of man second to the needs of all other creatures." Man, he says, is "as much a 'part of nature'" as the other creatures. But what he means is that human needs must be put *ahead* of the needs of all other creatures, as we see when he equates a hydroelectric or an irrigation dam with a beaver dam. His solution to the problem of hunger is therefore remarkably unencumbered by moral, cultural, or ecological considerations: "We'll turn to science and technology for the answer—we'll modify the environment." Thus, with a shrug, he sets agriculture free of ecology. But we are left with an awesome ecological question: How can humans, who are creatures, hope to survive in a world in which other creatures perish? Or how much can we "modify" the environment before we fatally "modify" ourselves? Here is Mr. Butz's answer: "The challenge to agriculture and science is to find the right application of technology to modify the environment in a way that will benefit both man and the rest of nature." One can only agree, pointing out, however, that the applications of technology so far advocated and defended by Mr. Butz have notoriously failed to do so and that his colleagues and constituents in the "agribusiness" system have so far failed notoriously even to consider the advisability of doing so.

The second assumption is, of course, closely related to the first. Mr. Butz begins with the term "survival" and a most dramatic issue he makes of it: "Backed in a corner with no job, no income, and an empty stomach churning from hunger, the most dedicated environmentalist will forget his fight for the seagull or the walrus. He will get down and scrap for survival like any other creature . . ." Of course he will—though he may even then remember that he and the sea gull and the walrus are all scrapping for survival in the same world and against the same abuses. But by the end of Mr. Butz's speech, without transition or warning, the term has changed: ". . . there can be little hope for mankind's continued *affluence* [my emphasis] unless we face up to the moral question of the need to limit our numbers. In the meantime, science and agriculture will have to buy the time for us to reach that solution." And with this shift of terms, "science and agriculture" have been nominated to do the work that can be done safely and adequately only by complex cultural changes leading to restraint of consumption and competent care of the earth. It is exactly this refusal to consider survival except as "continued affluence" that has brought our survival into doubt. There is, anyhow, only a fanciful connection between affluence and survival; we do not have to be as comfortable and extravagant as we are in order to survive. And there is no connection between affluence, as we understand it, and civilization. All that civilization requires is enough; it does not require extravagance. Until these distinctions are made, we cannot even begin to talk sensibly about the problem of hunger.

The fact is that Mr. Butz and his colleagues in the corporations and the universities do not know whether unorthodox technologies and methods will produce more food or less. The only information that they have, or that they acknowledge, is that which "proves" the efficacy of "agribusiness" technology. Where are the control plots which test the various organic systems of soil management? Where are the performance figures for present-day small farms using draft animals, small-scale machine technologies, and alternative energy sources? Where are the plots kept free of agricultural chemicals? If these exist, then they are the best-kept secrets of our time. But if they do not exist, whence comes the scientific authority of orthodox agriculture? Without appropriate controls, one has no proof; one does not, in any respectable sense, have an experiment.

HORSE-POWERED FARMS

Mr. Gavett, followed by Mr. Butz and others, bases an amazingly bitter attack against the use of horses upon a "projection." There was no need for so speculative a maneuver, for a number of horse-powered farms presently exist—not experiments or controls, but living examples, requiring only to be carefully observed. I have visited a number of farms powered either partially or exclusively by draft horses. I can offer only a few random figures having to do with these farms, and so what I have to say is offered as proof of nothing except their possibility. But the fact of their possibility suggests strongly that we ought to have thorough studies of their ecological and economic performance.

In the early spring of 1975 I visited three good Iowa farms, all of which made extensive use of horses. All three were farmed by older men, working for the most part alone, who farmed this way by conviction; who were thoughtful, indeed passionate, holdouts against the capital-intensive, highly mechanized farming of their neighbors; and who lived in the isolation of those who are "different." Of their financial condition, I can say only that from all visible signs they were better than solvent. Their homes were comfortable, their farm buildings well kept up, etc.

The first of these men farmed 120 acres. He had two teams, one of which was a young pair he was breaking for another horseman. He owned two thirty-year-old H Farmall tractors that he used mainly for the heavy work of plowing and disking; the rest of his work he did with the horses. Aside from the manure from his barn, he used no fertilizer. He did not use insecticides. Herbicides he used only selectively, for the control of thistles. In addition to the considerable saving of fuel, he mentioned two other benefits from his way of farming: he believed that he had less erosion than his neighbors and that his ground worked easier. His corn yielded an average of seventy bushels per acre.

The second of these farmers had 300 acres of excellent land surrounded by cash-grain farm "businesses" of the orthodox make. He did most of the farming with horses, keeping an ancient Farmall tractor to do only the heaviest field work and to provide stationary power. Except for plowing down "a little" fertilizer, he used no chemicals. His fields were fertilized with manure, and tilled in rotation from corn to beans to corn again to oats to hay. His corn yield ran to about seventy-five bushels to the acre. The economy of this farm was carefully diversified. Among his other enterprises, the farmer had a dairy herd of six cows, which he milked by hand.

The third farm was similar to the second in size and in the combination of horse power with an old tractor (a 1946 WC Allis-Chalmers) used for power-takeoff work and for the heaviest work in the field. This farmer said that he had "never used a pound of fertilizer." He owned twelve horses, one a Percheron stud. The income from this horse-breeding operation was paying *all* the operating expenses of the farm. I neglected to ask this farmer what his corn yield was. But I did ask him how his economic situation compared to that of his neighbors. He said that he couldn't say, but that they often called him over "to buy something that they ought to keep."

My visits to these farms involved long distance and short time, and so were necessarily far too hasty. And it was too early in the season to get a fair look at the condition of the fields. For those reasons my information is not nearly so complete as I now wish it were. I am able to fill out the impression somewhat by quoting a letter from an Iowa agriculture student who spent much more time on the second of these farms than I did. His visit, like mine, was before the growing season, and again the facts are scanty. But his description is much more detailed than mine, and a context is given in which the *meaning* of such a farm is made plain.

"At this time of the year 95% of the land has either been cut (soybeans) or plowed up (corn). The few guys who have any livestock at all have it on an enormous scale and they are among the few who have let fences remain. So the majority don't plow to the fence row; they plow to the culvert's edge. Row crop cultivation is done on such a large scale here that they must fall plow so that they can be timely in planting the vast acreages come spring. But what I saw on the south and west sides of fields in the culverts were snow drifts that were black to dark grey—each layer of snow has a layer of topsoil on top. You see this beside every field without exception that has been fall plowed. There's no protection from the wind in this flat country and if they don't get early freezes with heavy snow the land is vulnerable. There are many abandoned farm buildings with the ground cultivated within a few feet of them. I often saw good stock barns, the likes of which you rarely see in the southeast, with the south or east end cut out, and all they hold are large tractors and implements. You don't have to travel far to see whole square miles of land with no farm houses or outbuildings, plowed up north to south and east to west—right up to the culverts. And in the midst of this land, where farmers are no less dependent on Shell Oil Co. and John Deere than they are on the weather, stands _____'s place; honestly, to see it is to believe that it's an oasis in the midst of a desert. I knew from a mile and a half down the road that it was his place. His milking shorthorns were out gleaning corn. The fields were well fenced, the buildings being used for the purposes intended. His rotation is the old Iowa standard: 60 acres of corn (and some sorghum), 30 of oats, 30 of hay (clover), 30 of soybeans all on the home place. He has another 120 acres on a neighboring section. He keeps 40 shorthorns, five* of

*One had evidently been turned dry since my visit.

which he milks by hand. He fattens about twenty hogs, keeps 200 chickens by which he's able to sell eggs to his neighbors."

It will be observed that the use of horses is not just a means of doing work, a kind of power added to a farm from outside as petroleum or electricity is added to it. The use of horses is a means that *belongs* to the farm; it is a way of farming; it is, as Maurice Telleen points out, invariably accompanied or followed by a set of practices that belong together. If made to belong to the land by good care and good sense, horses tend to preserve its health. With horses come pastures and hay fields, because the horses must eat. And if one is going to grow forage for horses, then one finds it natural and economical to grow it also for other animals. From the growing of forage and the diversification of animal species, there follow naturally the principles of diversification and rotation of field crops. Having animals, one has manure, and so manure is used instead of commercial chemical fertilizers. And the use of manure, the conservation of humus, and the practices of rotation and diversification tend to work against diseases, insects, and weeds, and so one uses few or no pesticides. It is a way of farming that involves year-round use of the land by animals, plants, and the farm people—in contrast to the "corn, beans, and Florida" rotation of orthodox cash-grain farmers. Moreover, the farmer who farms with horses is not likely to be an expander. His way of farming tends to confine him to a limited acreage near home. He therefore concentrates his attention and, instead of getting more, takes good care of what he has—sows cover crops, guards against erosion, etc.

What we have here is a description of a permanent, settled, careful, largely independent agriculture that uses the land more efficiently, at least in the sense that it uses it more months in the year and more conservatively, than the orthodox agriculture. The defenders of the orthodoxy will immediately point out that the corn yields I have cited are extremely low and that we would run great risks should we reduce all yields to that level. This point must be taken seriously—not just by people on my side of the argument, but by all students and scholars of agriculture, for what is required is a definitive, scientifically sound answer. In the absence of such an answer, there are still a couple of points that need to be made.

First, it must be emphasized that all three of these farmers are older men whose children have left the farm and who are working for the most part alone. We must therefore consider that they may lack the energy, help, and motivation to push themselves and their fields toward maximum production. Second, these farms are survivals of an old way—a good way, when well followed, but not necessarily the best. The pressures of surviving, of keeping their inherited values intact in an increasingly alien atmosphere, have undoubtedly kept these farmers from being as innovative as they might have been in kinder circumstances. Particularly suggestive is the possibility of grafting the soil management methods of the more advanced organic farmers upon the traditional structures and skills of the old horse-powered farming.

But offsetting the smallness of these yields is their relative independence of economic and political conditions. Such yields are attainable on these farms year after year, *whatever* the availability of credit or of petroleum products—something that cannot be said of the much larger yields of orthodox farms, which depend absolutely on credit and on "purchased inputs" from the oil industries. The horse teams will go to the fields no matter what is happening on Wall Street or in the capitals of the Middle East. Seventy bushels of corn per acre is only half as good a yield as 140 bushels, true enough. But then it is infinitely preferable to no bushels at all.

THE AMISH

My final example of an exemplary marginal agriculture is that of the Amish. Nothing, I think, is more peculiarly characteristic of the agricultural orthodoxy—as of American society in general—than its inability to see the Amish for what they are. Oh, it *sees* them, all right. It sees them as quaint, picturesque, old-fashioned, backward, unprogressive, strange, extreme, different, perhaps slightly subversive. And that "sight" is perfect blindness. What is not seen is that the Amish are a community in the full sense of the word; they may well be the last surviving white community of any considerable size in this country. And for this there are reasons. It is especially the reasons that we do not want to see, for these reasons invalidate most of the assumptions and ambitions by which we proudly characterize ourselves as "modern."

My knowledge of the Amish, as of the other farmers I have discussed, is by no means thorough or detailed enough to satisfy the demands of strict scholarship. And I shall not pretend to be "objective" about them. I admire and respect them deeply, with few reservations; in many ways I envy them. In addition to reading several published accounts of Amish culture and agriculture, I am able to speak to some extent from experience. I have looked carefully at Amish farming in Iowa, Pennsylvania, Indiana, and Ohio, and these travels have involved some personal contacts.

What, then, are the reasons that the Amish have been able to survive as a community—or, it might be more correct to say, as a closely bound fellowship of many communities? I think that there are three primary reasons, from which spring many others.

First, the Amish communities are, at their center, religious. They are bound together not just by various worldly necessities, but by spiritual authority. Theirs is, moreover, a religion unusually attentive to its effects and obligations in this world. Whereas most contemporary sects of Christianity have tended to specialize in the interests of the spirit, leaving aside the issues of the use of the world, the Amish have not secularized their earthly life. They have not hesitated to see communal and agricultural implications in their religious principles, and these implications directly influence their behavior. The "goal" of Amish culture is not just the welfare of the spirit, but a larger harmony "among God, nature, family, and community."*

Second, the Amish have severely restricted the growth of institutions among themselves, and so they are not victimized, as we so frequently are, by organizations set up ostensibly to

*The Amish have two considerable problems, now, which trouble this ideal of harmony. They are having more children than, in present economic conditions, they can provide farms for, and so some of their young people are taking town jobs. And where coal underlies their land, some are permitting the strip-miners to come in. Reclamation was better than usual on the Amish farms I saw that had been stripped, and the land was going back into pasture. Some Amishmen nevertheless feel the practice to be wrong.

"serve" them. Though they pay the required deferences to our institutions, they accept few of the benefits, and so remain, in perhaps the most important respects, free of them. They do not become dependent on them and so maintain their integrity. As far as I know, the only institutions in our sense that the Amish have started are their schools—and this, by our standards, for a strange reason: to *keep* the responsibility for educating their children and so, in consequence, to keep their children. Amish ministers and bishops are chosen by lot, after fasting and prayer (as Mathias was chosen), and so they do not have a professional, a paid, an economically dependent, or an ambitious clergy. Their religious services are held in barns or homes; their charities are not organized or abstract but are usually in direct response to observed needs. And so they do not have a church building or a building fund or church functionaries or administrators. There is little distinction between the church and its members.

There are, one may as well say, only two Amish institutions: the family and the community. And these institutions fulfill directly, humanly, simply, and quietly nearly all the functions that we have delegated to our obtrusive, inhuman, indifferent, clumsy, expensive institutions. Family and community serve as insurance, welfare, social security, public safety. Indeed, they serve as, and replace, government. The simple living together of relatives and neighbors makes unnecessary to them our obsession with "security."

Third, the Amish are the truest geniuses of technology, for they understand the necessity of limiting it, and they know *how* to limit it. They have refused to see "technological innovation as an end in itself." And so their "religiously enforced family and community values are safeguarded against the social costs of changes which in their estimation did more harm than good to the community as a whole." Whereas our society tends to conceive of community as a loose political-economic mechanism of mutually competing producers, suppliers, and consumers, the Amish think of "the community as a whole"—that is, as all of the people, or perhaps, considering the excellence both of their neighborliness and their husbandry, as all the people and their land together. If the community is whole, then it is healthy, at once earthly and holy. The wholeness or health of the community is their standard. And by this standard they have been required to limit their technology.

By living well without such "necessities" as automobiles, tractors, electrical power, and telephones, the Amish prove them unnecessary and so give the lie to our "economy." And by these restraints they have kept their health, for by them they have kept themselves at home and have, for the most part, kept their children at home. They have not the knowledge of experts, which is by definition a homeless or rootless knowledge—the knowledge, in Sir Albert Howard's words, of people who cannot "take their own advice before offering it to other people"—and which is, as such, dangerous. They do not use knowledge to prey upon one another.

The healthy results of these restraints are readily visible to anyone who so much as drives an automobile through such an Amish community as the one in Holmes County, Ohio. Unlike so much of the best farmland, which has become a kind of agricultural desert, the Amish landscape in Holmes County is vibrantly populated with both people and animals. Busy people are seen everywhere. All the houses are lived in. All the buildings are in use. Fences and buildings are in excellent repair. And there are signs of a thriving and thrifty home life: vegetable gardens, flower gardens, fruit trees, grape arbors, berry vines, beehives, bird houses. People are making careful, comely, dignified work of the essential tasks defined by modern values as "drudgery." And because they have thought of the well-being of all the people, all are busy. There is a use for everyone. The Amish do not have the abandoned children, cast-off old people, criminals, indigents, and vagrants whom we have "freed from drudgery."

And these people practice a way of farming capable of taking exquisite care of the land. In the fall of 1976 I stood on a hillside that had been used and cared for by three generations of Amish farmers. It was steep land of the sort more often than not worn out under the old American agriculture and simply unusable by the new. This hillside had been cropped in alternating strips of corn and sod. The corn crop, which was excellent, had been cut with a binder and shocked. The farmer and his sons had carried the bundles off the plowed ground—eight rows up the hill, eight rows down—and shocked them on the sod strips, so as to get the cover crop sowed as soon as possible. By orthodox standards, this work was demeaning drudgery. By the standard of the health of the field, it was simply necessary, and so it had been done. When I was there the cover crop was coming up to safeguard the ground over the winter. I looked for marks of erosion. There were none. It is possible, I think, to say that this is a Christian agriculture, formed upon the understanding that it is sinful for people to misuse or destroy what they did not make. The Creation is a unique, irreplaceable gift, therefore to be used with humility, respect, and skill.

And so, though Amish agriculture is not modern or progressive, it is by no means ignorant or unintelligent. By the correct standard, it is much more sophisticated than orthodox agriculture. The Amish were among the first to understand the uses of rotation, manure, and legumes. They keep a balance between livestock and crops. They benefit from exchanges of labor and other forms of neighborliness. In lieu of massive consumption of fossil fuels and electricity, they make the fullest possible use of energies available on the farm—of the wind, of draft animals, and, of course, of their own bodies. Their technological restraints are balanced, quite naturally, by inventiveness. The Amish are good mechanics, and they have displayed much ingenuity in, among other things, the adaptation of tractor implements for use with teams of two to eight or more horses.

Another observer of Amish farming wrote a letter to the editor of the *Draft Horse Journal*, who published excerpts in the issue of Autumn 1976. The editor noted that "the writer of this letter owns no horses, has no vested interest in the horse business, that I know of." The following paragraphs are taken from that letter.

"My farming consists of just under a hundred acres of rather heavy, low lying land. At one time our family did something or other on three farms, of which one has gotten completely covered by houses and another has been sold to an Old Order Amishman. We still call the third one home.

"When the one farm was sold to the Amishman I forgot to tell him that one particular field was too heavy and low to grow alfalfa. By experience, I knew it wouldn't work. For a couple of years he had it in other crops, then he went to alfalfa. This embarrassed me because I knew I should have cautioned him on this. But he had a marvelous stand and a very heavy yield. It was puzzling. I puzzled over it for years but am now very persuaded of the why and wherefore.

"With our tractors we kept the soil rather permanently compacted because it was necessary to get on the land as soon as surface moisture conditions permitted. And the tire patterns pretty well rolled the entire area in the course of repeated passage. With his horses, this just didn't happen. And in the course of a couple of winters the deep frost had corrected my tractor compaction mistakes. Soil structure improved. Root penetration was facilitated. Water holding capacity as well as internal drainage both benefited, and the alfalfa flourished.

"The Amish will not argue the point because they don't think anyone is interested, but will say that a farm 'works' easier after a couple of years of horse farming. This compaction problem has to be the explanation. The resulting improved soil structure, allowing for better root penetration, is probably the reason they can get similar yields with less chemical fertilizer than their mechanized neighbors.

"But back to that heavy land alfalfa for a moment. The crop was disgustingly rank and lodged. Mowing it would be a problem. So I paid a visit, and . . . was flabbergasted to notice the farmer slowed down and mowed right through. . . . Then I realized that something very nice was being demonstrated; since the sickle drive was independent from ground travel (his horse drawn mower was equipped with an engine to drive the sickle . . .) this 'horse farmer' had sickle-cycle-to-ground-speed control that no tractor farmer could have unless he had a hydrostatic drive tractor! It gave him a control flexibility that I had never experienced and left me feeling sort of humble. I came away wondering which of us had the better technology . . . and I'm still not sure."

And then this writer addresses himself to the standard orthodox argument that we cannot feed draft horses without starving humans.

"Of course you have to feed draft animals. But this does not necessarily mean you'll have less to sell per acre farmed. In my observation, good horse farmers seem to have about as much to market per acre as the rest of us. Certainly they manage to nourish their animals very well, too. Closer examination would probably show the animals are at least partly nourished on what is wasted on fully mechanized farms. I'm not speaking just of corn fodder, either. Who else hand gathers the ears the picker missed now-a-days? . . . Does the man with the 4 row combine? And could he gather the shelling loss if he would? Hardly.

"Well, I'm not very impressed by the statistics that prove we'd starve if farming went back to animal power. In many sections of the country that is exactly what happens when a farm, or a group of farms, comes into the hands of an Amishman. The farming has gone from tractors back to horses on hundreds of farms in my part of the United States without any noticeable reduction in agricultural output. Any suggestion that our county produces less now than 20 years ago would seem outrageous to all the people I know."

In support of these impressions of the general good health of Amish agriculture, some more specific information is available in an article entitled "Agricultural Alternatives" in the March-April 1972 issue of *CBNS Notes*, published by the Center for the Biology of Natural Systems. I take the following quotations from that article:

"Amish attitudes toward fertility maintenance are amazingly varied. . . . Preliminary analysis of the data shows three distinct patterns. The traditional pattern consists of crop rotation which includes nitrogen-fixing legumes, heavy application of manure to at least 1/5 of the farm in any growing season and lime and rock phosphate to one of the fields in the rotation every year. With the aid of hybrid seed corn some farmers estimate their yields at 90 to 100 bushels per acre although some estimates fall as low as 70.

"A second pattern is the conventional with Amish modifications. The soil will be tested for its acidity and for its phosphorous and potash balance. A county agent or a fertilizer dealer will then make a recommendation for lime and fertilizer application in relation to specific cropping plans. This can include the use of anhydrous ammonia as a cheap source of nitrogen for corn. It is observed that Amish operators who follow this pattern 'factor in' the effects of their crop rotation and the availability of manure and thus apply fertilizer less heavily per acre. . . .

"The third pattern is the use of organic fertilizers. . . . Some of them tend to be costly in terms of additional yield per acre but a minority of the Amish farm operators are very enthusiastic users. The appeal is on the basis of a claim to a more nutritious quality of feed grain which in turn leads to healthier livestock, healthier soil and eventually healthier humans. Interestingly, several of the operators interviewed adopted a program of organic fertility maintenance after having been on a conventional program of commercial fertilizer for several years."

As for the effects of this agriculture, the article offers evidence which suggests that, ecologically and economically, the Amish methods are sounder than the orthodox. Water pollution from Amish fields was far less: "One of the comparative samples taken in March, 1971 showed concentrations of nitrate nitrogen of 12.1 ppm and 8.9 in the tile of conventional farms in Douglas County [Illinois]. The Amish tile had a concentration of 4.6 ppm (corn). A comparison in May showed a concentration of 26.6 ppm (corn) and 10.9 ppm (beans) in the conventional farms and 4.6 ppm in the tiles of the Amish farmer."

And the Amish, whose farms in 1965 averaged only 76.55 acres, were prospering financially during a time when many of the smaller orthodox farmers (with far larger holdings) were being "squeezed out": "Our best single indicator of economic viability is bank data which compares 88 Amish bank accounts in 1964 with those same accounts in 1971. During these years the Amish accounts showed an increase in net worth from $2,379,000 to $4,045,000."

Since the Amish are manifestly excellent farmers, and are so complexly successful in other ways, one wonders why they have been ignored by the officials and the scholars of agriculture —especially since their technology and methods are so well suited to land not even farmable by orthodox methods and to farmers not able to survive in the orthodox economy. I have been able to think of only two answers, aside from the conventional contempt for anything small: first, the Amish are a thrifty people, hence poor consumers of "purchased inputs" from the "agribusiness" industries; and, second, they are living disproof of some of the fundamental assumptions of the orthodoxy.

PRODUCTION AND REPRODUCTION

To these exemplary forms of unorthodox agriculture, we may add the new work in urban homesteading, aquaculture, solar green-houses, alternative energy sources, small technology, organic pest control, etc., by the New Alchemy Institute, the Farallones Institute, Rodale Press, and others, as well as the various farmers' and consumers' cooperatives that have been started in the past few years both as strategies of health and as protests against the "agribusiness" juggernaut. Together, these have restored a sense of possibility, both cultural and agricultural, that has been nearly obliterated by the ambitions of agriculture specialists and businessmen. They make possible a vision of an agriculture many times more versatile and diverse than the orthodox, hence many times more responsive to the demands of good husbandry, to local conditions, and to human needs.

For the orthodox obsession with production, profit, and expansion, this healthier agriculture would substitute a more complex consciousness, the terms of which would be ecological integrity, nutrition, technological appropriateness, social stability, skill, quality, thrift, diversity, decentralization, independence, usufruct. Or, put more simply, it would replace the concern for production with a concern for reproduction. Production, some would say, is the male principle in isolation from the female principle. Thus isolated, the male principle wants to exert itself absolutely; it wants to "do everything at once" —which is, of course, what doomsday will do. But reproduction, which is the male and the female principles in union, is nurturing, patient, resigned to the pace of seasons and lives, respectful of the nature of things. Production's tendency is to go "all out"; it always aims to set a new record. Reproduction is more conservative and more modest; its aim is not to happen once, but to happen again and again and again, and so it seeks a balance between saving and spending. At their best, farmers have always had this ancient purpose of reproduction. Without it, they make their art as sterile as mining.

There would, of course, be no need for a different vision of agriculture if the one we had were demonstrably working in the best long-term interests of the people and the land, or even if it were generally believed to be doing so. In fact, there are a great many people who do *not* believe that it is doing so, and their number is growing. And so the last agricultural margin remaining to be noticed is a political one: the people who feel that they are being victimized by orthodox agriculture and whose dissatisfaction is either ignored or held in contempt.

There are, first, many people—ex-farmers, heirs of farmers, and would-be farmers—who want to farm but are prevented from doing so by high land costs, taxes, inheritance taxes, and interest rates. And these economic barriers, which exclude the small operator, directly favor not just the survival, but also the expansion, of the big operator. This is not a necessary result of "the way things are." It is the calculated effect of a deliberate policy to allow the big to grow bigger at the expense of the small. In addition, there are many farmers of the same kinds who are presently farming, but whose survival is in doubt for the same reasons.

Second, there is a rapidly increasing number of consumers who wish to buy food that is nutritionally whole and uncontaminated by pesticides and other toxic chemical residues. And these people would prefer not to pay the exorbitant food prices required by long-distance transportation, processing, packaging, and advertising, all of which result from "agribusiness" control of food.

PUBLIC REMEDIES

And so we come to the question of what, in a public or governmental sense, ought to be done. Any criticism of an established way, if it is to be valid, must have as its standard not only a need, but a better way. It must show that a better way is desirable, and it must give examples to show that it is possible. I have produced the argument and the examples—not definitively, I am sure, but sufficiently to provide an agenda for the further work that is necessary.

It remains for me to suggest public changes that are necessary to bring the better way to realization. This is the most fearful part of my task, for what I have described at such length here is a big problem, and it is the overwhelming tendency of our time to assume that a big problem calls for a big solution.

I do not believe in the efficacy of big solutions. I believe that they not only tend to prolong and complicate the problems they are meant to solve, but that they cause new problems. On the other hand, if the solution is small, obvious, simple, and cheap, then it may quickly and permanently solve the immediate problem and many others as well. For example, if a city-dweller walks or rides a bicycle to work, he has found the simplest solution to his transportation problem—and at the same time he is reducing pollution, reducing the waste of natural resources, reducing the public expenditure for traffic control, saving his money, and improving his health. The same ramifying pattern of solutions attends all skills and strategies of economic independence: gardening, cooking, household maintenance, etc. To turn an agricultural problem over to the developers, promoters, and salesmen of industrial technology is not to ask for a solution; it is to ask for more industrial technology and for a bigger bureaucracy to handle the resulting problems of social upset, unemployment, ill health, urban sprawl, and overcrowding. Whatever their claims to "objectivity," these people will not examine the problem and apply the most fitting solution; they will reverse that procedure and define the problem to fit the solution in which their ambitions and their livelihoods have been invested. They are thriving on the problem and so can have little interest in solving it.

And so the first necessary public change is simply a withdrawal of confidence from the league of specialists, officials, and corporation executives who for at least a generation have had almost exclusive charge of the problem and who have enormously enriched and empowered themselves by making it worse.

Second, as a people, we must learn again to think of human energy, *our* energy, not as something to be saved, but as something to be used and to be enjoyed in use. We must understand that our strength is, first of all, strength of body, and that this strength cannot thrive except in useful, decent, satisfying, comely work. There is no such thing as a reservoir of bodily energy. By saving it—as our ideals of labor-saving and luxury bid us to do—we simply waste it, and waste much else along with it.

Third, we must see again, as I think the founders of our

government saw, that the most appropriate governmental powers are negative—those, that is, that protect the small and weak from the great and powerful, *not* those by which the government becomes the profligate, ineffectual parent of the small and weak after it has permitted the great and powerful to make them helpless. The governmental power that can be used most effectively to assure an equitable distribution of property, which alone can give some measure of strength and independence to ordinary citizens, is that of taxation. As our present economy clearly shows, the small can survive only if the great are restrained. And there is nothing undemocratic or anti-libertarian about restraining them. To assume that ordinary citizens can compete successfully with people of wealth and with corporations, as our government presently tends to do, is simply to abandon the ordinary citizens. Restraint by taxation is the smallest, most obvious, simplest, and cheapest answer. This is not my idea. It is Thomas Jefferson's. Writing to Reverend James Madison on October 28, 1785, Jefferson spoke of the desirability of freehold tenure of property. And then he said "Another means of silently lessening the inequality of property is to exempt all from taxation below a certain point, and to tax the higher portions of property in geometric progression as they rise. The earth is given as a common stock for man to labor and live on. If for the encouragement of industry [he means, of course, mainly agriculture] we allow it to be appropriated, we must take care that employment be provided to those excluded from the appropriation. If we do not, the fundamental right to labor the earth returns to the unemployed . . . it is not too soon to provide by every possible means that as few as possible shall be without a little portion of land. The small landholders are the most precious part of a state. . . ." It would, of course, be necessary to consider how much land in any region ought to constitute a living for a family.

Fourth, considering that the price of farmland has now been driven up by urban pressures and speculation until farmers often cannot afford to own it, low-interest loans ought to be made available to people wishing to buy family-size farms. This would probably need to be only a temporary or transitional measure.

Fifth, there should be a system of production and price controls that would tend to adjust production both to need and to the carrying capacities of farms. One purpose of this would be to curb the extreme fluctuations of supply, which work in the long run to the disadvantage of small producers. Another would be the elimination of the phenomenon of "harvest-time depressed prices"—which, in practice, means that the price of grain is low when it is in the hands of the wrong people (small farmers who cannot afford storage) and high when it is in the hands of the right people (big farmers and "agribusiness" corporations).

Sixth, there should be a program to promote local self-sufficiency in food. The cheapest, freshest food is that which is produced closest to home and is not delayed for processing. This should work toward the most direct dealing between farmers and merchants and farmers and consumers. Much might be done by the promotion of growers' and consumers' cooperatives.

Seventh, every town and city should be required to operate an organic-waste depot where sewage, garbage, waste paper, and the like would be composted and given or sold at cost to farmers. Every truck bringing a load of produce to town should go home with a load of compost. This would greatly improve the health of both the rivers and the fields and it would lower the cost of food.

Eighth, there should be a strenuous review of all sanitation laws governing the production of food, and those that are unnecessary should be eliminated. Sanitation laws have almost invariably worked against the small producer, destroying his markets or prohibitively increasing the cost of production. If we are as technologically adept as we claim to be, then it is inexcusable that we do not have, for instance, an acceptable, inexpensive technology for small dairies. And there is no reason, given the necessary collecting points, that we should not have markets for small quantities of other foods. If we are serious about increasing food production, then we must make room for the small producer. Moreover, decency and common sense require us to learn if it is necessary for cleanliness invariably to be expensive.

Ninth, we should encourage the greatest possible technological and genetic diversity, in conformation to local need, as opposed to the present dangerous uniformity in both categories. This diversity should be the primary goal of the land-grant schools. To this end, they should be *required*, as the Hatch Act instructs, "to assure agriculture a position in research equal to that of industry." These schools, and their professors individually, should be forbidden to accept work on assignment from any corporation or other outside interest that might wish to market any resulting product. (This, of course, would not apply to professors working on their own time outside the university.)

Tenth, to de-specialize the interests of the colleges of agriculture—that is, to shift their loyalty from "agribusiness" and industry back to the farmers—two other measures might be useful: (1) The faculties should be opened, on a part-time basis, to farmers, just as faculties of medicine and law are opened to doctors and lawyers; and (2) faculty members could be paid half their salary in cash and given the use of a boundary of college farmland the potential annual income from which would be equivalent to the other half. In both instances, the professor would be in a position to "take his own advice before offering it to other people." And much good might be expected from that. Professors might again become people of experience rather than experts. They might again be able to apply their learning to the small problems of ordinary people and to recommend means and methods not profitable to the suppliers of "purchased inputs."

Eleventh, we must address ourselves seriously, and not a little fearfully, to the problem of human scale. What is it? How do we stay within it? What sort of technology enhances our humanity? What sort reduces it? The reason is simply that we cannot live except within limits, and these limits are of many kinds: spatial, material, moral, spiritual. The world has room for many people who are content to live as humans, but only for a relative few intent upon living as giants or as gods.

Twelfth, having exploited "relativism" until, as a people, we have no deeply believed reasons for doing anything, we must now ask ourselves if there is not, after all, an absolute good by which we must measure ourselves and for which we must work. That absolute good, I think, is health—not in the merely hygienic sense of personal health, but the health, the wholeness, finally the holiness, of Creation, of which our personal health is only a share.

THE NECESSITY OF MARGINS

In Michigan in the fall of 1975, a fire-retarding chemical known as PBB was mistaken for a trace mineral and mixed into a large order of livestock feed. This feed was sent to four Michigan mills run by the Farm Bureau, and from there it went to farms and to the stock troughs. The resulting contamination of meat, milk, and eggs produced a disaster which is still continuing after three-and-a-half years and the limits of which are not known. The immediate and most noticeable result was a state program to destroy contaminated animals and food products. This did away with "about 1.5 million chickens, 29,000 head of cattle, 5,920 hogs, 1,470 sheep, 2,600 lb. of butter, 18,000 lb. of cheese, 34,000 lb. of dry milk products, and 5 million eggs." But many people were also affected, some seriously, and the long-term effects on human health are not known.

This was a tragedy—personal and economic, private and public—caused by one error that "may have been as simple as pulling the wrong lever." And we must recognize that, both in its carelessness and in its magnitude, this tragedy is characteristic of an agriculture, indeed of a culture, without margins. In a highly centralized and industrialized food-supply system there can be no small disaster. Whether it be a production "error" or a corn blight, the disaster is not foreseen until it exists; it is not recognized until it is widespread. By contrast, a highly diversified, small-farm agriculture combined with local marketing is literally crisscrossed with margins, and these margins work both to allow and encourage care and to contain damage.

But such an agriculture would do more than provide us with protective margins. In reducing industrial uniformity it would give us a new sense of our real unity, our common sharing in the good of health. It is a rule, apparently, that whatever is divided must compete. We have been wrong to believe that competition invariably results in the triumph of the best. Divided, body and soul, man and woman, producer and consumer, nature and technology, city and country are thrown into competition with one another. And none of these competitions is ever resolved in the triumph of one competitor, but only in the exhaustion of both.

For our healing we have on our side one great force: the power of Creation, with good care, with kindly use, to heal itself.

Afterword to the Third Edition

In *The Unsettling of America* I argue that industrial agriculture and the assumptions on which it rests are wrong, root and branch; I argue that this kind of agriculture grows out of the worst of human history and the worst of human nature. From my own point of view, the happiest fate of my labors would have been disproof. I would have been much relieved if somebody had proved me wrong, or if events had shown that I need not have worried. For this book certainly was written out of worry. It was written, in fact, out of the belief that we were living under the rule of an ideology that was destroying our land, our communities, and our culture—as we still are.

My argument, as I saw it twenty years ago, was addressed to leaders in the schools, the governments, and other places, who presumably were interested in argument as a way of approaching certain kinds of truths. The years since its publication have demonstrated, among other things, my naiveté. The argument set forth in this book, though it has been much and sometimes vehemently disagreed with, has never been answered, let alone disproved.

For this, surely the paramount reason is that events have continued to confirm my argument at every point. The enormous productivity of industrial agriculture cannot be denied, but neither can its enormous ecological, economic, and human costs, which are bound eventually to damage its productivity. This book's tragedy is that it is true.

Moreover, those who disagree with what I wrote are mainly those against whom I wrote it: adherents of the industrial program, which is too powerful, too rich, and too preoccupied with conquest to be diverted by anybody's mere argument. They simply are not obliged to care whether or not they may be wrong.

Another reason my book has received no vigorous counterargument, I fear, is that in centers of learning and power argument itself has become virtually obsolete, a lost art. Public discourse of all kinds now tends to pattern itself either upon the arts of advertisement and propaganda (that is, the arts of persuasion without argument, which lead to reasonless and even unconscious acquiescence) or upon the allegedly objective or value-free demonstrations of science.

When I was writing this book I still supposed, for example, that the land-grant universities, if confronted by an argument against their governing assumptions, would either have to produce a stronger counterargument or change their assumptions. That supposition may have been naive, but it was nonetheless one that I had every right to make, for the pursuit of truth by argument and counterargument is a major part of our cultural tradition from the Gospels and the Platonic dialogues to every county courthouse today.

The response to this book has shown, instead, that the universities are not interested in the pursuit of truth by argument. They are interested in preserving the conclusion of an old argument that for the most part they no longer bother to make: namely, that the world and all its creatures are machines. The organization of the modern university—and of modern intellectual life—rests upon this argument. Perhaps this line of thought began in metaphor, but now the likeness has become identity. It is assumed, as my friend Gene Logsdon puts it, that biology and mechanics are the same thing, and that other things don't matter. Here are some smaller assumptions that derive from and help to preserve the larger one:

1. If the world and all its creatures are machines, then the world and all its creatures are entirely comprehensible, manipulable, and controllable by humans.
2. The humans who have this power are experts.
3. Experts are made by education.
4. Education only happens in schools.
5. Experts are smarter than other people.
6. Thinking is best done by experts in offices and laboratories.
7. People who *do* work cannot be trusted to think about it.
8. People who work would prefer not to work.
9. Human workers are inefficient machines, encumbered by extraneous needs and desires, and they should be replaced by more efficient machines or by chemicals.
10. In general, the human machine is better at consumption than production.

11. A farm is or ought to be a factory in which plant and animal machines serve the economic machine in the most efficient way.
12. Efficiency has nothing to do with human or biological needs and desires.
13. Farm bankruptcy increases agricultural efficiency.
14. All farmers actually dislike farming and are secretly glad when they go bankrupt, because that gets them out of the sticks and into the bright lights where they have a chance to become experts.
15. Conventional agricultural science (like all conventional science) is disinterested and objective and serves no interest other than the advancement of human knowledge.

And so on.

Eventually this mechanistic line of thought brings us to the doctrine that whatever happens is inevitable. Actually, this stark determinism is altered in general use to a doctrine that is even more contemptible: every *bad* thing that happens is inevitable. For every good thing that happens there are mobs of claimers of credit. Every good and perfect gift comes from politicians, scientists, researchers, governments, and corporations. Evils, however, are inevitable; there is just no use in trying to choose against them. Thus all industrial comforts and labor-saving devices are the result only of human ingenuity and determination (not to mention the charity and altruism that have so conspicuously distinguished the industrial subspecies for the past two centuries), but the consequent pollution, land destruction, and social upheaval have been "inevitable."

Thus President Clinton (for whom I voted) could tell an audience of "farmers and agricultural organization leaders" in Billings, Montana, on June 1, 1995, that the American farm population now is "dramatically lower, obviously, than it was a generation ago. And that was inevitable because of the increasing productivity of agriculture." (See assumptions 9–13 above.)

That is to say that what happened happened because it had to happen. Thus the apologists for the ruin of agricultural lands, economies, and communities have shown always that they did nothing to stop it because there was nothing they could have done to stop it. (It's just progress, folks. Be glad your children won't suffer the drudgery and degradation of farm ownership.) The president also said that he wants to save the family farm, which is "alive and well" in Montana. He said he believes that we have "bottomed out in the shrinking of the farm sector." He said he wants to help young farmers. He spoke of the need to make American agriculture "competitive with people around the world." He praised our huge volume of agricultural exports. All of this had been said countless times before, and all of it sinks beneath that weighty adjective *inevitable.* If an utterly brainless and destructive agricultural economy has been inevitable for half a century, why should it now suddenly cease to be inevitable?

The president's remarks in Montana provide evidence enough that our national conversation about agriculture, at the "upper levels" of policy and research, is frozen solid. The people up there have not had a new or divergent thought in two generations. Furthermore, they do not wish to think a new or divergent thought; the old thoughts have suited their careers and their pocketbooks well enough. When threatening ideas or even threatening statistics or experimental evidence are introduced, those upper-level experts just freeze them into the ice and skate over the top of them.

But if the publication of *The Unsettling of America* and subsequent events have shown me that throwing a rock into a frozen river does not make a ripple, they have also shown that beneath the ice the waters are strongly flowing and stirred up and full of nutrients. Beneath the clichés of official science and policy, our national conversation about agriculture is more vigorous and exciting now than it has been since the 1930s.

This book has not had the happy fate of being proved wrong, but it has had the next-happiest fate of belonging to a growing effort to think again about the issues of American land use and to start the changes that are needed.

My book does not stand alone. It occurs in a lineage of works, influences, and exemplars that it acknowledges, and that I have more fully acknowledged in writings since. And it belongs to a company of present-day works and exemplars joined by a common commitment and a common aim: books by Marty Strange, Gene Logsdon, and Wes Jackson, among others; organizations such as the Land Institute, the Center for Rural Affairs, the Land Stewardship Project, Tilth, and the

E. F. Schumacher Society; the several conservation organizations; a rapidly increasing number of organizations interested in the local marketing of local products; and thousands of farmers and gardeners. I am much encouraged by the knowledge that if this book (and my other books) suddenly disappeared from print and from memory, its advocacy and its hope would continue undiminished.

The people to whom this book belongs are thinking about agriculture and other land-based enterprises in a way radically different from the way the people at the upper levels of policy and research are thinking. They think so differently, I believe, because their motives are different. Their thinking does not begin with a set of predetermining ideas but rather with particular places, people, needs, and desires. This book's friends and allies began to think and to work not because they had careers to make or ideologies to serve but because they loved certain places, people, possibilities, and ways that they could not indifferently see destroyed.

What we are working for, I think, is an authentic settlement and inhabitation of our country. We would like to see all human work lovingly adapted to the nature of the places where it is done and to the real needs of the people by whom and for whom it is done. We do not believe that any violence to places, to people, or to other creatures is "inevitable." We believe that the industrial ideology is wrong because it obscures and disrupts this necessary work of local adaptation or home making.

I do not believe that this effort will lead to perfection; the best agriculture and economics we can imagine will not return us to Eden; we will carry with us into whatever we do the weaknesses and limits inherent in our nature. But our imperfections argue more strongly than any hope of perfection for the adoption of ways and aims that can lead us beyond the selfishness that is institutionalized in the present economic system. To suggest that the health of places and communities might be the indispensable standard of economic behavior is finally to ask how a mere human, whose years are like the grass that is cut down in the evening, can justify on his or her own behalf the permanent destruction of anything.

Our effort to make something comely and enduring of our life on this earth will last as long as our species, I am confident of that; it is, after all, an ancient effort. I am confident that our present effort here in the United States can grow and accomplish much without the help of the upper levels of research and policy. It does worry me, however, that the people working in various ways to protect places and communities and ways of life now make up a sizable constituency that is virtually unclaimed and unrepresented. The dangers in this are obvious enough to anybody willing to look. Our government has shown considerable enthusiasm for "leveling the playing field" in the interest of international corporations. Its enthusiasm for leveling the playing field in the interest of local economies and local ecosystems remains to be demonstrated.

Wendell Berry
August 1995

FROM

THE GIFT OF GOOD LAND

(1981)

Horse-Drawn Tools and the Doctrine of Labor Saving

FIVE YEARS AGO, when we enlarged our farm from about twelve acres to about fifty, we saw that we had come to the limits of the equipment we had on hand: mainly a rotary tiller and a Gravely walking tractor; we had been borrowing a tractor and mower to clip our few acres of pasture. Now we would have perhaps twenty-five acres of pasture, three acres of hay, and the garden; and we would also be clearing some land and dragging the cut trees out for firewood. I thought for a while of buying a second-hand 8N Ford tractor, but decided finally to buy a team of horses instead.

I have several reasons for being glad that I did. One reason is that it started me thinking more particularly and carefully than before about the development of agricultural technology. I had learned to use a team when I was a boy, and then had learned to use the tractor equipment that replaced virtually all the horse and mule teams in this part of the country after World War II. Now I was turning around, as if in the middle of my own history, and taking up the old way again.

Buying and borrowing, I gathered up the equipment I needed to get started: wagon, manure spreader, mowing machine, disk, a one-row cultivating plow for the garden. Most of these machines had been sitting idle for years. I put them back into working shape, and started using them. That was 1973. In the years since, I have bought a number of other horse-drawn tools, for myself and other people. My own outfit now includes a breaking plow, a two-horse riding cultivator, and a grain drill.

As I have repaired these old machines and used them, I have seen how well designed and durable they are, and what good work they do. When the manufacturers modified them for use with tractors, they did not much improve either the machines or the quality of their work. (It is necessary, of course, to note some exceptions. Some horsemen, for instance, would argue that alfalfa sod is best plowed with a tractor. And one must also except such tools as hay conditioners and chisel plows that came after the development of horse-drawn tools had ceased. We do not know what innovations, refinements, and improvements would have come if it had continued.) At the peak of their development, the old horse tools were excellent. The coming of the tractor made it possible for a farmer to do more work, but not better. And there comes a point, as we know, when *more* begins to imply *worse*. The mechanization of farming passed that point long ago—probably, or so I will argue, when it passed from horse power to tractor power.

The increase of power has made it possible for one worker to crop an enormous acreage, but for this "efficiency" the country has paid a high price. From 1946 to 1976, because fewer people were needed, the farm population declined from thirty million to nine million; the rapid movement of these millions into the cities greatly aggravated that complex of problems which we now call the "urban crisis," and the land is suffering for want of the care of those absent families. The coming of a tool, then, can be a cultural event of great influence and power. Once that is understood, it is no longer possible to be simple-minded about technological progress. It is no longer possible to ask, What is a good tool? without asking at the same time, How *well* does it work? and, What is its influence?

One could say, as a rule of thumb, that a good tool is one that makes it possible to work faster *and* better than before. When companies quit making them, the horse-drawn tools fulfilled both requirements. Consider, for example, the International High Gear No. 9 mowing machine. This is a horse-drawn mower that certainly improved on everything that came before it, from the scythe to previous machines in the International line. Up to that point, to cut fast and to cut well were two aspects of the same problem. Past that point the speed of the work could be increased, but not the quality.

I own one of these mowers. I have used it in my hayfield at the same time that a neighbor mowed there with a tractor mower; I have gone from my own freshly cut hayfield into others just mowed by tractors; and I can say unhesitatingly that, though the tractors do faster work, they do not do it better. The same is substantially true, I think, of other tools: plows, cultivators, harrows, grain drills, seeders, spreaders, etc. Through the development of the standard horse-drawn equipment, quality and speed increased together; after that, the principal increase has been in speed.

Moreover, as the speed has increased, care has tended to decline. For this, one's eyes can furnish ample evidence. But we have it also by the testimony of the equipment manufacturers themselves. Here, for example, is a quote from the public relations paper of one of the largest companies: "Today we have multi-row planters that slap in a crop in a hurry, putting down seed, fertilizer, insecticide and herbicide in one quick swipe across the field."

But good work and good workmanship cannot be accomplished by "slaps" and "swipes." Such language seems to be derived from the he-man vocabulary of TV westerns, not from any known principles of good agriculture. What does the language of good agricultural workmanship sound like? Here is the voice of an old-time English farmworker and horseman, Harry Groom, as quoted in George Ewart Evans's *The Horse in the Furrow*: "It's all rush today. You hear a young chap say in the pub: 'I done thirty acres today.' But it ain't messed over, let alone done. You take the rolling, for instance. Two mile an hour is fast enough for a roll or a harrow. With a roll, the slower the better. If you roll fast, the clods are not broken up, they're just pressed in further. Speed is everything now; just jump on the tractor and way across the field as if it's a dirt-track. You see it when a farmer takes over a new farm: he goes in and plants straight-way, right out of the book. But if one of the old farmers took a new farm, and you walked round the land with him and asked him: 'What are you going to plant here and here?' he'd look at you some queer; because he wouldn't plant nothing much at first. He'd wait a bit and see what the land was like: he'd *prove* the land first. A good practical man would hold on for a few weeks, and get the feel of the land under his feet. He'd walk on it and feel it through his boots and see if it was in good heart, before he planted anything: he'd sow only when he knew what the land was fit for."

Granted that there is always plenty of room to disagree about farming methods, there is still no way to deny that in the first quotation we have a description of careless farming, and in the second a description of a way of farming as careful —as knowing, skillful, and loving—as any other kind of high workmanship. The difference between the two is simply that the second considers where and how the machine is used, whereas the first considers only the machine. The first is the point of view of a man high up in the air-conditioned cab of a tractor described as "a beast that eats acres." The second is that of a man who has worked close to the ground in the open air of the field, who has studied the condition of the ground as he drove over it, and who has cared and thought about it.

If we had tools thirty-five years ago that made it possible to do farm work both faster and better than before, then why did we choose to go ahead and make them no longer better, but just bigger and bigger and faster and faster? It was, I think, because we were already allowing the wrong people to give the wrong answers to questions raised by the improved horse-drawn machines. Those machines, like the ones that followed them, were *labor savers.* They may seem old-timey in comparison to today's "acre eaters," but when they came on the market they greatly increased the amount of work that one worker could do in a day. And so they confronted us with a critical question: How would we define labor saving?

We defined it, or allowed it to be defined for us by the corporations and the specialists, as if it involved no human considerations at all, as if the labor to be "saved" were not human labor. We decided, in the language of some experts, to look on technology as a "substitute for labor." Which means that we did not intend to "save" labor at all, but to *replace* it, and to *displace* the people who once supplied it. We never asked what should be done with the "saved" labor; we let the "labor market" take care of that. Nor did we ask the larger questions of what values we should place on people and their work and on the land. It appears that we abandoned ourselves unquestioningly to a course of technological evolution, which would value the development of machines far above the development of people.

And so it becomes clear that, by itself, my rule-of-thumb definition of a good tool (one that permits a worker to work both better and faster) does not go far enough. Even such a tool can cause bad results if its use is not directed by a benign and healthy social purpose. The coming of a tool, then, is not just a cultural event; it is also an historical crossroad—a point at which people must choose between two possibilities: to become more intensive or more extensive; to use the tool for quality or for quantity, for care or for speed.

In speaking of this as a choice, I am obviously assuming that the evolution of technology is *not* unquestionable or uncontrollable; that "progress" and the "labor market" do *not* represent anything so unyielding as natural law, but are aspects of an economy; and that any economy is in some sense a "managed" economy, managed by an intention to distribute the benefits of work, land, and materials in a certain way. (The present agricultural economy, for instance, is slanted to give the greater portion of these benefits to the "agribusiness" corporations. If this were not so, the recent farmers' strike would have been an "agribusiness" strike as well.) If those assumptions are correct, we are at liberty to do a little historical supposing, not meant, of course, to "change history" or "rewrite it," but to clarify somewhat this question of technological choice.

Suppose, then, that in 1945 we had valued the human life of farms and farm communities 1 percent more than we valued "economic growth" and technological progress. And suppose we had espoused the health of homes, farms, towns, and cities with anything like the resolve and energy with which we built the "military-industrial complex." Suppose, in other words, that we had really meant what, all that time, most of us and most of our leaders were saying, and that we had really tried to live by the traditional values to which we gave lip service.

Then, it seems to me, we might have accepted certain mechanical and economic limits. We might have used the improved horse-drawn tools, or even the small tractor equipment that followed, not to displace workers and decrease care and skill, but to intensify production, improve maintenance, increase care and skill, and widen the margins of leisure, pleasure, and community life. We might, in other words, by limiting technology to a human or a democratic scale, have been able to use the saved labor *in the same places where we saved it*.

It is important to remember that "labor" is a very crude, industrial term, fitted to the huge economic structures, the dehumanized technology, and the abstract social organization of urban-industrial society. In such circumstances, "labor" means little more than the sum of two human quantities, human energy plus human time, which we identify as "man-hours." But the nearer home we put "labor" to work, and the smaller and more familiar we make its circumstances, the more we enlarge and complicate and enhance its meaning. At work in a factory, workers are only workers, "units of production" expending "man-hours" at a task set for them by strangers. At work in their own communities, on their own farms or in their own households or shops, workers are *never* only workers, but rather persons, relatives, and neighbors. They work *for* those they work *among* and *with*. Moreover, workers tend to be independent in inverse proportion to the size of the circumstance in which they work. That is, the work of factory workers is ruled by the factory, whereas the work of housewives, small craftsmen, or small farmers is ruled by their own morality, skill, and intelligence. And so, when workers work independently and at home, the society as a whole may lose something in the way of organizational efficiency and economies of scale. But it begins to *gain* values not so readily quantifiable in the fulfilled humanity of the workers, who then bring to their work not just contracted quantities of "man-hours," but qualities such as independence, skill, intelligence, judgment, pride, respect, loyalty, love, reverence.

To put the matter in concrete terms, if the farm communities had been able to use the best horse-drawn tools to save labor in the true sense, then they might have used the saved time and energy, first of all, for leisure—something that technological progress has given to farmers. Second, they might have used it to improve their farms: to enrich the soil, prevent erosion, conserve water, put up better and more permanent fences and buildings; to practice forestry and its dependent crafts and economies; to plant orchards, vineyards, gardens of bush fruits; to plant market gardens; to improve pasture, breeding, husbandry, and the subsidiary enterprises of a local, small-herd livestock economy; to enlarge, diversify, and deepen the economies of households and homesteads. Third, they might have used it to expand and improve the specialized crafts necessary to the health and beauty of communities: carpentry, masonry, leatherwork, cabinetwork, metalwork, pottery, etc. Fourth, they might have used it to improve the homelife and the home instruction of children, thereby preventing the hardships and expenses now placed on schools, courts, and jails.

It is probable also that, if we *had* followed such a course, we would have averted or greatly ameliorated the present shortages of energy and employment. The cities would be much less crowded; the rates of crime and welfare dependency would be much lower; the standards of industrial production would probably be higher. And farmers might have avoided their present crippling dependence on money lenders.

I am aware that all this is exactly the sort of thinking that the technological determinists will dismiss as nostalgic or wishful. I mean it, however, not as a recommendation that we "return to the past," but as a criticism of the past; and my criticism is based on the assumption that we had in the past, and that we have now, a *choice* about how we should use technology and what we should use it for. As I understand it, this choice depends absolutely on our willingness to limit our desires as well as the scale and kind of technology we use to satisfy them. Without that willingness, there is no choice; we must simply abandon ourselves to whatever the technologists may discover to be possible.

The technological determinists, of course, do not accept that such a choice exists—undoubtedly because they resent the moral limits on their work that such a choice implies. They speak romantically of "man's destiny" to go on to bigger and more sophisticated machines. Or they take the opposite course and speak the tooth-and-claw language of Darwinism. Ex-secretary of agriculture Earl Butz speaks, for instance, of "Butz's Law of Economics" which is "Adapt or Die."

I am, I think, as enthusiastic about the principle of adaptation as Mr. Butz. We differ only on the question of what should be adapted. He believes that we should adapt to the machines, that humans should be forced to conform to technological conditions or standards. I believe that the machines should be adapted to us—to serve our *human* needs as our history, our heritage, and our most generous hopes have defined them.

Solving for Pattern

OUR DILEMMA in agriculture now is that the industrial methods that have so spectacularly solved some of the problems of food production have been accompanied by "side effects" so damaging as to threaten the survival of farming. Perhaps the best clue to the nature and the gravity of this dilemma is that it is not limited to agriculture. My immediate concern here is with the irony of agricultural methods that destroy, first, the health of the soil and, finally, the health of human communities. But I could just as easily be talking about sanitation systems that pollute, school systems that graduate illiterate students, medical cures that cause disease, or nuclear armaments that explode in the midst of the people they are meant to protect. This is a kind of surprise that is characteristic of our time: the cure proves incurable; security results in the evacuation of a neighborhood or a town. It is only when it is understood that our agricultural dilemma is characteristic not of our agriculture but of our time that we can begin to understand why these surprises happen, and to work out standards of judgment that may prevent them.

To the problems of farming, then, as to other problems of our time, there appear to be three kinds of solutions:

There is, first, the solution that causes a ramifying series of new problems, the only limiting criterion being, apparently, that the new problems should arise beyond the purview of the expertise that produced the solution—as, in agriculture, industrial solutions to the problem of production have invariably caused problems of maintenance, conservation, economics, community health, etc., etc.

If, for example, beef cattle are fed in large feed lots, within the boundaries of the feeding operation itself a certain factory-like order and efficiency can be achieved. But even within those boundaries that mechanical order immediately produces biological disorder, for we know that health problems and dependence on drugs will be greater among cattle so confined than among cattle on pasture.

And beyond those boundaries, the problems multiply. Pen feeding of cattle in large numbers involves, first, a manure-removal problem, which becomes at some point a health problem for the animals themselves, for the local watershed, and for the adjoining ecosystems and human communities. If the manure is disposed of without returning it to the soil that produced the feed, a serious problem of soil fertility is involved. But we know too that large concentrations of animals in feed lots in one place tend to be associated with, and to promote, large cash-grain monocultures in other places. These monocultures tend to be accompanied by a whole set of specifically agricultural problems: soil erosion, soil compaction, epidemic infestations of pests, weeds, and disease. But they are also accompanied by a set of agricultural-economic problems (dependence on purchased technology; dependence on purchased fuels, fertilizers, and poisons; dependence on credit)—and by a set of community problems, beginning with depopulation and the removal of sources, services, and markets to more and more distant towns. And these are, so to speak, only the first circle of the bad effects of a bad solution. With a little care, their branchings can be traced on into nature, into the life of the cities, and into the cultural and economic life of the nation.

The second kind of solution is that which immediately worsens the problem it is intended to solve, causing a hellish symbiosis in which problem and solution reciprocally enlarge one another in a sequence that, so far as its own logic is concerned, is limitless—as when the problem of soil compaction is "solved" by a bigger tractor, which further compacts the soil, which makes a need for a still bigger tractor, and so on and on. There is an identical symbiosis between coal-fired power plants and air conditioners. It is characteristic of such solutions that no one prospers by them but the suppliers of fuel and equipment.

These two kinds of solutions are obviously bad. They always serve one good at the expense of another or of several others, and I believe that if all their effects were ever to be accounted for they would be seen to involve, too frequently if not invariably, a net loss to nature, agriculture, and the human commonwealth.

Such solutions always involve a definition of the problem that is either false or so narrow as to be virtually false. To define an agricultural problem as if it were solely a problem of agriculture—or solely a problem of production or technology or economics—is simply to misunderstand the problem, either inadvertently or deliberately, either for profit or because of a prevalent fashion of thought. The whole problem must be solved, not just some handily identifiable and simplifiable aspect of it.

Both kinds of bad solutions leave their problems unsolved. Bigger tractors do not solve the problem of soil compaction any more than air conditioners solve the problem of air pollution. Nor does the large confinement-feeding operation solve the problem of food production; it is, rather, a way calculated to allow large-scale ambition and greed to profit from food production. The real problem of food production occurs within a complex, mutually influential relationship of soil, plants, animals, and people. A real solution to that problem will therefore be ecologically, agriculturally, and culturally healthful.

Perhaps it is not until health is set down as the aim that we come in sight of the third kind of solution: that which causes a ramifying series of solutions—as when meat animals are fed on the farm where the feed is raised, and where the feed is raised to be fed to the animals that are on the farm. Even so rudimentary a description implies a concern for pattern, for quality, which necessarily complicates the concern for production. The farmer has put plants and animals into a relationship of mutual dependence, and must perforce be concerned for balance or symmetry, a reciprocating connection in the pattern of the farm that is biological, not industrial, and that involves solutions to problems of fertility, soil husbandry, economics, sanitation—the whole complex of problems whose proper solutions add up to *health*: the health of the soil, of plants and animals, of farm and farmer, of farm family and farm community, all involved in the same internested, interlocking pattern—or pattern of patterns.

A bad solution is bad, then, because it acts destructively upon the larger patterns in which it is contained. It acts destructively upon those patterns, most likely, because it is formed in ignorance or disregard of them. A bad solution solves for a single purpose or goal, such as increased production. And it is typical of such solutions that they achieve stupendous increases in production at exorbitant biological and social costs.

A good solution is good because it is in harmony with those larger patterns—and this harmony will, I think, be found to have the nature of analogy. A bad solution acts within the larger pattern the way a disease or addiction acts within the body. A good solution acts within the larger pattern the way a healthy organ acts within the body. But it must at once be understood that a healthy organ does not—as the mechanistic or industrial mind would like to say—"give" health to the body, is not exploited for the body's health, but is *a part* of its health. The health of organ and organism is the same, just as the health of organism and ecosystem is the same. And these structures of organ, organism, and ecosystem—as John Todd has so ably understood—belong to a series of analogical integrities that begins with the organelle and ends with the biosphere.

It would be next to useless, of course, to talk about the possibility of good solutions if none existed in proof and in practice. A part of our work at *The New Farm* has been to locate and understand those farmers whose work is competently responsive to the requirements of health. Representative of these farmers, and among them remarkable for the thoroughness of his intelligence, is Earl F. Spencer, who has a 250-acre dairy farm near Palatine Bridge, New York.

Before 1972, Earl Spencer was following a "conventional" plan which would build his herd to 120 cows. According to this plan, he would eventually buy all the grain he fed, and he was already using as much as 30 tons per year of commercial fertilizer. But in 1972, when he had increased his herd to 70 cows, wet weather reduced his harvest by about half. The choice was clear: he had either to buy half his yearly feed supply, or sell half his herd.

He chose to sell half his herd—a very unconventional choice, which in itself required a lot of independent intelligence. But character and intelligence of an even more respectable order were involved in the next step, which was to understand that the initial decision implied a profound change in the pattern of the farm and of his life and assumptions as a farmer. With his herd now reduced by half, he saw that before the sale he had been overstocked, and had been abusing his land. On his 120 acres of tillable land, he had been growing 60 acres of corn and 60 of alfalfa. On most of his fields, he was growing corn three years in succession. The consequences of this he now saw as symptoms, and saw that they were serious: heavy dependence on purchased supplies, deteriorating soil structure, declining quantities of organic matter, increasing erosion, yield reductions despite continued large applications of fertilizer. In addition, because of his heavy feeding of concentrates, his cows were having serious digestive and other health problems.

He began to ask fundamental questions about the nature of the creatures and the land he was dealing with, and to ask if he could not bring about some sort of balance between their needs and his own. His conclusion was that "to be in balance with nature is to be successful." His farm, he says, had been going in a "dead run"; now he would slow it to a "walk."

From his crucial decision to reduce his herd, then, several other practical measures have followed:

1. A five-year plan (extended to eight years) to phase out entirely his use of purchased fertilizers.
2. A plan, involving construction of a concrete manure pit, to increase and improve his use of manure.
3. Better husbandry of cropland, more frequent rotation, better timing.
4. The gradual reduction of grain in the feed ration, and the concurrent increase of roughage—which has, to date, reduced the dependence on grain by half, from about 6000 pounds per cow to about 3000 pounds.
5. A breeding program which selects "for more efficient roughage conversion."

The most tangible results are that the costs of production have been "dramatically" reduced, and that per cow production has increased by 1500 to 2000 pounds. But the health of the whole farm has improved. There is a moral satisfaction in this, of which Earl Spencer is fully aware. But he is also aware that the satisfaction is not *purely* moral, for the good results are also practical and economic: "We have half the animals we had before and are feeding half as much grain to those remaining, so we now need to plant corn only two years in a row. Less corn means less plowing, less fuel for growing and harvesting, and less wear on the most expensive equipment." Veterinary bills have been reduced also. And in 1981, if the schedule holds, he will buy no commercial fertilizer at all.

From the work of Earl Spencer and other exemplary farmers, and from the understanding of destructive farming practices, it is possible to devise a set of critical standards for agriculture. I am aware that the list of standards which follows must be to some extent provisional, but am nevertheless confident that it will work to distinguish between healthy and unhealthy farms, as well as between the oversimplified minds that solve problems for some X such as profit or quantity of production, and those minds, sufficiently complex, that solve for health or quality or coherence of pattern. To me, the validity of these standards seems inherent in their general applicability. They will serve the making of sewer systems or households as readily as they will serve the making of farms:

1. A good solution accepts given limits, using so far as possible what is at hand. The farther-fetched the solution, the less it should be trusted. Granted that a farm can be too small, it is nevertheless true that enlarging scale is a deceptive solution; it solves one problem by acquiring another or several others.

2. A good solution accepts also the limitation of discipline. Agricultural problems should receive solutions that are agricultural, not technological or economic.

3. A good solution improves the balances, symmetries, or harmonies within a pattern—it is a qualitative solution—rather than enlarging or complicating some part of a pattern at the expense or in neglect of the rest.

4. A good solution solves more than one problem, and it does not make new problems. I am talking about health as opposed to almost any cure, coherence of pattern as opposed to almost any solution produced piecemeal or in isolation. The return of organic wastes to the soil may, at first glance, appear to be a good solution *per se*. But that is not invariably or necessarily true. It is true only if the wastes are returned to the right place at the right time in the pattern of the farm, if the

waste does not contain toxic materials, if the quantity is not too great, and if not too much energy or money is expended in transporting it.

5. A good solution will satisfy a whole range of criteria; it will be good in all respects. A farm that has found correct agricultural solutions to its problems will be fertile, productive, healthful, conservative, beautiful, pleasant to live on. This standard obviously must be qualified to the extent that the pattern of the life of a farm will be adversely affected by distortions in any of the larger patterns that contain it. It is hard, for instance, for the economy of a farm to maintain its health in a national industrial economy in which farm earnings are apt to be low and expenses high. But it is apparently true, even in such an economy, that the farmers most apt to survive are those who do not go too far out of agriculture into either industry or banking—and who, moreover, live like farmers, not like businessmen. This seems especially true for the smaller farmers.

6. A good solution embodies a clear distinction between biological order and mechanical order, between farming and industry. Farmers who fail to make this distinction are ideal customers of the equipment companies, but they often fail to understand that the real strength of a farm is in the soil.

7. Good solutions have wide margins, so that the failure of one solution does not imply the impossibility of another. Industrial agriculture tends to put its eggs into fewer and fewer baskets, and to make "going for broke" its only way of going. But to grow grain should not make it impossible to pasture livestock, and to have a lot of power should not make it impossible to use only a little.

8. A good solution always answers the question, How much is enough? Industrial solutions have always rested on the assumption that enough is all you can get. But that destroys agriculture, as it destroys nature and culture. The good health of a farm implies a limit of scale, because it implies a limit of attention, and because such a limit is invariably implied by any pattern. You destroy a square, for example, by enlarging one angle or lengthening one side. And in any sort of work there is a point past which more quantity necessarily implies less quality. In some kinds of industrial agriculture, such as cash grain farming, it is possible (to borrow an insight from Professor

Timothy Taylor) to think of technology as a substitute for skill. But even in such farming that possibility is illusory; the illusion can be maintained only so long as the consequences can be ignored. The illusion is much shorter lived when animals are included in the farm pattern, because the husbandry of animals is so insistently a human skill. A healthy farm incorporates a pattern that a single human mind can comprehend, make, maintain, vary in response to circumstances, and pay steady attention to. That this limit is obviously variable from one farmer and farm to another does not mean that it does not exist.

9. A good solution should be cheap, and it should not enrich one person by the distress or impoverishment of another. In agriculture, so-called "inputs" are, from a different point of view, outputs—*expenses.* In all things, I think, but especially in an agriculture struggling to survive in an industrial economy, any solution that calls for an expenditure to a manufacturer should be held in suspicion—not rejected necessarily, but *as a rule* mistrusted.

10. Good solutions exist only in proof, and are not to be expected from absentee owners or absentee experts. Problems must be solved in work and in place, with particular knowledge, fidelity, and care, by people who will suffer the consequences of their mistakes. There is no theoretical or ideal *practice.* Practical advice or direction from people who have no practice may have some value, but its value is questionable and is limited. The divisions of capital, management, and labor, characteristic of an industrial system, are therefore utterly alien to the health of farming—as they probably also are to the health of manufacturing. The good health of a farm depends on the farmer's mind; the good health of his mind has its dependence, and its proof, in physical work. The good farmer's mind and his body —his management and his labor—work together as intimately as his heart and his lungs. And the capital of a well-farmed farm by definition includes the farmer, mind and body both. Farmer and farm are one thing, an organism.

11. Once the farmer's mind, his body, and his farm are understood as a single organism, and once it is understood that the question of the endurance of this organism is a question about the sufficiency and integrity of a pattern, then the word *organic* can be usefully admitted into this series of standards.

It is a word that I have been defining all along, though I have not used it. An organic farm, properly speaking, is not one that uses certain methods and substances and avoids others; it is a farm whose structure is formed in imitation of the structure of a natural system; it has the integrity, the independence, and the benign dependence of an organism. Sir Albert Howard said that a good farm is an analogue of the forest which "manures itself." A farm that imports too much fertility, even as feed or manure, is in this sense as inorganic as a farm that exports too much or that imports chemical fertilizer.

12. The introduction of the term *organic* permits me to say more plainly and usefully some things that I have said or implied earlier. In an organism, what is good for one part is good for another. What is good for the mind is good for the body; what is good for the arm is good for the heart. We know that sometimes a part may be sacrificed for the whole; a life may be saved by the amputation of an arm. But we also know that such remedies are desperate, irreversible, and destructive; it is impossible to improve the body by amputation. And such remedies do not imply a safe logic. As *tendencies* they are fatal: you cannot save your arm by the sacrifice of your life.

Perhaps most of us who know local histories of agriculture know of fields that in hard times have been sacrificed to save a farm, and we know that though such a thing is possible it is dangerous. The danger is worse when topsoil is sacrificed for the sake of a crop. And if we understand the farm as an organism, we see that it is impossible to sacrifice the health of the soil to improve the health of plants, or to sacrifice the health of plants to improve the health of animals, or to sacrifice the health of animals to improve the health of people. In a biological pattern—as in the pattern of a community—the exploitive means and motives of industrial economics are immediately destructive and ultimately suicidal.

13. It is the nature of any organic pattern to be contained within a larger one. And so a good solution in one pattern preserves the integrity of the pattern that contains it. A good agricultural solution, for example, would not pollute or erode a watershed. What is good for the water is good for the ground, what is good for the ground is good for plants, what is good for plants is good for animals, what is good for

animals is good for people, what is good for people is good for the air, what is good for the air is good for the water. And vice versa.

14. But we must not forget that those human solutions that we may call organic are not natural. We are talking about organic *artifacts*, organic only by imitation or analogy. Our ability to make such artifacts depends on virtues that are specifically human: accurate memory, observation, insight, imagination, inventiveness, reverence, devotion, fidelity, restraint. Restraint —for us, now—above all: the ability to accept and live within limits; to resist changes that are merely novel or fashionable; to resist greed and pride; to resist the temptation to "solve" problems by ignoring them, accepting them as "trade-offs," or bequeathing them to posterity. A good solution, then, must be in harmony with good character, cultural value, and moral law.

Family Work

FOR THOSE of us who have wished to raise our food and our children at home, it is easy enough to state the ideal. Growing our own food, unlike buying it, is a complex activity, and it affects deeply the shape and value of our lives. We like the thought that the outdoor work that improves our health should produce food of excellent quality that, in turn, also improves and safeguards our health. We like no less the thought that the home production of food can improve the quality of family life. Not only do we intend to give our children better food than we can buy for them at the store, or than they will buy for themselves from vending machines or burger joints, we also know that growing and preparing food at home can provide family work—work for everybody. And by thus elaborating household chores and obligations, we hope to strengthen the bonds of interest, loyalty, affection, and cooperation that keep families together.

Forty years ago, for most of our people, whether they lived in the country or in town, this was less an ideal than a necessity, enforced both by tradition and by need. As is often so, it was only after family life and family work became (allegedly) unnecessary that we began to think of them as "ideals."

As ideals, they are threatened; as they have become (even allegedly) unnecessary, they have become by the same token less possible. I do not mean to imply that *I* think the ideal is any less valuable than it ever was, or that it is—in reality, in the long run—less necessary. Nor do I think that less possible means impossible.

I do think that the ideal is more difficult now than it was. We are trying to uphold it now mainly by will, without much help from necessity, and with no help at all from custom or public value. For most people now do seem to think that family life and family work are unnecessary, and this thought has been institutionalized in our economy and in our public values. Never before has private life been so preyed upon by public life. How can we preserve family life—if by that we mean, as I think we must, *home* life—when our attention is so forcibly drawn away from home?

We know the causes well enough.

Automobiles and several decades of supposedly cheap fuel have put longer and longer distances between home and work, household and daily needs.

TV and other media have learned to suggest with increasing subtlety and callousness—especially, and most wickedly, to children—that it is better to consume than to produce, to buy than to grow or to make, to "go out" than to stay home. If you have a TV, your children will be subjected almost from the cradle to an overwhelming insinuation that all worth experiencing is somewhere else and that all worth having must be bought. The purpose is blatantly to supplant the joy and beauty of health with cosmetics, clothes, cars, and ready-made desserts. There is clearly too narrow a limit on how much money can be made from health, but the profitability of disease—especially disease of spirit or character—has so far, for profiteers, no visible limit.

Another cause, and one that seems particularly regrettable, is public education. The idea that the public should be educated is altogether salutary, and since we insist on making this education compulsory we ought, in reason, to reconcile ourselves to the likelihood that it will be mainly poor. I am not nearly so much concerned about its quality as I am about its *length.* My impression is that the chief, if unadmitted, purpose of the school system is to keep children away from home as much as possible. Parents want their children kept out of their hair; education is merely a by-product, not overly prized. In many places, thanks to school consolidation, two hours or more of travel time have been added to the school day. For my own children the regular school day from the first grade—counting from the time they went to catch the bus until they came home—was nine hours. An extracurricular activity would lengthen the day to eleven hours or more. This is not education, but a form of incarceration. Why should anyone be surprised if, under these circumstances, children should become "disruptive" or even "ineducable"?

If public education is to have any meaning or value at all,

then public education *must* be supplemented by home education. I know this from my own experience as a college teacher. What can you teach a student whose entire education has been public, whose daily family life for twenty years has consisted of four or five hours of TV, who has never read a book for pleasure or even *seen* a book so read; whose only work has been schoolwork, who has never learned to perform any essential task? Not much, so far as I could tell.

We can see clearly enough at least a couple of solutions.

We can get rid of the television set. As soon as we see that the TV cord is a vacuum line, pumping life and meaning out of the household, we can unplug it. What a grand and neglected privilege it is to be shed of the glibness, the gleeful idiocy, the idiotic gravity, the unctuous or lubricious greed of those public faces and voices!

And we can try to make our homes centers of attention and interest. Getting rid of the TV, we understand, is not just a practical act, but also a symbolical one: we thus turn our backs on the invitation to consume; we shut out the racket of consumption. The ensuing silence is an invitation to our homes, to our own places and lives, to come into being. And we begin to recognize a truth disguised or denied by TV and all that it speaks and stands for: no life and no place is destitute; all have possibilities of productivity and pleasure, rest and work, solitude and conviviality that belong particularly to themselves. These possibilities exist everywhere, in the country or in the city, it makes no difference. All that is necessary is the time and the inner quietness to look for them, the sense to recognize them, and the grace to welcome them. They are now most often lived out in home gardens and kitchens, libraries, and workrooms. But they are beginning to be worked out, too, in little parks, in vacant lots, in neighborhood streets. Where we live is also a place where our interest and our effort can be. But they can't be there by the means and modes of consumption. If we consume nothing but what we buy, we are living in "the economy," in "television land," not at home. It is productivity that rights the balance, and brings us home. Any way at all of joining and using the air and light and weather of your own place—even if it is only a window box, even if it is only an opened window—is a making and a having that you cannot get from TV or government or school.

That local productivity, however small, is a gift. If we are parents we cannot help but see it as a gift to our children—and the *best* of gifts. How will it be received?

Well, not ideally. Sometimes it will be received gratefully enough. But sometimes indifferently, and sometimes resentfully.

According to my observation, one of the likeliest results of a wholesome diet of home-raised, home-cooked food is a heightened relish for cokes and hot dogs. And if you "deprive" your children of TV at home, they are going to watch it with something like rapture away from home. And obligations, jobs, and chores at home will almost certainly cause your child to wish, sometimes at least, to be somewhere else, watching TV.

Because, of course, parents are not the only ones raising their children. They are being raised also by their schools and by their friends and by the parents of their friends. Some of this outside raising is good, some is not. It is, anyhow, unavoidable.

What this means, I think, is about what it has always meant. Children, no matter how nurtured at home, must be risked to the world. And parenthood is not an exact science, but a vexed privilege and a blessed trial, absolutely necessary and not altogether possible.

If your children spurn your healthful meals in favor of those concocted by some reincarnation of Col. Sanders, Long John Silver, or the Royal Family of Burger; if they flee from books to a friend's house to watch TV, if your old-fashioned notions and ways embarrass them in front of their friends—does that mean you are a failure?

It may. And what parent has not considered that possibility? I know, at least, that I have considered it—and have wailed and gnashed my teeth, found fault, laid blame, preached and ranted. In weaker moments, I have even blamed myself.

But I have thought, too, that the term of human judgment is longer than parenthood, that the upbringing we give our children is not just for their childhood but for all their lives. And it is surely the *duty* of the older generation to be embarrassingly old-fashioned, for the claims of the "newness" of any younger generation are mostly frivolous. The young are born to the human condition more than to their time, and they face mainly the same trials and obligations as their elders have faced.

The real failure is to give in. If we make our house a household instead of a motel, provide healthy nourishment for mind and body, enforce moral distinctions and restraints, teach essential skills and disciplines and require their use, there is no certainty that we are providing our children a "better life" that they will embrace wholeheartedly during childhood. But we are providing them a choice that they may make intelligently as adults.

A Few Words for Motherhood

IT IS the season of motherhood again, and we are preoccupied with the pregnant and the unborn. When birth is imminent, especially with a ewe or a mare, we are at the barn the last thing before we go to bed, at least once in the middle of the night, and well before daylight in the morning. It is a sort of joke here that we have almost never had anything born in the middle of the night. And yet somebody must get up and go out anyway. With motherhood, you don't argue probabilities.

I set the alarm, but always wake up before it goes off. Some part of the mind is given to the barn, these times, and you can't put it to sleep. For a few minutes after I wake up, I lie there wondering where I will get the will and the energy to drag myself out of bed again. Anxiety takes care of that: maybe the ewe has started into labor, and is in trouble. But it isn't just anxiety. It is curiosity too, and the eagerness for new life that goes with motherhood. I want to see what nature and breeding and care and the passage of time have led to. If I open the barn door and hear a little bleat coming out of the darkness, I will be glad to be awake. My liking for that always returns with a force that surprises me.

These are bad times for motherhood—a kind of biological drudgery, some say, using up women who could do better things. Thoreau may have been the first to assert that people should not belong to farm animals, but the idea is now established doctrine with many farmers—and it has received amendments to the effect that people should not belong to children, or to each other. But we all have to belong to something, if only to the idea that we should not belong to anything. We all have to be used up by something. And though I will never be a mother, I am glad to be used up by motherhood and what it leads to, just as—most of the time—I gladly belong to my wife, my children, and several head of cattle, sheep, and horses. What better way to be used up? How else to be a farmer?

There are good arguments against female animals that need help in giving birth; I know what they are, and have gone over them many times. And yet—if the ordeal is not too painful or too long, and if it succeeds—I always wind up a little grateful to the ones that need help. Then I get to take part, get to go through the process another time, and I invariably come away from it feeling instructed and awed and pleased.

My wife and son and I find the heifer in a far corner of the field. In maybe two hours of labor she has managed to give birth to one small foot. We know how it has been with her. Time and again she has lain down and heaved at her burden, and got up and turned and smelled the ground. She is a heifer —how does she know that something is supposed to *be* there?

It takes some doing even for the three of us to get her into the barn. Her orders are to be alone, and she does all in her power to obey. But finally we shut the door behind her and get her into a stall. She isn't wild; once she is confined it isn't even necessary to tie her. I wash in a bucket of icy water and soap my right hand and forearm. She is quiet now. And so are we humans—worried, and excited too, for if there is a chance for failure here, there is also a chance for success.

I loop a bale string onto the calf's exposed foot, knot the string short around a stick which my son then holds. I press my hand gently into the birth canal until I find the second foot and then, a little further on, a nose. I loop a string around the second foot, fasten on another stick for a handhold. And then we pull. The heifer stands and pulls against us for a few seconds, then gives up and goes down. We brace ourselves the best we can into our work, pulling as the heifer pushes. Finally the head comes, and then, more easily, the rest.

We clear the calf's nose, help him to breathe, and then, because the heifer has not yet stood up, we lay him on the bedding in front of her. And what always seems to me the miracle of it begins. She has never calved before. If she ever saw another cow calve, she paid little attention. She has, as we humans say, no education and no experience. And yet she recognizes the calf as her own, and knows what to do for it. Some heifers don't, but most do, as this one does. Even before she gets up, she begins to lick it about the nose and face with loud, vigorous swipes of her tongue. And all the while she utters a kind of moan, meant to comfort, encourage, and reassure—or so I understand it.

How does she know so much? How did all this come about? Instinct. Evolution. I know those words. I understand the logic of the survival of the fittest: good mothering instincts have survived because bad mothers lost their calves: the good traits triumphed, the bad perished. But how come some are fit in the first place? What prepared in the mind of the first cow or ewe or mare—or, for that matter, in the mind of the first human mother—this intricate, careful, passionate welcome to the newborn? I don't know. I don't think anybody does. I distrust any mortal who claims to know. We call these animals dumb brutes, and so far as we can tell they are more or less dumb, and there are certainly times when those of us who live with them will seem to find evidence that they are plenty stupid. And yet, they are indisputably allied with intelligence more articulate and more refined than is to be found in any obstetrics textbook. What is one to make of it? Here is a dumb brute lying in dung and straw, licking her calf, and as always I am feeling honored to be associated with her.

The heifer has stood up now, and the calf is trying to stand, wobbling up onto its hind feet and knees, only to be knocked over by an exuberant caress of its mother's tongue. We have involved ourselves too much in this story by now to leave before the end, but we have our chores to finish too, and so to hasten things I lend a hand.

I help the calf onto his feet and maneuver him over to the heifer's flank. I am not supposed to be there, but her calf is, and so she accepts, or at least permits, my help. In these situations it sometimes seems to me that animals know that help is needed, and that they accept it with some kind of understanding. The thought moves me, but I am never sure, any more than I am sure what the cow means by the low moans she makes as the calf at last begins to nurse. To me, they sound like praise and encouragement—but how would I know?

Always when I hear that little smacking as the calf takes hold of the tit and swallows its first milk, I feel a pressure of laughter under my ribs. I am not sure what that means either. It certainly affirms more than the saved money value of the calf and the continued availability of beef. We all three feel it. We look at each other and grin with relief and satisfaction. Life is on its legs again, and we exult.

A Talent for Necessity

IN THE DAYS when the Southdown ram was king of the sheep pastures and the show-ring, Henry Besuden of Vinewood Farm in Clark County, Kentucky, was perhaps the premier breeder and showman of Southdown sheep in the United States. The list of his winnings at major shows would be too long to put down here, but the character of his achievement can be indicated by his success in showing carload lots of fat lambs in the Chicago International Livestock Exposition. Starting in 1946, he sent eighteen carloads to the International, and won the competition twelve times. "I had 'em fat," he says, remembering. "I had 'em good." Such was the esteem and demand for his stock among fellow breeders that in 1954 he sold a yearling ram for $1200, then a record price for a Southdown.

One would imagine that such accomplishments must have rested on the very best of Bluegrass farmland. But the truth, nearly opposite to that, is much more interesting. "If I'd inherited good land," Henry Besuden says, "I'd probably have been just another Bluegrass farmer."

What he inherited, in fact, was 632 acres of rolling land, fairly steep in places, thin soiled even originally, and by the time he got it, worn out, "corned to death." His grandfather would rent the land out to corn, two hundred acres at a time, and not even get up to see where it would be planted—even though "it was understood to be the rule that renters ruined land." By the time Henry Besuden was eight years old both his mother and father were dead, and the land was farmed by tenants under the trusteeship of a Cincinnati bank. When the farm came to him in 1927, it was heavily encumbered by debt and covered with gullies, some of which were deep enough to hide a standing man.

And so Mr. Besuden began his life as a farmer with the odds against him. But his predicament became his education and, finally, his triumph. "I was lucky," he told Grant Cannon of *The Farm Quarterly* in 1951. "I found that I had some talent for doing the things I *had* to do. I *had* to improve the farm or starve to death; and I *had* to go into the sheep business because sheep were the only animals that could have lived off the farm."

Now seventy-six years old and not in the best of health, Mr. Besuden has not owned a sheep for several years, but he speaks of them with exact remembrance and exacting intelligence; he is one of the best talkers I have had the luck to listen to. How did he get started with sheep? "I was told they'd eat weeds and briars," he says, looking sideways through pipesmoke to see if I get the connection, for the connection between sheep and land is the critical one for him. The history of his sheep and the history of his farm are one history, and it is his own.

Having only talent and necessity—and unusual energy and determination—Mr. Besuden set about the restoration of his ravaged fields. There was no Soil Conservation Service then, but a young man in his predicament was bound to get plenty of advice. To check erosion he first tried building rock dams across the gullies. That wasn't satisfactory; the dams did catch some dirt, but then the fields were marred by half buried rock walls that interfered with work. He tried huge windrows of weeds and brush to the same purpose, but that was not satisfactory either.

Some of the worst gullies he eventually had to fill with a bulldozer. But his main erosion-stopping tool turned out, strangely enough, to be the plow; the tool that in the wrong hands had nearly ruined the farm, in the right hands healed it. Starting at the edge of a gulley he would run a backfurrow up one side and down the other, continuing to plow until he had completed a sizable land. And then he would start at the gulley again, turning the furrows inward as before. He repeated this process until what had been a ditch had become a saucer, so that the runoff, rather than concentrating its force in an abrasive torrent, would be shallowly dispersed over as wide an area as possible. This, as he knew, had been the method of the renters to prepare the gullied land for yet another crop of corn. For them, it had been a temporary remedy; he made it a permanent one.

Nowadays Kentucky fescue 31 would be the grass to sow on such places, but fescue was not available then. Mr. Besuden used small grains, timothy, sweet clover, Korean lespedeza. He used mulches, and he did not overlook the usefulness of what he knew for certain would grow on his land—weeds: "Briars are a good thing for a little hollow." In places he planted thickets of black locust—a native leguminous tree that would serve four purposes: hold the land, encourage grass to grow, provide shade for livestock, and produce posts. But his highest praise is given to the sweet clover which he calls "the best land builder I've ever run into. It'll open up clay, and throw a lot of nitrogen into the ground." The grass would come then, and the real healing would start.

Once the land was in grass, his policy generally was to leave it in grass. Only the best-laying, least vulnerable land was broken for tobacco, the region's major money crop then as now. Even today, I noticed, he sees that his fields are plowed very conservatively. The plowlands are small and carefully placed, leaving out thin places and waterways.

The basic work of restoration continued for twenty-three years. By 1950 the scars were grassed over, and the land was supporting one of the great Southdown flocks of the time. But it was not healed. What was there is gone, and Henry Besuden knows that it will be a long time building back. "'Tain't in good shape, yet," he told an interviewer in 1978.

And so if Mr. Besuden built a reputation as one of the best of livestock showmen, the focus of his interest was nevertheless not the show-ring but the farm. It would be true, it seems, to say that he became a master sheepman and shepherd as one of the ways of becoming a master farmer. For this reason, his standards of quality were never frivolous or freakish, as show-ring standards have sometimes been accused of being, but insistently practical. He never forgot that the purpose of a sheep is to produce a living for the farmer and to put good meat on the table: "When they asked me, 'What do you consider a perfect lamb?' I said, 'One a farmer can make money on!' The foundation has to be the commercial flock." And he wrote in praise of the Southdown ram that "he paid his rent."

But it was perhaps even more characteristic of him to write in 1945 that "one very important thing is that sheep are land builders," and to plead for their continued inclusion in farm livestock programs. He had seen the handwriting on the wall: the new emphasis on row cropping and "production" which in the years after World War II would radically alter the balance of crops and animals on farms, and which, as he feared, would help to destroy the sheep business in his own state. (In 1947, Mr. Besuden's county of Clark had twenty-four breeding flocks of Southdowns, and 30,000 head of grade ewes. That is more than remain now in the whole state of Kentucky.) What he called for instead—and events are rapidly proving him right—was "a long-time program of land building" by which he meant a way of farming based on grass and forage crops, which would build up and maintain reserves of fertility. And in that kind of farming, he was prepared to insist, because he knew, sheep would have an important place.

"I think," he wrote in his series of columns, "Sheep Sense," published in *The Sheepman* in 1945 and 1946, "the fertilizing effect of sheep on the farm has never received the attention it deserves. As one who has had to farm poor land where the least amount of fertilizer shows up plainly, I have noticed that on land often thought too poor for cattle the sheep do well and in time benefit the crops and grass to such an extent that other stock can then be carried. I have seldom seen sheep bed down for the night on anything but high land, and their droppings are evenly scattered on the pasture while grazing, so that no vegetation is killed."

What he wanted was "a way of farming compatible with nature"; this was the constant theme of his work, and he followed it faithfully, both in his pleasure in the lives and events of nature, and in his practical solutions to the problems of farming and soil husbandry. He was never too busy to appreciate, and to praise, the spiritual by-products, as he called them, of farm life. Nor was he too busy to attend to the smallest needs of his land. At one time, for example, he built "two small houses on skids," each of which would hold twenty-five bales of hay. These could be pulled to places where the soil was thin, where the hay would be fed out, and then moved on to other such places. (In the spring they could be used to raise chickens.)

"It's good to have Nature working for you," he says. "She works for a minimum wage." But in reading his "Sheep Sense" columns, one realizes that he not only did not separate the spiritual from the practical, but insisted that they cannot be separated: "This thing of soil conservation involves more than laying out a few terraces and diversion ditches and sowing to grass and legumes, it also involves the heart of the man managing the land. If he loves his soil he will save it." Once, he says, he thought of numbering his fields, but decided against it—"That didn't seem fair to them"—for each has its own character and potential.

As a rule, he would have 400 head of ewes in two flocks—a flock of registered Southdowns and a flock of "Western" commercial ewes. After lambing, he would be running something in the neighborhood of 1000 head. To handle so many sheep on a diversified farm required a great deal of care, and Mr. Besuden's system of management, worked out with thorough understanding and attention to detail, is worth the interest and reflection of any raiser of livestock.

It was a system intended, first of all, to get the maximum use of forage. This rested on what he understands to be a sound principle of livestock farming and soil conservation, but it was forced upon him by the poor quality of his land. He had to keep row cropping to a minimum, and if that meant buying grain, then he would buy it. But he did not buy much. He usually fed, he told me, one pound of corn per ewe per day for sixty days. But in "Sheep Sense" for December 1945, he wrote: "One-half pound grain with three pounds legume hay should do the job, starting with the hay and adding the grain later." He creep-fed his early lambs, but took them off grain as soon as pasture was available. In "Sheep Sense," March 1946, he stated flatly that "creep-feeding after good grass arrives does not pay."

Grain, then, he considered not a diet, but a supplement, almost an emergency ration, to assure health and growth in the flock during the time when he had no pasture. It must be remembered that he was talking about a kind of sheep bred to make efficient use of pasture and hay, and that the market then favored that kind. In the decades following World War II, cheap energy and cheap grain allowed interest to shift to the larger breeds of sheep and larger slaughter lambs that must be grain fed. But now with the cost of energy rising, pushing up the cost of grain, and the human consumption of grain rising with the increase of population, Henry Besuden's sentence of a generation ago resounds with good sense: "Due to the shortage of grain throughout the world, the sheep farmer needs to study the possibilities of grass fattening."

Those, anyhow, were the possibilities that *he* was studying. And the management of pasture, the management of sheep *on* pasture, was his art.

In the fall he would select certain pastures close to the barn to be used for late grazing. This is what is now called "stock-piling"—which, he points out, is only a new word for old common sense. It was sometimes possible, in favorable years, to keep the ewes on grass all through December, feeding "very little hay" and "a small amount of grain." Sometimes he sowed rye early to provide late fall pasture and so extend the grazing season.

His ewes were bred to lamb in January and February. He fed good clover or alfalfa hay, and from about the middle of January to about the middle of March he gave the ewes their sixty daily rations of grain. In mid March the grain feeding ended, and ewes and lambs went out on early pasture of rye which had been sown as a cover crop on the last year's tobacco patches. "A sack of Balboa rye sown in the early fall," he wrote, "is worth several sacks of feed fed in the spring and is much cheaper." From the rye they went to the clover fields where tobacco had grown two years before. From the clover they were moved onto the grass pastures. The market lambs were sold straight off the pastures, at eighty to eighty-five pounds, starting the first of May.

After fescue became available, Mr. Besuden made extensive use of it in his pastures. But he feels that this grass, though an excellent land conserver, is not nutritious or palatable enough to make the best sheep pasture, and so he took pains to diversify his fescue stands with timothy and legumes. His favorite pasture legume is Korean lespedeza, though he joins in the fairly common complaint that it is less vigorous and productive now than it used to be. He has also used red clover, alsike, ladino, and birdsfoot trefoil. He says that he had trouble getting his ewes with lamb in the first heat when they were bred on clover pastures, but that he never had this trouble on lespedeza.

His pastures were regularly reseeded to legumes, usually in March, the sheep tramping in the seed, and he found this method of "renovation" to be as good as any. The pastures were clipped twice during the growing season, sometimes oftener, to keep the growth vigorous and uniform.

The key to efficient management of sheep on pasture is paying attention, and it was important to Mr. Besuden that he should be on horseback among his sheep in the early mornings. The sheep would be out of the shade then, grazing, and he could study their condition and the condition of the field. He speaks of the "bloom" of a pasture, referring to a certain freshness of appearance made by new, tender growth sprigging up through the old. When that bloom is gone, he thinks, the sheep should be moved. The move from a stale pasture to a fresh one can lengthen the grazing time by as much as two hours a day. He believes also that lambs do best when the flock is not too large. That is because sheep tend to bunch together when grazing, the least vigorous lambs coming last and having to feed on grass mouthed over and rejected by the others. He saw to it that his pastures were amply provided with shade, and he knew that the shade needed to be well placed: "I think the best lamb-growing pastures I have are the ones where the shade is close to the water. I have seen times during July and August when sheep would not leave the shade and go to water if the shade and water happened to be at opposite ends of a large field."

The crisis of the shepherd's year, of course, is lambing time. That is the time that the year's work stands or falls by. And because it usually takes place in cold weather, the success of lambing is almost as dependent on the shepherd's facilities as on his knowledge. The lambing barn at Vinewood is an instructive embodiment of Mr. Besuden's understanding of his work and his gift for order. He gives a good description of it himself in one of his columns:

"Practically all the lambing here at Vinewood in recent years has been in a barn especially made for the purpose, shiplap (tongue groove) boxing with a low loft and a window in each bent. The east end of the barn [away from the prevailing winds] is rarely ever closed, a gate being used. Often in extremely cold weather the temperature can be raised fifteen or twenty degrees by the heat from the sheep. Some thirty feet out in the front and extending the width of the barn [is] a heavy layer of rock. . . . This prevents the muddy place that often appears at the barn door and . . . pulls at the sheep as they walk through it, causing slipped lambs. Also at the entrance . . . a locust post is half embedded across the door. This serves as a protection in case of dogs trying to dig under the door or gate and helps to hold the bedding in the barn as the sheep go out. Any kind of a sill that is too high or causes the heavy-in-lamb ewes to jump or strain to cross is too risky."

The barn is admirably laid out, with pens, chutes, and gates to permit the feeding, handling, sorting, and loading of a large number of sheep with the least trouble. There were lambing pens for forty ewes. There was also a small room with pens that could be heated by a stove. Above each pen was a red wooden "button" that could be turned down to indicate that a ewe was near to lambing or for any other reason in need of close attention. These were used when Mr. Besuden had an experienced helper to share the nighttime duty with him. "They saved a lot of cold midnight talk," he says.

But experienced help was not always available, and then he would have to work through the days and nights of lambing alone. Staying awake would get to be a problem. Sometimes, sitting beside one of the pens, waiting for a ewe to lamb, he would tie a string from one of her hind legs to his wrist. When her labor pains came and she began to shift around, she would tug the string and he would wake up and tend to her.

And so the talent for what he "had to do" was in large measure the ability to bear the good outcome in mind: to envision, in spite of rocks and gullies, the good health of the fields; to foresee in the pregnant ewes and the advancing seasons a good crop of lambs. And it was the ability to carry in his head for nearly half a century the ideal character and pattern of the Southdown, and to measure his animals relentlessly against it—an ability, rare enough, that marked him as a master stockman.

He told me a story that suggests very well the distinction and the effect of that ability. On one of his trips to the International he competed against a western sheepman who had selected his carload of fifty fat lambs out of ten thousand head.

After the Vinewood carload had won the class, this gentleman came up and asked: "How many did you pick yours from, Mr. Besuden?"

"About seventy-five."

"Well," the western breeder said, "I guess it's better to have the right seventy-five than the wrong ten thousand."

But the ability to recognize the right seventy-five is worthless by itself. Just as necessary is the ability to do the work and to pay attention. To pay attention, above all—that is another of the persistent themes of Mr. Besuden's talk and of his life. He is convinced that paying attention pays, and this sets him apart from the mechanized "modern" farmers who are pushed to accept more responsibility than they can properly meet, and to work at freeway speeds. He wrote in his column of the importance of "little things done on time." He said that they paid, but he knew that people did them for more than pay.

He told me also about a farmer who wouldn't scrape the manure off his shoes until he came to a spot that was bare of grass. "That's what I mean," he said. "You have to keep it on your mind."

Seven Amish Farms

IN TYPICAL Midwestern farming country the distances between inhabited houses are stretching out as bigger farmers buy out their smaller neighbors in order to "stay in." The signs of this "movement" and its consequent specialization are everywhere: good houses standing empty, going to ruin; good stock barns going to ruin; pasture fences fallen down or gone; machines too large for available doorways left in the weather; windbreaks and woodlots gone down before the bulldozers; small schoolhouses and churches deserted or filled with grain.

In the latter part of March this country shows little life. Field after field lies under the dead stalks of last year's corn and soybeans, or lies broken for the next crop; one may drive many miles between fields that are either sodded or planted in winter grain. If the weather is wet, the country will seem virtually deserted. If the ground is dry enough to support their wheels, there will be tractors at work, huge machines with glassed cabs, rolling into the distances of fields larger than whole farms used to be, as solitary as seaborne ships.

The difference between such country and the Amish farmlands in northeast Indiana seems almost as great as that between a desert and an oasis. And it is the *same* difference. In the Amish country there is a great deal more life: more natural life, more agricultural life, more human life. Because the farms are small—most of them containing well under a hundred acres—the Amish neighborhoods are more thickly populated than most rural areas, and you see more people at work. And because the Amish are diversified farmers, their plowed croplands are interspersed with pastures and hayfields and often with woodlots. It is a varied, interesting, healthy looking farm country, pleasant to drive through. When we were there, on the twentieth and twenty-first of last March, the spring plowing had just started, and so you could still see everywhere the annual covering of stable manure on the fields, and the teams of Belgians or Percherons still coming out from the barns with loaded spreaders.

Our host, those days, was William J. Yoder, a widely respected breeder of Belgian horses, an able farmer and carpenter, and a most generous and enjoyable companion. He is a vigorous man, strenuously involved in the work of his farm and in the life of his family and community. From the look of him and the look of his place, you know that he has not just done a lot of work in his time, but has done it well, learned from it, mastered the necessary disciplines. He speaks with heavy stress on certain words—the emphasis of conviction, but also of pleasure, for he enjoys the talk that goes on among people interested in horses and in farming. But unlike many people who enjoy talking, he speaks with care. Bill was born in this community, has lived there all his life, and he has grandchildren who will probably live there all their lives. He belongs there, then, root and branch, and he knows the history and the quality of many of the farms. On the two days, we visited farms belonging to Bill himself, four of his sons, and two of his sons-in-law.

The Amish farms tend to divide up between established ones, which are prosperous looking and well maintained, and run-down, abused, or neglected ones, on which young farmers are getting started. Young Amish farmers *are* still getting started, in spite of inflation, speculators' prices, and usurious interest rates. My impression is that the proportion of young farmers buying farms is significantly greater among the Amish than among conventional farmers.

Bill Yoder's own eighty-acre farm is among the established ones. I had been there in the fall of 1975 and had not forgotten its aspect of cleanness and good order, its well-kept white buildings, neat lawns, and garden plots. Bill has owned the place for twenty-six years. Before he bought it, it had been rented and row cropped, with the usual result: it was nearly played out. "The buildings," he says, "were nothing," and there were no fences. The first year, the place produced five loads (maybe five tons) of hay, "and that was mostly sorrel." The only healthy plants on it were the spurts of grass and clover that grew out of the previous year's manure piles. The corn crop that first year "might have been thirty bushels an acre," all nubbins. The sandy soil blew in every strong wind, and when he plowed the fields his horses' feet sank into "quicksand potholes" that the share uncovered.

The remedy has been a set of farming practices traditional among the Amish since the seventeenth century: diversification, rotation of crops, use of manure, seeding of legumes. These practices began when the Anabaptist sects were disfranchised in their European homelands and forced to the use of poor soil. We saw them still working to restore farmed-out soils in Indiana. One thing these practices do is build humus in the soil, and humus does several things: increases fertility, improves soil structure, improves both water-holding capacity and drainage. "No humus, you're in trouble," Bill says.

After his rotations were established and the land had begun to be properly manured, the potholes disappeared, and the soil quit blowing. "There's something in it now—there's some substance there." Now the farm produces abundant crops of corn, oats, wheat, and alfalfa. Oats now yield 90–100 bushels per acre. The corn averages 100–125 bushels per acre, and the ears are long, thick, and well filled.

Bill's rotation begins and ends with alfalfa. Every fall he puts in a new seeding of alfalfa with his wheat; every spring he plows down an old stand of alfalfa, "no matter how good it is." From alfalfa he goes to corn for two years, planting thirty acres, twenty-five for ear corn and five for silage. After the second year of corn, he sows oats in the spring, wheat and alfalfa in the fall. In the fourth year the wheat is harvested; the alfalfa then comes on and remains through the fifth and sixth years. Two cuttings of alfalfa are taken each year. After curing in the field, the hay is hauled to the barn, chopped, and blown into the loft. The third cutting is pastured.

Unlike cow manure, which is heavy and chunky, horse manure is light and breaks up well coming out of the spreader; it interferes less with the growth of small seedlings and is less likely to be picked up by a hay rake. On Bill's place, horse manure is used on the fall seedings of wheat and alfalfa, on the young alfalfa after the wheat harvest, and both years on the established alfalfa stands. The cow manure goes on the corn ground both years. He usually has about 350 eighty-bushel spreader loads of manure, and each year he covers the whole farm—cropland, hayland, and pasture.

With such an abundance of manure there obviously is no *dependence* on chemical fertilizers, but Bill uses some as a "starter" on his corn and oats. On corn he applies 125 pounds of nitrogen in the row. On oats he uses 200–250 pounds of 16-16-16, 20-20-20, or 24-24-24. He routinely spreads two tons of lime to the acre on the ground being prepared for wheat.

His out-of-pocket costs per acre of corn last year were as follows:

Seed (planted at a rate of seven acres per bushel)	$ 7.00
Fertilizer	7.75
Herbicide (custom applied, first year only)	16.40

That comes to a total of $31.15 per acre—or, if the corn makes only a hundred bushels per acre, a little over $0.31 per bushel. In the second year his per acre cost is $14.75, less than $0.15 per bushel, bringing the two-year average to $22.95 per acre or about $0.23 per bushel.

The herbicide is used because, extra horses being on the farm during the winter, Bill has to buy eighty to a hundred tons of hay, and in that way brings in weed seed. He had no weed problem until he started buying hay. Even though he uses the herbicide, he still cultivates his corn three times.

His cost per acre of oats came to $33.00 ($12.00 for seed and $21.00 for fertilizer)—or, at ninety bushels per acre, about $0.37 per bushel.

Of Bill's eighty acres, sixty-two are tillable. He has ten acres of permanent pasture, and seven or eight of woodland, which produced the lumber for all the building he has done on the place. In addition, for $500 a year he rents an adjoining eighty acres of "hill and woods pasture" which provides summer grazing for twenty heifers; and on another neighboring farm he rents varying amounts of cropland.

All the field work is done with horses, and this, of course, comes virtually free—a by-product of the horse-breeding enterprise. Bill has an ancient Model D John Deere tractor that he uses for belt power.

At the time of our visit, there were twenty-two head of horses on the place. But that number was unusually low, for Bill aims to keep "around thirty head." He has a band of excellent brood mares and three stallions, plus young stock of assorted ages. Since October 1 of last year, he had sold eighteen head of registered Belgian horses. In the winters he operates a "urine line," collecting "pregnant mare urine," which is sold to a pharmaceutical company for the extraction of various hormones. For this purpose he boards a good many mares belonging to neighbors; that is why he must buy the extra hay that causes his weed problem. (Horses are so numerous on this farm because they are one of its money-making enterprises. If horses were used only for work on this farm, four good geldings would be enough.)

One bad result of the dramatic rise in draft horse prices over the last eight or ten years is that it has tended to focus attention on such characteristics as size and color to the neglect of less obvious qualities such as good feet. To me, foot quality seems a critical issue. A good horse with bad feet is good for nothing but decoration, and at sales and shows there are far too many flawed feet disguised by plastic wood and black shoe polish. And so I was pleased to see that every horse on Bill Yoder's place had sound, strong-walled, correctly shaped feet. They were good horses all around, but their other qualities were well founded; they stood on good feet, and this speaks of the thoroughness of his judgment and also of his honesty.

Though he is a master horseman, and the draft horse business is more lucrative now than ever in its history, Bill does not specialize in horses, and that is perhaps the clearest indication of his integrity as a farmer. Whatever may be the dependability of the horse economy, on this farm it rests upon a diversified agricultural economy that is sound.

He was milking five Holstein cows; he had fifteen Holstein heifers that he had raised to sell; and he had just marketed thirty finished hogs, which is the number that he usually has on fence. All the animals had been well wintered—Bill quotes his father approvingly: "Well wintered is half summered"—and were in excellent condition. Another saying of his father's that Bill likes to quote—"Keep the horses on the side of the fence the feed is on"—has obviously been obeyed here. The feeding is careful, the feed is good, and it is abundant. Though it was almost spring, there were ample surpluses in the hayloft and in the corn cribs.

Other signs of the farm's good health were three sizable garden plots, and newly pruned grapevines and raspberry canes. The gardener of the family is Mrs. Yoder. Though most of the children are now gone from home, Bill says that she still grows as much garden stuff as she ever did.

All seven of the Yoders' sons live in the community. Floyd, the youngest, is still at home. Harley has a house on nearly three acres, works in town, and returns in the afternoons to his own shop where he works as a farrier. Henry, who also works in town, lives with Harley and his wife. The other four sons are now settled on farms that they are in the process of paying for. Richard has eighty acres, Orla eighty, Mel fifty-seven, and Wilbur eighty. Two sons-in-law also living in the community are Perry Bontrager, who owns ninety-five acres, and Ervin Mast, who owns sixty-five. Counting Bill's eighty acres, the seven families are living on 537 acres. Of the seven farms, only Mel's is entirely tillable, the acreages in woods or permanent pasture varying from five to twenty-six.

These young men have all taken over run-down farms, on which they are establishing rotations and soil husbandry practices that, being traditional, more or less resemble Bill's. It seemed generally agreed that after three years of this treatment the land would grow corn, as Perry Bontrager said, "like anywhere else."

These are good farmers, capable of the intelligent planning, sound judgment, and hard work that good farming requires. Abused land heals and flourishes in their care. None of them expressed a wish to own more land; all, I believe, feel that what they have will be enough—when it is paid for. The big problems are high land prices and high interest rates, the latter apparently being the worst.

The answer, for Bill's sons so far, has been town work. All of them, after leaving home, have worked for Redman Industries, a manufacturer of mobile homes in Topeka. They do piecework, starting at seven in the morning and quitting at two in the afternoon, using the rest of the day for farming or other work. This, Bill thinks, is now "the only way" to get started farming. Even so, there is "a lot of debt" in the community —"more than ever."

With a start in factory work, with family help, with government and bank loans, with extraordinary industry and perseverance, with highly developed farming skills, it is still possible for young Amish families to own a small farm that will eventually support them. But there is more strain in that effort now than there used to be, and more than there should be. When the burden of usurious interest becomes too great, these young men are finding it necessary to make temporary returns to their town jobs.

The only one who spoke of his income was Mel, who owns fifty-seven acres, which, he says, *will be* enough. He and his family milk six Holsteins. He had nine mares on the urine line last winter, seven of which belonged to him. And he had twelve brood sows. Last year his gross income was $43,000. Of this, $12,000 came from hogs, $7,000 from his milk cows, the rest from his horses and the sale of his wheat. After his production costs, but *before* payment of interest, he netted $22,000. In order to cope with the interest payments, Mel was preparing to return to work in town.

These little Amish farms thus become the measure both of "conventional" American agriculture and of the cultural meaning of the national industrial economy.

To begin with, these farms give the lie direct to that false god of "agribusiness": the so-called economy of scale. The small farm is not an anachronism, is not unproductive, is not unprofitable. Among the Amish, it is still thriving, and is still the economic foundation of what John A. Hostetler (in *Amish Society*, third edition) rightly calls "a healthy culture." Though they do not produce the "record-breaking yields" so touted by the "agribusiness" establishment, these farms are nevertheless highly productive. And if they are not likely to make their owners rich (never an Amish goal), they can certainly be said to be sufficiently profitable. The economy of scale has helped corporations and banks, not farmers and farm communities. It has been an economy of dispossession and waste—plutocratic, if not in aim, then certainly in result.

What these Amish farms suggest, on the contrary, is that in farming there is inevitably a scale that is suitable both to the productive capacity of the land and to the abilities of the farmer; and that agricultural problems are to be properly solved, not in expansion, but in management, diversity, balance, order, responsible maintenance, good character, and in the sensible limitation of investment and overhead. (Bill makes a careful distinction between "healthy" and "unhealthy" debt, a "healthy debt" being "one you can hope to pay off in a reasonable way.")

Most significant, perhaps, is that while conventional agriculture, blindly following the tendency of any industry to exhaust its sources, has made soil erosion a national catastrophe, these Amish farms conserve the land and improve it in use.

And what is one to think of a national economy that drives such obviously able and valuable farmers to factory work? What value does such an economy impose upon thrift, effort, skill, good husbandry, family and community health?

In spite of the unrelenting destructiveness of the larger economy, the Amish—as Hostetler points out with acknowledged surprise and respect—have almost doubled in population in the last twenty years. The doubling of a population is, of course, no significant achievement. What is significant is that these agricultural communities have doubled their population *and yet remained agricultural communities* during a time when conventional farmers have failed by the millions. This alone would seem to call for a careful look at Amish ways of farming. That those ways have, during the same time, been ignored by the colleges and the agencies of agriculture must rank as a prime intellectual wonder.

Amish farming has been so ignored, I think, because it involves a complicated structure that is at once biological and cultural, rather than industrial or economic. I suspect that anyone who might attempt an accounting of the economy of an Amish farm would soon find himself dealing with virtually unaccountable values, expenses, and benefits. He would be dealing with biological forces and processes not always measurable, with spiritual and community values not quantifiable; at certain points he would be dealing with mysteries—and he would be finding that these unaccountables and inscrutables have results, among others, that are economic. Hardly an appropriate study for the "science" of agricultural economics.

The economy of conventional agriculture or "agribusiness"

is remarkable for the simplicity of its arithmetic. It involves a manipulation of quantities that are all entirely accountable. List your costs (land, equipment, fuel, fertilizer, pesticides, herbicides, wages), add them up, subtract them from your earnings, or subtract your earnings from them, and you have the result.

Suppose, on the other hand, that you have an eighty-acre farm that is not a "food factory" but your home, your given portion of Creation which you are morally and spiritually obliged "to dress and to keep." Suppose you farm, not for wealth, but to maintain the integrity and the practical supports of your family and community. Suppose that, the farm being small enough, you farm it with family work and work exchanged with neighbors. Suppose you have six Belgian brood mares that you use for field work. Suppose that you also have milk cows and hogs, and that you raise a variety of grain and hay crops in rotation. What happens to your accounting then?

To start with, several of the costs of conventional farming are greatly diminished or done away with. Equipment, fertilizer, chemicals all cost much less. Fuel becomes feed, but you have the mares and are feeding them anyway; the work ration for a brood mare is not a lot more costly than a maintenance ration. And the horses, like the rest of the livestock, are making manure. Figure that in, and figure, if you can, the value of the difference between manure and chemical fertilizer. You can probably get an estimate of the value of the nitrogen fixed by your alfalfa, but how will you quantify the value to the soil of its residues and deep roots? Try to compute the value of humus in the soil—in improved drainage, improved drought resistance, improved tilth, improved health. Wages, if you pay your children, will still be among your costs. But compute the difference between paying your children and paying "labor." Work exchanged with neighbors can be reduced to "man-hours" and assigned a dollar value. But compute the difference between a neighbor and "labor." Compute the value of a family or a community to any one of its members. We may, as we must, grant that among the values of family and community there is economic value—but what is it?

In the Louisville *Courier-Journal* of April 5, 1981, the Mobil Oil Corporation ran an advertisement which was yet another celebration of "scientific agriculture." American farming, the

Mobil people are of course happy to say, "requires *more petroleum products than almost any other industry.* A gallon of gasoline to produce a single bushel of corn, for example. . . ." This, they say, enables "each American farmer to feed sixty-seven people." And they say that this is "a-maizing."

Well, it certainly is! And the chances are good that an agriculture totally dependent on the petroleum industry is not yet as amazing as it is going to be. But one thing that is already sufficiently amazing is that a bushel of corn produced by the burning of one gallon of gasoline has already cost more than *six times* as much as a bushel of corn grown by Bill Yoder. How does Bill Yoder escape what may justly be called the petroleum tax on agriculture? He does so by a series of substitutions: of horses for tractors, of feed for fuel, of manure for fertilizer, of sound agricultural methods and patterns for the exploitive methods and patterns of industry. But he has done more than that—or, rather, he and his people and their tradition have done more. They have substituted themselves, their families, and their communities for petroleum. The Amish use little petroleum—and need little—because they have those other things.

I do not think that we can make sense of Amish farming until we see it, until we become willing to see it, as belonging essentially to the Amish practice of Christianity, which instructs that one's neighbors are to be loved as oneself. To farmers who give priority to the maintenance of their community, the economy of scale (that is, the economy of *large* scale, of "growth") can make no sense, for it requires the ruination and displacement of neighbors. A farm cannot be increased except by the decrease of a neighborhood. What the interest of the community proposes is invariably an economy of *proper* scale. A whole set of agricultural proprieties must be observed: of farm size, of methods, of tools, of energy sources, of plant and animal species. Community interest also requires charity, neighborliness, the care and instruction of the young, respect for the old; thus it assures its integrity and survival. Above all, it requires good stewardship of the land, for the community, as the Amish have always understood, is no better than its land. "If treated violently or exploited selfishly," John Hostetler writes, the land "will yield poorly." There could be no better statement of the meaning of the *practice* and the practicality of charity. Except to the insane narrow-mindedness of industrial economics, selfishness does not pay.

The Amish have steadfastly subordinated economic value to the values of religion and community. What is too readily overlooked by a secular, exploitive society is that their ways of doing this are not "empty gestures" and are not "backward." In the first place, these ways have kept the communities intact through many varieties of hard times. In the second place, they conserve the land. In the third place, they yield economic benefits. The community, the religious fellowship, has many kinds of value, and among them is economic value. It is the result of the practice of neighborliness, and of the practice of stewardship. What moved me most, what I liked best, in those days we spent with Bill Yoder was the sense of the continuity of the community in his dealings with his children and in their dealings with their children.

Bill has helped his sons financially so far as he has been able. He has helped them with his work. He has helped them by sharing what he has—lending a stallion, say, at breeding time, or lending a team. And he helps them by buying good pieces of equipment that come up for sale. "If he ever gets any money," he says of one of the boys, for whom he has bought an implement, "he'll pay me for it. If he don't, he'll just use it." He has been their teacher, and he remains their advisor. But he does not stand before them as a domineering patriarch or "authority figure." He seems to speak, rather, as a representative of family and community experience. In their respect for him, his sons respect their tradition. They are glad for his help, advice, and example, but there is nothing servile in this. It seems to be given and taken in a kind of familial friendship, respect going both ways.

Everywhere we went, when school was not in session, the children were at the barns, helping with the work, watching, listening, learning to farm in the way it is best learned. Wilbur told us that his eleven-year-old son had cultivated twenty-three acres of corn last year with a team and a riding cultivator. That reminded Bill of the way he taught Wilbur to do the same job.

Wilbur was little then, and he loved to sit in his father's lap and drive the team while Bill worked the cultivator. If Wilbur could drive, Bill thought, he could do the rest of it. So he got off and shortened the stirrups so the boy could reach them with his feet. Wilbur started the team, and within a few steps began plowing up the corn.

"Whoa!" he said.

And Bill, who was walking behind him, said, "Come up!"

And it went that way for a little bit:

"Whoa!"

"Come up!"

And then Wilbur started to cry, and Bill said:

"Don't cry! Go ahead!"

The Gift of Good Land

"Dream not of other Worlds . . ."
Paradise Lost VIII, 175

My PURPOSE here is double. I want, first, to attempt a Biblical argument for ecological and agricultural responsibility. Second, I want to examine some of the practical implications of such an argument. I am prompted to the first of these tasks partly because of its importance in our unresolved conflict about how we should use the world. That those who affirm the divinity of the Creator should come to the rescue of His creature is a logical consistency of great potential force.

The second task is obviously related to the first, but my motive here is somewhat more personal. I wish to deal directly at last with my own long held belief that Christianity, as usually presented by its organizations, is not *earthly* enough—that a valid spiritual life, in this world, must have a practice and a practicality—it must have a material result. (I am well aware that in this belief I am not alone.) What I shall be working toward is some sort of practical understanding of what Arthur O. Lovejoy called the "this-worldly" aspect of Biblical thought. I want to see if there is not at least implicit in the Judeo-Christian heritage a doctrine such as that the Buddhists call "right livelihood" or "right occupation."

Some of the reluctance to make a forthright Biblical argument against the industrial rape of the natural world seems to come from the suspicion that this rape originates with the Bible, that Christianity cannot cure what, in effect, it has caused. Judging from conversations I have had, the best known spokesman for this view is Professor Lynn White, Jr., whose essay, "The Historical Roots of Our Ecologic Crisis" has been widely published.

Professor White asserts that it is a "Christian axiom that nature has no reason for existence save to serve man." He seems to base his argument on one Biblical passage, Genesis 1:28, in which Adam and Eve are instructed to "subdue" the earth. "Man," says Professor White, "named all the animals, thus establishing his dominance over them." There is no doubt that Adam's superiority over the rest of Creation was represented, if not established, by this act of naming; he *was* given dominance. But that this dominance was meant to be tyrannical, or that "subdue" meant to destroy, is by no means a necessary inference. Indeed, it might be argued that the correct understanding of this "dominance" is given in Genesis 2:15, which says that Adam and Eve were put into the Garden "to dress it and to keep it."

But these early verses of Genesis can give us only limited help. The instruction in Genesis 1:28 was, after all, given to Adam and Eve in the time of their innocence, and it seems certain that the word "subdue" would have had a different intent and sense for them at that time than it could have for them, or for us, after the Fall.

It is tempting to quarrel at length with various statements in Professor White's essay, but he has made that unnecessary by giving us two sentences that define both his problem and my task. He writes, first, that "God planned all of this [the Creation] explicitly for man's benefit and rule: no item in the physical creation had any purpose save to serve man's purposes." And a few sentences later he says: "Christianity . . . insisted that it is God's will that man exploit nature for his *proper* ends" [My emphasis].

It is certainly possible that there might be a critical difference between "man's purposes" and "man's *proper* ends." And one's belief or disbelief in that difference, and one's seriousness about the issue of propriety, will tell a great deal about one's understanding of the Judeo-Christian tradition.

I do not mean to imply that I see no involvement between that tradition and the abuse of nature. I know very well that Christians have not only been often indifferent to such abuse, but have often condoned it and often perpetrated it. That is not the issue. The issue is whether or not the Bible explicitly or implicitly defines a *proper* human use of Creation or the natural world. Proper use, as opposed to improper use, or abuse, is a matter of great complexity, and to find it adequately treated it is necessary to turn to a more complex story than that of Adam and Eve.

The story of the giving of the Promised Land to the Israelites is more serviceable than the story of the giving of the

Garden of Eden, because the Promised Land is a divine gift to a *fallen* people. For that reason the giving is more problematical, and the receiving is more conditional and more difficult. In the Bible's long working out of the understanding of this gift, we may find the beginning—and, by implication, the end—of the definition of an ecological discipline.

The effort to make sense of this story involves considerable difficulty because the tribes of Israel, though they see the Promised Land as a gift to them from God, are also obliged to take it by force from its established inhabitants. And so a lot of the "divine sanction" by which they act sounds like the sort of rationalization that invariably accompanies nationalistic aggression and theft. It is impossible to ignore the similarities to the westward movement of the American frontier. The Israelites were following their own doctrine of "manifest destiny," which for them, as for us, disallowed any human standing to their opponents. In Canaan, as in America, the conquerors acted upon the broadest possible definition of idolatry and the narrowest possible definition of justice. They conquered with the same ferocity and with the same genocidal intent.

But for all these similarities, there is a significant difference. Whereas the greed and violence of the American frontier produced an ethic of greed and violence that justified American industrialization, the ferocity of the conquest of Canaan was accompanied from the beginning by the working out of an ethical system antithetical to it—and antithetical, for that matter, to the American conquest with which I have compared it. The difficulty but also the wonder of the story of the Promised Land is that, there, the primordial and still continuing dark story of human rapaciousness begins to be accompanied by a vein of light which, however improbably and uncertainly, still accompanies us. This light originates in the idea of the land as a gift—not a free or a deserved gift, but a gift given upon certain rigorous conditions.

It is a gift because the people who are to possess it did not create it. It is accompanied by careful warnings and demonstrations of the folly of saying that "My power and the might of mine hand hath gotten me this wealth" (Deuteronomy 8:17). Thus, deeply implicated in the very definition of this gift is a specific warning against *hubris* which is the great ecological sin, just as it is the great sin of politics. People are not gods.

They must not act like gods or assume godly authority. If they do, terrible retributions are in store. In this warning we have the root of the idea of propriety, of *proper* human purposes and ends. We must not use the world as though we created it ourselves.

The Promised Land is not a permanent gift. It is "given," but only for a time, and only for so long as it is properly used. It is stated unequivocally, and repeated again and again, that "the heaven and the heaven of heavens is the Lord's thy God, the earth also, with all that therein is" (Deuteronomy 10:14). What is given is not ownership, but a sort of tenancy, the right of habitation and use: "The land shall not be sold forever: for the land is mine; for ye are strangers and sojourners with me" (Leviticus 25:23).

In token of His landlordship, God required a sabbath for the land, which was to be left fallow every seventh year; and a sabbath of sabbaths every fiftieth year, a "year of jubilee," during which not only would the fields lie fallow, but the land would be returned to its original owners, as if to free it of the taint of trade and the conceit of human ownership. But beyond their agricultural and social intent, these sabbaths ritualize an observance of the limits of "my power and the might of mine hand"—the limits of human control. Looking at their fallowed fields, the people are to be reminded that the land is theirs only by gift; it exists in its own right, and does not begin or end with any human purpose.

The Promised Land, moreover, is "a land which the Lord thy God careth for: the eyes of the Lord thy God are always upon it" (Deuteronomy 11:12). And this care promises a re-possession by the true landlord, and a fulfillment not in the power of its human inhabitants: "as truly as I live, all the earth shall be filled with the glory of the Lord" (Numbers 14:21)—a promise recalled by St. Paul in Romans 8:21: "the creature [the Creation] itself also shall be delivered from the bondage of corruption into the glorious liberty of the children of God."

Finally, and most difficult, the good land is not given as a reward. It is made clear that the people chosen for this gift do not deserve it, for they are "a stiff-necked people" who have been wicked and faithless. To such a people such a gift can be given only as a moral predicament: having failed to deserve it beforehand, they must prove worthy of it afterwards; they must use it well, or they will not continue long in it.

How are they to prove worthy?

First of all, they must be faithful, grateful, and humble; they must remember that the land is a gift: "When thou hast eaten and art full, then thou shalt bless the Lord thy God for the good land which he hath given thee." (Deuteronomy 8:10).

Second, they must be neighborly. They must be just, kind to one another, generous to strangers, honest in trading, etc. These are social virtues, but, as they invariably do, they have ecological and agricultural implications. For the land is described as an "inheritance"; the community is understood to exist not just in space, but also in time. One lives in the neighborhood, not just of those who now live "next door," but of the dead who have bequeathed the land to the living, and of the unborn to whom the living will in turn bequeath it. But we can have no direct behavioral connection to those who are not yet alive. The only neighborly thing we can do for them is to preserve their inheritance: we must take care, among other things, of the land, which is never a possession, but an inheritance to the living, as it will be to the unborn.

And so the third thing the possessors of the land must do to be worthy of it is to practice good husbandry. The story of the Promised Land has a good deal to say on this subject, and yet its account is rather fragmentary. We must depend heavily on implication. For sake of brevity, let us consider just two verses (Deuteronomy 22:6–7):

> If a bird's nest chance to be before thee in the way in any tree, or on the ground, whether they be young ones, or eggs, and the dam sitting upon the young, or upon the eggs, thou shalt not take the dam with the young:
> But thou shalt in any wise let the dam go, and take the young to thee; that it may be well with thee, and that thou mayest prolong thy days.

This, obviously, is a perfect paradigm of ecological and agricultural discipline, in which the idea of inheritance is necessarily paramount. The inflexible rule is that the source must be preserved. You may take the young, but you must save the breeding stock. You may eat the harvest, but you must save seed, and you must preserve the fertility of the fields.

What we are talking about is an elaborate understanding of charity. It is so elaborate because of the perception, implicit here, explicit in the New Testament, that charity by its nature cannot be selective—that it is, so to speak, out of human control. It cannot be selective because between any two humans, or any two creatures, all Creation exists as a bond. Charity cannot be just human, any more than it can be just Jewish or just Samaritan. Once begun, wherever it begins, it cannot stop until it includes all Creation, for all creatures are parts of a whole upon which each is dependent, and it is a contradiction to love your neighbor and despise the great inheritance on which his life depends. Charity even for one person does not make sense except in terms of an effort to love all Creation in response to the Creator's love for it.

And how is this charity answerable to "man's purposes"? It is not, any more than the Creation itself is. Professor White's contention that the Bible proposes any such thing is, so far as I can see, simply wrong. It is not allowable to love the Creation according to the purposes one has for it, any more than it is allowable to love one's neighbor in order to borrow his tools. The wild ass and the unicorn are said in the Book of Job (39:5–12) to be "free," precisely in the sense that they are not subject or serviceable to human purposes. The same point —though it is not the main point of that passage—is made in the Sermon on the Mount in reference to "the fowls of the air" and "the lilies of the field." Faced with this problem in Book VIII of *Paradise Lost*, Milton scrupulously observes the same reticence. Adam asks about "celestial Motions," and Raphael refuses to explain, making the ultimate mysteriousness of Creation a test of intellectual propriety and humility:

> . . . for the Heav'n's wide Circuit, let it speak
> The Maker's high magnificence, who built
> So spacious, and his Line stretcht out so far;
> That Man may know he dwells not in his own;
> An Edifice too large for him to fill,
> Lodg'd in a small partition, and the rest
> Ordain'd for uses to his Lord best known.
>
> *(lines 100–106)*

The Creator's love for the Creation is mysterious precisely because it does not conform to human purposes. The wild ass and the wild lilies are loved by God for their own sake and yet they are part of a pattern that we must love because it includes us. This is a pattern that humans can understand well enough to respect and preserve, though they cannot "control" it or hope to understand it completely. The mysterious and the practical, the Heavenly and the earthly, are thus joined. Charity is a theological virtue and is prompted, no doubt, by a theological emotion, but it is also a practical virtue because it must be practiced. The requirements of this complex charity cannot be fulfilled by smiling in abstract beneficence on our neighbors and on the scenery. It must come to acts, which must come from skills. Real charity calls for the study of agriculture, soil husbandry, engineering, architecture, mining, manufacturing, transportation, the making of monuments and pictures, songs and stories. It calls not just for skills but for the study and criticism of skills, because in all of them a choice must be made: they can be used either charitably or uncharitably.

How can you love your neighbor if you don't know how to build or mend a fence, how to keep your filth out of his water supply and your poison out of his air; or if you do not produce anything and so have nothing to offer, or do not take care of yourself and so become a burden? How can you be a neighbor without *applying* principle—without bringing virtue to a practical issue? How will you practice virtue without skill?

The ability to be good is not the ability to do nothing. It is not negative or passive. It is the ability to do something well —to do good work for good reasons. In order to be good you have to know how—and this knowing is vast, complex, humble and humbling; it is of the mind and of the hands, of neither alone.

The divine mandate to use the world justly and charitably, then, defines every person's moral predicament as that of a steward. But this predicament is hopeless and meaningless unless it produces an appropriate discipline: stewardship. And stewardship is hopeless and meaningless unless it involves long-term courage, perseverance, devotion, and skill. This skill is not to be confused with any accomplishment or grace of spirit or of intellect. It has to do with everyday proprieties in the practical use and care of created things—with "right livelihood."

If "the earth is the Lord's" and we are His stewards, then obviously some livelihoods are "right" and some are not. Is there, for instance, any such thing as a Christian strip mine? A Christian atomic bomb? A Christian nuclear power plant or radioactive waste dump? What might be the design of a Christian transportation or sewer system? Does not Christianity imply limitations on the scale of technology, architecture, and land holding? Is it Christian to profit or otherwise benefit from violence? Is there not, in Christian ethics, an implied requirement of practical separation from a destructive or wasteful economy? Do not Christian values require the enactment of a distinction between an organization and a community?

It is impossible to understand, much less to answer, such questions except in reference to issues of practical skill, because they all have to do with distinctions between kinds of action. These questions, moreover, are intransigently personal, for they ask, ultimately, how each livelihood and each life will be taken from the world, and what each will cost in terms of the livelihoods and lives of others. Organizations and even communities cannot hope to answer such questions until individuals have begun to answer them.

But here we must acknowledge one inadequacy of Judeo-Christian tradition. At least in its most prominent and best known examples, this tradition does not provide us with a precise enough understanding of the commonplace issues of livelihood. There are two reasons for this.

One is the "otherworldly philosophy" that, according to Lovejoy, "has, in one form or another, been the dominant official philosophy of the larger part of civilized mankind through most of its history. . . . The greater number of the subtler speculative minds and of the great religious teachers have . . . been engaged in weaning man's thought or his affections, or both, from . . . Nature." The connection here is plain.

The second reason is that the Judeo-Christian tradition as we have it in its art and literature, including the Bible, is so strongly heroic. The poets and storytellers in this tradition have tended to be interested in the extraordinary actions of "great men"—actions unique in grandeur, such as may occur only once in the history of the world. These extraordinary actions do indeed bear a universal significance, but they cannot very well serve as examples of ordinary behavior. Ordinary behavior belongs to a different dramatic mode, a different understanding of action, even a different understanding of virtue. The drama of heroism raises above all the issue of physical and moral courage: Does the hero have, in extreme circumstances, the courage to obey —to perform the task, the sacrifice, the resistance, the pilgrimage that he is called on to perform? The drama of ordinary or daily behavior also raises the issue of courage, but it raises at the same time the issue of skill; and, because ordinary behavior lasts so much longer than heroic action, it raises in a more complex and difficult way the issue of perseverance. It may, in some ways, be easier to be Samson than to be a good husband or wife day after day for fifty years.

These heroic works are meant to be (among other things) instructive and inspiring to ordinary people in ordinary life, and they are, grandly and deeply so. But there are two issues that they are prohibited by their nature from raising: the issue of life-long devotion and perseverance in unheroic tasks, and the issue of good workmanship or "right livelihood."

It can be argued, I believe, that until fairly recently there was simply no need for attention to such issues, for there existed yeoman or peasant or artisan classes: these were the people who did the work of feeding and clothing and housing, and who were responsible for the necessary skills, disciplines, and restraints. As long as those earth-keeping classes and their traditions were strong, there was at least the hope that the world would be well used. But probably the most revolutionary accomplishment of the industrial revolution was to destroy the traditional livelihoods and so break down the cultural lineage of those classes.

The industrial revolution has held in contempt not only the "obsolete skills" of those classes, but the concern for quality, for responsible workmanship and good work, that supported their skills. For the principle of good work it substituted a secularized version of the heroic tradition: the ambition to be a "pioneer" of science or technology, to make a "breakthrough" that will "save the world" from some "crisis" (which now is usually the result of some previous "breakthrough").

The best example we have of this kind of hero, I am afraid, is the fallen Satan of *Paradise Lost*—Milton undoubtedly having observed in his time the prototypes of industrial heroism. This is a hero who instigates and influences the actions of others, but does not act himself. His heroism is of the mind only—escaped as far as possible, not only from divine rule, from its place in the order of creation or the Chain of Being, but also from the influence of material creation:

> A mind not to be chang'd by Place or Time.
> The mind is its own place, and in itself
> Can make a Heav'n of Hell, a Hell of Heav'n.
> *(Book I, lines 253–255)*

This would-be heroism is guilty of two evils that are prerequisite to its very identity: *hubris* and abstraction. The industrial hero supposes that "mine own *mind* hath saved me"—and moreover that it may save the world. Implicit in this is the assumption that one's mind is one's own, and that it may choose its own place in the order of things; one usurps divine authority, and thus, in classic style, becomes the author of results that one can neither foresee nor control.

And because this mind is understood only as a cause, its primary works are necessarily abstract. We should remind ourselves that materialism in the sense of the love of material things is not in itself an evil. As C. S. Lewis pointed out, God too loves material things; He invented them. The Devil's work is abstraction—not the love of material things, but the love of their quantities—which, of course, is why "David's heart smote him after that he had numbered the people" (II Samuel 24:10). It is not the lover of material things but the abstractionist who defends long-term damage for short-term gain, or who calculates the "acceptability" of industrial damage to ecological or human health, or who counts dead bodies on the battlefield. The true lover of material things does not think in this way, but is answerable instead to the paradox of the parable of the lost sheep: that each is more precious than all.

But perhaps we cannot understand this secular heroic mind until we understand its opposite: the mind obedient and in place. And for that we can look again at Raphael's warning in Book VIII of *Paradise Lost*:

> . . . apt the Mind or Fancy is to rove
> Uncheckt, and of her roving is no end;
> Till warn'd, or by experience taught, she learn

That not to know at large of things remote
From use, obscure and subtle, but to know
That which before us lies in daily life,
Is the prime Wisdom; what is more, is fume,
Or emptiness, or fond impertinence,
And renders us in things that most concern
Unpractic'd, unprepar'd, and still to seek.
Therefore from this high pitch let us descend
A lower flight, and speak of things at hand
Useful . . .

(lines 188–200)

In its immediate sense this is a warning against thought that is theoretical or speculative (and therefore abstract), but in its broader sense it is a warning against disobedience—the eating of the forbidden fruit, an act of *hubris*, which Satan justifies by a compellingly reasonable theory and which Eve undertakes as a speculation.

A typical example of the conduct of industrial heroism is to be found in the present rush of experts to "solve the problem of world hunger"—which is rarely defined except as a "world problem" known, in industrial heroic jargon, as "the world food problematique." As is characteristic of industrial heroism, the professed intention here is entirely salutary: nobody should starve. The trouble is that "world hunger" is not a problem that can be solved by a "world solution." Except in a very limited sense, it is not an industrial problem, and industrial attempts to solve it—such as the "Green Revolution" and "Food for Peace"—have often had grotesque and destructive results. "The problem of world hunger" cannot be solved until it is understood and dealt with by local people as a multitude of local problems of ecology, agriculture, and culture.

The most necessary thing in agriculture, for instance, is not to invent new technologies or methods, not to achieve "breakthroughs," but to determine what tools and methods are appropriate to specific people, places, and needs, and to apply them correctly. Application (which the heroic approach ignores) is the crux, because no two farms or farmers are alike; no two fields are alike. Just the changing shape or topography of the land makes for differences of the most formidable kind.

Abstractions never cross these boundaries without either ceasing to be abstractions or doing damage. And prefabricated industrial methods and technologies *are* abstractions. The bigger and more expensive, the more heroic, they are, the harder they are to apply considerately and conservingly.

Application is the most important work, but also the most modest, complex, difficult, and long—and so it goes against the grain of industrial heroism. It destroys forever the notions that the world can be thought of (by humans) as a whole and that humans can "save" it as a whole—notions we can well do without, for they prevent us from understanding our problems and from growing up.

To use knowledge and tools in a particular place with good long-term results is not heroic. It is not a grand action visible for a long distance or a long time. It is a small action, but more complex and difficult, more skillful and responsible, more whole and enduring, than most grand actions. It comes of a willingness to devote oneself to work that perhaps only the eye of Heaven will see in its full intricacy and excellence. Perhaps the real work, like real prayer and real charity, must be done in secret.

The great study of stewardship, then, is "to know / That which before us lies in daily life" and to be practiced and prepared "in things that most concern." The angel is talking about good work, which is to talk about skill. In the loss of skill we lose stewardship; in losing stewardship we lose fellowship; we become outcasts from the great neighborhood of Creation. It is possible—as our experience in *this* good land shows—to exile ourselves from Creation, and to ally ourselves with the principle of destruction—which is, ultimately, the principle of nonentity. It is to be willing *in general* for beings to not-be. And once we have allied ourselves with that principle, we are foolish to think that we can control the results. The "regulation" of abominations is a modern governmental exercise that never succeeds. If we are willing to pollute the air—to harm the elegant creature known as the atmosphere—by that token we are willing to harm all creatures that breathe, ourselves and our children among them. There is no begging off or "trading off." You cannot affirm the power plant and condemn the smokestack, or affirm the smoke and condemn the cough.

That is not to suggest that we can live harmlessly, or strictly at our own expense; we depend upon other creatures and survive by their deaths. To live, we must daily break the body and shed the blood of Creation. When we do this knowingly, lovingly, skillfully, reverently, it is a sacrament. When we do it ignorantly, greedily, clumsily, destructively, it is a desecration. In such desecration we condemn ourselves to spiritual and moral loneliness, and others to want.

FROM

STANDING BY WORDS

(1983)

Standing by Words

> "He said, and stood . . ."
> *Paradise Regained*, IV, 561.

TWO EPIDEMIC illnesses of our time—upon both of which virtual industries of cures have been founded—are the disintegration of communities and the disintegration of persons. That these two are related (that private loneliness, for instance, will necessarily accompany public confusion) is clear enough. And I take for granted that most people have explored in themselves and their surroundings some of the intricacies of the practical causes and effects; most of us, for example, have understood that the results are usually bad when people act in social or moral isolation, and also when, because of such isolation, they fail to act.

What seems not so well understood, because not so much examined, is the relation between these disintegrations and the disintegration of language. My impression is that we have seen, for perhaps a hundred and fifty years, a gradual increase in language that is either meaningless or destructive of meaning. And I believe that this increasing unreliability of language parallels the increasing disintegration, over the same period, of persons and communities.

My concern is for the *accountability* of language—hence, for the accountability of the users of language. To deal with this matter I will use a pair of economic concepts: *internal accounting*, which considers costs and benefits in reference only to the interest of the money-making enterprise itself; and *external accounting*, which considers the costs and benefits to the "larger community." By altering the application of these terms a little, any statement may be said to account well or poorly for what is going on inside the speaker, or outside him, or both.

It will be found, I believe, that the accounting will be poor—incomprehensible or unreliable—if it attempts to be purely internal or purely external. One of the primary obligations of language is to connect and balance the two kinds of accounting.

And so, in trying to understand the degeneracy of language, it is necessary to examine, not one kind of unaccountability, but two complementary kinds. There is language that is diminished by subjectivity, which ends in meaninglessness. But that kind of language rarely exists alone (or so I believe), but is accompanied, in a complex relationship of both cause and effect, by a language diminished by objectivity, or so-called objectivity (inordinate or irresponsible ambition), which ends in confusion.

My standpoint here is defined by the assumption that no statement is complete or comprehensible in itself, that in order for a statement to be complete and comprehensible three conditions are required:

1. It must designate its object precisely.

2. Its speaker must stand by it: must believe it, be accountable for it, be willing to act on it.

3. This relation of speaker, word, and object must be conventional; the community must know what it is.

These are still the common assumptions of private conversations. In our ordinary dealings with each other, we take for granted that we cannot understand what is said if we cannot assume the accountability of the speaker, the accuracy of his speech, and mutual agreement on the structures of language and the meanings of words. We assume, in short, that language is communal, and that its purpose is to tell the truth.

That these common assumptions are becoming increasingly uncommon, particularly in the discourse of specialists of various sorts, is readily evident to anyone looking for evidence. How far they have passed from favor among specialists of language, to use the handiest example, is probably implicit in the existence of such specialists; one could hardly become a language specialist (a "scientist" of language) so long as one adhered to the old assumptions.

But the influence of these specialists is, of course, not confined to the boundaries of their specializations. They write textbooks for people who are not specialists of language, but who are apt to become specialists of other kinds. The general purpose of at least some of these specialists, and its conformability to the purposes of specialists of other kinds, is readily suggested by a couple of recently published textbooks for freshman English.

One of these, *The Contemporary Writer*, by W. Ross Winterowd, contains a chapter on language, the main purpose of which is to convince the student of the illegitimate tyranny of any kind of prescriptive grammar and of the absurdity of judging language "on the basis of extra-linguistic considerations." This chapter proposes four rules that completely overturn all the old common assumptions:

1. "Languages apparently do not become better or worse in any sense. They simply change."

2. "Language is arbitrary."

3. "Rightness and wrongness are determined . . . by the purpose for which the language is being used, by the audience at which it is directed, and by the situation in which the use is taking place."

4. ". . . a grammar of a language is a description of that language, nothing more and nothing less."

And these rules have a pair of corollaries that Mr. Winterowd states plainly. One is that "you [the freshman student] have a more or less complete mastery of the English language. . . ." The other is that art—specifically, here, the literary art—is "the highest expression of the human need to play, of the desire to escape from the world of reality into the world of fantasy."

The second of these texts, *Rhetoric: Discovery and Change*, by Richard E. Young, Alton L. Becker, and Kenneth L. Pike, takes the standardless functionalism of Mr. Winterowd's understanding of language and applies it to the use of language. "The ethical dimension of the art of rhetoric," these authors say, is in "the attempt to reduce another's sense of threat in the effort to reach the goal of cooperation and mutual benefit. . . ." They distinguish between evaluative writing and descriptive writing, preferring the latter because evaluative writing tends to cause people "to become defensive," whereas "a description . . . does not make judgments. . . ." When, however, a writer "must make judgments, he can make them in a way that minimizes the reader's sense of threat." Among other things, "he can acknowledge the personal element in his judgment. . . . There is a subtle but important difference between saying 'I don't like it' and 'It's bad.'"

The authors equate evaluation—functionally, at least—with dogmatism: "The problem with dogmatism is that, like evaluation, it forces the reader to take sides." And finally they recommend a variety of writing which they call "provisional" because it "focuses on the process of enquiry itself and acknowledges the tentative nature of conclusions. . . . Provisional writing implies that more than one reasonable conclusion is possible."

The first of these books attempts to make the study of language an "objective" science by eliminating from that study all extra-linguistic values and the issue of quality. Mr. Winterowd asserts that "the language grows according to its own dynamics." He does not say, apparently because he does not believe, that its dynamics include the influence of the best practice. There is no "best." Anyone who speaks English is a "master" of the language. And the writers once acknowledged as masters of English are removed from "the world of reality" to the "world of fantasy," where they lose their force within the dynamics of the growth of language. Their works are reduced to the feckless status of "experiences": "we are much more interested in the imaginative statement of the message . . . than we are in the message. . . ." Mr. Winterowd's linguistic "science" thus views language as an organism that has evolved without reference to habitat. Its growth has been "arbitrary," without any principle of selectivity.

Against Mr. Winterowd's definition of literature, it will be instructive to place a definition by Gary Snyder, who says of poetry that it is "a tool, a net or trap to catch and present; a sharp edge; a medicine, or the little awl that unties knots." It will be quickly observed that this sentence enormously complicates Mr. Winterowd's simplistic statement-message dichotomy. What Mr. Winterowd means by "message" is an "idea" written in the dullest possible prose. His book is glib, and glibness is an inescapable doom of language without standards. One of the great practical uses of literary disciplines, of course, is to resist glibness—to slow language down and make it thoughtful. This accounts for the influence of verse, in its formal aspect, within the dynamics of the growth of language: verse checks the merely impulsive flow of speech, subjects it to another pulse, to measure, to extra-linguistic considerations; by inducing the hesitations of difficulty, it admits into language the influence of the Muse and of musing.

The three authors of the second book attempt to found an ethics of rhetoric on the idea expressed in one of Mr. Winterowd's rules: "Rightness and wrongness are determined" by purpose, audience, and situation. This idea apparently derives from, though it significantly reduces, the ancient artistic concern for propriety or decorum. A part of this concern was indeed the fittingness of the work to its occasion: that is, one would not write an elegy in the meter of a drinking song—though that is putting it too plainly, for the sense of occasion exercised an influence both broad and subtle on form, diction, syntax, small points of grammar and prosody—everything. But occasion, as I understand it, was invariably second in importance to the subject. It is only the modern specialist who departs from this. The specialist poet, for instance, degrades the subject to "subject matter" or raw material, so that the subject exists for the poem's sake, is *subjected* to the poem, in the same way as industrial specialists see trees or ore-bearing rocks as raw material subjected to their manufactured end-products. Quantity thus begins to dominate the work of the specialist poet at its source. Like an industrialist, he is interested in the subjects of the world for the sake of what they can be made to produce. He mines his experience for subject matter. The first aim of the propriety of the old poets, by contrast, was to make the language true to its subject—to see that it told the truth. That is why they invoked the Muse. The truth the poet chose as his subject was perceived as *superior* to his powers—and, by clear implication, to his occasion and purpose. But the aim of truth-telling is not stated in either of these textbooks. The second, in fact, makes an "ethical" aim of avoiding the issue, for, as the authors say, coining a formidable truth: "Truth has become increasingly elusive and men are driven to embrace conflicting ideologies."

This sort of talk about language, it seems to me, is fundamentally impractical. It does not propose as an outcome any fidelity between words and speakers or words and things or words and acts. It leads instead to muteness and paralysis. So far as I can tell, it is unlikely that one could speak at all, in even the most casual conversation, without some informing sense of what would be best to say—that is, without some sort of *standard*. And I do not believe that it is possible to act on the basis of a "tentative" or "provisional" conclusion. We may know that we are forming a conclusion on the basis of provisional or insufficient knowledge—that is a part of what we understand as the tragedy of our condition. But we must act, nevertheless, on the basis of *final* conclusions, because we know that actions, occurring in time, are irrevocable. That is another part of our tragedy. People who make a conventional agreement that all conclusions are provisional—a convention almost invariably implied by academic uses of the word "objectivity" —characteristically talk but do not act. Or they do not act deliberately, though time and materiality carry them into action of a sort, willy-nilly.

And there are times, according to the only reliable ethics we have, when one is required to tell the truth, whatever the urgings of purpose, audience, and situation. Ethics requires this because, in the terms of the practical realities of our lives, the truth is safer than falsehood. To ignore this is simply to put language at the service of purpose—*any* purpose. It is, in terms of the most urgent realities of our own time, to abet a dangerous confusion between public responsibility and public relations. Remote as these theories of language are from practical contexts, they are nevertheless serviceable to expedient practices.

In affirming that there is a necessary and indispensable connection between language and truth, and therefore between language and deeds, I have certain precedents in mind. I begin with the Christian idea of the Incarnate Word, the Word entering the world as flesh, and inevitably therefore as action —which leads logically enough to the insistence in the Epistle of James that faith without works is dead:

> For if any be a hearer of the word, and not a doer, he is like unto a man beholding his natural face in a glass:
> For he beholdeth himself, and goeth his way, and straightway forgetteth what manner of man he was.

I also have in mind the Confucian insistence on sincerity (precision) and on fidelity between speaker and word as essentials of political health: "Honesty is the treasure of states." I have returned to Ezra Pound's observation that Confucius "collected *The Odes* to keep his followers from abstract discussion. That is, *The Odes* give particular instances. They do not lead to exaggerations of dogma."

And I have remembered Thoreau's sentence: "Where shall we look for standard English, but to the words of a standard man?"

The idea of standing by one's word, of words precisely designating things, of deeds faithful to words, is probably native to our understanding. Indeed, it seems doubtful that we could understand anything without that idea.

But in order to discover what makes language that can be understood, stood by, and acted on, it is necessary to return to my borrowed concepts of internal and external accounting. And it will be useful to add two further precedents.

In *Mind and Nature*, Gregory Bateson writes that "'things' . . . can only enter the world of communication and meaning by their names, their qualities and their attributes (i.e., by reports of their internal and external relations and interactions)."

And Gary Snyder, in a remarkably practical or practicable or practice-able definition of where he takes his stand, makes the poet responsible for "possibilities opening both inward and outward."

There can be little doubt, I think, that any accounting that is *purely* internal will be incomprehensible. If the connection between inward and outward is broken—if, for instance, the experience of a single human does not resonate within the common experience of humanity—then language fails. In *The Family Reunion*, Harry says: "I talk in general terms / Because the particular has no language." But he speaks, too, in despair, having no hope that his general terms can communicate the particular burden of his experience. We readily identify this loneliness of personal experience as "modern." Many poems of our century have this loneliness, this failure of speech, as a subject; many more exhibit it as a symptom.

But it begins at least as far back as Shelley, in such lines as these from "Stanzas Written in Dejection, Near Naples":

> Alas! I have nor hope nor health,
> Nor peace within nor calm around,
> Nor that content surpassing wealth
>
> The sage in meditation found,
> And walked with inward glory crowned—
> Nor fame, nor power, nor love, nor leisure.
> . . . .
> I could lie down like a tired child,
> And weep away the life of care
> Which I have borne and must bear,
> Till death like sleep might steal on me . . .

This too is an example of particular experience concealing itself in "general terms"—though here the failure, if it was suspected, is not acknowledged. The generality of the language does not objectify it, but seals it in its subjectivity. In reading this—as, I think, in reading a great many poems of our own time—we sooner or later realize that we are reading a "complaint" that we do not credit or understand. If we fail to realize this, it is because we have departed from the text of the poem, summoning particularities of our own experience in support of Shelley's general assertions. The fact remains that Shelley's poem doesn't tell us what he is complaining about; his lines fail to "create the object [here, the experience] which they contemplate." The poem has forsaken its story. This failure is implicitly conceded by the editors of *The Norton Anthology of English Literature*, who felt it necessary to provide the following footnote:

> Shelley's first wife, Harriet, had drowned herself; Clara, his baby daughter by Mary Shelley, had just died; and Shelley himself was plagued by ill health, pain, financial worries, and the sense that he had failed as a poet.

But I think the poem itself calls attention to the failure by its easy descent into self-pity, finally asserting that "I am one / Whom men love not. . . ." Language that becomes too subjective lacks currency, to use another economic metaphor; it will not pass. Self-pity, like self-praise, will not pass. The powers of language are used illegitimately, to impose, rather than to elicit, the desired response.

Shelley is not writing gibberish here. It is possible to imagine that someone who does not dislike this poem may see in it a certain beauty. But it is the sickly beauty of generalized emotionalism. For once precision is abandoned as a linguistic or literary virtue, vague generalization is one of the two remaining possibilities, gibberish being the second. And without precise accounting, leading to responsible action, there is no escape from the Shelleyan rhythm of exaltation and despair—ideal passions culminating in real disasters.

It is true, in a sense, that "the particular has no language"—that at least in public writing, and in speech passing between strangers, there may only be degrees of generalization. But there are, I think, two kinds of precision that are particular and particularizing. There is, first, the precision in the speech of people who share the same knowledge of place and history and work. This is the precision of direct reference or designation. It sounds like this: "How about letting me borrow your tall jack?" Or: "The old hollow beech blew down last night." Or, beginning a story, "Do you remember that time . . . ?" I would call this community speech. Its words have the power of pointing to things visible either to eyesight or to memory. Where it is not much corrupted by public or media speech, this community speech is wonderfully vital. Because it so often works designatively it *has* to be precise, and its precisions are formed by persistent testing against its objects.

This community speech, unconsciously taught and learned, in which words live in the presence of their objects, is the very root and foundation of language. It is the source, the unconscious inheritance that is carried, both with and without schooling, into consciousness—but never *all* the way, and so it remains rich, mysterious, and enlivening. Cut off from this source, language becomes a paltry work of conscious purpose, at the service and the mercy of expedient aims. Theories such as those underlying the two textbooks I have discussed seem to be attempts to detach language from its source in communal experience, by making it arbitrary in origin and provisional in use. And this may be a "realistic" way of "accepting" the degradation of community life. The task, I think, is hopeless, and it shows the extremes of futility that academic specialization can lead to. If one wishes to promote the life of language, one must promote the life of the community—a discipline many times more trying, difficult, and long than that of linguistics, but having at least the virtue of hopefulness. It escapes the despair always implicit in specializations: the cultivation of discrete parts without respect or responsibility for the whole.

The other sort of precision—the sort available to public speech or writing as well as to community speech—is a precision that comes of tension either between a statement and a prepared context or, within a single statement, between more or less conflicting feelings or ideas. Shelley's complaint is incomprehensible not just because it is set in "general terms," but because the generalities are too simple. One doesn't credit the emotion of the poem because it is too purely mournful. We are—conventionally, maybe, but also properly—unprepared to believe without overpowering evidence that things are *all* bad. Self-pity may deal in such absolutes of feeling, but we don't deal with other people in the manner of self-pity.

Another general complaint about mortality is given in Act V, Scene 2, of *King Lear*, when Edgar says to Gloucester: "Men must endure / Their going hence, even as their coming hither." Out of context this statement is even more general than Shelley's. It is, unlike Shelley's, deeply moving because it is tensely poised within a narrative context that makes it precise. We know exactly, for instance, what is meant by that "must": a responsible performance is required until death. But the complaint is followed immediately by a statement of another kind, forcing the speech of the play back into its action: "Ripeness is all. Come on."

Almost the same thing is done in a single line of Robert Herrick, in the tension between the complaint of mortality and the jaunty metric:

> Out of the world he must, who once comes in . . .

Here the very statement of inevitable death sings its acceptability. How would you divide, there, the "statement of the message" from "the message"?

And see how the tension between contradictory thoughts particularizes the feeling in these three lines by John Dryden:

> Old as I am, for Ladies Love unfit,
> The Pow'r of Beauty I remember yet,
> Which once inflam'd my Soul, and still inspires my Wit.

These last three examples immediately receive our belief and sympathy because they satisfy our sense of the complexity, the cross-graining, of real experience. In them, an inward possibility is made to open outward. Internal accounting has made itself externally accountable.

Shelley's poem, on the other hand, exemplifies the solitude of inward experience that continues with us, both in and out of poetry. I don't pretend to understand all the causes and effects of this, but I will offer the opinion that one of its chief causes is a simplistic idea of "freedom," which also continues with us, and is also to be found in Shelley. At the end of Act III of *Prometheus Unbound*, we are given this vision of a liberated humanity:

> The loathsome mask has fallen, the man remains
> Sceptreless, free, uncircumscribed, but man
> Equal, unclassed, tribeless, and nationless,
> Exempt from awe, worship, degree, the king
> Over himself . . .

This passage, like the one from the "Stanzas Written in Dejection," is vague enough, and for the same reason; as the first hastened to emotional absolutes, this hastens to an absolute idea. It is less a vision of a free man than a vision of a definition of a free man. But Shelley apparently did not notice that this headlong scramble of adjectives, though it may produce one of the possible definitions of a free man, also defines a lonely one, unattached and displaced. This free man is described as loving, and love is an emotion highly esteemed by Shelley. But it is, like his misery, a "free" emotion, detached and absolute. In this same passage of *Prometheus Unbound*, he calls it "the nepenthe, love"—love forgetful, or inducing forgetfulness, of grief or pain.

Shelley thought himself, particularly in *Prometheus Unbound*, a follower of Milton—an assumption based on a misunderstanding of *Paradise Lost*. And so it is instructive in two ways to set beside Shelley's definition of freedom this one by Milton:

> To be free is precisely the same thing as to be pious, wise, just, and temperate, careful of one's own, abstinent from what is another's, and thence, in fine, magnanimous and brave.

And Milton's definition, like the lines previously quoted from Shakespeare, Herrick, and Dryden, derives its precision from tension: he defines freedom in terms of responsibilities. And it is only this tension that can suggest the possibility of *living* (for any length of time) in freedom—just as it is the tension between love and pain that suggests the possibility of carrying love into acts. Shelley's freedom, defined in terms of freedom, gives us only this from a "Chorus of Spirits" in Act IV, Scene I, of *Prometheus Unbound*:

> Our task is done,
> We are free to dive, or soar, or run;
> Beyond and around,
> Or within the bound
> Which clips the world with darkness round.
>
> We'll pass the eyes
> Of the starry skies
> Into the hoar deep to colonize . . .

Which, as we will see, has more in common with the technological romanticism of Buckminster Fuller than with anything in Milton.

In supposed opposition to this remote subjectivity of internal accounting, our age has developed a stance or state of mind which it calls "objective," and which produces a kind of accounting supposed to be external—that is, free from personal biases and considerations. This objective mentality, within the safe confines of its various specialized disciplines, operates with great precision and confidence. It follows tested and trusted procedures and uses a professional language that an outsider must assume to be a very exact code. When this language is used by its accustomed speakers on their accustomed ground, even when one does not understand it, it clearly voices the implication of a marvelously precise control over objective reality. It is only when it is overheard in confrontation with failure that this implication falters, and the adequacy of this sort of language comes into doubt.

The transcribed conversations of the members of the Nuclear Regulatory Commission during the crisis of Three Mile Island provide a valuable exhibit of the limitations of one of these objective languages. At one point, for example, the commissioners received a call from Roger Mattson, Nuclear Reactor Regulation chief of systems safety. He said, among other things, the following:

> That bubble will be 5,000 cubic feet. The available volume in the upper head and the candy canes, that's the hot legs, is on the order of 2,000 cubic feet total. I get 3,000 excess cubic feet of noncondensibles. I've got a horse race. . . . We have got every systems engineer we can find . . . thinking the problem: how the hell do we get the noncondensibles out of there, do we win the horse race or do we lose the horse race.

At another time the commissioners were working to "engineer a press release," of which "The focus . . . has to be reassuring. . . ." Commissioner Ahearne apparently felt that it was a bit *too* reassuring, and he would have liked to *suggest* the possibility of a bad outcome, apparently a meltdown. He said:

> I think it would be technically a lot better if you said —something about there's a possibility—it's small, but it could lead to serious problems.

And, a few sentences later, Commissioner Kennedy told him:

> Well I understand what you're saying. . . . You could put a little sentence in right there . . . to say, were this —in the unlikely event that this occurred, increased temperatures would result and possible further fuel damage.

What is remarkable, and frightening, about this language is its inability to admit what it is talking about. Because these specialists have routinely eliminated themselves, as such and as representative human beings, from consideration, according to the prescribed "objectivity" of their discipline, they cannot bring themselves to acknowledge to each other, much less to the public, that their problem involves an extreme danger to a lot of people. Their subject, as bearers of a public trust, is this danger, and it can be nothing else. It is a technical problem least of all. And yet when their language approaches this subject, it either diminishes it, or dissolves into confusions of both syntax and purpose. Mr. Mattson speaks clearly and coherently enough so long as numbers and the jargon of "candy canes" and "hot legs" are adequate to his purpose. But as soon as he tries to communicate his sense of the human urgency of the problem, his language collapses into a kind of rant around the metaphor of "a horse race," a metaphor that works, not to reveal, but to obscure his meaning. And the two commissioners, struggling with their obligation to inform the public of the possibility of a disaster, find themselves virtually languageless—without the necessary words and with only the shambles of a syntax. They cannot say what they are talking about. And so their obligation to *inform* becomes a tongue-tied—and therefore surely futile—effort to *reassure*. Public responsibility becomes public relations, apparently, for want of a language adequately responsive to its subject.

So inept is the speech of these commissioners that we must deliberately remind ourselves that they are not stupid and are probably not amoral. They are highly trained, intelligent, worried men, whose understanding of language is by now to a considerable extent a public one. They are atomic scientists whose criteria of language are identical to those of at least some linguistic scientists. They determine the correctness of their statement to the press exactly according to Mr. Winterowd's rule: by their purpose, audience, and situation. Their language is governed by the ethical aim prescribed by the three authors of *Rhetoric: Discovery and Change*: they wish above all to speak in such a way as to "reduce another's sense of threat." But the result was not "cooperation and mutual benefit"; it was incoherence and dishonesty, leading to public suspicion, distrust, and fear. It is beneficial, surely, to "reduce another's sense of threat" only if there is no threat.

The commissioners speak a language that is diminished by

inordinate ambition: the taking of more power than can be responsibly or beneficently held. It is perhaps a law of human nature that such ambition always produces a confusion of tongues:

> And they said, Go to, let us build us a city and a tower, whose top may reach unto heaven; and let us make us a name, lest we be scattered
>
> . . . .
>
> And the Lord said . . . now nothing will be restrained from them, which they have imagined to do.
>
> Go to, let us go down, and there confound their language, that they may not understand one another's speech.

The professed aim is to bring people together—usually for the implicit, though unstated, purpose of subjecting them to some public power or project. Why else would rulers seek to "unify" people? The idea is to cause them to speak the same language—meaning either that they will agree with the government or be quiet, as in communist and fascist states, or that they will politely ignore their disagreements or disagree "provisionally," as in American universities. But the result—though power may survive for a while in spite of it—is confusion and dispersal. Real language, real discourse are destroyed. People lose understanding of each other, are divided and scattered. Speech of whatever kind begins to resemble the speech of drunkenness or madness.

What this dialogue of the Nuclear Regulatory Commissioners causes one to suspect—and I believe the suspicion is confirmed by every other such exhibit I have seen—is that there is simply no such thing as an accounting that is *purely* external. The notion that external accounting can be accomplished by "objectivity" is an illusion. Apparently the only way to free the accounting of what is internal to people, or subjective, is to make it internal to (that is subject to) some other entity or structure just as limiting, or more so—as the commissioners attempted to deal with a possible public catastrophe in terms either of nuclear technology or of public relations. The only thing really externalized by such accounting is a bad result that one does not wish to pay for.

And so external accounting, alone, is only another form of internal accounting. The only difference is that this "objective" accounting does pretty effectively rule out personal considerations *of a certain kind.* (It does *not* rule out the personal desire for wealth, power, or for intellectual certainty.) Otherwise, the talk of the commissioners and the lines from "Stanzas Written in Dejection" are equally and similarly incomprehensible. The languages of both are obviously troubled, we recognize the words, and learn something about the occasions, but we cannot learn from the language itself exactly what the trouble is. The commissioners' language cannot define the problem of their public responsibility, and Shelley's does not develop what I suppose should be called the narrative context of his emotion, which therefore remains incommunicable.

Moreover, these two sorts of accounting, so long as they remain discrete, both work to keep the problem abstract, all in the mind. They are both, in different ways, internal to the mind. The real occasions of the problems are not admitted into consideration. In Shelley's poem, this may be caused by a despairing acceptance of loneliness. In the Nuclear Regulatory Commission deliberations it is caused, I think, by fear; the commissioners take refuge in the impersonality of technological procedures. They cannot bear to acknowledge considerations and feelings that might break the insulating spell of their "objective" dispassion.

Or, to put it another way, their language and their way of thought make it possible for them to think of the crisis only as a technical event or problem. Even a meltdown is fairly understandable and predictable within the terms of their expertise. What is unthinkable is the evacuation of a massively populated region. It is the disorder, confusion, and uncertainty of that exodus that they cannot face. In dealing with the unstudied failure mode, the commissioners' minds do not have to leave their meeting room. It is an *internal* problem. The other, the human, possibility, if they were really to deal with it, would send them shouting into the streets. Even worse, perhaps, from the point of view of their discipline, it would force them to face the absurdity of the idea of "emergency planning"—the idea, in other words, of a controlled catastrophe. They would have to admit, against all the claims of professional standing and "job security," that the only way to control the danger of a nuclear power plant is not to build it. That is to say, if they had a language strong and fine enough to consider *all* the considerations, it would tend to force them out of the confines of "objective" thought and into action, out of solitude into community.

It is the *purity* of objective thought that finally seduces and destroys it. The same thing happens, it seems to me, to the subjective mind. For certain emotions, especially the extremely subjective ones of self-pity and self-love, isolation holds a strong enticement: it offers to keep them pure and neat, aloof from the disorderliness and the mundane obligations of the human common ground.

The only way, so far as I can see, to achieve an accounting that is verifiably and reliably external is to admit the internal, the personal, as an appropriate, necessary consideration. If the Nuclear Regulatory Commissioners, for example, had spoken a good common English, instead of the languages of their specialization and of public relations, then they might have spoken of their personal anxiety and bewilderment, and so brought into consideration what they had in common with the people whose health and lives they were responsible for. They might, in short, have sympathized openly with those people—and so have understood the probably unbearable burden of their public trusteeship.

To be bound within the confines of either the internal or the external way of accounting is to be diseased. To hold the two in balance is to validate both kinds, and to have health. I am not using these terms "disease" and "health" according to any clinical definitions, but am speaking simply from my own observation that when my awareness of how I feel overpowers my awareness of where I am and who is there with me, I am sick, diseased. This can be appropriately extended to say that if what I think obscures my sense of whereabouts and company, I am diseased. And the converse is also true: I am diseased if I become so aware of my surroundings that my own inward life is obscured, as if I should so fix upon the value of some mineral in the ground as to forget that the world is God's work and my home.

But still another example is necessary, and other terms.

*

In an article entitled "The Evolution and Future of American Animal Agriculture," G. W. Salisbury and R. G. Hart consider the transformation of American agriculture "from an art form into a science." The difference, they say, is that the art of agriculture is concerned "only" with the "how . . . of farming," whereas the science is interested in the "whys."

As an example—or, as they say, a "reference index"—of this change, the authors use the modern history of milk production: the effort, from their point of view entirely successful, "to change the dairy cow from the family companion animal she became after domestication and through all of man's subsequent history into an appropriate manufacturing unit of the twentieth century for the efficient transformation of unprocessed feed into food for man."

The authors produce "two observations" about this change, and these constitute their entire justification of it:

> First, the total cow population was reduced in the period 1944 through 1975 by 67 percent, but second, the yield per cow during the same period increased by 60 percent. In practical terms, the research that yielded such dramatic gains produced a savings for the American public as a whole of approximately 50 billion pounds of total digestible nitrogen per year in the production of a relatively constant level of milk.

The authors proceed to work this out in dollar values, and to say that the quantity of saved dollars finally "gets to the point that people simply do not believe it." And later they say that, in making this change, "The major disciplines were genetics, reproduction, and nutrition."

This is obviously a prime example of internal accounting in the economic sense. The external account is not fully renderable; the context of the accounting is vast, some quantities are not known, and some of the costs are not quantifiable. However, there can be no question that the externalized costs are large. The net gain is not, as these authors imply, identical with the gross. And the industrialization of milk production is a part of a much larger enterprise that may finally produce a highly visible, if not entirely computable, net loss.

At least two further observations are necessary:

1. The period, 1944–1975, also saw a drastic decrease in the number of dairies, by reason of the very change cited by Salisbury and Hart. The smaller—invariably the smaller—dairies were forced out because of the comparative "inefficiency" of their "manufacturing units." Their failure was part of a major population shift, which seriously disrupted the life both of the country communities and of the cities, broke down traditional community forms, and so on.

2. The industrialization of agriculture, of which the industrialization of milk production is a part, has caused serious problems that even agricultural specialists are beginning to recognize: soil erosion, soil compaction, chemical poisoning and pollution, energy shortages, several kinds of money troubles, obliteration of plant and animal species, disruption of soil biology.

The human, the agricultural, and the ecological costs are all obviously great. Some of them have begun to force their way into the accounts and are straining the economy. Others are, and are likely to remain, external to all ledgers.

The passages I have quoted from Professors Salisbury and Hart provide a very neat demonstration of the shift from a balanced internal-external accounting (the dairy cow as "family companion animal") to a so-called "objective" accounting (the dairy cow as "appropriate manufacturing unit of the twentieth century"), which is, in fact, internal to an extremely limited definition of agricultural progress.

The discarded language, oddly phrased though it is, comes close to a kind of accountability: the internal (family) and the external (cow) are joined by a moral connection (companionship). A proof of its accountability is that this statement can be the basis of moral behavior: "Be good to the cow, for she is our companion."

The preferred phrase—"appropriate manufacturing unit of the twentieth century"—has nothing of this accountability. One can say, of course: "Be good to the cow, for she is productive (or expensive)." But that could be said of a machine; it takes no account of the cow as a living, much less a fellow, creature. But the phrase is equally unaccountable as language. "Appropriate" to what? Though the authors write "appropriate . . . of the twentieth century," they may mean "appropriate . . . *to* the twentieth century." But are there no families and no needs for companionship with animals in the twentieth century? Or perhaps they mean "appropriate . . . for the efficient transformation of unprocessed feed into food for man." But the problem remains. Who is this "man"? Someone, perhaps, who needs no companionship with family or animals? We are constrained to suppose so, for "objectivity" has apparently eliminated "family" and "companion" as terms subject to personal bias—perhaps as "merely sentimental." By the terms of this "objective" accounting, then, "man" is a creature who needs to eat, and who is for some unspecified reason more important than a cow. But for a reader who considers himself a "man" by any broader definition, this language is virtually meaningless. Because the terms of personal bias (that is, the terms of *value*) have been eliminated, the terms of judgment ("appropriate" and "efficient") mean nothing. The authors' conditions would be just as well satisfied if the man produced the milk and the cow bought it, or if a machine produced it and a machine bought it.

Sense, and the possibility of sense, break down here because too much that clearly belongs in has been left out. Like the Nuclear Regulatory Commissioners, Salisbury and Hart have eliminated themselves as representative human beings, and they go on to eliminate the cow as a representative animal—all "interests" are thus removed from the computation. They scrupulously pluck out the representative or symbolic terms in order to achieve a pristinely "objective" accounting of the performance of a "unit." And so we are astonished to discover, at the end of this process, that they have complacently allowed the dollar to stand as representative of *all* value. What announced itself as a statement about animal agriculture has become, by way of several obscure changes of subject, a crudely simplified statement about industrial economics. This is not, in any respectable sense, language, or thought, or even computation. Like the textbooks I have discussed, and like the dialogue of the Nuclear Regulatory Commission, it is a pretentious and dangerous deception, forgiveable only insofar as it may involve self-deception.

*

If we are to begin to make a reliable account of it, this recent history of milk production must be seen as occurring within a system of nested systems: the individual human within the family within the community within agriculture within nature:

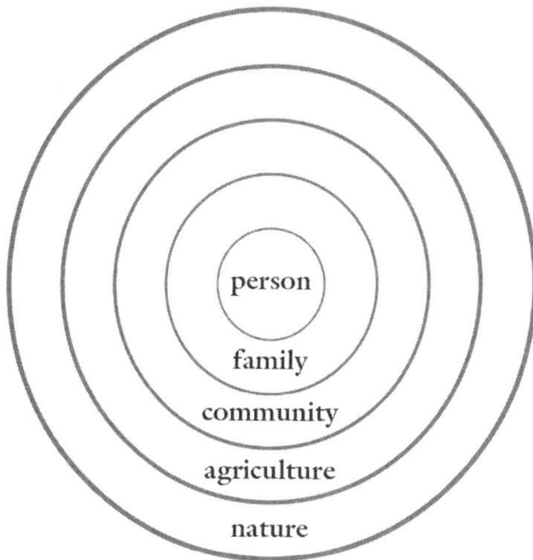

So long as the smaller systems are enclosed within the larger, and so long as all are connected by complex patterns of interdependency, as we know they are, then whatever affects one system will affect the others.

It seems that this system of systems is safe so long as each system is controlled by the next larger one. If at any point the hierarchy is reversed, and the smaller begins to control the larger, then the destruction of the entire system of systems begins. This system of systems is perhaps an updated, ecological version of the Great Chain of Being. That is, it may bring us back to a hierarchical structure not too different from the one that underlies *Paradise Lost*—a theory of the form of Creation which is at the same time a moral form, and which is violated by the "disobedience" or *hubris* of attempting to rise and take power above one's proper place.

But the sketch I have made of the system of systems is much too crude, for the connections between systems, insofar as this is a human structure, are not "given" or unconscious or automatic, but involve disciplines. Persons are joined to families, families to communities, etc., by disciplines that must be deliberately made, remembered, taught, learned, and practiced.

The system of systems begins to disintegrate when the hierarchy is reversed because that begins the disintegration of the connecting disciplines. Disciplines, typically, degenerate into professions, professions into careers. The accounting of Salisbury and Hart is defective because it upsets the hierarchies and so, perhaps unwittingly, fails to consider all the necessary considerations. Their "reference index" does occur within a system of systems—but a drastically abbreviated one, which involves a serious distortion:

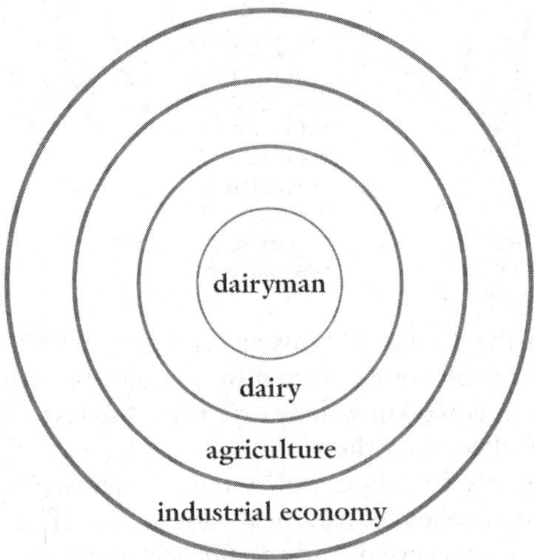

Two things are wrong with this. First, too much has been left out; the claims of family, community, and nature are all ignored. Second, the outer circle is too much within the interest of the inner. The dairyman is not *necessarily* under the control of simple greed—but this structure supplies no hint of a reason why he should not be.

The system of systems, as I first described it, involves three different kinds of interests:

1. The ontogenetic. This is self-interest and is at the center.
2. The phylogenetic. This is the interest that we would call

"humanistic." It reaches through family and community and into agriculture. But it does not reach far enough into agriculture because, by its own terms, it cannot.

3. The ecogenetic. This is the interest of the whole "household" in which life is lived. (I don't know whether I invented this term or not. If I did, I apologize.)

These terms give us another way to characterize the flaw in the accounting of Salisbury and Hart. Their abbreviated system of systems fails either to assemble enough facts, to account fully for the meaning of the facts, or to provide any standard of judgment, because the ontogenetic interest is both internal and external to it.

The system of systems, as I first sketched it, has this vulnerability: that the higher interests can be controlled or exploited by the lower interests simply by leaving things out—a procedure just as available to ignorance as to the highest cunning of "applied science." And given even the most generous motives, ignorance is always going to be involved.

There is no reliable standard for behavior anywhere within the system of systems except truth. Lesser standards produce destruction—as, for example, the standards of public relations make gibberish of language.

The trouble, obviously, is that we do not know much of the truth. In particular, we know comparatively little ecogenetic truth.

And so yet another term has to be introduced. The system of systems has to be controlled from above and outside. There has to be a religious interest of some kind above the ecogenetic. It will be sufficient to my purpose to say simply that the system of systems is enclosed within mystery, in which some truth can be known, but never all truth.

Neither the known truth nor the mystery is internal to any system. And here, however paradoxical it may seem, we begin to see a possibility of reliable accounting and of responsible behavior. The appropriateness of words or deeds can be determined only in reference to the whole "household" in which they occur. But this whole, as such, cannot enter into the accounting. (If it could, then the only necessary language would be mathematics, and the only necessary discipline would be military.) It can only come in as mystery: a factor of X which stands not for the unknown but the unknowable. This is an X that cannot be solved—which may be thought a disadvantage by some; its advantage is that, once it has been let into the account, it cannot easily be ignored. You cannot leave anything out of mystery, because by definition everything is always in it.

The practical use of religion, then, is to keep the accounting in as large a context as possible—to see, in fact, that the account is never "closed." Religion forces the accountant to reckon with mystery—the unsolvable X that keeps the debit and credit or cost and benefit columns open so that no "profit" can ever be safely declared. It forces the accounting outside of every enclosure that it might be internal to. Practically, this X means that all "answers" must be worked out within a limit of humility and restraint, so that the initiative to act would always imply a knowing acceptance of accountability for the results. The establishment and maintenance of this limit seems to me the ultimate empirical problem—the real "frontier" of science, or at least of the definition of the possibility of a *moral* science. It would place science under the rule of the old concern for propriety, correct proportion, proper scale—from which, in modern times, even the arts have been "liberated." That is, it would return to all work, artistic or scientific, the possibility of an external standard of quality. The quality of work or of a made thing would be determined by how conservingly it fitted into the system of systems. Judgment could then begin to articulate what is already obvious: that some work preserves the household of life, and some work destroys it. And thus a real liberation could take place: life and work could go free of those "professional standards" (and professional languages) that are invariably destructive of quality, because they always work as sheep's clothing for various kinds of ontogenetic motives. It is because of these professional standards that the industries and governments, while *talking* of the "betterment of the human condition," can *act* to enrich and empower themselves.

The connections within the system of systems are *practical* connections. The practicality consists in the realization that—despite the blandishments of the various short-circuited "professional" languages—you cannot speak or act in your own best interest without espousing and serving a higher interest.

It is not knowledge that enforces this realization, but the humbling awareness of the insufficiency of knowledge, of mystery.

Applying, then, the standard of ecogenetic health to the work of Salisbury and Hart, we get a third way to describe its failure: it makes a principle of replacing the complex concern for quality ("how") with the drastically simplifying concern for quantity. Thus motive is entirely "liberated" from method: any way is good so long as it increases per unit production. But everything except production is diminished in the process, and Salisbury and Hart do not have a way of accounting for, hence no ability even to recognize, these diminishments. All that has been diminished could have been protected by a lively interest in the question of "how," which Salisbury and Hart—like the interests they are accounting for—rule out at the beginning. They were nevertheless working under what I take to be a rule: When you subtract quality from quantity, the gross result is not a net gain.

And so a reliable account is personal at the beginning and religious at the end. This does not mean that a reliable account includes the whole system of systems, for no account can do that. It does mean that the account is made in precise reference to the system of systems—which is another way of saying that it is made in respect for it. Without this respect for the larger structures, the accounting shrinks into the confines of some smaller structure and becomes specialized, partial, and destructive.

It is this sort of external accounting that deals with connections and thus inevitably raises the issue of quality. Which, I take it, is always the same as the issue of propriety: how appropriate is the tool to the work, the work to the need, the need to other needs and the needs of others, and to the health of the household or community of all creatures?

And this kind of accounting gives us the great structures of poetry—as in Homer, Dante, and Milton. It is these great structures, I think, that carry us into the sense of being, in Gary Snyder's phrase, "at one with each other." They teach us to imagine the life that is divided from us by difference or enmity: as Homer imagined the "enemy" hero, Hector; as Dante, on his pilgrimage to Heaven, imagined the damned; as

Milton, in his awed study of the meaning of obedience, epitomized sympathetically in his Satan the disobedient personality. And as, now, ecological insight proposes again a poetry with the power to imagine the lives of animals and plants and streams and stones. And this imagining is eminently proprietous, fitting to the claims and privileges of the great household.

Unlike the problems of quantity, the problems of propriety are never "solved," but are ceaselessly challenging and interesting. This is the antidote to the romance of big technological solutions. Life would be interesting—there would be exciting work to do—even if there were no nuclear power plants or "agri-industries" or space adventures. The elaborations of elegance are at least as fascinating, and more various, more democratic, more healthy, more practical—though less glamorous—than elaborations of power.

Without this ultimate reference to the system of systems, and this ultimate concern for quality, any rendering of account falls into the service of a kind of tyranny: it accompanies, and in one way or another invariably enables, the taking of power, from people first and last, but also from all other created things.

In this degenerative accounting, language is almost without the power of designation because it is used conscientiously to refer to nothing in particular. Attention rests upon percentages, categories, abstract functions. The reference drifts inevitably toward the merely provisional. It is not language that the user will very likely be required to stand by or to act on, for it does not define any personal ground for standing or acting. Its only practical utility is to support with "expert opinion" a vast, impersonal technological action already begun. And it works directly against the conventionality, the community life, of language, for it holds in contempt, not only all particular grounds of private fidelity and action, but the common ground of human experience, memory, and understanding from which language rises and on which meaning is shaped. It is a tyrannical language: tyrannese.

Do people come consciously to such language and to such implicit purpose? I hope not. I do not think so. It seems likely to me that, first, a certain kind of confusion must occur. It is a confusion about the human place in the universe, and it has been produced by diligent "educational" labors. This confusion is almost invariably founded on some romantic proposition about the "high destiny of man" or "unlimited human horizons." For an example I turn to R. Buckminster Fuller, here defending the "cosmic realism" of space colonization:

> Conceptualizing realistically about humans as passengers on board 8,000-mile diameter Spaceship Earth traveling around the Sun at 60,000 miles an hour while flying formation with the Moon, which formation involves the 365 revolutions per each Sun circuit, and recalling that humans have always been born naked, helpless and ignorant though superbly equipped cerebrally, and endowed with hunger, thirst, curiosity and procreative instincts, it has been logical for humans to employ their minds' progressive discoveries of the cosmic principles governing all physical interattractions, interactions, reactions and intertransformings, and to use those principles in progressively organizing, to humanity's increasing advantage, the complex of cosmic principles interacting locally to produce their initial environment which most probably was that of a verdant south seas coral atoll—built by the coral on a volcano risen from ocean bottom ergo unoccupied by any animals, having only fish and birds as well as fruits, nuts and coconut milk.

That is a single sentence. I call attention not only to the vagueness and oversimplification of its generalities, but, more important, to the weakness of its grammar and the shapelessness and aimlessness of its syntax. The subject is an "it" of very tentative reference, buried in the middle of the sentence—an "it," moreover, that cannot possibly be the subject of the two complicated participial constructions that precede it. The sentence, then, begins with a dangling modifier half its length. On the other end, it peters out in a description of the biology of a coral atoll, the pertinence of which is never articulated. In general, the sentence is a labyrinth of syntactical confusions impossible to map. When we reflect that "sentence" means, literally, "a way of thinking" (Latin: *sententia*) and that it comes from the Latin *sentire*, to feel, we realize that the concepts of sentence and sentence structure are not merely grammatical or merely academic—not negligible in any sense. A sentence is both the opportunity and the limit of thought—what we have to think with, and what we have to think in. It is, moreover, a *feelable* thought, a thought that impresses its sense not just on our understanding, but on our hearing, our sense of rhythm and proportion. It is a pattern of felt sense.

A sentence that is completely shapeless is therefore a loss of thought, an act of self-abandonment to incoherence. And indeed Mr. Fuller shows himself here a man who conceives a sentence, not as a pattern of thought apprehensible to sense, but merely as a clot of abstract concepts. In such a syntactical clot, words and concepts will necessarily tend to function abstractly rather than referentially. It is the statement of a man for whom words have replaced things, and who has therefore ceased to think particularly about any thing.

The idea buried in all those words is, so far as I can tell, a simple one: humans are born earth-bound, ignorant, and vulnerable, but intelligent; by their intelligence they lift themselves up from their primitive origin and move to fulfill their destiny. As we learn in later sentences, this destiny is universal in scope, limitless ("ever larger"), and humans are to approach it by larger and larger technology. The end is not stated, obviously, because it is not envisioned, because it is not envisionable. It seems to me that the aimlessness, the limitlessness, of Mr. Fuller's idea produces the aimlessness and shapelessness of his sentence.

By contrast, consider another view of the human place in the universe, not a simple view or simply stated, but nevertheless comely, orderly, and clear:

> There wanted yet the Master work, the end
> Of all yet done: A Creature who not prone
> And Brute as other Creatures, but endu'd
> With Sanctity of Reason, might erect
> His Stature, and upright with Front serene
> Govern the rest, self-knowing, and from thence
> Magnanimous to correspond with Heav'n,
> But grateful to acknowledge whence his good
> Descends, thither with heart and voice and eyes
> Directed in Devotion to adore
> And worship God Supreme . . .

These lines of Milton immediately suggest what is wrong, first, with Mr. Fuller's sentence, and then with the examples of tyrannese that preceded it. They all assume that the human prerogative is unlimited, that we *must* do whatever we have the power to do. Specifically, what is lacking is the idea that humans have a place in Creation and that this place is limited by responsibility on the one hand and by humility on the other—or, in Milton's terms, by magnanimity and devotion. Without this precision of definition, this setting of bounds or ends to thought, we cannot mean, or say what we mean, or mean what we say; we cannot stand by our words because we cannot utter words that can be stood by; we cannot speak of our own actions as persons, or even as communities, but only of the actions of percentages, large organizations, concepts, historical trends, or the impersonal "forces" of destiny or evolution.

Or let us consider another pair of statements. The first, again from Buckminster Fuller, following the one just quoted, elaborates his theme of technological destiny:

> First the humans developed fish catching and carving tools, then rafts, dug-out canoes and paddles and then sailing outrigger canoes.
>
> Reaching the greater islands and the mainland they developed animal skin, grass and leafwoven clothing and skin tents. They gradually entered safely into geographical areas where they would previously have perished. Slowly they learned to tame, then breed, cows, bullocks, water buffalo, horses and elephants. Next they developed oxen, then horse-drawn vehicles, then horseless vehicles, then ships of the sky. Then employing rocketry and packaging up the essential life-supporting environmental constituents of the biosphere they made sorties away from their mothership Earth and finally ferried over to their Sun orbiting-companion, the Moon.

The other is from William Faulkner's story, "The Bear." Isaac McCaslin is speaking of his relinquishment of the ownership of land:

> He created the earth, made it and looked at it and said it was all right, and then He made man. He made the earth first and peopled it with dumb creatures, and then

> He created man to be His overseer on the earth and to hold suzerainty over the earth and the animals on it in His name, not to hold for himself and his descendants inviolable title forever, generation after generation, to the oblongs and squares of the earth, but to hold the earth mutual and intact in the communal anonymity of brotherhood, and all the fee He asked was pity and humility and sufferance and endurance and the sweat of his face for bread.

The only continuity recognized by Mr. Fuller is that of technological development, which is in fact not a continuity at all, for, as he sees it, it does not proceed by building on the past but by outmoding and replacing it. And if any other human concern accompanied the development from canoe to spaceship, it is either not manifest to Mr. Fuller, or he does not think it important enough to mention.

The passage from Faulkner, on the other hand, cannot be understood except in terms of the historical and cultural continuity that produced it. It awakens our memory of Genesis and *Paradise Lost*, as *Paradise Lost* awakens our memory of Genesis. In each of these the human place in Creation is described as a moral circumstance, and this circumstance is understood each time, it seems to me, with a deeper sense of crisis, as history has proved humanity more and more the exploiter and destroyer of Creation rather than its devout suzerain or steward. Milton knew of the conquests of Africa and the Americas, the brutality of which had outraged the humane minds of Europe and provided occasion to raise again the question of the human place in Creation; the devils of *Paradise Lost* are, among other things, conquistadores. (They are also the most expedient of politicians and technologists: ". . . by strength / They measure all. . . .") Faulkner's Isaac McCaslin, a white Mississippian of our own time, speaks not just with Milton's passion as a moral witness, but with the anguish of a man who inherits directly the guilt of the conqueror, the history of expropriation, despoliation, and slavery.

It is of the greatest importance to see how steadfastly this thrust of tradition, from Genesis to Milton to Faulkner, works toward the definition of personal place and condition, responsibility and action. And one feels the potency of this tradition to reach past the negativity of Isaac McCaslin's too simple relinquishment toward the definition of the atoning and renewing work that each person must do. Mr. Fuller's vision, by contrast, proposes that we have ahead of us only the next technological "breakthrough"—which, now that we have "progressed" to the scale of spaceships, is not work for persons or communities but for governments and corporations. We have in these two statements an open conflict between unlimited technology and traditional value. It is foolish to think that these are compatible. Value and technology can meet only on the ground of restraint.

The technological determinists have tyrannical attitudes, and speak tyrannese, at least partly because their assumptions cannot produce a moral or a responsible definition of the human place in Creation. Because they assume that the human place is any place, they are necessarily confused about where they belong.

Where does this confusion come from? I think it comes from the specialization and abstraction of intellect, separating it from responsibility and humility, magnanimity and devotion, and thus giving it an importance that, in the order of things and in its own nature, it does not and cannot have. The specialized intellectual assumes, in other words, that intelligence is all in the mind. For illustration, I turn again to *Paradise Lost*, where Satan, fallen, boasts in "heroic" defiance that he has

> A mind not to be chang'd by Place or Time.
> The mind is its own place, and in itself
> Can make a Heav'n of Hell, a Hell of Heav'n,
> What matter where, if I be still the same . . .

I do not know where one could find a better motto for the modernist or technological experiment, which assumes that we can fulfill a high human destiny anywhere, any way, so long as we can keep up the momentum of innovation; that the mind is "its own place" even within ecological degradation, pollution, poverty, hatred, and violence.

What we know, on the contrary, is that in any culture that could be called healthy or sane we find a much richer, larger concept of intelligence. We find, first, some way of acknowledging in action the existence of "higher intelligence." And we find that the human mind, in such a culture, is invariably strongly *placed*, in reference to other minds in the community and in cultural memory and tradition, and in reference to earthly localities and landmarks. Intelligence survives both by internal coherence and external pattern; it is both inside and outside the mind. People are born both with and into intelligence. What is thought refers precisely to what is thought about. It is this outside intelligence that we are now ignoring and consequently destroying.

As industrial technology advances and enlarges, and in the process assumes greater social, economic, and political force, it carries people away from where they belong by history, culture, deeds, association, and affection. And it destroys the landmarks by which they might return. Often it destroys the nature or the character of the places they have left. The very possibility of a practical connection between thought and the world is thus destroyed. Culture is driven into the mind, where it cannot be preserved. Displaced memory, for instance, is hard to keep in mind, harder to hand down. The little that survives is attenuated—without practical force. That is why the Jews, in Babylon, wept when they remembered Zion. The mere memory of a place cannot preserve it, nor apart from the place itself can it long survive in the mind. "How shall we sing the Lord's song in a strange land?"

The enlargement of industrial technology is thus analogous to war. It continually requires the movement of knowledge and responsibility away from home. It thrives upon the disintegration of homes, the subjugation of homelands. It requires that people cease to cooperate directly to fulfill local needs from local sources and begin instead to deal with each other always across the rift that divides producer and consumer, and always competitively. The idea of the independence of individual farms, shops, communities, and households is anathema to industrial technologists. The rush to nuclear energy and the growth of the space colony idea are powered by the industrial will to cut off the possibility of a small-scale energy technology—which is to say the possibility of small-scale personal and community acts. The corporate producers and their sycophants in the universities and the government will do virtually anything (or so they have obliged us to assume) to keep people from acquiring necessities in any way except by *buying* them.

Industrial technology and its aspirations enlarge along a line described by changes of verb tense: I need this tool; I will need this tool; I would need this tool. The conditional verb rests by nature upon *ifs.* The ifs of technological rationalization (*if* there were sufficient demand, money, knowledge, energy, power) act as wedges between history and futurity, inside and outside, value and desire, and ultimately between people and the earth and between one person and another.

By such shifts in the tenses of thought (as sometimes also by the substitution of the indefinite for the definite article) it is possible to impair or destroy the power of language to designate, to shift the focus of reference from what is outside the mind to what is inside it. And thus what already exists is devalued, subjugated, or destroyed for the sake of what *might* exist. The modern cult of planners and "futurologists" has thus achieved a startling resemblance to Swift's "academy of projectors":

> In these colleges, the professors contrive new rules and methods of agriculture and building, and new instruments and tools for all trades and manufactures; whereby, as they undertake, one man shall do the work of ten: a palace may be built in a week, of materials so durable, as to last for ever without repairing. All the fruits of the earth shall come to maturity, at whatever season we think fit to chuse, and encrease an hundred fold more than they do at present, with innumerable other happy proposals. The only inconvenience is, that none of these projects are yet brought to perfection; and, in the mean time, the whole country lies miserably waste. . . .

People who are willing to follow technology wherever it leads are necessarily willing to follow it away from home, off the earth, and outside the sphere of human definition, meaning, and responsibility. One has to suppose that this would be all right if they did it only for themselves and if they accepted the terms of their technological romanticism absolutely—that is, if they would depart absolutely from all that they propose to supersede, never to return. But past a certain scale, as C. S. Lewis wrote, the person who makes a technological choice does not choose for himself alone, but for others; past a certain scale, he chooses for *all* others. Past a certain scale, if the break with the past is great enough, he chooses for the past, and if the effects are lasting enough he chooses for the future. He makes, then, a choice that can neither be chosen against nor unchosen. Past a certain scale, there is no dissent from a technological choice.

People speaking out of this technological willingness cannot speak precisely, for what they are talking about does not yet exist. They cannot mean what they say because their words are avowedly speculative. They cannot stand by their words because they are talking about, if not *in*, the future, where they are not standing and cannot stand until long after they have spoken. All the grand and perfect dreams of the technologists are happening in the future, but nobody is there.

What can turn us from this deserted future, back into the sphere of our being, the great dance that joins us to our home, to each other and to other creatures, to the dead and the unborn? I think it is love. I am perforce aware how baldly and embarrassingly that word now lies on the page—for we have learned at once to overuse it, abuse it, and hold it in suspicion. But I do not mean any kind of abstract love, which is probably a contradiction in terms, but particular love for particular things, places, creatures, and people, requiring stands and acts, showing its successes or failures in practical or tangible effects. And it implies a responsibility just as particular, not grim or merely dutiful, but rising out of generosity. I think that this sort of love defines the effective range of human intelligence, the range within which its works can be dependably beneficent. Only the action that is moved by love for the good at hand has the hope of being responsible and generous. Desire for the future produces words that cannot be stood by. But love makes language exact, because one loves only what one knows. One cannot love the future or anything in it, for nothing is known there. And one cannot unselfishly make a future for someone else. Love for the future is self-love—love for the present self, projected and magnified into the future, and it is an irremediable loneliness.

Because love is not abstract, it does not lead to trends or

percentages or general behavior. It leads, on the contrary, to the perception that there is no such thing as general behavior. There is no abstract action. Love proposes the work of settled households and communities, whose innovations come about in response to immediate needs and immediate conditions, as opposed to the work of governments and corporations, whose innovations are produced out of the implicitly limitless desire for future power or profit. This difference is the unacknowledged cultural break in Mr. Fuller's evolutionary series: oxen, horse-drawn vehicles, horseless vehicles, ships of the sky. Between horse-drawn vehicles and horseless vehicles, human life disconnected itself from local sources; energy started to flow away from home. A biological limit was overrun, and with it the deepest human propriety.

Or, to shift the terms, love defines the difference between the "global village" which is a technological and a totalitarian ideal, directly suited to the purposes of centralized governments and corporations, and the Taoist village-as-globe, where the people live frugally and at peace, pleased with the good qualities of necessary things, so satisfied where they are that they live and die without visiting the next village, though they can hear its dogs bark and its roosters crow.

We might conjecture and argue a long time about the meaning and even the habitability of such a village. But one thing, I think, is certain: it would not be a linguistic no-man's-land in which words and things, words and deeds, words and people failed to stand in reliable connection or fidelity to one another. People and other creatures would be known by their names and histories, not by their numbers or percentages. History would be handed down in songs and stories, not reduced to evolutionary or technological trends. Generalizations would exist, of course, but they would be distilled from experience, not "projected" from statistics. They would sound, says Lao Tzu, this way:

> "Alert as a winter-farer on an icy stream,"
> "Wary as a man in ambush,"
> "Considerate as a welcome guest,"
> "Selfless as melting ice,"
> "Green as an uncut tree,"
> "Open as a valley . . ."

I come, in conclusion, to the difference between "projecting" the future and making a promise. The "projecting" of "futurologists" *uses* the future as the safest possible context for whatever is desired; it binds one only to selfish interest. But making a promise binds one *to someone else's future*. If the promise is serious enough, one is brought to it by love, and in awe and fear. Fear, awe, and love bind us to no selfish aims, but to each other. And they enforce a speech more exact, more clarifying, and more binding than any speech that can be used to sell or advocate some "future." For when we promise in love and awe and fear there is a certain kind of mobility that we give up. We give up the romanticism of progress, that is always shifting its terms to fit its occasions. We are speaking where we stand, and we shall stand afterwards in the presence of what we have said.

Poetry and Marriage: The Use of Old Forms

THE MEANING of marriage begins in the giving of words. We cannot join ourselves to one another without giving our word. And this must be an unconditional giving, for in joining ourselves to one another we join ourselves to the unknown. We can join one another *only* by joining the unknown. We must not be misled by the procedures of experimental thought: in life, in the world, we are never given two known results to choose between, but only *one* result that we choose without knowing what it is.

Marriage rests upon the immutable *givens* that compose it: words, bodies, characters, histories, places. Some wishes cannot succeed; some victories cannot be won; some loneliness is incorrigible. But there is relief and freedom in knowing what is real; these givens come to us out of the perennial reality of the world, like the terrain we live on. One does not care for this ground to make it a different place, or to make it perfect, but to make it inhabitable and to make it better. To flee from its realities is only to arrive at them unprepared.

Because the condition of marriage is worldly and its meaning communal, no one party to it can be solely in charge. What you alone think it ought to be, it is not going to be. Where you alone think you want it to go, it is not going to go. It is going where the two of you—and marriage, time, life, history, and the world—will take it. You do not know the road; you have committed your life to a way.

In marriage as in poetry, the given word implies the acceptance of a form that is never entirely of one's own making. When understood seriously enough, a form is a way of accepting and of living within the limits of creaturely life. We live only one life, and die only one death. A marriage cannot include everybody, because the reach of responsibility is short. A poem cannot be about everything, for the reach of attention and insight is short.

There are two aspects to these forms. The first is the *way* of making or acting or doing, which is to some extent technical. That is to say that definitions—settings of limits—are involved. The names *poetry* and *marriage* are given only to certain things, not to anything or to everything. Poetry is made of words; it is expected to keep a certain fidelity to everyday speech and a certain fidelity to music; if it is unspeakable or unmusical, it is not poetry. Marriage is the mutual promise of a man and a woman to live together, to love and help each other, in mutual fidelity, until death. It is understood that these definitions cannot be altered to suit convenience or circumstance, any more than we can call a rabbit a squirrel because we preferred to see a squirrel. Poetry of the traditionally formed sort, for instance, does not propose that its difficulties should be solved by skipping or forcing a rhyme or by mutilating syntax or by writing prose. Marriage does not invite one to solve one's quarrel with one's wife by marrying a more compliant woman. Certain limits, in short, are prescribed—imposed *before* the beginning.

The second aspect of these forms is an opening, a generosity, toward possibility. The forms acknowledge that good is possible; they hope for it, await it, and prepare its welcome—though they dare not *require* it. These two aspects are inseparable. To forsake the way is to forsake the possibility. To give up the form is to abandon the hope.

A certain awesome futurity, then, is the inescapable condition of word-giving—as it is, in fact, of all speech—for we speak into no future that we know, much less into one that we desire, but into one that is unknown. But that it is unknown requires us to be generous toward it, and requires our generosity to be full and unconditional. The unknown is the mercy and it may be the redemption of the known. The given word may come to appear to be wrong, or wrongly given. But the unknown still lies ahead of it, and so who is finally to say? If time has apparently proved it wrong, more time may prove it right. As growth has called it into question, further growth may reaffirm it.

These forms are artificial; if they exist they have to be made. Sexual love is natural, but marriage is not. The impulse to sing is natural, but language and the forms of song are not.

These forms are also initially arbitrary, because at the outset they can always be argued against. Until the wedding vows are said, the argument that one might find a better spouse has standing because there is no argument or evidence that can be produced against it; statistical probability would seem to support it: given the great number of theoretically possible choices, one *might* make a better choice. The vows answer that argument simply by cloture: the marriage now exists beyond all possibility of objection. A vow, Beatrice says, in *Paradiso* V, "is never canceled save by being kept . . . ," and must not be changed or broken on one's own judgment, though her stricture does not apply to vows that are foolish or perverse like those of Jephthah and Agamemnon. (She is acknowledging, let me emphasize, that some vows *ought* to be broken. Undoubtedly, some marriages are wrong, some divorces right. But it must also be understood, I think, that the possibility of breaking a vow can tell us nothing of what is meant by making and keeping one. Divorce is the contradiction of marriage, not one of its proposed results.)

Similarly, before the poem, there is no necessity governing the choice of form. Why *The Faerie Queene* should have been written in Spenserian stanzas is an unanswerable and probably a pointless question. All we can do is acknowledge that it *is* so written and go on to questions that are answerable.

The forms, then, are arbitrary *before* they are entered upon. Afterward, they have the same undoubtable existence and standing as the forms of an elm or a river. A poet such as Spenser evidently entered upon the form of a poem as solemnly as he entered upon any other cultural form—that of public service, say, or that of marriage. He understood it both as enablement and as constraint, and he meant not to break it, for in keeping the form he did not merely obey an arbitrarily imposed technical requirement but maintained his place in his cultural lineage, as both inheritor and bequeather, which saved him from loneliness and enabled him to mean—as witness his filial apostrophe to Chaucer in *The Faerie Queen*:

through infusion sweete
Of thine owne spirit, which doth in me survive
I follow here the footing of thy feete,
That with thy meaning so I may the rather meete.

(Though my concern here is with *making*, I should point out that the reader of a poem also participates in its form and in the community it makes, precisely as Spenser says: by following the footing of the poet's feet.)

Arbitrary in the choosing, these forms, once chosen and kept, are not arbitrary, but become inseparable from our definition as human beings. But the decision to break them is again arbitrary, for it cannot be based on any sufficient argument or evidence. And this decision does not return us to the state (perfectly inhabitable and respectable) that we were in before we chose the form but throws us into a state of formlessness, which is painful and dangerous, and which we can escape only by a return to form. The choice of a form is rightly solemn, both because of the gravity of the responsibility to keep it and because of the danger of failing to keep it. This choice does not present itself in any way that we may safely take less than seriously. To have a life or a place or a poem that is formless—into which anything at all may, or may not, enter—is to be condemned, at best, to bewilderment.

It is often assumed, as if under the influence of the promises of advertisements, that need or desire, ambition or inspiration may proceed straight to satisfaction. But this is false, contrary to the nature of form and the nature of discipline, as it is to common experience. It may sometimes happen by chance, but it does not dependably happen by chance. When it happens by luck, it will generally be found to be the luck of the well prepared. These impulses dependably come to fruition only by encountering the resistances of form, by being balked, baffled, forced to turn back and start again. They come to fruition by error and correction. Form is the means by which error is recognized and the means by which correctness is recognized.

There are, it seems, two Muses: the Muse of Inspiration, who gives us inarticulate visions and desires, and the Muse of Realization, who returns again and again to say, "It is yet more difficult than you thought." This is the muse of form.

The first muse is the one mainly listened to in a cheap-energy civilization, in which "economic health" depends on the assumption that everything desirable lies within easy reach of anyone. To hear the second muse one must move outside the cheap-energy enclosure. It is the willingness to hear the second muse that keeps us cheerful in our work. To hear only the first is to live in the bitterness of disappointment.

It is true that any form can be applied with a stupid rigidity. It can be falsely exclusive, consigning all that it cannot contain "to cold oblivion," as Shelley wrongly believed marriage was supposed to do. A set verse form can, of course, be used like a cookie cutter or a shovel, including and excluding arbitrarily by its own rule. But a set form can be used also to summon into a poem, or into a life, its unforeseen belongings, and thus is not rigid but freeing—an invocation to unknown possibility.

Form, crudely or stupidly used, may indeed be inimical to freedom. Well used, it may be the means of earning freedom, the price of admission or permission, the enablement to be free. But the connection may be even closer, more active and interesting, than that; it may be that form, strictly kept, *enforces* freedom. The form can be fulfilled only by a kind of abandonment to hope and to possibility, to unexpected gifts. The argument for freedom is not an argument against form. Form, like topsoil (which is intricately formal), empowers time to do good.

Properly used, a verse form, like a marriage, creates impasses, which the will and present understanding can solve only arbitrarily and superficially. These halts and difficulties do not ask for immediate remedy; we fail them by making emergencies of them. They ask, rather, for patience, forbearance, inspiration—the gifts and graces of time, circumstance, and faith. They are, perhaps, the true occasions of the poem: occasions for surpassing what we know or have reason to expect. They are points of growth, like the axils of leaves. Writing in a set form, rightly understood, is anything but force and predetermination. One puts down the first line of the pattern *in trust* that life and language are abundant enough to complete it. Rightly understood, a set form prescribes its restraint to the poet, not to the subject.

Marriage too is an attempt to rhyme, to bring two different lives—within the one life of their troth and household—periodically into agreement or consent. The two lives stray apart necessarily, and by consent come together again: to "feel together," to "be of the same mind." Difficult virtues are again necessary. And failure, permanent failure, is possible. But it is this possibility of failure, together with the formal bounds, that turns us back from fantasy, wishful thinking, and self-pity into the real terms and occasions of our lives.

It may be, then, that form serves us best when it works as an obstruction to baffle us and deflect our intended course. It may be that when we no longer know what to do we have come to our real work and that when we no longer know which way to go we have begun our real journey. The mind that is not baffled is not employed. The impeded stream is the one that sings.

In this way the keeping of the form instructs us. We had been prepared to learn what we had the poor power to expect. But fidelity to the form has driven us beyond expectation. The world, the truth, is more abounding, more delightful, more demanding than we thought. What appeared for a time perhaps to be mere dutifulness, that dried skull, suddenly breaks open in sweetness—and we are not where we thought we were, nowhere that we could have expected to be. It was expectation that would have kept us where we were.

In *Taking the Path of Zen*, Robert Aitken Roshi says: "It is not unusual to find true resonance with a so-called advanced koan in just a single dokusan, though often more time is necessary, and sometimes one gets stuck, and must stay there for a while." That necessity to "stay there for a while" is the gist of the meaning of form. Forms join us to time, to the consequences and fruitions of our own passing. The Zen student, the poet, the husband, the wife—none knows with certainty what he or she is staying for, but all know the likelihood that they will be staying "a while": to find out what they are staying for. And it is the faith of all of these disciplines that they will not stay to find that they should not have stayed.

That faith has nothing to do with what is usually called optimism. As the traditional marriage ceremony insists, not everything that we stay to find out will make us happy. The faith, rather, is that by staying, and only by staying, we will learn something of the truth, that the truth is good to know, and that it is always both different and larger than we thought.

The exploiter and the Shelleyan romantic (who are often the same person) are always in flight from consequence, the troubles of duration. The religious disciple, the husband and wife, the poet, like the true husbandman, accept the duration and effort, even the struggle, of formal commitment. They must come prepared to stay; if they mean to stay they will have to work, and they must learn the difference between good work and bad—which is to say that the capability of good work must be handed down from old to young.

The filial piety of Spenser, following the footing of the feet of Chaucer, has its opposite in the character of Michael in Eliot's *The Elder Statesman*, who longs to be fatherless in a present "freed" from the past, so that his name will be, as we say, "just a word":

> I simply want to lead a life of my own,
> According to my own ideas of good and bad,
> Of right and wrong. I want to go far away
> To some country where no one has heard the name of
> Claverton . . .

We will find that dream of escape in Shelley, who might be surprised—though *we* should not be—to discover that what Michael wants to do with his freedom is

> to go abroad
> As a partner in some interesting business.
> . . . .
> Import and export,
> With opportunity of profits both ways.

Part of the nature of a form seems to be that it is communal—that it can be bequeathed and inherited, that it can be taught, not as an instance (a relic), but as a way still usable. Both its validity and its availability depend upon our common understanding that we humans are all fundamentally alike.

Forms are broken, usually, on the authority of the opposing principle that we are all fundamentally or essentially different. Each individual, each experience, each life is assumed to be unique—hence, each individual should be "free" to express or fulfill his or her unique self in a way appropriately unique.

Both the communal and the individual emphases can be carried to extremes, and the extremity of each is loneliness. One can be lonely in the totalitarian crowd, in which no difference is perceived or tolerated; and one can be lonely in the difference or uniqueness of individuality in which community is repudiated.

The whole range of possibilities can be exemplified within language itself. It is possible to speak a language so commonized by generality or jargon or slang that one's own mind and life virtually disappear into it. And it is possible to speak a language made so personal by contrivance, affectation, or slovenliness that one makes no sense. Between the two are the Confucian principles, dear to Pound, of fidelity ("the man . . . standing by his word") and sincerity (precise speech: words that can be stood by).

This word-keeping, standing by one's word, is a double fidelity: to the community and to oneself. It is the willingness of difference both to represent itself and to account for itself. The individual is thereby at once free and a member. To break one's word in order to be "free" of it, on the other hand, is to make and enforce a damning equation between freedom and loneliness.

The freedom that depends upon or results in the breaking of words and the breaking of forms comes, I think, from a faith in the individual intelligence, in "genius," as opposed to a faith in the community or in culture. Belief in culture calls for the same disciplines as belief in marriage. It calls, indeed, for *more* patience and *more* faith, for it requires dedication to work and hope of more than a lifetime. This work, as I understand it, consists of the accumulation of local knowledge *in place*, generation after generation, children learning the visions and failures, stories and songs, names, ways, and skills of their elders, so that the costs of individual trial-and-error learning can be lived with and repaid, and the community thus enabled to preserve both itself and its natural place and neighborhood.

I do not mean that the rebellions of genius are necessarily bad; they may be both inevitable and indispensable. But they *can* be bad; they are dangerous, and it is prudent to understand what their dangers are. We may be properly suspicious of them when they are carried out in the name of "personal freedom" or "personal fulfillment," and when their program is a general rebellion against the imperfect past.

"Freedom" and "fulfillment" are often coupled together in the idea that one must be free in order to fulfill oneself. Behind the divorce epidemic—as behind the epidemics of venereal disease and accidental parenthood among the young—is the unappeasable demon of "sexual fulfillment." Behind the epidemic of poetic illiteracy is the conviction that one must free oneself from poetic tradition (by not learning it) in order to fulfill one's ambition. The question here is whether we want "fulfill" to mean more than "satisfy"—as it must if we take seriously even its literal sense: to fill full. Sexual desire is an appetite, and so, in a different way, is ambition; they can be fulfilled only temporarily and soon have to be refulfilled. Though we would like it to be, "sexual fulfillment" is really no more exalted an idea than "hunger fulfillment." Ambition fulfillment, as we still understand when talking of politics or war, is a dangerous enterprise—for a mental appetite may be larger and more rapacious than a physical one, may require to be refulfilled more frequently and is less likely to be twice appeased by the same refulfillment.

If we mean "fulfill" in its larger sense—to discharge or put to rest, as when one fulfills an obligation or a promise—then we see that what we fulfill are not appetites but forms. Marriage is a form of sexual love which allows its fulfillment in both senses: in satisfaction and in responsibility for its consequences, and it sets a term to this responsibility—"until death"—at which it may be said to be fulfilled. The form of a particular poem similarly allows a valid fulfillment of poetic ambition. And the particularity of these fulfillments, working as they do to define complex obligations to discipline, to community, to tradition, to forebears and successors, mitigates the dangers of self-renewing appetites. The seriousness of this formality can hardly be overstated, for in the fulfillment of form is death—as the marriage vows instruct us. Fulfillment "bears it out even to the edge of doom." Past fulfillment, desire is at an end; there is no need to return. Great cost is obviously involved. But to stop short of such fulfillment is to make love "time's fool," and to fix a value upon a worth unknown.

Outrage and rebellion against the past are undoubtedly human necessities, but they are limited necessities, and they probably should be limited to youth. Things are obviously wrong with the past; young people have the clarity to see them and the energy to rebel against them. But as a general principle, such rebellion is destructive, for it keeps us from seeing that the past, unsatisfactory as it is, is the source of nearly all our good. Maturity sees that the past is not to be rejected, destroyed, or replaced, but rather that it is to be judged and corrected, that the work of judgment and correction is endless, and that it necessarily involves one's *own* past.

The industrial economy has made a general principle of the youthful antipathy to the past, and the modern world abounds with heralds of "a better future" and with debunkers happy to point out that Yeats was "silly like us" or that Thomas Jefferson may have had a Negro slave as a mistress—and so we are disencumbered of the burden of great lives, set free to be as cynical or desperate as we please. Cultural forms, it is held, should change apace to keep up with technology. Sexual discipline should be replaced by the chemicals, devices, and procedures of "birth control," and poetry must hasten to accept the influence of typewriter or computer.

It can be better argued that cultural forms ought to change by analogy with biological forms. I assume that they *do* change in that way, and by the same necessity to respond to changes of circumstance. It is necessary, nevertheless, to recognize a difference in kinds of cultural change: there is change by necessity, or adaptation; and there is contrived change, or novelty. The first is the work of species or communities or lineages of descent, occurring usually by slow increments over a long time. The second is the work of individual minds, and it happens, or is intended to happen, by fiat. Individual attempts to change cultural form—as to make a new kind of marriage or family or community—are nearly always shallow or foolish and are frequently totalitarian. The assumption that it can easily be otherwise comes from the faith in genius.

To adopt a communal form with the idea of changing or discarding it according to individual judgment is hopeless, the despair and death of meaning. To keep the form is an act of faith in possibility, not of the form, but of the life that is given to it; the form is a question addressed to life and time, which only life and time can answer.

Individual genius, then, goes astray when it proposes to do the work of community. We rightly follow its promptings, on the other hand, when it can point out correctly that *we* have gone astray—when forms have become rigid or empty, when we have forgot their use or their meaning. We then follow our genius or our geniuses back to reverence, to truth, or to nature. This alternation is one of the long rhythms of our history.

But the faith in genius and the rebellions of genius, at the times when these are necessary, should lead to the renewal of forms, not to their destruction. No individual can justifiably destroy anything of communal value on behalf of the community. Though individual geniuses have often enough assumed otherwise, there is no reason to grant them special privileges or exemptions. No artistic or scientific genius is justified in abusing nature or culture.

The analogy I have been working with here is most readily apparent if we think of marriage and poetic forms as *set* forms—that is, forms that in a sense *precede* the content, that are in a sense *prescriptive*. These set forms are indispensable, I believe, because they accommodate and serve that part of our life which is cyclic, drawing minds and lives back repeatedly through the same patterns, as each year moves through the same four seasons in the same order.

My remaining problem is to see how so-called free verse may fit into my scheme. It *has* to be fitted in if I am to respect my scheme, and if I acknowledge, as I certainly do, that much free verse is poetry. There is some danger of becoming cute or precious in carrying this analogy out to such length, and yet I am working on the assumption that the analogy is valid.

One analogue of free verse, I think, must be courtship, a way of accommodating our minds to something new, of finding out what it is that we are doing or about to do. The encounter between the English language and the subjects and objects of a new continent strange to it and to its poetic tradition seems to have required this of a succession of American poets from Whitman's time until now. Who can help feeling in the early *Leaves of Grass* a kind of falling in love? The lines must reach out impulsively, become capacious and tensile, to include in their full stance and particularity the images of American experience:

> The negro that drives the long dray of the stone-yard,
> steady and tall he stands pois'd on one leg on the
> string-piece . . .

Never mind that these loving inventories have occurred in

poetry before. This is new, this confrontation with a continent needing to be realized, and we grant Whitman his liberty and his exultation; we feel ourselves free and exultant with him; we willingly forgive the absurdities that occasionally jeopardize his exuberance—his ostentatious French phrases, for instance, or his industrial optimism.

But if Whitman's work is the prime example of the freeing of verse, it is an example of something else too, for at its best Whitman's line, as we will see if we try to shorten or lengthen it, is a form. He set his line free only to make it into a *kind* of line that we recognize anywhere we see it—a new power, a new music, added to poetry, which can be learned and used. Theoretically, I suppose, any line can be written a different way, but I don't think that we are tempted to imagine this line as anything but what it is:

Seasons pursuing each other the plougher ploughs, the
mower mows, and the winter-grain falls in the ground . . .

And such newness does not destroy the old set forms, but renews them in renewing our understanding of what a line of verse is, our sense of its properties of duration and coherence, beginning and end. The term "organic," when applied to free or "open" poetic forms, should alert us to the nature of all form, traditional or new, for the organic forms of nature, like the ballad stanza or the stanza of Spenser, are principles that are repeatable and recognizable through a series of variations. This recognizability within difference suggests the proper relation of abstraction and particularity—and suggests, moreover, that the right function of abstraction is to give appropriate clarity and distinction to the particular. The form, recognizable from verse to verse, shapes and measures what is said, makes it musical.

So it is with Whitman's line. So it is, I believe, with the line of every writer of free verse worth reading. At some point the poet ceases to be an experimenter or innovator and begins a life's work. *Leaves of Grass* answers for itself the questions that stood before *The Divine Comedy* or *The Canterbury Tales*: What can be done in this way? What *must* be done in this way? Once a way is chosen with enough seriousness, the analogy shifts from courtship to marriage.

The work of poetic form is coherence, joining things that need to be joined, as marriage joins them—in words by which a man or a woman can stand, words confirmable in acts. Forms join, and this is why forms tend to be analogues of each other and to resonate with each other. Forms join the diverse things that they contain; they join their contents to their context; they join us to themselves; they join us to each other; they join writers and readers; they join the generations together, the young and the old, the living and the dead. Thus, for a couple, marriage is an entrance into a timeless community. So, for a poet (or a reader), is the mastery of poetic form. Joining the form, we join all that the form has joined.

FROM

HOME ECONOMICS

(1987)

Getting Along with Nature

THE DEFENDERS of nature and wilderness—like their enemies the defenders of the industrial economy—sometimes sound as if the natural and the human were two separate estates, radically different and radically divided. The defenders of nature and wilderness sometimes seem to feel that they must oppose any human encroachment whatsoever, just as the industrialists often apparently feel that they must make the human encroachment absolute or, as they say, "complete the conquest of nature." But there is danger in this opposition, and it can be best dealt with by realizing that these pure and separate categories are pure ideas and do not otherwise exist.

Pure nature, anyhow, is not good for humans to live in, and humans do not want to live in it—or not for very long. Any exposure to the elements that lasts more than a few hours will remind us of the desirability of the basic human amenities: clothing, shelter, cooked food, the company of kinfolk and friends—perhaps even of hot baths and music and books.

It is equally true that a condition that is *purely* human is not good for people to live in, and people do not want to live for very long in it. Obviously, the more artificial a human environment becomes, the more the word "natural" becomes a term of value. It can be argued, indeed, that the conservation movement, as we know it today, is largely a product of the industrial revolution. The people who want clean air, clear streams, and wild forests, prairies, and deserts are the people who no longer have them.

People cannot live apart from nature; that is the first principle of the conservationists. And yet, people cannot live in nature without changing it. But this is true of *all* creatures; they depend upon nature, and they change it. What we call nature is, in a sense, the sum of the changes made by all the various creatures and natural forces in their intricate actions and influences upon each other and upon their places. Because of the woodpeckers, nature is different from what it would be without them. It is different also because of the borers and ants that live in tree trunks, and because of the bacteria that live in the soil under the trees. The making of these differences is the making of the world.

Some of the changes made by wild creatures we would call beneficent: beavers are famous for making ponds that turn into fertile meadows; trees and prairie grasses build soil. But sometimes, too, we would call natural changes destructive. According to early witnesses, for instance, large areas around Kentucky salt licks were severely trampled and eroded by the great herds of hoofed animals that gathered there. The buffalo "streets" through hilly country were so hollowed out by hoofwear and erosion that they remain visible almost two centuries after the disappearance of the buffalo. And so it can hardly be expected that humans would not change nature. Humans, like all other creatures, must make a difference; otherwise, they cannot live. But unlike other creatures, humans must make a choice as to the kind and scale of the difference they make. If they choose to make too small a difference, they diminish their humanity. If they choose to make too great a difference, they diminish nature, and narrow their subsequent choices; ultimately, they diminish or destroy themselves. Nature, then, is not only our source but also our limit and measure. Or, as the poet Edmund Spenser put it almost four hundred years ago, Nature, who is the "greatest goddesse," acts as a sort of earthly lieutenant of God, and Spenser represents her as both a mother and judge. Her jurisdiction is over the relations between the creatures; she deals "Right to all . . . indifferently," for she is "the equall mother" of all "And knittest each to each, as brother unto brother." Thus, in Spenser, the natural principles of fecundity and order are pointedly linked with the principle of justice, which we may be a little surprised to see that he attributes also to nature. And yet in his insistence on an "indifferent" natural justice, resting on the "brotherhood" of *all* creatures, not just of humans, Spenser would now be said to be on sound ecological footing.

In nature we know that wild creatures sometimes exhaust their vital sources and suffer the natural remedy: drastic population reductions. If lynxes eat too many snowshoe rabbits —which they are said to do repeatedly—then the lynxes starve down to the carrying capacity of their habitat. It is the carrying capacity of the lynx's habitat, not the carrying capacity of the lynx's stomach, that determines the prosperity of lynxes. Similarly, if humans use up too much soil—which they have often done and are doing—then they will starve down to the carrying capacity of *their* habitat. This is nature's "indifferent" justice. As Spenser saw in the sixteenth century, and as we must learn to see now, there is no appeal from this justice. In the hereafter, the Lord may forgive our wrongs against nature, but on earth, so far as we know, He does not overturn her decisions.

One of the differences between humans and lynxes is that humans can see that the principle of balance operates between lynxes and snowshoe rabbits, as between humans and topsoil; another difference, we hope, is that humans have the sense to act on their understanding. We can see, too, that a stable balance is preferable to a balance that tilts back and forth like a seesaw, dumping a surplus of creatures alternately from either end. To say this is to renew the question of whether or not the human relationship with nature is necessarily an adversary relationship, and it is to suggest that the answer is not simple.

But in dealing with this question and in trying to do justice to the presumed complexity of the answer, we are up against an American convention of simple opposition to nature that is deeply established both in our minds and in our ways. We have opposed the primeval forests of the East and the primeval prairies and deserts of the West, we have opposed man-eating beasts and crop-eating insects, sheep-eating coyotes and chicken-eating hawks. In our lawns and gardens and fields, we oppose what we call weeds. And yet more and more of us are beginning to see that this opposition is ultimately destructive even of ourselves, that it does not explain many things that need explaining—in short, that it is untrue.

If our proper relation to nature is not opposition, then what is it? This question becomes complicated and difficult for us because none of us, as I have said, wants to live in a "pure" primeval forest or in a "pure" primeval prairie; we do not want to be eaten by grizzly bears; if we are gardeners, we have a legitimate quarrel with weeds; if, in Kentucky, we are trying to improve our pastures, we are likely to be enemies of the nodding thistle. But, do what we will, we remain under the spell of the primeval forests and prairies that we have cut down and broken; we turn repeatedly and with love to the thought of them and to their surviving remnants. We find ourselves attracted to the grizzly bears, too, and know that they and other great, dangerous animals remain alive in our imaginations as they have been all through human time. Though we cut down the nodding thistles, we acknowledge their beauty and are glad to think that there must be some place where they belong. (They may, in fact, not always be out of place in pastures; if, as seems evident, overgrazing makes an ideal seedbed for these plants, then we must understand them as a part of nature's strategy to protect the ground against abuse by animals.) Even the ugliest garden weeds earn affection from us when we consider how faithfully they perform an indispensable duty in covering the bare ground and in building humus. The weeds, too, are involved in the business of fertility.

We know, then, that the conflict between the human and the natural estates really exists and that it is to some extent necessary. But we are learning, or relearning, something else, too, that frightens us: namely, that this conflict often occurs at the expense of *both* estates. It is not only possible but altogether probable that by diminishing nature we diminish ourselves, and vice versa.

The conflict comes to light most suggestively, perhaps, when advocates for the two sides throw themselves into absolute conflict where no absolute difference can exist. An example of this is the battle between defenders of coyotes and defenders of sheep, in which the coyote-defenders may find it easy to forget that the sheep ranchers are human beings with some authentic complaints against coyotes, and the sheep-defenders find it easy to sound as if they advocate the total eradication of both coyotes and conservationists. Such conflicts—like the old one between hawk-defenders and chicken-defenders—tend to occur between people who use nature indirectly and people who use it directly. It is a dangerous mistake, I think, for either side to pursue such a quarrel on the assumption that victory would be a desirable result.

The fact is that people need both coyotes and sheep, need a world in which both kinds of life are possible. Outside the heat of conflict, conservationists probably know that a sheep is one of the best devices for making coarse foliage humanly edible and that wool is ecologically better than the synthetic fibers, just as most shepherds will be aware that wild nature is of value to them and not lacking in interest and pleasure.

The usefulness of coyotes is, of course, much harder to define than the usefulness of sheep. Coyote fur is not a likely substitute for wool, and, except as a last resort, most people don't want to eat coyotes. The difficulty lies in the difference between what is ours and what is nature's: What is ours is ours because it is directly useful. Coyotes are useful *indirectly*, as part of the health of nature, from which we and our sheep alike must live and take our health. The fact, moreover, may be that sheep and coyotes need each other, at least in the sense that neither would prosper in a place totally unfit for the other.

This sort of conflict, then, does not suggest the possibility of victory so much as it suggests the possibility of a compromise—some kind of peace, even an alliance, between the domestic and the wild. We know that such an alliance is necessary. Most conservationists now take for granted that humans thrive best in ecological health and that the test or sign of this health is the survival of a diversity of wild creatures. We know, too, that we cannot imagine ourselves apart from those necessary survivals of our own wildness that we call our instincts. And we know that we cannot have a healthy agriculture apart from the teeming wilderness in the topsoil, in which worms, bacteria, and other wild creatures are carrying on the fundamental work of decomposition, humus making, water storage, and drainage. "In wildness is the preservation of the world," as Thoreau said, may be a spiritual truth, but it is also a practical fact.

On the other hand, we must not fail to consider the opposite proposition—that, so long at least as humans are in the world, in human culture is the preservation of wildness—which is equally, and more demandingly, true. If wildness is to survive, then *we* must preserve it. We must preserve it by public act, by law, by institutionalizing wildernesses in some places. But such preservation is probably not enough. I have heard Wes Jackson of the Land Institute say, rightly I think, that if we cannot preserve our farmland, we cannot preserve the wilderness. That said, it becomes obvious that if we cannot preserve our cities, we cannot preserve the wilderness. This can be demonstrated practically by saying that the same attitudes that destroy wildness in the topsoil will finally destroy it everywhere; or by saying that if *everyone* has to go to a designated public wilderness for the necessary contact with wildness, then our parks will be no more natural than our cities.

But I am trying to say something more fundamental than that. What I am aiming at—because a lot of evidence seems to point this way—is the probability that nature and human culture, wildness and domesticity, are not opposed but are interdependent. Authentic experience of either will reveal the need of one for the other. In fact, examples from both past and present prove that a human economy and wildness can exist together not only in compatibility but to their mutual benefit.

One of the best examples I have come upon recently is the story of two Sonora Desert oases in Gary Nabhan's book, *The Desert Smells Like Rain*. The first of these oases, A'al Waipia, in Arizona, is dying because the park service, intending to preserve the natural integrity of the place as a bird sanctuary for tourists, removed the Papago Indians who had lived and farmed there. The place was naturally purer after the Indians were gone, but the oasis also began to shrink as the irrigation ditches silted up. As Mr. Nabhan puts it, "an odd thing is happening to their 'natural' bird sanctuary. They are losing the heterogeneity of the habitat, and with it, the birds. The old trees are dying. . . . These riparian trees are essential for the breeding habitat of certain birds. Summer annual seed plants are conspicuously absent. . . . Without the soil disturbance associated with plowing and flood irrigation, these natural foods for birds and rodents no longer germinate."

The other oasis, Ki:towak, in old Mexico, still thrives because a Papago village is still there, still farming. The village's oldest man, Luis Nolia, is the caretaker of the oasis, cleaning the springs and ditches, farming, planting trees: "Luis . . . blesses the oasis," Mr. Nabhan says, "for his work keeps it healthy." An ornithologist who accompanied Mr. Nabhan found twice as many species of birds at the farmed oasis as he found at the bird sanctuary, a fact that Mr. Nabhan's Papago friend, Remedio, explained in this way: "That's because those birds, they come where the people are. When the people live and work in a place, and plant their seeds and water their trees, the birds go live with them. They like those places, there's plenty to eat and that's when we are friends to them."

Another example, from my own experience, is suggestive in a somewhat different way. At the end of July 1981, while I was using a team of horses to mow a small triangular hillside pasture that is bordered on two sides by trees, I was suddenly aware of wings close below me. It was a young red-tailed hawk, who flew up into a walnut tree. I mowed on to the turn and stopped the team. The hawk then glided to the ground not twenty feet away. I got off the mower, stood and watched, even spoke, and the hawk showed no fear. I could see every feather distinctly, claw and beak and eye, the creamy down of the breast. Only when I took a step toward him, separating myself from the team and mower, did he fly. While I mowed three or four rounds, he stayed near, perched in trees or standing erect and watchful on the ground. Once, when I stopped to watch him, he was clearly watching me, stooping to see under the leaves that screened me from him. Again, when I could not find him, I stooped, saying to myself, "This is what he did to look at me," and as I did so I saw him looking at me.

Why had he come? To catch mice? Had he seen me scare one out of the grass? Or was it curiosity?

A human, of course, cannot speak with authority of the motives of hawks. I am aware of the possibility of explaining the episode merely by the hawk's youth and inexperience. And yet it does not happen often or dependably that one is approached so closely by a hawk of any age. I feel safe in making a couple of assumptions. The first is that the hawk came because of the conjunction of the small pasture and its wooded borders, of open hunting ground and the security of trees. This is the phenomenon of edge or margin that we know to be one of the powerful attractions of a diversified landscape, both to wildlife and to humans. The human eye itself seems drawn to such margins, hungering for the difference made in the countryside by a hedgy fencerow, a stream, or a grove of trees. And we know that these margins are biologically rich, the meeting of two kinds of habitat. But another difference also is important here: the difference between a large pasture and a small one, or, to use Wes Jackson's terms, the difference between a field and a patch. The pasture I was mowing was a patch—small, intimate, nowhere distant from its edges.

My second assumption is that the hawk was emboldened

to come so near because, though he obviously recognized me as a man, I was there with the team of horses, with whom he familiarly and confidently shared the world.

I am saying, in other words, that this little visit between the hawk and me happened because the kind and scale of my farm, my way of farming, and my technology *allowed* it to happen. If I had been driving a tractor in a hundred-acre cornfield, it would not have happened.

In some circles I would certainly be asked if one can or should be serious about such an encounter, if it has any value. And though I cannot produce any hard evidence, I would unhesitatingly answer yes. Such encounters involve another margin —the one between domesticity and wildness—that attracts us irresistibly; they are among the best rewards of outdoor work and among the reasons for loving to farm. When the scale of farming grows so great and obtrusive as to forbid them, the *life* of farming is impoverished.

But perhaps we do find hard evidence of a sort when we consider that *all* of us—the hawk, the horses, and I—were there for our benefit and, to some extent, for our *mutual* benefit: The horses live from the pasture and maintain it with their work, grazing, and manure; the team and I together furnish hunting ground to the hawk; the hawk serves us by controlling the field-mouse population.

These meetings of the human and the natural estates, the domestic and the wild, occur invisibly, of course, in any well-farmed field. The wilderness of a healthy soil, too complex for human comprehension, can yet be husbanded, can benefit from human care, and can deliver incalculable benefits in return. Mutuality of interest and reward is a possibility that can reach to any city backyard, garden, and park, but in any place under human dominance—which is, now, virtually everyplace —it is a possibility that is *both* natural and cultural. If humans want wildness to be possible, then they have to make it possible. If balance is the ruling principle and a stable balance the goal, then, for humans, attaining this goal requires a consciously chosen and deliberately made partnership with nature.

In other words, we can be true to nature only by being true to human nature—to our animal nature as well as to cultural patterns and restraints that keep us from acting like animals. When humans act like animals, they become the most dangerous of animals to themselves and other humans, and this is because of another critical difference between humans and animals: Whereas animals are usually restrained by the limits of physical appetites, humans have mental appetites that can be far more gross and capacious than physical ones. Only humans squander and hoard, murder and pillage because of notions.

The work by which good human and natural possibilities are preserved is complex and difficult, and it probably cannot be accomplished by raw intelligence and information. It requires knowledge, skills, and restraints, some of which must come from our past. In the hurry of technological progress, we have replaced some tools and methods that worked with some that do not work. But we also need culture-borne instructions about who or what humans are and how and on what assumptions they should act. The Chain of Being, for instance—which gave humans a place between animals and angels in the order of Creation—is an old idea that has not been replaced by any adequate new one. It was simply rejected, and the lack of it leaves us without a definition.

Lacking that ancient definition, or any such definition, we do not know at what point to restrain or deny ourselves. We do not know how ambitious to be, what or how much we may safely desire, when or where to stop. I knew a barber once who refused to give a discount to a bald client, explaining that his artistry consisted, not in the cutting off, but in the knowing when to stop. He spoke, I think, as a true artist and a true human. The lack of such knowledge is extremely dangerous in and to an individual. But ignorance of when to stop is a modern epidemic; it is the basis of "industrial progress" and "economic growth." The most obvious practical result of this ignorance is a critical disproportion of scale between the scale of human enterprises and their sources in nature.

The scale of the energy industry, for example, is too big, as is the scale of the transportation industry. The scale of agriculture, from a technological or economic point of view, is too big, but from a demographic point of view, the scale is too small. When there are enough people on the land to use it but not enough to husband it, then the wildness of the soil that we call fertility begins to diminish, and the soil itself begins to flee from us in water and wind.

If the human economy is to be fitted into the natural economy in such a way that both may thrive, the human economy must be built to proper scale. It is possible to talk at great length about the difference between proper and improper scale. It may be enough to say here that that difference is *suggested* by the difference between amplified and unamplified music in the countryside, or the difference between the sound of a motorboat and the sound of oarlocks. A proper human sound, we may say, is one that allows other sounds to be heard. A properly scaled human economy or technology allows a diversity of other creatures to thrive.

"The proper scale," a friend wrote to me, "confers freedom and simplicity . . . . and doubtless leads to long life and health." I think that it also confers joy. The renewal of our partnership with nature, the rejoining of our works to their proper places in the natural order, reshaped to their proper scale, implies the reenjoyment both of nature and of human domesticity. Though our task will be difficult, we will greatly mistake its nature if we see it as grim, or if we suppose that it must always be necessary to suffer at work in order to enjoy ourselves in places specializing in "recreation."

Once we grant the possibility of a proper human scale, we see that we have made a radical change of assumptions and values. We realize that we are less interested in technological "breakthroughs" than in technological elegance. Of a new tool or method we will no longer ask: Is it fast? Is it powerful? Is it a labor saver? How many workers will it replace? We will ask instead: Can we (and our children) afford it? Is it fitting to our real needs? Is it becoming to us? Is it unhealthy or ugly? And though we may keep a certain interest in innovation and in what we may become, we will renew our interest in what we have been, realizing that conservationists must necessarily conserve *both* inheritances, the natural and the cultural.

To argue the necessity of wildness to, and in, the human economy is by no means to argue against the necessity of wilderness. The survival of wilderness—of places that we do not change, where we allow the existence even of creatures we perceive as dangerous—is necessary. Our sanity probably requires it. Whether we go to those places or not, we need to know that they exist. And I would argue that we do not need just the great public wildernesses, but millions of small private or semi-private ones. Every farm should have one; wildernesses can occupy corners of factory grounds and city lots—places where nature is given a free hand, where no human work is done, where people go only as guests. These places function, I think, whether we intend them to or not, as sacred groves—places we respect and leave alone, not because we understand well what goes on there, but because we do not.

We go to wilderness places to be restored, to be instructed in the natural economies of fertility and healing, to admire what we cannot make. Sometimes, as we find to our surprise, we go to be chastened or corrected. And we go in order to return with renewed knowledge by which to judge the health of our human economy and our dwelling places. As we return from our visits to the wilderness, it is sometimes possible to imagine a series of fitting and decent transitions from wild nature to the human community and its supports: from forest to woodlot to the "two-story agriculture" of tree crops and pasture to orchard to meadow to grainfield to garden to household to neighborhood to village to city—so that even when we reached the city we would not be entirely beyond the influence of the nature of that place.

What I have been implying is that I think there is a bad reason to go to the wilderness. We must not go there to escape the ugliness and the dangers of the present human economy. We must not let ourselves feel that to go there is to escape. In the first place, such an escape is now illusory. In the second place, if, even as conservationists, we see the human and the natural economies as necessarily opposite or opposed, we subscribe to the very opposition that threatens to destroy them both. The wild and the domestic now often seem isolated values, estranged from one another. And yet these are not exclusive polarities like good and evil. There can be continuity between them, and there must be.

What we find, if we weight the balance too much in favor of the domestic, is that we involve ourselves in dangers both personal and public. Not the least of these dangers is dependence on distant sources of money and materials. Farmers are in deep trouble now because they have become too dependent on corporations and banks. They have been using methods and species that enforce this dependence. But such a dependence is not safe, either for farmers or for agriculture. It is not safe for urban consumers. Ultimately, as we are beginning to see, it is not safe for banks and corporations—which, though they have evidently not thought so, are dependent upon farmers. Our farms are endangered because—like the interstate highways or modern hospitals or modern universities—they cannot be inexpensively used. To be usable at all they require great expense.

When the human estate becomes so precarious, our only recourse is to move it back toward the estate of nature. We undoubtedly need better plant and animal species than nature provided us. But we are beginning to see that they can be too much better—too dependent on us and on "the economy," too expensive. In farm animals, for instance, we want good commercial quality, but we can see that the ability to produce meat or milk can actually be a threat to the farmer and to the animal if not accompanied by qualities we would call natural: thriftiness, hardiness, physical vigor, resistance to disease and parasites, ability to breed and give birth without assistance, strong mothering instincts. These natural qualities decrease care, work, and worry; they also decrease the costs of production. They save feed and time; they make diseases and cures exceptional rather than routine.

We need crop and forage species of high productive ability also, but we do not need species that will not produce at all without expensive fertilizers and chemicals. Contrary to the premise of agribusiness advertisements and of most expert advice, farmers do not thrive by production or by "skimming" a large "cash flow." They cannot solve their problems merely by increasing production or income. They thrive, like all other creatures, according to the difference between their income and their expenses.

One of the strangest characteristics of the industrial economy is the ability to increase production again and again without ever noticing—or without acknowledging—the *costs* of production. That one Holstein cow should produce 50,000 pounds of milk in a year may appear to be marvelous—a miracle of modern science. But what if her productivity is dependent upon the consumption of a huge amount of grain (about a bushel a day), and therefore upon the availability of cheap petroleum? What if she is too valuable (and too delicate) to be allowed outdoors in the rain? What if the proliferation of her kind will again drastically reduce the number of dairy farms and farmers? Or, to use a more obvious example, can we afford a bushel of grain at a cost of five to twenty bushels of topsoil lost to erosion?

"It is good to have Nature working for you," said Henry Besuden, the dean of American Southdown breeders. "She works for a minimum wage." That is true. She works at times for almost nothing, requiring only that we respect her work and give her a chance, as when she maintains—indeed, improves—the fertility and productivity of a pasture by the natural succession of clover and grass or when she improves a clay soil for us by means of the roots of a grass sod. She works for us by preserving health or wholeness, which for all our ingenuity we cannot make. If we fail to respect her health, she deals out her justice by withdrawing her protection against disease—which we *can* make, and do.

To make this continuity between the natural and the human, we have only two sources of instruction: nature herself and our cultural tradition. If we listen only to the apologists for the industrial economy, who respect neither nature nor culture, we get the idea that it is somehow our goodness that makes us so destructive: The air is unfit to breathe, the water is unfit to drink, the soil is washing away, the cities are violent and the countryside neglected, all because we are intelligent, enterprising, industrious, and generous, concerned only to feed the hungry and to "make a better future for our children." Respect for nature causes us to doubt this, and our cultural tradition confirms and illuminates our doubt: No good thing is destroyed by goodness; good things are destroyed by wickedness. We may identify that insight as Biblical, but it is taken for granted by both the Greek and the Biblical lineages of our culture, from Homer and Moses to William Blake. Since the start of the industrial revolution, there have been voices urging that this inheritance may be safely replaced by intelligence, information, energy, and money. No idea, I believe, could be more dangerous.

Two Economies

SOME TIME AGO, in a conversation with Wes Jackson in which we were laboring to define the causes of the modern ruination of farmland, we finally got around to the money economy. I said that an economy based on energy would be more benign because it would be more comprehensive.

Wes would not agree. "An energy economy still wouldn't be comprehensive enough."

"Well," I said, "then what kind of economy *would* be comprehensive enough?"

He hesitated a moment, and then, grinning, said, "The Kingdom of God."

I assume that Wes used the term because he found it, at that point in our conversation, indispensable; I assume so because, in my pondering over its occurrence at that point, I have found it indispensable myself. For the thing that troubles us about the industrial economy is exactly that it is not comprehensive enough, that, moreover, it tends to destroy what it does not comprehend, and that it is *dependent* upon much that it does not comprehend. In attempting to criticize such an economy, we naturally pose against it an economy that does not leave anything out, and we can say without presuming too much that the first principle of the Kingdom of God is that it includes everything; in it, the fall of every sparrow is a significant event. We are in it whether we know it or not and whether we wish to be or not. Another principle, both ecological and traditional, is that everything in the Kingdom of God is joined both to it and to everything else that is in it; that is to say, the Kingdom of God is orderly. A third principle is that humans do not and can never know either all the creatures that the Kingdom of God contains or the whole pattern or order by which it contains them.

The suitability of the Kingdom of God as, so to speak, a place name is partly owing to the fact that it still means pretty much what it has always meant. Because, I think, of the embarrassment that the phrase has increasingly caused among the educated, it has not been much tainted or tampered with by the disinterested processes of academic thought; it is a phrase that comes to us with its cultural strings still attached. To say that we live in the Kingdom of God is both to suggest the difficulty of our condition and to imply a fairly complete set of culture-borne instructions for living in it. These instructions are not always explicitly ecological, but it can be argued that they are always implicitly so, for all of them rest ultimately on the assumptions that I have given as the second and third principles of the Kingdom of God: that we live within order and that this order is both greater and more intricate than we can know. The difficulty of our predicament, then, is made clear if we add a fourth principle: Though we cannot produce a complete or even an adequate description of this order, severe penalties are in store for us if we presume upon it or violate it.

I am not dealing, of course, with perceptions that are only Biblical. The ancient Greeks, according to Aubrey de Sélincourt, saw "a continuing moral pattern in the vicissitudes of human fortune," a pattern "formed from the belief that men, as men, are subject to certain limitations imposed by a Power —call it Fate or God—which they cannot fully comprehend, and that any attempt to transcend those limitations is met by inevitable punishment." The Greek name for the pride that attempts to transcend human limitations was *hubris*, and hubris was the cause of what the Greeks understood as tragedy.

Nearly the same sense of *necessary* human limitation is implied in the Old Testament's repeated remonstrances against too great a human confidence in the power of "mine own hand." Gideon's army against the Midianites, for example, was reduced from thirty-two thousand to three hundred expressly to prevent the Israelites from saying, "Mine own hand hath saved me." A similar purpose was served by the institution of the Sabbath, when, by not working, the Israelites were meant to see the limited efficacy of their work and thus to understand their true dependence.

Though I hope that my insistence on the usefulness of the term, the Kingdom of God, will be understood, I must acknowledge that the term is local, in the sense that it is fully available only to those whose languages are involved in Western or Biblical tradition. A person of Eastern heritage, for example, might speak of the totality of all creation, visible and invisible, as "the Tao." I am well aware also that many people would not willingly use either term, or any such term. For these reasons, I do not want to make a statement that is specially or exclusively Biblical, and so I would like now to introduce a more culturally neutral term for that economy that I have been calling the Kingdom of God. Sometimes, in thinking about it, I have called it the Great Economy, which is the name I am going to make do with here—though I will remain under the personal necessity of Biblical reference. And that, I think, must be one of my points: We can name it whatever we wish, but we cannot define it except by way of a religious tradition. The Great Economy, like the Tao or the Kingdom of God, is both known and unknown, visible and invisible, comprehensible and mysterious. It is, thus, the ultimate condition of our experience and of the practical questions rising from our experience, and it imposes on our consideration of those questions an extremity of seriousness and an extremity of humility.

I am assuming that the Great Economy, whatever we may name it, is indeed—and in ways that are, to some extent, practical—an economy: It includes principles and patterns by which values or powers or necessities are parceled out and exchanged. But if the Great Economy comprehends humans and thus cannot be fully comprehended by them, then it is also not an economy in which humans can participate directly. What this suggests, in fact, is that humans can live in the Great Economy only with great uneasiness, subject to powers and laws that they can understand only in part. There is no human accounting for the Great Economy. This obviously is a description of the circumstance of religion, the circumstance that *causes* religion. De Sélincourt states the problem succinctly: "Religion in every age is concerned with the vast and fluctuant regions of experience which knowledge cannot penetrate, the regions which a man knows, or feels, to stretch away beyond the narrow, closed circle of what he can *manage* by the use of his wits."

If there is no denying our dependence on the Great Economy, there is also no denying our need for a little economy—a narrow circle within which things are manageable by the use of our wits. I don't think Wes Jackson was denying this need when he invoked the Kingdom of God as the complete economy; rather, he was, I think, insisting upon a priority that is both proper and practical. If he had a text in mind, it must have been the sixth chapter of Matthew, in which, after speaking of God's care for nature, the fowls of the air and the lilies of the field, Jesus says: "Therefore take no thought, saying, What shall we eat? or, What shall we drink? or, Wherewithal shall we be clothed? . . . But seek ye first the kingdom of God, and his righteousness; and all these things shall be added unto you."

There is an attitude that sees in this text a denial of the value of *any* economy of this world, but this attitude makes the text useless and meaningless to humans who must live in this world. These verses make usable sense only if we read them as a statement of considerable practical import about the real nature of worldly economy. If this passage meant for us to seek *only* the Kingdom of God, it would have the odd result of making good people not only feckless but also dependent upon bad people busy with quite other seekings. It says, rather, to seek the Kingdom of God *first*; that is, it gives an obviously necessary priority to the Great Economy over any little economy made within it. The passage also clearly includes nature within the Great Economy, and it affirms the goodness, indeed the sanctity, of natural creatures.

The fowls of the air and the lilies of the field live within the Great Economy entirely by nature, whereas humans, though entirely dependent upon it, must live in it partly by artifice. The birds can live in the Great Economy only as birds, the flowers only as flowers, the humans only as humans. The humans, unlike the wild creatures, may choose not to live in it —or, rather, since no creature can escape it, they may choose to *act* as if they do not, or they may choose to try to live in it on their own terms. If humans choose to live in the Great Economy on *its* terms, then they must live in harmony with it, maintaining it in trust and learning to consider the lives of the wild creatures.

Certain economic restrictions are clearly implied, and these restrictions have mainly to do with the economics of futurity. We know from other passages in the Gospels that a certain preparedness or provisioning for the future is required of us.

It may be that such preparedness is part of our obligation to today, and for *that* reason we need "take no thought for the morrow." But it is clear that such preparations can be carried too far, that we can provide too much for the future. The sin of "a certain rich man" in the twelfth chapter of Luke is that he has "much goods laid up for many years" and thus believes that he can "eat, drink, and be merry." The offense seems to be that he has stored up too much and in the process has belittled the future, for he has reduced it to the size of his own hopes and expectations. He is prepared for a future in which he will be prosperous, not for one in which he will be dead. We know from our own experience that it is possible to live in the present in such a way as to diminish the future practically as well as spiritually. By laying up "much goods" in the present—and, in the process, *using up* such goods as topsoil, fossil fuel, and fossil water—we incur a debt to the future that we cannot repay. That is, we diminish the future by deeds that we call "use" but that the future will call "theft." We may say, then, that we seek the Kingdom of God, in part, by our economic behavior, and we fail to find it if that behavior is wrong.

If we read Matthew 6:24–34 as a teaching that is *both* practical and spiritual, as I think we must, then we must see it as prescribing the terms of a kind of little economy or human economy. Since I am deriving it here from a Christian text, we could call it a Christian economy. But we need not call it that. A Buddhist might look at the working principles of the economy I am talking about and call it a Buddhist economy. E. F. Schumacher, in fact, says that the aim of "Buddhist economics" is "to obtain the maximum of well-being with the minimum of consumption," which I think is partly the sense of Matthew 6:24–34. Or we could call this economy (from Matthew 6:28) a "considerate" economy or, simply, a good economy. Whatever the name, the human economy, if it is to be a good economy, must fit harmoniously within and must correspond to the Great Economy; in certain important ways, it must be an analogue of the Great Economy.

A fifth principle of the Great Economy that must now be added to the previous four is that *we* cannot foresee an end to it: The same basic stuff is going to be shifting from one form to another, so far as we know, forever. From a human point of view, this is a rather heartless endurance. As cynics sometimes point out, conservation is always working, for what is lost or wasted in one place always turns up someplace else. Thus, soil erosion in Iowa involves no loss because the soil is conserved in the Gulf of Mexico. Such people like to point out that soil erosion is as "natural" as birdsong. And so it is, though these people neglect to observe that soil conservation is also natural, and that, before the advent of farming, nature alone worked effectively to keep Iowa topsoil in Iowa. But to say that soil erosion is natural is only a way of saying that there are some things that the Great Economy cannot do for humans. Only a little economy, only a good human economy, can define for us the value of keeping the topsoil where it is.

A good human economy, that is, defines and values human goods, and, like the Great Economy, it conserves and protects its goods. It proposes to endure. Like the Great Economy, a good human economy does not propose for itself a term to be set by humans. That termlessness, with all its implied human limits and restraints, is a human good.

The difference between the Great Economy and any human economy is pretty much the difference between the goose that laid the golden egg and the golden egg. For the goose to have value as a layer of golden eggs, she must be a live goose and therefore joined to the life cycle, which means that she is joined to all manner of things, patterns, and processes that sooner or later surpass human comprehension. The golden egg, on the other hand, can be fully valued by humans according to kind, weight, and measure—but it will not hatch, and it cannot be eaten. To make the value of the egg *fully* accountable, then, we must make it "golden," must remove it from life. But if in our valuation of it, we wish to consider its relation to the goose, we have to undertake a different kind of accounting, more exacting if less exact. That is, if we wish to value the egg in such a way as to preserve the goose that laid it, we find that we must behave, not scientifically, but humanely; we must understand ourselves as humans as fully as our traditional knowledge of ourselves permits. We participate in our little human economy to a considerable extent, that is, by factual knowledge, calculation, and manipulation; our participation in the Great Economy also requires those things, but requires as well humility, sympathy, forbearance, generosity, imagination.

Another critical difference, implicit in the foregoing, is that, though a human economy can evaluate, distribute, use, and preserve things of value, it cannot make value. Value can originate only in the Great Economy. It is true enough that humans can add value to natural things: We may transform trees into boards, and transform boards into chairs, adding value at each transformation. In a good human economy, these transformations would be made by good work, which would be properly valued and the workers properly rewarded. But a good human economy would recognize at the same time that it was dealing all along with materials and powers that it did not make. It did not make trees, and it did not make the intelligence and talents of the human workers. What the humans have added at every step is artificial, made by art, and though the value of art is critical to human life, it is a secondary value.

When humans presume to originate value, they make value that is first abstract and then false, tyrannical, and destructive of real value. Money value, for instance, can be said to be true only when it justly and stably represents the value of necessary goods, such as clothing, food, and shelter, which originate ultimately in the Great Economy. Humans can originate money value in the abstract, but only by inflation and usury, which falsify the value of necessary things and damage their natural and human sources. Inflation and usury and the damages that follow can be understood, perhaps, as retributions for the presumption that humans can make value.

We may say, then, that a human economy originates, manages, and distributes secondary or added values but that, if it is to last long, it must also manage in such a way as to make continuously available those values that are primary or given, the secondary values having mainly to do with husbandry and trusteeship. A little economy is obliged to receive them gratefully and to use them in such a way as not to diminish them. We might make a long list of things that we would have to describe as primary values, which come directly into the little economy from the Great, but the one I want to talk about, because it is the one with which we have the most intimate working relationship, is the topsoil.

We cannot speak of topsoil, indeed we cannot know what it is, without acknowledging at the outset that we cannot make it. We can care for it (or not), we can even, as we say, "build" it, but we can do so only by assenting to, preserving, and perhaps collaborating in its own processes. To those processes themselves we have nothing to contribute. We cannot make topsoil, and we cannot make any substitute for it; we cannot do what it does. It is apparently impossible to make an adequate description of topsoil in the sort of language that we have come to call "scientific." For, although any soil sample can be reduced to its inert quantities, a handful of the real thing has life in it; it is full of living creatures. And if we try to describe the behavior of that life we will see that it is doing something that, if we are not careful, we will call "unearthly": It is making life out of death. Not so very long ago, had we known about it what we know now, we would probably have called it "miraculous." In a time when death is looked upon with almost universal enmity, it is hard to believe that the land we live on and the lives we live are the gifts of death. Yet that is so, and it is the topsoil that makes it so. In fact, in talking about topsoil, it is hard to avoid the language of religion. When, in "This Compost," Whitman says, "The resurrection of the wheat appears with pale visage out of its graves," he is speaking in the Christian tradition, and yet he is describing what happens, with language that is entirely accurate and appropriate. And when at last he says of the earth that "It gives such divine materials to men," we feel that the propriety of the words comes not from convention but from the actuality of the uncanny transformation that his poem has required us to imagine, as if in obedience to the summons to "consider the lilies of the field."

Even in its functions that may seem, to mechanists, to be mechanical, the topsoil behaves complexly and wonderfully. A healthy topsoil, for instance, has at once the ability to hold water and to drain well. When we speak of the health of a watershed, these abilities are what we are talking about, and the word "health," which we do use in speaking of watersheds, warns us that we are not speaking merely of mechanics. A healthy soil is made by the life dying into it and by the life living in it, and to its double ability to drain and retain water we are complexly indebted, for it not only gives us good crops but also erosion control as well as *both* flood control and a constant water supply.

Obviously, topsoil, not energy or money, is the critical quantity in agriculture. And topsoil *is* a quantity; we need it in quantities. We now need more of it than we have; we need to help it to make more of itself. But it is a most peculiar quantity, for it is inseparable from quality. Topsoil is by definition *good* soil, and it can be preserved in human use only by good care. When humans see it as a mere quantity, they tend to make it that; they destroy the life in it, and they begin to measure in inches and feet and tons how much of it they have "lost."

When we see the topsoil as the foundation of that household of living creatures and their nonliving supports that we now call an "ecosystem" but which some of us understand better as a "neighborhood," we find ourselves in debt for other benefits that baffle our mechanical logic and defy our measures. For example, one of the principles of an ecosystem is that diversity increases capacity—or, to put it another way, that complications of form or pattern can increase greatly within quantitative limits. I suppose that this may be true only up to a point, but I suppose also that that point is far beyond the human capacity to understand or diagram the pattern.

On a farm put together on a sound ecological pattern, the same principle holds. Henry Besuden, the great farmer and shepherd of Clark County, Kentucky, compares the small sheep flock to the two spoons of sugar that can be added to a brimful cup of coffee, which then becomes "more palatable [but] doesn't run over. You can stock your farm to the limit with other livestock and still add a small flock of sheep." He says this, characteristically, after rejecting the efforts of sheep specialists to get beyond "the natural physical limits of the ewe" by breeding out of season in order to get three lamb crops in two years or by striving for "litters" of lambs rather than nature's optimum of twins. Rather than chafe at "natural physical limits," he would turn to nature's elegant way of enriching herself *within* her physical limits by diversification, by complication of pattern. Rather than strain the productive capacity of the ewe, he would, without strain, enlarge the productive capacity of the farm—a healthier, safer, and cheaper procedure. Like many of the better traditional farmers, Henry Besuden is suspicious of "the measure of land in length and width," for he would be mindful as well of "the depth and quality."

A small flock of ewes, fitted properly into a farm's pattern, virtually disappears into the farm and does it good, just as it virtually disappears into the time and energy economy of a farm family and does it good. And, properly fitted into the farm's pattern, the small flock virtually disappears from the debit side of the farm's accounts but shows up plainly on the credit side. This "disappearance" is possible, not to the extent that the farm is a human artifact, a belonging of the human economy, but to the extent that it remains, by its obedience to natural principle, a belonging of the Great Economy.

A little economy may be said to be good insofar as it perceives the excellence of these benefits and husbands and preserves them. It is by holding up this standard of goodness that we can best see what is wrong with the industrial economy. For the industrial economy does not see itself as a little economy; it sees itself as the *only* economy. It makes itself thus exclusive by the simple expedient of valuing only what it can use—that is, only what it can regard as "raw material" to be transformed mechanically into something else. What it cannot use, it characteristically describes as "useless," "worthless," "random," or "wild," and gives it some such name as "chaos," "disorder," or "waste"—and thus ruins it or cheapens it in preparation for eventual use. That western deserts or eastern mountains were once perceived as "useless" made it easy to dignify them by the "use" of strip mining. Once we acknowledge the existence of the Great Economy, however, we are astonished and frightened to see how much modern enterprise is the work of hubris, occurring outside the human boundary established by ancient tradition. The industrial economy is based on invasion and pillage of the Great Economy.

The weakness of the industrial economy is clearly revealed when it imposes its terms upon agriculture, for its terms cannot define those natural principles that are most vital to the life and longevity of farms. Even if the industrial economists could afford to do so, they could not describe the dependence of agriculture upon nature. If asked to consider the lilies of the field or told that the wheat is resurrected out of its graves, the agricultural industrialist would reply that "my engineer's mind inclines less toward the poetic and philosophical, and more toward the practical and possible," unable even to suspect that such a division of mind induces blindness to possibilities of the utmost practical concern.

That good topsoil both drains and retains water, that diversity increases capacity, are facts similarly alien to industrial logic. Industrialists see retention and drainage as different and opposite functions, and they would promote one at the expense of the other, just as, diversity being inimical to industrial procedure, they would commit themselves to the forlorn expedient of enlarging capacity by increasing area. They are thus encumbered by dependence on mechanical solutions that can work only by isolating and oversimplifying problems. Industrialists are condemned to proceed by devices. To facilitate water retention, they must resort to a specialized water-holding device such as a terrace or a dam; to facilitate drainage, they must use drain tile, or a ditch, or a "subsoiler." It is possible, I know, to argue that this analysis is too general and to produce exceptions, but I do not think it deniable that the discipline of soil conservation is now principally that of the engineer, not that of the farmer or soil husband—that it is now a matter of digging in the earth, not of enriching it.

I do not mean to say that the devices of engineering are always inappropriate; they have their place, not least in the restoration of land abused by the devices of engineering. My point is that, to facilitate both water retention and drainage in the same place, we must improve the soil, which is not a mechanical device but, among other things, a graveyard, a place of resurrection, and a community of living creatures. Devices may sometimes help, but only up to a point, for soil is improved by what humans do not do as well as by what they do. The proprieties of soil husbandry require acts that are much more complex than industrial acts, for these acts are conditioned by the ability *not* to act, by forbearance or self-restraint, sympathy or generosity. The industrial act is simply prescribed by thought, but the act of soil building is also *limited* by thought. We build soil by knowing what to do but also by knowing what not to do and by knowing when to stop. Both kinds of knowledge are necessary because invariably, at some point, the reach of human comprehension becomes too short, and at that point the work of the human economy must end in absolute deference to the working of the Great Economy. This, I take it, is the practical significance of the idea of the Sabbath.

To push our work beyond that point, invading the Great Economy, is to become guilty of hubris, of presuming to be greater than we are. We cannot do what the topsoil does, any more than we can do what God does or what a swallow does. We can fly, but only as humans—very crudely, noisily, and clumsily. We can dispose of corpses and garbage, but we cannot, by our devices, turn them into fertility and new life. And we are discovering, to our great uneasiness, that we cannot dispose at all of some of our so-called wastes that are toxic or radioactive. We can appropriate and in some fashion use godly powers, but we cannot use them safely, and we cannot control the results. That is to say that the human condition remains for us what it was for Homer and the authors of the Bible. Now that we have brought such enormous powers to our aid (we hope), it seems more necessary than ever to observe how inexorably the human condition still contains us. We only do what humans can do, and our machines, however they may appear to enlarge our possibilities, are invariably infected with our limitations. Sometimes, in enlarging our possibilities, they narrow our limits and leave us more powerful but less content, less safe, and less free. The mechanical means by which we propose to escape the human condition only extend it; thinking to transcend our definition as fallen creatures, we have only colonized more and more territory eastward of Eden.

II

Like the rich man of the parable, the industrialist thinks to escape the persistent obligations of the human condition by means of "much goods laid up for many years"—by means, in other words, of quantities: resources, supplies, stockpiles, funds, reserves. But this is a grossly oversimplifying dream and, thus, a dangerous one. All the great natural goods that empower agriculture, some of which I have discussed, have to do with quantities, but they have to do also with qualities, and they involve principles that are not static but active; they have to do with formal processes. The topsoil exists as such because it is ceaselessly transforming death into life, ceaselessly supplying food and water to all that lives in it and from it; otherwise, "All flesh shall perish together, and man shall turn again unto dust." If we are to live well on and from our land, we must live by faith in the ceaselessness of these processes and by faith in our own willingness and ability to collaborate with them. Christ's prayer for "daily bread" is an affirmation of such faith, just as it is a repudiation of faith in "much goods laid up." Our life and livelihood are the gift of the topsoil and of our willingness and ability to care for it, to grow good wheat, to make good bread; they do not derive from stockpiles of raw materials or accumulations of purchasing power.

The industrial economy can define potentiality, even the potentiality of the living topsoil, only as a *fund*, and thus it must accept impoverishment as the inescapable condition of abundance. The invariable mode of its relation both to nature and to human culture is that of mining: withdrawal from a limited fund until that fund is exhausted. It removes natural fertility and human workmanship from the land, just as it removes nourishment and human workmanship from bread. Thus the land is reduced to abstract marketable quantities of length and width, and bread to merchandise that is high in money value but low in food value. "Our bread," Guy Davenport once said to me, "is more obscene than our movies."

But the industrial use of *any* "resource" implies its exhaustion. It is for this reason that the industrial economy has been accompanied by an ever-increasing hurry of research and exploration, the motive of which is not "free enterprise" or "the spirit of free inquiry," as industrial scientists and apologists would have us believe, but the desperation that naturally and logically accompanies gluttony.

One of the favorite words of the industrial economy is "control": We want "to keep things under control"; we wish (or so we say) to "control" inflation and erosion; we have a discipline known as "crowd control"; we believe in "controlled growth" and "controlled development," in "traffic control" and "self-control." But, because we are always setting out to control something that we refuse to limit, we have made control a permanent and a helpless enterprise. If we will not limit causes, there can be no controlling of effects. What is to be the fate of self-control in an economy that encourages and rewards unlimited selfishness?

More than anything else, we would like to "control the forces of nature," refusing at the same time to impose any limit on human nature. We assume that such control and such freedom are our "rights," which seems to ensure that our means of control (of nature and of all else that we see as alien) will be violent. It is startling to recognize the extent to which the industrial economy depends upon controlled explosions—in mines, in weapons, in the cylinders of engines, in the economic pattern known as "boom and bust." This dependence is the result of a progress that can be argued for, but those who argue for it must recognize that, in all these means, good ends are served by a destructive principle, an association that is difficult to control if it is not limited; moreover, they must recognize that our failure to limit this association has raised the specter of uncontrollable explosion. Nuclear holocaust, if it comes, will be the final detonation of an explosive economy.

An explosive economy, then, is not only an economy that is dependent upon explosions but also one that sets no limits on itself. Any little economy that sees itself as unlimited is obviously self-blinded. It does not see its real relation of dependence and obligation to the Great Economy; in fact, it does not see that there *is* a Great Economy. Instead, it calls the Great Economy "raw material" or "natural resources" or "nature" and proceeds with the business of putting it "under control."

But "control" is a word more than ordinarily revealing here, for its root meaning is to roll against, in the sense of a little wheel turning in opposition. The principle of control, then, involves necessarily the principle of division: One thing may turn against another thing only by being divided from it. This mechanical division and turning in opposition William Blake understood as evil, and he spoke of "Satanic wheels" and "Satanic mills": "wheel without wheel, with cogs tyrannic / Moving by compulsion each other." By "wheel without wheel," Blake meant wheel outside of wheel, one wheel communicating motion to the other in the manner of two cogwheels, the point being that one wheel can turn another wheel outside itself only in a direction opposite to its own. This, I suppose, is acceptable enough as a mechanism. It becomes "Satanic" when it becomes a ruling metaphor and is used to describe and to organize fundamental relationships. Against the Satanic "wheel without wheel," Blake set the wheels of Eden, which "Wheel within wheel in freedom revolve, in harmony and peace." This is the "wheel in the middle of a wheel" of Ezekiel's vision, and it is an image of harmony. That the relation of these wheels is not mechanical we know from Ezekiel 1:21: "the spirit of the living creature was in the wheels." The wheels of opposition oppose the spirit of the living creature.

What had happened, as Blake saw accurately and feared justifiably, was a fundamental shift in the relation of humankind to the rest of creation. Sometime between, say, Pope's verses on the Chain of Being in *An Essay on Man* and Blake's "London," the dominant minds had begun to see the human race, not as a part or a member of Creation, but as outside it and opposed to it. The industrial revolution was only a part of this change, but it is true that, when the wheels of the industrial revolution began to revolve, they turned against nature, which became the name for all of Creation thought to be below humanity, as well as, incidentally, against all once thought to be above humanity. Perhaps this would have been safe enough if nature—that is, if all the rest of Creation—had been, as proposed, passively subject to human purpose.

Of course, it never has been. As Blake foresaw, and as we now know, what we turn against must turn against us. Blake's image of the cogwheels turning in relentless opposition is terrifyingly apt, for in our vaunted war against nature, nature fights back. The earth may answer our pinches and pokes "only with spring," as e. e. cummings said, but if we pinch and poke too much, she can answer also with flood or drouth, with catastrophic soil erosion, with plague and famine. Many of the occurrences that we call "acts of God" or "accidents of nature" are simply forthright natural responses to human provocations. Not always; I do not mean to imply here that, by living in harmony with nature, we can be free of floods and storms and drouths and earthquakes and volcanic eruptions; I am only pointing out, as many others have done, that, by living in opposition to nature, we can *cause* natural calamities of which we would otherwise be free.

*

The problem seems to be that a human economy cannot prescribe the terms of its own success. In a time when we wish to believe that humans are the sole authors of the truth, that truth is relative, and that value judgments are all subjective, it is hard to say that a human economy can be wrong, and yet we have good, sound, practical reasons for saying so. It is indeed possible for a human economy to be wrong—not relatively wrong, in the sense of being "out of adjustment," or unfair according to some human definition of fairness, or weak according to the definition of its own purposes—but wrong absolutely and according to practical measures. Of course, if we see the human economy as the *only* economy, we will see its errors as political failures, and we will continue to talk about "recovery." It is only when we think of the little human economy in relation to the Great Economy that we begin to understand our errors for what they are and to see the qualitative meanings of our quantitative measures. If we see the industrial economy in terms of the Great Economy, then we begin to see industrial wastes and losses not as "trade-offs" or "necessary risks" but as costs that, like all costs, are chargeable to somebody, sometime.

That we can prescribe the terms of our own success, that we can live outside or in ignorance of the Great Economy are the greatest errors. They condemn us to a life without a standard, wavering in inescapable bewilderment from paltry self-satisfaction to paltry self-dissatisfaction. But since we have no place to live but in the Great Economy, whether or not we know that and act accordingly is the critical question, not about economy merely, but about human life itself.

It is possible to make a little economy, such as our present one, that is so short-sighted and in which accounting is of so short a term as to give the impression that vices are necessary and practically justifiable. When we make our economy a little wheel turning in opposition to what we call "nature," then we set up competitiveness as the ruling principle in our explanation of reality and in our understanding of economy; we make of it, willy-nilly, a virtue. But competitiveness, as a ruling principle and a virtue, imposes a logic that is extremely difficult, perhaps impossible, to control. That logic explains why our cars and our clothes are shoddily made, why our "wastes" are toxic, and why our "defensive" weapons are suicidal; it explains why it is so difficult for us to draw a line between "free enterprise" and crime. If our economic ideal is maximum profit with minimum responsibility, why should we be surprised to find our corporations so frequently in court and robbery on the increase? Why should we be surprised to find that medicine has become an exploitive industry, profitable in direct proportion to its hurry and its mechanical indifference? People who pay for shoddy products or careless services and people who are robbed outright are equally victims of theft, the only difference being that the robbers outright are not guilty of fraud.

If, on the other hand, we see ourselves as living within the Great Economy, under the necessity of making our little human economy within it, according to its terms, the smaller wheel turning in sympathy with the greater, receiving its being and its motion from it, then we see that the traditional virtues are necessary and are practically justifiable. Then, because in the Great Economy *all* transactions count and the account is never "closed," the ideal changes. We see that we cannot *afford* maximum profit or power with minimum responsibility because, in the Great Economy, the loser's losses finally afflict the winner. Now the ideal must be "the maximum of well-being with the minimum of consumption," which both defines and requires neighborly love. Competitiveness cannot be the ruling principle, for the Great Economy is not a "side" that we can join nor are there such "sides" within it. Thus, it is not the "sum of its parts" but a *membership* of parts inextricably joined to each other, indebted to each other, receiving significance and worth from each other and from the whole. One is obliged to "consider the lilies of the field," not because they are lilies or because they are exemplary, but because they are fellow members and because, as fellow members, we and the lilies are in certain critical ways alike.

To say that within the Great Economy the virtues are necessary and practically justifiable is at once to remove them from that specialized, sanctimonious, condescending practice of virtuousness that is humorless, pointless, and intolerable to its beneficiaries. For a human, the good choice in the Great Economy is to see its membership as a neighborhood and oneself as a neighbor within it. I am sure that virtues count in a neighborhood—to "love thy neighbor as thyself" requires the help of all seven of them—but I am equally sure that in a neighborhood the virtues cannot be practiced as such. Temperance has no appearance or action of its own, nor does justice, prudence, fortitude, faith, hope, or charity. They can only be employed on occasions. "He who would do good to another," William Blake said, "must do it in Minute Particulars." To help each other, that is, we must go beyond the coldhearted charity of the "general good" and get down to work where we are:

Labour well the Minute Particulars, attend to the Little-ones,
And those who are in misery cannot remain so long
If we do but our duty: labour well the teeming Earth.

It is the Great Economy, not any little economy, that invests minute particulars with high and final importance. In the Great Economy, each part stands for the whole and is joined to it; the whole is present in the part and is its health. The industrial economy, by contrast, is always striving and failing to make fragments (pieces that *it* has broken) *add up* to an ever-fugitive wholeness.

Work that is authentically placed and understood within the Great Economy moves virtue toward virtuosity—that is, toward skill or technical competence. There is no use in helping our neighbors with their work if we do not know how to work. When the virtues are rightly practiced within the Great Economy, we do not call them virtues; we call them good farming, good forestry, good carpentry, good husbandry, good weaving and sewing, good homemaking, good parenthood, good neighborhood, and so on. The general principles are submerged in the particularities of their engagement with the world. Lao Tzu saw the appearance of the virtues as such, in the abstract, as indicative of their loss:

When people lost sight of the way to live
Came codes of love and honesty . . .
When differences weakened family ties
Came benevolent fathers and dutiful sons;
And when lands were disrupted and misgoverned
Came ministers commended as loyal.

And these lines might be read as an elaboration of the warning against the *appearances* of goodness at the beginning of the sixth chapter of Matthew.

The work of the small economy, when it is understandingly placed within the Great Economy, minutely particularizes the virtues and carries principle into practice; to the extent that it does so, it escapes specialization. The industrial economy requires the extreme specialization of work—the separation of work from its results—because it subsists upon divisions of interest and must deny the fundamental kinships of producer and consumer; seller and buyer; owner and worker; worker, work, and product; parent material and product; nature and artifice; thoughts, words, and deeds. Divided from those kinships, specialized artists and scientists identify themselves as "observers" or "objective observers"—that is, as outsiders without responsibility or involvement. But the industrialized arts and sciences are false, their division is a lie, for there is no specialization of results.

There is no "outside" to the Great Economy, no escape into either specialization or generality, no "time off." Even insignificance is no escape, for in the membership of the Great Economy everything signifies; whatever we do counts. If we do not serve what coheres and endures, we serve what disintegrates and destroys. We can *presume* that we are outside the membership that includes us, but that presumption only damages the membership—and ourselves, of course, along with it.

In the industrial economy, the arts and the sciences are specialized "professions," each having its own language, speaking to none of the others. But the Great Economy proposes arts and sciences of membership: ways of doing and ways of knowing that cannot be divided from each other or within themselves and that speak the common language of the communities where they are practiced.

The Loss of the University

THE PREDICAMENT of literature within the university is not fundamentally different from the predicament of any other discipline, which is not fundamentally different from the predicament of language. That is, the various disciplines have ceased to speak to each other; they have become too specialized, and this overspecialization, this separation, of the disciplines has been enabled and enforced by the specialization of their languages. As a result, the modern university has grown, not according to any unifying principle, like an expanding universe, but according to the principle of miscellaneous accretion, like a furniture storage business.

I assume that there is a degree of specialization that is unavoidable because concentration involves a narrowing of attention; we can only do one thing at a time. I assume further that there is a degree of specialization that is desirable because good work depends upon sustained practice. If we want the best work to be done in teaching or writing or stone masonry or farming, then we must arrange for that work to be done by proven master workers, people who are prepared for the work by long and excellent practice.

But to assume that there is a degree of specialization that is proper is at the same time to assume that there is a degree that is improper. The impropriety begins, I think, when the various kinds of workers come to be divided and cease to speak to one another. In this division they become makers of *parts* of things. This is the impropriety of industrial organization, of which Eric Gill wrote, "Skill in making . . . degenerates into mere dexterity, i.e. skill in doing, when the workman . . . ceases to be concerned for the thing made or . . . has no longer any responsibility for the thing made and has therefore lost the knowledge of what it is that he is making. . . . The factory hand can only know what he is *doing*. What is being made is no concern of his." Part of the problem in universities now (or part of the cause of the problem) is this loss of concern for the thing made and, back of that, I think, the loss of agreement on what the thing is that is being made.

The thing being made in a university is humanity. Given the current influence of universities, this is merely inevitable. But what universities, at least the public-supported ones, are *mandated* to make or to help to make is human beings in the fullest sense of those words—not just trained workers or knowledgeable citizens but responsible heirs and members of human culture. If the proper work of the university is only to equip people to fulfill private ambitions, then how do we justify public support? If it is only to prepare citizens to fulfill public responsibilities, then how do we justify the teaching of arts and sciences? The common denominator has to be larger than either career preparation or preparation for citizenship. Underlying the idea of a university—the bringing together, the combining into one, of all the disciplines—is the idea that good work and good citizenship are the inevitable by-products of the making of a good—that is, a fully developed—human being. This, as I understand it, is the definition of the name *university.*

In order to be concerned for the thing made, in order even to know what it is making, the university as a whole must speak the same language as all of its students and all of its graduates. There must, in other words, be a common tongue. Without a common tongue, a university not only loses concern for the thing made; it loses its own unity. Furthermore, when the departments of a university become so specialized that they can speak neither to each other nor to the students and graduates of other departments, then that university is displaced. As an institution, it no longer knows where it is, and therefore it cannot know either its responsibilities to its place or the effects of its irresponsibility. This too often is the practical meaning of "academic freedom": The teacher feels free to teach and learn, make and think, without concern for the thing made.

For example, it is still perfectly acceptable in land-grant universities for agricultural researchers to apply themselves to the development of more productive dairy cows without considering at all the fact that this development necessarily involves the failure of many thousands of dairies and dairy farmers—that it has already done so and will inevitably continue to do so. The researcher feels at liberty to justify such work merely on the basis of the ratio between the "production unit" and the volume of production. And such work is permitted to continue, I suspect, because it is reported in language that is unreadable and probably unintelligible to nearly everybody in the university, to nearly everybody who milks cows, and to nearly everybody who drinks milk. That a modern university might provide a forum in which such researchers might be required to defend their work before colleagues in, say, philosophy or history or literature is, at present, not likely, nor is it likely, at present, that the departments of philosophy, history, or literature could produce many colleagues able or willing to be interested in the ethics of agricultural research.

Language is at the heart of the problem. To profess, after all, is "to confess before"—to confess, I assume, before all who live within the neighborhood or under the influence of the confessor. But to confess before one's neighbors and clients in a language that few of them can understand is not to confess at all. The specialized professional language is thus not merely a contradiction in terms; it is a cheat and a hiding place; it may, indeed, be an ambush. At the very root of the idea of profession and professorship is the imperative to speak plainly in the common tongue.

That the common tongue should become the exclusive specialty of a department in a university is therefore a tragedy, and not just for the university and its worldly place; it is a tragedy for the common tongue. It means that the common tongue, so far as the university is concerned, *ceases* to be the common tongue; it becomes merely one tongue within a confusion of tongues. Our language and literature cease to be seen as occurring in the world, and begin to be seen as occurring within their university department and within themselves. Literature ceases to be the meeting ground of all readers of the common tongue and becomes only the occasion of a deafening clatter *about* literature. Teachers and students read the great songs and stories to learn *about* them, not to learn *from* them. The *texts* are tracked as by the passing of an army of ants, but the power of songs and stories to affect life is still little acknowledged, apparently because it is little felt.

The specialist approach, of course, is partly justifiable; in both speech and literature, language does occur within itself. It echoes within itself, reverberating endlessly like a voice echoing within a cave, and speaking in answer to its echo, and the answer again echoing. It must do this; its nature, in part, is to do this.

But its nature also is to turn outward to the world, to strike its worldly objects cleanly and cease to echo—to achieve a kind of rest and silence in them. The professionalization of language and of language study makes the cave inescapable; one strives without rest in the interior clamor.

The silence in which words return to their objects, touch them, and come to rest is not the silence of the plugged ear. It is the world's silence, such as occurs after the first hard freeze of autumn, when the weeks-long singing of the crickets is suddenly stopped, and when, by a blessedly recurring accident, all machine noises have stopped for the moment, too. It is a silence that must be prepared for and waited for; it requires a silence of one's own.

The reverberations of language within itself are, finally, mere noise, no better or worse than the noise of accumulated facts that grate aimlessly against each other in think tanks and other hollow places. Facts, like words, are not things but verbal tokens or signs of things that finally must be carried back to the things they stand for to be verified. This carrying back is not specialist work but an act generally human, though only properly humbled and quieted humans can do it. It is an act that at once enlarges and shapes, frees and limits us.

It is necessary, for example, that the word *tree* evoke memories that are both personal and cultural. In order to understand fully what a tree is, we must remember much of our experience with trees and much that we have heard and read about them. We destroy those memories by reducing trees to facts, by thinking of *tree* as a mere word, or by treating our memory of trees as "cultural history." When we call a tree a tree, we are not isolated among words and facts but are at once in the company of the tree itself and surrounded by ancestral voices calling out to us all that trees have been and meant. This is simply the condition of being human in this world, and there is nothing that art and science can do about it, except get used to it. But, of course, only specialized "professional" arts and sciences would propose or wish to do something about it.

This necessity for words and facts to return to their objects in the world describes one of the boundaries of a university, one of the boundaries of book learning anywhere, and it describes the need for humility, restraint, exacting discipline, and high standards within that boundary.

Beside every effort of making, which is necessarily narrow, there must be an effort of judgment, of criticism, which must be as broad as possible. That is, every made thing must be submitted to these questions: What is the quality of this thing as a human artifact, as an addition to the world of made and of created things? How suitable is it to the needs of human and natural neighborhoods?

It must, of course, sooner or later be submitted as well to the special question: How good is this poem or this farm or this hospital as such? For it to have a human value, it obviously must be well made; it must meet the specialized, technical criteria; it must be *good* as such. But the question of its quality as such is not interesting—in the long run it is probably not even askable—unless we ask it under the rule of the more general questions. If we are disposed to judge apart from the larger questions, if we judge, as well as make, as specialists, then a good forger has as valid a claim to our respect as a good artist.

These two problems, how to make and how to judge, are the business of education. But education has tended increasingly to ignore the doubleness of its obligation. It has concerned itself more and more exclusively with the problem of how to make, narrowing the issue of judgment virtually to the terms of the made thing itself. But the thing made by education now is not a fully developed human being; it is a specialist, a careerist, a graduate. In industrial education, the thing *finally* made is of no concern to the makers.

In some instances, this is because the specialized "fields" have grown so complicated within themselves that the curriculum leaves no time for the broad and basic studies that would inform judgment. In other instances, one feels that there is a potentially embarrassing conflict between judgment broadly informed and the specialized career for which the student is being prepared; teachers of advertising techniques, for example, could ill afford for their students to realize that they are learning the arts of lying and seduction. In all instances, this narrowing is justified by the improbable assumption that young students, before they know anything else, know what they need to learn.

If the disintegration of the university begins in its specialist ideology, it is enforced by a commercial compulsion to satisfy the customer. Since the student is now so much a free agent in determining his or her education, the department administrators and the faculty members must necessarily be preoccupied with the problem of how to keep enrollments up. Something obviously must be done to keep the classes filled; otherwise, the students will wander off to more attractive courses or to courses more directly useful to their proposed careers. Under such circumstances it is inevitable that requirements will be lightened, standards lowered, grades inflated, and instruction narrowed to the supposed requirements of some supposed career opportunity.

Dr. Johnson told Mrs. Thrale that his cousin, Cornelius Ford, "advised him to study the Principles of every thing, that a general Acquaintance with Life might be the Consequence of his Enquiries—Learn said he the leading Precognita of all things . . . grasp the Trunk hard only, and you will shake all the Branches." The soundness of this advice seems indisputable, and the metaphor entirely apt. From the trunk it is possible to "branch out." One can begin with a trunk and develop a single branch or any number of branches; although it may be possible to begin with a branch and develop a trunk, that is neither so probable nor so promising. The modern university, at any rate, more and more resembles a loose collection of lopped branches waving about randomly in the air. "Modern knowledge is departmentalized," H. J. Massingham wrote in 1943, "while the essence of culture is initiation into wholeness, so that all the divisions of knowledge are considered as the branches of one tree, the Tree of Life whose roots went deep into earth and whose top was in heaven."

This Tree, for many hundreds of years, seems to have come almost naturally to mind when we have sought to describe the form of knowledge. In Western tradition, it is at least as old as Genesis, and the form it gives us for all that we know is organic, unified, comprehensive, connective—and moral. The tree, at the beginning, was two trees: the tree of life and the tree of the knowledge of good and evil. Later, in our understanding of them, the two trees seem to have become one, or each seems to stand for the other—for in the world after the Fall, how can the two be separated? To know life is to know good and evil; to prepare young people for life is to prepare them to know the difference between good and evil. If we represent knowledge as a tree, we know that things that are divided are yet connected. We know that to observe the divisions and ignore the connections is to destroy the tree. The history of modern education may be the history of the loss of this image, and of its replacement by the pattern of the industrial machine, which subsists upon division—and by industrial economics ("publish or perish"), which is meaningless apart from division.

The need for broadly informed human judgment nevertheless remains, and this need requires inescapably an education that is broad and basic. In the face of this need, which is *both* private and public, "career preparation" is an improper use of public money, since "career preparation" serves merely private ends; it is also a waste of the student's time, since "career preparation" is best and most properly acquired in apprenticeships under the supervision of employers. The proper subject for a school, for example, is how to speak and write well, not how to be a "public speaker" or a "broadcaster" or a "creative writer" or a "technical writer" or a journalist or a practitioner of "business English." If one can speak and write well, then, given the need, one can make a speech or write an article or a story or a business letter. If one cannot speak or write well, then the tricks of a trade will be no help.

The work that should, and that can, unify a university is that of deciding what a student should be required to learn—what studies, that is, constitute the trunk of the tree of a person's education. "Career preparation," which has given so much practical support to academic specialization (and so many rewards to academic specialists), seems to have destroyed interest in this question. But the question exists and the failure to answer it (or even to ask it) imposes severe penalties on teachers, students, and the public alike. The penalties imposed on students and graduates by their failure to get a broad, basic education are, I think, obvious enough. The public penalties are also obvious if we consider, for instance, the number of certified expert speakers and writers who do not speak or write well, who do not know that they speak or write poorly, and who apparently do not care whether or not they speak or write honestly.

The penalties that this failure imposes on teachers are not so obvious, mainly, I suppose, because so far the penalties have been obscured by rewards. The penalties for teachers are the same as those for students and the public, plus one more: The failure to decide what students should be required to learn keeps the teacher from functioning as, and perhaps from becoming, a responsible adult.

There is no one to teach young people but older people, and so the older people must do it. That they do not know enough to do it, that they have never been smart enough or experienced enough or good enough to do it, does not matter. They must do it because there is no one else to do it. This is simply the elemental trial—some would say the elemental tragedy—of human life: the necessity to proceed on the basis merely of the knowledge that is available, the necessity to postpone until too late the question of the sufficiency and the truth of that knowledge.

There is, then, an inescapable component of trial and error in human education; some things that are taught will be wrong because fallible humans are the teachers. But the reason for education, its constant effort and discipline, is surely to reduce the young person's dependence on trial and error as far as possible. For it *can* be reduced. One should not have to learn everything, or the basic things, by trial and error. A child should not have to learn the danger of heat by falling into the fire. A student should not have to learn the penalties of illiteracy by being illiterate or the value of a good education by the "object lesson" of a poor one.

Teachers, moreover, are not providing "career preparation" so much as they are "preparing young people for life." This statement is not the result of educational doctrine; it is simply the fact of the matter. To prepare young people for life, teachers must dispense knowledge and enlighten ignorance, just as supposed. But ignorance is not only the affliction that teaching seeks to cure; it is also the condition, the predicament, in which teaching is done, for teachers do not know the life or the lives for which their students are being prepared.

This condition gives the lie to the claims for "career preparation," since students may not *have* the careers for which they have been prepared: The "job market" may be overfilled; the requirements for this or that career may change; the student may change, or the world may. The teacher, preparing the student for a life necessarily unknown to them both, has no excusable choice but to help the student to "grasp the Trunk."

Yet the arguments for "career preparation" continue to be made and to grow in ambition. On August 23, 1983, for example, the Associated Press announced that "the head of the Texas school board wants to require sixth-graders to choose career 'tracks' that will point them toward jobs." Thus, twelve-year-old children would be "free to choose" the kind of life they wish to live. They would even be free to change "career tracks," though, according to the article, such a change would involve the penalty of a delayed graduation.

But these are free choices granted to children not prepared or ready to make them. The idea, in reality, is to impose adult choices on children, and these "choices" mask the most vicious sort of economic determinism. This idea of education as "career track" diminishes everything it touches: education, teaching, childhood, the future. And such a thing could not be contemplated for sixth-graders, obviously, if it had not already been instituted in the undergraduate programs of colleges and universities.

To require or expect or even allow young people to choose courses of study and careers that they do not yet know anything about is not, as is claimed, a grant of freedom. It is a severe limitation upon freedom. It means, in practice, that when the student has finished school and is faced then, appropriately, with the need to choose a career, he or she is prepared to choose only one. At that point, the student stands in need of a freedom of choice uselessly granted years before and forfeited in that grant.

The responsibility to decide what to teach the young is an adult responsibility. When adults transfer this responsibility to the young, whether they do it by indifference or as a grant of freedom, they trap themselves in a kind of childishness. In that failure to accept responsibility, the teacher's own learning and character are disemployed, and, in the contemporary industrialized education system, they are easily replaced by bureaucratic and methodological procedures, "job market" specifications, and tests graded by machines.

This question of what all young people should be expected to learn is now little discussed. The reason, apparently, is the tacit belief that now, with the demands of specialization so numerous and varied, such a question would be extremely hard, if not impossible, to answer. And yet this question appears to be as much within the reach of reason and common sense as any other. It cannot be denied, to begin with, that all the disciplines rest upon the knowledge of letters and the knowledge of numbers. Some rest more on letters than numbers, some more on numbers than letters, but it is surely true to say that people without knowledge of both letters and numbers are not prepared to learn much else. From there, one can proceed confidently to say that history, literature, philosophy, and foreign languages rest principally on the knowledge of letters and carry it forward, and that biology, chemistry, and physics rest on the knowledge of numbers and carry it forward. This provides us with a description of a probably adequate "core curriculum"—one that would prepare a student well both to choose a direction of further study and to go in that direction. An equally obvious need, then, is to eliminate from the curriculum all courses without content—that is, all courses in methodologies and technologies that could, and should, be learned in apprenticeships.

Besides the innate human imperfections already mentioned, other painful problems are involved in expecting and requiring students to choose the course of their own education. These problems have to do mainly with the diversity of gifts and abilities: that is, some people are not talented in some kinds of work or study; some, moreover, who are poor in one discipline may be excellent in another. Why should such people be forced into situations in which they must see themselves as poor workers or as failures?

The question is not a comfortable one, and I do not believe that it can or should be comfortably answered. There is pain in the requirement to risk failure and pain in the failure that may result from that requirement. But failure is a possibility; in varying degrees for all of us, it is inescapable. The argument for removing the possibility of failure from schoolwork is therefore necessarily specious. The wrong is not in subjecting students to the possibility of failure or in calling their failures failures; the wrong is in the teacher's inability to see that failure in school is not necessarily synonymous with and does not necessarily lead to failure in the world. The wrong is in the failure to see or respect the boundaries between the school and the world. When those are not understood and respected, then the school, the school career, the diploma are all surrounded by such a spurious and modish dignity that failure in school *is* failure in the world. It is for this reason that it is so easy to give education a money value and to sell it to consumers in job lots.

It is a fact that some people with able minds do not fit well into schools and are not properly valued by schoolish standards and tests. If such people fail in a school, their failure should be so called; a school's worth and integrity depend upon its willingness to call things by their right names. But, by the same token, a failure in school is no more than that; it does not necessarily imply or cause failure in the world, any more than it implies or causes stupidity. It is not rare for the judgment of the world to overturn the judgment of schools. There are other tests for human abilities than those given in schools, and there are some that cannot be given in schools. My own life has happened to acquaint me with several people who did not attend high school but who have been more knowledgeable in their "field" and who have had better things to say about matters of general importance than most of the doctors of philosophy I have known. This is not an "anti-intellectual" statement; it is a statement of what I take to be fact, and it means only that the uses of schools are limited—another fact, which schools prepare us to learn by surprise.

Another necessary consideration is that low expectations and standards in universities encourage the lowering of expectations and standards in the high schools and elementary schools. If the universities raise their expectations and standards, the high schools and elementary schools will raise theirs; they will have to. On the other hand, if the universities teach high school courses because the students are not prepared for university courses, then they simply relieve the high schools of their duty and in the process make themselves unable to do their own duty. Once the school stoops to meet the student, the standards of judgment begin to topple at all levels. As standards are lowered—as they cease to be the measure of the students and come to be measured by them—it becomes manifestly less possible for students to fail. But for the same reason it becomes less possible for them to learn and for teachers to teach.

The question, then, is what is to determine the pattern of education. Shall we shape a university education according to the previous schooling of the students, which we suppose has made them unfit to meet high expectations and standards, and to the supposed needs of students in some future still dark to us all? Or shall we shape it according to the nature and demands of the "leading Precognita of all things"—that is, according to the essential subjects of study? If we shape education to fit the students, then we clearly can maintain no standards; we will lose the subjects and eventually will lose the students as well. If we shape it to the subjects, then we will save both the subjects and the students. The inescapable purpose of education must be to preserve and pass on the essential human means—the thoughts and words and works and ways and standards and hopes without which we are not human. To preserve these things and to pass them on is to prepare students for life.

That such work cannot be done without high standards ought not to have to be said. There are necessarily increasing degrees of complexity in the studies as students rise through the grades and the years, and yet the standards remain the same. The first-graders, that is, must read and write in simple sentences, but they read and write, even so, in the language of the King James Bible, of Shakespeare and Johnson, of Thoreau, Whitman, Dickinson, and Twain. The grade-schooler and the graduate student must study the same American history, and there is no excuse for falsifying it in order to make it elementary.

Moreover, if standards are to be upheld, they cannot be specialized, professionalized, or departmented. Only common standards can be upheld—standards that are held and upheld in common by the whole community. When, in a university, for instance, English composition is made the responsibility exclusively of the English department, or of the subdepartment of freshman English, then the quality of the work in composition courses declines and the standards decline. This happens necessarily and for an obvious reason: If students' writing is graded according to form and quality in composition class but according only to "content" in, say, history class and if in other classes students are not required to write at all, then the message to the students is clear: namely, that the form and quality of their writing matters only in composition class, which is to say that it matters very little indeed. High standards of composition can be upheld only if they are upheld everywhere in the university.

Not only must the standards be held and upheld in common but they must also be applied fairly—that is, there must be no conditions with respect to persons or groups. There must be no discrimination for or against any person for any reason. The quality of the individual performer is the issue, not the category of the performer. The aim is to recognize, reward, and promote good work. Special pleading for "disadvantaged" groups—whether disadvantaged by history, economics, or education—can only make it increasingly difficult for members of that group to do good work and have it recognized.

If the university faculties have failed to answer the question of the internal placement of the knowledges of the arts and sciences with respect to each other and to the university as a whole, they have, it seems to me, also failed to ask the question of the external placement of these knowledges with respect to truth and to the world. This, of course, is a dangerous question, and I raise it with appropriate fear. The danger is that such questions should be *settled* by any institution whatever; these questions are the proper business of the people in the institutions, not of the institutions as such. I am arguing here against the specialist absorption in career and procedure that destroys what I take to be the indispensable interest in the question of the truth of what is taught and learned, as well as the equally indispensable interest in the fate and the use of knowledge in the world.

I would be frightened to hear that some university had

suddenly taken a lively interest in the question of what is true and was in the process of answering it, perhaps by a faculty vote. But I am equally frightened by the fashionable lack of interest in the question among university teachers individually. I am more frightened when this disinterest, under the alias of "objectivity," is given the status of a public virtue.

Objectivity, in practice, means that one studies or teaches one's subject *as such*, without concern for its relation to other subjects or to the world—that is, without concern for its truth. If one is concerned, if one cares, about the truth or falsity of anything, one cannot be objective: one is glad if it is true and sorry if it is false; one believes it if it is judged to be true and disbelieves it if it is judged to be false. Moreover, the truth or falsity of some things cannot be objectively demonstrated, but must be determined by feeling and appearance, intuition and experience. And this work of judgment cannot take place at all with respect to one thing or one subject alone. The issue of truth rises out of the comparison of one thing with another, out of the study of the relations and influences between one thing and another and between one thing and many others.

Thus, if teachers aspire to the academic virtue of objectivity, they must teach as if their subject has nothing to do with anything beyond itself. The teacher of literature, for example, must propose the study of poems as relics left by people who, unlike our highly favored modern selves, believed in things not subject to measurable proof; religious poetry, that is, may be taught as having to do with matters once believed but not believable. The poetry is to be learned *about*; to learn *from* it would be an embarrassing betrayal of objectivity.

That this is more than a matter of classroom technique is made sufficiently evident in the current fracas over the teaching of the Bible in public schools. Judge Jackson Kiser of the federal district court in Bristol, Virginia, recently ruled that it would be constitutional to teach the Bible to public school students if the course is offered as an elective and "taught in an objective manner with no attempt made to indoctrinate the children as to either the truth or falsity of the biblical materials." James J. Kilpatrick, who discussed this ruling approvingly in one of his columns, suggested that the Bible might be taught "as Shakespeare is taught" and that this would be good because "the Bible is a rich lode of allusion, example and quotation." He warned that "The line that divides propaganda from instruction is a wavering line drawn on shifting sands," and he concluded by asserting that "Whatever else the Bible may be, the Bible is in fact literature. The trick is to teach it that way."

The interesting question here is not whether young English-speakers should know the Bible—they obviously should—but whether a book that so directly offers itself to our belief or disbelief can be taught "as literature." It clearly cannot be so taught except by ignoring "whatever else [it] may be," which is a very substantial part of it. The question, then, is whether it can be adequately or usefully taught as something less than it is. The fact is that the writers of the Bible did not think that they were writing what Judge Kiser and Mr. Kilpatrick call "literature." They thought they were writing the truth, which they expected to be believed by some and disbelieved by others. It is conceivable that the Bible could be well taught by a teacher who believes that it is true, by a teacher who believes that it is untrue, or by a teacher who believes that it is partly true. That it could be well taught by a teacher uninterested in the question of its truth is not conceivable. That a lively interest in the Bible could be maintained through several generations of teachers uninterested in the question of its truth is also not conceivable.

Obviously, this issue of the Bible in the public schools cannot be resolved by federal court decisions that prescribe teaching methods. It can only be settled in terms of the freedom of teachers to teach as they believe and in terms of the relation of teachers and schools to their local communities. It may be that in this controversy we are seeing the breakdown of the public school system, as an inevitable consequence of the breakdown of local communities. It is hard to believe that this can be remedied in courts of law.

My point, anyhow, is that we could not consider teaching the Bible "as literature" if we were not already teaching literature "as literature"—as if we do not care, as if it does not matter, whether or not it is true. The causes of this are undoubtedly numerous, but prominent among them is a kind of shame among teachers of literature and other "humanities" that their truths are not objectively provable as are the truths of science. There is now an embarrassment about any statement that depends for confirmation upon experience or imagination or feeling or faith, and this embarrassment has produced an overwhelming impulse to treat such statements merely as artifacts, cultural relics, bits of historical evidence, or things of "aesthetic value." We will study, record, analyze, criticize, and appreciate. But we will not believe; we will not, in the full sense, know.

The result is a stance of "critical objectivity" that causes many teachers, historians, and critics of literature to sound—not like mathematicians or chemists: their methodology does not permit that yet—but like ethologists, students of the behavior of a species to which they do not belong, in whose history and fate they have no part, their aim being, not to know anything for themselves, but to "advance knowledge." This may be said to work, as a textual mechanics, but it is not an approach by which one may know any great work of literature. That route is simply closed to people interested in what "they" thought "then"; it is closed to people who think that "Dante's world" or "Shakespeare's world" is far removed and completely alienated from "our world"; and it is closed to the viewers of poetic devices, emotional effects, and esthetic values.

The great distraction behind the modern fate of literature, I think, is expressed in Coleridge's statement that his endeavor in *Lyrical Ballads* was "to transfer from our inward nature . . . . a semblance of truth sufficient to procure for these shadows of imagination that willing suspension of disbelief for the moment, which constitutes poetic faith." That is a sentence full of quakes and tremors. Is our inward nature true only by semblance? What is the difference, in a work of art, between truth and "semblance of truth"? What must be the result of separating "poetic faith" from faith of any other kind and then of making "poetic faith" dependent upon will?

The gist of the problem is in that adjective *willing*, which implies the superiority of the believer to what is believed. This implication, I am convinced, is simply untrue. Belief precedes will. One either believes or one does not, and, if one believes, then one willingly believes. If one disbelieves, even unwillingly, all the will in the world cannot make one believe. Belief is involuntary, as is the Ancient Mariner's recognition of the beauty and sanctity of the water snakes:

> A spring of love gushed from my heart,
> And I blessed them unaware . . .

This involuntary belief is the only approach to the great writings. One may, assuredly, not believe, and we must, of course, grant unbelievers the right to read and comment as unbelievers, for disbelief is a legitimate response, because it is a possible one. We must be aware of the possibility that belief may be false, and of the need to awaken from false belief; "one need not step into belief as into an abyss." But we must be aware also that to disbelieve is to remain, in an important sense, outside the work. When we are *in* the work, we are long past the possibility of any debate with ourselves about whether or not to be willing to believe. When we are *in* the work, we simply *know* that great Odysseus has come home, that Dante is in the presence of the celestial rose, that Cordelia, though her father carries her in his arms, is dead. If we know these things, we are apt to know too that Mary Magdalene mistook the risen Christ for the gardener—and are thus eligible to be taken lightly by objective scholars, and to be corrected by a federal judge.

We and these works meet in imagination; by imagination we know their truth. In imagination, there is no specifically or exclusively "poetic faith," just as there is no faith that is specifically or exclusively religious. Belief is the same wherever it happens, and its terms are invariably set by the imagination. One believes, that is, because one *sees*, not because one is informed. That is why, four hundred years after Copernicus, we still say, "The sun is rising."

When we read the ballad of Sir Patrick Spens we know that the knight and his men have drowned because "Thair hats they swam aboone," not because we have confirmed the event by the study of historical documents. And if our assent is forced also by the ballad of Thomas Rhymer, far stranger than that of Sir Patrick, what are we to say? Must we go, believing, into the poem, and then return from it in disbelief because we find the story in no official record, have read of no such thing in the newspaper, and know nothing like it in our own experience? Or must we live with the poem, with our awareness of its power over us, as a piece of evidence that reality may be larger than we thought?

"Does that mean," I am asked, "that it's not possible for us to read Homer properly because we don't believe in the Greek gods?" I can only answer that I suspect that a proper reading of Homer will *result* in some manner of belief in his gods. How else explain their survival in the works of Christian writers into our own time? This survival has its apotheosis, it seems to me, in C. S. Lewis's novel, *That Hideous Strength*, at the end of which the Greek planetary deities reappear on earth as angels. Lewis wrote as a Christian who had read Homer, but he had read, obviously, as a man whose imagination was not encumbered with any such clinical apparatus as the willing suspension of disbelief. As such a reader, though he was a Christian, his reading had told him that the pagan gods retained a certain authority and commanded a certain assent. Like many of his forebears in English literary tradition, he yearned toward them. Their triumphant return, at the end of *That Hideous Strength*, as members of the heavenly hierarchy of Christianity, is not altogether a surprise. It is a profound resolution, not only in the novel itself, but in the history of English literature. One hears the ghosts of Spenser and Milton sighing with relief.

Questions of the authenticity of imaginings invite answers, and yet may remain unanswered. For the imagination is not always subject to immediate proof or demonstration. It is often subject only to the slow and partial authentication of experience. It is subject, that is, to a practical, though not an exact, validation, and it is subject to correction. For a work of imagination to endure through time, it must prove valid, and it must survive correction. It is correctable by experience, by critical judgment, and by further works of imagination.

To say that a work of imagination is subject to correction is, of course, to imply that there is no "world of imagination" as distinct from or opposed to the "real world." The imagination is *in* the world, is at work in it, is necessary to it, and is correctable by it. This correcting of imagination by experience is inescapable, necessary, and endless, as is the correcting of experience by imagination. This is the great general work of criticism to which we all are called. It is not literary criticism any more than it is historical or agricultural or biological criticism, but it must nevertheless be a fundamental part of the work of literary criticism, as it must be of criticisms of all other kinds. One of the most profound of human needs is for the truth of imagination to prove itself in every life and place in the world, and for the truth of the world's lives and places to be proved in imagination.

This need takes us as far as possible from the argument for works of imagination, human artifacts, as special cases, privileged somehow to offer themselves to the world on their own terms. It is this argument and the consequent abandonment of the general criticism that have permitted the universities to organize themselves on the industrial principle, as if faculties and students and all that they might teach and learn are no more than parts of a machine, the purpose of which they have, in general, not bothered to define, much less to question. And largely through the agency of the universities, this principle and this metaphor now dominate our relation to nature and to one another.

If, for the sake of its own health, a university must be interested in the question of the truth of what it teaches, then, for the sake of the world's health, it must be interested in the fate of that truth and the uses made of it in the world. It must want to know where its graduates live, where they work, and what they do. Do they return home with their knowledge to enhance and protect the life of their neighborhoods? Do they join the "upwardly mobile" professional force now exploiting and destroying local communities, both human and natural, all over the country? Has the work of the university, over the last generation, increased or decreased literacy and knowledge of the classics? Has it increased or decreased the general understanding of the sciences? Has it increased or decreased pollution and soil erosion? Has it increased or decreased the ability and the willingness of public servants to tell the truth? Such questions are not, of course, precisely answerable. Questions about influence never are. But they are askable, and the asking, should we choose to ask, would be a unifying and a shaping force.

Preserving Wildness

THE ARGUMENT over the proper relation of humanity to nature is becoming, as the sixties used to say, polarized. And the result, as before, is bad talk on both sides. At one extreme are those who sound as if they are entirely in favor of nature; they assume that there is no necessary disjuncture or difference between the human estate and the estate of nature, that human good is in some simple way the same as natural good. They believe, at least in principle, that the biosphere is an egalitarian system, in which all creatures, including humans, are equal in value and have an equal right to live and flourish. These people tend to stand aloof from the issue of the proper human use of nature. Indeed, they have begun to use "stewardship" (meaning the responsible use of nature) as a term of denigration.

At the other extreme are the nature conquerors, who have no patience with an old-fashioned outdoor farm, let alone a wilderness. These people divide all reality into two parts: human good, which they define as profit, comfort, and security; and everything else, which they understand as a stockpile of "natural resources" or "raw materials," which will sooner or later be transformed into human good. The aims of these militant tinkerers invariably manage to be at once unimpeachable and suspect. They wish earnestly, for example, to solve what they call "the problem of hunger"—if it can be done glamorously, comfortably, and profitably. They believe that the ability to do something is the reason to do it. According to a recent press release from the University of Illinois College of Agriculture, researchers there are looking forward to "food production without either farmers or farms." (This is perhaps the first explicit acknowledgment of the program that has been implicit in the work of the land-grant universities for forty or fifty years.)

If I had to choose, I would join the nature extremists against the technology extremists, but this choice seems poor, even assuming that it is possible. I would prefer to stay in the middle, not to avoid taking sides, but because I think the middle *is* a side, as well as the real location of the problem.

*

The middle, of course, is always rather roomy and bewildering territory, and so I should state plainly the assumptions that define the ground on which I intend to stand:

1. We live in a wilderness, in which we and our works occupy a tiny space and play a tiny part. We exist under its dispensation and by its tolerance.
2. This wilderness, the universe, is *somewhat* hospitable to us, but it is also absolutely dangerous to us (it is going to kill us, sooner or later), and we are absolutely dependent upon it.
3. That we depend upon what we are endangered by is a problem not solvable by "problem solving." It does not have what the nature romantic or the technocrat would regard as a solution. We are not going back to the Garden of Eden, nor are we going to manufacture an Industrial Paradise.
4. There does exist a possibility that we can live more or less in harmony with our native wilderness; I am betting my life that such a harmony is possible. But I do not believe that it can be achieved simply or easily or that it can ever be perfect, and I am certain that it can never be made, once and for all, but is the forever unfinished lifework of our species.
5. It is not possible (at least, not for very long) for humans to intend their own good specifically or exclusively. We cannot intend our good, in the long run, without intending the good of our place—which means, ultimately, the good of the world.
6. To use or not to use nature is not a choice that is available to us; we can live only at the expense of other lives. Our choice has rather to do with how and how much to use. This is not a choice that can be decided satisfactorily in principle or in theory; it is a choice intransigently impractical. That is, it must be worked out in local practice because, by necessity, the practice will vary somewhat from one locality to another. There is, thus, no *practical* way that we can intend the good of the world; practice can only be local.
7. If there is no escape from the human use of nature, then human good cannot be simply synonymous with natural good.

What these assumptions describe, of course, is the human predicament. It is a spiritual predicament, for it requires us to be properly humble and grateful; time and again, it asks us to be still and wait. But it is also a practical problem, for it requires us to *do* things.

In going to work on this problem it is a mistake to proceed on the basis of an assumed division or divisibility between nature and humanity, or wildness and domesticity. But it is also a mistake to assume that there is no difference between the natural and the human. If these things could be divided, our life would be far simpler and easier than it is, just as it would be if they were not different. Our problem, exactly, is that the human and the natural are indivisible, and yet are different.

The indivisibility of wildness and domesticity, even within the fabric of human life itself, is easy enough to demonstrate. Our bodily life, to begin at the nearest place, is half wild. Perhaps it is more than half wild, for it is dependent upon reflexes, instincts, and appetites that we do not cause or intend and that we cannot, or had better not, stop. We live, partly, because we are domestic creatures—that is, we participate in our human economy to the extent that we "make a living"; we are able, with variable success, to discipline our appetites and instincts in order to produce this artifact, this human living. And yet it is equally true that we breathe and our hearts beat and we survive as a species because we are wild.

The same is true of a healthy human economy as it branches upward out of the soil. The topsoil, to the extent that it is fertile, is wild; it is a dark wilderness, ultimately unknowable, teeming with wildlife. A forest or a crop, no matter how intentionally husbanded by human foresters or farmers, will be found to be healthy precisely to the extent that it is wild—able to collaborate with earth, air, light, and water in the way common to plants before humans walked the earth. We know from experience that we can increase our domestic demands upon plants so far that we force them into kinds of failure that wild plants do not experience.

Breeders of domestic animals, likewise, know that, when a breeding program is too much governed by human intention, by economic considerations, or by fashion, uselessness is the result. Size or productivity, for instance, will be gained at the cost of health, vigor, or reproductive ability. In other words, so-called domestic animals must remain half wild, or more than half, because they are creatures of nature. Humans are intelligent enough to select for a type of creature; they are not intelligent enough to *make* a creature. Their efforts to make an entirely domestic animal, like their efforts to make an entirely domestic human, are doomed to failure because they do not have and undoubtedly are never going to have the full set of production standards for the making of creatures. From a human point of view, then, creature making is wild. The effort to make plants, animals, and humans ever more governable by human intentions is continuing with more determination and more violence than ever, but that does not mean that it is nearer to success. It means only that we are increasing the violence and the magnitude of the expectable reactions.

To be divided against nature, against wildness, then, is a human disaster because it is to be divided against ourselves. It confines our identity as creatures entirely within the bounds of our own understanding, which is invariably a mistake because it is invariably reductive. It reduces our largeness, our mystery, to a petty and sickly comprehensibility.

But to say that we are not divided and not dividable from nature is not to say that there is no difference between us and the other creatures. Human nature partakes of nature, participates in it, is dependent on it, and yet is different from it. We feel the difference as discomfort or difficulty or danger. Nature is not easy to live with. It is hard to have rain on your cut hay, or floodwater over your cropland, or coyotes in your sheep; it is hard when nature does not respect your intentions, and she never does exactly respect them. Moreover, such problems belong to all of us, to the human lot. Humans who do not experience them are exempt only because they are paying (or underpaying) other humans such as farmers to deal with nature on their behalf. Further, it is not just agriculture-dependent humanity that has had to put up with natural dangers and frustrations; these have been the lot of hunting and gathering societies also, and the wild creatures do not always live comfortably or easily with nature either.

But humans differ most from other creatures in the extent to which they must be *made* what they are—that is, in the extent to which they are artifacts of their culture. It is true that what we might as well call culture does go into the making of some birds and animals, but this teaching is so much less than the teaching that makes a human as to be almost a different thing. To take a creature who is biologically a human and to make him or her fully human is a task that requires many years (some of us sometimes fear that it requires more than a lifetime), and this long effort of human making is necessary, I think, because of our power. In the hierarchy of power among the earth's creatures, we are at the top, and we have been growing stronger for a long time. We are now, to ourselves, incomprehensibly powerful, capable of doing more damage than floods, storms, volcanoes, and earthquakes. And so it is more important than ever that we should have cultures capable of making us into humans—creatures capable of prudence, justice, fortitude, temperance, and the other virtues. For our history reveals that, stripped of the restraints, disciplines, and ameliorations of culture, humans are not "natural," not "thinking animals" or "naked apes," but monsters—indiscriminate and insatiable killers and destroyers. We differ from other creatures, partly, in our susceptibility to monstrosity. It is perhaps for this reason that, in the wake of the great wars of our century, we have seen poets such as T. S. Eliot, Ezra Pound, and David Jones making an effort to reweave the tattered garment of culture and to reestablish the cultural tasks, which are, as Pound put it, "To know the histories / to know good from evil / And know whom to trust." And we see, if we follow Pound a little further, that the recovery of culture involves, leads to, or is the recovery of nature:

the trees rise
and there is a wide sward between them
. . . myrrh and olibanum on the altar stone
giving perfume,
and where was nothing
now is furry assemblage
and in the boughs now are voices . . .

In the recovery of culture *and* nature is the knowledge of how to farm well, how to preserve, harvest, and replenish the forests, how to make, build, and use, return and restore. In this *double* recovery, which is the recovery of our humanity, is the hope that the domestic and the wild can exist together in lasting harmony.

This doubleness of allegiance and responsibility, difficult as it always is, confusing as it sometimes is, apparently is inescapable. A culture that does not measure itself by nature, by an understanding of its debts to nature, becomes destructive of nature and thus of itself. A culture that does not measure itself by its own best work and the best work of other cultures (the determination of which is its unending task) becomes destructive of itself and thus of nature.

Harmony is one phase, the good phase, of the inescapable dialogue between culture and nature. In this phase, humans consciously and conscientiously ask of their work: Is this good for us? Is this good for our place? And the questioning and answering in this phase is minutely particular: It can occur only with reference to particular artifacts, events, places, ecosystems, and neighborhoods. When the cultural side of the dialogue becomes too theoretical or abstract, the other phase, the bad one, begins. Then the conscious, responsible questions are not asked; acts begin to be committed and things to be made on their own terms for their own sakes, culture deteriorates, and nature retaliates.

The awareness that we are slowly growing into now is that the earthly wildness that we are so complexly dependent upon is at our mercy. It has become, in a sense, our artifact because it can only survive by a human understanding and forbearance that we now must make. The only thing we have to preserve nature with is culture; the only thing we have to preserve wildness with is domesticity.

To me, this means simply that we are not safe in assuming that we can preserve wildness by making wilderness preserves. Those of us who see that wildness and wilderness need to be preserved are going to have to understand the dependence of these things upon our domestic economy and our domestic behavior. If we do not have an economy capable of valuing in particular terms the durable good of localities and communities, then we are not going to be able to preserve anything. We are going to have to see that, if we want our forests to last, then we must make wood products that last, for our forests are more threatened by shoddy workmanship than by clear-cutting or by fire. Good workmanship—that is, careful, considerate, and loving work—requires us to think considerately of the whole process, natural and cultural, involved in the making of wooden artifacts, because the good worker does not share the industrial contempt for "raw material." The good worker loves the board before it becomes a table, loves the tree before it yields the board, loves the forest before it gives up the tree. The good worker understands that a badly made artifact is both an insult to its user and a danger to its source. We could say, then, that good forestry begins with the respectful husbanding of the forest that we call stewardship and ends with well-made tables and chairs and houses, just as good agriculture begins with stewardship of the fields and ends with good meals.

In other words, conservation is going to prove increasingly futile and increasingly meaningless if its proscriptions are not answered positively by an economy that rewards and enforces good use. I would call this a loving economy, for it would strive to place a proper value on all the materials of the world, in all their metamorphoses from soil and water, air and light to the finished goods of our towns and households, and I think that the only effective motive for this would be a particularizing love for local things, rising out of local knowledge and local allegiance.

Our present economy, by contrast, does not account for affection at all, which is to say that it does not account for value. It is simply a description of the career of money as it preys upon both nature and human society. Apparently because our age is so manifestly unconcerned for the life of the spirit, many people conclude that it places an undue value on material things. But that cannot be so, for people who valued material things would take care of them and would care for the sources of them. We could argue that an age that *properly* valued and cared for material things would be an age properly spiritual. In my part of the country, the Shakers, "unworldly" as they were, were the true materialists, for they truly valued materials.

And they valued them in the only way that such things *can* be valued in practice: by good workmanship, both elegant and sound. The so-called materialism of our own time is, by contrast, at once indifferent to spiritual concerns and insatiably destructive of the material world. And I would call our economy, not materialistic, but abstract, intent upon the subversion of both spirit and matter by abstractions of value and of power. In such an economy, it is impossible to value anything that one *has*. What one has (house or job, spouse or car) is only valuable insofar as it can be exchanged for what one believes that one wants—a limitless economic process based upon boundless dissatisfaction.

Now that the practical processes of industrial civilization have become so threatening to humanity and to nature, it is easy for us, or for some of us, to see that practicality needs to be made subject to spiritual values and spiritual measures. But we must not forget that it is also necessary for spirituality to be responsive to practical questions. For human beings the spiritual and the practical are, and should be, inseparable. Alone, practicality becomes dangerous; spirituality, alone, becomes feeble and pointless. Alone, either becomes dull. Each is the other's discipline, in a sense, and in good work the two are joined.

"The dignity of toil is undermined when its necessity is gone," Kathleen Raine says, and she is right. It is an insight that we dare not ignore, and I would emphasize that it applies to *all* toil. What is not needed is frivolous. Everything depends on our right relation to necessity—and therefore on our right definition of necessity. In defining our necessity, we must be careful to discount the subsidies, the unrepaid borrowings, from nature that have so far sustained industrial civilization: the "cheap" fossil fuels and ores; the forests that have been cut down and not replanted; the virgin soils of much of the world, whose fertility has not been replenished.

And so, though I am trying to unspecialize the idea and the job of preserving wildness, I am not against wilderness preservation. I am only pointing out, as the Reagan administration has done, that the wildernesses we are trying to preserve are standing squarely in the way of our present economy, and that the wildernesses cannot survive if our economy does not change.

The reason to preserve wilderness is that we need it. We need wilderness of all kinds, large and small, public and private. We need to go now and again into places where our work is disallowed, where our hopes and plans have no standing. We need to come into the presence of the unqualified and mysterious formality of Creation. And I would agree with Edward Abbey that we need as well some tracts of what he calls "absolute wilderness," which "through general agreement none of us enters at all."

We need wilderness also because wildness—nature—is one of our indispensable studies. We need to understand it as our source and preserver, as an essential measure of our history and behavior, and as the ultimate definer of our possibilities. There are, I think, three questions that must be asked with respect to a human economy in any given place:

1. What is here?
2. What will nature permit us to do here?
3. What will nature help us to do here?

The second and third questions are obviously the ones that would define agendas of practical research and of work. If we do not work with and within natural tolerances, then we will not be permitted to work for long. It is plain enough, for example, that if we use soil fertility faster than nature can replenish it, we are proposing an end that we do not desire. And to ignore the possibility of help from nature makes farming, for example, too expensive for farmers—as we are seeing. It may make life too expensive for humans.

But the second and third questions are ruled by the first. They cannot be answered—they cannot intelligently be asked—until the first has been answered. And yet the first question has not been answered, or asked, so far as I know, in the whole history of the American economy. All the great changes, from the Indian wars and the opening of agricultural frontiers to the inauguration of genetic engineering, have been made without a backward look and in ignorance of whereabouts. Our response to the forest and the prairie that covered our present fields was to get them out of the way as soon as possible. And the obstructive human populations of Indians and "inefficient" or small farmers have been dealt with in the same spirit. We have never known what we were doing because we have never known what we were *un*doing. We cannot know what we are doing until we know what nature would be doing if we were doing nothing. And that is why we need small native wildernesses widely dispersed over the countryside as well as large ones in spectacular places.

However, to say that wilderness and wildness are indispensable to us, indivisible from us, is not to say that we can find sufficient standards for our life and work in nature. To suggest that, for humans, there is a simple equation between "natural" and "good" is to fall prey immediately to the cynics who love to point out that, after all, "everything is natural." They are, of course, correct. Nature provides bountifully for her children, but, as we would now say, she is also extremely permissive. If her children want to destroy one another entirely or to commit suicide, that is all right with her. There is nothing, after all, more natural than the extinction of species; the extinction of *all* species, we must assume, would also be perfectly natural.

Clearly, if we want to argue for the existence of the world as we know it, we will have to find some way of qualifying and supplementing this relentless criterion of "natural." Perhaps we can do so only by a reaffirmation of a lesser kind of naturalness —that of self-interest. Certainly human self-interest has much wickedness to answer for, and we are living in just fear of it; nevertheless, we must take care not to condemn it absolutely. After all, we value this passing work of nature that we call "the natural world," with its graceful plenty of animals and plants, precisely because *we* need it and love it and want it for a home.

We are creatures obviously subordinate to nature, dependent upon a wild world that we did not make. And yet we are joined to that larger nature by our own nature, a part of which is our self-interest. A common complaint nowadays is that humans think the world is "anthropocentric," or human-centered. I understand the complaint; the assumptions of so-called anthropocentrism often result in gross and dangerous insubordination. And yet I don't know how the human species can avoid some version of self-centeredness; I don't know how any species can. An earthworm, I think, is living in an earthworm-centered world; the thrush who eats the earthworm is living in a thrush-centered world; the hawk who eats the thrush is living in a hawk-centered world. Each creature, that is, does what is necessary in its own behalf, and is domestic in its own *domus* or home.

Humans differ from earthworms, thrushes, and hawks in their capacity to do more—in modern times, a great deal more—in their own behalf than is necessary. Moreover, the vast majority of humans in the industrial nations are guilty of this extravagance. One of the oldest human arguments is over the question of how much is necessary. How much must humans do in their own behalf in order to be fully human? The number and variety of the answers ought to notify us that we never have known for sure, and yet we have the disquieting suspicion that, almost always, the honest answer has been "less."

We have no way to work at this question, it seems to me, except by perceiving that, in order to have the world, we must share it, both with each other and with other creatures, which is immediately complicated by the further perception that, in order to live in the world, we must use it somewhat at the expense of other creatures. We must acknowledge both the centrality and the limits of our self-interest. One can hardly imagine a tougher situation.

But in the recognition of the difficulty of our situation is a kind of relief, for it makes us give up the hope that a solution can be found in a simple preference for humanity over nature or nature over humanity. The only solutions we have ahead of us will need to be worked for and worked out. They will have to be practical solutions, resulting in good local practice. There is work to do that can be done.

As we undertake this work, perhaps the greatest immediate danger lies in our dislike of ourselves as a species. This is an understandable dislike—we are justly afraid of ourselves—but we are nevertheless obliged to think and act out of a proper self-interest and a genuine self-respect as human beings. Otherwise, we will allow our dislike and fear of ourselves to justify further abuses of one another and the world. We must come to terms with the fact that it is not natural to be disloyal to one's own kind.

For these reasons, there is great danger in the perception that "there are too many people," whatever truth may be in it, for this is a premise from which it is too likely that somebody, sooner or later, will proceed to a determination of *who* are the surplus. If we conclude that there are too many, it is hard to avoid the further conclusion that there are some we do not need. But how many do we need, and which ones? Which ones, now apparently unnecessary, may turn out later to be indispensable? We do not know; it is a part of our mystery, our wildness, that we do not know.

I would argue that, at least for us in the United States, the conclusion that "there are too many people" is premature, not because I know that there are *not* too many people, but because I do not think we are prepared to come to such a conclusion. I grant that questions about population size need to be asked, but they are not the *first* questions that need to be asked.

The "population problem," initially, should be examined as a problem, not of quantity, but of pattern. Before we conclude that we have too many people, we must ask if we have people who are misused, people who are misplaced, or people who are abusing the places they have. The facts of most immediate importance may be, not how many we are, but where we are and what we are doing. At any rate, the attempt to solve our problems by reducing our numbers may be a distraction from the overriding population statistic of our time: that *one* human with a nuclear bomb and the will to use it is 100 percent too many. I would argue that it is not human fecundity that is overcrowding the world so much as technological multipliers of the power of individual humans. The worst disease of the world now is probably the ideology of technological heroism, according to which more and more people willingly cause large-scale effects that they do not foresee and that they cannot control. This is the ideology of the professional class of the industrial nations—a class whose allegiance to communities and places has been dissolved by their economic motives and by their educations. These are people who will go anywhere and jeopardize anything in order to assure the success of their careers.

We may or may not have room for more people, but it is certain that we do not have more room for technological heroics. We do not need any more thousand-dollar solutions to ten-dollar problems or million-dollar solutions to thousand-dollar problems—or multibillion-dollar solutions where there was never a problem at all. We have no way to compute the inhabitability of our places; we cannot weigh or measure the pleasures we take in them; we cannot say how many dollars domestic tranquillity is worth. And yet we must now learn to bear in mind the memory of communities destroyed, disfigured, or made desolate by technological events, as well as the memory of families dispossessed, displaced, and impoverished by "labor-saving" machines. The issue of human obsolescence may be more urgent for us now than the issue of human population.

The population issue thus leads directly to the issue of proportion and scale. What is the proper amount of power for a human to use? What are the proper limits of human enterprise? How may these proprieties be determined? Such questions may seem inordinately difficult, but that is because we have gone too long without asking them. One of the fundamental assumptions of industrial economics has been that such questions are outmoded and that we need never ask them again. The failure of that assumption now requires us to reconsider the claims of wildness and to renew our understanding of the old ideas of propriety and harmony.

When we propose that humans should learn to behave properly with respect to nature so as to place their domestic economy harmoniously upon and within the sustaining and surrounding wilderness, then we make possible a sort of landscape criticism. Then we can see that it is not primarily the number of people inhabiting a landscape that determines the propriety of the ratio and the relation between human domesticity and wildness, but it is the way the people divide the landscape and use it. We can see that it is the landscape of monoculture in which both nature and humanity are most at risk. We feel the human fragility of the huge one-class housing development, just as we feel the natural fragility of the huge one-crop field.

Looking at the monocultures of industrial civilization, we yearn with a kind of homesickness for the humanness and the naturalness of a highly diversified, multipurpose landscape, democratically divided, with many margins. The margins are of the utmost importance. They are the divisions between holdings, as well as between kinds of work and kinds of land. These margins—lanes, streamsides, wooded fencerows, and the like—are always freeholds of wildness, where limits are set on human intention. Such places are hospitable to the wild lives of plants and animals and to the wild play of human children. They enact, within the bounds of human domesticity itself, a human courtesy toward the wild that is one of the best safeguards of designated tracts of true wilderness. This is the landscape of harmony, safer far for life of all kinds than the landscape of monoculture. And we should not neglect to notice that, whereas the monocultural landscape is totalitarian in tendency, the landscape of harmony is democratic and free.

A Good Farmer of the Old School

AT THE 1982 Draft Horse Sale in Columbus, Ohio, Maury Telleen summoned me over to the group of horsemen with whom he was talking: "Come here," he said, "I want you to hear this." One of those horsemen was Lancie Clippinger, and what Maury wanted me to hear was the story of Lancie's corn crop of the year before.

The story, which Lancie obligingly told again, was as interesting to me as Maury had expected it to be. Lancie, that year, had planted forty acres of corn; he had also bred forty gilts that he had raised so that their pigs would be ready to feed when the corn would be ripe. The gilts produced 360 pigs, an average of nine per head. When the corn was ready for harvest, Lancie divided off a strip of the field with an electric fence and turned in the 360 shoats. After the shoats had fed on that strip for a while, Lancie opened a new strip for them. He then picked the strip where they had just fed. In that way, he fattened his 360 shoats and also harvested all the corn he needed for his other stock.

The shoats brought $40,000. Lancie's expenses had been for seed corn, 275 pounds of fertilizer per acre, and one quart per acre of herbicide. He did not say what the total costs amounted to, but it was clear enough that his net income from the forty acres of corn had been high, in a year when the corn itself would have brought perhaps two dollars a bushel.

At the end of the story, I remember, Lancie and Maury had a conversation that went about like this:

"Do you farrow your sows in a farrowing house?"

"No."

"Oh, you do it in huts, then?"

"No, I have a field I turn them out in. It has plenty of shade and water. And I see them every day."

Here was an intelligent man, obviously, who knew the value of doing his own thinking and paying attention, who understood clearly that the profit is in the difference between costs and earnings, and who proceeded directly to minimize his costs. In a time when hog farmers often spend many thousands of dollars on highly specialized housing and equipment, Lancie's "hog operation" consisted almost entirely of hogs. His principal outlays otherwise were for the farm itself and for fencing. But what struck me most, I think, was the way he had employed nature and the hogs themselves to his own advantage. The bred sows needed plenty of shade, water, and room for exercise; Lancie provided those things, and nature did the rest. He also supplied his own care and attention, which came free; they did not have to be purchased at an inflated cost from an industrial supplier. And then, instead of harvesting his corn mechanically, hauling it, storing it, grinding it, and hauling it to his shoats, he let the shoats harvest and grind it for themselves. He had the use of the whole hog, whereas in a "confinement operation," the hog's feet, teeth, and eyes have virtually no use and produce no profit.

At the next Columbus Sale, I hunted Lancie up, and again we spent a long time talking. We talked about draft horses, of course, but also about milk cows and dairying. And that part of our conversation interested me about as much as the hog story had the year before. What so impressed me was Lancie's belief that there is a limit to the number of cows that a dairy farmer can manage well; he thought the maximum number to be about twenty-five: "If a fellow milks twenty-five cows, he'll *see* them all." If he milks more than that, Lancie said, even though he may touch them all, he will not *see* them all. As in Lancie's account of his corn crop and the 360 shoats, the emphasis here was on the importance of seeing, of paying attention. That this is important economically, he made clear in something he said to me later: "You can take care of twenty or twenty-five cows and do it right. More, you're overlooking things that cost you money." It is necessary, Lancie thinks, to limit the scale of operation, not only in dairying, but in all other enterprises on the farm because proper scale permits a correct balance between work and care. The distinction he was making, it seemed to me, was between work, as it has been understood traditionally on the farm, and processing, as it is understood in industry.

Those two conversations stayed in my mind, proving useful many times in my effort to understand the troubles developing in our agricultural economy. I knew that Lancie Clippinger was one of the best farmers of the old school, and I promised myself that I would visit him at his farm, which I was finally able to do in October 1985.

The farm is on somewhat rolling land, surrounded by woodlots and brushy fencerows, so that it has a little of the feeling of a large forest clearing. There are 175 acres, of which about 135 are cropped; the rest is in permanent pasture and woods. Although conveniently close to the state road, the farm is at the end of a lane, set off to itself. It is pretty and quiet, a pleasant place to live and to farm, as well as to visit. Lancie and his wife, Verna Bell, bought the place and moved there in the fall of 1971.

When my wife and I drove into the yard, Kathy, one of Lancie's granddaughters, who had evidently been watching for us, came out of the house to meet us. She took us out through the barn lot to a granary where Lancie, his son Keith, and Sherri, another granddaughter, were sacking some oats. We waited, talking with Kathy, while they finished the job, and then we went with Lancie and Keith to look at the horses.

Lancie keeps only geldings, buying them at sales as weanlings, raising and breaking them, selling them, and then replacing them with new colts. When we were there, he had nine head: a pair of black Percherons, a handsome crossbred bay with black mane and tail, and six Belgians. Though he prefers Percherons, he does not specialize; at the sales, his only aim is to buy "colts that look like they'll grow into good big horses." He wants them big because the big ones bring the best prices, but, like nearly all draft horse people who use their horses, he would rather have smaller ones—fifteen hundred pounds or so—if he were keeping them only to work.

The horses he led out for us were in prime condition, and he had been right about them: They had, sure enough, grown into good big ones. These horses may be destined for pulling contests and show hitches, but while they are at Lancie's they put in a lot of time at farm work—they work their way through school, you might say. Like so many farmers of his time, Lancie once made the change from horses to tractors, but with him this did not last long. He was without horses "for a little while" in the seventies, and after that he began to use them again. Now he uses the horses for "just about everything" except cutting and baling his hay and picking his corn. Last spring he used his big tractor only two days. The last time he went to use it, it wouldn't start, and he left it sitting in the shed; it was still sitting there at the time of our visit.

Part of the justification for the return to the use of horses is economic. When he was doing all his work with tractors, Lancie's fuel bill was $6,000 a year; now it is about $2,000. Since the horses themselves are a profit-making enterprise on this farm, the $4,000 they save on fuel is money in the bank. But the economic reason is not the only one: "Pleasure," Lancie says, "is a big part of it." At the year's end, his bank account will show a difference that the horses have made, but day by day his reason for working them is that he *likes* to.

He does not need nine horses in order to do his farming. He has so many because he needs to keep replacements on hand for the horses he sells. He aims, he says, to sell "two or three or four horses every year." To farm his 175 acres, he needs only four good geldings, although he would probably like to keep five, in case he needed a spare. With four horses on his grain drill, he can plant fifteen or twenty acres in a day. He uses four horses also on an eight-foot tandem disk and a springtooth harrow, and he can plant twelve or fifteen acres of corn a day "and not half try."

In plowing, he goes by the old rule of thumb that you can plow an acre per horse per day, provided the horses are in hard condition. "If you start at seven in the morning and stay there the way you ought to," he says, "you can plow three acres a day with three horses." That is what he does, and he does it with a walking plow because, he says, it is easier to walk than to ride. That, of course, is hardly a popular opinion, and Lancie is amused by the surprise it sometimes causes.

One spring, he says, after he had started plowing, he ordered some lime. When the trucker brought the first load, he stopped by the house to ask where to spread it. Mrs. Clippinger told him that Lancie was plowing, and pointed out to the field where Lancie could be seen walking in the furrow behind his plow and team. The trucker was astonished: "Even the *Amish* ride!"

In 1936, Lancie remembers, he plowed a hundred acres, sixty of them in sod, with two horses, Bob and Joe. Together, that team weighed about thirty-five hundred pounds. They were blacks. Lancie had been logging with them before he started plowing, and they were in good shape, ready to go. They plowed two acres a day, six days a week, for nearly nine weeks. It is the sort of thing, one guesses, that could have been done only because all the conditions were right: a strong young man, a tough team, a good season. "Looked like, back then, there wasn't any bad weather," Lancie says, laughing. "You could work all the time."

This farmer's extensive use of live horsepower is possible because his farm is the right size for it and because a sensible rotation of crops both reduces the acreage to be plowed each year and distributes the other field work so that not too much needs to be done at any one time. Of the farm's 135 arable acres, approximately fifty-five will be in corn, forty in oats, and forty in alfalfa. Each of the crops will be grown on the same land two years in order to avoid buying alfalfa seed every year.

The two-year-old alfalfa, turned under, supplies enough nitrogen for the first year of corn. In the second year, the corn crop receives a little commercial nitrogen. The routine application of fertilizer on the corn is 275 pounds per acre of 10-10-20, drilled into the row with the planter. The oats are fertilized at the same rate as the corn, while the alfalfa field, because Lancie sells quite a bit of hay, receives 600 pounds per acre of 3-14-42 in two applications every year. The land is limed at a rate of two tons per acre every time it is plowed. Otherwise, for fertilization Lancie depends on manure from his cattle and horses. "That's what counts," he says. It counts because it pays but does not cost. He usually has enough manure to cover his corn ground every year.

This system of management has not only maintained the productive capacity of the farm but has greatly improved it. Fourteen years ago, when Lancie began on it, the place was farmed out. The previous farmer had plowed it all and planted it all in corn year after year. When the farm sold in the fall of 1971, the corn crop, which was still standing, was bought by a neighboring farmer, who found it not worth picking. Lancie plowed it under the next spring. In order to have a corn crop that first year, he used 900 pounds of fertilizer to the acre—300 pounds of nitrogen and 600 of "straight analysis." After that, when his rotations and other restorative practices had been established, he went to his present rate of 275 pounds of 10-10-20. The resulting rates of production speak well for good care: The corn has made 150 bushels per acre, Lancie says, "for a long time"; this year his oats made 109 bushels per acre, and he also harvested 11,000 fifty-pound bales of alfalfa hay from a forty-acre field (a per-acre yield of about seven tons) and sold 4,800 bales for $12,000.

In addition to seed and fertilizer, Lancie purchases some insecticide and herbicide. This year his alfalfa was sprayed once for weevils, and he used a half-pint of 2-4-D per acre on his corn. The 2-4-D, he says, would not have been necessary if he had cultivated four times instead of twice. Using the chemical saved two cultivations that would have interfered with hay harvest.

What is most significant about Lancie's management of his crops is that it gives his farm a degree of independence that is unusual in these times. The farm, first of all, is ordered and used according to its own nature and carrying capacity, not according to the dictates of farm policy, expert advice, or fluctuations of the economy. The possibility of solving one's economic problems by production alone is not, in Lancie's opinion, a good possibility. If you are losing money on the corn you produce, he points out, the more you produce the more you lose. That so many farmers continue to compensate for low grain prices by increasing production, at great cost to their farms and to themselves, is a sort of wonder to him. "The cheaper it is, the more they plow," he says. "I don't know what they mean." His own farm, by contrast, grows approximately the same acreages of the same crops every year, not because that is what the economy supposedly demands, but because that is what the land can produce at the least cost for the longest time.

Since the farm itself is so much the source of its own fertility and operating energy, Lancie's use of purchased supplies can be minimal, selective, and nonaddictive. Because his cropping pattern and system of management are sound, Lancie can buy these things to suit his convenience. His total expense for 2-4-D for his corn this year, for example, was fifty-six dollars—a very small price to pay in order to have his hands and his mind free at haying time. The point, I think, is that he had a choice: He could choose to do what made the most sense. A further point is that he can quit using chemicals and purchased fertilizer if it ever makes economic sense to do so. As a farmer, he is not addicted to these things.

The conventional industrial farmer, on the other hand, is too often the prisoner of his own technology and methods and has no choice but to continue to do as he has done, whatever the disadvantages. A farmer who has no fences cannot turn hogs in to harvest his corn when prices are low. A farmer who has invested heavily in a farrowing house and all the equipment that goes with it is stuck with that investment. If, for some reason, it ceases to be profitable for him to produce feeder pigs, he still has the farrowing house, which is good for little else, and perhaps a debt on it as well. Thus, mental paralysis and economic slavery can be instituted on a farm by the farmer's technological choices.

One of the main results of Lancie Clippinger's independence is versatility, enabling him to take advantage quickly of opportunities as they appear. Because he has invested in no expensive specialized equipment, he can change his ways to suit his wishes or his circumstances. That he did well raising and finishing shoats one year does not mean that he must continue to raise them. Last year, for instance, he thought there was money to be made on skinny sows. He bought sixty-two at $100 a head, turned them into his cornfield, and, while they ate, he picked. "We all worked together," he says. The sows did a nearly perfect job of gleaning the field, and they brought $200 a head when he sold them.

There is a direct economic payoff in this freedom of choice: It pays to be able to choose to substitute a team of horses for a tractor, or manure for fertilizer, or cultivation for herbicides. When you cultivate a field of corn, as Lancie says, "you're selling your labor"; in other words, you ensure a relation between production and consumption that is proper because it makes sound economic sense. If the farmer does not achieve that proper relation on his farm, he will be a victim. When Lancie prepares his ground with plow and harrow and cultivates his crop instead of buying chemicals, he is a producer, not a consumer; he is selling his labor, not buying an expensive substitute for labor. Moreover, when he does this with a team of horses instead of buying fuel, he is selling his team's labor, not paying for an expensive substitute. When he uses his own corn, oats, and hay to replace petroleum, he is selling those feeds for a far higher return than he could get on the market. He and his horses are functioning, in effect, as solar converters, making usable and profitable the free sunlight that falls onto the farm. They are producing at home the energy, weed control, and fertility that other farmers are going broke trying to pay for.

The industrial farmer consumes more than he produces and is a captive consumer of the suppliers who have prospered by the ruination of such farmers. So far as the national economy is concerned, this kind of farmer exists only to provide cheap food and to enrich the agribusiness corporations, at his own expense.

Sometimes Lancie's intelligent methods and his habit of paying attention yield unexpected dividends. The year after he hogged down the forty acres of corn with the 360 shoats, the field was covered with an excellent stand of alsike clover. "It was pretty," Lancie says, but he didn't know where it came from. He asked around in the neighborhood and discovered that the field had been in alsike seventeen years before. The seed had lain in the ground all that time, waiting for conditions to be right, and somehow the hogs had made them right. Thus, that year's very profitable corn harvest, which had been so well planned, resulted in a valuable gift that nobody had planned—or could have planned. There is no recipe, so far as I know, for making such a thing happen. Obviously, though, a certain eligibility is required. It happened on Lancie's farm undoubtedly because he is the kind of farmer he is. If he had been plowing the whole farm every year and planting it all in corn, as his predecessor had, such a thing would not have happened.

It is care, obviously, that makes the difference. The farm gives gifts because it is given a chance to do so; it is not overcropped or overused. One of Lancie's kindnesses to his farm is his regular rotation of his crops; another is his keeping of livestock, which gives him not only the advantages I have already described but also permits him to make appropriate use of land not suited to row cropping. Like many farms in the allegedly flat corn belt, Lancie's farm includes some land that should be kept permanently grassed, and on his farm, unlike many, it *is* kept permanently grassed. He can afford this because he can make good use of it that way, without damaging it, for these thirty or so acres give him five hundred bales of bluegrass hay early in the year and, after that, months of pasture, at the cost only of a second clipping. The crop on that land does not need to be planted or cultivated, and it is harvested by the animals; it is therefore the cheapest feed on the place.

Lancie Clippinger is as much in the business of growing crops and making money on them as any other farmer. But he is also in the business of making sense—making sense, that is, for himself, not for the oil, chemical, and equipment companies, or for the banks. He is taking his own advice, and his advice comes from his experience and the experience of farmers like him, not from experts who are not farmers. For those reasons, Lancie Clippinger is doing all right. He is farming well and earning a living by it in a time when many farmers are farming poorly and making money for everybody but themselves.

"I don't know what they mean," he says. "You'd think some in the bunch would use their heads a *little* bit."

FROM

WHAT ARE PEOPLE FOR?

(1990)

Damage

I

I have a steep wooded hillside that I wanted to be able to pasture occasionally, but it had no permanent water supply.

About halfway to the top of the slope there is a narrow bench, on which I thought I could make a small pond. I hired a man with a bulldozer to dig one. He cleared away the trees and then formed the pond, cutting into the hill on the upper side, piling the loosened dirt in a curving earthwork on the lower.

The pond appeared to be a success. Before the bulldozer quit work, water had already begun to seep in. Soon there was enough to support a few head of stock. To heal the exposed ground, I fertilized it and sowed it with grass and clover.

We had an extremely wet fall and winter, with the usual freezing and thawing. The ground grew heavy with water, and soft. The earthwork slumped; a large slice of the woods floor on the upper side slipped down into the pond.

The trouble was the familiar one: too much power, too little knowledge. The fault was mine.

I *was* careful to get expert advice. But this only exemplifies what I already knew. No expert knows everything about every place, not even everything about any place. If one's knowledge of one's whereabouts is insufficient, if one's judgment is unsound, then expert advice is of little use.

II

In general, I have used my farm carefully. It could be said, I think, that I have improved it more than I have damaged it.

My aim has been to go against its history and to repair the damage of other people. But now a part of its damage is my own.

The pond was a modest piece of work, and so the damage is not extensive. In the course of time and nature it will heal.

And yet there *is* damage—to my place, and to me. I have carried out, before my own eyes and against my intention, a part of the modern tragedy: I have made a lasting flaw in the face of the earth, for no lasting good.

Until that wound in the hillside, my place, is healed, there will be something impaired in my mind. My peace is damaged. I will not be able to forget it.

III

It used to be that I could think of art as a refuge from such troubles. From the imperfections of life, one could take refuge in the perfections of art. One could read a good poem—or better, write one.

Art was what was truly permanent, therefore what truly mattered. The rest was "but a spume that plays / Upon a ghostly paradigm of things."

I am no longer able to think that way. That is because I now live in my subject. My subject is my place in the world, and I live in my place.

There is a sense in which I no longer "go to work." If I live in my place, which is my subject, then I am "at" my work even when I am not working. It is "my" work because I cannot escape it.

If I live in my subject, then writing about it cannot "free" me of it or "get it out of my system." When I am finished writing, I can only return to what I have been writing about.

While I have been writing about it, time will have changed it. Over longer stretches of time, *I* will change it. Ultimately, it will be changed by what I write, inasmuch as I, who change my subject, am changed by what I write about it.

If I have damaged my subject, then I have damaged my art. What aspired to be whole has met damage face to face, and has come away wounded. And so it loses interest both in the anesthetic and in the purely esthetic.

It accepts the clarification of pain, and concerns itself with healing. It cultivates the scar that is the course of time and nature over damage: the landmark and mindmark that is the notation of a limit.

To lose the scar of knowledge is to renew the wound.

An art that heals and protects its subject is a geography of scars.

IV

"You never know what is enough unless you know what is more than enough."

I used to think of Blake's sentence as a justification of youthful excess. By now I know that it describes the peculiar condemnation of our species. When the road of excess has reached the palace of wisdom it is a healed wound, a long scar.

Culture preserves the map and the records of past journeys so that no generation will permanently destroy the route.

The more local and settled the culture, the better it stays put, the less the damage. It is the foreigner whose road of excess leads to a desert.

Blake gives the just proportion or control in another proverb: "No bird soars too high, if he soars with his own wings." Only when our acts are empowered with more than bodily strength do we need to think of limits.

It was no thought or word that called culture into being, but a tool or a weapon. After the stone axe we needed song and story to remember innocence, to record effect—and so to describe the limits, to say what can be done without damage.

The use only of our bodies for work or love or pleasure, or even for combat, sets us free again in the wilderness, and we exult.

But a man with a machine and inadequate culture—such as I was when I made my pond—is a pestilence. He shakes more than he can hold.

Wallace Stegner and the Great Community

IN THE SPRING of 1958 I received notice that I had been given a Stegner Fellowship for the next school year at Stanford University. I assumed, properly enough, that I was going to be dealing with people my own age who would know a great deal more than I did. Clearly, I needed to prepare myself.

My wife and new daughter and I were staying in Lexington, Kentucky, that summer, and I decided that I would read the works of Wallace Stegner. The only thing I had read by him up to then was *Field Guide to the Western Birds*, published by Ballantine in a collection called *New Short Novels 2*. In those days I assumed that anything that interested me had already interested everybody else, and so I was surprised to find that the university library yielded only his first novel, *Remembering Laughter*, and his second book of stories, *The City of the Living*. I found a copy of the first short story collection, *The Women on the Wall*, in a used book store.

My reading of those three books is still clear and immediate to me, perhaps because my uneasiness about going to Stanford had sharpened my wits. I thought *Remembering Laughter* a perfect little novel, clean and swift and assured, and I can still feel the weight of the disaster in it. As the would-be author of a first novel myself, I envied it and was intimidated by it. I saw plainly that this man, when he was perhaps my age, had known how to write, and that he had known how much better than I did.

Because I had no ambition to write short stories, I read *The Women on the Wall* and *The City of the Living* more dispassionately. By then I had read a good many short stories. Besides the ones scattered in textbooks and anthologies, I had read Hemingway's in the Modern Library collection and Faulkner's *Collected Stories*. But I had read those mainly, I think, as a reader—without asking how they had been made. *The Women on the Wall* and *The City of the Living* were the first collections of stories by one writer that I read asking that question, seeing how an able workman made use of a form.

And so Wallace Stegner became my teacher before I ever laid eyes on him, and he was already teaching me in a way that I have come to see as characteristic of him: by bestowing a kindness that implied an expectation, and by setting an example.

It has been twenty-seven years since 1958, and I am no longer certain what I expected him to be. All I am sure of is that I did not expect him to be as he was when he took over the writing seminar at the start of the winter quarter. Perhaps, since he taught at a great university and had written many books, I expected him to be magisterial. Perhaps I expected him to display the indisputable field marks of Literary Genius. Or perhaps I expected a dogmatist of some sort, for I remember that he had a reputation with some former writing students still around as a stickler for grammar—as if *grammar* had anything to do with *art*.

The man himself, it turned out, was something altogether different, and a great deal better. When I was asked to write about him as a teacher, I had the feeling that I was expected to tell anecdotes of memorable classroom performances. But that is not what I have to tell. As a teacher, as I knew him, he had none of the performer in him. When he spoke to the class, the class felt spoken to. You did not feel that he was glancing at himself.

What struck me first about him was the way he looked. He was, and still is, an extraordinarily handsome man. And unlike many writers and professors, he dressed with care and looked good in his clothes. I don't mean that he was a dandy; he was as far from that as he was from being a performer. His was a neatness, one felt, that had to do with respect, for himself and for others.

But just as striking, just as unexpected, was the way he understated his role as teacher. I think that he is naturally a reticent man, not given to self-revelation or self-advertisement, and for years I assumed that his reticence in the classroom was merely a part of his character. But as I have come to know him better, have read him more, and have thought more about him, I have changed my mind. I am sure that character had something to do with it, but it seems to me now that his stance and manner in the classroom were the results of an accurate and generous understanding of his situation as teacher and of ours as students. The best explanation may be in these sentences from his essay "The Book and the Great Community":

> Thought is neither instant nor noisy. . . . It thrives best in solitude, in quiet, and in the company of the past, the great community of recorded human experience. That recorded experience is essential whether one hopes to reassert some aspect of it, or attack it.

The community here is that of "recorded human experience," not the Pantheon of Great Writers. It is immense and diverse, more like the Library of Congress than the Harvard Five-Foot Shelf. But it does include the great writers. It is bewildering both in its amplitude and in the eminence of some of its members. A teacher leading his students to the entrance to that community, as would-be contributors to it, must know that both he and they are coming into the possibility of error. The teacher may make mistakes about the students; the students may make much more serious ones about themselves. He is leading them, moreover, to a community, not to some singular stump or rostrum from which he will declare the Truth.

I think, then, that when Mr. Stegner sat with us at the long table in the Jones Room of the Stanford Library, he felt that a certain modesty and a certain discretion were in order. He did not speak as though he confused any utterance of his with the earthquake that one afternoon gently shook us while we talked.

There must have been about twenty of us in the class, and among its members were Ernest Gaines, Ken Kesey, and Nancy Packer. Nancy, I think, was the oldest of us. She had read a great deal, talked well, and was useful to us all. But as a group we had a good deal to say, and Mr. Stegner would let us say it. He would read a piece of somebody's work and then sit back, sometimes with a cigar, listening attentively while we had our say.

One sunny afternoon he read us a piece of work of his own —a chapter, I think, from *All the Little Live Things*—and then graciously paid attention to all we said to him about it. I remember that I made an extensive comment myself, and I am tempted to wish I could remember what I said. Probably I should be glad to have forgotten. I would be glad, anyhow, to know that he has forgotten.

He had his say too, of course. He commended generously where commendation was due. He was a good encourager when encouragement was needed. And, more important, he was a willing dealer with problems, even little problems, even —yes—problems of grammar. That is, he gave good practical help.

He did not adopt the pose or use the tactics of the authoritarian teacher, and yet one felt his authority. I am not talking, of course, about institutional authority, which must be asserted to be felt, and which he did not assert. We felt in him, rather, the authority of authentic membership in the great community, of one who had thought and worked in solitude, in quiet, in the company of the past—an authority that would be destroyed in being asserted.

It was this implicit authority in him that most impressed me during my time at Stanford. That and an expectation implicit in his dealings with me, that perhaps was not conscious with him, but which I felt keenly. At one time in the midst of my fellowship year I thought (mistakenly, it turned out) that I had finished the novel I was working on, and I could not see clearly what I ought to do next. For a while, paralyzed in my confusion, I was not able to write anything. And I remember both my embarrassment, for I felt that Mr. Stegner expected me to be at work, and how paltry, in light of his expectation, my excuses appeared in my own eyes. That was, maybe, my definitive encounter with the hardest truth that a writer—or any other worker—can learn: that, to all practical purposes, excuses are not available.

Mr. Stegner's teaching, then, as I have come to see it, was as important for what it did not do as for what it did. One thing he did not do was encourage us too much. If we wrote well, he said so. But he did not abet any suspicions we may have had that we were highly accomplished writers or that we were ever going to be highly accomplished. In this, I think, he respected the past, for he was better acquainted with the art of writing than we were and knew better than we did how much we had to learn. But he was being respectful of us too. He did not want to mislead us or help us to mislead ourselves. He did not say what he had not considered or did not mean. He did not deal in greetings at the beginnings of great careers.

He did not pontificate or indoctrinate or evangelize. We were not expected to become Stegnerians. None of us could have doubted that he wanted us to know and think and write as well as we possibly could. But no specific recipe or best way was recommended. The emphasis was on workmanship. What we were asked to be concerned with was the job of work at hand, what one or another of us had done or attempted to do. Our teacher was a writer, he too was at work on what he had chosen to do; he would help us if he could.

And so what I began by calling reticence—at some risk, for it is not a fashionable virtue now—finally declares itself as courtesy toward both past and future: courtesy toward the art of writing, which needs to be carefully learned and generously passed on; and courtesy toward us, who as young writers needed all the help we could get, but needed also to be left to our own ways.

In his conversations with Richard Etulain, there is a passage in which Mr. Stegner names several of his old students, speaks of their accomplishments, and then says, "I try not to take credit for any of that." In the mouths of some people, that statement would not be trustworthy; in the mouths of some it would contradict itself. Coming from Mr. Stegner, it is trustworthy, for in fact he has not been a taker of credit. The fellows *have* been left to their ways. They have come, benefited as they were able, and left free of obligation.

My bet, however, is that, by the fellows and other students, Wallace Stegner is given a great deal more credit than he would be comfortable in taking. The first fellowships were given in 1946. By 1971, when Mr. Stegner retired from teaching, there had been about a hundred of them, and the list of fellows contains a good many names that do it credit: Evan S. Connell, Dan Jacobson, Hannah Green, Eugene Burdick, Edward Abbey, Thomas McGuane, Ken Gangemi, Tillie Olsen, Robert Stone, Jim Houston, Larry McMurtry, Ernest Gaines, James Baker Hall, Gurney Norman, Ed McClanahan, Peter Beagle, Nancy Huddleston Packer, Max Apple, Blanche Boyd, Judith Rascoe, Max Crawford, Robin White, Merrill Joan Gerber,

Charlotte Painter, Al Young, Raymond Carver, William Wiegand, William Kittredge.

My bet is that every one of those people owes something to Wallace Stegner. My further bet is that most of them, maybe years after their participation in the seminar, have been surprised by some recognition of what and how much they owe. And yet I know that a teacher is right in trying not to take credit for his students. The fellows, one assumes, must have learned something from attending the seminars given by Mr. Stegner, Richard Scowcraft, and other teachers. But they must have learned something too from coming to California and living there for a year. They must have learned something from one another, and from other people they came to know while they were there. They must have learned something from the books they read while they were there. How would you sort all that out, for the hundred fellows and the hundreds of others who sat in the Jones Room from 1946 to 1971, and make a precise estimate of the influence of one teacher? Obviously, you cannot. As thousands of faculty committees have found out, "teacher evaluation" is a hopeless business. There is, thank God, no teacher-meter, and there never is going to be one. A teacher's major contribution may pop out anonymously in the life of some ex-student's grandchild. A teacher, finally, has nothing to go on but faith, a student nothing to offer in return but testimony.

What I have to testify is that, although I do not think that Mr. Stegner thinks of himself as my teacher, my awareness of him as a teacher has grown over the years, and I think myself more than ever his student. And this has to do with the changes that have happened in me since my time at Stanford.

In those days I assumed, as my education had prepared me to assume, that I was going to follow a literary career that would lead me far from home. I assumed that I would teach (and write) in a university in a large city, and in fact I did so for a while. And then my family and I returned to my native county in Kentucky. That is, I became, in both habitat and subject, what is called a "regional writer."

There are, of course, problems in being a regional writer. If one is regional only in subject, then there is a temptation —and an abundance of precedent—for becoming a sort of industrialist of letters, mining one's province for whatever can be got out of it in the way of "raw material" for stories and novels. I would argue that it has been possible for such writers to write so exploitively, condescendingly, and contemptuously of their regions and their people as virtually to prepare the way for worse exploitation by their colleagues in other industries: if it's a god-forsaken boondocks full of ignorant hillbillies, or a god-forsaken desert populated by a few culturally deprived ranchers, why *not* strip-mine it?

On the other hand, if one both lives and writes in one's region, one becomes aware of good reasons to be more watchful and more careful. It was not until I began the struggle to live and write in my region that I began to be aware of Wallace Stegner as a writer struggling to live and write in *his* region —something I would never have done, I think, had I chosen to live in San Francisco or New York. Or at least his struggle would not have meant so much, would not have been so instructive and reassuring to me, had I lived in one of those places. In *Wolf Willow* and then in *The Sound of Mountain Water*, I saw him dealing head-on with the problems of the history and the literary history of the West, the challenges and flaws and threats of those histories, and the possible meanings and uses of his own history as a westerner. And of course, once one sees this effort in some of his work, one begins to see it in the rest. One sees him becoming a new kind of writer: one who not only writes about his region but also does his best to protect it, by writing and in other ways, from its would-be exploiters and destroyers.

As a regional writer he seems to me exemplary. He has worked strenuously to know his region. He has been not just a student of its history, but one of its historians. There is an instructive humility in his studentship as a historian of the West. It is hard to imagine Hemingway researching and writing a history of Michigan or Africa; to him, as to many writers now, history was immediate experience. To Mr. Stegner, it is also memory. He has the care and the scrupulousness of one who understands remembering as a duty, and who therefore understands historical insight and honesty as duties. He has endeavored to understand the differences of his region from other regions and also from its own pipe dreams and fantasies of itself. He has never condescended to his region—an impossibility, since he has so profoundly understood himself as a part of it. He has not dealt in the quaint, the fantastical, or the picturesque. And, above all, he has written well. He is a highly accomplished and an extremely versatile writer. One of the pleasures in reading him is to see how many kinds of writing he has done well. He does not allow suppositions about the homeliness or provinciality of his subject to qualify or limit his own powers. He writes as intelligently about cowboys as about historians and literary critics.

He is a re-readable writer. One reads out of interest or curiosity to see how the story or the argument or the explanation will play itself out. One can re-read, I think, only to be surprised. If re-reading does not yield surprises, one does not continue. Not long ago, when I re-read "Genesis," the long story in *Wolf Willow*, I was surprised by the excellence of the artistry, its clarity and crispness, the cleanness of movement. It is a story by a man who learned about work and hardship, weather and country, from experience, but who has also thought well about them, and has read well. More recently, re-reading some of the essays, I was surprised by the way they spoke, amplifying themselves, to the time and experience added to me since I read them last.

He is, then, a regional writer who has escaped the evils of regional-ism. And he has escaped, as well, the evils of idiosyncrasy. He has been a writer whose work has not exhausted literary possibility, but has made it, opened it up. Younger writers of the West, I would think, would find it possible to build on his work directly. And younger writers of other regions, I know, will find it illustrative of possibilities, a measure of excellence, and a source of comfort.

I have been using the adjective "regional" as if I am not aware of the antipathy attached to it in some literary circles. I am using it that way because I feel strongly the need for the sort of regional identification and commitment that is exemplified in Mr. Stegner's work. But I am, of course, inescapably aware of the scorn with which that identification and commitment are frequently met. Mr. Stegner himself cites a typical instance in his conversations with Richard Etulain: the *New York Times*, he says, called him "the Dean of Western Writers"—and got his name wrong. The adjective "western," as all regional writers will understand, would have been dismissive even if the name had been correctly given. This is the regionalism of New York, which will use the West, indeed depend on it, but not care for it. And this regionalism is opposed and corrected by writing that is authentically and faithfully regional.

In the face of this metropolitan regionalism which has so far been under no constraint to see itself as such, and which has condescended to and exploited all other regions, Mr. Stegner has taken his stand as a westerner and has produced work that seems to me not only excellent, but indispensable as well. And what I most respect in him is that in all the years of his effort he has not stooped to the self-promotion endemic to the literary regionalism of New York. He has represented himself solely by his writings and his acts of citizenship. In a self-exploiting, world-exploiting age, this is a high and admirable accomplishment in itself.

Writer and Region

I FIRST READ *Huckleberry Finn* when I was a young boy. My great-grandmother's copy was in the bookcase in my grandparents' living room in Port Royal, Kentucky. It was the Webster edition, with E. W. Kemble's illustrations. My mother may have told me that it was a classic, but I did not *know* that it was, for I had no understanding of that category, and I did not read books because they were classics. I don't remember starting to read *Huckleberry Finn*, or how many times I read it; I can only testify that it is a book that is, to me, literally familiar: involved in my family life.

I can say too that I "got a lot out of it." From early in my childhood I was not what was known as a good boy. My badness was that I was headstrong and did not respond positively to institutions. School and Sunday school and church were prisons to me. I loved being out of them, and I did not behave well in them. *Huckleberry Finn* gave me a comforting sense of precedent, and it refined my awareness of the open, outdoor world that my "badness" tended toward.

That is to say that *Huckleberry Finn* made my boyhood imaginable to me in a way that it otherwise would not have been. And later, it helped to make my grandfather's boyhood in Port Royal imaginable to me. Still later, when I had come to some knowledge of literature and history, I saw that that old green book had, fairly early, made imaginable to me my family's life as inhabitants of the great river system to which we, like Mark Twain, belonged. The world my grandfather had grown up in, in the eighties and nineties, was not greatly changed from the world of Mark Twain's boyhood in the thirties and forties. And the vestiges of that world had not entirely passed away by the time of my own boyhood in the thirties and forties of the next century.

My point is that *Huckleberry Finn* is about a world I know, or knew, which it both taught me about and taught me to imagine. That it did this before I could have known that it was doing so, and certainly before anybody told me to expect it to do so, suggests its greatness to me more forcibly than any critical assessment I have ever read. It is called a great American book; I think of it, because I have so experienced it, as a transfiguring regional book.

As a boy resentful of enclosures, I think I felt immediately the great beauty, the great liberation, at first so fearful to him, of the passage in Chapter 1 when Huck, in a movement that happens over and over in his book, escapes the strictures of the evangelical Miss Watson and, before he even leaves the house, comes into the presence of the country:

> By-and-by they fetched the niggers in and had prayers, and then everybody was off to bed. I went up to my room with a piece of candle and put it on the table. Then I set down in a chair by the window and tried to think of something cheerful, but it warn't no use. I felt so lonesome I most wished I was dead. The stars was shining, and the leaves rustled in the woods ever so mournful; and I heard an owl, away off, who-whooing about somebody that was dead, and a whippoorwill and a dog crying about somebody that was going to die; and the wind was trying to whisper something to me and I couldn't make out what it was, and so it made the cold shivers run over me.

It is a fearful liberation because the country, so recently settled by white people, is already both haunted and threatened. But the liberation is nevertheless authentic, both for Huck and for the place and the people he speaks for. In the building and summoning rhythm of his catalog of the night sounds, in the sudden realization (his and ours) of the equality of his voice to his subject, we feel a young intelligence breaking the confines of convention and expectation to confront the world itself: the night, the woods, and eventually the river and all it would lead to.

By now we can see the kinship, in this respect, between Huck's voice and earlier ones to the east. We feel the same sort of outbreak as we read:

> When I wrote the following pages, or rather the bulk of them, I lived alone, in the woods, a mile from any neighbor, in a house which I had built myself, on the shore of Walden Pond, in Concord, Massachusetts, and earned my living by the work of my hands only.

That was thirty years before *Huckleberry Finn*. The voice is certainly more cultivated, more adult, more reticent, but the compulsion to get *out* is the same.

And a year after that we hear:

> I loaf and invite my soul,
> I lean and loaf at ease . . . observing a spear of summer grass.

And we literally *see* the outbreak here as Whitman's line grows long and prehensile to include the objects and acts of a country's life that had not been included in verse before.

But Huck's voice is both fresher and historically more improbable than those. There is something miraculous about it. It is not Mark Twain's voice. It is the voice, we can only say, of a great genius named Huckleberry Finn, who inhabited a somewhat lesser genius named Mark Twain, who inhabited a frustrated businessman named Samuel Clemens. And Huck speaks of and for and as his place, the gathering place of the continent's inland waters. His is a voice governed always by the need to flow, to move outward.

It seems miraculous also that this voice should have risen suddenly out of the practice of "comic journalism," a genre amusing enough sometimes, but extremely limited, now hard to read and impossible to need. It was this way of writing that gave us what I understand as regional*ism*: work that is ostentatiously provincial, condescending, and exploitive. That *Huckleberry Finn* starts from there is evident from its first paragraph. The wonder is that within three pages the genius of the book is fully revealed, and it is a regional genius that for 220 pages (in the Library of America edition) remains untainted by regionalism. The voice is sublimely confident of its own adequacy to its own necessities, its eloquence. Throughout those pages the book never condescends to its characters or its subject; it never glances over its shoulder at literary opinion; it never fears for its reputation in any "center of culture." It reposes, like Eliot's Chinese jar, moving and still, at the center of its own occasion.

I should add too that the outbreak or upwelling of this voice, impulsive and freedom-bent as it is, is not disorderly. The freeing of Huck's voice is not a feat of power. The voice is enabled by an economy and a sense of pace that are infallible, and innately formal.

*

That the book fails toward the end (in the sixty-seven pages, to be exact, that follow the reappearance of Tom Sawyer) is pretty generally acknowledged. It does not fail exactly into the vice that is called regionalism, though its failure may have influenced or licensed the regionalism that followed; rather, it fails into a curious frivolity. It has been all along a story of escape. A runaway slave is an escaper, and Huck is deeply implicated, finally by his deliberate choice, in Jim's escape; but he is making his own escape as well, from Miss Watson's indoor piety. After Tom reenters the story, these authentic escapes culminate in a bogus one: the freeing of a slave who, as Tom knows, has already been freed. It is as though Mark Twain has recovered authorship of the book from Huck Finn at this point—only to discover that he does not know how to write it.

Then occurs the wounding and recovery of Tom and the surprising entrance of his Aunt Polly who, true to her character, clears things up in no time—a delightful scene; there is wonderful writing in the book right through to the end. But Mark Twain is not yet done with his theme of escape. The book ends with Huck's determination to "light out for the Territory" to escape being adopted and "sivilized" by Tom's Aunt Sally. And here, I think, we are left face-to-face with a flaw in Mark Twain's character that is also a flaw in our national character, a flaw in our history, and a flaw in much of our literature.

As I have said, Huck's point about Miss Watson is well taken and well made. There is an extremity, an enclosure, of conventional piety and propriety that needs to be escaped, and a part of the business of young people is to escape it. But this point, having been made once, does not need to be made again. In the last sentence, Huck is made to suggest a virtual identity between Miss Watson and Aunt Sally. But the two women are not at all alike. Aunt Sally is a sweet, motherly, entirely affectionate woman, from whom there is little need to escape because she has no aptitude for confinement. The only time she succeeds in confining Huck, she does so by *trusting* him. And so when the book says, "Aunt Sally she's going to adopt me and sivilize me and I can't stand it. I been there before," one can only conclude that it is not Huck talking about Aunt Sally, but Mark Twain talking, still, about the oppressive female piety of Miss Watson.

Something is badly awry here. At the end of this great book we are asked to believe—or to believe that Huck believes—that there are no choices between the "civilization" represented by pious slave owners like Miss Watson or lethal "gentlemen" like Colonel Sherburn and lighting out for the Territory. This hopeless polarity marks the exit of Mark Twain's highest imagination from his work. Afterwards we get *Pudd'nhead Wilson*, a fine book, but inferior to *Huckleberry Finn*, and then the inconsolable grief, bitterness, and despair of the last years.

It is arguable, I think, that our country's culture is still suspended as if at the end of *Huckleberry Finn*, assuming that its only choices are either a deadly "civilization" of piety and violence or an escape into some "Territory" where we may remain free of adulthood and community obligation. We want to be free; we want to have rights; we want to have power; we do not yet want much to do with responsibility. We have imagined the great and estimable freedom of boyhood, of which Huck Finn remains the finest spokesman. We have imagined the bachelorhoods of nature and genius and power: the contemplative, the artist, the hunter, the cowboy, the general, the president—lives dedicated and solitary in the Territory of individuality. But boyhood and bachelorhood have remained our norms of "liberation," for women as well as men. We have hardly begun to imagine the coming to responsibility that is the meaning, and the liberation, of growing up. We have hardly begun to imagine community life, and the tragedy that is at the heart of community life.

Mark Twain's avowed preference for boyhood, as the time of truthfulness, is well known. Beyond boyhood, he glimpsed the possibility of bachelorhood, an escape to "the Territory," where individual freedom and integrity might be maintained—and so, perhaps, he imagined Pudd'nhead Wilson, a solitary genius devoted to truth and justice, standing apart in the preserve of cynical honesty.

He also imagined Aunt Polly and Aunt Sally. They, I think, are the true grown-ups of the Mississippi novels. They have their faults, of course, which are the faults of their time and place, but mainly they are decent people, responsible members of the community, faithful to duties, capable of love, trust, and long-suffering, willing to care for orphan children. The characters of both women are affectionately drawn; Mark Twain evidently was moved by them. And yet he made no acknowledgment of their worth. He insists on regarding them as dampeners of youthful high spirits, and in the end he refuses to distinguish them at all from the objectionable Miss Watson.

There is, then, something stunted in *Huckleberry Finn*. I have hated to think so—for a long time I tried consciously *not* to think so—but it is so. What is stunted is the growth of Huck's character. When Mark Twain replaces Huck as author, he does so apparently to make sure that Huck remains a boy. Huck's growing up, which through the crisis of his fidelity to Jim ("All right, then, I'll *go* to hell") has been central to the drama of the book, is suddenly thwarted first by the Tomfoolery of Jim's "evasion" and then by Huck's planned escape to the "Territory." The real "evasion" of the last chapters is Huck's, or Mark Twain's, evasion of the community responsibility that would have been a natural and expectable next step after his declaration of loyalty to his friend. Mark Twain's failure or inability to imagine this possibility was a disaster for his finest character, Huck, whom we next see not as a grown man, but as partner in another boyish evasion, a fantastical balloon excursion to the Pyramids.

I am supposing, then, that *Huckleberry Finn* fails in failing to imagine a responsible, adult community life. And I am supposing further that this is the failure of Mark Twain's life, and of our life, so far, as a society.

Community life, as I suggested earlier, is tragic, and it is so because it involves unremittingly the need to survive mortality, partiality, and evil. Because Huck Finn and Mark Twain so clung to boyhood, and to the boy's vision of free bachelorhood, neither could enter community life as I am attempting to understand it. A boy can experience grief and horror, but he cannot experience that fulfillment and catharsis of grief, fear, and pity that we call tragedy and still remain a boy. Nor can he experience tragedy in solitude or as a stranger, for tragedy is experienceable only in the context of a beloved community. The fulfillment and catharsis that Aristotle described as the communal result of tragic drama is an artificial enactment of the way a mature community survives tragedy in fact. The community wisdom of tragic drama is in the implicit understanding that no community can survive that cannot survive the worst. Tragic drama attests to the community's need to survive the worst that it knows, or imagines, can happen.

In his own life, Mark Twain experienced deep grief over the deaths of loved ones, and also severe financial losses. But these experiences seem to have had the effect of isolating him, rather than binding him to a community. Great personal loss, moreover, is not much dealt with in those Mississippi books that are most native to his imagination: *Tom Sawyer*, *Huckleberry Finn*, and *Life on the Mississippi*. The only such event that I remember in those books is the story, in *Life on the Mississippi*, of his brother Henry's death after an explosion on the steamboat *Pennsylvania*. Twain's account of this is extremely moving, but it is peculiar in that he represents himself—though his mother, a brother, and a sister still lived—as Henry's *only* mourner. No other family member is mentioned.

What is wanting, apparently, is the tragic imagination that, through communal form or ceremony, permits great loss to be recognized, suffered, and borne, and that makes possible some sort of consolation and renewal. What is wanting is the return to the beloved community, or to the possibility of one. That would return us to a renewed and corrected awareness of our partiality and mortality, but also to healing and to joy in a renewed awareness of our love and hope for one another. Without that return we may know innocence and horror and grief, but not tragedy and joy, not consolation or forgiveness or redemption. There is grief and horror in Mark Twain's life and work, but not the tragic imagination or the imagined tragedy that finally delivers from grief and horror.

He seems instead to have gone deeper and deeper into grief and horror as his losses accumulated, and deeper into outrage as he continued to meditate on the injustices and cruelties of history. At the same time he withdrew further and further from community and the imagining of community, until at last his Hadleyburg—such a village as he had earlier written about critically enough, but with sympathy and good humor too—is used merely as a target. It receives an anonymous and indiscriminate retribution for its greed and self-righteousness—evils that community life has always had to oppose, correct, ignore, indulge, or forgive in order to survive. All observers of communities have been aware of such evils, Huck Finn having been one of the acutest of them, but now it is as if Huck has been replaced by Colonel Sherburn. "The Man that Corrupted Hadleyburg" is based on the devastating assumption that people are no better than their faults. In old age, Mark Twain had become obsessed with "the damned human race" and the malevolence of God—ideas that were severely isolating and, ultimately, self-indulgent. He was finally incapable of that magnanimity that is the most difficult and the most necessary: forgiveness of human nature and human circumstance. Given human nature and human circumstance, our only relief is in this forgiveness, which then restores us to community and its ancient cycle of loss and grief, hope and joy.

And so it seems to me that Mark Twain's example remains crucial for us, for both its virtues and its faults. He taught American writers to be writers by teaching them to be *regional* writers. The great gift of *Huckleberry Finn*, in itself and to us, is its ability to be regional without being provincial. The provincial is always self-conscious. It is the conscious sentimentalization of or condescension to or apology for a province, what I earlier called regionalism. At its most acute, it is the fear of provinciality. There is, as I said, none of that in the first thirty-two chapters of *Huckleberry Finn*. (In the final eleven chapters it is there in the person of Tom Sawyer, who is a self-made provincial.) Mark Twain apparently knew, or he had the grace to trust Huck to know, that *every* writer is a regional writer, even those who write about a fashionable region such as New York City. The value of this insight, embodied as it is in a great voice and a great tale, is simply unreckonable. If he had done nothing else, that would have made him indispensable to us.

But his faults are our own, just as much as his virtues. There are two chief faults and they are related: the yen to escape to the Territory, and retribution against the life that one has escaped or wishes to escape. Mark Twain was new, for his place, in his virtue. In his faults he was old, a spokesman for tendencies already long established in our history.

That these tendencies remain well established among us ought to be clear enough. Wallace Stegner had them in mind when he wrote in *The Sound of Mountain Water*:

> For many, the whole process of intellectual and literary growth is a movement, not through or beyond, but away from the people and society they know best, the faiths they still at bottom accept, the little raw provincial world for which they keep an apologetic affection.

Mr. Stegner's "away from" indicates, of course, an escape to the Territory—and there are many kinds of Territory to escape to. The Territory that hinterland writers have escaped to has almost always been first of all that of some metropolis or "center of culture." This is not inevitably dangerous; great cities are probably necessary to the life of the arts, and all of us who have gone to them have benefited. But once one has reached the city, other Territories open up, and some of these *are* dangerous. There is, first, the Territory of retribution against one's origins. In our country, this is not just a Territory, but virtually a literary genre. From the sophisticated, cosmopolitan city, one's old home begins to look like a "little raw provincial world." One begins to deplore "small town gossip" and "the suffocating proprieties of small town life"—forgetting that gossip occurs only among people who know one another and that propriety is a dead issue only among strangers. The danger is not just in the falsification, the false generalization, that necessarily attends a *distant* scorn or anger, but also in the loss of the subject or the vision of community life and in the very questionable exemption that scorners and avengers customarily issue to themselves.

And so there is the Territory of self-righteousness. It is easy to assume that we do not participate in what we are not in the presence of. But if we are members of a society, we participate, willy-nilly, in its evils. Not to know this is obviously to be in error, but it is also to neglect some of the most necessary and the most interesting work. How do we reduce our dependency on what is wrong? The answer to that question will necessarily be practical; the wrong will be correctable by practice and by practical standards. Another name for self-righteousness is economic and political unconsciousness.

There is also the Territory of historical self-righteousness: if *we* had lived south of the Ohio in 1830, *we* would not have owned slaves; if *we* had lived on the frontier, *we* would have killed no Indians, violated no treaties, stolen no land. The probability is overwhelming that if we had belonged to the generations we deplore, we too would have behaved deplorably. The probability is overwhelming that we *belong* to a generation that will be found by its successors to have behaved deplorably. Not to know that is, again, to be in error and to neglect essential work, and some of this work, as before, is work of the imagination. How can we imagine our situation or our history if we think we are superior to it?

Then there is the Territory of despair, where it is assumed that what is objectionable is "inevitable," and so again the essential work is neglected. How can we have something better if we do not imagine it? How can we imagine it if we do not hope for it? How can we hope for it if we do not attempt to realize it?

There is the Territory of the national or the global point of view, in which one does not pay attention to anything in particular.

Akin to that is the Territory of abstraction, a regionalism of the mind. This Territory originally belonged to philosophers, mathematicians, economists, tank thinkers, and the like, but now some claims are being staked out in it for literature. At a meeting in honor of *The Southern Review*, held in the fall of 1985 at Baton Rouge, one of the needs identified, according to an article in the *New York Times Book Review*, was "to redefine Southernness without resort to geography." If the participants all agreed on any one thing, the article concluded,

> it is perhaps that accepted definitions of regionalism have been unnecessarily self-limiting up to now. The gradual disappearance of the traditional, material South does not mean that Southernness is disappearing, any more than blackness is threatened by integration, or sacredness by secularization. If anything, these metaregions . . . , based as they are upon values, achieve distinction in direct proportion to the homogenization of the physical world. By coming to terms with a concept of regionalism that is no longer based on geographical or material considerations, *The Southern Review* is sidestepping those forces that would organize the world around an unnatural consensus.

Parts of that statement are not comprehensible. Blackness, I would think, *would* be threatened by integration, and sacredness by secularization. Dilution, at least, is certainly implied in both instances. We might as well say that fire is a state of mind and thus is not threatened by water. And how might blackness and sacredness, which have never been regions, be "metaregions"? And is the natural world subject to limitless homogenization? There are, after all, southern species of plants and animals that will not thrive in the north, and vice versa.

This "metaregion," this region "without resort to geography," is a map without a territory, which is to say a map impossible to correct, a map subject to become fantastical and silly like that Southern chivalry-of-the-mind that Mark Twain so properly condemned. How this "metaregion" could resist homogenization and "unnatural consensus" is not clear. At any rate, it abandons the real region to the homogenizers: You just homogenize all you want to, and we will sit here being Southern in our minds.

Similar to the Territory of abstraction is the Territory of artistic primacy or autonomy, in which it is assumed that no value is inherent in subjects but that value is conferred upon subjects by the art and the attention of the artist. The subjects of the world are only "raw material." As William Matthews writes in a recent article, "A poet beginning to make something needs raw material, something to transform." For Marianne Moore, he says,

> subject matter is not in itself important, except that it gives her the opportunity to speak about something that engages her passions. What is important instead is what she can discover to say.

And he concludes:

> It is not, of course, the subject that is or isn't dull, but the quality of attention we do or do not pay to it, and the strength of our will to transform. Dull subjects are those we have failed.

This assumes that for the animals and humans who are not fine artists, who have discovered nothing to say, the world is dull, which is not true. It assumes also that attention is of interest in itself, which is not true either. In fact, attention is of value only insofar as it is paid in the proper discharge of an obligation. To pay attention is to come into the presence of a subject. In one of its root senses, it is to "stretch toward" a subject, in a kind of aspiration. We speak of "paying attention" because of a correct perception that attention is *owed*—that without our attention and our attending, our subjects, including ourselves, are endangered.

Mr. Matthews's trivializing of subjects in the interest of poetry industrializes the art. He is talking about an art oriented exclusively to production, like coal mining. Like an industrial entrepreneur, he regards the places and creatures and experiences of the world as "raw material," valueless until exploited.

The test of imagination, ultimately, is not the territory of art or the territory of the mind, but the territory underfoot. That is not to say that there is no territory of art or of the mind, only that it is not a separate territory. It is not exempt either from the principles above it or from the country below it. It is a territory, then, that is subject to correction—by, among other things, paying attention. To remove it from the possibility of correction is finally to destroy art and thought, and the territory underfoot as well.

Memory, for instance, must be a pattern upon the actual country, not a cluster of relics in a museum or a written history. What Barry Lopez speaks of as a sort of invisible landscape of communal association and usage must serve the visible as a guide and as a protector; the visible landscape must verify and correct the invisible. Alone, the invisible landscape becomes false, sentimental, and useless, just as the visible landscape, alone, becomes a strange land, threatening to humans and vulnerable to human abuse.

To assume that the context of literature is "the literary world" is, I believe, simply wrong. That its real habitat is the household and the community—that it can and does affect, even in practical ways, the life of a place—may not be recognized by most theorists and critics for a while yet. But they will finally come to it, because finally they will have to. And when they do, they will renew the study of literature and restore it to importance.

*

Emerson in "The American Scholar," worrying about the increasing specialization of human enterprises, thought that the individual, to be whole, "must sometimes return from his own labor to embrace all the other laborers"—a solution that he acknowledged to be impossible. The result, he saw, was that "man is thus metamorphosed into a thing, into many things." The solution that he apparently did think possible was a return out of specialization and separateness to the human definition, so that a thinker or scholar would not be a "mere thinker," a thinking specialist, but "Man Thinking." But this return is not meant to be a retreat into abstraction, for Emerson understood "Man Thinking" as a thinker committed to action: "Action is with the scholar subordinate, but it is essential. Without it, he is not yet man."

And action, of course, implies place and community. There can be disembodied thought, but not disembodied action. Action—embodied thought—requires local and communal reference. To act, in short, is to live. Living "is a total act. Thinking is a partial act." And one does not live alone. Living is a communal act, whether or not its communality is acknowledged. And so Emerson writes:

> I grasp the hands of those next me, and take my place in the ring to suffer and to work, taught by an instinct, that so shall the dumb abyss be vocal with speech.

Emerson's spiritual heroism can sometimes be questionable or tiresome, but he can also write splendidly accurate, exacting sentences, and that is one of them. We see how it legislates against what we now call "groupiness." Neighborhood is a given condition, not a contrived one; he is not talking about a "planned community" or a "network," but about the necessary interdependence of those who are "next" each other. We see how it invokes dance, acting in concert, as a metaphor of almost limitless reference. We see how the phrase "to suffer and to work" refuses sentimentalization. We see how common work, common suffering, and a common willingness to join and belong are understood as the conditions that make speech possible in "the dumb abyss" in which we are divided.

This leads us, probably, to as good a definition of the beloved community as we can hope for: common experience and common effort on a common ground to which one willingly belongs. The life of such a community has been very little regarded in American literature. Our writers have been much more concerned with the individual who is misunderstood or mistreated by a community that is in no sense beloved, as in *The Scarlet Letter*. From Thoreau to Hemingway and his successors, a great deal of sympathy and interest has been given to the individual as pariah or gadfly or exile. In Faulkner, a community is the subject, but it is a community disintegrating, as it was doomed to do by the original sins of land greed, violent honor, and slavery. There are in Faulkner some characters who keep alive the hope of community, or at least the fundamental decencies on which community depends, and in Faulkner, as in Mark Twain, these are chiefly women: Dilsey, Lena Grove, the properly outraged Mrs. Littlejohn.

The one American book I know that is about a beloved community—a settled, established white American community with a sustaining common culture, and mostly beneficent toward both its members and its place—is Sarah Orne Jewett's *The Country of the Pointed Firs.* The community that the book describes, the coastal village of Dunnet, Maine, and the neighboring islands and back country, is an endangered species on the book's own evidence: many of its characters are old and childless, without heirs or successors—and with the Twentieth Century ahead of it, it could not last. But though we see it in its last days, we see it whole.

We see it whole, I think, because we see it both in its time and in its timelessness. The centerpiece of the book, the Bowden family reunion, is described in the particularity of a present act, but it is perceived also—as such an event must be—as a reenactment; to see is to remember:

> There was a wide path mowed for us across the field, and, as we moved along, the birds flew up out of the thick second crop of clover, and the bees hummed as if it still were June. There was a flashing of white gulls over the water where the fleet of boats rode the low waves together in the cove, swaying their small masts as if they kept time to our steps. The plash of the water could be heard faintly, yet still be heard; we might have been a company of ancient Greeks.

Thus, though it precisely renders its place and time, the book never subsides into the flimsy contemporaneity of "local color." The narrator of the book is one who departs and returns, and her returns are homecomings—to herself as well as to the place:

> The first salt wind from the east, the first sight of a lighthouse set boldly on its outer rock, the flash of a gull, the waiting procession of seaward-bound firs on an island, made me feel solid and definite again, instead of a poor incoherent being. Life was resumed, and anxious living blew away as if it had not been. I could not breathe deep enough or long enough. It was a return to happiness.

Anyone acquainted with the sentimentalities of American regionalism will look on that word "happiness" with suspicion. But here it is not sentimental, for the work and suffering of the community are fully faced and acknowledged. The narrator's return is not to an idyll of the boondocks; it is a re-entrance into Emerson's "ring." The community is happy in that it has survived its remembered tragedies, has re-shaped itself coherently around its known losses, has included kindly its eccentrics, invalids, oddities, and even its one would-be exile. The wonderful heroine of the book, and its emblem, Mrs. Elmira Todd, a childless widow, who in her youth "had loved one who was far above her," is a healer—a grower, gatherer, and dispenser of medicinal herbs.

She is also a dispenser of intelligent talk about her kinfolk and neighbors. More than any book I know, this one makes its way by conversation, engrossing exchanges of talk in which Mrs. Todd and many others reveal to the narrator their life and history and geography. And perhaps the great cultural insight of the book is stated by Mrs. Todd:

> Conversation's got to have some root in the past, or else you've got to explain every remark you make, an' it wears a person out.

The conversation wells up out of memory, and in a sense *is* the community, the presence of its past and its hope, speaking in the dumb abyss.

An Argument for Diversity

Elegant solutions will be predicated upon the uniqueness of place. JOHN TODD

I LIVE IN and have known all my life the northern corner of Henry County, Kentucky. The country here is narrowly creased and folded; it is a varied landscape whose main features are these:

1. A rolling upland of which some of the soil is good and some, because of abuse, is less so. This upland is well suited to mixed farming, which was, in fact, traditional to it, but which is less diversified now than it was a generation ago. Some row-cropping is possible here, but even the best-lying ridges are vulnerable to erosion and probably not more than ten percent should be broken in any year. It is a kind of land that needs grass and grazing animals, and it is excellent for this use.

2. Wooded bluffs where the upland breaks over into the valleys of the creeks and the Kentucky River. Along with virtually all of this region, most of these bluffs have been cleared and cropped at one time or another. They should never have been cropped, and because of their extreme vulnerability to erosion they should be logged only with the greatest skill and care. These bluffs are now generally forested, though not many old-growth stands remain.

3. Slopes of gentler declivity below the bluffs and elsewhere. Some of these slopes are grassed and, with close care, are maintainable as pasture. Until World War II they were periodically cropped, in a version of slash-and-burn agriculture that resulted in serious damage by erosion. Now much of this land is covered with trees thirty or forty years old.

4. Finally, there are the creek and river bottoms, some of which are subject to flooding. Much of this land is suitable for intensive row-cropping, which, under the regime of industrial agriculture, has sometimes been too intensive.

Within these four general divisions this country is extremely diverse. To familiarity and experience, the landscape divides into many small facets or aspects differentiated by the kind or quality of soil and by slope, exposure, drainage, rockiness, and so on. In the two centuries during which European races have occupied this part of the country, the best of the land has sometimes been well used, under the influence of good times and good intentions. But virtually none of it has escaped ill use under the influence of bad times or ignorance, need or greed. Some of it—the steeper, more marginal areas—never has been well used. Of virtually all this land it may be said that the national economy has prescribed ways of use but not ways of care. It is now impossible to imagine any immediate way that most of this land might receive excellent care. The economy, as it now is, prescribes plunder of the landowners and abuse of the land.

The connection of the American economy to this place—in comparison, say, to the connection of the American economy to just about any university—has been unregarding and ungenerous. Indeed, the connection has been almost entirely exploitive—and it has never been more exploitive than it is now. Increasingly, from the beginning, most of the money made on the products of this place has been made in other places. Increasingly the ablest young people of this place have gone away to receive a college education, which has given them a "professional status" too often understood as a license to become the predators of such places as this one that they came from. The destruction of the human community, the local economy, and the natural health of such a place as this is now looked upon not as a "trade-off," a possibly regrettable "price of progress," but as a good, virtually a national goal.

Recently I heard, on an early-morning radio program, a university economist explaining the benefits of off-farm work for farm women: that these women are increasingly employed off the farm, she said, has made them "full partners" in the farm's economy. Never mind that this is a symptom of economic desperation and great unhappiness on the farm. And never mind the value, which was more than economic, of these women's previous contribution *on* the farm to the farm family's life and economy—in what was, many of them would have said, a full partnership. *Now* they are "earning forty-five percent of total family income"; *now* they are playing "a major role." The forty-five percent and the "major role" are allowed to defray all other costs. That the farm family now furnishes labor and (by its increased consumption) income to the economy that is destroying it is seen simply as an improvement. Thus the abstract and extremely tentative value of money is thoughtlessly allowed to replace the particular and fundamental values of the lives of household and community. Obviously, we need to stop thinking about the economic functions of individuals for a while, and try to learn to think of the economic functions of communities and households. We need to try to understand the long-term economies of places—places, that is, that are considered as dwelling places for humans and their fellow creatures, not as exploitable resources.

What happens when farm people take up "off-farm work"? The immediate result is that they must be replaced by chemicals and machines and other purchases from an economy adverse and antipathetic to farming, which means that the remaining farmers are put under yet greater economic pressure to abuse their land. If under the pressure of an adverse economy the soil erodes, soil and water and air are poisoned, the woodlands are wastefully logged, and everything not producing an immediate economic return is neglected, that is apparently understood by most of the society as merely the normal cost of production.

This means, among other things, that the land and its human communities are not being thought about in places of study and leadership, and this failure to think is causing damage. But if one lives in a country place, and if one loves it, one must think about it. Under present circumstances, it is not easy to imagine what might be a proper human economy for the country I have just described. And yet, if one loves it, one must make the attempt; if one loves it, the attempt is irresistible.

Two facts are immediately apparent. One is that the present local economy, based like the economies of most rural places exclusively on the export of raw materials, is ruinous. Another is that the influence of a complex, aggressive national economy upon a simple, passive local economy will also be ruinous. In a varied and versatile countryside, fragile in its composition and extremely susceptible to abuse, requiring close human care and elaborate human skills, able to produce and needing to produce a great variety of products from its soils, what is needed, obviously, is a highly diversified local economy.

We should be producing the fullest variety of foods to be consumed locally, in the countryside itself and in nearby towns and cities: meats, grains, table vegetables, fruits and nuts, dairy products, poultry and eggs. We should be harvesting a sustainable yield of fish from our ponds and streams. Our woodlands, managed for continuous yields, selectively and carefully logged, should be yielding a variety of timber for a variety of purposes: firewood, fence posts, lumber for building, fine woods for furniture makers.

And we should be adding value locally to these local products. What is needed is not the large factory so dear to the hearts of government "developers." To set our whole population to making computers or automobiles would be as gross an error as to use the whole countryside for growing corn or Christmas trees or pulpwood; it would discount everything we have to offer as a community and a place; it would despise our talents and capacities as individuals.

We need, instead, a system of decentralized, small-scale industries to transform the products of our fields and woodlands and streams: small creameries, cheese factories, canneries, grain mills, saw mills, furniture factories, and the like. By "small" I mean simply a size that would not be destructive of the appearance, the health, and the quiet of the countryside. If a factory began to "grow" or to be noisy at night or on Sunday, that would mean that another such factory was needed somewhere else. If waste should occur at any point, that would indicate the need for an enterprise of some other sort. If poison or pollution resulted from any enterprise, that would be understood as an indication that something was absolutely wrong, and a correction would be made. Small scale, of course, makes such changes and corrections more thinkable and more possible than does large scale.

I realize that, by now, my argument has crossed a boundary line of which everyone in our "realistic" society is keenly aware. I will be perceived to have crossed over into "utopianism" or fantasy. Unless I take measures to prevent it, I am going to hear somebody say, "All that would be very nice, if it were possible. Can't you be realistic?"

Well, let me take measures to prevent it. I am not, I admit, optimistic about the success of this kind of thought. Otherwise, my intention, above all, is to be realistic; I wish to be practical. The question here is simply that of convention. Do I want to be realistic according to the conventions of the industrial economy and the military state, or according to what I know of reality? To me, an economy that sees the life of a community or a place as expendable, and reckons its value only in terms of money, is not acceptable because it is *not* realistic. I am thinking as I believe we must think if we wish to discuss the *best* uses of people, places, and things, and if we wish to give affection some standing in our thoughts.

If we wish to make the best use of people, places, and things, then we are going to have to deal with a law that reads about like this: as the quality of use increases, the scale of use (that is, the size of operations) will decline, the tools will become simpler, and the methods and the skills will become more complex. That is a difficult law for us to believe, because we have assumed otherwise for a long time, and yet our experience overwhelmingly suggests that it *is* a law, and that the penalties for disobeying it are severe.

I am making a plea for diversity not only because diversity exists and is pleasant, but also because it is necessary and we need more of it. For an example, let me return to the countryside I described at the beginning of this essay. From birth, I have been familiar with this place and have heard it talked about and thought about. For the last twenty-five years I have been increasingly involved in the use and improvement of a little part of it. As a result of some failures and some successes, I have learned some things about it. I am certain, however, that I do not know the best way to use this land. Nor do I believe that anyone else does. I no longer expect to live to see it come to its best use. But I am beginning to see what is needed, and everywhere the need is for diversity. This is the need of every American rural landscape that I am acquainted with. We need a greater range of species and varieties of plants and animals, of human skills and methods, so that the use may be fitted ever more sensitively and elegantly to the place. Our places, in short, are asking us questions, some of them urgent questions, and we do not have the answers.

The answers, if they are to come and if they are to work, must be developed in the presence of the user and the land; they must be developed to some degree *by* the user *on* the land. The present practice of handing down from on high policies and technologies developed without consideration of the nature and the needs of the land and the people has not worked, and it cannot work. Good agriculture and forestry cannot be "invented" by self-styled smart people in the offices and laboratories of a centralized economy and then sold at the highest possible profit to the supposedly dumb country people. That is not the way good land use comes about. And it does not matter how the methodologies so developed and handed down are labeled; whether "industrial" or "conventional" or "organic" or "sustainable," the professional or professorial condescension that is blind to the primacy of the union between individual people and individual places is ruinous. The challenge to the would-be scientists of an ecologically sane agriculture, as David Ehrenfeld has written, is "to provide unique and particular answers to questions about a farmer's unique and particular land." The proper goal, he adds, is *not* merely to "substitute the cult of the benevolent ecologist for the cult of the benevolent sales representative."

The question of what a beloved country is to be used for quickly becomes inseparable from the questions of who is to use it or who is to prescribe its uses, and what will be the ways of using it. If we speak simply of the use of "a country," then only the first question is asked, and it is asked only by its would-be users. It is not until we speak of "a beloved country" —a particular country, particularly loved—that the question about ways of use will arise. It arises because, loving our country, we see where we are, and we see that present ways of use are not adequate. They are not adequate because such local cultures and economies as we once had have been stunted or destroyed. As a nation, we have attempted to substitute the *concepts* of "land use," "agribusiness," "development," and the like for the *culture* of stewardship and husbandry. And this change is not a result merely of economic pressures and adverse social values; it comes also from the state of affairs in our educational system, especially in our universities.

It is readily evident, once affection is allowed into the

discussion of "land use," that the life of the mind, as presently constituted in the universities, is of no help. The sciences are of no help, indeed are destructive, because they work, by principle, outside the demands, checks, and corrections of affection. The problem with this "scientific objectivity" becomes immediately clear when science undertakes to "apply" itself to land use. The problem simply is that land users are using people, places, and things that cannot be well used without affection. To be well used, creatures and places must be used sympathetically, just as they must be known sympathetically to be well known. The economist to whom it is of no concern whether or not a family loves its farm will almost inevitably aid and abet the destruction of family farming. The "animal scientist" to whom it is of no concern whether or not animals suffer will almost inevitably aid and abet the destruction of the decent old ideal of animal husbandry and, as a consequence, increase the suffering of animals. I hope that my country may be delivered from the remote, cold abstractions of university science.

But "the humanities," as presently constituted in the universities, are of no help either, and indeed, with respect to the use of a beloved country, they too have been destructive. (The closer I have come to using the term "humanities," the less satisfactory it has seemed to me; by it I mean everything that is not a "science," another unsatisfactory term.) The humanities have been destructive not because they have been misapplied, but because they have been so frequently understood by their academic stewards as not applicable. The scientific ideals of objectivity and specialization have now crept into the humanities and made themselves at home. This has happened, I think, because the humanities have come to be infected with a suspicion of their uselessness or worthlessness in the face of the provability or workability or profitability of the applied sciences. The conviction is now widespread, for instance, that "a work of art" has no purpose but to be itself. Or if it is allowed that a poem, for instance, has a meaning, then it is a meaning peculiar to its author, its time, or its convention. A poem, in short, is a relic as soon as it is composed; it can be taught, but it cannot teach. The issue of its truth and pertinence is not raised because literary study is conducted with about the same anxiety for "control" as is scientific study. The context of a poem is its text, or the context of its history and criticism as a text. I have not, of course, read all the books or sat in all the classrooms, but my impression is that not much importance is attached to the question of the truth of poems. My impression is that "Comus," for example, is not often taught as an argument with a history and a sequel, with the gravest importance for us in our dilemma now. My impression is that the great works are taught less and less as Ananda Coomaraswamy said they should be: with the recognition "that nothing will have been accomplished unless men's lives are affected and their values changed by what we have to show." My impression is that in the humanities as in the sciences the world is increasingly disallowed as a context. I hope that my country may be delivered from the objectivity of the humanities.

Without a beloved country as context, the arts and the sciences become oriented to the careers of their practitioners, and the intellectual life to intellectual (and bureaucratic) procedures. And so in the universities we see forming an intellectual elite more and more exclusively accomplished in procedures such as promotion, technological innovation, publication, and grant-getting. The context of a beloved country, moreover, implies an academic standard that is not inflatable or deflatable. The standard—the physical, intellectual, political, ecological, economic, and spiritual health of the country—cannot be too high; it will be as high, simply, as we have the love, the vision, and the courage to make it.

I would like my country to be seen and known with an attentiveness that is schooled and skilled. I would like it to be loved with a minutely particular affection and loyalty. I would like the work in it to be practical and loving and respectful and forbearing. In order for these things to happen, the sciences and the humanities are going to have to come together again in the presence of the practical problems of individual places, and of local knowledge and local love in individual people—people able to see, know, think, feel, and act coherently and well without the modern instinct of deference to the "outside expert."

What should the sciences have to say to a citizen in search of the criteria by which to determine the best use of a beloved place or countryside, or of the technical or moral means by which to limit that use to its best use? What should the humanities have to say to a scientist—or, for that matter, a citizen—in search of the cultural instructions that might effectively govern the use of a beloved place? These questions or such questions could reunite the sciences and the humanities. That a scientist and an artist can speak and work together in response to such questions I know from my own experience. All that is necessary is a mutuality of concern and a mutual willingness to speak common English. When friends speak across these divisions or out of their "departments," in mutual concern for a beloved country, then it is clear that these diverse disciplines are not "competing interests," as the university structure and academic folklore suggest, but interests with legitimate claims on all minds. It is only when the country becomes an abstraction, a prize of conquest, that these interests compete—though, of course, when that has happened *all* interests compete.

But in order to assure that a beloved country might be lovingly used, the sciences and the humanities will have to do more than mend their divorce at "the university level"; they will also have to mend their divorce from the common culture, by which I do not mean the "popular culture," but rather the low and local wisdom that is now either relegated to the compartments of anthropology or folklore or "oral history," or not attended to at all.

Some time ago, after I had given a lecture at a college in Ohio, a gentleman came up and introduced himself to me as a fellow Kentuckian.

"Where in Kentucky are you from?" I asked.

"Oh, a little place you probably never heard of—North Middletown."

"I *have* heard of North Middletown," I said. "It was the home of my father's great friend John W. Jones."

"Well, John W. Jones was my uncle."

I told him then of my father's and my own respect for Mr. Jones.

"I want to tell you a story about Uncle John," he said. And he told me this:

When his Uncle John was president of the bank in North Middletown, his policy was to give a loan to any graduate of the North Middletown high school who wanted to go to college and needed the money. This practice caused great consternation to the bank examiners who came and found those unsecured loans on the books and no justification for them except Mr. Jones's conviction that it was right to make them.

As it turned out, it was right in more than principle, for in the many years that Mr. Jones was president of the bank, making those "unsound loans," *all* of the loans were repaid; he never lost a dime on a one of them.

I do not mean to raise here the question of the invariable goodness of a college education, which I doubt. My point in telling this story is that Mr. Jones was acting from a kind of knowledge, inestimably valuable and probably indispensable, that comes out of common culture and that cannot be taught as a part of the formal curriculum of a school. The students whose education he enabled were not taught this knowledge at the colleges they attended. What he knew—and this involved his knowledge of himself, his tradition, his community, and everybody in it—was that trust, in the circumstances then present, could beget trustworthiness. This is the kind of knowledge, obviously, that is fundamental to the possibility of community life and to certain good possibilities in the characters of people. Though I don't believe that it can be taught and learned in a university, I think that it should be known about and respected in a university, and I don't know where, in the sciences and the humanities as presently constituted, students would be led to suspect, much less to honor, its existence. It is certainly no part of banking or of economics as now taught and practiced. It is a part of community life, which most scientists ignore in their professional pursuits, and which most people in the humanities seem to regard as belonging to a past now useless or lost or dispensed with.

Let me give another, more fundamental example. My brother, who is a lawyer, recently had as a client an elderly man named Bennie Yeary who had farmed for many years a farm of about three hundred acres of hilly and partly forested land. His farm and the road to his house had been damaged by a power company.

Seeking to determine the value of the land, my brother asked him if he had ever logged his woodlands. Mr. Yeary answered, "Yes, sir, since 1944. . . . I have never robbed [the land]. I have always just cut a little out where I thought it needed it. I have got as much timber right now, I am satisfied, . . . as I had when I started mill runs here in '44."

That we should not rob the land is a principle to be found readily enough in the literary culture. That it came into literature out of the common culture is suggested by the fact that it is commonly phrased in this way by people who have not inherited the literary culture. That we should not rob the land, anyhow, is a principle that can be learned from books. But the ways of living on the land so as not to rob it probably cannot be learned from books, and this is made clear by a further exchange between my brother and Mr. Yeary.

They came to the question of what was involved in the damage to the road, and the old farmer said that the power company had destroyed thirteen or fourteen water breaks. A water break is a low mound of rock and earth built to divert the water out of a hilly road. It is a means of preventing erosion both of the roadbed and of the land alongside it, one of the ways of living on the land without robbing it.

"How long . . . had it been since you had those water breaks constructed in there?"

"I had been working on them . . . off and on, for about twelve years, putting them water breaks in. I hauled rocks out of my fields . . . and I would dig out, bury these rocks down, and take the sledgehammer and beat rock in here and make this water break."

The way to make a farm road that will not rob the land cannot be learned from books, then, because the long use of such a road is a part of the proper way of making it, and because the use and improvement of the road are intimately involved with the use and improvement of the place. It is of the utmost importance that the rocks to make the water breaks were hauled from the fields. Mr. Yeary's solution did not, like the typical industrial solution, involve the making of a problem, or a series of problems, elsewhere. It involved the making of a solution elsewhere: the same work that improved the road improved the fields. Such work requires not only correct principles, skill, and industry, but a knowledge of local particulars and many years; it involves slow, small adjustments in response to questions asked by a particular place. And this is true in general of the patterns and structures of a proper human use of a beloved country, as examination of the traditional landscapes of the Old World will readily show: they were made by use as much as by skill.

This implication of use in the making of essential artifacts and the maintenance of the landscape—which are to so large an extent the making and the maintenance of culture—brings us to the inescapable final step in an argument for diversity: the realization that without a diversity of people we cannot maintain a diversity of anything else. By a diversity of people I do not mean a diversity of specialists, but a diversity of people elegantly suited to live in their places and to bring them to their best use, whether the use is that of uselessness, as in a place left wild, or that of the highest sustainable productivity. The most abundant diversity of creatures and ways cannot be maintained in preserves, zoos, museums, and the like, but only in the occupations and the pleasures of an appropriately diversified human economy.

The proper ways of using a beloved country are "humanities," I think, and are as complex, difficult, interesting, and worthy as any of the rest. But they defy the present intellectual and academic categories. They are *both* science and art, knowing and doing. Indispensable as these ways are to the success of human life, they have no place and no standing in the present structures of our intellectual life. The purpose, indeed, of the present structures of our intellectual life has been to educate them out of existence. I think I know where in any university my brother's client, Mr. Yeary, would be laughed at or ignored or tape-recorded or classified. I don't know where he would be appropriately honored. The scientific disciplines certainly do not honor him, and the "humane" ones almost as certainly do not. We would have to go some distance back in the literary tradition—back to Thomas Hardy at least, and before Hardy to Wordsworth—to find the due respect paid to such a person. He *has* been educated almost out of existence, and yet an understanding of his importance and worth would renew the life of the mind in this country, in the university and out.

The Pleasures of Eating

MANY TIMES, after I have finished a lecture on the decline of American farming and rural life, someone in the audience has asked, "What can city people do?"

"Eat responsibly," I have usually answered. Of course, I have tried to explain what I meant by that, but afterwards I have invariably felt that there was more to be said than I had been able to say. Now I would like to attempt a better explanation.

I begin with the proposition that eating is an agricultural act. Eating ends the annual drama of the food economy that begins with planting and birth. Most eaters, however, are no longer aware that this is true. They think of food as an agricultural product, perhaps, but they do not think of themselves as participants in agriculture. They think of themselves as "consumers." If they think beyond that, they recognize that they are passive consumers. They buy what they want—or what they have been persuaded to want—within the limits of what they can get. They pay, mostly without protest, what they are charged. And they mostly ignore certain critical questions about the quality and the cost of what they are sold: How fresh is it? How pure or clean is it, how free of dangerous chemicals? How far was it transported, and what did transportation add to the cost? How much did manufacturing or packaging or advertising add to the cost? When the food product has been manufactured or "processed" or "precooked," how has that affected its quality or price or nutritional value?

Most urban shoppers would tell you that food is produced on farms. But most of them do not know what farms, or what kinds of farms, or where the farms are, or what knowledge or skills are involved in farming. They apparently have little doubt that farms will continue to produce, but they do not know how or over what obstacles. For them, then, food is pretty much an abstract idea—something they do not know or imagine—until it appears on the grocery shelf or on the table.

The specialization of production induces specialization of consumption. Patrons of the entertainment industry, for example, entertain themselves less and less and have become more and more passively dependent on commercial suppliers. This is certainly true also of patrons of the food industry, who have tended more and more to be *mere* consumers—passive, uncritical, and dependent. Indeed, this sort of consumption may be said to be one of the chief goals of industrial production. The food industrialists have by now persuaded millions of consumers to prefer food that is already prepared. They will grow, deliver, and cook your food for you and (just like your mother) beg you to eat it. That they do not yet offer to insert it, prechewed, into your mouth is only because they have found no profitable way to do so. We may rest assured that they would be glad to find such a way. The ideal industrial food consumer would be strapped to a table with a tube running from the food factory directly into his or her stomach.

Perhaps I exaggerate, but not by much. The industrial eater is, in fact, one who does not know that eating is an agricultural act, who no longer knows or imagines the connections between eating and the land, and who is therefore necessarily passive and uncritical—in short, a victim. When food, in the minds of eaters, is no longer associated with farming and with the land, then the eaters are suffering a kind of cultural amnesia that is misleading and dangerous. The current version of the "dream home" of the future involves "effortless" shopping from a list of available goods on a television monitor and heating precooked food by remote control. Of course, this implies and depends on a perfect ignorance of the history of the food that is consumed. It requires that the citizenry should give up their hereditary and sensible aversion to buying a pig in a poke. It wishes to make the selling of pigs in pokes an honorable and glamorous activity. The dreamer in this dream home will perforce know nothing about the kind or quality of this food, or where it came from, or how it was produced and prepared, or what ingredients, additives, and residues it contains—unless, that is, the dreamer undertakes a close and constant study of the food industry, in which case he or she might as well wake up and play an active and responsible part in the economy of food.

There is, then, a politics of food that, like any politics, involves our freedom. We still (sometimes) remember that we cannot be free if our minds and voices are controlled by someone else. But we have neglected to understand that we cannot be free if our food and its sources are controlled by someone else. The condition of the passive consumer of food is not a democratic condition. One reason to eat responsibly is to live free.

But if there is a food politics, there are also a food esthetics and a food ethics, neither of which is dissociated from politics. Like industrial sex, industrial eating has become a degraded, poor, and paltry thing. Our kitchens and other eating places more and more resemble filling stations, as our homes more and more resemble motels. "Life is not very interesting," we seem to have decided. "Let its satisfactions be minimal, perfunctory, and fast." We hurry through our meals to go to work and hurry through our work in order to "recreate" ourselves in the evenings and on weekends and vacations. And then we hurry, with the greatest possible speed and noise and violence, through our recreation—for what? To eat the billionth hamburger at some fast-food joint hellbent on increasing the "quality" of our life? And all this is carried out in a remarkable obliviousness to the causes and effects, the possibilities and the purposes, of the life of the body in this world.

One will find this obliviousness represented in virgin purity in the advertisements of the food industry, in which food wears as much makeup as the actors. If one gained one's whole knowledge of food from these advertisements (as some presumably do), one would not know that the various edibles were ever living creatures, or that they all come from the soil, or that they were produced by work. The passive American consumer, sitting down to a meal of pre-prepared or fast food, confronts a platter covered with inert, anonymous substances that have been processed, dyed, breaded, sauced, gravied, ground, pulped, strained, blended, prettified, and sanitized beyond resemblance to any part of any creature that ever lived. The products of nature and agriculture have been made, to all appearances, the products of industry. Both eater and eaten are thus in exile from biological reality. And the result is a kind of solitude, unprecedented in human experience, in which the eater may think of eating as, first, a purely commercial transaction between him and a supplier and then as a purely appetitive transaction between him and his food.

And this peculiar specialization of the act of eating is, again, of obvious benefit to the food industry, which has good reasons to obscure the connection between food and farming. It would not do for the consumer to know that the hamburger she is eating came from a steer who spent much of his life standing deep in his own excrement in a feedlot, helping to pollute the local streams, or that the calf that yielded the veal cutlet on her plate spent its life in a box in which it did not have room to turn around. And, though her sympathy for the slaw might be less tender, she should not be encouraged to meditate on the hygienic and biological implications of mile-square fields of cabbage, for vegetables grown in huge monocultures are dependent on toxic chemicals—just as animals in close confinement are dependent on antibiotics and other drugs.

The consumer, that is to say, must be kept from discovering that, in the food industry—as in any other industry—the overriding concerns are not quality and health, but volume and price. For decades now the entire industrial food economy, from the large farms and feedlots to the chains of supermarkets and fast-food restaurants, has been obsessed with volume. It has relentlessly increased scale in order to increase volume in order (presumably) to reduce costs. But as scale increases, diversity declines; as diversity declines, so does health; as health declines, the dependence on drugs and chemicals necessarily increases. As capital replaces labor, it does so by substituting machines, drugs, and chemicals for human workers and for the natural health and fertility of the soil. The food is produced by any means or any shortcut that will increase profits. And the business of the cosmeticians of advertising is to persuade the consumer that food so produced is good, tasty, healthful, and a guarantee of marital fidelity and long life.

It is possible, then, to be liberated from the husbandry and wifery of the old household food economy. But one can be thus liberated only by entering a trap (unless one sees ignorance and helplessness as the signs of privilege, as many people apparently do). The trap is the ideal of industrialism: a walled city surrounded by valves that let merchandise in but no consciousness out. How does one escape this trap? Only voluntarily, the same way that one went in: by restoring one's consciousness of what is involved in eating; by reclaiming responsibility for one's own part in the food economy. One might begin with the illuminating principle of Sir Albert Howard's *The Soil and Health*, that we should understand "the whole problem of health in soil, plant, animal, and man as one great subject." Eaters, that is, must understand that eating takes place inescapably in the world, that it is inescapably an agricultural act, and that how we eat determines, to a considerable extent, how the world is used. This is a simple way of describing a relationship that is inexpressibly complex. To eat responsibly is to understand and enact, so far as one can, this complex relationship. What can one do? Here is a list, probably not definitive:

1. Participate in food production to the extent that you can. If you have a yard or even just a porch box or a pot in a sunny window, grow something to eat in it. Make a little compost of your kitchen scraps and use it for fertilizer. Only by growing some food for yourself can you become acquainted with the beautiful energy cycle that revolves from soil to seed to flower to fruit to food to offal to decay, and around again. You will be fully responsible for any food that you grow for yourself, and you will know all about it. You will appreciate it fully, having known it all its life.

2. Prepare your own food. This means reviving in your own mind and life the arts of kitchen and household. This should enable you to eat more cheaply, and it will give you a measure of "quality control": you will have some reliable knowledge of what has been added to the food you eat.

3. Learn the origins of the food you buy, and buy the food that is produced closest to your home. The idea that every locality should be, as much as possible, the source of its own food makes several kinds of sense. The locally produced food supply is the most secure, the freshest, and the easiest for local consumers to know about and to influence.

4. Whenever possible, deal directly with a local farmer, gardener, or orchardist. All the reasons listed for the previous suggestion apply here. In addition, by such dealing you eliminate the whole pack of merchants, transporters, processors, packagers, and advertisers who thrive at the expense of both producers and consumers.

5. Learn, in self-defense, as much as you can of the economy and technology of industrial food production. What is

added to food that is not food, and what do you pay for these additions?

6. Learn what is involved in the *best* farming and gardening.

7. Learn as much as you can, by direct observation and experience if possible, of the life histories of the food species.

The last suggestion seems particularly important to me. Many people are now as much estranged from the lives of domestic plants and animals (except for flowers and dogs and cats) as they are from the lives of the wild ones. This is regrettable, for these domestic creatures are in diverse ways attractive; there is much pleasure in knowing them. And farming, animal husbandry, horticulture, and gardening, at their best, are complex and comely arts; there is much pleasure in knowing them, too.

It follows that there is great *dis*pleasure in knowing about a food economy that degrades and abuses those arts and those plants and animals and the soil from which they come. For anyone who does know something of the modern history of food, eating away from home can be a chore. My own inclination is to eat seafood instead of red meat or poultry when I am traveling. Though I am by no means a vegetarian, I dislike the thought that some animal has been made miserable in order to feed me. If I am going to eat meat, I want it to be from an animal that has lived a pleasant, uncrowded life outdoors, on bountiful pasture, with good water nearby and trees for shade. And I am getting almost as fussy about food plants. I like to eat vegetables and fruits that I know have lived happily and healthily in good soil, not the products of the huge, bechemicaled factory-fields that I have seen, for example, in the Central Valley of California. The industrial farm is said to have been patterned on the factory production line. In practice, it looks more like a concentration camp.

The pleasure of eating should be an *extensive* pleasure, not that of the mere gourmet. People who know the garden in which their vegetables have grown and know that the garden is healthy will remember the beauty of the growing plants, perhaps in the dewy first light of morning when gardens are at their best. Such a memory involves itself with the food and is one of the pleasures of eating. The knowledge of the good health of the garden relieves and frees and comforts the eater. The same goes for eating meat. The thought of the good pasture and of the calf contentedly grazing flavors the steak. Some, I know, will think it bloodthirsty or worse to eat a fellow creature you have known all its life. On the contrary, I think it means that you eat with understanding and with gratitude. A significant part of the pleasure of eating is in one's accurate consciousness of the lives and the world from which food comes. The pleasure of eating, then, may be the best available standard of our health. And this pleasure, I think, is pretty fully available to the urban consumer who will make the necessary effort.

I mentioned earlier the politics, esthetics, and ethics of food. But to speak of the pleasure of eating is to go beyond those categories. Eating with the fullest pleasure—pleasure, that is, that does not depend on ignorance—is perhaps the profoundest enactment of our connection with the world. In this pleasure we experience and celebrate our dependence and our gratitude, for we are living from mystery, from creatures we did not make and powers we cannot comprehend. When I think of the meaning of food, I always remember these lines by the poet William Carlos Williams, which seem to me merely honest:

There is nothing to eat,
  seek it where you will,
    but of the body of the Lord.
The blessed plants
  and the sea, yield it
    to the imagination
intact.

The Work of Local Culture

FOR MANY YEARS, my walks have taken me down an old fencerow in a wooded hollow on what was once my grandfather's farm. A battered galvanized bucket is hanging on a fence post near the head of the hollow, and I never go by it without stopping to look inside. For what is going on in that bucket is the most momentous thing I know, the greatest miracle that I have ever heard of: it is making earth. The old bucket has hung there through many autumns, and the leaves have fallen around it and some have fallen into it. Rain and snow have fallen into it, and the fallen leaves have held the moisture and so have rotted. Nuts have fallen into it, or been carried into it by squirrels; mice and squirrels have eaten the meat of the nuts and left the shells; they and other animals have left their droppings; insects have flown into the bucket and died and decayed; birds have scratched in it and left their droppings or perhaps a feather or two. This slow work of growth and death, gravity and decay, which is the chief work of the world, has by now produced in the bottom of the bucket several inches of black humus. I look into that bucket with fascination because I am a farmer of sorts and an artist of sorts, and I recognize there an artistry and a farming far superior to mine, or to that of any human. I have seen the same process at work on the tops of boulders in a forest, and it has been at work immemorially over most of the land surface of the world. All creatures die into it, and they live by it.

The old bucket started out a far better one than you can buy now. I think it has been hanging on that post for something like fifty years. I think so because I remember hearing, when I was just a small boy, a story about a bucket that must have been this one. Several of my grandfather's black hired hands went out on an early spring day to burn a tobacco plant bed, and they took along some eggs to boil to eat with their dinner. When dinner time came and they looked around for something to boil the eggs in, they could find only an old bucket that at one time had been filled with tar. The boiling water softened the residue of tar, and one of the eggs came out of the water black. The hands made much sport of seeing who would have to eat the black egg, welcoming their laughter in the midst of their day's work. The man who had to eat the black egg was Floyd Scott, whom I remember well. Dry scales of tar still adhere to the inside of the bucket.

However small a landmark the old bucket is, it is not trivial. It is one of the signs by which I know my country and myself. And to me it is irresistibly suggestive in the way it collects leaves and other woodland sheddings as they fall through time. It collects stories, too, as they fall through time. It is irresistibly metaphorical. It is doing in a passive way what a human community must do actively and thoughtfully. A human community, too, must collect leaves and stories, and turn them to account. It must build soil, and build that memory of itself—in lore and story and song—that will be its culture. These two kinds of accumulation, of local soil and local culture, are intimately related.

In the woods, the bucket is no metaphor; it simply reveals what is always happening in the woods, if the woods is let alone. Of course, in most places in my part of the country, the human community did not leave the woods alone. It felled the trees and replaced them with pastures and crops. But this did not revoke the law of the woods, which is that the ground must be protected by a cover of vegetation and that the growth of the years must return—or be returned—to the ground to rot and build soil. A good local culture, in one of its most important functions, is a collection of the memories, ways, and skills necessary for the observance, within the bounds of domesticity, of this natural law. If the local culture cannot preserve and improve the local soil, then, as both reason and history inform us, the local community will decay and perish, and the work of soil building will be resumed by nature.

A human community, then, if it is to last long, must exert a sort of centripetal force, holding local soil and local memory in place. Practically speaking, human society has no work more important than this. Once we have acknowledged this principle, we can only be alarmed at the extent to which it has been ignored. For although our present society does generate a centripetal force of great power, this is not a local force, but one centered almost exclusively in our great commercial and industrial cities, which have drawn irresistibly into themselves both the products of the countryside and the people and talents of the country communities.

There is, as one assumes there must be, a countervailing or centrifugal force that also operates in our society, but this returns to the countryside not the residue of the land's growth to re-fertilize the fields, not the learning and experience of the greater world ready to go to work locally, and not—or not often—even a just monetary compensation. What are returned, instead, are overpriced manufactured goods, pollution in various forms, and garbage. A landfill on the edge of my own rural county in Kentucky, for example, daily receives about eighty truckloads of garbage. Fifty to sixty of these loads come from cities in New York, New Jersey, and Pennsylvania. Thus, the end result of the phenomenal modern productivity of the countryside is a debased countryside, which becomes daily less pleasant, and which will inevitably become less productive.

The cities, which have imposed this inversion of forces on the country, have been unable to preserve themselves from it. The typical modern city is surrounded by a circle of affluent suburbs, eating its way outward, like ringworm, leaving the so-called inner city desolate, filthy, ugly, and dangerous.

My walks in the hills and hollows around my home have inevitably produced in my mind the awareness that I live in a diminished country. The country has been and is being reduced by the great centralizing process that is our national economy. As I walk, I am always reminded of the slow, patient building of soil in the woods. And I am reminded of the events and companions of my life—for my walks, after so long, are cultural events. But under the trees and in the fields I see also the gullies and scars, healed or healing or fresh, left by careless logging and bad farming. I see the crumbling stone walls and the wire fences that have been rusting out ever since the 1930s. In the returning woods growth of the hollows, I see the sagging and the fallen barns, the empty and ruining houses, the houseless chimneys and foundations. As I look at this evidence of human life poorly founded, played out, and gone, I try to recover some understanding, some vision, of what this country was at the beginning: the great oaks and beeches and hickories, walnuts and maples, lindens and ashes, tulip poplars, standing in beauty and dignity now unimaginable, the black soil of their making, also no longer imaginable, lying deep at their feet—an incalculable birthright sold for money, most of which we did not receive. Most of the money made on the products of this place has gone to fill the pockets of people in distant cities who did not produce the products.

If my walks take me along the roads and streams, I see also the trash and the junk, carelessly manufactured and carelessly thrown away, the glass and the broken glass and the plastic and the aluminum that will lie here longer than the lifetime of trees—longer than the lifetime of our species, perhaps. And I know that this also is what we have to show for our participation in the American economy, for most of the money made on these things too has been made elsewhere.

It would be somewhat more pleasant for country people if they could blame all this on city people. But the old opposition of country versus city—though still true, and truer than ever economically, for the country is more than ever the colony of the city—is far too simple to explain our problem. For country people more and more live like city people, and so connive in their own ruin. More and more country people, like city people, allow their economic and social standards to be set by television and salesmen and outside experts. Our garbage mingles with New Jersey garbage in our local landfill, and it would be hard to tell which is which.

As local community decays along with local economy, a vast amnesia settles over the countryside. As the exposed and disregarded soil departs with the rains, so local knowledge and local memory move away to the cities or are forgotten under the influence of homogenized salestalk, entertainment, and education. This loss of local knowledge and local memory—that is, of local culture—has been ignored, or written off as one of the cheaper "prices of progress," or made the business of folklorists. Nevertheless, local culture has a value, and part of its value is economic. This can be demonstrated readily enough.

For example, when a community loses its memory, its members no longer know one another. How can they know one another if they have forgotten or have never learned one another's stories? If they do not know one another's stories, how can they know whether or not to trust one another? People who do not trust one another do not help one another, and moreover they fear one another. And this is our predicament now. Because of a general distrust and suspicion, we not only lose one another's help and companionship, but we are all now living in jeopardy of being sued.

We don't trust our "public servants" because we know that they don't respect us. They don't respect us, as we understand, because they don't know us; they don't know our stories. They expect us to sue them if they make mistakes, and so they must insure themselves, at great expense to them and to us. Doctors in a country community must send their patients to specialists in the city, not necessarily because they believe that they are wrong in their diagnoses, but because they know that they are not infallible and they must protect themselves against lawsuits, at great expense to us.

The government of my home county, which has a population of about ten thousand people, pays an annual liability insurance premium of about $34,000. Add to this the liability premiums that are paid by every professional person who is "at risk" in the county, and you get some idea of the load we are carrying. Several decent family livelihoods are annually paid out of the county to insurance companies for a service that is only negative and provisional.

All of this money is lost to us by the failure of community. A good community, as we know, insures itself by trust, by good faith and good will, by mutual help. A good community, in other words, is a good local economy. It depends on itself for many of its essential needs and is thus shaped, so to speak, from the inside—unlike most modern populations that depend on distant purchases for almost everything and are thus shaped from the outside by the purposes and the influence of salesmen.

I was walking one Sunday afternoon several years ago with an older friend. We went by the ruining log house that had belonged to his grandparents and great-grandparents. The house stirred my friend's memory, and he told how the oldtime people used to visit each other in the evenings, especially in the long evenings of winter. There used to be a sort of institution in our part of the country known as "sitting till bedtime." After supper, when they weren't too tired, neighbors would walk across the fields to visit each other. They popped corn, my friend said, and ate apples and talked. They told each other stories. They told each other stories, as I knew myself, that they all had heard before. Sometimes they told stories about each other, about themselves, living again in their own memories and thus keeping their memories alive. Among the hearers of these stories were always the children. When bedtime came, the visitors lit their lanterns and went home. My friend talked about this, and thought about it, and then he said, "They had everything but money."

They were poor, as country people have often been, but they had each other, they had their local economy in which they helped each other, they had each other's comfort when they needed it, and they had their stories, their history together in that place. To have everything but money is to have much. And most people of the present can only marvel to think of neighbors entertaining themselves for a whole evening without a single imported pleasure and without listening to a single minute of sales talk.

Most of the descendants of those people have now moved away, partly because of the cultural and economic failures that I mentioned earlier, and most of them no longer sit in the evenings and talk to anyone. Most of them now sit until bedtime watching TV, submitting every few minutes to a sales talk. The message of both the TV programs and the sales talks is that the watchers should spend whatever is necessary to be like everybody else.

By television and other public means, we are encouraged to believe that we are far advanced beyond sitting till bedtime with the neighbors on a Kentucky ridgetop, and indeed beyond anything we ever were before. But if, for example, there should occur a forty-eight-hour power failure, we would find ourselves in much more backward circumstances than our ancestors. What, for starters, would we do for entertainment? Tell each other stories? But most of us no longer talk with each other, much less tell each other stories. We tell our stories now mostly to doctors or lawyers or psychiatrists or insurance adjusters or the police, not to our neighbors for their (and our) entertainment. The stories that now entertain us are made up for us in New York or Los Angeles or other centers of such commerce.

But a forty-eight-hour power failure would involve almost unimaginable deprivations. It would be difficult to travel, especially in cities. Most of the essential work could not be done. Our windowless modern schools and other such buildings that depend on air conditioning could not be used. Refrigeration would be impossible; food would spoil. It would be difficult or impossible to prepare meals. If it was winter, heating systems would fail. At the end of forty-eight hours many of us would be hungry.

Such a calamity (and it is a modest one among those that our time has made possible) would thus reveal how far most of us are now living from our cultural and economic sources, and how extensively we have destroyed the foundations of local life. It would show us how far we have strayed from the locally centered life of such neighborhoods as the one my friend described—a life based to a considerable extent on what we now call solar energy, which is decentralized, democratic, clean, and free. If we note that much of the difference we are talking about can be accounted for as an increasing dependence on energy sources that are centralized, undemocratic, filthy, and expensive, we will have completed a sort of historical parable.

How has this happened? There are many reasons for it. One of the chief reasons is that everywhere in our country the local succession of the generations has been broken. We can trace this change through a series of stories that we may think of as cultural landmarks.

Throughout most of our literature, the normal thing was for the generations to succeed one another in place. The memorable stories occurred when this succession failed or became difficult or was somehow threatened. The norm is given in Psalm 128, in which this succession is seen as one of the rewards of righteousness: "Thou shalt see thy children's children, and peace upon Israel."

The longing for this result seems to have been universal. It presides also over *The Odyssey*, in which Odysseus's desire to return home is certainly regarded as normal. And this story is also much concerned with the psychology of family succession. Telemachus, Odysseus's son, comes of age in preparing for the return of his long-absent father; and it seems almost that Odysseus is enabled to return home by his son's achievement of enough manhood to go in search of him. Long after the return of both father and son, Odysseus's life will complete itself, as we know from Teiresias's prophecy in Book XI, much in the spirit of Psalm 128:

A seaborne death
soft as this hand of mist will come upon you
when you are wearied out with sick old age,
your country folk in blessed peace around you.

The Bible makes much of what it sees as the normal succession, in such stories as those of Abraham, Isaac, and Jacob, or of David and Solomon, in which the son completes the work or the destiny of the father. The parable of the prodigal son is prepared for by such Old Testament stories as that of Jacob, who errs, wanders, returns, is forgiven, and takes his place in the family lineage.

Shakespeare was concerned throughout his working life with the theme of the separation and rejoining of parents and children. It is there at the beginning in *The Comedy of Errors*, and he is still thinking about it when he gets to *King Lear* and *Pericles* and *The Tempest.* When Lear walks onstage with Cordelia dead in his arms, the theme of return is fulfilled, only this time in the way of tragedy.

Wordsworth's poem "Michael," written in 1800, is in the same line of descent. It is the story of a prodigal son, and return is still understood as the norm; before the boy's departure, he and his father make a "covenant" that he will return home and carry on his father's life as a shepherd on their ancestral pastures. But the ancient theme here has two significant differences: the son leaves home for an economic reason, and he does not return. Old Michael, the father, was long ago "bound / In surety for his brother's son." This nephew has failed in his business, and Michael is "summoned to discharge the forfeiture." Rather than do this by selling a portion of their patrimony, the aged parents decide that they must send their son to work for another kinsman in the city in order to earn the necessary money. The country people all are poor; there is no money to be earned at home. When the son has cleared the debt from the land, he will return to it to "possess it, free as the wind / That passes over it." But the son goes to the city, is corrupted by it, eventually commits a crime, and is forced "to seek a hiding place beyond the seas."

"Michael" is a sort of cultural watershed. It carries on the theme of return that goes back to the beginnings of Western culture, but that return now is only a desire and a memory; in the poem it fails to happen. Because of that failure, we see in "Michael" not just a local story of the Lake District of England, which it is, but the story of rural families in the industrial nations from Wordsworth's time until today. The children go to the cities, for reasons imposed by the external economy, and they do not return; eventually the parents die and the family land, like Michael's, is sold to a stranger. By now it has happened millions of times.

And by now the transformation of the ancient story is nearly complete. Our society, on the whole, has forgotten or repudiated the theme of return. Young people still grow up in rural families and go off to the cities, not to return. But now it is felt that this is what they *should* do. Now the norm is to leave and not return. And this applies as much to urban families as to rural ones. In the present urban economy the parent-child succession is possible only among the economically privileged. The children of industrial underlings are not likely to succeed their parents at work, and there is no reason for them to wish to do so. We are not going to have an industrial "Michael" in which it is perceived as tragic that a son fails to succeed his father on an assembly line.

According to the new norm, the child's destiny is not to succeed the parents, but to outmode them; succession has given way to supersession. And this norm is institutionalized not in great communal stories, but in the education system. The schools are no longer oriented to a cultural inheritance that it is their duty to pass on unimpaired, but to the career, which is to say the future, of the child. The orientation is thus necessarily theoretical, speculative, and mercenary. The child is not educated to return home and be of use to the place and community; he or she is educated to *leave* home and earn money in a provisional future that has nothing to do with place or community. And parents with children in school are likely to find themselves immediately separated from their children, and made useless to them, by the intervention of new educational techniques, technologies, methods, and languages. School systems innovate as compulsively and as eagerly as factories. It is no wonder that, under these circumstances, "educators" tend to look upon the parents as a bad influence and wish to take the children away from home as early as possible. And many parents, in truth, are now finding their children an encumbrance at home, where there is no useful work for them to do, and are glad enough to turn them over to the state for the use of the future. The extent to which this order of things is now dominant is suggested by a recent magazine article on the discovery of what purports to be a new idea:

> The idea that a parent can be a teacher at home has caught the attention of educators. . . . Parents don't have to be graduates of Harvard or Yale to help their kids learn and achieve.

Thus the home as a place where a child can learn becomes an *idea* of the professional "educator," who retains control of the idea. The home, as the article makes clear, is not to be a place where children may learn on their own, but a place where they are taught by parents according to the instructions of professional "educators." In fact, the Home and School Institute, Inc., of Washington, D.C. (known, of course, as "the HSI") has been "founded to show . . . how to involve families in their kids' educations."

In such ways as this, the nuclei of home and community have been invaded by the organizations, just as have the nuclei of cells and atoms. And we must be careful to see that the old cultural centers of home and community were made vulnerable to this invasion by their failure as economies. If there is no household or community economy, then family members and neighbors are no longer useful to one another. When people are no longer useful to one another, then the centripetal force of family and community fails, and people fall into dependence on exterior economies and organizations. The hegemony of professionals and professionalism erects itself on local failure, and from then on the locality exists merely as a market for consumer goods and as a source of "raw material," human and natural. The local schools no longer serve the local community; they serve the government's economy and the economy's government. Unlike the local community, the government and the economy cannot be served with affection, but only with professional zeal or professional boredom. Professionalism means more interest in salaries and less interest in what used to be known as disciplines. And so we arrive at the idea, endlessly reiterated in the news media, that education can be improved by bigger salaries for teachers—which may be true, but education cannot be improved, as the proponents too often imply, by bigger salaries alone. There must also be love of learning and of the cultural tradition and of excellence—and this love cannot exist, because it makes no sense, apart from the love of a place and a community. Without this love, education is only the importation into a local community of centrally prescribed "career preparation" designed to facilitate the export of young careerists.

Our children are educated, then, to leave home, not to stay home, and the costs of this education have been far too little acknowledged. One of the costs is psychological, and the other is at once cultural and ecological.

The natural or normal course of human growing up must begin with some sort of rebellion against one's parents, for it is clearly impossible to grow up if one remains a child. But the child, in the process of rebellion and of achieving the emotional and economic independence that rebellion ought to lead to, finally comes to understand the parents as fellow humans and fellow sufferers, and in some manner returns to them as their friend, forgiven and forgiving the inevitable wrongs of family life. That is the old norm.

The new norm, according to which the child leaves home as a student and never lives at home again, interrupts the old course of coming of age at the point of rebellion, so that the child is apt to remain stalled in adolescence, never achieving any kind of reconciliation or friendship with the parents. Of course, such a return and reconciliation cannot be achieved without the recognition of mutual practical need. In the present economy, however, where individual dependences are so much exterior to both household and community, family members often have no practical need or use for one another. Hence the frequent futility of attempts at a purely psychological or emotional reconciliation.

And this interposition of rebellion and then of geographical and occupational distance between parents and children may account for the peculiar emotional intensity that our society attaches to innovation. We appear to hate whatever went before, very much as an adolescent hates parental rule, and to look on its obsolescence as a kind of vengeance. Thus we may explain industry's obsessive emphasis on "this year's model," or the preoccupation of the professional "educators" with theoretical and methodological innovation. Similarly, in modern literature we have had for many years an emphasis on "originality" and "the anxiety of influence" (an adolescent critical theory), as opposed, say, to Spenser's filial admiration for Chaucer, or Dante's for Virgil.

But if the new norm interrupts the development of the relation between children and parents, that same interruption, ramifying through a community, destroys the continuity and so the integrity of local life. As the children depart, generation after generation, the place loses its memory of itself, which is its history and its culture. And the local history, if it survives at all, loses its place. It does no good for historians, folklorists, and anthropologists to collect the songs and the stories and the lore that make up local culture and store them in books and archives. They cannot collect and store—because they cannot know—the pattern of reminding that can survive only in the living human community in its place. It is this pattern that is the life of local culture and that brings it usefully or pleasurably to mind. Apart from its local landmarks and occasions, the local culture may be the subject of curiosity or of study, but it is also dead.

The loss of local culture is, in part, a practical loss and an economic one. For one thing, such a culture contains, and conveys to succeeding generations, the history of the use of the place and the knowledge of how the place may be lived in and used. For another, the pattern of reminding implies affection for the place and respect for it, and so, finally, the local culture will carry the knowledge of how the place may be well and lovingly used, and also the implicit command to use it *only* well and lovingly. The only true and effective "operator's manual for spaceship earth" is not a book that any human will ever write; it is hundreds of thousands of local cultures.

Lacking an authentic local culture, a place is open to exploitation, and ultimately destruction, from the center. Recently, for example, I heard the dean of a prominent college of agriculture interviewed on the radio. What have we learned, he was asked, from last summer's drouth? And he replied that "we" need to breed more drouth resistance into plants, and that "we" need a government "safety net" for farmers. He might have said that farmers need to re-examine their farms and their circumstances in light of the drouth, and to think again on such subjects as diversification, scale, and the mutual helpfulness of neighbors. But he did not say that. To him, the drouth was merely an opportunity for agribusiness corporations and the government, by which the farmers and rural communities could only become more dependent on the economy that is destroying them. This is as good an example as any of the centralized thinking of a centralized economy—to which the only effective answer that I know is a strong local community with a strong local economy and a strong local culture.

For a long time now, the prevailing assumption has been that if the nation is all right, then all the localities within it will be all right also. I see little reason to believe that this is true. At present, in fact, both the nation and the national economy are living at the expense of localities and local communities—as all small-town and country people have reason to know. In rural America, which is in many ways a colony of what the government and the corporations think of as the nation, most of us have experienced the losses that I have been talking about: the departure of young people, of soil and other so-called natural resources, and of local memory. We feel ourselves crowded more and more into a dimensionless present, in which the past is forgotten and the future, even in our most optimistic "projections," is forbidding and fearful. Who can desire a future that is determined entirely by the purposes of the most wealthy and the most powerful, and by the capacities of machines?

Two questions, then, remain: Is a change for the better possible? And who has the power to make such a change? I still believe that a change for the better is possible, but I confess that my belief is partly hope and partly faith. No one who hopes for improvement should fail to see and respect the signs that we may be approaching some sort of historical waterfall, past which we will not, by changing our minds, be able to change anything else. We know that at any time an ecological or a technological or a political event that we will have allowed may remove from us the power to make change and leave us with the mere necessity to submit to it. Beyond that, the two questions are one: the possibility of change depends on the existence of people who have the power to change.

Does this power reside at present in the national government? That seems to me extremely doubtful. To anyone who has read the papers during the recent presidential campaign, it must be clear that at the highest level of government there is, properly speaking, no political discussion. Are the corporations likely to help us? We know, from long experience, that the corporations will assume no responsibility that is not forcibly imposed upon them by government. The record of the corporations is written too plainly in verifiable damage to permit us to expect much from them. May we look for help to the universities? Well, the universities are more and more the servants of government and the corporations.

Most urban people evidently assume that all is well. They live too far from the exploited and endangered sources of their economy to need to assume otherwise. Some urban people are becoming disturbed about the contamination of air, water, and food, and that is promising, but there are not enough of them yet to make much difference. There is enough trouble in the "inner cities" to make them likely places of change, and evidently change is in them, but it is desperate and destructive change. As if to perfect their exploitation by other people, the people of the "inner cities" are destroying both themselves and their places.

My feeling is that if improvement is going to begin anywhere, it will have to begin out in the country and in the country towns. This is not because of any intrinsic virtue that can be ascribed to rural people, but because of their circumstances. Rural people are living, and have lived for a long time, at the site of the trouble. They see all around them, every day, the marks and scars of an exploitive national economy. They have much reason, by now, to know how little real help is to be expected from somewhere else. They still have, moreover, the remnants of local memory and local community. And in rural communities there are still farms and small businesses that can be changed according to the will and the desire of individual people.

In this difficult time of failed public expectations, when thoughtful people wonder where to look for hope, I keep returning in my own mind to the thought of the renewal of the rural communities. I know that one revived rural community would be more convincing and more encouraging than all the government and university programs of the last fifty years, and I think that it could be the beginning of the renewal of our country, for the renewal of rural communities ultimately implies the renewal of urban ones. But to be authentic, a true encouragement and a true beginning, this would have to be a revival accomplished mainly by the community itself. It would have to be done not from the outside by the instruction of visiting experts, but from the inside by the ancient rule of neighborliness, by the love of precious things, and by the wish to be at home.

Why I Am Not Going to Buy a Computer

LIKE ALMOST EVERYBODY ELSE, I am hooked to the energy corporations, which I do not admire. I hope to become less hooked to them. In my work, I try to be as little hooked to them as possible. As a farmer, I do almost all of my work with horses. As a writer, I work with a pencil or a pen and a piece of paper.

My wife types my work on a Royal standard typewriter bought new in 1956 and as good now as it was then. As she types, she sees things that are wrong and marks them with small checks in the margins. She is my best critic because she is the one most familiar with my habitual errors and weaknesses. She also understands, sometimes better than I do, what *ought* to be said. We have, I think, a literary cottage industry that works well and pleasantly. I do not see anything wrong with it.

A number of people, by now, have told me that I could greatly improve things by buying a computer. My answer is that I am not going to do it. I have several reasons, and they are good ones.

The first is the one I mentioned at the beginning. I would hate to think that my work as a writer could not be done without a direct dependence on strip-mined coal. How could I write conscientiously against the rape of nature if I were, in the act of writing, implicated in the rape? For the same reason, it matters to me that my writing is done in the daytime, without electric light.

I do not admire the computer manufacturers a great deal more than I admire the energy industries. I have seen their advertisements, attempting to seduce struggling or failing farmers into the belief that they can solve their problems by buying yet another piece of expensive equipment. I am familiar with their propaganda campaigns that have put computers into public schools in need of books. That computers are expected to become as common as TV sets in "the future" does not impress me or matter to me. I do not own a TV set. I do not see that computers are bringing us one step nearer to anything that does matter to me: peace, economic justice, ecological health, political honesty, family and community stability, good work.

What would a computer cost me? More money, for one thing, than I can afford, and more than I wish to pay to people whom I do not admire. But the cost would not be just monetary. It is well understood that technological innovation always requires the discarding of the "old model"—the "old model" in this case being not just our old Royal standard, but my wife, my critic, my closest reader, my fellow worker. Thus (and I think this is typical of present-day technological innovation), what would be superseded would be not only something, but somebody. In order to be technologically up-to-date as a writer, I would have to sacrifice an association that I am dependent upon and that I treasure.

My final and perhaps my best reason for not owning a computer is that I do not wish to fool myself. I disbelieve, and therefore strongly resent, the assertion that I or anybody else could write better or more easily with a computer than with a pencil. I do not see why I should not be as scientific about this as the next fellow: when somebody has used a computer to write work that is demonstrably better than Dante's, and when this better is demonstrably attributable to the use of a computer, then I will speak of computers with a more respectful tone of voice, though I still will not buy one.

To make myself as plain as I can, I should give my standards for technological innovation in my own work. They are as follows:

1. The new tool should be cheaper than the one it replaces.

2. It should be at least as small in scale as the one it replaces.

3. It should do work that is clearly and demonstrably better than the one it replaces.

4. It should use less energy than the one it replaces.

5. If possible, it should use some form of solar energy, such as that of the body.

6. It should be repairable by a person of ordinary intelligence, provided that he or she has the necessary tools.

7. It should be purchasable and repairable as near to home as possible.

8. It should come from a small, privately owned shop or store that will take it back for maintenance and repair.

9. It should not replace or disrupt anything good that already exists, and this includes family and community relationships.

1987

After the foregoing essay, first published in the *New England Review and Bread Loaf Quarterly*, was reprinted in *Harper's*, the *Harper's* editors published the following letters in response and permitted me a reply. W. B.

LETTERS

Wendell Berry provides writers enslaved by the computer with a handy alternative: Wife—a low-tech energy-saving device. Drop a pile of handwritten notes on Wife and you get back a finished manuscript, edited while it was typed. What computer can do that? Wife meets all of Berry's uncompromising standards for technological innovation: she's cheap, repairable near home, and good for the family structure. Best of all, Wife is politically correct because she breaks a writer's "direct dependence on strip-mined coal."

History teaches us that Wife can also be used to beat rugs and wash clothes by hand, thus eliminating the need for the vacuum cleaner and washing machine, two more nasty machines that threaten the act of writing.

Gordon Inkeles
Miranda, Calif.

I have no quarrel with Berry because he prefers to write with pencil and paper; that is his choice. But he implies that I and others are somehow impure because we choose to write on a computer. I do not admire the energy corporations, either. Their shortcoming is not that they produce electricity but how they go about it. They are poorly managed because they are blind to long-term consequences. To solve this problem, wouldn't it make more sense to correct the precise error they are making rather than simply ignore their product? I would be happy to join Berry in a protest against strip mining, but

I intend to keep plugging this computer into the wall with a clear conscience.

James Rhoads
Battle Creek, Mich.

I enjoyed reading Berry's declaration of intent never to buy a personal computer in the same way that I enjoy reading about the belief systems of unfamiliar tribal cultures. I tried to imagine a tool that would meet Berry's criteria for superiority to his old manual typewriter. The clear winner is the quill pen. It is cheaper, smaller, more energy-efficient, human-powered, easily repaired, and non-disruptive of existing relationships.

Berry also requires that this tool must be "clearly and demonstrably better" than the one it replaces. But surely we all recognize by now that "better" is in the mind of the beholder. To the quill pen aficionado, the benefits obtained from elegant calligraphy might well outweigh all others.

I have no particular desire to see Berry use a word processor; if he doesn't like computers, that's fine with me. However, I do object to his portrayal of this reluctance as a moral virtue. Many of us have found that computers can be an invaluable tool in the fight to protect our environment. In addition to helping me write, my personal computer gives me access to up-to-the-minute reports on the workings of the EPA and the nuclear industry. I participate in electronic bulletin boards on which environmental activists discuss strategy and warn each other about urgent legislative issues. Perhaps Berry feels that the Sierra Club should eschew modern printing technology, which is highly wasteful of energy, in favor of having its members hand-copy the club's magazines and other mailings each month?

Nathaniel S. Borenstein
Pittsburgh, Pa.

The value of a computer to a writer is that it is a tool not for generating ideas but for typing and editing words. It is cheaper than a secretary (or a wife!) and arguably more fuel-efficient. And it enables spouses who are not inclined to provide free labor more time to concentrate on *their* own work.

We should support alternatives both to coal-generated electricity and to IBM-style technocracy. But I am reluctant to entertain alternatives that presuppose the traditional subservience of one class to another. Let the PCs come and the wives and servants go seek more meaningful work.

Toby Koosman
Knoxville, Tenn.

Berry asks how he could write conscientiously against the rape of nature if in the act of writing on a computer he was implicated in the rape. I find it ironic that a writer who sees the underlying connectedness of things would allow his diatribe against computers to be published in a magazine that carries ads for the National Rural Electric Cooperative Association, Marlboro, Phillips Petroleum, McDonnell Douglas, and yes, even Smith-Corona. If Berry rests comfortably at night, he must be using sleeping pills.

Bradley C. Johnson
Grand Forks, N.D.

WENDELL BERRY REPLIES:

The foregoing letters surprised me with the intensity of the feelings they expressed. According to the writers' testimony, there is nothing wrong with their computers; they are utterly satisfied with them and all that they stand for. My correspondents are certain that I am wrong and that I am, moreover, on the losing side, a side already relegated to the dustbin of history. And yet they grow huffy and condescending over my tiny dissent. What are they so anxious about?

I can only conclude that I have scratched the skin of a technological fundamentalism that, like other fundamentalisms, wishes to monopolize a whole society and, therefore, cannot tolerate the smallest difference of opinion. At the slightest hint of a threat to their complacency, they repeat, like a chorus of toads, the notes sounded by their leaders in industry. The past was gloomy, drudgery-ridden, servile, meaningless, and slow. The present, thanks only to purchasable products, is meaningful, bright, lively, centralized, and fast. The future, thanks only to more purchasable products, is going to be even better. Thus consumers become salesmen, and the world is made safer for corporations.

I am also surprised by the meanness with which two of these writers refer to my wife. In order to imply that I am a tyrant, they suggest by both direct statement and innuendo that she is subservient, characterless, and stupid—a mere "device" easily forced to provide meaningless "free labor." I understand that it is impossible to make an adequate public defense of one's private life, and so I will only point out that there are a number of kinder possibilities that my critics have disdained to imagine: that my wife may do this work because she wants to and likes to; that she may find some use and some meaning in it; that she may not work for nothing. These gentlemen obviously think themselves feminists of the most correct and principled sort, and yet they do not hesitate to stereotype and insult, on the basis of one fact, a woman they do not know. They are audacious and irresponsible gossips.

In his letter, Bradley C. Johnson rushes past the possibility of sense in what I said in my essay by implying that I am or ought to be a fanatic. That I am a person of this century and am implicated in many practices that I regret is fully acknowledged at the beginning of my essay. I did not say that I proposed to end forthwith all my involvement in harmful technology, for I do not know how to do that. I said merely that I want to limit such involvement, and to a certain extent I do know how to do that. If some technology does damage to the world—as two of the above letters seem to agree that it does—then why is it not reasonable, and indeed moral, to try to limit one's use of that technology? *Of course*, I think that I am right to do this.

I would not think so, obviously, if I agreed with Nathaniel S. Borenstein that "'better' is in the mind of the beholder." But if he truly believes this, I do not see why he bothers with his personal computer's "up-to-the-minute reports on the workings of the EPA and the nuclear industry" or why he wishes to be warned about "urgent legislative issues." According to his system, the "better" in a bureaucratic, industrial, or legislative mind is as good as the "better" in his. His mind apparently is being subverted by an objective standard of some sort, and he had better look out.

Borenstein does not say what he does after his computer has drummed him awake. I assume from his letter that he must send donations to conservation organizations and letters to officials. Like James Rhoads, at any rate, he has a clear conscience. But this is what is wrong with the conservation movement. It has a clear conscience. The guilty are always other people, and the wrong is always somewhere else. That is why Borenstein finds his "electronic bulletin board" so handy. To the conservation movement, it is only production that causes environmental degradation; the consumption that supports the production is rarely acknowledged to be at fault. The ideal of the run-of-the-mill conservationist is to impose restraints upon production without limiting consumption or burdening the consciences of consumers.

But virtually all of our consumption now is extravagant, and virtually all of it consumes the world. It is not beside the point that most electrical power comes from strip-mined coal. The history of the exploitation of the Appalachian coal fields is long, and it is available to readers. I do not see how anyone can read it and plug in any appliance with a clear conscience. If Rhoads can do so, that does not mean that his conscience is clear; it means that his conscience is not working.

To the extent that we consume, in our present circumstances, we are guilty. To the extent that we guilty consumers are conservationists, we are absurd. But what can we do? Must we go on writing letters to politicians and donating to conservation organizations until the majority of our fellow citizens agree with us? Or can we do something directly to solve our share of the problem?

I am a conservationist. I believe wholeheartedly in putting pressure on the politicians and in maintaining the conservation organizations. But I wrote my little essay partly in distrust of centralization. I don't think that the government and the conservation organizations alone will ever make us a conserving society. Why do I need a centralized computer system to alert me to environmental crises? That I live every hour of every day in an environmental crisis I know from all my senses. Why then is not my first duty to reduce, so far as I can, my own consumption?

Finally, it seems to me that none of my correspondents recognizes the innovativeness of my essay. If the use of a computer is a new idea, then a newer idea is not to use one.

Feminism, the Body, and the Machine

SOME TIME AGO *Harper's* reprinted a short essay of mine in which I gave some of my reasons for refusing to buy a computer. Until that time, the vast numbers of people who disagree with my writings had mostly ignored them. An unusual number of people, however, neglected to ignore my insensitivity to the wonders of computer enhancement. Some of us, it seems, would be better off if we would just realize that this is already the best of all possible worlds, and is going to get even better if we will just buy the right equipment.

Harper's published only five of the letters the editors received in response to my essay, and they published only negative letters. But of the twenty letters received by the *Harper's* editors, who forwarded copies to me, three were favorable. This I look upon as extremely gratifying. If these letters may be taken as a fair sample, then one in seven of *Harper's* readers agreed with me. If I had guessed beforehand, I would have guessed that my supporters would have been fewer than one in a thousand. And so I suppose, after further reflection, that my surprise at the intensity of the attacks on me is mistaken. There are more of us than I thought. Maybe there is even a "significant number" of us.

Only one of the negative letters seemed to me to have much intelligence in it. That one was from R. N. Neff of Arlington, Virginia, who scored a direct hit: "Not to be obtuse, but being willing to bare my illiterate soul for all to see, is there indeed a 'work demonstrably better than Dante's' . . . which was written on a Royal standard typewriter?" I like this retort so well that I am tempted to count it a favorable response, raising the total to four. The rest of the negative replies, like the five published ones, were more feeling than intelligent. Some of them, indeed, might be fairly described as exclamatory.

One of the letter writers described me as "a fool" and "doubly a fool," but fortunately misspelled my name, leaving me a speck of hope that I am not the "Wendell Barry" he was talking about. Two others accused me of self-righteousness, by which they seem to have meant that they think they are righter than I think I am. And another accused me of being more concerned about my own moral purity than with "any ecological effect," thereby making the sort of razor-sharp philosophical distinction that could cause a person to be elected president.

But most of my attackers deal in feelings either feminist or technological, or both. The feelings expressed seem to be representative of what the state of public feeling currently permits to be felt, and of what public rhetoric currently permits to be said. The feelings, that is, are similar enough, from one letter to another, to be thought representative, and as representative letters they have an interest greater than the quarrel that occasioned them.

Without exception, the feminist letters accuse me of exploiting my wife, and they do not scruple to allow the most insulting implications of their indictment to fall upon my wife. They fail entirely to see that my essay does not give any support to their accusation—or if they see it, they do not care. My essay, in fact, does not characterize my wife beyond saying that she types my manuscripts and tells me what she thinks about them. It does not say what her motives are, how much work she does, or whether or how she is paid. Aside from saying that she is my wife and that I value the help she gives me with my work, it says nothing about our marriage. It says nothing about our economy.

There is no way, then, to escape the conclusion that my wife and I are subjected in these letters to a condemnation by category. My offense is that I am a man who receives some help from his wife; my wife's offense is that she is a woman who does some work for her husband—which work, according to her critics and mine, makes her a drudge, exploited by a conventional subservience. And my detractors have, as I say, no evidence to support any of this. Their accusation rests on a syllogism of the flimsiest sort: my wife helps me in my work, some wives who have helped their husbands in their work have been exploited, therefore my wife is exploited.

This, of course, outrages justice to about the same extent that it insults intelligence. Any respectable system of justice exists in part as a protection against such accusations. In a just society nobody is expected to plead guilty to a general indictment, because in a just society nobody can be convicted on a general indictment. What is required for a just conviction is a particular accusation that can be *proved*. My accusers have made no such accusation against me.

That feminists or any other advocates of human liberty and dignity should resort to insult and injustice is regrettable. It is equally regrettable that all of the feminist attacks on my essay implicitly deny the validity of two decent and probably necessary possibilities: marriage as a state of mutual help, and the household as an economy.

Marriage, in what is evidently its most popular version, is now on the one hand an intimate "relationship" involving (ideally) two successful careerists in the same bed, and on the other hand a sort of private political system in which rights and interests must be constantly asserted and defended. Marriage, in other words, has now taken the form of divorce: a prolonged and impassioned negotiation as to how things shall be divided. During their understandably temporary association, the "married" couple will typically consume a large quantity of merchandise and a large portion of each other.

The modern household is the place where the consumptive couple do their consuming. Nothing productive is done there. Such work as is done there is done at the expense of the resident couple or family, and to the profit of suppliers of energy and household technology. For entertainment, the inmates consume television or purchase other consumable diversion elsewhere.

There are, however, still some married couples who understand themselves as belonging to their marriage, to each other, and to their children. What they have they have in common, and so, to them, helping each other does not seem merely to damage their ability to compete against each other. To them, "mine" is not so powerful or necessary a pronoun as "ours."

This sort of marriage usually has at its heart a household that is to some extent productive. The couple, that is, makes around itself a household economy that involves the work of both wife and husband, that gives them a measure of economic independence and self-protection, a measure of self-employment, a measure of freedom, as well as a common ground and a common satisfaction. Such a household economy may employ the disciplines and skills of housewifery, of carpentry and other trades of building and maintenance, of gardening and other branches of subsistence agriculture, and even of woodlot management and woodcutting. It may also involve a "cottage industry" of some kind, such as a small literary enterprise.

It is obvious how much skill and industry either partner may put into such a household and what a good economic result such work may have, and yet it is a kind of work now frequently held in contempt. Men in general were the first to hold it in contempt as they departed from it for the sake of the professional salary or the hourly wage, and now it is held in contempt by such feminists as those who attacked my essay. Thus farm wives who help to run the kind of household economy that I have described are apt to be asked by feminists, and with great condescension, "But what do you *do*?" By this they invariably mean that there is something better to do than to make one's marriage and household, and by better they invariably mean "employment outside the home."

I know that I am in dangerous territory, and so I had better be plain: what I have to say about marriage and household I mean to apply to men as much as to women. I do not believe that there is anything better to do than to make one's marriage and household, whether one is a man or a woman. I do not believe that "employment outside the home" is as valuable or important or satisfying as employment at home, for either men or women. It is clear to me from my experience as a teacher, for example, that children need an ordinary daily association with *both* parents. They need to see their parents at work; they need, at first, to play at the work they see their parents doing, and then they need to work with their parents. It does not matter so much that this working together should be what is called "quality time," but it matters a great deal that the work done should have the dignity of economic value.

I should say too that I understand how fortunate I have been in being able to do an appreciable part of my work at home. I know that in many marriages both husband and wife are now finding it necessary to work away from home. This issue, of course, is troubled by the question of what is meant by "necessary," but it is true that a family living that not so long ago was ordinarily supplied by one job now routinely requires two or more. My interest is not to quarrel with individuals, men or women, who work away from home, but rather to ask why we should consider this general working away from home to be a desirable state of things, either for people or for marriage, for our society or for our country.

If I had written in my essay that my wife worked as a typist and editor for a publisher, doing the same work that she does for me, no feminists, I daresay, would have written to *Harper's* to attack me for exploiting her—even though, for all they knew, I might have forced her to do such work in order to keep me in gambling money. It would have been assumed as a matter of course that if she had a job away from home she was a "liberated woman," possessed of a dignity that no home could confer upon her.

As I have said before, I understand that one cannot construct an adequate public defense of a private life. Anything that I might say here about my marriage would be immediately (and rightly) suspect on the ground that it would be only *my* testimony. But for the sake of argument, let us suppose that whatever work my wife does, as a member of our marriage and household, she does both as a full economic partner and as her own boss, and let us suppose that the economy we have is adequate to our needs. Why, granting that supposition, should anyone assume that my wife would increase her freedom or dignity or satisfaction by becoming the employee of a boss, who would be in turn also a corporate underling and in no sense a partner?

Why would any woman who would refuse, properly, to take the marital vow of obedience (on the ground, presumably, that subservience to a mere human being is beneath human dignity) then regard as "liberating" a job that puts her under the authority of a boss (man or woman) whose authority specifically requires and expects obedience? It is easy enough to see why women came to object to the role of Blondie, a mostly decorative custodian of a degraded, consumptive modern household, preoccupied with clothes, shopping, gossip, and outwitting her husband. But are we to assume that one may fittingly cease to be Blondie by becoming Dagwood? Is the life of a corporate underling—even acknowledging that corporate underlings are well paid—an acceptable end to our quest for human dignity and worth? It is clear enough by now that one does not cease to be an underling by reaching "the top." Corporate life is composed only of lower underlings and higher underlings. Bosses are everywhere, and all the bosses are underlings. This is invariably revealed when the time comes for accepting responsibility for something unpleasant, such as the Exxon fiasco in Prince William Sound, for which certain lower underlings are blamed but no higher underling is responsible. The underlings at the top, like telephone operators, have authority and power, but no responsibility.

And the oppressiveness of some of this office work defies belief. Edward Mendelson (in the *New Republic*, February 22, 1988) speaks of "the office worker whose computer keystrokes are monitored by the central computer in the personnel office, and who will be fired if the keystrokes-per-minute figure doesn't match the corporate quota." (Mr. Mendelson does not say what form of drudgery this worker is being saved from.) And what are we to say of the diversely skilled country housewife who now bores the same six holes day after day on an assembly line? What higher form of womanhood or humanity is she evolving toward?

How, I am asking, can women improve themselves by submitting to the same specialization, degradation, trivialization, and tyrannization of work that men have submitted to? And that question is made legitimate by another: How have men improved themselves by submitting to it? The answer is that men have not, and women cannot, improve themselves by submitting to it.

Women have complained, justly, about the behavior of "macho" men. But despite their he-man pretensions and their captivation by masculine heroes of sports, war, and the Old West, most men are now entirely accustomed to obeying and currying the favor of their bosses. Because of this, of course, they hate their jobs—they mutter, "Thank God it's Friday" and "Pretty good for Monday"—but they do as they are told. They are more compliant than most housewives have been. Their characters combine feudal submissiveness with modern helplessness. They have accepted almost without protest, and often with relief, their dispossession of any usable property and, with that, their loss of economic independence and their consequent subordination to bosses. They have submitted to the destruction of the household economy and thus of the household, to the loss of home employment and self-employment, to the disintegration of their families and communities, to the desecration and pillage of their country, and they have continued abjectly to believe, obey, and vote for the people who have most eagerly abetted this ruin and who have most profited from it. These men, moreover, are helpless to do anything for themselves or anyone else without money, and so for money they do whatever they are told. They know that their ability to be useful is precisely defined by their willingness to be somebody else's tool. Is it any wonder that they talk tough and worship athletes and cowboys? Is it any wonder that some of them are violent?

It is clear that women cannot justly be excluded from the daily fracas by which the industrial economy divides the spoils of society and nature, but their inclusion is a poor justice and no reason for applause. The enterprise is as devastating with women in it as it was before. There is no sign that women are exerting a "civilizing influence" upon it. To have an equal part in our juggernaut of national vandalism is to be a vandal. To call this vandalism "liberation" is to prolong, and even ratify, a dangerous confusion that was once principally masculine.

A broader, deeper criticism is necessary. The problem is not just the exploitation of women by men. A greater problem is that women and men alike are consenting to an economy that exploits women and men and everything else.

Another decent possibility my critics implicitly deny is that of work as a gift. Not one of them supposed that my wife may be a consulting engineer who helps me in her spare time out of the goodness of her heart; instead they suppose that she is "a household drudge." But what appears to infuriate them the most is their supposition that she works for nothing. They assume—and this is the orthodox assumption of the industrial economy—that the only help worth giving is not given at all, but sold. Love, friendship, neighborliness, compassion, duty—what are they? We are realists. We will be most happy to receive your check.

*

The various reductions I have been describing are fairly directly the results of the ongoing revolution of applied science known as "technological progress." This revolution has provided the means by which both the productive and the consumptive capacities of people could be detached from household and community and made to serve other people's purely economic ends. It has provided as well a glamor of newness, ease, and affluence that made it seductive even to those who suffered most from it. In its more recent history especially, this revolution has been successful in putting unheard-of quantities of consumer goods and services within the reach of ordinary people. But the technical means of this popular "affluence" has at the same time made possible the gathering of the real property and the real power of the country into fewer and fewer hands.

Some people would like to think that this long sequence of industrial innovations has changed human life and even human nature in fundamental ways. Perhaps it has—but, arguably, almost always for the worse. I know that "technological progress" can be defended, but I observe that the defenses are invariably quantitative—catalogs of statistics on the ownership of automobiles and television sets, for example, or on the increase of life expectancy—and I see that these statistics are always kept carefully apart from the related statistics of soil loss, pollution, social disintegration, and so forth. That is to say, there is never an effort to determine the *net* result of this progress. The voice of its defenders is not that of the responsible bookkeeper, but that of the propagandist or salesman, who says that the net gain is more than 100 percent—that the thing we have bought has perfectly replaced everything it has cost, and added a great deal more: "You just can't lose!" We thus have got rich by spending, just as the advertisers have told us we would, and the best of all possible worlds is getting better every day.

The statistics of life expectancy are favorites of the industrial apologists, because they are perhaps the hardest to argue with. Nevertheless, this emphasis on longevity is an excellent example of the way the isolated aims of the industrial mind reduce and distort human life, and also the way statistics corrupt the truth. A long life has indeed always been thought desirable; everything that is alive apparently wishes to continue to live.

But until our own time, that sentence would have been qualified: long life is desirable and everything wishes to live *up to a point.* Past a certain point, and in certain conditions, death becomes preferable to life. Moreover, it was generally agreed that a good life was preferable to one that was merely long, and that the goodness of a life could not be determined by its length. The statisticians of longevity ignore good in both its senses; they do not ask if the prolonged life is virtuous, or if it is satisfactory. If the life is that of a vicious criminal, or if it is inched out in a veritable hell of captivity within the medical industry, no matter—both become statistics to "prove" the good luck of living in our time.

But in general, apart from its own highly specialized standards of quantity and efficiency, "technological progress" has produced a social and ecological decline. Industrial war, except by the most fanatically narrow standards, is worse than war used to be. Industrial agriculture, except by the standards of quantity and mechanical efficiency, diminishes everything it affects. Industrial workmanship is certainly worse than traditional workmanship, and is getting shoddier every day. After forty-odd years, the evidence is everywhere that television, far from proving a great tool of education, is a tool of stupefaction and disintegration. Industrial education has abandoned the old duty of passing on the cultural and intellectual inheritance in favor of baby-sitting and career preparation.

After several generations of "technological progress," in fact, we have become a people who *cannot* think about anything important. How far down in the natural order do we have to go to find creatures who raise their young as indifferently as industrial humans now do? Even the English sparrows do not let loose into the streets young sparrows who have no notion of their identity or their adult responsibilities. When else in history would you find "educated" people who know more about sports than about the history of their country, or uneducated people who do not know the stories of their families and communities?

To ask a still more obvious question, what is the purpose of this technological progress? What higher aim do we think it is serving? Surely the aim cannot be the integrity or happiness of our families, which we have made subordinate to the education system, the television industry, and the consumer economy. Surely it cannot be the integrity or health of our communities, which we esteem even less than we esteem our families. Surely it cannot be love of our country, for we are far more concerned about the desecration of the flag than we are about the desecration of our land. Surely it cannot be the love of God, which counts for at least as little in the daily order of business as the love of family, community, and country.

The higher aims of "technological progress" are money and ease. And this exalted greed for money and ease is disguised and justified by an obscure, cultish faith in "the future." We do as we do, we say, "for the sake of the future" or "to make a better future for our children." How we can hope to make a good future by doing badly in the present, we do not say. We cannot think about the future, of course, for the future does not exist: the existence of the future is an article of faith. We can be assured only that, if there is to be a future, the good of it is already implicit in the good things of the present. We do not need to plan or devise a "world of the future"; if we take care of the world of the present, the future will have received full justice from us. A good future is implicit in the soils, forests, grasslands, marshes, deserts, mountains, rivers, lakes, and oceans that we have now, and in the good things of human culture that we have now; the only valid "futurology" available to us is to take care of those things. We have no need to contrive and dabble at "the future of the human race"; we have the same pressing need that we have always had—to love, care for, and teach our children.

And so the question of the desirability of adopting any technological innovation is a question with two possible answers—not one, as has been commonly assumed. If one's motives are money, ease, and haste to arrive in a technologically determined future, then the answer is foregone, and there is, in fact, no question, and no thought. If one's motive is the love of family, community, country, and God, then one will have to think, and one may have to decide that the proposed innovation is undesirable.

The question of how to end or reduce dependence on some of the technological innovations already adopted is a baffling one. At least, it baffles me. I have not been able to see, for example, how people living in the country, where there is no public transportation, can give up their automobiles without becoming less useful to each other. And this is because, owing largely to the influence of the automobile, we live too far from each other, and from the things we need, to be able to get about by any other means. Of course, you *could* do without an automobile, but to do so you would have to disconnect yourself from many obligations. Nothing I have so far been able to think about this problem has satisfied me.

But if we have paid attention to the influence of the automobile on country communities, we know that the desirability of technological innovation is an issue that requires thinking about, and we should have acquired some ability to think about it. Thus if I am partly a writer, and I am offered an expensive machine to help me write, I ought to ask whether or not such a machine is desirable.

I should ask, in the first place, whether or not I wish to purchase a solution to a problem that I do not have. I acknowledge that, as a writer, I need a lot of help. And I have received an abundance of the best of help from my wife, from other members of my family, from friends, from teachers, from editors, and sometimes from readers. These people have helped me out of love or friendship, and perhaps in exchange for some help that I have given them. I suppose I should leave open the possibility that I need more help than I am getting, but I would certainly be ungrateful and greedy to think so.

But a computer, I am told, offers a kind of help that you can't get from other humans; a computer will help you to write faster, easier, and more. For a while, it seemed to me that every university professor I met told me this. Do I, then, want to write faster, easier, and more? No. My standards are not speed, ease, and quantity. I have already left behind too much evidence that, writing with a pencil, I have written too fast, too easily, and too much. I would like to be a *better* writer, and for that I need help from other humans, not a machine.

The professors who recommended speed, ease, and quantity to me were, of course, quoting the standards of their universities. The chief concern of the industrial system, which is to say the present university system, is to cheapen work by increasing volume. But implicit in the professors' recommendation was the idea that one needs to be up with the times. The pace-setting academic intellectuals have lately had a great hankering to be up with the times. They don't worry about keeping up with the Joneses: as intellectuals, they know that they are supposed to be Nonconformists and Independent Thinkers living at the Cutting Edge of Human Thought. And so they are all a-dither to keep up with the times—which means adopting the latest technological innovations as soon as the Joneses do.

Do I wish to keep up with the times? No.

My wish simply is to live my life as fully as I can. In both our work and our leisure, I think, we should be so employed. And in our time this means that we must save ourselves from the products that we are asked to buy in order, ultimately, to replace ourselves.

The danger most immediately to be feared in "technological progress" is the degradation and obsolescence of the body. Implicit in the technological revolution from the beginning has been a new version of an old dualism, one always destructive, and now more destructive than ever. For many centuries there have been people who looked upon the body, as upon the natural world, as an encumbrance of the soul, and so have hated the body, as they have hated the natural world, and longed to be free of it. They have seen the body as intolerably imperfect by spiritual standards. More recently, since the beginning of the technological revolution, more and more people have looked upon the body, along with the rest of the natural creation, as intolerably imperfect by mechanical standards. They see the body as an encumbrance of the mind—the mind, that is, as reduced to a set of mechanical ideas that can be implemented in machines—and so they hate it and long to be free of it. The body has limits that the machine does not have; therefore, remove the body from the machine so that the machine can continue as an unlimited idea.

It is odd that simply because of its "sexual freedom" our time should be considered extraordinarily physical. In fact, our "sexual revolution" is mostly an industrial phenomenon, in which the body is used as an idea of pleasure or a pleasure machine with the aim of "freeing" natural pleasure from natural consequence. Like any other industrial enterprise, industrial sexuality seeks to conquer nature by exploiting it and ignoring the consequences, by denying any connection between nature and spirit or body and soul, and by evading social responsibility. The spiritual, physical, and economic costs of this "freedom" are immense, and are characteristically belittled or ignored. The diseases of sexual irresponsibility are regarded as a technological problem and an affront to liberty. Industrial sex, characteristically, establishes its freeness and goodness by an industrial accounting, dutifully toting up numbers of "sexual partners," orgasms, and so on, with the inevitable industrial implication that the body is somehow a limit on the idea of sex, which will be a great deal more abundant as soon as it can be done by robots.

This hatred of the body and of the body's life in the natural world, always inherent in the technological revolution (and sometimes explicitly and vengefully so), is of concern to an artist because art, like sexual love, is of the body. Like sexual love, art is of the mind and spirit also, but it is made with the body and it appeals to the senses. To reduce or shortcut the intimacy of the body's involvement in the making of a work of art (that is, of any artifice, anything made by art) inevitably risks reducing the work of art and the art itself. In addition to the reasons I gave previously, which I still believe are good reasons, I am not going to use a computer because I don't want to diminish or distort my bodily involvement in my work. I don't want to deny myself the *pleasure* of bodily involvement in my work, for that pleasure seems to me to be the sign of an indispensable integrity.

At first glance, writing may seem not nearly so much an art of the body as, say, dancing or gardening or carpentry. And yet language is the most intimately physical of all the artistic means. We have it palpably in our mouths; it is our *langue*, our tongue. Writing it, we shape it with our hands. Reading aloud what we have written—as we must do, if we are writing carefully—our language passes in at the eyes, out at the mouth, in at the ears; the words are immersed and steeped in the senses of the body before they make sense in the mind. They *cannot* make sense in the mind until they have made sense in the body.

Does shaping one's words with one's own hand impart character and quality to them, as does speaking them with one's own tongue to the satisfaction of one's own ear? There is no way to prove that it does. On the other hand, there is no way to prove that it does not, and I believe that it does.

The act of writing language down is not so insistently tangible an act as the act of building a house or playing the violin. But to the extent that it is tangible, I love the tangibility of it. The computer apologists, it seems to me, have greatly underrated the value of the handwritten manuscript as an artifact. I don't mean that a writer should be a fine calligrapher and write for exhibition, but rather that handwriting has a valuable influence on the work so written. I am certainly no calligrapher, but my handwritten pages have a homemade, handmade look to them that both pleases me in itself and suggests the possibility of ready correction. It looks hospitable to improvement. As the longhand is transformed into typescript and then into galley proofs and the printed page, it seems increasingly to resist improvement. More and more spunk is required to mar the clean, final-looking lines of type. I have the notion—again not provable—that the longer I keep a piece of work in longhand, the better it will be.

To me, also, there is a significant difference between ready correction and easy correction. Much is made of the ease of correction in computer work, owing to the insubstantiality of the light-image on the screen; one presses a button and the old version disappears, to be replaced by the new. But because of the substantiality of paper and the consequent difficulty involved, one does not handwrite or typewrite a new page every time a correction is made. A handwritten or typewritten page therefore is usually to some degree a palimpsest; it contains parts and relics of its own history—erasures, passages crossed out, interlineations—suggesting that there is something to go back to as well as something to go forward to. The light-text on the computer screen, by contrast, is an artifact typical of what can only be called the industrial present, a present absolute. A computer destroys the sense of historical succession, just as do other forms of mechanization. The well-crafted table or cabinet embodies the memory of (because it embodies respect for) the tree it was made of and the forest in which the tree stood. The work of certain potters embodies the memory that the clay was dug from the earth. Certain farms contain hospitably the remnants and reminders of the forest or prairie that preceded them. It is possible even for towns and cities to remember farms and forests or prairies. All good human work remembers its history. The best writing, even when printed, is full of intimations that it is the present version of earlier versions of itself, and that its maker inherited the work and the ways of earlier makers. It thus keeps, even in print, a suggestion of the quality of the handwritten page; it is a palimpsest.

Something of this undoubtedly carries over into industrial products. The plastic Clorox jug has a shape and a loop for the forefinger that recalls the stoneware jug that went before it. But something vital is missing. It embodies no memory of its source or sources in the earth or of any human hand involved in its shaping. Or look at a large factory or a power plant or an airport, and see if you can imagine—even if you know—what was there before. In such things the materials of the world have entered a kind of orphanhood.

It would be uncharitable and foolish of me to suggest that nothing good will ever be written on a computer. Some of my best friends have computers. I have only said that a computer cannot help you to write *better*, and I stand by that. (In fact, I know a publisher who says that under the influence of computers—or of the immaculate copy that computers produce—many writers are now writing worse.) But I do say that in using computers writers are flirting with a radical separation of mind and body, the elimination of the work of the body from the work of the mind. The text on the computer screen, and the computer printout too, has a sterile, untouched, factory-made look, like that of a plastic whistle or a new car. The body does not do work like that. The body *characterizes* everything it touches. What it makes it traces over with the marks of its pulses and breathings, its excitements, hesitations, flaws, and mistakes. On its good work, it leaves the marks of skill, care, and love persisting through hesitations, flaws, and mistakes. And to those of us who love and honor the life of the body in this world, these marks are precious things, necessities of life.

But writing is of the body in yet another way. It is preeminently a walker's art. It can be done on foot and at large. The beauty of its traditional equipment is simplicity. And cheapness. Going off to the woods, I take a pencil and some paper (*any* paper—a small notebook, an old envelope, a piece of a feed sack), and I am as well equipped for my work as the president of IBM. I am also free, for the time being at least, of everything that IBM is hooked to. My thoughts will not be coming to me from the power structure or the power grid, but from another direction and way entirely. My mind is free to go with my feet.

I know that there are some people, perhaps many, to whom you cannot appeal on behalf of the body. To them, disembodiment is a goal, and they long for the realm of pure mind—or pure machine; the difference is negligible. Their departure from their bodies, obviously, is much to be desired, but the rest of us had better be warned: they are going to cause a lot of dangerous commotion on their way out.

Some of my critics were happy to say that my refusal to use a computer would not do any good. I have argued, and am convinced, that it will at least do *me* some good, and that it may involve me in the preservation of some cultural goods. But what they meant was real, practical, public good. They meant that the materials and energy I save by not buying a computer will not be "significant." They meant that no individual's restraint in the use of technology or energy will be "significant." That is true.

But each one of us, by "insignificant" individual abuse of the world, contributes to a general abuse that is devastating. And if I were one of thousands or millions of people who could afford a piece of equipment, even one for which they had a conceivable "need," and yet did not buy it, *that* would be "significant." Why, then, should I hesitate for even a moment to be one, even the first one, of that "significant" number? Thoreau gave the definitive reply to the folly of "significant numbers" a long time ago: Why should anybody wait to do what is right until everybody does it? It is not "significant" to love your own children or to eat your own dinner, either. But normal humans will not wait to love or eat until it is mandated by an act of Congress.

One of my correspondents asked where one is to draw the line. That question returns me to the bewilderment I mentioned earlier: I am unsure where the line ought to be drawn, or how to draw it. But it is an intelligent question, worth losing some sleep over.

I know how to draw the line only where it is easy to draw. It is easy—it is even a luxury—to deny oneself the use of a television set, and I zealously practice that form of self-denial. Every time I see television (at other people's houses), I am more inclined to congratulate myself on my deprivation. I have no doubt, as I have said, that I am better off without a computer. I joyfully deny myself a motorboat, a camping van, an off-road vehicle, and every other kind of recreational machinery. I have, and want, no "second home." I suffer very comfortably the lack of colas, TV dinners, and other counterfeit foods and beverages.

I am, however, still in bondage to the automobile industry and the energy companies, which have nothing to recommend them except our dependence on them. I still fly on airplanes, which have nothing to recommend them but speed; they are inconvenient, uncomfortable, undependable, ugly, stinky, and scary. I still cut my wood with a chainsaw, which has nothing to recommend it but speed, and has all the faults of an airplane, except it does not fly.

It is plain to me that the line ought to be drawn without fail wherever it can be drawn easily. And it ought to be easy (though many do not find it so) to refuse to buy what one does not need. If you are already solving your problem with the equipment you have—a pencil, say—why solve it with something more expensive and more damaging? If you don't have a problem, why pay for a solution? If you love the freedom and elegance of simple tools, why encumber yourself with something complicated?

And yet, if we are ever again to have a world fit and pleasant for little children, we are surely going to have to draw the line where it is *not* easily drawn. We are going to have to learn to give up things that we have learned (in only a few years, after all) to "need." I am not an optimist; I am afraid that I won't live long enough to escape my bondage to the machines. Nevertheless, on every day left to me I will search my mind and circumstances for the means of escape. And I am not without hope. I knew a man who, in the age of chainsaws, went right on cutting his wood with a handsaw and an axe. He was a healthier and a saner man than I am. I shall let his memory trouble my thoughts.

Word and Flesh

TOWARD THE END of *As You Like It*, Orlando says: "I can live no longer by thinking." He is ready to marry Rosalind. It is time for incarnation. Having thought too much, he is at one of the limits of human experience, or of human sanity. If his love does put on flesh, we know he must sooner or later arrive at the opposite limit, at which he will say, "I can live no longer without thinking." Thought—even consciousness—seems to live between these limits: the abstract and the particular, the word and the flesh.

All public movements of thought quickly produce a language that works as a code, useless to the extent that it is abstract. It is readily evident, for example, that you can't conduct a relationship with another person in terms of the rhetoric of the civil rights movement or the women's movement—as useful as those rhetorics may initially have been to personal relationships.

The same is true of the environment movement. The favorite adjective of this movement now seems to be "planetary." This word is used, properly enough, to refer to the interdependence of places, and to the recognition, which is desirable and growing, that no place on the earth can be completely healthy until all places are.

But the word "planetary" also refers to an abstract anxiety or an abstract passion that is desperate and useless exactly to the extent that it is abstract. How, after all, can anybody—any particular body—do anything to heal a planet? The suggestion that anybody could do so is preposterous. The heroes of abstraction keep galloping in on their white horses to save the planet—and they keep falling off in front of the grandstand.

What we need, obviously, is a more intelligent—which is to say, a more accurate—description of the problem. The description of a problem as planetary arouses a motivation for which, of necessity, there is no employment. The adjective "planetary" describes a problem in such a way that it cannot be solved. In fact, though we now have serious problems nearly everywhere on the planet, we have no problem that can accurately be described as planetary. And, short of the total annihilation of the human race, there is no planetary solution.

There are also no national, state, or county problems, and no national, state, or county solutions. That will-o'-the-wisp, the large-scale solution to the large-scale problem, which is so dear to governments, universities, and corporations, serves mostly to distract people from the small, private problems that they may, in fact, have the power to solve.

The problems, if we describe them accurately, are all private and small. Or they are so initially.

The problems are our lives. In the "developed" countries, at least, the large problems occur because all of us are living either partly wrong or almost entirely wrong. It was not just the greed of corporate shareholders and the hubris of corporate executives that put the fate of Prince William Sound into one ship; it was also our demand that energy be cheap and plentiful.

The economies of our communities and households are wrong. The answers to the human problems of ecology are to be found in economy. And the answers to the problems of economy are to be found in culture and in character. To fail to see this is to go on dividing the world falsely between guilty producers and innocent consumers.

The planetary versions—the heroic versions—of our problems have attracted great intelligence. But these problems, as they are caused and suffered in our lives, our households, and our communities, have attracted very little intelligence.

There are some notable exceptions. A few people have learned to do a few things better. But it is discouraging to reflect that, though we have been talking about most of our problems for decades, we are still mainly *talking* about them. The civil rights movement has not given us better communities. The women's movement has not given us better marriages or better households. The environment movement has not changed our parasitic relationship to nature.

We have failed to produce new examples of good home and community economies, and we have nearly completed the destruction of the examples we once had. Without examples, we are left with theory and the bureaucracy and meddling that come with theory. We change our principles, our thoughts, and our words, but these are changes made in the air. Our lives go on unchanged.

For the most part, the subcultures, the countercultures, the dissenters, and the opponents continue mindlessly—or perhaps just helplessly—to follow the pattern of the dominant society in its extravagance, its wastefulness, its dependencies, and its addictions. The old problem remains: How do you get intelligence *out* of an institution or an organization?

My small community in Kentucky has lived and dwindled for at least a century under the influence of four kinds of organizations: governments, corporations, schools, and churches—all of which are distant (either actually or in interest), centralized, and consequently abstract in their concerns.

Governments and corporations (except for employees) have no presence in our community at all, which is perhaps fortunate for us, but we nevertheless feel the indifference or the contempt of governments and corporations for communities such as ours.

We have had no school of our own for nearly thirty years. The school system takes our young people, prepares them for "the world of tomorrow"—which it does not expect to take place in any rural area—and gives back "expert" (that is, extremely generalized) ideas.

The church is present in the town. We have two churches. But both have been used by their denominations, for almost a century, to provide training and income for student ministers, who do not stay long enough even to become disillusioned.

For a long time, then, the minds that have most influenced our town have not been *of* the town and so have not tried even to perceive, much less to honor, the good possibilities that are there. They have not wondered on what terms a good and conserving life might be lived there. In this my community is not unique but is like almost every other neighborhood in our country and in the "developed" world.

The question that *must* be addressed, therefore, is not how to care for the planet, but how to care for each of the planet's millions of human and natural neighborhoods, each of its millions of small pieces and parcels of land, each one of which is in some precious way different from all the others. Our understandable wish to preserve the planet must somehow be reduced to the scale of our competence—that is, to the wish to preserve all of its humble households and neighborhoods.

What can accomplish this reduction? I will say again, without overweening hope but with certainty nonetheless, that only love can do it. Only love can bring intelligence out of the institutions and organizations, where it aggrandizes itself, into the presence of the work that must be done.

Love is never abstract. It does not adhere to the universe or the planet or the nation or the institution or the profession, but to the singular sparrows of the street, the lilies of the field, "the least of these my brethren." Love is not, by its own desire, heroic. It is heroic only when compelled to be. It exists by its willingness to be anonymous, humble, and unrewarded.

The older love becomes, the more clearly it understands its involvement in partiality, imperfection, suffering, and mortality. Even so, it longs for incarnation. It can live no longer by thinking.

And yet to put on flesh and do the flesh's work, it must think.

In his essay on Kipling, George Orwell wrote: "All left-wing parties in the highly industrialized countries are at bottom a sham, because they make it their business to fight against something which they do not really wish to destroy. They have internationalist aims, and at the same time they struggle to keep up a standard of life with which those aims are incompatible. We all live by robbing Asiatic coolies, and those of us who are 'enlightened' all maintain that those coolies ought to be set free; but our standard of living, and hence our 'enlightenment,' demands that the robbery shall continue."

This statement of Orwell's is clearly applicable to our situation now; all we need to do is change a few nouns. The religion and the environmentalism of the highly industrialized countries are at bottom a sham, because they make it their business to fight against something that they do not really wish to destroy. We all live by robbing nature, but our standard of living demands that the robbery shall continue.

We must achieve the character and acquire the skills to live much poorer than we do. We must waste less. We must do more for ourselves and each other. It is either that or continue merely to think and talk about changes that we are inviting catastrophe to make.

The great obstacle is simply this: the conviction that we cannot change because we are dependent on what is wrong. But that is the addict's excuse, and we know that it will not do.

How dependent, in fact, are we? How dependent are our neighborhoods and communities? How might our dependences be reduced? To answer these questions will require better thoughts and better deeds than we have been capable of so far.

We must have the sense and the courage, for example, to see that the ability to transport food for hundreds or thousands of miles does not necessarily mean that we are well off. It means that the food supply is more vulnerable and more costly than a local food supply would be. It means that consumers do not control or influence the healthfulness of their food supply and that they are at the mercy of the people who have the control and influence. It means that, in eating, people are using large quantities of petroleum that other people in another time are almost certain to need.

Our most serious problem, perhaps, is that we have become a nation of fantasists. We believe, apparently, in the infinite availability of finite resources. We persist in land-use methods that reduce the potentially infinite power of soil fertility to a finite quantity, which we then proceed to waste as if it were an infinite quantity. We have an economy that depends not on the quality and quantity of necessary goods and services but on the moods of a few stockbrokers. We believe that democratic freedom can be preserved by people ignorant of the history of democracy and indifferent to the responsibilities of freedom.

Our leaders have been for many years as oblivious to the realities and dangers of their time as were George III and Lord North. They believe that the difference between war and peace is still the overriding political difference—when, in fact, the difference has diminished to the point of insignificance. How would you describe the difference between modern war and modern industry—between, say, bombing and strip mining, or between chemical warfare and chemical manufacturing? The difference seems to be only that in war the victimization of humans is directly intentional and in industry it is "accepted" as a "trade-off."

Were the catastrophes of Love Canal, Bhopal, Chernobyl, and the *Exxon Valdez* episodes of war or of peace? They were, in fact, peacetime acts of aggression, intentional to the extent that the risks were known and ignored.

We are involved unremittingly in a war not against "foreign enemies," but against the world, against our freedom, and indeed against our existence. Our so-called industrial accidents should be looked upon as revenges of Nature. We forget that Nature is necessarily party to all our enterprises and that she imposes conditions of her own.

Now she is plainly saying to us: "If you put the fates of whole communities or cities or regions or ecosystems at risk in single ships or factories or power plants, then I will furnish the drunk or the fool or the imbecile who will make the necessary small mistake."

Nature as Measure

I LIVE IN a part of the country that at one time a good farmer could take some pleasure in looking at. When I first became aware of it, in the 1940s, the better land, at least, was generally well farmed. The farms were mostly small and were highly diversified, producing cattle, sheep, and hogs, tobacco, corn, and the small grains; nearly all the farmers milked a few cows for home use and to market milk or cream. Nearly every farm household maintained a garden, kept a flock of poultry, and fattened its own meat hogs. There was also an extensive "support system" for agriculture: every community had its blacksmith shop, shops that repaired harness and machinery, and stores that dealt in farm equipment and supplies.

Now the country is not well farmed, and driving through it has become a depressing experience. Some good small farmers remain, and their farms stand out in the landscape like jewels. But they are few and far between, and they are getting fewer every year. The buildings and other improvements of the old farming are everywhere in decay or have vanished altogether. The produce of the country is increasingly specialized. The small dairies are gone. Most of the sheep flocks are gone, and so are most of the enterprises of the old household economy. There is less livestock and more cash-grain farming. When cash-grain farming comes in, the fences go, the livestock goes, erosion increases, and the fields become weedy.

Like the farm land, the farm communities are declining and eroding. The farmers who are still farming do not farm with as much skill as they did forty years ago, and there are not nearly so many farmers farming as there were forty years ago. As the old have died, they have not been replaced; as the young come of age, they leave farming or leave the community. And as the land and the people deteriorate, so necessarily must the support system. None of the small rural towns is thriving as it did forty years ago. The proprietors of small businesses give up or die and are not replaced. As the farm trade declines, farm equipment franchises are revoked. The remaining farmers must drive longer and longer distances for machines and parts and repairs.

Looking at the country now, one cannot escape the conclusion that there are no longer enough people on the land to farm it well and to take proper care of it. A further and more ominous conclusion is that there is no longer a considerable number of people knowledgeable enough to look at the country and see that it is not properly cared for—though the face of the country is now everywhere marked by the agony of our enterprise of self-destruction.

And suddenly in this wasting countryside there is talk of raising production quotas on Burley tobacco by twenty-four percent, and tobacco growers are coming under pressure from the manufacturers to decrease their use of chemicals. Everyone I have talked to is doubtful that we have enough people left in farming to meet the increased demand for either quantity or quality, and doubtful that we still have the barnroom to house the increased acreage. In other words, the demand going up has met the culture coming down. No one can be optimistic about the results.

Tobacco, I know, is not a food, but it comes from the same resources of land and people that food comes from, and this emerging dilemma in the production of tobacco can only foreshadow a similar dilemma in the production of food. At every point in our food economy, present conditions remaining, we must expect to come to a time when demand (for quantity or quality) going up will meet the culture coming down. The fact is that we have nearly destroyed American farming, and in the process have nearly destroyed our country.

How has this happened? It has happened because of the application to farming of far too simple a standard. For many years, as a nation, we have asked our land only to produce, and we have asked our farmers only to produce. We have believed that this single economic standard not only guaranteed good performance but also preserved the ultimate truth and rightness of our aims. We have bought unconditionally the economists' line that competition and innovation would solve all problems, and that we would finally accomplish a technological end-run around biological reality and the human condition.

Competition and innovation have indeed solved, for the time being, the problem of production. But the solution has been extravagant, thoughtless, and far too expensive. We have been winning, to our inestimable loss, a competition against our own land and our own people. At present, what we have to show for this "victory" is a surplus of food. But this is a surplus achieved by the ruin of its sources, and it has been used, by apologists for our present economy, to disguise the damage by which it was produced. Food, clearly, is the most important economic product—except when there is a surplus. When there is a surplus, according to our present economic assumptions, food is the *least* important product. The surplus becomes famous as evidence to consumers that they have nothing to worry about, that there is no problem, that present economic assumptions are correct.

But our present economic assumptions are failing in agriculture, and to those having eyes to see the evidence is everywhere, in the cities as well as in the countryside. The singular demand for production has been unable to acknowledge the importance of the sources of production in nature and in human culture. Of course agriculture must be productive; that is a requirement as urgent as it is obvious. But urgent as it is, it is not the *first* requirement; there are two more requirements equally important and equally urgent. One is that if agriculture is to remain productive, it must preserve the land, and the fertility and ecological health of the land; the land, that is, must be used *well.* A further requirement, therefore, is that if the land is to be used well, the people who use it must know it well, must be highly motivated to use it well, must know how to use it well, must have time to use it well, and must be able to afford to use it well. Nothing that has happened in the agricultural revolution of the last fifty years has disproved or invalidated these requirements, though everything that has happened has ignored or defied them.

In light of the necessity that the farm land and the farm people should thrive while producing, we can see that the single standard of productivity has failed.

Now we must learn to replace that standard by one that is more comprehensive: the standard of nature. The effort to do this is not new. It was begun early in this century by Liberty Hyde Bailey of the Cornell University College of Agriculture, by F. H. King of the University of Wisconsin College of Agriculture and the United States Department of Agriculture, by J. Russell Smith, professor of economic geography at Columbia University, by the British agricultural scientist Sir Albert Howard, and by others; and it has continued into our own time in the work of such scientists as John Todd, Wes Jackson, and others. The standard of nature is not so simple or so easy a standard as the standard of productivity. The term "nature" is not so definite or stable a concept as the weights and measures of productivity. But we know what we mean when we say that the first settlers in any American place recognized that place's agricultural potential "by its nature"—that is, by the depth and quality of its soil, the kind and quality of its native vegetation, and so on. And we know what we mean when we say that all too often we have proceeded to ignore the nature of our places in farming them. By returning to "the nature of the place" as standard, we acknowledge the necessary limits of our own intentions. Farming cannot take place except in nature; therefore, if nature does not thrive, farming cannot thrive. But we know too that nature includes us. It is not a place into which we reach from some safe standpoint outside it. We are in it and are a part of it while we use it. If it does not thrive, we cannot thrive. The appropriate measure of farming then is the world's health and our health, and this is inescapably *one* measure.

But the oneness of this measure is far different from the singularity of the standard of productivity that we have been using; it is far more complex. One of its concerns, one of the inevitable natural measures, is productivity; but it is also concerned for the health of all the creatures belonging to a given place, from the creatures of the soil and water to the humans and other creatures of the land surface to the birds of the air. The use of nature as measure proposes an atonement between ourselves and our world, between economy and ecology, between the domestic and the wild. Or it proposes a conscious and careful recognition of the interdependence between ourselves and nature that in fact has always existed and, if we are to live, must always exist.

Industrial agriculture, built according to the single standard of productivity, has dealt with nature, including human nature, in the manner of a monologist or an orator. It has not asked for anything, or waited to hear any response. It has told nature what it wanted, and in various clever ways has taken what it wanted. And since it proposed no limit on its wants, exhaustion has been its inevitable and foreseeable result. This, clearly, is a dictatorial or totalitarian form of behavior, and it is as totalitarian in its use of people as it is in its use of nature. Its connections to the world and to humans and the other creatures become more and more abstract, as its economy, its authority, and its power become more and more centralized.

On the other hand, an agriculture using nature, including human nature, as its measure would approach the world in the manner of a conversationalist. It would not impose its vision and its demands upon a world that it conceives of as a stockpile of raw material, inert and indifferent to any use that may be made of it. It would not proceed directly or soon to some supposedly ideal state of things. It *would* proceed directly and soon to serious thought about our condition and our predicament. On all farms, farmers would undertake to know responsibly where they are and to "consult the genius of the place." They would ask what nature would be doing there if no one were farming there. They would ask what nature would permit them to do there, and what they could do there with the least harm to the place and to their natural and human neighbors. And they would ask what nature would *help* them to do there. And after each asking, knowing that nature will respond, they would attend carefully to her response. The use of the place would necessarily change, and the response of the place to that use would necessarily change the user. The conversation itself would thus assume a kind of creaturely life, binding the place and its inhabitants together, changing and growing to no end, no final accomplishment, that can be conceived or foreseen.

Farming in this way, though it certainly would proceed by desire, is not visionary in the political or utopian sense. In a conversation, you always expect a reply. And if you honor the other party to the conversation, if you honor the *otherness* of the other party, you understand that you must not expect always to receive a reply that you foresee or a reply that you will like. A conversation is immitigably two-sided and always to some degree mysterious; it requires faith.

For a long time now we have understood ourselves as traveling toward some sort of industrial paradise, some new Eden conceived and constructed entirely by human ingenuity. And we have thought ourselves free to use and abuse nature in any way that might further this enterprise. Now we face overwhelming evidence that we are not smart enough to recover Eden by assault, and that nature does not tolerate or excuse our abuses. If, in spite of the evidence against us, we are finding it hard to relinquish our old ambition, we are also seeing more clearly every day how that ambition has reduced and enslaved us. We see how everything—the whole world—is belittled by the idea that all creation is moving or ought to move toward an end that some body, some human body, has thought up. To be free of that end and that ambition would be a delightful and precious thing. Once free of it, we might again go about our work and our lives with a seriousness and pleasure denied to us when we merely submit to a fate already determined by gigantic politics, economics, and technology.

Such freedom is implicit in the adoption of nature as the measure of economic life. The reunion of nature and economy proposes a necessary democracy, for neither economy nor nature can be abstract in practice. When we adopt nature as measure, we require practice that is locally knowledgeable. The particular farm, that is, must not be treated as any farm. And the particular knowledge of particular places is beyond the competence of any centralized power or authority. Farming by the measure of nature, which is to say the nature of the particular place, means that farmers must tend farms that they know and love, farms small enough to know and love, using tools and methods that they know and love, in the company of neighbors that they know and love.

In recent years, our society has been required to think again of the issues of the use and abuse of human beings. We understand, for instance, that the inability to distinguish between a particular woman and any woman is a condition predisposing to abuse. It is time that we learn to apply the same understanding to our country. The inability to distinguish between a farm and any farm is a condition predisposing to abuse, and abuse has been the result. Rape, indeed, has been the result, and we have seen that we are not exempt from the damage we have inflicted. Now we must think of marriage.

CHRONOLOGY

NOTE ON THE TEXTS

NOTES

INDEX

Chronology

1934 Born Wendell Erdman Berry on August 5 in Henry County, Kentucky, the first child of John Marshall and Virginia Erdman Perry Berry. Father, born November 8, 1900, near Lacie, Kentucky, earned a bachelor's degree at Georgetown College in central Kentucky in 1922, and then worked as secretary to Congressman Virgil Chapman while earning a law degree from The George Washington University in Washington, D.C. He graduated in 1927 and later that year opened a private legal practice in Henry County. Mother born November 6, 1907, in Port Royal, in Henry County. She attended Randolph-Macon Woman's College in Lynchburg, Virginia, a Methodist women's college tied to the all-male Randolph-Macon College in Ashland, Virginia, graduating with a degree in English in 1930. Both families can trace roots back to early years of Henry County's history; both families also owned slaves before the Civil War (at time of Berry's birth, there are many children of both slaves and slaveholders in the community). Parents married on July 14, 1933.

1935 Brother John Marshall Jr. born October 13.

1936 Family moves to New Castle, Kentucky, where father practices law and helps with paternal grandfather's farm, the "Home Place."

1938 Sister Mary Jo born March 18.

1939 Sister Martha Frances born July 22.

1940 Enters New Castle Elementary School in Henry County. Over the next few years, mother introduces Berry to books about King Arthur's knights and Robin Hood, as well as *The Swiss Family Robinson*, *Treasure Island*, and *The Yearling*. Later will read Mary O'Hara's *My Friend Flicka*, *Thunderhead*, and *Green Grass of Wyoming*.

1941 The Burley Tobacco Growers Cooperative Association, representing burley tobacco farmers in Kentucky, Missouri, Indiana, Ohio, and West Virginia with an average farm size of 150 acres, which had been founded in 1921, is revived under the New Deal, establishing parity prices and production control as protections for farmers. Father is one of the principal authors of the program and serves as counsel and vice-president and later president.

1942 Uncle Morgan Perry serves in the navy medical corps during World War II, and because of his knowledge of agriculture, is placed in charge of the garden of the naval hospital at Pearl Harbor.

1944 On July 3, uncle Wendell Holmes Berry, with whom Berry was extremely close, is shot at a defunct lead mine property, where he had been working with a crew salvaging lumber and roofing, after a quarrel with Floyd Martin, one of the owners of the property, and dies the same day.

1948 Finishes the eighth grade at the New Castle School. Enters Millersburg Military Institute in Millersburg, Kentucky, at the same time as his brother, John, Jr. Berry will remember, "It was very confining and military, of course. I never liked the military part of it. I did have some good teachers there, and I began there to read more seriously than I had before. I began to have a favorite subject: literature." Father speaks before Congress on behalf of the Tobacco Program. In high school, begins tentatively to write poems.

1952 Graduates from Millersburg Military Institute. In the fall, enters University of Kentucky, in Lexington, Kentucky. Co-edits freshman magazine, *Green Pen*.

1953 Takes "Introduction to Literature" from Thomas Stroup. Begins bringing Stroup poems for criticism. Later takes classes from Hollis Summers and Robert Hazel. Reads T. S. Eliot and Ezra Pound. In May, publishes first essay, "The Wings of the Future," in *Green Pen*, about the survival of society in the light of two world wars and the entrance of the U.S. into the Korean conflict; he writes that it is the "American debt to tomorrow" to ensure that a democratic form of government succeeds over Soviet-style communism. During a composition class, the instructor Robert D. Jacobs introduces Berry to the term "agrarian" and to the 1930s manifesto of the twelve Southern Agrarian writers (who include Allen Tate, John Crowe Ransom, and Robert Penn Warren), *I'll Take My Stand*, a work that will be significant (both in influence and in reaction) to the development of Berry's thought.

1954 Publishes first poem, "Spring," and first short story, "Summer Crop," in the spring issue of *Stylus*, the university literary magazine. Becomes editor of *Stylus*. At a writers' conference at Morehead State College in Kentucky meets writer James Still, whose work will become an influence (Berry will later write, "His short stories, I think, are as near to perfect as any writing I know"). In fall, meets James Baker Hall, who will become a close friend, in a creative writing class taught by Hollis Summers, for which Berry writes story "The Brothers" (later part of *Nathan Coulter*).

1955 Wins *Stylus*'s Dantzler Award for short story "The Brothers." In fall, meets Tanya Amyx (born April 30, 1936, in Berkeley, California), a French and music major and daughter of University of Kentucky art professor Clifford Amyx and textile artist Dee Amyx. "The first time I ever saw Tanya," Berry would later remark, "she was standing by [a] wooden newel post in Miller Hall at the University of Kentucky. Years later, they started to remodel the place. I went over and said, 'Look. When you tear that post out, I want it.'" (The post is now in the Berrys' home.) With fellow student Edward M. Coffman, travels to Kenyon College to meet John Crowe Ransom.

1956 Wins *Stylus*'s Dantzler Award for short story "The Chestnut Stud." Meets fellow student Gurney Norman. On May 16, submits term paper on *I'll Take My Stand*: "The Regional Context: A Consideration of the Southern Agrarians as Southerners." Graduates from the University of Kentucky with a bachelor's degree in English. In summer, Berry and James Baker Hall join literature seminars at Indiana University School of Letters; Berry takes course in Joyce and Yeats from Richard Ellman, and course in modern poetry from Karl Shapiro. "The Brothers" wins first prize in the *Carolina Quarterly*'s 6th Annual Fiction Contest and is published in its summer issue. In the fall, enters the master's program in English at the University of Kentucky.

1957 In February, poem "Rain Crow" is published in *Poetry* magazine; in the next five years, his work will be published in *Poetry* five times. Meets writer Ed McClanahan during a graduate course. Completes MA in English. "Elegy" wins the Farquhar Award for Poetry from the University of Kentucky. Two short stories, "Whippoorwills" and "Apples," are published in *Coraddi: Arts*

Forum 1957. Marries Tanya Amyx, May 29. In summer, they live in a cabin built by Berry's great-uncle Curran Mathews, called "the Camp," on the Kentucky River near Port Royal, which was used as a family retreat from the 1920s. In the fall, they move to an apartment in Georgetown, Kentucky; Berry begins teaching freshman composition and sophomore literature at Georgetown College, father's alma mater, while Tanya continues to study at the University of Kentucky.

1958 Daughter Mary Dee born May 10 in Lexington. Awarded a Wallace Stegner Fellowship at Stanford University; in August, family moves to Mill Valley, California, to live while he studies creative writing at Stanford University under Stegner (about whom Berry will later write, "My debt to him is probably greater than I know") and Richard Scowcroft; fellow members of the fiction seminars include Ernest J. Gaines (also a Stegner fellow), Ken Kesey, and Nancy Packer. Lives with family in small house owned by Tanya's aunt and uncle, Ann and Dick O'Hanlon, which would later become part of the O'Hanlon Center. Works on novel *Nathan Coulter*: "I had about half of it written when I went out there, and I just continued to work on it." Introduced by Holman Hamilton, one of Berry's history professors at the University of Kentucky, to Craig Wylie of Houghton Mifflin, to whom Berry submits the manuscript of *Nathan Coulter*. Houghton Mifflin purchases an option on the book for $250.

1959 Accepts appointment as Edward H. Jones Lecturer in Creative Writing at Stanford University for one year.

1960 In January, begins writing second novel, *A Place on Earth*. First novel, *Nathan Coulter*, published in April by Houghton Mifflin. In spring, family moves back to Kentucky, living on the "Home Place" in Henry County, and also farming with Berry's friend and neighbor Owen Flood.

1961 Receives a Guggenheim Foundation Fellowship. Will later write of Henry County that summer: "I began to understand that so long as I did not know the place fully, or even adequately, I belonged to it only partially. That summer I began to see, however dimly, that one of my ambitions, perhaps my governing ambition, was to belong fully to this place, to belong as the thrushes and the herons and the muskrats belonged, to be altogether at home here." In August, leaves with family for a year in Tuscany and southern France to study and write: "the land in Tuscany has had about 2,000 years of good care. And it looked like it when I was there. . . . The sight of that changed my mind about what was possible in land use."

1962 Meets critic and translator Wallace Fowlie when both stay at La Napoule Arts Foundation near Cannes. Awarded Vachel Lindsay Prize by *Poetry* magazine. Returns to Kentucky from Europe in July. Son Pryor Clifford (Den) born August 19 in Lexington. In the fall, moves with family to New York to take up post as assistant professor of English and director of freshman writing at New York University's University Heights campus in the Bronx. Rents apartment in New Rochelle, New York. Meets Bobbie Ann Mason.

1963 Family moves to loft at 277 Greenwich Street, New York City, downstairs from poet Denise Levertov (whom Berry had met the previous fall, although they had corresponded since 1958, when Levertov wrote Berry about a poem he had published in *Poetry*) and writer Mitchell Goodman. Tanya arranges playdates for daughter Mary and a fellow student at St. Luke's School, the daughter of James Parks Morton, later dean of St. John the Divine. Spends summer in Kentucky. Dismantles the Camp, which has flooded several times over the years, and rebuilds it higher on the bank of the river. Works on novel *A Place on Earth*. Reads Harry Caudill's *Night Comes to the Cumberlands*, an important influence ("It showed me what it might mean to be a responsible Kentucky writer living in Kentucky, and it affected me deeply"). Returns to New York for fall semester. Meets poet Donald Hall, who becomes friend and correspondent, at a cocktail party on Riverside Drive in Manhattan. Offered and accepts position as professor of English at the University of Kentucky to begin in the fall of 1964, where he will teach creative writing and other courses.

1964 In May, visits Levertov and Goodman's Maine farm, where he reads in typescript Hayden Carruth's poetry collection *North Winter* (Berry will later begin a correspondence with Carruth that will last until Carruth's death in 2008). Leaves New York for Kentucky soon after. A poem about Kennedy's assassination, *November Twenty Six Nineteen Hundred Sixty Three* (with drawings by Ben Shahn) published by George Braziller in May. Begins commuting weekly from Lexington (where family lives) to the Camp to write. Meets Kentucky artist Harlan Hubbard and his wife Anna by chance while on a canoe trip on the Ohio River; later writes that their homestead "differed from Thoreau's economy radically in some respects, and also advanced and improved upon it. The main differences were that, whereas Thoreau's was a bachelor's economy, the Hubbards' was that of a married couple." First poetry collection *The Broken Ground* published by Harcourt, Brace & World in September (a *New York Times* review will begin: "The quiet but sure and melodious voice of Wendell Berry makes 'The Broken Ground' an immediate pleasure"). In November, purchases the Lanes Landing property as a weekend place, a twelve-acre hillside farm adjacent to the Camp and bordering his maternal grandfather's farm on the west side of the Kentucky River. Berry's editor at Harcourt, Dan Wickenden, introduces him to the work of organic farming pioneer Sir Albert Howard.

1965 Awarded a Rockefeller Foundation Fellowship. On July 4, moves with family to Lanes Landing Farm, where they will continue to live and farm year round. "I remember a moment—in 1965, or a little after—when I realized that I didn't have to be a writer; there were other kinds of work also that required artistry and offered satisfaction. From here, looking back, I can see what a defining moment that was. I had, in effect, decided not to be a 'professional' writer, but instead, in the literal sense, an amateur: I would work for love." Over time, expands the farm until it consists of about 117 acres in two tracts; at first, raises only food for the family. In July, sees first strip mine above Hardburly, Kentucky, while visiting writer Gurney Norman; Norman introduces him to Kentucky lawyer, writer, and environmentalist Harry Caudill and Caudill's wife Anne, who will become lifelong friends, at a meeting of the Appalachian Group to Save the Land and People. Will one day write, "Gurney has been my Virgil on the upper end (the mountain end) of the Kentucky River watershed, where he is at home. I'm at home on the lower end."

1967 In summer, makes first notes for a work that will become *The Unsettling of America* after reading about the report of President Johnson's "special commission on federal food and fiber policies" that defined the nation's greatest agriculture problem as a surplus of farmers. Novel *A Place on Earth* published by Harcourt, Brace & World in September. Receives Bess Hoken Prize from *Poetry* magazine.

Becomes president of the Cumberland Chapter of the Sierra Club in Kentucky. Begins work with farmers and landowners to oppose a dam across the Red River Gorge that would destroy the gorge's unique topography and ecosystem. In December, visits Trappist monk Thomas Merton at Abbey of Our Lady of Gethsemani near Bardstown, Kentucky, with Ralph Eugene Meatyard and his wife Madelyn, Tanya, and Denise Levertov.

1968 On February 10, delivers speech, "A Statement Against the War in Vietnam," during the Kentucky Conference on War and the Draft in Lexington, Kentucky: "I have come to the realization that I can no longer imagine a war that I would believe to be either useful or necessary. I would be against any war." In fall, takes leave from University of Kentucky to become visiting professor of creative writing at Stanford University during fall 1968 and winter 1969 quarters; lives with family in Menlo Park. Colleagues at Stanford include Wallace Stegner, Ed McClanahan, and Ken Fields. Meets writer John Haines when he gives a reading at Stanford. During the Christmas holiday, writes *The Hidden Wound*, an extended meditation on race and memories of two older black people who were his friends during his childhood. Poetry collection *Openings* published by Harcourt, Brace & World in October; *The New York Times* reviewer calls it a book "to win the respect of anyone who cares about contemporary verse."

1969 Returns with family to Kentucky. Essay collection *The Long-Legged House*, of which the title essay is about the history of the Camp, published by Harcourt, Brace & World in April; other essays concern place and belonging, agriculture, community, small farming, the Vietnam War, and consumerism. The same month, poetry collection *Findings* published by The Prairie Press. Wins first prize in the Borestone Mountain Poetry Awards. Receives grant from the National Endowment for the Arts.

1970 On March 7, speaks at the march at the state capitol in Frankfort, Kentucky, to protest the war in Vietnam; his speech notes government disdain for the will of the people and the huge expense of the war: "With the very earth dying under our feet, with the air full of pollution; we are spending billions of dollars and thousands of lives to assure the success of tyranny in Vietnam." At the first Earth Day celebration at the University of Kentucky, delivers speech "Think Little." Publishes essay "The Regional Motive" in the autumn issue of *The Southern Review*, a response to the Southern Agrarians; in it, he is critical of those Agrarians who had moved to northern universities, saying that it "invalidated their thinking, and reduced their effort to the level of an academic exercise." (Receives letter from Allen Tate in response, and when the essay is included in *A Continuous Harmony* in 1972, Berry appends an apologetic footnote; he later writes that "whatever the amount of truth in that statement, and there is some, it is also a piece of smartassery.") Poetry collection *Farming: A Hand Book* published by Harcourt Brace Jovanovich in September. Meets writer and farmer Gene Logsdon, who becomes a friend and ally, after Logsdon reads *Farming: A Hand Book*. *The Hidden Wound* published by Houghton Mifflin in September.

1971 *The Unforeseen Wilderness: An Essay on Kentucky's Red River Gorge* published by University Press of Kentucky in April, with photographs by Ralph Eugene Meatyard; it is a response to the plan to dam the Red River, which Berry had opposed since 1967. (Congress will withdraw funds for the project in 1976. In 1993 the river is designated by Congress in the National Wild and Scenic Rivers System Act.) Elected Distinguished Professor of the Year by the University of Kentucky. Receives the Arts and Letters Literary Award from the American Academy of Arts and Letters.

1972 With his brother John Berry Jr. and the Save Our Land Committee, helps in the successful opposition to the Jefferson County Air Board's plan for an international jetport in Henry and Shelby Counties. In the fall, takes a one-year sabbatical from the University of Kentucky; his courses are covered by Ed McClanahan. Essay collection *A Continuous Harmony: Essays Cultural and Agricultural* published by Harcourt Brace Jovanovich in September. Meets bookseller and future publisher Jack Shoemaker when Shoemaker visits Port Royal after reading *The Long-Legged House*.

1973 *The Country of Marriage* published by Harcourt Brace Jovanovich in February, his fifth poetry collection, with thirty-five poems, including the title poem, "Prayer After Eating," and "Manifesto: The Mad Farmer Liberation

Front." Buys neighboring forty-acre farm that had been sold to a developer, bulldozed, and graveled; begins process of repair and restoration. Invites writer and photographer James Baker Hall to take photographs during a tobacco harvest (they will be published with an essay by Berry in 2004). Begins correspondence with poet and essayist Gary Snyder, and after a few months Snyder travels to Port Royal to visit Lanes Landing. In November, brother elected to the Kentucky senate, in which he will serve until 1981.

1974 Serves as Elliston Poetry Lecturer at the University of Cincinnati for winter 1974. Novel *The Memory of Old Jack* published by Harcourt Brace Jovanovich in February (*The New York Times Book Review* calls it "a slab of rich Americana" and *Library Journal* says the novel is "worthy of a place among the best pieces of prose written by American writers of this century"). Meets Maurice Telleen at a draft horse sale in Waverly, Iowa, whose *Draft Horse Journal* will publish many of Berry's stories. Gives speech on July 1 at "Agriculture for a Small Planet Symposium" in Spokane, Washington, which will form the basis for the first chapters of *The Unsettling of America*: "I was asked to talk about 'Labor Intensive Micro-Systems Agriculture.' That's not my language, and it's not the sort of language I wish to use because it's the way people speak when they don't want to be understood by most people. I'm not sure what to make of these particular phrases, but they seem to suggest a very methodological or technological approach to agriculture. Part of my purpose here is to suggest that any such approach will necessarily be too simple." Mentions critically Secretary of Agriculture Earl Butz's "adapt or die" policy. Speech helps encourage the Tilth movement, a regional network of organic farmers in the Northwest (still active today). Poetry collection *An Eastward Look* published by Sand Dollar Press in the fall, Berry's first publication with editor Jack Shoemaker. Wins the Emily Clark Balch Prize from *The Virginia Quarterly Review*.

1975 *The Memory of Old Jack* wins the Friends of American Writers Award. A poetry pamphlet, *Horses*, published by Larkspur Press in Kentucky in April; another, *To What Listens*, published by Best Cellar Press. Poetry collection *Sayings and Doings* published by Gnomon Press in December.

1977 Serves as writer-in-residence for the winter quarter at Centre College in Danville, Kentucky. Poetry collection *Clearing* published by Harcourt Brace Jovanovich in March. *The Unsettling of America: Culture and Agriculture* published by Sierra Club Books in August and dedicated to Maurice Telleen, the editor of *Draft Horse Journal*. Donald Hall reviews both volumes for *The New York Times*, writing, "Berry is a prophet of our healing, a utopian poet-legislator like William Blake." Resigns from the University of Kentucky to become contributing editor at Rodale Press, including its magazines *Organic Gardening* and *The New Farm*.

1978 With Tanya, buys first flock (six ewes and one buck) of Border Cheviot sheep, a Scottish breed, which they will raise for the next four decades. In November, debates Secretary of Agriculture Earl Butz on the agricultural crisis at Manchester University, North Manchester, Indiana: "As I see it, the farmer standing in his field is not simply a component of a production machine. He stands where lots of cultural lines cross. The traditional farmer, that is the farmer who first fed himself off his farm and then fed other people, who farmed with his family, who passed the land on down to people who knew it and had the best reasons to take care of it—that farmer stood at the convergence of traditional values, our values: independence, thrift, stewardship, private property, political liberties, family, marriage, parenthood, neighborhood—values that decline as that farmer is replaced by a technologist whose only standard is efficiency."

1979 On June 3, opposes and with eighty-nine others is arrested during nonviolent protests against the construction of the Marble Hill nuclear power plant on the Ohio River near Madison, Indiana. The company eventually abandons the effort to finish it. Poetry collection *The Gift of Gravity* published by Deerfield Press.

1980 Fired from Rodale Press: "I think this was because I was more for small farmers than I was for organic farmers. Also, I don't think I've ever been a very good employee." Ernest J. Gaines visits the Berrys in Kentucky. Begins working with editor Jack Shoemaker, who had left bookselling to cofound North Point Press in Berkeley, California, in 1979, with William Turnbull. (Most of Berry's books will henceforward be published with Shoemaker.) Poetry collection

A Part published in October. Joins with brother and other residents of Henry County along with the group Kentuckians for the Commonwealth in a successful effort to oppose the building of a hazardous chemical waste incinerator in the county. Publishes poetry pamphlet *The Salad.* On November 11, writes to Wes Jackson, founder of The Land Institute, after reading Jackson's book, *New Roots for Agriculture.* Visits Jackson in December to write an article about The Land Institute for *The New Farm* magazine. The two begin a long correspondence and friendship.

1981 *Recollected Essays: 1965–1980* published in August. Essay collection *The Gift of Good Land: Further Essays Cultural and Agricultural*, dedicated to Gene Logsdon, published in November. Both volumes are reviewed in *The Washington Post* by Larry Woiwode, who calls them "reference works of the body and soul." Granddaughter Katie Jean Smith born, December 16.

1982 Poetry collection *The Wheel* published in October.

1983 Speaking at Oberlin College in February, meets David and Elsie Kline, an Amish couple from Holmes County who had read *The Unsettling of America* and came to hear him speak. Kline (with Meatyard, Telleen, and Jackson) is one of the only four agrarian friends Berry has from outside of Henry County for the next decade. Essay collection *Standing by Words* published in October, which *The Christian Science Monitor* review calls "nothing short of splendid." Publishes substantially revised and shortened version of 1967 novel *A Place on Earth* in March; in a new introduction, writes that the original book was "clumsy, overwritten, wasteful."

1985 Co-edits with Wes Jackson and Bruce Coleman *Meeting the Expectations of the Land: Essays in Sustainable Agriculture and Stewardship*, published by North Point in February; Berry's contribution is the essay "Whose Head is the Farmer Using? Whose Head is Using the Farmer?" Granddaughter Virginia Dee Smith born, March 6. *Collected Poems 1957–1982* published in May; *The New York Times* reviewer writes that Berry "can be said to have returned American poetry to a Wordsworthian clarity of purpose." Publishes substantially revised version of *Nathan Coulter* in May. Helps found Community Farm Alliance, a group of Kentucky farmers, bankers, businessmen, and clergy members organized to help small farmers shift from tobacco to other products. Begins correspondence with writer John Haines.

1986 *The Wild Birds: Six Stories of the Port William Membership* published in March. Receives honorary doctorate from the University of Kentucky. In November, delivers lecture "Preserving Wildness" at the first Temenos Academy Conference in Devon, England.

1987 Returns to the English Department of the University of Kentucky, now teaching courses for "future teachers and farmers or anyone preparing to work in practical and comprehensive ways with young minds or with nature"; courses include "Composition for Teachers," aimed at public school English teachers; "Readings in Agriculture," which assigns Spenser, Milton, Shakespeare, Pope, and Wordsworth, as well as Sir Albert Howard, J. Russell Smith, Wes Jackson, and Gene Logsdon; and "The Pastoral." Publishes *Home Economics: Fourteen Essays* in June; the *Christian Science Monitor*, in reviewing it, calls Berry "*the* prophetic American voice of our day." Serves as writer-in-residence for the interim term at Bucknell University. Receives the American Academy of Arts and Letters Jean Stein Award and the Kentucky Governor's Milner Award in the Arts. In the fall, "Why I Am Not Going to Buy a Computer" published in the *New England Review and Bread Loaf Quarterly* and reprinted in *Harper's*; the essay inspires many critical responses. *Sabbaths: Poems* published in September, the first of what will become a sustained project of poems on the themes of work and rest, fields and woods. *The Landscape of Harmony: Two Essays on Wildness and Community*, including the lecture "Preserving Wildness," published by Five Seasons Press, United Kingdom, in September. Poetry collection *Some Differences* published by Confluence Press in December.

1988 Novel *Remembering* published in October and reviewed favorably in the *Los Angeles Times*.

1989 Receives Lannan Foundation Award for nonfiction. Poetry collection *Traveling at Home* published in November. Delivers the Blazer Lecture on the life and work of Kentucky artist and author Harlan Hubbard at the University of Kentucky.

1990 Granddaughter Tanya Christine Smith born February 19.

Essay collection *What Are People For?*, dedicated to Gurney Norman, published in March; it includes the oft-repeated line, "Eating is an agricultural act." (Berry later calls this quote, as repeated without context, "an oversimplification he is now damned sorry to have written.") Bill McKibben publishes a major profile of Berry in *The New York Review of Books*, writing that "wherever we live, however we do so, we desperately need a prophet of responsibility; and although the days of the prophets seem past to many of us, Berry may be the closest to one we have. But, fortunately, he is also a poet of responsibility. He makes one believe that the good life may not only be harder than we're used to but sweeter as well." Inducted into the Fellowship of Southern Writers. Berry's Blazer lecture, *Harlan Hubbard: Life and Work*, published by University Press of Kentucky in November.

1991 *Standing on Earth: Selected Essays* published in the United Kingdom by Brian Keeble's Golgonooza Press in April. Publishes poetry collection *Sabbaths 1987* with Larkspur Press in October. Father dies October 31. North Point Press closes, and Berry follows Shoemaker to Pantheon.

1992 Receives the Victory of Spirit Ethics Award from the University of Louisville and the Louisville Community Foundation. From late August to late September, teaches a course at Schumacher College in South Devon, England, on "Nature as Teacher: The Lineage of Writings Which Link Culture and Agriculture"; writers assigned include Shakespeare, Sir Albert Howard, and Wes Jackson. On September 8, meets Charles, Prince of Wales. "It was a much easier meeting than I expected, for he is intelligent, considerate, and talks and listens well. He is an ally, is deeply concerned about what is happening to rural life and to civilization," Berry wrote in a letter to Wes Jackson. *Fidelity: Five Stories* published in October to favorable reviews, including in *The New York Times* and *The New Criterion*.

1993 Granddaughter Emily Rose Berry born, May 2. *Sex, Economy, Freedom and Community: Eight Essays* published in October; *The New York Times* writes that the book "in its eight essays distills the author's radically conservative views (in the good senses of both words)." Receives The Orion Society's John Hay Award for nature writing. Quits his job at the University of Kentucky.

1994 Receives T. S. Eliot Award for creative writing from the

Ingersoll Foundation. Poetry collection *Entries* published in May. *Watch With Me and Six Other Stories of the Yet-Remembered Ptolemy Proudfoot and His Wife, Miss Minnie, Née Quinch* published in August. Berry follows his editor to Counterpoint Press, which Shoemaker cofounds in Washington, D.C., with Frank H. Pearl.

1995 Grandson Marshall Amyx Berry born June 22. In October, long poem *The Farm* (one of the *Sabbath* series) published by Larkspur Press and the essay collection *Another Turn of the Crank* published by Counterpoint.

1996 Co-authors *Three on Community* with Gary Snyder and Carole Koda, published by Limberlost Press in April. Receives the Harry M. Caudill Conservationist Award from the Cumberland Chapter of the Sierra Club. Novel *A World Lost* published in October.

1997 Mother dies, January 3. Father-in-law Clifford Amyx dies, July 30. *Two More Stories of the Port William Membership* published by Gnomon Press. Receives the Lyndhurst Prize for individuals making a significant contribution to the arts. Speaks at a benefit for The Garden Project in San Francisco, November 10.

1998 Gives keynote address, "In Distrust of Movements," at the Northeast Organic Farming Association conference. *A Timbered Choir: The Sabbath Poems 1979–1997* published in April. *The Selected Poems of Wendell Berry* published in October.

1999 Receives the Thomas Merton Award from the Thomas Merton Center for Peace and Social Justice in Pittsburgh. On November 10, joins Gary Snyder and Jack Shoemaker in reading and conversation at a Lannan Foundation event in Santa Fe, New Mexico.

2000 Wins Poets' Prize for *The Selected Poems of Wendell Berry.* In May publishes *Life Is a Miracle: An Essay Against Modern Superstition*; Bill McKibben, reviewing it for *The Washington Monthly*, writes, "it's hard to imagine that the millennium has seen a more important book than this slim volume from our finest essayist." Novel *Jayber Crow* published in September; *The New York Times* reviewer writes that "by the end this melancholy barber has won both our attention and our hearts." In October, teaches one-week course at Schumacher College in England on "Community, Sustainability and Globalisation."

2001 In response to the events following 9/11, writes "Thoughts in the Presence of Fear," which is published on October 30 by *The Land Report*; in December it is joined by two other essays, "The Idea of a Local Economy" and "In Distrust of Movements," and published by The Orion Society in paperback as *In the Presence of Fear: Three Essays for a Changed World*. Publishes *Sonata at Payne Hollow: A Play* with Larkspur Press.

2002 *The Art of the Commonplace: The Agrarian Essays of Wendell Berry*, edited and introduced by Norman Wirzba, published in April. Berry follows his editor to Shoemaker & Hoard, which Shoemaker cofounds with Trish Hoard.

2003 On February 9, essay "A Citizen's Response to the National Security Strategy of the United States," critical of the Bush administration's post-9/11 strategy, published as a full-page advertisement in *The New York Times*. Essay collection *Citizenship Papers* published in August. *Citizen's Dissent: Security, Morality, and Leadership in an Age of Terror: Essays*, co-authored with David James Duncan, published by The Orion Society. Receives Lifetime Achievement Award from the Cathedral Heritage Foundation in Louisville, Kentucky.

2004 *That Distant Land: The Collected Stories* published in February. Receives the Eli M. Oboler Memorial Award in Orlando in June, shared by David James Duncan, awarded by the American Library Association Intellectual Freedom Round Table, for *Citizen's Dissent*. Mother-in-law Dee Rice Amyx dies, July 3. Awarded the Charity Randall Citation from the International Poetry Forum, as well as the Soil and Water Conservation Society Honor Award from the Kentucky Bluegrass Chapter for dedicated efforts toward the preservation of unique rural and cultural resources. Writes essay for *Tobacco Harvest: An Elegy*, to accompany photographs by James Baker Hall; published by University Press of Kentucky in September. Novel *Hannah Coulter* published in September. Poetry collection *Sabbaths 2002* published by Larkspur Press.

2005 Receives O. Henry Prize for short story "The Hurt Man." *Given: Poems* published in May. Inducted into the University of Kentucky Arts and Sciences Hall of Fame. *Blessed Are the Peacemakers: Christ's Teachings about Love, Compassion & Forgiveness*, selections of the gospels compiled and introduced by Berry, published in October. *The Way of Ignorance and Other Essays* published in October. Essay "Not a Vision of our Future, But of Ourselves," about the earth's destruction due to coal mining, included in *Missing Mountains: We Went to the Mountaintop but It Wasn't There*, published by Wind Publications in October. Receives Lifetime Achievement Award from the Conference on Christianity and Literature, December 29.

2006 In August, speaks at "One Thing to Do About Food: A Forum" with Alice Waters, Eric Schlosser, Marion Nestle, Peter Singer, and others. In October, gives the keynote address for the thirtieth anniversary of The Land Institute in Salina, Kansas. Novel *Andy Catlett: Early Travels* published in November.

2007 Visits Ernest J. Gaines in Louisiana. Shoemaker and Charlie Winton acquire both Counterpoint Press and Soft Skull Press; merge both with Shoemaker & Hoard to re-form Counterpoint.

2008 On February 14, at the Kentuckians for the Commonwealth's annual I Love Mountains Day, delivers speech at the Kentucky state capitol in Frankfort calling for nonviolent resistance and civil disobedience to protest mountaintop removal in Kentucky. In May, awarded honorary doctorate by Duke University. *The Mad Farmer Poems* published by Counterpoint in November. Children's book *Whitefoot: A Story from the Center of the World* published in December. Receives the Cynthia Pratt Laughlin Medal from the Garden Club of America.

2009 On January 4, publishes an op-ed in *The New York Times* with Wes Jackson, "A 50-Year Farm Bill," calling for legislation that would address soil loss, pollution, fossil fuels, and the preservation of rural communities. On January 23, releases public statement against the death penalty through the Kentucky Coalition to Abolish the Death Penalty: "As I am made deeply uncomfortable by the taking of a human life before birth, I am also made deeply uncomfortable by the taking of a human life after birth." On March 2, joins protests in Washington, D.C., against mountaintop removal and burning of fossil fuels. On April 3, awarded the Fellowship of Southern Writers' Cleanth Brooks Medal for Excellence in Southern Letters. Essay collection *Bringing It to the Table: On Farming and Food* with an introduction by Michael Pollan, published in August. Poetry collection *Leavings* published in October. In November, joins public letter signed by forty Kentucky writers to Kentucky governor and attorney general asking for a moratorium on the death penalty. On December 20, removes personal papers on loan to the University of Kentucky Archives as a result of the University's too close relationship with the coal industry. (In August 2012, those papers will be donated to the Kentucky Historical Society in Frankfort.)

2010 Receives O. Henry Prize for short story "Stand By Me." Essay collections *Imagination in Place* (including reflections on Wallace Stegner, Gurney Norman, Hayden Carruth, Donald Hall, Jane Kenyon, John Haines, James Still, Gary Snyder, and Kathleen Raine) published in January, and *What Matters? Economics for a Renewed Commonwealth*, with a foreword by Herman E. Daly, published in May.

2011 In mid-February, joins nineteen other activists protesting mountaintop removal in eastern Kentucky by occupying the office of Governor Steve Beshear in Frankfort for a weekend. The governor agrees only to tour coal communities to see the damage. *The Poetry of William Carlos Williams of Rutherford*, a personal tribute to the poet, published in February and is reviewed in *The New York Review of Books*. On March 2, awarded National Humanities Medal by President Barack Obama at the White House. On May 4, speaks at the Future of Food conference at Georgetown University: "the future of food is not distinguishable from the future of the land, which is indistinguishable, in turn, from the future of human care." Daughter Mary Berry establishes The Berry Center in New Castle, Kentucky, to continue the work of Berry and his father and brother toward healthy and sustainable agriculture.

2012 Great-granddaughter Charlcye Anne Johnson born February 7. Receives O. Henry Prize for short story "Nothing Living Lives Alone." *New Collected Poems* published in March; *Christian Science Monitor* hails it as "a welcome complement to the rest of his work." Receives the Steward of God's Creation Award at the National Cathedral in Washington, D.C., April 22. Delivers the 41st Jefferson Lecture, the highest honor the federal government confers for distinguished intellectual achievement in the humanities, at the John F. Kennedy Center for the Performing Arts on April 23: "We cannot know the whole truth, which belongs to God alone, but our task nevertheless is to seek to know what is true. And if we offend gravely enough against what we know to be true, as by failing badly enough to deal affectionately and responsibly with our land and our neighbors, truth will retaliate with ugliness, poverty, and disease." *It All Turns on Affection: The Jefferson Lecture and Other Essays* published in September. *A Place in Time: Twenty Stories of the Port William Membership* published in October. On October 17, receives the James Beard Foundation Leadership Award for efforts toward a healthier, safer, and more sustainable food system, in New York City. On October 30, receives the inaugural Green Cross Award, given by the Bishop of California, for increasing cultural engagement in environmental issues.

2013 In April, elected into the American Academy of Arts and Sciences. The Berry Center and St. Catharine College in Springfield, Kentucky, establish The Berry Farming Program, an undergraduate, multidisciplinary degree inspired by Wes Jackson and Berry's philosophy and work, and designed for students from generational farm families. The program accepts its first students in the fall (St. Catharine College will close in 2016, and the Berry Farming Program will move to partnership with Sterling College in Vermont). *This Day: Sabbath Poems Collected & New 1979–2013* published in October. On October 17, awarded the Freedom Medal from the Roosevelt Institute's Four Freedoms Awards in New York City. Daughter Mary marries Steve Smith, October 26.

2014 *Distant Neighbors: The Selected Letters of Wendell Berry and Gary Snyder*, edited by Chad Wriglesworth, published in June with reviews in the *Los Angeles Review of Books*, *Commonweal*, and other publications. *Terrapin and Other Poems*, with illustrations by Tom Pohrt, published in November. Poetry collection *Roots to the Earth*, with woodcuts by Wesley Bates, published by Larkspur Press.

2015 On January 28, inducted into the Kentucky Writers Hall of Fame, the first living writer to be so honored. *Our Only World: Ten Essays* published in February. Poetry collection *Sabbaths 2013* published by Larkspur Press.

2016 Receives O. Henry Prize for short story "Dismemberment." On March 17, receives the Ivan Sandrof Lifetime Achievement Award from the National Book Critics Circle in New York City, presented by Nick Offerman. *A Small Porch: Sabbath Poems 2014 and 2015* together with *"The Presence of Nature in the Natural World: A Long Conversation"* published in June. On September 25, delivers the Strachan Donnelley Lecture on Conservation and Restoration at the 2016 Prairie Festival in Salina, Kansas, celebrating the fortieth anniversary of The Land Institute and the tenure and retirement of Wes Jackson as president. Brother John M. Berry Jr. dies, October 27. Speaks in conversation with Wes Jackson, and moderated by Mary Berry, at the 36th Annual Schumacher Lectures, October 22. Great-granddaughter Wendy Jean Johnson born, October 29. In December, speaks with Eric Schlosser at the twentieth anniversary of the Center for a Livable Future at Johns Hopkins University.

2017 *The World Ending Fire: The Essential Wendell Berry*, selected by Paul Kingsnorth, published by Allen Lane in the United Kingdom in January. Contributes to *Letters to a Young Farmer: On Food, Farming, and Our Future* by Stone Barns Center for Food and Agriculture, published by Princeton Architectural Press. *The Art of Loading Brush* published by Counterpoint Press in October.

Note on the Texts

This volume—the first of a two-volume set, *What I Stand On: The Collected Essays of Wendell Berry 1969–2017*—presents a selection of Berry's nonfiction from the first half of his career. It contains the whole of the author's long argumentative book *The Unsettling of America* (1977) and thirty-two essays culled from eight other books by Berry published from 1969 to 1990: *The Long-Legged House* (1969), *The Hidden Wound* (1970), *A Continuous Harmony* (1972), *Recollected Essays: 1965–1980* (1981), *The Gift of Good Land* (1981), *Standing by Words* (1983), *Home Economics* (1987), and *What Are People For?* (1990). The contents were chosen by the author and his longtime editor and publisher, Jack Shoemaker, and are arranged in chronological order by first book publication. In accord with Shoemaker and Berry's wishes, and as described below, this volume prints the texts of the essays from the most recent Counterpoint Press printings, which reflect corrections made by the author. The companion volume presents selections from books published from 1993 to 2017.

After the publication of *The Unsettling of America* (1977) by Sierra Club Books, Berry began to look for a new publisher. In 1980 he embarked on his long collaboration with Jack Shoemaker, then a thirty-four-year-old editor who, with William Turnbull, had recently cofounded North Point Press in Berkeley, California. From that point forward most of Berry's books would be edited by Shoemaker. Berry followed his editor first to Pantheon when North Point closed in 1991; then to Counterpoint Press, which Shoemaker cofounded with Frank H. Pearl in Washington, D.C., in 1994; then to Shoemaker & Hoard in 2002, which Shoemaker cofounded with Trish Hoard; and finally back to Counterpoint, which merged with Shoemaker & Hoard when Shoemaker and Charlie Winton purchased both Counterpoint Press and Soft Skull Press in 2007. None of the books from which selections have been made for this volume appeared in British editions.

The Long-Legged House was published by Harcourt, Brace & World on April 9, 1969; the most recent Counterpoint Press paperback printing, the source for the texts printed here, was published in 2012. The first publication of the essays selected from *The Long-Legged House* is given below.

The Rise. *The Rise* (Lexington: The University of Kentucky Library Press, 1968).

The Long-Legged House. *The Long-Legged House* (New York: Harcourt, Brace & World, 1969).
A Native Hill. *The Hudson Review*, Winter 1968–1969.

The Hidden Wound was published by Houghton Mifflin Co. on September 14, 1970; the most recent Counterpoint Press paperback printing, the source for the text of the excerpt printed here (Chapters 4 through 8), was published in 2010. These chapters first appeared in the 1970 Houghton Mifflin edition of *The Hidden Wound*. They were later published, in a slightly different version, as "Nick and Aunt Georgie," in *Recollected Essays: 1965–1980* (1981).

A Continuous Harmony was published by Harcourt Brace Jovanovich on September 20, 1972; the most recent Counterpoint Press paperback printing, the source for the texts printed here, was published in 2012. The first publication of the essays selected from *A Continuous Harmony* is given below.
Think Little. *Whole Earth Catalog*, September 1970.
Discipline and Hope. *A Continuous Harmony* (New York: Harcourt Brace Jovanovich, 1972). (A portion of the essay was delivered at the University of Kentucky as the Distinguished Professor Lecture on November 17, 1971.)
In Defense of Literacy. *A Continuous Harmony* (New York: Harcourt Brace Jovanovich, 1972).

The Unsettling of America: Culture and Agriculture was published by Sierra Club Books in San Francisco on August 27, 1977. Parts of *The Unsettling of America* appeared previously in slightly different form in *The Nation* and *CoEvolution Quarterly*. Sierra Club Books issued a revised second edition in paperback on August 12, 1986, and ten years later, on March 1, 1996, a revised third edition. The text of *The Unsettling of America* used in this volume is that of the most recent Counterpoint Press paperback printing, published September 15, 2015, which includes the author's Preface to the Second Edition (1986) and his Afterword to the Third Edition (1995).

Recollected Essays: 1965–1980 was published by North Point Press in August 1981. While "The Making of a Marginal Farm" was selected from *Recollected Essays: 1965–1980*, the text of the essay used in this volume is taken from *The World-Ending Fire: The Essential Wendell Berry*, published by Counterpoint Press in 2018. "The Making of a Marginal Farm" was first published in *The Smithsonian*, August 1980.

The Gift of Good Land was published by North Point Press on November 30, 1981; the most recent Counterpoint Press paperback printing, the source for the texts printed here, was published in 2012. The first publication of the essays selected from *The Gift of Good Land* is given below.

Horse-Drawn Tools and the Doctrine of Labor Saving. Lane de Moll and Gigi Coe, eds., *Stepping Stones: Appropriate Technology and Beyond* (New York: Schocken Books, 1978).

Solving for Pattern. *The New Farm*, January 1981.

Family Work. Roger B. Yepsen, Jr., ed., *Home Food Systems* (Emmaus, PA: Rodale Press, 1981).

A Few Words for Motherhood. *The New Farm*, May/June 1981.

A Talent for Necessity. *The New Farm*, February 1981.

Seven Amish Farms. *The Gift of Good Land* (San Francisco: North Point Press, 1981).

The Gift of Good Land. *Sierra Club Bulletin*, November/December 1979.

Standing by Words was published by North Point Press on October 30, 1983; the most recent Counterpoint Press paperback printing, the source for the texts printed here, was published in 2011. The first publication of the essays selected from *Standing by Words* is given below.

Standing by Words. *The Hudson Review*, Winter 1980/1981. (An earlier version of this essay was delivered as a speech at the annual meeting of the Lindisfarne Fellows, Summer 1978.)

Poetry and Marriage: The Use of Old Forms. *CoEvolutionary Quarterly*, Winter 1982.

Home Economics was published by North Point Press on June 24, 1987; the most recent Counterpoint Press paperback printing, the source for the texts printed here, was published in 2009. The first publication of the essays selected from *Home Economics* is given below.

Getting Along with Nature. *The Land Report*, Summer 1982.

Two Economies. *Review and Expositor*, May 1984.

The Loss of the University. Gerald Graff and Reginald Gibbons, eds., *Criticism in the University* (Evanston, Ill: Northwestern University Press, 1985).

Preserving Wildness. Speech delivered at the Temenos Conference at Dartington Hall, in November 1986, first published in *Wilderness*, Spring 1987.

A Good Farmer of the Old School. *Draft Horse Journal*, Spring 1986.

What Are People For? was published by North Point Press on April 1, 1990; the most recent Counterpoint Press paperback printing, the source for the texts printed here, was published in 2010. The first publication of the essays selected from *What Are People For?* is given below.

Damage. *The American Poetry Review*, May–June 1975.

Wallace Stegner and the Great Community. *South Dakota Review*, Winter 1985.
Writer and Region. *The Hudson Review*, Spring 1987.
An Argument for Diversity. *The Hudson Review*, Winter 1980.
The Pleasures of Eating. *What Are People For?* (San Francisco: North Point Press, 1990).
The Work of Local Culture. An Iowa Humanities Lecture delivered at Harlan Community High School, Harlan, Iowa, November 13, 1988, and published after the occasion as a pamphlet by the Iowa Humanities Board, 1988.
Why I Am Not Going to Buy a Computer. *New England Review and Bread Loaf Quarterly*, Autumn 1987, reprinted in *Harper's* in May 1987 with the letters and response.
Feminism, the Body, and the Machine. *What Are People For?* (San Francisco: North Point Press, 1990).
Word and Flesh. *Whole Earth Review*, Spring 1990.
Nature as Measure. *What Are People For?* (San Francisco: North Point Press, 1990).

This volume presents the texts of the printings chosen for inclusion here, but it does not attempt to reproduce features of their typographic design. Original citations have been altered to conform to Library of America practices; however, original footnotes that take the form of personal observation or argument have been preserved verbatim. In a few instances quotations have been corrected with the author's approval. The texts are presented without change, except for the correction of typographical errors and minor changes necessitated by the formatting of notes. Spelling, punctuation, and capitalization are often expressive features, and they are not altered, even when inconsistent or irregular. The following is a list of typographical errors corrected, cited by page and line number: 16.12, W.O.A.; 25.16, and end; 26.27, simple; 37.24, And how; 38.20, 'tis; 40.11, now the; 44.20, camp; 45.33, content; 53.12, though; 63.15, camp; 71.20, Ever; 73.39, of Kentucky; 77.21, to more; 100.34, father; 161.2 and 5, Castenada; 210.14, "destiny,"; 213.11, furnance; 291.21, [California}; 314.18, clams.; 380.8, THE LAND GRANT-COLLEGES; 384.2, Allis-Chalmer; 408.12, suceeded; 409.16, *a*; 446.28, WD Allis-Chalmer; 463.15, with.; 475.2, principle; 500.24, farm, It; 538.15, best"'; 548.40, Commissioners; 580.23, occuring; 625.35, specialists) seem; 653.3, principle; 699.28, asked; 704.26, on, a; 718.19, sociey,.

Notes

In the notes below, the reference numbers denote page and line of this volume (the line count includes headings). No note is made for material included in standard desk-reference books. Quotations from Shakespeare are keyed to *The Riverside Shakespeare*, ed. G. Blakemore Evans (Boston: Houghton Mifflin, 1974). Biblical quotations are keyed to the King James Version. Original citations have been altered to conform to Library of America practices; however, original footnotes that take the form of personal observation or argument have been preserved verbatim. For references to other studies and further information than is included in the Chronology, see *Wendell Berry: Life and Work*, ed. Jason Peters (Lexington: The University Press of Kentucky, 2007); and *Conversations with Wendell Berry*, ed. Morris Allen Grubbs (Jackson: University Press of Mississippi, 2007).

From THE LONG-LEGGED HOUSE

8.26 George Caleb Bingham's painting of trappers] American painter George Caleb Bingham (1811–1879) portrayed frontier life along the Missouri and Mississippi Rivers. His *Fur Traders Descending the Missouri* (1845)—originally entitled *French Trader and His Half-Breed Son*—was followed by the similar but often seen as less skillful painting *The Trappers' Return* (1851).

36.29–30 Thoreau's idea of a hypaethral book] "Hypaethral" means roofless, open to the sky. Writing in his *Journal* on June 29, 1851, Henry David Thoreau (1817–1862) characterized *A Week on the Concord and Merrimack Rivers* (1849) as "a hypaethral or unroofed book, lying open under the ether and permeated by it, open to all weathers, not easy to be kept on a shelf."

36.33–34 a long poem . . . titled "Diagon,"] Published in the February 1959 issue of *Poetry*, when Berry was twenty-four.

36.40 "Upon Appleton House, to my Lord *Fairfax*."] A country-house poem written by Marvel in 1651 for his patron Thomas, Lord Fairfax, while the poet was a tutor at Appleton House, the Fairfax Yorkshire estate.

53.40 *A Place on Earth*] Berry's second novel, published in 1967 (revised 1983).

64.30–32 "It requires . . . a day."] Thoreau, "Life Without Principle," published in the *Atlantic Monthly Magazine* (October 1863).

65.17–19 "For an inheritance . . . not be seen."] René Char (1907–1988), *Hypnos Waking: Poetry and Prose by René Char*, selected and translated by Jackson Matthews (1956).

73.21–22 the autobiography of Rev. Jacob Young] One of many itinerant Methodist preachers in the early American republic, Jacob Young (1776–1859) traveled Kentucky's Salt River Circuit. His *Autobiography of a Pioneer: Or, The Nativity, Experience, Travels, and Ministerial Labors of Rev. Jacob Young, with Incidents, Observations, and Reflections* was published in 1857.

86.9–10 the music of the spheres] The idea that the celestial bodies—the sun, the moon, the planets—resonated with harmonious sound dates back to antiquity. According to Aristotle in *On the Heavens*, the Pythagoreans supposed the music of the spheres could not be heard because, having become habituated to it, we could not distinguish it from silence.

92.29 Solomon in all his glory . . . one of these.] Matthew 6:29.

97.38–39 like Valéry's sycamore] An allusion to the poem "Au Platane" [The Plane Tree] by French poet Paul Valéry (1871–1945), published in *Charmes* (1922). The American plane tree is also known as the sycamore.

From THE HIDDEN WOUND

110.25 "Get along home . . . along home!"] "Cindy," a popular Southern folk song, possibly African American in origin, dating back to the nineteenth century.

113.12–13 Witch of Endor] In 1 Samuel 28:3–25, the witch of Endor summons the prophet Samuel from the dead, at King Saul's request.

113.24 the Back to Africa movement.] Originating in the nineteenth century, the Back to Africa movement gained new life and momentum in the 1920s through the efforts of the Jamaican-born social activist Marcus Garvey (1887–1940).

117.9–10 Dr. Bell's Pine Tar Honey] A patented elixir manufactured by E. E. Sutherland Medicine Company of Paducah, Kentucky, from 1894 to 1921. It advertised itself as a remedy for coughs, colds, croup, whooping cough, and lung soreness: "Like the sun's rays through a cloud comes Dr. Bell's Pine-Tar-Honey to the weak and weary cough-worn lungs."

122.33 Joe Louis] World heavyweight boxing champion from 1937 to 1949, nicknamed the "Brown Bomber" (1914–1981).

From A CONTINUOUS HARMONY

136.13 My Lai] My Lai Massacre of approximately 500 Vietnamese civilians by U.S. troops during the Vietnam War, on March 16, 1968.

137.14 Frankfort] Frankfort, Kentucky, home to the state government of the Commonwealth of Kentucky.

138.31–34 "chief way . . . consumers be few. . . ."] See Ezra Pound's translation of *The Great Digest* (1928) in his anthology *Confucius* (1951).

141.3–4 The lotus-eaters] In Homer's *Odyssey*, Book IX, a people living in stupefied indolence.

142.16 who turns up the air conditioner] President Nixon, who liked to read by the fireside in the White House, sometimes ran the air conditioner in summer months while burning a fire in the fireplace.

143.1–2 Louis Bromfield] American novelist and conservationist (1896–1956); Bromfield and his family spent more than a decade living in France.

143.5–9 F. H. King . . . cultivated land."] Franklin Hiram King (1848–1911), American agricultural scientist and inventor of the cylindrical-tower storage silo. See the Introduction to his book *Farmers of Forty Centuries: Or Permanent Agriculture in China, Korea and Japan* (1911), drawn from observations made during his extensive travels to Asia.

144.27–30 "It is the story . . . green things. . . ."] See Chapter 1 of John G. Neihardt's *Black Elk Speaks* (1932), based on the author's conversations with the Oglala Lakota medicine man.

144.31–35 "I saw . . . was holy."] *Black Elk Speaks*, Chapter 3.

149.13–16 "We are fatalists . . . other way."] Edward Dahlberg, "Thoreau and *Walden*," collected in *Do These Bones Live* (1941), later retitled *Can These Bones Live* (1960).

152.27–28 Sir Albert Howard . . . *An Agricultural Testament*] Albert Howard (1873–1947), British botanist and agriculturist. Howard's publications drew on extensive practical research in India, where, working as an agricultural advisor, he came to see the wisdom of traditional Indian farming practices. *An Agricultural Testament* (1940) addresses the proper management of soil fertility. The quotation concerning "the wheel of life" comes from Chapter II.

154.1–2 F. H. King's *Farmers of Forty Centuries*] See note 143.5–9.

155.39 John Collier] Advocate of indigenous rights and cultural pluralism (1884–1968) whose appointment as commissioner for the Bureau of Indian Affairs by Franklin D. Roosevelt in 1933 would alter federal assimilationist policies toward Native Americans. His book *The Indians of the Americas* appeared in 1947.

156.4–5 Chief Rekayi . . . refusing to leave his ancestral home] Chief Rekayi Tangwena (1910–1984) and the people of Kaerezi, on Zimbabwe's eastern border (then Rhodesia), resisted the colonial government's evictions from their ancestral lands in the late 1960s and early 1970s.

157.10–12 "are tied . . . lasting bonds."] Thomas Jefferson, letter to John Jay, August 23, 1785.

157.12–13 ". . . legislators cannot invent . . . subdividing property. . . ."] Thomas Jefferson, letter to James Madison, October 28, 1785.

157.14–16 ". . . it is not . . . a state."] Thomas Jefferson, letter to James Madison, October 28, 1785.

157.18–19 "natural aristocracy . . . government of society,"] Thomas Jefferson, letter to John Adams, October 28, 1813.

157.38–40 ". . . that the producers . . . the consumers temperate. . . ."] See Ezra Pound's translation of *The Great Digest* (1928) in his anthology *Confucius.*

160.16 Paul Radin] American anthropologist and ethnographer (1883–1959).

160.36 the Native American Church] A pan-Native American religion, also known as Peyotism, combining traditional Native American beliefs and Christianity. Meetings involve the ingestion of peyote, a hallucinogen containing mescaline.

161.2–3 *The Teachings of Don Juan: A Yaqui Way of Knowledge*] Published in 1968, the first in a series of popular books by anthropologist Carlos Castaneda (1925–1998) based on the teaching of Yaqui shaman don Juan Matus. Critics of Castaneda's work have questioned whether or not the shaman don Juan existed.

166.33 "the police of meaningless labor."] Thoreau, "Life Without Principle" (1863).

167.31–37 Thoreau in his *Journal* . . . music and poetry."] November 20, 1851.

168.11–12 the Twelve Southerners] The Twelve Southerners, also known as the Southern Agrarians, were a circle of Southern writers and academics formed in the 1920s and united in the cause of preserving an agrarian way of life against the forces of industrialism. Among the group's most prominent members were John Crowe Ransom, Robert Penn Warren, Donald Davidson, and Allen Tate. Each of the Twelve Southerners contributed an essay to *I'll Take My Stand: The South and Agrarian Tradition* (1930), sometimes described as an "agrarian manifesto."

172.23 Sir Albert Howard] See note 152.27–28.

172.27–31 "I think I have told you . . . to see."] John G. Neihardt, *Black Elk Speaks* (1932), Chapter 18. See also note 144.27–30.

172.32–33 "No ideas but in things."] William Carlos Williams (1883–1963), *Paterson*, Book I (1946).

174.23–25 "by the alacrity . . . its way."] Thoreau, "Civil Disobedience" (1849), published initially under the title "Resistance to Civil Government" in Elizabeth Peabody's *Aesthetic Papers.*

176.2–11 The *I Ching* says . . . the good."] See hexagram 43 of the *I Ching*, one of the *Wujing* (Five Classics) of Confucianism. Berry quotes from the English translation by Cary F. Baynes (1950), based on the German translation by Richard Wilhelm.

176.11–12 "if not obtained . . . do not last."] The Confucian *Analects*, 4:5, translation by Ezra Pound (1950), in his anthology *Confucius.*

176.13–15 "Ye have heard . . . not evil . . ."] Matthew 5:38–39.

176.16–17 "As soon as . . . based on."] Ken Kesey in conversation with Berry.

176.17–22 In 1931, Judge Lusk . . . non-existent."] After a jury had found three Communist Party members in Chattanooga guilty of violating the Sedition Statute in March 1931, Lusk granted the defendants a new trial.

176.23 the University of Emily's Run] Emily's Run, a creek in Henry County, Kentucky.

178.19–20 J. Russell Smith's . . . *Tree Crops*] *Tree Crops: A Permanent Agriculture* (1929) describes the traditional use of tree crops and how trees such as the carob, honey locust, mulberry, chestnut, and black walnut can be an important source for food, soil conservation, and sustainable agriculture.

181.16–17 "Take therefore . . . of itself."] Matthew 6:34.

182.4–16 "Everything the Power . . . our children."] *Black Elk Speaks*, Chapter 17.

182.23–25 "the sacred hoop . . . as starlight. . . .] *Black Elk Speaks*, Chapter 3.

183.11–14 "The Six Grandfathers . . . make happy."] *Black Elk Speaks*, Chapter 16.

183.35–37 "medicine seems . . . more disease."] Leon R. Kass, *Toward a More Natural Science: Biology and Human Affairs* (1985), Chapter 1.

184.3–6 "Our father . . . for them."] Paul Radin, *The Road of Life and Death: A Ritual of the American Indian* (1945), Prologue.

186.5–10 Justus von Liebig wrote . . . of Africa."] See Letter XII in *Letters on Modern Agriculture* (1859) by the German agricultural chemist Justus Freiherr von Liebig (1803–1873).

187.6–7 "Art," A. R. Ammons says . . . unconscious event. . . ."] In his long poem "Essay on Poetics" (1970), reprinted in his *Collected Poems: 1951–1971* (1972).

187.18–20 "We have constructed . . . returning curve. . . ."] Thoreau, *Walden* (1854), Chapter 4.

192.3–18 ". . . though I recognize . . . American citizen."] Robert E. Lee, letter to Anne R. Marshall, April 20, 1861.

193.3–4 ". . . all things whatsoever . . . to them. . . ."] Matthew 7:12.

193.12–16 ". . . whosoever shall smite . . . persecute you. . . ."] Matthew 5:39–44.

193.19–20 "An eye . . . a tooth"] Matthew 5:38.

194.28–30 "a trumpet . . . the pitchers."] Judges 7:16.

201.17–18 "neither the day nor the hour. . . ."] Matthew 25:13.

201.25–27 "archer, when he misses . . . in himself."] *The Unwobbling Pivot*, one of the Four Books (*Sìshū*) of Confucianism. Berry quotes from Ezra Pound's translation (1947) in his anthology *Confucius*.

202.3–5 Thomas Merton . . . work well."] Merton, American Trappist monk of the Abbey of Our Lady of Gethsemani, in Nelson County, Kentucky (1915–1968), whose numerous books include *The Seven Storey Mountain* (1948). He had a deep and abiding interest in Shakers and their work. Not far from Gethsemani is the Pleasant Hill Shaker Village in Harrodsburg, Kentucky, which Merton visited at least twice. Merton's comment about Shakers and craftsmanship was made in conversation with Berry.

205.26 "Read not the Times. Read the Eternities,"] H. D. Thoreau, "Life without Principle" (1863).

205.27 "literature is news that STAYS news."] Ezra Pound, *ABC of Reading* (1939), Chapter 2.

205.31–37 Men are . . . of the heart . . .] Edwin Muir, "The Island," first collected in *One Foot in Eden* (1956) and reprinted in *Collected Poems* (1965).

From RECOLLECTED ESSAYS

212.12 a Gravely "walking tractor"] A multipurpose two-wheeled walking tractor, also known as a "walk behind," manufactured by Gravely Tractor, Inc.

212.12–13 an old Farmall A] A compact row-crop tractor with a four-cylinder engine, manufactured by International Harvester from 1939 to 1947. In 1947 Harvester introduced the Farmall Super A, manufactured until 1954.

THE UNSETTLING OF AMERICA

220.1 *For Maurice Telleen*] American farmer, writer, founder of *The Draft Horse Journal*, and one of Berry's close personal friends (1928–2011). See Berry's tribute to Telleen in *It All Turns on Affection* (2012).

223.13 former Secretary of Agriculture Earl L. Butz] Butz (1909–2008) was Secretary of Agriculture from 1971 to 1976 under Richard Nixon and Gerald Ford.

223.29–30 an article in the *Louisville Courier-Journal*] "Surplus of People Called Biggest Farm Problem," the Louisville *Courier-Journal* (July 24, 1967).

229.1–2 Who so hath . . . hath taken.] Montaigne, "Of Coaches," *The Essayes of Michael Lord of Montaigne*, John Florio translation (1603).

230.1–7 *So many goodly citties . . . base conquest.*] Montaigne, "Of Coaches," *The Essayes of Michael Lord of Montaigne*, John Florio translation (1603).

232.13 El Dorado] Spanish: literally "the golden one," referring initially to a (mythical) king of the ancient indigenous Muisca culture of Colombia, and later synonymous with a lost kingdom of gold. The legend of El Dorado led many Spanish conquistadors (and other Europeans) to South America in search of riches.

233.25–40 "affected every aspect . . . the red man."] Bernard DeVoto, *The Course of Empire* (1952).

234.24–29 "More than four-fifths . . . adverse interest."] DeVoto, *The Course of Empire.*

234.31–33 "The interest . . . northern Mexico . . ."] DeVoto, *The Course of Empire.*

235.9 Tenochtitlán] Capital city of the Aztec Empire, located on an island on Lake Texcoco (present-day Mexico City).

236.39–40 our recent secretary of agriculture remarked . . . "Food is a weapon."] Earl L. Butz, see note 221.13. Butz's comment appeared in "What to Do: Costly Choices," *Time* magazine (November 11, 1974): "Food is a weapon. It is now one of the principal tools in our negotiating kit." See also Butz's speech to the Advertising Council, Washington, D.C., June 24, 1974: ". . . food was . . . a major weapon in achieving an honorable peace in Vietnam."

237.1–2 a secretary of defense . . . "palatable" levels of devastation.] James R. Schlesinger (1929–2014), Secretary of Defense from 1973 to 1975 under Richard Nixon and Gerald Ford, advocated making plans for limited nuclear war a part of U.S. military strategy.

237.24–32 Men are made . . . the rose.] Edwin Muir, "The Island," collected in *One Foot in Eden* (1956).

239.2 "Snirt"] Dirty snow.

239.16–17 "a farmer . . . about $320,000"] Roy Reed, "'Paper Rich'—The Farmer's Plaint," the Louisville *Courier-Journal* (February 18, 1976).

240.5–6 "If a man . . . about him. . . ."] Ezra Pound, Canto XIII, *The Cantos.*

242.11 Homestead Act of 1862] Legislative action signed into law by Lincoln that granted land to private citizens (up to 160 acres), encouraging settlement of the West.

243.5–12 Only a man . . . Dynasties pass.] Thomas Hardy, "In a Time of 'The Breaking of Nations,'" first published in the *Saturday Review* (January 1916).

244.1–3 . . . *wanting good government . . . disciplined themselves . . .*] See Ezra Pound's translation of *The Unwobbling Pivot* (1947) in his anthology *Confucius*.

245.3–4 William Rood in the *Los Angeles Times*] William Rood, "Environment Groups Invest in Polluters," the *Los Angeles Times* (July 29, 1975).

245.8 The Sierra Club] National conservation organization founded in 1872 by Scottish American naturalist John Muir (1838–1914).

247.3–4 methods of transportation . . . bridge."] Ivan Illich, *Tools for Conviviality* (1973). Illich (1926–2002) was an Austrian philosopher and Roman Catholic priest who was a fierce critic of modern institutions, including public schooling and medicine.

257.1–2 David Budbill] American poet and playwright (1940–2016).

259.22–25 Sir Albert Howard suggests . . . balance] See Howard's Introduction to *An Agricultural Testament* (1940).

261.34–36 "feeds himself . . . food production"] Richard E. Bell (1934–2015), Assistant Secretary of Agriculture, International Affairs and Commodity Programs from 1975 to 1977, "Meeting World Food Needs," speech to the Fertilizer Institute Annual Meeting, Chicago, February 3, 1976.

261.39–40 a recent assistant secretary of agriculture] Richard E. Bell.

262.2–5 former Secretary of Agriculture Butz could say . . . farm goods,"] In "Agriculture—200 Years After," a speech to the National Council of Farmer Cooperatives Annual Meeting, Washington, D.C., January 15, 1976. See also note 223.13.

263.20–21 one of these speeches . . . by former Assistant Secretary Richard E. Bell] Bell's comments were made at a speech before the Fall Convention of the National Farm Broadcasters Association, Kansas City, Kansas, November 15, 1975.

270.21–22 what Maurice Telleen calls . . . hedonism."] In a conversation with Berry.

271.39–272.3 Dr. John Nicolai . . . needed to be eliminated.] James R. Russell, "Dairy farming in Kentucky registers sharp decline," the Louisville *Courier-Journal* (April 22, 1974).

276.22–24 "that dreary precept . . . after truth."] Howard, *The Soil and Health* (1947), Chapter 6.

276.25–29 "The natural universe . . . group of experts."] *The Soil and Health*, Chapter 6.

276.30–31 "I could not . . . other people."] *The Soil and Health*, Introduction.

276.32–33 "wide chasm . . . in the field."] *The Soil and Health*, Introduction.

276.37–38 "for treating . . . one great subject."] *The Soil and Health*, Introduction.

277.5–6 "Death supersedes . . . dead and decayed."] *The Soil and Health*, Chapter 2.

283.39–284.15 Carl Sauer wrote . . . European nations."] Carl Sauer, *Northern Mists* (1968). An influential American geographer, Sauer (1889–1975) explored in his many writings the relationship between human cultures and physical environments. *Northern Mists* examines the challenges faced by pre-Columbian explorers in North America.

291.19–20 former Agriculture Secretary Clifford Hardin] Hardin (1915–2010) was Secretary of Agriculture from 1969 to 1971 under Richard Nixon.

296.5 Freedom City] Experimental 400-acre commune near Greenville, Mississippi, intended to aid African Americans who had lost their farm jobs as a result of mechanization and the introduction of chemical herbicides. Freedom City foundered in the mid-1970s.

309.4 F. M. Esfandiary] Fereidoun M. Esfandiary (1930–2000), American futurist, transhumanist, and author of several books including *Are You Transhuman?: Monitoring and Stimulating Your Personal Rate of Growth in a Rapidly Changing World* (1989), who would change his name to FM-2030, signifying in part his belief that by 2030 science and technology would overcome death. Esfandiary was vitrified after his death.

309.6–7 an article entitled "Homo Sapiens, the Manna Maker,"] F. M. Esfandiary, "Homo Sapiens, the Manna Maker," *The New York Times* (August 9, 1975).

311.13–14 "All these things . . . worship me."] Matthew 4:9.

312.3 "Energy . . . Eternal Delight."] William Blake, *The Marriage of Heaven and Hell* (1790–93), plate 4.

312.9–10 "Energy is the only life . . ."] *The Marriage of Heaven and Hell*, plate 4.

312.28–29 F. H. King's *Farmers of Forty Centuries*] See note 143.5–9.

312.31–32 Sir Albert Howard . . . the "Wheel of Life."] See note 152.27–28. In *The Soil and Health* (1947), Howard writes: "[Stability in nature] dominates by means of an ever-recurring cycle, a cycle which, repeating itself silently and ceaselessly, ensures the continuation of living matter. This cycle is constituted of the successive and repeated processes of birth, growth, maturity, death, and decay. An eastern religion calls this cycle the Wheel of Life and no better name could be given to it. The revolutions of this Wheel never falter and are perfect. Death supersedes life and life rises again from what is dead and decayed."

320.10–17 According to Lauren Soth . . . Indiana."] Lauren Soth, "Politics and Agribuz: The Operations of Dr. Butz," the *Nation* (October 26, 1974).

323.38 The digging stick] An agricultural tool for digging, a sharpened stick.

327.3–5 Can we . . . from its use"?] Ivan Illich, *Energy and Equity* (1974). See also note 247.3–4.

328.1–3 *But just stop . . . own food.*] James E. Bostic, Jr., "Rural America: Where the Action Is," speech to the Farmers Home Administration State Conference, Raleigh, North Carolina, April 16, 1976.

328.6–7 *Find the shortest . . . the mouth.*] Richard Deats, "A Conversation with Lanza del Vasto," *Fellowship* (September 1975).

330.11–18 The crows . . . for sight.] *King Lear*, IV.vi.13–20.

330.24–25 a Bedlamite] An insane person, literally a resident or ex-patient of Bedlam, as London's Bethlem Royal Hospital came to be known.

331.8–10 You ever-gentle gods . . . you please.] *King Lear*, IV.vi.217–19.

331.15–16 "'Twixt two extremes . . . and grief . . ."] *King Lear*, V.iii.199.

332.19 Lookout Mountain] Located at the border between Tennessee and Georgia, a popular tourist destination. From the rock formations known as "Rock City," it is possible, the tourism industry claims, to see seven states: Tennessee, Kentucky, Virginia, South Carolina, North Carolina, Georgia, and Alabama.

333.22–25 Out of . . . speechless caravan.] Hart Crane, "Proem: To Brooklyn Bridge," lines 17–18, *The Bridge* (1930).

335.22–23 "Man has no Body distinct from his Soul . . ."] Blake, *The Marriage of Heaven and Hell* (1790–93), plate 4.

342.20–30 I went by the field . . . armed man.] Proverbs 24:30–34.

356.26–30 Maybe the bride-bed . . . a single light . . .] W. B. Yeats, "Solomon and the Witch," collected in *Michael Robartes and the Dancer* (1921).

358.36–37 "revel and rest softly, side by side,"] *The Odyssey*, Book V, line 236, translation by Robert Fitzgerald (1961). All quotations of Homer are from the Fitzgerald translation, and citations are keyed to the 1998 edition.

359.4–11 If you could . . . grace and form?] *The Odyssey*, Book V, lines 215–222.

359.13–17 My quiet Penélopê . . . for home . . .] *The Odyssey*, Book V, lines 225–229.

361.20–24 grew like a pillar . . . a bedpost . . .] *The Odyssey*, Book XXIII, lines 217–223.

361.27–32 Now from his breast . . . went down . . .] *The Odyssey*, Book XXIII, lines 259–264.

362.23–24 "The government of the state . . . family order."] See Ezra Pound's translation of *The Great Digest* (1928) in his anthology *Confucius*.

363.6–10 Odysseus found . . . the brambles . . .] *The Odyssey*, Book XXIV, lines 251–255.

363.31–32 "beat their swords . . . learn war any more,"] Isaiah 2:4.

363.38 "in blessed peace."] *The Odyssey*, Book XI, line 151.

365.17–19 To learn to preserve . . . study the forest.] See the Introduction to Albert Howard's *An Agricultural Testament*: "The main characteristic of Nature's farming can therefore be summed up in a few words. Mother earth never attempts to farm without live stock; she always raises mixed crops; great pains are taken to preserve the soil and to prevent erosion; the mixed vegetable and animal wastes are converted into humus; there is no waste; the processes of growth and the processes of decay balance one another; ample provision is made to maintain large reserves of fertility; the greatest care is taken to store the rainfall; both plants and animals are left to protect themselves against disease."

367.8–10 "so that perhaps acres . . . that danger. . . ."] See "The Danger," by American poet and novelist Millen Brand (1906–1980), collected in *Local Lives* (1975), a commonplace history of the Pennsylvania Dutch.

367.37–368.1 a recent *National Geographic* article] Sabrina Michaud and Roland Michaud, "Trek to Lofty Hunza—and Beyond," *National Geographic* (November 1975).

375.20–27 ". . . the ploughman . . . oxen's work . . ."] See *The Horse in the Furrow* (1960), Chapter 2, by George Ewart Evans (1909–1988), an Anglo Welsh oral historian.

377.7–10 "that the mass . . . grace of God."] Thomas Jefferson, letter to Roger C. Weightman, June 24, 1826.

377.17–21 "Cultivators of the earth . . . lasting bonds."] Thomas Jefferson, letter to John Jay, August 23, 1785.

377.25–27 "I consider . . . generally overturned."] Thomas Jefferson, letter to John Jay, August 23, 1785.

378.2–7 ". . . I do most anxiously wish . . . distrustful superintendence."] Thomas Jefferson, letter to Mann Page, August 30, 1795.

378.10–16 ". . . the long succession . . . the Atlantic . . ."] Thomas Jefferson, letter to James Madison, February 17, 1826.

378.17 JUSTIN MORRILL] Justin Smith Morrill (1810–1898), a representative and senator from Vermont, best known for his sponsorship of the Land-Grant College Act of 1862 and the Agricultural College Act of 1890.

378.20–23 "an amount . . . in Congress . . ."] Morrill Act, U.S.C., Title 7, Section 301.

378.24–29 "to the endowment . . . professions in life."] Morrill Act, U.S.C., Title 7, Section 304.

378.32–34 "a sound . . . prosperity and security."] Hatch Act, U.S.C., Title 7, Section 361b.

378.35–38 "It is also . . . the economy."] Hatch Act, U.S.C., Title 7, Section 361b.

379.2–9 "It shall be . . . rural life . . ."] Hatch Act, U.S.C., Title 7, Section 361b.

379.11–14 "In order to aid . . . the same . . ."] Smith-Lever Act, U.S.C., Title 7, Section 341.

379.19–20 Morrill's statement . . . written "apparently in 1874."] William Belmont Parker, *The Life and Public Services of Justin Smith Morrill* (1924). Morrill's autobiographical memorandum, found among his papers after his death, enumerates five reasons for writing the Land-Grant College Act.

380.19–20 "natural aristocracy" of "virtue and talents"] Thomas Jefferson, letter to John Adams, October 28, 1813: "For I agree with you that there is a natural aristocracy among men. The grounds of this are virtue and talents."

380.34 "degrees of genius."] Thomas Jefferson, letter to John Page, August 30, 1795: "I do most anxiously wish to see the highest degrees of education given to the higher degrees of genius, and to all degrees of it, so much as may enable them to read & understand what is going on in the world, and to keep their part of it going on right: for nothing can keep it right but their own vigilant and distrustful superintendence."

383.34–35 Jim Hightower and Susan DeMarco give . . . the central argument] Jim Hightower and Susan DeMarco of the Agribusiness Accountability Project, statements before the Senate Subcommittee on Migratory Labor, June 19, 1972. Berry's source is Jim Hightower, *Hard Tomatoes, Hard Times: The Failure of the Land Grant College Complex* (1972).

384.40–385.1 An article in the *Louisville Courier-Journal*] Phil Norman, "UK Agriculture College touches many lives," the Louisville *Courier-Journal* (October 10, 1976).

385.40–386.3 "to assume . . . better citizenship."] House Committee on Agriculture, U.S. Congress, "Cooperative Agricultural Extension Work," *Report No. 110*, 63rd Congress, 2nd Session (1913).

386.22–23 "corporate control . . . agriculture production"] Sarah Shaver Hughes, "Agricultural Surpluses and American Foreign Policy 1952–60," an unpublished master's thesis, University of Wisconsin (1964). Quoted in Darryl McLeod, "Urban-Rural Food Alliances: A Perspective on Recent Community Food Organizing," *Radical Agriculture*, edited by Richard Merrill (1976).

386.24–25 then Secretary of Agriculture Ezra Taft Benson] Ezra Taft Benson (1899–1994) served as Secretary of Agriculture from 1953 to 1961, under Dwight D. Eisenhower.

391.27 Gresham's Law] Economic principle that "bad money drives out good."

393.25 a textbook of rhetoric] W. Ross Winterowd, *The Contemporary Writer* (1975).

396.32 F. H. King] See note 143.5–9.

401.27–30 "mechanized agriculture . . . systems are."] Sterling Wortman, "Food and Agriculture," *Scientific American* (September 1976).

407.6–7 Professor Philip M. Raup . . . testified] Philip M. Raup, "Needed Research into the Effects of Large Scale Farm and Business Firms on Rural America," testimony before the Subcommittee on Monopoly of the United States Senate Small Business Committee, March 1, 1972.

411.28–29 an unpublished paper by Professor Stephen B. Brush] "Andean Culture and Agriculture: Perspectives on Development," presented at the International Hill Land Symposium, West Virginia University, Morgantown, West Virginia, October 5, 1976.

416.15–16 "the regional specialization . . . individual specialization."] Maurice Telleen, personal correspondence with Berry.

424.19 ladino] A variety of white clover.

424.20 Korean lespedeza] A very good annual clover for pasture and hay.

424.24 harrow] An agricultural instrument consisting of spikes, teeth, or discs, used specifically for the purpose of breaking up and smoothing the surface of the soil.

425.29–426.2 "In Kentucky . . . horses or mules."] Thomas P. Cooper, comments in University of Kentucky College of Agriculture Circular No. 306 (November 1937).

427.4 Percheron horses] A breed of tall heavy draft horses, usually gray or black.

430.27–28 "Except a corn . . . much fruit."] John 12:24.

432.2 chisel plow] A plow that allows reduced tillage, loosening but not inverting soil.

432.37 Charolais cows] A breed of large white beef cattle, developed in France.

434.16 bu. per acre] Bushels per acre.

435.29–32 a report published in 1975 and entitled *A Comparison . . . Pesticides.*] William Lockeretz, Robert Klepper, Barry Commoner, Michael Gertler, Sarah Fast, Daniel O'Leary, and Roger Blobaum, "A Comparison of the Production, Economic Returns, and Energy Intensiveness of Corn Belt Farms That Do and Do Not Use Inorganic Fertilizers and Pesticides," Center for the Biology of Natural Systems at Washington University (1975).

435.32–33 Barry Commoner] American cellular biologist and pioneer in the modern environmental movement (1917–2012). Berry quotes from Chapter 7 of Commoner's *The Poverty of Power: Energy and the Economic Crisis* (1976).

437.25 BTU] British thermal unit, the amount of heat necessary to raise the temperature of one pound of water by one degree Fahrenheit.

438.5 vetch] Any of several legumes of the genus *Vicia*, grown for fodder or green manure.

439.34 in a speech by . . . Butz himself.] Earl L. Butz, speech to the American Chemical Society Meeting, New York, April 8, 1976. Subsequent quotations of Butz refer to this speech.

440.6–7 As the *Draft Horse Journal* noted in an editorial] *Draft Horse Journal* (Winter 1976).

441.9–10 Morrison's *Feeds and Feeding*] First published in 1898, a guide to the feeding, care, and management of livestock that underwent numerous revisions and editions through the mid-twentieth century.

446.5 H Farmall tractors] H Farmall tractor, a row-crop tractor manufactured by International Harvester from 1939 to 1953.

446.27–28 a 1946 WC Allis-Chalmers] A nimble row-crop tractor manufactured by Allis-Chalmers from 1948 to 1953.

446.30–31 Percheron stud] See note 427.4.

448.7 as Maurice Telleen points out] In a conversation with Berry. See note 220.1.

450.29 "among God . . . community."] "Agricultural Alternatives," *CBNS Notes*, Center for the Biology of Natural Systems, Washington University (March–April 1972).

451.29–33 "technological innovation . . . as a whole."] "Agricultural Alternatives," *CBNS Notes.*

452.12–13 "take their own . . . other people."] Albert Howard, *The Soil and Health: A Study of Organic Agriculture* (1947), Introduction.

453.27–28 a letter to the editor of the *Draft Horse Journal*] *Draft Horse Journal* (Autumn 1976).

457.5–6 the New Alchemy Institute, the Farallones Institute, Rodale Press] The New Alchemy Institute (1969–91), located on a former dairy farm on Cape Cod, was a research center focused on decreasing human dependence on fossil fuels through ecological approaches such as organic agriculture, aquaculture, and bioshelter; the Farallones Institute, in Berkeley, California, founded in 1972, was a community of Northern California scientists, designers, and horticulturalists dedicated to the development of ecologically integrated living design; Rodale Press, in Emmaus, Pennsylvania, founded in 1930, was a family-owned independent publisher of health and wellness books and magazines, until its sale in 2018 to Hearst Communications.

463.15–18 "about 1.5 million chickens . . . 5 million eggs."] "Michigan Episode," *The Shepherd* (July 1976).

463.21–22 "may have been . . . the wrong lever."] "Michigan Episode," *The Shepherd*.

468.38 Marty Strange, Gene Logsdon, and Wes Jackson] Marty Strange (b. 1947), environmentalist, author of *Family Farming: A New Economic Vision* (1988), and cofounder of the Center for Rural Affairs in Nebraska, championing the sustainable agriculture movement; Gene Logsdon (1931–2016), American farmer and author of numerous farm-related books including *The Contrary Farmer* (1994) and *Letter to a Young Farmer* (2017), with a foreword by Wendell Berry; Wes Jackson (b. 1936), American plant geneticist and advocate for sustainable practices, author of several books including *New Roots for Agriculture* (1980) and *Becoming Native to This Place* (1994), and cofounder of The Land Institute in Salina, Kansas, dedicated to the development of perennial crops.

468.39–469.1 the Land Institute, the Center for Rural Affairs, the Land Stewardship Project, Tilth, and the E. F. Schumacher Society] The Land Institute, see note 468.38; the Center for Rural Affairs, see note 468.38; the Land Stewardship Project, a Minnesota-based organization dedicated to promoting an ethic of farmland stewardship; Tilth, the Tilth Alliance, a sustainable agricultural movement and organization in the Pacific Northwest inspired by a speech made by Berry at the "Agriculture for a Small Planet" symposium in Spokane, Washington, July 1, 1974; E. F. Schumacher Society, now the Schumacher Center for a New Economics, an organization based in Great Barrington, Massachusetts, working to build a just and sustainable global economy.

From THE GIFT OF GOOD LAND

473.6 a Gravely walking tractor] See note 212.12.

473.28 breaking plow . . . cultivator . . . grain drill.] A ground-breaking plow with a moldboard; cultivator, a secondary tillage plow often used for weeding; grain drill, an implement for planting small grains.

473.36 hay conditioners and chisel plows] Hay conditioner, tool that crimps newly cut hay to promote faster drying; chisel plow, see note 432.2.

475.17–18 Harry Groom, as quoted in . . . George Ewart Evans's *The Horse in the Furrow*] *The Horse in the Furrow* (1960), Chapter 3. See also note 375.20–27.

479.28 Earl Butz] See note 223.13.

483.15 John Todd] A Canadian ecological designer (b. 1939) whose design innovations address challenges of food production, waste treatment, and environmental repair through ecosystem technologies that emulate nature's patterns and strategies. Todd was one of the founders of Cape Cod's New Alchemy Institute (1969–91), a research center dedicated to ecological design solutions.

483.20 *The New Farm*] A magazine devoted to organic farming, published by the Rodale Press from 1979 to 1994. Berry served as a contributing editor to Rodale in the late 1970s.

486.40–487.1 an insight from . . . Timothy Taylor] Timothy H. Taylor (1918–2010), a faculty member in the College of Agriculture at University of Kentucky whose expertise was forage production and grassland ecology, and a friend of Wendell Berry.

488.7–8 a good farm . . . "manures itself."] Albert Howard, *An Agricultural Testament* (1940), Introduction.

495.24–25 Thoreau . . . the first to assert that people should not belong to farm animals] In the first chapter of *Walden* (1854), Thoreau writes: "I see young men, my townsmen, whose misfortune it is to have inherited farms, houses, barns, cattle, and farming tools; for these are more easily acquired than got rid of. Better if they had been born in the open pasture and sucked by a wolf, that they might have seen with clearer eyes field they were called to labor in. Who made them serfs of the soil? Why should they eat their sixty acres, when man is condemned to eat only his peck of dirt? Why should they begin digging their graves as soon as they are born?"

498.2 the Southdown ram] Southdown sheep, a hornless breed of sheep prized for its fleece and meat and used for the improvement of other breeds.

498.3 Henry Besuden] Henry Carlisle Besuden (1904–1985), sheep breeder, conservationist, and the writer of the column "Sheep Sense" for *The Sheepman* magazine, 1945–47.

498.35–36 *The Farm Quarterly*] Published from 1946 to 1972, a large-format popular farm magazine that included photographs and human-interest stories.

499.37 Kentucky fescue 31] A low-maintenance fescue that tolerates heat and drought.

499.39 Korean lespedeza] See note 424.20.

507.36 Belgians or Percherons] Belgian, a breed of heavy draft horse; Percheron, see note 427.4.

508.37–38 all nubbins] Nubbin, a small and stunted ear of corn.

509.22 silage] Fermented fodder, chopped and stored, for cattle, sheep, and other cud-chewing farm animals.

510.33 Model D John Deere tractor] The first tractor built by John Deere, manufactured from 1923 until 1953.

512.9 a farrier] A craftsman who specializes in shoeing horses.

513.27 John A. Hostetler] American writer and scholar of Amish and Hutterite societies (1918–2001). His *Amish Society* (1963) saw an expanded fourth edition in 1993.

519.19–21 what Arthur O. Lovejoy . . . Biblical thought.] Arthur Oncken Lovejoy (1873–1962), American philosopher and historian of ideas. In *The Great Chain of Being: A Study of the History of an Idea* (1936, 1963), Lovejoy writes: "through the Middle Ages there were at least kept alive, in an age of which the official doctrine was predominantly otherworldly, certain roots of an essentially 'this-worldly' philosophy: the assumption that there is a true and intrinsic multiplicity in the divine nature, that is to say, in the world of Ideas . . . that the world of temporal and sensible experience is thus good, and the supreme manifestation of the divine."

519.29–30 Lynn White, Jr. . . . "The Historical Roots of Our Ecologic Crisis"] Delivered as an address at the annual meeting of the American Association for the Advancement of Science in Washington, D.C., in December 1966, prior to its publication in *Science* (March 10, 1967).

519.34 Genesis 1:28] "And God blessed them, and God said unto them, Be fruitful, and multiply, and replenish the earth, and subdue it: and have dominion over the fish of the sea, and over the fowl of the air, and over every living thing that moveth upon the earth."

522.17 "year of jubilee,"] Leviticus 25:10.

522.22–23 "my power . . . mine hand"] Deuteronomy 8:17.

524.25 "the fowls of the air" and "the lilies of the field."] Matthew 6:26, 6:28.

526.25–31 "otherworldly philosophy" . . . Nature."] Lovejoy, *The Great Chain of Being*, Chapter II.

528.20–21 As C. S. Lewis . . . material things] In *Mere Christianity* (1952), Book II, Chapter 5, Lewis writes: "There is no good trying to be more spiritual than God. God never meant man to be a purely spiritual creature. That is why He uses material things like bread and wine to put the new life into us. We may think this rather crude and unspiritual. God does not: He invented eating. He likes matter. He invented it."

From STANDING BY WORDS

538.26–27 Against Mr. Winterowd's . . . a definition by Gary Snyder] See Gary Snyder's "Poetry, Community, & Climax," *Field 20* (Spring 1979). Snyder (b. 1930), American poet of the Pacific Rim, essayist, and environmental activist, whose many works include the 1974 Pulitzer Prize–winning poetry collection *Turtle Island*. Berry and Snyder have been friends and correspondents since the 1970s. *Distant Neighbors* (2015), edited by Chad Wriglesworth, is a selection of their correspondence.

540.34–37 For if any . . . he was.] James 1:23–24.

541.3 "Honesty is the treasure of states."] See Ezra Pound's translation of *The Great Digest* (1928) in his anthology *Confucius*.

541.4–7 Pound's observation . . . exaggerations of dogma."] See Pound's translator's note ("Procedure") in his translation of the Confucian *Analects* (1950) in his anthology *Confucius*.

541.8–10 "Where shall . . . a standard man?"] Thoreau, *A Week on the Concord and Merrimack Rivers* (1849), the "Sunday" chapter.

541.26–27 "possibilities . . . inward and outward."] Gary Snyder, "Poetry, Community, & Climax," *Field 20* (Spring 1979).

541.33–34 "I talk . . . no language."] T. S. Eliot, *The Family Reunion*, Part I, Scene I, a verse play, first performed in 1939.

542.25–26 "create the object . . . they contemplate."] T. S. Eliot, "John Dryden" (1921), reprinted in *Selected Essays of T. S. Eliot* (1950).

542.30–33 Shelley's first wife . . . a poet.] *The Norton Anthology of English Literature* (1962), Volume 2, edited by M. H. Abrams et al.

544.24–25 "Men must endure . . . coming hither."] *King Lear*, V.ii.9–10.

544.36 Out of the world he must, who once comes in . . .] Robert Herrick, "None Free from Fault," collected in *Hesperides* (1648).

545.3–5 Old as I am . . . my Wit.] The opening lines of John Dryden's "Cymon and Iphigenia, from Boccace," collected in *Fables Ancient and Modern* (1700).

546.4–7 To be free . . . magnanimous and brave.] John Milton, *The Second Defense of the People of England*, published originally in Latin in 1654, in which the English poet offers legal justification for Parliament's execution of King Charles. It follows and continues Milton's first defense, published three years earlier.

546.25–26 technological romanticism of Buckminster Fuller] Richard Buckminster Fuller (1895–1983), American inventor, futurist, and self-described "design scientist," believed that through technological innovation humans would one day transform the planet, feeding and housing themselves in workless leisure.

547.7–9 transcribed conversations . . . of the Nuclear Regulatory Commission . . . Three Mile Island] Berry, in an original note, acknowledges his indebtedness, for a sampling and commentary on the Nuclear Regulatory Commission transcripts, to Paul Trachtman, "Phenomena, comment, and notes," *Smithsonian* (July 1979). Through mechanical failures and human errors, the Three Mile Island nuclear power station, near Harrisburg, Pennsylvania, had a partial meltdown of its nuclear core on March 28, 1979, releasing radioactive gases into the atmosphere.

549.5–13 And they . . . another's speech.] Genesis 11:4–6.

552.2–4 In an article . . . transformation of American agriculture] G. W. Salisbury and R. G. Hart, "The Evolution and Future of American Animal Culture," *Perspective in Biology and Medicine* (Spring 1979).

559.34–35 in Gary Snyder's phrase, "at one with each other."] Gary Snyder, "Poetry, Community, & Climax," *Field 20* (Spring 1979).

561.6–24 Conceptualizing realistically . . . coconut milk.] See Buckminster Fuller's comments in the anthology *Space Colonies* (1977), edited by Stewart Brand, which gathers in book form the debate on space colonization that took place in the 1970s in the pages of *CoEvolution Quarterly*.

562.30–40 There wanted . . . God Supreme . . .] Milton, *Paradise Lost*, VII.505–515.

563.19–33 First the humans . . . the Moon.] Buckminster Fuller, comments in *Space Colonies*.

563.37–564.9 He created . . . for bread.] William Faulkner, "The Bear," *Go Down, Moses* (1942).

565.27–30 A mind . . . still the same . . .] Milton, *Paradise Lost*, I.253–256.

567.19–32 "academy of projectors . . . miserably waste. . . .] Jonathan Swift, *Gulliver's Travels* (1726), Part III, Chapter IV. Among the many impractical experiments conducted by Lagado's Academy of Projectors are a project to extract sunshine from cucumbers, an operation to reduce human waste to its original food, and another to turn ice into gunpowder.

568.1–4 But past a certain scale, as C. S. Lewis wrote . . . *all* others.] See Lewis's *Abolition of Man* (1943): "Man's conquest of Nature, if the dreams of some scientific planners are realized, means the rule of a few hundreds of men over billions upon billions of men. . . . Each new power won *by* man is a power *over* man as well. . . . For the power of Man to make himself what he pleases means, as we have seen, the power of some men to make other men what *they* please."

569.18–22 the Taoist village-as-globe . . . roosters crow.] *Tao Te Ching*, section 80.

569.35–40 "Alert as a winter-farer . . . a valley . . ."] *Tao Te Ching*, section 15, Witter Bynner translation (1944).

573.34–37 through infusion sweete . . . rather meete.] Edmund Spenser, *The Faerie Queen*, IV, canto ii, stanza 34.

575.2–4 consigning all . . . "to cold oblivion," as Shelley wrongly believed marriage was supposed to do.] Percy Bysshe Shelley, "Epipsychidion" (1821), lines 149–154.

577.9–13 I simply want . . . Claverton . . .] T. S. Eliot, *The Elder Statesman*, Act II, a verse play, first performed in 1958.

577.17–21 to go abroad . . . both ways.] Eliot, *The Elder Statesman*, Act II.

578.7–9 the Confucian principles, dear to Pound . . . stood by).] See Pound's glossary to his translation of *The Great Digest* (1928) in his anthology *Confucius.*

581.36–38 The negro . . . string-piece . . .] Walt Whitman, "Song of Myself," section 13, first published as one of twelve untitled poems in the 1855 edition of *Leaves of Grass.*

582.16–17 Seasons pursuing . . . the ground . . .] Whitman, "Song of Myself," section 15.

From HOME ECONOMICS

588.21–28 as the poet Edmund Spenser put it . . . brother unto brother."] *The Faerie Queen*, VII, canto vii, stanza 14.

591.26 "In wildness . . . the world,"] Thoreau, "Walking," delivered at the Concord Lyceum on April 23, 1851, and printed in the June 1862 issue of the *Atlantic Monthly* shortly after the author's death.

591.34–35 Wes Jackson of the Land Institute] See note 468.38.

596.14–16 "The proper scale," a friend wrote to me . . . life and health."] Perhaps a reference to a postcard from Jane Kenyon, according to Berry.

601.16–22 The ancient Greeks . . . inevitable punishment."] Aubrey de Sélincourt, *The World of Herodotus* (1982).

601.30–31 "Mine own hand hath saved me."] Judges 7:2.

602.32–36 "Religion in every age . . . his wits."] De Sélincourt, *The World of Herodotus.*

603.6–10 "Therefore take . . . unto you."] Matthew 6:31, 6:33.

604.2–3 "take no thought for the morrow."] Matthew 6:34.

604.28–30 the aim of "Buddhist economics . . . minimum of consumption,"] E. F. Schumacher, *Small Is Beautiful* (1973), Part I, Chapter 4.

608.26–609.4 Henry Besuden . . . depth and quality."] From Henry Besuden's speech to the International Stockmen's School, San Antonio, Texas, January 2–6, 1983.

610.2–4 "my engineer's mind . . . practical and possible,"] Gordon Millar, "Agriculture: Is Small Really Beautiful?," *World Research INK* (January 1978).

612.5–6 "All flesh . . . unto dust."] Job 34:15.

612.25–26 "Our bread," Guy Davenport . . . our movies."] Personal remark made by Davenport to Berry. Guy Mattison Davenport (1927–2005), American fiction writer, essayist, translator, teacher, and longtime resident of Lexington, Kentucky. Davenport began teaching at the University of Kentucky in 1964, the same year Berry joined the faculty as a creative writing instructor.

613.35 "Satanic wheels"] William Blake, *Jerusalem* (1804–20), plate 12, line 44, and plate 13, line 37.

613.35–36 "Satanic mills"] Blake, *Milton* (1810), Preface.

613.36–37 "wheel without wheel . . . each other."] *Jerusalem*, plate 15, lines 18–19.

614.4–5 "Wheel within wheel . . . harmony and peace."] *Jerusalem*, plate 15, line 20.

614.6 "wheel in the middle of a wheel"] Ezekiel 1:16.

614.30 "only with spring," as e. e. cummings said] In his poem "O sweet spontaneous," first published in *The Dial* in 1920 and collected in *Tulips and Chimneys* (1923).

617.5–6 "He who would . . . Minute Particulars."] *Jerusalem*, plate 55, line 60.

617.9–11 Labour well . . . teeming Earth.] *Jerusalem*, plate 55, lines 51–53.

617.31–36 When people . . . as loyal.] *Tao Te Ching*, section 18, Witter Bynner translation (1944).

617.37–39 warning . . . sixth chapter of Matthew.] Matthew 6:1: "Take heed that ye do not your alms before men, to be seen of them: otherwise ye have no reward of your Father which is in heaven."

619.28–34 "Skill in making . . . concern of his."] Eric Gill, "What Is Art?," *A Holy Tradition of Working: Passages from the Writings of Eric Gill* (1983), edited by Brian Keeble. Gill (1882–1940), an English sculptor and type designer.

624.18–23 Dr. Johnson . . . all the Branches."] W. Jackson Bate, *Samuel Johnson* (1975), Chapter 4.

624.30–35 "Modern knowledge . . . in heaven."] H. J. Massingham, *The Tree of Life* (1943).

627.12–14 "the head . . . toward jobs."] "Texas School Board Chief Wants Sixth-Graders to Pick Job 'Tracks,'" Louisville *Courier-Journal* (August 13, 1983).

632.32–633.6 Judge Jackson Kiser . . . that way."] James J. Kilpatrick, "Plan to Teach the Bible as Literature May Wind up in the Supreme Court," Louisville *Courier-Journal* (September 15, 1983).

634.24–28 Coleridge's statement . . . poetic faith."] Samuel Taylor Coleridge, *Biographia Literaria* (1817), Chapter 14.

635.3–4 A spring of love . . . them unaware . . .] Coleridge, "The Rime of the Ancient Mariner," Part IV, stanza 14, published in *Lyrical Ballads* (1798).

635.10–11 "one need . . . an abyss."] Harry Mason, in a letter to Berry.

638.26–27 According to a recent press release] Larry Green, "Family Farm Threatened as Technology Raises Productivity: Science Reshaping Agriculture's Future," *Los Angeles Times* (July 14, 1985).

642.26–27 "To know . . . whom to trust."] The opening lines of Pound's Canto LXXXV, *The Cantos.*

642.30–36 the trees rise . . . are voices . . .] Pound, Canto XC, *The Cantos.*

645.37–38 pointing out, as the Reagan administration has done] When James Watt (b. 1938) was appointed Secretary of the Interior in 1981 by Ronald Reagan, he pledged: "We will mine more, drill more, cut more timber to use our resources rather than simply keep them locked up."

646.8–10 I would agree with Edward Abbey that we need . . . "absolute wilderness,"] Edward Abbey, *Desert Solitaire* (1968), chapter entitled "Polemic: Industrial Tourism and the National Parks."

652.2–3 Maury Telleen] See note 220.1.

656.20–21 10-10-20] A fertilizer grade. The numbers refer to the primary nutrients needed by plant life: nitrogen, phosphorus, and potash. 10-10-20 fertilizer contains 10 percent nitrogen, 10 percent phosphorus, and 20 percent potash.

659.18 alsike clover] Typically used for hay, pasture, and soil improvement. Alsike produces white or pale pink flowers.

From WHAT ARE PEOPLE FOR?

665.13–14 "You never . . . enough."] One of the "Proverbs of Hell" in Blake's *The Marriage of Heaven and Hell* (1790–93), plate 7.

665.25 "No bird . . . own wings."] *The Marriage of Heaven and Hell*, plate 7. Another of the "Proverbs of Hell."

667.1 *Wallace Stegner*] Stegner (1909–1993) was an American novelist, short story writer, historian, environmentalist, and Berry's friend as well as his teacher. He won the Pulitzer Prize for his novel *Angle of Repose* (1971) and the National Book Award for the novel that followed, *The Spectator Bird* (1977).

669.11–12 the Harvard Five-Foot Shelf.] In 1909 P. F. Collier published its highly successful fifty-volume library of the world's classics, edited by retired Harvard University president Charles W. Eliot, who promoted the idea that a five-foot shelf "would hold books enough to afford a good substitute for a liberal education to anyone who would read them with devotion, even if he could spare but fifteen minutes a day for reading."

676.5 Webster edition, with E. W. Kemble's illustrations.] Samuel Clemens, who wrote under the pen name Mark Twain, started and personally financed the publishing firm Charles L. Webster & Co. *Adventures of Huckleberry Finn* (1884), illustrated by E. W. Kemble, was one of the firm's first publications. Despite notable successes, the firm went bankrupt in 1894, after only ten years of operation.

677.35–39 When I wrote . . . my hands only.] The opening sentence of Thoreau's *Walden* (1854).

678.5–7 I loaf . . . summer grass.] Walt Whitman, "Song of Myself," section 1.

678.34–35 like Eliot's Chinese jar] In "Burnt Norton," the first of his *Four Quartets* (1943), Eliot asserts, "Only by the form, the pattern, / Can words or music reach / The stillness, as a Chinese jar still / Moves perpetually in its stillness."

681.37–38 The fulfillment . . . Aristotle described] In his *Poetics*, Aristotle says the purpose of dramatic tragedy is to bring about catharsis in theatergoers, to arouse sensations such as pity and fear, thereby cleansing them of these passions.

685.22–23 according . . . *Times Book Review*] Mark K. Stengel, "Modernism on the Mississippi: The Southern Review 1935–85," *The New York Times Book Review* (November 24, 1985).

686.22–23 As William Matthews writes in a recent article] William Matthews, "Dull Subjects," *New England Review and Bread Loaf Quarterly* (Winter 1985).

691.2–3 *Elegant solutions . . . uniqueness of place.*] John Todd, "Tomorrow Is Our Permanent Address," published in *The Book of the New Alchemists*, edited by Nancy Jack Todd (1977). See note 483.15.

698.5 "Comus,"] A poem by English poet John Milton, whose actual title is "A Mask Presented at Ludlow Castle," performed on Michaelmas, 1634, before John Egerton, Earl of Bridgewater, at Ludlow Castle and published anonymously in 1637.

698.9–11 "that nothing . . . to show."] Ananda K. Coomaraswamy, "Why Exhibit Works of Art?," collected in his book of essays *Christian and Oriental Philosophy of Art* (1956).

705.17–18 billionth hamburger] On November 20, 1984, in a highly publicized event, McDonald's served its 50 billionth hamburger.

707.3–4 "the whole problem . . . one great subject."] Albert Howard, *The Soil and Health: A Study of Organic Agriculture*, Introduction, originally published under the title *Farming and Gardening for Health or Disease* (1945) before its 1947 reissue. See note 152.27–28.

709.21–27 There is nothing . . . intact.] William Carlos Williams, "The Host," collected in *The Desert Music and Other Poems* (1954).

717.7–8 in the spirit of Psalm 128] Psalms 128:3–6: "Thy wife shall be as a fruitful vine by the sides of thine house: thy children like olive plants round about thy table. . . . and thou shalt see the good of Jerusalem all the days of thy life. Yea, thou shalt see thy children's children, and peace upon Israel."

717.9–12 A seaborne death . . . around you.] *The Odyssey*, Book XI, lines 148–151, translation by Robert Fitzgerald.

717.35 "bound / In surety for his brother's son."] Wordsworth, "Michael," lines 210–211, first published in the 1800 edition of Wordsworth and Coleridge's *Lyrical Ballads.*

717.36–37 "summoned to discharge the forfeiture."] Wordsworth, "Michael," line 215.

718.3–4 "possess it . . . over it."] Wordsworth, "Michael," lines 246–247.

718.5–6 "to seek . . . the seas."] Wordsworth, "Michael," line 449.

719.14 a recent magazine article] Marianne Merrill Moates, "Learning . . . Every Day," *Creative Ideas for Living* (July/August 1988).

721.15 "the anxiety of influence"] A term employed by the American critic Harold Bloom (b. 1930) to characterize the poetic tradition: more particularly the anxiety aroused by predecessors as an individual poet attempts to make his or her own lasting contribution. This idea was first outlined at length in Bloom's book *The Anxiety of Influence: A Theory of Poetry* (1973).

737.39–40 cease to be Blondie . . . becoming Dagwood?] Blondie (Boopadoop) Bumstead and her husband Dagwood Bumstead, the principal characters in the still-running syndicated comic strip *Blondie* created by American cartoonist Chic Young (1901–1973).

738.8–9 the Exxon fiasco in Prince William Sound] On March 24, 1989, the oil tanker *Exxon Valdez*, bound for California, ran aground on Alaska's Bligh Reef, spilling nearly 11 million gallons of crude oil into Prince William Sound.

748.33–36 Thoreau . . . until everybody does it?] An allusion to H. D. Thoreau's essay "Civil Disobedience" (1849), in which he writes: "The only obligation which I have a right to assume is to do at any time what I think right."

751.2–3 "I can live no longer by thinking."] *As You Like It*, V.ii.50.

752.15 Prince William Sound] See note 738.8–9.

754.11 the singular sparrows of the street] Cf. Matthew 10:29.

754.11 the lilies of the field] Matthew 6:28.

754.12 "the least of these my brethren."] Matthew 25:40.

754.21–30 "All left-wing parties . . . shall continue."] George Orwell, "Rudyard Kipling" (1942), reprinted in *Collected Essays* (1961).

755.32–34 oblivious . . . as were George III and Lord North.] King George III of Great Britain and Ireland (1738–1820) and his prime minister Lord North (1732–1792) lost the American colonies.

756.4–5 Love Canal, Bhopal, Chernobyl, and the *Exxon Valdez*] Love Canal, near Niagara Falls, New York, the site of a chemical waste dump in the 1940s and 1950s that forced the evacuation and resettlement of more than a thousand families in the Love Canal neighborhood, after President Jimmy Carter (twice) declared a state of emergency in the area, in 1978 and 1981; Bhopal, the capital city in Madhya Pradesh, India, which gained international attention in 1984 when a pesticide-manufacturing plant leaked toxic gas, killing thousands; Chernobyl, an uninhabitable city in northern Ukraine, synonymous with the nuclear disaster there in 1986, when the Chernobyl nuclear power station emitted large amounts of radiation into the atmosphere; *Exxon Valdez*, see note 738.8–9.

758.12 Burley tobacco] A light air-cured tobacco used in cigarettes, once the "king" of crops in Kentucky.

760.1–8 Liberty Hyde Bailey . . . F. H. King . . . J. Russell Smith . . . Sir Albert Howard . . . John Todd, Wes Jackson] Liberty Hyde Bailey (1858–1954), American botanist, horticulturalist, and cofounder of the American Society for Horticultural Science; F. H. King, see note 143.5–9; Joseph Russell Smith (1874–1966), American economic geographer, conservationist, and writer whose books include *Tree Crops: A Permanent Agriculture* (1929); Sir Albert Howard, see note 152.27–28; John Todd, see note 483.15; Wes Jackson (b. 1936), see note 468.38.

Index

This book is set in 10 point ITC Galliard, a face designed for digital composition by Matthew Carter and based on the sixteenth-century face Granjon. The paper is acid-free lightweight opaque that will not turn yellow or brittle with age. The binding is sewn, which allows the book to open easily and lie flat. The binding board is covered in Brillianta, a woven rayon cloth made by Van Heek–Scholco Textielfabrieken, Holland.
Composition by Dedicated Book Services.
Printing and binding by LSC Communications.
Designed by Bruce Campbell.

THE LIBRARY OF AMERICA SERIES

Library of America fosters appreciation of America's literary heritage by publishing, and keeping permanently in print, authoritative editions of America's best and most significant writing. An independent nonprofit organization, it was founded in 1979 with seed funding from the National Endowment for the Humanities and the Ford Foundation.

1. Herman Melville: Typee, Omoo, Mardi
2. Nathaniel Hawthorne: Tales & Sketches
3. Walt Whitman: Poetry & Prose
4. Harriet Beecher Stowe: Three Novels
5. Mark Twain: Mississippi Writings
6. Jack London: Novels & Stories
7. Jack London: Novels & Social Writings
8. William Dean Howells: Novels 1875–1886
9. Herman Melville: Redburn, White-Jacket, Moby-Dick
10. Nathaniel Hawthorne: Collected Novels
11 & 12. Francis Parkman: France and England in North America
13. Henry James: Novels 1871–1880
14. Henry Adams: Novels, Mont Saint Michel, The Education
15. Ralph Waldo Emerson: Essays & Lectures
16. Washington Irving: History, Tales & Sketches
17. Thomas Jefferson: Writings
18. Stephen Crane: Prose & Poetry
19. Edgar Allan Poe: Poetry & Tales
20. Edgar Allan Poe: Essays & Reviews
21. Mark Twain: The Innocents Abroad, Roughing It
22 & 23. Henry James: Literary Criticism
24. Herman Melville: Pierre, Israel Potter, The Confidence-Man, Tales & Billy Budd
25. William Faulkner: Novels 1930–1935
26 & 27. James Fenimore Cooper: The Leatherstocking Tales
28. Henry David Thoreau: A Week, Walden, The Maine Woods, Cape Cod
29. Henry James: Novels 1881–1886
30. Edith Wharton: Novels
31 & 32. Henry Adams: History of the U.S. during the Administrations of Jefferson & Madison
33. Frank Norris: Novels & Essays
34. W.E.B. Du Bois: Writings
35. Willa Cather: Early Novels & Stories
36. Theodore Dreiser: Sister Carrie, Jennie Gerhardt, Twelve Men
37. Benjamin Franklin: Writings (2 vols.)
38. William James: Writings 1902–1910
39. Flannery O'Connor: Collected Works
40, 41, & 42. Eugene O'Neill: Complete Plays
43. Henry James: Novels 1886–1890
44. William Dean Howells: Novels 1886–1888
45 & 46. Abraham Lincoln: Speeches & Writings
47. Edith Wharton: Novellas & Other Writings
48. William Faulkner: Novels 1936–1940
49. Willa Cather: Later Novels
50. Ulysses S. Grant: Memoirs & Selected Letters
51. William Tecumseh Sherman: Memoirs
52. Washington Irving: Bracebridge Hall, Tales of a Traveller, The Alhambra
53. Francis Parkman: The Oregon Trail, The Conspiracy of Pontiac
54. James Fenimore Cooper: Sea Tales
55 & 56. Richard Wright: Works
57. Willa Cather: Stories, Poems, & Other Writings
58. William James: Writings 1878–1899
59. Sinclair Lewis: Main Street & Babbitt
60 & 61. Mark Twain: Collected Tales, Sketches, Speeches, & Essays
62 & 63. The Debate on the Constitution
64 & 65. Henry James: Collected Travel Writings
66 & 67. American Poetry: The Nineteenth Century
68. Frederick Douglass: Autobiographies
69. Sarah Orne Jewett: Novels & Stories
70. Ralph Waldo Emerson: Collected Poems & Translations
71. Mark Twain: Historical Romances
72. John Steinbeck: Novels & Stories 1932–1937
73. William Faulkner: Novels 1942–1954
74 & 75. Zora Neale Hurston: Novels, Stories, & Other Writings
76. Thomas Paine: Collected Writings
77 & 78. Reporting World War II: American Journalism
79 & 80. Raymond Chandler: Novels, Stories, & Other Writings

81. Robert Frost: Collected Poems, Prose, & Plays
82 & 83. Henry James: Complete Stories 1892–1910
84. William Bartram: Travels & Other Writings
85. John Dos Passos: U.S.A.
86. John Steinbeck: The Grapes of Wrath & Other Writings 1936–1941
87, 88, & 89. Vladimir Nabokov: Novels & Other Writings
90. James Thurber: Writings & Drawings
91. George Washington: Writings
92. John Muir: Nature Writings
93. Nathanael West: Novels & Other Writings
94 & 95. Crime Novels: American Noir of the 1930s, 40s, & 50s
96. Wallace Stevens: Collected Poetry & Prose
97. James Baldwin: Early Novels & Stories
98. James Baldwin: Collected Essays
99 & 100. Gertrude Stein: Writings
101 & 102. Eudora Welty: Novels, Stories, & Other Writings
103. Charles Brockden Brown: Three Gothic Novels
104 & 105. Reporting Vietnam: American Journalism
106 & 107. Henry James: Complete Stories 1874–1891
108. American Sermons
109. James Madison: Writings
110. Dashiell Hammett: Complete Novels
111. Henry James: Complete Stories 1864–1874
112. William Faulkner: Novels 1957–1962
113. John James Audubon: Writings & Drawings
114. Slave Narratives
115 & 116. American Poetry: The Twentieth Century
117. F. Scott Fitzgerald: Novels & Stories 1920–1922
118. Henry Wadsworth Longfellow: Poems & Other Writings
119 & 120. Tennessee Williams: Collected Plays
121 & 122. Edith Wharton: Collected Stories
123. The American Revolution: Writings from the War of Independence
124. Henry David Thoreau: Collected Essays & Poems
125. Dashiell Hammett: Crime Stories & Other Writings
126 & 127. Dawn Powell: Novels
128. Carson McCullers: Complete Novels
129. Alexander Hamilton: Writings
130. Mark Twain: The Gilded Age & Later Novels
131. Charles W. Chesnutt: Stories, Novels, & Essays
132. John Steinbeck: Novels 1942–1952
133. Sinclair Lewis: Arrowsmith, Elmer Gantry, Dodsworth
134 & 135. Paul Bowles: Novels, Stories, & Other Writings
136. Kate Chopin: Complete Novels & Stories
137 & 138. Reporting Civil Rights: American Journalism
139. Henry James: Novels 1896–1899
140. Theodore Dreiser: An American Tragedy
141. Saul Bellow: Novels 1944–1953
142. John Dos Passos: Novels 1920–1925
143. John Dos Passos: Travel Books & Other Writings
144. Ezra Pound: Poems & Translations
145. James Weldon Johnson: Writings
146. Washington Irving: Three Western Narratives
147. Alexis de Tocqueville: Democracy in America
148. James T. Farrell: Studs Lonigan Trilogy
149, 150, & 151. Isaac Bashevis Singer: Collected Stories
152. Kaufman & Co.: Broadway Comedies
153. Theodore Roosevelt: Rough Riders, An Autobiography
154. Theodore Roosevelt: Letters & Speeches
155. H. P. Lovecraft: Tales
156. Louisa May Alcott: Little Women, Little Men, Jo's Boys
157. Philip Roth: Novels & Stories 1959–1962
158. Philip Roth: Novels 1967–1972
159. James Agee: Let Us Now Praise Famous Men, A Death in the Family, Shorter Fiction
160. James Agee: Film Writing & Selected Journalism
161. Richard Henry Dana Jr.: Two Years Before the Mast & Other Voyages
162. Henry James: Novels 1901–1902
163. Arthur Miller: Plays 1944–1961
164. William Faulkner: Novels 1926–1929
165. Philip Roth: Novels 1973–1977

166 & 167. American Speeches: Political Oratory
168. Hart Crane: Complete Poems & Selected Letters
169. Saul Bellow: Novels 1956–1964
170. John Steinbeck: Travels with Charley & Later Novels
171. Capt. John Smith: Writings with Other Narratives
172. Thornton Wilder: Collected Plays & Writings on Theater
173. Philip K. Dick: Four Novels of the 1960s
174. Jack Kerouac: Road Novels 1957–1960
175. Philip Roth: Zuckerman Bound
176 & 177. Edmund Wilson: Literary Essays & Reviews
178. American Poetry: The 17th & 18th Centuries
179. William Maxwell: Early Novels & Stories
180. Elizabeth Bishop: Poems, Prose, & Letters
181. A. J. Liebling: World War II Writings
182. American Earth: Environmental Writing Since Thoreau
183. Philip K. Dick: Five Novels of the 1960s & 70s
184. William Maxwell: Later Novels & Stories
185. Philip Roth: Novels & Other Narratives 1986–1991
186. Katherine Anne Porter: Collected Stories & Other Writings
187. John Ashbery: Collected Poems 1956–1987
188 & 189. John Cheever: Complete Novels & Collected Stories
190. Lafcadio Hearn: American Writings
191. A. J. Liebling: The Sweet Science & Other Writings
192. The Lincoln Anthology
193. Philip K. Dick: VALIS & Later Novels
194. Thornton Wilder: The Bridge of San Luis Rey & Other Novels 1926–1948
195. Raymond Carver: Collected Stories
196 & 197. American Fantastic Tales
198. John Marshall: Writings
199. The Mark Twain Anthology
200. Mark Twain: A Tramp Abroad, Following the Equator, Other Travels
201 & 202. Ralph Waldo Emerson: Selected Journals
203. The American Stage: Writing on Theater
204. Shirley Jackson: Novels & Stories
205. Philip Roth: Novels 1993–1995
206 & 207. H. L. Mencken: Prejudices
208. John Kenneth Galbraith: The Affluent Society & Other Writings 1952–1967
209. Saul Bellow: Novels 1970–1982
210 & 211. Lynd Ward: Six Novels in Woodcuts
212. The Civil War: The First Year
213 & 214. John Adams: Revolutionary Writings
215. Henry James: Novels 1903–1911
216. Kurt Vonnegut: Novels & Stories 1963–1973
217 & 218. Harlem Renaissance Novels
219. Ambrose Bierce: The Devil's Dictionary, Tales, & Memoirs
220. Philip Roth: The American Trilogy 1997–2000
221. The Civil War: The Second Year
222. Barbara W. Tuchman: The Guns of August, The Proud Tower
223. Arthur Miller: Plays 1964–1982
224. Thornton Wilder: The Eighth Day, Theophilus North, Autobiographical Writings
225. David Goodis: Five Noir Novels of the 1940s & 50s
226. Kurt Vonnegut: Novels & Stories 1950–1962
227 & 228. American Science Fiction: Nine Novels of the 1950s
229 & 230. Laura Ingalls Wilder: The Little House Books
231. Jack Kerouac: Collected Poems
232. The War of 1812
233. American Antislavery Writings
234. The Civil War: The Third Year
235. Sherwood Anderson: Collected Stories
236. Philip Roth: Novels 2001–2007
237. Philip Roth: Nemeses
238. Aldo Leopold: A Sand County Almanac & Other Writings
239. May Swenson: Collected Poems
240 & 241. W. S. Merwin: Collected Poems
242 & 243. John Updike: Collected Stories
244. Ring Lardner: Stories & Other Writings
245. Jonathan Edwards: Writings from the Great Awakening
246. Susan Sontag: Essays of the 1960s & 70s

247. William Wells Brown: Clotel & Other Writings
248 & 249. Bernard Malamud: Novels & Stories of the 1940s, 50s, & 60s
250. The Civil War: The Final Year
251. Shakespeare in America
252. Kurt Vonnegut: Novels 1976–1985
253 & 254. American Musicals 1927–1969
255. Elmore Leonard: Four Novels of the 1970s
256. Louisa May Alcott: Work, Eight Cousins, Rose in Bloom, Stories & Other Writings
257. H. L. Mencken: The Days Trilogy
258. Virgil Thomson: Music Chronicles 1940–1954
259. Art in America 1945–1970
260. Saul Bellow: Novels 1984–2000
261. Arthur Miller: Plays 1987–2004
262. Jack Kerouac: Visions of Cody, Visions of Gerard, Big Sur
263. Reinhold Niebuhr: Major Works on Religion & Politics
264. Ross Macdonald: Four Novels of the 1950s
265 & 266. The American Revolution: Writings from the Pamphlet Debate
267. Elmore Leonard: Four Novels of the 1980s
268 & 269. Women Crime Writers: Suspense Novels of the 1940s & 50s
270. Frederick Law Olmsted: Writings on Landscape, Culture, & Society
271. Edith Wharton: Four Novels of the 1920s
272. James Baldwin: Later Novels
273. Kurt Vonnegut: Novels 1987–1997
274. Henry James: Autobiographies
275. Abigail Adams: Letters
276. John Adams: Writings from the New Nation 1784–1826
277. Virgil Thomson: The State of Music & Other Writings
278. War No More: American Antiwar & Peace Writing
279. Ross Macdonald: Three Novels of the Early 1960s
280. Elmore Leonard: Four Later Novels
281. Ursula K. Le Guin: The Complete Orsinia
282. John O'Hara: Stories
283. The Unknown Kerouac: Rare, Unpublished & Newly Translated Writings
284. Albert Murray: Collected Essays & Memoirs
285 & 286. Loren Eiseley: Collected Essays on Evolution, Nature, & the Cosmos
287. Carson McCullers: Stories, Plays & Other Writings
288. Jane Bowles: Collected Writings
289. World War I and America: Told by the Americans Who Lived It
290 & 291. Mary McCarthy: The Complete Fiction
292. Susan Sontag: Later Essays
293 & 294. John Quincy Adams: Diaries
295. Ross Macdonald: Four Later Novels
296 & 297. Ursula K. Le Guin: The Hainish Novels & Stories
298 & 299. Peter Taylor: The Complete Stories
300. Philip Roth: Why Write? Collected Nonfiction 1960–2013
301. John Ashbery: Collected Poems 1991–2000
302. Wendell Berry: Port William Novels & Stories: The Civil War to World War II
303. Reconstruction: Voices from America's First Great Struggle for Racial Equality
304. Albert Murray: Collected Novels & Poems
305 & 306. Norman Mailer: The Sixties
307. Rachel Carson: Silent Spring & Other Writings on the Environment
308. Elmore Leonard: Westerns
309 & 310. Madeleine L'Engle: The Kairos Novels
311. John Updike: Novels 1959–1965
312. James Fenimore Cooper: Two Novels of the American Revolution
313. John O'Hara: Four Novels of the 1930s
314. Ann Petry: The Street, The Narrows
315. Ursula K. Le Guin: Always Coming Home
316 & 317. Wendell Berry: Collected Essays
318. Cornelius Ryan: The Longest Day, A Bridge Too Far